International Geomorphology 1986

From left to right, Dr. Nicholas Short, Professor A. Zhivago, Professor Denys Brunsden, His Grace the Duke of Devonshire and Professor Ian Douglas.

Some participants at the Conference.

International Geomorphology 1986

Proceedings of the First International Conference on Geomorphology

PART II

Edited on behalf of the
British Geomorphological Research Group

by

V. Gardiner
Department of Geography
University of Leicester

and sectional editors

M. G. Anderson	K. S. Richards
A. P. Carr	D. E. Sugden
I. Douglas	M. A. Summerfield
J. C. Doornkamp	M. M. Sweeting
R. I. Ferguson	M. F. Thomas
M. J. Kirkby	J. R. G.Townshend
J. Lewin	D. E. Walling
J. McManus	A. Warren
A. C. Millington	P. Worsley
R. P. C. Morgan	

A Wiley–Interscience Publication

JOHN WILEY & SONS

CHICHESTER • NEW YORK • BRISBANE • TORONTO • SINGAPORE

Library of Congress Cataloging-in-Publication Data:

International Conference on Geomorphology (1st : 1985 :
University of Manchester)
International geomorphology, 1986.

'Held at the University of Manchester from September 15–21, 1985'—Pref.
Includes index.
1. Geomorphology—Congresses. I. Gardiner, V.
II. British Geomorphological Research Group.
III. Title.
GB400.2.I58 1985 551 86-18465

ISBN 0 471 91471 1 (Part I)
ISBN 0 471 91472 X (Part II)
ISBN 0 471 90869 X (Set)

British Library Cataloguing in Publication Data:

International Conference on Geomorphology
(1st : 1985 : University of Manchester)
International geomorphology 1986.
1. Geomorphology
I. Title II. British Geomorphological
Research Group III. Gardiner, V.
551.4 GB401.5

ISBN 0 471 91471 1 (Part I)
ISBN 0 471 91472 X (Part II)
ISBN 0 471 90869 X (Set)

Printed in Great Britain

List of Contributors

A. D. ABRAHAMS, *Department of Geography, State University of New York at Buffalo, Faculty of Social Sciences, Buffalo, New York 14260, USA*

D. ADAMSON, *Department of Geography, Monash University, Victoria 3168, Australia*

F. AHNERT, *Geographisches Institut der Rhein.-Westf. Techn. Hochschule Aachen, 5100 Aachen, Templergraben 55, West Germany*

NANSHAN AI, *Department of Geography, Lanzhou University, Lanzhou, People's Republic of China*

M. ALLARD, *Centre d'études Nordiques, Université Laval, Sainte-Foy, Quebec G1K 7P4, Canada*

B. AMBROISE, *Lab. de Geographie Physique, RCP 741, CNRS, 3 rue de l'Argonne, F67083 Strasbourg, France*

J. F. ARAYA-VERGARA, *Department of Geography, University of Chile, Marcoleta, Santiago 250, Chile*

Y. BATTIAU-QUENEY, *Department of Geography, University of Lille 1, Villeneuve 59655, France*

J. BAXTER

M. T. BENAZZOUZ, *University Constantine, 7 Cité Bouhalilaid, Constantine, Wilaya, Algeria*

A. BILLARD, *Laboratory of Physical Geography, CNRS, Paris UA 0141, France*

R. W. BLAIR JNR., *Department of Geology, Fort Lewis College, Durango, Colorado 81301, USA*

J. A. BLANCO

R. P. BOURMAN, *School of Applied Science, South Australian College of Advanced Education, Underdale Site, Holbrook Road, Underdale, South Australia, Australia*

M. D. CAMPBELL, *Department of Geography, University College Swansea, Singleton Park, Swansea SA2 8PP, UK*

M. CANTANO

V. CASTELLANI, *Universita La Sapienza, Roma, Italy*

J. F. CERVANTES-BORJA, *Instituto de Geografia, Universidad Nacional, Autonoma de Mexico, 04510 Mexico DF, Mexico*

XI-QING CHEN

S. CHEN, *Chongqing Institute of Architecture and Engineering, People's Republic of China*

S. CICCACCI

C. M. CLAPPERTON, *Department of Geography, University of Aberdeen, Old Aberdeen, Aberdeen AB9 2UF, UK*

H. J. COOKE, *University of Botswana, Gaborone 0022, Botswana*

C. COSANDEY, *Physical Geography Laboratory, 1 Place Austide Briand, Meudon 92190, France*

HEZHI DAI, *Chongqing, Sichuan, People's Republic of China*

A. B. DE VILLIERS, *Department of Geography, P.U. for C.H.E., Potchefstroom 2520, Republic of South Africa*

J. C. DOHRENWEND, *US Geological Survey, MS 901, 345 Middlefield Road, Menlo Park, CA 94025, USA*

G. R. DOUGLAS, *Mentaskolinn vid Hamrahlid, Reykjavik, Iceland*

W. DRAGONI, *CNR–IRPI, Localita Madonna Alta, 06100 Perugia, Italy*

E. DROZDOWSKI, *Department of Lowland Geomorphology, Polish Academy of Sciences, Kopernika 19, Torun 87100, Poland*

ZI-RONG DU, *Southwest China Teachers University, Beibei, Chongqing, Sichuan, People's Republic of China*

F. S. EBISEMIJU, *Department of Geography, Obafemi Awolowo University, Ado Ekiti, Ondo State, Nigeria*

I. S. EVANS, *Department of Geography, Durham University, Science Laboratories, South Road, Durham DH1 3LE, UK*

R. I. FERGUSON, *Department of Environmental Science, University of Stirling, Stirling FK9 4LA, UK*

H. J. FINNEY, *Silsoe College, Silsoe, Bedford MK45 4DT, UK*

C. T. FOSTER, *Department of Geography, University of Liverpool, Roxby Building, Liverpool L69 3BX, UK*

P. FREDI, *Inst. Geologia e Paleontologia, Universita Degli Studi di Roma, Piazzale Aldo Moro 5, Roma 00100, Italy*

V. GARDINER, *Department of Geography, University of Leicester, University Road, Leicester LE1 7RH, UK*

A. F. GELLATLY, *Department of Geography, The University of Sheffield, Sheffield S10 2TN, UK*

D. GILLIESON, *Department of Geography, The Australian National University, Canberra ACT, GPO 4, Australia*

J. E. GORDON, *Nature Conservancy Council, Pearl House, Bartholomew Street, Newbury, Berks. RG14 5LS, UK*

P. GORECKI

H. GU, *Chongqing Institute of Communication, Chongqing, People's Republic of China*

M. J. HAIGH, *Department of Social Studies, Oxford Polytechnic, Oxford OX3 0BP, UK*

A. HALL, *Fettes College, Edinburgh EH4 1QX, UK*

CAIHUA HE, *Department of Geography, Guizhou Teachers' University, Guizhou, People's Republic of China*

J. HEAD

M. R. HENDRICKS, *Instituut voor aardwetenschappen, Free University of Amsterdam, Postbus 7161, 1007 MC Amsterdam, The Netherlands*

C. G. HIGGINS, *Department of Geology, University of California, Davis, California 95616, USA*

G. HOPE

J. K. HOWARD

YUNLIN HUANG, *Institute of Geography, Academia Sinica, Beijing, People's Republic of China*

ZHENGUO HUANG, *Guangzhou Institute of Geography, Yellow Flower Hill, Guangzhou, People's Republic of China*

P. HUGHES

L. K. JEJE, *Department of Geography, University of Ife, Ile Ife, Nigeria*

A. R. JONES, *Department of Geography, University of Reading, Whiteknights, PO Box 227, Reading RG6 2AB, UK*

C. A. JONES

L. KASPRKAK

R. B. KING, *Land Resources Development, Tolworth Tower, Surbiton, Surrey KT6 7DY, UK*

M. J. KIRKBY, *School of Geography, The University of Leeds, Leeds LS2 9JT, UK*

S. KOZARSKI, *Quaternary Research Institute, Adam Mickiewicz University, ul. Fredry 10, Poznan 61701, Poland*

M. LASSILA, *Department of Geography, University of Umea, PO 187, Umea, Sweden*

R. LEVESQUE, *Ventre d'études Nordiques, Université Laval, Sainte-Foy, Quebec G1K 7P4, Canada*

FENGHUA LI

XINGGUO LI

KONGHONG LI, *Guangzhou Institute of Geography, Yellow Flower Hill, Guangzhou, People's Republic of China*

PINGRI LI, *Guangzhou Institute of Geography, Yellow Flower Hill, Guangzhou, People's Republic of China*

SHUPEI LI

JUNSHU LIN, *Institute of Geography, Academia Sinica, Beijing, People's Republic of China*

ZECHUN LIU, *Department of Geography, Nanjing Normal University, Nanjing, People's Republic of China*

TUNGSHENG LIU

SHUZHENG LIU, *Chengdu Institute of Geology, Academia Sinica, Chengdu, Sichuan, People's Republic of China*

YAORU LU, *The Institute of Hydrogeology and Engineering Geology, Ministry of Geology, Zhengding, Hebei, People's Republic of China*

E. LUPIA PALMIERI, *Dipartimento di scienze della terra, Universita degli studi di Roma, Piazzale Aldo Moro 5, Roma 00100, Italy*

M. E. MARKER, *Department of Geography, University of Fort Hare, Private Bag X1314, Alice, Ciskei, South Africa*

M. A. MARQUES, *Department of Geomorphology, University of Barcelona, Grand Via 585, Barcelona 08007, Spain*

J. J. MCALISTER, *Department of Geography, University of Belfast, Belfast BT9 6BB, Northern Ireland, UK*

L. D. MCFADDEN, *Department of Geology, University of New Mexico, Albuquerque, NM 87131, USA*

J. P. McGreevy, *Public Record Office, Belfast, Northern Ireland, UK*

J.-P. Metailie, *Institut de Geographie, Université Toulouse, 109 bis rue Vauquelie, Toulouse 31058, France*

M. Meza-Sanchez, *Instituto de Geografia, Universidad Nacional, Autonoma de Mexico, 04510 Mexico DF, Mexico*

A. C. Millington, *Department of Geography, University of Reading, Whiteknights, PO Box 227, Reading RG6 2AB, UK*

M. R. Mohammad, *Department of Geology, Faculty of Science, Alexandria University, Moharram-Beh, Alexandria, Egypt*

E. Molina, *Dept. Geomorfologia, Facultad de Ciencias, Universidad Salamanca, Spain*

R. P. C. Morgan, *Silsoe College, Silsoe, Bedford MK45 4DT, UK*

Guichun Mu, *Chongqing, Sichuan, People's Republic of China*

H. M. Mushala

T. Muxart

R. H. Neale, *School of Geography, The University of Leeds, Leeds LS2 9JT, UK*

R. S. Pandey, *Govind Ballabh Pant Social Science Institute, 80 Tagore Town, Allahabad, India*

A. J. Parsons, *Department of Geography, University of Keele, Keele, Staffs. ST5 5BG, UK*

E. Pellitero

N. R. Peters, *Department of Geography, University of Hull, Hull HU6 7RX, UK*

J. Poesen, *Laboratory for Experimental Geomorphology, Redingenstraat 16 bis, Leuven B3000, Belgium*

F. Pugliese, *Dipartimento di Scienze della Terra, Universita degli studi di Roma ''La Sapienza'', Piazzale Aldo Moro 5, 00100 Roma, Italy*

X. QIAN, *Chongqing University, People's Republic of China*

SHANWEN QIU

R. K. RAI, *Department of Geography, North-Eastern Hill University, Shillong, Meghalaya, India*

J. H. REYNHARDT, *Department of Geography, Vista University, Mamelodi Campus, Private Bag X03, Mamelodi 0100, South Africa*

D. A. ROBINSON, *The Geography Laboratory, The University of Sussex, Arts Building, Falmer, Brighton BN1 9QN, UK*

J. ROCA, *Department of Geography, University of Barcelona, Grand Via 585, Barcelona 08007, Spain*

K. ROGNER, *Universitat Paderborn N4 110, Warburger Str. 100, Paderborn D4790, West Germany*

L. RUSSELL, *Department of Geography, University of Fort Hare, Private Bag X1314, Alice 5700, Ciskei, South Africa*

G. B. RYDOUT

F. SALVINI

C. J. SCHOUTEN, *Department of Geography, State University of Utrecht, PO Box 80.115, 3508 TC, Utrecht, The Netherlands*

M. K. SEGUIN, *Centre d'études Nordiques, Université Laval, Sainte-Foy, Quebec G1K 7P4, Canada*

E. SEYHAN, *Instituut voor aardwetenschappen, Free University of Amsterdam, Postbus 7161, 1007 MC, Amsterdam, The Netherlands*

R. A. SHAKESBY, *Department of Geography, University College Swansea, Singleton Park, Swansea SA2 8PP, UK*

H. S. SHARMA, *Department of Geography, University of Rajasthan, Jaipur, Rajasthan, India*

A. N. SHARPLEY, *Water Quality, Watershed Laboratory, PO Box 1480, Oklahoma 74702, USA*

P. A. Shaw

I. Simmers, *Instituut voor aardwetenschappen, Free University of Amsterdam, Postbus 7161, 1007 MC, Amsterdam, The Netherlands*

S. Singh, *Department of Geography, University of Allahabad, Allahabad-211 002, India*

R. Sjoberg

B. J. Smith, *Department of Geography, The Queen's University of Belfast, Belfast BT7 1NN, Northern Ireland, UK*

S. J. Smith

L. Starkel, *Inst. Geografii PAN, ul. Sw. Jana 22, 31-018 Krakow, Poland*

D. E. Sugden, *Department of Geography, University of Aberdeen, Old Aberdeen AB9 2UF, UK*

Xiulan Sui

M. Sullivan, *Geography/Demography Department, The University of Papua New Guinea, PO Box 320, Papua New Guinea*

M. A. Summerfield, *Department of Geography, University of Edinburgh, Drummond Street, Edinburgh EH8 9XP, UK*

Jianzhong Sun, *Department of Hydrogeology and Engineering Geology, Xian Geological College, Xian City, People's Republic of China*

M. M. Sweeting, *School of Geography, University of Oxford, Mansfield Road, Oxford OX1 3TB, UK*

M. F. Thomas, *Department of Environmental Science, University of Stirling, Stirling FK9 4LA, UK*

J. R. G. Townshend, *NERC Unit for Thematic Information Systems, Department of Geography, University of Reading, Whiteknights, PO Box 227, Reading RG6 2AB, UK*

B. D. Turrin, *U.S. Geological Survey, MS 901, 345 Middlefield Road, Menlo Park, CA 94025, USA*

C. VERGNOLLE, *CIMA-LA 366 CNRS, Institut de Geographie, Université de Toulouse — Le Mirail, 5 Allées Antonio Machado, 31 058 Toulouse, France*

L. VIVAS, *School of Geography, Universidad de los Andes, Merida, Venezuela*

D. VIVILLE, *Lab. de Geographie Physique, RCP 741 CNRS, 3 rue de l'Argonne, F67083 Strasbourg, France*

R. P. D. WALSH, *Department of Geography, University College Swansea, Singleton Park, Swansea SA2 8PP, UK*

A. WARREN, *Department of Geography, University College London, 26 Bedford Way, London WC1H 0AP, UK*

S. G. WELLS, *Department of Geology, University of New Mexico, Albuquerque, New Mexico 87131, USA*

W. B. WHALLEY, *Department of Geography, The Queen's University of Belfast, Belfast BT7 1NN, Northern Ireland, UK*

M. A. J. WILLIAMS, *Geography Department, Monash University, Clayton, Victoria, Australia 3168*

R. B. G. WILLIAMS, *The Geography Laboratory, The University of Sussex, Arts Building, Falmer, Brighton BN1 9QN, UK*

J. R. WILLIAMS

P. WORSLEY, *Department of Geography, University of Nottingham, University Park, Nottingham NG7 2RD, UK*

YUMEI XIA

YOULIN YANG, *Institute of Desert Research, Academia Sinica, Lanzhou, 14 Dong Gang West Road, Lanzhou, Gansu, People's Republic of China*

HUAI-JEN YANG

DAOXIAN YUAN, *Institute of Karst Geology, Ministry of Geology and Mineral Resources, Guilin, Guangxi, People's Republic of China*

BAOYIN YUAN

T. ZAN, *Department of Geography, Lanzhou University, Lanzhou, People's Republic of China*

YAOGUANG ZHANG, *Institute of Geography, Academia Sinica, Beijing, People's Republic of China*

ZHONQYING ZHANG, *Guangzhou Institute of Geography, Yellow Flower Hill, Guangzhou, People's Republic of China*

A. V. ZHIVAGO, *P.P. Shirsov Institute for Oceanology, Academy of Sciences USSR, 23 Krasikova Street, Moscow V-218, USSR*

XIANGHAO ZHONG, *Chengdu Institute of Geography, Academia Sinica, Chengdu, Sichuan, People's Republic of China*

JINGHU ZHU, *Department of Geography, Harbin Normal University, Harbin, People's Republic of China*

ZHENDA ZHU, *Institute of Desert Research, Academia Sinica, Lanzhou, 14 Dong Gang West Road, Lanzhou, Gansu, People's Republic of China*

H. ZILLIACUS, *Department of Geography, University of Helsinki, Hallitushata 11-13, 00100 Helsinki, Finland*

BENGONG ZOU, *Institute of Desert Research, Academia Sinica, 14 Dong Gang West Road, Lanzhou, People's Republic of China*

Contents

APPLIED GEOMORPHOLOGY, LAND CONSERVATION AND RESOURCE EVALUATION

WEATHERING

GLACIAL GEOMORPHOLOGY

ARID LAND AND AEOLIAN GEOMORPHOLOGY

CONTENTS OF PART I

Preface

Geomorphology is a truly international science. Its subject matter is independent of international boundaries, political systems or racial groupings. Its practitioners form an informal international fraternity of scientists sharing a common interest in landforms, their processes and evolution. Despite the difficulties of language, ideas are exchanged throughout the world and collaboration, to varying degrees, exists between geomorphologists of many nations. However there is still much potential for formalising existing international links and developing new ones. During the 1970s much progress was made by the British Geomorphological Research Group towards fostering international geomorphology. Two conferences were held in London, with a considerable number of overseas participants, and joint meetings were organised with geomorphological groups in other countries, notably Germany and Poland. The culmination of all of this effort was the suggestion that the time was ripe, in the mid-1980s, for a major international conference in geomorphology, independent of its parent disciplines, and with truly international participation. Manchester was chosen as the venue, organising and local committees were set up, and thus was the scene set for the conference from which these Proceedings originated.

The time was also appropriate in other ways. The practical significance of geomorphological processes had been tragically demonstrated during preceding years by several major natural catastrophes and by land management and conservation problems in Africa. The significance of global geomorphology in human terms was underlined during the period of the conference itself by events associated with earthquakes in Mexico. Increasing financial pressures had forced geomorphologists in some countries to examine potentials for increased collaboration, and, finally, financial incentives for international collaboration were slowly but surely becoming available from both governments and international bodies such as the EEC.

The objectives of the Conference, which was held at the University of Manchester from September 15–21, 1985, were in general terms to provide an international forum in which work on all aspects of geomorphology could be reported and discussed, to emphasise the role of geomorphology in environmental management and the Developing World, and to foster further international collaboration in the discipline. Great efforts were made to inform all geomorphologists of the conference, in all countries and in all disciplines, and to obtain funds which could be used to provide bursaries for prospective participants in less fortunate situations.

There was an almost overwhelming response to the call for papers and the Organising Committee had, with great regret, to conclude that it was just not possible to include all worthy papers in the programme. It was an extremely difficult task to select from amongst the many excellent abstracts submitted. The subjects suggested for the Conference were divided into 26 symposia, addressing virtually all of geomorphology. Some papers were presented verbally, others as posters, but all manuscripts arising from them have been considered equally in preparing these Proceedings. One-page abstracts of all papers, however presented, were included in a Conference Abstracts volume which was circulated to participants on registration. This did much to aid understanding of ideas, and much to stimulate discussion.

Manuscripts submitted for consideration for publication have been refereed in a manner similar to that employed for periodical papers, and about 20 per cent were excluded as a result of this process. The accepted papers have been organised into nineteen major thematic groups, in most cases reflecting the symposia of the Conference. Very broadly, volume one of these Proceedings contains process geomorphology and volume two contains symposia having a more significant chronological and climato-genetic component. Symposia on aspects of applied geomorphology have been distributed between both volumes. Papers from associated workshops have not, however, been included in these Proceedings, but will form special issues of periodicals. Earth Surface Processes and Landforms will contain a selection of papers on loess, and two supplementary volumes of *Zeitschrift für Geomorphologie* will carry papers resulting from the Neotectonics and Morphotectonics workshop and Laterites workshop. Finally, the papers from the Drumlin workshop have been published in a separate book.

The format of these Proceedings is worthy of mention. It was felt essential to produce the Proceedings within as short a time and at as low a cost as possible. A camera-ready format was therefore chosen. Some camera-ready manuscripts were prepared by authors, but to ensure speed and uniformity about one third of the papers were done in Leicester, and many more were processed by sectional editors. The adoption of this procedure inevitably results in a slightly less uniform set of papers in terms of conventions and general appearance, but we feel that this slight sacrifice is worthwhile under the circumstances. It is a great pleasure to offer thanks for the work of all the contributors in preparing their contributions, the translations from French manuscripts carried out by Professor Ian Douglas, the help so freely given in many ways by sectional editors and their Institutes, and the assistance rendered by all technical and clerical staff of the Department of Geography at Leicester University, particularly Carol Deacon for dealing with an enormous amount of correspondence. Finally, I would like to give special thanks to Mr and Mrs Pigden of Lutterworth Wordscan, who word-processed a quarter of a million words on geomorphology with remarkable efficiency and care to detail, and to my family for having put up with international geomorphology for so long.

It is clear that the First International Conference on Geomorphology was unique in the history of geomorphology. Virtually every country in which geomorphology is practised was represented, and for five days in Manchester geomorphologists were to be seen making almost impossible decisions concerning which of the many excellent concurrent sessions to attend. Particularly noteworthy was the recognition of the event's significance by China, resulting in one of the largest delegations of Chinese scientists to attend a conference in Europe on any subject.

Perhaps even more important to geomorphology than the results and ideas conveyed in formal sessions are the many contacts and friendships made. It is the British Geomorphological Research Group's hope that the First International Conference on Geomorphology will be the first of many, and that it will long be remembered for the productive friendships which will develop. It is my hope that these Proceedings will provide not only a set of papers reporting geomorphological research and encapsulating international geomorphology in 1986, but also a reminder of the spirit of international friendship in which the conference was conducted.

Foreword

PRELIMINARY

For many years the British Geomorphological Research Group cherished the belief that the world's geomorphologists would welcome an opportunity to meet together, to exchange views and to build firm friendships that might enhance our subject, contribute to the solution of international geomorphological problems and lead to profitable research projects. At Manchester on September 15–21, 1985, these dreams became a reality as 675 delegates from 51 countries met in an informal but highly organised and historic conference.

The BGRG had previously practised for this meeting by organising two smaller international conferences at the Joint School of Geography of King's College and the London School of Economics and Political Science, on 'Geomorphology, Present Problems and Future Prospects' (London, 5–9 April 1976) and on 'Mega Geomorphology' (London, 26–30 March 1981). In addition it had opened correspondence links with other organised groups around the world and arranged several exchange field visits, particularly with our Dutch and German colleagues, in order to build firm support for future activities. The Group also placed considerable emphasis on its publications and on building its international membership (147 by 1981), so that it would be able to contribute financially, in a substantial way, to the funding of a future conference. The launching of the journal *Earth Surface Processes* in 1978 firmly set the BGRG on the international stage and our acceptance as a Study Group of the Institute of British Geographers and a specialist group of the Geological Society of London ensured that our activities would be interdisciplinary in nature.

The first circular for the conference was issued in January 1983. At the outset it was determined by the BGRG that every effort would be made to make our efforts of value to developing countries and to facilitate attendance by as many people as possible from areas where travel funds are difficult or impossible to obtain. Accordingly we announced the special theme of the conference as:

'Geomorphology, Resources, Environment and the Developing World'

The invitation listed three objectives:

(1) To provide an international forum for geomorphology at which we may discuss the developments and future research problems of our subject.
(2) To discuss in plenary session the important question of how we may apply knowledge and techniques of geomorphology to problems of development, resource and environmental conservation and management, hazard alleviation and other problems involved in the use of the earth as the human habitat.
(3) To establish a framework for future international cooperation, an international newsletter and a basis for further meetings at regular (4-year?) intervals.

An organising committee was established consisting of Ian Douglas (Manchester), Denys Brunsden (King's College, London), Ed Derbyshire (Keele, later Leicester), Vince Gardiner (Leicester), Keith Richards (Hull, later Cambridge), and Michael Thomas (Stirling), later to be joined by the current secretary and treasurer of the BGRG, Alan Werritty and Ian Reid and members of the local organising committee. This committee was able, under the firm hand of Ian Douglas, to settle all the basic problems involved in policy decisions, excursions, publications, funding, and programme details with a minimum of meetings or cost and issued the Second Circular in January 1984. The response to the First Circular had been overwhelming, with over 600 people from 65 countries expressing an interest. With confidence, a programme was announced that included plenary sessions, paper sessions, poster presentations, associated meetings of IGU, INQUA, the Glaciological Society, QRA/IGU Periglacial Commission field meeting, field excursions, technical workshops, social events, accompanying persons programme, exhibitions and publications. A local committee consisting of David Collins, Brian Kear, Chuff Johnson, Tom Spencer, Adrian Harvey, Wilf Theakstone, John Gunn, Donald Warth, Jim Petch and Eric Isaac was established. Conference Secretaries, Meena Shah and Charmaine Arnold-Reed were put hard at work. Excursions or workshops were organised by Marjorie Sweeting, Adrian Harvey, Peter Worsley, Frank Oldfield, A. Thompson, Tom Spencer, Jan Hooke, John Gunn, E. M. Driscoll, Peter James, Chuff Johnson, Jim Petch, Eric Isaac, Roy Morgan, Malcolm Newson, Ed Derbyshire, Brian Whalley, Peter Bull, Julian Orford, John Dearing, R. Thompson, A. J. Parsons and Mike Kirkby. In the event they did not all come to fruition but it was obvious at this stage that the whole geomorphological community was willing this conference to success.

The team was, of course, eventually joined by innumerable helpers from the Manchester department, other universities, the BGRG, postgraduates and even the families of the organisers. I cannot name them all but I am sure that every participant will remember their efforts with gratitude. Geomorphology owes them a great debt for creating such an efficient and happy atmosphere.

THE CONFERENCE

Objective (1): An International Forum

The conference was opened on Sunday 15th September by His Grace, the Duke of Devonshire who emphasised our role in the improvement of the world's environment, the conservation of its resources and the mitigation of its hazards. He congratulated us on our efforts to improve international understanding and goodwill. He was accompanied on the platform by Ian Douglas as Conference Organiser, Professor A. Zhivago from the USSR Academy of Sciences who made a presentation of a fine oceanographic atlas, Dr Nicholas Short from NASA/Goddard Space Flight Center who made a presentation on 'Regional Landforms from Space — a perspective on world landforms as seen from Landsat imagery', and Denys Brunsden acting as chairman of the BGRG and the conference.

There were major conference lectures on each evening. Professor H. Th. Verstappen, 'Geomorphology, Resources, Environment and the Developing World'; Professor Victor R. Baker gave the *Zeitschrift für Geomorphologie* lecture on 'A New Global Geomorphology'; Professor J. H. Walker on 'Potential for International Collaboration in Geomorphological Research'; Professor J. Tricart on 'Geomorphology and Development'; and Professor R. U. Cooke who delivered the annual Frost lecture on 'The Use of Geomorphology'. The last was preceded by the presentation of the David Linton Award to Professor Leszek Starkel, a distinguished Polish colleague and long term friend and visitor to the BGRG.

Tom Spencer edited a huge volume containing Abstracts of the over 700 papers submitted to the Conference, to provide a unique record of the state of the subject in 1985. The present volume now presents over 170 of these submissions in full. The statistics of the abstracts are impressive. Sixty countries are represented, but the fact that 70 per cent came from ten countries and 40 per cent from three countries suggests how much more work there is to be done by future international collaboration. The breakdown into the most numerous contributions per country and by subject (Table 1) gives a quick guide to our present preoccupations. The emergence of Chinese geomorphology and the dominance of applied aspects are particularly noteworthy but even this breakdown does not portray the full list of conference themes which fell under 31 separate headings.

Did we achieve the objective of our international forum? We must leave the memory of Manchester, the Abstracts volume and the present Proceedings to speak for themselves. A personal view is an enthusiastic affirmative, but with qualified knowledge that there were missing friends, that despite all efforts many developing countries were under-represented, and that there is a potential for a huge growth in the future.

It is also important to note that there were several special publications on the occasion of the conference. These include:

Table 1. Origin (a) and Subject (b) of Abstracts submitted at Manchester

(a)	Country of Abstract	No. of Abstracts	%
	UK	125	18
	People's Republic of China	81	12
	USA	78	11
	Canada	44	6
	France	38	5
	USSR	26	4
	Australia	22	3
	Italy	22	3
	India	21	3
	Japan	21	3
(b)	Subject Area	No. of Abstracts	%
	Applied conservation	64	9
	Glacial	54	8
	Basin sediment systems	50	7
	Applied resources	47	7
	Long term landform evolution	45	7
	River channel dynamics	42	6
	Neotectonics and morphotectonics	41	6
	Mass movements	41	6
	Quaternary studies	37	5
	Theoretical	36	5
	Coastal	32	4

(Other fields strongly represented: Laterite (13), Less (10), Karst (9), Remote Sensing (9), Aeolian (6), Periglacial (6), Slopes (5), and Government (5)).

R. H. Johnson (ed.) (1985) *The Geomorphology of North-West England*. Manchester University Press, 421pp.

S. Kosarski (ed.) *Quaestiones Geographicae*, Special Publication 1, Poznan, Poland.

M. Pecsi (ed.) *Environmental and Dynamical Geomorphology*, Budapest, Hungary.

Wang Nailand, Shi Yafent *et al. Geomorphology in China*, Beijing, China.

D. Barsch, H. Liedtke, K. Moller, G. Stablein. *Geomorphological Mapping in the Federal Republic of Germany*. Berlin, Germany.

M. Panizza (ed.) *Italian Research in Physical Geography and Geomorphology: an overview*. Modena, Italy.

R. S. Hayden (ed.) *Global Mega-Geomorphology*. NASA Conf. Publ. 2312, Washington, USA.

In addition the accompanying workshops have generated several volumes of material that will be published in international journals. There were also two special lectures, the Edward Arnold Lecture by Olav Slaymaker on 'The erosion–sediment and solute yield–landform evolution problem in British Columbia's Coastal Mountains' and the John Wiley & Sons Lecture by L. Jijun, Tang Lingyu and Feng Zhaodan on 'Late Quaternary monsoon patterns on the loess plateau of China'.

Objective (2): The Applications of Geomorphology to the Use of the Earth as a Human Habitat

The success of this objective has already been answered by the listing in Table 1 which clearly shows the dominance of applied aspects in the Abstracts submitted. The two plenary sessions by Verstappen and Tricart provided the setting for fruitful discussion and although it cannot be pretended that we solved any of the world's problems, we were made aware of the magnitude and variety of the tasks before the subject.

Paper sessions covered geomorphology and environmental management, hazard assessment, river management, erosion prediction, mineral exploration, planning, urban environment, techniques for application, land resource evaluation, anthropogenic landforms, land conservation, engineering geomorphology in fluvial, karst in coastal situations, slope stability and foundations. The George Allen & Unwin Lecture was delivered by Peter Fookes on 'Geomorphology and Civil Engineering'.

For those of us who believe that a healthy subject needs a strong applied wing this was a particularly gratifying response.

Objective (3): Future International Collaboration

Three business meetings discussed the future of international collaboration in geomorphology. For the historical record these were:

Tuesday 17 September 1985	National Delegates	*c.* 45 participants
Wednesday 18 September 1985	Plenary Session	*c.* 500 participants
Thursday 19 September 1985	Working Committee	6 participants

At the first meeting 34 countries were represented, at the second 51 countries. The discussion was based on a survey carried out by Dr D. Sugden (reported elsewhere in this volume) of international views on the establishment of an International Geomorphological Organisation. The Plenary Session was introduced by Professor H. J. Walker who read a paper as 'Potentials for International Collaboration in Geomorphological Research' (in this volume).

At these historic meetings, held in a friendly, informative and productive setting, the following conclusions were reached:

1. All delegates agreed that there is an important need for future international cooperation at all levels.
2. That our links with other international organisations and subject disciplines should be strengthened and affiliations developed.

3. That geomorphology needs a separate organisation in order to strengthen its negotiating position within science, education, politics and government.
4. Slow, careful and considered progress was emphasised.

The following recommendations were accepted:

1. A large national geomorphological organisation should be asked to organise the Second International Geomorphological Conference in 4–5 years time. The date of 1989 was accepted.
2. An international committee should be set up to investigate and solve the problems associated with choosing the correct organisational form. The committee should produce positive proposals to be presented at the Second International Geomorphological Conference. Professor D. Brunsden was asked to nominate a committee of not more than 10 persons.
3. That an international newsletter be produced to report on matters of international interest.

These recommendations were implemented in the following ways:

1. A formal invitation was given by and accepted from the West German delegation to hold the Second International Conference on Geomorphology in 1989 at Frankfurt. The organiser is Professor A. Semmel. Initial correspondence should be addressed to Professor Dr D. Barsch, Geographisches Institut, Im Neueuheimer Feld, 69, Heidelberg; or Professor H. Bremer, Geographisches Institut, Albertus-Magnus-Platz, D-500 Köln. We must also record that invitations were also offered by Italian, Australia–New Zealand and Japanese delegates to whom we are very grateful.
1. The following working party was established:

   Working Committee: Denys Brunsden (Chairman), Jesse Walker (Secretary), Hanna Bremer, Jane Soons, Theo Verstappen, Wang Nailing and Stepan Kozarski.

   Advisory Associates: V. Baker (Arizona), A. Condrero (Spain), A. Faniran (Nigeria), J. C. Nansen (Australia), S. Okuda (Japan), D. St Onge (Canada), M. Panizza (Italy), M. Pecsi (Hungary), J. de Ploey (Belgium), R. K. Rai (India), A. Rapp (Sweden), J. Tricart (France), A. Zhivago (USSR).

   Corresponding Members (List still open for nominations from countries): M. T. Benazzouz (Algeria), H. Fischer (Austria), L. Peeters and A. Pissart (Belgium), J. P. Queiroz Neto (Brazil), A. Heginbottom and O. Slaymaker (Canada), J. F. Araya (Chile), Shi Yafeng (China), M. Hermelin (Columbia), M. Stankoviansky (Czechoslovakia), J. Kruger (Denmark), D. Barsch (Fed. Republic of Germany), M. Seppala (Finland), H. Maroukian (Greece), E. Koster (Holland), C. L. So (Hong Kong), A. S. K. Nair (India), A. Yair (Israel), T. Mizuyama (Japan), D. E. Kapula (Kenya), A. Lopez (Mexico),

M. E. Sullivan (Papua New Guinea), C. A. Crelhol (Portugal), I. Ichim (Romania), P. P. Wong (Singapore), B. P. Moon (South Africa), P. Lopez-Bermudez (Spain), C. M. Madduma-Bandara (Sri Lanka), A. Rapp (Sweden), O. Erol and M. Karabiyikoglu (Turkey), K. J. Gregory (UK), A. A. Abrahams (USA), L. Vivas (Venezuela), I. Gams (Yugoslavia).

3. An international newsletter has been established that gives all details of these meetings as well as questionnaires for a World Dictionary of Geomorphologists. Copies obtainable from Professor H. J. Walker (Louisiana).

These successes are having many further consequences. Future newsletters are to be published in the journals *Zeitschrift für Geomorphologie* and *Earth Surface Processes and Landforms*. Letters of interest or national committees have been nominated for the USSR, Belgium and China and France has held meetings to establish an Association Française de Géomorphologie. It seems certain that the lasting influence of Manchester will be a strong international geomorphological community.

CONCLUSION

These successes and this volume have only been possible because of the hard work and financial support of many individuals. I have tried to list as many as possible in this report but I am aware that many people must remain unmentioned. We must acknowledge:

The British Geomorphological Research Group
The Royal Society
The Royal Geographical Society
The Geological Society of London
The Institute of British Geographers
The US Army European Research Office
The US Army Research, Development and Standardisation Group, UK
The UNESCO Major International Project on Research and Training leading to the Integrated Management of the Coastal Systems (COMAR)
The Department of Education and Science, UK
The British Council
The Great Britain–China Educational Trust
Geomorphological Services Ltd
Zeitschrift für Geomorphologie
George Allen & Unwin
Methuen & Co. Ltd
Edward Arnold (Publishers) Ltd
John Wiley & Sons Ltd

Longman Group Ltd
British Airways
University of Manchester
Mr Phil Blinkhorn
The Cooperative Bank

who sponsored the conference and made it possible to use nearly £15,000 to entertain visitors from the Developing World. The receptions given by the Department of Geography, the BGRG, Methuen & Co. Ltd, and the University of Manchester at the Whitworth Art Gallery, the rather festive conference dinner in the Haworth Room of the University, the distinguished guest speakers Sir George Bishop CB, OBE, President of the Royal Geographical Society, Professor Hanna Bremer (University of Cologne) and Professor Liu Tungsheng, Secretary General, China Association of Science and Technology, Academia Sinica, Beijing and the friendly atmosphere generated by all the participants have left warm memories.

May I thank, on your behalf, all the helpers at the conference; the Department of Geography at the University of Manchester; Vince Gardiner and the session coordinators and editors for their painstaking work on this volume; and above all Professor Ian Douglas for this remarkable achievement. See you in Frankfurt?

Denys Brunsden
Chairman BGRG, 1985

THEORETICAL GEOMORPHOLOGY

International Geomorphology 1986 Part II
Edited by V. Gardiner

PAPERS ON THEORETICAL GEOMORPHOLOGY: INTRODUCTION

Mike Kirkby

School of Geography, University of Leeds,
Leeds, LS2 9JT, England

Theory may be described as any abstract explanation of phenomena. Low level theories are essentially derived from particular data sets, and only exceptionally have application beyond that data set: such theories are generally derived using statistical methods, most commonly varieties of multiple regression. Higher level theories are less closely related to particular data, and are intended to have wider applicability. Greater generality is generally obtained at the expense of precise application to any individual case, even where quantitative forecasts are made. The power of a theory is not a precise measure, but can be considered as describing their applicability, allowing for this trade-off between range and reliability. Two theories might thus be equally powerful where one gives very reliable forecasts for one site, and the other gives general guidance over a very wide range of sites and conditions. In either case a good theory should preferably be based on general physical, chemical or biological principles.

Twelve papers have been published from those presented orally and through posters in the 'Theoretical Geomorphology' session at the Manchester conference. They represent some variety in their range of applicability, although the largest group (6) are concerned with the analysis of drainage basin data. Three papers are concerned with broader issues related to morphometric analysis, and the remaining three show some variety in style and topic.

Within the group of papers on drainage basins, two are concerned with internal correlations. Ebisemiju has analysed data for 30 variables to compare two lithologies in south eastern Nigeria. De Villiers found that only 16 of his original 59 variables were needed to describe the variability within 167 second order catchments. Two papers by Ciccacci and co-workers describe analysis of catchment data from Italy. The first forecasts suspended sediment yield for twenty catchments over a wide span of conditions: drainage density is found to be the best single predictor. The second paper examines channel directions around Monti Sabatini in relation to fracture trends, using stream order as an indicator for age of the streams, and hence fractures. Sharma's paper on Rajasthan in north west India explores the relationships between climate and morphometric properties. The final drainage basin paper, by Parsons, uses data from both Britain and Australia to look at changes in slope form down the length of a valley. The data are analyzed in terms of changing slope-base boundary conditions.

Among the more general papers on morphometry, Evans' on 'The morphometry of specific landforms' is the most general. It offers guidance about both conceptual stages of an analysis, and the kinds of measures needed to describe a landscape, but avoids explicit advice on the use of particular indices. Hendriks **et al** examine the problem of determining grid scales for hydrological regionalization schemes, comparing the different criteria commonly used, for a data set from Luxembourg. An attempt is made to reduce the degree of subjectivity in choosing grid size. Abrahams' paper is concerned with differences from the random network model to meet the constraints of space filling, particularly where adjacent tributaries join the main stream on the same side. He argues that more discrepancies will be demonstrated as appropriate tests are devised and refined.

Ahnert's paper aims to simplify the analysis of magnitude frequency relationships by approximating many climatic variables to an exponential distribution. He demonstrates that this provides a good fit for daily and short-period rainfall, frost frequency and wind velocities. Kirkby and Neale propose a generalized soil erosion model, combining monthly water balances, a vegetation growth model and existing empirical estimators for sediment yield. This is allowed to stabilize towards an equilibrium natural vegetation cover, giving seasonal variations in both cover and erosion rates. The final paper, by Gu **et al**, applies Prigogine's work on open system thermodynamics to dissipative structures within a conceptual river system. They propose a maximum-minimum criterion for energy dissipation in a river, which also provides criteria for judging whether a river is in equilibrium.

The overall balance of these papers shows that the state of theory in geomorphology is far from satisfactory. Although multivariate studies are useful as a forecasting basis within their data range, they rarely transcend it or lead to more general principles about, for example, the way in which drainage basins and networks evolve or function. There is a pressing need for such theories which, even if initially incomplete, could lead to a better understanding of the impact of changes in land use or climate. The greatest current progress in theory is in understanding process mechanics, and evidence of this progress is reported within other sessions of the conference. Our efforts are however much more ill-formed and tentative in re-applying process understanding to the development of theories about landscape form, and it is in this area that almost everything must still be done.

International Geomorphology 1986 Part II
Edited by V. Gardiner

ENVIRONMENTAL CONSTRAINTS ON THE INTERDEPENDENCE OF DRAINAGE BASIN MORPHOMETRIC PROPERTIES

F. S. Ebisemiju

Department of Geography, Ondo State University,
Ado-Ekiti, Ondo State, Nigeria.

ABSTRACT

Analysis of the intercorrelation structures of 30 morphometric properties of monolithologic mature drainage basins from two contrasting environments within the Mamu River Basin in south-eastern Nigeria supports the hypothesis that the nature of an environment is of considerable importance in determining the pattern and sensitivity of the interaction amongst morphometric variables. Although the underlying dimensions of drainage basin morphology and their defining parameters are similar in both the sandstone and clay shale environments, the interaction amongst the relief/slope variables is relatively stronger in the sandstone environment. The interactions between basin size and drainage texture as well as between the drainage texture and relief/slope parameters are stronger in the clay shale basin group. The observed differences are attributed to variations in the dominant runoff processes that control stream channel initiation and basin development, and in the extent of floodplain development in the two environments. The lumping together of observations from the two environments produces some important differences in the sensitivity of the links between the variables. It is concluded that the differences between findings reported in the literature are attributable more to environmental constraints than to differences in the mix of variables. It is argued, therefore, that the findings of applied morphometric intercorrelation studies can be valid only if the observations are from a homogeneous environment.

INTRODUCTION

Investigations carried out in different parts of the world have revealed significant climatic and lithological controls on the morphology of drainage basins (see, for example, Chorley, 1957; Chorley and Morgan, 1962; Morisawa, 1962; Ebisemiju, 1976a). Also many of the linear, areal and relief aspects of drainage systems have been shown to be strongly interdependent. Attempts have been made to examine the intercorrelation structure of morphometric parameters, identify their underlying dimensions and reduce the battery of intercorrelated parameters to a few irreducible elements that adequately simulate drainage basin morphology (Mather and Doornkamp, 1970; Gardiner, 1978; Ebisemiju, 1979a, 1979b). An examination of the findings, however, reveals differences in the number, identity and

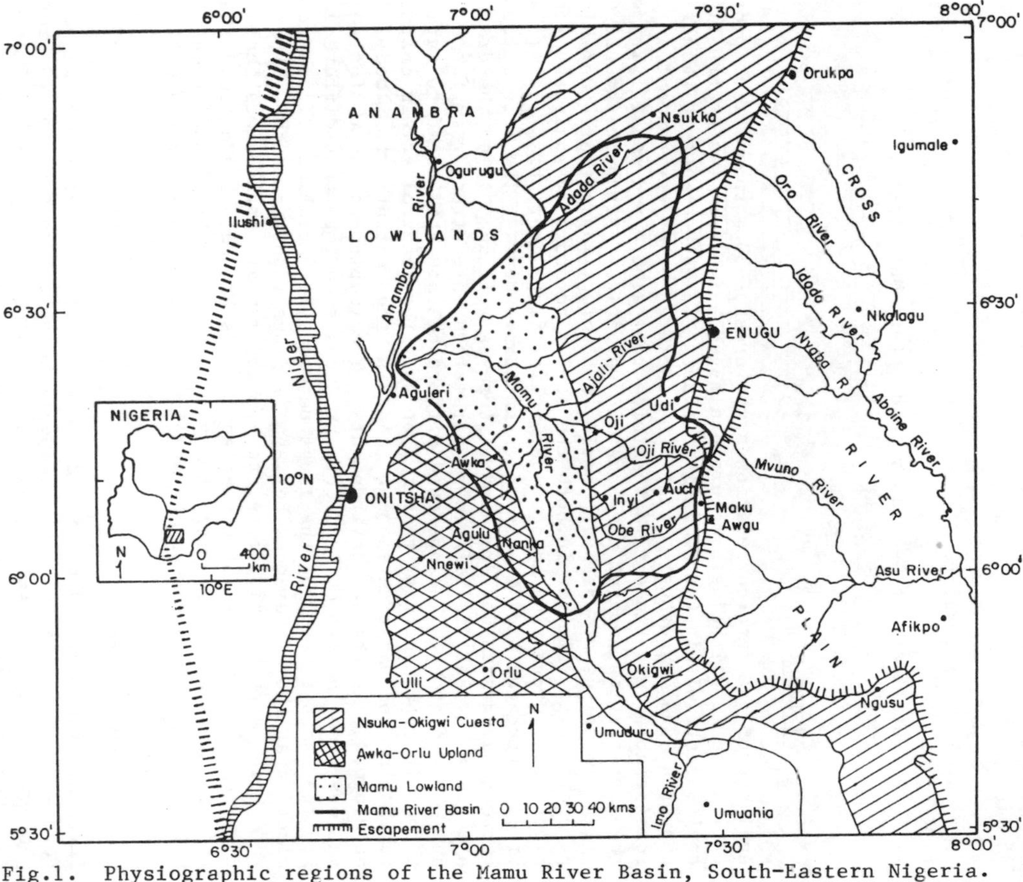

Fig.1. Physiographic regions of the Mamu River Basin, South-Eastern Nigeria.

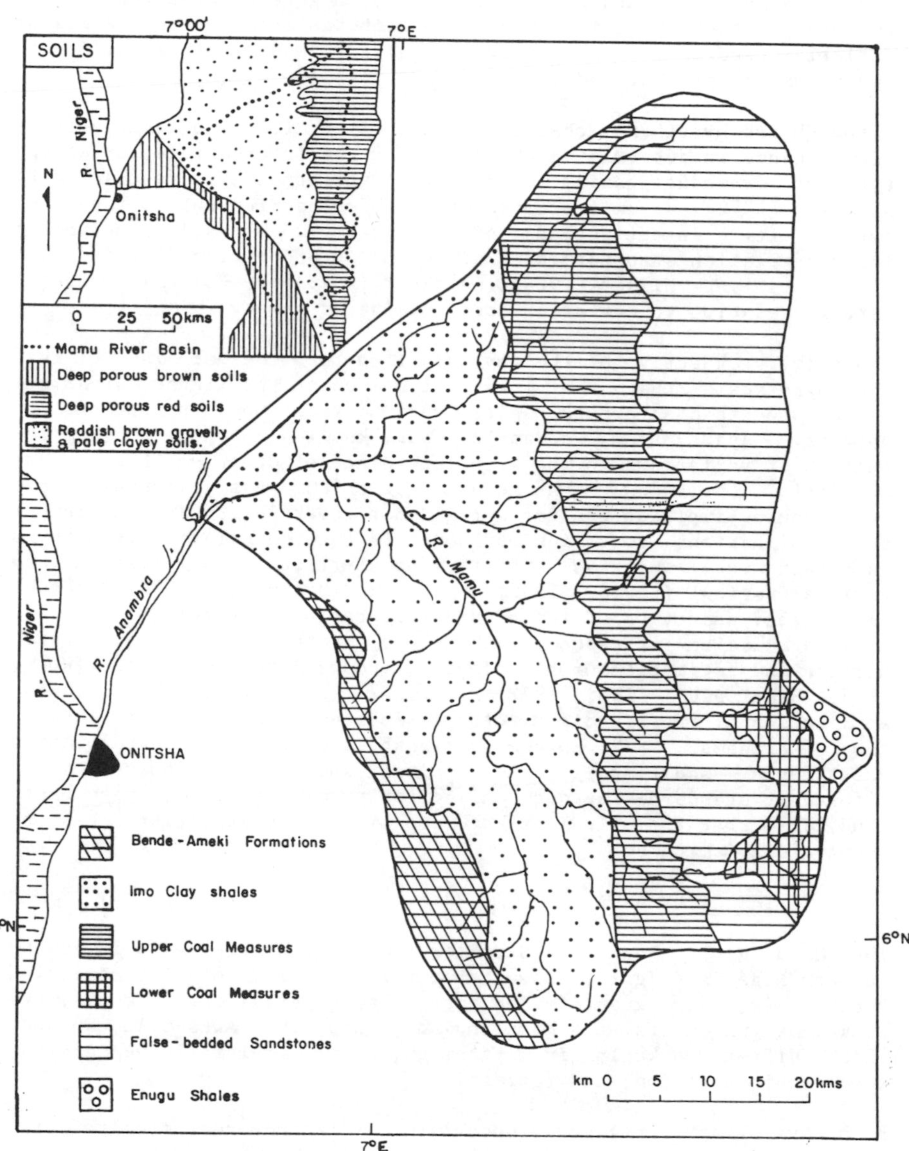

Fig.2. Geological formations and major soil groups, Mamu River Basin, South-Eastern Nigeria.

relative importance of the underlying dimensions and their defining variables, as well as in the interaction patterns of some morphometric variables. Gardiner (1978, p.425), suggested that the "differences between studies may be attributable to either the mix of variables used or to the geomorphological conditions specific to individual studies or indeed, to both".

Although the hypothesis that the nature of an area is of considerable significance in determining the nature of the intercorrelation structure of drainage basin morphometric variables was postulated by Abrahams (1972), no empirical study has been carried out to impugn or support it. Investigations of the pattern of variations in the intercorrelation structure of drainage basin properties carried out by Onesti and Miller (1974), Gardiner (1978) and Ebisemiju (1985a) were in relation to different order basins or sizes.

The primary objective of the investigation reported here is to test the hypothesis that the interaction amongst the linear, areal and relief aspects of drainage basins varies from one environment to another. Apart from advancing our knowledge about the spatial pattern of variations in the interdependence of drainage basin morphometric attributes, the results of this investigation will have far-reaching implications for applied morphometric studies in view of the applications to which morphometric intercorrelation structures have been put, such as multivariate terrain classification and regionalisation (Lewis, 1969; Mather and Doornkamp, 1970; Blake et al., 1973), selection of representative basins (Ebisemiju, 1979b), the formulation of reduced rank multiple regression models for prediction and estimation of hydrological and geomorphic events (Wong, 1963; Haan and Allen, 1972; Rice, 1973; Ebisemiju, 1976a), in the application of the ergodic hypothesis in studies of landscape evolution (Abrahams, 1972; Ebisemiju, 1985b) and in environmental impact assessments. Additionally, the findings should contribute towards efforts at standardising data collection procedures in drainage basin studies so that the findings of empirical investigations can be meaningfully compared.

THE STUDY AREA

The Mamu River Basin, a Strahler order-7 drainage basin in southeastern Nigeria (Fig.1), is suitable for this kind of investigation because it can be subdivided into three physical environments, the Nsukka-Okigwi Cuesta, the Mamu Lowland, and the Awka-Orlu Upland, which differ markedly in lithology, soil, rainfall, vegetation, topographic position, and terrain.

Each physiographic region is underlain by distinct geological formations (sandstones and clay shales) which have been weathered to produce equally distinct regolith (Fig.2). The Awka-Orlu Upland is associated with the Bende-Ameki Formation, the Mamu Lowland with Imo Clay Shales, while the Nsukka-Okigwi Cuesta is underlain by sandstone formations (Upper Coal Measures, False-bedded Sandstones, Lower Coal Measures, and Awgu Sandstones (Fig.2). The geological evolution of the area and the structure and lithology of the formations have been

TABLE 1. Lithology and topography of the Mamu River Basin, South-Eastern Nigeria.

Period	Age	Formation	Lithology	Topography
TERTIARY	Miocene	Bende-Ameki Formation	Coarse sandstones, white to greenish clayey sandstones, grey-green argillaceous sandstones, shales, sandy shales, clays, mudstones and thin limestones.	Dissected plateau and ridge; active gully erosion and mass wasting. (Awka-Orlu Upland)
TERTIARY	Lower Eocene	Imo Clay Shales	Clay-shales, with intra-formational sandbodies (sandstone units which appear at several horizons)	Undulating to nearly level plain with low ridges (less than 120m above sea level). - Mamu Lowland
TERTIARY / CRETACEOUS	Palaeocene / Maes-trichtian/ Campanian and (?) Santonian	Upper Coal Measures	White to grey, coarse to medium-grained sandstones, carbonaceous shales; sandy shales; subordinate coals; limestones. Predominantly siltstones.	Maturely-dissected dipslope of the Nsukka-Okigwi Cuesta. Forms residual hills on the Nsukka-Udi Plateau
CRETACEOUS	Maes-trichtian/ Campanian and (?) Santonian	False-bedded Sandstones	Medium to coarse-grained cross-bedded sandstones sometimes poorly consolidated, with subordinate white and pale grey shale bands. Very permeable, coarse friable sandstone.	Nsukka-Udi Plateau, with residual hills and numerous dry valleys.
CRETACEOUS	Maes-trichtian/ Campanian and (?) Santonian	Lower Coal Measures	Fine to coarse-grained, white to grey sandstones; shaly sandstones and sandy shales; subordinate carbonaceous shales and coals. Resistant siltstones and sandstones.	Dissected and rugged scarp slope of the Nsukka-Okigwi Cuesta.

studied in great detail (see, for example, Wilson, 1925; Wilson and Bain, 1928; Simpson, 1954; Iloeje, 1961a; Jungerius, 1964a). The lithology of each formation and the associated terrain characteristics are detailed in Table 1. Most importantly, no flexures have been found in any of the geological formations, and only a few small faults have been located in the sandstone formations; consequently, there is no structural control on the evolution of drainage networks and surface forms. Differences in the morphology of the drainage basins and in the interdependence of their morphometric parameters, therefore, are attributable primarily to variations in lithology, soil, and the nature of geomorphic processes.

The hot humid climate has favoured rapid disintegration of these sedimentary rocks, resulting in the formation of a mantle of soils and earth materials whose textural, geotechnical, and mineralogical properties are controlled by the lithological composition of the parent rocks (Obihara, 1961; Ofomata, 1964; Jungerius, 1964b, 1965; Jungerius and Levelt, 1964) hence the boundaries of the major soil groups coincide approximately with those of the geological formations (Fig.2). The sandstone formations which underlie the Nsukka-Okigwi Cuesta and the sandstone and shale formations of the Awka-Orlu Upland give rise to very deep, porous ferrallitic soils known as Acid Sands (Obihara, 1961). The mantle of these coarse sandy materials is so thick that the parent rocks are rarely exposed except along the sides of deeply incised valleys and scarps. On the other hand, hydromorphic soils with gravelly and loamy topsoils overlying a clay substratum are associated with the clay shales of the Mamu Lowland. Prolonged subaerial erosion has, however, stripped away much of the weathered materials leaving a relatively thin veneer of regolith over the underlying soft impervious clay shales. Undecomposed clay shale is often encountered within a depth of 2m (Jungerius, 1964b), and the soils are generally waterlogged in the wet season. These differences in the properties of bedrocks and weathered mantles in the two regions are likely to induce spatial variations in the manner in which water accumulates on the surface, to generate the conditions necessary for channel initiation, growth, and integration.

The dissected and rugged upland topography of the Cuesta contrasts markedly with the low undulating to nearly level plain of the Mamu Lowland with its low ridges of terrace sands, wide shallow valleys and relatively lower drainage density (Tables 1 and 2); see also Iloeje, 1961a, 1961b; Ofomata, 1967; Uzoroh, 1968). The western rim of the Mamu River Basin is formed by the much-eroded scarp of the Awka-Orlu Upland, an area of active gully erosion and mass wasting, especially earth slumping and slippage (Ofomata, 1964, 1967).

The vegetation of the Nsukka-Okigwi Cuesta and the Awka-Orlu Upland is derived savanna dominated by grassland and scrub, except along river courses and around settlements where the primeval moist evergreen forest is still found. The Mamu Lowland, on the other hand, carries a denser savanna woodland and forest vegetation.

Although the three physiographic regions fall within the same climatic belt (Koppen's Awi climate) with a uniformly high mean monthly temperature of about 27°C and a seasonal distribution of rainfall (25mm to 340mm per month) which totals over 1,500mm a year, the number of rainy days per year varies significantly within the drainage basin from a mean value of 51 in the Mamu Lowland to 65 in the Awka-Orlu Upland and 158 in the Nsukka-Okigwi Cuesta. The violent storms of short duration, characteristic of the entire area, occur therefore more frequently in the cuesta and are also more erosive here because of the sparser vegetation cover.

The drainage basins are selected from the Mamu Lowland (Clay Shales) and the Nsukka-Okigwi Cuesta (Sandstones). Analysis of variance reveals significant differences at the 0.01 level of significance in

TABLE 2. Drainage basin morphometric properties analysed, mean values and transformations.

Morphometric Property	Symbol	Mean Values: Sand-stone	Mean Values: Clay Shale	Transformations: Sand-stone	Transformations: Clay Shale	Transformations: All
Basin area (km^2)	A	2.669	7.380	log	log	normal
Basin perimeter (km)	P	6.926	11.282	log	log	normal
Basin length (km)	Lb	2.648	4.328	log	log	normal
Basin circularity ratio	Rc	0.682	0.711	log	normal	log
Number of exterior links	ne	10	12	log	log	log
No. of 2nd. order stream segments	N2	3	4	log	log	log
Total number of stream segments	ΣN	14	17	log	log	normal
Bifurcation ratio (1:2)	Rb	3.702	3.324	log	log	log
Total length of ext. links (km)	le	4.247	7.737	log	log	log
Total length of int. links (km)	li	3.167	6.211	log	log	log
Total stream length (km)	ΣL	7.414	13.948	log	log	log
Length of the mainstream (km)	Lm	2.811	4.904	log	log	normal
Average length of ext. links (km)	$\bar{l}e$	0.423	0.675	log	log	log
Average length of int. links (km)	$\bar{l}i$	0.352	0.565	log	log	log
Average length of 2nd. order stream segments (km)	$\bar{L}2$	0.658	0.979	log	log	log
Link length ratio ($\bar{l}e$: $\bar{l}i$)	λR	1.202	1.195	log	sq.rt.	sq.rt.
Stream length ratio (2:1)	RL	1.534	1.492	log	normal	log
Drainage texture ratio	Rdt	1.366	0.815	log	log	log
Total drainage density (km^{-1})	Dd	2.906	2.020	normal	normal	log
Density of exterior links (km^{-1})	De	1.674	1.126	normal	normal	log
Total stream segment frequency (km^{-2})	Fss	6	3	sq.rt.	normal	log
Frequency of ext. links (km^{-2})	Fe	4	2	sq.rt.	sq.rt.	log
Drainage intensity	DI	18.419	7.448	log	sq.rt.	log
Length of overland flow (m)	Lg	179.2	264.2	log	log	log
Basin relative relief (m)	Hb	135.4	52.9	sq.rt.	log	log
Main channel fall (m)	Hcm	119.8	44.6	sq.rt.	log	log
Relief ratio	Rh	0.052	0.013	sq.rt.	log	sq.rt.
Average ground slope (sin.)	θg	0.192	0.013	sq.rt.	log	sq.rt.
Mean max. valley-side slope (sin.)	θm	0.338	0.066	normal	normal	log
Mean stream channel slope (sin.)	θcm	0.041	0.010	normal	log	log

the mean values of most of the morphometric parameters of the basins in the two lithological and physiographic regions (Table 1). In view of the marked contrasts in the lithology, soils, vegetation, topographic position, and morphology of the two sets of drainage basins, it is likely that the intercorrelation structures of their morphometric parameters are also significantly different.

Although in this paper the two groups of catchments are named after their respective lithologies, the preceding analyses clearly show that they can be equally differentiated on the basis of spatial scale and other environmental factors such as topographic positions (upland and lowland basins), and vegetation type (savanna and forest basins). These other factors, no doubt, confound the lithological differences in varying degrees. This research, therefore, does not focus on the isolated effect of lithology on morphometric interaction patterns; rather, it seeks to produce reduced rank models of the intercorrelation structures of drainage basin morphology in the two regions and explain the observed similarities and differences in terms of the particular ways in which the environmental factors operate and are combined to determine form-form interaction patterns in each region. The observed differences in morphometric interaction patterns, therefore, are functions of lithology as well as catchment position, vegetation, and rainfall intensity.

DATA COLLECTION AND ANALYSIS

The drainage network on the 1:50,000-scale topographical maps was augmented by drainage channels identified from interpretation of contour patterns following the system outlined by Morisawa (1957). These were checked by both a stereoscopic examination of 1:40,000-scale aerial photographs and field survey to produce the complete drainage network which is not significantly different from the analysis based upon contour criteria (Ebisemiju, 1976b). The Strahler hierarchical ordering scheme was adopted in classifying the streams and their drainage basins (Strahler, 1952).

One hundred and eight monolithologic third-order drainage basins were selected randomly, 62 from the sandstones of the upland environment, and 46 from the clay shales of the lowland environment; these represent 78% of all monolithologic third-order basins in each geological formation. Our interest is on the pattern of variations in the intercorrelation structure of basin morphometric properties, so it is important that all major linear, areal and relief attributes of drainage basins are quantified and that the morphometric parameters that have featured prominently in most geomorphological and hydrological investigations are included for analysis. However, variables which attempt to summarize in one index the gross morphology of drainage basins, such as the ruggedness number, geometry number, and the texture/slope product were excluded because the focus is on the interaction amongst the elements of the system. Thirty parameters, covering a wide range of basin properties, were measured (Table 2). Lengths and areas were measured with an opisometer and compensating polar planimeter respectively, while the average ground slope of each basin was estimated from the 1:50,000-scale topographical maps by the Wentworth method (Wentworth, 1930).

For each of the two data sets (sandstone and clay shale basin groups) as well as for the combined data the distribution of each variable was tested for normality using Snedecor's (1956) test for symmetry and kurtosis. Most of the variables have varying degrees of skewness and kurtosis. For these variables, logarithmic and square root transformations produced distributions that are not significantly different from the normal at the 0.01 level (Table 2). Each data matrix was then subjected to principal components analysis, and those components having eigenvalues greater than 1.0 were rotated using the Varimax method in order to produce an orthogonal transformation of the components and simplify interpretation. The loadings of the variables on the factors are presented in Table 3, while the rotated factor solutions are summarised in Table 4. These and the simple correlation matrices (Fig.3) are examined for differences and similarities in the intercorrelation structures of the morphometric variables.

Comparative analysis. Table 4 reveals that the total variances accounted for by the six factors and the underlying dimensions of drainage basin morphology are similar for both groups of drainage basins in spite of the significant differences in their lithological characteristics, soil properties, vegetation cover type and density, terrain and topographic position. Similar underlying dimensions - texture of dissection, stream network size, relief/slope, link length ratio, basin shape and mean length of exterior links - have been reported for the third-order drainage basins of the Bruce Vale River Basin, Barbados and the Great River Valley Basin in Tobago (Ebisemiju, 1979a). Even when the two sets of basins were lumped together for factor analysis, the same underlying dimensions emerged (Tables 3 and 4). These findings suggest that the morphology of drainage basins can be adequately described by the defining variables of the six factors, viz: total drainage density, total stream length, basin relief, link length ratio, circularity ratio, and the mean length of exterior links.

Another similarity is in the interaction of the relief/slope variables. In both environments, this set of variables have their highest loadings on Factor III which, therefore, identifies a relief/slope dimension. However, the correlation matrices (Fig.3) reveal that the strength of the interaction amongst these slope parameters is relatively weaker in the clay shale environment where the coefficient of correlation of average ground slope and mean stream channel gradient is 0.759 compared to 0.941 in the sandstone environment. This is attributable to differences in floodplain deposition in the two regions. The extent of floodplain or valley fill deposition has been shown to be a major constraint on the interaction between stream channel and valley side slopes (Richards, 1977). The rugged terrain of the maturely dissected zones of the Nsukka-Okigwi Cuesta is characterised by short, steep rectilinear valleyside slopes which merge abruptly with the basal stream channels. Floodplain or valley fill deposition is generally absent.

TABLE 3. Rotated factor loading matrices of the morphometric properties of third-order drainage basins, Mamu River Basin.

Morphometric Properties	FACTORS																	
	Sandstone Basins						Clay Shale Basins						All Basins					
	I	II	III	IV	V	VI	I	II	III	IV	V	VI	I	II	III	IV	V	VI
A	52	80	-13	-18	-08	18	74	60	-40	-13	-02	-22	72	54	-35	17	00	-02
P	48	78	-08	-16	-32	17	71	57	-36	12	-03	-22	71	55	-30	17	-20	00
Lb	47	71	-06	-13	-38	04	70	43	-40	07	-09	-03	70	49	-31	10	-31	06
Rc	-08	-29	-12	01	94	-03	-10	-07	-14	03	97	05	02	-05	-20	-01	96	-08
ne	-40	88	-09	21	-02	02	-16	87	-13	04	-23	05	-17	97	-07	01	-02	16
N2	-34	79	-25	23	05	-07	-07	89	-11	10	31	05	-05	93	-18	-09	02	-31
ΣN	-40	88	-13	21	00	00	-15	90	-12	06	11	05	-15	98	-10	-01	-01	05
Rb	-11	14	33	-06	-14	10	-17	10	-04	10	-07	-01	-24	04	23	20	-09	92
le	10	93	-02	-31	-01	00	21	93	-05	04	13	13	37	80	-26	02	01	07
li	31	86	20	-08	00	10	31	87	19	12	-10	-02	47	52	-10	68	01	-07
ΣL	07	95	-01	-19	-11	18	18	94	-11	06	-05	-18	46	77	-26	17	-06	02
Lm	37	73	00	-18	-39	07	60	50	-30	10	-02	-11	64	53	-28	-12	-29	09
Īe	47	16	12	-86	01	04	55	24	30	04	07	-76	71	11	-29	00	04	-06
Īi	55	32	18	-09	00	53	47	40	19	72	10	-21	46	10	22	60	10	00
Ī2	51	13	34	-21	-03	18	53	48	20	-38	-05	-16	54	50	01	-20	00	11
λR	-53	12	08	-12	19	-80	-55	11	02	-78	22	-24	-38	11	13	-80	17	03
RL	48	08	34	11	35	23	56	42	16	33	-12	20	78	-37	39	04	-03	17
Rdt	-90	08	-03	05	06	-17	-72	23	27	-18	13	13	-82	13	26	-21	13	12
Dd	-96	06	25	02	-02	-05	-95	00	33	-13	-05	02	-83	01	38	-10	-09	09
De	-90	06	13	-18	17	-26	-91	11	16	-30	14	-16	-87	05	29	-29	05	12
Fss	-89	13	03	37	08	-19	-89	-03	31	-17	03	27	-89	-07	33	-19	00	05
Fe	-90	-06	06	37	07	-17	-88	-02	35	-17	11	27	-94	-03	20	-17	-01	11
DI	-94	-07	11	26	05	-15	-92	-01	28	-23	18	11	-89	-05	35	16	-03	06
Lg	90	-16	-20	08	03	09	93	-01	-31	15	-05	-02	82	00	-37	11	10	-03
Hb	09	15	94	-08	-13	14	21	17	95	03	-04	-03	-26	-10	90	07	-13	11
Hcm	07	08	94	-07	-08	15	18	16	92	00	-10	-13	-17	-11	94	08	-04	01
Rh	-14	-25	93	-02	10	11	-32	-12	92	00	-05	-08	-45	-28	83	03	05	06
θg	-18	-17	90	-02	-04	03	-38	-19	71	10	-27	-18	-49	-27	76	-07	-05	15
θm	-15	08	92	-03	-01	-02	-46	-22	84	06	-02	-14	-50	-24	78	-06	-06	11
θcm	-09	-36	87	10	06	12	-47	-03	83	05	00	-06	-47	-35	78	00	00	00
Eigenvalue	8.72	7.56	5.75	1.62	1.57	1.40	9.87	7.61	5.95	1.76	1.40	1.20	10.63	6.19	5.86	1.33	1.26	1.15
% of total variance	31.28	27.13	20.63	5.81	5.63	5.02	33.57	25.89	20.34	5.99	4.76	4.08	37.97	22.11	20.93	6.39	4.50	4.14

TABLE 4. Summary of Varimax rotation solutions.

Factor	Sandstone Basins: Diagnostic Variables*	Sandstone Basins: Identity	Sandstone Basins: % Var.	Clay Shale Basin: Diagnostic Variables*	Clay Shale Basin: Identity	Clay Shale Basin: % Var.	All Basins: Diagnostic Variables*	All Basins: Identity	All Basins: % Var.
I	Dd, DI, Lg, Rdt, De, Fe, Fss	Texture of dissection	31.28	Dd, Lg, DI, De, Fss, Fe, A, Rdt, P, Lb	Texture of dissection	33.57	Fe, DI, Fss, De, Dd, Lg, Rdt, RL, A, P, $\overline{le}$, Lb	Texture of dissection	37.97
II	ΣL, le, ΣN, ne, li, A, N2, P, Lm, Lb	Stream network size	27.13	ΣL, le, ΣN, N2, ne, li	Stream network size	25.89	ΣN, ne, N2, le, ΣL	Stream network size	22.11
III	Hb, Hcm, Rh, θm, θg, θcm	Relief/ Slope	20.63	Hb, Hcm, Rh, θm, θcm, θg	Relief/ Slope	20.34	Hcm, Hb, Rh, θm, θcm, θg	Relief/ Slope	20.93
IV	λR	Link length ratio	5.02	λR, $\overline{li}$	Link length ratio	5.99	λR	Link length ratio	6.89
V	Rc	Basin shape	5.63	Rc	Basin shape	4.76	Rc	Basin shape	4.50
VI	$\overline{le}$	Mean length of exterior links	5.81	$\overline{le}$	Mean length of exterior links	4.08	Rb	Bifurcation ratio	4.14

* Variables with high loadings (70 and above) are arranged in descending order of importance, the first being the factor-defining variable.

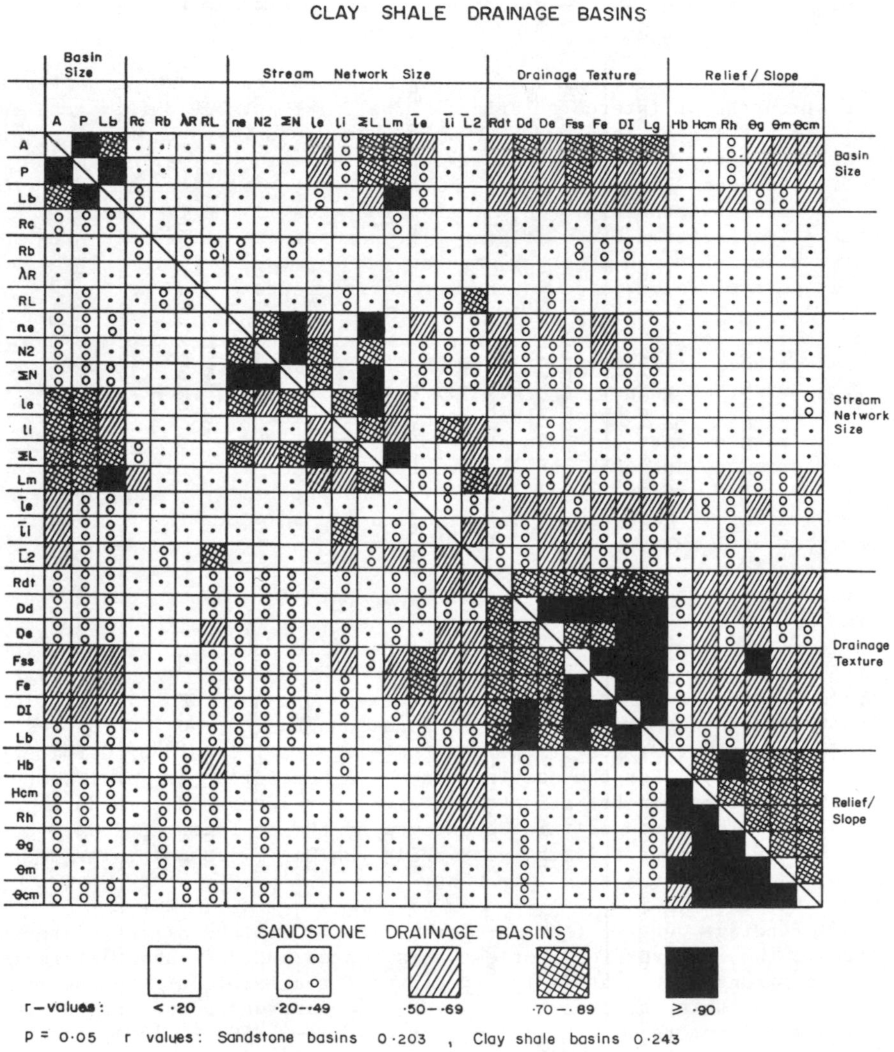

CLAY SHALE DRAINAGE BASINS
Basin Size
Stream Network Size
Drainage Texture
Relief / Slope
A P Lb Rc Rb λR RL ne N2 ΣN le li ΣL Lm l̄e l̄i L̄2 Rdt Dd De Fss Fe DI Lg Hb Hcm Rh θg θm θcm
Basin Size
Stream Network Size
Drainage Texture
Relief/ Slope
SANDSTONE DRAINAGE BASINS
r-values:
< ·20
·20--·49
·50--·69
·70--·89
≥ ·90
p = 0·05 r values: Sandstone basins 0·203 , Clay shale basins 0·243

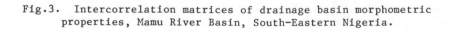

Fig.3. Intercorrelation matrices of drainage basin morphometric properties, Mamu River Basin, South-Eastern Nigeria.

Therefore, a strong link exists between valley-side slope and the slope of the basal stream channel. In the gently undulating to flat terrain of the Mamu Lowland with long gentle valley-side slopes and wide shallow valleys, however, floodplain deposition is found adjacent to even some second order stream segments. Alluvial terraces and ridges of terrace sands are also widespread on the floodplains. The interdependence of the slope variables in the Mamu Lowland basins, therefore, is not as strong as in the dissected zones of the Cuesta.

The analyses also reveal other important differences in the patterns and strengths of intercorrelations. The first notable difference is in the pattern of loading of the variables describing basin size (A, P, Lb). In the sandstone basin group, this set of variables have their highest loadings on the stream network size factor. In the clay shale environment, however, they load heavily on the texture of dissection factor. These loading patterns suggest differences in the interaction of basin size parameters with stream network size and drainage texture variates in the two environments.

The nature of the relationship between drainage texture and basin area has been controversial (Gregory and Walling, 1973; Pethick, 1975, 1978; Gardiner et al., 1978; Ferguson, 1978; Gerrard, 1978; Richards, 1978). Pethick (1975) argued that the observed dependence of drainage density on basin area is due to the bias of using similar order drainage networks and the lumping together of drainage basins with a wide range of environmental characteristics, and concluded that drainage density and basin area are independent in homogeneous environments but that the relationship is strong and negative in heterogeneous environments. Gardiner et al., (1977), however, contended that "an inverse relationship between drainage density and area may exist even when the sample is derived from basins which are uniform in character" (p.118). The data from the Mamu river basin, with its strikingly contrasting environments, do not support Pethick's hypothesis. The relationship between drainage density and area is strong and negative (-0.489 for sandstone basins; -0.750 for clay shale basins), thus confirming the postulate put forward by Gardiner, et al., (1977). For the combined data - a heterogeneous environment - the relationship between the two areal parameters is even stronger (-0.795). The stronger interdependence of the drainage texture and basin size parameters in the clay shale basins (Fig.3) explains their high loadings on the same component.

These findings suggest that, at least for a given drainage basin order, the interaction between basin area and drainage texture is always strong and negative irrespective of the nature of the environment. In this study area, however, the interdependence of the two sets of parameters is considerably stronger (-0.750, -0.489), in the clay shale environment. It seems, therefore, that the physical principles that govern the interaction of basin size and drainage texture parameters vary between the two environments. Unfortunately we cannot presently explain the observed differences in the interdependence of these sets of plainmetric variables.

Another major difference in the intercorrelation structures of drainage basin morphometry in the two environments is in the interaction of ground slope and drainage texture parameters. Although the rotated factor solutions suggest that ground slope and drainage texture are orthogonal dimensions of basin morphology in both the sandstone and clay shale environments, an examination of the simple correlation matrices reveals moderately strong interdependence of the two sets of variables in the clay shale environment (0.514) but very weak association in the sandstone environment (0.161).

Much controversy also exists among fluvial geomorphologists on the role of ground slope in controlling texture of dissection in fluvially eroded landscapes. Haggett and Chorley (1969, p.150) first put forward the hypothesis that relief or slope may affect drainage density only when the surface is bare and of low resistivity, but that this effect will be masked in vegetated landscapes; the erosion mechanics, however, were not discussed. Dunne (1980) outlined the physical principles and erosion mechanics governing the initiation, growth and integration of stream channel networks by the various types of flow. Abrahams (1980) applied these principles to resolve the confusion over the role of ground slope in controlling drainage texture by demonstrating that "the effectiveness of ground slope in this role appears to vary directly with the proportion of the landscape subject to overland flow" (p.80), thus implicitly lending support to Haggett and Chorley's hypothesis. The deep soils, high infiltration capacities, and dense vegetation cover of the sandstone drainage basins inhibit widespread occurrence of overland flow in the cuesta landscape. Under these conditions, subsurface flow and saturation overland flow are the processes involved in runoff generation and stream channel initiation. Since the proportion of the landscape subject to overland flow is small and the depth of overland flow within the partial areas is not proportional to source-area size, drainage texture should correlate poorly with ground slope. Indeed, the relative permeability of the arenaceous soils derived from the sandstone formations was found to account for 69% of the variation in drainage density with ground slope contributing an extra 4% (Ebisemiju, 1979c).

The moderately strong interdependence of drainage texture and ground slope in the clay shale environment (0.514) suggests that overland flow occurs over a large proportion of the landscape and controls stream channel initiation. This is characteristic of landscapes with shallow soils, low infiltration capacities, and sparse vegetation covers (Abrahams, 1980). Topographic, geologic, and pedologic conditions in the forested Mamu Lowland are conducive to the widespread occurrence of overland flow. In this landscape of low relief and gentle slopes, an impervious layer of undecomposed clay shales underlies, at a depth of about 2 metres, a shallow layer of hydromorphic soils with equally low infiltration capacities. Under these conditions, overland flow occurs rapidly during rainstorms and covers a large proportion of the landscape. Similar conditions exist in the forested clayey drainage basins studied by Bonell, et al., (1984) in Luxembourg. In such landscapes, the "depth of overland flow is proportional to source-area size . . . and hence drainage texture is positively correlated with ground slope" (Abrahams, 1980, p.80). The apparent orthogonality of the two sets of variables in the clay shale basin group is due to the stronger within-group interdependence of the variables in each set rather than any lack of between-group interactions (Fig.3). Perhaps, the interaction between ground slope and drainage texture parameters could have been stronger in this environment but for the high hydraulic resistance of the forest floor and dense grass cover.

This observed difference between reality and the Varimax rotation solutions once again underscores the need to look beyond rotated factor solutions if causal relationships are to be identified (Carson, 1966; Davies, 1971; Daultrey, 1976). This analysis, therefore, reveals a difference in the control of relief/slope on drainage texture in the sandstone and clay shale environments, the interaction between the two sets of variables being stronger in the clay shale environment.

IMPLICATIONS OF THE FINDINGS

Although the above analyses reveal similarities in the underlying dimensions of drainage basin morphology in the two contrasting environments, significant differences in the interdependence of their morphometric parameters are also highlighted. In view of these differences, drainage basins with a wide range of environmental conditions should not be lumped together in studies aimed at highlighting the interdependence of morphometric attributes. The factor solution obtained from the lumping together of data from the sandstone and clay shale environments (Tables 3 and 4) is different in some important respects. For example, average length of exterior links which is an orthogonal dimension of drainage basin morphology in both the sandstone and clay shale environments (Factor VI in both) has its highest loading on the texture of dissection factor (I) in the heterogeneous data set. More importantly, morphometric variables from three orthogonal morphological dimensions of the sandstone basins (I, II, VI) and two of the clay shale basins (I, VI) have their highest loadings on the texture factor. Also, the defining variables for each of the three major factors (I, II, III) are different in the heterogeneous data set (Table 4).

This analysis suggests that different environments produce some variations in the interdependence of drainage basin morphometric properties because the physical principles and the runoff and erosion mechanics governing stream channel initiation, growth, and integration and the development of drainage basin systems vary from one environment to another. When observations from such different environments are lumped together, each region loses its individuality, the differences between regions are masked, and significantly different interaction patterns may emerge. The differences between findings reported in the literature, therefore, are attributable more to environmental constraints than differences in the mix of variables. The findings reported by Mather and Doornkamp (1970) and Gardiner (1978) should be interpreted, therefore, with caution since the drainage basins analysed in each case are from a heterogeneous environment. For example, the orthogonality of the total drainage density and total stream frequency of third-order basins in North-West Devon (Gardiner, 1978) and the low loadings of total stream length on the basin size and stream network size components as well as on all other components in the Uganda study (Mather and Doornkamp, 1970) raise some questions about the experimental design of both researches, and are probably a consequence of the heterogeneity of the environments. Any multivariate terrain classification and regionalisation, selection of representative basins, formulation of reduced rank multiple regression models, studies of drainage basin evolution and other applied morphometric studies based on the inter-correlation structure of such a heterogeneous data, therefore, cannot be optimal.

It is more useful to consider different homogeneous environments, compare the intercorrelation structures of their drainage basin morphometry, explain the observed similarities and differences, and then formulate a general model of drainage basin morphometric interaction founded upon physical principles of runoff and erosion mechanics.

REFERENCES

Abrahams, A. D., 1972. Factor analysis of drainage properties: evidence for stream abstraction accompanying the degradation of relief. Water Resour. Res., 8, 624-633.

Abrahams, A. D., 1980. Channel link density and ground slope. Annals Assoc. Am.Geog., 70, 80-93.

Blake, G. J., Cook, A. D., and Greenhall, D. H., 1973. The use of principal component factor analysis to establish the uniformity of a hydrological region in Northland, New Zealand, in Symposium on the results of research on representative and experimental basins, pp183-200. IASH-UNESCO, publ.96.

Bonell, M., Hendricks, M. R., Imeson, A. C., and Hazelhoff, L., 1984. The generation of storm runoff in a forested clayey drainage basin in Luxembourg. J. Hydrol., 71, 53-77.

Carson, M., 1966. Some problems with the use of correlation techniques in morphometric studies. Br. Geomorph. Res. Group Occ. Paper 4, 149-166.

Chorley, R. J., 1957. Climate and morphometry. J. Geol., 65, 628-668.

Chorley, R. J., and Morgan, M. A., 1962. Comparison of morphometric features: Unaka Mountains Tennessee and North Carolina, and Dartmoor, England. Geol. Soc. Am. Bull., 73, 17-34.

Daultrey, S., 1976. Principal Components Analysis. Concepts and Techniques in Modern Geography, No.8, Geo Abstracts, Norwich.

Davies, W. K. D., 1971. Varimax and the destruction of generality. Area, 3, 112-118.

Dunne, T., 1980. Formation and controls of channel networks. Prog. in Phys. Geog., 4, 211-239.

Ebisemiju, F. S., 1976a. The Structure of the Interrelationships of Drainage Basin Characteristics. Unpublished Ph.D. Thesis, University of Ibadan, Nigeria.

Ebisemiju, F. S., 1976b. Morphometric work with Nigerian topographical maps. Nigerian Geog. J., 19, 65-77.
Ebisemiju, F. S., 1979a. A reduced rank model of drainage basin morphology. Geogr. Ann., 61A, 103-112.
Ebisemiju, F. S., 1979b. An objective criterion for the selection of representative basins. Water Resour. Res., 15, 148-159.
Ebisemiju, F. S., 1979c. Analysis of drainage density and similar parameters in relation to soil and vegetation characteristics. Nigerian Geog. J., 22, 1, 33-43.
Ebisemiju, F. S., 1985a. Spatial scale and drainage basin morphometric interaction. Catena, in press.
Ebisemiju, F. S., 1985b. A re-evaluation of Hack's general equation expressing drainage basin shape. Nigerian Geog. J., 28, in press.
Ferguson, R. I., 1978. Drainage density - area relationship: comment. Area, 10, 350-352.
Gardiner, V., 1978. Redundancy and spatial organization of drainage basin form indices: an empirical investigation of data from North-West Devon. Trans. Inst. Br. Geog. New Series, 3, 416-431.
Gardiner, V., Gregory, K. J., and Walling, D. E., 1977. Further notes on the drainage density - basin area relationship. Area, 9, 117-121.
Gardiner, V., Gregory, K. J., and Walling, D. E., 1978. Further notes on the drainage density-area relationship: comment. Area, 10, 354-355.
Gerrard, A. J., 1978. Drainage density - area relationship: comment. Area, 10, 5, 352-353.
Gregory, K. J., and Walling, D. E., 1973. Drainage Basin Form and Process: A Geomorphological Approach. Edward Arnold, London.
Haan, C. T., and Allen, O. M., 1972. Comparison of multiple regression and principal component regression for predicting water yields in Kentucky. Water Resour. Res., 8, 1593-1596.
Haggett, P., and Chorley, R. J., 1969. Network Analysis in Geography. Edward Arnold, London.
Iloeje, N. P., 1961a. The structure and relief of the Nsukka-Okigwi cuesta. Nigerian Geog. J., 4, 21-39.
Iloeje, N. P., 1961b. The geomorphology of the Nsukka-Okigwi cuesta. The Nigerian Scientist, 1, 13-23.
Jungerius, P. D., 1964a. The Upper Coal Measure Cuesta in eastern Nigeria. Zeit. für Geomorph. Suppl., 5, 167-176.
Jungerius, P. D., 1964b. The soils of eastern Nigeria. Publication Service Geologique du Luxembourg, XIV, 185-198.
Jungerius, P. D., 1965. Some aspects of the geomorphological significance of soil texture in eastern Nigeria. Zeit. für Geomorph. Suppl., 9, 332-345.
Jungerius, P. D., and Levelt, Th., 1964. The clay mineralogy of some soils on sedimentary rocks in eastern Nigeria. Soil Sci., 97, 89-95.
Lewis, L. A., 1969. Analysis of surficial landform properties - the regionalisation of Indiana into units of morphometric similarity. Proc. Indiana Aca. Sci., 78, 317-328.
Mather, P., and Doornkamp, J. C., 1970. Multivariate analysis in geography, with particular reference to drainage basin morphometry. Trans. Inst. Br. Geogr., 51, 163-187.

Morisawa, M. E., 1957. Accuracy of determination of stream lengths from topographic maps. Trans. Am. Geophys. Union, 38, 86-88.

Morisawa, M. E., 1962. Quantitative geomorphology of some watersheds in the Appalachian Plateau. Geol. Soc. Am. Bull., 73, 1025-1046.

Obihara, C. E., 1961. The acid sands of Eastern Nigeria. Nigerian Scientist, 1, 57-64.

Ofomata, G. E. K., 1964. Soil erosion in the Enugu region of Nigeria. African Soils, 9, 289-348.

Ofomata, G. E. K., 1967. Some observations on relief and erosion in Eastern Nigeria. Rev. de Geom. Dyn., 17, 21-29.

Onesti, L. J., and Miller, T. K., 1974. Patterns of variation in a fluvial system. Water Resour. Res., 10, 1178-1186.

Pethick, J. S., 1975. A note on the drainage density - basin area relationship. Area, 7, 217-222.

Pethick, J. S., 1978. Drainage density - basin area relationship: comment. Area, 10, 249.

Rice, R. M., 1973. Factor analysis for the interpretation of basin physiography, in Symposium on the Results of Research on Representative and Experimental Basins, pp253-268. IASH-UNESCO, publ.96.

Richards, K. S., 1977. Slope form and basal stream relationships: some further comments. Earth Surface Processes, 2, 87-92.

Richards, K. S., 1978. Yet more notes on the drainage density - basin area relationship. Area, 10, 344-348.

Simpson, A., 1954. The Nigerian coalfield: the geology of parts of Onitsha, Owerri and Benue Prov. Geol. Surv. Nigeria, Bull., 24

Snedecor, G. W., 1956. Statistical Methods. The Iowa State College Press, Ames, Iowa.

Strahler, A. N., 1950. Equilibrium theory of erosional slopes approached by frequency analysis. Amer. J. Sci., 248, 673-696, 800-814.

Strahler, A. N., 1952. Hypsometric (area-altitude) analysis of erosional topography. Geol. Soc. Amer. Bull., 63, 1117-1142.

Uzoroh, E. J., 1968. Erosional Development of the Mamu River Basin, Eastern Nigeria. Unpublished M.Sc. Thesis, Dept. of Geog., University of Ibadan, Nigeria.

Wentworth, C. K., 1930. A simplified method of determining the average slope of land surfaces. Amer. J. Sci., 20, 184-194.

Wilson, R. C., 1925. The Geology of the eastern railway. Geol. Surv. Nigeria, Bull., 8.

Wilson, R. C., and Bain, A. D. N., 1928. The Nigerian Coalfield: Section II, Parts of Onitsha and Owerri Provinces. Geol. Surv. Nigeria, Bull., 12.

Wong, S. T., 1963. A multivariate statistical model for predicting mean annual flood in New England. Ann. Assoc. Am. Geog., 53, 298-311.

International Geomorphology 1986 Part II
Edited by V. Gardiner

A MULTIVARIATE EVALUATION OF A GROUP OF DRAINAGE BASIN VARIABLES - A SOUTH AFRICAN CASE STUDY

A.B. de Villiers

Dept. of Geography, Potchefstroom University for CHE, Potchefstroom, 2520, RSA

ABSTRACT

The aim of this paper is to evaluate a group of morphological variables that can be used in the description of the morphology of drainage basins. The variables used in this study, comprise two groups namely a group of previously defined and a group of newly defined variables.

Data used in this study was obtained from 167 second-order drainage basins (Strahler system of ordering). The data was submitted to factor and multiple linear regression analysis.

The results obtained in this study has led to the conclusion that only 16 variables (out of the original 59 variables defined) need to be used to describe the morphology of this drainage basin.

INTRODUCTION

The proliferation of variables used in the quantitative evaluation of drainage basins has led to a certain amount of confusion as to the relative importance, applicability and usefulness of the individual variables. The list of 59 variables shown in Table 1 comprises two groups of variables namely of a previously defined group, that has been extensively used in similar geomorphological studies, as well as of a newly defined group of variables depicting aspects of soils, geology and vegetation. The aspects of geology and soils defined as new variables includes areas covered by the different soil and rock types per drainage basin, as well as indexes of erosion as defined by researchers in the respective fields. As for the vegetation component, the areas covered by three distinct ground cover components per drainage area were used as variables.

TABLE 1 . List of 59 variables used

VARIABLE	SYMBOL	REFERENCE
Stream length	L_u	Horton, 1945; Strahler, 1975
Basin area	A_u	Horton, 1945
Basin length	L_b	Horton, 1945
Basin width	B_r	Horton, 1945
Basin perimeter	P	Smith, 1950
Height where stream originates	X	De Villiers, 1981
Height of basin mouth	Z_o	Doornkamp, King, 1971
Height of highest point on watershed	Z_b	Morisawa, 1968
Basin circularity	R_c	Miller, 1953
Basin elongation	R_e	Schumm, 1956
Total height difference in stream	h	Swanepoel, 1980
Stream gradient	GRAD	Horton, 1945
Total basin relief	R_b	Doornkamp, King, 1971
Relief ratio	R_h	Schumm, 1956
Topographic factor	T	Potter, 1953
Drainage density	D_u	Horton, 1945
Lemniscate ratio	K	Chorley et al, 1957

TABLE 1 (Continued)

VARIABLE	SYMBOL	REFERENCE
Individual areas covered by each of five slope classes	M1-M5	De Villiers, 1981
Aspect of slopes	Asp	De Villiers, 1981
Individual areas covered by each of the geological series	G1-G6	De Villiers, 1981
Basin erodibility index	YPI	De Villiers, 1981
Individual areas covered by each of 20 soil associations	B1-B20	De Villiers, 1981
Clay ratio - subsoil	K1/B	Van Zyl, 1973
Exchangeable sodium %-subsoil	UNB	Harmse, 1973
Clay ratio - topsoil	K1/A	Van Zyl, 1973
Exchangeable sodium %-topsoil	UNA	Harmse, 1973
Silt-clay ratio subsoil	SKB	Van Zyl, 1973
Silt-clay ratio topsoil	SKA	Van Zyl, 1973
Basal grass cover component	J1	De Villiers, 1981
Crown cover component shrubs (low)	J2	De Villiers, 1981
Crown cover component schrubs (tall)	J3	De Villiers, 1981

STUDY AREA

The data for this study was obtained from the Grootspruit basin, a sixth order (Strahler method of ordering) drainage basin in the Eastern Orange Free State.

This basin is underlain by a diversity of rock types of different ages, these rocks include both sedimentary and igneous types. The basin is bordered by a mountain range in the west, the north and north east. In the central portion of the basin some erosion remnants are to be found as well as a rather extensive undulating plain covered partly by recent alluvial materials. Twenty one different soil associations are also present in the area. The area is well suited for an analysis of this nature because a diversity of different landforms is formed within the basin.

Because the basin is rather small the macro-climate was found to be homogeneous. The average daily maximum temperature showed a range of 0,5°C for the area as a whole, while the average daily minimum temperature differed by only 0,7°C. The yearly average rainfall is 800 mm in the high areas and between 700 mm and 800 mm for the central portion of the basin. It is believed that with more accurate and longer term records from more weather stations a rather greater range in temperature in the mountainous areas would be observed. With the present data available, however, the climate has to be considered as homogeneous (De Villiers, 1981). The seasonal climatic range is typical of the eastern Highveld regions of Southern Africa with warm, moist summers (thunderstorm activity is pronounced), and cool to cold dry winters with frost at night.

The natural vegetation of the area is dominated by different species of grass with a short shrub component in disturbed areas. In the mountain areas some low trees and shrubs are to be found on protected slopes and in some valleys. Cultivation is restricted to the flat to undulating central portion of the basin. Extensive grazing by cattle, and to a certain extent sheep, is the main agricultural activity for the rest of the area.

DATA COLLECTION

The data used in this study was obtained for the 167 second- order basins that form part of this basin. The drainage lines on the existing 1:50 000 topo-cadastral maps of the region were checked with the aid of aerial photographs at a scale of 1:20 000 and followed by systematic field checks. Missing drainage lines were drawn in on the base maps before the ordering was was done.

All linear measurements were obtained with the aid of a opisometer and areas were measured with a polar planimeter. Heights were obtained by contour interpolation and from spot heights on the existing maps. The values of variables not obtained by direct measurements were calculated with the aid of a computer.

The erodibility of rocks and soils was calculated by obtaining field samples and laboratory analysis of the different samples. The methods used are discussed in detail by De Villiers (1981). The variables for depicting the vegetation were obtained from a map compiled by Van Ginkel and Swart (1971).

DATA ANALYSIS

The data were analyzed by means of factor and multiple linear regression analysis. These methods were chosen because it has already been demonstrated by Mather and Doornkamp (1970), Doornkamp and King (1971), and Mather (1976) that these methods, if properly applied, could be used for investigations of this nature. The programmes used for this analysis were those developed by the Health Sciences Computing Facility (HSCF), University of California, Los Angeles (Copyright, 1977 by the Regents of the University). These so-called BMD Programmes are described by Brown (1977). The BMDP1R (multiple linear regression), and BMDP4M (factor analysis) packages were used in this study.

Some of the results used in this section are based on earlier work by De Villiers (1981) while the bulk of the analytical work, concerning the regression analysis constitutes new analysis.

FACTOR ANALYSIS

In this analysis the VARIMAX-orthogonal rotation method for the factor axes was used. This was done to ensure a standard end product (Mather and Doornkamp, 1970, 167). Initially a factor analysis with all the variables for all the basins was done. This analysis yielded fourteen factors that explained 73% of the total variance in the data set. The grouping of certain variables on some of the factor axes proved to be extremely difficult to interpret on a geomorphologically sound basis. As this analysis involved 59 variables, it makes the interpretation of results extremely complex so it was decided to follow a split-and-merge procedure in an effort to establish more interpretable factors that explain larger amounts of variance in the original data. The original variables were split into data groups depicting planimetric basin measurements, slopes and heights, geology, soils and vegetation. Each of these groups constituted an independent analysis. This procedure invariably yielded factors with high levels of explained variance. For example in the analysis of planimetric basin measurements (9 variables), four factors comprising 89,7% of the total variance were obtained. This resulted in a clearer picture of the underlying order in the different data sets.

The merging procedure involved analysis, firstly, of pairs of data sets and then progressively larger combinations of the different data sets. In all twenty seven different combinations were analysed. As could be expected larger data sets yielded more factors to account for acceptable levels of variation in the original data.

The results of the different analyses were used to choose the variables listed in Table 2. Variables were chosen on the basis of their factor loadings on the different factor axes in the 27 analyses performed. If for example two variables, that depict different measures of soil erodibility, showed high loadings on different factor axes, in any of the performed analyses, both would be included in the list of prime variables.

Before it was decided to exclude a variable its performance was checked in the analysis where only a single group of variables was analysed. This was done to ensure that no variable that contributed significantly to different factors in the original single data set was excluded. This procedure has however one disadvantage, namely that it could lead to the inclusion of highly correlated variables into the list of prime variables.

It is fully appreciated that this splitting and merging procedure might not be absolutely statistically justifiable, but it is not statistical optimality that is sought, but rather geomorphic interpretation.

Table 2 also shows the result of factor analyses. The sorted rotated factor loadings pattern (varimax orthogonal rotation) on each variable for each factor is shown. Factor loadings smaller than 0,250 are shown as zero's. The percentage of variation by each individual factor is shown at the bottom of this table. The result was obtained from a factor analysis using only the listed variables.

The factor loadings pattern enables one to place the variables into different groups. Factor 1 includes variables which could be called measures of drainage basin size and drainage network, with the notable exception of T (topographic factor). The reason why this variable is included into this factor group is not at all clear. Variables classified on the second factor axis includes all variables connected with heights and or differences in height in the basins. Factor group 3 includes those that can be used to depict the erodibility of soils, while the basin erodibility index (YPI) is classed in group 5 with basin circularity (R_c) in group 6. Drainage density (D_u) is also significant in as much as it is classified in a separate factor group (group 4), although it also has a factor loading in group one.

In all, 6 factors are generated to replace 19 variables. These 6 factors represent nearly 85% of the total variance of the 19 variables. It is however, of interest to note that the 19 variables "chosen" as prime variables can be grouped into five main classes namely basin size, drainage network, heights of basins, erodibility of soils, and a erodibility index for the basin.

TABLE 2. Sorted rotated factor loadings pattern.

VARIABLE		FACTORS 1	2	3	4	5	6
Basin perimeter	(P)	0,941	0	0	0	0	0
Basin area	(A_u)	0,907	0	0	0	0	0
Stream length	(L_u)	0,901	0	0	0	0	0
Basin length	(L_b)	0,887	0	0	0	0	0
Topographic factor	(T)	0,803	0	0	0	0	0
Basin width	(B_r)	0,712	0	0	0	0	0
Basin relief	(R_b)	0	-0,917	0	0	0	0
Highest point	(Z_b)	0	0,911	0	0	0	0
Basin mouth	(Z_o)	0	0,989	0	0	0	0
Relief ratio	(R_h)	0	0,850	0	0	0	0
Stream origin	(X)	0	0,712	0	0	0	0
Gradient	(GRAD)	0,382	-0,501	0	0	0	0
Exchangeable Sodium-A	(UNA)	0	0	0,931	0	0	0
Clay ratio-A	(K1/A)	0	0	0,909	0	0	0
Exchangeable Sodium-B	(UNB)	0	0	0,766	0	0	0
Silt/Clay-A	(SKA)	0	0	0,737	0	0	0
Drainage density	(D_u)	-0,392	0	0	0,592	0	0
Erodibility index	(YPI)	0	0	0	0	0,751	0
Basin circu= larity	(R_c)	0,297	0	0	0	0	0,552
	% Variation	37,95	30,75	9,70	2,54	2,21	1,83

Further careful investigation of the percentages of variation explained by the different factors in Table 2 has led to the conclusion that the total amount of variation explained by the variables grouped into the last three factors (6,58%), in relation to the amount of variation explained by the first three factors (78,40%), is very low. It could very well be argued with the above in mind, that these variables need not be used to explain the morphology of this basin. It should however be mentioned that the basin erodibility index had a rather small, but significant squared multiple correlation coefficient of 20,7% with basin size in the original correlation matrix.

The correlation matrix computed for the variables contained in the first three factors of Table 2 is given in Table 3. The interrelationships between the variables contained in this matrix were further investigated in an effort to find single representative variables out of these different factor groups. This was done with the aid of multiple linear regression.

MULTIPLE LINEAR REGRESSION

Thirty-one of the more important regression equations between the independent and dependent variables defined in this process of investigation is shown in Table 4. The correlation coefficients between the individual independent variables and their dependent variables, in each of these cases, have been examined together with the multiple correlation coefficient of the regression in the following way:

In the multiple linear regression between P (dependent) and L_b and A_u (independent) variables the R^2 = 91,73% (Table 4). The correlation coefficients (calculated from the data in Table 3) between P and L_b = 83,53% and between P and A_u = 84,09%. This means that if P has to be predicted by A_u alone, 84,09% of the variance in P can be explained by A_u. However if P can be predicted by both A_u and L_b, only 7,64% more of the variance in the values of P can be predicted (91,73 - 84,09 = 7,6%). Similarly if P can be predicted by L_b, 83,53% of the variance in P can be explained by L_b. A gain of 8,2% in the variance in P is found if the value of A_u is also known. The same process of investigation was followed with firstly defining A_u as dependent and L_b and P as independent variables, and then by taking L_b as dependent and the other two as independent variables.

All the cases shown in Table 4 (as well as another 45 not shown in this table) were examined in this way. The conclusions drawn from these investigations are that within the variable group of basin size, depending on the availability of data, and the purpose of the study, virtually any one of the variables could be used as a prime variable. It would seem though that it would be advisable to use either P (basin perimeter) or A_u (basin size). In the basin height group, total basin relief (R_b) and, or the relief ratio (R_h) seems to be the best within-group predictors, while stream length (L_u) could be used as a measure of the drainage component.

TABLE 3. Correlation matrix for 16 variables

	P	A_u	L_u	L_b	T	B_r	R_b	Z_b	Z_o	R_h	X	GRAD	UNA	$K\ell/A$	UNB	SKA
P	1															
A_u	0,917	1														
L_u	0,925	0,868	1													
L_b	0,914	0,827	0,904	1												
T	0,728	0,620	0,808	0,709	1											
B_r	0,644	0,694	0,581	0,618	0,431	1										
R_b	0,166	0,083	0,191	0,161	0,613	0,093	1									
Z_b	-0,032	-0,104	-0,006	-0,037	0,430	0,070	0,905	1								
Z_o	-0,322	-0,349	-0,304	-0,324	-0,031	-0,290	0,412	0,761	1							
R_h	-0,485	-0,430	-0,450	-0,518	-0,092	-0,341	0,568	0,688	0,606	1						
X	-0,049	-0,115	-0,011	-0,062	0,467	-0,105	0,768	0,926	0,809	0,630	1					
GRAD	-0,377	-0,322	-0,436	-0,386	-0,138	-0,242	0,355	0,519	0,630	0,678	0,630	1				
UNA	0,260	0,255	0,258	0,263	-0,072	0,139	-0,544	-0,688	-0,599	-0,621	-0,635	-0,621	1			
$K\ell/A$	0,212	0,190	0,201	0,201	-0,101	0,088	-0,523	-0,608	-0,503	-0,582	-0,583	-0,582	0,919	1		
UNB	0,329	0,345	0,293	0,286	-0,607	0,236	-0,518	-0,692	-0,690	-0,631	-0,703	-0,631	0,684	0,733	1	
SKA	-0,146	0,114	-0,119	-0,163	0,110	0,046	0,344	0,378	0,284	0,459	0,405	0,459	-0,504	-0,597	-0,381	1

TABLE 4. Regression equations found by multiple linear regression

			R^2 (%)
P	=	$2,26 + 1,17\ L_u + 0,68\ A_u$	90,83
A_u	=	$-1,69 + 0,21\ L_u + 0,54\ P$	84,39
L_u	=	$-0,22 + 0,08\ A_u + 0,36\ P$	85,81
P	=	$1,85 + 0,75\ A_u + 1,17\ L_b$	91,73
A_u	=	$-1,77 - 0,11\ L_b + 0,67\ P$	84,18
L_b	=	$-0,07 - 0,04\ A_u + 0,41\ P$	83,54
P	=	$1,17 + 1,12\ L_u + 1,01\ L_b$	88,82
L_u	=	$-0,37 + 0,38\ L_b + 0,27\ P$	87,68
L_b	=	$0,15 + 0,38\ L_u + 0,22\ P$	85,90
Z_b	=	$1628,30 + 1,07\ R_b + 496,24\ R_h$	86,29
R_b	=	$-1085,52 + 0,69\ Z_b - 141,35\ R_h$	82,37
R_h	=	$-0,79 + 0,001\ Z_b - 0,001\ R_b$	48,94
T	=	$257,98 + 2,05\ R_b - 1864,32\ R_h$	66,19
R_b	=	$-24,47 + 865,35\ R_h + 0,03\ T$	76,9
R_h	=	$0,08 + 0,001\ R_b - 0,001\ T$	63,32
B_r	=	$0,49 + 0,06\ L_u + 0,27\ L_b$	38,5
L_b	=	$0,36 + 0,77\ L_u + 0,27\ B_r$	83,89
L_u	=	$0,12 + 0,94\ L_b + 0,07\ B_r$	81,85
GRAD	=	$2,16 - 0,61\ L_u + 33,31\ R_h$	48,18
L_u	=	$3,29 - 0,07$ GRAD $- 4,24\ R_h$	23,44
R_h	=	$0,11 - 0,01\ L_u + 0,01$ GRAD	48,95
Z_b	=	$0,06 + 1,0\ Z_o + 1,0\ R_b$	100,00
Z_o	=	$0,09 + Z_b - 0,99\ R_b$	100,00
R_b	=	$0,49 - 1,0\ Z_o + 0,99\ Z_b$	100,00
UNA	=	$-29,33 + 0,64$ Kℓ/A $+ 0,69$ SKA	85,25
UNB	=	$330,23 + 1,06$ UNA $- 0,71$ SKA	45,51
SKA	=	$132,15 - 0,00$ UND $- 0,05$ UNA	27,23
Kℓ/A	=	$395,97 + 1,25$ UNA $- 2,85$ SKA	87,26
UNA	=	$61,63 + 0,01$ UNB $+ 0,60$ Kℓ/A	84,99
Kℓ/A	=	$-20,33 + 0,18$ UNB $+ 1,20$ UNA	86,95
UNB	=	$220,49 + 0,75$ Kℓ/A $+ 0,06$ UNA	52,53

In the erosion group of variables the exchangeable sodium % of the top soil (UNA) could be used to predict the clay ratio of the topsoil (K1/A), while it seems important to retain the variables for the silt-clay ratio of the topsoil (SKA) and the exchangeable sodium % of the underlying horizon (UNB).

Some of the between-variable group correlations that were examined is also shown in Table 4. In this table it can be seen that very high multiple correlation coefficients (%) are attained in cases where for example basin size and drainage measurements are used in dependent and independent variables. The best correlation coefficients being where perimeter (P), stream length (L_u) and basin size (A_u) were used (R^2 = 90,83%). Another series of cor= relations were also computed using variables out of three or more variable groups. Disappointingly, however, none of these showed multiple correlation coefficients of more than approximately 30%.

CONCLUSION

It is suggested that the original 59 variables defined could by factor analysis be succesfully reduced to the 19 variables listed in Table 2. These 19 variables could be further reduced to 16 (Table 3) with the loss of only 6,58% of the total (84,98%) variance explained by the defined 19 variables. Depending largely on the type of analysis, the data available, and the purpose of a study it could be possible to reduce the amount of variables, within the different data groups, to for example one as is the case with drainage network, or to three as is the case with the soil erodibility class. It needs to be stressed however that the selection of the prime varibales to be used in a particular study needs very careful thought and insight into the exact aims of the particular study.

ACKNOWLEDGEMENTS

I would like to thank the following persons for their contributions:

Prof. J.S. le Roux of the University of the Orange Free State for his leadership in the original study. Prof. H.S. Steyn and Mrs. S. Uys of the Statistical Consultant Service (Potchefstroom University) for their contribution regarding the statistical methods used in the study.

REFERENCES

Brown, M.S., 1977. *BMDP-77- Biomedical computer programmes. P-series*. University California Press, Berkley, California, p. 880.

Chorley, R.J., Malm, D.E.G., Pogorzelsky, A., 1957. A new standard for estimating drainage basin shape. *American J. Sci.*, 255, 138-144.

De Villiers, A.B. , 1981 . *'n Kwantitatiewe analise van sekere morfologiese eienskappe in die Grootspruitopvanggebied in die Oranje-Vrystaat.* Ph.D. proefskrif, Fakulteit Natuurwetenskappe, Departement Aardrykskunde, U.O.V.S. p. 231.

Doornkamp, J.C., King, C.A.M. , 1971 . *Numerical analysis in Geomorphology - an introduction.* Arnold, London, p. 372.

Harmse, H.J. von M. , 1973 . Die invloed van die tipe geadsorbeerde katione en die grondoplossing op die erodeerbaarheid van gronde , *Annale - 5de jaarlikse konvensie van die S.A. Instituut van Siviele Ingenieurs,* Afdeling 7.04, 1-11.

Horton, R.E. , 1945 . Erosional development of streams and their basins: hydrophysical approach to quantitative morphology . *Geol. Soc. Am. Bull.*, 56, 275-370.

Mather, P.M. ,1976 . *Computational methods of multivariate analysis in Physical Geography,* Wiley, London, p. 532.

Mather, P.M., Doornkamp, J.C. , 1970 . Multivariate analysis in Geography . *Trans. Inst. British Geographers,* 51, 163-187.

Miller, V.C. , 1953 . Quantitative study of drainage basin characteristics in the Clinch Mountain area, Virginia and Tennesee . *Columbia University, Deparment of Geography. Technical Report,* 3.

Morisawa, M. , 1968 . *Streams: their dynamics and morphology.* McGraw-Hill, New York, p. 175.

Potter, W.D. , 1953 . Rainfall and topographic factors that affect runoff . *Am. Geographical Union Transactions,* 33, 67-7.

Schumm, S.A. , 1956 . The evaluation of drainage systems and slopes in badlands at Perth Amboy, New Jersey . *Geol. Soc. Am. Bull.*, 67, 597-646.

Smith, K.G. , 1950 . Standards for grading texture of erosional topography . *Am. J. Sci.*, 248, 655-668.

Swanepoel, A.J. , 1980 . *'n Kwantitatiewe geomorfologiese analise van 'n deel van die Vredefortkoepel.* M.Sc.-verhandeling, Departement Geografie, Potchefstroomse Universiteit vir CHO. p. 135.

Van Ginkel, B., Swart, J.S., 1971 . *Die plantegroei van die Grootspruitopvanggebied.* Bo-Oranjerivier Opvanggebied-projek, Kaart - Dept. L.T.D., O.V.S.

Van Zyl, D.J.A. , 1973 , Gronderosie deur water - 'n fisiese proses . *Annale - 5de jaarlikse konvensie van die S.A. Instituut van Siviele Ingenieurs,* Afdeling 7.01, 1-16.

International Geomorphology 1986 Part II
Edited by V. Gardiner

INDIRECT EVALUATION OF EROSION ENTITY IN DRAINAGE BASINS THROUGH GEOMORPHIC, CLIMATIC AND HYDROLOGICAL PARAMETERS

S. Ciccacci, P. Fredi, E. Lupia Palmieri and F. Pugliese

Dipartimento di Scienze della Terra
Universita' "La Sapienza", Roma, Italy

ABSTRACT

The regression equations connecting suspended sediment yield by streams with some parameters - expressing the main climatic, hydrologic and morphologic characters which control erosion intensity in drainage basins - were improved by extending the quantitative geomorphic studies already started by the authors on drainage networks.

To have a sample which is representative of the different physiograpic conditions of Italy, twenty drainage basins were choosen. The values of annual suspended sediment yield(Tu,ton/sq.km/year) measured at the gauging stations were used as index of erosional processes acting within the drainage basins examined. The main factors affecting erosion intensity were translated into the following synthetic parameters: two pluviometric regime indices, mean annual discharge, drainage density and some parameters expressing the topological structure of drainage networks.

The correlation between Tu values and those of the above mentioned parameters provided some relations of this type:

$\log Tu = ax + b$ and $\log Tu = ax + by + c$; a, b, c being the numerical coefficients and x and y two of the parameters examined. High values of determination coefficients were observed, explaining more than 95% of Tu variations. Drainage density resulted to be the best predictor of Tu. The practical utility of these regressions is evident. Although indirectly and approximately, they allow the evaluation of Tu entity in drainage basins lacking suspended sediment monitoring stations. As Tu values are the most widespread data which can evaluate - although limitedly - erosion processes in drainage basins, the relations obtained can give information also about the intensity of these processes and can lead to the compilation of "erosion index" maps expressed in quantitative terms.

INTRODUCTION

Systematic and widespread direct measurements of erosion in experimental catchments are rare and often incomplete; for this reason many attempts were made by authors from different countries to identify empirical equations in which physiographic variables are employed as predictors of erosional process entity.

It is known that stream load can partially and approximately express the amount of the erosional processes acting within drainage basins. For this expression to be more precise it would be necessary to know the total amount of stream load. However, bed load and dissolved load are rarely measured and with different techniques; consequently most of the studies concerning the indirect evaluation of erosion entity in drainage basins are based on suspended sediment yield data. In Italy the only extensively available data about stream load refer exclusively to suspended sediment yield which is measured at 116 gauging stations controlled by the Hydrographic Office of the Ministry of Works.

Previous researches carried out on some drainage basins of Italy allowed the identification of significant relationships between suspended sediment yield and some climatic and geomorphic parameters which express the main erosion controlling features of drainage basins and networks (Ciccacci et al., 1977, 1979, 1981). The aim of this paper is to improve the regression equations already obtained by extending the quantitative geomorphic studies to a higher number of drainage basins in order to have a more representative sample of the various physiographic conditions of Italy.

METHODS

Twenty drainage basins of Italy were considered (Fig.1); their location and their main lithological characteristics are listed in Table I.

The choice of the drainage basins to examine was unfortunately conditioned by the lack or limited working of gauging stations, which often makes suspended sediment data deficient or relevant to a short recording period.

The mean annual suspended sediment yield, Tu, (tons/sq.km/year) was assumed as index of the erosion processes acting within drainage basins; Tu values express the specific degradation and allow the comparison among basins different in size, as they are independent from the total area of catchments.

The choice of the indices to be compared with Tu was guided by the results of previous researches (Ciccacci et al., 1981) which had shown that p^2/P and $P \times \sigma$, among climatic parameters, and drainage density (D), index and density of hierarchical anomaly (Δa and ga respectively) among geomorphic parameters were the most significantly related to Tu values; moreover the mean annual

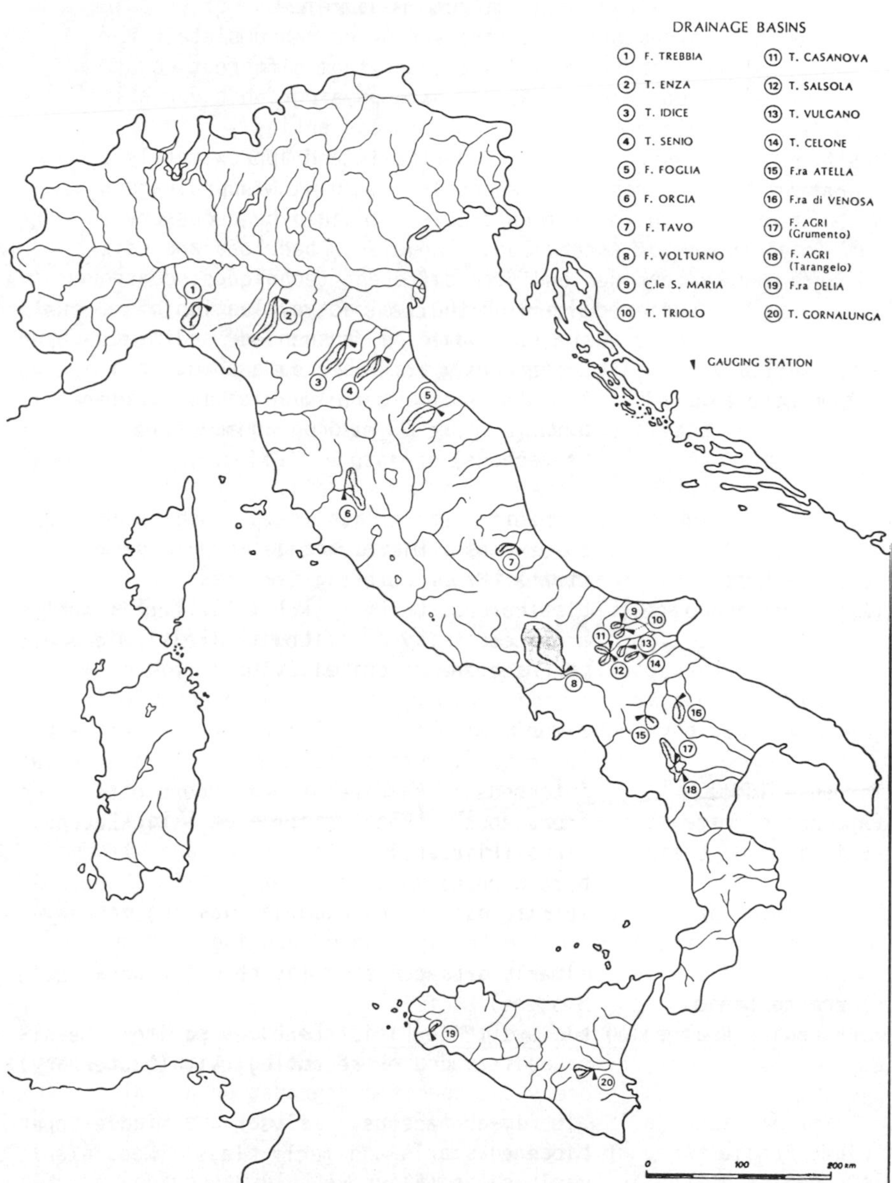

Fig. 1. Geographical location of the twenty drainage basins examined.

TABLE I. Location and main lithological characters of drainage basins examined

BASIN (Location)	LITHOTYPES a)Prevailing b)Subordinate
Fiume Trebbia (Appennino Ligure)	a)marly-arenaceous flysch (Upper Cretaceous); shales with olistostromes (Middle-Upper Cretaceous); schisty shales with siliceous limestones and sandstones (Upper Cretaceous) b)marls and marly limestones (Lower Cretaceous) sandstones (Paleocene).
Torrente Enza (Appennino Emiliano)	a)chaotic shales with calcareous interbeds, calcareous microbreccias and arenaceous strata (Upper Jurassic-Eocene) b)calcareous-marly flysch (Paleocene); marly-arenaceous flysch (Lower-Middle Miocene); terraced and recent alluvium (Quaternary).
Torrente Idice (Appennino Tosco-Emiliano)	a)marly-arenaceous flysch (Upper Cretaceous-Paleocene); chaotic shales with calcareous interbeds, calcareous microbreccias and arenaceous strata (Upper Jurassic-Eocene); clays (Pliocene) b)calcareous marls and sandy marls (Miocene), sandstones (Miocene), alluvium (Quaternary).
Torrente Senio (Appennino Romagnolo)	a)marly-arenaceous flysch (Lower-Middle Miocene) b)clays (Plio-Pleistocene),evaporites (Messinian),terraced or recent alluvium(Quaternary)
Fiume Foglia (Appennino Marchigiano)	a)marly-arenaceous flysch (Middle-Upper Miocene), marls and marly clays (Messinian); marly clays (Pliocene) b)calcareous-marly flysch (Eocene); chaotic shales (Cretaceous-Eocene);evaporites (Messinian).

follows TABLE I.

Fiume Orcia (Appennino Toscano)	a)clays and sandy clays (Plio-Pleistocene); calcareous-marly flysch (Upper Cretaceous-Eocene) b)limestones (Upper Triassic-Cretaceous); alluvium (Quaternary
Fiume Tavo (Appennino Abruzzese)	a)arenaceous-marly flysch (Upper Miocene) b)limestones(Triassic-Cretaceous);sandy marls (Lower-Middle Miocene).
Fiume Volturno (Appennino Campano)	a)limestones (Triassic-Lower Miocene); arenaceous-marly flysch (Miocene) b)marls and marly limestones (Oligocene-Miocene),tuff and pyroclastic flows (Pleistocene); fluvio-lacustrine deposits (Pleistocene); recent alluvium (Holocene).
Canale S. Maria (Monti della Daunia Puglia)	a)clays and sandy clays (Pliocene); terraced alluvium (Quaternary) b)marly-arenaceous flysch (Lower-Middle Miocene).
Torrente Triolo (Monti della Daunia Puglia)	a)terraced alluvium (Quaternary); clays and sandy clays (Pliocene) b)variegated shales(Cretaceous-Eocene);marly-arenaceous flysch (Lower-Middle Miocene).
Torrente Casanova (Monti della Daunia Puglia)	a)marly-arenaceous flysch (Lower-Middle Miocene b)variegated shales (Cretaceous-Eocene),clays (Pliocene) fluvial sands and conglomerates (Quaternary).
Torrente Salsola (Monti della Daunia Puglia)	a)sandy and marly clays (Pliocene); marly-arenaceous flysch (Lower-Middle Miocene) b)variegated shales (Cretaceous-Eocene) alluvium (Quaternary).
Torrente Vulgano (Monti della Daunia Puglia)	a)terraced and recent alluvium (Quaternary) b)variegated shales(Cretaceous-Eocene),marly-arenaceous flysch (Lower-Middle Miocene).

follows TABLE I.

Torrente Celone (Monti della Daunia Puglia)	a)marly-arenaceous flysch (Lower-Middle Miocene); sands and marly clays (Pliocene) b)alluvium(Quaternary);sandy clays (Pliocene)
Fiumara di Venosa (Appennino Lucano)	a)sands and conglomerates (Calabrian);fluvio-lacustrine deposits and tuffites (Quaternary) b)variegated shales(Cretaceous-Eocene);marly-arenaceous flysch (Lower-Middle Miocene).
Fiumara di Atella (Appenino Lucano)	a)variegated shales (Cretaceous-Eocene); arenaceous-marly flysch (Upper Oligocene-Lower Miocene) b)tuffs and tuffites (Pleistocene); conglomerates, sands and clays (Pliocene); alluvium (Holocene).
Fiume Agri (Grumento) (Appennino Lucano)	a)limestones, dolomitic limestones and chert limestones (Triassic-Upper Cretaceous) b)argillophyllitic flysch (Lower-Middle Cretaceous); marly calcarenites, argillites and sandy limestones (Middle Cretaceous); shales (Middle-Upper Cretaceous); alluvium (Quaternary).
Fiume Agri (Tarangelo) (Appennino Lucano)	a)limestones, dolomitic limestones and chert limestones (Triassic-Upper Cretaceous);marly-arenaceous flysch(Middle Miocene);arenaceous-calcareous flysch (Paleocene-Eocene) b)argillophyllitic flysch (Lower-Middle Cretaceous);shales (Middle-Upper Cretaceous); marly calcarenites,argillites and sandy limestones (Middle Cretaceous); terraced fluvio-lacustrine deposits (Pleistocene).
Fiumara Delia (Sicilia occidentale)	a)clays (Middle Miocene), evaporites (Messinian), marls (Lower Pliocene) b)alluvium and litoraneous deposits (Quaternary).
Fiume Gornalunga (Sicilia orientale)	a)clays and sandy clays (Pliocene);variegated shales (Cretaceous-Eocene); marls (Messinian) b)evaporites and diatomites (Messinian);alluvium (Quaternary).

discharge (Q, in m^3/sec) was added.
Both the climatic parameters express the pluviometric regime; p^2/P is the well known Fournier's parameter (Fournier, 1949, 1960) and $P \times \sigma$ is a parameter proposed by the authors where P is the total annual precipitation and σ is the standard deviation of the precipitation of the twelve months of a given year from their relevant average (Ciccacci et al., 1977).
In this phase of the research the discharge values were also considered among the parameters to be compared with Tu. For each basin these values refer to the same recording period as Tu and are measured at the same gauging station.
To the end of this inquiry, the drainage density (D) is certainly one of the most significant parameters because it expresses many of the factors which control the amount of suspended sediment yield. In this investigation the drainage density values were obtained by measuring the length of all the identifiable paths for channelled runoff. The drainage network to be measured was drawn from topographic maps(scale 1:25,000) and was integrated through both the examination of aerial photographs and the field survey. In the calculation of D all the endoreic areas were excluded as well as all the strictly artificial canals almost lacking in suspended load.
Finally, the index (Δa) and the density (ga) of hierarchical anomaly were considered (Avena et al., 1967; Avena and Lupia Palmieri 1969); they express the organization of drainage networks which depends on the lithological, tectonic, climatic and vegetation conditions of the basins. To define these parameters it is necessary to explain first the number of hierarchical anomaly (Ga): it represents the smallest number of 1st order streams necessary to make a drainage network perfectly conservative. The ratio between the value of the hierarchical anomaly (Ga) and the number of 1st order channels really present in the drainage network (N1) provides the index of hierarchical anomaly ($\Delta a = Ga/N1$); the ratio between Ga and the drainage basin area (A, in sq.km) provides the density of hierarchical anomaly ($ga = Ga/A$). These two parameters express exactly the organization degree of drainage networks, as they take into account the hierarchically anomalous influences as well as the weight of their anomaly; moreover Δa and ga are not depending on basin area, therefore they allow the comparison among different drainage basins. The values of Tu and of the climatic, hydrologic and geomorphic parameters calculated for the twenty basins examined are listed in Table II.
The following step of the inquiry was the identification of the relationships between the dependent variable Tu and one or more of the independent variables (i.e. climatic, hydrologic and geomorphic parameters).
To this end bivariate and multiple regression analyses were

TABLE II. Climatic, hydrologic and geomorphic parameters of the drainage basins examined

Basins	Area sq.km	Years of recording	Tu (tons/sq.km/yr)	p^2/P	$\frac{P \times \sigma}{10^4}$	Q (cu.m/sec)	D	Δa	ga
F. Trebbia	226	5	1523	81.52	22.41	10.215	4.90	1.59	19.35
T. Enza	670	11	2500	38.31	7.17	12.596	5.62	1.18	18.80
T. Idice	397	8	2397	33.01	4.90	5.042	5.61	0.86	30.27
T. Senio	269	23	868	42.22	5.96	3.255	4.17	1.32	15.10
F. Foglia	603	9	1957	43.50	5.96	7.819	4.99	1.39	15.03
F. Orcia	580	20	1309	35.50	3.78	4.060	4.80	1.34	17.60
F. Tavo	109	7	307	40.42	5.85	3.919	3.14	1.38	8.75
F. Volturno	2015	24	697	67.44	12.52	31.837	3.50	1.62	17.50
Canale S. Maria	60	8	230	31.74	2.78	0.168	2.21	1.14	3.00
T. Triolo	54	8	363	28.65	2.97	0.200	2.60	0.91	2.80
T. Casanova	52	8	185	26.58	2.89	0.208	2.70	0.36	1.40
T. Salsola	43	8	228	24.93	2.57	0.185	2.60	1.15	4.60
T. Vulgano	94	8	251	25.72	3.14	0.385	2.44	0.94	3.00
T. Celone	86	8	248	23.23	3.02	0.550	2.80	0.76	3.10
F.ra di Venosa	261	13	225	23.13	2.44	1.070	2.76	1.08	2.90
F.ra di Atella	158	21	513	30.61	3.97	1.643	3.65	1.17	6.09
F.Agri (Grumento)	278	6	231	51.30	7.37	7.061	2.42	1.10	4.30
F.Agri (Tarangelo)	507	6	340	48.85	7.59	10.444	3.06	1.40	10.80
F.ra Delia	140	16	261	33.39	3.62	0.695	2.59	0.57	2.05
F. Gornalunga	232	5	1107	35.45	3.67	0.557	4.40	1.13	0.73

performed.

RESULTS AND DISCUSSION

Mean annual suspended sediment yield was plotted first against each independent variable in order to determine general relationships. Simple bivariate regression analysis pointed out that a significant relationship exists between the logarithm of the dependent variable Tu and the independent variable D. The resulting regression equation and the determination coefficient (r^2) were:

$$\log Tu = 1.52390 + 0.33708\ D \qquad r^2 = 0.95679 \qquad (1)$$

The regression line is showed in Figure 2a; the per cent differences between measured and calculated values of Tu are listed in Table III. Analysis of variance (ANOVA f test) evidenced that the variances of the two sets of data are the same at the $\alpha = 0.05$ level of significance.

As already observed in previous works (Ciccacci et al., 1977 and 1979) less significant relations were evidenced by the bivariate analysis between Tu and the other parameters. However mention must be made for the climatic parameters; if they failed in explaining Tu variability among different basins, they proved to be good predictors of annual variation of Tu for a given drainage basin.

The very high correlation between suspended sediment load (tons/sq.km/year) and drainage density shows that D can be considered a very good predictor of Tu values. The fact that drainage density explains much of Tu variability can be easily justified. This geomorphic parameter synthetically expresses many of the factors which control the erosion entity in drainage basins and as a consequence the amount of suspended load by watercourses.

Drainage density is strongly depending on climatic conditions, moreover it is tied to the type and density of the vegetation cover and it can be partially modified in response to human activity. D values are also a function of the lithological and structural characters of outcropping rocks: more precisely they are inversely proportional to the rock permeability and proportional to their fracturing degree. Besides, if sedimentary lithotypes crop out, D can exhaustively express also the erodibility of outcropping rocks; in particular , other conditions being equal, the higher the rock erodibility is, the higher are the values of drainage density. Taking into account that the basins examined are mainly emplaced on sedimentary terrigenous lithotypes, it is easily understood why a very high correlation exists between Tu - which expresses the specific degradation of drainage basins - and D which synthesizes the degree of rock erodibility.

Although D can explain more than 95% of Tu variations, it is to underline that in some cases the difference between the measured values of Tu and those calculated by equation (1) is rather high

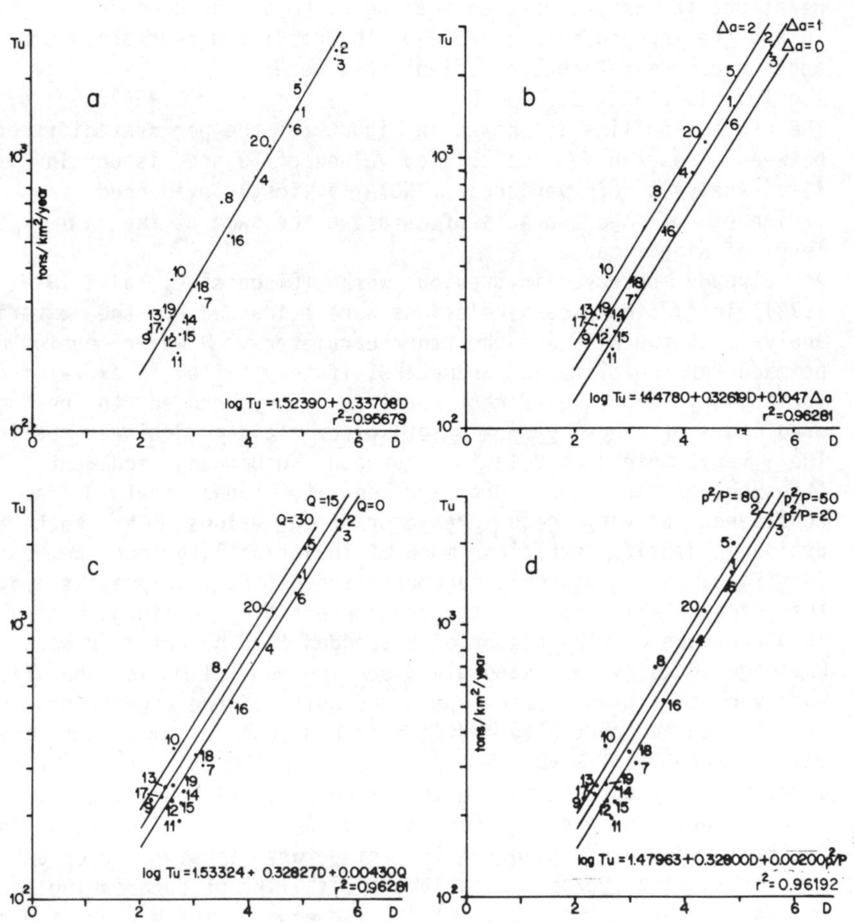

Fig. 2. Simple bivariate and multiple regressions relating Tu values with D (a), D and **Δ**a (b), D and Q (c), D and p^2/P (d).

TABLE III. Per cent differences between measured Tu (Tu m) and Tu values calculated (Tu c) by regressions (1), (2), (3) and (4)

Relations		(1)		(2)		(3)		(4)	
Basins	Tu m	Tu c	diff. %	Tu c	diff. %	Tu c	diff. %	Tu c	diff. %
F. Trebbia	1523	1498	1.64	1618	6.24	1534	0.72	1778	16.74
T. Enza	2500	2620	4.80	2523	0.92	2707	8.28	2510	0.40
T. Idice	2397	2600	8.47	2322	3.13	2492	3.96	2431	1.42
T. Senio	868	850	2.07	878	1.15	824	5.07	855	1.50
F. Foglia	1957	1607	17.88	1652	15.59	1603	18.09	1597	18.40
F. Orcia	1309	1386	5.88	1415	8.10	1338	2.22	1334	1.91
F. Tavo	307	382	24.43	411	33.88	381	24.10	389	26.71
F. Volturno	697	505	27.55	569	18.36	659	5.45	579	16.93
Canale S.Maria	230	186	19.13	193	16.09	182	20.87	185	19.57
T. Triolo	363	251	30.85	245	32.51	244	32.78	245	32.51
T. Casanova	185	272	47.03	232	25.41	263	42.16	262	41.62
T. Salsola	228	251	10.09	259	13.60	244	7.02	241	5.70
T. Vulgano	251	222	11.55	219	12.75	217	13.55	214	14.74
T. Celone	248	294	18.55	275	10.89	285	14.92	278	12.10
F.ra di Venosa	225	285	26.67	288	28.00	278	23.56	270	20.00
F.ra di Atella	513	568	10.72	573	11.70	548	6.82	547	6.63
F.Agri (Grum.)	231	219	5.19	224	3.03	228	1.30	238	3.03
F.Agri (Tar.)	340	359	5.59	389	14.41	383	12.65	381	12.06
F.ra Delia	261	249	4.60	224	14.18	243	6.90	249	4.60
F. Gornalunga	1107	1016	8.22	997	9.94	955	13.73	986	10.93
	Mean % diff.		14.55		13.99		13.21		13.37

(Table III).

In order to determine if two of the computed parameters could provide a better explanation of Tu variation than D alone, the multiple regression analysis was performed. The contribution made by each variable other than D was tested by using partial correlation coefficients computed for the remaining variables with D held constant. Table IV shows that if the effects of D are controlled the highest partial correlation ceofficient is between Tu and Δa (0.35), followed by the coefficient between Tu and Q and between Tu and p^2/P; very slight are the relations between Tu and $Px\sigma$ and ga respectively.

TABLE IV. Partial correlation coefficients (D constant)

	Q	Δa	ga	p^2/P	$Px\sigma$
Tu	0.24	0.35	0.06	0.24	0.14

The following step of the analysis was then the inclusion of the variable Δa which - after D - is the one which contributes most to the unexplained variation of Tu. The result was this regression equation:

$$\log Tu = 1.44780 + 0.32619\ D + 0.10247\ \Delta a \qquad r^2 = 0.96281 \qquad (2)$$

The determination coefficient of this regression is higher than that of regression (1); the regression line is shown in Figure 2b.

The sample size (20 basins) is perhaps too small to allow the inclusion of a third independent variable, which will be done when more basins will have been analyzed.

Although the partial correlation coefficients between Tu and Q and Tu and p^2/P, holding D constant, were lower as respect to the coefficients between Tu and Δa, these two variables show a less marked correlation with D than ga (Table V).

TABLE V. Correlation matrix of independent variables

	D	Q	Δa	ga	p^2/P	Px
D	1	0.47	0.63	0.92	0.59	0.57
Q		1	0.66	0.59	0.78	0.63
Δa			1	0.71	0.72	0.65
ga				1	0.67	0.62
p^2/P					1	0.95
Px						1

For this reason the multiple regressions were computed also by adding to D the parameters Q and p^2/P in turn; the following equations were obtained:

$$\log Tu = 1.53324 + 0.32827\ D + 0.00430\ Q \qquad r^2 = 0.96281 \qquad (3)$$

$\log Tu = 1.47963 + 0.32800\ D + 0.00200\ p^2/P \quad r^2 = 0.96192$ (4)

The regression lines are shown in Figure 2c and 2d respectively.

Per cent differences between Tu values measured at the gauging stations and Tu values calcalated through multiple regression equations are shown in Table III. The analysis of variance (ANOVA f test) evidenced that the variances of the two sets of data are the same at the $\alpha = 0.05$ level of significance.

The examination of Table III suggests that the per cent differences between calculated and measured values of Tu range from 0.40 to 47.03 with a mean value that is about 14% for all the relations. The highest values generally refer to watercourses having a very irregular regime ("fiumare"); in these cases Tu value of a single year is often enough to modify significantly the mean value of the whole recording period, especially when few years of observation are available. Lower per cent differences generally refer to drainage basins larger in size and characterized by a more regular regime.

The comparison between these multiple regressions and those obtained in a previous phase of the inquiry (Ciccacci et al., 1981) for a smaller sample of basins showed that the degree of significance of the corresponding new regressions is still very high; however some observations must be made. The increase of the sample size showed that - at least for the 20 basins examined - there is a very strong correlation between D and ga (Tab.V); in such a case the latter parameter loses a good deal of its significance in explaining Tu variations. For this reason,if the hierarchical parameters are used, the multiple regression (2) where Δa is present must be preferred to the one relating Tu, D and ga ($1.58355 + 0.30694\ D + 0.00482\ ga$, $r^2 = 0.95877$). The relation (4), where p^2/P is present showed a value of r higher than in previous work (Ciccacci et al., 1981); this can be explained considering that the higher number of basins examined and their various geographic distribution resulted in a more marked variability of climatic conditions which the computed parameters succeeded in quantifying. Furthermore, the regression (4) is more significant than the one where $Px\sigma$ is used ($\log Tu = 1.52320 + 0.33091\ D + 0.00394\ Px\sigma$, $r^2 = 0.95865$). At present, then, the regression (4) seems to be preferred when the climatic parameters are considered. Finally it is to underline the good results obtained by adding to D the parameter Q which had not been used previously.

Concluding,the r^2 of the simple regression between Tu and D was only slightly improved by adding the other variables, which can be partially explained with some colinearity of the independent variables (Tab. V); therefore it seems that other variables do not add too much to the regression (1). However, it is noteworthy considering that if the adding of other variables does not increase significantly the r^2 value, in some cases it decreases the per cent differences between calculated and measured values of Tu (Tab. III). For example, in the case of F. Volturno the per cent difference between Tum and Tuc decreases from 27.55 to 5.45 if Tu is predicted through regression (3) instead of (1), thus suggesting that in this basin the combined effect of D and Q accounts for more of the Tu variability than D alone. Analogously, if Δa is added to D the per cent difference between Tum and Tuc for the Torrente Casanova is reduced from 47.03 to 25.41.

Finally from a practical point of view it is to underline that the regressions (1) and (2) are always applicable while the utilization of regressions (3) and (4) is conditioned by the availability of discharge and precipitation data. Furthermore, regressions (1) and (2) afford more detailed information, as they can be used to predict the potential supply of partial basisns to the total Tu of a given basin, while regressions (3) and (4) are not utilizable for this purpose. In particular regression (3) can not be used because the discharge of tributary is difficulty known and the (4) is not advisable because the value of p^2/P of partial basins, if computable, rarely differs greatly from the mean value of the whole basin they belong to.

FINAL REMARKS

The regression analysis showed that Tu values are significantly related to some climatic and geomorphic parameters; hence the determination of these easily computable parameters allows the prediction of Tu values.Beyond their purely scientific meaning, the regression equations obtained have also an applied interest. For example they can be useful to partly evaluate the potential contribution of watercourses to the beach nourishment. Moreover, in the lack of experimental data, the indirect evaluation of Tu can allow the estimation of sedimentation in reservoirs; an attempt in this direction was already made for a reservoir in Central Italy (Ciccacci et al., 1983) and the calculated Tu provided a value of sedimentation rate consistent with the data obtained in different subsequent bathimetric surveys.

It is to underline that for drainage basins mainly emplaced on terrigenous lithotypes, like those considered in this research, the computation of Tu through the found relations can afford a realistic measure of erosional processes which supply sediments to rivers.

Moreover, the areal variation of erosion intensity within a given basin can be evidenced by predicting Tu through the relations applicable to partial basins; this approach was followed in previous works (Ciccacci et al., 1981; Biasini et al., 1983) and the result was the elaboration of maps where the variability of erosion entity within two Italian drainage basins is shown.

Certainly these maps of "erosion index" compiled following the method proposed by the authors, give a partial view of erosional processes acting within drainage basins, but they have the merit of being expressed in quantitative terms and can represent a useful integration for geomorphic maps and for maps of slope instability. Finally, they can be a reference mark for unreplaceable experimental studies which aim at the direct determination of soil erosion.

ACKNOWLEDGEMENTS

This study was funded by Ministero della Pubblica Istruzione (Ministry of Education) and by C.N.R. (National Council of Researches), Research Program: Quantitative Geomorphology of Italian Drainage Basins (E. Lupia Palmieri, Principal Investigator).

REFERENCES

Avena, G.C., Giuliano, G., and Lupia Palmieri, E.,1967. Sulla valutazione quantitativa della gerarchizzazione ed evoluzione dei reticoli fluviali. Boll. Soc. Geol. It., 86, 781-796.

Avena, G.C., and Lupia Palmieri, E., 1969. Analisi geomorfica quantitativa, in Idrogeologia dell'Alto bacino del Liri (Appennino centrale). Geol. Rom., VIII, 319-378.

Biasini, A., Buonasorte, G., Fredi, P.,and Lupia Palmieri, E.,1983. Bacino dell'Ovito (Pietrasecca), carte tematiche.Istituto di Geologia e Paleontologia Univ. Roma, Multigrafica Ed.

Ciccacci, S., D'Alessandro, L., and Fredi, P., 1983. Sulla valutazione indiretta dell'interrimento nei bacini lacustri: il lago artificiale di Scandarello (Rieti). Atti XXIII Congr. Geogr. It., 2 (3), 37-52.

Ciccacci, S., Fredi, P., and Lupia Palmieri, E., 1977. Rapporti fra trasporto solido e parametri climatici e geomorfici in alcuni bacini idrografici italiani.Atti Conv. Misura del trasporto solido al fondo nei corsi d'acqua . C.N.R., Firenze, C-4.1/C-4.16.

Ciccacci, S., Fredi, P. , and Lupia Palmieri, E., 1979. Quantitative expression of climatic and geomorphic factors affecting erosional processes: indirect determination of the amount of erosion in drainage basins in Italy. An approach. Polish-Italian seminar : Superficial mass movements in mountain regions , Szymbark, 76-89.

Ciccacci, S., Fredi, P., Lupia Palmieri, E.,and Pugliese, F.,1981. Contributo dell'analisi geomorfica quantitativa alla valutazione dell'entita' dell'erosione nei bacini fluviali. Boll. Soc. Geol. It., 99, 455-516.

Fournier, F., 1960. Climat et erosion: la relation entre

l'erosion du sol par l'eau et les precipitations atmospheriques, Press Univ. de France, Paris, 201pp..

Fournier, F., 1960a. Debit solide des cours d'eau. Essai d'estimation de la perte en terre subie par l'ensamble du globe terrestre. Inter. Assoc. Sci. Hydrol., 53, 19-22.

Horton, R.E., 1945. Erosional development of streams and their drainage basins: hydrophysical approach to quantitative morphology. Geol. Soc. Am. Bull., 56, 275-370.

Lupia Palmieri, E., 1983. Il problema della valutazione dell'entita' dell'erosione nei bacini fluviali. Atti XXIII Congr. Geogr. It., 2 (1), 143-176.

Ministero Lavori Pubblici, 1924-1978. Annali Idrologici, parte II, sezioni: Bari, Bologna, Catanzaro, Genova, Napoli, Palermo, Parma, Pescara, Pisa.

Servizio Geologico d'Italia. Carta Geologica d'Italia, scala 1:100,000. Roma.

International Geomorphology 1986 Part II
Edited by V. Gardiner

AN APPROACH TO THE QUANTITATIVE ANALYSIS OF THE RELATIONS BETWEEN DRAINAGE PATTERN AND FRACTURE TREND

S. Ciccacci, P. Fredi, E. Lupia Palmieri and F. Salvini

Dipartimento di Scienze della Terra
Universita' "La Sapienza", Roma, Italy

ABSTRACT

To single out the possible relations between drainage pattern and tectonics in the volcanic area of the Monti Sabatini (Northern Latium) a methodology was set up which allows the automatic identification of preferential directions of stream channels treated as statistical set.

The drainage network of the study area was traced and rectified by a semiautomatic process. Eventually the stream order, the geographical coordinates of the extremes, the azimuth and the length of each of the 20,131 rectilinear segments so obtained were memorized.

Then a sophisticated analysis of azimuthal distribution was processed by cumulative length and cumulative number for the whole network and for the rectilinear segments of each stream order. The same analysis was also processed for partial areas to investigate the areal variations of drainage net preferential orientations.

Successively the preferential directions of field surveyed fractures were identified through the same kind of analysis; some main fracture domains were found in the Monti Sabatini area; they are : E-W, N-S, NE and WNW domains.

The comparison between the identified drainage network and fracture domains showed that their main orientations are consistent. The marked prevalence of N-S domain in the lower stream orders (in particular in the 1st and 2nd) could lead to the hypothesis that the corresponding N-S fracture domain might represent the most recent fracture direction in the Monti Sabatini area. On the other hand the E-W domain, that represents the fracture main orientation, is present only in higher stream orders; therefore this fracture direction might reflect the reactivation of tectonics which were preexisting to the volcanic activity.

INTRODUCTION

Within a wide research program concerning the geological and structural arrangement of the volcanic area of the Monti Sabatini (20 km North of Rome, Figure 1) the azimuthal systems of field surveyed fractures and the drainage network pattern were analyzed. The aim of this analysis was the identification of the possible relations existing in this area between the orientation of channels of different orders and that of the fracture systems.

Analogous researches evidenced the existence of a connection between these two elements which is often very complex (Bannister and Arbor, 1980; Scheidegger, 1980; Lanzhou and Scheidegger, 1981; Pohn, 1983). The peculiar geo-structural arrangement of the Monti Sabatini area makes the comparison between drainage network pattern and fracture trend particularly interesting. This area underwent a complex tectonic evolution and was recently covered by a pyroclastic mantle, which surely reduced the dissection of the previous landscape. Moreover, this pyroclastic cover is intersected with fracture systems which are subsequent to the volcanic activity. It is likely that the development of the drainage network was controlled not only by the new morphology and by the difference of lithology, but also by the orientation of fracture systems.

STUDY AREA

The Monti Sabatini area is located in the Latium region (Central Italy), about 20 km North of Rome and it is part of the "Provincia Romana", a K-alkaline volcanic province. This volcanic province is related to the intense NW-SE trending extensional tectonics of the eastern margin of the Tyrrhenian Sea basin (1.0 M. Yr. to Recent). The products of this volcanic complex are extensively present over an area of about 1,500 square kilometers and are not related to a central edifice but to different volcanic features in different sectors. The western sector is characterized by several pyroclastic flows connected with NNW-SSE fractures and their age ranges between 1.0 and 0.1 M.Yr.; subordinately acid lava domes are present. The main volcanic edifices are located in the eastern sector of the volcanic complex; the oldest activity began to the East and progressively migrated to the West along an E-W alignment. This activity produced mainly pyroclastic flows - the main one being larger than 10 cubic kilometers - and to a minor extent a series of scoriae cones and lava flows (Locardi and Sommavilla, 1974; De Rita et al., 1983).

The main morphological feature of this area is the Recent volcano-tectonic depression of the Lake of Bracciano, in the center of the study area. From a morphological point of view the sectors to the

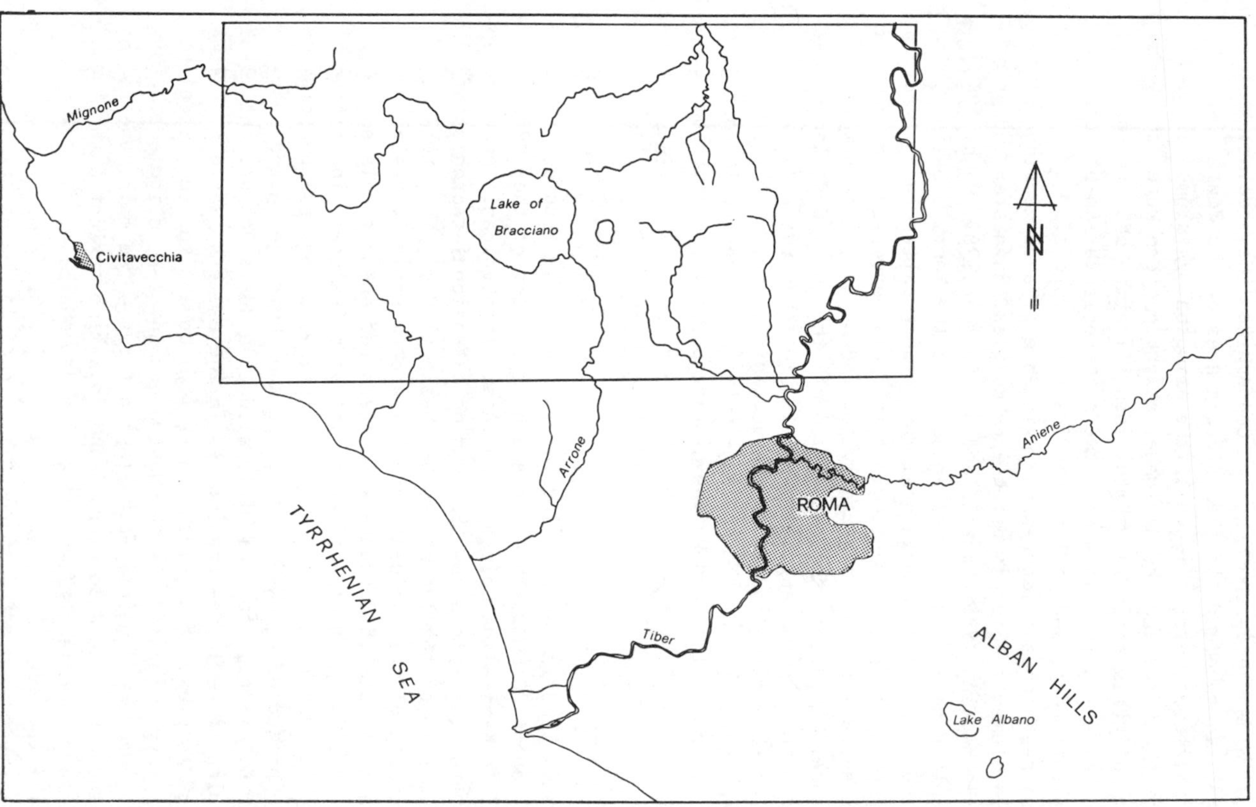

Fig. 1. Location of the study area.

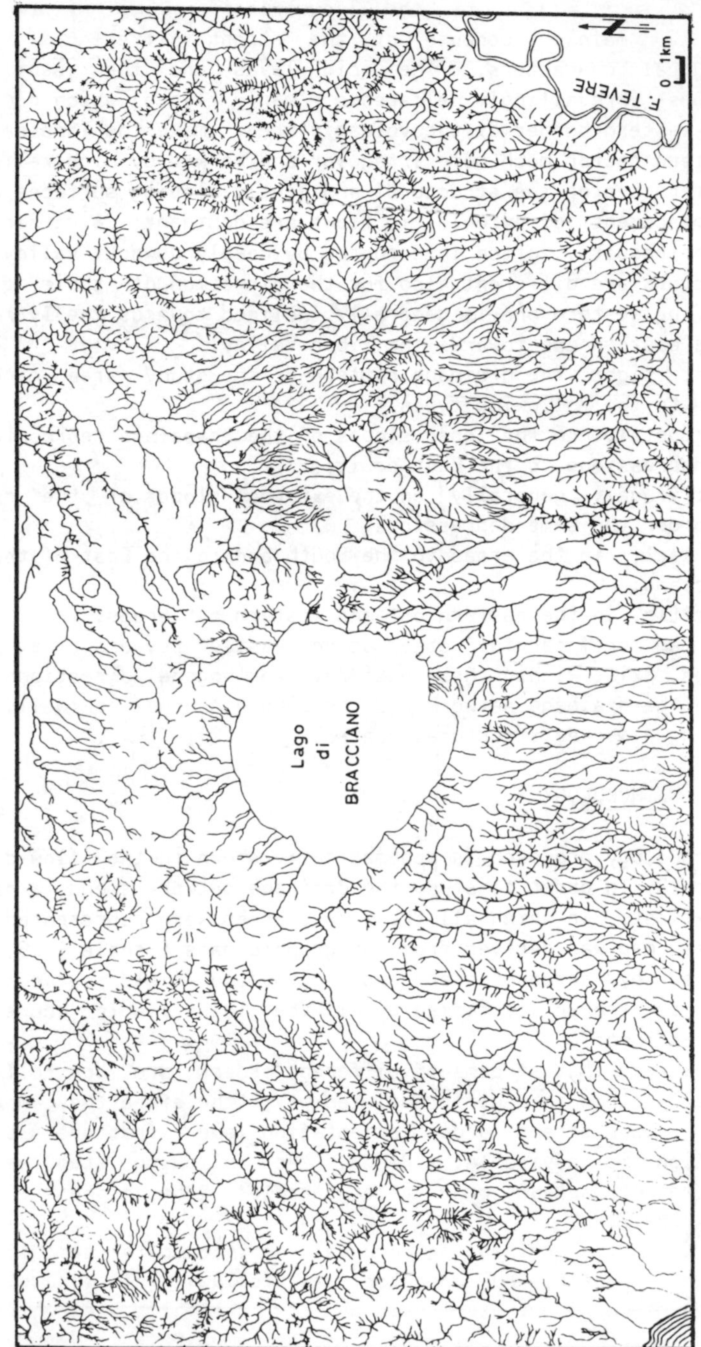

Fig. 2. Drainage network of the Monti Sabatini area.

West and to the East of the lacustrine depression are quite different, mainly because of the strong differences in the lithological characters. The acid lavas and the sedimentary lithotypes (prevailingly calcareous - marly flysch) which crop out at the western border of the volcanic area make the landscape of the western sector much more irregular and marked by high relief. In the eastern sector the structural and gently sloping surfaces due to the emplacement of pyroclastic flows are dominant; a rather flat landscape results, often interrupted by deeply downcut valleys. The two sectors are different also for the direction of their overall slope; the western one is generally inclined towards the Tyrrhenian Sea, the eastern one towards the Tiber Valley.

At first sight the drainage pattern of the study area might be ascribed to the following types:

a) centripetal, on the inner slopes of the Bracciano depression and of the depressions to the East of it;

b) centrifugal, generally on the external slopes of the reliefs which border the same depressions;

c) parallel, in the areas to the South and to the East of the Lake of Bracciano;

d) dendritic, in the easternmost and westernmost areas.

However a more careful observation easily reveals that many anomalies exist within these patterns: orthogonal directions often concern the drainage network and the single channel as well, thus suggesting a marked structural control (Fig. 2).

METHODOLOGY

The methodology proposed is schematically shown in the flow-diagram of Figure 3. In order to study the drainage network trend, a map was traced first, in which all the watercourses were represented (Fig. 2). This map was derived from topographic maps (scale 1:25,000) integrated by aerophotograph examination and field surveys; the network so traced is formed by all the identifiable courses of channelled runoff.

It is likely that streams of different orders are controlled by fracture systems in a differentiated way; therefore the comparison between each stream order orientation and fracture systems orientation is advisable. Hence, the following step was the definition of stream order after Strahler; the highest order resulted the 7th.

Successively the drainage network was rectified by a semiautomatic digitizing process: an operator was following continously the watercourses with the electronic pen of a graphic tablet while the computer was densely sampling a series of coordinates. Whenever the azimuth of two adjacent rectilinear segments differred in less than

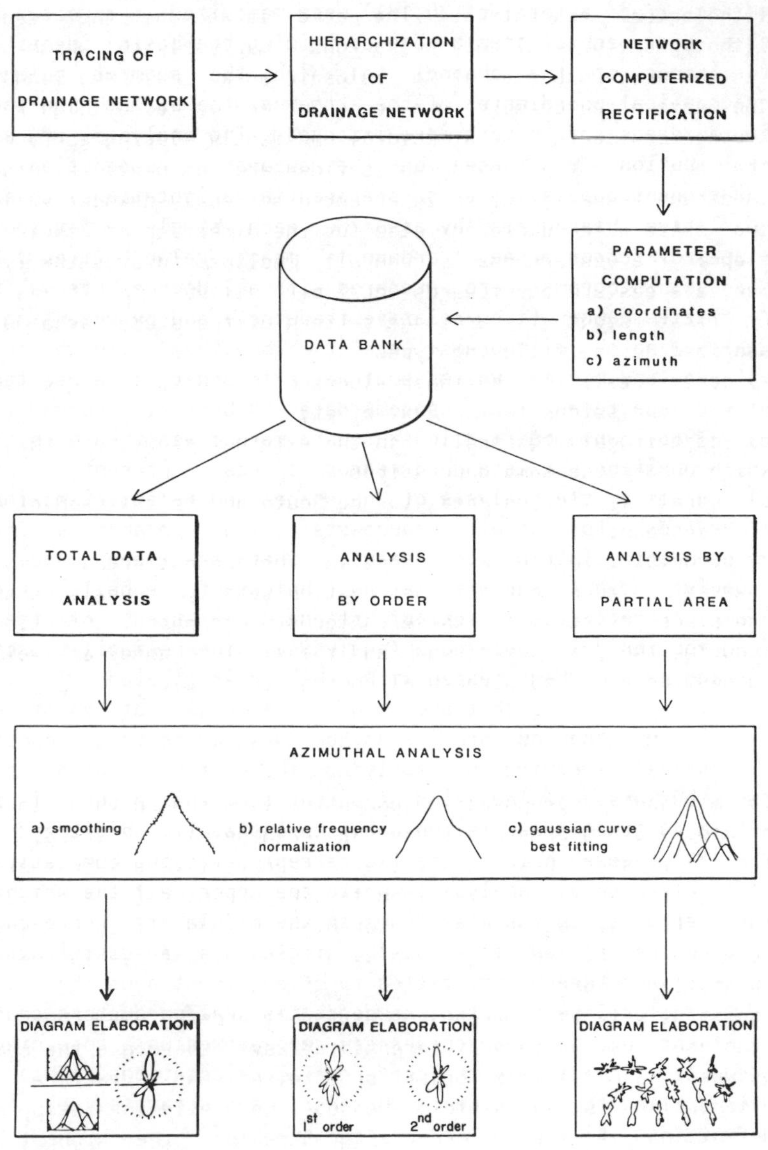

Fig. 3. Flow diagram schematically showing the performed methodology.

two degrees they were joined to form a single longer segment. The rectilinear segments obtained in such a way and longer than 100 meters (for a total of 20,131) were memorized in a data bank in which for each of them a code containing the basin identification, the order of the channel to which the segment belongs, the geographical coordinates of the extremes, the azimuth and the length in hundreds of meters were inserted. The analysis of azimuthal distribution was based on the automatic recognition of the independent population (i.e. preferential orientations) in azimuthal cumulative histograms by assuming each single distribution, a standard error one, around a mean value (i.e. Gaussian curves).Histograms were prepared with a 2 degree interval both by cumulative number (i.e. relative frequency) and by weighting data by their length. Histograms were than smoothed with a 10 degree interval and minimal values subtracted in order to reduce the effect of very scattered (i.e. random) data. A best fit program (Wise, et al., 1985) by modified least square method was then used to detect the position of standard error curves present in the new histograms by the analysis of the shapes and relative maxima, and afterwards to compute pseudo-statistical parameters for each independent distribution (peak), that are mean value (mode), relative height and width at half height of the bell curve (this last parameter is a sort of "standard deviation" of the single distributions in polymodal analysis). The reliability of these fitting were then tested with Chi Square, Student's T and F distribution tests that showed positive results at 95% of level of comparison; the authors are anyhow well aware of the questionable statistical meaning of applying these tests with polymodal distributions. The kind of output from this program is shown in Figure 4, that presents the analysis of total digitized segments (20,131). Lower half of the figure represents the cumulative number (i.e. frequency) analysis, while the upper half the weighted one (see above). In the histograms in the middle are represented the smoothed normalized data (dots), single Gaussian distribution (bell curves) and theoretical fitted functions (continous lines). In the left side are the computed pseudo-statistical parameters and on the right the same Gaussian curves are represented on a rose diagram, to visualize immediately the preferential orientations. This analysis was applied to all segments together, and separately by order of stream (Fig. 5).

In order to investigate areal variability of the preferential orientations the same type of analysis was repeatedly performed by selecting data through segment coordinates by circular areas placed on an hexagonal centered grid. The side of the grid was 10 km and the radius of the circle was computed in order to have a total of 50% of overlap among one circle and the surrounding ones. The result

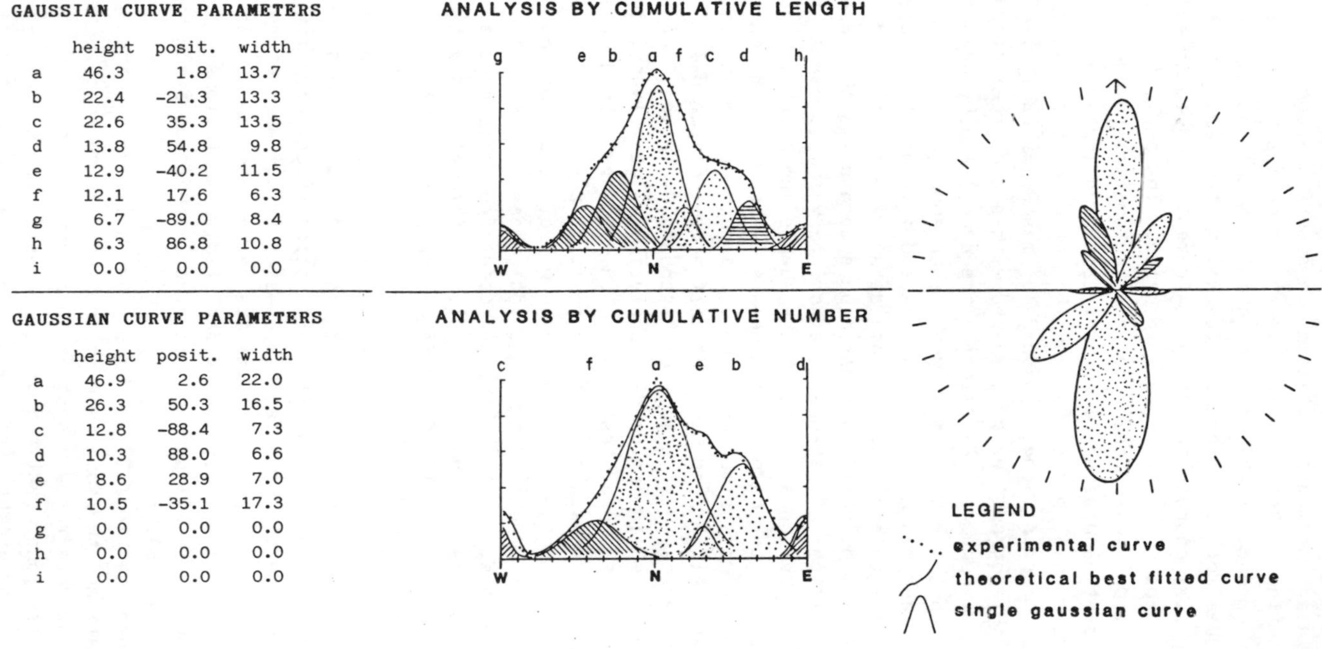

GAUSSIAN CURVE PARAMETERS

	height	posit.	width
a	46.3	1.8	13.7
b	22.4	-21.3	13.3
c	22.6	35.3	13.5
d	13.8	54.8	9.8
e	12.9	-40.2	11.5
f	12.1	17.6	6.3
g	6.7	-89.0	8.4
h	6.3	86.8	10.8
i	0.0	0.0	0.0

GAUSSIAN CURVE PARAMETERS

	height	posit.	width
a	46.9	2.6	22.0
b	26.3	50.3	16.5
c	12.8	-88.4	7.3
d	10.3	88.0	6.6
e	8.6	28.9	7.0
f	10.5	-35.1	17.3
g	0.0	0.0	0.0
h	0.0	0.0	0.0
i	0.0	0.0	0.0

Fig . 4. Azimuthal analysis by cumulative length (upper half) and by cumulative number (lower half) of all the rectilinear segments (20,131) collected in the Monti Sabatini area.

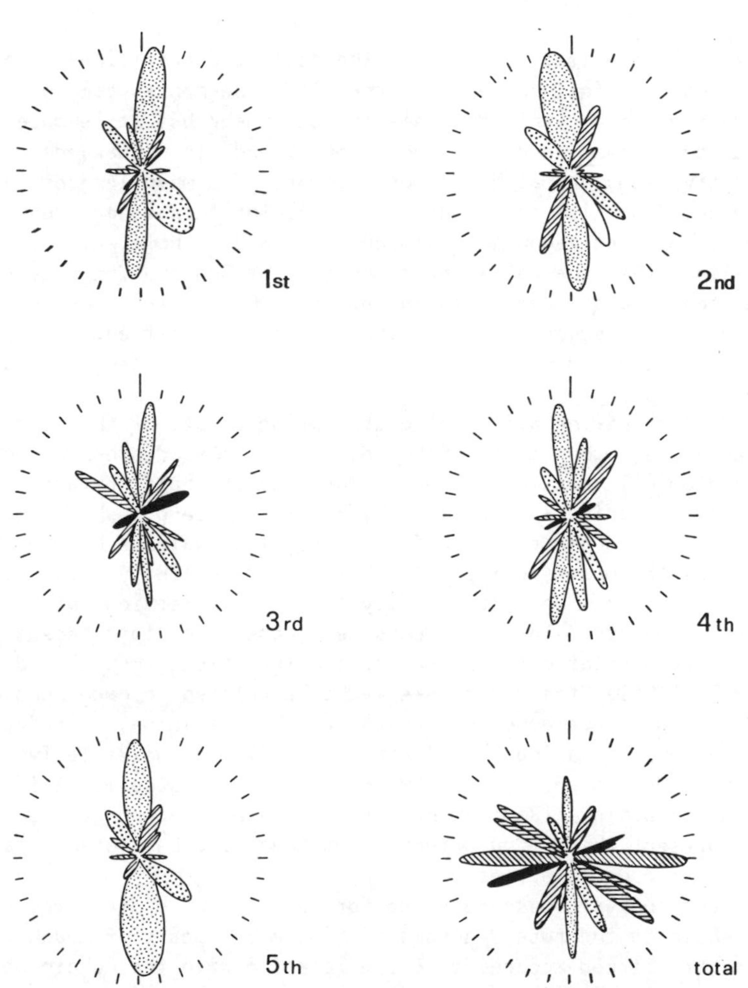

Fig. 5. Rose diagrams showing the preferential orientations of stream orders from 1st to 5th and of the whole drainage network.

was then presented by placing a rose diagram representing the fitted functions in the center of each circle. Figure 6 shows the result of this procedure for total data.

RESULTS

In Table I some statistics about the collected rectilinear segments are shown. The number of rectified segments decreases with increasing order; this decrease is not linear but it is more marked among the lower orders. The same trend is observed for the cumulative length, although more linear. The mean lengths tend to increase from 1st to 3rd order, while for the higher orders mean lengths do not show any particular trend; however, it is to underline that the rectilinear segments of 3rd and 7th orders seem to be too long as respect to the others. This observation about mean lengths might suggest the hypothesis, to be verified, that these channel orders may be controlled by fractures longer than the others.

Figure 4 shows the azimuthal distribution of all of the rectilinear segments in the Monti Sabatini area. As discussed in the methodological section, the figure upper half refers to the azimuthal distribution weighted by cumulative length while the lower half refers to the relative frequency. The analysis by cumulative length is the most significant, because it is likely that regional tectonic trends are more easily found in connection with longer stream segments. The main peaks were named "domains" because they are surely related to preferential orientations of the drainage network, while the minor ones might be related to data scattering. The marked prevalence of N-S (N 2° E) azimuthal preferential orientation or "azimuthal domain" is easily recognizable.Two minor domains, NE (N 35° E) and NNW (N 21° W), are also identifiable in the distribution. Beside these domains small secondary peaks are also present which show orientations that will become important in the higher order analyses.

The same analysis was processed for each stream order. The results are shown in the rose diagrams of Figure 5, peaks of each stream order are marked according to the relation with the domain observed in the total analysis.

Table II shows the azimuthal values of the identified preferential orientations or azimuthal domains. The rose diagram of the 1st stream order (Fig. 5) shows the prevalence of the N-S (N 9° E) domain, while NNW (N 21° W) and NE (N 46° E) domains assume a greater relative importance than in the total analysis. It is self-evident that the 1st order streams greatly influence the total data analysis, as they represent 61% of data set. The NNE peak is noticed only in the frequency analysis (lower half) while the

TABLE I. Statistics of stream rectilinear segments

Order	Number	Cum. length (km)	% number	% length	av. length (km)
Total	20131	4161.24	100	100	.207
1st	12273	2303.69	61	55	.188
2nd	4156	930.96	21	22	.224
3rd	1886	493.82	9	12	.262
4th	1078	252.93	5	6	.235
5th	459	111.39	2	3	.243
6th	231	50.11	1	1	.217
7th	45	17.84	1	1	.396

TABLE II. Main preferential orientations (azimuthal domains) of rectified drainage network

Preferential orientation	Stream order					
	1st	2nd	3rd	4th	5th	total
N-S	N9°E	N6°W	N7°E N10°W	N12°E N5°W	N3°E	N2°E
NE	N46°E	N39°E	N31°E	N31°E	N28°E	N35°E
ENE	--	--	--	N56°E	N62°E N53°E	--
E-W	--	--	--	--	E-W	
WNW	--	--	--	N65°W	N70°W	--
NNW	N21°W	N36°W	N34°W	N34°W	N26°W	N21°W

weighted one is included in N-S domain, this being related to particular variations of rectilinear segment lengths with azimuths.
The 2nd order rose diagram is similar to the 1st order one, but with a counterclockwise rotation of about 10 degrees (Fig. 5). The 3rd order rose diagram shows the presence of the three domains already cited (N-S, NNW and NE). The N-S domain is present as a double peak (N 7° E and N 10° W) which repeats the orientations of N-S domains observed in the 1st and 2nd order analysis; the direction relatable to 1st order N-S domain is prevailing between the two. Three minor peaks are present which rise to importance in higher stream orders.
The 4th order rose diagram (Fig. 5) shows a more equilibrated relative importance of the previously identified domains; moreover two new domains appear: WNW (N 65° W) and ENE (N 56° E). The N-S domain is represented by a double peak as in the previous order, but with reversed relative importance (cf. 3rd order analysis). The more dispersed distribution observed might well be related to poor control (number of data = 459). Because of even smaller data sets, for higher orders they were not analyzed.
As explained earlier, the same azimuthal analysis was performed by partial areas for the whole drainage network and for each stream order from 1st to 4th; higher orders were discarded because of data inadequacy. The aim was to the areal variability of the preferential orientations. Figure 6 shows the areal azimuthal distribution of all the linear segments. With rare exceptions, the N-S azimuthal domain is always present, but it prevails in the central and southern sectors of the study area. The NE domain is more concentrated in the south-western sector, while it is often absent in the others; for the NNW domain the situation is practically reversed: it is present in the easternmost sector and generally tends to disappear when the NE domain appears. The ENE domain is observable in central and northern sectors, while it is absent in the southern one. Much more limited is the presence of the WNW domain which practically appears only in the easternmost sector. Finally, the E-W domain is clearly present in the north-western sector.
To find out the possible relations between drainage pattern and field surveyed fracture trend, the preferential fracture direction found through a fracture station net (1,054 local systems on a total of 427 stations) were processed in a way similar to the rectilinear segments azimuthal analysis (Salvini, in press). This time the two analyses were progressed by number (i.e. simple relative frequency, lower diagrams in all figures) and by taking into account the intensity of each fracture system (upper diagrams), that is a weighting parameter according to the expression:

$$Is = If \times (P \times 0.1)^{1/2}$$

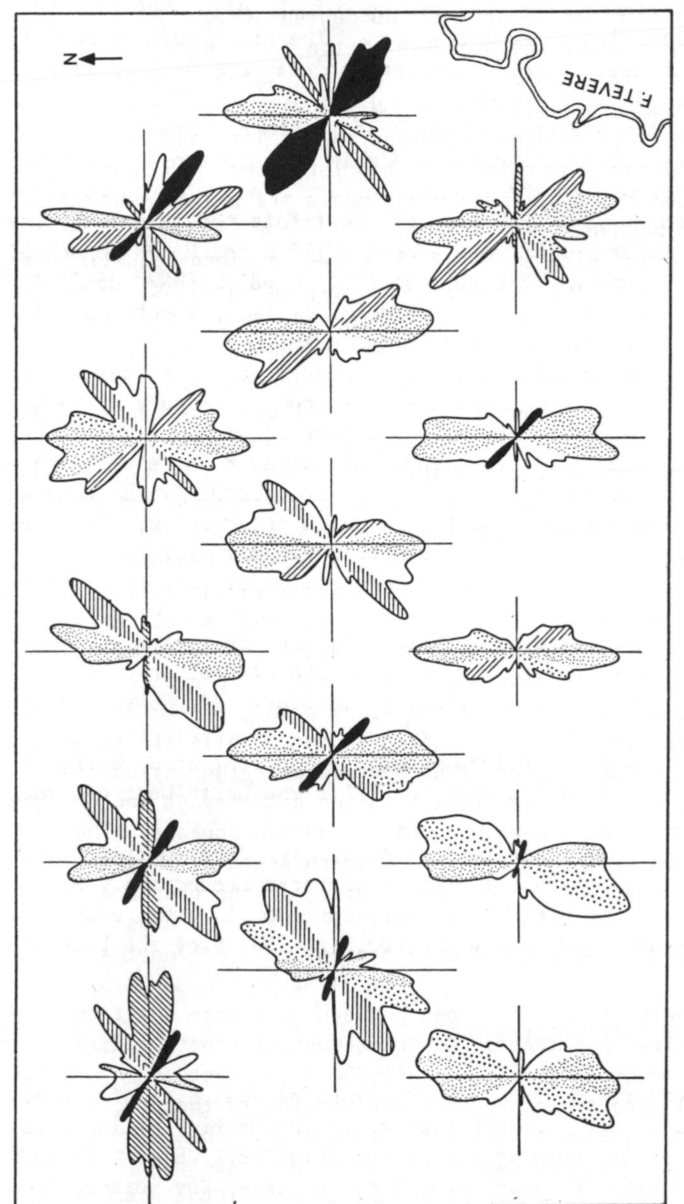

Fig. 6. Azimuthal analysis by partial areas. The rose diagrams evidence the areal variation of the whole drainage network orientations.

where: Is = intensity of the system; If = fracture intensity of the station; P = percentage of data within the station that belong to the system.

This adopted weighting factor (according to Salvini, 1984) affords a better interpretation of fracture data. Therefore this type of analysis will be used in the comparison of fracture with drainage pattern.

Figure 7 presents the analysis of total of fracture systems (1,054) and shows the possibility to identify 4 main orientations of systems or "fracture domains" in the Monti Sabatini area. The directions about N 85° E and N 89° W (E-W domain) constitute the main peak in both analyses; other peaks are around N 11° E and N 9° W (N-S domain), N 44° E and N 32° E (NE domain), N 68° W (WNW domain); moreover the direction N 23° W (NNW domain) is present but it appears only in the analysis by cumulative number.

The results of the areal analysis, progressed with the same parameters as for the drainage network, is shown in Figure 8. Peaks were identified according to the main fracture domains.

An E-W domain is immediately noticeable almost everywhere with the exception of the area South of the Lake of Bracciano. This domain presents two preferential azimuths: N 87° E and N 84° W. In the eastern sector the N 84° W is prevalent while in the western one the N 87° E prevails; in the northern sector the two directions appear always together.

The NE domain shows a consistent direction; in the central-western sector it appears concentrated around the Lake of Bracciano and is clearly present in the north-western boundary of Rome (south sector).

The N-S domain shows an evident pattern concentrated along a diagonal of the figure with smaller peaks to the North-West of the Lake of Bracciano.

The WNW domain appears widely scattered in five preferential orientations; it is very well defined in the area around the Lake of Bracciano. Finally the NNW domain is present in all the examined area, with particular evidence to the North and South of the Lake of Bracciano.

DISCUSSION AND CONCLUSIONS

Although the comparison between the azimuthal analysis of total rectilinear segments and fracture systems in the Monti Sabatini area suggests an apparently unsatisfying correlation (cf. Fig. 4 and 7), many interesting considerations can be drawn by discussing Figure 9 which shows in a series of azimuthal diagrams the variation of the domains by increasing stream order and at the same time compares them with fracture domains.

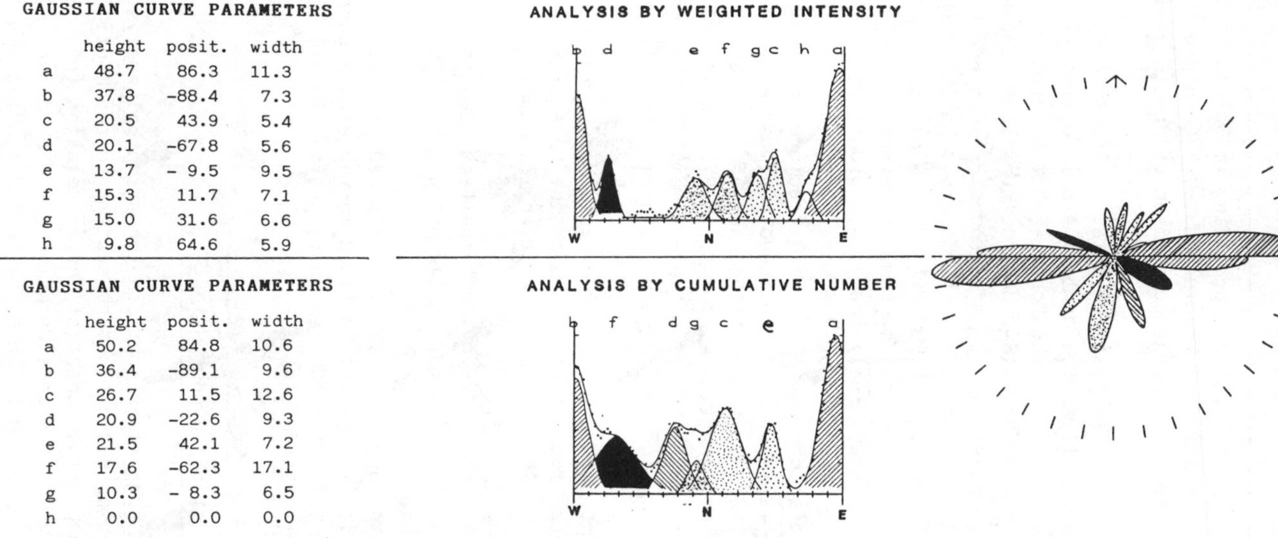

GAUSSIAN CURVE PARAMETERS

	height	posit.	width
a	48.7	86.3	11.3
b	37.8	-88.4	7.3
c	20.5	43.9	5.4
d	20.1	-67.8	5.6
e	13.7	- 9.5	9.5
f	15.3	11.7	7.1
g	15.0	31.6	6.6
h	9.8	64.6	5.9

GAUSSIAN CURVE PARAMETERS

	height	posit.	width
a	50.2	84.8	10.6
b	36.4	-89.1	9.6
c	26.7	11.5	12.6
d	20.9	-22.6	9.3
e	21.5	42.1	7.2
f	17.6	-62.3	17.1
g	10.3	- 8.3	6.5
h	0.0	0.0	0.0

Fig. 7. Azimuthal analysis by weighted local intensity (upper half) and by cumulative number (lower half) of fractures surveyed in the Monti Sabatini area.

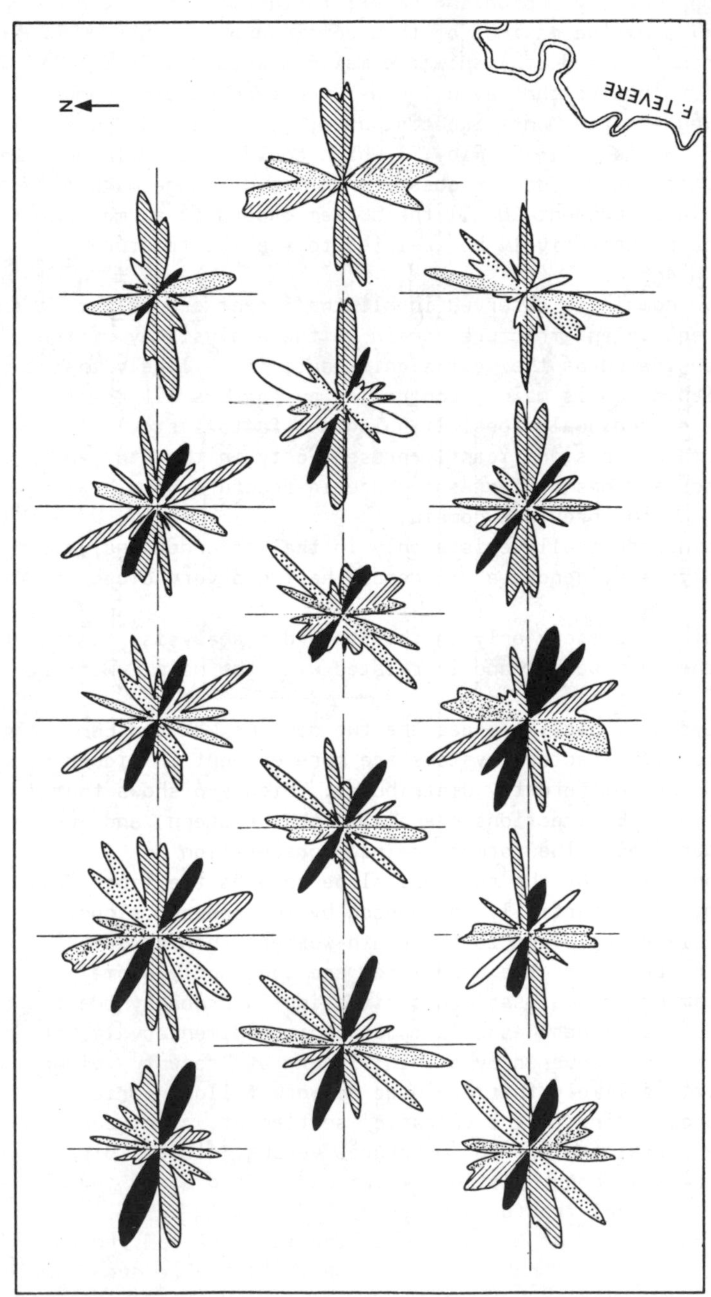

Fig. 8. Azimuthal analysis of fracture systems by partial areas.

The N-S domain is present in all the stream orders with decreasing relative importance by increasing order; furthermore the even orders (2nd and 4th) show the maximum of this domain around N 5° W, while the odd ones (3rd and 5th) show the maximum around N 8° E. It is interesting to notice that even the N-S fracture domain shows the same doubling in the Monti Sabatini area; in fact two peaks are present (N 9° W and N 11° E, Fig. 7) which could have influenced the double direction of N-S domain observed in the drainage network.

The NE domain is present in all the stream orders as a medium-size peak and could tentatively be related to the NE fracture domain (N 32° E and N 44° E).

Also the NNW domain is observed in all the stream orders but there is not any equivalent fracture domain in the analysis by intensity which was considered as the most significant; it is likely that this drainage orientation is mainly controlled by factors different from fractures (i.e. regional slope, lithological factors etc.).

The ENE domain is significantly present only in the 4th and 5th order analyses and has no correspondence in fracture data, with the same meaning of the previous domain.

The WNW domain practically exists only in the 4th order analysis but fracture analysis evidenced a WNW domain having a very close azimuth (N 68° W).

The E-W domain is present only in the 5th order analysis, where it represents the main peak, and is related with the main E-W fracture domain (N 87° E).

In the analysis by partial areas the two domains NNW and ENE which do not relate with fracture systems are more evident in higher order analyses and are differently distributed. Figure 6 shows that the NNW and the ENE directions prevail in the eastern and western sectors respectively. The former drainage orientation might be controlled by the regional slope towards the Tiber Valley, the latter might be partially influenced by the seaward slope.

The areal analyses relevant to the south-western, north-western and north-eastern corners of the study area are constantly anomalous in all the stream order analyses. This situation is probably due to the fact that in these zones Mesozoic to Cenozoic sedimentary lithotypes crop out or are covered by a thin layer of recent volcanics; therefore it is likely that drainage network follows orientations which are preexisting to the volcanic settlement. The comparison between the areal analyses of fractures versus stream directions does not show any good relation which can be explained by minor significance of local fracture analysis.

The marked prevalence of N-S domain in the lower stream orders (in particular in the 1st and 2nd) with respect to the other domains could lead to the hypothesis that the corresponding N-S fracture domain could represent the most recent fracture direction in the

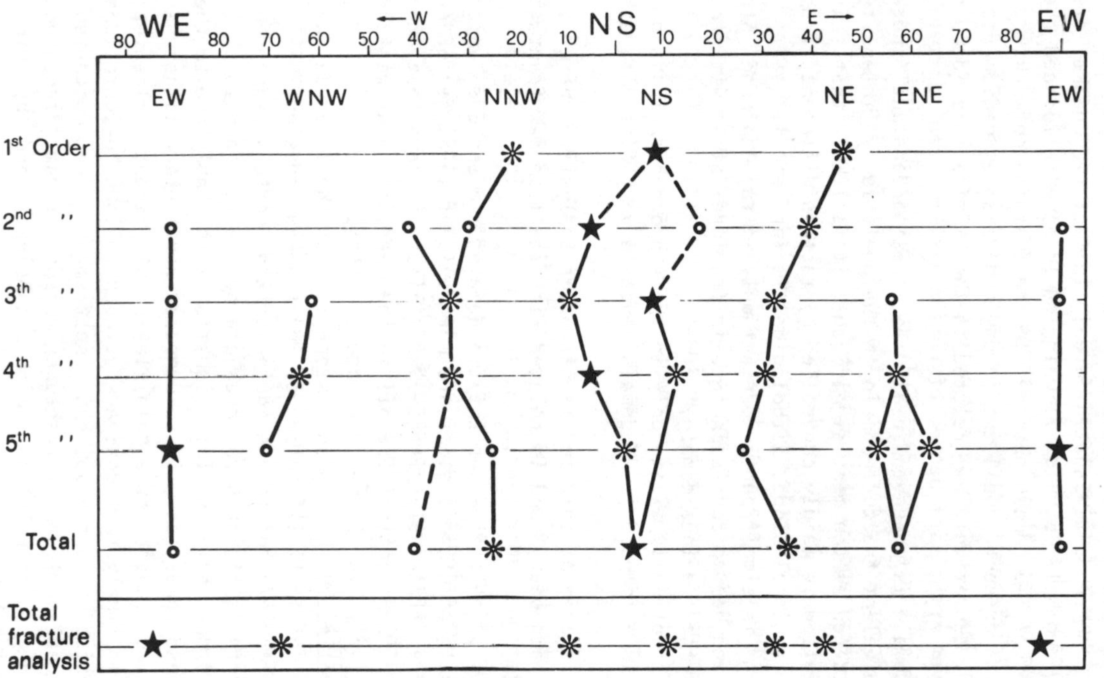

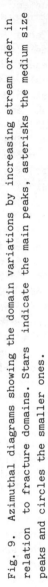

Fig. 9. Azimuthal diagrams showing the domain variations by increasing stream order in relation to fracture domains. Stars indicate the main peaks, asterisks the medium size peaks and circles the smaller ones.

Monti Sabatini area. On the other hand the E-W domain, that represents the fracture main orientation, compares to a drainage network domain that is present only in higher orders; therefore this fracture direction could reflect the reactivation of tectonics which were preexisting to the volcanic activity. Although these considerations require further verification, they do not contrast with the geological and structural indications of central Italy.

Concluding it can be said that the drainage orientation in this area is controlled by different factors. The main dislocations which led the volcanic complex formation surely determined a marked structural control. Moreover, the influence of regional slope can not be disregarded even if its role is difficult to be properly defined; in fact regional slopes are determined not only by the Tyrrhenian Sea (to the South-West) and by the valley of the River Tiber (to the South-East), but also by the emplacement of the volcanic cover, which followed a very complex pattern because of the lack of a central edifice. The complexity of this problem suggests the need of more detailed studies. - which are now being carried on - on the influence of regional slopes on the drainage orientation. Finally, the results of this research demonstrated that the drainage orientation of the Monti Sabatini area is strongly controlled by recent tectonics.

The methodology proposed in this paper proved adequate in the identification of the relations existing between drainage and fracture orientations. Moreover, on the basis of the relations observed in the study area, the hypothesis was made that the analysis of drainage orientation can help to determine the relative age of fractures; this last hypothesis however needs further tests on other areas with different geological conditions.

ACKNOWLEDGEMENTS

This study was funded by the Ministero della Pubblica Istruzione (Ministry of Education), National Project: Morphotectonics (E. Lupia Palmieri, Principal Investigator).

The authors thank M. Albano and M. Salvati for having prepared all the drawings.

REFERENCES

Bannister,E., and Arbor, N., 1980. Joint and drainage orientation of SW Pennsylvania. Zeit. für Geom.,N.F. Bd., 24 (3), 273-286.

Biasini, A., Buonasorte, G., Ciccacci, S., Fredi, P., and Lupia Palmieri, E., Geomorphological features of Monti Sabatini, in Geology of the volcanic complex of Monti Sabatini (in press).

De Rita, D., Funiciello, R., Rossi, U., and Sposato, A., 1983. Structure and evolution of the Sacrofano and Baccano Caldera. Journ. of Volcan. and Geoth. Res., 17, 219-236.

Diamant, E., Polishook, B., and Flexer, A., 1983. A positive correlation between joint trends and drainage patterns (underground storage project, Israel). Bull. Intern. Assoc. of Engin. Geol., 28, 103-107.

Lanzhou, N.S., and Scheidegger, A.E., 1981. Valley trends in Tibet. Zeit. fur Geom. N.F. Bd., 25 (2), 203-212.

Locardi, E., and Sommavilla, E., 1974. I vulcani sabatini nell'evoluzione della struttura regionale. Mem. Soc. Geol. It.,13 (2), 455-468.

Pohn, H.A., 1983. The relationship of joints and stream drainage in flat laying rocks of south-central New York and northern Pennsylvania. Zeit. fur Geom., 27 (3), 375-384.

Salvini, F. Le deformazioni fragili: individuazione dei sistemi, distribuzione areale e rapporto con le principali unita' litostratigrafiche, in Geology of the volcanic complex of Monti Sabatini (in press).

Scheidegger, A.E., 1980. The orientation of valleys trends in Ontario. Zeit. fur Geom. N.F. Bd., 24 (1), 19-30.

Wise,D.U., Funiciello, R., Parotto, M., and Salvini, F., 1985. Topographic lineament swarms: Clues to their origin from domain analysis of Italy. Geol. Soc. of Am. Bull., 96, 952-967.

International Geomorphology 1986 Part II
Edited by V. Gardiner

CLIMATE AND DRAINAGE BASIN MORPHOMETRIC PROPERTIES - A CASE STUDY OF RAJASTHAN

H. S. Sharma

Department of Geography, University of Rajasthan, Jaipur, Rajasthan, India

ABSTRACT

Morphometric properties were measured for 116 drainage basins from various climatic zones in Rajasthan. Drainage density increases from arid to semi-arid regions, and decreases in humid regions; it is also affected by rock type. Stream frequency behaves similarly. Dissection is maximum where rejuvenation has occurred, or in areas of higher rainfall, and texture ratio increases with increasing rainfall from arid to semi-arid zones, and decreases into the humid zones. Stream sinuosity is also at a maximum in the semi-arid regions.

INTRODUCTION

The validity of climatic geomorphology depends on its ability to establish a relationship between climate and process and to show that regional variations in climatic factors cause regional differences in landforms. The significance of climate in relation to process must also be evaluated in order to make this exercise more meaningful than a simple descriptive morphogenetic study. Although much information is now available concerning the variation of drainage density in relation to climate (Abrahams, 1972, 1980; Chorley, 1958, 1969; Derbyshire, 1973, 1976; Gardiner, 1971, 1974, 1982; Gregory, 1976; Gregory and Gardiner, 1975; Madduma Bandara, 1974; Wilson, 1973), very little attention has been directed towards the impact of climate on other aspects of fluvial morphometry. In India sufficient data are available to evaluate the geomorphic characteristics of drainage basins. However, very little work has been done on the relationships between climate and morphometry or process. Undoubtedly, such studies play a significant role in watershed management, especially in a water deficit state like Rajasthan. Keeping this in view, an attempt has been made in this paper to study the relationship between climate and morphometric properties of drainage basins in Rajasthan.

GEOGRAPHICAL LOCATION OF THE AREA

The state of Rajasthan covers an area of 342,261 square kilometres and lies between Latitudes 23°03' and 30°10' (north) and Longitudes 69°30' and 78°27' (east). It is bounded on the west and north-west by Pakistan, on the north and north-east by Haryana and Uttar Pradesh and on the south-east and south-west by Medhya Pradesh and Gujarat states resprectively.

Geologically, it is in the north-western part of the Indian shield which has not been affected by tectonic orogenic movements since the Mesozoic. However, western Rajasthan has witnessed various phases of marine transgression and regression up to the Tertiary period, while north and south-eastern Rajasthan have experienced cymatogenic upwarping. The evidences of recent uplift are incised meanders, ravines and gullies along the Chambal and its tributaries (Sharma, 1968).

Climatically, Rajasthan lies within a region which may be described as having arid (Ed), semi-arid (Dd), transitional (between Ed and Dd) and sub-humid (Cw) zones. The normal annual rainfall distribution and its variability are shown in Fig.1 from which it is inferred that rainfall decreases in the region from south and south-east to north-west, whereas coefficient of variability shows a reverse trend (Gupta and Singh, 1981). The area under study receives rainfall ranging from 100mm to 1650mm. The rainfall variability is high and evaporation losses are very high. Almost 90 per cent of the rainfall occurs between the months of June and September. During the remaining months of the year rainfall is light and sporadic, and several consecutive rainless months, and perhaps years, are not uncommon in arid parts of the state. Annual and daily ranges of temperature are high and there are marked seasonal contrasts in humidity, wind speed and radiation.

DATA

Drainage networks have long been considered to embrace several types of components, such as perennial, intermittent and ephemeral streams (Gregory, 1976). In this case study of Rajasthan an expanded network (intermittent streams) has mainly been considered. These stream channels are shown as continuous lines on the Survey of India topographical maps. It is unrealistic to standardize drainage density and stream frequency only on the basis of the permanent perennial drainage network, because most of the landforming processes are more active during conditions of high runoff, when water is flowing in the intermittent channels as well as those of base flow.

In order to show the relationship between climate and morphometric properties, 116 basins (54 third order basins and 62 random order basins) were randomly selected from various climatic zones of Rajasthan (Fig.2) and the values were derived from recent 1:50,000 topographical maps based on aerial photographs. The selection of sample basins has further been based on various rock types and morphological units within climatic regions. The morphometric properties of total number of stream segments, stream length, drainage density, stream frequency, dissection index, texture ratio, and stream sinuosity were measured and correlated with mean annual rainfall. The distribution of 116 sampled basins in this regional analysis is biased towards the eastern part of Rajasthan. The lack of sufficient well identified stream channels and rainfall data in the western part of Rajasthan precludes more extensive random sampling. The 116 basins correspond to rainfall measurement stations. Where particular rainfall data were not available, the equivalent value for the nearest station in the same climatic region has been used. Mean annual rainfall was selected as the sole independent variable for two reasons. Firstly, good estimates of mean annual rainfall for each sample basin could be obtained. Secondly, most of the rainfall (90 per cent) in this sub-tropical climatic region is concentrated in three months (July to September), during which rainfall intensity remains high. For want of reliable intensity data, this analysis has been carried out taking only the mean annual rainfall into account.

The spatial patterns of drainage density, stream frequency and dissection index have also been analysed. The data in this case were obtained from 1:253,440 topographical maps and supplemented with Landsat images. The data were measured by grid methods proposed by workers such as Gardiner (1971). The values so computed were later reduced to the 1:1,000,000 scale.

MORPHOMETRIC PROPERTIES

Stream Number. Quantitative methods to analyse morphometric characteristics of fluvial landforms were intitiated by Horton (1945). The first law of Horton is the law of stream numbers. Tables 1 and 2 provide the Rajasthan data for third order and random order basins respectively. The number of streams decreases with increase in stream order in all basins irrespective of climatic control (Sharma, 1983). However, looking at the number of stream segments of the first order, basins of humid climatic regions tend to have a higher number of stream segments than arid regions. Within a climatic zone, however, the number of stream segments of different orders does vary from basin to basin, which may be attributed to the character of the underlying geology.

Stream Length. The second law of Horton concerns stream length. Total stream length (Tables 1 and 2) again varies both within and between regions. For example, the value of total stream length for basin number 16 which falls in the sub-humid zone is recorded as the highest at 65km; that for basin 35 which falls in the semi-arid region is 57.15km; total stream length values tend to be smallest in the arid and transitional zones.

Drainage Density. Drainage density is the total length of stream per unit area. It is the result of the response to the controls of climate and catchment characteristics. It is also a main factor influencing runoff and sediment response of a drainage basin. Previous studies have indicated that drainage density increases through arid and semi-arid climates and decreases through humid continental and sub-tropical areas to values found in humid temperate areas (Gregory and Gardiner, 1975).

TABLE 1. Morphometric data for third order basins.

Climatic region	Basin number	Total no. of stream segments	Total stream length (km)	Drainage density (km/km)	Stream frequency (No./km)	Texture ratio (Nu/Pu)	Mean annual rainfall (mm)
Humid	1	23	30.50	1.28	0.96	0.92	985.7
	2	27	27.20	1.23	1.22	1.33	1128.4
	3	69	47.75	5.41	7.82	3.28	732.6
	4	39	24.85	3.25	5.10	2.67	929.2
	5	83	55.95	3.37	5.01	3.35	1017.9
	6	27	17.75	5.59	11.67	5.03	638.1
	7	42	26.70	3.65	5.75	3.28	638.1
Sub-humid	8	22	22.45	1.72	1.68	1.10	1008.8
	9	44	27.50	3.28	5.25	3.20	709.5
	10	33	18.50	1.95	3.49	2.38	758.6
	11	43	20.25	1.54	3.28	2.68	779.4
	12	15	42.00	1.25	0.86	0.96	781.6
	13	29	24.00	0.85	0.53	0.62	781.6
	14	19	27.50	0.99	0.79	0.88	709.5
	15	84	18.00	1.08	0.42	0.27	1008.8
	16	84	65.00	2.15	2.55	2.80	1008.8
	17	77	31.50	1.70	1.24	1.24	709.8
	18	69	25.00	2.58	2.31	1.75	1008.8
	19	21	11.87	0.42	0.75	0.41	754.0
	20	22	7.75	0.69	1.98	2.66	754.0
	21	21	6.87	0.86	2.63	1.60	754.0
	22	28	14.00	0.50	1.01	0.73	754.0
	23	32	26.14	0.37	0.46	-	758.6
	24	19	23.94	0.52	0.41	-	758.6
	25	28	10.72	0.87	2.27	0.77	758.6
	26	10	8.12	0.51	0.64	1.30	758.6
	27	24	12.28	0.70	1.51	-	758.6
	28	11.75	11.75	0.34	0.32	-	758.6
Semi-arid	29	33	26.95	2.43	2.98	2.06	748.5
	30	38	25.00	4.34	6.60	3.36	504.7
	31	15	15.00	1.38	1.33	0.92	513.6
	32	29	19.25	2.71	4.09	2.48	887.6
	33	53	24.00	3.24	7.16	4.18	645.4
	34	37	31.50	1.93	2.26	1.62	639.7
	35	24	57.25	0.58	0.24	0.45	656.8
	36	22	21.50	0.94	0.83	0.95	748.5
	37	72	33.00	2.24	2.10	1.16	748.5
	38	31	25.00	0.95	1.18	1.47	660.4
	39	29	28.00	0.88	0.92	1.45	704.9
Transitional zone	40	11	16.55	1.26	0.84	0.66	378.2
	41	33	19.00	3.28	5.69	3.00	418.8
	42	22	17.40	3.18	4.02	1.76	472.1
	43	12	14.80	0.95	0.77	0.60	378.2
	44	25	14.00	2.50	4.47	2.32	560.6
Arid	45	34	22.50	2.58	3.89	2.12	264.8
	46	10	13.25	2.25	1.70	0.76	270.1
	47	29	23.80	1.94	2.37	1.65	366.0

[continuation of Table 1]

Climatic region	Basin number	Total no. of stream segments	Total stream length (km)	Drainage density (km/km^2)	Stream frequency (No./km^2)	Texture ratio (Nu/Pu)	Mean annual rainfall (mm)
Arid	48	36	15.75	3.28	7.50	3.39	286.8
(cont.)	49	14	10.95	1.80	2.30	1.13	207.6
	50	14	7.00	3.09	6.19	2.33	178.5
	51	26	19.25	2.08	2.81	2.16	182.6
	52	23	18.05	1.30	1.66	1.42	182.5
	53	22	19.25	1.29	1.47	1.22	207.6
	54	32	21.00	3.81	5.81	2.66	314.0

TABLE 2. Morphometric data for random order basins.

Climatic region	Basin number	Total no. of stream segments	Total stream length (km)	Drainage density (km/km^2)	Stream frequency (No./km^2)	Texture ratio (Nu/Pu)	Mean annual rainfall (mm)
Humid	1	400	234.00	2.83	4.84	9.85	856.1
	2	193	212.25	3.14	5.40	6.20	732.6
	3	196	147.95	2.72	3.60	5.60	1017.9
	4	375	262.20	2.50	3.58	8.28	1044.7
	5	66	84.75	3.82	2.98	3.34	739.4
	6	159	68.70	2.58	5.97	6.29	834.0
	7	95	92.75	1.56	1.59	2.62	1089.3
	8	35	36.25	1.98	1.91	1.97	903.6
	9	134	66.80	4.43	8.90	5.25	1630.0
Sub-humid	10	400	184.20	4.93	10.70	16.00	537.7
	11	100	116.70	0.92	0.79	2.04	781.6
	12	289	210.55	2.05	2.81	5.47	709.5
	13	76	81.25	1.24	1.16	2.08	758.6
	14	120	65.75	1.37	2.51	4.21	754.0
	15	25	52.00	2.41	3.34	2.72	638.1
	16	231	114.50	2.43	2.08	3.67	537.7
	17	156	233.00	1.52	1.40	3.42	638.1
	18	31	24.50	1.72	1.35	0.58	638.1
	19	37	47.00	3.18	2.16	2.70	638.1
	20	7	52.00	0.91	0.65	0.92	781.6
	21	23	83.00	2.06	2.08	2.58	1008.8
	22	88	96.75	1.35	1.17	2.33	709.5
	23	134	111.00	2.21	3.11	3.46	1008.8
	24	105	110.50	1.61	1.28	2.24	709.5
	25	28	121.00	1.64	1.89	2.72	748.5
	26	40	11.62	0.69	2.39	3.50	754.0
	27	332	119.07	0.97	2.70	-	754.0
	28	32	19.21	0.53	0.86	-	758.6
	29	34	15.50	0.84	1.85	1.60	758.6
	30	81	21.00	1.13	4.37	1.37	758.6

[continuation of Table 2]

Climatic region	Basin number	Total no. of stream segments	Total stream length (km)	Drainage density (km/km)	Stream frequency (No./km)	Texture ratio (Nu/Pu)	Mean annual rainfall (mm)
	31	197	53.00	0.59	2.18	2.25	758.6
	32	66	45.00	0.71	1.04	-	758.6
	33	26	18.08	0.39	0.56	-	758.6
	34	47	26.46	0.50	0.90	-	758.6
	35	44	62.37	0.49	0.43	-	758.6
Semi-arid	36	88	99.75	1.04	0.92	2.12	748.5
	37	57	59.20	2.18	2.10	2.15	468.5
	38	78	56.50	0.94	1.30	2.31	536.9
	39	83	89.50	1.29	1.20	2.15	651.3
	40	252	128.75	2.88	5.63	6.76	618.5
	41	105	98.00	4.02	4.31	4.00	572.0
	42	188	97.50	1.00	1.93	4.82	644.4
	43	113	87.00	1.05	1.36	4.10	644.4
	44	57	44.50	1.10	1.41	2.53	660.4
	45	265	193.00	1.00	1.38	3.48	708.6
	46	76	52.00	1.09	1.60	3.70	708.6
	47	104	70.50	1.13	1.66	2.81	708.6
	48	108	87.00	1.44	1.79	3.60	660.4
	49	76	83.50	0.84	0.77	2.71	631.3
	50	128	175.00	1.76	1.29	4.12	661.8
	51	230	197.00	1.78	2.08	5.97	708.6
	52	21	21.50	0.94	0.92	1.68	702.9
Transitional zone	53	48	37.00	1.66	2.16	2.31	418.8
	54	103	87.25	2.66	3.14	4.03	472.1
	55	48	55.75	1.56	1.34	1.62	472.1
	56	87	49.70	3.30	5.78	4.72	460.6
Arid	57	60	35.60	3.61	6.10	4.21	434.4
	58	98	59.00	2.07	3.44	3.65	336.0
	59	65	53.75	2.08	2.52	2.81	434.4
	60	27	45.25	0.88	0.52	0.63	270.1
	61	30	23.00	1.54	2.01	1.73	207.6
	62	49	32.25	3.50	5.32	3.26	264.8

Nu - number of stream segments. Pu - length of basin perimeter in kilometres

The plot of drainage density against rainfall for third order basins (Fig.3) shows that drainage density increases from arid towards semi-arid regions and decreases in humid regions. The points of sub-humid regions appear to represent a wide range of drainage density values, for rainfalls between 700mm-800mm and about 1000mm. Drainage density attains its maximum value in the rainfall zone between 600mm and 700mm. The results are further supported by correlation coefficient values of each climatic zone (Table 3).

TABLE 3. Correlation coefficients.

	Humid region	Sub-humid region	Semi-arid region	Transi- tional	Arid region
Rainfall vs. Drainage density					
Third order basins	-0.81	+0.22	-0.17	-0.24	+0.40
Random order basins	+0.03	-0.32	-0.44	+0.35	+0.40
Rainfall vs. Stream frequency					
Third order basins	-0.81	+0.01	-0.16	+0.67	+0.62
Random order basins	+0.51	-0.02	-0.09	+0.13	+0.46

Correlation coefficient values are insignificant for all except the humid region

When the drainage density of random order basins is plotted against rainfall (Fig.4) the pattern appears to be the same as that observed for third order basins, but correlation coefficient values are very low.

The significance of correlation coefficients was also tested and the values have been found to be insignificant in all the climatic zones, except the humid climatic zone. These results, therefore, suggest that the hypotheses of earlier workers do not apply in the case of Rajasthan at the small basin scale. On the other hand, analysis of the drainage density map of Rajasthan (Fig.5) in relation to rainfall (Fig.1) shows that drainage density is related to mean annual rainfall, but that the relationship is a complex one substantially affected by rock type.

Stream Frequency. Stream frequency, plotted against mean annual rainfall (Figs.6 and 7) shows the same trend as that of drainage density. The correlation coefficients of various climatic zones are given in Table 3. The stream frequency map (Fig.8) also shows a very similar pattern to that of drainage density. The variation in stream frequency in different climatic zones is because of local factors such as geological structure and vegetation cover.

Dissection Index. The dissection index (Number of contour crenulations/Area of basin in square kilometres) is an index of the degree of dissection in the area (Fig.9). The value of dissection is very low in the north-western part of Rajasthan. It is higher in the southern and south-west parts of Aravallis. The dissection index does not show any particular relationship with the rainfall pattern of Rajasthan, but it is evident that the dissection is maximum in the areas which have either undergone stream rejuvenation, as in the lower course of the Chambal, or have higher rainfall. Maximum dissection indices are found in south Aravallis, perhaps because of severe fluvial erosion and higher rainfall in the Mount Abu zone.

TABLE 4. Sinuosity index and climate.

Stream segment	Channel length in km (X)	Direct length in km (Y)	Sinuosity index (X/Y) (Leopold, 1964)
Arid region (The Luni Basin)			
I	31.00	21.00	1.47
II	0.25	7.50	1.10
III	10.00	8.00	1.25
IV	27.75	21.50	1.25
V	20.50	18.50	1.10
VI	8.00	7.25	1.10
VII	13.00	8.50	1.52
VIII	16.00	13.00	1.23
IX	26.00	23.00	1.13
X	33.25	29.00	1.14
Total	139.75	157.25	12.33
Mean	19.37	15.72	1.23
Semi-arid region (The Gambhir Basin)			
I	27.00	15.50	1.74
II	46.50	28.25	1.64
III	41.25	27.00	1.52
IV	12.50	7.50	1.66
V	23.25	20.50	1.13
VI	12.00	8.75	1.37
VII	36.00	26.00	1.38
VIII	35.50	25.50	1.39
IX	20.25	16.00	1.26
X	31.00	23.50	1.31
XI	9.50	6.50	1.46
Total	294.75	205.00	15.86
Mean	26.79	18.63	1.44
Sub-humid region (The Berach Basin)			
I	25.00	19.00	1.31
II	34.00	27.50	1.23
III	21.00	19.00	1.10
IV	9.50	9.00	1.05
V	29.50	25.50	1.15
VI	32.00	29.50	1.08
VII	10.00	9.50	1.05
VIII	33.50	30.55	1.09
IX	4.00	3.50	1.14
Total	198.50	173.00	10.20
Mean	22.05	19.22	1.13

[continuation of Table 4]

Stream segment	Channel length in km (X)	Direct length in km (Y)	Sinuosity index (X/Y) (Leopold, 1964)
Semi-humid region (The Mej Basin)			
I	1.65	2.25	1.36
II	1.10	1.30	1.19
III	2.25	3.00	1.33
IV	0.75	1.85	2.45
V	0.60	1.50	2.50
VI	1.00	2.35	2.35
VII	1.00	2.60	2.60
VIII	1.19	2.12	1.95
	-		-
Total	9.54	16.97	15.73
Mean	1.19	2.13	1.96
	-		-
Humid region (The Chambal and Mahi Basins)			
I	44.50	24.00	1.85
II	28.00	19.25	1.45
III	33.25	25.75	1.29
IV	40.25	28.25	1.42
V	31.00	26.75	1.15
VI	26.50	24.25	1.09
VII	24.50	22.00	1.11
VIII	33.50	28.50	1.17
IX	38.00	22.50	1.68
X	42.00	32.50	1.29
XI	31.50	25.75	1.22
XII	50.75	30.50	1.66
	-		-
Total	423.75	310.00	16.38
Mean	35.31	25.83	1.36

Texture Ratio. Texture ratio is the combined result of drainage density and stream frequency. The length and spacing of streams collectively form a texture of drainage. The texture ratios of random order and third order basins in different climatic zones are presented in Tables 1 and 2. The values of texture ratio of third order basins have been plotted against rainfall in Fig.10, which shows that texture ratio increases with increase in rainfall from arid to semi-arid zones and decreases with increase in rainfall in humid zones.

Stream Sinuosity Index. Stream sinuosity index is one of the parameters used to differentiate various types of land units. It is commonly expressed quantitatively as the ratio between the direct length and the channel length, or as the percentage increase in channel length per unit direct length. Leopold, Wolman and Miller

(1964) have maintained that a stream with sinuosity over 1.5 may be considered as meandering, and below 1.5 as non-meandering. In nature all streams show significant departures from a straight line course and hence possess some degree of sinuosity. There are several factors, such as hydraulics, lithology, topography and climate, which are responsible for the variation in the geometry of a meandering channel. However, it has been established that sinuosity index increases with the advancement of the cycle of erosion. When direct length and channel length are plotted for different climatic regions, a distinct pattern emerges signifying the role of climate in stream sinuosity irrespective of the stage of the cycle of erosion.

In the present study an attempt has been made to correlate sinuosity index with rainfall in the different climatic regions of Rajasthan. Measurements were taken at fifty sites in different river basins in different climatic zones of Rajasthan. The sinuosity index was calculated according to Leopold's method. The data used in this paper (Table 4) were measured from 1:50,000 Survey of India Toposheets.

The sinuosity index increases from arid to semi-humid with the exception of the Berach basin, and then decreases in the humid region. The table further indicates that channel length tends to increase from the arid to semi-arid, and then decreases in sub-humid regions, and again increases in the humid region, while direct length increases from arid to humid regions.

CONCLUSIONS

The analysis of 116 small basins of Rajasthan demonstrates that drainage networks vary substantially with climate, and particularly in relation to rainfall. Additional data are required to examine more thoroughly the hypothesis of whether drainage density, stream frequency, texture ratio and sinuosity index truly increase through arid to semi-arid zones and then decrease in the humid climatic zone. However, it can be ascertained from the little data available that the above hypothesis does not apply in the case of Rajasthan at the small basin scale, although at the meso scale (of the whole of Rajasthan) morphometric properties like drainage density and stream frequency are related to mean annual rainfall.

Acknowledgements

The author is thankful to Professor Moonis Raza, presently Vice-Chancellor, University of Delhi, Delhi, India, for his guidance in the preparation of this study; the University Grants Commission, New Delhi, for the financial assistance to carry out the research under the Career Award Scheme; and to Mr. R. K. Singh for secretarial assistance.

REFERENCES

Abrahams, A. D., 1972. Drainage densities and sediment yields in Eastern Australia. Australian Geographical Studies, 10, 19-41.

Abrahams, A. D., 1980. Channel link density and ground slope. Annals, Association of American Geographers, 70, 80-93.

Chorley, R. J., 1956. Climate and morphometry. Journal of Geology, 65, 628-38.

Chorley, R. J. (Ed.), 1969. Water, earth and man. Methuen, London.

Derbyshire, E., (Ed.), 1973. Climatic geomorphology. Macmillan, London.

Derbyshire, E., (Ed.), 1976. Geomorphology and climate. Wiley, London.

Gardiner, V., 1971. A drainage density map of Dartmoor. Reports and Transactions of the Devonshire Association for the Advancement of Science, 103, 167-80.

Gardiner, V., 1974. Landform and land classification in North West Devon. Reports and Transactions of the Devonshire Association for the Advancement of Science, 106,141-53.

Gardiner, V., 1982. The impact of climate on fluvial systems, in Perspectives in geomorphology, Vol. 2, (Ed. Sharma), pp19-40. Concept Publishing Company, New Delhi.

Gregory, K. J., 1976. Drainage networks and climate, in Geomorphology and climate, (Ed. Derbyshire), pp289-315. Wiley, London.

Gregory, K. J., and Gardiner, V., 1975. Drainage density and climate. Zeitschrift für Geomorphologie, 19, 287-98.

Gupta, B. L., and Singh Nityanand, 1981. Some findings of rainfall characteristics in Rajasthan. Geographical Review, 43, 326-341.

Horton, R. E., 1945. Erosional development of streams and their drainage basins: a hydrophysical approach to quantitative morphology. Bulletin Geological Society of America, 56, 275-370.

Leopold, L. B., Wolman, M. G., and Miller, J. P., 1964. Fluvial processes in geomorphology. W. H. Freeman, San Francisco.

Madduma Bandara, C. M., 1974. Drainage density and effective precipitation. Journal of Hydrology, 21, 187-90.

Sharma, H. S., 1968. Genesis of ravines of the Lower Chambal Valley, India. Selected Papers, 1, 114-118, 21st International Geographical Congress, India.

Sharma, H. S., 1983. Morphogenetic Regionalization of Rajasthan. Unpublished Project Report submitted to University Grants Commission under Career Award.

Wilson, L., 1973. Relationships between geomorphic processes and modern climates as a method in Palaeoclimatology, in Climatic geomorphology), (Ed. Derbyshire), pp269-84. Macmillan, London.

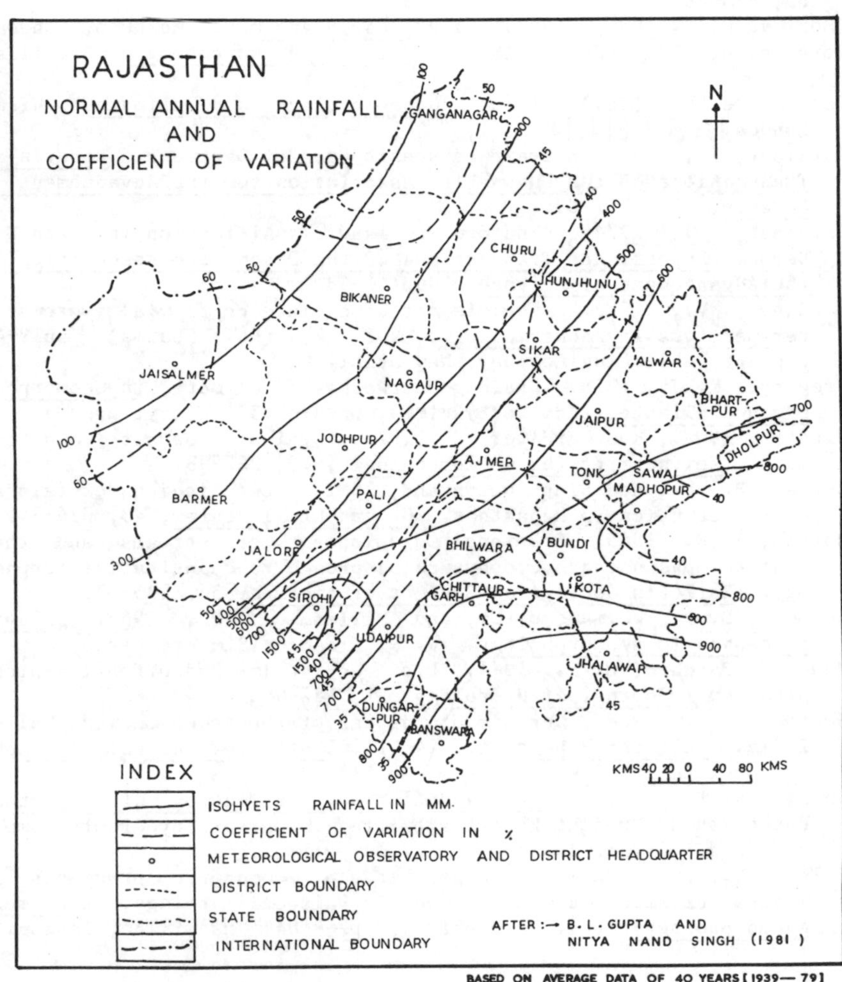

Fig.1. Rajasthan : normal annual rainfall and coefficient of variation, based on data from 1939-1979.

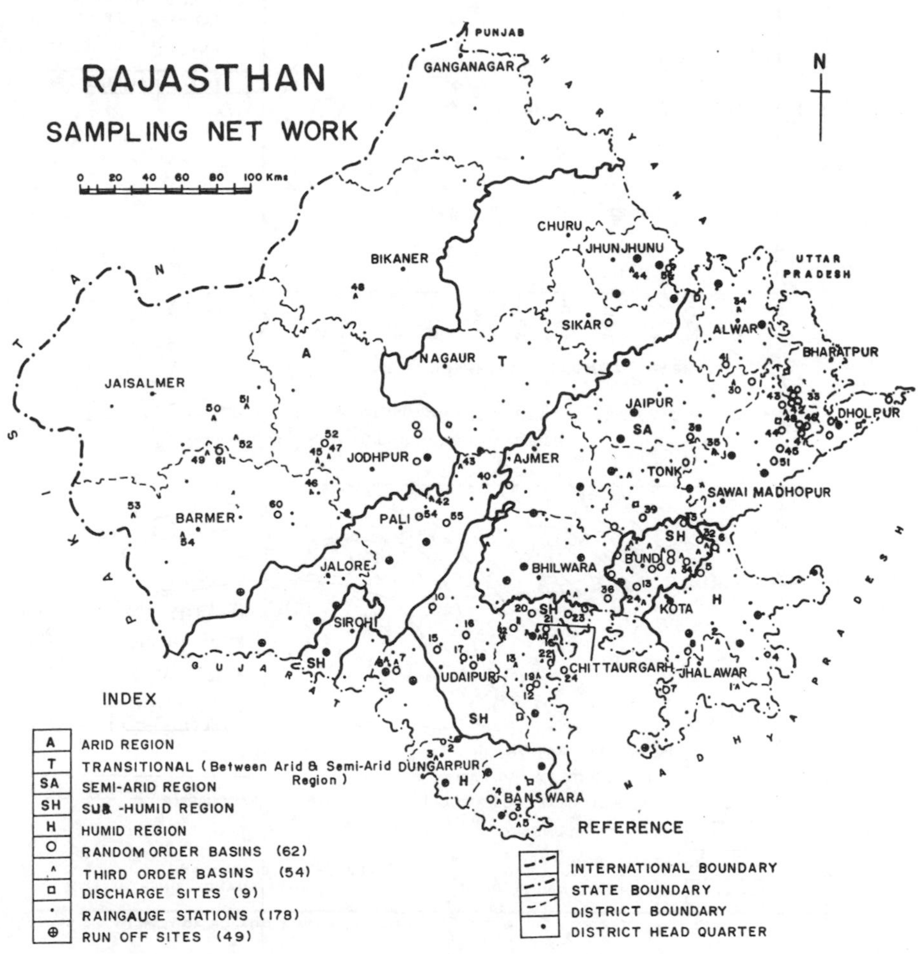

Fig.2. Rajasthan : sampling network.

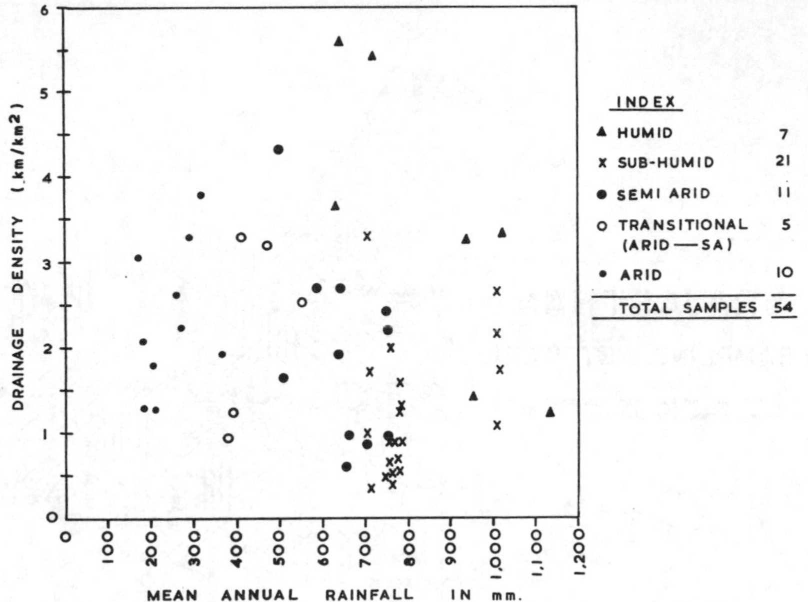

Fig.3. Drainage density and mean annual rainfall, third order basins.

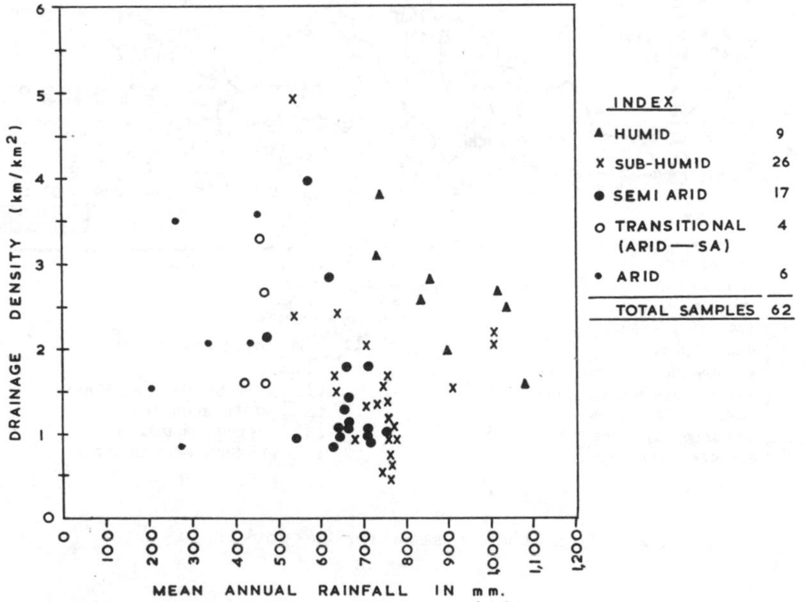

Fig.4. Drainage density and mean annual rainfall, random order basins.

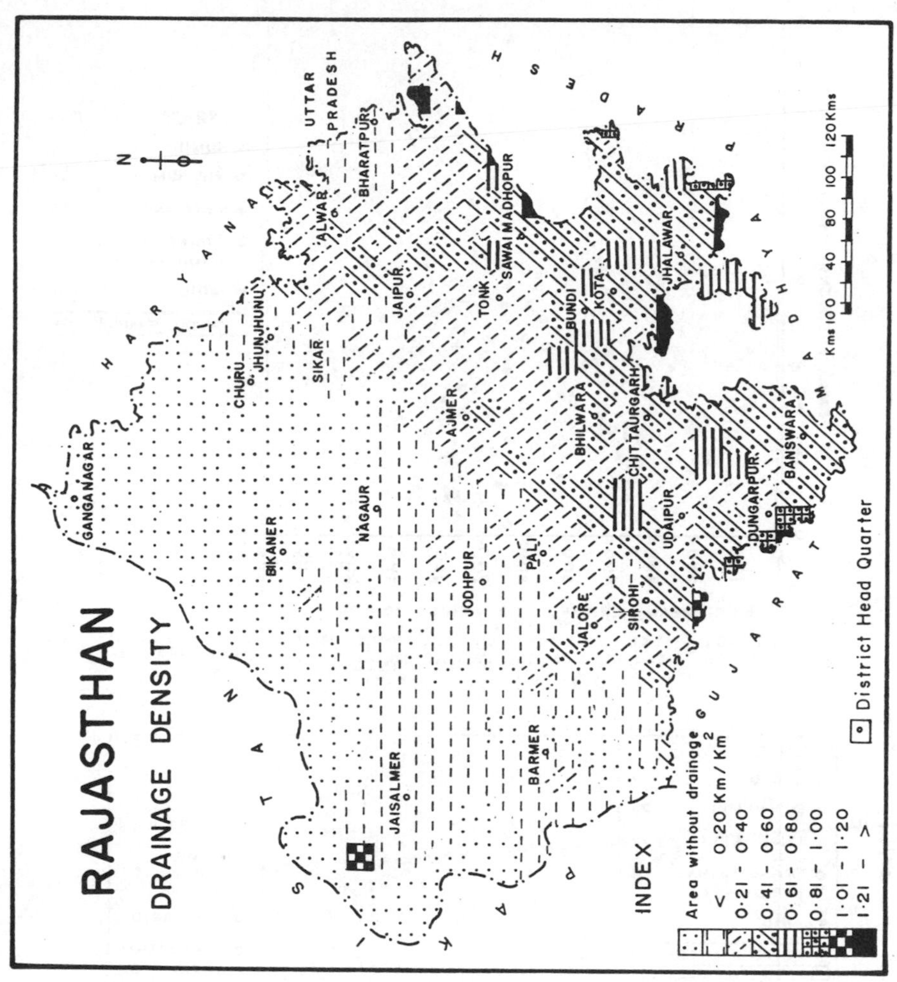

Fig.5. Rajasthan : drainage density.

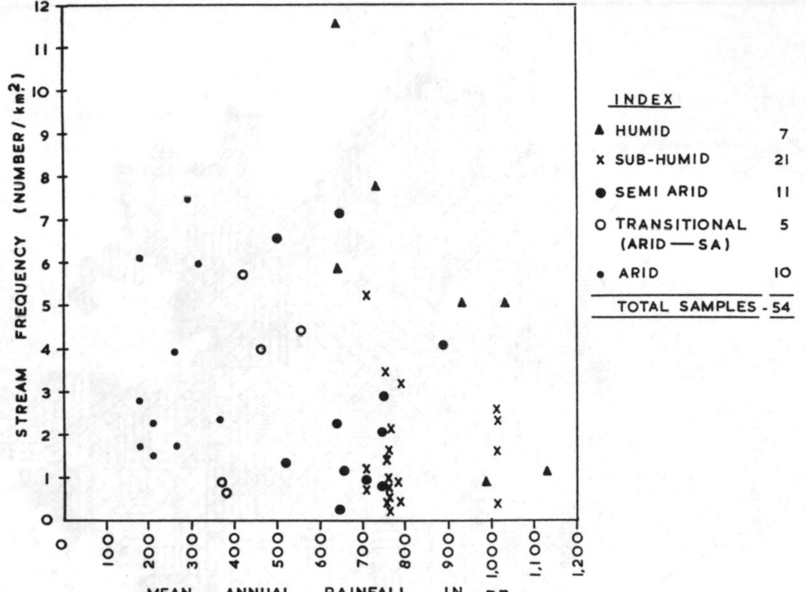

Fig.6. Stream frequency and mean annual rainfall, third order basins.

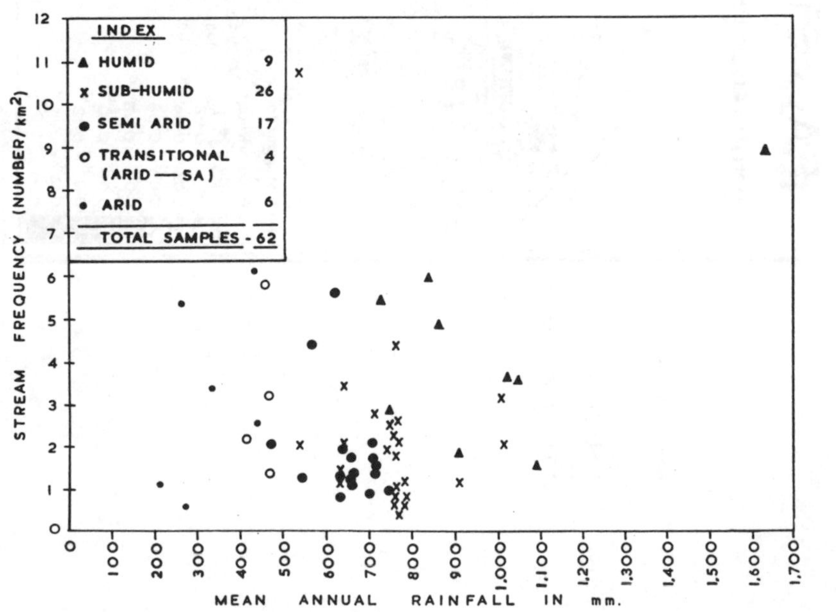

Fig.7. Stream frequency and mean annual rainfall, random order basins.

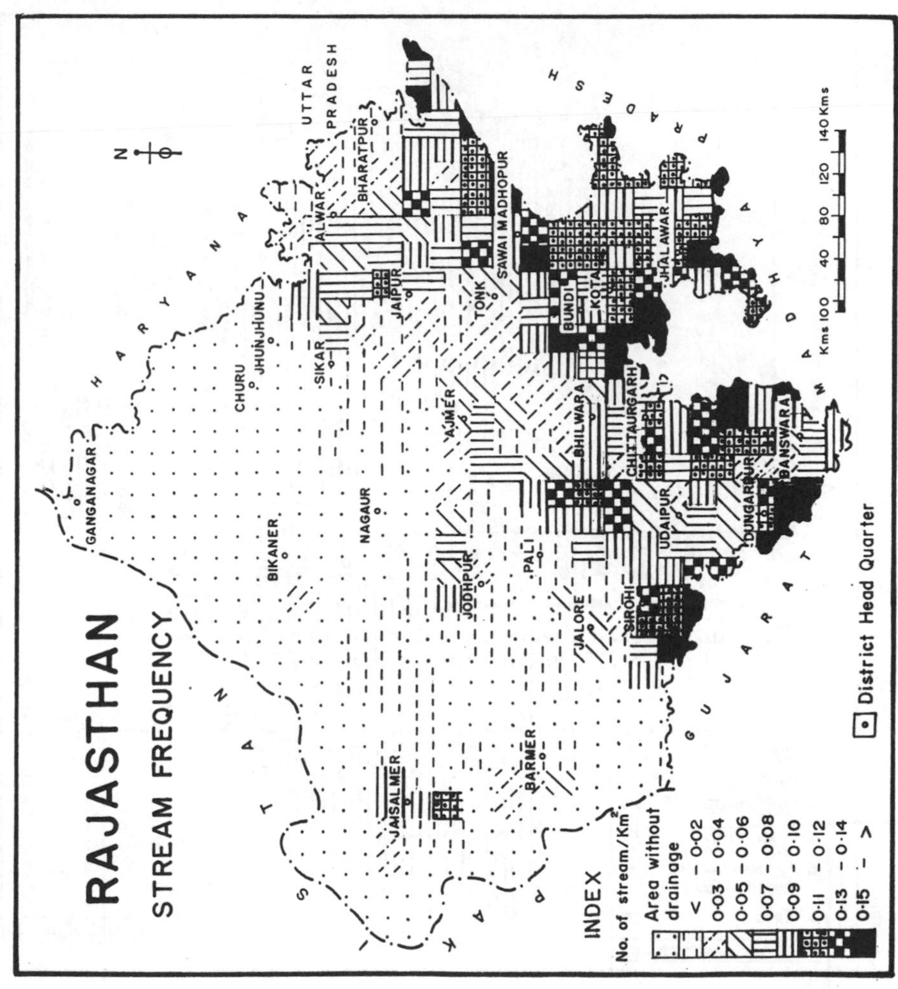

Fig.8. Rajasthan : stream frequency.

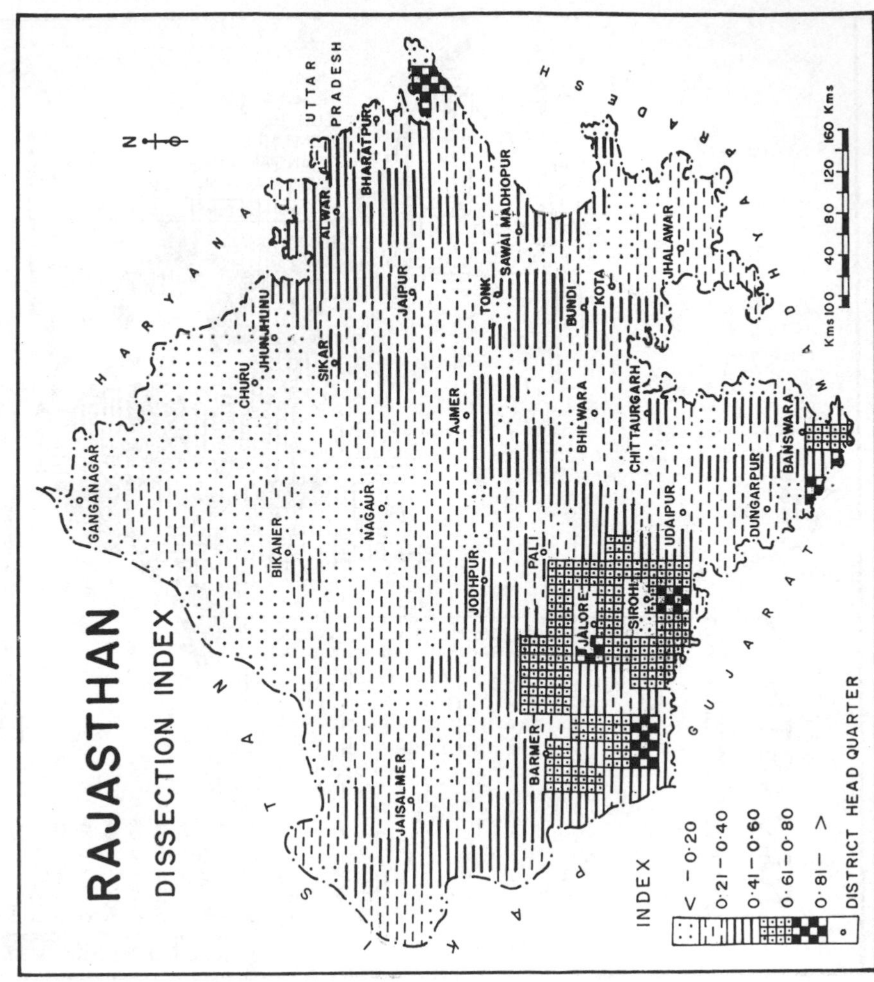

Fig.9. Rajasthan : dissection index.

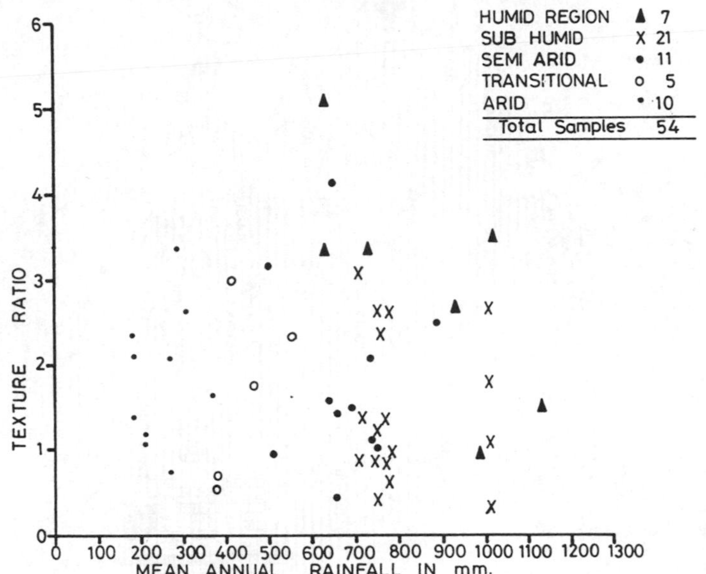

Fig.10. Texture ratio and mean annual rainfall, third order basins.

International Geomorphology 1986 Part II
Edited by V. Gardiner

PROCESS, FORM AND BOUNDARY CONDITIONS ALONG VALLEY-SIDE SLOPES

A. J. Parsons,

Department of Geography, University of Keele,
Keele, Staffs.

ABSTRACT

Differences along a valley side in the position of the basal stream may be regarded as differences in the local boundary conditions of the valley side. Studies from three areas show that there is a downvalley increase in maximum slope along valley sides. In one of these areas this increase is attributed to the coincident downvalley increase in relief accompanying stream incision. But in the other two study areas such a downvalley increase in relief appears to be inadequate to explain the observed increase in maximum valley-side slope. It is argued that a downvalley increase in valley-floor width and undercutting by the basal stream is the likely cause of slope steepening. The observed downvalley changes in maximum slope may be regarded as the realisation of a response to downvalley changes in slope boundary conditions which result from both vertical and lateral changes in the position of the basal stream. That hillslope form varies in response to varying boundary conditions suggests that both the spatial and temporal changes in boundary conditions that occur in natural landscapes need to be considered if the complexity of naturally-occurring hillslope form is to be understood.

INTRODUCTION

The form of a valley side is a response to the operation of geomorphic processes. Along the length of any valley side it is common to observe substantial differences among hillslope profiles measured perpendicular to the valley-side contours. These differences in form reflect variation in local factors controlling the operation of geomorphic processes. Such local factors include geological inhomogeneities, variations in hillslope plan curvature, and differences in the position of the basal stream. Differences along a valley side in the position of the basal stream may be regarded as differences in local boundary conditions. Depending upon the degree of incision of the basal stream and upon its lateral movement across the valley floor, the lower boundary of the valley side will alter both vertically and horizontally from place to place along the valley side. This paper is concerned with differences among hillslope profiles measured along valley sides in relation to local boundary conditions. More specifically, it examines variation in maximum slope measured on hillslope profiles and the extent to which these variations in maximum slope can be related to variations in boundary conditions induced by the basal stream.

Previous work on the variation in valley-side slope (both maximum and mean) in relation to basal streams has considered this variation either across valleys (Melton, 1960) or among drainage basins of different order (Strahler, 1950; Carter and Chorley, 1961; Arnett, 1971; Richards, 1977). Melton, in a study of valley asymmetry, concluded that asymmetric basal corrasion was, in the majority of instances that he studied, the most probable cause of differences in steepness of opposing valley sides. Studies of the variation in valley-side slope with stream order have reached two apparently contradictory conclusions. Strahler (1950), taking data from nine widely spaced sites across the United States, found a positive correlation between mean maximum valley-side slope at each site and channel gradient, and argued that this relationship indicated an adjustment between erosion from the slopes and debris removed by the streams. On the other hand, Carter and Chorley (1961) and Arnett (1971), both working in relatively small areas, found that the mean maximum slope for basins of a given order increased with increase in basin order (up to order 4 in Carter's and Chorley's study and up to order 5 in Arnett's study), suggesting that as order (and hence discharge) increases so maximum valley-side slope increases. Carter and Chorley noted the difference between their result and Strahler's and suggested that it was due partly to a difference in scale of study and partly to the particular character of their study area (an expanding stream network). Richards (1977) pointed out that the apparent contradiction between Strahler's results, on the one hand, and Carter's and Chorley's and Arnett's, on the other, can be separately accounted for by adjustment of stream gradient to the supply of sediment from valley sides, in the former case, and downstream increases in relief, in the latter case. Richards argued that the increase in discharge that accompanies the increase in stream order results in a greater depth of incision and that "the result is, for simple geometrical reasons, gradual steepening of mean slope angle" (p.88). The increase in valley-side slope with increase in stream order is due simply to an increase in relief. Among basins of a given low order (and particularly first order) in which floodplain development has not occurred to such an extent that the sensitivity of the slope-stream relationship has been weakened stream gradient is adjusted to the supply of sediment which, in turn, depends on valley-side gradient, and this leads to a positive relationship between the two gradients.

Although these studies provide some insight into the relationship between the gradients of valley sides and properties of their basal streams, and suggest some causative mechanisms, they do not deal directly with the issue of differences in maximum slope along one valley side within a drainage basin. Richards' model for increase in mean valley-side slope with stream order as a function of increasing depth of incision may, however, be applied at a within-basin level. In a landscape in which stream incision is still active it may be expected that there will be an increase in relief downvalley within a drainage basin. Under such conditions valley-side slopes must become longer downvalley and hence the likelihood of slope failure must increase downvalley. Such slope failures, although tending to reduce average slope, typically result in short sections inclined at steeper angles than before. Hence the increase in relief downvalley is likely to lead to an increase in maximum valley-side slope, i.e. a downward change in the lower boundary of a valley side with respect to the upper one is likely to cause an increase in maximum valley-side slope. Such a situation is most likely to be found where stream incision into a relatively level surface is taking place and where this incision has not progressed so far as to remove all of the pre-existing low-angle surface. In landscapes of this type the elevation of the divides is determined, primarily, by the gradient of the pre-existing surface whereas that of the valley floors is determined by the amount of stream incision. Exmoor in south-west England is such a landscape and thus provides a suitable locality in which to test the validity of Richards' model at the within-basin level.

EXMOOR STUDY AREA

Exmoor is an upland area reaching a maximum elevation of 433 m developed on Devonian slates, siltstones and sandstones that have a regional dip to the south. These rocks were eroded, principally in Tertiary times, and a low-angle surface, cutting across geological structures, is thought to have formerly existed. Present-day drainage is incising into this surface (now tilted to the south) but extensive remnants of it survive at elevations up to 400 m. Within the upper parts of this area, where the vegetation consists of heather and grass moorland, rough grazing and improved pasture, suitable valley sides on which to test Richards' model were sought. The requirements for these valley sides were that they should (i) be in valleys of low order; (ii) have divides that do not exhibit a significant decrease in elevation in the downvalley direction; (iii) be relatively straight in plan; (iv) be free of any tributaries entering across them; and (v) be long enough for several, widely spaced profiles to be measured down them.

TABLE 1. Summary characteristics of valley sides in the three study areas

Study area	Exmoor	South Downs	New South Wales Bathurst	New South Wales Braidwood	New South Wales Uralla
Mean length (metres)	429.1	245.8	121.7	100.0	263.3
Mean slope (degrees)	8.6	9.1	8.1	9.3	1.5

Four valley sides were identified which satisfied all of these conditions and a fifth valley side (Pennycombe Water) which met all of the conditions except (iv) was included in the sample (Fig. 1).

Summary characteristics of these valley sides, determined from the field measurements, are presented in Table 1. Within the selected valley sides, a systematic sample of hillslope profiles was taken.

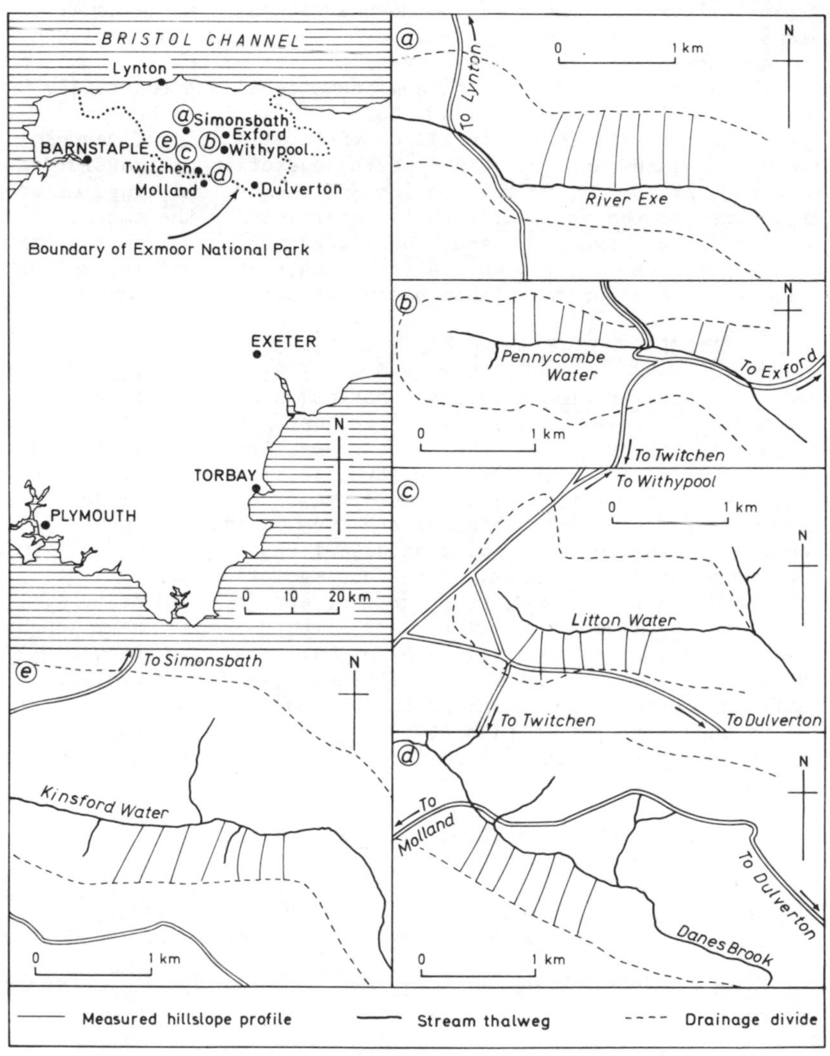

Fig. 1 Exmoor study area

Commencing at a randomly selected point well downstream of the valley head, profile locations were identified at 200 m intervals along the profile sampling line (Young, 1974, p.17) until 7 profiles had been located. Profiles were measured from these points using an Abney level and tape adopting a 5 m measured length between survey stations. Profiles were terminated at the downslope end at the stream bank and at the upslope end where a consistent reverse slope was recorded. The only exception to these general rules is the set of profiles measured along Pennycombe Water. Here, because of the tributary stream, the interval between profiles 5 and 6 was 900 m (Fig. 1).

Table 2. Characteristics of Exmoor hillslope profiles

Drainage basin	Profile no.	Profile length (m)	Profile relief (m)	Slope of maximum segment (°)
R. Exe	1	515.0	44.6	16.3
	2	480.0	48.6	18.6
	3	525.0	50.2	20.7
	4	560.0	58.6	24.6
	5	590.0	61.7	20.5
	6	610.0	65.5	19.4
	7	597.0	67.3	20.8
Pennycombe Water	1	395.0	46.2	17.4
	2	395.0	53.8	20.7
	3	325.0	49.1	19.2
	4	233.0	43.8	18.2
	5	172.0	38.2	18.4
	6	349.9	70.7	20.3
	7	365.0	79.6	22.7
Litton Water	1	413.5	35.3	8.5
	2	449.1	38.1	11.1
	3	338.0	37.6	11.3
	4	314.0	42.6	12.3
	5	355.0	46.4	14.9
	6	355.0	52.6	14.6
	7	380.0	52.6	14.5
Danes Brook	1	418.4	58.3	15.0
	2	385.0	59.7	15.7
	3	370.0	57.0	21.9
	4	423.7	63.2	22.9
	5	505.0	66.0	17.6
	6	505.0	70.7	22.2
	7	464.0	71.6	24.2
Kinsford Water	1	475.0	93.8	24.5
	2	425.0	90.9	24.0
	3	465.0	89.9	25.2
	4	595.0	94.4	22.0
	5	490.0	85.1	18.9
	6	395.0	89.5	25.4
	7	395.0	92.9	27.9

From the field measurements best segments (Young, 1971) were identified adopting the recommended value of 10 per cent for V_{amax} and from the best segments profile the maximum slope was identified for each profile. This value was arbitrarily defined as the slope of the steepest segment provided that it extended for a minimum of 30 m. This length was selected as being long enough to overcome the effects of minor irregularities on the profile but short enough for its slope to be affected by any head scars resulting from significant mass movement processes. Where such a segment did not exist adjacent segments were amalgamated and their weighted mean slope determined, thus ensuring that the defined maximum slope still extended for a minimum of 30 m. The maximum segments on these profiles are in all cases found on the lower part of a convex profile but, with very few exceptions, they are separated from the stream channel by a lower-angle footslope. None of the maximum segments appears to be associated with presently active slope failure. Data on length, relief and maximum slope of each profile (grouped by valley side) are presented in Table 2.

Rather than examine each valley side separately for downvalley changes in maximum slope it is preferable to consider all five valley sides together. Both because of the small size of each sample (7) and the inherent variability of valley-side profile form (Parsons, 1982), statistically significant results might not be obtained even if the data do support the hypothesis of systematic change in value in the downvalley direction. However, in their raw state the data from the five valley sides are not comparable and need to be standardised. This was achieved by determining the mean maximum slope for each valley side and subtracting this value from the maximum slope of each profile measured on that valley side. The data were then available for analysis as a single data set of 35 observations of differences from a zero mean. Because of the small number of profiles taken from each valley side the data were not additionally standardised to unit variance.

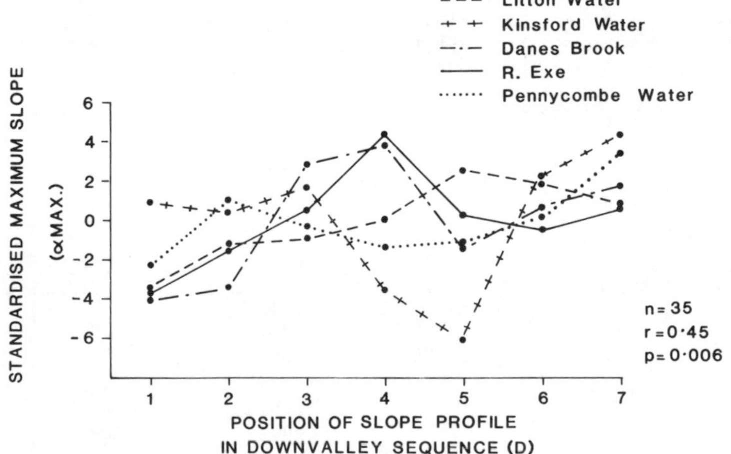

Fig. 2 Scatterplots of α_{max} against D for Exmoor data

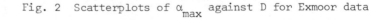

Plots of the standardised value (α_{max}) for each profile against its position in the downvalley sequence (D) are shown for each valley

side in Figure 2. Taking the data set as a whole, the regression of α_{max} on D is significant at the 0.001 probability level. According to the hypothesis developed from Richards' model it is to be expected that this relationship between α_{max} and D is to be explained as a function of increasing relief in the downvalley direction. Accordingly, both the regression of α_{max} on standardised relief (H) and the partial correlation of α_{max} with D controlling for H were determined (Table 3). Results of this analysis would appear to confirm that in this landscape the downvalley increase in maximum slope can be attributed to the coincident increase in relief. There is a strong correlation between H and D as was predicted from Richards' model so that the step-wise multiple regression of α_{max} on both variables showed little increase in explained variance over the regression of maximum slope on H, and D failed to enter the equation as a significant independent variable ($p = 0.24$). The actual amount of variance in α_{max} explained by H is relatively small (23 per cent). This is to be expected because H is only one of several local factors likely to influence α_{max}. What is important is that, despite the influence of these other factors, a significant relationship is found to exist.

TABLE 3. Percentages of explained variance in α_{max} by simple linear regression on D and on H, correlation coefficients between D and H, and correlation correlation coefficients between α_{max} and D controlling for H

Study area	Exmoor	South Downs	New South Wales
Regression of α_{max} on D	$R^2 = 21\%$ ***	$R^2 = 36\%$ **	$R^2 = 30\%$ ***
Regression of α_{max} on H	$R^2 = 23\%$ **	$R^2 = 13\%$ n.s.	$R^2 = 30\%$ ***
Correlation of H with D	R = .66 ***	R = .44 *	R = .39 **
Partial correlation of α_{max} with D controlling for H	R = .21 n.s.	R = .52 **	R = .43 **

*** Significant at the 0.1% level
** Significant at the 1% level
* Significant at the 5% level
n.s. not significant

SOUTH DOWNS STUDY AREA

The results of the study so far are in agreement with the contention that downvalley changes in maximum slope result from downvalley increases in relief. In order to test the validity of this model further it was applied to data taken from a study of valley sides in the South Downs (Parsons, 1982). This landscape is one in which increase in relief downvalley is much less pronounced. Any low-angle surface which may have existed prior to stream incision no longer remains. The divides between adjacent valleys are relatively sharp and the watersheds are well-defined. Processes of erosion would appear to have operated effectively as far as the divides so that they, too, slope downvalley. Although Parsons did identify some evidence for a downvalley increase in maximum slope in this environment, it can be accounted for within the framework of increases in relief downvalley if either it is less than that found in the Exmoor study, or, if a substantial increase in maximum slope does exist, it is interpreted as an inheritance from that which may have developed earlier when valley dissection had proceeded less far and a greater increase in relief downvalley may have existed, or it may be attributed to some other factor.

TABLE 4. Characteristics of South Downs hillslope profiles

Drainage basin	Profile no.	Profile length (m)	Profile relief (m)	Slope of maximum segment (°)
	1	225.0	25.6	9.6
	2	280.0	31.9	8.3
	3	155.0	22.5	16.1
1	4	105.0	23.5	20.3
	5	180.0	45.4	21.3
	6	195.0	47.2	20.6
	1	335.0	46.9	11.4
	2	255.0	44.2	22.2
2	3	210.0	42.0	20.7
	4	215.0	46.2	21.9
	5	287.0	50.7	19.5
	1	345.0	46.8	15.0
	2	260.0	35.3	16.6
	3	200.0	26.7	13.3
3	4	145.0	24.3	17.9
	5	150.0	28.3	18.3
	6	200.0	40.3	17.7
	1	100.0	6.2	5.8
	2	245.0	21.1	9.3
4	3	313.0	53.5	18.2
	4	505.0	51.6	10.4
	5	500.0	55.2	9.0

Within the area studied by Parsons (1982) four valley sides meet the requirements for selection described in the previous section. These valley sides are long enough for either five or six slope profiles spaced at 200 m to be measured on them (Fig. 3). Summary characteristics of these four valley sides, derived from the profile measurements, are presented in Table 1 and details of length, relief and maximum slope of each profile are given in Table 4. Maximum segments on the profiles are all found on the lower part of convex profiles above a lower-angle footslope. None is associated with presently active slope failure.

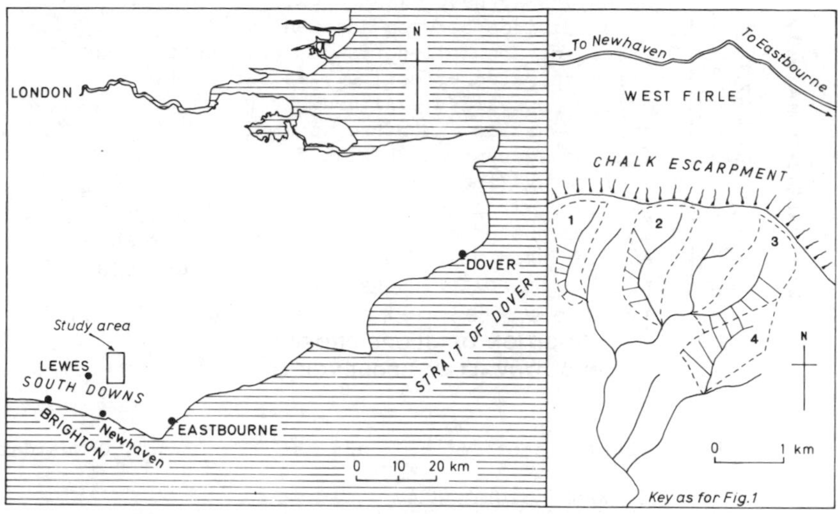

Fig. 3 South Downs study area

The same analysis was carried out on the profiles from these valley sides as on the Exmoor data. The plots of α_{max} against D for each valley side are shown in Figure 4. Again, taking the data as a whole the regression of α_{max} on D is significant ($p = 0.003$), but important differences are apparent in the statistical tests (Table 3). Although there is still a significant increase in relief downvalley it is position in the downvalley sequence and not relief which appears to be the parameter controlling α_{max}. The correlation of α_{max} with H is not significant at the 0.05 probability level whereas the partial correlation of α_{max} with D controlling for H is significant. In the step-wise multiple regression H failed to enter the equation as a significant independent factor ($p = 0.56$). The indication is either that some other factor is affecting maximum valley-side slope or that the effect of relief on valley-side slope persists long after the relief difference itself has largely disappeared.

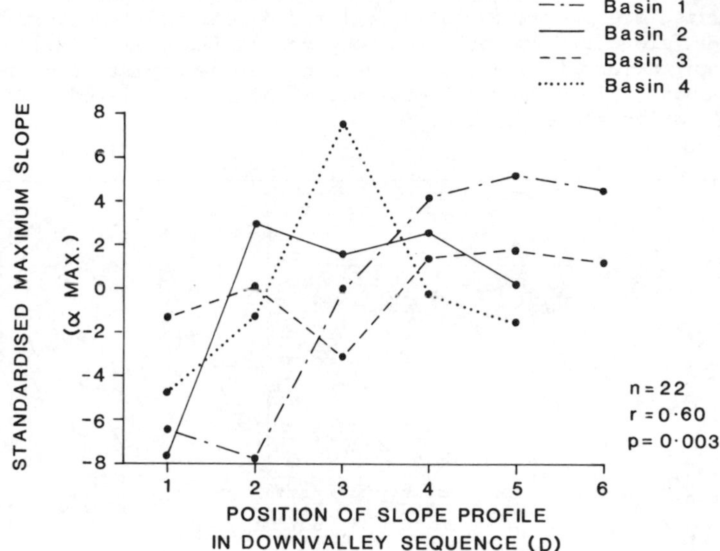

Fig. 4 Scatterplots of α_{max} against D for South Downs data

NEW SOUTH WALES STUDY AREA

In order to explore these two hypotheses further, data from a third area were examined. These data were taken from a study of long segments on low-angle hillslopes (Abrahams and Parsons, 1977). They come from three localities (hereafter referred to as Bathurst, Braidwood and Uralla) in eastern New South Wales, Australia, and are suitable for the present study because the landscapes from which they come have features in common with both previously studied areas. Like the Exmoor valley sides, those from New South Wales have divides which do not slope significantly downvalley, but like the South Downs ones, the divides are sharp and the watersheds are well-defined. These three localities are similar to each other inasmuch as their landscapes have each developed on an eroded granitic batholith with a weathered mantle but they differ in hillslope form (Table 1). Although the valley sides meet the conditions for selection previously defined, the data collected are not strictly comparable to those from the other two study areas in that the profiles were not equally spaced along the valley sides (Fig. 5). They were located at sites where the hillslopes were almost straight in plan (i.e. $+500>R_h>-500$, see Young, 1972, p.176). There was a minimum spacing between any two profiles of approximately 50 m but no maximum was set. In practice, the maximum spacing between any two adjacent profiles was approximately 400 m. From Bathurst data one valley side of 500 m extent was used on which 9 profiles were measured. From Braidwood data two valley sides of 1000 m and 500 m extent were used on which 6 and 8 profiles, respectively, were measured. From Uralla data two valley sides of 200 m and 600 m extent were used on which 5 and 10 profiles, respectively, were measured. The maximum segments all lie above a lower-angle footslope and none is associated with any active slope failure. Data on each profile are presented in Table 5.

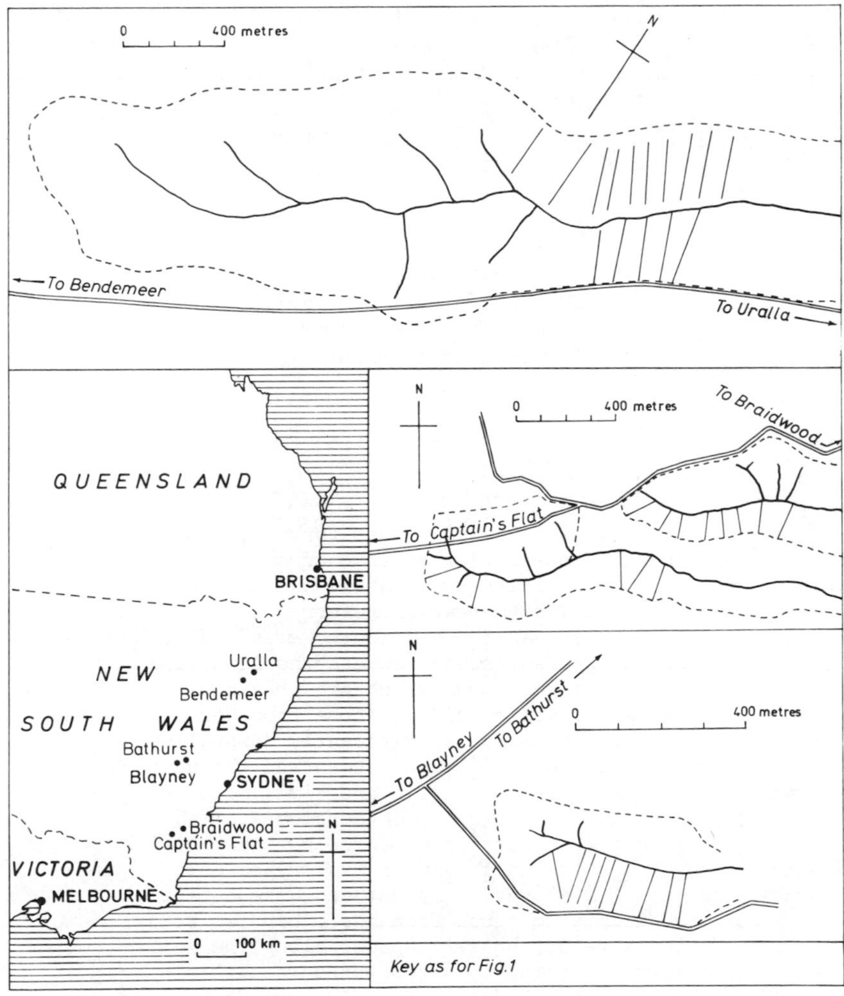

Fig. 5 New South Wales study area

It would be expected that if relief were the only factor causing a downvalley increase in maximum valley side slope then in this area, because there is a downvalley increase in relief due to the unequal downvalley gradients of the divides and the valley floors, the

TABLE 5. Characteristics of New South Wales hillslope profiles

Drainage basin	Profile no.	Profile length (m)	Profile relief (m)	Slope of maximum segment (°)
Bathurst	1	80.0	8.7	7.0
	2	145.0	17.7	7.7
	3	145.0	20.5	8.2
	4	135.0	19.9	9.9
	5	120.0	17.1	9.3
	6	115.0	16.5	7.3
	7	110.0	15.3	7.2
	8	120.0	16.8	8.3
	9	125.0	16.1	8.2
Braidwood south basin	1	95.0	14.7	8.9
	2	75.0	9.8	7.5
	3	100.0	8.0	4.6
	4	100.0	15.1	8.7
	5	155.0	20.4	7.6
	6	120.0	17.8	8.6
Braidwood north basin	1	110.0	20.1	10.5
	2	100.0	16.7	9.6
	3	90.0	15.3	9.8
	4	100.0	17.0	9.8
	5	85.0	15.2	10.3
	6	80.0	15.8	11.4
	7	90.0	17.2	11.0
	8	100.0	21.6	12.5
Uralla north side	1	140.0	1.8	1.1
	2	285.0	4.9	1.7
	3	285.0	6.2	3.2
	4	300.0	8.6	3.7
	5	300.0	7.5	2.6
	6	320.0	10.1	3.2
	7	295.0	9.3	4.1
	8	265.0	8.9	3.9
	9	280.0	8.6	3.4
	10	305.0	8.8	3.9
Uralla south side	1	200.0	5.2	2.9
	2	200.0	5.4	3.0
	3	215.0	5.5	3.5
	4	270.0	7.2	2.4
	5	290.0	8.3	2.6

landscape would show a pattern of statistical relationships similar to that for Exmoor. In particular, the partial correlation of α_{max} with D controlling for H should not be significant. If, on the other hand, a second factor affects maximum valley-side slope in landscapes of sharp divides, the analysis should show a significant partial correlation of α_{max} with D when H is controlled for.

Undertaking the same analysis for this study area as for the other two, significant results were again obtained (Fig. 6 and Table 3). These results suggest that the latter proposition is more likely to be correct; namely that downvalley increase in relief is one factor which causes an increase in maximum slope but that a second factor also operates. Both variables (D and H) appear to affect α_{max} independently and more or less equally. As a consequence, in the step-wise multiple regression both variables entered as significant independent variables in the following equation:

$$\alpha_{max} = 0.14H + 0.17D - 0.75.$$

This equation explains 43 per cent of the variance in α_{max}, the highest value obtained in any of the three study areas.

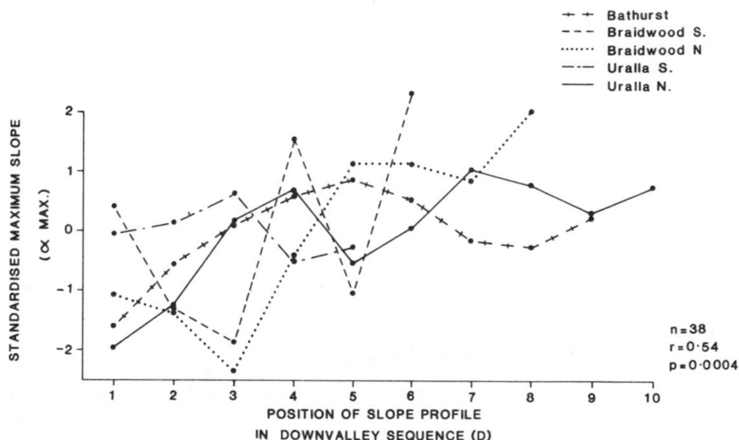

Fig. 6 Scatterplots of α_{max} against D for New South Wales data

DISCUSSION

This analysis has shown that a downvalley increase in maximum valley-side slope can be identified within drainage basins (of low order) as well as at the larger scale investigated by Carter and Chorley (1961) and Arnett (1971). However it has also shown that the model proposed by Richards (1977) to explain this increase is inadequate at the within basin level. In particular, it has suggested that in landscapes of sharp divides some other factor may operate to cause valley sides to steepen downvalley.

Where opposing valley sides meet at narrow divides changes in valley-side slope may result from the lowering of the divides or the migration of the base of the valley sides. Whereas the former can only lead to a reduction in slope the latter can lead to either a reduction or an increase (Fig. 7). That increases are common suggests that downvalley migration of the valley side towards the divides is taking place, presumably by a process of valley-floor widening. As streams begin to undercut the valley sides they can be expected to induce mass movements leading to increases in maximum slope. It follows that the divide lowering which may also be occurring downvalley is less significant than the downvalley increase in valley-floor width in determining maximum valley-side slope.

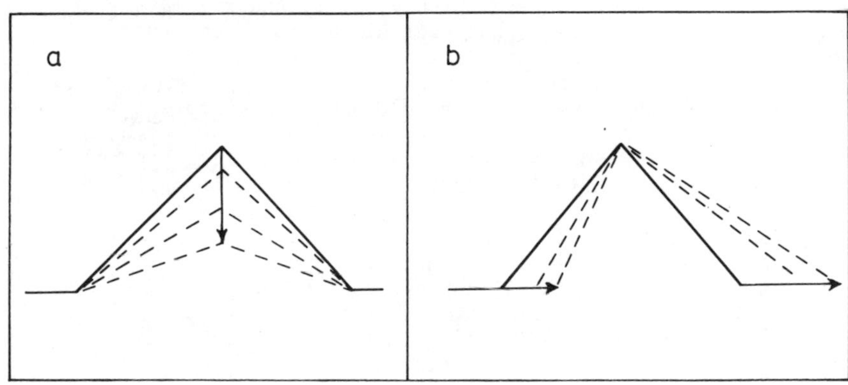

Fig. 7 (a) Lowering of the divide causes a reduction in slope
(b) Lateral migration of the base of the hillslope causes either a reduction in slope (migration away from the divide) or an increase in slope (migration towards the divide)

The results of this study are important for understanding the relationship between valley-side slope form and boundary conditions and they may usefully be compared with those of Armstrong (1980), and Melton's (1960) study of asymmetry. Using a simulation model of valley side development, Armstrong obtained downvalley increases in valley-side slope which he attributed to changes in basal conditions as a result of stream incision. The downvalley increases found in the present study have likewise been attributed to changes in basal conditions, though in this case resulting from both lateral and vertical stream incision. Armstrong considered the increase in slope steepness downvalley to be an equilibrium response to changing a boundary condition, specifically lowering of the lower boundary. From the present study it can be argued that a downvalley increase in maximum slope may be the equilibrium response both to a vertical shift in the position of the lower boundary as a result of stream incision (as in Armstrong's case) and to a lateral shift in the position of the lower boundary (as in Melton's case), though in the present study it seems likely that the lateral shift is not just a local phenomenon but a result of general valley widening. The equilibrium response appears to be similar in both cases but there is a change in its cause.

In some earlier studies (e.g. Palmer, 1956) downvalley sequences have been interpreted as temporal sequences. Armstrong (1980) argued that such an approach is invalid because a downvalley sequence is no more than a realisation of the response to downvalley changes in basal conditions. Using spatial sequences to study hillslope evolution has been largely superseded by more controlled process-response models (e.g. Kirkby, 1971; Ahnert, 1976) in which an important parameter is that controlling the boundary conditions. Armstrong (1982) has recently pointed out that Kirkby's model is very sensitive to the boundary conditions and choosing ones different from those used by Kirkby produces a different outcome. The result of the present empirical study suggests that boundary conditions vary in natural landscapes and that hillslope form responds to these variations as well as to variations in process. It is likely that in addition to varying from place to place in the landscape, boundary conditions at any one place change through time. Attempts to model the evolution of hillslopes need to give close attention not only to the links between form and process but also to hillslope boundary conditions and the manner in which they may change under natural conditions. If boundary conditions are treated in an unrealistic manner the evolutionary hillslope sequence produced by a process-response model may have little relevance for natural landscapes. Likewise, studies of the evolution of hillslopes on anthropogenic landscapes where boundary conditions may be very different (e.g. Goodman and Haigh, 1981) cannot be easily extrapolated to naturally-occurring hillslopes. At present, little appears to be known of the manner in which boundary conditions vary throughout the landscape or of the ways in which they change through time. If the complexity of naturally-occurring hillslopes is to be understood serious attention may need to be given to both these types of change in boundary conditions.

ACKNOWLEDGEMENTS

I am grateful to Susan Parsons for her assistance in obtaining the profile data of the Exmoor and South Downs study areas, and to Athol Abrahams and Lorraine Oak for providing much of the New South Wales profile data and the background material for Figure 5. The University of Keele provided a grant towards the cost of fieldwork on Exmoor.

REFERENCES

Ahnert, F.,1976. Brief description of a comprehensive three-dimensional model of landform development . Zeit für Geomorphologie Supplementband 25, 29-49

Arnett, R.R.,1971. Slope form and geomorphological processes: an Australian example . Inst. Br. Geogr. Spec. Pubn.,3, 81-92

Armstrong, A.C.,1980. Simulated slope development sequences in a three-dimensional context . Earth Surface Processes, 5, 256-270

Armstrong, A.C.,1982. A comment on the continuity equation model of slope profile development and its boundary conditions . Earth Surface Processes and Landforms,7, 283-284

Carter, C.S.,and Chorley, R.J.,1961. Early slope development in an expanding stream system . Geol. Mag.,98, 117-130

Goodman, J.M.,and Haigh, M.J.,1981. Slope evolution on abandoned spoil banks eastern Oklahoma . Physical Geography, 2, 160-173

Kirkby, M.J.,1971. Hillslope process-response models based on the continuity equation . Inst. Br. Geogr. Spec. Pubn.,3, 15-30

Melton, M.A.,1960. Intravalley variation in slope angles related to microclimate and erosional environment . Bull. Geol. Soc. Am., 71, 133-144

Palmer, J.,1956. Tor formation at the Bridestones in north-east Yorkshire and its significance in relation to problems of valley-side development and regional glaciation . Trans. Inst. Br. Geogr.,22, 55-71

Parsons, A.J.,1982. Slope profile variability in first order drainage basins . Earth Surface Processes and Landforms,7, 71-78

Richards, K.S.,1977. Slope form and basal stream relationships: some further comments . Earth Surface Processes,2, 87-95

Strahler, A.N.,1950. Equilibrium theory of erosional slopes approached by frequency distribution analysis . American J. Sci., 248, 673-696 and 800-814

Young, A.,1971. Slope profile analysis: the system of best units . Inst. Br. Geogr. Spec. Pubn.,3, 1-13

Young, A.,1972. Slopes. Oliver & Boyd, Edinburgh, p.288

Young, A.,1974. Slope profile survey . Br. Geomorph. Res. Group Tech. Bull. 11

International Geomorphology 1986 Part II
Edited by V. Gardiner

THE MORPHOMETRY OF SPECIFIC LANDFORMS

Ian S. Evans

Department of Geography,
University of Durham

ABSTRACT

Whatever the landform, nine stages in morphometric analysis can be recognised: conceptualising, defining, delimiting, measuring, calculating derived measures, assessing frequency distributions, mapping, interrelating and assessing meaning. Measures are needed to cover (a) position; (b) direction; (c) size; (d) gradient; (e) shape; (f) context. Many landforms are due to 'linear' geomorphic agents and these tend to possess bilateral symmetry: others have radial symmetry or none. Summits and 'foci' provide geomorphic points of reference. The morphometric measures chosen should facilitate interpretation in process, chronologic and material terms.

Relationships between shape and size provide some constraints for speculations on landform development, but there are notorious pitfalls in interpretation. Advice is offered on these points, and on use of units, ratios, transformations, clustering, and multivariate relationships, with examples from work on drumlins and cirques. The increasing availability of altitude matrices makes it feasible to measure distributions of altitude, gradient, aspect, profile and plan convexity within a landform. Summaries of these may replace many of the ratios currently in use, thus providing a convergence between specific and general geomorphometry. This 'distributional' approach forms a third alternative to the more common use of lists of indices and the fitting of equations to landform surfaces.

INTRODUCTION

The land surface can be analysed either as a whole with techniques appropriate to any rough surface, or subdivided into meaningful units labelled 'landforms' which are then analysed by methods appropriate to each landform type. Engineers have concentrated on the former, generating statistics for arbitrary patches of the land surface such as map sheets: this 'general geomorphometry' approach includes spectral analysis (Pike & Rozema, 1975) and fractal analysis (Mandelbrot, 1977). Geomorphologists on the other hand have invested much effort in the recognition of landforms of distinct types: river terraces, river channels, pingos, sand dunes and others have been mapped, measured and analysed, and most morphometry by geomorphologists has been devoted to specific landforms. This has the advantage that indices are defined to suit the landform in question, but the disadvantage that cross-fertilization between studies of different landforms has been limited: researchers have not always chosen the most suitable indices, still less have they used the most suitable forms of analysis.

In this paper I suggest that there are common themes in specific geomorphometry such that if a new landform type was recognised tomorrow, some accumulated experience could be brought to bear upon its analysis. Researchers are invited to use the following as a checklist, while adopting or developing definitions which are appropriate to their landform type, its symmetry, outline and unity or subdivision. Whatever the landform type, nine stages in morpho-metric analysis can be recognised: conceptualising, defining, delimiting and measuring the landform type; calculating derived measures, assessing frequency distributions, mapping and interrelating the measures; and interpreting the results.

STAGES IN A MORPHOMETRIC ANALYSIS

(i) Specific morphometry is founded on the conceptualisation of landform types, examples of which can be distinguished from their surroundings, including neighbouring examples. Such conceptualisation is now quite advanced for those landforms such as cirques and (active) pingos, which can be related to specific (even if vaguely understood) processes. It is less clear, and less generally agreed, for components of subaerial slopes, despite the widely quoted 'nine-unit land surface model' of Dalrymple, Blong and Conacher (1968). Sauchyn and Gardner (1983) used morphometric results to help conceptualise rock-face basins, distinguishing chutes, funnels and open cirques. Landforms such as tors, bornhardts, inselbergs, castle-kopjes and perched blocks overlap and grade into each other confusingly, as defined by different geomorphologists.

(ii) Hence there is a need to adopt an operational definition sufficiently precise to help distinguish the landform from all others. Dictionary definitions are rarely adequate, being encumbered with vague attributions such as what the form 'may have' or 'sometimes has'. Examples (most obviously of cirques, of barchan dunes and of river meanders) are confusingly described in textbooks as 'typical', when they are in fact unusually well-developed and clear-cut.

If there is not great similarity in the sets of landforms defined by different researchers, comparisons and the cumulation of reliable results will be impossible. More attention needs to be paid to marginal cases and those which have been considered and rejected: reasons for rejection may be more revealing than reasons for acceptance (Table 1). Full consideration of marginal and rejected cases increases the workload, and morphometric testing of operational definitions has been rare. Since it is rare for one landform type to be obviously different from others (e.g. with all cases being twice as steep, twice as large or twice as elongate), much more testing of definitions and distinctions is required.

TABLE 1. Marginal cases: Cumbrian cirques.

Exclusion is on the basis of all attributes. W, N etc refer to the Wainwright divisions of the Cumbrian Lake District Fells.

marginal cirques (included) :			non-cirques (excluded) :		
		POOREST HEADWALLS; maximum gradients			
Grisedale Tarn	E	44°	Gavel Moss	E	40°
Haycock East	W	44°	Tongue Gill Head	W	40°
Nine Gills	N	47°	Slades Head	N	41°
Bayscar Slack	W	47°	Yewgrove Gill	FE	41°
Thack Bottom	E	47°	The Bog, Wythburn	C	46°
Scoat Tarn	W	48°			
		POOREST FLOORS; minimum gradients			
Bowscale West (Ling Thrang Crags)	N	19°	Tongue Gill, Rosthwaite	NW	22°
Heck Cove	FE	18°	Deepdale Crag	E	22°
Over Cove	FE	18°	Gasgale Crag East	NW	19°
Pillar Cove	W	17°			
		POOREST PLAN CLOSURES			
Boathow Crag, Ennerdale Water	W	20°	Seatallan	W	29°
Dove Crag	E	36°	Band Knotts, Kentmere	FE	42°
Red Screes, Clough Head	E	46°	Bowderbeck Head	NW	40°
		IRREGULAR			
Gladgrove Gill, Kentmere	FE		Glencoyne (broader)	E	
Dove Crag	E		Houndshope Cove, Dovedale	E	
		UNROUNDED			
Southerndale	N		Hayeswater Gill	FE	
Angle Crags	C		Gatesgarthdale Beck	W/NW	
			Caiston Glen, Kirkstone	E	
			Greendale Tarn, Wasdale	W	

(iii) Testing requires that each case be delineated or delimited from the surrounding land surface. Sometimes landforms are mapped with incomplete limits, e.g. an arc on the upstream side of a cirque or landslide scar, or without limits, e.g. the axes of drumlins or the crests of dunes, or the centre points of pingos or volcanoes. Lack of a complete outline prevents measurement of size and inhibits measurement of shape. Even if delimitation appears (in part) arbitrary, specific geomorphometry is conceptually based on discontinuities; if delimitation of landforms cannot be completed, we must resort instead to general geomorphometry.

(iv) Measurements of the delimited cases can be taken either in the field, or from remote sensing images, or from maps, or from a description held in computer-readable form. For a particular problem or area, choice is usually constrained strongly by what is available and feasible.

At present, field measurement or photogrammetry from large-scale air photos (Sauchyn and Gardner, 1983) remains desirable for accurate detail. However, a certain amount of generalisation is desirable, and this may be provided by reliable contour maps such as the new 1:10,000 Ordnance Survey maps of Britain with photogrammetric contours at intervals of 5 or 10m, depending on relief. Cirque floor and wall gradients measured from these maps are probably more reliable than those measured in the field (at least, by me). Critiques by Clayton (1953) and by Rose and Letzer (1975) of morphometry from earlier, provisional 1:25,000 maps, with many contours interpolated, do not necessarily apply to such modern maps. Of course, studies of small features such as earth hummocks or solution hollows require field measurement or special photogrammetry.

(v) Directly measured attributes such as length or width are usually supplemented by calculated indices derived from them, e.g. the width/length ratio or the difference between maximum and minimum gradient. Ratios should be defined to have statistically manageable properties: those varying between zero and one are easier to handle than those between one and infinity. Hence it is better to divide minimum gradient by maximum gradient than vice versa.

Measures are required for (a) position (usually x, y, z); (b) direction or orientation (except for radially symmetric landforms); (c) size; (d) gradient; (e) shape; and (f) context (density, spacing, contiguity, pattern).(See also Gardiner's examples, 1983 p 4) . Usually several indices are required under each of these headings, especially if several different elements are distinguished within the landform. In biological morphometry the emphasis is on size and shape, and sometimes density and pattern. In geomorphology, because all landforms have an areal extent, position, direction and context are important; and because most landforms have a vertical extent, gradient assumes great importance and is considered separately, not just as one component of shape. Shape indices may be ratios between size indices, or additional measures (Jarvis, 1981): one or more may be adopted as a measure of 'degree of development' of the landform.

Each attribute is either dimensionless or in specific units, and morphometrists should be sensitive to this: for particular interpretations, either dimensional or dimensionless indices may be appropriate. Sets of variables in the same units may be analysed more meaningfully than mixed sets (see viii below). (a) may be expressed in m (or km); (b) in degrees clockwise from north; (c) in m, m^2 or m^3; (d) and (e) are dimensionless, though gradient may be expressed in degrees or radians. Directional (azimuthal, orientational) measurements are 'on the circle' and must be summarized by vector techniques (Evans, 1977) rather than the linear statistics used for 'polar' variables; Jauhiainen's (1975) analysis seems suspect on this basis.

(vi) All attributes (measurements and derived indices) have frequency distributions over a set of cases. Whether the set is a population, or a probability sample, the frequency distributions should be checked for symmetry and in particular for the absence of long tails or outliers, before any further analysis: even the calculation of mean values may be distorted by outlying values. Since almost all size measures have positively skewed distributions (many small, few large cases), the question of transformation is not a peripheral issue: only on the appropriate measurement scale can the position of extreme values be properly assessed (Evans, 1983).

Square root and cube root transformations are valuable in converting variables expressed in m^2 (area) or m^3 (volume) to m (length), thus reducing the number of units involved prior to multivariate analysis. Square root and especially logarithmic transformations have been particularly useful for morphometric variables (Gardiner, 1973), at least for size variables, but a further check that the transformed frequency distribution has the desired qualities is essential. Approximation to the Gaussian or so-called 'normal' model (once emphasized by Gardiner and Gardiner, 1978) is not essential, as Gardiner (p.c.1985) now agrees : anyway, most morphometric variables are positive or bounded and values cannot tend toward both minus and plus infinity. Rectangular distributions are no problem. What matters is that tails of extremely high or low values or both should not be so extensive as to distort correlation, regression and other analyses. Different transformations are required for closed ratios (bounded on both sides : Evans and Jones, 1981): the angular transformation rectifies mild skewness, while the logit transformation has a stronger effect.

If a distribution is essentially unimodal, then an average size may be attributed to the landform. If this is real, then contra Mandelbrot (1977) we are dealing with a scale-specific, non-fractal component of the land surface: I suggest that glacial cirques fall into this category (Derbyshire and Evans,1976). Some features have increasing numbers of smaller cases: erosional summit magnitudes, for example, follow a gamma distribution and the mode is always at the smallest magnitude determinable. Careful attention should therefore be paid to the lower limb of positively skewed size distributions; is the distribution truncated in reality, or by the resolution of the data available, or by the subconscious decisions of a researcher whose mental model of the landform is scale-specific?

Finally, a distribution may be multi-modal. There may be valid reasons for this e.g. in structural control or in the existence of process thresholds, providing Wilson's (1972) scale hierarchy of aeolian bedforms. Alternatively, different types of landform may have been mixed together. Discovery of the latter may be a valuable step forward, but the modes must be clearly distinct and the gap between them statistically significant. Difficult though this concept is, it can only be assessed if a complete population or a probability sample has been analysed. A complete population is treated as a sample since it is one outcome of the relevant processes. Even statistical significance, though necessary, is not sufficient where measurements may be autocorrelated because of an underlying control. An east-west ridge may support a set of north-facing cirques, and a north-south ridge an east-facing set, but this does not mean that cirque aspect is inherently bimodal: if a similar northwest-southeast ridge were present, it would probably produce northeast-facing cirques. The history of altimetric analysis (reviewed in Ollier, 1963, Clarke, 1966 and Richards, 1981) suggests that geomorphologists are predisposed to over-interpret multimodality.

(vii) Morphometric analyses are rarely convincing (to geographers) unless <u>maps</u> of the landforms analysed are presented. Usually this leads on to analyses of the spatial distribution of landforms (density, spacing, pattern) and of their attributes (especially altitude, size, and shape). Since there is considerable local scatter, spatial distributions of attributes may require generalisation by trend surface analysis (Unwin, 1975).

(viii) <u>Interrelation</u> of different attributes is commonly a major aim of morphometric studies: how does shape relate to size, how do both relate to position, and how do different shape attributes and different size attributes interrelate? Correlation, regression and multivariate analysis are used to compress a long list of attributes into a smaller number of compound dimensions, or to select the 'most important' attributes which can represent the list (Evans, 1984). Here the units and type of attribute are important: to include a ratio and any one of its component parts in the same Principal Components Analysis (P.C.A.) or numerical classification is to court confusion. Correlation of ratios with <u>any</u> elements in common can be misleading (Evans and Jones, 1981): for example, there is an inbuilt negative relation between length and (width/length), especially if length varies more than width.

It may be advisable to perform separate multivariate analyses for each set of attributes, defined in particular units (metres, degrees or dimensionless). This permits covariance-based P.C.A., which may be more informative than P.C.A. based on standardised variables of diverse types, i.e. correlation-based P.C.A. Information on the variability of attributes should not be ignored; probably a dimension which varies between 100 and 900m <u>should</u> be given more weight than one varying between 250 and 300m.

It is tempting to apply standard techniques such as regression. When interrelating different size attributes or different shape attributes, however, neither variable is truly 'independent' or dependent. For example, Lancaster (1981) relates the width and height of linear dunes by the regression equation

$$W = 532 + 2.87H$$

That is, a width of 532m is predicted for dunes of zero height. Yet judging from the scatter, and the correlation coefficient of only +0.59, a regression of H on W would predict finite height for zero width. The regression is probably misleading: in these cases, common in morphometry, more use should be made of the reduced major axis (Mark and Church, 1977) or of other general trend lines. We need to test whether the axis passes through zero, or whether any zero intercept is significant or substantial. This may lead on to allometric (power function) analysis (Church and Mark, 1980; Olyphant, 1981) . Other special techniques, suitable for azimuthal (directional or orientational) data, were reviewed by Mardia (1972, 1975).

(ix) Interpretation includes assessing the value of different attributes in genetic or chronologic terms. For erosional landforms such as cirques there is a temptation to use size as a surrogate for age (Derbyshire and Evans, 1976), since development of these concave forms involves enlargement unless neighbouring forms enlarge faster. However, topographic sites for cirque initiation vary and a broad range of sizes may be expected from an early stage. Landforms such as landslide scars may be created catastrophically with a broad range of sizes, and it would be difficult to defend landslide size as an indicator of age. Despite these difficulties, relations between shape and size are important to an understanding of how landforms vary, and they do provide some constraints to speculations on landform development.

Drumlins and other bedforms may either enlarge or reduce over time, as they adjust to changing force fields. Even if made of bedrock, they may enlarge or reduce as erosion progresses to lower levels. Clearly there is a fundamental distinction between such forms, stabilised by negative feedbacks, and forms such as cirques and rock basins where positive feedback is important.

The concept of convergence or equifinality has been repeated so often that it has become a cliché. Of course forms similar in some attributes may have been produced by different processes: scientific method (e.g. Chamberlin, 1890) suggests a search for further attributes (possibly material composition rather than form) which will provide the necessary distinctions. Evans (1972) showed how discrimination between different hypotheses for glacial valley asymmetry could be achieved by refining and extending the morphometric analysis.

Interpretation also involves suggestions, for future work, of improvements in any of the above stages. If the morphometric attributes used fail to discriminate between clear cases and marginal cases, or between the landform type and related but distinct types, it is desirable to develop more sensitive methods. Taken as a set, morphometric measures should express the 'degree of development' of a landform, which may be interpreted either in chronologic terms (stage), in material resistance terms (structure), or in terms of variations in effective force (process). Measures which facilitate such interpretation, such as profile or plan closure for cirques, are to be preferred over others such as width/length ratio, which varies little for cirques. Measures which summarize aspects of the broad form, such as height/length, or height/square root of area, are preferred over those which emphasise irrelevant details; length of perimeter should be avoided because it is influenced by minor crenulations and hence is more sensitive to measurement resolution than are other size measurements.

This discussion has emphasised the nine stages which are particularly important for morphometry. Some further considerations are important in any scientific study. Samples should be of a size adequate to test the questions of interest. The more subdivisions are to be made, and the more attributes are to be interrelated (e.g. in a factor analysis or a multiple regression), the larger the overall sample needs to be, or the research effort is wasted. Absence of a significant result has often been misinterpreted as a negative result, when the sample size is so inadequate that significance cannot be expected.

Sampling should be on a probability basis, either random or systematic or some compromise between these: even in the 1980's, this is fairly rare in geomorphology, and greater effort must be made to overcome pressures (see below) of convenience and feasibility. Where definition and delimitation of landforms is as time-consuming as their measurement, it may be best to analyse a whole population within a defined study area. This in effect is cluster sampling, and the choice of study area becomes critical. For example, landforms which appear clustered in a broad area may appear more regular than random in a small area . For a particular problem, 'natural' study areas may be defined by mountain ranges (for cirques), drainage basins (for river terraces, or channel forms) or morphological regions (for drumlins or sand dunes). When overall tendencies are important (e.g. in cirque aspect or drumlin orientation), such areas are probably preferable to map sheets or grid squares. Actual study areas are almost always chosen for convenience, e.g. accessibility, map availability or previous work: hence researchers should consider the respects in which their area is distinctive, and those in which it is similar to other areas with landforms of the type under study.

VARIABLES AND SYMMETRY

The selection of attributes for a given landform type depends on the type of symmetry expected, and on whether our concept of the form includes internal discontinuities. Symmetry may be radial (volcanoes, craters, pingos), bilateral directed (drumlins, cirques, yardangs, landslides, barchans) or absent (kames, spits). Also, more complex combinations of symmetry are found in river channels. Bilaterally symmetric (i.e. axial) directed forms are particularly common because of the importance of directed geomorphic agents, with forces operating in a dominant direction. Even if this direction varies over time, landforms are adapted to the last morphogenetically significant direction or directions.

TABLE 2. CIRQUE MORPHOMETRY: MEASURED ATTRIBUTES
(+D = applicable also to drumlins)

a POSITION :

V1 Identification number (+D)

V2 Easting (+D) V3 Northing (+D) (100m)
(4 digit National Grid in G.B.)

altitudes: (m)

V4	Lowest (+D)	V7	Max. crest (+D)
V5	Modal floor	V8	Max. above (draining into cirque)
V6	Max. floor	V9	Crest on median axis (+D)

b DIRECTION: aspects (0-360 degrees)

V10 Median axis, outward, through middle of threshold, leaving ½ cirque area on left, ½ on right (+D)

V11 Headwall, visually weighted for gradient & area

c SIZE: (m)

V12 Length of median axis (+D)

V13 Width : max., perpendicular to median axis (+D)

V14 Max. headwall height, crest to floor, along a single slope line

d GRADIENT : (degrees)

V15 Max. headwall, over 30m vertically

V16 Min. floor, over 10m vertically (0° if large lake or bog)

e SHAPE : closure

V17 Plan closure : change in azimuth along mid-height contour (degrees)

V18 Lake? 1: major rock basin lake
2: major, possibly solely moraine-dammed
3: major peat bog, former lake
4: minor lake(s) or peat bog(s)
5: drift-covered floor, may conceal a basin
6: outsloping bedrock, no basin

V19 Number of cols over 30m deep in crest

(FACTOR)

V20 Geology (local coding scheme) (materials, for drumlins)

TABLE 3. CIRQUE MORPHOMETRY:
ATTRIBUTES CALCULATED IN COMPUTER

c SIZE

V21 Median axial height, V9-V4 (+D) (m)
(Crest median altitude - Lowest altitude)

V22 'Volume', V12 x V13 x V21 x 10^{-6}
(Length x width x height) (+D) ($m^3 \times 10^{-6}$)

V23 Floor altitude range, V6-V4 (m)
(Max. floor altitude - lowest altitude)

d GRADIENT (degrees)

V24 Axial gradient, arctan V21/V12
(arctangent height/length)

e SHAPE : closure (degrees)

V25 Profile closure, V15-V16
(Max. headwall gradient - min. floor gradient)

V26 Concavity, V17 + 4x V25
(Plan closure + 4x Profile closure)

(other shape)

V27 Plan /profile closure ratio, V17/V25
(Plan closure/profile closure)

V28 Width/length, V13/V12 (+D)

f CONTEXT

(Such variables could be calculated from grid references, V2 and V3)

We might expect transferability of variables between landforms with comparable symmetry, regardless of the relative importance of erosion and deposition (which often remains uncertain, as for drumlins). For example, the list of cirque descriptors in Tables 2 and 3 can be applied in its entirety to landslides if for 'headwall' we read 'scar' and for 'floor' we read 'depositional zone'. V19 (cols) is not very relevant, but it is applicable. This transferability is because both types of landform are axial, directed downslope, and eroded into a slope.

Only some of these variables can be applied to drumlins. Variables 5, 6, 11, 14, 15, 16, 18 and 23 do not apply because there is no subdivision of drumlins comparable to the division of cirques into floors and headwalls: but for gradient, stoss, lee and maximum side gradients can be distinguished. Variables 8, 17, 19, 25, 26 and 27 do not apply because drumlins are not eroded into slopes; their higher contours should close through 360 degrees. For V20, bedrock geology would be replaced by material composition. Variables 9 and 24 also seem irrelevant to drumlins. And for V10, the axis should be a crest axis, ideally coinciding with the long axis, and its direction should be measured from steep end to gentler end. Work with drumlins has in fact dealt mainly with length, width, height and their ratios (Mills, 1980), plus direction, spacing and pattern (Rose and Letzer, 1975: Jauhiainen, 1975).

For bilaterally-symmetric, directed landforms there may be a useful reference point such as the summit of a drumlin, yardang or barchan, the sink of a doline, the middle of a cirque threshold, or the mouth (downstream end) of a drainage basin. The axis will usually pass through this point, and this helps us choose between the different definitions of length as listed by Gardiner (1975). In radially symmetric landforms, the centre of symmetry will be the reference point and it may be useful to analyse variation as a function of polar coordinates, ρ, θ. Subdivision of landforms, for example into outer slopes, crater walls, central depressions and possibly central peaks, makes further indices possible. Even radially symmetric landforms like pingos commonly have some elongation, and length may be measured as well as width and height (Stager, 1956).

THREE APPROACHES

So far, we have assumed that landforms will be summarised by overall measurements and indices. Though most common, the use of indices is only one of three approaches to the numerical expression of landforms. The second is the use of equations fitted to the whole form, or part of it, and the third, introduced here, is the use of distributions of 'point' variables.

The traditional use of indices involves a series of measurement operations on each landform. If these are defined without reference to each other, the analysis may become disjointed: there is also the pitfall of overlapping indices - some authors have even put length/width and width/length into one analysis, commenting on the strong correlation between them ! (The correlation coefficient does not quite reach -1.0 because a reciprocal relationship, though perfect, is curvilinear.) Many of the correlations are obvious: it is not surprising that different size variables are well correlated (Fig. 1), still less different altitude variables (Fig. 2). We can make these clusters of inter- correlated variables larger or smaller simply by adding or deleting measures of altitude or size: hence the magnitude of an eigenvalue or factor is not of great importance, though it has often been overemphasised. Likewise if we measure drainage density in three ways (length of stream, number of stream segments, and number of contour crenulations, each per unit area), it will be represented by a cluster of these three strongly interrelated variables.

Analysis of relations within clusters serves mainly the modest purpose of deciding which variables are really needed. For example, from Fig. 2 and a related variance-mode principal component analysis, we might decide that cirque altitude could be summarised by one of the 'low' variables and one of the 'high', but if pressed, reduction to a single altitude variable would not sacrifice too much information. From Fig. 1 it was concluded that area was redundant, since it could be predicted from length x width, both of which can be measured much more rapidly than can area.

Relations between size, shape, gradient and position are more interesting so long as they are not inbuilt ratio correlations such as overall gradient (height/length) versus length. Mosimann (1970) showed that it is extremely difficult to separate 'size' and

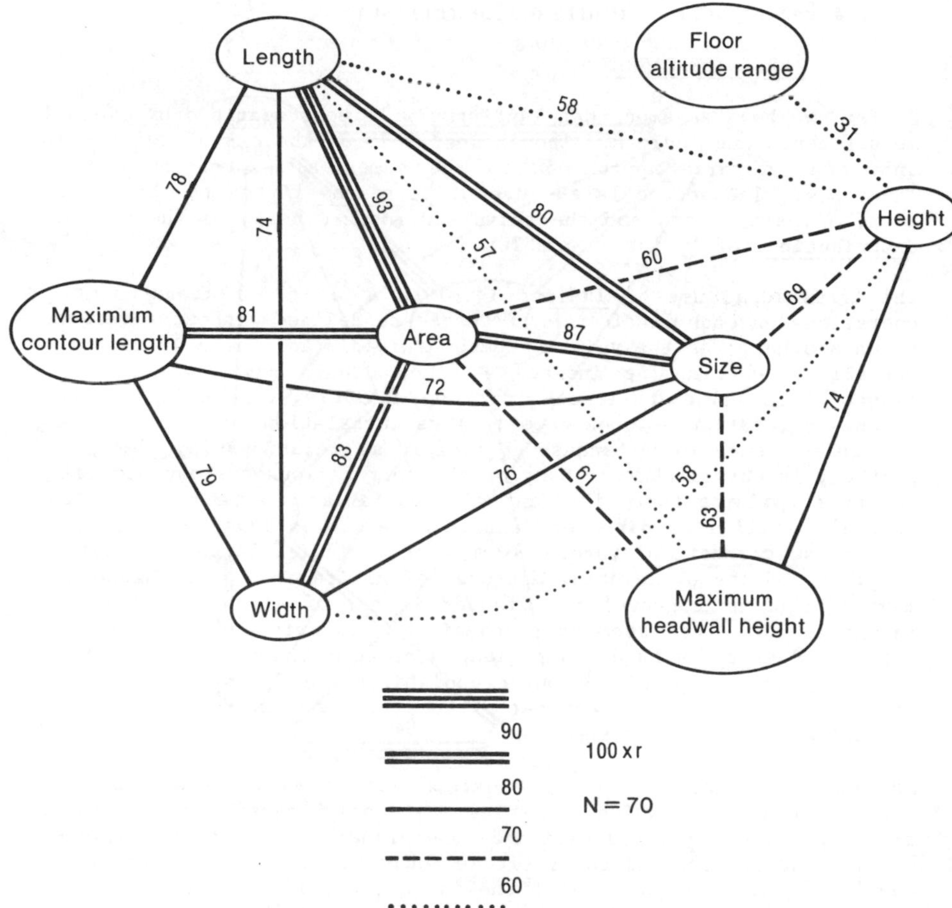

Fig. 1. A correlation structure for eight size variables, measured for 70 glacial cirques west of the Keswick-Ambleside road. Square root transformations were applied throughout. Area was not measured for further cirques because of its predictability from length and width. 69% of this information (variance) can be represented in the first principal component.

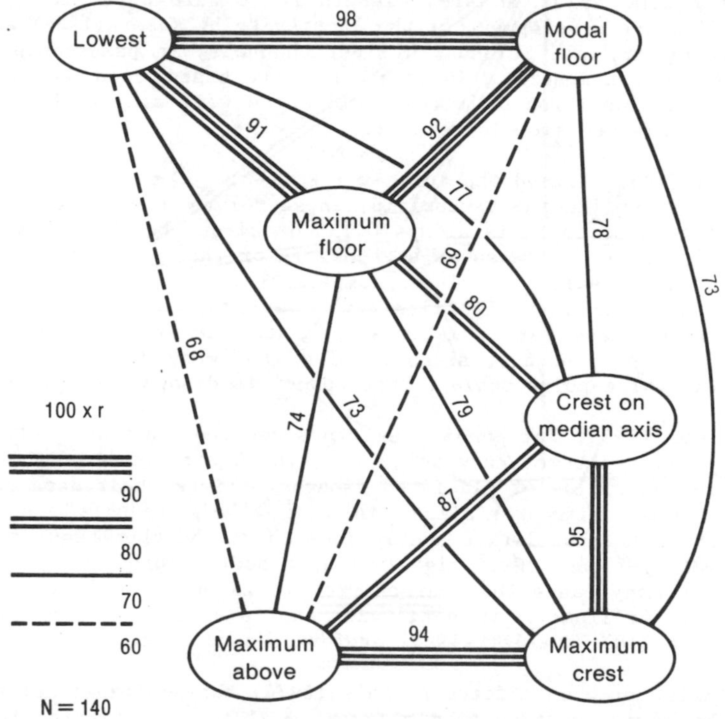

Fig. 2. A correlation structure for six altitude variables, measured for 140 glacial cirques (subject to revision) in the English Lake District, Cumbria. 85% of this information can be incorporated in the first principal component. Varimax rotation suggests two dimensions, one with 48% (all 'floor' variables load over 0.83) and one with 47% (all 'high' variables load over 0.82).

'shape': dividing x by length only produces a 'length-independent' index if the relationship of x to length is a simple linear one through the origin, and if x and length have similar variability. One possibility, apparently not yet used in geomorphometry, would be to analyse 'shape' (including gradient) as a set of residuals from regressions between size variables.

The use of equations , e.g. the lemniscate loop (Chorley, 1959), to represent a landform invokes a mathematical model of the landform, implying greater thought and more sophisticated measurement than the use of indices. Yet this sophistication may be more apparent than real. Chorley did not fit loops to the outlines of drumlins: he calculated k from area and length, so that statistically the outcome is the same as if an area: length ratio were used. The useful quantitative development of the lemniscate by Komar (1984) extended it to streamlined islands in river channels, in palaeochannels and in Martian channels. Komar's work was based upon width and downstream position of maximum width, as well as length and area, but it remained two-dimensional.

Haynes (1968) fitted the curve $y = k\,(1-x)\,e^{-x}$ to cirque centre-line profiles: unlike the paraboloid, these curves provide a rock basin, but one whose depth is always 0.135 of cirque height. The 'depth', k (which actually measures steepness), of these curves was assigned one of three values by visual comparison of centreline profiles with three model curves. Profiles could be enlarged or reduced for the comparison, and were sometimes truncated at an arbitrary point. This fitting procedure seems to offer too many degrees of freedom, and the three-point scale for steepness is disappointingly coarse.

Svensson (1959, for one valley cross-section) and Graf (1970, for sixty) fitted power curves $y = ax^b$, which are parabolae if b is 2, to cross-profiles of glaciated troughs. Since their data points on each valley side are almost in line, their results are very sensitive to speculative assumptions about the thickness of drift in the valley-floor. Graf rightly sought bedrock profiles (1970, p 308), but "in many cases the bedrock profile is obscured in whole or in part ... Filling ... with alluvium and talus reduced the form ratio" (1970, p 310). Wheeler (1984) prefers a quadratic.

A similar problem affects Aniya's (1974) three-dimensional extension of the power equation to $z = (X/a)^k + (Y/b)^h$ for fitting to cirque form. The 21 to 181 points used per cirque by Aniya and Welch (1981) were only 46 to 76% of the points measured, since at other points the bedrock was buried under ice or moraine. The thickness of the latter is thus estimated from equations fitted mainly to the exposed rock walls. Residual errors range from 7 to 29m, but though this may seem reasonable, only two Antarctic cirques out of their thirteen are well approximated by a paraboloid. This model is descriptive rather than process-based, since different processes are operative on cirque floors and headwalls. Segments with different equations might be more appropriate: cirque headwalls tend to a limiting gradient related to the bulk strength of jointed rock.

Thus it is clear that despite its apparent sophistication, equation fitting may be inaccurate, even simplistic and misleading: which is not to say that we should give up trying! Few authors compare the fit of different equations; if they did, they would find it difficult to discriminate between, say, a power function and an exponential, over a limited range of values. It is best if equations can be fitted to a dense grid of point altitudes (see below), giving a three-dimensional model with a specified residual error. Fitting an equation to a digitised outline would be a somewhat less adequate approach for drumlins, but irrelevant for cirques since their two-dimensional plan form is of little significance, depending mainly on the surrounding topography.

A third and new possibility is proposed here. As digital terrain models (altitude matrices) become more accurate and more available, it is becoming easier to measure altitude and the surface derivatives gradient, aspect, profile convexity and plan convexity at a sample of points over the landform (Evans, 1980, 1981). These 'point' variables are more clearly related to geomorphic processes than are indices. Their frequency distributions and relationships can provide a much fuller summary of surface form than can the selected extremes and key points of the indices approach, or the simplifying assumptions of the equation-fitting approach. This 'distribution' approach amounts to the application to specific landforms of some general geomorphometric techniques previously applied to map sheets (Evans, 1984) or drainage basins. It provides the 'battery of measures' required by Jarvis (1981) to describe the geometrical complexity of the three-dimensional land surface. It permits comparisons both between landforms of a given type, and between different types of landform, providing a potential convergence between specific and general geomorphometry.

DISTRIBUTION APPROACH: SOME RESULTS

Preliminary indications are that summary measures for individual landforms are more distinct and easier to interpret than the corresponding values for map sheets or large drainage basins. Using modern photogrammetric 1:10,000 maps of Cumbria, England, I have produced altitude matrices with a grid mesh of 50m by manual interpolation to the nearest metre from contours at 10m intervals for cirques, and at 5m intervals for drumlins.

Ten cirques were selected at random from some 150 defined for the Lake District by the author: two of these contained 'cirques within cirques', giving a set of twelve. Sampling was not possible for drumlins since only a few suitable maps were available, and the population has not been defined. Seven drumlins which cut about four contours were selected; this set is therefore biased toward higher drumlins. Five of them are on sheet NY62SE just north of Appleby and the other two (larger and with gentler slopes) are on sheet NY35SW southwest of Carlisle. After outlining the landforms the 50m grid was superimposed, and altitude was interpolated at each point within the outline and at (Queen's case) neighbours outside. The latter provide a peripheral 'halo' of points required to calculate gradient and convexity at internal points. Except for one 'cirque-in-cirque' which covers only 35 data points, the cirques cover 79 to 476 points (0.20 to 1.19 km^2) and the drumlins 43 to 216 points (0.11 to 0.54 km^2).

The most conspicuous common feature of the drumlins (Table 4) is the dominance of convexity in plan: mean values are 16 to 59 degrees/100m and most distributions are positively skewed, with high standard deviations of 50 to 130. Slope profiles might be expected to be similarly convex, but as the drumlins stand today this is roughly balanced by concavity low on the slopes. Gradients average 2.4 to 6 degrees and show little skew: altitude standard deviations are 4 to 9m. Drumlin orientation makes vector strengths modulo 180° moderate or strong (37 to 82%),i.e. slope aspect is bimodal . Both plan and (especially) profile convexity increase with altitude, but the expected decline in gradient with altitude is not consistently found; the top is always gentle but in some drumlins the gentle ends and lower sides are more conspicuous. Taken together these general characteristics are useful in distinguishing drumlins from other landforms; these and further characteristics are useful for describing variations between drumlins. A fuller account of this analysis is given in Evans (1986).

The cirques on the other hand have much greater mean gradients (20 to 34°), with little skewness and relatively low variability in gradient within the more poorly developed examples (Table 4). Altitude is unskewed or positively skewed, with high standard deviations. Profile convexities are positively skewed: that is, many gentle concavities are balanced by relatively few, sharp convexities (mainly around the cirque crest). The latter might be omitted if the peripheral 'halo' of points were excluded. Vector strength is stronger modulo 360°, showing direction rather than orientation, but aspect is often bimodal with fewer observations near the vector mean than on either side.

The strongest correlation for a cirque is usually the increase of gradient with altitude. Profile convexity also increases with altitude, and plan convexity does so weakly but consistently. Consistency within these admittedly small data sets suggests that this approach will be valuable so long as accurate high-resolution data sets are available.

In conclusion, note that like the indices approach, the equation approach and the distribution approach still involve the nine stages defined above. Landforms similar in their symmetry, subdivision and topographic position may be fitted by similar types of equation, and are likely to involve similar distributional attributes. The use of indices, especially the indiscriminate multivariate analysis of mixed units and ratios, may be what has given morphometry a 'bad name' in some quarters; but it will continue because its speed is useful where many landforms are to be measured and where morphometry is a minor part of a broader study. Equations are not necessarily better; they are needed in modelling but require a process rationale

TABLE 4. Summary of Morphometric Distribution Statistics.
For seven separate analyses of individual Cumbrian drumlins, the mean, median and maximum of the seven statistics are given: likewise for twelve Cumbrian cirques.

	SEVEN DRUMLINS			TWELVE CIRQUES			UNITS
Altitude	min	median	max	min	median	max	
St. deviation	4.0	5.3	9.1	44.9	70.0	102.8	m
Skew	-.57	.04	.29	-.17	.21	.94	-
Gradient							
Mean	2.43	4.30	5.91	19.6	29.2	33.8	deg
St. deviation	1.42	1.84	2.26	4.7	9.6	15.4	deg
Skew	-.40	.21	3.06	-.68	.01	.22	-
Profile convexity							
Mean	-1.06	-.04	.78	-6.14	-.37	5.32	deg/100m
St. deviation	3.59	6.83	7.77	11.8	18.2	30.4	deg/100m
Skew	-.84	.40	1.57	.30	1.44	3.18	-
Kurtosis	-.67	-.18	3.96	.11	2.44	14.48	-
Plan Convexity							
Mean	16	41	59	-25	-12	-2	deg/100m
St. deviation	50	110	130	38	52	78	deg/100m
Skew	-.90	2.89	7.87	-1.82	.24	3.21	-
Kurtosis	.22	13.38	84.97	-.41	2.63	38.19	-
Vector strengths (gradient weighted)							
360 deg.	.05	.22	.89	.40	.72	.91	-
180 deg.	.10	.21	.82	.09	.25	.72	-
90 deg.	.18	.27	.47	.06	.41	.54	-
Correlations							
ALT:GRAD	-.52	-.07	.42	.14	.60	.86	-
ALT:PROFC	.38	.70	.75	.33	.54	.68	-
ALT:PLANC	.23	.40	.53	.08	.24	.40	-
PROFC:PLANC	.09	.21	.29	.03	.18	.39	-
GRAD:PROFC	-.25	-.01	.28	-.38	-.02	.18	-
GRAD:PLANC	-.57	-.30	.27	-.00	.20	.48	-

and careful calibration, preferably against gridded altitude data. With increasing altitude matrix availability, increasing computer power to manipulate these data, and systems and programs which are easier to use, we may expect an increase in the importance of the 'distribution' approach, combined with rigorous equation-fitting.

The need for careful data analysis is illustrated by Carson and Lapointe's (1983) demonstration of the inadequacy of the 'sine-generated curve' and other symmetrical models of river channel plan form. The present paper emphasises non-fluvial landforms, and even so a comprehensive review cannot be provided here. I hope that sufficient comparisons have been made that those contemplating morphometric work on one landform will be persuaded of the value of consulting work on other landforms.

ACKNOWLEDGEMENTS

I am very grateful to Nick Cox and Vince Gardiner for comments on this paper, to David Hume for drawing the figures, and to Elizabeth Pearson and Catherine Reed for typing.

REFERENCES

Aniya,M.,1974. Model for cirque morphology . Geographical Review of Japan, 47, 776-784.

Aniya, M.,and Welch, R.,1981. Morphometric analyses of Antarctic cirques from photogrammetric measurements . Geografiska Annaler, 63A, 41-53.

Carson, M.A.,and Lapointe, M.F.,1983. The inherent asymmetry of river meander planform . Journal of Geology, 91, 41-55.

Chamberlin, T.C.,1890. The method of multiple working hypotheses Science (old series),15, 92 . Reprinted 1965, 148, 754-759 . See also 1897 Journal of Geology, 5, 837-848 and 1931, 39, 155-165.

Chorley, R.J.,1959. The shape of drumlins . Journal of Glaciology, 3, 339-344.

Church, M.,and Mark, D.M.,1980. On size and scale in geomorphology. Progress in Physical Geography, 4, 342-390

Clarke, J.I.,1966. Morphometry from maps , in Essays in geomorphology, G.H. Dury (ed.), 235-274. Heinemann, London.

Clayton, K.M.,1953. A note on 25-foot contours on O.S. 1:25,000 maps . Geography, 38, 77-83.

Dalrymple, J.B., Blong, R.J.,and Conacher, A.J.,1968. A hypothetical nine-unit landsurface model . Zeitschrift für Geomorphologie, N. F., 12, 60-76.

Derbyshire, E.,and Evans, I.S.,1976. The climatic factor in cirque variation , in Geomorphology and climate, E. Derbyshire (ed.), ch.15, 447-494. Wiley, New York and London.

Evans, I.S.,1972. Inferring process from form; the asymmetry of glaciated mountains , in W.P. Adams and F.M. Helleiner (eds.), International Geography 1972, University of Toronto Press, v.1, 17-19.

Evans, I.S.,1977. World-wide variations in the direction and concentration of cirque and glacier aspects . Geografiska Annaler 59A, 151-175.

Evans, I.S., 1980. An integrated system of terrain analysis and slope mapping. Zeitschrift für Geomorphologie, N.F. Supplementband, 36, 274-295.

Evans, I.S., 1981. General geomorphometry, in Geomorphological Techniques, A.Goudie et al. (eds.), 31-37. George Allen and Unwin, London.

Evans, I.S., 1983. Univariate analysis; presenting and summarizing single variables, in A Census user's handbook, D. Rhind (ed.), ch.4, 115-149. Methuen, London.

Evans, I.S., 1984. Correlation structures and factor analysis in the investigation of data dimensionality: statistical properties of the Wessex land surface, England, Proceedings of the International Symposium on Spatial Data Handling, Aug. 20-24, 1984, Zürich, Geographisches Institut, Universität Zürich-Irchel, v.1, 98-116.

Evans, I.S., 1986. A new approach to drumlin morphometry, Drumlins: 1985 symposium, J.Menzies and J.Rose (eds.), A.A. Balkema.

Evans, I.S., and Jones, K., 1981. Ratios and closed number systems, in Quantitative geography: a British view, N. Wrigley and R.J. Bennett (eds.), ch.12, 123-134. Routledge and Kegan Paul, London.

Gardiner, V., 1973. Univariate distributional characteristics of some morphometric variables. Geografiska Annaler 54A, 147-153.

Gardiner, V., 1975. Drainage basin morphometry, British Geomorphological Research Group Technical Bulletin, 14. Geo Abstracts, Norwich.

Gardiner, V., 1983. The relevance of geomorphometry to studies of Quaternary morphogenesis, in Studies in Quaternary Geomorphology; Proceedings 6th. British-Polish seminar. D.J. Briggs and R.S. Waters (eds.), 1-18, Geo Books, Norwich.

Gardiner, V., and Gardiner, G., 1978. Analysis of frequency distributions, CATMOG 19, p.68. Geo Abstracts, Norwich.

Graf, W.L., 1970. The geomorphology of the glacial valley cross-section. Arctic and Alpine Research, 2, 303-312.

Haynes, V.M., 1968. The influence of glacial erosion and rock structure on corries in Scotland. Geografiska Annaler, 50A, 221-234.

Jarvis, R.S., 1981. Specific geomorphometry, in Geomorphological Techniques, A. Goudie et al. (eds.), ch. 2.5, 42-46. George Allen & Unwin, London.

Jauhiainen, E., 1975. Morphometric analysis of drumlin fields in northern Central Europe. Boreas, 4, 219-230.

Komar, P.D., 1984. The lemniscate loop - comparisons with the shapes of streamlined landforms. Journal of Geology, 92, 133-145.

Lancaster, N., 1981. Aspects of the morphometry of linear dunes of the Namib Desert. South African Journal of Science, 77, 366-368.

Mandelbrot, B.B., 1977. Fractals: form, chance and dimension. W.H. Freeman, San Francisco, p 365. See also 1982 The fractal geometry of nature, p 460.

Mardia, K.V., 1972. Statistics of directional data. Academic Press, London.

Mardia, K.V., 1975. Statistics of directional data Journal, Royal Statistical Society, B 37, 349-393.

Mark, D.M., and Church, M., 1977. On the misuse of regression in earth science . Mathematical Geology, 9, 63-75.

Mills, H.H., 1980. An analysis of drumlin form in the northeastern and north-central United States . Geological Society of America Bulletin, 91, Part I 637-639, Part II 2214-2289

Mosimann, J.E., 1970. Size allometry; size and shape variables with characterizations of the log-normal and generalised gamma distributions . Journal, American Statistical Association, 65, 930-945.

Ollier, C.D., 1963. Contour map accuracy and analysis. Australian Geographical Studies, 1, 96-99.

Olyphant, G.A., 1981. Allometry and cirque evolution. Geological Society ofAmerica Bulletin, 92, Part I, 679-685.

Pike, R.J., and Rozema, W.J., 1975. Spectral analysis of landforms. Annals, Association American Geographers, 65, 499-514.

Richards, K.S., 1981. Geomorphometry and geochronology , in Geomorphological Techniques. A. Goudie et al. (eds), ch. 2.4, 38-41. George Allen and Unwin, London.

Rose, J., and Letzer, J.M., 1975. Drumlin measurements: a test of the reliability of data derived from 1:25,000 scale topographic maps. Geological Magazine, 112, 361-371.

Sauchyn, D.J., and Gardner, J.S., 1983. Morphometry of open rock basins, Kananaskis area, Canadian Rocky Mountains . Canadian Journal of Earth Science, 20, 409-419.

Stager, J.K., 1956. Progress report on the characteristics and distribution of pingos east of the Mackenzie Delta . Canadian Geographer, 7, 13-20.

Svensson, H., 1959. Is the cross section of a glacial valley a parabola? Journal of Glaciology, 3, 362-363.

Unwin, D.J., 1975. An introduction to trend surface analysis , CATMOG, 5, Geo Abstracts, Norwich.

Wheeler, D.A., 1984. Use of parabolas to describe the cross-sections of glaciated valleys . Earth Surface Processes & Landforms, 9, 391-394.

Wilson, I.G., 1972. Aeolian bedforms - their development and origins . Sedimentology, 19, 173-210

International Geomorphology 1986 Part II
Edited by V. Gardiner

DETERMINATION OF GRID SCALES OF LANDSCAPE VARIABLES FOR HYDROLOGICAL REGIONALISATION

M.R. Hendriks, E. Seyhan and I. Simmers

Institute of Earth Sciences, Free University,
1007 MC Amsterdam (The Netherlands)

ABSTRACT

For four areas in East-Luxembourg a comparative study has been completed to determine grid scales to be used for hydrological regionalisation. Grid scales required to estimate mean catchment values for classifying basins and for hydrological prediction by statistical modelling have been determined using (a) an empirical method, (b) prejudged confidence limits, (c) Cramér's equation, (d) Kendall and Stuart's equation, (e) frequency analysis and (f) autocorrelation analysis. Autocorrelation analysis stands out as an explicit method which takes the areal distribution of landscape variable-values into account. For each landscape variable a different grid scale can be produced; by weighting the importance of landscape variables for a specific 'problem' the *'overall' grid scale* to be used for hydrological regionalisation can hence be determined. In this way a considerable reduction can be achieved in the subjective element normally encountered in literature when it comes to determining grid scales for hydrological regionalisation.

INTRODUCTION AND RATIONALE

Hydrological regionalisation is concerned with extending records in space as opposed to generating them in time and has for a number of years been used as a standard tool to facilitate extrapolation from sites at which records have been collected to others at which data are required but unavailable (Riggs, 1973; Mosley, 1981). The need for such information has escalated dramatically in recent years to the point where the procedure has become a standard prerequisite for adequate solution of innumerable water-resources management problems in a spectrum of climatic regions and geographical areas (Simmers, 1984).

The Australian studies reported by Laut et al. (1982) represent a significant development in the field of landscape classification and regionalisation. The strategy adopted was to provide a standardised hydrologically-oriented landscape description and classification and to use this to typify or classify sub-basins. Grid cells formed the basic data collection units and each grid-cell data set included *continuous* (e.g., altitude, relief and slope), *disordered multistate* (e.g., lithology, land use) and *non-exclusive ordered multistate* (e.g., aspect) types of attributes. A numerical classification technique was adopted to compute a data set dissimilarity matrix and resultant classification of grid cells, with sub-basins classified according to their grid-cell composition. In their studies Laut et al. (1982) used grid cells of 4 km^2 as basic data collection units stating that this provided a useful compromise between the high costs of establishing and using a data set based on 1 km^2 cells and the loss of spatial information associated with using larger units.

A considerable reduction in the subjective element normally encountered in literature when it comes to determining grid scales can be achieved, when the grid size is directly related to the purpose for which regionalisation is to be carried out. Regionalisation of hydrological data can be shown to be bound by two fundamental issues : the *'problem'* to be resolved and the *scale* at which a solution is required. Each 'problem' can be thought of as having its own set of *'problem'-related variables* and each variable as having its own *sampling scale* or *grid scale*. By weighting the importance of landscape variables for a specific 'problem' the *'overall' grid scale* to be used for hydrological regionalisation can thus be determined. Since sub-basins are classified according to their grid-cell composition when using the Laut et al. (1982)-model, sub-basins of different sizes can hence be classified according to the scale at which a solution is required.

The use of geometric grid cells has a number of distinct advantages: (a) units can be quantitatively derived according to the required precision of contained data, (b) units can be of any size, directly related to the purpose for which regionalisation is to be carried out (the 'problem' to be resolved), (c) information stored in grid cells lends itself to solution of 'problems' occurring at different scales, (d) a grid-cell configuration allows attachment of other (more sophisticated) physically based grid-models for the solution of 'problems' of a distributed nature, and (e) grid-cell information is easily stored on computer, forming a data-bank for resolution of future 'problems'.

An experimental configuration as indicated above will be used to classify catchments and to predict flood parameters by statistical modelling in the Luxembourg sandstone and Keuper areas (Hendriks, in prep). Reservations concerning underlying linear assumptions in statistical models are recognized and the appropriateness of the model for predicting flood parameters remains to be tested. Alternatively, the grid cell classification may be used for the allocation of unit areas within a simulation model.

APPROACH

In order to compare the suitability of different methods for the determination of grid scales four areas were selected in East-Luxembourg (Figure 1 and Table I). The areas used in the analyses are the indicated rectangular shapes, except in the case of the empirical approach where the enclosed catchments were used.

Topographical maps at a scale of 1 : 10,000 formed the basis for di-

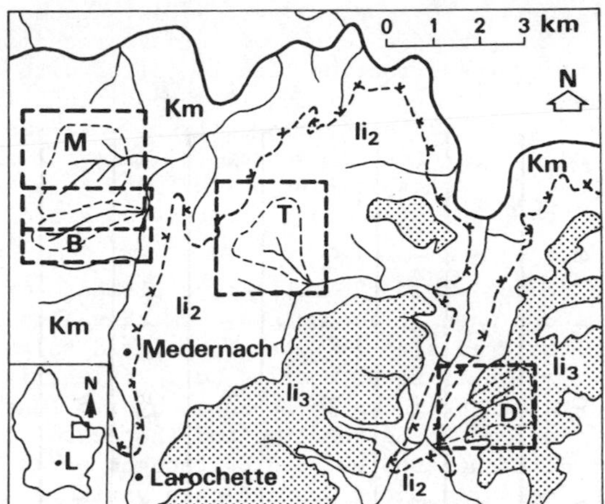

Figure 1 Location of the selected catchments/areas (B = Brücherbaach, D = Dosbach, M = Moserbach, T = Tollbach; L = Luxembourg-city; Km = Keuper marls, li_2 = exposed Luxembourg sandstone, li_3 = Luxembourg sandstone covered by Ariëten marls; -✕- = Keuper formation-Luxembourg sandstone boundary)

gitising contourlines at 5 m altitude intervals (digitising equidistance = 5 m). Using the program ISOTOGRID (Hazelhoff, 1983) and La Grange interpolation, variable values with an arbitrarily chosen grid spacing in nature of 50 m were derived for the first three of the following four variables :

(1) *Altitude* : elevation above mean sea level in meters.

(2) *Slope* : slope of the land surface in degrees. The variables altitude and slope represent *continuous* variables.

(3) *Aspect* : Aspect is treated as a *non-exclusive ordered multistate* variable and is the compass direction of slope in a downhill direction in degrees, North being defined as 0° or 360°. In order to give this variable some physical meaning (in connection with insolation and evapo(transpi)ration) the values are classified into five groups scoring 1 to 5 as suggested by the Tennessee Valley Authority (1964) (Figure 2a), with slopes less than 10 % (4.5^o) receiving an intermediate aspect-class value of 3.

For analyses requiring discrete (yes or no; one or zero) data aspect has been divided into two classes representing values from NW to and including SE and from SE to and including NW in a clockwise direction (Figure 2b).

(4) *Forest* : This fourth variable has been determined by hand using 1 : 10,000 land use maps established in the field. The variable forest represents one of the states of the *disordered multistate* variable land use. The percentage of forested area in rounded fifths per grid has been determined as a 'within grid' description with a grid spacing in nature of 100 m. For analyses needing

Catchment/area	Main formations	Altitude Range (m)	Area (km^2)	Land use % F	% P	% C	% U
Brücherbaach	Keuper-marls	238-350 (c)	c : 1.77	10	80	8	2
			r : 4.26	11	80	7	2
Dosbach	Arieten-marls (c, r) Luxembourg sandstone (c, r) Psiloten-marls (c, r) Keuper-marls (r)	210-343 (c)	c : 1.98	40	47	11	2
			r : 3.69	53	35	10	2
Moserbach	Keuper-marls	255-339 (c)	c : 3.30	48	46	5	1
			r : 6.76	45	47	6	2
Tollbach	Luxembourg sandstone (c, r) Psiloten-marls (r) Keuper-marls (r)	285-421 (c)	c : 2.54	87	0	13	0
			r : 6.00	75	11	13	1

TABLE I : Brief description of the selected areas (see Figure 1) (c = catchment; r = rectangular area; F = forest; P = pasture and grassland; C = cropland; U = urban areas and roads)

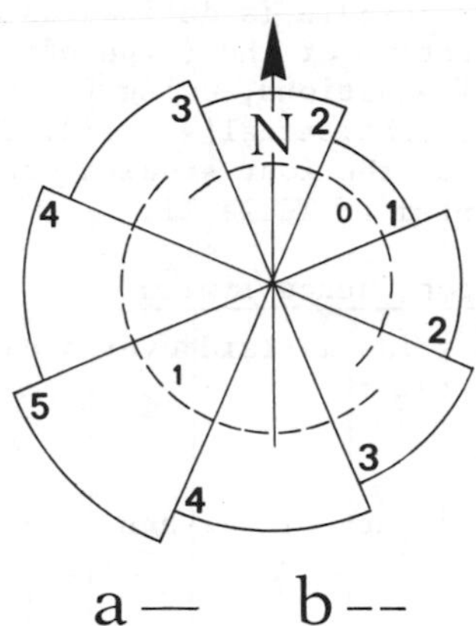

Figure 2 Classification of aspect into (a) five (TVA, 1964) and (b) two (discrete) groups

discrete data a subdivision has been made into grids with less than 50 % forested area and those with 50 % or more.

In all analyses the grid variable-values are treated as centroids describing a surrounding square grid with a width equal to the spacing used, thereby providing an equal number of variable values and grids.

The present paper critically examines several of the available methods for determining grid scales to be used for hydrological regionalisation and considers only a limited number of variables. The following methods are evaluated :

(1) Empirical approach
(2) Prejudged confidence limits
(3) Cramér's approach
(4) Kendall and Stuart's approach
(5) Frequency analysis
(6) Autocorrelation analysis

THEORETICAL CONSIDERATIONS AND ANALYSES

1. Empirical Approach

Typical of this approach are the empirical equations derived by Enderlein et al. (1982) :

$$0.05\sqrt{A_c} \leq D_n \leq 0.1\sqrt{A_c} \quad (1)$$

$$0.05M\sqrt{A_c} \leq D_m \leq 0.1M\sqrt{A_c} \qquad (2)$$

The symbols are listed in Appendix I.

In these expressions grid scale is defined as a function of catchment area only, irrespective of the (type of) landscape variable involved. Other empirical equations, using for example local relief, may also be found to be statistically valid. The resulting grid scales in nature (D_n) for the four studied catchments when using equation (1) are presented in Table II.

2. Prejudged confidence limits

If a network of square grids is laid over a catchment area one can define the grid scale from :

$$D_n = \left|\sqrt{A/n}\right| \qquad (3)$$

Statistically n is determined by (Seyhan, 1983) as :

$$n = (ts/c)^2 \qquad (4)$$

Combining equations (3) and (4) gives :

$$D_n = \left|\frac{c\sqrt{A}}{ts}\right| \qquad (5)$$

Standard deviations of the variables altitude and slope for the four rectangular shaped areas using 50 by 50 m grids are presented in Table III. The values of D_n using equation (5), with c = 5 m (altitude) or 0.5^o (slope), $\alpha = 0.05$ and the above mentioned values for s, are again given in Table II. It is obvious that the method assumes normally distributed data and that it is subjective since the 'unknown' values of c, α and s need to be prejudged. Furthermore, the method is not suitable for determining grid scales of landscape variables such as aspect (class) and (percentage of) forested area, since the choice of values for c, α and s are meaningless. A lognormal version of equation (5) is provided by Montgomery and Hart (1971).

3. Cramér's approach

For discrete distributed data Cramér (1948) shows that :

$$\delta \geq \frac{n}{n+t^2}\left[p^* + \frac{t^2}{2n} \pm t\sqrt{\frac{p^*q^*}{n} + \frac{t^2}{4n^2}}\right] \qquad (6)$$

Solving equation (6) for n gives (Gottschalk and Krassovskaia, 1980):

$$n \geq t^2\left[\frac{2p^*q^*}{\delta^2} - 1 \pm \sqrt{\left(\frac{2p^*q^*}{\delta^2} - 1\right)^2 + \frac{1}{\delta^2} - 1}\right] \qquad (7)$$

Combining equations (7) and (3) produces :

$$D_n \leq \left|\frac{\sqrt{A}}{t\sqrt{\frac{2p^*q^*}{\delta^2} - 1 \pm \sqrt{\left(\frac{2p^*q^*}{\delta^2} - 1\right)^2 + \frac{1}{\delta^2} - 1}}}\right| \qquad (8)$$

This equation has been used to determine the grid scales for the

Variable	Area	Grid scale D_n (m) (1)	(2)	(3)	(4)	(5)	(6)	(7)
Altitude	Brücherbaach	67-133	220	-	308	300	300	560
	Dosbach	70-141	141	-	197	300	200	335
	Moserbach	91-182	320	-	461	300	200	590
	Tollbach	80-159	201	-	281	200	300	395
Slope	Brücherbaach	67-133	151	-	212	200	100	180
	Dosbach	70-141	59	-	83	800(?)	100	135
	Moserbach	91-182	203	-	284	400	300	100
	Tollbach	80-159	107	-	149	$\leq$50	200	150
Aspect	Brücherbaach	67-133	-	106	-	-	-	60
	Dosbach	70-141	-	182	-	-	-	55
	Moserbach	91-182	-	136	-	-	-	55
	Tollbach	80-159	-	135	-	-	-	90
Forest	Brücherbaach	67-133	-	171	-	-	-	60
	Dosbach	70-141	-	99	-	-	-	220
	Moserbach	91-182	-	117	-	-	-	300
	Tollbach	80-159	-	125	-	-	-	250

TABLE II : Grid scales determined by using :

(1) an empirical method (Enderlein et al. (1982), giving maximum and minimum grid scales)
(2) prejudged confidence limits ($\alpha = 0.05$; $c = 5$ m (altitude), $c = 0.5^\circ$ (slope))
(3) Cramér's approach ($\alpha = 0.05$; $\delta = 0.1$)
(4) Kendall and Stuart's approach ($\alpha = 0.05$; $\delta = 10$ m (altitude), $\delta = 1^\circ$ (slope))
(5) Frequency analysis
(6) Similar analysis to (5) using the arithmetic mean
(7) Autocorrelation analysis

variables slope and forest presented as discrete (yes or no; one or zero) data, with the empirical probability p^* taken equal to the number of zeros as a fraction of the total number of grid points ($0 \leq p^* \leq 1$). Since p^* and q^* ($= 1-p^*$) only occur linked as p^*q^* in equation (8) they are interchangeable. As such, p^* can equally well be taken to represent either the number of zero scores or the number of one scores. For example : The Moserbach-rectangle has an area of 6.76 km^2, covered for the variable aspect by 2703 grids and an equal number of grid points with a 50 m spacing in nature. When classifying aspect into two classes (see Figure 2) : 1083 points fall into class 0 and 1620 into class 1, therefore p = 1083/2703 = 0.401. For $\delta = 0.1$ and $\alpha = 0.05$: n : 365 (equation (7)) and $D_n = 136$ m (equation (3) or (8)). The values of D_n using equation (8) with $\delta = 0.1$ and $\alpha = 0.05$ are presented in Table II. Since the data should have a discrete probability distribution and since the technique is subjective in the sense that δ and α need to be prejudged, the use of equation (8) is limited.

4. Kendall and Stuart's Approach

For continuous data Kendall and Stuart (1958) state that

$$n \geqq \frac{4\sigma^2 t z^2}{\delta^2} \tag{9}$$

Combining equations (9) and (3) with z = 1 gives :

$$D_n \leqq \left| \frac{\delta\sqrt{A}}{2\sigma\sqrt{t}} \right| \tag{10}$$

By considering for large samples that the population standard deviation (σ) is equal to the sample standard deviation (s) when using 50 by 50 m grids (see Table III), D_n values for the variables altitude and slope (with $\delta = 10$ m (altitude) or 1° (slope) and $\alpha = 0.05$) have been determined for the four rectangular shaped areas and are presented in Table II. The method assumes normally distributed data and is subjective since the 'unknown' values of δ, α and σ must again be prejudged. The method is not suitable for determining grid scales of landscape variables such as aspect (class) and (percentage of) forested area, since the choice of values for δ, α and σ are meaningless.

5. Frequency Analysis

Using Weibull's (1939) equation :

$$P = \frac{m}{n+1} \tag{11}$$

frequency analyses for the variables altitude and slope were completed for the four rectangular areas using different grid spacings of 50 m and 100-1200 m in 100 m intervals.

From the frequency plots established for these different spacings mean values (x_m) were determined (at P = 0.5). Arithmetic mean values (x_a) for the different grid spacings were also separately determined. The derived values for the two variables are plotted against grid spacing in Figures 3 and 4.
Proceeding from large to small grid spacings, the largest spacing after which no significant change in mean value occurs, corresponding to the asymptotic value of x_m or x_a, is the determined grid scale for the area and variable involved. These are shown in Figures 3 and 4 by small arrows and are listed in Table II. The outcome of the frequency analysis using x_m for slopes in the Dosbach area (determined grid scale : 800 m) does not appear to be sensible knowing the terrain and also when compared with that derived from x_a (100 m).

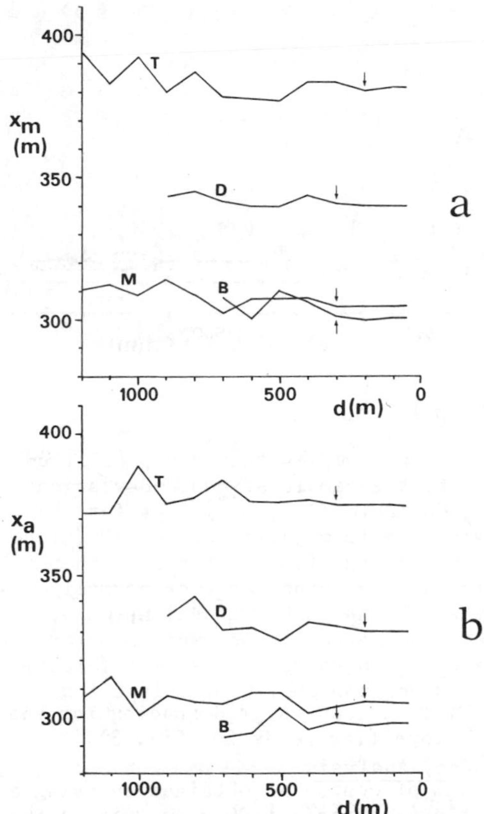

Figure 3 Plot of mean vs. grid spacing for the variable altitude (x_m = mean value from frequency plots; x_a = arithmetic mean; d = grid spacing). B = Brücherbaach (n = 1705), D = Dosbach (n=1476), M = Moserbach (n = 2703), T = Tollbach (n = 2401) (n = number of variable values for d = 50 m).

Because of the small number of possible variable values for aspect (1 to 5) and forest (0 to 5) values of x_a and x_m remain quite constant independent of the grid spacing and no significant change in mean values occurs. This is particularly evident for values of x_m. Mean values using either frequency analysis or x_a have been determined from the original input data matrix at 1, 2, 4, 6, etc. times the grid spacing used (50 m in this study for altitude and slope). This influences the precision with which grid scales are determined; the values given in column (5) and (6) of Table II must therefore be interpreted as minima, a grid scale of, for example, 100 m (= 2*50 m) meaning 100 $\leq$ grid scale < 200 m (= 4*50 m).

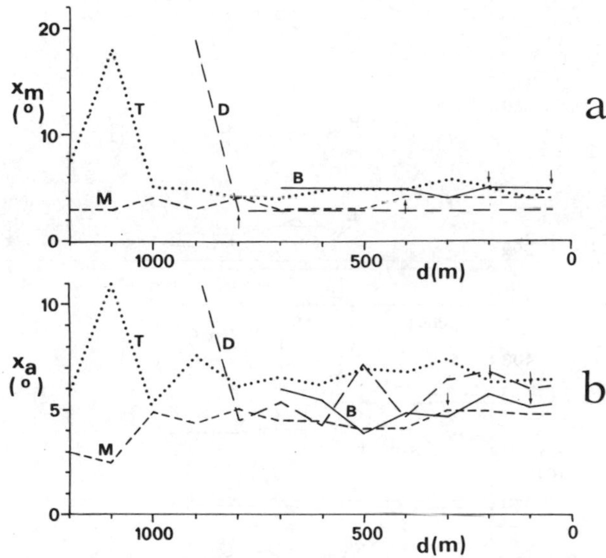

Figure 4 Plot of mean vs grid spacing for the variable slope (legend as for fig. 3).

Greater precision can of course be obtained by using a smaller grid scale (e.g. 5 m), a grid scale of 100 m (= 20*5 m) then giving 100 ≤ grid scale < 110 m (= 22*5 m). However, computer costs for generating such a fine grid network become excessive. The nature of this present study did not warrant a rerun using a finer grid network.

6. Autocorrelation analysis

Autocorrelation analysis is an extension of conventional correlation analysis and is generally used for analysing time series. Autocorrelation coefficients for a single variable in a one-dimensional series are given by :

$$r_k = \frac{\sum_{i=1}^{n-k} x_i \, x_{i+k}}{\sqrt{\sum_{i=1}^{n-k} x_i^2 \sum_{i=1}^{n-k} x^2_{i+k}}} \qquad (12)$$

For a two-dimensional series equation (12) can be rewritten in three ways :

N-S autocorrelation :

$$r_k = \frac{\sum_{i=1}^{nc}\left(\sum_{j=1}^{nr-k} x_{i,j}\, x_{i,j+k}\right)}{\sqrt{\left[\sum_{i=1}^{nc}\left(\sum_{j=1}^{nr-k} x_{i,j}^2\right)\right]\left[\sum_{i=1}^{nc}\left(\sum_{j=1}^{nr-k} x_{i,j+k}^2\right)\right]}} \tag{13}$$

E-W autocorrelation :

$$r_k = \frac{\sum_{j=1}^{nr}\left(\sum_{i=1}^{nc-k} x_{i,j}\, x_{i+k,j}\right)}{\sqrt{\left[\sum_{j=1}^{nr}\left(\sum_{i=1}^{nc-k} x_{i,j}^2\right)\right]\left[\sum_{j=1}^{nr}\left(\sum_{i=1}^{nc-k} x_{i+k,j}^2\right)\right]}} \tag{14}$$

Averaged autocorrelation :

$$r_k = \frac{\sum_{i=1}^{nc}\left(\sum_{j=1}^{nr} x_{i,j}\, y_{i,j}\right)}{\sqrt{\left[\sum_{i=1}^{nc}\left(\sum_{j=1}^{nr} x_{i,j}^2\right)\right]\left[\sum_{i=1}^{nc}\left(\sum_{j=1}^{nr} y_{i,j}^2\right)\right]}} \tag{15}$$

The N-S and E-W autocorrelation coefficients are calculated by pairing the deviations from the series mean of the variable in respectively a north-south and east-west direction. For averaged autocorrelation the deviation from the series mean of the variable is paired with the deviations from the series mean of its surrounding grid points lying at distance k. In the analyses the deviations from the series mean have been measured with the standard deviation as unity, that is : standardised data have been used. Because of the nature of the data (distance series) and the purpose of analysis (determination of sampling- or grid scales) it is important to remain in touch with the reality of the original data as much as possible. Consequently, no removal of (deterministic) trend has been carried out. Autocorrelation coefficients were calculated using the program AUTOMAT (Waardenburg, 1984) and the respective correlograms plotted for the standardised landscape variables altitude, slope, aspect (5 classes, see Figure 2) and forest (in rounded fifths per grid) for all the rectangular shaped areas. The optimum sampling scale (or grid scale) can be determined by analogy with the theory of turbulency (Monin and Yaglom, 1971; Bradshaw, 1976), by integrating the autocorrelation function (Gottschalk and Krasovskaia, 1980) :

$$D_1 = \int_{o}^{\infty} \rho(\tau)\,d\tau \tag{16}$$

This is illustrated graphically in Figure 5, where the shaded area under the autocorrelation function is equal to the area of a rectangle with height unity (r_k = 1) and base equal to the required grid scale value.

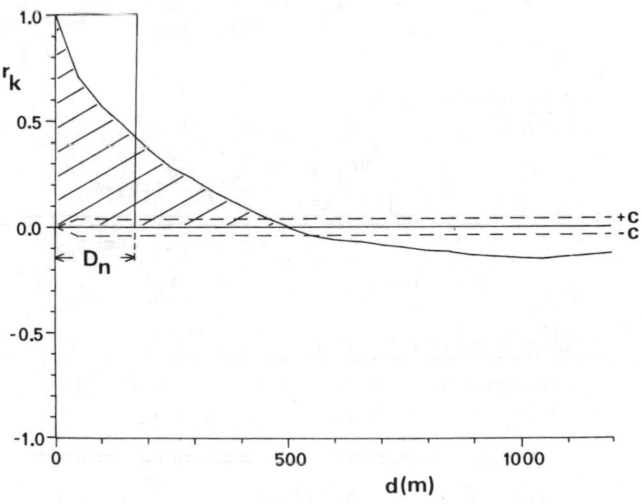

Figure 5 Determination of grid scale by integration of the autocorrelation function (equation (16)) for the variable slope (Tollbach area) using circular autocorrelation (equation (15)) (r_k autocorrelation coefficient; d = grid spacing (m); c = 95 % confidence limit; D_n = grid scale (m)).

In order to obtain the required grid scale in meters (D_n) one simply multiplies D_1 by the grid spacing (m) taken in developing the input matrix. The resulting D_n values are shown in Table III. Because the number of pairs in the calculation of autocorrelation coefficients decreases as the lag increases, n/2 has been taken as the maximum lag in the analyses (n = nr, nc, minimum of nc, nr for respectively equations (13), (14), (15)). The computer program AUTOMAT calculates the area under the autocorrelation function (D_1) until r_k equals zero. Biased estimates of D_1 and D_n occur when values of r_k remain larger than zero; this occurs when a deterministic trend exists in the data. These biased estimates are given the superscript * in Table III. In the situation when (for k = 0 to n/2 lags) r_k remains larger than +c (c = 95 % confidence limit; see Figure 5) values of D_n are given with a greater than sign (>) in Table III. When r_k remains larger than zero but at some point does not significantly differ from zero ($0 < r_k \leq + c$) exact values of D_n are given with only the superscript *; these estimates are only slghtly higher than those that would have been obtained if the abscissa had been raised by +c and the area under the autocorrelation function determined accord-

Variable	Area.	x_a	s	A	B	C	D
Altitude	Brücherbaach	297.31	23.97	> 635*	560	> 595*	560
(m)	Dosbach	329.55	34.80	> 670*	335	> 465*	335
	Moserbach	304.56	20.14	> 860*	590	> 730*	590
	Tollbach	374.32	31.15	395	470	435	395
Slope (°)	Brücherbaach	5.27	3.48	180*	230*	> 195*	180*
	Dosbach	6.11	8.27	> 315*	135	> 225*	135
	Moserbach	4.82	3.27	100	215	200*	100
	Tollbach	6.48	5.86	150	190	170	150
Aspect	Brücherbaach	2.91	0.72	60	145	85	60
(5 class-	Dosbach	3.07	0.41	55	90	80	55
es)	Moserbach	2.92	0.70	55	175*	100	55
	Tollbach	3.04	0.83	90	140	115	90
Forest	Brücherbaach	0.42	1.16	60	360	> 220*	60
(rounded	Dosbach	2.59	2.28	> 390*	220	> 300*	220
fifths)	Moserbach	2.16	2.32	300	> 650*	440*	300
	Tollbach	3.81	1.90	> 520*	250	390*	250

TABLE III : Grid scales determined by autocorrelation analysis

x_a = arithmetic mean (grid spacing : 50 m, except for forest : 100 m)
s = standard deviation (grid spacing : 50 m, except for forest : 100 m)
A = grid scale determined by N-S autocorrelation (m)
B = grid scale determined by E-W autocorrelation (m)
C = grid scale determined by averaged autocorrelation (m)
D = minimum of A, B and C (m)
* = biased estimation

ingly.

Table III shows that averaged autocorrelation always gives a value of D_n that is intermediate to the outcome from the other two modes of autocorrelation. Since E-W and N-S autocorrelation do not produce the same value of D_n it is logical - when square grids are to be used in the regionalisation model (as with Laut et al., 1982) - to assign the minimum grid scale in one of these perpendicular directions as the determined grid scale for the variable and area involved. These values of D_n are presented in Table II (and in Table III under D).

Altitude and slope data (with a grid spacing of 50 m) have been tested for Gaussian normality using (a) Kolmogorov-Smirnov, (b) Chi-square, (c) Kurtosis and (d) Skewness tests. For those found to be non-normally distributed on any one of these tests, log, square root and square transformations were carried out with the transformed data again tested for normality. Even after transformations normality was only accepted by all four tests for the variable height (non-transformed) in the Brücherbaach and Moserbach area; in nearly all cases normality was accepted by Chi-square testing ($\alpha = 0.05$, nr. of classes = 15). Grid scales determined by autocorrelation analysis using original, non-normally distributed, standardised data and transformed, normally distributed, standardised data (using Chi-square) were not found to be significantly different (see Table IV : Dosbach area). The effects of using an input data matrix with 100 m grid spacing (instead of 50 m) and/or a different digitising equidistance (= altitude interval of digitised contours : 25 m instead of 5 m) are shown in Table IV for one area only (Tollbach). The effect of using a larger grid spacing is negligible (provided autocorrelation coefficients are based on sufficient data points). Similar results were also obtained when using a smaller grid spacing of 30 m for the variables altitude and slope in the Tollbach area (6889 grid points), so that for these cases at least no evidence was found of the integral scale being dependent on the grid spacing used. The use of a larger equidistance can be expected to result in less reliable grid data. However, not all the landscape variables show the same sensitivity : going from 25 m to 5 m made remarkably little difference for the variables altitude and slope as can be seen in Table IV.

Table IV also shows the effect on autocorrelation of a different spatial distribution of variable values (using the same values) for the variable forest in the Moserbach area. In part of this area forest appears as a continuous cover and the determined grid scale for the whole area is large (300 m). If forest had been distributed more randomly over the area autocorrelation would produce a smaller grid scale (of about 40 m) to be used for hydrological regionalisation purposes.

DISCUSSION AND CONCLUSIONS

Table II shows the grid scales of chosen landscape variables determined by the various methods; it is obvious that differences occur when using different methods.

The first method, an *empirical approach* gives grid scales as a

Area	Variable	x_a	s	A	B	C	D
Dosbach	Altitude	329.55	34.80	> 670*	335	> 465*	335
(a)	$(altitude)^2$	n.d.	n.d.	> 660*	335	> 465*	335
	Slope	6.11	8.27	> 315*	135	> 225*	135
	$(slope)^{0.5}$	2.05	1.37	> 290*	130	> 215*	130
Tollbach	Altitude (1)	373.84	31.59	408	477	447	408
(b)	Altitude (2)	374.32	31.15	395	470	435	395
	Altitude (3)	374.95	30.45	400	470	420	400
	Altitude (4)	372.96	29.06	390	470	420	390
	Slope (1)	6.55	5.96	150	189	171	150
	Slope (2)	6.48	5.86	150	190	170	150
	Slope (3)	6.51	5.96	160	200	180	160
	Slope (4)	5.05	5.21	170	210	190	170
	Aspect (2)	3.04	0.83	90	140	115	90
	Aspect (3)	3.03	0.81	100	150	120	100
	Aspect (4)	2.99	0.73	120	150	130	120
Moserbach	Forest (5)	2.27	2.35	300	> 650*	440*	300
(b)	Forest (6)	2.16	2.32	300	> 650*	440*	300
(c)	Forest (7)	2.27	2.35	80	40	75	40

TABLE IV : Effect on the determined grid scale of (a) using transformed, normally distributed data, (b) a different grid spacing and/or digitising equidistance and, (c) a different areal distribution of forest.

x_a, s , *, A, B, C, D : see TABLE III.

(1) = grid spacing of 30 m, digitising equidistance: 5 m
(2) = grid spacing of 50 m, digitising equidistance: 5 m
(3) = grid spacing of 100 m, digitising equidistance: 5 m
(4) = grid spacing of 100 m, digitising equidistance: 25 m
(5) = grid spacing of 50 m
(6) = grid spacing of 100 m
(7) = generated data with the same variable-values as in Moserbach, forest (4), but randomly distributed over the area
(n.d. = not determined)

function of catchment area only and the derived values are cautiously low. Because the determined grid scales are independent of the landscape variable involved, this method cannot be used in problem-oriented regionalisation, as outlined in the Introduction and Rationale.

In order to compare grid scales calculated by the method of *prejudged confidence limits* and *Kendall and Stuart's approach*, equal values of α, $\delta = 2c$ and $\sigma = s$ have been used. As can be expected, the same fluctuations in grid scale with variable and area are produced by both methods; Table II further shows that the method of prejudged confidence limits produces more conservative estimates of grid scale than Kendall and Stuart's approach. Grid scales determined by *Cramér's approach* can be used in addition to either of these last two methods, thereby covering the range of variables involved in this study. However all these methods are subjective since statistical parameters need to be prejudged and, of added importance, the areal distribution of variable values is not taken into account.

Frequency analysis does not involve any subjective decision other than determining the point where no significant change occurs in mean value (at $P = 0.5$) with decreasing grid spacing. In this respect it is better than the previous methods. However, the method was found to be applicable only to continuous variables (height, slope), and even then the results often did not appear to be sensible; similar analysis using the arithmetic mean gave more stable results, but can also only be used for continuous variables.

Autocorrelation analysis, together with determination of the grid scale by integration, can be used for all types of variables : continuous, disordered multistate and non-exclusive ordered multistate. The method takes the areal distribution of landscape variable-values into account and can be said to be at least explicit (that is : the same results will be obtained by different people if they make the same choice of procedure), if not totally objective. In analysing spatial data one must be aware of the danger of spatial data being self-similar so that the determined grid scale (integral scale) may depend on the scale of the measurements. However, for the variables altitude and slope in the Tollbach area no evidence of this was found when comparing grid scales derived from 30-, 50- and 100 m input data matrices. The stationarity requirement in autocorrelation analysis presents both practical and methodological problems. Because of the nature of the data (distance series) and the purpose of analysis (determination of sampling- or grid scales) it is desirable to remain in touch with the reality of the original data as much as possible. The data have therefore only been standardised and removal of trend has not been attempted. Since square grids are to be used in the regionalisation model the minimum grid scale in one of two perpendicular directions is logically assigned as the determined grid scale. Because in most cases one direction is less influenced by (deterministic) trend (this is logically always the direction giving the minimum grid scale), the outcome of the analyses were not seriously affected.

In this study a number of methods determining grid scales to be used for hydrological regionalisation have been tested using only a limited number of variables. Further research on the use of autocorrelation analysis for a number of other variables and variable combinations (e.g., permeability in connection with geology, soils and land use) and the effect of using different scales on grid-model output is continuing (Hendriks, in prep.).

In conclusion, looking at the results presented in Table II and favouring autocorrelation analysis for determining D_n : the grid scale for altitude is 480 m, slope 130 m, aspect 70 m, and forest 220 m. These are the mean grid scales (weighted for area, as rounded 10 m-values) for the four studied areas and may be taken as representative of areas lying in the Luxembourg sandstone and Keuper formations. The *'overall' grid scale* to be used for hydrological regionalisation can be determined by weighting the importance of landscape variables for a specific hydrological 'problem' (factor analysis). In this way a considerable reduction can be achieved in the subjective element normally encountered in literature when it comes to determining grid scales for hydrological regionalisation.

ACKNOWLEDGEMENTS

We wish to thank Drs. Frank Waardenburg for his help with computer programming. For this study, computer facilities of both the Laboratory of Physical Geography and Soil Science, University of Amsterdam and the Institute of Earth Sciences, Free University, Amsterdam were used.

The investigations were supported (in part) by the Netherlands Foundation for Earth Science Research (AWON) with financial aid from the Netherlands Organisation for the Advancement of Pure Research (Z.W.O.).

REFERENCES

Bradshaw, P.,1976 . Turbulence. Topics in Applied Physics Volume 12 . Springer Verlag Berlin-Heidelberg.

Cramér, H.,1948 . Mathematical methods of statistics . Princeton University Press, Princeton.

Enderlein, R., Glugla, G.,and Eyrich, A.,1982 . Grid method for calculating groundwater recharge from geographical and climatological characteristics. UNESCO Studies and Reports in Hydrology No. 32. Application of results from representative and experimental basins, 163-176.

Gottschalk, L.,and Krasovskaia, I.,1980 . Synthesis, processing and display of comprehensive hydrologic information . SMHI Rapporter No. RHO 22.

Hazelhoff, L.,1983 . Verwerking van hoogtelijnenkaarten . Mimeogr. Rep., Phys. Geogr. Soil Sci. Lab., Univ. of Amsterdam, Amsterdam.

Hendriks, M.R.,in prep. . Regionalisation of hydrological data . Ph.D. thesis, Inst. of Earth Sci., Free Univ., Amsterdam.

Kendall, M.G.,and Stuart, A.,1958 . The advanced theory of statistics. Vol. 1, Charles Griffin, London.

Laut, P., Austin, M.P. Body, D.N., Faith, D.P., Goodspeed, M.J.,and Paine, T.,1982 . Hydrologic classification of sub-basins in the Macleay Valley, New South Wales . CSIRO Div. Water and Land Resources, Canberra, Tech. Memo. 82/13.

Monin, A.S.,and Yaglom, A.M.,1971 . Statistical fluid mechanics : mechanics of turbulence . English ed. updated, augm. and rev. by the authors, ed. by Lumley, J.L., M.I.T. Press, Cambridge, Mass.

Montgomery, H.A.C.,and Hart, I.C.,1971 . The planning of sampling programmes with particular reference to river management . Water Pollution Research Laboratory Report No. 1253, Stevenage, England.

Mosley, M.P.,1981 . Delimitation of New Zealand hydrologic regions . J. Hydrology, 49, 173-192.

Riggs, H.C.,1973 . Regional analyses of streamflow characteristics . Techniques of water resources investigations, Book 4, ch. B3. US Geological Survey, Washington, D.C.

Seyhan, E.,1983 . Application of statistical methods to hydrology . Publ. Inst. of Earth Sci., Free Univ., Amsterdam.

Simmers, I.,1984 . A systematic problem-oriented approach to hydrological data regionalisation . J. Hydrology, 73 : 71-87.

Tennessee Valley Authority (TVA), 1964 . Bradshaw Creek Elk River - A pilot study in area-stream factor correlation . Knoxville , Office of Tributary Area Development, Research Paper 4.

Waardenburg, F.D.E.,1984 . Statistical programs . Mimeogr. Rep., Inst. of Earth Sci., Free Univ., Amsterdam.

Weibull, W.,1939 . A statistical theory of the strength of materials vol. 151 . Ing. Vetenskaps Akad. Handl., Stockholm.

APPENDIX I List of symbols :

A = area of rectangle covering all of the catchment (m^2)

A_c = catchment area (m^2)

α = level of significance

c = confidence limit

d = grid spacing (m)

D_1 = grid scale (expressed as lag-value)

D_m = grid scale on the map (m)

D_n = grid scale in nature (m)

δ = length of confidence interval

i = 1, 2, 3, ..., nc

j = 1, 2, 3, ..., nr

k = lag (k = 1, 2, 3, ...)

l = number of grid variable-values that surround $x_{i,j}$ at distance k (l = 4, 3 or 2)

m = rank number

M = map scale

n = number of variable values

nc = number of columns in the data matrix

nr = number of rows in the data matrix

p = probability of occurrence

p^* = empirical probability

q^* = $1-p^*$

r_k = autocorrelation coefficient for lag k

$\rho(\tau)$ = autocorrelation function

s = sample standard deviation

σ = population standard deviation

t = critical value (two tailed test) corresponding to a given level of significance (α)

τ = integration parameter

x_a = arithmetic mean

x_i = $x_1, x_2, x_3, \ldots, x_n$ as deviations from the series mean

$x_{i,j}$ = observed values as deviations from the series mean

x_m = mean value at P = 0.5

$y_{i,j}$ = $(x_{i,j+k} + x_{i,j-k} + x_{i+k,j} + x_{i-k,j}) / l$

z = coefficient; for z = 1 the mean value is estimated

International Geomorphology 1986 Part II
Edited by V. Gardiner

CHANNEL NETWORK TOPOLOGY: REGULAR OR RANDOM?

Athol D. Abrahams

State University of New York at Buffalo,
Buffalo, New York 14260, U.S.A.

ABSTRACT

Although the random topology model contains no implication for the way in which channel networks develop, it clearly implies that at a point in time the branching (or merging) pattern is completely random. Early direct and indirect tests of the model were generally supportive. However, these tests suffered from several deficiencies that made them insensitive to regularities (non-random configurations) in network topology. These deficiencies included the preferential testing of small networks which conform to the random topology model more closely than do large ones, a loss of topologic information in testing all but the smallest networks, the spatial aggregation of networks with compensating topological biases, and a bias in the usual method of statistically testing the model that favors its acceptance. Recent less-conventional tests have identified numerous regularities in network topology caused by the spatial requirements of tributaries and the need for their basins to fit together within larger basins; the size, sinuosity, and migration rate of valley bends; and the length, steepness, and plan form of valley sides. It is concluded that the extent to which channel network topology appears to be random depends on the way in which it is examined. The illusion that channel networks branch randomly can be traced to the multiple and complex controls of network topology and to deficiencies in the early tests of the random topology model. As more discriminating tests have been developed, progressively more regularities have been found, and this trend seems likely to continue.

INTRODUCTION

In 1966 the study of channel networks was revolutionized with the publication of a paper by Shreve (1966) in which he introduced the random topology (RT) model. This model rests on the postulate that in the absence of environmental controls, all topologically distinct channel networks (TDCNs) of a given magnitude (i.e., with a given number of sources) are equally likely. Although the RT model contains no implication for the manner in which channel networks develop, it clearly implies that at a point in time the branching (or merging) pattern is completely random.

During the ensuing decade there were numerous tests of the RT model (reviewed by Abrahams (1984a)) that for the most part supported the model, even in the presence of strong environmental controls (e.g., Mock, 1976). As a result of these tests, it became widely accepted by the mid-1970s that the model was a good approximation to natural networks and, hence, that such networks branched more or less randomly (e.g., Werritty, 1972; Smart, 1972; Shreve, 1975).

DIRECT AND INDIRECT TESTS OF THE RT MODEL

The tests of the RT model that were critical in establishing its credibility were both direct and indirect. The indirect tests involved quantitatively predicting known aspects of drainage basin morphometry. Shreve (1966) demonstrated that the RT model predicts Horton's (1945) law of stream numbers, and subsequent research showed that when a second postulate dealing with link lengths and areas is added to the model, the extended model is able to predict many general characteristics of drainage basin morphometry (e.g., Shreve, 1975; Smart, 1978). Although these predictions lent considerable credence to the model, they are relatively insensitive tests owing to the high degree of aggregation or averaging of topological information in the predicted quantities.

The direct tests of the RT model compared observed and expected frequencies of TDCNs or groups of TDCNs in samples of networks of a given magnitude (M). Many researchers tested the model in this fashion and generally obtained favorable results (e.g., Smart, 1969; Krumbein and Shreve, 1970; Werritty, 1972). Although direct tests appear on the surface to be the best method of evaluating the RT model, in fact they suffer from several deficiencies.

1. Most direct tests of the RT model have been performed on small networks, partly because data are most easily collected for such networks and partly because the factorial increase in the number of TDCNs with M has restricted tests of TDCNs to networks with $M \leq 5$. However, because small networks seem to conform more closely to the RT model than do large ones (Abrahams, 1984a), the overall effect of these tests has been to give an overly favorable view of the model.

2. When networks of higher magnitudes have been tested, it has been necessary to group the TDCNs in order to obtain manageable classes. Such grouping, which Abrahams and Mark (1986) termed topological aggregation, has resulted in a dramatic loss of information (Werritty, 1972; Jarvis and Werritty, 1975), so that only the most pronounced departures from the RT model, such as the variation in the proportion of tributary source links with ground slope (Smart, 1969; Abrahams, 1977), have been detected.

3. Large samples of networks must be selected from large areas. Several studies have shown that local environmental inhomogeneities within such areas can produce systematic deviations from the

RT model in subsamples of networks that cancel out one another when the subsamples are combined and tested (Krumbein and Shreve, 1970; Abrahams, 1975b; Abrahams and Flint, 1983). Abrahams and Mark (1986) coined the term spatial aggregation for this phenomenon. Spatial aggregation almost certainly plays a major role in obscuring topological biases, even in networks with uniform environments (Flint, 1980).

4. Direct tests of the RT model have always employed the 0.05 level of significance. However, as exercises in model confirmation, these tests should have used a larger significance level. The fact that they did not means that the tests have been biased in favor of the model (Abrahams and Mark, 1986). Unfortunately, in order to calculate the appropriate significance level, it is first necessary to specify an alternative hypothesis. This is rarely possible in situations where goodness-of-fit tests are employed, and has never been done in direct tests of the RT model. Abrahams and Mark (1986) repeated all published tests of TDCNs and ambilateral classes at significance levels of 0.10, 0.20, and 0.30. Only 7 (15 percent) of the 46 tests resulted in rejection of the RT model at the 0.05 level, but the number increased to 21 (46 percent) at the 0.30 level. It is therefore clear that the proportion of tests accepting the RT model is very sensitive to the choice of significance level. When somewhat higher and more realistic levels than 0.05 are employed, this proportion drops sharply.

Thus both the direct and indirect tests have problems of one sort or another. The effect of these problems was to obscure many of the limitations of the RT model. As a consequence, for more than a decade, geomorphologists harbored the impression that the model was a better representation of natural networks than it actually is, and that channel networks have a more or less random branching pattern. However, with the development of less-conventional methods of testing the model in the late 1970s and early 1980s, many regularities (i.e., non-random configurations) in the branching pattern of channel networks in uniform environments have been recognized. These tests and the regularities they have revealed are reviewed in the following section.

LESS-CONVENTIONAL TESTS OF THE RT MODEL

Channel Network Elongation. In a study of 39 mature drainage basins selected from the literature and 35 mature basins selected from Australian topographic maps, all with $M \geq 50$, Abrahams (1977) discovered a direct relationship between a measure of relative relief and the percentage of exterior links in a basin that are tributary source (TS) links (i.e., exterior links that join interior links downstream) (Fig. 1). This relationship indicates that the RT model best approximates channel networks in basins with moderate to low relative relief. Basins with high relative relief typically contain more TS links than are predicted by the model, whereas basins with very low relative relief contain fewer TS links than predicted.

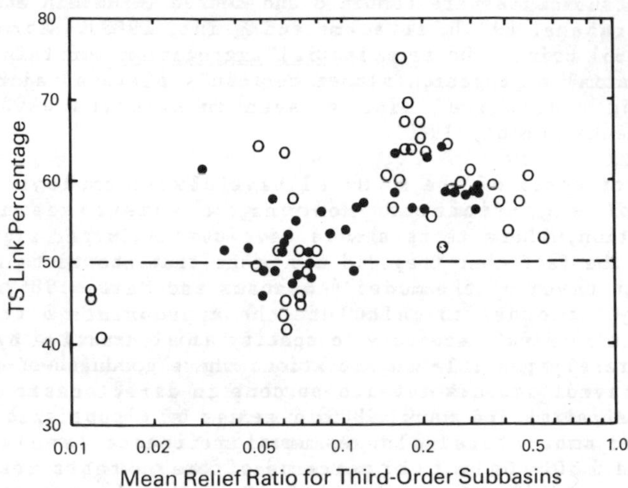

Fig. 1. Plot of TS link percentage (%TS) against an index of relative relief for 39 drainage basins selected from the literature (open circles) and 35 Australian drainage basins selected from topographic maps (solid circles). In topologically random populations of networks with M $\geq$ 50, %TS ≅ 50 percent.

Smart (1978) tested the RT model in a wide-ranging study of 30 channel networks in eastern Kentucky. After comparing the observed and predicted diameter distributions for channel networks of various M, he concluded that there is a persistent tendency for the observed networks to be more elongate than predicted. Smart drew attention to the similarities between his findings and those of Abrahams (1977) and concluded "that network elongation generally increases with increasing relative relief (or slope) and that the Kentucky data mark a specific point in the relationship" (Smart, 1978, p. 166).

Tributary Arrangements Along Main Streams: Effect of Space Filling. Some of the most convincing evidence for regularities in channel network topology has come from studies of the arrangement of tributaries along main streams. This type of study was initiated by James and Krumbein (1969) who viewed every stream in a channel network as a main stream receiving tributaries, except where it joins a larger stream downstream and itself becomes a tributary. Working in the Middle Fork (Rockcastle Creek) channel network in eastern Kentucky, they found that 60.4 percent of main stream links are trans links (i.e., links bounded by tributaries entering from opposite sides), and that there are significantly more very short trans links than very short cis links (i.e., links bounded by tributaries entering from the same side). These findings are generally attributed to the fact that tributaries on the same side of a main stream are prevented from developing close to one another by their surrounding drainage basins, whereas tributaries on opposite sides of a main stream are subject to no such constraint (Horton, 1945, p. 344-345; Smart and Wallis, 1971; Abrahams, 1975a; Flint, 1980; Abrahams and Flint, 1983).

The role of space filling in controlling the locations of tributaries was taken up by Flint (1980), who concluded that distinctive tributary arrangements obtain along the lower, middle, and upper reaches of main streams because these locations are subject to different sets of space-filling constraints. In a study of 45 dendritic channel networks from various parts of the United States, Flint found that 65.6 percent of first, 57.3 percent of second, and 56.7 percent of third tributaries upstream from main stream outlets are obtuse (Fig. 2) (Abrahams and Flint, 1983), whereas the expectable percentage is 50 percent according to the RT model. He ascribed this excess of obtuse tributaries to "basin asymmetry," by which he meant the unequal amounts of land on either side of main streams near their outlets, and to the unequal spatial constraints this asymmetry imposes on tributary development. Abrahams (1980) expressed basin asymmetry in terms of unequal semidivide angles (Fig. 2), and showed that these angles are a function of channel and valley-side slopes in the vicinity of the junction.

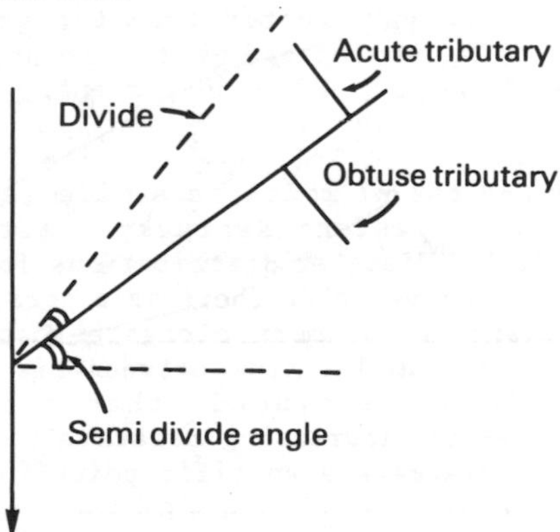

Fig. 2. Schematic diagram of a fork illustrating terminology.

Along the middle reaches of main streams, Flint found that the occurrence of cis and trans links is strongly influenced by the size and hence the spatial requirements of the tributary at the downstream end of each link. He recorded a higher proportion of trans links upstream of large tributaries than of small ones and suggested that these proportions simply average out to yield trans link percentages in the order of the 60.4 percent reported by James and Krumbein (1969). In contrast, along the very upper

TABLE 1. Variation in percentage of obtuse tributaries with tributary number and magnitude

Tributary number	Percentage of obtuse tributaries Sample size	
	M < 4*	M ≥ 4
1 - 3	58.50 2260	68.69 99
4 - 7	52.32 1227	65.98 97
8 - 15	50.20 990	60.76 158
≥ 16	50.67 1048	54.87 226

* M denotes tributary magnitude

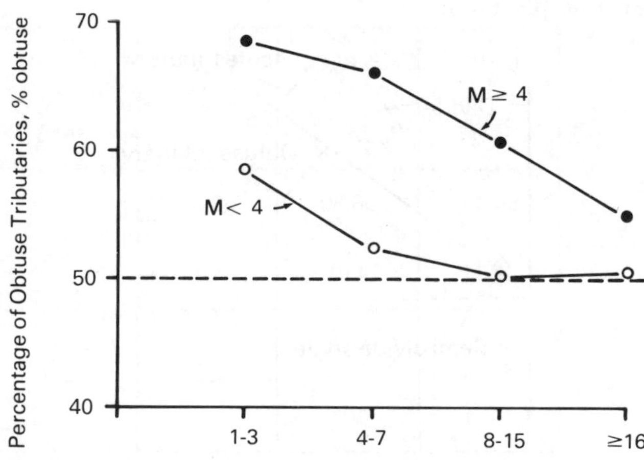

Fig. 3. Plot of percentage of obtuse tributaries against the number of the tributary counting upstream from the outlet for small (M < 4) and large (M ≥ 4) tributaries at 6105 junctions in Kentucky and West Virginia.

reaches of main streams, cis links outnumber trans links. Flint attributed this situation to the narrowness of headwater basins. In such basins the distance between the main stream and each divide is commonly so short that any irregularity in either the shape of the basin or the position of the stream that reduces this distance is likely to inhibit tributary development on one side of the stream and so favor the formation of cis links between tributaries on the other side.

Abrahams and Updegraph (in preparation) have extended Flint's (1980) work on the influence of unequal semidivide angles at the outlets of main streams on the arrangement of tributaries along such streams. These workers examined 6105 junctions in the Rockhouse Fork, Middle Fork, and Coldwater Fork (Rockcastle Creek) channel networks, eastern Kentucky, and in the Pigeon Creek channel network, western West Virginia. In the first of two analyses they distinguished between small tributaries with magnitudes $M < 4$ and large ones with $M \geq 4$. Then for each group of tributaries they investigated how the percentage of obtuse tributaries (%obtuse) varies with the number of a tributary (T_N) counting upstream from the outlet. Their results, which are summarized in Table 1 and Fig.3, show (1) that %obtuse is greater for large tributaries than for small ones for all T_N; and (2) that whereas small tributaries are preferentially obtuse as far upstream as $T_N = 3$, large tributaries are preferentially obtuse over the full length of main streams.

Magnitude of Main Stream, M_D	Magnitude of Tributary, M: 1	2-3	4-10	≥11
≥11	51.81 1548	53.22 590	57.06 333	63.09 149
4-10	54.45 1381	56.59 410	69.39 98	—
2-3	55.27 1491	64.76 105	—	—

Fig. 4. Percentage of obtuse tributaries as a function of the magnitude of the tributary and the magnitude of the main stream. The second number in each cell is the sample size.

The second analysis performed by Abrahams and Updegraph was concerned with whether %obtuse varies with either the magnitude of the tributary (M) or the magnitude of the main stream (M_D) at its confluence with the tributary. Fig. 4 shows that for small tributaries joining large main streams, %obtuse is close to 50 percent; but as M increases or M_D decreases, %obtuse increases. Thus %obtuse is at a maximum for tributaries joining main streams only slightly larger than themselves.

Taken together, the results of the two analyses suggest that large tributaries exhibit a stronger tendency than do small ones to form on the obtuse side of main streams. In the immediate vicinity of main stream outlets, this is probably true of all large tributaries regardless of their size relative to the main stream they join, because large tributaries are more sensitive than small ones to the spatial constraints imposed by unequal semidivide angles. However, where $T_N > 3$, large tributaries that join main streams only slightly larger than themselves are more likely than other large tributaries to be obtuse. The reason for this is that where a tributary is almost as large as the main stream it joins, it typically extends to the boundary of the main stream's drainage basin. Consequently, the side on which such a tributary develops is very sensitive to the relative amounts of space available on either side of the main stream. Abrahams and Updegraph found that there is a tendency for main streams to curve upstream -- that is, toward the larger stream that they join at their downstream end. This tendency restricts the amount of space available on the acute side of main streams and favors the development of large tributaries on their obtuse side.

These analyses go much further than Flint's in identifying regularities in channel network topology caused by the spatial requirements of tributaries and the differential availability of space on the obtuse and acute sides of main streams. The differential availability of space is a consequence of unequal semidivide angles at the outlets of main streams and of the upstream curvature of these streams. Flint (1980) viewed the variation in semidivide angles as a function of the need for tributary basins to fit together within larger basins, whereas Schumm (1956) suggested that the upstream curvature of tributaries (subbasin main streams) in the Perth Amboy badlands channel network was due to the hydrophilic behavior of headward-growing channels during the initial development of the channel network. It is unclear whether this explanation applies to full-sized dendritic channel networks, as in most cases their origin is unknown.

<u>Tributary Arrangements Along Main Streams: Effect of Valley Winding</u>. Along the middle reaches of main streams, James and Krumbein (1969) and Flint (1980) showed that the spatial requirements of tributaries give rise to regularities in the arrangement of tributaries, and Jarvis and Sham (1981) demonstrated that these requirements also produce regularities in the sequence of tributary sizes, with large tributaries tending to be separated by clusters of small ones. Most streams have valleys that wind to

some extent, and Abrahams (1984b, 1984c) has shown in two recent studies that valley winding can have a profound effect on the arrangement and sizes of tributaries. Both studies were based on a sample of 40 winding streams and valleys from the eastern United States and eastern Australia.

In the first study, Abrahams (1984b) employed causal analysis to gain some insight into the factors that control the percentage of tributaries on the concave (out) side of bends (%CVT) and the percentage of trans links (%trans) along the 40 winding streams and valleys. He found that 81 percent of the variation in %CVT could be accounted for by the mean number of tributaries per bend ($\bar{N}_T$), valley sinuosity (P), and the average bend migration rate (M_R), and that 72 percent of the variation in %trans could be explained by these three variables plus %CVT. A causal diagram summarizing the relationships between these variables is presented in Fig. 5.

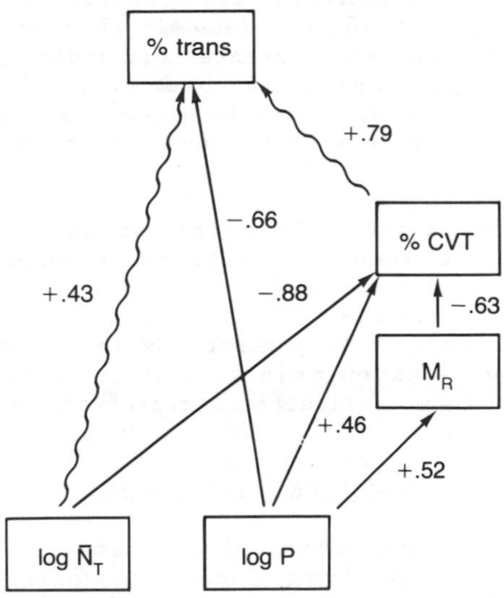

Fig. 5. Causal diagram indicating the results of a partial correlation analysis of the factors influencing the percentage of trans links and the percentage of tributaries on the concave side of bends along winding streams and valleys. The straight arrows denote linear relations, whereas the wavy arrows denote curvilinear relations. The associated numbers are the appropriate partial correlation coefficients. The notation is explained in the text.

In an attempt to sort out the complex web of causal connections that underlies the pattern of tributary development along the middle reaches of winding streams and valleys, Abrahams identified four independent and semi-independent factors that regulate or control this pattern: (1) the spatial requirements of tributaries, (2) valley sinuosity, (3) mean valley bend length, and (4) mean rate of bend migration. Factors 2, 3, and 4 represent valley winding. But because the effect of factors 2 and 3 on tributary development is dependent on factor 1, all four factors may be viewed as contributing to the effect of valley winding on tributary development. In the causal analysis of %CVT and %trans (Fig. 5), factor 2 is represented by P, factor 4 by M_R, and factors 1 and 3 by $\bar{N}_T$.

In the second study, Abrahams (1984c) investigated whether valley winding affects small ($M \leq 5$) and large tributaries ($M > 5$) differently. He calculated the percentages of small tributaries (%SCVT) and large tributaries (%LCVT) on the concave side of bends along each of the 40 reaches investigated previously. A paired t test revealed that $\overline{\%LCVT}$ is significantly larger than $\overline{\%SCVT}$, indicating that large tributaries are more likely than small ones to be located on the outside of bends. This was attributed to large tributaries experiencing greater difficulty than small ones forming in the limited amount of space on the inside of bends, especially where valley sinuosity is high. This result shows that the size of tributaries as well as their arrangement is regulated by valley winding.

The foregoing discussion of tributary arrangements along main streams leaves little doubt that these arrangements possess a large non-random or deterministic component. However, the ultimate test of the deterministic or non-random character of tributary development would be to predict the locations of individual tributaries along a given main stream. A recent study of the locations of magnitude 1 tributaries represents an initial effort in this direction.

LOCATIONS OF MAGNITUDE 1 TRIBUTARIES

This study (Abrahams and Motsay, in preparation) was performed along a reach of the Cox River, located approximately 90 km west of Sydney, Australia. Along this reach, which is part of a longer reach of the same river studied by Abrahams (1984b, 1984c), the stream channel and valley axis more or less coincide. The landscape is maturely dissected, with the river being incised 500-700 m below the adjacent ridge crests. As a result of this considerable amount of incision, the river's valley sides are typically 0.75-1.0 km long, and the magnitude 1 tributaries (mean length of 470 m) are developed on these valley sides and do not extend beyond the river valley. Thus it seems reasonable to assume that most, if not all, of the magnitude 1 tributaries formed in response to the incision of the Cox River and are closely adjusted to the general configuration of the valley sides.

Increments of 2.5 mm on the map (= 40 m on the ground) were marked off along the study reach. Working separately along each side of the river, a systematic sample was collected of every tenth increment that did not receive a tributary. The sampling continued in an upstream direction until this "non-tributary" sample contained 100 increments. All 61 increments along the sampled reach that received a magnitude 1 tributary were included in the "tributary" sample. The following variables were measured for each increment in the two samples:

1. The length of overland flow (L) from the divide down the valley side or magnitude 1 tributary to the increment. This variable was included as a measure of length of the area contributing runoff to the base of the valley side.

2. The height difference (H) between the increment and the divide. H was then divided by L to obtain a measure of mean valley-side slope (S).

3. The curvature or angle of deflection (θ) of the main stream at the increment (see Abrahams, 1984b, Fig. 13). θ is a surrogate for valley-side plan form, and was recorded as positive for the concave side and negative for the convex side of bends so that it varies directly with the degree of concentration of runoff at the base of the valley side.

4. The distance (D_T) to the nearest tributary upstream or downstream along the same side of the main stream. D_T is a crude measure of the availability of space as determined by the proximity of another tributary.

TABLE 2. Discriminant analysis of tributary and non-tributary main stream increments

Discriminating variables	Rao's V	Change in Rao's V	Probability (α = 0.05)*	Standardized discriminant function coefficients
log L	192.72	192.72	< 0.0001	0.9121
θ	199.84	7.12	0.008	0.3542
S	204.54	4.70	0.030	-0.1660
log D_T	208.49	3.95	0.047	0.1757

* H_o: The change in Rao's V is zero in the population.

A stepwise discriminant analysis was then performed to ascertain whether or not a clear distinction could be drawn between tributary and non-tributary increments on the basis of their values of L, θ, D_T, and S. Rao's V, which is a generalized measure of the distance between the groups, was used as the criterion for controlling the entry of variables. As Table 2 shows, all four variables entered the discriminant function, which has a canonical correlation of 0.753 and correctly predicts the group membership of 91.8 percent of the tributary and 90 percent of the non-tributary increments. This is an important result for two reasons. First, it indicates that the locations of magnitude 1 tributaries along the Cox River are determined almost entirely by the proximity of the nearest tributary and the length, slope, and plan curvature of the valley side. Second, it means that the derived discriminant function may be used to predict the locations of other individual magnitude 1 tributaries along this river with a success rate of about 90 percent.

This kind of analysis cannot be applied meaningfully to less deeply incised main streams where some magnitude 1 tributaries extend beyond the main stream valley. In these circumstances, L will obviously be longer for tributary than non-tributary increments, making a discriminant analysis pointless. Still, it is clear that at least along deeply incised main streams with long valley sides, like the Cox River, the locations of magnitude 1 tributaries are very largely controlled by the configuration of the valley sides and the proximity of competing tributaries. Given that magnitude 1 tributaries comprise about half the tributaries along main streams, this conclusion suggests that there is a large deterministic or non-random component in the development of tributaries along main streams and, hence, in the topology of the channel networks containing these streams.

SUMMARY AND CONCLUSION

Implicit in the RT model is that channel networks free of environmental controls branch (or merge) randomly. Early direct and indirect tests generally supported the model and, hence, the notion of random branching. However, subsequent less-conventional tests, including the above discriminant analysis, have demonstrated that there are pronounced regularities in network topology. These regularities have been linked to a variety of factors that control network topology, including the upstream curvature and spatial requirements of tributaries and the need for their basins to fit together within larger basins; the size, sinuosity, and migration rate of valley bends; and the length, steepness, and plan form of valley sides.

Given the extent of these regularities, it is intriguing that for more than a decade the RT model appeared to work so well and gave rise to the notion that channel networks branch randomly. The early success of the model can be attributed to two factors: (1) the deficiencies in the direct and indirect tests, which made these tests insensitive to the regularities in network topology and permitted most of these regularities to go undetected; and (2) the multiple controls of network topology, which combine to produce complex branching patterns. This complexity was mistaken for randomness by early investigators, who found support for this point of view in the results of the direct and indirect tests.

In conclusion, it is evident that the extent to which channel network topology appears to be random depends on the way in which it is examined. When the complex branching patterns of channel networks were examined by the early direct and indirect tests, they appeared to be random. However, as more discriminating ways of analyzing network topology have been developed in recent years, progressively more regularities have been discovered, and this trend seems likely to continue. These regularities clearly indicate that channel networks do not branch randomly. But this does not mean that there is no random element in network topology, only that it is smaller than implied by the RT model.

REFERENCES

Abrahams, A. D., 1975a. Initial bifurcation processes in natural channel networks. Geology, 3, 307-308.

Abrahams, A. D., 1975b. Topologically random channel networks in the presence of environmental controls. Bulletin of the Geological Society of America, 86, 1459-1462.

Abrahams, A. D., 1977. The factor of relief in the evolution of channel networks in mature drainage basins. American Journal of Science, 277, 626-645.

Abrahams, A. D., 1980. Divide Angles and their relation to interior link lengths in natural channel networks. Geographical Analysis, 12, 157-171.

Abrahams, A. D., 1984a. Channel networks: a geomorphological perspective. Water Resources Research, 20, 161-188.

Abrahams, A. D., 1984b. Tributary development along winding streams and valleys. American Journal of Science, 284, 863-892.

Abrahams, A. D., 1984c. The development of tributaries of different sizes along winding streams and valleys. Water Resources Research, 20, 1791-1796.

Abrahams, A. D., and Flint, J. J., 1983. Geological controls on the topological properties of some trellis channel networks. Bulletin of the Geological Society of America, 94, 80-91.

Abrahams, A. D., and Mark, D. M., 1986. The random topology model of channel networks: bias in statistical tests. The Professional Geographer, in press.

Flint, J. J., 1980. Tributary arrangements in fluvial systems. American Journal of Science, 280, 26-45.

Horton, R. E., 1945. Erosional development of streams and their drainage basins; hydrophysical approach to quantitative morphology. Bulletin of the Geological Society of America, 56, 275-370.

Howard, A. D., 1972. Problems of interpretation of simulation models of geologic processes, in Quantitative Geomorphology: Some Aspects and Applications (Ed. M. E. Morisawa), pp 63-82. Publications in Geomorphology, Binghamton, N.Y.

James, W. R., and Krumbein, W. C., 1969. Frequency distribution of stream link lengths. Journal of Geology, 77, 544-565.

Jarvis, R. S., and Sham, C. H., 1981. Drainage network structure and the diameter-magnitude relation. Water Resources Research, 17, 1019-1027.

Jarvis, R. S., and Werritty, A., 1975. Some comments on testing random topology stream network models. Water Resources Research, 11, 309-318.

Krumbein, W. C., and Shreve, R. L., 1970. Some statistical properties of dendritic channel networks, Office of Naval Research Task 389-150, Technical Report 13, Department of Geological Sciences, Northwestern University, Evanston, Ill., 117 pp.

Leopold, L. B., and Langbein, W. B., 1963. Association and indeterminancy in geomorphology, in The Fabric of Geology (Ed. C. C. Albritton), pp 184-192, Addison-Wesley, Reading, Mass.

Mock, S. J., 1976. Topological properties of some trellis pattern channel networks, CRREL Report 76-46, U.S. Cold Regions Research and Engineering Laboratory, Hanover, N.H., 54 pp.

Schumm, S. A., 1956. Evolution of drainage systems and slopes in badlands at Perth Amboy, New Jersey. Bulletin of the Geological Society of America, 67, 597-646.

Schumm, S. A., 1977. The Fluvial System, Wiley, New York, 338 pp.

Shreve, R. L., 1966. Statistical law of stream numbers. Journal of Geology, 74, 17-37.

Shreve, R. L., 1975. The probabilistic-topologic approach to drainage-basin geomorphology. Geology, 3, 527-529.

Smart, J. S., 1969. Topological properties of channel networks. Bulletin of the Geological Society of America, 80, 1757-1774.

Smart, J. S., 1972. Channel networks. Advances in Hydroscience, 8, 305-346.

Smart, J. S., 1978. The analysis of drainage network composition. Earth Surface Processes, 3, 129-170.

Smart, J. S., and Wallis, J. R., 1971. Cis and trans links in natural channel networks. Water Resources Research, 7, 1346-1348.

Werritty, A., 1972. The topology of stream networks, in Spatial Analysis in Geomorphology (Ed. R. J. Chorley), pp 167-196, Methuen, London.

International Geomorphology 1986 Part II
Edited by V. Gardiner

AN APPROACH TO THE IDENTIFICATION OF MORPHOCLIMATES

Frank Ahnert

Dept. of Geography, RWTH Aachen
Federal Republic of Germany

ABSTRACT

One of the requirements for progress in climatic geomorphology is a more specific yet standardized and widely applicable identification of morphoclimatic conditions. An approach to this goal is seen in the systematic application of magnitude-frequency analysis to rainfall and other morphoclimatic phenomena; the magnitudes of relevant meteorological events are subjected to regression analysis as functions of the decadic logarithm of their recurrence interval, expressed in years. Thus the constant (Y) of the regression equation indicates the magnitude of the 1-year event, the sum of the constant (Y) and the regression coefficient (A) the magnitude of the 10-year event, and Y+2A, the magnitude of the 100-year event. Y and A combinedly make up the magnitude-frequency index MFI, written (Y;A). Isoline maps of Y and A show the spatial differentiation of magnitudes and their frequencies. The method is applied to daily rainfall, to short duration rainfall, duration of rainless periods, severity of frosts, freeze/thaw cycles and wind velocities. It appears that this approach can serve as the core of a process oriented morphoclimatology.

INTRODUCTION

Up to the present it has been customary in geomorphology and in other earth sciences to introduce climatic factors in the form of data supplied directly by general climatology - that is, mean annual or monthly values, mean frequencies and mean extremes. Examples are the much-copied diagrams of Peltier (1950) which relate many exogenic processes to mean annual temperature and mean annual rainfall. Such data may be reasonably adequate for broad generalizations concerning processes that continue over a long time, as for example chemical weathering of bedrock under a thick soil cover. For most other exogenic geomorphological processes, however, such a mean-value approach is inadequate because these processes consist of discontinuous events of varying magnitudes and frequencies. Accordingly, it is necessary to characterize the morphoclimatic conditions for such processes in terms of the magnitudes and frequencies of the climatic (or rather meteorological) events that cause them - for example, high intensity rainfall, high winds and occurrences of frost.

This has long been known and practised in fluvial hydrology; magnitude-frequency analysis has been used for decades to assess the probabilities of river floods. Wolman and Miller (1960) have transferred this concept from hydrology to fluvial geomorphology, and Hershfield (1961) has applied it to extreme rainfall events.

Apart from analyses of stream flow and fluvial sediment transport, however, the magnitude-frequency approach has been used only rarely in geomorphology. Instead, process research based on field measurements has frequently included the ad-hoc measurement of the rainfall at the field sites, in an effort to determine directly the rainfall energy available locally for individual geomorphological process events (see, for example, the discussion by Douglas, 1976, pp. 271-272). While such measurements are very useful for the explanation of local short-term process interactions, they can neither be readily extrapolated over longer periods of time nor be automatically viewed as representative of other areas. Such measurements seem, therefore, appropriate mainly for investigations of the physics of geomorphological process systems, in particular for field experiments (cf. Ahnert, 1980), but not for long- term process studies over larger areas.

Between the general-climatological mean-value approach to morphoclimate on one hand and the short-term on-site measurement on the other, there is a gap that might be bridged by an analysis of official (and usually published) meteorological data with the goal of assessing the frequencies of geomorphologically relevant meteorological events of varying magnitudes over longer periods of time and, if required, over larger areas than is possible by ad-hoc measurements. Such a method would serve especially well for research in developing countries where there are meteorological records available, where the need to apply them as specifically as possible to soil erosion and other environmental problems is great and where there are few if any financial and man-power resources available to carry out special field measurement programs over long periods of time.

With that purpose in mind this paper was written. In particular, the aim is to arrive at a simple numerical designation (an "index") from which the entire magnitude-frequency distribution of geomorphologically relevant meteorological events may easily be recognized or reconstructed. The paper is concerned mainly with diurnal rainfall events but examples will show that the same concept also applies to other kinds of data.

THE METHOD

The procedure is quite similar to the usual method of magnitude-frequency analysis. It differs only in two respects:

- instead of the customary annual maximum series of data, the partial duration series is used;
- and instead of using a Gumbel plot (Gumbel, 1958, Chow, 1964, Chptr. 8, Zeller, Geiger and Röthlisberger, 1983), a semilogarithmic plot is used, with a semilogarithmic regression equation.

The principle of this approach was already successfully applied in an earlier study of the morphoclimate of south-central Kenya (Ahnert, 1982, 22-28). Raw material are the data of meteorological events from specific stations, for example, daily rainfall data as shown in Figure 1 (note: in this paper the words "daily" and "diurnal" both refer to the 24-hour day as a time unit). Figure 1 also illustrates the main reason for the magnitude-frequency approach: all three stations have about the same annual rainfall, but differ with respect to the magnitudes and frequencies of daily rainfall amounts; the latter are much more important as inputs to geomorphological processes than the annual sums or the mean annual values which have been used as morphoclimatic indicators even in the very recent literature.

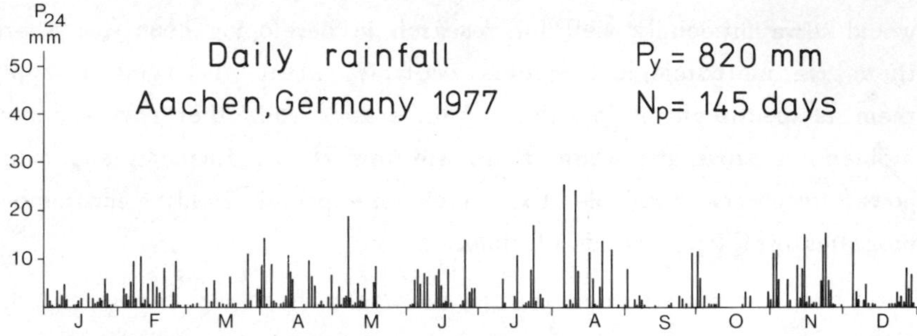

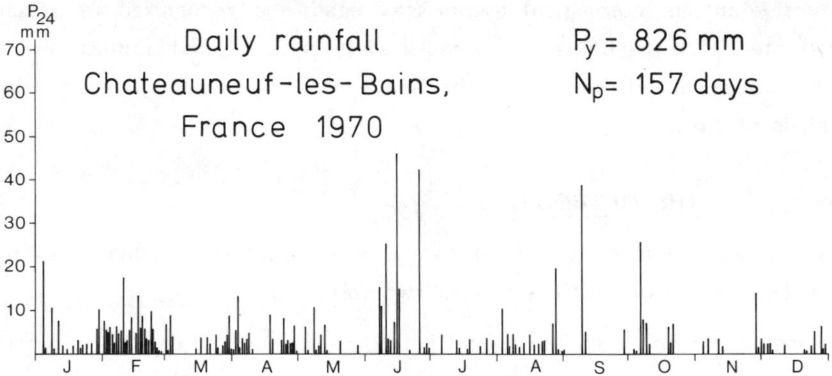

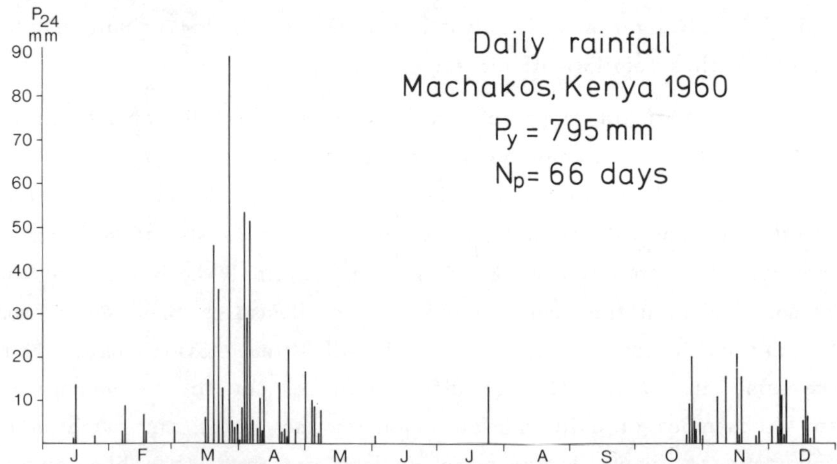

Fig. 1. Daily rainfall data of three stations with similar annual rainfall P_y. N_p= Number of days with rain. (Sources: Official meteorological data)

Over a sufficient length of record (recommended: at least five years in most climates, longer in arid climates), the data are ranked (the highest values gets the rank = 1, the second highest the rank = 2, etc.) and the recurrence interval RI of each data value is determined according to

$$RI = \frac{N+1}{rank} \qquad (1),$$

where N is the total number of time units in the record. The data are then plotted as a function of the decadic logarithm of the recurrence interval; and the regression equation

$$P_{24} = Y+A \log_{10} RI_y \qquad (2)$$

is determined, in which P_{24} is the daily rainfall amount and R_y is the recurrence interval expressed in years. The meaning of the constant Y and of the coefficient or gradient A will be explained below.

For the three stations of Figure 1 and five years of data, the plots and regression equations are shown in Figures 2-4. At Machakos, Kenya (Fig. 2), both the constant Y and the coefficient A of the regression equation are high; at Chateauneuf-les- Bains (Fig. 3), located in the Massif Central of France, both Y and A are appreciably lower, and at Aachen in West Germany (Fig. 4) they are lower still. The maximum values for Machakos and Chateauneuf represent very rare events which happened to occur in the five-year records used. It is advisable to exclude such outliers from the regression procedure because they would unduly change the position of the regression line.

One can now estimate the expected daily rainfall amount for any recurrence interval and the expected recurrence interval for any daily rainfall amount - either directly on the diagram of the plotted data or by means of the regression equation. In particular, the constant Y and the coefficient A of the semilogarithmic regression equation (2) allow not only an easy reconstruction of the regression line and thus of the entire magnitude-frequency distribution for a given station; they also yield by simple addition the magnitudes of events at several important recurrence intervals for:

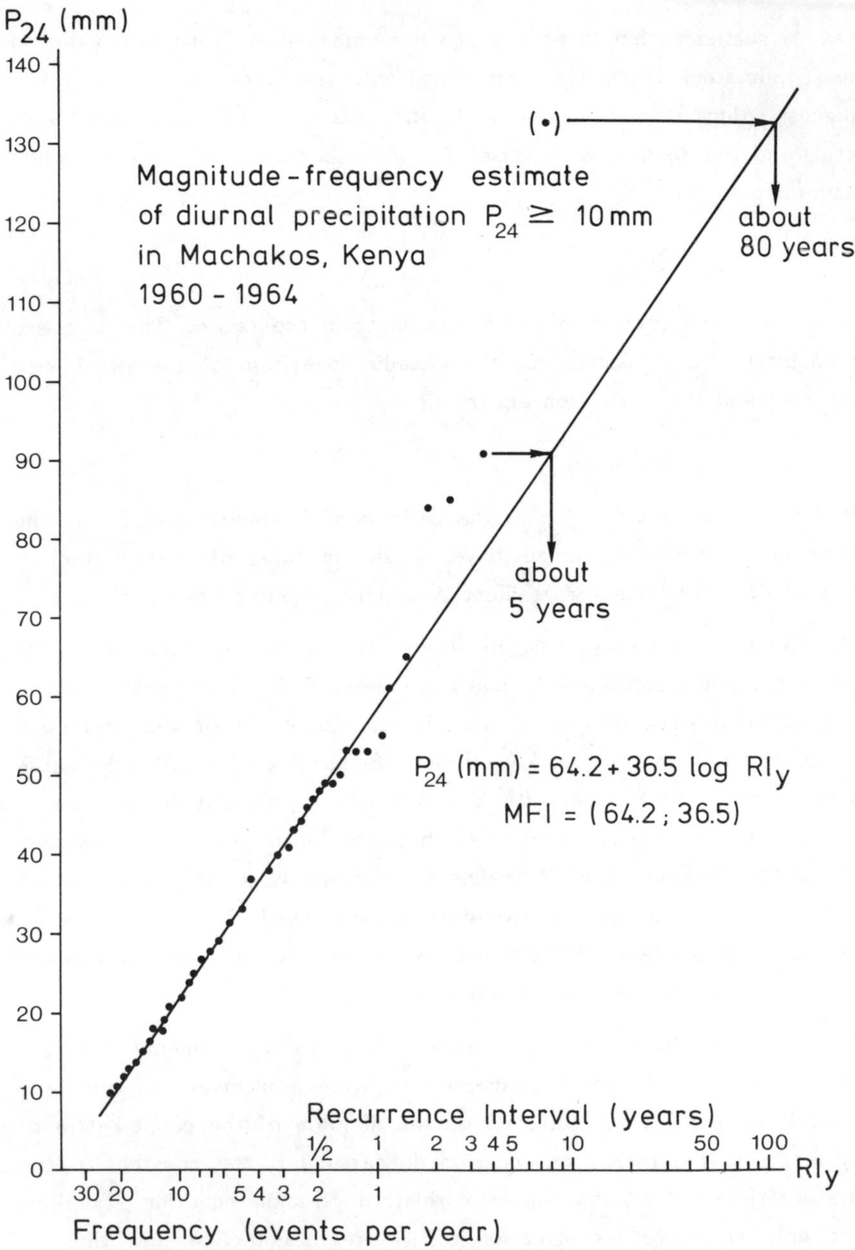

Fig. 2. Magnitude-frequency estimate of rainfall at Machakos, Kenya. (Source of data: Archive of East African Meteorological Department, Nairobi)

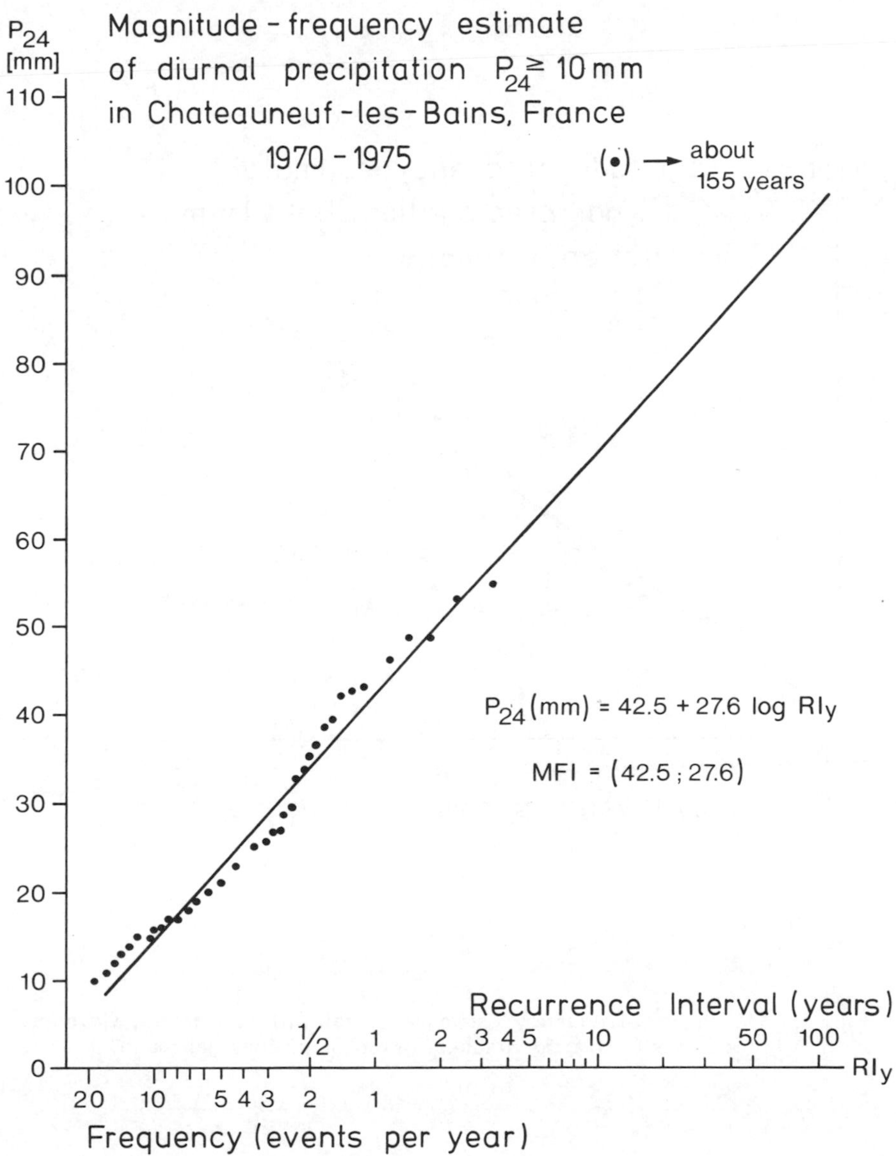

Fig. 3. Magnitude-frequency estimate of rainfall at Chateauneuf-les-Bains, France. (Source of data: Commission Météorologique du Puy-de-Dôme)

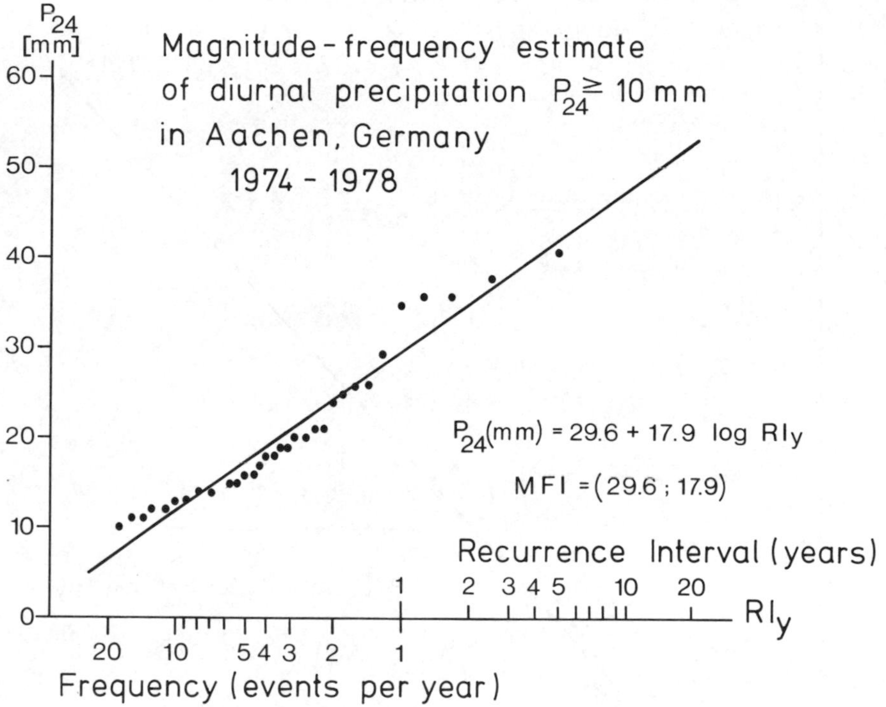

Fig. 4. Magnitude-frequency estimate of rainfall at Aachen, Germany (Source of data: Deutsches Meteorologisches Jahrbuch)

- the constant Y of equation (2) is identical with the magnitude of the rainfall event P_{24} that has a recurrence interval RI_y = 1 year ($\log_{10} RI_y = 0.0$),
- the sum Y+A is equal to the magnitude of the rainfall event P_{24} that has a recurrence interval RI_y = 10 years ($\log_{10} RI_y$=1.0),
- the sum Y+2A is equal to the magnitude of the rainfall event P_{24} with a recurrence interval RI_y = 100 years ($\log_{10} RI_y$=2.0).

Y and A are therefore combined as a convenient two-number index, the

magnitude-frequency index or MFI, which identifies the characteristic recurrence interval of events of various magnitudes at the given station. The MFI is written (Y;A).

For example, the magnitude-frequency distribution of daily rainfall amounts at Machakos, Kenya, is well described by (64.2; 36.5), at Chateauneuf-les-Bains, France by (42.5; 27.6) and at Aachen, Germany by (29.6; 17.9) (cf. Figures 2-4). Accordingly, in Aachen one may expect that a daily rainfall of 29.6 mm would be reached or exceeded once a year, a daily rainfall of 29.6+17.9 = 47.5 mm once every ten years and a daily rainfall of 29.6+2x17.9 = 65.4 mm once every one hundred years. Figure 5 shows the magnitude-frequency graphs and MFI values for rainfall events with durations of 15 minutes, 30 minutes, one hour and three hours as well as for daily rainfall at Nairobi, Kenya.

COMPARISON WITH THE GUMBEL-TYPE METHOD

The semilogarithmic method described in the preceding section offers several distinct advantages over the Gumbel method:

1. The plot of data points is made on ordinary graph paper rather than on special "Gumbel paper" which is not easily obtained in most parts of the world.
2. The convenience and immediate interpretability of the magnitude-frequency index MFI cannot be matched by numerical indicators derived from the Gumbel method.
3. The Gumbel method has usually been applied to annual maximum series of data, which means that the second highest value of any

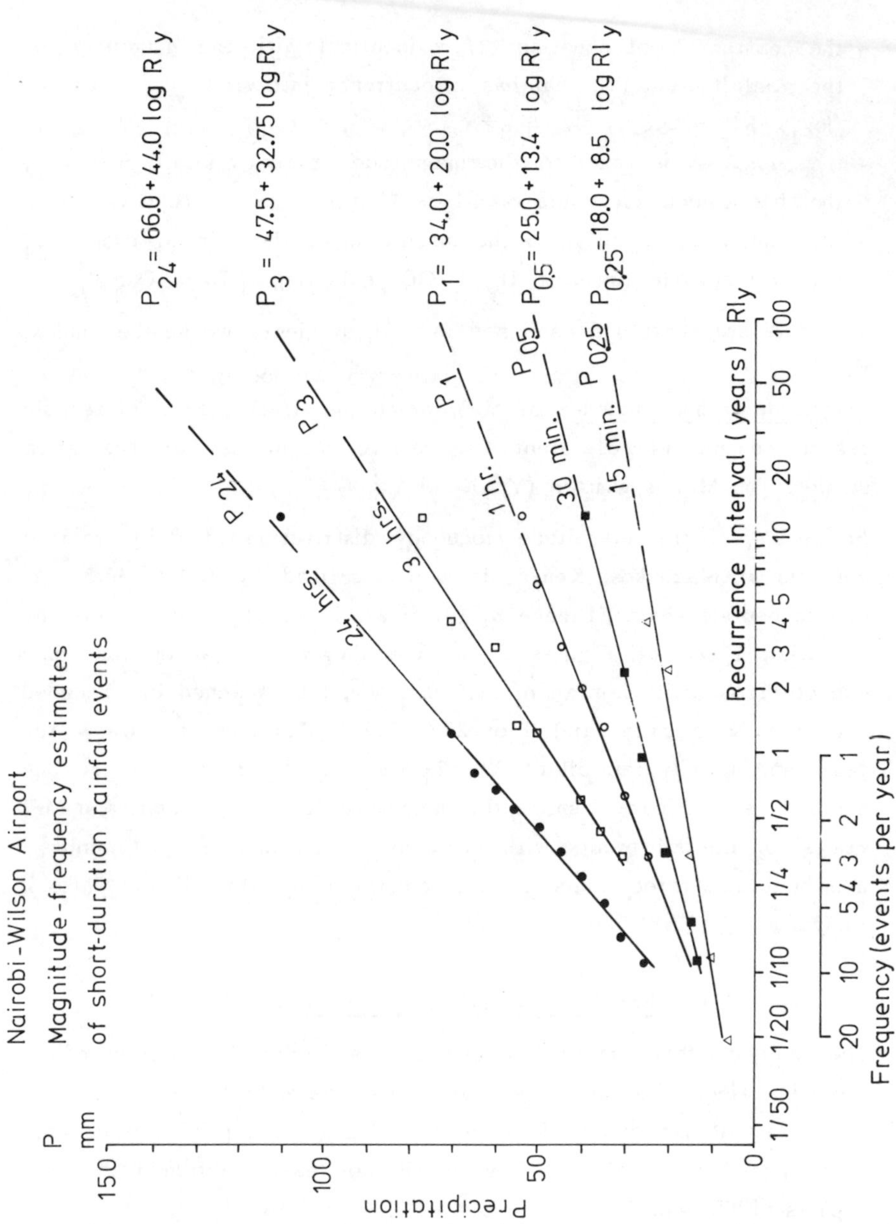

Fig. 5. Magnitude-frequency estimates of short-duration rainfall events (15 minutes, 30 minutes, one hour, three hours and 24 hour total rainfall) at Nairobi, Kenya (Source of data: Lawes, 1974)

year is not included in the data set even though it may be higher than the maximum for other years. This is adequate for engineers who merely need to consider the probable magnitude of the maximum event that might endanger a structure during its expected time span of suitable existence; it is less useful, however, in geomorphology where all events above a certain threshold contribute to landform changes. For this reason, the partial duration series is used here rather than the annual maximum series.

It seems therefore that the semilogarithmic method provides an easier key to the geomorphologically relevant magnitude-frequency information than does the Gumbel method. In order to be equally applicable, however, its qualities as an estimator and probabilistic predictor of events must match those of the Gumbel approach.

This question was tested with pluviograph data that were provided by Lawes (1974) in a study of short-duration intensities of rainfall at stations in East Africa. Seventeen of the stations lie in Kenya; their pluviograph records are between eight years and thirty-four years long, with a mean length of about 18 years. As a part of this study, Lawes has estimated also the magnitudes of 25-year, 50-year and 100-year rainfall events with durations of 15 minutes, 30 minutes, 1 hour and 3 hours by means of the Gumbel method. For comparison, a magnitude-frequency analysis of the same set of pluviograph data was carried out with the semilogarithmic method and the magnitudes of the 100-year rainfall events were estimated for the various durations by means of equation (2), i.e. by adding Y+2A. Figure 6 compares these semilogarithmic estimates with the Gumbel estimates made by Lawes; the difference is so small that it can be ignored - especially if one considers that these estimates are extrapolations made for a time that is about three to twelve times as long as the length of the data records. Thus the accuracy of both methods appears to be about equal.

CARTOGRAPHIC REPRESENTATION OF THE MFI

The magnitude-frequency index serves not only to characterize individual weather stations. It can be applied also regionally by means of isoline maps of the two MFI components Y and A. The isolines are

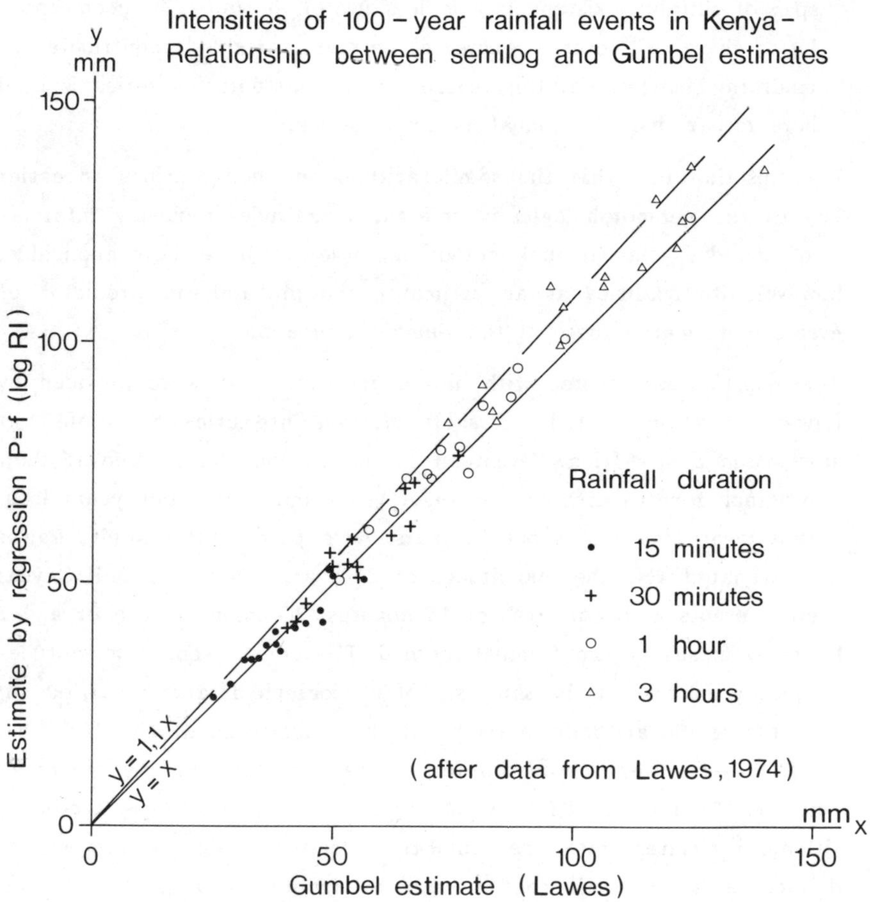

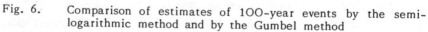

Fig. 6. Comparison of estimates of 100-year events by the semi-logarithmic method and by the Gumbel method

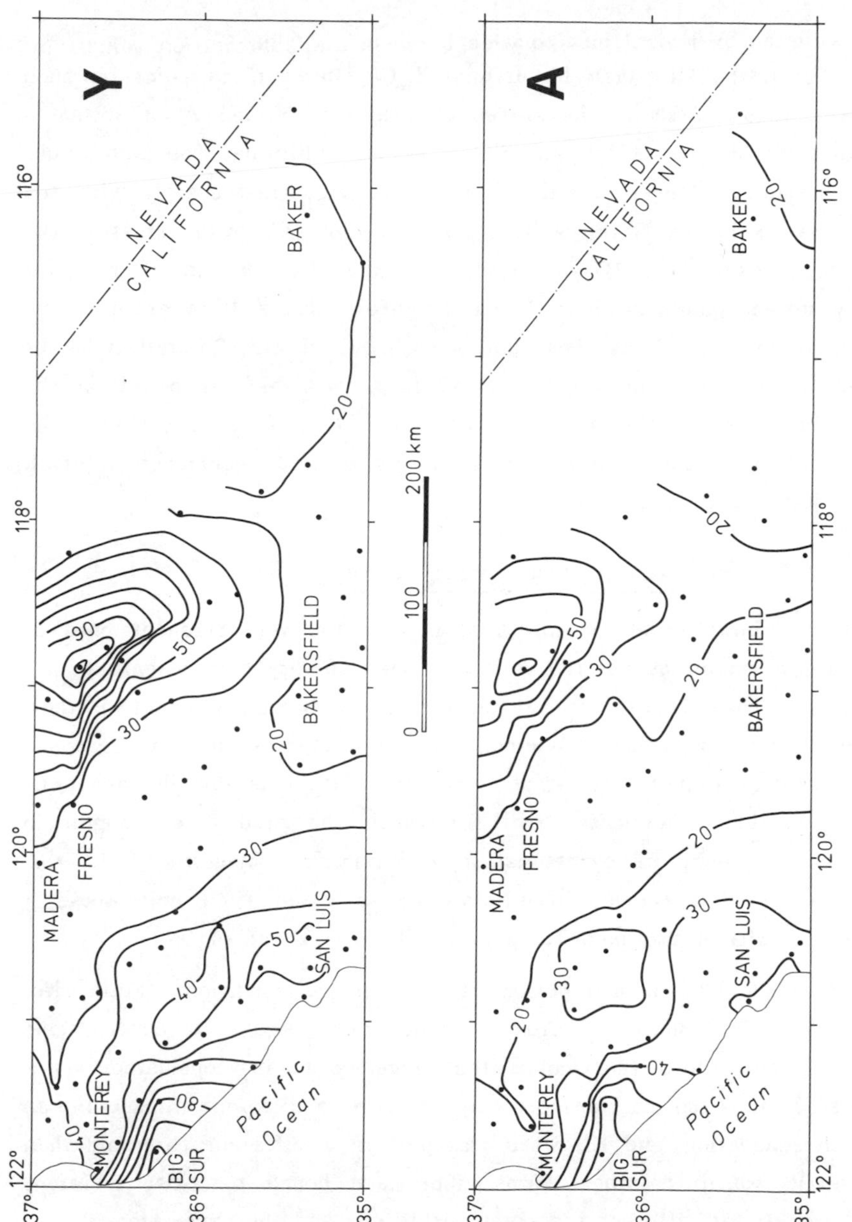

Fig. 7. Isoline maps of the constant Y and the coefficient A of the magnitude-frequency estimating equation in a belt between 35° and 37° N lat. across the State of California. (Source of data: Climatic data 1961-197O, U.S. Weather Bureau)

constructed by logical interpolation between available station values. In this way, the MFI may be estimated for any surface point on such maps. Figure 7 shows the spatial variations of Y and A in an east-west belt between 35° N and 37° N across California. The high values in the Coast Ranges and the Sierra Nevada contrast sharply with the low values in the Great Valley and with the still lower values in the Mojave Desert. The 10-year events obtained by addition of Y+A for any surface points in Figure 7 match quite well the 10 year maximum of 24-hour rainfall of these points as shown in the Rainfall Atlas of the United States by Hershfield (1961, p. 99) which is based on the Gumbel method. The advantage of Figure 7 is that it contains the information about rainfall events of any desired recurrence interval, not just the 10-year values.

THE "MOST EFFECTIVE EVENT"

Of all rainfall events in an area, neither the very frequent ones of very low magnitude nor the very rare ones of very high magnitude are likely to have the greatest cumulative geomorphological effect. Rather, following a suggestion of Wolman and Miller (1960), one may reasonably assume that the greatest cumulative effect is brought about by that particular intensity level for which the product of magnitude times frequency of events is at a maximum. "Magnitude" in this context is the amount of rainfall per selected time unit above a certain critical threshold T_R, that is, $(P-T_R)$ instead of P.

The threshold T_R is a function of the surface conditions (grain size, presence or absence of crusts, perviousness, vegetation cover, slope angle), but also a function of the processes to be considered. On a given surface, splash transport may set in at a different threshold than wash denudation, and rill wash transport at a different threshold than interrill wash; in other words, there are bound to exist different thresholds for different processes within one and the same area.

Figure 8 shows the product of magnitude times frequency as a function of the recurrence interval RI_y, for some selected stations. Both coordinate axes are logarithmic. Next to the left end of each curve the magnitude-frequency index (Y;A) of that particular station is given. In the absence of more detailed local information, the threshold $T_R = 25$ mm was used for all stations; 25 mm per day is approximately the rainfall at which runoff may be expected in several parts of the world (cf. Ahnert, 1982, p. 24-25). The curves in Figure 8 are very similar in shape, but their position varies within the coordinate field according to the values of the MFI. In areas of high rainfall (for example, Big Sur, Calif.) the product of magnitude and frequency has a high maximum value; furthermore, this high maximum is coupled with a short recurrence interval. In areas of low rainfall (for example, Baker, Calif.), the maximum of the product is low and its recurrence interval is measured in years or tens of years. This means that in arid regions there is not only a lower intensity of this "most effective rainfall event" than in humid regions but the occurrences of this event are also less frequent. Taken by itself, this result would indicate that landforms shaped by runoff events evolve more slowly, and relic landforms consequently are preserved longer, in arid regions than in humid ones. Of course such a generalization disregards the different surface conditions (e.g. vegetation cover, grain sizes of surface material) in different climates; however, the effects of these differences may neutralize one another to some extent.

In Figure 8 the recurrence interval for which the product of magnitude times frequency has its maximum is obtained graphically; however, it may also be calculated directly by differentiation. The frequency F of a rainfall event of a given magnitude $(P-T_R)$ is the inverse of the recurrence interval:

$$F = RI_y^{-1} \qquad (3);$$

the product of magnitude - as defined by equation (2) - and frequency is then

$$(P-T_R)F = \left[(Y-T_R)+A \log_{10} RI_y\right] RI_y^{-1} \qquad (4).$$

This product has its maximum value if its first derivative is equal to zero, i.e. if

$$\frac{d\left[(P-T_R)F\right]}{dRI_y} = \left[A \log_{10} e-(Y-T_R)-A \log_{10} RI_y\right] RI_y^{-2} = 0 \qquad (5),$$

which is true for

$$\log_{10} RI_y = \log_{10} e - \frac{(Y-T_R)}{A} \qquad (6)$$

(e is the base of natural logarithms).

The recurrence interval RI_{yE} of the most effective rainfall event P_E is then

$$RI_{yE} = 10^{-\frac{(Y-T_R)}{A}} \cdot e \qquad (7).$$

The mathematical structure of this functional relationship is similar to that derived by Carson and Kirkby (1972, pp. 215-217) for the peak value of soil wash under conditions of particular daily rainfall and infiltration/evaporation losses; indeed the shape of their wash rate curves (1972, p. 217) very closely resemble the shape of the curves in Figure 8.

In Figure 9 the radiating lines are lines of constant RI_{yE} as a function of both $(P-T_R)$ and A. For any given value of A, the recurrence interval of the most effective event (RI_{yE}) increases with decreasing values of $(P-T_R)$. For any given value of $(P-T_R) > 0$, it increases with increasing values of A, but for any given value of $(P-T_R) < 0$, it decreases with increasing A. At $(P-T_R) = 0$, RI_{yE} = const. = e, i.e. 2.72 years, for all values of A. Although in Figure 9 the threshold also has been set at T_R = 25 mm, its actual numerical value has no bearing upon the distribution of the RI_{yE} - isolines; for this purpose, the designation of the abscissa could simply be "$Y-T_R$". However, T_R = 25 mm does influence the locations of the data points that have also been plotted in Figure 9 according to their MFI values Y-25 and A. The data points make it directly possible to read off the RI_{yE} values for each station, assuming a threshold of 25 mm. The three non-Californian stations Machakos, Chateauneuf and Aachen fit quite well into

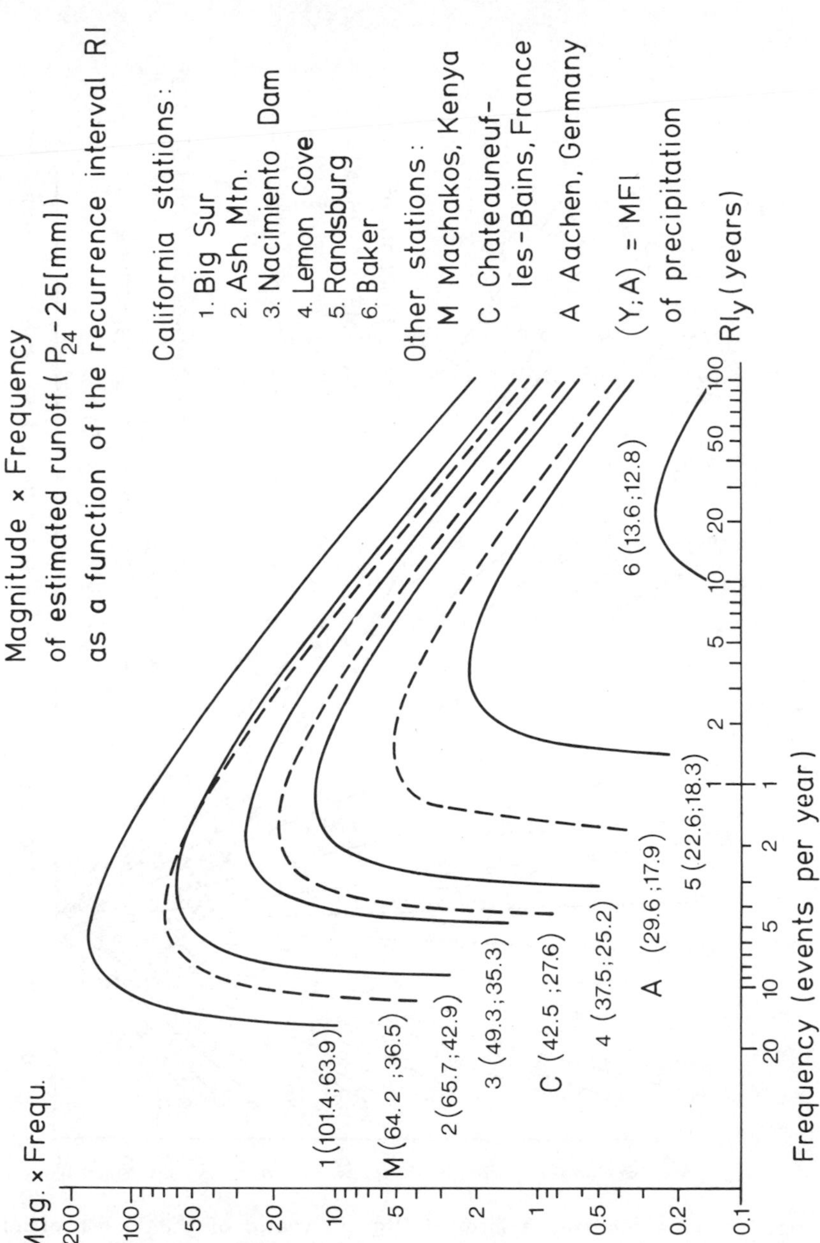

Fig. 8. The product of magnitude times frequency as a function of the recurrence interval. For explanation see text. (Sources: see Figures 2,3,4 and 7)

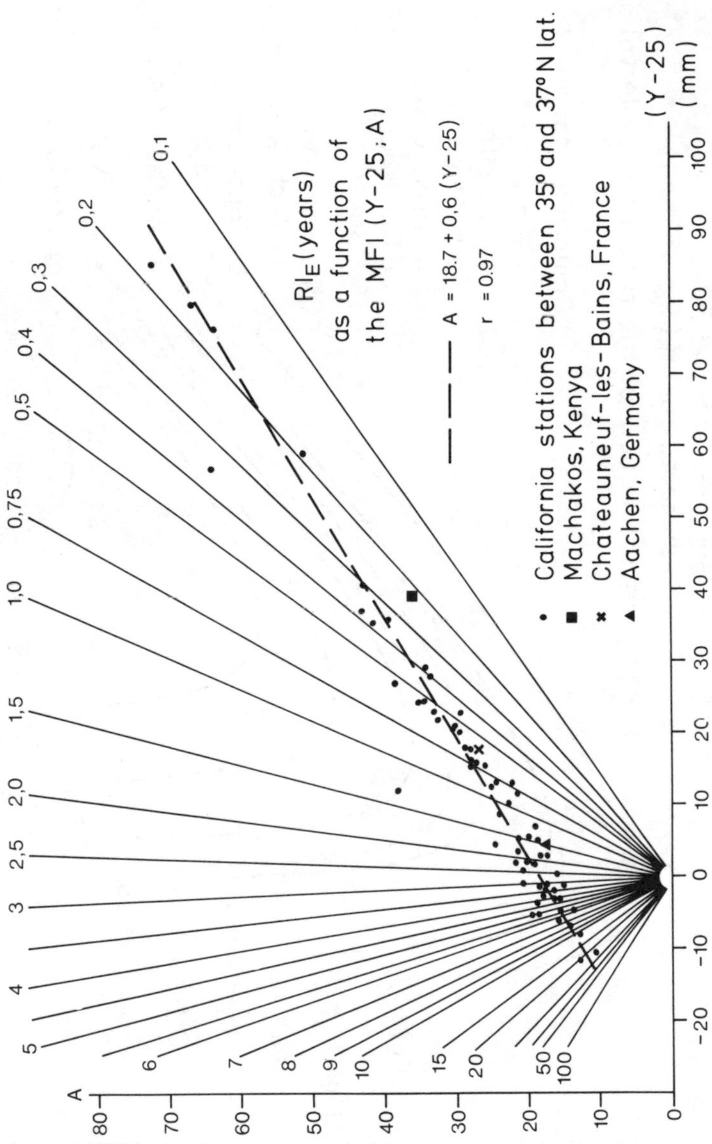

Fig. 9. The recurrence interval RI_E (in years) of the "most efficient runoff event" as a function of the MFI-components Y-25 mm (25 mm per day is the arbitrary runoff threshold T_R) and A. The dashed regression line shows A=f(Y-25 mm) for the Californian stations shown. (Sources: same as for Figure 7)

the same distribution pattern but have somewhat lower A-values than California stations of the same Y-value.

The regression line A=f(Y-25) in Figure 9 indicates that over the entire range of stations shown, the values of A tend to stay near 0.6 Y; the two components of the MFI appear therefore to be interdependent. This relationship will need further analysis. If it remains reasonably consistent in all other climatic regions, then a single number might suffice for the magnitude-frequency index instead of the two numbers used now.

EXAMPLES OF APPLICATIONS TO OTHER SETS OF DATA

The magnitude-frequency index also describes other sets of data rather well. Figure 10 shows the lengths of rainless periods as a function of their recurrence intervals in Cologne. Here the data base (Table 1) is a count of the number of rainless periods of varying lengths that occurred within the years under consideration. These periods are then ranked and treated further according to the method described above. The MFI in this case (14.5; 10.8) states that one may expect a rainless period of at least 14-15 consecutive days once every year and of 14.5+10.8 or about 25 days once in 10 years. Obviously such information is useful in hydrology and agriculture.

Another example is the occurrence of frost of varying severity in Munich (Table 2 and Figure 11); the MFI (-18.5; -8.6) indicates that each year the temperature is likely to drop at least to a minimum of -18.5° C and that in a ten-year period a minimum of -27° C may occur. Together with frost severity, the frequency of freeze/thaw cycles also is of geomorphological interest, particularly as a factor in mechanical weathering. The data for Munich (Table 3) are well described by the regression equation of Figure 12, except for one low outlier (52 cycles in 1977) which was excluded from the regression procedure.

In studies of aeolian erosion, transport and deposition as well as in studies of marine or lacustrine shore processes by wave action, the magnitude-frequency distribution of wind speeds becomes an important

TABLE 1. Recurrence intervals RI_y of rainless periods of varying lengths at Cologne aiport, Germany, 1969-1978

Duration D (days)	Number of occurrences	Ranks	Mean Rank	RI_y (years)	log RI_y
24	1	1	1	11	1.041
22	1	2	2	5.5	0.740
21	3	3-5	4	2.75	0.439
19	1	6	6	1.83	0.263
17	1	7	7	1.57	0.196
16	1	8	8	1.38	0.138
15	4	9-12	10.5	1.05	0.020
14	1	13	13	0.85	-0.073
13	3	14-16	15	0.73	-0.135
12	2	17-18	17.5	0.63	-0.202
11	5	19-23	21	0.52	-0.281
10	9	24-32	28	0.39	-0.406
9	3	33-35	34	0.32	-0.490
8	7	36-42	39	0.28	-0.550
7	15	43-57	50	0.22	-0.658
6	14	58-71	64.5	0.17	-0.768
5	23	72-94	83	0.13	-0.878

Source: Deutsches Meteorologisches Jahrbuch

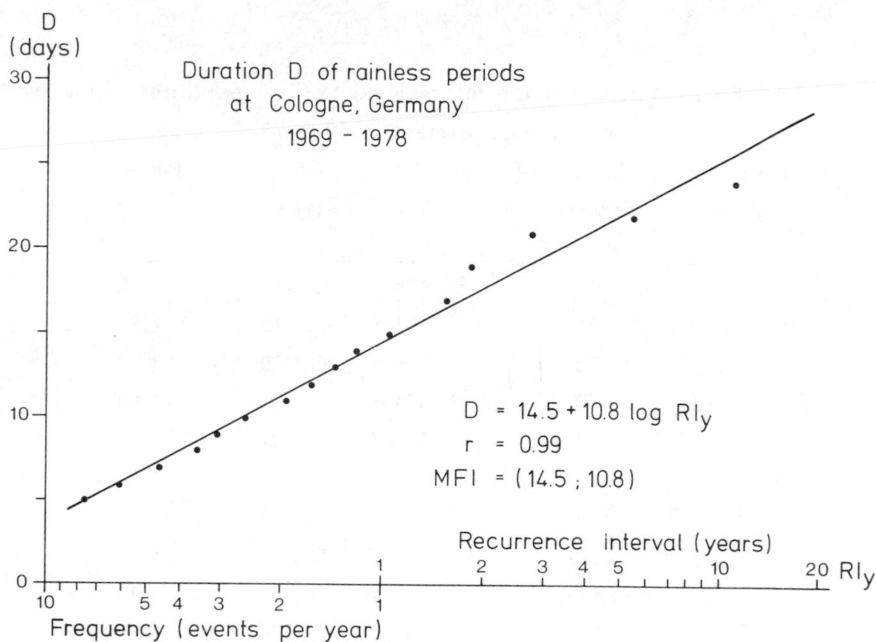

Fig. 10. Durations of rainless periods, Cologne. (cf. Table 1)

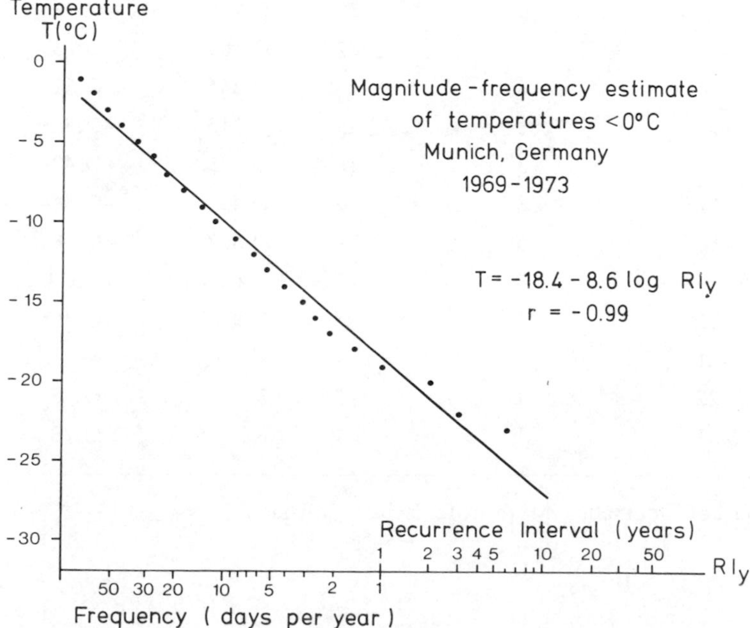

Fig. 11. Occurrences of frost of varying severity, Munich. (cf. Table 2)

TABLE 2. Recurrence intervals RI_y of temperatures below 0° C at Munich, Germany 1969-1973

Temperature T (° C)	Number of occurrences (days)	Mean Rank	RI_y (years)	log RI_y
-1	93	476	0.013	-1.90
-2	64	397.5	0.015	-1.82
-3	72	329.5	0.018	-1.74
-4	58	264.5	0.023	-1.64
-5	52	209.5	0.029	-1.54
-6	38	164.5	0.036	-1.44
-7	24	132.5	0.045	-1.34
-8	30	106.5	0.056	-1.25
-9	20	81.5	0.074	-1.13
-10	16	63.5	0.094	-1.03
-11	14	48.5	0.124	-0.91
-12	6	38.5	0.156	-0.81
-13	7	32	0.188	-0.73
-14	8	24.5	0.245	-0.61
-15	3	19	0.315	-0.50
-16	3	16	0.375	-0.43
-17	3	13	0.461	-0.34
-18	5	9	0.667	-0.18
-19	3	5	1.200	0.01
-20	1	3	2.000	0.30
-21	-	-	-	-
-22	1	2	3	0.48
-23	1	1	6	0.78

Source: Deutsches Meteorologisches Jahrbuch

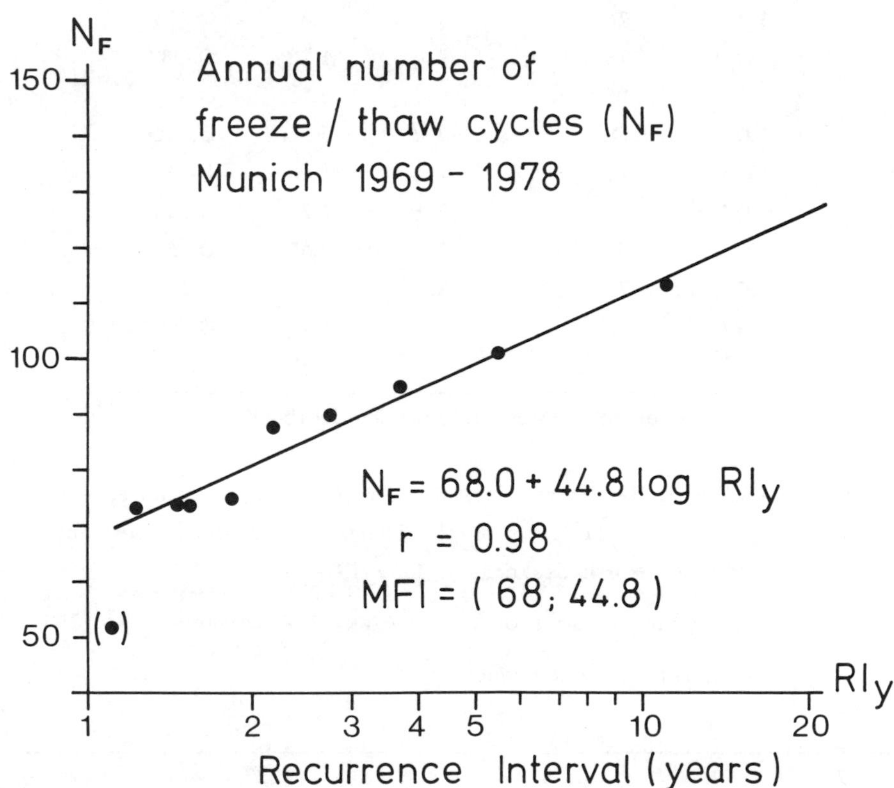

Fig. 12. Annual frequency of freeze/thaw cycles in Munich. (cf. Table 3)

TABLE 3. Recurrence intervals RI_y of the annual number of freeze-thaw cycles in Munich, Germany 1969-1978

Year	Number of freeze-thaw cycles N_F	Rank	RI_y (years)	log RI_y
1969	75	6	1.83	0.263
1970	73	9	1.22	0.087
1971	90	4	2.75	0.439
1972	101	2	5.50	0.740
1973	113	1	11.0	1.041
1974	88	5	2.20	0.342
1975	95	3	3.67	0.564
1976	74	7.5	1.47	0.167
1977	52	10	1.1	0.041
1978	74	7.5	1.47	0.167

Source: Deutsches Meteorologisches Jahrbuch

TABLE 4. Recurrence intervals RI_d (in days) of wind speeds 11.2 m/s on the island of Norderney (Germany North Sea coast), 1972-1974

Beaufort	Minimum V(m/s)	Days of occurrence	Rank	Recurrence interval RI_d (days)	log RI_d
9	21.0	8	8	137.1	2.14
8	17.4	20	28	39.2	1.59
7	14.3	84	112	9.8	0.99
6	11.2	169	281	3.9	0.59

Source: Deutsches Meteorologisches Jahrbuch

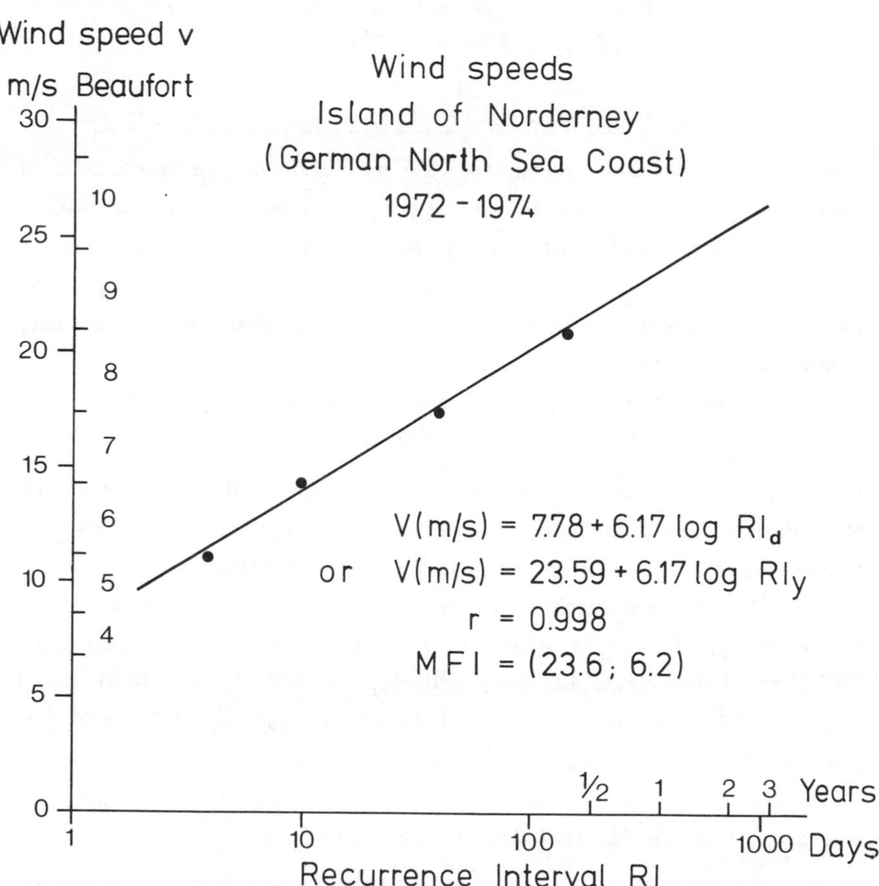

Fig. 13. Wind speeds at the island of Norderney. (cf. Table 4)

factor. Table 4 and Figure 13 show them for the east Frisian island of Norderney; winds of 23-24 m/s (force 9 on the Beaufort scale) are probable every year and of 29-30 m/s (force 11) once in 10 years. For specific purposes one could make also separate magnitude-frequency analyses for different wind directions.

EXTRAPOLATION LIMITS AND A NOTE OF CAUTION

Theoretically one could extrapolate the prediction of the magnitudes of extreme events by means of the pertinent regression equation or of the MFI to recurrence intervals of any length. However, in practice there are limits which are set by common sense. The longer the recurrence interval, the greater is the chance that it will in fact contain climatic changes that are not foreseeable or, in the case of long extrapolation into the past, not estimable from the short data record on which the regression equation is based.

In the case of frequency data (i.e., number of occurrences per year), an additional limit is posed by the limited length of the time unit used (year) and by the limited length of the climatic cycles (seasons, day/night) that exist within this time unit at most places on earth. For example, if one extrapolates the annual number of freeze/thaw cycles at Munich (Fig. 12) over millenia, the MFI (68.0; 44.8) would indicate 202 cycles per year as a 1000-year value and 247 cycles per year as a 10,000-year value. The absurdity of such long-term projections is obvious in this case, quite apart from the fact that the length of record is for too short for such estimates.

The most obvious limitation of the reliability of any value obtained by extrapolation is set by the length of the data record used. Of course the data should reach at least over several years - the more the better. The examples in this paper are based on relatively short data records in order to test the method under unfavorable conditions. In practice, a 10-year record of data should suffice, however, to identify the MFI and the "most effective event" with reasonable accuracy. Figure 14 illustrates the effect of record length upon the magnitude-frequency regression lines of diurnal rainfall at Aachen: the twenty-

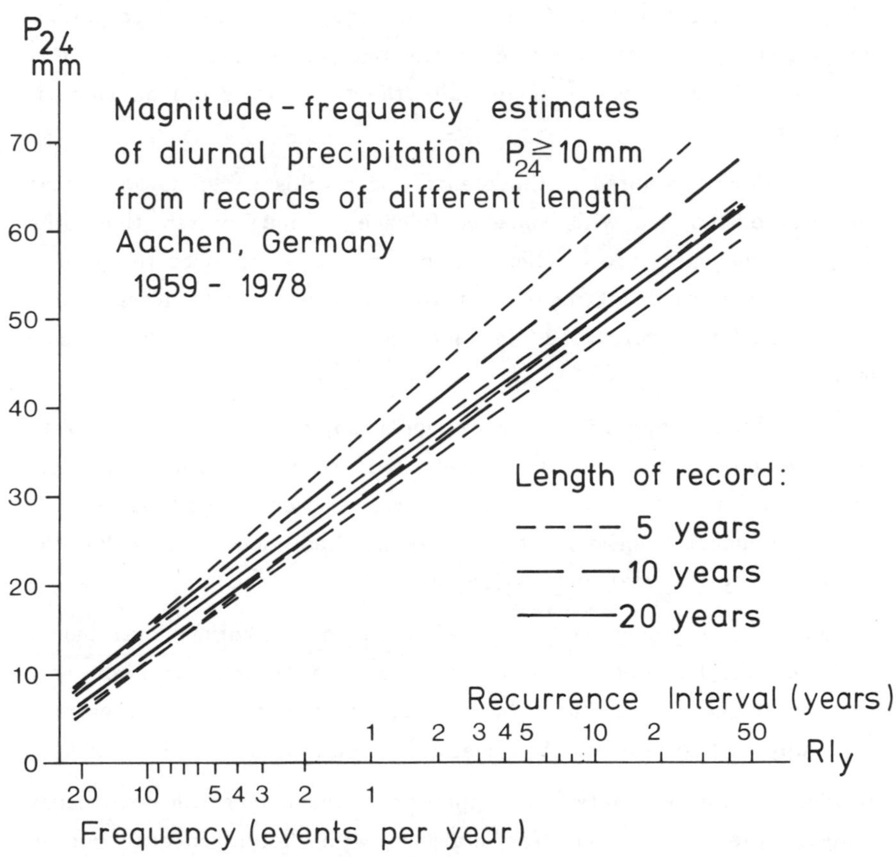

Fig. 14. Magnitude-frequency estimating lines for 5-year, 10-year and 20-year records of diurnal rainfall greater than 10 mm at Aachen. Note the close bundling of all but the two uppermost lines

year record 1959-1978 was divided into two ten-year records and four five-year records, and the MFI was calculated for each of these. The twenty-year MFI is (32.1; 18.4), the two ten-year MFI's are (35.3; 19.9) and (30.6; 18.2), the four five-year MFI's are (32.2; 18.4), (38.3; 22.2), (31.2; 19.2) and (29.6; 17.9). Only one of the five-year periods yields a magnitude-frequency estimate that differs appreciably from the others; it influences of course also the ten-year record of which it is a part.

The experience gathered in the preparation of this paper suggests that one may extrapolate with some confidence to roughly ten times the length of the data record. This does not preclude the occurrence of a single extraordinarily extreme event that happens to take place within the extrapolation period but in fact has a much longer recurrence interval.

As a reasonable upper limit for extrapolation, one century is suggested. During 100 years, it is unlikely that climatic changes are big enough to affect the magnitudes of extreme events significantly; and for most annual frequencies the 100-year predictions stay within the bounds set by the year and its seasons.

Perhaps the Gumbel method is better suited for predictions over longer time periods; however, it seems that such predictions are not very meaningful anyway, so that the simple, convenient method presented here fully suffices for morphoclimatic purposes.

A special note of caution is appropriate when magnitude-frequency analyses are carried out for stations with heterogeneous rainfall distributions. In the southeastern U.S., for example, high diurnal rainfall amounts may come either from short but very severe thunderstorms or from day-long hurricane rains. The magnitude-frequency distributions and hence the recurrence intervals of these two types of rainfall events probably differ considerably. Their individual effects could be discerned, however, if one compares the magnitude-frequency index of diurnal precipitation with that of 1-hour or 3-hour precipitation.

CONCLUSION

The examples demonstrate that semilogarithmic magnitude-frequency analysis, and with it the magnitude-frequency index (MFI), is a useful tool for the characterization not only of morphoclimates but also of certain climatic aspects that are relevant to botany, agriculture, tourism, civil engineering and other fields, whenever the climatic "factor" in the original meaning of that word consists of distinct events of varying magnitudes and frequencies.

Of course this method cannot directly identify climatic traits that are not events - for example, seasonality. To some extent, one might overcome this by stratifying the data, that is, by calculating the MFI separately for each season or even for each month of the year. Another aspect is the successive occurrence of events within a shorter time. Heavy rain often does not fall on isolated days but on several days in succession, which affects the soil water budget as well as the runoff to be expected. The assessment of such conditions requires additional information to complement the MFI.

The approach presented here is seen as a possible first step towards a comprehensive morphoclimatology which is an applied climatology for geomorphological purposes; its concepts, its terms, its variables are not merely taken over from general climatology but are specifically developed to fit the needs of geomorphology. Its data base, however, must as far as possible be the publicly available meteorological record in order to insure the possibility of standardization, of long-term coverage and of world-wide application.

ACKNOWLEDGEMENTS

The author thanks Dr. A. Werrity for helpful discussion of this subject matter and Mr. W. Berndt for assistance in the compilation of the data.

REFERENCES

Ahnert, F., 198O. A note on measurements and experiments in geomorphology. Z. Geomorph. Suppl. 35, 1-1O

Ahnert, F., 1982. Untersuchungen über das Morphoklima und die Morphologie des Inselberggebiets von Machakos, Kenia. Catena Suppl. 2, 1-72

Carson, M.A., and Kirkby, M.J., 1972. Hillslope form and process. Cambridge

Chow, V.T. (ed.), 1964. Handbook of applied hydrology. McGraw-Hill, New York

Commission Météorologique du Puy-de-Dôme, - Bulletin Météorologique. Clermont-Ferrand

Deutscher Wetterdienst, - Deutsches Meteorologisches Jahrbuch. Published annually. Offenbach

Douglas, I., 1976. Erosion rates and climate: geomorphological implications. In: Derbyshire, E. (ed.), Geomorphology and climate. John Wiley, London - New York - Sidney - Toronto. 27O-287

Gumbel, E.J., 1958. Statistics of extremes. Columbia University Press, New York

Hershfield, D., 1961. Rainfall Atlas of the United States. U.S. Government Printing Office, Washington, D.C.

Lawes, E.F., 1974. An analysis of short duration rainfall intensities. East African Meteorological Dept., Technical Memorandum No. 23. Nairobi

Peltier, L.C., 195O. The geographical cycle in periglacial regions as it is related to climatic geomorphology. Ann. Ass. Am. Geogr., 4O, 214-236

Wolman, M.G., and Miller, J.P., 196O. Magnitude and frequency of forces in geomorphic processes. J. Geol. 68, 54-74

Zeller, J., Geiger, H., and Röthlisberger, G., 1983. Starkniederschläge des schweizerischen Alpen- und Alpenrandgebietes. Vol. 6: Ausgewählte Stationen von Jura, Mittelland und grenznahem Ausland. Eidgenössische Anstalt für das forstliche Versuchswesen. Birmendorf

International Geomorphology 1986 Part II
Edited by V. Gardiner

A SOIL EROSION MODEL INCORPORATING SEASONAL FACTORS

M.J. Kirkby and R.H. Neale

School of Geography, University of Leeds, England

ABSTRACT

A model has been developed for estimating monthly overland flow and sub-surface flow, together with dynamic changes in soil moisture storage, living vegetation and organic soil biomasses and vegetation cover. It draws on previous work on a catchment hydrological model (TOPMODEL) based on an exponential store; combined with a vegetation growth model synthesized from existing ecological sources. The model forecasts hydrological and vegetation status, for any topographic location within a catchment, and uses these forecasts to provide estimates of soil erosion by splash and wash processes. It is seen both as a conceptual advance, and as a compromise in complexity and physical realism between fully distributed catchment models and the Universal Soil Loss Equation.

When monthly means of rainfall totals, numbers of rain-days and temperature are applied repeatedly, sites under natural vegetation take about 50 model years to reach a stable annual cycle. Response times in response to drastic changes in land use or climate are similar. In seasonal climates, the stable cycle shows that erosion and overland flow tend to be greatest at the beginning of the rainy season, whereas vegetation and organic soil biomasses lag behind the rainfall peak. The model reproduces broad global distributions of vegetation and organic soil, providing significant internal validation. The semi-arid maximum and temperate minimum in erosion rates, as mean annual rainfall is varied, are clearly shown in the simulations. Where the annual rainfall distribution is varied in range and season of maximum, the temperate minimum is reduced in importance as the rainfall becomes more seasonal, and this effect is greatest for winter rainfall peaks. Highly seasonal Mediterranean climates are thus forecast as showing a continuous increase in erosion with increasing rainfall.

The proposed model is currently cast in the form of a global model, which can provide default parameter values from a minimum of local data. Where local data for climate and/or vegetation are more complete, the model may be adapted to replace default values as far as available.

REVIEW

Soil erosion is a widespread hazard which is often exacerbated by man. Erosion by water is thought to be most rapid under semi-arid climates, where rainfall may be intense and vegetation sparse. Where existing natural or semi-natural vegetation is reduced even seasonally the increase in erosion may be striking, especially in areas of high rainfall intensity. For example, in southeastern USA, stripping of natural vegetation cover has been shown to increase erosion by a factor of 10^3 to 10^4 times. Such increases can be catastrophic in areas where conservation practises are not applied, perhaps most seriously where marginally semi-arid lands are subjected to over-grazing or unsuitable cultivation, as in much of the Sahel. Although the interaction of vegetation cover with erosion is imperfectly understood, accelerated erosion is commonly preceded by a progressive degradation of vegetation cover and loss of organic matter in the soil. Erosion losses are commonly greatest at times of year when rainfall is intense and vegetation relatively sparse. For these reasons, vegetation is seen as an essential component of any effective model for erosion forecasting. This paper presents an approach towards such a model, which is intended to provide low cost forecasts of net erosion, taking into account both hydrological and vegetation factors on a monthly basis. Relevant existing approaches are through the Universal Soil Loss equation, through distributed models for catchment hydrology and through vegetation models. These are briefly reviewed below.

The most widespread model for forecasting soil erosion at present is the Universal Soil Loss Equation (referred to as USLE below). It was originally proposed by Wischmeier and Smith in 1958, although it has been extensively revised since. Its formulation is a simple multiplication of factors for rainfall erosivity, soil erodibility, slope length, slope gradient, cropping management and erosion control practise; derived from extensive empirical experimentation using standard runoff plots of 1.83m width and 22.13m length (0.01 acres area) on a 9% gradient (for non topographic factors). A large amount of such data is available, predominantly for the USA, but with substantial work elsewhere. Although well parameterized for certain regions and soil types, the USLE generally requires considerable additional plot data for new areas. Its success even as a forecasting model is therefore limited. This is because the physical basis for USLE is extremely weak. Two examples illustrate this point, related to its hydrological and topographic behaviour.

First there is strong evidence that the dominant process in serious soil erosion events is rill erosion, and that this is driven by overland flow through hydraulic tractive stress. In any realistic hydrological model, overland flow is obtained by subtraction of an infiltration or storage capacity from incident rainfall. Because this is a subtractive process, no multiplicative model such as USLE can give a realistic forecast of the hydrological processes. Furthermore the soil infiltration or storage capacity depends on previous moisture history, soil composition and vegetation cover; all of which change with time and season.

A second example of failure of the simple multiplicative form of the USLE lies in the topographic response. Along the length of a hillslope, sediment load at a point is controlled by the load delivered from upslope, consequent surplus transporting capacity and the erodibility of surface soil aggregates. Erosion loss or aggradation at any site may only be forecast reliably by routing this sediment downslope, and calculating loss as rate of increase of load. Again a simple multiplicative model of gradient and slope length factors, or even a combined factor as in some later USLE variants, cannot give a proper physical representation of local rates of erosion and deposition on a hillside.

Although there has been substantial further work on soil erosion, primarily sponsored by the US Department of Agriculture (e.g. Foster, 1977, 1982), the large part of it related to the USLE has almost entirely failed to come to terms with its fundamental limitations: that it cannot provide useful forecasts of hillslope flow generation and that it fails to distinguish between the soil and surface properties which are separately relevant to hydrology, hydraulic resistance and aggregate transport. By simply embodying generalized regressions over time and space without any view of the physical basis of the erosion processes, USLE cannot realistically forecast the frequency, severity and spatial distribution of the relatively limited number of erosion events at a given site, especially under changing conditions. It is therefore an insensitive tool at best.

The second area of active research into erosion forecasting has been concerned with the development of distributed hydrological models to give detailed patterns of overland flow and sediment yield during the course of individual storm events (e.g. Alonso **et al**, 1978; Li, 1979). This group of approaches to erosion forecasting, although fruitful, requires very large amounts of field data to parameterize the models, and very substantial computer run times to forecast erosion rates for a significant period of,say, a decade. It is therefore not, at present, thought to be readily applicable to more generalised forecasting, particularly at the low cost per unit area required for many problem areas in the world's semi arid lands. It may also be criticized for not giving sufficient weight to recent developments in partial area models of hillslope hydrology. Despite these reservations, the application of hydrological principles to soil erosion forecasting is seen as an important step in the direction of removing the objections to USLE stated above.

The third area of relevant models is concerned with vegetation biomass and morphology. Despite its importance in controlling erosion, vegetation is normally only included in current erosion models as one or more parameters, not as an interacting component of the model. Vegetation models may either describe the distribution of gross morphological types/ plant assemblages as an empirical function of climate (e.g. Box, 1981); or may forecast the productivity or growth rate of individual species or communities. The latter type of model is most relevant to the needs of erosion forecasting, because it allows seasonal variations in plant community and therefore hydrological and erosion responses. Recent advances in general vegetation production ecology have been based largely on the detailed case studies of production and biomass in representative terrestrial biomes, conducted under the International Biological Programme (IBP). Global regressions of the resulting data on climatic variables have been attempted (e.g. Lieth and Whittaker, 1975; Lieth and Box, 1977), and have met with some success despite the range of species responses and non climatic controls influencing growth.

Improved understanding of plant metabolism in general and the physiology of certain well studied crop species in particular has led to the development of specific growth models for crops or other individual species which realistically simulate physiological and growth responses to climate in the growing season (e.g. Hanks and Rasmussen, 1982). However most are only applicable over a very restricted range of conditions. These limitations also generally apply to empirical yield models developed for agronomic forecasting.

Because vegetation models have been developed primarily by botanists, ecologists and agronomists, their hydrological implications have not generally been considered. Vegetation acts through intercepting precipitation and influencing total evapo-transpiration and water balance. It also influences crusting of the soil surface and soil moisture retention, both of which profoundly influence infiltration capacity. Forecasting the nature of the vegetation cover at a site, and its development over the year in response to seasonal elements of the climate, or the weather of a particular year, is seen as perhaps the most significant need in improving the hydrological basis of soil erosion forecasts.

STRUCTURE OF PROPOSED MODEL

The model presented here is conceived as part of a larger model, with somewhat more ambitious aims. Figure 1 outlines the main sub-systems considered in the model. Features in parentheses or with thinner arrows and/or box margins in the figure are intended to be part of the final model, but have not been included in the version reported in detail here. The most important interactions controlling erosion transport on partially vegetated surfaces are thought to be (1) the environmental and hydrological control of plant and soil biomass, (2) the control of soil hydrological properties by soil biomass and (3) the control of the various erosion processes by overland flow (for wash) and by the combined action of vegetation cover and rainfall intensity (for splash). Also of considerable importance, although not yet incorporated into the model are (4) the detailed mechanics of erosion transport processes, including their distribution and interactions downslope, and (5) the factors influencing soil erodibility and its dynamic adjustment in response to biological and erosional processes. It may be seen that the current model is intended to advance soil erosion forecasts chiefly through its improved handling of hillslope hydrology and of vegetation interacting with it. In this section, each sub-system of the current model is discussed in some detail.

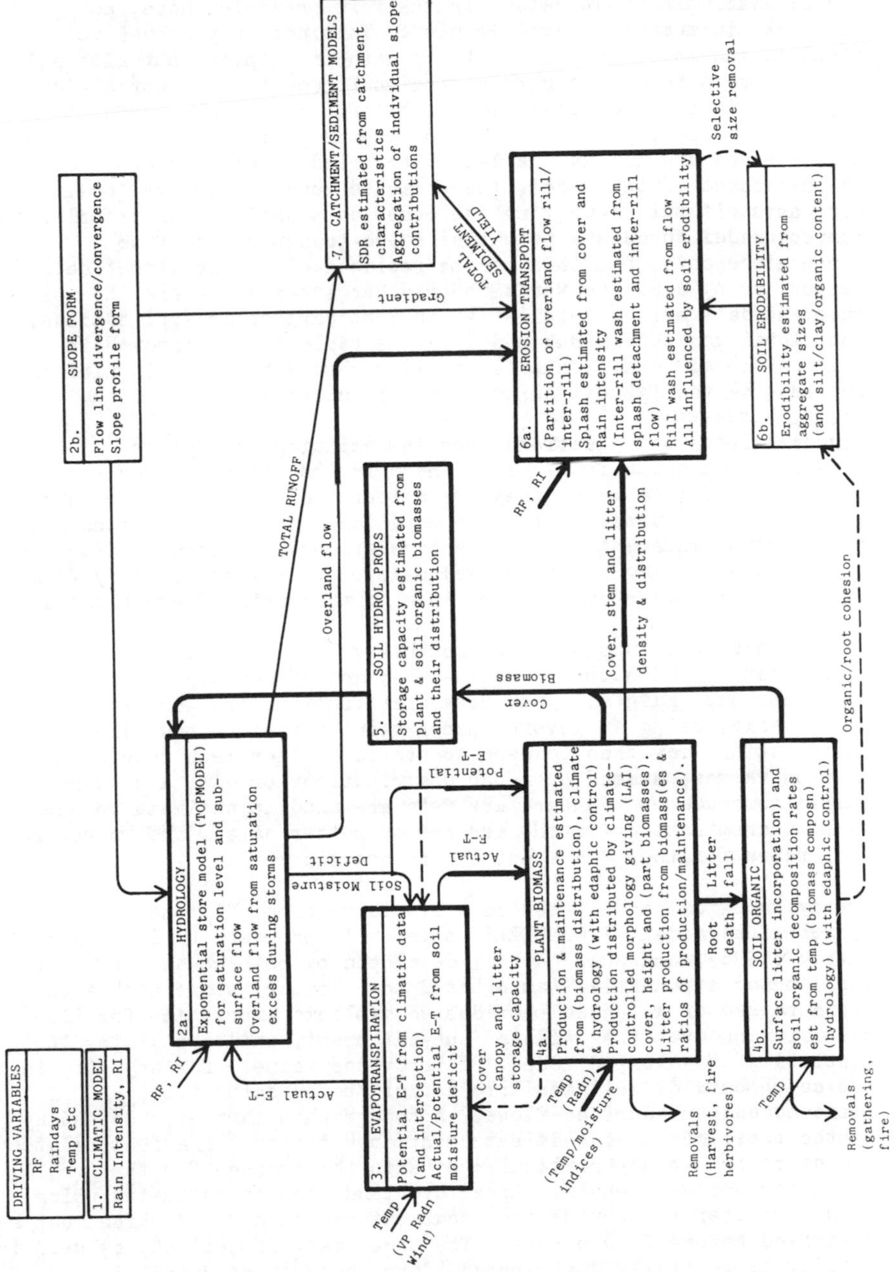

FIG.1 Systems diagram showing the main processes and interactions included in the current erosion model (Heavy lines); and in prospective versions of it (Light lines and parentheses).

The approach adopted in the model is intended to be flexible in its use of available input data. In the form presented here, an absolute minimum data level is used. This not only serves to simplify the model to bring out its main principles, but also allows it to be used here to generate erosion forecasts on a world wide scale. This scale of operation is helpful in defining default parameter values, which can be used in any particular context where local values are not available. Where good local data are available on any aspect of the model, these can be used to improve local forecasts without re-writing the model as a whole. For example, the current model forecasts potential evapo-transpiration from temperature data, but this can be replaced either by direct data values, or by a Penman Monteith or other sub-model where the relevant data exist. Similarly, in most particular applications, the global vegetation sub-model can be replaced or improved in the light of local data. This flexibility should be borne in mind in considering the 'global default' model outlined here.

Figure 2 lists the equations used in computing the system relationships shown in Figure 1 and described below. The sequence shows the order in which they are calculated, and the arrows show the logical precursor(s) to each equation. These expressions give the current implementation of each relationship discussed for the global model, using parameter values either derived directly from other studies, or modified from them on the basis of empirical data.

Minimum climatic inputs are of monthly values for total precipitation and number of rain-days; and of monthly mean temperatures. Rainfall intensity distributions are generated on a daily basis, using an inverse exponential distribution. These distributions may then be used to provide either mean runoff and erosion values, summed over the distribution, or else a stochastic rainfall sequence. Temperature data are used to estimate potential evapo-transpiration (in the absence of better data), and to control unit rates for the decomposition of organic soil.

The hydrological model used is a simple version of a one-store hydrological model, 'TOPMODEL', originally proposed by Beven and Kirkby (1978), and subsequently developed by both co-authors. It is based on an exponential (as opposed to a linear) store with a scale depth parameter **m**, which provides an explicit expression for flow distribution over the relevant sub-catchment, while retaining the simplicity of a lumped model. It provides values, for any site in a sub-catchment, of soil mositure deficit below saturation, and of overland and sub-surface flows. A topographic parameter depending on the ratio of local unit catchment area to local gradient relates the hydrology at any particular site to the average for the sub-catchment as a whole. Local overland flow is estimated using this parameter to provide the local saturation deficit which must be satisfied before flow occurs. The model may, if desired, be used in a fully or partially distributed form where great detail is required. In its original version, flows were forecast over time increments of 1/2 to 3 hours, but an aggregated version has been

Kirkby and Neale: Soil erosion model

(1) M = MBASE + 5 (MH(i-1) (Soil storage parameter in mm)

(2) RJ = 1 M^2 + exp [ST(i-1)/M] (Saturated soil flow)

(3) RC = 5 [1 - C(i - 1)] - C(i - 1) [ST (i - 1) + M ASRAT] (Critical storage to initiate overland flow in mm)

(4) OF = RF(i)/ {exp [RC/Rϕ(i) (1 + M/Rϕ (i))] }(Overland flow in mm)

(5) PE = 0.12 $[T(i) + 6]^2$ (Potential evapo-transpiration mm)

(6) 0 < = AE = ST(i - 1) + RF(i - 1) - OF + 7M < = PE (Actual E - T)

(7) F = RF(i) - AE - OF if > = 0: otherwise 0 (mm Infiltration)

(8) DS = M ln [RJ/F + (1 - RJ/F) exp (- F/M)] if f> o
= M ln (1 + RJ/M if F = o (Increased deficit over month)

(9) ST(i) = ST(i - 1) - DS (Updated soil water storage in mm)

(10) SF = F - DS (Total sub-surface flow for month in mm)

(11) C(i) = (PCOV C(i - 1) + (1 - PCOV) AE/PE (Vegetation cover %)

(12) GPP = 0.0033 {AE -3 [1 - C(i)] RF(i)/RO(i)} (Gross production)

(13) RES = 0.0005 exp [- T(i)/12] MV(i - 1)(Respiration etc)

(14) LF = MV (i – 1)/{ 72.8 ln [MV(i - 1) + 1.4] }(Leaf fall etc)

(15) MV(i) = MV (i - 1) + GPP - RES - LF (1 + VCROP) (Updated living biomass in kg m^{-2})

(16) DEC = MH (i - 1) exp [0.12 T(i) - 5.75] (Organic soil decomposition)

(17) MH(i) = MH(i - 1) + LF (1 - HCROP) - DEC (Updated organic soil biomass in kg m^{-2})

(18) TANϕ = 0.2 + 0.5 DRAT (Surface roughness)

(19) SPL = 0.00084 [i - C(i)] GRAD/[LEN GRSIZE $(TAN\phi)^2$] (μm yr^{-1})

(20) WASH = 8 x 10^{-6} OF^2 LEN GRAD/(GRSIZE TANϕ) (μm yr^{-1})

Additional notation in order of use above

MBASE	Basal value of M for mineral soil
RF	Mean precipitation for month
T	Mean temperature for month
i,i=1	Array indices referring to current and previous month
λ	Parameter linking saturated hydraulic conductivity at surface to M. λ should be inversely proportional to sub-catchment mean of topographic ratio a/GRAD. λ = 8 (mm month)$^{-1}$ currently.
ASRAT	Ratio of local value of topographic parameter a/GRAD to average for sub-catchment
a	Area drained at point per unit contour length
GRAD	Local surface gradient (tangent slope)
R_O	Mean rainfall per rain-day for month
PCOV	Persistence of cover parameter (currently 0.75 to indicate 75% surviving next month if AE = 0
VCROP	Proportion of annual leaf fall which is cropped from plant, in addition to any removal of normally shed material
HCROP	Proportion of annual leaf fall cropped, and therefore not added to litter
DRAT	Ratio of grain sizes of Roughness elements (D_{84}) to Mean size, DM
DM	Mean grain size of surface aggregates
LEN	Length of slope plot for erosion estimates

Figure 2. List of equations used in current model, in order of calculation, showing logical precursor(s) of each

developed to give monthly flows (and their distributions) in response to monthly rainfall inputs.

TOPMODEL provides an estimate of overland flow, by summing, over their frequency distribution for each month, the rainfall in excess of that required to fill soil water storage. This expression (4) includes a correction term for the change in sub-surface flow during the storm, and a term to correct the average sub-catchment values to the local site of interest. Rainfall, less overland flow and actual evapo-tranpiration, is then used to estimate the change of soil water storage and total sub-surface flow for the month (equation 7 to 10), using the relationship (8) from TOPMODEL to integrate the net change.

Potential evapo-transpiration is currently (though not necessarily) estimated in equation (5) from monthly temperatures. This expression approximates to Thornthwaite's estimates, but is simpler in form and provides a better forecast over the whole range. Actual evapotranspiration is constrained to lie between zero and the potential value. Within this range, it falls as a linear function of soil moisture deficit, reaching zero for a rooting constant which is related, through the soil hydrological parameter **m**, to the organic soil biomass. This estimate is made in equation (6).

On the global scale, net productivity of the plant community is calculated as the difference between a gross, or unstressed productivity which is proportional to actual evapo-transpiration less an allowance for bare ground evaporation (equation 12); and a temperature dependent loss term representing respiration etc. (equation 13). The gross productivity term is based on data from Larcher (1975). The combined effect of these two terms is to produce a net primary productivity which closely follows the equation (based on a different estimator of actual evapo-tranpiration) of Lieth and Box (1977). It differs in order to provide the best available estimator for the published data of Whittaker and Likens (1975) from gross environmental parameters. This may be replaced by a more detailed model, like those for crop species, or by direct data at a local scale.

In assessing plant growth, net productivity is set against losses which may collectively be termed 'leaf fall', though including death of all parts of the plant, at a total rate which increases with plant biomass (equation 14). The increase is initially more or less linear, and decreases in rate for communities with large standing biomass, corresponding to their greater proportion of woody and other non-shedding tissues. The form used follows the data of Whittaker and Likens (1975) and O'Neill and De Angelis (1981). Allowance can be made for cropping of parts of the living plant, for example through grazing; and for removal of some of the litter, for example for firewood. Both of these losses are scaled to the leaf fall rate, as a measure of sustainable production. They influence the living plant biomass (equation 15) and the addition of material to the organic soil (equation 17). One important element of the global erosion model is that the estimates of vegetation biomass obtained from it can be compared with a large body of ecological literature, which provides significant independent validation both of the vegetation model itself and also of the hydrological model which is used to 'grow' it through controlling evapo-transpiration.

The organic soil biomass receives material from 'leaf fall', and loses material through decomposition to CO_2 and inorganic soil nutrients. The former is controlled by the plants as described above. The latter (equation 16) is directly proportional to soil biomass and to a rate constant which is thought to depend mainly on temperature, through its control on the population density of decomposer organisms. The temperature dependence of the rate constant gives values which are in reasonable agreement with data by Singh and Gupta (1977). In the current model, good soil aeration is assumed, so that no account is taken of the considerable depression of decomposition rates by waterlogged conditions. Low soil moisture, litter composition and its pH are also known to influence decomposition rates. These features could readily be included, either by reducing parameter values or in response to forecast soil water deficits. Neither has any distinction yet been made between litter layers and organic material admixed with the mineral soil.

A third property of the vegetation cover which is also estimated independently is the total crown cover. For a non-seasonal climate, it is estimated simply as the ratio of actual to potential evapo-transpiration. The basis for using an estimate of this kind is the assumption that plant root systems spread out to intercept available rainfall over the whole of the ground surface, whereas their crown cover is restricted, in one of their most direct strategies for limiting actual transpiration losses. For a seasonal climate, the ratio of actual to potential evapo-transpiration is used in a first order Markov model (equation 11) with a persistence of **PCOV** = 0.75. It then provides a crude estimate of cover, as it changes gradually over the year, which is used in the model to forecast exposure of the soil to splash and crusting processes. The ratio of mean plant biomass to mean cover may also be treated as a rough indicator of height for the plant community, though not of its seasonal variations, helping to give a better picture of the types of community modelled, although this measure is not used in the model. The calculated cover may also be used to provide an approximate estimate of Leaf Area Index (LAI), if this is needed by the Penman Monteith or other evapo-transpiration sub-model. On an assumption of stochastic leaf overlap with a Poisson distribution, LAI is equal to -ln(proportion bare area). This expression is not however used in the model at present.

Soil hydrological properties provide the key feedback from vegetation characteristics to the hydrological model. At a global level, no assumptions have been made about mineral soil differences, even though some consistent differences and trends with climate are known to exist. Soil parameters, particularly the key parameter **m** of the subsurface hydrological store, are instead forecast solely from the vegetation and organic biomass values, principally the latter. The basis in reason for this linkage is that soil organic matter provides one of the major reservoirs of readily reversible soil water storage. High soil organic biomass therefore bestows a good soil structure for moisture storage and a high surface permeability: qualities which are known to have a profound influence on the parameter **m** from experience with TOPMODEL. This linkage is used in two ways. First to forecast the parameter **m** as a linear function (equation 1) of the organic soil biomass (from the previous month), together with a constant term representing the mineral soil. The second expression of this linkage, in equation (2), forecasts the monthly sub-surface runoff at the start-of-month soil water level. This term, which is used in calculating the total sub-surface runoff for the month (equation 8), includes an estimate of saturated hydraulic conductivity at the soil surface, which is taken to be directly proportional to the parameter **m**.

Vegetation also influences hydrological forecasts through a term in equation (3) which represents crusting and sealing of the soil surface. The available storage capacity which must be filled before overland flow can begin is assumed to be equal to the local saturation deficit in the proportion of the area covered by vegetation, and to a low constant value (of 2-5mm) for the bare part of the area. Under high canopies, this term may need correcting to allow full or partial crusting beneath the canopy.

The core of the current model is concerned with forecasting hillslope hydrology, and its interaction with vegetation cover. At present the forecast of erosion losses is relatively crude, and is intended to illustrate the implications of the hydrological forecasts rather than to provide a final basis for forecasting erosion, which will be the subject of further work.

The wash erosion forecast uses an estimate of overland flow derived from the hydrological sub-model. This estimate, either for an individual storm or as a monthly mean on the basis of the rainfall distribution, has considerable indirect dependence on vegetation, particularly through the moisture retention of the organic soil, but also through the influence of bare area on crusting of the soil surface. The values generated are based on empirical estimates previously used by Kirkby (1976, 1980), but do not yet include any detailed analysis of the relative significance of rill and inter-rill contributions such as those monitored by Meyer **et al.** (1975). The forecast of equation (20) nominally relates to average denudation from a 10 metre long convex slope, falling from a divide. It may readily be adjusted to other configurations by adjusting the sediment transport rate constant, which is proportional to slope length squared multiplied by basal gradient; and by adjusting the hydrological parameter which relates the local site to the average for the sub-catchment (**ASRAT** in equation 3).

Splash erosion forecasts are related to local gradient and to storm rainfall (equation 19). The process is assumed to be effective only where the ground is bare of vegetation, although there is scope for revising this assumption where there is known to be little canopy cover within 5-10 metres of the ground. Both splash and wash expressions contain a correction for the relative roughness of the surface in terms of the spread of surface stone sizes (equation 18), but there is not at present any forecast of progressive armouring as erosion proceeds.

The set of equations in figure 2 is calculated in order from (1) to (20) for each month, and values for next month calculated on the basis of the previous month's storage, vegetation etc. The thrust of the model is currently concerned with the interaction of hydrology with vegetation in response to climate. There is no doubt that there is a second major set of interactions concerned with hydrological and strength properties of the mineral soil which merits similar exploration.

MODEL PERFORMANCE

The equations listed in Figure 2 form the core of a computer simulation model, which calculates values of saturation deficit, flows and erosion based on averages over the distributions of rainfall for each month of the year. It is currently implemented on the University of Leeds AMDAHL mainframe computer in the FORTRAN 77 language, although its requirements are well within the scope of most microcomputers. The model, in its current 'global default' form, uses monthly data for precipitation, number of rain-days and mean temperature as its sole explicit inputs, although there are also a number of implicit inputs representing the constants in the governing equations.

To initialize the model, arbitrary values may be taken for living and organic soil biomasses. Over a wide range of assumed values, the simulation converges on a stable annual cycle of values if the monthly climatic means are input repeatedly. The time for effective convergence depends on the initial values chosen and their difference from the final values. A period of twenty to eighty simulation years is generally needed to reach stability. The runs reported here reflect the values of biomass, mean erosion etc. for the stable pattern obtained in this way. Simulations may however continue from this stable state to examine a historical climatic sequence, either as represented by mean values or for the particular sequence of storms. They may also examine the response to a stochastic sequence of storms generated from the mean monthly distributions, and so gain some insight into the expected variability in erosion rates over time.

The performance of the model is illustrated in a series of figures (3 to 7) which illustrate its behaviour for the set of parameter values shown in Figure 2. Figure 3 shows the course of the main variables of interest for Almeria, a fairly arid Mediterranean climate, that is with maximum rainfall in the winter (coolest) months. (a) shows the march of rainfall, temperature and forecast overland flow: most of the remaining rainfall contributes to the actual evapo-transpiration. It may be seen that the proportion, as well as the total amount of overland flow is greatest in the months of greatest (and most intense) rainfall, and that there is more overland flow in the autumn, when vegeation is thin, than at similar rainfall levels in the spring.

In Figure (3b), it may be seen that the forecast vegetation biomass is greatest in the spring, at the end of the rainy season, and least in the autumn, growing in the winter rainy season which is warm enough not to inhibit growth. The vegetation of southern Spain consists largely of low perennial shrubs, so that the persistence of a substantial proportion of the biomass through the summer is an accurate reflection of the true conditions. The organic soil biomass peaks even later, in May/June, depending as it does on leaf fall from the spring growth to sustain it. (c) shows the forecast for the erosion components. Wash erosion peaks strongly in the winter months in association with overland flow, and more or less in phase with it, and has a very pronounced minimum during the summer months. For splash the variation over the year, while present, is less extreme, and there is some tendency for relatively greater erosion in the autumn than the spring, when rain acts on a minimum vegetation cover. The overall pattern shown in Figure 3 is of erosion and overland flow tending to peak a little in advance of the rainfall maximum, and for vegetation and organic soil biomasses to lag behind the rainfall peak. The effect of the strongly seasonal climate is to provide a time of year when vegetation is sparse and rainfall intense, so that total erosion is notably greater than for a climate in which the same rainfall is equably distributed over the twelve months of the year.

Figures 4 to 6 all refer to a range of mean annual temperatures and rainfalls; with superimposed variations in the degree of seasonality of rainfall and temperature, and in the mean rain per rain-day, allowing the diagrams to show many commonly ocurring combinations of global climates with summer rainfall maxima. The diagrams are thus intended to show a reasonably realistic pattern of world-wide variations in climate.

Figure 4 shows the forecast values for equilibrium biomass of living plants (a) and organic soil (b). It may be seen to forecast the broad pattern of world ecotypes. The range of values for standing biomass follows mean values reported by Whittaker and Likens (1975), with a maximum of 45 kg m^{-2} in the humid tropics, declining to 0.15 kg m^{-2} for tundra, and virtually to zero in extreme desert conditions. Within this overall range, estimates for boreal forest are on the low side, and for savanna somewhat too high (although savanna ecosystems almost certainly reflect many factors beside direct hydrological control). Soil organic matter peaks in cool-temperate and boreal areas, conforming well to the data of Basilevich and Rodin (1975). Both living and organic soil biomass distributions show a primary dependence on temperature at high rainfalls, when growth is limited by potential evapo-transpiration; and a joint dependence on rainfall and temperature for semi-arid climates, tending towards zero values for very hot, dry climates. The forecast behaviour thus provides a fair summary of world-wide values, especially when it is noted that no assumptions are made about vegetation form (e.g. grass, shrubs or trees etc) or

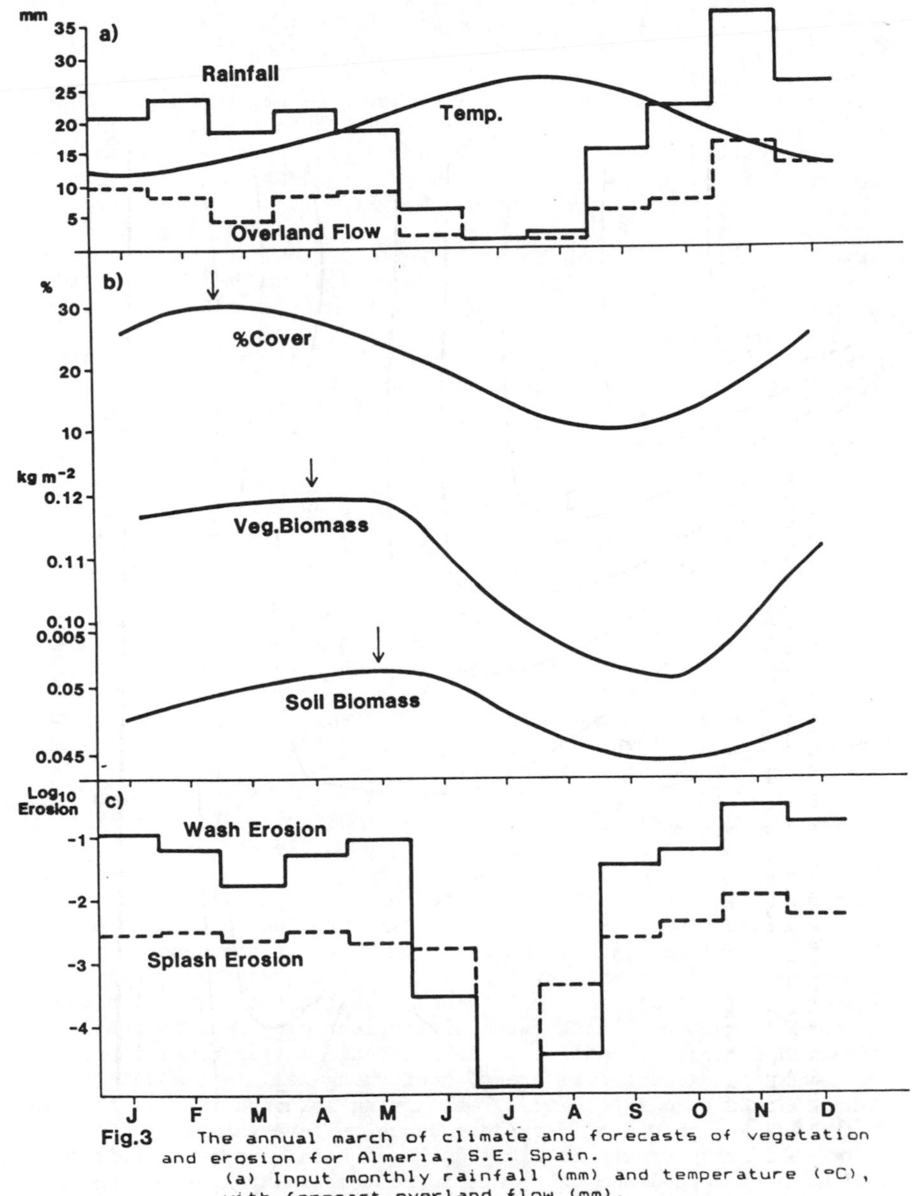

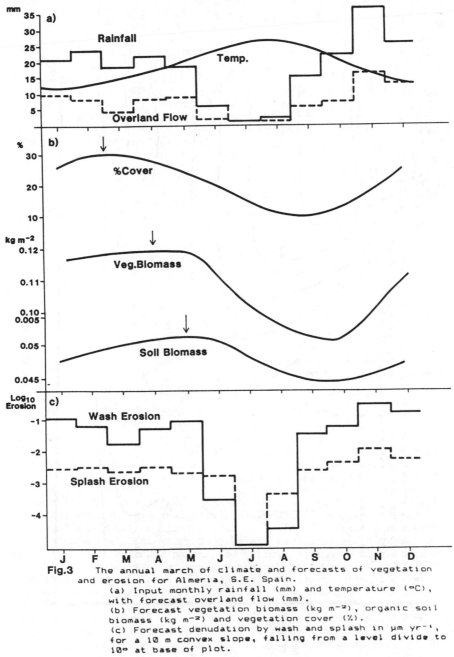

Fig.3 The annual march of climate and forecasts of vegetation and erosion for Almeria, S.E. Spain.

(a) Input monthly rainfall (mm) and temperature (°C), with forecast overland flow (mm).

(b) Forecast vegetation biomass (kg m^{-2}), organic soil biomass (kg m^{-2}) and vegetation cover (%).

(c) Forecast denudation by wash and splash in μm yr^{-1}, for a 10 m convex slope, falling from a level divide to 10° at base of plot.

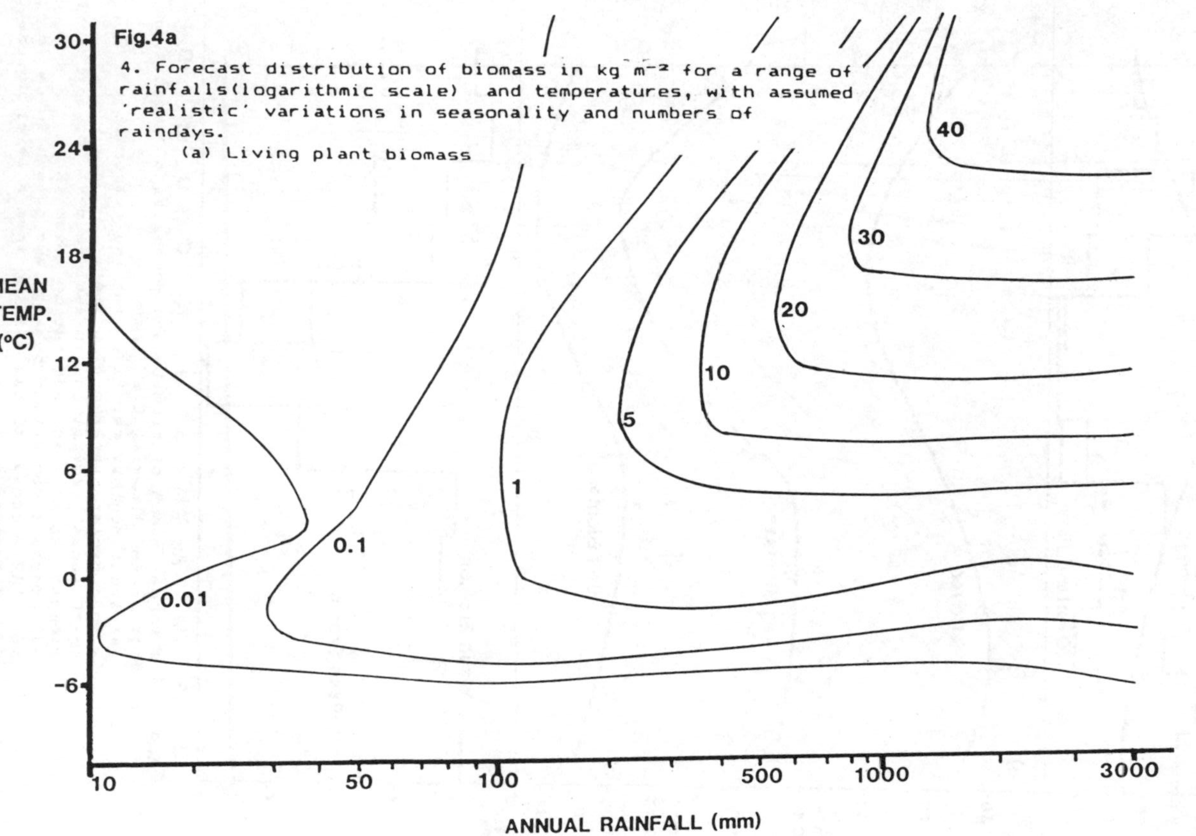

Fig.4a

4. Forecast distribution of biomass in kg m^{-2} for a range of rainfalls(logarithmic scale) and temperatures, with assumed 'realistic' variations in seasonality and numbers of raindays.

(a) Living plant biomass

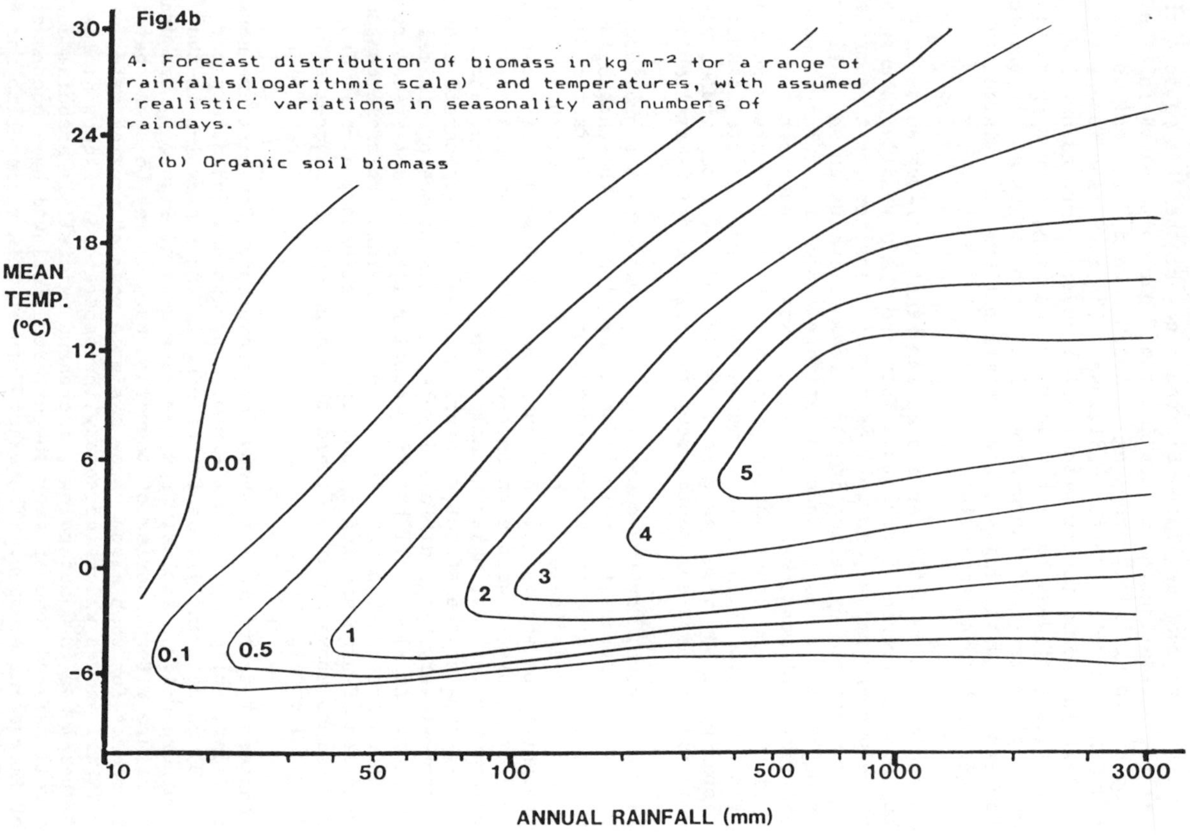

Fig.4b

4. Forecast distribution of biomass in kg m^{-2} for a range of rainfalls(logarithmic scale) and temperatures, with assumed 'realistic' variations in seasonality and numbers of raindays.

(b) Organic soil biomass

deciduous/ evergreen habit. The model can therefore be considerably improved if parameter values are allowed to take even such broad categories of local cover into account. If estimates of plant 'height' are made as living biomass / cover, then the pattern is of tallest vegetation (more than 40 kg m^{-2} per unit cover) for the tropical rain forest, dropping to about 2 kg m^{-2} for hot deserts, and 0.2 kg m^{-2} for tundra conditions. The resulting forecast pattern is one of tundra with a total cover of ground hugging vegetation; of deserts with widely spaced large shrubs or cacti; and of complete cover of increasing height, from temperate to tropical forests. As with biomass, the overall pattern is satisfactory, though with local anomalies.

These vegetation estimates are an important half-way house in building up erosion forecasts. They provide a readily checked set of values and ranges, as well as representing important controls on erosion rates. By validating the gross estimates shown in Figure 4, we may have some confidence in estimates both of regular seasonal change and of responses to particular weather sequences or agricultural and conservation practices which influence vegetation type or cover. With suitable parameter changes, adaptations of the global model may also serve as a stable model for particular plant communities where no special purpose growth model is available.

Figure 5 shows, for the same range of climates as the previous figure, the forecast levels of total erosion for the chosen 10°, 10 m slope. The area for which denudation by splash is faster than by wash is also indicated. The forecasts for low-temperature areas are thought to be too large. Although these cold areas are known to be subject to severe wash erosion because of their sparse vegetation cover, the forecast makes no allowance for snow or ice, and so over-estimates the effect of overland flow. The temperate zone, except at very high rainfall levels, is seen as one of minimal erosion. At almost all but the lowest rainfalls, a temperature transect for erosion from this forecast, holding rainfall constant, consists of high erosion in sub-freezing climates; a rapid drop to low values above about 5°C; and then a progressive increase to high tropical erosion rates.

Constant temperature transects for the erosion forecast are more familiar, following the pattern of Langbein and Schumm's paper (1958) on sediment yields for catchments and reservoirs in the southern USA. Examples, from the same data as Figure 5, are drawn in Figure 6 for a series of temperatures. It may be seen that forecasts for wash alone show a simple peak of erosion rate, at rainfall values which increase with temperature. Except at low temperatures (6°C and below), increased rainfall is associated with a well defined zone of very low erosion, followed by a second zone of increasing erosion. The exact form of this rising limb for high rainfalls is strongly influenced by the hydraulic conductivity of the near-surface soils; high conductivities giving a more subdued rise. The addition of splash somewhat confuses the pattern of total erosion at low temperatures. Splash gives the curve a 'shoulder' on the right of the semi-arid wash erosion peak, which at lower

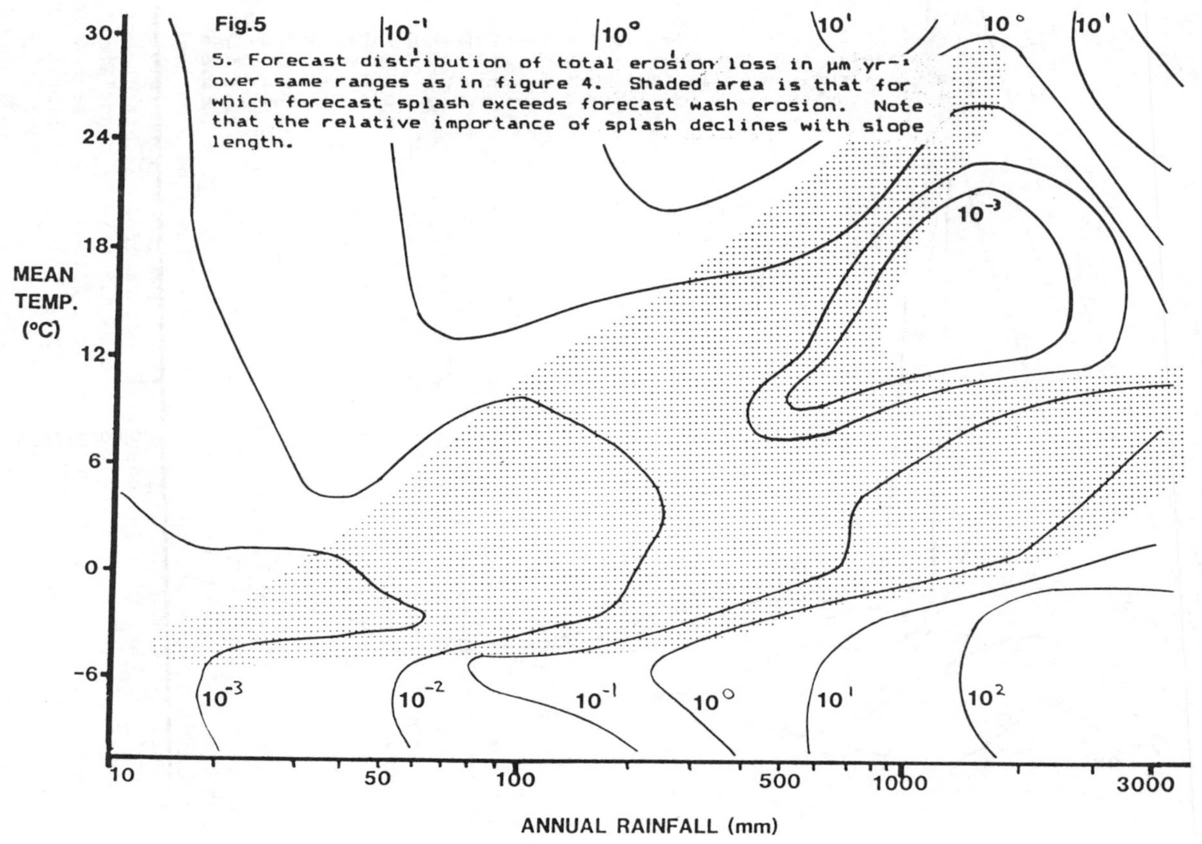

Fig.5

5. Forecast distribution of total erosion loss in μm yr⁻¹ over same ranges as in figure 4. Shaded area is that for which forecast splash exceeds forecast wash erosion. Note that the relative importance of splash declines with slope length.

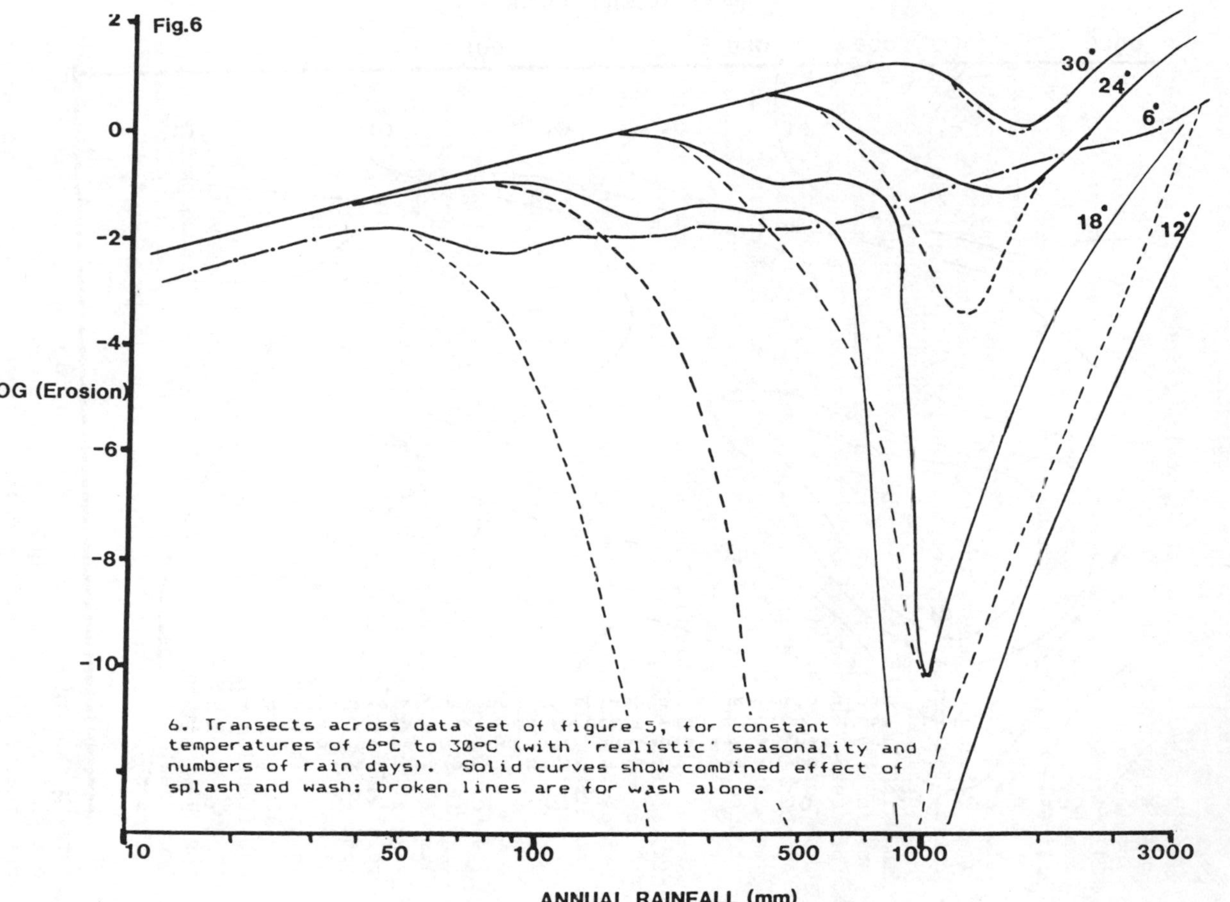

6. Transects across data set of figure 5, for constant temperatures of 6°C to 30°C (with 'realistic' seasonality and numbers of rain days). Solid curves show combined effect of splash and wash: broken lines are for wash alone.

temperatures gives a secondary peak of total erosion. At high temperatures, the influence of splash is proportionately less.

In Figure 7, the effect of seasonality has been explicitly separated out in the forecasts of erosion. Five curves are shown here, all for the variation of wash and splash erosion with rainfall along a 21°C transect. The basic curve is for zero seasonality, with every month at equal rainfall and temperature. The 50% curves have rainfall ranging as a sine wave from 50% of the annual mean to 150% of the mean; and temperature varying sinusoidally over 5°C either side of the mean. On the +50% curve, these changes are in phase, so that rainfall peaks in the summer (i.e. hot season). On the -50% curve, the changes are out of phase, giving winter rainfall (i.e. a Mediterranean climate). The other two curves are for +/-90% changes in rainfall, with a temperature fluctuation of +/-10°C. Increasing seasonality can be seen to have a strong influence on the erosion forecasts. Greater seasonality always increases erosion, as has been noted above, but its greatest impact is in reducing the depth of the temperate erosion minimum. This effect is seen to be a great deal more marked for the case of winter rainfall minima. For the strongest seasonality (-90% curve), which is for a climatic transect similar to that along the Spanish coast from Almeria to Gibraltar, the temperate minimum of erosion has almost completely disappeared. It is clear that, at the very least, more attention should be paid to seasonality in collating erosion data, and that Fournier's (1960) expression for seasonality should be given due credit, but should not be the last word on the subject.

CONCLUSIONS AND PROSPECT

The forecasts for soil erosion, overland flow, living biomass and organic soil have been explored here to demonstrate the relevance of a model which in primarily concerned with the relationship between vegetation and slope hydrology. At present many of the relationships, especially those used to estimate soil hydrological properties from organic soil biomass, are experimental, without direct evidence for their form. It is clear too, that other inter-relationships, especially those including the mineral soil, are also important; and that the present model fails for sub-freezing regimes. Despite these shortcomings, the model presented is well founded in empirical data at many points, and in our understanding of detailed hydrological and geomorphological processes. It represents one important conceptual step towards a better physical basis for erosion forecasting.

At the global scale of the model presented, there has not been time here to explore all the dimensions of interest. There is, for example, scope for attempting to place cropping and conservation practice on a more rational basis. There is also scope for using the estimates of sub-surface flow, in conjunction with vegetation and soil organic matter, to generate a parallel model for solution rates, taking account of the influence of vegetation not only in controlling hydrology, but also in controlling soil CO_2 levels.

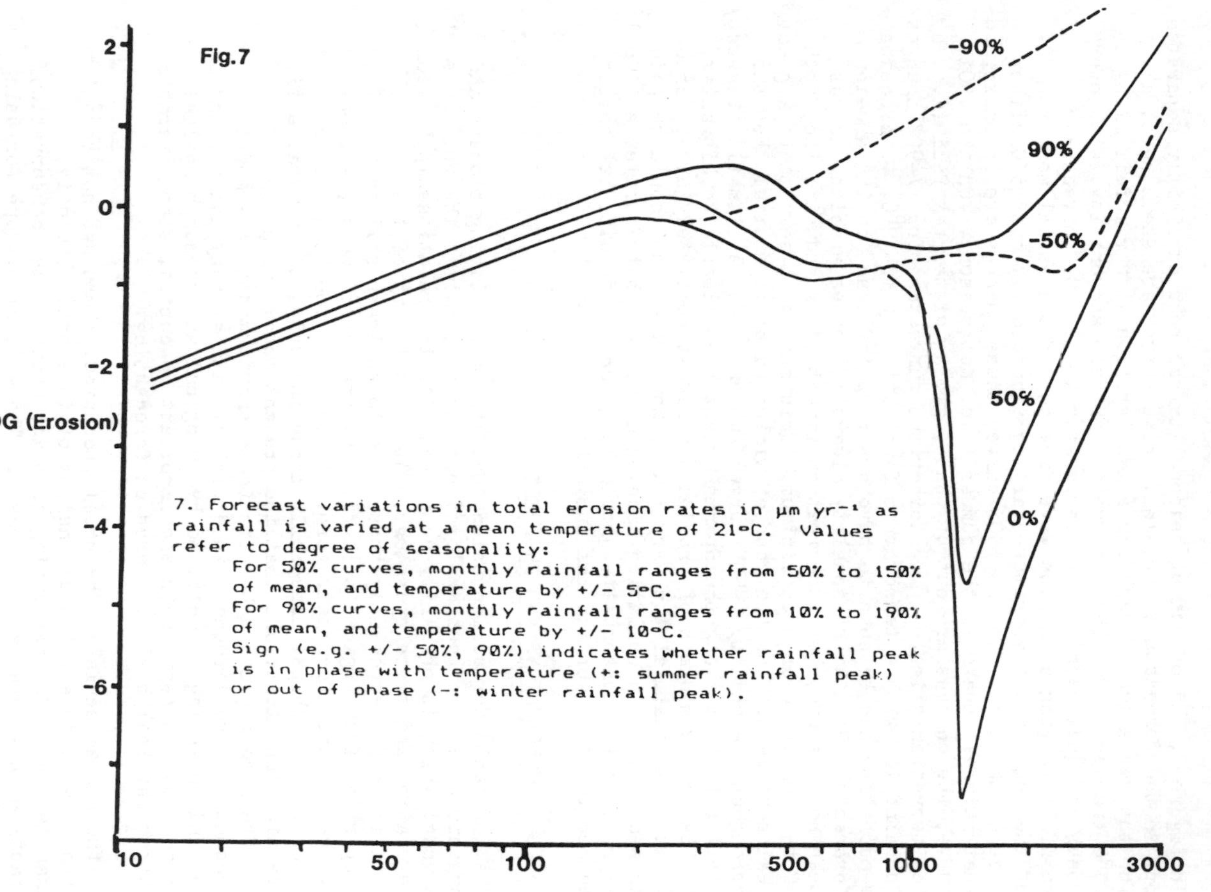

7. Forecast variations in total erosion rates in $\mu m \ yr^{-1}$ as rainfall is varied at a mean temperature of 21°C. Values refer to degree of seasonality:
For 50% curves, monthly rainfall ranges from 50% to 150% of mean, and temperature by +/- 5°C.
For 90% curves, monthly rainfall ranges from 10% to 190% of mean, and temperature by +/- 10°C.
Sign (e.g. +/- 50%, 90%) indicates whether rainfall peak is in phase with temperature (+: summer rainfall peak) or out of phase (-: winter rainfall peak).

The prospect for better detailed erosion forecasting is seen to rest on the global model presented here, in the sense that it provides a default basis for any parameters which are not better defined in each local context. Many parameters are however known in any local forecasting context, generally including the type of vegetation cover. The overall framework is still thought to be highly relevant however, in allowing forecasts to respond, through vegetation growth (for the known species of the site) to the weather of particular years or runs of years; or to changes in land use or conservation practice.

Current soil conservation practice aims at limiting erosion at a level which, while sustainable in the short term, cannot be replaced by natural rates of bedrock and soil weathering. To improve our control of the land for agriculture without degrading our soil resources in this way, we must aim at a better knowledge of the whole erosional system, and not simply those parts which are seen as amenable to straightforward physical analysis. The role of vegetation as a major control on erosion is unquestioned. We must intensify our study of erosional systems, going beyond the normally accepted bounds of physical hydrology by including the vegetation as a dynamic part of that system.

REFERENCES

Alonso, C.Y. **et al**, 1978. Field test of a distributed sediment yield model, in Verification of mathematical and physical models in hydraulic engineering. American Society of Civil Engineering, Hydraulics Division.

Basilevich, N.I., and Rodin, L.Y., 1971. Geographical regularities in productivity and circulation of chemical elements in the earth's main vegetation types, in Soviet Geography, (Rev. and Trans.) American Geographical Society, New York.

Beven, K.J., and Kirkby, M.J., 1978. A physically based, variable contributing area model of basin hydrology. Hydrological Sciences Bulletin, 24(1), 43-69.

Box, E.L., 1981. Macro-climate and plant forms. An introduction to predictive modelling in phytogeography. De Junk Pub.

Foster, G.R. (Editor), 1977. Soil erosion: prediction and control. Soil Conservation Society of America.

Foster, G.R.,1982. Modelling the erosion process, in Hydrologic modelling of small watersheds (ed. C.T. Haan), American Society of Agricultural Engineering, Monograph 5.

Fournier, F., 1960. Climat et erosion: la relation entre l'erosion du sol par l'eau et les precipitations atmospheriques. Presses Universitaire de France, Paris, 201pp.

Hanks, R.J., and Rasmussen,V.P., 1982. Predicting crop production as related to plant water stores. Advances in Agronomy, 35, 193-215.

Langbein, W.B., and Schumm, S.A., 1958. Yield of sediment in relation to mean annual precipitation. Transactions, American Geophysical Union, 39, 1076-1084.

Larcher, W., 1975. Physiological plant ecology. Springer Verlag.

Li, R.M., 1979. Water and sediment routing from watersheds, in Modelling in Rivers (edited H.W. Shen). John Wiley, 9.1-9.88

Lieth, H.,and Box, E.O., 1977. The gross primary productivity of land vegetation: a first attempt. Tropical Ecology, 18, 109-115.

Lieth, H.,and Whittaker, R.H. (Editors), 1975. The primary production of the biosphere.

O, Neill, R.V.,and De Angelis, D.L., 1981. Comparative productivity and biomass relations of forest ecosystems, in Dynamic properties of forest systems (ed. D.E. Reichle), IBP Publication 23.

Singh, J.S.,and Gupta, S.R., 1977. Plant decomposition and soil respiration in terrestrial ecosystems. Botanical Review, 43, 449-528.

Whittaker, R.H., and Likens, G.E., 1975. The biosphere and man, in The primary production of the biosphere (ed. H. Lieth and R.H. Whittaker).

Wischmeier, W.H.,and Smith, D.D., 1958. Rainfall energy and its relationship to soil loss. Transactions, American Geophysical Union, 39, 285-291.

International Geomorphology 1986 Part II
Edited by V. Gardiner

RIVER-GEOMORPHOLOGIC PROCESSES AND DISSIPATIVE STRUCTURE

H. Gu,
Chongqing Institute of Communication, China

S. Chen,
Chongqing Institute of Architecture and Engineering, China

X. Qian,
Chongqing University, China

N. Ai and T. Zan,
Lanzhou University, China

ABSTRACT

A river system is an assemblage of stream flow and environment in a certain region (or space) with a certain structure and function; and is an open system far from any equilibrium state. The rivers exchange mass, energy and information with their environment constantly. By dissipation of energy and non-linear interaction, the river-geomorphologic system forms and keeps its macroscopic order structure in time and space; described as its dissipative structure. In this paper, the authors study the mainly antagonistic actions between force and resistance, dynamic conditions and mass balance. They propose the dissipative structure theory of mass and energy dissipation from the point of view of non-equilibrium thermodynamics and a minimum-maximum rate of energy dissipation. According to the river's phase relations and the river-bed and its evolution, we discuss the appropriate relationships between river self-organisation and environmental changes.

THE RIVER SYSTEM AND ITS DISSIPATIVE STRUCTURE

A river is an open system, which contains a series of complex mechanical, physical, chemical, biological, and even societal processes. Generally speaking the system is related to the hydrological cycle, which involves climate, vegetation cover, soil and groundwater. It should be pointed out that, however a river system is defined, it always has the stream valley as its environment, and constantly exchanges mass and energy with it. By dissipating energy and through non-linear dynamic mechanisms, the river system forms and keeps its macroscopic order structure in time which is described here as its "dissipative structure". The evolution of a river is the evolution of this structure.

In an open system, negative entropy flow leads to the formation and maintenance of order structures. The entropy change of a river system is shown in Fig. 1. According to the thermodynamics of irreversible processes, we have:

$$dS = d_iS + d_eS \qquad (1)$$

where d_eS expresses entropy flow from the environment, it can be considered as the change of natural and artificial environmental

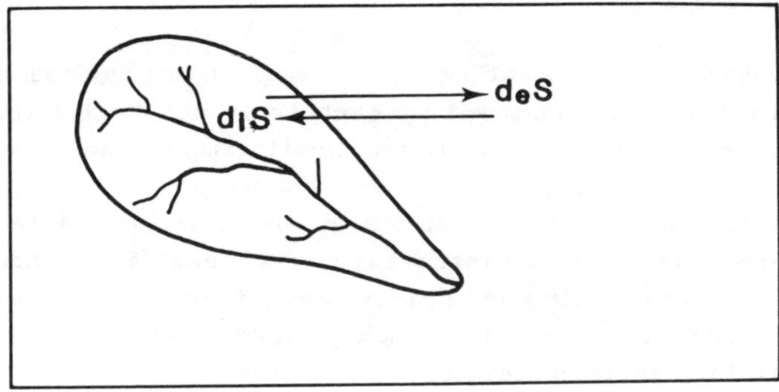

Fig. 1. Entropy change of a river system

conditions such as climate, vegetation cover, soil and irrigation works etc. d_eS can be positive or negative or zero, determined by the direction and intensity of the above changes, including important man-made effects related to river basin development. d_iS is the entropy production in the interior of the river system. It results from stream flow, fluctuation of water levels, mass diffusion, sediment transportation and erosion. In all cases, we have:

$$d_i S > 0 \qquad (2)$$

because the river system undergoes irreversible changes.

The river is a dynamic system. The bigger d_iS is, the more complex its processes are, and the farther from an equilibrium state it is. The entropy production d_iS occurs, even under equilibrium conditions (that is $dS = 0$), because of the irreversible processes going on. For this equilibrium case, the environment supplies enough negative entropy flow to the river system. The river system generates a related d_iS, and

$$d_iS = -d_eS > 0 \qquad (3)$$

If the river system loses stability due to some disturbance, d_iS increases or decreases. In the former case the river develops towards a more ordered state, and in the latter case the river tends to degrade towards an equilibrium state, and finally to disappear.

Entropy effects are a simple measure of dissipation quantities. In the river system, the mass, energy and information from its valley maintain its development. The relationship between supply and demand, and the difference between them lead river systems to develop or degrade, and control their direction of change.

The above analysis shows that the river is an open system far from equilibrium, and that the interactions between its elements are non-linear. Within the system there are many stochastic phenomena with fluctuating rates. All such phenomena satisfy the basic conditions for forming dissipative structures.

FLUVIAL PROCESSES

A river system is a subsystem of a larger river system. It contains stream flow, sediment and all types of geomorphological patterns resulted from sediment erosion and deposition under varying conditions.

Fluvial Processes

Under the actions of stream flow, the time and space changes in topography are considered as fluvial processes. Sediment transport by the stream flow is its main characteristic, and the evolution of the river course is its concentrated expression.

In a big and complex river system, the river geomorphology is the comprehensive result of the actions of all factors such as climate, vegetation cover, soil, geologic structure and human activity. The fluvial processes $G_m(t)$ can be expressed as following:

$$G_m(t) = \begin{pmatrix} C(t) \\ P(t) \\ S(t) \\ G(t) \\ M(t) \end{pmatrix} \qquad (4)$$

where: C(t), P(t), S(t), G(t) and M(t) express climate, vegetation cover, soil geologic structure and human activity processes respectively.

The river geomorphologic pattern $G_m t$ may alternatively be expressed by the set junction:

$$G_m(t) = C(t) \cap P(t) \cap S(t) \cap G(t) \cap M(t) \qquad (5)$$

Forces and Resistance

The energy of fluvial processes is provided by gravity and climate. Obviously the forces which act on the ground surface and cause erosion and mass transportation, are mainly due to gravity, tensile stresses and pressure in the water, the dynamic forces of stream flow, collision forces exerted by water drops, expansion forces resulting from the changes of water and water temperature, and diffusion forces. Carson and Kirkby(1972) have estimated their quantitative levels and effects (Table 1).

The data in Table 1 show that in a river system with slopes and river courses, the action of stream flow is always very important both for energy dissipation and for sediment transport.

The concentrated expression of the resistance to erosion and mass transportation is the intrinsic shear strength of rock or soil mass. It is composed of plane friction forces, cohesion, effective normal force and adhesion, and is in opposition to gravity and hydraulic shear stresses.

TABLE 1. The rates of forces and their effects for transporting sediment in geomorphological systems

Force	Total work (Jm^{-2} y^{-1})	Total work for transporting silt (Jm^{-2} y^{-1})	Efficiency (%)
Gravity	1-100	1-100	100
Stream flow	10^5-10^6	10-100	0.01
Collisions of	2000	0.02	0.002
Waterdrops	500	0.5	0.1
Stream	2000	0.4	0.02
Expansion:			
freezing	$5x10^7$	0.1	$2.5x10^{-7}$
temp	$2x10^8$	0.2	$1.5x10^{-9}$

The opposition and combination of forces and resistances form the three basic phases of river geomorphology:

(a) Force > Resistance : erosion
(b) Force = Resistance : transportation
(c) Force < Resistance : deposition

On the basis of erosion mechanics, according to the principle of antagonism proposed by Scheidegger (1961) we think of river geomorphology as the result of the opposing action of forces and resistances. Different balances between them result in different geomorphologic patterns. It should also be pointed out that there are both endogenous tectonics and exogenous stresses (eg drop impact, stream flow); and the resistances also consist of both endogenous and exogenous stresses.

Dynamic conditions and mass conditions

From the above analysis, we conclude that fluvial geomorphology is the result of the comprehensive actions of water flow and sediment transport. After making a "Pansystems analysis" (Ai and Gu, 1984) of actions of internal and external stresses for developing rivers, we have obtained the basic conditions for forming and maintaining fluvial processes as follows:

(a) Dynamic conditions: Stream flow is the main expression of energy dissipation, sediment production and its transportation.

(b) Mass condition: Rivers mould varying geomorphological forms in their sediment which generates self-organization.

The principle of extreme values of energy dissipation

(a) The minimum rate of energy dissipation: C. T. Yang and C. C. Song (Yang and Song, 1979; Yang, 1981) proposed the theory of minimum rate of energy dissipation, resulting from comprehensive analysis of much field and laboratory data for rivers. The main concept is as follows:

Firstly, if constrained by fixed banks and bed, all stream flow satisfies continuity equations, boundary conditions, and the condition that the energy dissipation by stream flow is a minimum.

Secondly, under the conditions of free and plastic modification of banks and bed, if the energy dissipation of stream flow deviates from its minimum the stream flow will adjust itself to a state in which the energy dissipation is again a minimum. The stream flow is then stable. That is, the energy dissipation rate of stable flows under the conditions of a fixed boundary is a minimum, as in the first case.

The nature of the principle is that in a river system of any kind, the forces leading the water to flow and the primary forces eroding the riverbed always tend to satisfy a minimum rate of energy dissipation.

(b) The maximum rate of energy dissipation

In opposition to the above theory, Professor Huang Wanli (1981) proposed a theory of maximum rate of energy dissipation. He considers that when any system consisting of solid, liquid, gas and other critical continuous media undergoes change, all their particles produce a maximum rate of energy transformation in the whole in order to form a stress field, strain field or velocity field and pressure field. From this we can conclude the second law of liquid dynamics. That is, when a liquid, or liquid with solid particles in the system changes under given initial and boundary conditions, the distributions of density, velocity and pressure at any time always tends to satisfy a maximum rate of energy dissipation in the whole system.

The theory of dissipative structure of energy dissipation

The theory of minimum rate of energy dissipation has been widely used in practice, and has proven practical application. However, since the theory is inferred under the conditions of fixed boundary and with silt, its application is limited. In addition, the field evidence in support of the theory lacks rigorous logic, and there are natural phenomena which deviate from the theory. Further study thus shows that the theory is not always valid.

The theory of maximum rate of energy dissipation may be demonstrated by many natural phenomena. Professor Huang derived the theory by calculus of variations, but there is no way to verify the theory by experiment.

After analysis and study of the principles of extreme values, we present the theory of dissipative structure for energy dissipation in a river geomorphologic system.

Order comes out of non-equilibrium. This is the basic point of view for the developments of non-equilibrium thermodynamics initiated by the Brussels School. When the system is in an unstable state far from non-equilibrium, not only may a fluctuation making the system deviate from the present state decrease, but also can be amplified to form new and more ordered state: the dissipative structure. Therefore, we can say that the evolution of a river system is the evolution of its dissipative structure. In fact, if we want to study river evolution we can study its structures, and then use them as a basis for classifying the rivers. Obviously, the dissipative structure is different from an equilibrium, or "dead" structure.

According to thermodynamic theory, a process responds to one or more forms. In the fluvial system, the relationship between forces and flows is generally non-linear. From the branch theory of modern mathematics we know that the equations of states of the fluvial system generally have multiple solutions. Some of them may be stable, and others unstable. A real fluvial system evolves through a sequence of fluctuations. The real processes may or may not satisfy a maximum rate of energy dissipation, determined by the initial and boundary conditions and the values of the fluctuations. If the relationship between the forces and flows is linear, the system must be in a near-equilibrium state and the processes in the system satisfy a minimum rate of energy dissipation. Hitherto, we conclude that the principles of maximum and minimum rates of energy dissipation are valid in certain situations respectively. The general theory, encompassing both, should be the theory of dissipative structure for energy dissipation which the authors present in this paper, and which is derived below.

The energy for transporting sediment in unit stream width is equal to its potential. Thus, the rate of potential dissipation per unit stream is a function of silt. From general physics, we have:

$$y' = \frac{dy}{dt} = \frac{dx}{dt} = -x\frac{dy}{dx} = vs \qquad (6)$$

where: y = potential loss of unit stream within a given river reach

t = time

x = length of the river reach under study

v = average velocity and

s = water energy slope

From the above simple formula, we infer the change of y': for a given river reach. It is clearly greater for a flood period than for a low water period; and greater for erosion than for deposition. For a whole river, y' is greater up-stream than down-stream, greater for straight channels than for meander bends, greater for rapidly flowing tributaries than for slow flowing tributaries. For example, we may consider the effect of flood stage on the water energy slope. From the curvilinear relationship between water and its fluxes it can be seen that for both V = f(H) and S = f(H), increasing water level gives rise to increases in V and S, so that $f'(H) > 0$. As water level goes down:

$$V = f(H) \text{ and } S = f(H) \text{ decreases, so that } f'(H) < 0.$$

Therefore, at the peak we have:

$$V_p = V_{max},\ S_p = S_{max}$$

At the minimum flow, we similarly have:

$$V_v = V_{min}, \; S_v = S_{min}$$

Thus, from (6) we get:

$$y'_p = Y_{min}, \; Y'_v = Y_{min}$$

where V_p, S_p and Y'_p express flow velocity, slope and the rate of energy dissipation at the peak respectively; and V_v, S_v and Y'_v express those for the minimum flow respectively.

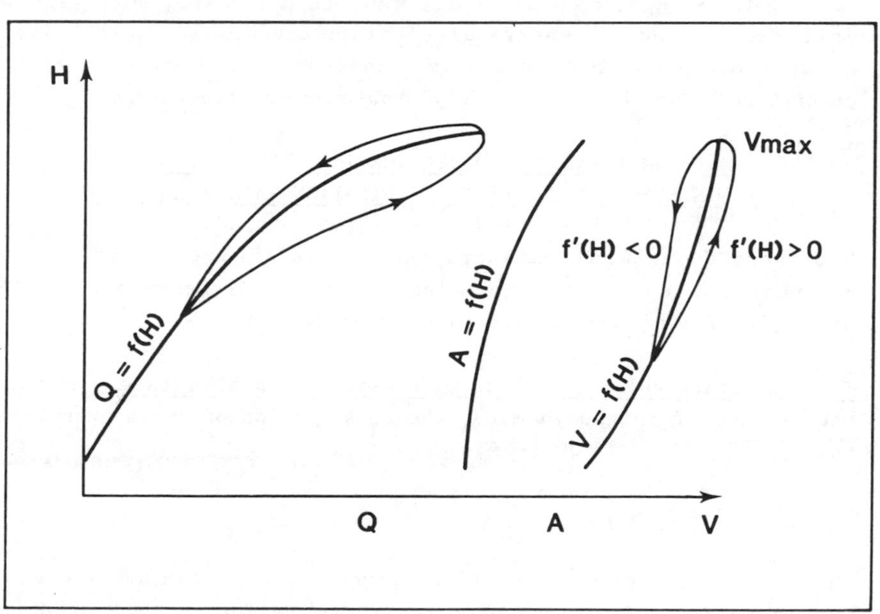

Fig. 2. The curve relating water level and sediment flux, H

The above analysis shows that there exist objective processes $Y_{min} \rightarrow Y_{max} \rightarrow Y_{min}$ describing the rate of energy dissipation. This is in accordance with the above discussion. We consider it as the principle of maximum-minimum rates of energy dissipation.

Now, we shall give the criterion for the evolution of the fluvial system:

(a) If $d_iS < - d_eS$, then the total entropy change $dS=d_eS+d_iS>0$.

Under this condition, the negative entropy flow from the environment is enough to maintain the irreversible processes of river systems. The river geomorphologic pattern therefore develops towards a more ordered structure.

(b) If $d_iS > -d_cS$, then $ds = d_i S + d_e S > 0$. In this case the environment cannot supply negative entropy flow to the river system. The rate of energy dissipation decreases and the fluvial patterns degrade. In consequence the order structure tends to vanish.

(c) If $d_iS = -d_eS$, $dS = d_iS + d_eS = 0$. In this case, both stream valley ecosystem and energy dissipation are stable. The river geomorphologic processes are in a constant state and have great stability. They have an ideally stable order structure.

A VARIATIONAL DERIVATION FOR THE PRINCIPLE OF MINIMUM-MAXIMUM ENERGY DISSIPATION RATE

Professor Huang Wanli has proposed a variational proof on the maximum rate law of energy dissipation in (4). Here we derive the principle of minimum-maximum energy dissipation rate.

The stationary value principle of energy dissipation rate. For an open system of thermodynamics, the conservation of energy requires that

$$\dot{E} = \dot{E}_c + \dot{E}_d \qquad (7)$$

where $\dot{E}$ - Total rate of input of energy to the system at time t
$\dot{E}_c$ - Rate of storage of internal energy for the system at t
$\dot{E}_d$ - Rate of energy dissipation for the system at t

and $\dot{E}$ is non-negative. The irreversible processes of the system require that:

$$\dot{E}_d \geqslant 0 \qquad (8)$$

Now consider the variation of the system from a real state to other conceivable states, assuming t is constant in this process. From

(7) we have

$$\delta \dot{E} = \delta \dot{E}_c + \delta \dot{E}_d \qquad (9)$$

Assuming the input energy rate E keeps constant for the systems considered, then we have:

$$\delta \dot{E} = \delta \dot{E}_c + \delta \dot{E}_d = 0 \qquad (10)$$

In (9), $\delta \dot{E}$ corresponds to a certain increment of input energy rate added to the system, and would lead to increments in both $\delta \dot{E}_c$ and $\delta \dot{E}_d$. In other words, $\delta \dot{E}_c$ and $\delta \dot{E}_d$ should have same sign. Therefore, from (10) we obtain

$$\delta \dot{E}_c = 0 \qquad (11)$$

$$\delta \dot{E}_d = 0 \qquad (12)$$

(12) is the stationary value condition for energy dissipation rate. It can be stated as follows: of all the possible systems with same input energy rate, the one for which the energy dissipation rate assumes a stationary value is the real system. (11) shows that the energy storage rate assumes a stationary value too.

<u>The principle of minimum maximum energy dissipation rate</u>. Let us study the stationary value conditions (11) and (12). As mentioned above, $\dot{E} \geqslant 0$ and $\dot{E}_d \geqslant 0$, and (10) is true for any two neighbouring coneivable states of the system, ie $\delta \dot{E}_c = -\delta \dot{E}_d$. But $\delta \dot{E}_c = 0$, $\delta \dot{E}_d = 0$.

If we may assume that $\dot{E}_c > 0$ for any system considered, we can show that $\dot{E}_d = \max$ for the real system.

In Fig 3a, the point A represents the real state and ECABD represent conceivable variations in $\dot{E}_d$ for fixed $\dot{E} > 0$. $\dot{E}_d = \max$ means that the arbitrary directions of variation shown by ECABD must be convex down. Otherwise, if the direction of variation were concave, then, owing to the arbitrariness of variation, we could always find a way, say, E'C'AB'D', for which $\dot{E}_c < 0$.

But this is in contradiction with the assumption above. Therefore, $\dot{E}_d = \max$ must be true.

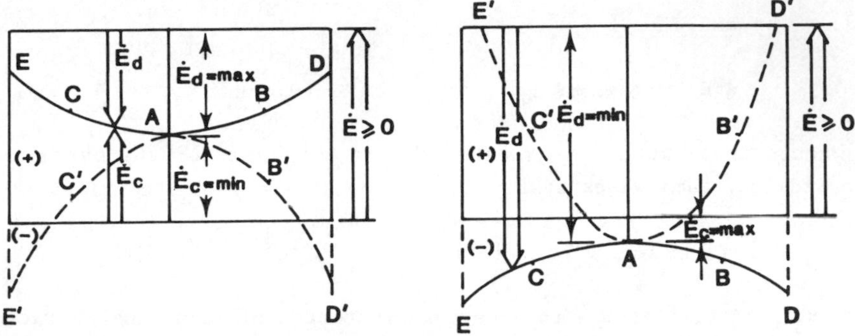

Fig. 3. The principle of minimum-maximum energy dissipation rate.

Similarly, if we assume $\dot{E}_c < 0$, for all arbitrary systems considered, we can show that $\dot{E}_d$ = min for the real system. As shown in Fig 3b, the point A represents the real state. $\dot{E}_d$ = min means that the arbitrary direction of variation ECABD must be concave down. Otherwise, if the direction were convex, then, owing to the arbitrariness of variation we could always find a way, say, the assumption mentioned, so $\dot{E}_d$ = min must be true.

These results can be stated as follows:

In all possible systems with the same input energy rate $\dot{E}$ ($\geqslant 0$), the energy dissipation rate for the real system $\dot{E}_d$ is a maximum, considered among all possible systems for which the energy storage rate $\dot{E}_c \geqslant 0$. The energy dissipation rate $\dot{E}_d$, for all possible real system with $\dot{E}_c \leqslant 0$ is similarly a minimum. This principle is called the principle of minimum-maximum energy dissipation rate.

We can also see that, $\dot{E}_d$ must be constant for all the possible systems with zero energy storage rate $\dot{E}_c$. These systems show a constant rate of energy dissipation. It is also clear that, $\dot{E}_d$ = max corresponds to $\dot{E}_c$ = min, and $\dot{E}_d$ = min to $\dot{E}_c$ = max. This provides a complementary principle of maximum-minimum rates for the storage of internal energy.

By considering the different cases for energy store rate $\dot{E}_c$, we have formulated a principle which combines the principle of minimum with the principle of maximum energy dissipation rate. In practice, this fact shows that the total entropy of the system ds = $d_eS + D_iS$ is the appropriate measure combining the two principles.

SUMMARY

(i) River systems meet the basic conditions to form dissipative structures.

a. The river system is an open system with the valley as its environment, and exchange mass, energy and information with them.

b. River systems are far from equilibrium states.

c. In the interior of the river system there are nonlinear relations between its elements, such as water flow, silt and riverbed etc. They may produce synergetic effects.

d. In the system there are stochastic fluctuations in water levels.

(ii) The maximum and minimum rates of energy dissipation are both valid in special situations respectively. The most general theory of evolution of river system should be the dissipative structure theory proposed by the authors.

(iii) The complexity of the environment, especially change in water levels, leads to stochastic changes of local fluvial patterns in space and time, and determines the systematics of the river system and the average geomorphologic patterns over a period.

(iv) The ecosystem equilibrium in river valley directly effects the change of water levels, and is expressed in the stability of dissipative structures for the river geomorphology.

REFERENCES

Ai, N.,and Gu, H., 1984. Pansystems Analysis of the Action of Endogenetic and Exogemetic Forces in Formation of River Network. Journal of Chongqing University, 4.

Carson, M. A., and Kirkby, M. J., 1972. Hillslope Form and Process. Cambridge University Press.

Gu. H., 1984. The Sediment Source and River Geomorphologic Processes of the Jialin River. Journal of Chongqing Jiaotong Institute, Vol 3, 14.

Huang, W., 1981. Maximum rate law of energy dissipation of continuous medium dynamics. Journal of Quinghua University, Vol 21, 1.

Nicolis, G.,and Prigogine, I., 1977. Self-Organization in Non-equilibrium System

Prigogine, I., 1978. Time Structure and Fluctuation. Science, 201, 4538

Prigogine, I., 1979. From being to becoming

Scheidegger, A. E., 1961. Theoretical Geomorphology. Prentice Hall; Springer Verlag, 333p.

Yang, C. T., 1981. Theory of Minimum Rate of Energy Dissipation and Its Application, "A Compilation of the Lectures on River Sedimentation" (Edited by Sedimentation Committee ChineseSociety of Hydraulic Engineering, June, 1981, Beijing, pp 159-194.

Yang, C. T., and Song, C. C., 1979. Theory of Minimum Rate of Energy Dissipation. Proc ASCE J. Hydr. Div., Hy 7, pp 769-784.

Zan, T., 1985. Entropy model of thermal stability of draft frozen wall. Lanzhou Institute of Glaciology and Geocryology, Academia Sinica.

REGIONAL LANDFORM STUDIES, PERIGLACIAL GEOMORPHOLOGY AND REMOTE SENSING

International Geomorphology 1986 Part II
Edited by V. Gardiner

REGIONAL LANDFORM STUDIES, PERIGLACIAL GEOMORPHOLOGY AND REMOTE SENSING : AN INTRODUCTION.

V. Gardiner

Department of Geography, University of Leicester, Leicester, LE1 7RH, U.K.

Although to some extent a pot-pourri, the thirteen papers presented in this section of the proceedings provide a microcosm of Geomorphology as a whole, with scales of investigation ranging from the mega-geomorphic and satellite-sensed to that of the individual slope profile, approaches ranging from the descriptive to the analytical, and attention being directed towards form and process, academic understanding and application. The papers fall into two categories, with those by Liu Shuzheng and Zhong Xianghao, Du Zi-rong, Liu Tung-sheng and Yuan Baoyin, Rai, and Singh and Pandey in the first one. All abstracts submitted to the organisers of the First International Conference on Geomorphology were allocated to sessions according to the limited evidence of their abstracts. However when full manuscripts became available it became clear that some papers really addressed several aspects of geomorphology in particular regions. The systematic classification adopted for the Conference structure was inevitably a rather arbitrary one for these papers, and it was felt appropriate to gather them together here.

The first three papers give accounts of landforms in specific parts of China, thus providing useful introductions to these areas for those not familiar with the geomorphology of China. The other two papers in this group illustrate the breadth of geomorphology in India. Rai synthesizes the evidence for rejuvenation of the Deccan Foreland, and illustrates the difficulties of correlating tectonic events on a sub-continental scale, particularly when much of the evidence is essentially subjective in nature. At a very different scale Singh and Pandey consider the evolution of slopes by examining detailed field evidence derived from measured slope profiles. The fuller integration of such diverse scales of study clearly represents a significant future goal for geomorphologists interpreting landscapes in these areas.

A second group of papers consists of papers presented at sessions for which only a very small number of manuscripts were submitted for publication, and which do not therefore warrant a separate section in the Proceedings. Zhu Jinghu describes periglacial landforms and deposits in part of China, and Allard et al. consider the classification and origin of some types of periglacial feature in northern

Quebec. The volcanoes of Yunnan Province, China, are described and evaluated as a natural resource by Mu Guichun and Dai Hezhi, and Zhivago describes the results of a survey of the bathymetry of part of the Gulf of Aden. This illustrates how geological, geophysical and geomorphological evidence must all be used in the interpretation of submarine landforms, and is an illustration of how fortunate are most geomorphologists, in that they can examine their subject matter directly! More conventional remote sensing in geomorphology is considered by the next two papers. Millington and Townshend examine the potential of satellite remote sensing for geomorphological investigations in general, and Jones evaluates the potential of one specific source of imagery for geomorphological mapping. The paper by Mohammad demonstrates how lineations visible on aerial photographs can be related to jointing, and hence slope evolution. Finally Cervantes-Borja and Meza-Sanchez give an effective demonstration of how studies of coastal landform evolution and contemporary dynamics can provide a basis for natural resource assessment and management.

International Geomorphology 1986 Part II
Edited by V. Gardiner

THE GEOMORPHOLOGY OF THE HENGDUAN MOUNTAINS, CHINA

Liu Shuzheng and Zhong Xianghao

Chengdu Institute of Geography
Academia Sinica, Chengdu, Sichuan, Chinas

ABSTRACT

The Hengduan Mountain Region is, structurally and geomorphologically, very complex. The most characteristic feature of the region is the parallel system of mountain ranges and river valleys which results from a series of north-south trending fault zones. During the Tertiary, an extensive erosion surface was developed. Recent neotectonic activity has resulted in differential uplift of fault bounded blocks. One result of this is that fragments of the Tertiary surface now occur at a variety of altitudes. The area was glaciated during the Pleistocene with three major ice advances. The post glacial climatic pattern is complex as a result of the rapid variations in altitude linked to the variations in longitude.

INTRODUCTION

A series of north-south oriented mountain ranges in Western Sichuan and Yunnan Province of south-west China, and in the adjacent area of eastern Tibet, have been given the collective name of the Hengduan Mountains. Literally translated, this means the "transversely cutting mountains" and refers to the fact that the mountain ranges cut east-west routeways. The exact boundaries of the Hengduan Mountains have not yet been defined but most earth scientists in China regard them as bordered on the east by the Qionglai Mountains, in the west by Baishula Mountains, by the Changdu and Ganzi mountains in the north and to the south by the Chinese-Burmese border. This corresponds to an area between 21° and 31°N and 98°-103°E (Fig. 1).

The area has a complex geological structure, a wide variety of geomorphological forms, a varied flora and fauna and a dramatic landscape. Geologically the Hengduan Mountains are located on the eastern wing of the linking zone between the South Asian and the Eurasian plates and between the Pacific and Palaeo-Mediterranean plates. The area was tectonically very active during the Quaternary; it was also glaciated during the Pleistocene with three main glaciations. The post glacial climatic pattern is complex, varying from tropical to peri-glacial, due to the interaction of latitude and altitude. The recent tectonic activity, the Pleistocene glaciation and the wide range of post-glacial climates have produced a unique combination of geomorphological landscapes, the basic features of which are outlined in this paper.

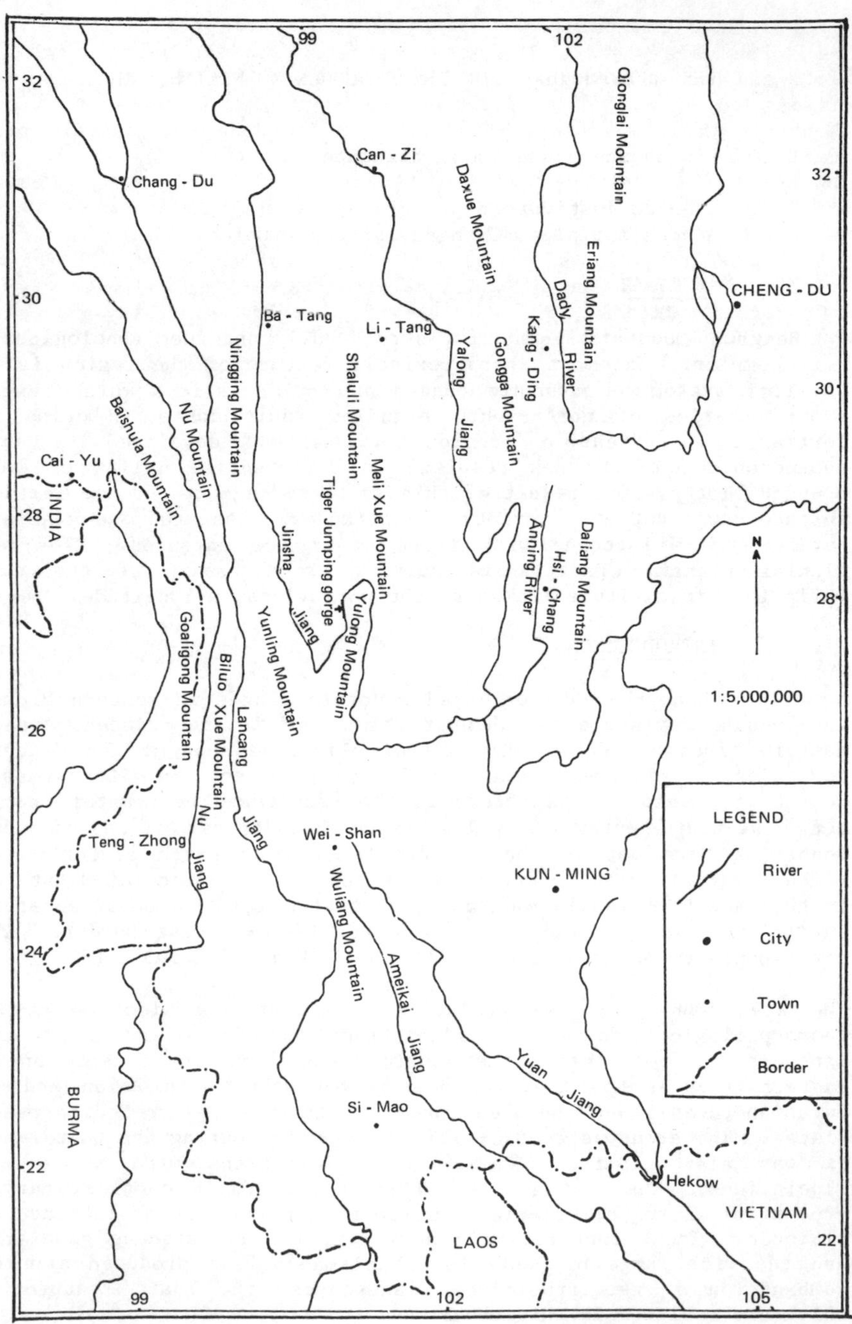

Fig. 1. Map of the Hengduan Mountains.

THE STEP-LIKE NATURE OF THE MOUNTAINS

The Hengduan Mountains are highest in the north and fall southwards in three main steps. The most northerly zone is located north of 28°N and has an average elevation of about 4500 m with the main peaks rising to over 5000 m. Among the more famous peaks are Gongga Mountain (7556 m), the highest peak in the Hengduan mountain block, Meli Xue Mountain (6740 m), Sigumang Mountain (6250 m), Queer Mountain (6740 m), Gongga Xue Mountain (6032 mm), Genie Mountain (6204 m), Baishula Mountain (6005 m), Xuebaoding Mountain (5588 m) and Bamang Xue Mountain (5137 m). All have permanent snowfields above 5000 m and a variety of valley glaciers. For example, an area of 360 km^2 around Gongga Mountain is covered by ice and snow with 159 glaciers, including valley glaciers, hanging glaciers and cirque glaciers. Data collected between 1930 and 1980 show that the glaciers are retreating but that the rate of retreat has declined during this century. Between 1930 and 1966 the average rate of retreat was c. 100 m per year and between 1966 and 1980, 40-50 m per year.

The second, central block of the Hengduan Mountains lies between 28° N and 25°30' N. It has a mean altitude of about 3500 m with peaks generally rising to about 4000 m, but with a few reaching over 5000 m, for example Yulong Xue Mountain (5596 m) and Haba Xue Mountain (5396 m). This mountain block contains many intermontane basins; these are mainly structural in origin. Some of the basins are large, for example the Xichang basin in the Anning River valley of Sichuan Province is 150 km long with an average width of 4 - 4.5 km and a maximum width of 12.3 km. This developed along the Anning River fault zone and has a fault lake with a surface area of 31 km^2 and a depth of 30 m.

The third, most southerly block of the mountains is located south of 25°30'N. It is mostly below 2500 m in altitude but a few peaks reach above 3000 m, for example Gaoliging Mountain (3374 m), Wuliang Mountain (3300 m) and Ailao Mountain (3138 m). The lowest point of the Hengduan Mountains, 76.4 m, is located at the confluence of the Yuanziang River and its tributary the NanXi river in Henkou County, Yunnan Province.

THE PARALLEL ARRANGEMENT OF THE MOUNTAIN RIDGES AND VALLEYS

One of the most striking, and important features of the Hengduan Mountains is the alternation of parallel moutain ranges and rivers. This pattern has resulted from a series of large, parallel, north-south oriented fault zones. The main north-south ranges, and rivers are, from west to east,

the Baishula - Gaolijong Mountains
the Nuziang River
the Nushan - Biluo Xue Mountains
the Lancangjiang River
the Ningjing - Yunling Mountains
the Jinshajiang River

the Shululi - Jiuguai - Yulong Mountains
the Yalongjiang River
the Maummangqi-da Xue Mountains
the Dadu River
The Qionglai - Daliang Mountain

The mountain ranges, and intervening valleys, are narrowest in an area in the west, demarcated by Cai-Yu and Li-Tang in the north and Teng-Zhong and Wei-Shan in the south (Fig. 1). To the north eastwards and south eastwards of this area the mountain ranges and valleys increase in width. Near 27°30'N four mountain ranges, and three valleys, occur in an east-west distance of only 65 km. The valleys here are very steep sided, with a classic V-shaped cross section, and there are a number of dramatic gorges. The "tiger jumping gorge" on the Jinshajiang River is the most famous: it is 16 km long and 60-80 km wide, and has a fall of 220 m".

THE VERTICAL ZONATION OF GEOMORPHOLOGICAL FEATURES AND PROCESSES

The altitudinal range, between the ridges and the valley floors, varies from 1000 to 3000 m (Table 1).

Table 1. Percentage of land in the Hengduan Mountains in relative relief zones.

Relative relief (m)	Frequency of occurrence	percent of total area	Accumulated per cent
0 - 499	34	3.3	3.3
500 - 999	155	15.2	18.5
1000 - 1499	196	19.3	37.8
1500 - 1999	222	21.8	59.6
2000 - 2499	182	17.9	77.5
2500 - 2999	139	13.7	91.2
3000 - 3499	61	6.0	97.2
3500 - 3999	20	2.0	99.2
4000 - 4500	6	0.6	99.8
>4500	3	0.2	100.0

Over some 73% of the area the altitudinal range, between valley floor and mountain summits, is between 1000 and 3000 m, and for 40% of the area is 2000 m (Fig. 2). A small part of the area has an altitudinal range of 4000-6000 m, for example the eastern slopes of Gongga mountain fall 6400 m in 29 km, and the eastern slopes of the Meli Xue Mountain fall 4760 m in 12 km.

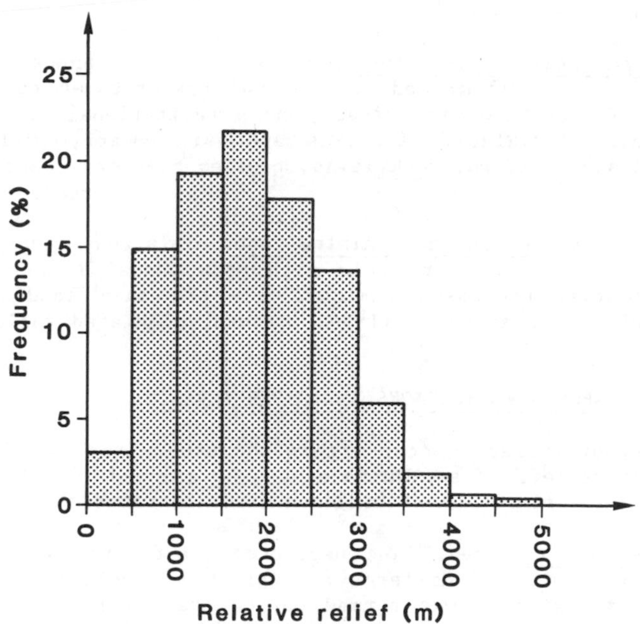

Fig. 2. Percentage of the Hengduan Mountains in relative relief classes.

These large variations in altitude produce a pronounced vertical, climatic zonation. Five main climatic zones can be recognised in the central (and northern) Hengduan Mountains:

subtropical zone	2000 to 2400 m
warm temperate zone	2000 (or 2400 m) to 2800 (or 3000 m)
cool temperate zone	2800 (or 3000 m) to 3700 (or 4200 m)
subfrigid zone	3700 (or 4200 m) to 4500 (or 5000 m)
frigid zone	5000 m

The dominant geomorphological agents and processes vary with climatic zone and have produced a variety of landforms. Four climatic-geomorphological zones are recognised:

The dry hot valley zone. This is characterised by intense physical weathering and by gravitational movements. Average annual temperature is generally greater than 15°C and mean annual rainfall is less than 800 mm. The main landforms are debris cones, alluvial cones, alluvial fans, landslips and, a variety of mudflow deposits.

The warm-cool, moist semi-mountain zone. This zone is found on the mid-slopes of the mountains. The mean annual rainfall is 1000-2000 mm and the mean annual temperature 15° - 0°C, depending on altitude. The dominant geomorphological agent is running water and there is intense erosion with the development of many deep gullies, for example on the eastern slopes of Gongga Mountain the density of gullies reaches 4.3 km/km^2.

The frigid moist sub-alpine zone. The average annual rainfall is generally 1000-1500 mm and mean annual temperatures range from 0 - 3°C. Freeze-thaw processes and gravitational movements are important. Periglacial features are very common, including stone circles, stone rivers, rockfields, boulder glaciers, debris flows and screes.

The permanent ice and snow, alpine zone. This zone occurs above 5000 m. The climate is very cold with mean annual temperatures below 0°C. Glacial processes dominate and the main landforms include pyramidal peaks, arêtes, cliffs, cirques, glaciated valleys, glacial hollows.

EROSIONAL SURFACES AND RIVER TERRACES

A variety of surfaces occur within the study area, which produces a step-like relief. These surfaces can be grouped into two main types - erosional surfaces (uplifted) and river terraces.

The erosional surfaces. The neotectonic activity was characterised by intermittent, and differential uplift of fault blocks. A broad denudational surface was formed during the Tertiary. This surface has been dissected and affected by differential uplift during the Quaternary. Remnants of the surface occur on mountain peaks and ridges, as plateau-like hills, and as benches on mountain sides. This can give a step-like appearance to the relief. Table 2 and Fig. 3 give the altitudes of the main surfaces.

Table 2. Altitudes of the main raised surfaces in the Hengduan Mountains.

Survey area	Latitude (°N)	Elevation altitude (m)
The Northern Shaluli mountain	31 - 30.5	4700 - 4600
The Ningging Mountain, Li-Tang	30.5 - 29.5	4500 - 4400
Aroung Qiang-Ning	31 - 30	4300 - 4200
Xiaozhong-Dian, Sanbi-Hai	30 - 28.5	4100 - 4000
Zhong-Dian, the Daliang Mountain	28.5 - 27.5	3800 - 3600
The Southern Yunling Mountain	28 - 27	3400 - 3200
Jian-Chuan, Li-Jiang	26.5 - 25	3100 - 3000
Dali county	26 - 25	2900 - 2800
Both sides of Cheng-Hai	26 - 27	2800 - 2400
Hui-Li, Hui-Dong	26 - 27	2200 - 2000
The Wuliang Mountain	24.5 - 23	2400 - 2200
Si-Mao, Yang-Jiang	22 - 23	1800 - 1700

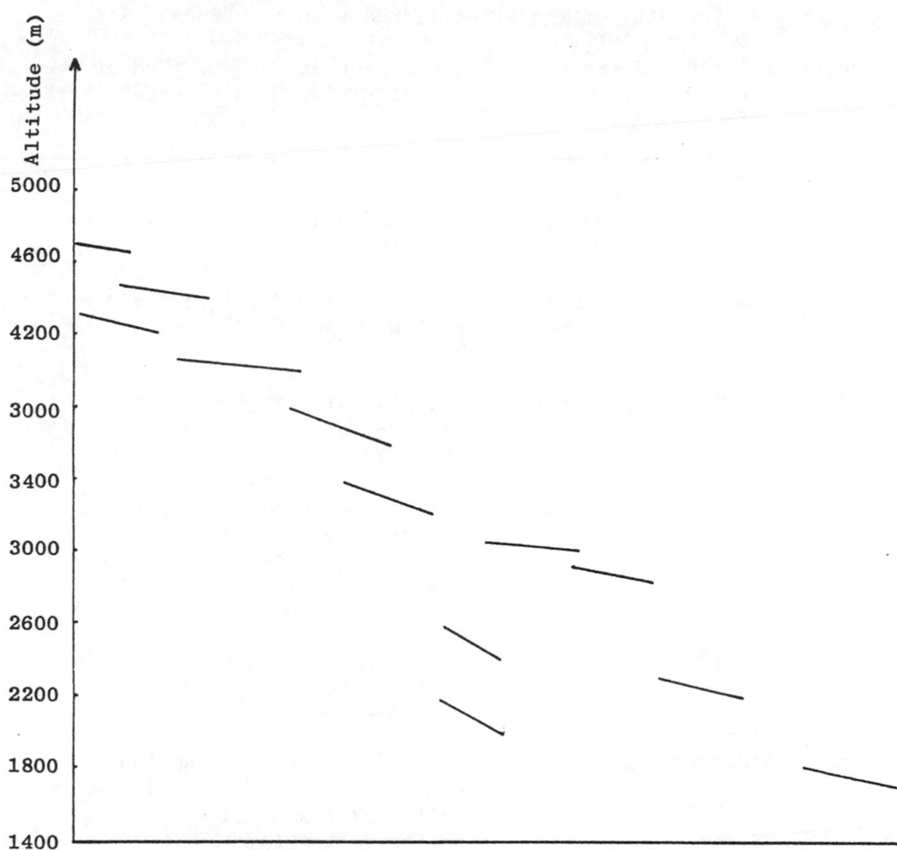

Fig. 3. Variation of the altitude of the main raised surfaces with latitude.

In the north of the Hengduan Mountains the surface forms a plateau with many small, gentle hills which rise 100-200 m above the valley floors. In the central and southern areas of the mountains the remnants of the surface are scattered and smaller. In some areas the surface can be identified by concordant altitudes of mountain peaks and ridges.

River terraces. Terraces are found in many of the river valleys but are best developed in the larger ones. Some valleys have a number of terraces at different altitudes, for example the Lancang River valley, near Changdu, (Table 3) has eight terraces and the Jingsha Jhang valley, near Dengke, six (Table 4).

These terraces result from adjustment following neotectonic movement during the Quaternary.

Table 3. Details of the River Terraces on the Lancang River.

Terrace	Relative altitude (m)	Terrace type	Constitutent materials	Width of the terrace surface
T1	5-6	depositional	Sand and gravel deposits	500-700 m
T2	8-10	depositional	Gravels below T2, yellow sub-clay upper T2	120-150 m
T3	20-25	depositional	Gravels below T2, silt sub-clay upper T2	250-300 m
T4	40-45	depositional	Material components are same as T1 & T2 Thickness of the gravel layer is more than 30 m	400-500 m and 700-1000 m long
T5	55-60	Rock	Gravel and debris of the upper rock, 1 m thick	
T6	110-120	depositional	Gravel with 1 m on the rock, and yellow silt sub-clay on the gravel layer	80-150 m

Table 4. Details of River Terraces on the Jingsha Jiang.

Terrace	T1	T2	T3	T4	T5	T6	T7	T8
Relative altitude (m)	9	25	55	72	105	130	150	180
Terrace type	depositional	rock	rock	rock	rock	rock	rock	rock

Radio-Carbon dating shows that the second and fourth terraces in the Anding River valley. a tributary of the Jingsha, were formed 25930-15995 and 42000 years ago respectively. The second terrace is generally between 1 and 3 km wide.

INTERMONTANE BASINS

The Hengduan Mountains contain a variety of intermontane basins; these are structural basins, karst basins and erosional basins. Structural basins are the most widespread and result from differential uplift along the fault zones. They are mainly distributed between 24°N and 30°N. Western Yunna contains some 450 basins larger than 1 km^2 and with a total area of 5000 km^2. Among the most famous are the Da-Li, Yong-Sheng, Dao-Cheng, Yan-Yuan and Hsi-Chang basins. These are generally 2-10 km wide and 10-30 km long and contain deposits thicker than 1000 m. The structural basins tend to occur at higher altitudes in the north and lower ones towards the south.

PLEISTOCENE GLACIAL FEATURES

The Hengduan Mountains contain a variety of glacial landforms dating from the Pleistocene glaciation. These include cirques, glaciated valleys, till plains, end moraines and glacial lakes. The cirques in western Sichuan can be divided into three groups according to their altitude; the first group lies between 4600 and 4700 m, the second between 4200 and 4300 m and the third between 4000 and 4200 m. The area of Gongga Mountain contains 247 cirques of Pleistocene age, in four altitudinal groups (Table 5).

Glaciated valleys occur widely between 1000 and 5000 m above sea level. The ratio of valley depth to width in these U-shaped valleys varies from 1:2.5 to 1:3.0 in the area of Gongga Mountain. Some of the glaciated valleys in this latter region are between 15 and 30 km long and 1.0 to 1.5 km wide. The Pleistocene tills in the Hengduan Mountains contain much more gravel and debris than fine material. The end moraines are well preserved and are widely distributed in valleys above 3000 m.

Table 5. Altitude of the Palaeocirques in the Gongga Mountain, the Hengduan Mountains (numbers in each class are bracketed).

Slope direction	First group	Second group	Third group	Fourth group
N	4830 (6)	4680 (46)	4450 (42)	4200 (7)
W	4920 (6)	4700 (43)	4540 (40)	-
S	4880 (6)	4640 (22)	4470 (36)	-
E	4830 (6)	4610 (2)	4410 (7)	4150 (8)

CONCLUSION

The interactions of geological structure and tectonic factors with the effects of the Pleistocene glaciation and present day variations in climate have produced in this area a great variety of landforms, which requires further detailed study.

International Geomorphology 1986 Part II
Edited by V. Gardiner

ON THE LANDFORMS OF THE BASIN OF SICHUAN, CHINA

Du Zi-rong

Southwest China Teacher's University,
Beibei, Sichuan, PRC

ABSTRACT

The Sichuan Basin is a syncline. Nine major landform types may be recognised within it, and at the coarsest scale, three major geomorphological regions. Geological structure is the most important factor in the development of landforms of the basin.

INTRODUCTION

This paper is a summary of fieldwork and analysis of landforms in Eastern Sichuan Province, People's Republic of China, which covers the whole Sichuan Basin and its surrounding mountains, an area of 227,073 square kilometres. Maps have been prepared of the entire region, depicting various geomorphological landform types and divisions. Three aspects are considered: structural characteristics, landform types and geomorphological regions.

STRUCTURE

Jurassic-Cretaceous red beds cover nearly the entire centre of the basin but Lower Palaeozoic and Pre-Cambrian rocks are found in the surrounding mountains. The topographic basin is therefore a structural basin or syncline in geological terms.

LANDFORM TYPES

The basin may be divided into nine main geomorphological types according to the combination of landforms present and their origins; each geomorphological type may then be subdivided into three groups according to the main geomorphological factors of geological structure, major or unique exogenetic agents, and rock type. In considering these factors one may be found to be dominant or all three may be of equal significance. Each group may be subdivided into several different kinds of landform, so each one, and its special characteristics, should be reflected on the map of geomorphological types. Thus the morphology and genesis are both considered in forming geomorphological types.

GEOMORPHOLOGICAL REGIONS

By detailed study of the landforms they have been regionalised at three levels, with three regions at the first, five at the second level and nineteen at the third. The Ghengdu Plain is a compound diluvial and alluvial fan plain to the west of the Longquan Shan fault, and has been subsiding since the Tertiary. To the east of the Hua-ying Shan basement fault there are some narrow but dominant elongated anticlines and synclines, each between ten and twenty kilometres in width. This area of parallel ranges and valleys is a unique geomorphological landscape in China. The well-known grand "Three gorges" on the River Yangtze are in this area. Between the two faults mentioned above there is a vast hilly area generally called the "Central Hills area" of the basin, with several gentle domes or short anticlines. It has been proven that the central part of the basin has a hard and stable basement of metamorphic rocks. Therefore the overlying strata cannot be folded to the same extent as in the eastern area, where the parallel ranges and valleys have developed so perfectly as a particular feature of the basin.

CONCLUSIONS

The formation and development of the Sichuan Basin has been controlled to a great degree by the geological structure; the external agents have played only a subsidiary role in modifying landforms.

International Geomorphology 1986 Part II
Edited by V. Gardiner

MEGAGEOMORPHIC FEATUES AND HISTORY OF THE CHINESE LOESS

Liu Tungsheng and Yuan Baoyin

Institute of Geology, Academia Sinica,
Beijing, China

ABSTRACT

Loess covers an area of approximately 440,000 km^2 in China, between latitudes 33° and 57°N and longitudes 75° and 125° E (Liu Tungsheng et al., 1985). The Huanghe, the Yellow River, the second largest river in China flows through the Loess Plateau, which is dissected by the river and its tributaries into a spectacular landscape, whether observed from space or earth. Soil erosion on the Loess Plateau supplies 16 x 10^9 tonnes of sediment to the Yellow River annually. The bed of the lower Yellow River aggrades about 10 cm per year creating a potential hazard for the inhabitants of the North China Plain. Geomorphological investigations are one of the most important means of understanding the devlopment of both the Loess and the North China Plain.

In the middle reaches of the Yellow River, loess forms a virtually continuous thick mantle, hence the name the Loess Plateau. We recommend that as a single geomorphological entity, because of its large extent and unique geomorphological history, the Loess Plateau be referred to as a megageomorphological unit.

THE MEGA-GEOMORPHOLOGY OF THE LOESS PLATEAU

Long ago, the inhabitants of the Loess Plateau gave various names to the different loess geomorphic features. The most widely used are "Yuan" - loess flat highlands; "Liang" - loess elongated ridges; "Mao" - loess round topped hillocks (Liu Tungsheng et al., 1962 (Fig. 1). These names have been used to discuss loess geomorphic features by Derbyshire (1983) in his paper to the BGRG Mega-geomorphology conference and in the forthcoming comprehensive NASA work "Regional Landforms from Space".

The Chinese people appreciate the international use of these terms. In China, people often take "Yuan" as the primary unbroken flat highland. After dissection, it breaks into the elongated ridges which could be called "Liang". By further transverse dissection the elongated "Liang" develop into isolated rounded hillocks called the loess "Mao". After careful study of the geomorphic history of these features it can be demonstrated that although they may be genetically related, sometimes each may develop independently of the other.

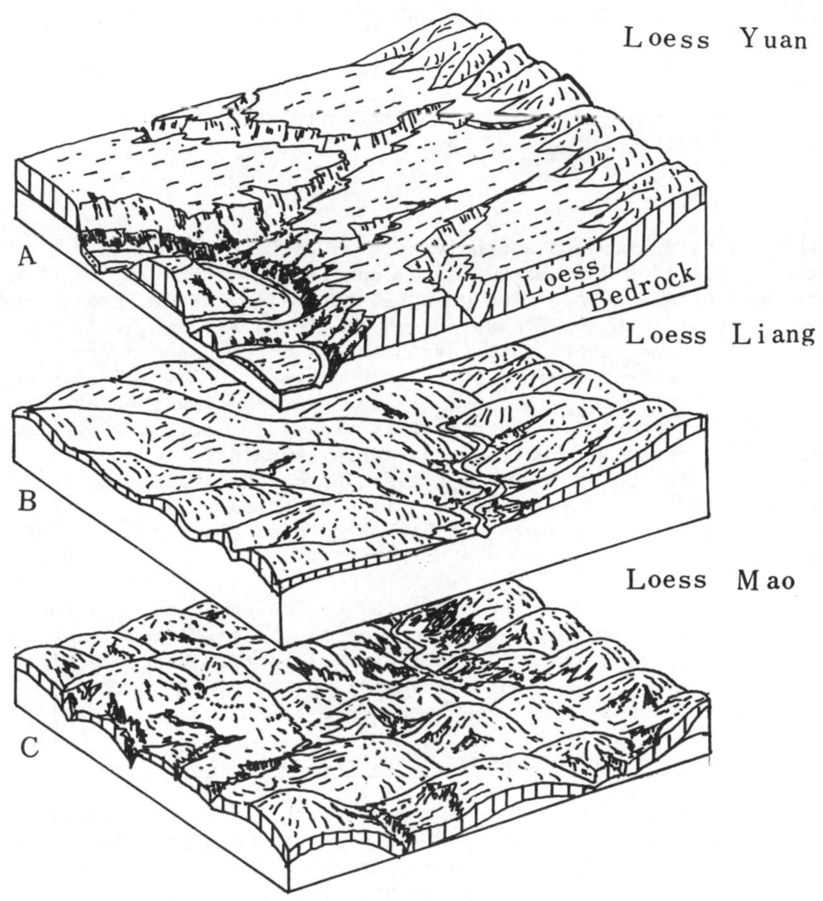

Fig. 1 Different loess geomorphic features in Loess Plateau

"Yuan", "Liang" and "Mao" are best developed in those regions where the loess thickness exceeds 50 metres. Most of them have a geological history older than 125,000 years BP. In China, Loess younger than 125,000 years BP usually forms a veneer less than 30 metres thick over mountain slopes, river terraces, alluvial fans and does not develop these geomorphological features.

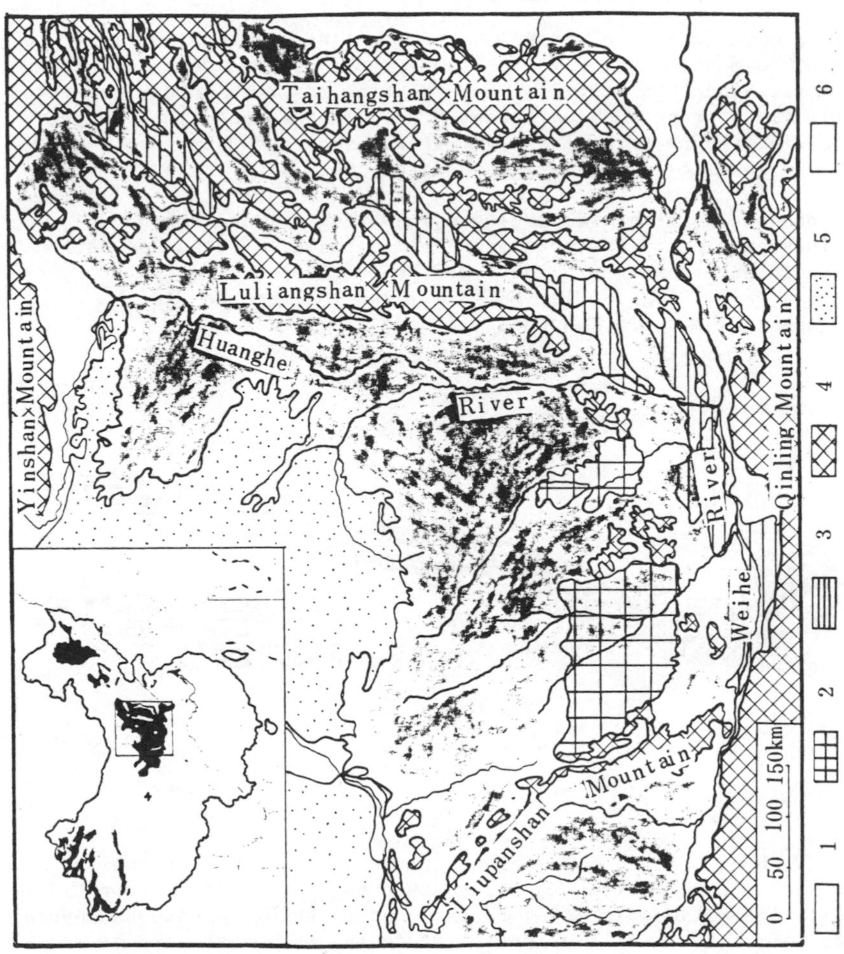

Fig. 2 Distribution of the different loess geomorphic features in Loess Plateau: 1) Loess "Liang" and "Mao" region; 2) Loess "yuan" region; 3) Loess Terrace region; 4) Bedrock; 5) Desert; 6) Alluvial Plain.

"Yuan", "Liang" and "Mao" occur mainly in the Loess Plateau, which is bounded in the west by Liupanshan mountain, in the east by Taihangshan mountain, in the south by the Tsungling mountain range and in the north by the Yinshan mountain, with a large part of deserts (Fig. 2).

STRATIGRAPHY OF THE LOESS

Before describing the morphology of the Loess Plateau, it is useful to mention briefly the stratigraphy of the loess in this region. Since the fifties Loess in China has been subdivided into Early Pleistocene Wucheng Loess, Middle Pleistocene Lishi Loess, Late Pleistocene Malan Loess (Wucheng, Lishi and Malan are names of type locality) (Liu Tungsheng et al. 1962).

Paleomagnetic stratigraphical studies using C_{14} and thermoluminescence dating methods of a well exposed, continuous section in Luochuan, Shanxi Province 220 km north of the famous ancient capital Xi'an (Fig. 3), may be summarized (Heller and Liu Tungsheng, 1984; Liu Tungsheng, 1985):

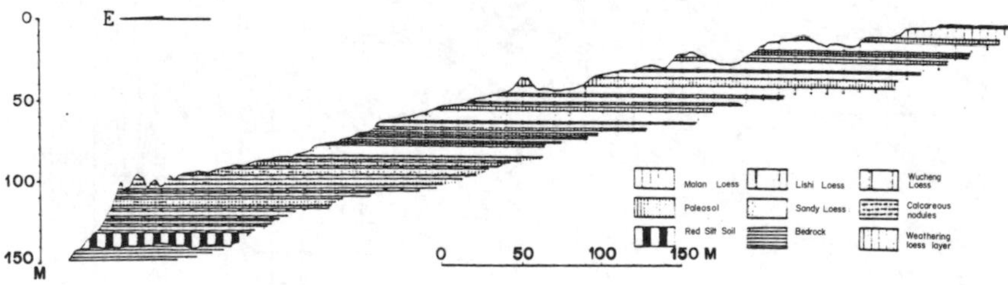

Fig. 3 Loess exposure at Heimugou, Luochuan

1) Wuchung Loess conformably overlying the upper Pliocene Red Silt Soil (more Ancient Loess) reaches 45 m of reddish compact loess with palaeosols often represented by thick calcareous nodule beds. Loess and soils alternate producing at least 9 cycles.

From a palaeomagnetic study, Matuyama/Gauss boundary is situated on top of the Red Silt Soil (Ancient Loess) about 2 m beneath the loess. Thus the loess in this region was deposited at the beginning of the Matuyama chron about 2.4 million years BP.

2) Lishi Loess

This unit conformably overlies the Wucheng Loess with a thickness of 75 m. It contains a series of 14 reddish to yellow loess units with cinnamon soils. The Brunhes/Matuyama Boundary (about 720,000 years BP) is located in the eighth loess layer (counted from the top downward). Among these palaeosols, the fifth palaeosol (counted from top downward) is a complex of 3 soils, the age of which has been calculated at approximately 560,000 years BP.

3) Malan Loess

This unit conformably overlies the Lishi Loess reaching approximately 10 m thick of light, greyish yellow sediment. The basal part of the Malan Loess has a thermoluminescence age of about 100,000 years BP.

4) Holocene Loess and Recent Loess

In the uppermost part of the Malan Loess there are up to 3 dark buried soils. Carbon 14 dating indicates ages from 7,000-9,000 years BP. Above this black soil, there is a layer of loess approximately 1 m thick with fragments of neolithic painted pottery. The charcoals in it yield C_{14} ages of 5,000-6,000 years BP.

From the calibrated age of the index layers, the thickness of the loess and palaeosols, the sedimentary rate of the loess layers is calculated, providing an estimated age for each loess and soil layer.

Based on this chronology of the loess-palaeosol sequence an attempted correlation with the deep sea core climatic curve has been made (Liu Tungsheng and Yuan Baoyin, 1982; Liu Tungsheng et al., 1985). This is now a move (Douglas, 1985) "to a more critical event-based interpretation of the stratigraphy of the deposits which correlate with the development of our erosion landforms".

BURIED PALAEOGEOMORPHOLOGY OF THE LOESS PLATEAU

Within the frame of the mountain ranges mentioned above, the Loess Plateau is situated on the cratonic basement, namely the "Ordos Platform" in bedrock geology. Since the early Mesozoic it has been an uplifted but stable region covered with terrestrial sediments. After Cretaceous planation, this large area evolved to a near horizontal erosional surface. This area was not subjected to Pleistocene continental glaciation. This Cretaceous to Tertiary planation surface extending over 240,000 km^2, accumulated deposits of Red Clay and Red Silt Soil in local depressions. It was overlain by a blanket of loess at least 50 metres thick. This cratonic basement together with the long duration of an arid or subarid environment in the Pleistocene provided a suitable environment for aeolian processes to construct the Loess Plateau mega-geomorphic feature (Liu Tungsheng et al., 1982; Wu Zirong et al., 1982).

Palaeolandforms in this ancient planation underneath the loess can be divided into the following forms: a) Low Rolling Hills: most of the underlying Mesozoic rocks are flat-lying sandstones and shales. Uplift and erosion has modified some of the planated surface into low rolling hills. Initial loess deposition on the basins and hills formed a blanket cover, which can be traced from the cover near rock outcrops and by the arched-expression of palaeosols overlying loess and hills of Mesozoic rocks.

The loess forms in this region are "Liang" and "Mao", but not "Yuan" (Fig.4):

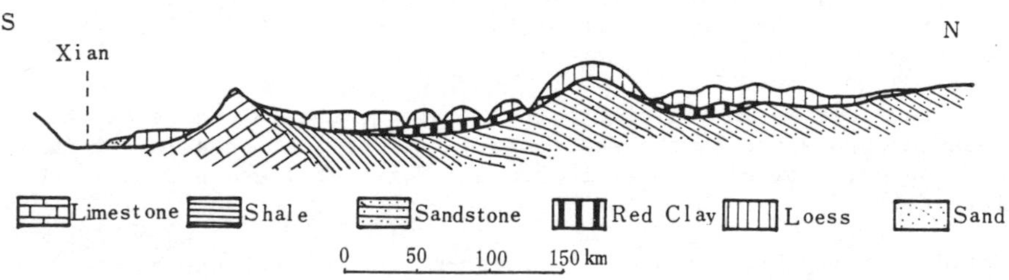

Fig 4 Loess "Ling", "Mao" and palaeolandforms underneath loess.

Buried landforms west of Liupanshan different from those to the east. Most of the hills buried underneath the loess in that region are Tertiary Red Beds which form a rugged hill region. This is another type of Buried landform which could be called Badland topography (E. Derbyshire, oral communication). (b) Basins and Inclined Plains: Within the Loess Plateau, the most interesting buried palaeogeomorphic features are the ancient basins. These basins usually are much more than 10 km in length and width. The central part is filled with Pliocene Red Clays and Red Silt Soil, above which loess is deposited as in Luochuan.

The loess and palaeosols are horizontal without the dips found in the "Liang", "Mao" region. In these palaeogeomorphic basins, the most complete succession of the loess-palaeosol sequences and "Yuan" landforms occurs(Fig. 5).

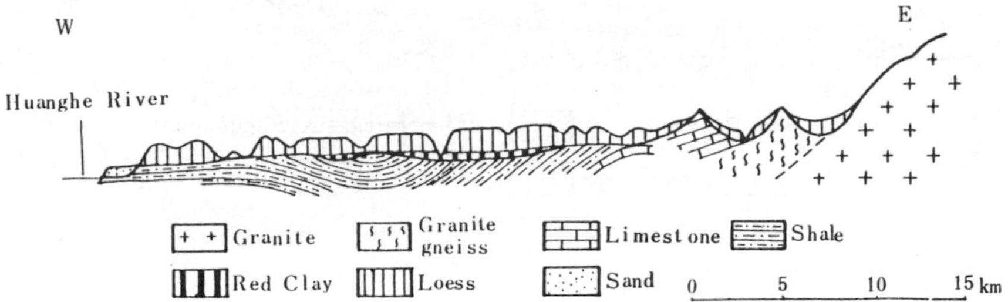

Fig. 5 Loess "Yuan" and Palaeolandforms underneath loess

(c) Limestone Karst: In the Northeastern part of the Loess Plateau, a large area was occupied by Palaeozoic basement limestones with well developed karst landforms. Loess buried these old karst regions and formed "Liang" and "Mao", with isolated limestone pinnacles protruding above the loess (Fig 6)

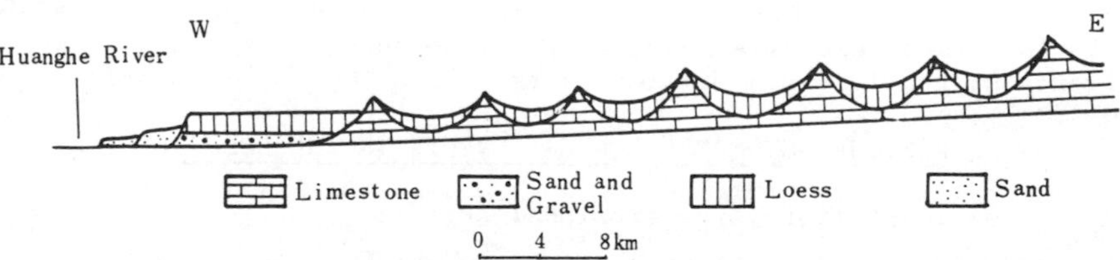

Fig 6 Loess buried under old karst regions and formed Loess "Liang" and "Mao".

(d) Wide River Terraces: Along some parts of the Yellow River and its large tributaries, loess covered ancient terraces with the flat surface known as "Yuan" (Fig. 7, Liu Tungsheng et al., 1965).

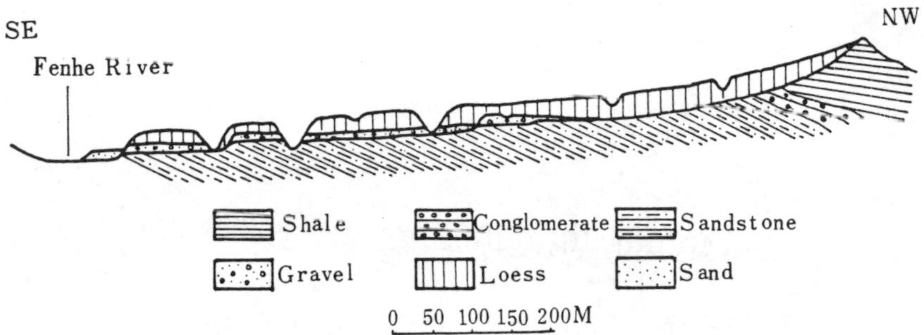

Fig. 7 Loess buried ancient river terraces and formed Loess "Yuan" (Loess Terrace).

Most of the remains of the ancient fossil man, such as the Lantian Man, Tali Man were found in regions with these buried landforms (Woo, 1964, 1966). From the foregoing, the following summary may be made:

1. The Loess Plateau as a whole is an aeolian depositional mega-geomorphologic unit. It has such second order varieties as "Yuan", "Liang" and "Mao" which are also depositional features.
2. Those depositional features were inherited from the Quaternary and Pre-Quaternary geomorphic features of the Cratonic unit since the Mesozoic age.
3. Available geochronological calibration of different layers of the loess and paleosols makes it possible to study the erosional events of the last 2.4 million years.

EROSIONAL GEOMORPHOLOGICAL EVENTS IN THE LOESS PLATEAU

1. Pre-2.4 million years erosional surface

River systems such as the ancient Yellow River and the Luochuan (Luohe River) are drainage lines which existed on the erosional surface predating the loess. The main course of Luohe river which now cuts into the Mesozoic sandstones has left a rock bench 80 metres high (Fig. 8). The consistency of the drainage pattern in the Loess Plateau showing regional uplift has been a major geologic process since 2.4 my BP. The main geomorphic features were now preserved as buried geomorphic features.

2. 2.4 my - 0.56 my deposition with some erosional events

Not adequately studied, episodic oscillations occurred with repeated climatic fluctuations from dry/cold to warm/wet, as demonstrated by the repeated loess and palaeosol layers. During this time, loess extended gradually to the north,south and east.

3. 560,000 yr BP erosion event

Strong erosion occurred in the "Liang", and "Mao" region while the river systems expanded. In depositional region at this time there developed the fifth palaeosol, a complex of three cinnamon type soils representing a prolonged period of climatic optimum in the Loess Plateau. Soil No 5 (S_5) was developed either horizontally or slightly inclined in "Yuan" region, usually expressed as 3 separate units in the "Liang" and "Mao" region. In many places above and below S_5 loesses are in unconformable contact with each other. It is the biggest erosional surface in the loess region. In this time river systems were extended and Mammals (including ancient man) flourished.

From the shape of the loess and soils above and beneath the fifth soil it can be observed that the lower parts are gently undulating while the upper parts may be steeply inclined. This is the period responsible for creating the modern expression of the "Yuan" and "Liang" and "Mao".

4. 200,000 yr BP and erosion event

The second palaeosol (S_2), sometimes composed of two soil layers, represents a prolonged warm wet period, occasionally separated by a short dry phase. This period produced wide valleys of about 30 metres in depth in "Yuan" and "Liang" regions. Later, down-cutting of V-shaped valleys usually developed along the wide valleys of 200,000 yr BP (Fig.9a).

5. 125,000 yr BP erosion event

The first palaeosol (S_1), has an estimted age of about 125,000 yr BP. This might be correlated with the last interglacial age.

Valleys were further deepened at this stage, producing many transverse gullies. (Fig.9b).

After the cutting of 125,000 yr BP the Malan Loess, which has the largest extension was deposited (Fig.9c).

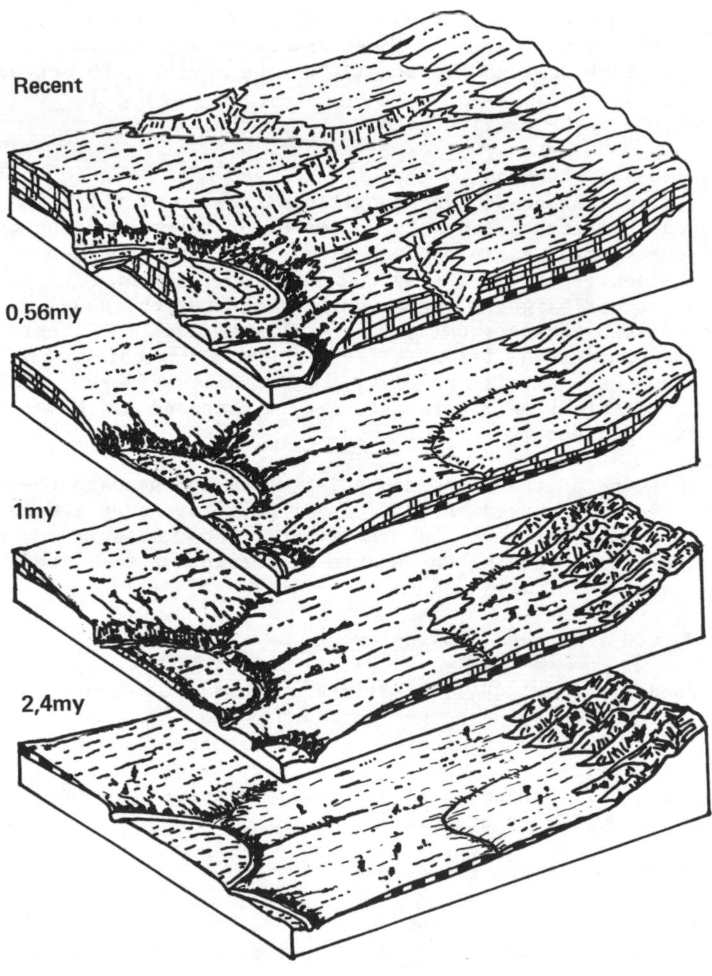

Fig. 8 Development of the geomorphic features of the "Yuan" Loess region.

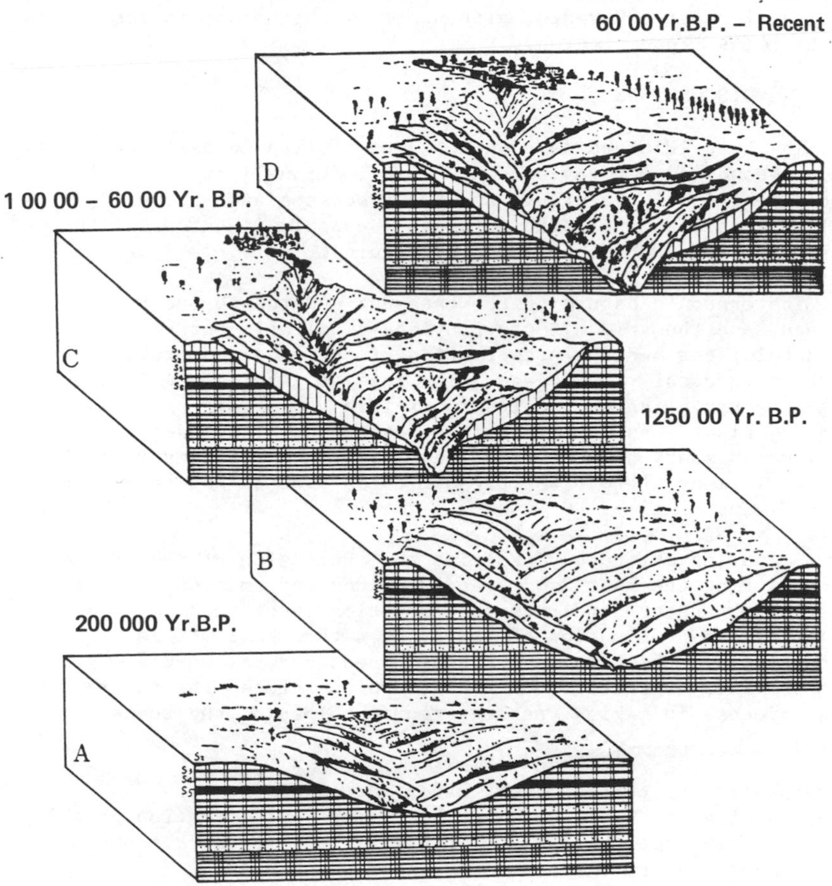

Fig. 9 Development of valleys in the "Yuan" Loess region of Luochuan

6. 5,000 yr BP erosion and post 5,000 yr erosion

A more recent erosion period observed from the "Yuan" region, might have occurred at the black soil period. Wherever Malan Loess has been dissected, it provides the modern densely dissected landforms of the Loess Plateau. (Fig. 9d).

In some places in the "Yuan" region, neolithic sites with fire places, kitchen midden, painted potteries, were discovered. C_{14} dates showed that they are 5,000 to 6,000 yr BP old. Some of these sites were cut by gully-head advance suggesting that the most recent gully extension of 50 metres occurred in 5,000 years, at about 1 m per 100 years, although this phenomenon is restricted in the loess "Yuan" region.

DISCUSSIONS

1. The Loess Plateau as a whole is an aeolian depositional unit. Its second order forms such as "Yuan", "Liang", and "Mao" are also depositional features. Aeolian processes have produced a uniform mantle at the earth surface. We would like to call the Loess Plateau in China a mega-geomorphologic unit.

2. The cratonic nature of the basement underneath the Loess Plateau, and the arid climate,provided a suitable environment for the development and the preservation of the mega-geomorphic unit which emerged only 2.4 my ago. The role of inheritance in these mega-geomorphic features helps to explain the origin of the "Yuan", "Liang" and "Mao". This is essential to understand the erosional history of the Loess Plateau, which is so important to the development of both the present Loess Plateau and the North China Plain.

3. A detailed chronology explains the succession of the erosional events during the last 2.4 my and provides information for further study of the nature and rate of dissection (neotectonic and climatic), identification of the process thresholds of aeolian deposition and the erosional events (loess and palaeosol). It also provides evidence to help distinguish between the natural and human process in making the mega-geomorphology of the Loess Plateau.

Geomorphologists through their study should develop an understanding of the process and history of the formation of this mega-geomorphology which inevitably will contribute to the development and protection of the whole environment.

REFERENCES

Derbyshire, E., 1983. On the morphology, sediments and origin of the Loess Plateau of central China, in Mega-geomorphology (eds Gardner and Scoging), pp 172-194. Oxford University Press, Oxford.

Douglas, I., 1985. Global megageomorphology. NASA Conference Publication 2312, 10-17.

Heller, F. and Liu Tungsheng, 1982. Magnetostratigraphical dating of loess deposits in China. Nature, 300, (5891).

Liu Tungsheng, An Zhisheng, and Yuan Baoyin, 1982. Aeolian Processes and Dust mantle (loess) in China. In: Quaternary Dust Mantles of China, New Zealand and Australia, pp 1-17. Australian National University Press. Canberra.

Liu Tungsheng, and Yuan Baoyin, 1982. Quaternary Climatic Fluctuation - A Correlation of Records in Loess with that of the Deep Sea Core V28-238. Research on Geology (1), Cultural Relics Publishing House.

Liu Tungsheng et. al., 1962. The Hungtu (Loess) of China. Acta Geologica Sinica, 42, 1-14, (in Chinese).

Liu Tungsheng et. al., 1965. Loess Deposits in China, Academia Press, (in Chinese).

Liu Tungsheng et. al., 1985. The Loess-Palaeosol Sequence in China and Climatic History. Episodes, 8, (1), 21-28.

Liu Tungsheng and Others, 1985. Loess and Environment, China Ocean Press.

Wang Yongyan and Song Hanliang, 1983. Rock desert, gravel desert, sand desert, loess, 178p, Shanxi People's Art Publishing House.

Woo Judang, 1964. Mandible of the Sinanthropus-type discovered at Lantian, Shensi-Sinanthropus lantianensis, Vertebrata Palasiatica, 8, 1-17. (in Chinese, English summary).

Wu Zirong, Yuan Baoyin, and Gao Fuqing, 1982. Geological Environment of Loess Deposits, In: Quaternary Dust Mantles of China, New Zealand and Australia, pp 19-20, The Australian National University Press, Canberra.

International Geomorphology 1986 Part II
Edited by V. Gardiner

EVIDENCES OF REJUVENATION OF THE DECCAN FORELAND, INDIA, WITH PARTICULAR REFERENCE TO THE MEGHALAYA PLATEAU

R. K. Rai

Geography Department, North-Eastern Hill University, Shillong, Meghalaya, India.

ABSTRACT

Synthesis of geomorphological studies from different parts of the Deccan Foreland, India, may enable greater understanding of uplift in the region. The Meghalaya Plateau is a detached part of this Foreland. Much evidence for uplift and rejuvenation has previously been reported from the Aravalli, Central Highlands and Chotanagpur/Baghelkhand Plateaux regions of the Foreland. In the Meghalaya Plateau five erosion surfaces occur, as evidenced by morphometric and field data. By considering the ages and causes of intermittent uplift throughout the Foreland it is tentatively concluded that the rejuvenation of the Deccan Foreland started with the Himalayan Orogeny and continued until the Recent period.

INTRODUCTION

The term Deccan Foreland refers to the northern zone of the Deccan plateau (22°0'-28°20'N and 72°15'-92°45'E) (Fig.1) facing the Himalayas. This zone seems to have acted as one of the forelands in the Himalayan Orogeny and was effected by immense orogenic activity (Wooldridge and Morgan, 1959). The collision of plates has a significant bearing on the palaeo-environment of regions such as the Himalayan region and the Deccan Foreland, drastically altering the morphogenetic conditions. It is relevant to seek evidence which may help to determine rifting times and hence opening up of the subduction zone along the southern margin of the Tethys sea. Advocates of the subcontinental collision against the Sino-Siberian landmass have thrown light on many interesting tectonic and geomorphological features (Sahni, 1981). Field studies have proved that in the central highland the Vindhyans and the Bundelkhand gneiss extend below the Ganges alluvium and that the northern margin of the old Gondwanaland was involved in the Himalayan mountain building. In the same manner the Archaean gneisses are also buried under the Brahmaputra alluvium to the north of Meghalaya Plateau.

Geomorphological study of different parts of the Deccan Foreland (Fig.1) could make it possible to develop a model applicable to the whole of the Foreland. Attempts to synthesise geomorphological characteristics of the Deccan Foreland which appear to be closely associated with the Himalayan Orogeny have been made (Kumar and Rai,

1980). The results obtained in the course of these studies were quite encouraging in understanding the uplift of the region.

A recent study of the Khasi and Jaintia Hills, Meghalaya, conducted by Rai and Panda (1981) has further added to knowledge of the geomorphological characteristics of this isolated part of the eastern extension of the Deccan Foreland, which now stands totally detached from the main Peninsular Plateau. The geological formations of the Meghalaya Plateau are, in general, similar to the rock formations of the Peninsular Plateau and it is thus regarded geologically as a part of the Peninsula, which has been cut off by the intervening spread of the Ganges and Brahmaputra alluvium. The gap between the Meghalaya Plateau and Rajmahal Hills is known as the Malda gap or Rajmahal-Garo gap.

The rocks of the Meghalaya Plateau resemble those of the Chotanagpur Plateau of Bihar and Bengal, and structural trends of the rocks of the two parts can be matched. Marine transgression affected the southern parts of the Meghalaya Plateau in the Cretaceous to early Eocene period and deposited marine sediments similar to those along the Coromandal coast of the Peninsular Plateau. The geological, geomorphological, and biotic evidence strongly supports the assumption that the Meghalaya Plateau is a continuation of the Chotanagpur Plateau of the Deccan Foreland.

The Deccan Foreland consists of ten physiographic regions (Fig.2) which have been grouped into four major geomorphological regions for convenience of discussion: (a) Aravalli region (b) Central Highland (c) Chotanagpur Plateau and Baghelkhand Plateau and (d) Meghalaya Plateau. The Aravalli region (Fig.2) is composed of Pre-Cambrian rocks; the Central Highland is mainly composed of the Vindhyan (Pre-Cambrian) group of sedimentary rocks and partly covered by the Deccan Traps (Mesozoic and Tertiary). The Chotanagpur and Baghelkhand Plateaux are composed of granites and gneisses belonging to the Archaean and Dharwar (Pre-Cambrian) rocks. The Meghalaya Plateau is composed of (a) Archaean gneisses with acid and basic intrusives (b) Shillong (Archaean) series rocks (c) Lower Gondwana (Permo-Carboniferous) rocks (d) Sylhet traps and (e) Cretaceous-Tertiary rocks. The Deccan Foreland is therefore mainly a Pre-Cambrian landmass which did not undergo any long period of marine transgression, and which has a long history of continuous erosion and degradation throughout the Palaeozoic and Mesozoic eras. Logically it should have eroded into a peneplain surface, but the highest elevations recorded in various regions, viz., Aravalli region, 1160m, Central Highland, 850m, Chotanagpur and Baghelkhand Plateaux, 1160m and Shillong Plateau 1960m, suggest rejuvenation (Kumar and Rai, 1972).

THE ARAVALLI REGION

Literature on the geomorphology of the Aravalli region is rather limited. However, publications of the Geological Survey of India (Heron, 1922, 1933, 1953) refer to certain characteristics of topography and landforms which are quite revealing. The Aravalli range might have reached the mature stage of peneplanation sometime in the Cretaceous period. The present ranges are the stumps of roots of the folded structure and are good examples of the reversal of topography. Dunn (1939) has recognised three erosional surfaces at elevations of 1200m, marked by the higher peaks, 600-680m around Udaipur region towards the end of the range, and 300-430m. An intensive study of the whole of the Aravalli region may throw further light on the erosion surfaces and the related geomorphological features.

The following evidence further supports the process of uplift and rejuvenation of the Aravalli region:

(1) The uniform sky line of the ranges.

(2) Resistant formations like quartzites and conglomerates stand as ranges while the phyllites and schists form the lowland.

(3) There are numerous examples of superimposed drainage in Udaipur region and in the lower Chambal Valley.

(4) Erosional scarps facing the northwest bank of the Chambal river.

(5) The dissected ravines of the Chambal and its tributaries. The studies conducted by Sharma (1980) in the lower Chambal valley reveal that due to regional rejuvenation the Chambal has entrenched into its own alluvium deposits. Once the river cuts a deeper valley the water from the adjoining bank rushes down to the level of the main river, and the high banks are dissected into ravines.

(6) Incised meanders have been identified in the Chambal and its tributaries.

(7) Sharma (1980) has traced erosion surfaces indicative of uplift in the lower Chambal valley at elevations of 480-600m, 360m and 180-305m.

THE CENTRAL HIGHLAND

The Central Highland is a triangular plateau between the Aravalli range, Chotanagpur - Baghelkhand Plateaux and the Narmada-Son axis; it comprises the undisturbed horizontal Vindhyan formations which remained a passive landmass throughout the Palaeozoic and Mesozoic eras. Nonetheless, study of the Sonar-Bearma basin by the author (Rai, 1980) revealed almost the same geomorphological characteristics as in other adjoining regions of the Deccan Foreland. After the emergence of the Central Highland as a landmass the major events in the history of the Sonar-Bearman basin were the formation of the Lameta beds near Jabalpur and volcanic activity of the Cretaceous period flooding the western Deccan Foreland, giving rise to the Deccan Traps. The basin falls on the border between the

Vindhyans and the Deccan Traps, which offered the opportunity to study not only the landforms developed on these two formations but the exhumed topography and superimposed drainage as well.

Morphometric analysis and good field evidence such as the presence of deep gorges along small streams, waterfalls, break in slopes, river terraces and the thick mantle of weathered material indicate the presence of five erosion surfaces in the region (Rai, 1978), at altitudes of 580m, 440 to 550m, 365 to 400m and 305m. It is very interesting to note that two surfaces, one the exhumed Vindhyan surface and the other on the Deccan Traps, are at the same altitude of 580m. The erosion surfaces corresponding to the Sonar-Bearma basin have also been identified in the Rewa Plateau by Dube (1968). He is of the opinion that the Bhander, the Panna and the Binjh scarps north of the Rewa Plateau, facing the Gangetic alluvium, are erosional scarps. The rejuvenation accelerated active erosion in the region.

CHOTANAGPUR AND BAGHELKHAND PLATEAUX

Dunn (1942) has identified four 'peneplains' at altitudes of 1,000m, 700m, 550m and 450m, in the Chotanagpur Plateau; these have been studied intensively by Chatterjee (1940, 1963), Bagchi and Sengupta (1958), Ahmad (1963), Singh (1957) and Varma (1958), who suggested the following evidence in support of rejuvenation:

(1) Well defined scarps between the erosional surfaces, which have been identified as either erosional or composite scarps.

(2) Zones of intense dissection, in Kodarma and Southeast of the Ranchi Plateau.

(3) Misfit streams in the Natarhat Plateau and Pat country of the Baghelkhand Plateau.

(4) The undulating surface of the Ranchi and Hazaribagh Plateau, identified as an uplifted peneplain with old remnants as monadnocks.

(5) Active erosion by streams with steep gradients, entrenched valley meanders, river terraces, boulder deposits and ungraded long profiles dotted with waterfalls (Verma, 1958) and (Khan, 1954).

(6) Presence of superimposed drainage, particularly over the Dalma hills.

MEGHALAYA PLATEAU

The analyses of geomorphological features of the Meghalaya Plateau by Singh (1968), Murthy (1968), Panda (1983) and Rai (1982) indicate that the most significant are erosion surfaces. Morphometric analysis of landforms and field evidence supported by palaeoclimatic evidence display erosion surfaces at different levels. The Meghalaya Plateau shows a stepped topography with maximum altitudes of about 1900m. To the southern part of the plateau the levelled base of the Cretaceous-Eocene sediments occurs. The evidence indicates that development of erosion surfaces has been closely associated with the Himalayan Orogeny. It appears that various phases of crustal movement were responsible for the development of salient relief features of the plateau, which were caused by dismemberment of Gondwanaland.

Morphometric analysis and field evidence indicate the presence of five erosion surfaces, as follows:

1. Gondwana surface 1500 - 1800m

2. Cretaceous surface 1200 - 1500m

3. Eocene surface 900 - 1200m

4. Pleistocene surface 600 - 900m

5. Pleistocene to Recent surface 300 - 600m

Analysis of superimposed profiles (Fig.3A) composite profiles (Fig.3B), projected profiles (Fig.3C), N-S cross section of plateau (Fig.3D), longitudinal profiles of selected rivers (Fig.4) and field evidence such as the presence of deep gorges, V-shaped valleys, waterfalls, river terraces, breaks in slope, the thick layer of weathered material and the uniform skyline of hills, support the presence of these five erosion surfaces. Analysis of the longitudinal profiles of the major rivers of the plateau show that the gradients are steep and the profiles are not smooth; the rivers are actively engaged in deepening their valleys. Numerous waterfalls, river terraces and springs in the region indicate the rejuvenation of the plateau.

AGES AND CAUSES OF INTERMITTENT UPLIFT

The erosional surfaces found in various parts of the Deccan Foreland, and their heights, are tabulated in Table 1. These observations made by different workers well support the concept of Recent rejuvenation of the Deccan Foreland. Bearing in mind the above arguments, one is faced with the questions of (a) periods of rejuvenation and (b) causes of intermittent uplifts.

The geological and geomorphological evidence in the Aravalli region, Central Highlands and Chotanagpur Plateau place the Great Boundary Fault and the faults astrides the Damodar valley in the Cretaceous period. For the rejuvenation of the Aravalli range Heron (1953) remarked that the range appears to have been peneplaned during the Mesozoic era, not much earlier than the Cretaceous. This might well have been the first planation since the range was uplifted in Pre-Vindhyan times. This surface may be taken as the Cretaceous

TABLE 1. Erosion surfaces in the Deccan Foreland.

Erosion surface	Height (m)	Region	Author
Gondwana	1500-1800	Meghalaya Plateau	Panda
Early Tertiary	1000	Natarhat Plateau	Chatterjee
Early Tertiary	1000	Chotanagpur Plateau	Dunn
Cretaceous	1200	Aravalli Region	Dunn
Bhander	580-750	Rewa Plateau	Dube
Pre-Cretaceous	480-600	Lower Chambal Valley	Sharma
Exhumed Vindhyan	580	Sonar-Bearma Basin	Rai
Cretaceous	1200-1500	Meghalaya Plateau	Panda
Middle-Late Tertiary	700	Chotanagpur Plateau	Dunn
Middle-Late Tertiary	700	Ranchi Plateau	Chatterjee
Late Eocene	580	Sonar-Bearma Basin	Rai
Panna	400-500	Rewa Plateau	Dube
Eocene	900-1200	Meghalaya Plateau	Panda
Late Tertiary	550	N.E.of Chotanagpur Plateau	Dunn
Late Tertiary	500	Basia Plateau	Chatterjee
Middle Miocene	440-530	Sonar-Bearma Basin	Rai
Middle Miocene	360-450	Lower Chambal Valley	Sharma
Rewa	300-350	Rewa Plateau	Dube
Pleistocene	600-900	Maghalaya Plateau	Panda
Pleistocene	365-395	Sonar-Bearma Basin	Rai
Pleistocene	450	N.E.of Chotanagpur Plateau	Dunn
Pleistocene	450	Simdega Plateau	Chatterjee
Pleistocene-Recent	300-600	Maghalaya Plateau	Panda
Pleistocene-Recent	305	Sonar Bearma Basin	Rai
Trans-Jamuna	100-150	Rewa Plateau	Dube
Pleistocene	100-305	Lower Chambal Valley	Sharma

surface. The second peneplain (600-680m) has been correlated by Heron (1953) with the Tertiary peneplain of the Chotanagpur Plateau.

About the peneplains of the Chotanagpur Plateau, Dunn (1939) is of the opinion that rejuvenation of the western highland (including adjoining Baghelkhand Plateau) occurred during the Cretaceous period, and that of the Ranchi Plateau in the later part of the Tertiary period.

While analysing the causes of earthquakes in Bihar Dunn (1939) advanced a mechanism of subsidence of the Indo-Gangetic trough and rising of the Himalayas and the Deccan Foreland. Wadia (1953), on the basis of studies of the Himalayas, has postulated three intermittent uplifts of the Himalayas. "The first of these was post Nummulitic i.e. towards the end of the Eocene culminating in the Oligocene ... greater intensity about the Middle Miocene. The last stage was mainly of post Pliocene age later than the deposition of the greater parts of Siwaliks and did not cease till after middle of Pleistocene". Dunn (1939) is of the opinion that the intermittent uplift of the Himalayas had corresponding periods of rejuvenation in the Chotanagpur Plateau, which may be correlated to the interruptions in the cycle of erosion. Rai (1978) is also of the opinion that the development of erosion surfaces of the Central Highland might have been associated with the Himalayan Orogeny. Dutta (1967), while analysing the palaeogeography of the Meghalaya Plateau, has postulated that the plateau seems to have been faulted and uplifted during the Tertiary.

The rock formations and geomorphological characteristics of the landforms of the region present good evidence of erosion, sedimentation, diastrophism, intrusion and movement of land and sea in the Meghalaya Plateau. The events of Himalayan uplift appear to have deeply influenced the landforms of the Meghalaya Plateau. The intervals between uplifts were long enough for the development of erosion surfaces in the region. The most important earth movement that is relevant to the formation of the Siwalik Himalayas and the Shillong plateau in the Miocene period is the development of a system of faults.

It appears that the oldest Gondwana surface was developed on the plateau and reached its peneplanation stage before the eruption of Sylhet traps. The tectonic history of the plateau began with the effusion of the Sylhet traps through fractures and faults. The southern block subsided and the northern block uplifted. Marine sediments of the Cretaceous period lie unconformably over the eroded surface in the south.

Murthy (1970) has also concluded that in the Shillong Plateau and Upper Assam vertical movement of fractured basement blocks has occurred since Eocene times. He is of the opinion that high seismicity of the region is an expression of neo-tectonic activity as a consequence of vertically dominant tectonics i.e. due to abrupt movement between adjacent basement blocks along faults. Ahmad (1965) seems to agree with this view and writes "Probably as the side effects of mighty Himalayan mountain movements the southern plateau of Bihar appears to have undergone three successive intermittent uplifts". It appears that all through the Deccan Foreland the zones of weakness i.e. the boundary fault east of the Aravalli range, faults along Narmada Son axis, along the Damodar valley, and the Dawki fault of Meghalaya Plateau have all been affected in this rejuvenation.

The above synthesis of the geomorphological evidences from widely scattered areas of the Deccan Foreland leads one to a possible conclusion that the Deccan Foreland has all been involved in diastrophic activity in the geological past. In the geological history of the Deccan Foreland the rejuvenation seems to have started from the same period as that of the Himalayan Orogeny and continued up to the Recent period. Such colossal activity may not have left the Foreland untouched.

Acknowledgement

The author expresses his sincere thanks to Dr. R. P. S. Pahuja, Senior Geologist, Geological Survey for India, for fruitful discussions.

REFERENCES

Ahmad, E., 1965. Bihar-Physical, Economic, Regional Geography. Ranchi University, Ranchi.

Bagchi, K., and Sengupta, A. K., 1958. Physiographic regions of the Ranchi Plateau, Bihar. Geog. Rev., India, 20, (Silver Jubilee number), 127-133.

Chatterjee, S. P., 1940. Gneissic topography of the Ranchi Plateau. Cal. Geog. Rev. 3, (2).

Chatterjee, S. P., 1963. Geomorphology of Ranchi Plateau. Fifty years of Science in India, Progress of Geography.

Dube, R. S., 1968. Erosion surfaces on the Rewa Plateau Madhya Pradesh, India. 21st International Geographical Congress, selected papers, 1, 35-42.

Dunn, J. A., 1939. Post Mesozoic Movements in the northern part of the Peninsula. Memoirs, Geological Survey of India, 73, 139-142.

Dunn, J. A., 1942. The economic geology and mineral resources of Bihar. Memoirs, Geological survey of India, 73, 6-14.

Dutta, S. K., 1967. Palaeogeography of Assam plateau. Bull. Directorate of Geology and Mining Deptt., Assam.

Heron, A. M., 1922. The Gwalior and Vindhyan System in South-Eastern Rajputana. Memoirs, Geological Survey of India, 45.

Heron, A. M., 1932. The Vindhyans of Rajputana. Memoirs, Geological Survey of India, 62.

Heron, A. M., 1953. Geology of Central Rajputana. Memoirs, Geological Survey of India, 29, 37-42.

Khan, P. K., 1954. Waterfalls of Chota Nagpur Plateau. Indian Geog. Journal, Madras, 29, (2/3).

Kumar, P., and Rai, R. K., 1972. Has Deccan Foreland rejuvenated? The Geographer. (The Aligarh Muslim University, Geographical Society, A.M.U., Aligarh), 19, 34-46.

Mittal, R. S., 1968. Physiographical and structural evolution of the Himalayas, Mountains and Rivers of India. National Committee for Geography, Calcutta.

Murthy, M. V. N., 1968. An outline in geomorphological evolution of the Assam Region. Procs. of the Pre-I.G.U., Cong. Symp. on Geomorphology and Plant Geography of North East India. Gauhati University, Gauhati, (Eds. Das and Rao) pp 10-15.

Murthy, M. V. N., 1970. Tectonic and mafic igneous activity in North East India in relation to upper mantle. Proc. II symp. on upper Mantle Project, pp 287-304.

Panda, P. C., 1983. Geomorphology and rural settlements in Khasi and Jaintia Hills, Meghalaya, Ph.D. thesis (unpublished), NEHU, Shillong.

Rai, R. K., 1978. The Study of erosion surfaces in the Sonar-Bearma basin and adjoining regions, Madhya Pradesh. Proc. of the symp. on "The Purana formations of Peninsular India". (Eds. Lakshmanan and West), pp 223-232.

Rai, R. K., 1980. Geomorphology of the Sonar-Bearma Basin. Concept Publishing Co., New Delhi.

Rai, R. K., 1982. Geomorphology and rural settlement in Meghalaya, in Perspectives in Geomorphology, Vol. 1, (Ed.Sharma), pp 107-118. Concept Publishing Company, New Delhi.

Rai, R. K., and Panda, P., 1981. Influence of Land forms in location and distribution of rural settlements in Khasi-Jaintia Hills, Meghalaya. National Geog., 16, 99-105.

Sahni, A., 1981. The timing of Fragmentation and collision of the Indian Plates : its bearing on Himalayan Geology. Himalayan Geology Seminar, Geological Survey of India, Miscellaneous Publication No.41, part IV, pp 235-242.

Sharma, H. S., 1980. The Physiography of Lower Chambal Valley and its Agricultural Development. Concept Publishing Co., New Delhi.

Singh, R. P., 1957. Landscape of Southeast Chota Nagpur plateau and its evolution. Nat. Jour. India, 137-168.

Singh, R. P., 1968. Geomorphology of Shillong Plateau. Procs. of the Pre-I.G.U. Cong. Symp. on Geomorphology and Plant Geography of North-East India. Gauhati University, Gauhati, (Eds. Das and Rao).

Varma, P., 1958. Ranchi Plateau : its Geomorphology and Human Settlements, Ph.D. thesis (unpublished), Allahabad University.

Wadia, D. N., 1953. Geology of India. Macmillan & Co. Ltd., London.

Wooldridge, S. W., and Morgan, R. S., 1959. An outline of Geomorphology : The Physical basis of Geography. Longmans, London.

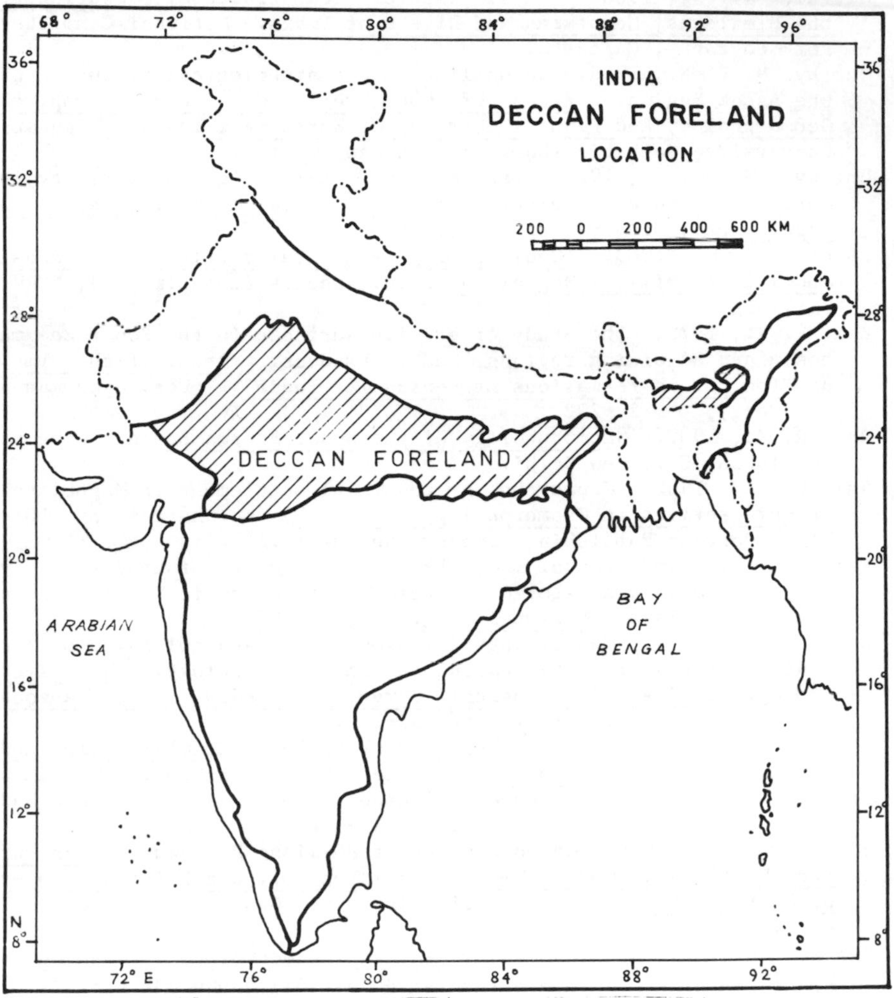

INDIA
DECCAN FORELAND
LOCATION

200 0 200 400 600 KM

DECCAN FORELAND

ARABIAN SEA

BAY OF BENGAL

Fig. 1. (Above). Location. Fig. 2. (Below). Physiographic regions

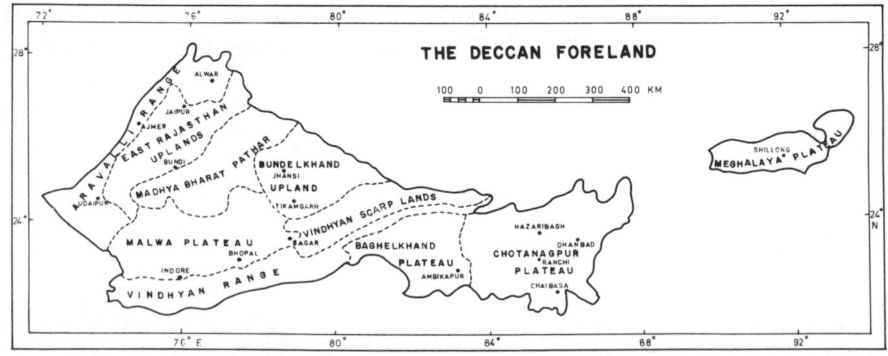

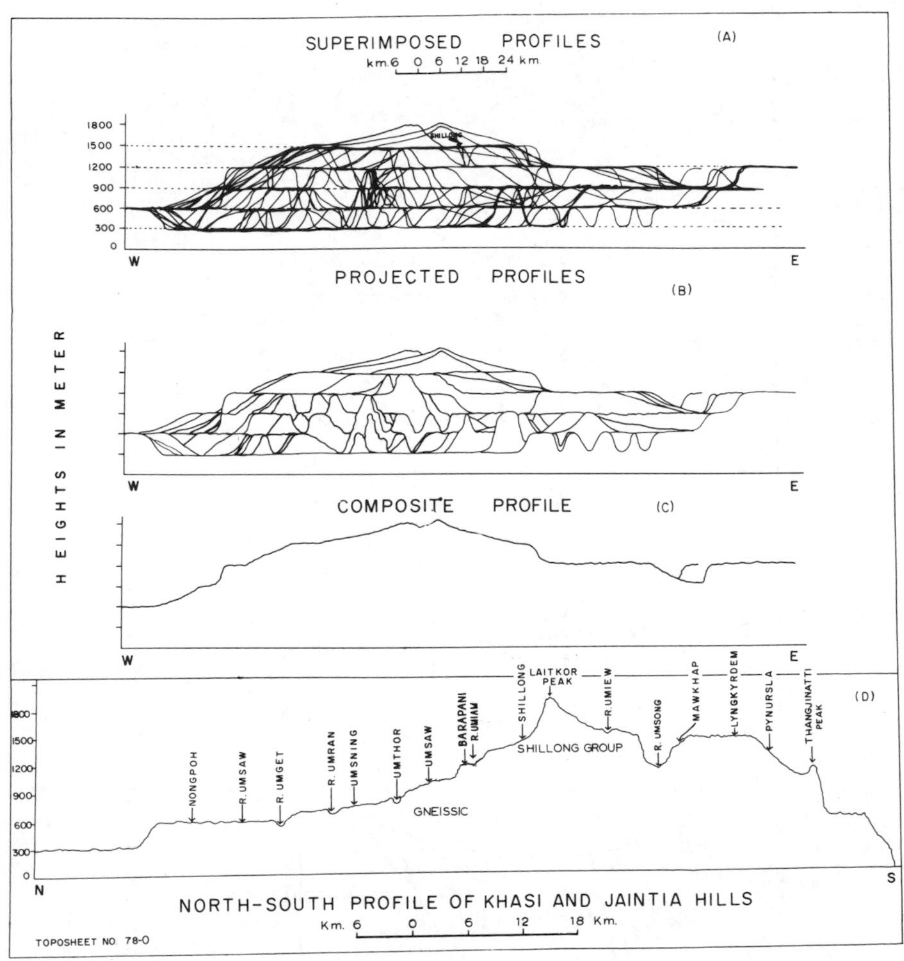

Fig. 3. Topographic profiles.

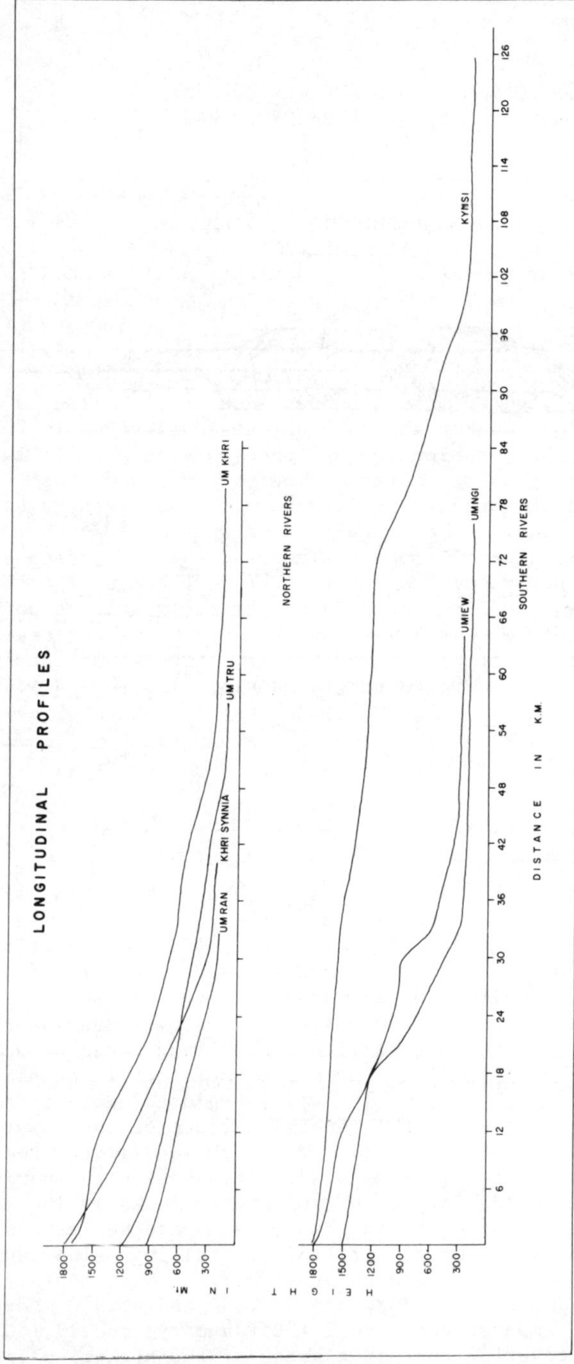

Fig. 4. Longitudinal profiles of selected rivers.

International Geomorphology 1986 Part II
Edited by V. Gardiner

MORPHOLOGICAL ANALYSIS AND DEVELOPMENT OF SLOPE PROFILES OVER BHANDER SCARPS, INDIA.

S. Singh
Department of Geography, University of Allahabad, India.
R. S. Pandey
G. P. Pant Social Science Institute, Allahabad, India.

ABSTRACT

Five sample profiles of the Bhander scarps, having a massive sandstone cap above weaker shales, have been surveyed in the field. Each profile is characterised by a higher percentage of length in concave elements than in convex elements. Maximum slope gradients (31.3% of the total profile lengths) are free-faces below limited summital convexities. The curvature ratio of less than 1.0 also confirms the dominance of concave elements. Very steep and cliff categories together occupy the greatest percentage of total length of each profile. Regolith and sand occupy the gentler slopes of the lower segments of the scarps whereas bedrock is characteristic of the steep and cliff slopes of the scarps. The sandstone free-face elements are subjected to mechanical disintegration and movement of disintegrated materials down the slope occurs because of gravity and rainwash during heavy rainstorms. The study shows parallel retreat of scarps and structural control of slope development.

THE REGION

The Bhandar Plateau of Madhya Pradesh, India, (24°3'29"N - 24°39'1"N and 80°16'30"E - 80°53'15"E) is located between Panna Plateau in the north-west and Rewa Plateau in the east (Fig.1). It is a well defined geomorphological unit, characterised by a wide flat-topped plateau surface (500-550m) having steep scarp faces on all sides except in the south-west. The plateau rises over the lower uplands where a few of the accordant summits resembling mesas project above the general surface (Fig.2). The margins of the higher plateau are highly indented scarps, having well developed embayments, rising abruptly about 350m above the general surface of the lower uplands (Singh and Pandey, 1982; 1983). The rocks underlying the plateau are Vindhyan sandstones, shales and limestones with horizontally bedded alternate bands of hard and soft rocks. The region is drained by the tributaries of the Tons, Satna and Ken rivers and has an average annual rainfall of 1137mm, falling mainly during the rainy season from 15th of June to September. Mean monthly maximum temperatures of January and June are 30.5°C and 45.3°C whereas mean monthly minimum temperatures are 2.4°C (January) and 23.1°C (June). The hilly tract covering the major areas of the higher plateau has

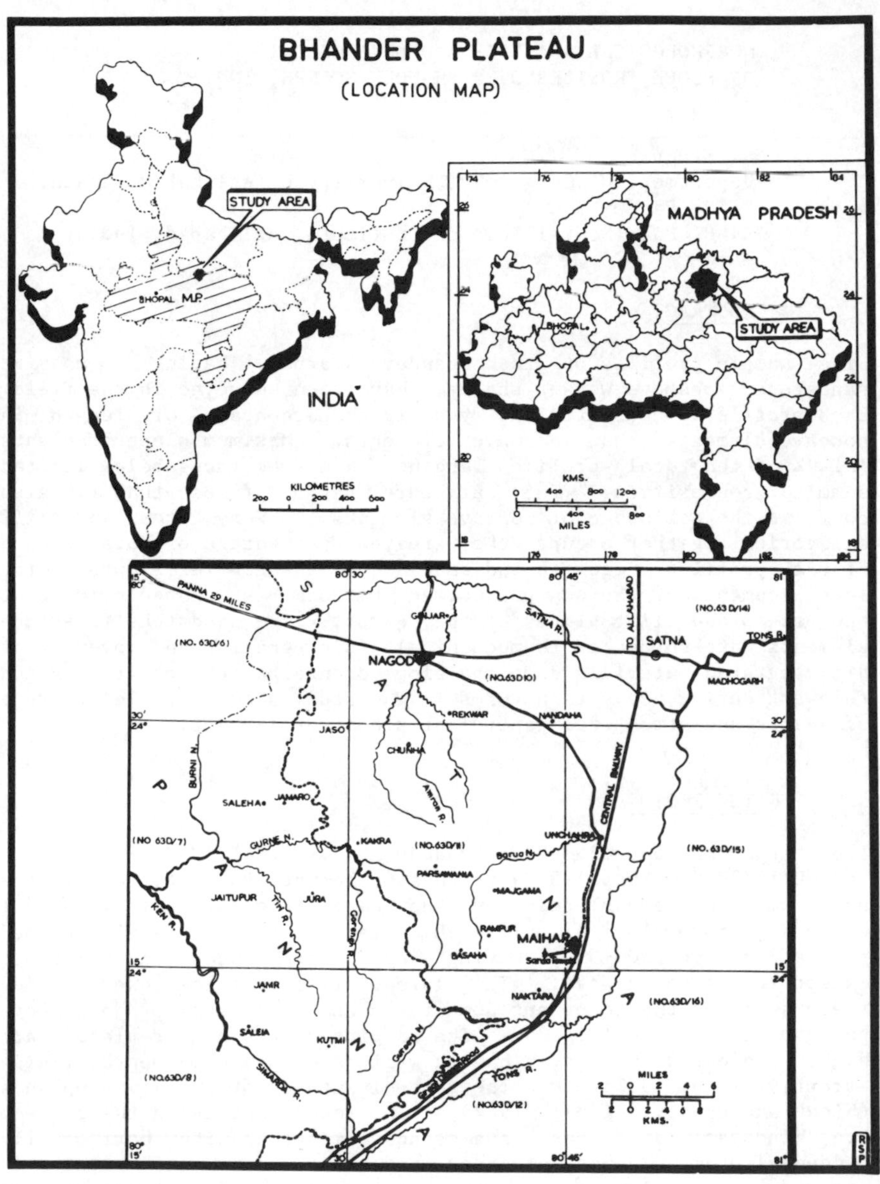

Fig.1. Location of the Bhander Plateau.

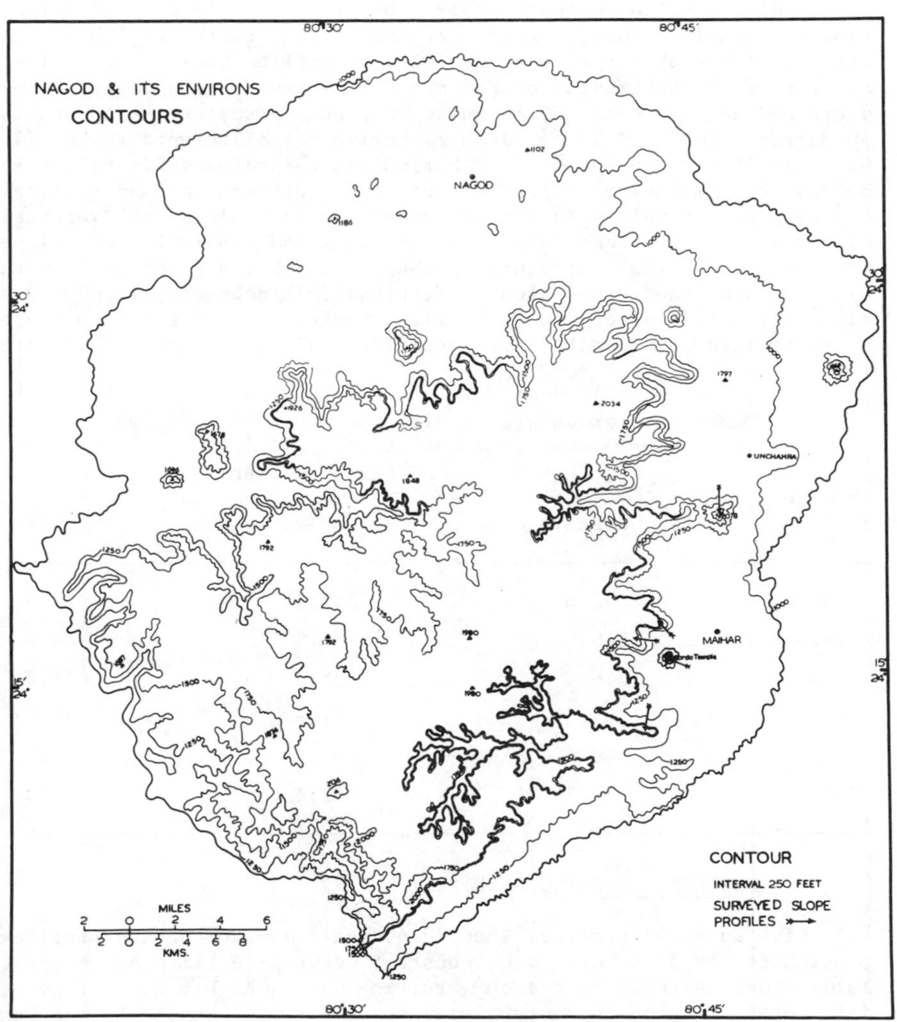

Fig.2. Topography and location of the surveyed profiles.

mixed open and dense forest cover whereas the lower segments of the scarps are characterised by open scrub.

METHODOLOGY

Five sample profiles in the localities of Rewa Fort Hill (R-1), Sharda Temple Hill (S-2), Ganesh Hill (G-3), Naktara Hill (N-4) and Kushla Hill (K-5) have been surveyed using abney level, measuring tape and ranging rods, from the foothills to the hill crest in straight lines as far as was possible. Profile stations were fixed at 20m or 10m horizontal distance. Slope angles were measured upslope and derived data of measured lengths, distances, slope angle in degrees (Ø), Cos Ø, Sin Ø, horizontal difference (D.Cos Ø), vertical difference (D.Sin Ø), horizontal co-ordinates (x values in metres or cumulative values of horizontal difference) and vertical co-ordinates (y values in metres or cumulative values of vertical difference) have been used for drawing and analysis of slope profiles. Profile curvature, the rate of change of angle with distance down the true slope, expressed in degrees per 100m, was also calculated, after Young (1972). Convex and concave slopes are characterised by positive and negative values of the curvature index.

TABLE 1. Percentage of Slope Segments of Profiles

Profile Number	Slope Segment Concave	Maximum	Convex
R-1	32.8	37.7	29.5
S-2	40.9	34.9	24.2
G-3	46.4	28.6	25.0
N-4	47.3	21.8	30.9
K-5	48.6	33.7	17.7
Total	43.2	31.3	25.5

RESULTS AND DISCUSSION

All five surveyed profiles show in general convexo-concave profiles punctuated by free-face and constant debris-rectilinear slopes. Table 1 demonstrates that each profile has a higher percentage of its total length in concave elements (over 30%) than in convex elements (below 30%), and thus the dominance of concavity over convexity on the escarpments is very much indicated. It is axiomatic to deduce that "the relative lengths of maximum units tend to decrease as the profile approaches towards a perfect convexo-concave form" (Kumar, 1981). The percentage of the total length occupied by the maximum segments (Table 1) ranges between 21.8 (profile N-4) and 37.3 (profile R-1) and on average the maximum segments occupy 31.3% of the total surveyed lengths of the profiles. The general pattern (Figs.3 and 4) is one of steeper profiles just below the summital convexity, characterised by scarps of free-face above and rectilinear segments below. It may be pointed out that markedly limited lengths of maximum segments in relation to convex and concave elements express slope-flattening, indicating slope decline and a late stage of slope development, but in the present case the fairly large extent of maximum segments (31.3%) suggests the parallel retreat of slopes where the slopes maintain their maximum angles. This process is still under operation adjacent to the escarpments on some of the residual flat-topped hills and mesas, but has ceased on some of the hills like Sharda Pole (not surveyed) near Naktara village, where scarps have disappeared giving the hill a well-developed profile of convexo-concave elements, this being the result of slope-flattening as a late stage of slope development.

TABLE 2. Profile characteristics.

Profile number	Aspect	Length of units in %		Angular change in units, in %		Index of curve form		Curvature ratio
		Convex Lx	Concave Lv	Convex Ax	Concave Av	Convex Cx=Lx/Ax	Concave Cv=Lv/Av	Cx/Cv
R-1	S	47.2	70.5	42.6	57.4	1.11	1.23	0.90
S-2	E	49.7	74.7	42.1	57.9	1.18	1.29	0.91
G-3	SE	48.2	72.6	40.5	59.5	1.19	1.22	0.98
N-4	NE	42.1	69.3	38.4	61.6	1.10	1.12	0.98
K-5	N	37.8	81.1	32.4	67.6	1.17	1.20	0.97

x = Convex v = Concave

It is apparent from Table 2 that for all five profiles Lv (length of concave units in percent) is larger than Lx (length of convex units in percent). Similarly curvature ratios of all five profiles (Table 2) fall below 1.00, confirming that concavity is most common. The near constancy of curvature ratios amongst the surveyed profiles (ranging between 0.90 and 0.98) indicates the generally similar properties of the escarpments of the study region.

Figs.3 and 4 indicate the convex and concave elements and segments of the slope units at a micro-level. The summital convexities in profiles K-5, N-4 and G-3 are 46, 40 and 45 degrees per 100 metres respectively. Profile K-5 has a summital convexity of 46° per 100m but, strikingly, has a limited basal minimum segment of 6° instead of a concavity, but it is succeeded by a concavity of 10° per 100m up-slope. The most pronounced segments are found just below the summital convexities in all surveyed profiles viz. 42° (R-1), 35° (S-2), 38° (G-3), 42° (N-4) and 40° (K-5). These indicate maximum segments of the profiles characterising free-face elements. In addition each profile has local irregularities, but on average the

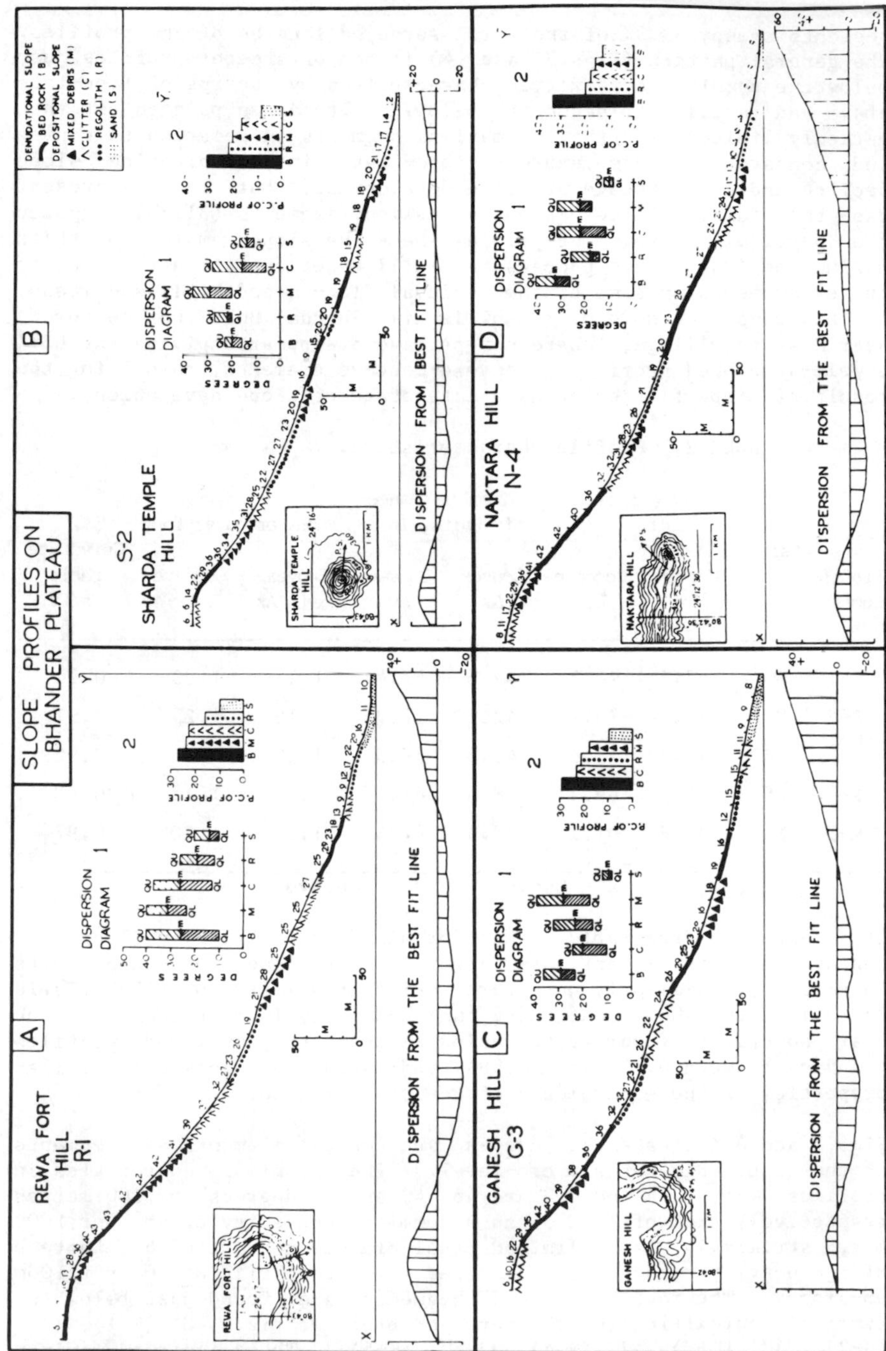

Fig.3. Slope profiles, dispersion diagrams and dispersions from the best fit lines.

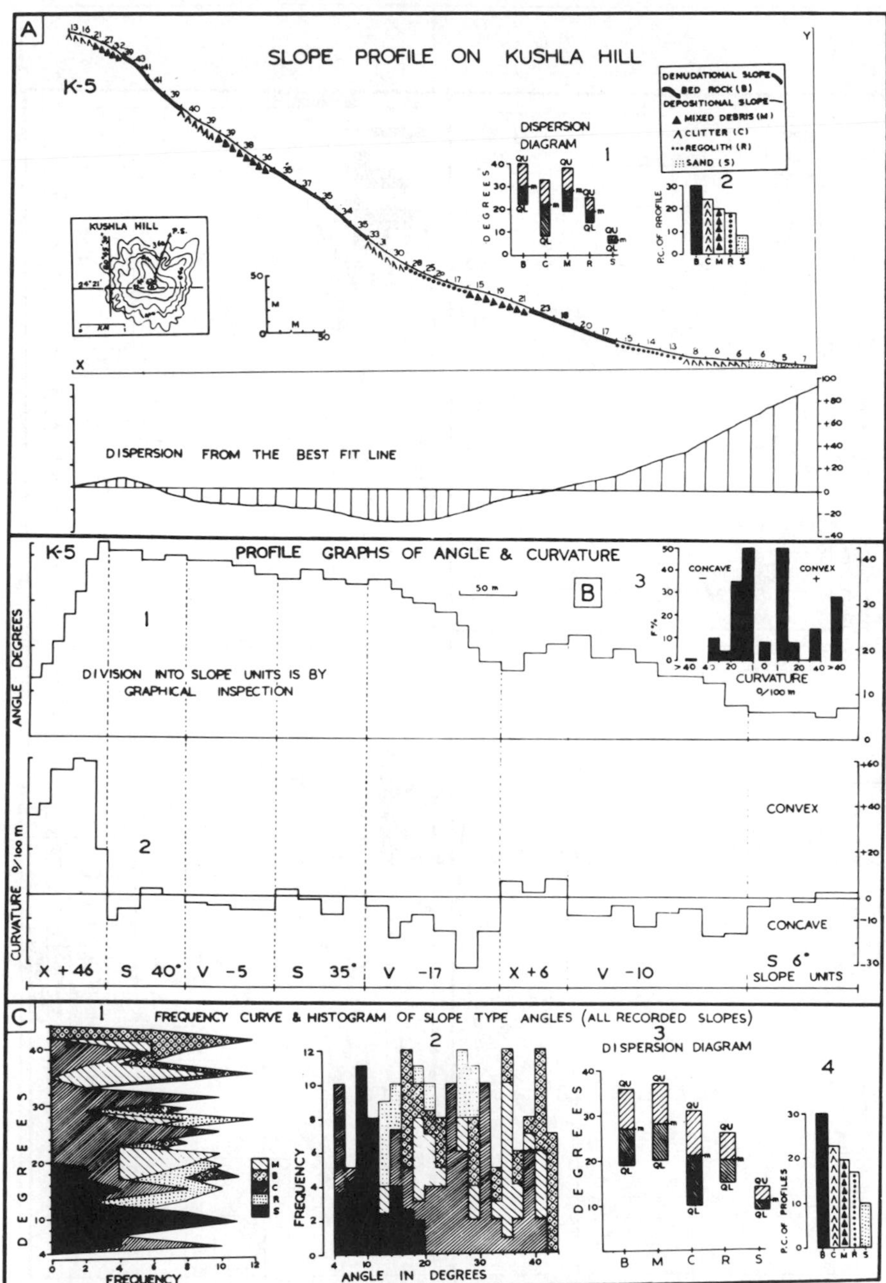

Fig.4. Slope profile, dispersion diagram, dispersion from the best fit line and frequency curve and histogram of slope type angle (all recorded slopes).

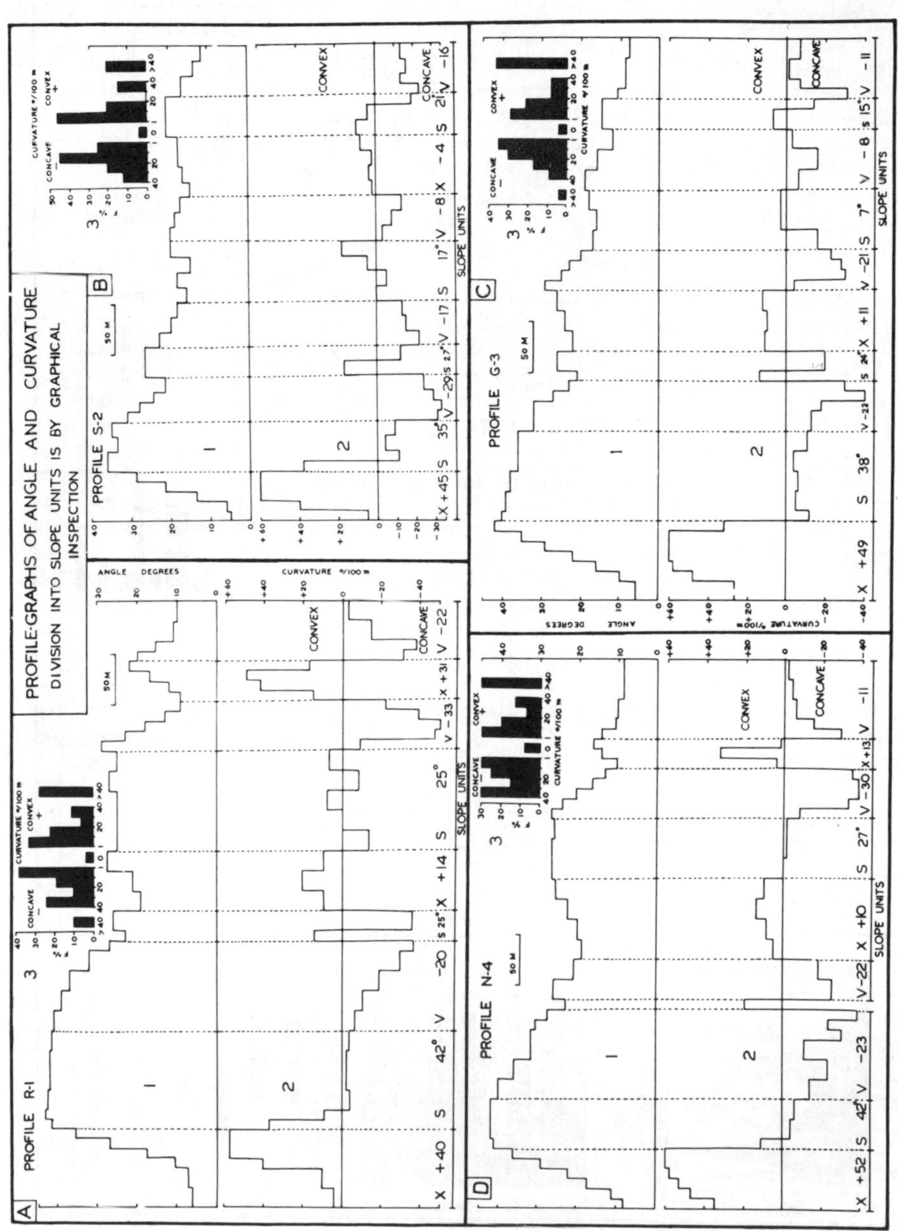

Fig.5. Profile graphs of angle and curvature.
Division into slope units by graphical inspection.

curvature of concavities in each profile increases up-slope, with minor exceptions, and the curvature of local convexities shows mixed trends.

The histograms (Figs. 4 and 5) of curvature indicate that for all five profiles the maximum frequency of the curvature of concave elements lies between 1°-20° per 100m. The curvature of convex elements is most frequently between 1°-10° per 100m in profiles R-1, S-2 and K-5, whereas profile N-4 shows a bimodal distribution with peaks in the categories of 1°-10° and 30°-49° per 100m. Profile G-3 has a maximum concentration of frequencies in the category of above 40° per 100m.

It may be concluded that the observation of Young that "the proportions of total ground length of the profile occupied by convexity, maximum segment and concavity differ substantially according to whether a free-face occurs. On slopes without free-face there is wide variation, but it is unusual for concavity to occupy more than 50%" (Young, 1972, p.158) holds good in the present study as all the surveyed profiles except S-2 (Sharda Temple Hill) are characterised by scarps with free-faces and have (Table 1) in total 43.2% of the ground length under concave elements. It may be further pointed out that a profile having a free-face is characterised by a short length of convex element; the free-face is succeeded by a rectilinear or slightly concave unit below. The development of segmented convexity over sandstone-shale structure in the present study conforms with the finding of Young in part of the Mato Grosso, Brazil, having sandstones under a moist Savanna climate (Young, 1970). Wherever the slope is steep (above 30°), the maximum segments have relatively long lengths. This is also true in the present study, as about 31% of total ground length of all the five profiles is in the maximum segment (Table 1).

Table 3 depicts the distribution of percentage relative length in six slope categories. On average more than 50% of the total length

TABLE 3. Distribution of slope lengths in gradient classes (%).

Profile number	Total length of profiles (m)	Gentle 0-6°	Moderate 6°-10°	Moderately steep 10°-16°	Steep 16°-20°	Very steep 20°-30°	Cliff over 30°	Total
R-1	610.00	3.28	13.11	9.84	8.20	34.42	31.15	100.00
S-2	475.00	4.21	-	17.87	42.10	19.02	16.80	100.00
G-3	560.00	1.82	10.91	18.18	12.73	25.45	30.91	100.00
N-4	550.00	-	20.00	9.09	15.09	23.10	32.72	100.00
K-5	740.00	12.16	5.41	12.16	14.86	10.81	44.59	100.00
All	2935.00	4.29	9.89	13.43	18.60	22.56	31.23	100.00

of all the surveyed profiles is in very steep and cliff categories because of the steep free-faces of the scarps along the plateau rims and some of the residual flat-topped hills or mesas. The rather anomalous profile of Sharda Temple Hill (S-2) results from the obliteration of scarps through Human activities. The positive curvature in the upper section and the negative curvature in the lower section of each profile, except for some deviation in the lower section of Kushla Hill (K-5), present an overall picture of the general pattern of slope profiles. Frequent repetition of positive and negative curvatures in the larger middle section of the profiles indicates alternate bands of sandstones and shales.

It is apparent from Figs. 3 and 4) (slopes categorised according to the presence or absence of debris, as bedrock, clitter, mixed debris, regolith or sand) that exposed bedrock is the most significant for all five surveyed profiles, where it varies from 26.23% to 32.85% of the total lengths, and the mean for all the profiles is 29.85%. Debris is unable to collect on the scarp faces and they undergo mechanical weathering; broken rock falls down to form clitter on the rectilinear or slightly concave sections just below the scarp free-face. On average clitter covers 20.79% of the total surveyed lengths of the five profiles.

Sand, the finest of the debris types, occupies the relatively gentle slopes, mainly on the lower sections of the profiles; it is practically absent from the steeper slopes of the scarp faces. On the other hand bare rock occurs mainly at the steep slopes of the scarps (generally above 30°), but local rises on the profiles also account for some rock outcrops, at slopes of over 18°. Regolith is generally found over the lower middle and lower sections of the profiles, and clitter is found over a wide range of slope angles.

TABLE 4. Upper (U) & lower quartile (L) & mean values ($\bar{x}$) of slopes for different cover types (degrees).

Profile No.	Bedrock			Clitter			Mixed debris			Regolith			Sand		
	$\bar{x}$	U	L	$\bar{x}$	U	L	$\bar{x}$	U	L	$\bar{x}$	U	L	$\bar{x}$	U	L
R-1	26	40	10	26	37	13	31	40	25	19	26	12	14	20	10
S-2	20	39	16	16	28	6	29	36	20	22	27	20	14	17	12
G-3	29	36	23	20	24	10	28	39	17	23	32	16	9	11	8
N-4	32	40	26	22	31	11	24	31	17	16	21	14	10	11	9
K-5	30	40	23	22	33	8	28	38	19	19	25	14	6	8	5
All	27	39	19	21	31	10	28	37	20	20	26	15	11	13	9

Table 4 gives the mean and quartile ranges of slope angles under the different debris types. The mean of bare rock sections range between 20° and 32° while the mean for all profiles stands at 27°.

The means for rock and mixed debris are very close (27° and 28°), while clitter (21°) and regolith (20°) also have virtually identical mean values; sand has the lowest mean value, of 11°. The interquartile range of bedrock is very wide (19°-39°) while that of sands is very limited (9°-13°). The moderate interquartile ranges of clitter and mixed debris are more or less identical (21° and 17° respectively). It is apparent that the steepest slopes have developed over bedrock and mixed debris, the blocks of sandstones outlined by joints being dislodged and removed either by gravity or wash during sudden cloudbursts, and then becoming further fragmented downslope.

SLOPE EVOLUTION

The level and gentle slopes in the northern lower plateau, locally known as the Nagod Plain, and in the Tons Plain region are the result of the penultimate stage of terrain development, due to lateral planation by left-bank feeders of the Satna and Tons rivers. The thickness of alluvium over such slopes varies from about 18m near the main Satna and Tons rivers to 8m near the toes of the hill slopes i.e. the northern and eastern scarps. The scarp zones or hill slopes are concave in profile with an upper free-face element. The concave section is covered with scree, with its maximum thickness at the base and thinning out up-slope. The free-face section is devoid of any scree. The evolution of hill slopes of scarp zones may be interpreted in terms of the slope replacement model of W. Penck (1924-25) and Fisher-Lehmann's process-response model (Fisher, 1866, Lehmann, 1933; 1934). The free-face section is subjected to severe physical and chemical weathering with the result that blocks of weathered rock fall down the slope. It is pertinent to point out that the sandstones in the alternate bands of sandstones and shales are well jointed and fine to medium grained, and are easily weathered, yielding a substantial volume of scree. The scree resulting from rock-fall from the free-face is transported down-slope due to gravity, and by surface wash and rills. Frequent showers during the rainy months (June to September) yield maximum runoff which transports the scree down-slope. Thus the free-face undergoes gradual parallel retreat, maintaining its maximum angle, and scree accumulation extending up-slope from the base of the hill replaces the free-face from below. The recession of the free-face has given birth to a rock core of convex plan over which rests the weathered scree derived from up-slope. Since there is no active erosion at the base of the hill sides and there is no substantive removal of materials from the slope base the debris slope is extending up-slope at the expense of the free-face. Wherever the recession of free-face segments on both sides of the hill-range has been particularly active the free-face has been eliminated and has been succeeded by a summital convexity, giving rise to a convexo-concave slope profile, as illustrated by Sharda Pole near Sharda Temple of Maihar. This hill exemplifies the complete elimination of scarps with free-faces. The surface wash has been so active that a major part of the previous scree has been removed and at certain places the former rock core of convex plan buried under the scree has been exposed and breached.

The observation of L. C. King (1951; 1953; 1962) whereby "slopes having scarp are 'normal' and if it is absent the 'active' element in slope retreat is gone, this applies particularly to humid temperate regions where slopes are generally relict, owing their form to evolution under periglacial conditions" may not be tenable, as examples of slopes with scarps experiencing active retreat and convexo-concave slopes having no scarp are found within a distance of about 8km, near Maihar. There are also other examples in the study region. A hill range between Naktara and Kunhwara villages, projecting out for a distance of about four kilometres from the main eastern rim of the plateau, has steep scarps on its northern and southern sides, the scarps are undergoing parallel retreat and are progressively narrowing the hill top (at present it is hardly 200m wide). In future the retreating scarps will meet and will thus be eliminated and succeeded by a summital convexity, resulting in a convexo-concave profile.

CONCLUSIONS

The present analysis suggests that the precipitous free-face element of the Bhander scarps has been maintained because of the presence of a massive sandstone capping lying over weaker shales. The Bhander scarps are characterised by four slope elements: summital convexity, free-face, rectilinear slope and basal concavity. The precipitous free-face section is subjected to mechanical disintegration assisted by oxidation and the disintegrated materials fall down-slope. The Bhander scarps are experiencing parallel retreat due to back wasting and the steepness of slope angles is maintained. Slope decline due to down wasting is active only over those residual hills where the cap rock of sandstone has been removed.

REFERENCES

Fisher, O., 1866. On the disintegration of a chalk cliff. Geological Magazine, 3, 354-356.

King, L. C., 1951. South African Scenery, 2nd edn. Oliver and Boyd, Edinburgh.

King, L. C., 1953. Canons of landscape evolution. Bulletin, Geological Society of America, 64, 721-751.

King, L. C., 1962. Morphology of the Earth. Oliver and Boyd, Edinburgh.

Kumar, A., 1981. The nature of slope profiles on some residual hills in the Jamalpur-Kiul Hills, Monghyr, India. Zeitschrift für Geomorphologie, 25, 391-399.

Lehmann, O., 1933. Morphologische Theorie der Verwitterung von Steinschlagwanden. Vierteljahrsschrift Naturforsch, Ges. Zurich, 78, 83-126.

Lehmann, O., 1934. Ueber die Morphologischen Folgen der Wandwitterung. Z. Geomorph. 8, 93-9.

Singh, S., and Pandey, R. S., 1982. A gemorphological study of slopes in the adjoining areas of Nagod. National Geographer, 17, 107-120.

Singh, S., and Pandey, R. S., 1983. Morphogenetic processes, landforms and environmental planning in Bhander plateau region, India. National Geographer, 18, 29-54.

Young, A., 1970. Slope form in part of the Mato Grosso, Brazil. Geographical Journal, 136, 383-392.

Young, A., 1972. Slopes, third impression (1978). Longman, London.

International Geomorphology 1986 Part II
Edited by V. Gardiner

PERIGLACIAL LANDFORMS IN THE NORTHERN PART OF THE GREATER XINGAN RANGES, CHINA.

Zhu Jinghu

Geography Department, Harbin Normal University,
Harbin, People's Republic of China.

ABSTRACT

The northern part of the Greater Xingan Ranges has an extensive range of fossil and contemporary periglacial landforms, which includes blockfields, blockslopes, tors, nivation hollows, cryoplanation surfaces, patterned ground, pingos, flat-floored and asymmetric valleys; gelifluction deposits, loess and periglacial involutions also occur. Four periglacial landform regions can be recognised, characterised by the intensity of development of periglacial landforms. Periglacial phases occurred in the Mid and Late Pleistocene and a subperiglacial phase occurred during the Holocene.

INTRODUCTION

The northern part of the Greater Xingan Ranges is at the southern edge of the high latitude periglacial zone of the Eurasian continent. This is the most important periglacial area in China, and has more than thirty types of periglacial landform, some of which were first found in North-east China. The area has experienced periglacial conditions for much of the Quaternary and so mid-Pleistocene, late Pleistocene and modern periglacial phenomena all occur in the area.

THE MAIN PERIGLACIAL LANDFORMS AND SEDIMENTS

Blockfields, blockslopes and tors. Extensive blockfields are developed on level mountain summits and on gentle hillslopes about 600-700m above sea level in the north of the area, whereas modern blockfields generally occur on cryoplanation surfaces 1300-1400m above sea level, and have a lesser extent. Blockfields are mainly made up of angular boulders of rhyolite, 30-50cm in diameter, and with some giant blocks.

Blockslopes are extensively distributed on convex hillslopes having gradients between 24° and 28°; their debris are nearly all frost-susceptible rhyolite. In the middle and upper parts of blockslopes stone-banked terraces occur. These are commonly 6-7m high, with the highest being 13m. In some regions stone pavements have developed.

Tors are found mainly on mountain summits or flanks. In most cases they occur singly but in a few cases they appear as a group. In the northern part of the area angular rhyolite and tuff tors have relative heights ranging from a few metres to more than 20m; in the south of the area tors of coarse granite have relative heights of up to 40m.

Geliflurtion deposits. These commonly spread down the gentle slopes, forming lobes, benches and terraces with steps ranging from 10cm to over 1m in height. Rock fragments, sand and clay occur in these unbedded deposits. At the bottom of dells there are normally two layers of gelifluction deposits beneath a metre of alluvial gravel. The upper layer is a yellow-dark brown gravel loam, approximately five metres thick. The lower layer is a yellow-white bouldery clay, about 2.5m thick. These layers represent two phases of gelifluction, the deposits being laid down in the dells formed by fluvial activity.

Nivation hollows and cryoplanation terraces. Nivation hollows are common at altitudes above 800-1000m. Their widths range from several hundred metres to a kilometre, depths from 80m to more than 300m, and slopes are from 3° to 10° in their lower parts. They are covered with boulders, sands and clay and at the outlet of the hollows pluvial-solifluction fans occur.

Cryoplanation terraces occur near the mountain summits. These have from three to seven steps, with heights from 10m to almost 40m. Terrace surfaces slope at gradients between 2° and 8° and the terrace scarps between 15° and 25°. Terraces are from 20m to 90m wide and the thickness of rock mantling them is less than 1m.

Stone circles, stone stripes, pingos, fossil ice-wedges and periglacial involutions. Sorted circles are extensively developed in this area. They occur mainly at the boundary between the floodplains and the slopes, where the ground has a slope between 3° and 5°. A few of these features occur on the high cryoplanation surfaces. The stone circles appear in groups, with the large sorted circles of granite boulders being over 8m in diameter. The smaller circles, of igneous rocks and some alluvial gravel, range from 3.8m to 4.6m in diameter.

Stone stripes occur on slopes with gradients between 15° and 30°, commonly being 30-50m long, and over 10m wide in the granite areas and 0.5 to 5m wide in other areas.

Pingos are found on the solifluction fans or on river valley bottoms near the headwaters. They are 1m to 5m high and are covered by turf and soil. Fossil pingos occur in the south of the area. For example, one in Ninth bog is pear-shaped, 400m long, 300m wide, 1-2m deep, and the bow-shaped bar of the pingo is 3 to 4.5m high. Deposits on its bottom consist of clay and silt.

Fossil ice wedges and periglacial involutions are extensively developed in the Pleistocene deposits on the east and west slopes of the Greater Xingan Ranges. In Pingtai and other places fossil ice wedges stretch into the underlying Baitu Shan formation. The upper parts of these ice wedges are 40-50cm wide, and they are 1.0-2.5m deep, with the deepest being 3.5m. The material in the wedges is the deposits of the lower part of the Pingtai Formation, with the debris generally being arranged in a vertical direction and with that at the ice wedge walls also being arranged directionally. In the lower part of the Pingtai Formation there are many periglacial involutions, with depths from 20cm to 2.7m.

Periglacial loess. This occurs mainly in the south-west of the area, on the mountain pass 90m above the river level and on river terraces at 6-7m, 20-25m and 70m above river level. The brown-yellow soil consists mainly of silt, with much calcium. It has small-scale stratification, with clayey lenticles and partings. In the upper parts there is debris and throughout the profile there are black mud lumps.

Flat-floored valleys and asymmetric valleys. Many small rivers have U-shaped valleys and small gradients. The river channels are very sinuous and some valley bottoms form string bogs. A series of east-west rivers which are tributaries of large rivers have asymmetric valleys. The north slope is steep and the south slope is more gentle. Differential moisture retention is the main factor in their evolution.

PERIGLACIAL LANDFORM REGIONS

The northern part of the Greater Xingan Ranges can be divided into four periglacial landform regions, according to climatic conditions, periglacial processes, periglacial landforms and their degree of development.

The most intensely developed region. The north-west part of the Greater Xingan Ranges has been subject to a semi-humid continental climate for a long time. The mean annual temperature is between -5.3°C and -8.7°C, the mean annual precipitation is 423-437mm. The amplitude of annual temperature variation can reach 74°C, and permafrost is extensive. Because of the cold climate, the large amplitude of temperature variation, the adequate moisture and frequent freeze-thaw alternations the main periglacial processes are frost heaving, thermo-thaw action and freeze-thaw weathering. The main periglacial landforms are stone circles, stone stripes, pingos, hummocks and thermokarst features.

The intensely developed region. The north-east part of the Ranges has semi-humid conditions with a mean annual temperature of -1.0°C to -5.0°C and a mean annual precipitation of 451-482mm. The amplitude of temperature variation is larger than in the first region. There is some snow every year and discontinuous permafrost is developed. Gelifluction and nivation are the most important processes, with frost heaving and frost weathering of secondary importance. Gelifluction terraces, gelifluction fans, blockstreams, nivation hollows and cryoplanation terraces are widespread.

The moderately developed region. The west slopes of the Greater Xingan Ranges experience cold and dry semi-arid continental climatic conditions. The mean annual temperature is -3.1°C, the mean annual precipitation 338m. Only sporadic permafrost exists in the dry soil and rocks. Fossil ice wedges, periglacial involutions and pingos have been formed by frost heaving, tors and stone stripes have formed by frost weathering, and the role of the wind is more significant than in the other regions. Thus there are ventifacts in the bottom of the valleys and on the river terraces, and sand dunes and periglacial loess also occur. In addition, dells, asymmetric valleys and string bogs are well developed in this region.

The weakly developed region. The east slope of the Ranges and the area fronting the mountains have warm and semi-humid climatic conditions. The mean annual temperature is 3.4°C and the mean annual precipitation 350-400mm. Because of the lower latitude, low altitude, higher temperatures and the destruction of vegetation permafrost is not well developed. Tors exist and there is a little periglacial loess. However some ancient periglacial phenomena, including fossil ice wedges, periglacial involutions, dells and asymmetric valleys, are very well developed.

PERIGLACIAL CHRONOLOGY

Two periglacial phases and a subperiglacial phase can be recognised on the basis of landform types, stratigraphy, clay mineralogy and pollen analysis.

Pingtai periglacial age. This is represented by fossil ice-wedges and periglacial involutions which occur extensively on the east slope of the Greater Xingan Ranges. Material which infilled fossil ice-wedges and which has been disturbed has been determined by thermoluminescence dating to date from about 538,000 ∓ 8,100 B.P. The pollen content indicates plant types typical of a cold and wet climate. In some profiles Pingtai strata are covered by Late Pleistocene periglacial loess, so the Pingtai periglacial age is dated about 400,000-600,000 B.P., in the Mid Pleistocene.

Hailaer periglacial phase. This is the most important periglacial phase and is represented by fossil ice-wedges and periglacial involutions which occur on the river terraces of the Hailaer River. Landforms associated with it are asymmetric valleys, nivation hollows, pingos, cryoplanation terraces, blockfields, stone stripes and circles, all of which are now inactive. By ^{14}C dating of fauna contained in Hailaer strata the phase has been dated to 15,200 ∓ 480 B.P., which ascribes it to the Late Pleistocene.

The Dabaishan subperiglacial phase. Periglacial processes did not cease during the Holocene, but were mostly concentrated in the north part of the mountain area, 1400-1500m above sea level. Modern periglacial phenomena such as blockfields, blockslopes, blockstreams and stone stripes are its representatives. These are 400-500m higher than similar periglacial phenomena from former periglacial phases, showing how altitude modifies latitudinal periglacial zones.

International Geomorphology 1986 Part II
Edited by V. Gardiner

PALSAS AND MINERAL PERMAFROST MOUNDS IN NORTHERN QUÉBEC

M. Allard, M.K. Seguin and R. Lévesque

Centre d'études nordiques,
Université Laval,
Sainte-Foy,
Québec, Canada,
G1K 7P4

ABSTRACT

Palsas and mineral permafrost mounds are widespread periglacial landforms characteristic of the discontinuous permafrost zone. Palsas are found dominantly in the forest tundra and mineral permafrost mounds occur mainly in the shrub tundra zone in somewhat colder environments, although exceptions and transitional features between the two landform types do exist. Almost all palsas have a mineral core within which most of the ice segregation and heaving have taken place. Post-glacial marine silts are the dominant material in which all these landforms are developed. Both palsas and mineral permafrost mounds can form either by the growth of single mounds or by dissection of large peat or permafrost plateaus. A general parabolic function curve has been established between feature size and permafrost thickness. Thermal and isotopic data indicate that the permafrost is in equilibrium with present climatic regime. However, the thermal equilibrium in each feature is controlled by vegetation height and density which control snow distribution around the sides. Peat stratigraphy and ^{14}C dates indicate that three generations of palsas and mineral permafrost mounds dominate amongst these landforms: from 1500 to 1050 BP, from 650 to 200 BP and very recent. For the last 150 years, degradation following active layer and slope processes has been dominant over incipience and growth of new features.

INTRODUCTION

Palsas, mineral permafrost mounds, frost-heaved bedrock mounds, icings, frost and naled blisters are all known to exist in the subarctic and arctic climate zones of Québec-Labrador. Until the contrary is eventually proven, only pingos are absent from this inventory. From that list, palsas and mineral permafrost mounds are of special interest for they are a major component of the northern Québec periglacial landscape. The term "mineral permafrost mound", used in this paper, applies to mounds with shape and size similar to those of palsas but made exclusively of mineral sediments; the terms mineral cryogenic mounds (Payette *et al.*, 1976; Payette and

Seguin, 1979; Lagarec, 1982), mineral palsas (Dionne, 1978; Pissart, 1983) and palsa-like landforms (Wrammer, 1972) have also been used for these features.

Since many of these landforms appear either in fens and bogs or are somehow related to the Holocene evolution of wetlands, indirect means to date them and to reconstruct their evolution are provided by using peat stratigraphy, pollen analysis and ^{14}C dating. The maximum age of some mounds and their growing conditions can also be deduced from post-glacial land uplift data for features that have developed on recently emerged terraces. The aim of this contribution is to present new data which, added to existing knowledge, will enable us to propose explanations for the inception and the evolution of these landforms. Whether their formation is linked to climate changes or not is also a question of great interest. Since landforms are to be named with terms related to both shape and genesis, terminological considerations will be further discussed in the conclusion.

PREVIOUS WORK AND PRESENT STATE OF KNOWLEDGE

After a comprehensive literature review, Washburn (1983a) concludes that palsas are "peaty permafrost mounds ranging from 0.5 to 10 m in height and exceeding 2 m in diameter, comprising (1) aggradation forms due to permafrost aggradation at an active layer/permafrost contact zone, and (2) similar-appearing degradation forms due to disintegration on an extensive peat deposits". This dual origin has been proposed by many investigators, including Schunke (1973), P'yvchenko (1969), Novikov and Usova (1979) and Lagarec (1980), the later in Northern Québec. Many authors also agree that palsas may occur in a large variety of shapes and that they are often genetically and spatially associated with peat plateaus (Wrammer, 1973; Hyvönen, 1972; Salmi, 1972: Åhman, 1977 and Zoltaï and Tarconai, 1975). Although the purely mineral frost mounds have not been widely accepted as real palsas, some authors mentioned their close kinship with them (Wrammer, 1972; Cailleux and Lagarec, 1977; Pissart and Gangloff, 1984), even to the point of coining the term "mineral palsa" (Dionne, 1978; Pissart, 1983; Gangloff and Pissart, 1983). The demonstrated dominance of palsas consisting of a peat cover over a frozen mineral core, as mentioned by French (1976) and by Brown and Kupsch (1974), has lead Åhman (1976) to consider peat as being only incidental in palsa formation and to use the term "minerogenic palsa".

As to the origin, it is now being increasingly accepted that palsa inception and growth are initiated after most of the peat comprised in the landform has accumulated (Zoltaï, 1972; Zoltaï and Tarnocai, 1975; Seppälä, 1983; Allard and Seguin, in press) although it has been for long believed that palsas form by the combined action of peat accumulation and ice segragation in the underlying mineral soil (French, 1976; Hamelin and Cailleux, 1969; Heim, 1976, Dever _et al_., 1984). Seppälä's (1982) recent experiment, following ideas expressed as soon as the turn of the century (Fries and Bergström,

1910), has stressed the importance of thinned snow cover for the inception of the frozen core. The insulating effect of peat drying upon heave for the preservation of the incipient permafrost has been discussed many times (see Zoltaï and Tarnocai, 1975). It has been recently suggested that segregation of ice in the ground heaving process is fed by suction of unfrozen soil water to the freezing front and, to some extent, within the frozen ground along temperature and water gradients in the soil (Pissart's cryosuction, 1983; see also Smith, 1985). It has been suggested that these landforms may be inherited from past climatic deterioration periods although it is also argued that changes in the local environment (e.g. snow distribution change, drainage variation, peat thickness increase) are sufficient to initiate them (Seppälä, 1982; Savoie and Gangloff, 1980).

REGIONAL AND SITE FACTORS IN NORTHERN QUEBEC

General distribution and the ecological zones. Nearly all palsas and permafrost mounds are grouped in scattered fields either within the forest tundra zone or in the shrub subzone of the tundra (Fig. 1) (Payette *et al*., 1976), although such landforms have been reported either in more severe climate zones such as the tundra (Savoie and Gangloff, 1980, see also Washburn, 1983b) or in a bog in the coastal forest tundra not far from the gulf of St-Lawrence River (Dionne, 1984). At a higher altitude (950 m), a small palsa was even found 100 km only northeast of Québec City in an alpine forest tundra zone (Payette, 1984). Lagarec (1982) mapped the distribution of palsas and mineral permafrost mounds east of Hudson Bay. In agreement with observations and more detailed local mapping by Seguin and Allard (1984), Lagarec's map shows that a close relationship exists between landform types distribution and ecological zones: mineral permafrost mounds are widespread in the shrub-tundra zone north of the treeline while peat landforms are found principally in the fens and the peat bogs of the forest tundra.

However, some exceptions and some transitional forms do exist. In the shrub tundra zone, some mounds have a small peat cap on their top; this peat cap is 1-1.5 m thick and measures about 2-3 m across. As one moves closer and closer to the treeline, peat cap size increases progressively and eventually makes a cover that blankets over 90% of the mound's surface. Along a 10 km long transect from out in shrub tundra to within the forest tundra, a geomorphological gradient from mineral permafrost mounds to palsas is thus evident. Mineral permafrost mounds also exist in the forest tundra where they can be found under three circumstances: 1 - as mineral mounds amongst a palsa field; they appear to be former palsas that have been stripped of their peat cover by erosion, 2 - on some river terraces where local drainage and regional geomorphology suggest that peat has never been present (Gangloff and Pissart, 1983) and 3 - along the shoreline, on the lower terrace just above the spring tide level; a situation where peat had no time to accumulate due to recent land emergence.

Hence mineral permafrost mounds, with the exception of those that might result from the erosion of the peat cover on palsas, are found in colder environments, for example north of the treeline, that is north of the isotherm of 10°C for the warmest month of the year and under mean annual temperatures at least below -5°C as is the case in the Nastapoca river area (Fig. 1). They also occur in colder sites within openings in the forest-tundra as well as under wind-blown and nearly snow-free sites such as unforested terraces and low terraces along the seashore. Palsas have a more southern distribution, in the range of -1°C to about -5°C mean annual temperature (Dionne, 1978, 1984).

Quaternary deposits. Palsas and mineral mounds have been reported and studied mainly from areas underlain by marine silty sediments deposited either in the post-glacial Tyrrell sea (Hudson Bay basin) (Hamelin and Cailleux, 1969; Lagarec, 1982; Brown, 1979) or d'Iberville sea (Ungava Bay basin) (Lagarec, 1976; Pissart and Gangloff, 1984). These fine sediments are a favourable material for ice segregation. However, palsas and mineral permafrost mounds have also been reported in thick marine sands (Seguin and Allard, 1984a) and we have found some that had developed in gravels, in till overlain by a peat layer and in a stony glaciomarine diamicton. With the exception of a few very small features (less than 60 cm high) found in a deep fen (more than 2 m of sedge peat), all palsas have a mineral core; this observation is in accordance with recent ones from other countries, for example Scandinavia (Åhman, 1976; Seppälä, 1983).

Characteristics of the permafrost. The permafrost contains principally ice lenses, but thin ice veins from one to five millimetres thick are also abundant and a few ice wedges have been found in a polygonal peat plateau. Most ice layers are less than 5-10 mm thick and form a reticulate network with the veins (Mackay, 1974); this network is apparent mostly in marine silts. Ice layers up to 30 cm thick have been found at some intervals both in sections or through coring. Some of these thicker ice layers are often found along the stratigraphic contact between the peat layer and the mineral substrate.

The surface of a high peat plateau in Sheldrake River area (Fig. 1) is patterned with ice-wedges polygons 14 m in diameter. The ground surface within the polygons (mostly hexagons) is nearly flat. The ice-wedges are 30 cm wide and their top is coincident with the permafrost table. Five to fifteen centimetre wide cracks dissect the peat surface above the wedges and contain collapsed peat from the sides, wind-blown peat dust and ponded water in summer.

Permafrost thickness varies with the size of palsas and mounds. Forty-six electrical resistivity soundings in mineral permafrost mounds in the Nastapoca River area indicate that permafrost thickness averages about 20 m with a minimum of 6 m and a maximum of 37 m. During that field season only mound widths were systematically measured. The relationship between mound width (W) and permafrost thickness (T) is expressed by the following equation:

$$T = 2.29822\ W^{0.454545} \times 0.993773^{W} \quad (1)$$

In Sheldrake River area, the same type of statistical analysis was carried out on palsas. Small ones were cored through (to a depth of 4 m) and permafrost thickness in the larger ones was measured with the help of electrical resistivity soundings (for methods, see Seguin, 1976; Seguin and Crépault, 1979; Seguin and Allard, 1984a and b). The length (L), width (W), height (H) and permafrost thickness (T) of 39 peat plateaus and palsas were measured. The best statistical relationship between permafrost thickness and landform size was obtained with volume (V) approximations simply defined by L x W x H (Blais, 1984):

$$T = 0.707719\ V^{0.304009} \times 0.999994^{V} \quad (2)$$

Both curves (Fig. 2) show a tendency for permafrost thickness to increase with feature size until it tends to a maximum although the fit is not very good for very large peat and mineral permafrost plateaus. The ratio between mean palsa height above the fen level and mean permafrost thickness is 0.37. Hence palsa height makes in general a little more than one third of permafrost thickness.

Climate and thermal regime. Permanent thermistor probes have been installed in one palsa and in three mineral permafrost mounds (Fig. 3. See also Seguin and Allard, 1984a, fig. 10). In the near-surface permafrost (down to 2 or 3 m deep) thermal fluctuations are in close agreement with the local climatic regime. For instance, in the palsa near Kuujjuarapik (Fig. 3) thaw almost reaches its maximum depth by the end of July; however, permafrost temperature below the active layer keeps going up until October (from -1.22°C to -1°C in 1982). On January 11th of the same winter, all the active layer had frozen back but the "heat wave" of the previous summer had just reached the base of the probe (-0.75°C). Cooling of the profile then goes on to April (-6.4°C at 185 cm) at which time warming starts from the surface.

Mean soil surface temperature is approximately -4.5°C in Kuujjuarapik, -5.8°C in Nastapoca, and between -2°C and -5°C in Manitounuk 1. These values are very close to recorded mean air temperatures and reflect the fact that palsas' and mounds' summits are snow-free in winter (Plate 1). The Manitounuk 2 site is a wooded (*Picea glauca*) mound; as the tree cover acts as a snow trap (Payette *et al.*, 1975), snow accumulates up to very near the summit and only a few square metres on top remain exposed in winter. Thus the -1.8°C mean permafrost temperature in that specific mound is about 3 degrees higher than mean annual air temperature due to thermal insulation provided by the snow layer.

Hence permafrost in palsas and cryogenic mounds maintains itself in a delicate thermal equilibrium with both present climatic regime and local terrain factors, principally snow cover distribution which is closely dependent on vegetation structure. Dever *et al.* (1984), working in the LG2 area (Fig. 1), have also shown with 3H determinations on lenticular ice that modern ice forms within palsas. These data on permafrost thermal and hydrologic regime are insufficient however to state that the landforms themselves are in equilibrium with present climate.

GEOMORPHOLOGICAL CHANGES

Morphological types; plateaus and single mounds. Palsa fields occur principally in two types of patterns in northern Québec: 1- in stripes: palsas are elongated transversally to the valley into which they occur and are roughly parallel to each other (Plate 2); 2- in complex patterns: there is no regular organisation and the landforms occur in various shapes such as sinuous, round, starshaped or horseshoe-shaped (plate 3). In both cases, a general summit concordance is perceptible although local doming or hollowing seem to have occurred. Both the elongated and parallel palsas and the complex forms are parts of patterns including plateaus and single palsas. Peat plateaus are about 3.3 m high and they average 75 m in length and 25 m in width; maximum dimensions are 300 m X 50 m X 8 m.

Mineral permafrost mounds also occur in patterns. Seguin and Allard (1984a), for instance, have described four permafrost landscape patterns in such landforms from the Nastapoca river area. There too, parallel and elongated mounds transverse to valleys (and the dominant wind) are important; they are the interfluves of an "en treillis" gulley system (Seguin and Allard, 1984a, fig. 2). Permafrost plateaus made of large heaved surfaces over 200 m across co-exist with individual mounds as small as 6 m in diameter.

Observations (Seguin and Allard, 1984a; Dionne, 1978, 1984; Lagarec, 1980, 1982), comparisons of old and recent air photographs (1948 and 1974 in Pissart and Gangloff, 1984) and repeated mapping (Payette *et al*., 1976; Samson, 1975) have demonstrated that peat plateaus and mineral permafrost plateaus are degrading and have been for a time affected by thermokarst processes that are dissecting them into individual mounds. One can actually see palsas or mounds neighbouring plateaus of which they were part and from which they most apparently were isolated by the melting of ground ice (Plate 4).

Active layer and slope processes. Due to the scarcity or absence of peat on mineral permafrost mounds, the active layer is deeper on them than on palsas. While active layer depth in peat is about 35-45 cm, it is 100-125 cm deep in silts (Fig. 3) and somewhat more in sands. In silty soils, a widespread network of mudboils extends on top of permafrost mounds and plateaus; on terrain with slopes as low as 1°, the mudboils tend to become elongated transverse to slope direction due to shearing induced by gelifluction. Active layer thawing on the steeper outer slopes of mounds induces thaw-slumping. Hence, as surface material is removed by thaw-slumping and gelifluction on the sides and the summit of each mound, new thermal conditions are set within the permafrost and degradation of the landform can continue.

Peat thickness on palsas varies from nothing where it has been eroded to a measured maximum of 2.7 m. This peat cover is being gradually destroyed; tension cracking due to doming and dessiccation has induced a network of polygons about 1-2 m in diameter in almost all peat covers. Wind erosion of peat is evident, principally in winter as blown dust is interstratified with snow layers in the snowbanks around the palsas. Rainsplash and sheetflow in summer contribute to erosion and thinning of the peat cover as small "peat flows" are visible on the sides of the features. When peat is less than 30-40 cm deep, summer thaw penetrates underneath in the mineral sediments. From then on, silt and fine sand surging up from the mineral substrate in mudboils disrupt the peat cover and solifluction helps to evacuate it. Polygonal peat blocks also gradually slide from the summits to the sides as they are carried over the slowly flowing substrate. Slope processes become similar to those of mineral permafrost mounds.

However, where deep peat occurs on top of the side slopes, these are steep and often vertical with overhanging peat blocks. As the exposed mineral substrate melts at the base of the slope, the overhanging blocks collapse. The fallen blocks eventually disintegrate in ponded water; when they no longer protect the slope foot, permafrost melting re-initiates the process again.

Evidences of recent degradation and slope backwasting. Peat plateaus are apparently being dissected along cracks. Although many cracks have been investigated in numerous areas, ice-wedges were found under one site only. However, melting along furrows and cracks, ice-filled or not, is the most probable mechanism for plateau dissection (Lagarec, 1980; P'yavchenko, 1969). Surface lowering and slope retreat is faster along these cracks (Plate 5). Ponds and fens expand in the gap between the isolated mounds and their parent peat plateaus so that each mound is really a permafrost island; confusion with high center polygons is easy to avoid (Zoltaï and Tarnocai, 1975).

Recent thaw slides which disrupted the vegetal cover are readily evident and most of the trees on palsas appear to have been disturbed by soil movements. Fifteen mature black spruces (Picea mariana) were sampled in Sheldrake River area, near the tree line, and tree ring counts were performed on reaction wood that has formed after each stem oscillation, a method used by Zoltaï (1975). On Fig. 4, it can be seen that, with the exception of only three individuals, all trees were born after 1870; this suggests a somewhat recent colonisation of the palsas. Since this date, tilting of trees has remained incessant and small peaks of slope activity in the years 1935-42, around 1953 and since 1969-70 suggest short periods of faster degradation.

Growing features. Some small palsas (4 to 10 m in diameter; 0.6 to 1.2 m in height) coexist here and there in fens amongst larger palsas and peat plateaus. In a few cases, the peaty bottom of shallow ponds has been recently heaved (see also Pissart, 1983). These small palsas "float" on the peat. When coring through the frozen peat (1.9 m thick) water rose in the drill-hole to the level of the standing water in the surrounding fen. When permafrost extends into the underlying silts in somewhat larger forms (1.5 m high), the palsas are "grounded" into the substrate. These observations are in accordance with Zoltaï's (1972) in western Canada and with modelling by Outcalt and Nelson (1984).

Some small, and probably incipient, mineral permafrost mounds were also found in silts in Kangiqsualujjuaq (Fig. 1) along the shoreline (Plate 6). Both small palsas and incipient mineral permafrost mounds occur in windswept areas almost devoid of snow-cover. Winter observations show this is especially the case along the shore. In the palsa fields, the small palsas are found in open space at some distance from the larger features which accumulate deep snowbanks on their sides. They are thus initiated in areas of very shallow snow cover; after they have grown up a few centimetres, they become exposed for the subsequent winters and permafrost can aggrade (Seppälä, 1982, 1983; Outcalt and Nelson, 1984).

THE AGE OF PALSAS AND MINERAL PERMAFROST MOUNDS

When segregation ice begins to accumulate in the ground, the bog surface heaves above the ground water level and a more xeric vegetation starts colonising the newly formed palsa. Kershaw and Gill (1979) have shown that remains of hydrophytic vegetation can be found in the peat cover up to very near the surface where a layer of volcanic ash provided a maximum age of 1200 years. On palsas in general, sedge and moss peat occur in the peat stratigraphy up to near the top of the sequence where it can be covered by forest peat (Zoltaï and Tarnocai, 1975). The surface heaving may also result simply in the cessation of peat accumulation (Savoie and Gangloff, 1980).

Allard and Seguin (in press) have studied on top of mineral permafrost mounds small peat remnants on what should have formerly been palsas. Two of these peat remains were themselves covered by slope debris indicating a complex Holocene history of repeated aggradation and degradation of permafrost landforms. The pollen composition of this sedge peat and ^{14}C dates indicate that for the time of peat accumulation until its burial by more recent sediments, that is from 5520 $\pm$ 160 BP (UQ-622) to 2300 $\pm$ 370 BP (UQ-632) in one case and from 4320 $\pm$ 120 BP (UQ-630) to 1560 $\pm$ 110 BP (UQ-629) BP in the other case, there was no permafrost at these particular sites. In this type of situation, the dates on the top of the peat layer are to be considered as maxima for the beginning of growth of these two mounds.

Richard (1981, p. 104) established the following sequence of local environmental changes using pollen analysis of palsa peat in Leaf River area (Figs 1 and 5); by 5000 BP, a shallow pond was occupying the site, probably since deglaciation that had taken place about one millenium earlier. A fen, with pools of open water, progressively invaded the pond until about 3800 BP; this was followed by a progressively dryer phase until 2965 BP, at which date palsa growth resulted in the near ending of peat accumulation. Heim (1976) has established in Ouiatchouan bog (Fig. 1) that a fen (Cyperacae dominated peat) was initiated by 4920 BP and replaced around 3500 BP by a bog (Sphagnum dominated peat with Ericacae); only the top 5 centimetres, dated as "modern", reflect drying of the surface and dense colonisation by shrubs (mainly Betula glandulosa). Fig. 5 shows schematically the peat stratigraphic sequence which has been established at six locations in northern Québec.

From these considerations, radiocarbon dating of summit peat from palsa (and mineral permafrost mounds when there are suitable peat remains) can provide an age estimate for mound inception. In such an attempt, five sources of error must be minimised:

1- Contamination by living rootlets from surface vegetation: after the modern litter is removed from the sampling site and the subsurface sample has been collected, rootlets and all apparently living matter is removed with tweezers.

2- Slow rate of peat accumulation: data from the above cited authors indicate that peat accumulation rates since the mid-Holocene have been slow (average 0.075 cm yr^{-1}; often as low as 0.01 cm yr^{-1}). Hence a sample 1 cm thick may represent 100 years of sedimentation and a thicker sample may yield a date with a standard error larger than the period represented by the sampled layers; the depth of sampling is also critical in order to obtain the best age estimate for the incipience of the mound. Samples must be sliced thin with a sharp knife.

3- Surface erosion of peat: since the original surface is eroded here and there, care must be taken in choosing the collecting site in order to get a sample really representative of the non-eroded, original peat.

4- Difficulty in identifying Sphagnum remains to the species level. This may cause an error since some species (e.g. S. fuscum) naturally tend to form mats a few centimetres above the bog water level. Whenever possible, macroremains must be checked to help in the dating of pertinent samples.

5- ^{14}C dating errors: a sufficient quantity of peat must be sampled to provide for cleaning, pre-treatment and sufficient CO_2 or C_6H_6 production for precise radiocarbon dating.

Since many of these potential errors are unavoidable to some extent, a statistical approach on many ^{14}C dates should help to compensate for errors on single dates. Figure 6 is a cumulative statistical weight histogram of 21 ^{14}C dates (Table 1). Statistical weights are calculated for each date on 10 years class interval from probability

TABLE 1. Radiocarbon dates of palsas and mineral permafrost mounds (+ two other maximum age determinations from emergence data)

Date	± 1 σ	Lab. no	Location (fig. 1)	Comments	Reference
2965	75	Beta-1081	Leaf River	palsa	Richard (1981)
2300	370	UQ-632	Nastapoca	Top of buried residual peat on mineral mound	Allard and Seguin (in press)
1910	80	I-13408	Sheldrake	palsa	Allard, unpub.
1700	70	UL-109	"	palsa	" "
1560	110	UQ-629	Nastapoca	idem UQ-632	idem UQ-632
1400	80	I-13132	"	top of buried peat bed in a permafrost plateau	" "
1350	80	I-13422	Sheldrake	palsa	Allard, unpub.
1260	60	UL-93	Kangiqsuallujjuaq	palsa	" "
1240	60	UL-91	"	palsa	" "
1200	80	I-13304	Sheldrake	palsa	" "
1100	70	UL-97	Kangiqsuallujjuaq	palsa plateau	" "
900	60	UQ-611	Koroc river	palsa, eroded surface: date older than palsa inception	Mathieu (1983)
600	60	UL-110	Nastapoca	palsa	Allard, unpub.
580	75	UQ-156	Killiniq	"	Savoie and Gangloff (1980)

Date	± 1 σ	Lab. no	Location (fig. 1)	Comments	Reference
550	60	UL-99	Kangiqsuallujjuaq	palsa	Allard, unpub.
470	90	QU-466	Leaf River	buried wood in mineral mound	Payette and Seguin (1979)
410	60	UQ-422	Koroc river	palsa	Mathieu (1983)
370	70	UL-103	Kangiqsuallujjuaq	mineral mound; dead *Sphagnum* heaved with sand	Allard, unpub.
350	60	UL-94	"	palsa	"
280	60	UL-112	Nastapoca	palsa	"
110	70	UL-102	Kangiqsuallujjuaq	palsa	"
modern	-	Lv-795	Ouiatchouan	palsa	Heim (1976)
modern	-	UQ-580	Koroc river (Labrador)	palsa	Mathieu (1983)
modern	-	UL-139	Ouiatchouan	palsa	Allard (unpub.)
modern	-	UL-141	"	palsa	"
Other age determinations from emergence data					
Younger than 300 years			Manitounuk	Mineral permafrost mound	Seguin and Allard (1984b)
Younger than 3000 years			Leaf River	" "	Payette and Seguin (1979)

tables for a normal distribution ($\pm$ 2 σ). Statistical weight of dates overlapping at $\pm$ 2 σ are summed to build up the histogram. Two major peaks show palsa inception periods from 1500 to 1050 BP and 650 to 200 BP. A number of "modern" dates (Table 1) also show that some palsas are of very recent inception.

DISCUSSION

Most palsas and mineral permafrost mounds have undergone some thermokarstic backwasting of their slopes. Summit downwasting by gelifluction and surface erosion has also been effective on many features. The overwhelming abundance of thermokarst ponds over mounds (Plates 2, 3 and 4) and the evident evolution from mineral and organic plateaus to separated mounds and palsas clearly indicate that permafrost was formerly more widespread than presently. Observation of active processes and dendrochronological data also indicate that degradation has been dominant for the last one hundred and fifty years. As the frozen mounds are permafrost islands in the discontinuous zone, each one has its own thermal regime which is regulated by snow cover distribution around its sides. Snow cover itself is largely dependent upon the vegetation structure. In a warmer climatic period, such as the one that extended from 1870 to 1960 (Kelly et al., 1982), tree and shrub population expansion occurred (Payette and Filion, 1985) and thermal conditions conducive to thermokarst degradation were set in the landforms as illustrated by the Manitounuk 2 case. Thinning of the permafrost goes on with mound shrinking.

Statistical compilation of ^{14}C dates however illustrates two major Holocene periods for palsa inception in northern Québec. These periods coincide very well with known cold intervals that Filion (1984) has identified by analysing periglacial dune stratigraphy. Gagnon (1983) has also shown that the treeline has receeded a few kilometres during the same periods. The two intervals are climatic deterioration peaks in a general cooling trend since the mid-Holocene (about 3500-3000 BP) (Richard, 1981; Short and Nichols, 1977; Heims, 1976; Filion, 1984). Forest and krummholz have generally regressed in northern Québec and their density has been reduced between 3000 and 2100 BP, 1800 and 1050 BP, 800 and 650 BP and 450 and 100 BP, while they have somewhat expanded in the intervening periods as well as in the last one hundred years (Payette and Gagnon, 1985).

However, modern dates and incipient features also show that local factors alone, amongst which snow distribution is dominant, can initiate a palsa or a frost mound during a relatively warm period. Hence, climate change is not an absolute prerequisite (Seppälä, 1982). However, a period of climatic deterioration increases the probability for mound initiation and growth because vegetal structures then become sparser, winters become of a more continental type with less snowfall, subfreezing temperatures get colder and thaw seasons get shorter or cooler.

Both palsas and mineral permafrost mounds result from the build up of segregation ice in mineral sediments. Both types of features also have a thermal regime which is in equilibrium with climatic conditions and controlled locally by snow distribution around them. ^{14}C data (Fig. 6 and Table 1) indicate that the older ones were generally formed during the same cold climatic periods. All these landforms can grow as single mounds or result from thermokarst dissection of mineral or peat plateaus. Only the peat cover makes a significant difference in appearance and is responsible for different slope processes on the mounds; peat thermal properties are also responsible for a more southern distribution for palsas. Hence, the term "mineral palsa" used by Dionne (1978), Pissart (1983) and Pissart and Gangloff (1984) for mineral permafrost mounds is not an illogical one and in many cases it corresponds to the original definition of palsas: "hummocks rising out of a bog with a core of ice" (Seppälä, 1972). However, in northern Québec, the problem with the term mineral palsa will arise when we will come to deal with mounds of more variable shape and size, in very variable sediments and associated with other features such as polygons. Furthermore, the word palsa has so evidently meant peaty frozen mounds for years to so many researchers that it is difficult to modify its general acceptation. These are the reasons for the use of the simple expression "permafrost mounds" ("Buttes de pergélisol" in French).

An identification problem between palsas and mineral permafrost mounds may be faced when dealing with transitional features which have a partly eroded peat cover. We suggest to start considering a feature as a mineral permafrost mound when the peat remnants have been reduced to the point that annual thawing and surface processes take place almost exclusively in mineral sediments rather than in peat. This criterion is easily assessed in the field.

REFERENCES

Åhman, R., 1976. The structure and morphology of minerogenic palsas in Northern Norway. Biuletyn Periglacjalny, 26, 25-31.

Åhman, R., 1977. Palsa i NordNorge. Medd. från Lunds Un., Geog. inst., Avhandlingar 78, 165 pp.

Allard, M.,and Seguin, M.K., in press. The Holocene evolution of permafrost near the tree-line, East of Hudson Bay, Northern Québec. Canadian Journal of Earth Sciences.

Blais, R., 1984. Modèles mathématiques pour l'évaluation de l'épaisseur du pergélisol des palses de la rivière Sheldrake. Université Laval, département de géographie, mémoire de baccalauréat, non publié, 69 pp.

Brown, R.J.E., 1979. Permafrost distribution in the southern part of the discontinuous zone in Québec and Labrador. Géographie physique et Quaternaire, 33, 279-290.

Brown, R.J.E.,and Kupch, W.O., 1974. Permafrost Terminology. National Research Council of Canada, Associate Committee on Geotechnical Research, Technical memorandum no 111, NRCC 14274, 62 pp.

Cailleux, A.,and Lagarec, D., 1977. Aspekte des Periglazials in Kanada. Nova Acta Leopoldina, Neve Folge Nr. 227, Bd. 47, pp 9-49.

Dever, L., Hillaire-Marcel, C., and Fontes, J. Ch., 1984. Composition isotopique, géochimie et génèse de la glace en lentilles (palsen) dans les tourbières du Nouveau-Québec (Canada). Journal of Hydrology, 71, 107-130.

Dionne, J.C., 1984. Palses et limites méridionales du pergélisol dans l'hémisphère nord: le cas de Blanc-Sablon, Québec. Géographie physique et Quaternaire, 38, 165-184.

Dionne, J.C., 1978. Formes et phénomènes périglaciaires en Jamésie, Québec subarctique. Géographie physique et Quaternaire, 32, 187-247.

Filion, L., 1984. A relationship between dunes, fire and climate recorded in the Holocene deposits of Québec. Nature, 309, 543-546.

French, H.M., 1976. The periglacial environment. Longman, London and New York, 309 pp.

Fries, I., and Bergström, E., 1910. Några iakttagelser ölver palsar och deras förëkomst i nordligaste Sverige. Geol. Fören. Stockholm Förh., 32, 195-205.

Gagnon, R., 1982. Fluctuations holocènes de la limite des forêts, Rivière aux Feuilles, Québec nordique: une analyse macro-fossile. Université Laval, Ph.D. Thesis, 108 pp.

Gangloff, P., and Pissart, A., 1983. Évolution géomorphologique et palses minérales près de Kuujjuaq (Fort-Chimo, Québec). Bulletin de la société géographique de Liège, 19, 119-132.

Hamelin, L.E., and Cailleux, A., 1969. Les palses dans le bassin de la Grande Rivière de la Baleine. Revue de géographie de Montréal, 23, 329-337.

Heim, J., 1976. Étude palynologique d'une palse de la région du golfe de Richmond (Nouveau-Québec, Canada). Cahiers de géographie de Québec, 20, 185-220.

Hyvönen, O., 1972. Palsojen morfologiasta ja esiintymisestä Fennoskandiassa. Terra, 84, 72-77.

Kelly, P.M., Jones, P.O., Sear, C.B., Gherry, B.S.G., and Tavakol, R.K., 1982. Variations in surface air temperatures: part 2. Arctic regions, 1881-1980. Monthly weather review, 110, 71-83.

Kershaw, G.P., and Gill, D., 1979. Growth and decay of palsas and peat plateaus in the Macmillan Pass-Tsichu River area, Northwest Territories, Canada. Canadian Journal of Earth Sciences, 16, 1362-1374.

Lagarec, D., 1976. Étude géomorphologique de palses dans la région de Chimo, Nouveau-Québec, Canada. Cahiers géologiques, 92, 153-163.

Lagarec, D., 1980. Étude géomorphologique de palses et autres buttes cryogènes en Hudsonie (Nouveau-Québec). Université Laval, Ph.D. Thesis, 308 pp.

Lagarec, D., 1982. Cryogenetic mounds as indicators of permafrost conditions, northern Québec. Proceedings 4th Canadian permafrost conference (Roger J.E. Brown memorial volume), pp 43-48.

Lundqvist, S., and Mattson, J.O., 1965. Studies on the thermal structure of a pals. Svensk Geografisk Årsbok, 41, 38-49.

Mackay, J.R., 1974. Reticulate ice veins in permafrost. Northern Canada. Canadian Geotechnical Journal, 11, 230-237.

Mathieu, C., 1983. Morphogénèse holocène des dunes et des palses de la basse vallée du Koroc (Nouveau-Québec). Université de Montréal, Département de géographie, master's thesis, 198 pp.

Novikov, S.M., and Usova, L.I., 1979. Nature and classification of palsa bogs. Soviet Hydrology, Selected papers 18, 109-113.
Outcalt, S.I., and Nelson, F., 1984. Computer simulation of buoyancy and snow-cover effects in palsa dynamics. Arctic and Alpine Research, 16, 259-263.
Payette, S., 1984. Un ilôt de pergélisol sur les hauts sommets de Charlevoix, Québec. Géographie physique et Quaternaire, 38, 305-307.
Payette, S., and Filion, L., 1985. White Spruce expansion at the tree-line and recent climatic change. Canadian Journal of Forest Research, 15, 241-251.
Payette, S., and Gagnon, R., 1985. Late Holocene deforestation and tree regeneration in the forest-tundra of Québec. Nature, 313, 570-572.
Payette, S., Ouzilleau, J., and Filion, L., 1975. Zonation des conditions d'enneigement en toundra forestière, Baie d'Hudson, Nouveau-Québec. Canadian Journal of Botany, 53, 1021-1030.
Payette, S., Samson, H., and Lagarec, D., 1976. The evolution of permafrost in the taïga and forest-tundra, western Québec-Labrador peninsula. Canadian Journal of Forest Research, 6, 203-220.
Payette, S., and Seguin, M.K., 1979. Les buttes minérales cryogènes dans les basses terres de la Rivière-aux-Feuilles, Nouveau-Québec. Géographie physique et Quaternaire, 33, 339-358.
Pissart, A., 1983. Pingos et palses: un essai de synthèse des connaissances actuelles, in Mesoformen des Reliefs im heutigen Periglazialraum. H. Poser et E. Schunke, edit. Abhandlungen der Akademie der Wissenschaften in Göttingen Mathematisch-Physilalische Klasse, Dritte Folge Nr. 35, 48-69.
Pissart, A., and Gangloff, P., 1984. Les palses minérales et organiques de la vallée de l'Aveneau, près de Kuujjuaq. Québec subarctique. Géographie physique et Quaternaire, 38, 217-228.
P'yavchenko, N.I., 1969. Swampy forests and bogs of Siberia (Zabolochennyye lesa i bolota Siberi) US Army Foreign Science and Technology Center technical translation, FSTC-HT-23-310-70, 215 pp.
Richard, P.S.H., 1981. Paléophytogéographie postglaciaire en Ungava par l'analyse pollinique. Paléo-Québec n° 13, 153 pp.
Salmi, M., 1972. Present developmental stages of palses in Finland. Proc. 4th Int. Peat Cong., Helsinki, 121-141.
Samson, H., 1975. Évolution du pergélisol en milieu tourbeux en relation avec le dynamisme de la végétation. Golfe de Richmond, Nouveau-Québec. Université Laval, Département de phytologie, M.Sc. thesis, 158 pp.
Savoie, L., and Gangloff, P., 1980. Analyse pollinique d'une palse au site archéologique de Vieux-Port-Burwell (Killiniq). Territoires du Nord-Ouest. Géographie physique et Quaternaire, 34, 301-321.
Schunke, E., 1973. Palsen und kryokarst in Zentral-Island. Nachrichten der Akademie der Wissenschaften in Gottingen II. Mathematisch Physikalische Klasse, 4, 65-102.
Seguin, M.K., 1976. Observations géophysiques sur le pergélisol des environs du lac Minto, Nouveau-Québec. Cahiers de géographie du Québec, 50, 327-346.

Seguin, M.K., and Allard, M., 1984a. Le pergélisol et les processus thermokarstiques de la région de la rivière Nastapoka, Nouveau-Québec. Géographie physique et Quaternaire, 38, 11-25.
Seguin, M.K., and Allard, M., 1984b. La répartition du pergélisol dans la région du détroit de Manitounuk; côte est de la mer d'Hudson, Canada. Canadian Journal of Earth Sciences, 21, 354-364.
Seguin, M.K., and Crépeault, J., 1979. Étude géophysique d'un champ de palses à Poste-de-la-Baleine, Nouveau-Québec. Géographie physique et Quaternaire, 33, 327-338.
Seppälä, M., 1972. The term palsa. Zeitschrift für Geomorphologie, 16, pp 463.
Seppälä, M., 1976. Seasonal thawing of a palsa at Enontekiö, Finnish Lapland, in 1974. Biuletyn Periglacjalny, 26, 17-24.
Seppälä, M., 1982. An experimental study of the formation of palsas. 4th Canadian permafrost conference (Roger J.E. Brown memorial volume), 36-42.
Seppälä, M., 1983a. Seasonal thawing of palsas in Finnish Lapland. In Permafrost: Fourth International Conference, Proceedings, National Academy Press, Washington, D.C., 1127-1131.
Seppälä, M., 1983b. Present-day periglacial phenomena in Northern Finland. Biuletyn Peryglacjalny, 29, 231-243.
Short, S.K., and Nichols, H., 1977. Holocene pollen diagrams from subarctic Labrador-Ungava: vegetation history and climatic change. Arctic and Alpine research, 9, 265-290.
Smith, M.W., 1985. Observations of soil freezing and frost heave at Inuvik, Northwest Territories, Canada. Canadian Journal of Earth Sciences, 22, 283-290.
Washburn, A.L., 1983a. What is a palsa? in Mesoformen des Reliefs im heutigen Periglazialraum. H. Poser and E. Schunke (eds) Abhandlungen der Akademie der Wissenchaften in Göttingen Mathematisch - Physikalische Klasse. Dritte folge Nr. 35, 34-37.
Washburn, A.L., 1983b. Palsas and continuous permafrost, in Permafrost Fourth International Conference, Proceedings, Washington, D.C., National Academy Press, 1372-1377.
Wrammer, P., 1972. Palslika bildningar i mineraljord. Några iakttegelser från Taavavuoma, Lapland (Palsa-like formations in mineral soil. Some observations from Taavanuoma, Swedish Lapland). Göteborgs Univ. Naturgeogr. Inst. Rapp. 1, 60 pp.
Wrammer, P., 1973. Palsmyrar i Taavavuoma, Lapland (Palsa bogs in Taavanuoma, Swedish Lapland). Göteborgs Univ. Naturgeogr. Inst. Rapp. 3, 140.
Zoltaï, S.C., 1972. Palsas and peat-plateaus in Central Manitoba and Saskatchewan. Canadian Journal of Forest Research, 2, 291-302.
Zoltaï, S.C., 1975. Tree-ring record of soil movements on permafrost. Arctic and Alpine Research, 4, 331-340.
Zoltaï, S.C., and Tarnocai, C., 1975. Perennially frozen peatlands in the Western Arctic and Subarctic of Canada. Canadian Journal of Earth Sciences, 12, 28-43.

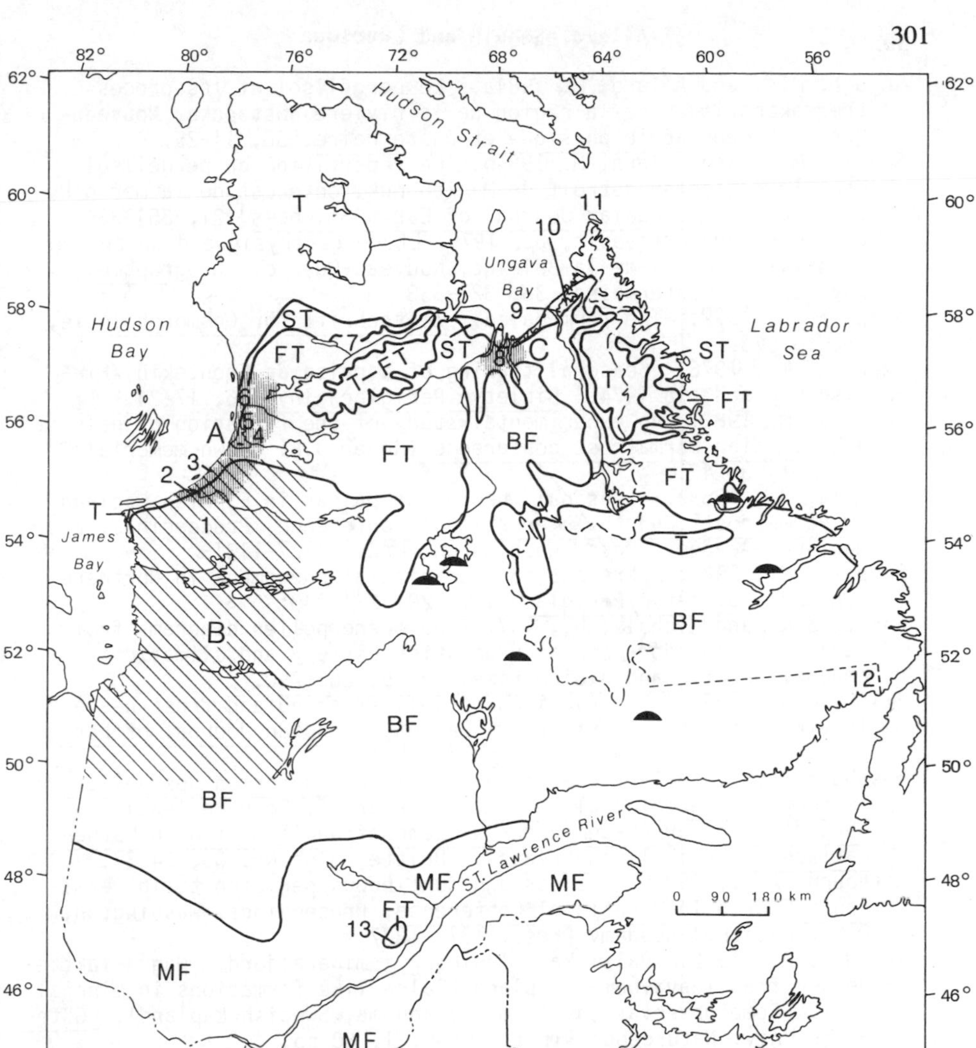

Fig. 1. Previous studies on palsas and mineral permafrost mounds in northern Québec. A: area studied by Lagarec (1982), B: Dionne (1978), C: Lagarec (1976); 1: LG2 (Dever et al. 1984), 2: Kuujjuarapik (Seguin and Crépault, 1979), 3: Manitounuk (Seguin and Allard, 1984b), 4: Ouiatchouan (Heim, 1976), 5: Rivière Sheldrake (Blais, 1984), 6: Rivière Nastapoca (Seguin and Allard, 1984a), 7: Rivière-aux Feuilles (Payette and Seguin, 1979), 8: Rivière Aveneau (Pissart and Gangloff, 1984), 9: Kangiqsuallujjuaq (Allard et al., unpub.), 10: rivière Koroc (Mathieu, 1983), 11: Killiniq (Savoie and Gangloff, 1980), 12: Blanc-Sablon (Dionne, 1984), 13: Charlevoix (Payette, 1984). Semicircles are reported palsa fields in Dionne (1984). T, tundra; ST, shrub-tundra; FT, forest-tundra; BF, boreal forest; MF, mixed forests.

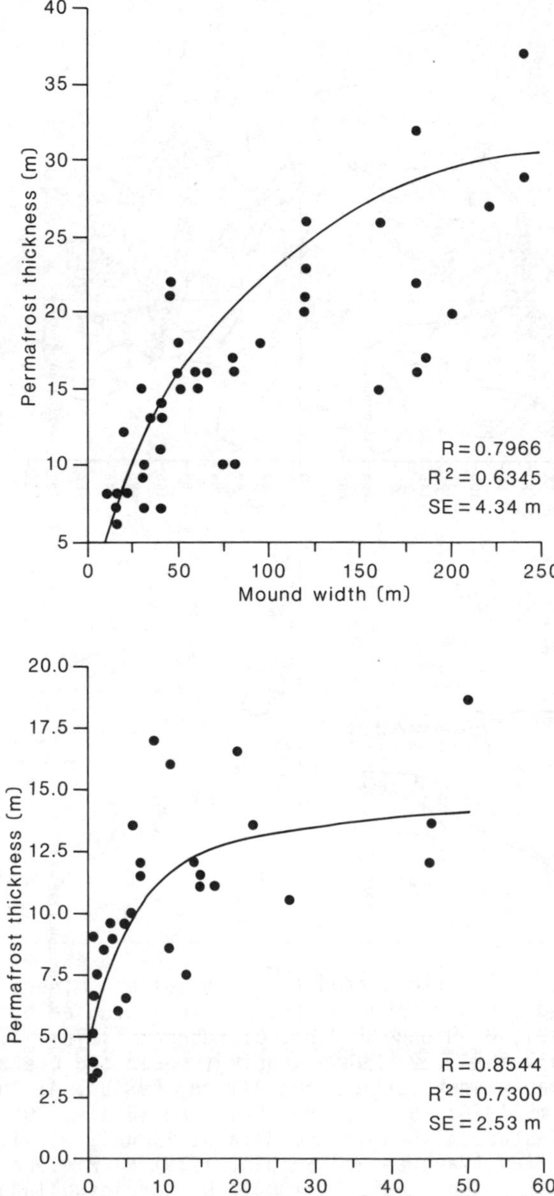

Fig. 2. Permafrost thickness vs width of mineral permafrost mounds (upper graph) and vs palsa volume (lower graph).

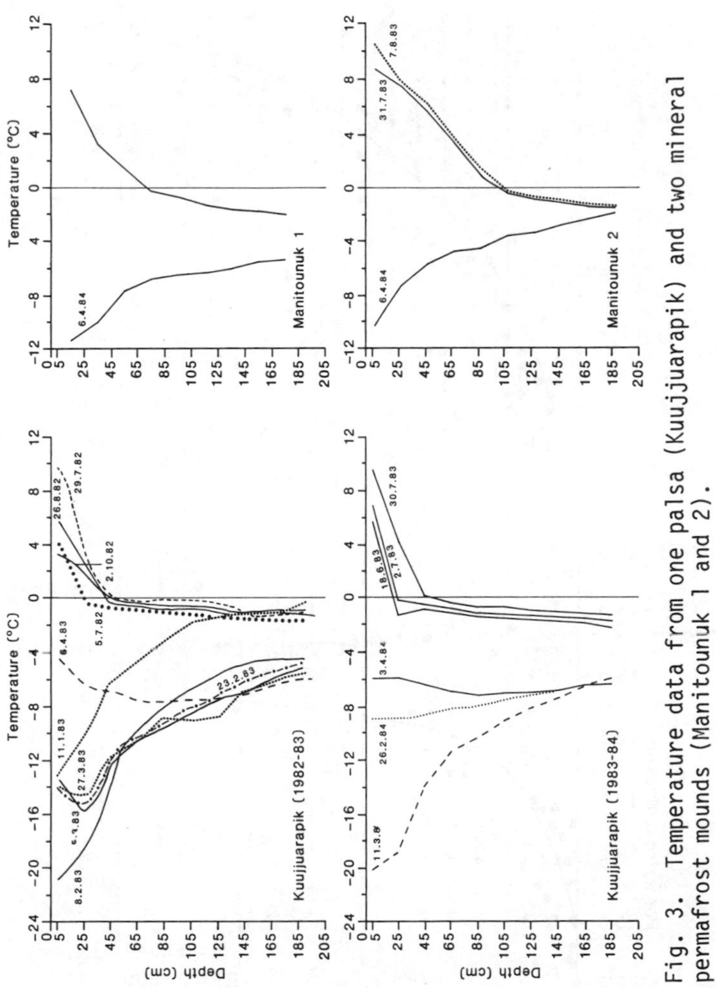

Fig. 3. Temperature data from one palsa (Kuujjuarapik) and two mineral permafrost mounds (Manitounuk 1 and 2).

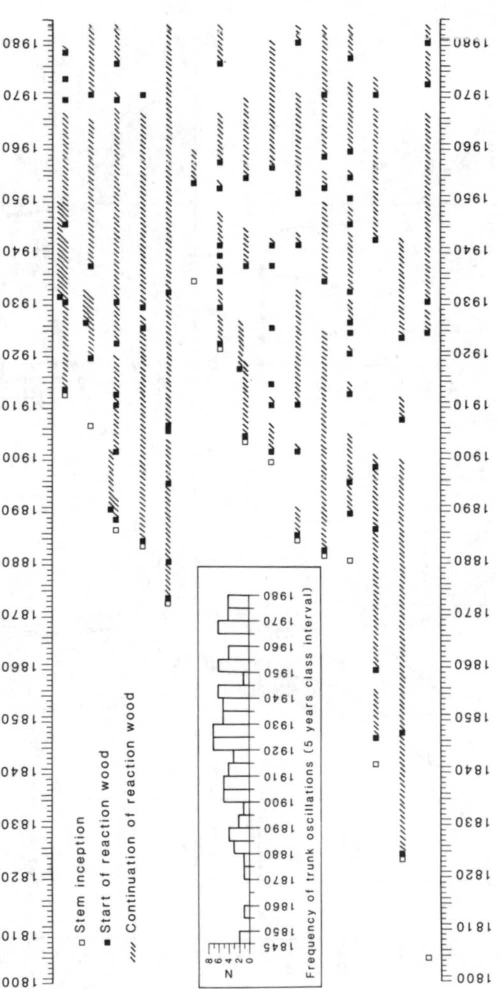

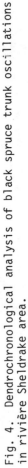

Fig. 4. Dendrochronological analysis of black spruce trunk oscillations in rivière Sheldrake area.

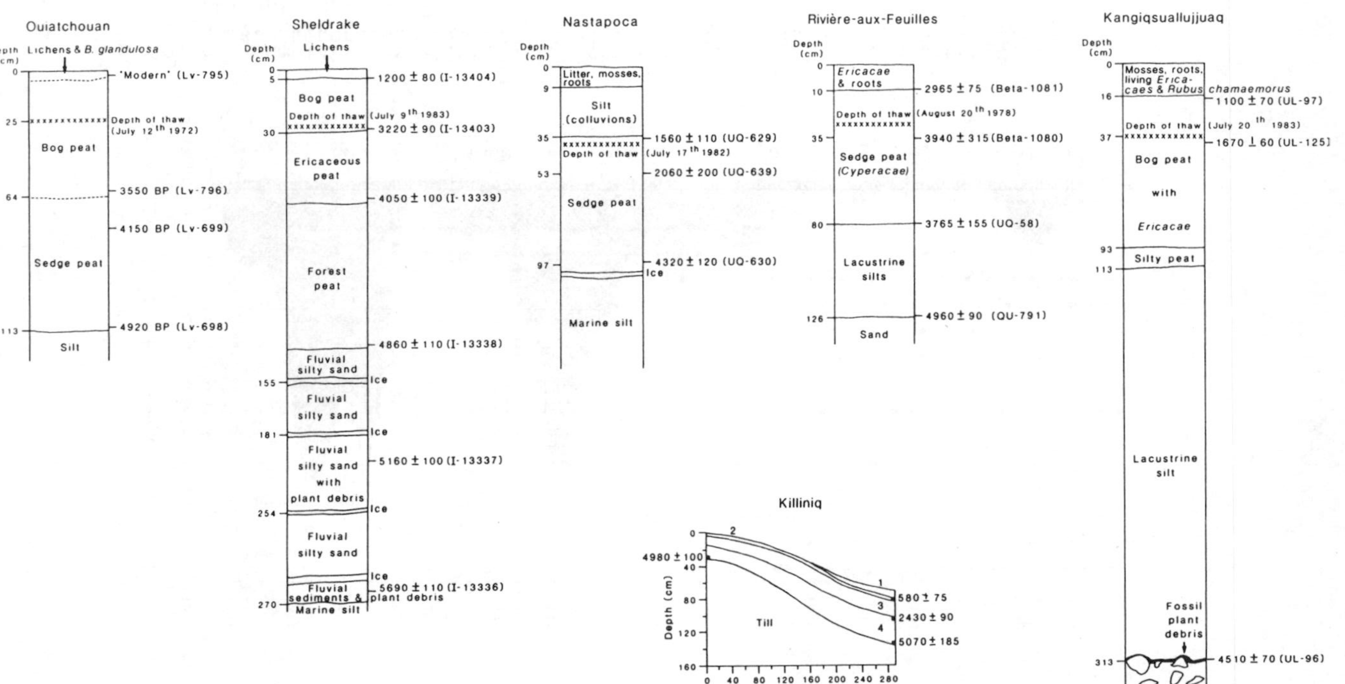

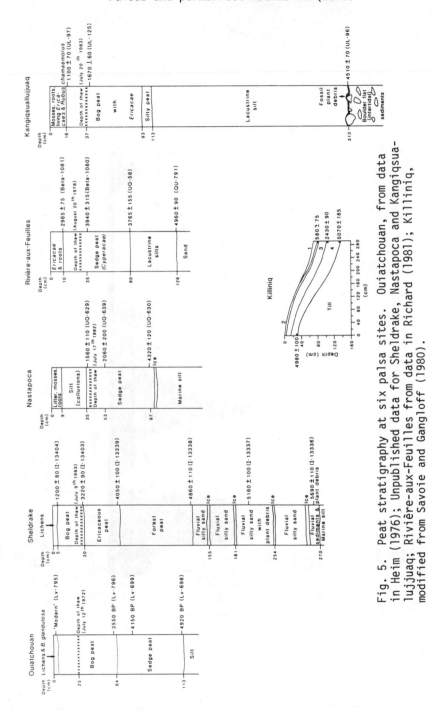

Fig. 5. Peat stratigraphy at six palsa sites. Ouiatchouan, from data in Heim (1976); Unpublished data for Sheldrake, Nastapoca and Kangiqsualujjuaq; Rivière-aux-Feuilles from data in Richard (1981); Killiniq, modified from Savoie and Gangloff (1980).

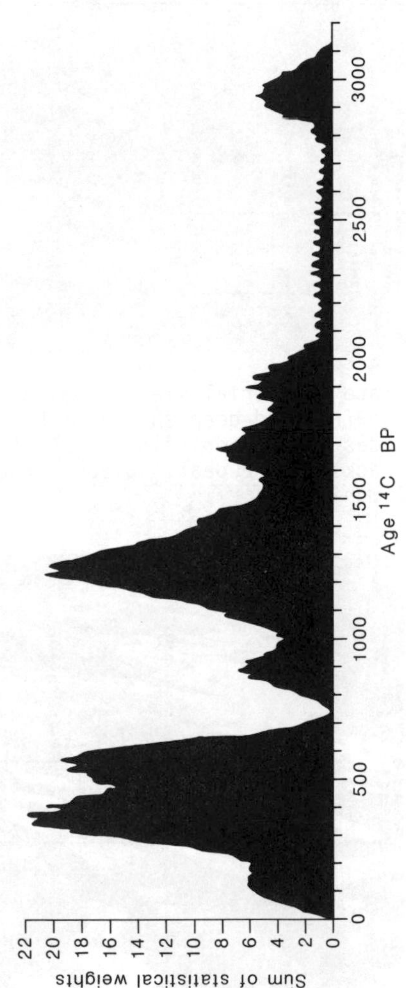

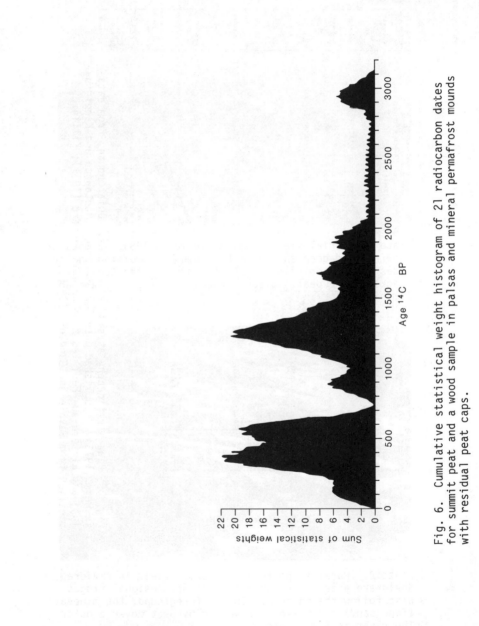

Fig. 6. Cumulative statistical weight histogram of 21 radiocarbon dates for summit peat and a wood sample in palsas and mineral permafrost mounds with residual peat caps.

Plate 1. Aerial view of palsas in march 1984; summits are snowfree and deep snowbanks have been accumulated on the sides by wind drifting. Note the snow-filled tension cracks in the peat. Black spruces in foreground are about 2 m high.

Plate 2. Parallel pattern in a palsa field in rivière Sheldrake area. Thermokarst ponds and residual ridges also follow the pattern. In the foreground, the mineral (fine sand) substrate is revealed by peat cover erosion. The river is 5 m across.

Plate 3. Complex pattern of peat plateaus, palsas and thermokarst ponds in rivière Sheldrake area. The river is 5 m across.

Plate 4. In foreground, a mineral permafrost plateau is being dissected into mounds. Shrubs colonize the hollows between the separating mounds. Resulting pattern is roughly parallel. Note the high turbidity of thermokarst ponds and the numerous frostboils on the mounds; they are in average 1 m in diameter.

Plate 5. Open crack (30 cm wide) in a polygonal peat plateau. Collapsed peat in the crack. Permafrost degradation and fen expansion (in background) occur along the crack network.

Plate 6. Two mineral permafrost mounds at the upper edge of a tidal marsh. The individual on the left side of picture provides a scale.

[illegible] and [illegible]
[illegible]
© [illegible] John Wiley & Sons Ltd

VOLCANOES IN TENGCHONG, YUNNAN PROVINCE CHINA

[illegible] Caichen and [illegible]

[illegible] Tengchong, China

ABSTRACT

[illegible] county is an area of recent volcanism. They occurred in four stages [illegible] to Holocene [illegible] eruptions [illegible] the [illegible] but the documentary evidence is not [illegible].

The area is rich in geothermal resources, having high [illegible] geothermal gradients. It is likely that any future eruptions will be [illegible] small [illegible] if indeed any occur. [illegible] exploration of [illegible] [illegible] [illegible]

INTRODUCTION

A group of volcanoes in Tengchong County, with frequent earthquakes and much geothermal activity, lies at the west foot of the Gaoligong [illegible] near the southwest border of China. Hydrothermal activity [illegible] in the geothermal fields located [illegible] the bases of young [illegible] volcanoes. Temperatures of some springs have exceeded [illegible] the local boiling point. Water and steam are boiling [illegible] underground, and hydrothermal explosions are frequent. This indicates that crustal movements and [illegible] development are [illegible] in this region.

VOLCANIC LANDFORMS

[illegible] volcanoes [illegible] the [illegible] between [illegible] the Hengduan ranges and the [illegible] of the [illegible]. The [illegible] [illegible] [illegible] [illegible] was [illegible] in the [illegible] and [illegible] [illegible] [illegible] [illegible] [illegible]

International Geomorphology 1986 Part II
Edited by V. Gardiner

VOLCANOES IN TENGCHONG, YUNNAN PROVINCE CHINA

Mu Guichun and Dai Hezhi

Chongqing, Sichuan, China

ABSTRACT

Tengchong County is an area of recent volcanism. This occurred in four stages, from the Pliocene to Holocene, with diminishing amounts of lava being erupted in each successive stage. The most recent eruptions may have occurred during the Ming dynasty, but the documentary evidence is not, in itself conclusive.

The area is rich in geothermal resources, having a high near-surface geothermal gradient. It is likely that any future eruptions will be on a very small scale, if indeed any occur. The rational exploitation of the geothermal resources of the area demands further research.

INTRODUCTION

A group of volcanoes in Tengchong County, with frequent earthquakes and much geothermal activity, lies at the west foot of the Gaoligong range, near the southwest border of China. Hydrothermal activity is strong in the geothermal fields located at the edge of recently erupted volcanoes. Temperatures of some springs have reached or exceeded the local boiling point. Water and steam are oozing from underground, and hydrothermal explosions are frequent. This indicates that crustal movements and geomorphic development are active in this region.

VOLCANIC LANDFORMS

Tengchong is located at the junction between the fault block belt of the Hengduan ranges and the folded belt of the Himalayas. The crust is quite unstable and fault activity is strong. Granite was intruded on a large scale in the Yanshan period (Mesozoic Era), and the region was later influenced strongly by the Himalayan movement (Cenozoic Era), resulting in faults striking S-N, NNE-SSW and NNW-SSE. The stages of volcanic eruption and the distribution of volcanoes are controlled by these faults, and surface hot springs and modern seismic activities occur along these belts.

Volcanoes in Tengchong have erupted frequently since the Pliocene; many are still morphologically intact. According to incomplete statistics, there are more than fifty volcanoes near Tengchong, including thirteen volcanic cones with intact craters and nine with calderas. A large number of volcanic cones have cratered peaks. The plugs formed by late lava can be found in some vents. Some cone walls have been eroded into gullies, but other cones still retain crater rims with steep inside walls. Much water has gathered in some calderas, forming lakes such as Lakes Qinghai and Beihai. There are also lava flows which covered the valleys and basins surrounding some volcanic cones, and several stages of lava platforms were formed. The volcanoes may be divided into two categories: according to the type of volcanic eruption, fissure and central eruptions.

Fissure Eruptions. These quiet eruptions occurred in the Pliocene. Basic lava flows, with low gas content and high mobility, formed compact, massive, dark basalt, with clear columnar jointing. This forms the bedrock of river terraces, mainly exposed in the valleys of the Rivers Daying and Longchuan.

Central Eruptions. Several of these eruptions occurred in Pleistocene and Holocene. They can be divided according to the order of eruption and lithological differences of the eruptive rocks into:

(1) Eruption of neutral and acid-neutral andesite and dacite, showing that the lava had a relatively high viscosity with poor mobility and higher gas content. Tiny crystals of plagioclase in the lava have a definite orientation, and a rhyotaxitic or banded structure exists. High mountains were formed, such as the Yubipo, Daliuchong Shan and Xiaoliuchong-shan near the city of Tengchong County. The volcanoes have been heavily modified, but traces of craters can still be seen on the tops of some cones, such as Gantang and Xiangtang of Gaoshansi.

(2) Eruption of basic or neutral-basic basalt and andesite basalt formed many stepped lava platforms and some lofty volcanic cones with craters, composed of interbedded and stratified lava, scoria and volcanic debris. Volcanic foam occurs at the top of volcanic cones, spindle-shaped or twisted bombs and lapilli occur near the volcanoes, and volcanic dust is found in lacustrine deposits near Tengchong. These volcanoes are mainly found at Mts. Daying, Dakong and Heikong in the north of Tengchong, and Mts. Laoguipuo and Maan in the southwest.

The two eruptive patterns mentioned above have the same origin, and their distributions are continuous.

STAGES AND TIMES OF VOLCANIC ERUPTION

Volcanic activity of Tengchong can be divided into four stages. The first, third and fourth may each be divided into two. The first stage occurred in the Pliocene, the second and third in the Pleistocene and the fourth in the Holocene. The criteria for division are: (1) The volcanic strata are interrupted, with sediments of fluvio-lacustrine or weathering facies on the denuded surface between two stages of volcanic eruption, and (2) There are two different volcanic rocks. Although similar in lithology it is clear that the later rock covered the first one.

The first stage. This consists of widely distributed basic lava which flowed freely in the valleys and basins. The first sub-stage is the basalt which was exposed at Mangbang in the valley of Longchuan River. A layer of grey-green epidotized augite-basalt was embedded in the coal seam of the River Zhoujia in the Pliocene Series, on which there was a layer of sandstone, sand-conglomerate and basaltic sand-conglomerate, and tuffaceous sand-conglomerate 80-280m thick. The uppermost layer is dark black dolerite. In the Lanlin coalmine located on the left side of the River Daying, the Tertiary is divided into six seams on the basis of the data offered by the Geological Department of Yunnan Province. The third seam is 170-200m thick and consists of sandy shale with a few sandstones, and on the bottom is basalt. The fourth and fifth seams are 240m thick and contain sandstone, sand-conglomerate and shale. The sixth is composed of andesitic basalt and basalt. The two layers of basalt were obviously formed by the eruptions of two phases. We do not have sufficient fieldwork and analysis concerning the low layer; it is treated as a second substage of the first stage here. The basic lava of the second substage of the first stage is widely distributed as bedrock, terrace basement or precipices in the valleys of the Rivers Longchuan and Daying, as the Xiaxingtang dolerite, whose absolute age is 7.2 million years. It is dark black and massive-compact, with a spheroidally weathered upper part; the lower part has columnar jointing. Outcrops in Langyian, Xingtang, Zhengyiguan and Muchangshan are covered by fluvial sand and conglomerate sediments.

The second stage. This is neutral and neutral-basic andesite, andesite-dacite and dacite lava, which is now mainly distributed in the north part of the Laifengshan, Gaoshanyubi Village, Daliuchongshan, Xiaoliuchongshan, Yujiadashan, Dujiapo and Liangdapotou, Bangeshan and Bapai in the southwest of the city. The lava is extensive, and the volcanoes are huge and have been heavily modified. Qinghai, Beihai, Gantang and Xiangtang might be calderas of this stage, in which water accumulated and lakes formed. Further study is necessary to decide whether the volcanoes of this stage should be subdivided.

The surface of the volcanic rocks has been weathered and a thick layer of brown-red soil formed, for example, the weathered red soil covered by the basalt flow of the late stage can be seen in the Ganzhezai, Hehuachi and Xiaozhuang.

The third stage. This is basalt and andesite-basalt and, according to lithological differences, may be subdivided into two. The first substage, in the middle-late part of the Pleistocene was olivine basalt, in Lujiapo-Xiaomipo, Taipingcun-Xiaozhuang-Longhuan-Languan, Maershan-Chenjiaxiang-Dieshuihe and Mt. Wuya in the extreme north. The olivine basalt can also be found in the Huiyao-Cheqiaotang. The outlines of the volcanic cones are still discernible but their craters have been damaged and the caprocks look like shields or mounds. An exception to this is the top of the low Maer Mountain where the original crater can be seen. The surface layer of the lava platform has been heavily weathered and covered by the regolith.

The second substage is andesite-basalt, at Nanlaifeng Mountain, Jiaoyiao and Bajiaoguan. Some damaged craters can be found. The volcanic cones consist of interbedded lava and volcanic breccia; gas bubbles in the lava were filled by zeolite and silicate secondary minerals. Elephas masindus molar fossils covered by the weathering soil of lava were found in Shuangpo of Nanlaifeng Mountain; these could date from the Pleistocene or Holocene, but most likely the Holocene. The eruption of andesite-basalt in Nanlaifeng Mountain is therefore of the late Pleistocene.

The fourth stage. This consisted of andesite-basalt. The volcanoes, with clear cones and craters, still appear whole. A variety of volcanic bombs can be found both inside and outside the craters. This stage is subdivided into two according to the relationship of lava flows to the underlying strata.

The first substage refers to eruption of andesite-basalt and olivine-basalt in the early Holocene. There are twelve truncated volcanic cones in Mazhan-Jiaoshan; for example, Heikong Mountain, with an elevation of 2,073m is 214m higher than the surrounding lava platforms, and on the top of the cone there is a funnel-shaped crater, 70m deep. The erupted lava, volcanic debris, slag and pumice were distributed regularly, and spindle-shaped volcanic bombs can be seen inside the craters. There is regolith on the surface of the volcanic cones. In addition, Laoguipo and Fugoshan in the west of the city were also formed.

There is a layer of fluvio-lacustrine sediments more than ten metres thick between the base of this volcanic eruption and the basalt of the first substage of the third stage; this is found in the vicinity of Xiangyangqiao, Longchuan River.

The second substage of the fourth stage refers to late volcanic activity in the Tengchong region. The lithologies are andesite-basalt and two pyroxene basalts. Four volcanoes, Mt. Daying (2,614.5m), Mt. Tieguo, Mt. Dakong and Mt. Maan are clearly whole, having typical volcano morphology, with cones, craters, rims, walls and plugs. The lava flow in this substage draped unconformably over the volcanic rocks of the first substage. The platforms found in this substage often have unique geomorphic structures. Loose and porous volcanic debris and slags as well as sediments of ancient stream channels, were buried under the surface lava flows. Shafts, sinkholes, talus, small depressions, stone embankments and long trenches were formed after the surfaces of the platforms subsided. The surfaces appear rather gentle in the distance, but in fact the ground is uneven. There are underground caverns and streams beneath the platforms.

In summary, the four stages of volcanic eruptions of Tengchong constitute a cycle of magma activity from basic to neutral-acid and then from basic to neutral-basic. The later the stage, the less the amount of volcanic lava erupted. For example, lava from Mt. Daying poured out along previous valleys, some of which reached the foot of the mountain; later flows only reached half way down the mountain.

The absolute ages of volcanic eruptions of Tengchong have not yet been determined. It is important to find out the date of the last volcanic eruption, as this concerns local people. Some works have pointed out that volcanic eruptions in Tengchong occurred in the Ming dynasty, according to "The Travel Notes of Xu Xiake". There is also an account of the volcanoes, earthquakes and hot springs in the local chronicles of the historical administrative government in Tengchong, but there is not enough evidence to ascertain the exact time of volcanic eruptions. We have analysed these historical documents but because of space limitations we do not quote and evaluate them here. However, we do not think it is reliable to conclude that volcanic eruptions occurred in Tengchong during the Ming dynasty, on only the evidence of these historical documents. In our opinion, the volcanoes of Tengchong are young. The original eruption occurred in the Pliocene, the eruption of the last stage was in the Holocene and the eruptions of the two stages between them occurred in the Pleistocene. The last eruption (the second substage of the fourth stage) might have occurred in historical times, even in the Ming dynasty, but it is necessary to find some reliable scientific evidence such as trees scorched by lava, pollen evidence, or historical traces and relics of the eruptions.

VOLCANOES AND GEOTHERMY

The volcanic region of Tengchong is rich in geothermal resources. According to incomplete statistics, there are 79 groups of springs and ten of them have a temperature of over 90°C. Each group has steam vents, boiling and hot springs, fountains, fumaroles and steaming ground. It has been preliminarily estimated that the natural thermal energy on the surface reaches 6.1×10^4 kilocalorie per second, equivalent to burning 270,000 tons of standard coal. Several marked high-temperature geothermal fields are on the edges of late volcanoes, such as Rehai field (1.6% of the total area of Tengchong County), located in the south of the last phase Maan Mountain volcano. The Rehai field has 5×10^4 kilocalorie per second natural thermal flow, equivalent to burning 210,000 tons standard coal per year. More than ten large fumaroles, with outlet temperatures over 94°C, can be seen in a ravine less than 200m long near Huangguaqing. Near Liuhuangtang there are some similar fumaroles whose temperatures reach the boiling point, and steaming ground. The strong steam flows form lofty steam columns in the air, and the hot water from around the fumaroles is boiling, so the field is steaming. Water temperatures of the boiling springs at Liuhuangtang and Reshuitang were, in January 1974, 96.6°C and 98.7°C respectively. The shallow layers of the ground have a great geothermal gradient, for example one drill hole 3.63m deep has a surface temperature of 94.1°C and at the bottom the temperature reaches 102.7°C - a geothermal gradient of 2.36°C per metre. A second drill hole, 1,500m from the first, with a depth of 4m, has a surface temperature of 94.8°C and geothermal gradient of 2.45°C per metre. According to records of the twelfth team of the Geological Department of Yunnan Province, the temperature at 12m from the surface reached 145°C and the geothermal gradient was 4.08°C per metre. It is apparent that there is a large amount of geothermal energy reserves, which provide favourable conditions for power station development.

The surface heat reflects the existence of high-energy geothermal heat flow. Large amounts of natural sulphur, alum and other minerals gather on the fumaroles and steaming ground; there is much water, nitrogen, hydrogen sulphide and methane in the released gases. The hot water contains volatile components derived from magma. In addition, hydrothermal explosions, ground roaring, ground groaning, eruptions of mud volcanoes and formation of sinter mounds frequently occur. These phenomena indicate that magma exists under or not far from the heat fields. This is the source of geothermal flows which has been continuously releasing much thermal energy for over 300 years. The granite of the Yanshan Period and the Tertiary granitic sand-conglomerate above the remaining magma are the reservoirs of geothermal energy, and the erupted lava becomes the roofs of the reservoirs. So the high temperature steam springs and hot springs are found around the volcanic lava, namely at the fractures of granite and sand-conglomerate. Moreover, seismic activity near Tengchong indicates magma-induced earthquakes, shown by frequent occurrence of microseismic activity, shallow earthquakes, and by frequent releases of crustal strain.

The active geological phenomena described above indicate that the volcanoes are now in a phase of slight activity, the after-vibrations of the last volcanic eruption. As for whether the volcanoes will erupt in future, we are not able to say "yes" or "no" on the basis of our present research. But the intensity of volcanic eruptions has diminished since the Pliocene and even if there is some magma remaining, the scale and intensity of eruptions will be much less than those of the latest stage. The heat flows provided by the after-vibrations of volcanic eruptions are considerable and relatively stable and can be exploited and used for centuries.

We have a superficial understanding of the volcanoes in Tengchong but do not have satisfactory answers to many problems. In order to carry out more rational exploitation of local geothermal resources, it is necessary to make further researches on the volcanoes in Tengchong.

International Geomorphology 1986 Part II
Edited by V. Gardiner

BOTTOM GEOMORPHOLOGY OF THE GULF OF ADEN, TADJURA RIFT ZONE

A. V. Zhivago

Institute of Oceanology,
USSR Academy of Sciences, Moscow, V-218, USSR.

ABSTRACT

A bathymetrical, geological and geomorphological survey of the Tadjura Rift zone, in the west of the Gulf of Aden, was carried out in 1984. The rift zone has the main characteristics of the Sheba Ridge previously described to the east, but at a reduced scale. The floor of the Gulf is interpreted as a young spreading structure, with two genetically distinct morphological provinces. A lower one includes the rift valley, volcanic "highs" and "lows", small fault scarps, vast basins and volcanic massifs. The upper one consists of scarps of rift mountains fringing the rift valley to north and south, with sediment-covered terraces.

INTRODUCTION

Complex geological-geomorphological studies, including direct visual observations of the bottom from the manned submersible "Pisces", were carried out during 1984 by the research vessel "Academic Mstislav Keldysh" in the Gulf of Aden, Tadjura rift zone. The work continued observations made by a British expedition in the 1960s, headed by A. C. Laughton on board the research vessels "Discovery" and "Owen"; this involved mainly the central and eastern parts of the gulf.

Laughton, Whitmarsh and Roberts interpreted most of the bathymetric, geological and geophysical data obtained by 1970 (Laughton, 1966; Laughton et al., 1970; Roberts and Whitmarsh, 1969). At this time the Sheba Ridge stretching along the gulf axis was considered as a link of the mid-oceanic ridge system. An assumption was made of an axial rift valley within its limits, marked by negative magnetic anomalies. Transform faults were mapped along which individual blocks of the Sheba ridge were subjected to left-side displacement. Uplifted margins with blocks striking SW-NE, parallel to transform faults, were revealed by Laughton as transverse submarine ridges.

The less-studied western part of the Gulf of Aden, known as the "Trench of Tadjura", was an object of study for the Institute of Oceanology, USSR Academy of Sciences. The rift zone appeared to preserve here, between 43° and 45°E, the main structural features of the Sheba Ridge but in somewhat reduced form. The general level of

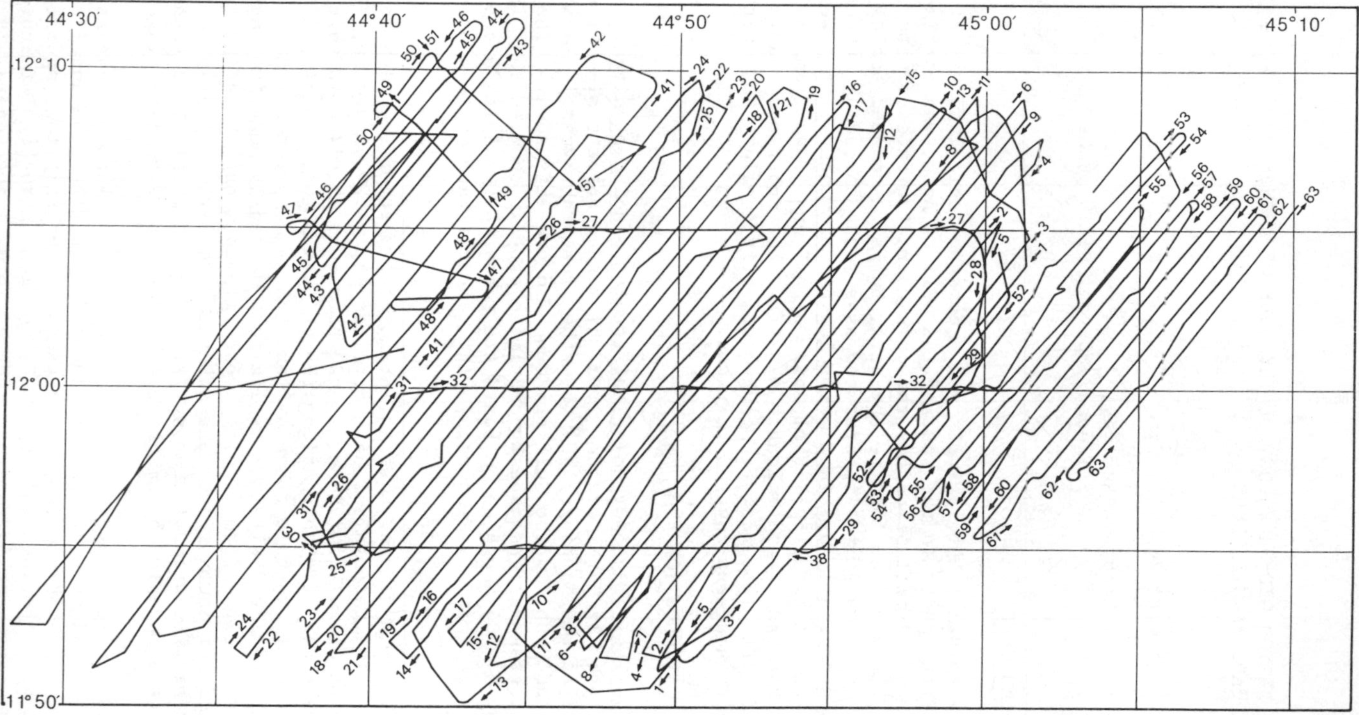

Fig.1. The echometric survey. Arrows indicate directions of the traverses, figures their numbers.

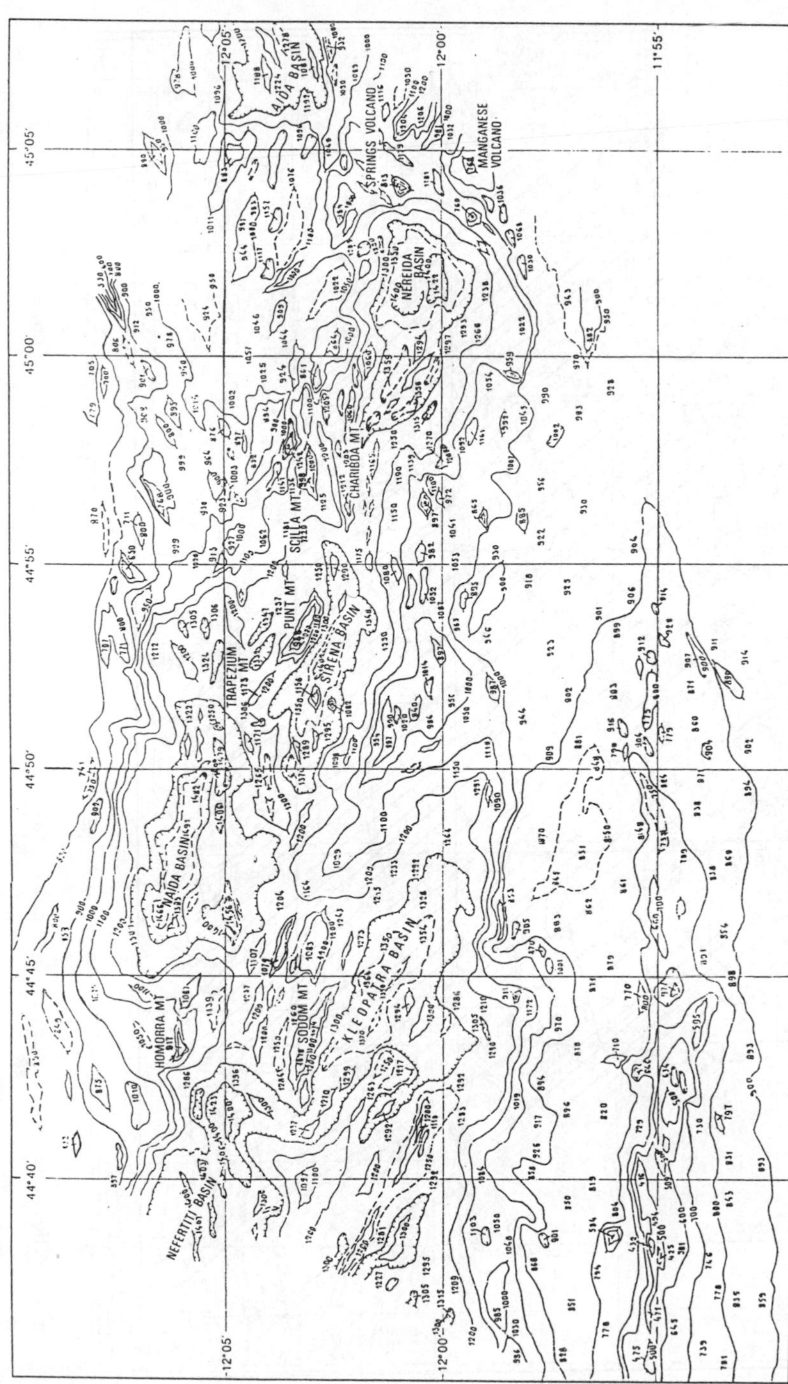

Fig.2. Bathymetry of the Tadjura rift zone.

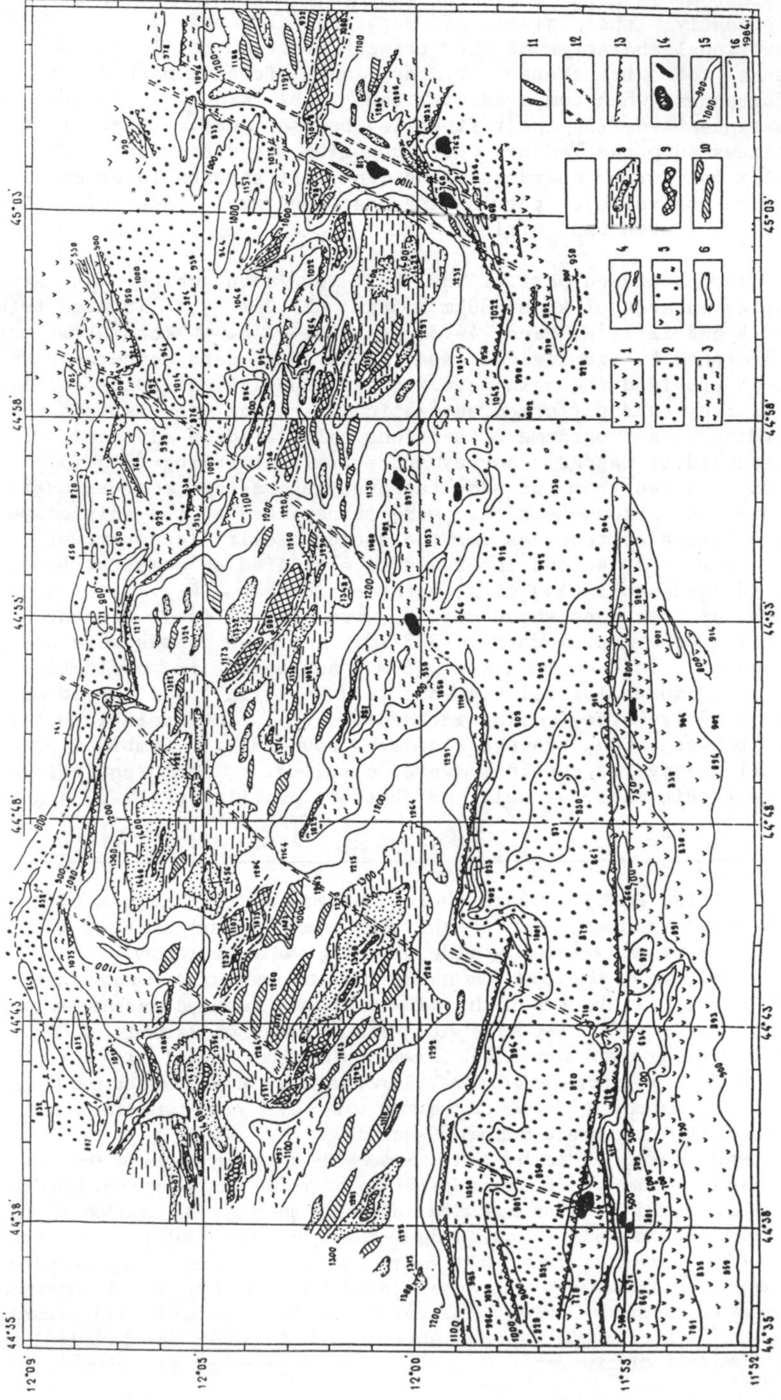
45°03'
44°58'
44°53'
44°48'
44°43'
44°38'
44°35'
12°09'
12°05'
12°00'
11°55'
11°52'
1
2
3
4
5
6
7
8
9
10
11
12
13
14
15
16

the bottom is more elevated whereas the relief amplitude and consequently the sizes of separate elements decrease. The transitional character of the "trench" bottom is indicated by the absence of transverse ridges and feebly marked transform disturbances when compared with the Sheba Ridge. To the west elevations bordering the rift valley become lower. The rift valley continues into the Tadjura Gulf, changing its direction. The whole complex of forms characterizing the rift appears later on land in the Ardukoba valley, terminating near the Azal lake of the Afar territory (north-east Africa).

The diamond-shaped part of the Gulf of Aden which was studied occupies an area of about 30km x 58km ($2310km^2$), located between 11°52'N and 12°11'N, and 44°35'E and 45°11'E. Echometric survey supplemented by geological and geophysical data concerning relief genesis provided the main data for geomorphological interpretations. The survey was performed automatically along the ship's course according to a programme of sections calculated in advance. It was accompanied by magnetic survey and seismic profiling (Fig.1). Depth values obtained from two simultaneously operating echo-sounders (a narrow- and a broad-beam one) were transmitted to the ship computing centre where corrections for the sound velocity in water were made. The depths at 20 second intervals were printed on computer tapes, to be used for compilation of a bathymetric map (Fig.2). The total length of sections whose depths were taken into account while compiling the map amounts to 2037km, and 6600 depths were analyzed and used. In compiling the map we succeeded in reflecting the structural-morphological elements of the rift zone by means of a pattern of isobaths, using geomorphological interpolation for the areas between the traverses of echo soundings. Combined profiles for all transverse sections were compiled. Geomorphological and physiographic maps, as well as bottom profiles (Figs.3, 4 and 5),

---

Fig.3 (FACING PAGE). Geomorphological map of the Tadjura rift zone.
I. The upper structural level of the rift zone - scarps and steps of the rift mountains of the early spreading cycle, bordering the median valley.
1. The crest zone of the upper scarp with little thickness of sedimentary cover (the upper terrace). 2. The step of the middle scarp with increased thicknesses of sedimentary cover. 3. The lower scarp with partial covering of sediments. 4. The raised edges of the scarps and steps. 5. Slopes of the scarps exposed aside from the axis of the valley. 6. Narrow longitudinal depressions on the surfaces of the steps.
II. Lower structural level of the rift zone - rift valley with the relief predetermined by recent manifestations of spreading.
7. General surface of the low bottom between the marginal scarps of the rift mountains (lower terrace). 8. Deeps inside the bottom of the valley. 9. Longitudinal prominences formed by recent extrusions of lavas, marking the location of the rift fissure. 10. Inner scarps (small ridges) formed by step faults to the north and south of the extrusive zone, with sharp crests and surfaces sloping away from the rift axis. 11. Ridges with transverse orientations, caused by tectonic movements in the zones of transform faults. 12. Zones of transform faults. 13. Escarpments along the main faults. 14. Big volcanoes. 15. Isobaths. 16. Structural limits (not coinciding with isobaths).

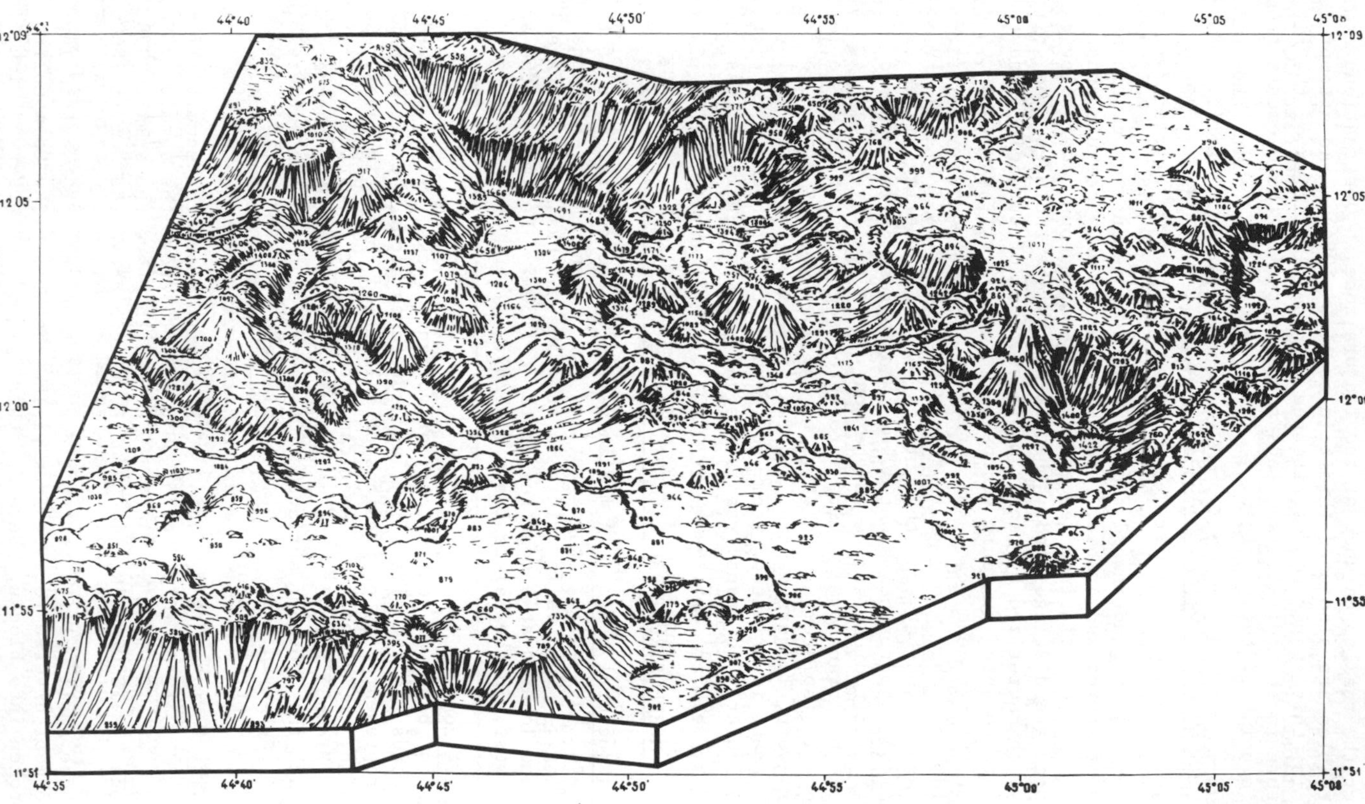

Fig.4. Physiographic map of the Tadjura rift zone.

reflect the real appearance of the Tadjura zone. Observations from submersibles are also used (Lisitzin et al., 1984). An attempt is made in this paper to describe the main peculiarities of the rift structure in the relief, and thus to characterize the elements from the viewpoint of their origin.

GEOMORPHOLOGICAL ANALYSIS

The floor of the Gulf of Aden is interpreted as a spreading structure causing northward and southward movement of continental blocks of Africa and Arabia, between which new oceanic crust is formed. Two distinct genetically homogeneous morphological provinces are distinguished: the lower, which includes the rift valley, characterized by recent manifestations of spreading, and the upper, represented by fault scarps and steps of rift mountains of an earlier (Pleistocene) cycle of spreading, fringing the rift valley. The width of the whole rift zone within the limits of the studied floor area, and between the crests of the bordering ridges (rift mountains), approaches about 30km. An impression is created of a general latitudinal strike in the zone. In reality though it is divided by transform faults into three segments subjected to left horizontal displacement relative to one another, by from 5 to 8km: Tadjura-west, Tadjura-centre, and Tadjura-east. The axial valley in each segment is oriented from south-east to north-west, strictly perpendicularly to transverse faults and at an angle of less than 40° to the general strike of the zone.

MORPHOSTRUCTURAL ELEMENTS OF THE RIFT ZONE

Lower structural stage. A median valley in the central part of the rift zone has relief characterized by recent manifestations of spreading. The valley width is from 5km to 16km. Its surface is rough and located at a mean sea depth of 1130m. Maximum depths are in the west, in the Naiada basin (1482m).

Extrusive zone.
A narrow band of extrusions of volcanic rocks, 1.5-2.5km wide, corresponds to the rift slit, and stretches along the valley axis. It is represented by protrusions of solidified basaltic lavas, 60-262m high, ("highs" of American authors), or by narrow slit-like basins ("lows") formed where melted masses from magmatic chambers do not reach the bottom surface. Observers on board the submersible "Pisces" surveyed four volcanic prominences on the valley bottom, mainly in Central Tadjura:

Trapezium	12°04 18'N,	44°42 00'E	(Depth above the summit 1173m)
Punt	12°03 10'N,	44°53 50'E	(Depth above the summit 988m)
Scilla	12°03 10'N,	44°57 50'E	(Depth above the summit 998m)
Charibda	12°02 70'N,	44°59 00'E	(Depth above the summit 1060m)

All these small uplifts of the floor are composed of basaltic lavas of the most diverse forms, from pillow and pipe-like to monolithic and broken into blocks. Several other "highs" not observed visually were echometrically surveyed.

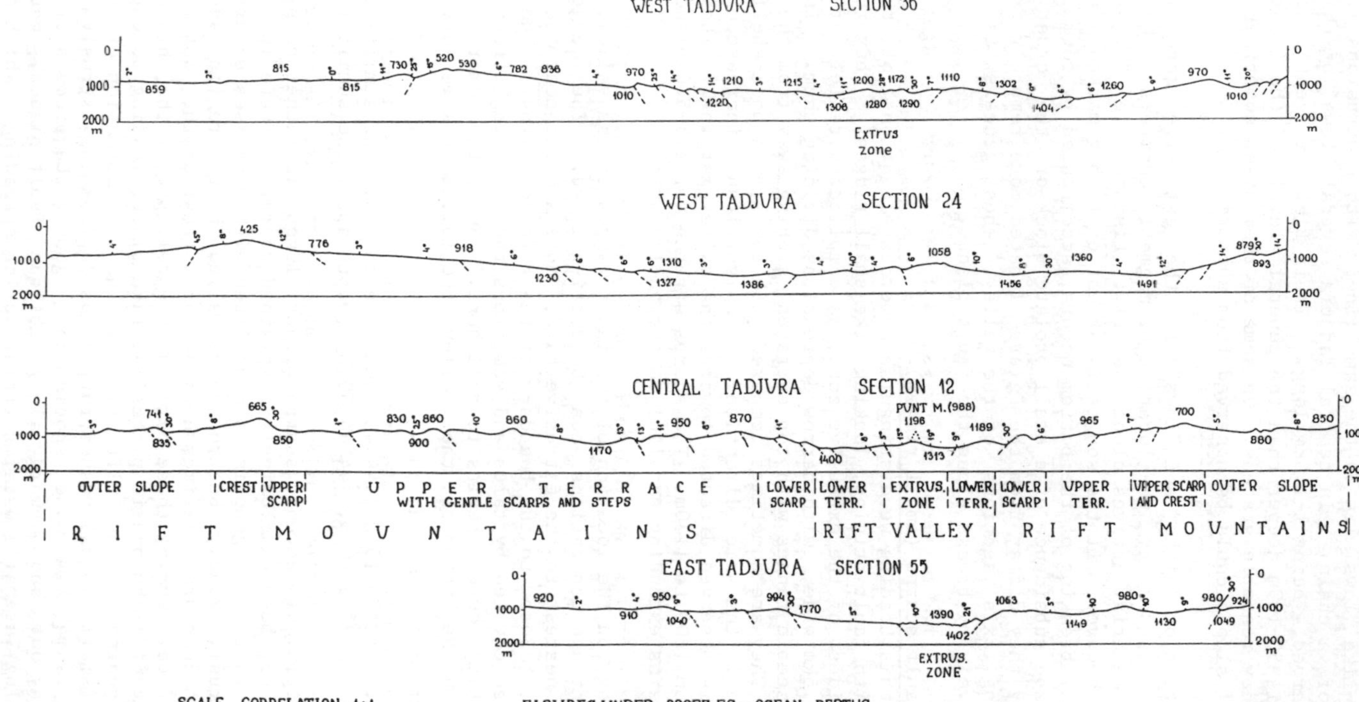

Fig.5. Typical transverse profiles of the Tadjura rift zone, compiled to true scale.

Fissure-like "lows" lie on the same line, with protrusions like the links of a chain. In central Tadjura several "lows" of this type are observed, being 1-1.5km long. Their flat or gently concave bottom is 50-100m deeper than the general level of the floor. In shape they resemble grabens. On some of them steep sometimes nearly vertical slopes could be observed from submersibles.

Alternation of "highs" and "lows" in the extrusive zone is most probably caused by the location of magmatic chambers. Where the melted material arrives along the rift fissure the "highs" are formed; however, if the source is absent or feebly marked a process of normal breaking of the bottom occurs, accompanied by formation of graben and subsidence of a small prolate block of the crust. In this way they differ from the "giars" located more remotely from the spreading axis sections of the valley, whose genesis is caused exclusively by breaks along the fault planes.

Inner fault scarps (small ridges).

Under this heading we distinguish numerous elevations of the sea floor with fault origins, looking like small ridges. They form, on both sides of the extrusive zone, a complicated relief of rises about 0.5km wide, with narrow depressions dividing them. In shape they resemble ranges with top surfaces sloping away from the valley axis. They are sometimes compared with books on a book-shelf, all inclined in the same direction. Insignificant increase of scarp height is observed directed towards the valley margins. Their tops vary from 1200 to 1130m. The scarps stretch for 2.5-5.0km and often form chains extending parallel to the valley axis.

Depending on the location of the inner ranges of the valley, their different age generations can be distinguished. The youngest are located nearest to the extrusive band, more ancient near the valley margins, close to the boundary of the rift mountains. The slopes are often covered by talus and sometimes the levelling effect of the sedimentary cover begins to tell here. Among the numerous fault scarps on the bottom of the valley the following were surveyed by sumbersibles:

Sodom	12°03 10'N,	44°43 70'E	(Depth above the summit 1070m)
Homorra	12°05 90'N,	44°44 50'E	(Depth above the summit 1087m)

The origin of tectonic scarps (inner ridges) is rather complicated and has not been conclusively explained so far. There is no doubt however that their genesis is associated with processes occurring in the extrusive zone. Due to the so-called "hungry regime" ascending melted masses are sometimes detained at a great depth, not reaching the surface. During these periods some collapse of the thin margins of adjacent lithospheric plates may take place, leading to expansion and deepening of the rift fissure and to formation of marginal scarps (small ranges) pushed further aside from the extrusive zone. At the same time one should not ignore the process of blocks squeezing out and upward along oblique fault planes, caused by continuing lateral crustal shifting during spreading, which creates significant compressive strains.

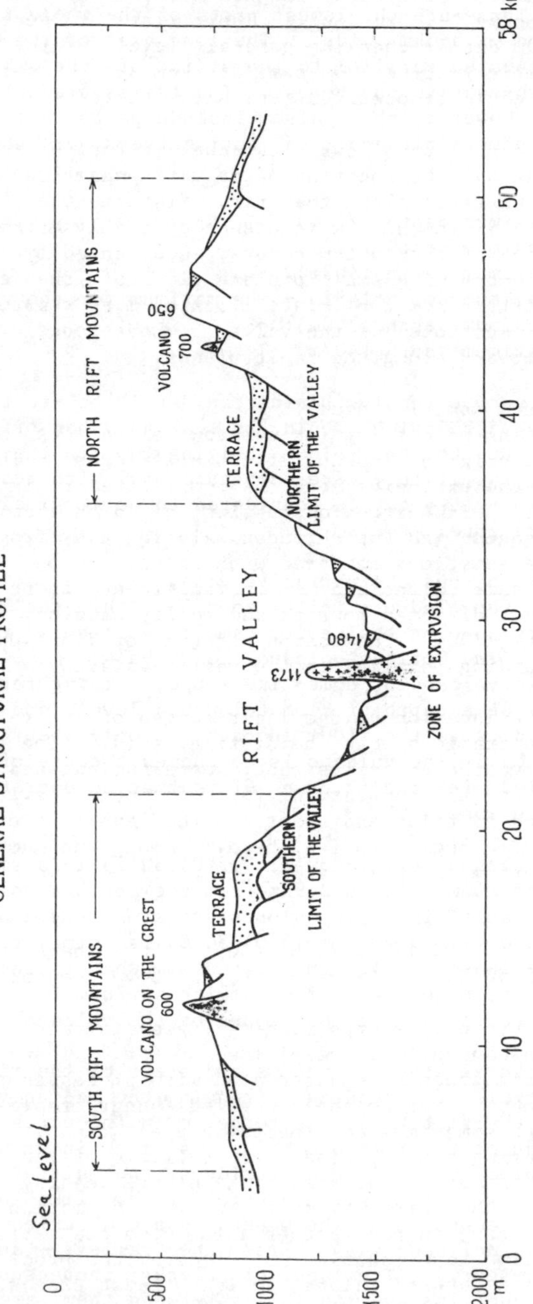

Fig.6. Summary structural-geomorphological profile of the Tadjura rift zcne.

Vast basins on the bottom of the rift valley

There were several vast depressions observed on the bottom of the rift valley. These are the lowest parts of the whole rift zone, 8-9.5km long and 1.5-3km wide. The long axes of the basins are generally orientated parallel to the strike of the whole valley, i.e. from south-east to north-west. The basins are outlined by the 1300m isobath, however they also include parts up to 1400m deep. The "Pisces" submersibles carried out surveys of the following basins:

Aida (maximum depth 1278m)
Nereida (maximum depth 1422m)
Sirena (maximum depth 1402m)
Naiada (maximum depth 1482m - the maximum for the whole valley)
Cleopatra (maximum depth 1390m)
Nefertiti (maximum depth 1407m)

The origins of such extensive basins can hardly be tectonics alone. The most likely assumption would be that they had originated in a "passive" way due to the floor bulging up in the neighbouring extrusive band, as if they were compensational with respect to the latter. Sometimes they may be considered to be marginal depressions of the extrusive zone.

Volcanic massifs beyond the extrusive zone.

In the east of the Tadjura rift zone, within the limits of the Keldysh fracture zone or beside it, small volcanoes rise from the valley bottom. The Springs volcano (12°00 90'N, 45°04 30'E) is a classical single volcano of dome-like shape, rising for 287m. Its acute summit is at a depth of 813m below sea level, and its width at the base, according to the 1100m isobath, is only about 0.5km. The slopes are 12-20°. The volcano is so named because of the thermal springs revealed in the course of submarine operations in the Keldysh fracture zone.

The Nameless volcano (11°59 30'N, 45°03 30'E) is a small mountain with a summit of 760m. It is 2.9km from the Springs volcano, on the same transform fault line. Its slopes are similarly about 15°. The Manganese volcano (11°59 40'N, 45°04 80'E) is similar in shape to the Nameless volcano. It is located somewhat to the east of the transform fracture zone, with its summit at 762m. Ferromanganese crusts occur here on the top and slopes of the volcano.

The upper structural stage - scarps and steps of the rift mountains at an earlier cycle of spreading, fringing the median valley. The depression of the rift valley is bordered in the north and south by the fault scarps of the rift mountains, similar in structure to the classical examples of mid-oceanic ridges, but differing in having much lower heights. The mountain relief is, like in the valley, caused by fault tectonics. The mountains are a series of blocks of oceanic crust inclined away from the valley, and gradually increasing in height. The age of blocks increases in the same direction. Mountains and the valley between them are a dynamic mega-structure within which new crust is continuously created, and the elevation of crust blocks along fault planes to the level of the crest of the rift mountains takes place. Transition from valley to mountains is caused only by, at some stage, uplift of one or several consecutive blocks, with greater rate and energy, which leads to creation of a scarp bordering the valley, i.e. the inner boundary of the rift mountains.

Northern mountains fringing the valley.

The lower marginal scarp of the rift mountains is 90-100m higher than the valley level. It is usually delineated by the 1000m and 1100m isobaths and looks like an escarpment with slopes of 15-20°. In certain parts the marginal scarp is of step-like structure, which indicates that here several blocks of crust are closely pressed together. The upper margins of the scarps are often elevated and from them "reverse" slopes may be observed, directed towards the outer margin of the whole construction. The latter may lead to narrow elongated depressions at the contact with the following higher scarp.

Above the marginal scarp between the 1000m and 800m isobaths a surface is located, with a general small inclination towards the rift valley. This terrace is formed by a combination of several scarps of approximately the same height, and is from 2.5-5.5km wide. Bottom sediments form here a few basins, in the subsidences between the scarps.

The upper scarp and the crest are the most elevated parts of the rift mountains, their distance from sea level being on average 750m. The scarp formation is caused by the same bulging up of one or several crust blocks as described above, but at a greater height when compared with the terrace blocks. These are the most ancient blocks, whose age is defined by linear magnetic anomaly two as two million years (The Matuyama epoch, Early Pleistocene). The crest of the rift mountains is most vividly manifested in Central Tadjura, with its sharp bottom ruggedness. It is a narrow line where bedrock is completely exposed and from where the "reverse" slope of the mountains towards the shelf of Arabia begins.

Southern mountains fringing the valley.

These are in many ways similar to the northern ones. The lower marginal scarp of the rift mountains, which is 100m high, is also located here between the 1000m and 1100m isobaths. However in the east of the zone some parts occur with inclinations of 30°, and the scarp edge has in some places longitudinal ranges, which are relics of extrusive volcanic structures shifted and elevated from the bottom in the process of spreading.

The terrace in the southern fringing mountains is particularly pronounced. Its sloping (3-5°) surface stretches between the 800m and 1000m isobaths as a plain covered with sediments 200m thick, completely concealing the roughness of the basement.

The upper scarp and crest of the mountains in the south of the rift zone are expressed as ranges with asymmetrical slopes. The slope between the 600m and 800m isobaths facing the valley is very steep, sometimes being up to 30°, with incision and dissection. Summits of the dome-like volcanoes are located along the line of tectonic disturbances of the crest. The linear magnetic anomaly two (Early Pleistocene) occurs here as it does in the north, i.e. the composition of the sea floor and stages of development appear to be mirror images of those of the northern fringing mountains. The spreading rate for the last two million years amounted to 1.85cm/year (complete opening).

GENERAL FEATURES OF THE MORPHO-STRUCTURE

The submarine relief of the Tadjura rift zone is exceptionally complicated and in some ways differs from classic examples of such structures on plate boundaries in the oceans. It is a young zone between the sub-continents of Arabia and Somali, which is at an early stage of its development. The divergence of sub-continents did not create the necessary conditions for formation of a typical ocean floor basin. The rift valley bottom elevation and the low level of fringing mountains are unusual for oceanic zones. The general amplitude of the relief hardly approaches 900m (Fig.6).

The principal features of the rift valley morphology are defined by dislocations of normal fault type. Volcanic processes are manifested only in the narrow zone of extrusion. However here, as a result of the insignificant spreading rate and the high position of the valley bottom, the magma cannot rise sufficiently high to form a high ridge as happens at mid-oceanic zones. Only short relief forms are generated, such as small prominences ("highs") alternating with subsidences ("lows") not filled by lavas. Their mapping presents a number of difficulties, but for this the manned submersibles operating on the floor of the Gulf of Aden appeared to be very useful. Specific features of the upper structural level are manifested in the general small height of the rift mountains and meandering outlines of the lower scarp, which are not usually observed in large rift systems.

The general structure of Tadjura is noted for the en echelon character of individual elements of the floor. However whereas in the ocean median ridges each echelon may be traced for large distances, here the extent does not usually exceed 8-10km. En echelon-like structure is caused by sinistral horizontal displacements of large blocks of the crust, by 2.5-4km, along transverse faults. In the relief, however, this is mainly manifested within the limits of the rift valley and hardly affects the mountains fringing it. The crests of the rift mountains are rectilinear, especially in the South. The zones of fault disturbances cross the median valley. Finally, the faults are not accompanied by high parallel ridges, as is so typical for the Sheba ridge in the eastern part of the Gulf of Aden. Additional studies of a wider bottom area are necessary to solve the problem of faults. The available materials allow us to conclude that typical well-developed transform faults were not observed in the rift zone of the Tadjura.

In the general geodynamic setting of the north-western part of the Indian ocean, the Tadjura zone, as well as the whole Gulf of Aden, may be considered to be an ocean which is being conceived between the spreading Arabian and Somalian lithospheric plates.

REFERENCES

Lisitsin, A. P., Bogdanov, Yu. A., Zonenshain, L. P., Kuzmin, M. N., and Sagalevich, A. M., 1984. The structure of the Tadjura Rift in the Gulf of Aden by the Submersible-born Data. DAN SSSR, 279, 1189-1193.

Laughton, A. S., 1966. The Gulf of Aden. Philosophical Transactions of the Royal Society of London, Series A, 259, 150-171.

Laughton, A. S., Whitmarsh, R. B., and Jones, M. T., 1970. The Evolution of the Gulf of Aden. Philosophical Transactions of the Royal Society of London, Series A, 267, 227-266.

Roberts, D. C., and Whitmarsh, R. B., 1969. A Bathymetric and Magnetic Survey of the Gulf of Tadjura. Earth and Planetary Science Letters, 5, 253-258. North-Holland Publishing Company, Amsterdam.

International Geomorphology 1986 Part II
Edited by V. Gardiner

THE POTENTIAL OF SATELLITE REMOTE SENSING FOR GEOMORPHOLOGICAL INVESTIGATIONS - AN OVERVIEW

A.C. Millington[a] and J.R.G. Townshend[b]

[a]Dept. of Geography and [b]NERC Unit for Thematic Information Services, University of Reading, Whiteknights, PO Box 227, Reading, Berkshire, RG6 2AB

ABSTRACT

Although satellite-based systems have not been designed to specifically gather data relevant to geomorphological phenomena many recent developments make such data relevant to geomorphologists. Particularly significant are improvements in spectral sensitivity and spatial resolution and the length of time over which data have been collected.

These three topics are analysed in the light of current geomorphological research interests and directions for future geomorphological investigations using remotely sensed data.

It is suggested that the quality of currently available remotely sensed data means that they are applicable to many areas of geomorphological research and that in the future more applicable data will become available to geomorphologists. The two main areas in which expansion of geomorphological remote sensing is likely to take place in the 1980s are in geomorphological monitoring and hazard prediction.

INTRODUCTION

Research into the applications of satellite remote sensing data has been vigourously conducted over the past fifteen years. Geomorphological applications have however significantly lagged behind those of most other disciplines - most notably geology and, even more so, ecologically-based subjects. This paper will review the potential of data from currently operational satellite remote sensing systems for geomorphological investigations. In doing so the reasons for the previous dearth of geomorphological work will be analysed and the future of geomorphological investigations using satellite data will be appraised.

Most of the satellite imagery that was available to geomorphologists until the early 1980's was generated by the Multi-Spectral Scanner (MSS) carried on-board the first three satellites in the LANDSAT Series - LANDSAT's 1, 2 and 3. Other satellite data have been acquired but used far less frequently. In order of use amongst geomorphologists the other types of satellite data that were used in the 1970s and early 1980s were microwave (radar) data from SEASAT,

Return Beam Vidicon (RBV) imagery from LANDSAT satellites and thermal infra-red data from the Heat Capacity Mapping Mission (HCMM). These different missions, the characteristics of the sensors and the uses of their data are described by many authors (e.g. Curran, 1985; Townshend,1981).

Geomorphological investigations utilising satellite data during the 1980's have focussed on imagery from the MSS and Thematic Mapper (TM) sensors carried on-board LANDSAT 4 and 5. In the latter half of the decade the HRV scanner carried on-board the French SPOT-satellites will be used with increasing frequency. In addition visible and microwave imagery has been acquired on an irregular basis from the Space Shuttle by the Large Format Camera (LFC) and the Shuttle Imaging Radars (SIR A and B) respectively.

This overview will concentrate on the spectral sensitivity and spatial resolution of the sensors and the resultant imagery. The applications of satellite remotely sensed data to current geomorphological research trends will be assessed.

SPECTRAL SENSITIVITY

The MSS, TM and SPOT-HRV sensors each sense the radiation in the electromagnetic spectrum as a number of discrete bands. The characteristics of each sensor are summarised in TABLE 1.

TABLE 1. Main sensor parameters for Landsat and SPOT satellites

Operational Parameters	Landsat MSS	Landsat TM	SPOT HRV
Launched	1972	1982	1986
Sensor type	line scanner	line scanner	linear array
Spectral Bands (nm)	4 500-600 5 600-700 6 700-800 7 800-1100	1 450-520 2 520-600 3 630-690 4 760-900 5 1550-1750 610400-12500 7 2080-2350	1 500-590 2 610-680 3 790-890 panchromatic 510-730 band
Spatial resolution (m)	79	30	10 (panchromatic) 20 (multispectral)
Temporal resolution (days)	18	16	26

The MSS scanner has four separate spectral bands. Two of these detect radiation in the green and red parts of the visible spectra and the other two in the near infra-red (NIR). This combination of visible and NIR bands has many ecological applications. The sensors on SPOT-1 and - 2 satellites will broadly mimic the MSS configuration and consequently they too may well be seen as a primarily ecological sensors. Despite the ecological bias of the MSS sensors' configuration, geologists have been able to effectively map sub-surface lithology and structure through nutrient and moisture stress patterns in the vegetation. Geomorphologists have however been slow to realise the potential of geobotanical remote sensing - particularly in well vegetated regions where it is often felt that satellite remote sensing has very limited potential. However in the African savanna zones dry season vegetation patterns have been used to detect ferricrete exposures and to map relict floodplain features (King, 1981; Millington, 1986; Vass, 1983), Fig.1.

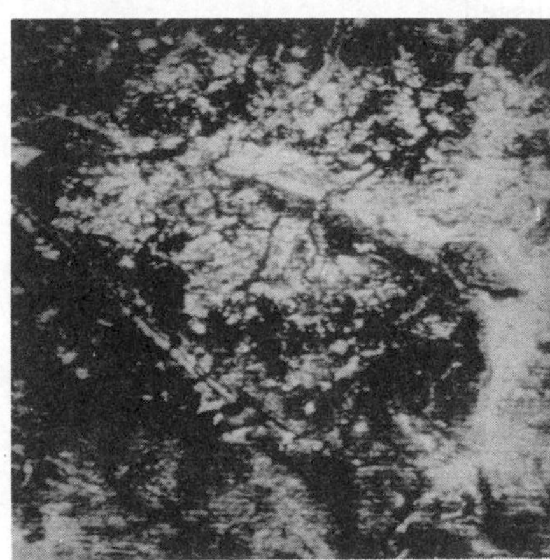

Fig.1 False Colour Composite (Bands 4,5 and 7) of a MSS subscene of northern Guinea. An eroded and dissected synclinal structure in the northern Fouta Djallan can easily be seen with rivers cutting through to the north. The geomorphological and geological features of the area are detected on the basis of differential vegetation responses resulting from varying soil moisture stresses, (Millington, 1986).

The inclusion of middle infra red (MIR) sensors, with spectral bandwidths of 1550-1750nm and 2080-2350nm, on the TM has greatly expanded the geomorphological potential of satellite imagery. The two MIR bands provide the capability to distinguish between different types of surficial materials through absorption peaks of iron, hydroxls, clay minerals and carbonates. These peaks are caused by the vibrations due to small displacements of atoms from equilibrium positions in the crystal lattices of minerals in surficial deposits. Although these are basically a feature of wavelengths greater than 2500nm the effect of overtones and combinations in the vibrations leads to absorption effects within the range of TM sensors. For example water in surficial materials leads to absorption bands at 1400 and 1900nm and a strong hydroxl fundamental at 2700nm causes absorption by hydroxl- bearing minerals. Clay minerals, amongst others, exhibit decreasing reflectance at wavelengths greater than

160nm and the ratio of wavelengths between 2100-2400nm and 1600nm can be used to detect clay-rich areas. Overtone bending and stretching vibrations for aluminium and magnesium combined with hydroxl ions within layered silicates creates narrow spectral absorption features in the 2100-2400nm range. At wavelengths of less than about 1200nm the vibrational effects are far less important than electronic effects in crystal lattices and it is these latter effects that provide information on surficial materials in unvegetated areas when using MSS imagery. Two electronic processes are important at these wavelengths - charge transfers between adjacent metal ions and crystal field effects. The former create absorption bands in the shorter wavelengths; for instance the fall off in the blue part of the spectrum is attributable to an iron-oxygen absorption band.

Electronic transitions caused by changing electron energy levels causes the commonly found iron absorption band between 850 and 920nm; however other transitional metal absorption bands attributable to copper, chromium and nickel are also important, (Hunt, 1980).

The division of the electromagnetic spectrum into discrete bands is an important advance of satellite imagery over aerial photography. Sampling the spectrum in discrete bands is important because it enables interactions between electromagnetic radiation and surficial features to be isolated and examined at specific wavelengths. Armed with a sound scientific knowledge of the nature of these interactions in certain parts of the spectrum it is possible to characterise surficial properties (Fig.2). The bandwidths of between 60-300nm on currently operational satellite remote sensing systems are still too broad for detecting absorption peaks in many minerals and for detecting edge shifts in stressed vegetation. Bandwidths to detect these phenomena need to be about 10-20nm and one of the current directions in sensor development is the specification and testing of narrow bandwidth imaging spectrometers, (Abrams and Goetz, 1986). Proposals exist for both shuttle and space station based imaging spectrometers (McElroy and Schneider, 1985a,b).

Middle IR information is particularly useful when combined with information from other parts of the electromagnetic spectrum. A number of computer algorithms have been developed to extract information from digital imagery by combining different spectral data. It is beyond the scope of this paper to deal with these here and useful summaries applicable to geomorphology can be found in Fabbri (1984), Gillespie (1980) and Moik(1980). Furthermore specific attention needs to be drawn to recent work by Jones (1986) who has evaluated a wide range of algorithms specifically for geomorphological mapping from TM imagery.

Thermal infra-red (TIR) imagery also needs to be examined in a geomorphological context. TIR data from the TM sensor has not been used with any regularity because of its coarse spatial resolution (120m) when compared to the other six TM bands (30m). Moreover this represents the finest resolution TIR data so far available from satellites. However the few geomorphological studies that have been carried out, mainly using airborne TIR imagery, indicate that a

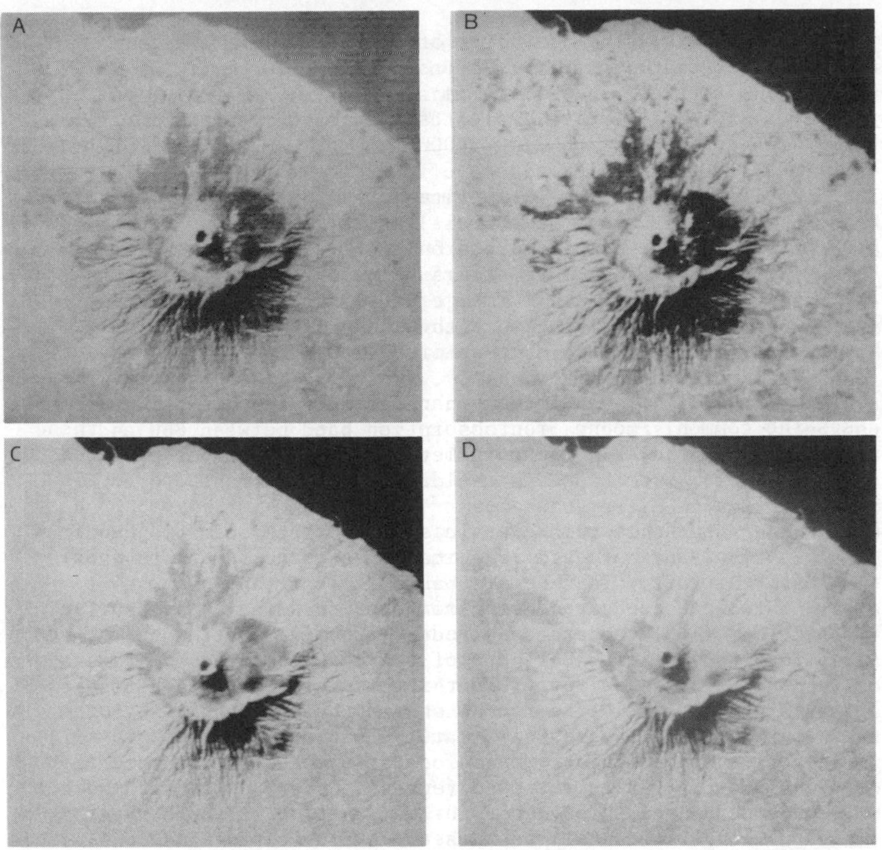

Fig.2 Mt. Vesuvius, S. Italy. Different spectral responses in four TM bands (a) 520-600nm; (b) 630-690nm; (c) 760-980nm (d) 1550-1750nm can be used to identify and map the different lava flows and crater features.

great potential exists for detecting shallow subsurface features such as paleochannels found under buried peat (Lynn, 1985) and in karst areas (Harvey et al., 1977; Kennie and Edmonds, 1986). TIR imagery will undoubtly provide much useful geomorphological information in the future. Nevertheless the fundamental relationships between thermal responses and different geomorphological features and surficial materials are poorly known, especially when compared to the visible, NIR and MIR wavelengths. Much more research is needed in this area before systematic TIR data can be routinely used for geomorphological investigations. A further research theme of great relevance to the geomorphological community concerning the TIR spectrum is the examination of the role of multispectral TIR imagery in the identification of diagnostic absorption bands, known as restrahlen features, in surficial materials. These have been related to the silica contents of surficial materials, (Kahle et al.,

1980; Lyon, 1965). The potential of TIR data therefore cannot be fully appreciated until it is collected on a systematic basis at resolutions far less than those employed by the TM sensor and further fundamental research into TIR interactions with surficial materials has been undertaken.

The other main part of the electromagnetic spectrum that provides important data for geomorphological investigations is the microwave area at wavelengths between 5 and 500mm. The geomorphological potential of active microwave imagery was first realised in the late 1960s and early 1970s when drainage networks were mapped under dense humid tropical vegetation in Colombia and Panama (MacDonald, 1969). Early workers felt that these examples showed the ability of active radar data to penetrate vegetation. Research since then has shown that these and similar affects are related to overall topographic roughness and moisture variations in vegetation. In fact in these case studies it was the river valleys that were being identified due to topographic dissection and the differences in moisture levels between trees in different catenary situations. Moreover, it has since been shown that radar is a useful tool in estimating soil moisture levels of bare and vegetated areas because of its ability to penetrate the soil (Ulaby et al., 1978, 1979). These early experiments were conducted using airborne sensors and as yet the potential of a fully operational satellite radar system has not been fully realised. The SEASAT satellite system launched in 1978 malfunctioned after only five months. Imagery obtained from SEASAT and by SIR-A and SIR-B, carried on-board the Space Shuttle, has again indicated the potential of active microwave systems in geomorphology. The entire area of radar remote sensing in geomorphology has recently been reviewed (Munday, 1984) and the future for regularly obtaining satellite radar information in the 1990s is very promising with satellites scheduled from Canada (RADARSAT), Japan (J-ERS-1), the USA (SIR-C and -D) as well as the European Space Agency's ERS-1. However to fully realise the potential of future satellite microwave systems there is a need to carry out research into the fundamental empirical relationships between radar parameters such as backscatter and the dielectric constants and surficial materials. Laboratory research to obtain such empirical relationships for various rock and soil types is currently being undertaken (Xiao Jin-Kai, 1986) but great difficulties will occur in transferring this to field situations because of the significance of surface roughness at these wavelengths. Empirical research in the microwave region therefore needs to be carried out in field situations rather than laboratories. Notwithstanding this problem the interpretation of satellite radar imagery will be severely restricted until such studies are undertaken.

SPATIAL RESOLUTION

Further justification for the use of TM imagery, particularly when compared to MSS imagery, can be gained from an analysis of their respective spatial resolutions.

The spatial resolution of the MSS is 79m for all bands; for the TM the spatial resolution of all bands, except Band 6 (TIR), is 30m. This represents a seven-fold areal improvement in ground resolution (Townshend, 1981). This improvement allows far more detail to be interpreted and mapped on the ground. Consequently many geomorphological phenomena are now readily identifiable on TM imagery which were previously poorly represented on MSS imagery (Millington and Townshend, 1984). These include linear features, often of fluvial origin, such as cutoffs, levees and sloughs; although a variety of types of mass movements, coastal features and larger gully systems can now also be mapped. A comparison between the MSS and TM imagery in a volcanically active area to the north-west of Naples indicates the effects of increased spatial resolution features on the identification of calderas and slope instability features, (Fig. 3).

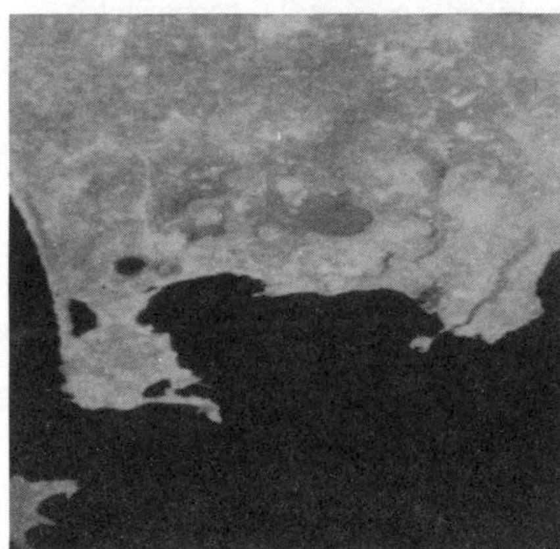

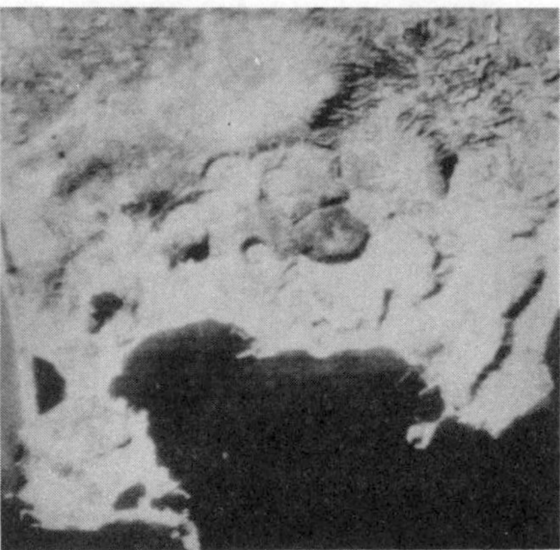

Fig.3 A comparison of FCC's of a Landsat 4 TM subscene of (right) and a Landsat 3 MSS subscene (left) of the northern Bay of Naples. The finer spatial resolution of geomorphological features is evident.

The increased spatial resolution of SPOT data will enable even finer detail to be mapped. The spatial resolutions for SPOT imagery are 20m in the multispectral mode and 10m in the panchromatic mode. This latter resolution will represent a 62-fold areal increase over MSS imagery.

A further advantage of SPOT data for the geomorphologist is its ability to provide stereoscopic imagery by using pairs of images taken in nadir and off-nadir positions. This is a different method for stereoscopic image production to that provided by conventional aerial photography, but for geomorphological research this should not prove problematical. Conventional overlapping stereoscopic imagery is provided by the LFC, carried on-board the Space Shuttle, in the form of high quality photography with spatial resolutions varying between approximately 10 and 20m (Millington and Moutsoulas, 1986).

The combination of fine spatial resolution imagery and the stereoscopic capability of SPOT-imagery will provide a product that can be used for detailed geomorphological mapping and could allow volumetric changes in geomorphological phenomena to be detected.

THE ROLE OF SATELLITE REMOTE SENSING IN GEOMORPHOLOGICAL MONITORING

Geomorphological research that has been undertaken using satellite remotely sensed data has so far almost entirely concentrated on the identification and mapping of specific features or assemblages of features. Feature identification and mapping is not, however, a major emphasis in geomorphology at the present time, except for geomorphological mapping. The research frontier in geomorphological remote sensing must therefore surely lie in matching current remote sensing technologies with the major research areas in geomorphology.

Ten areas of contemporary geomorphological research are defined in terms of the temporal and spatial resolution requirements of their data needs, Fig.4. These are compared to the data volume rates of current remote sensing systems which are limited by three factors:-

a) the data volume threshold line, this oblique line is taken as a function of the spatial and temporal resolutions and indicates the limits to change detection with multitemporal data sets,

b) the minimum spatial resolution line, this is a vertical line corresponding to the finest spatial resolution available and indicates the minimum size of feature identifiable from a single image,

c) the limit of archived remotely sensed data that can be used in change detection studies, this is indicated by the horizontal line in the upper part of the diagram.

Those areas of study falling in the lower left-hand sector (small catchment and related monitoring studies and aeolian landform monitoring) require more data than can be provided by current satellite systems. A further group of study areas (climate change studies, neotectonics, plate tectonics and large scale geomorphological studies) need comparative data over longer time periods than are currently available in archived data to detect geomorphological changes. Current geomorphological research areas such as coastal and estuarine studies, glaciology, change detection in alluvial landforms and duststorm monitoring can however take advantage of remotely sensed data. This does not mean that these areas of study should rely solely on remotely sensed data but rather data from remote sensing systems should be used, when appropriate, in combination with ground observations.

A further important point needs to be made here in relation to geomorphological research. Current research advances in sensor design and computer hardware and software mean that the data volume threshold line is constantly moving to smaller spatial, and finer temporal, resolutions thereby potentially incorporating more fields

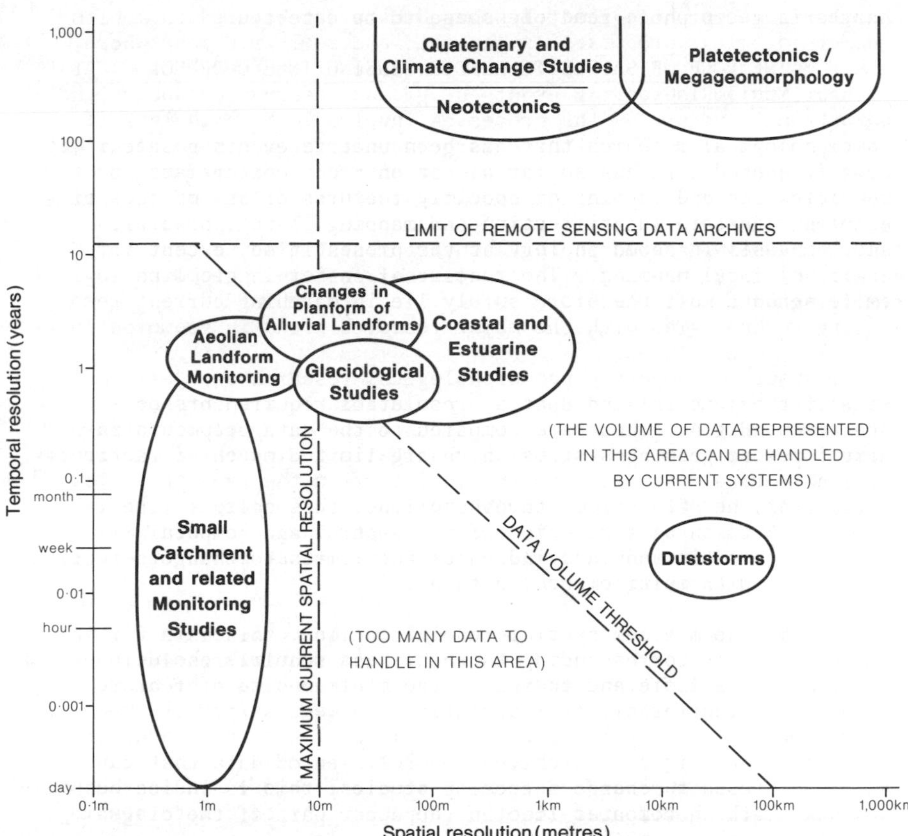

Fig.4 Geomorphological applications of satellite remote sensing in relation to spatial and temporal resolution and illustrating the factors limiting their suitability.

of geomorphological research.

The most innovative future use of remotely sensed data in geomorphology lies in monitoring geomorphological change and hazard prediction. This is because image acquisition is repetitive and an archive of Landsat imagery now exists back to 1971. However little research has been carried out in these areas so far. Yet advantages will definitely accrue to the geomorphological community in applying remotely sensed data to such studies in areas where monitoring is difficult due either to logistical problems or because of the nature of the processes involved. This role of remotely sensed data was well illustrated by research carried out on iceberg production monitoring on a monthly basis in western Greenland.

Another area in which remotely sensed data can be used to monitor geomorphological processes is the arid and semi-arid zone where field monitoring is very difficult because of the very low probability of monitoring events by ground observation due to the catastrophic nature of the processes involved. Nevertheless if a large enough area is monitored large magnitude events do occur with high frequency. Such large areas can only be monitored using remotely sensed data and consequently the probability of detecting geomorphological change is vastly increased. This approach is currently being tested in three environments in south-central Tunisia - large braided rivers, alluvial fans and salt lakes (Millington et al., 1986).

CONCLUSIONS

Satellite remote sensing data are now available with broader spectral and finer spatial resolutions that are appropriate for many types of geomorphological investigation. Imagery is currently available to almost all geomorphologists from the Landsat series of satellites, and fine resolution, stereoscopic imagery is available from the SPOT satellites. An expansion in available data will occur in the late 1980s and early 1990s with the launch of future SPOT missions and orbiting microwave satellites.

The maximum geomorphological potential can be achieved if digital imagery is computer enhanced. However this requires the development and testing of different computer algorithms and must be based on a sound understanding of the interactions between different types of electromagnetic radiation and surficial materials and features.

Although the past record of geomorphological remote sensing has been somewhat lacking compared to other applications of remote sensing, the time is now ripe for expansion; particularly into the application of satellite imagery to geomorphological monitoring and hazard prediction.

ACKNOWLEGEMENTS

We would like to thank Arwyn Jones for his help in preparing this paper and to Chris Holland, Erika Meller and Chris Howitt for technical assistance during its preparation.

REFERENCES

Abrams, M.J. and Goetz, A.F.H. 1986 Imaging Spectrometry: Past, Present, Future. Proc. 3rd Int. Coll. on Spectral Signatures of Objects in Remote Sensing. Les Arcs, France, ESA SP-247, 215-218

Curran, P.J. 1985 Principles of Remote Sensing, Longman, London

Fabbri, G. 1984, Image Processing of Geological Data. Van Nostrand Reinhold, New York

Gillespie, A.R., 1980. Digital Techniques of Image Enhancement, in Remote Sensing in Geology (Eds. Siegal and Gillespie), pp. 139-225, Wiley, New York

Harvey, E.J., Williams, J.H. and Dinkel, T.R. 1977 Application of Thermal Imagery and Aerial Photography to Hydrologic Studies of Karst Terrain in Missouri USGS Water Resources Division Rept., 78/005

Hunt, G.R. 1980 Electromagnetic Radiation: The Communication Link in Remote Sensing, in Remote Sensing in Geology (Eds. Siegal and Gillespie), pp. 5-46, Wiley, New York.

Jones, A.R. 1986 An evaluation of satellite Thematic Mapper in arid and semi-arid environments. (This Proceedings)

Kahle, A.B., Madura, D.P. and Soha, J.M. 1980 Middle infrared multispectral aircraft scanner data: analysis for geological applications Applied Optics, 19(14), 2279-2290

Kennie, T.J.M. and Edmonds, C.N. 1986 The Location of Potential Ground Subsidence and Collapse Features in Soluble Carbonate Rocks by Remote Sensing Techniques ASTM Int. Symp. on Geotechnical Applications of Remote Sensing, Florida (in press)

King, R.B. 1981 An evaluation of Landsat-RBV imagery for obtaining environmental data in Tanzania. Matching Remote Sensing Technologies and their Applications. Proc. 9th Tech. Conf. Remote Sensing Society London, 85-95

Lynn, D. 1985 The timeliness of thermal infrared data acquisition for soil and surficial material survey in humid temperate environments. Proc. 4th Symp. of ISSS Working Group, Remote Sensing for Soil Survey, Wangeningen and ITC, Enschede, Netherlands, March 1985.

Lyon, R. 1965 Analysis of rocks by spectral infrared emission (8 to 25 microns) Economic Geology 60, 715

McElroy, J.H. and Schneider, J.R. 1985a Earth Observations and The Polar Platform NOAA Tech. Rept., NESDIS 18, 16p.

McElroy, J.H. and Schneider, J.R. 1985b The Space Station Polar Platform: Integrating Research and Operational Missions, NOAA Tech. Rept. NESDIS 19, 19p

MacDonald, H. 1969 Geologic evaluation of radar imagery from Darien Province, Panama. Modern Geology 1, 1-63

Millington, A.C. 1986 Spectral Signatures of Land Cover Types in the Sahel for Geobotanical Modelling Proc. 3rd Int. Coll. on Spectral Signatures of Objects in Remote Sensing. Les Arcs, France, ESA SP-247, 485-89

Millington, A.C. and Moutsoulas, M. 1986 Evaluation of Large Format Camera imagery for geomorphological and geological mapping in Greece. Proc. Int. Symp. on Remote Sensing for Resources Development and Environmental Management. Enschede, (in press)

Millington, A.C. and Townshend, J.R.G. 1984 Remote sensing applications in African erosion and sedimentation studies. Challenges in African Hydrology and Water Resources. Proc. Harare Symposium, IAHS Publication 144, 373-384

Millington, A.C. Quarmby, N.Q., Jones A.R. and Townshend, J.R.G. 1986 Monitoring sediment transfer processes in Semi-arid Tunisia using multi-temporal Landsat TM imagery. Proc. Int. Symp. on Remote Sensing for Resources Development and Environmental Management, Enschede (in press)

Moik, J. 1980. Digital Processing of Remotely Sensed Images NASA SP-431, 330p.

Munday, T.J. 1984 Earth Science Applications of Radar Remote Sensing in Radar Remote Sensing over the Land Surface Remote Sensing Society, London

Townshend, J.R.G. 1981 The spatial resolving power of earth resources satellites Prog. phys. Geog., 5(1), 32-55

Townshend, J.R.G. (ed) 1981 Terrain Analysis and Remote Sensing George Allen and Unwin, London

Ulaby, F.T., Bativala, P.P. and Dobson, M.C. 1978 Microwave backscatter dependence on surface roughness, soil moisture and soil texture. Part 1 - bare soil IEEE Trans. on Geoscience Electronics, 16, 286-95

Ulaby, F.T., Bradley, G.A. and Dobson, M.C. 1979 Microwave backscatter dependence on surface roughness, soil moisture and soil texture, Part II - vegetation cover soil IEEE Trans. on Geoscience Electronics, 17, 33-40

Vass, P.A. 1983 A Landsat study of vegetation and seasonal livestock grazing in the southern Sudan. Remote Sensing for Rangeland Monitoring and Management. Proc. 9th Ann. Conf. Remote Sensing Society, Silsoe, 51-68

Xiao Jin-Kai, 1986 A study of Microwave Dielectric Properties of Minerals and Rocks. Proc. 3rd Int. Coll. on Spectral Signatures of Objects in Remote Sensing, Les Arcs, ESA SP-247, 293-296

International Geomorphology 1986 Part II
Edited by V. Gardiner

AN EVALUATION OF SATELLITE THEMATIC MAPPER IMAGERY FOR GEOMORPHOLOGICAL MAPPING IN ARID AND SEMI-ARID ENVIRONMENTS

A.R. Jones

Department of Geography, University of Reading, Whiteknights, Reading, Berks. U.K.

ABSTRACT

Test areas in southern Tunisia were used to evaluate the potential of satellite Thematic Mapper imagery for geomorphological mapping. Single band subscenes and standard false colour composites contain a great deal of geomorphological detail. However, much more information can be derived from the imagery if selective image processing techniques are applied to the digital data. The results demonstrate the applicability of suitably processed TM imagery for geomorphological investigations in arid and semi-arid environments.

INTRODUCTION

Geomorphological mapping is the depiction of landforms by showing their form, distribution, constituent materials, age and indicating the processes that led to their development. The applications of these maps are of considerable interest not only to geomorphologists but also to civil engineers, geologists and urban planners.

Traditionally, ground survey has been the primary source of information for geomorphological mapping, supplemented by information from topographic and other thematic maps. However, in recent years an increasing role in such surveys has been undertaken by remotely sensed imagery, mainly from aircraft. Numerous examples of such usage exists (e.g. Verstappen, 1977;Doornkamp *et al*, 1980). The advantages of aerial photographs include the excellent depiction of topography, rapid familiarisation with a study area and the availability of stereoscopic capability which provides an excellent base from which the geomorphologist can work.

With the advent of the space borne sensors, ranging from hand held cameras to line scanners and radars, earth scientists have been given a new perspective from which to study the earth's surface. Lillesand and Kieffer (1979) describe the advantages of satellite acquired imagery. These include synoptic scale coverage, multi-spectral and multidate capability and as a result, imagery from such sources are of great interest to geomorphologists. Studies of such diverse topics as Saharan aeolian dynamics (Mainquet, 1984), mapping of faults (Bailey *et al*, 1982) and mapping of tropical coastal phenomena (Sobur et al, 1978) demonstrate this. However, the use of such data for geomorphological mapping has been largely neglected. This is mainly due to an over reliance on inferior, standard photographic products as opposed to suitably processed digital data and unsuitable scanners. The development of the Thematic Mapper (TM) sensor, which was carried aboard Landsat 4 and 5 satellites, has enabled this situation to be remedied. This sensor was better suited to earth science applications than the earlier Multispectral Scanner because of its improved spatial and spectral resolutions (NASA, 1984).

This paper, by demonstrating the potential of TM imagery for geomorphological investigations, will show how selective image enhancement techniques of TM data can result in images which have an increased geomorphological information content. The resulting images will be better suited for geomorphological mapping purposes than standard products. The analysis in this work involved visual interpretations of digitally processed data using test sites in the arid and semi-arid region of southern Tunisia. The principal emphasis will be on the advantages of spectral qualities of the data and its processing for geomorphological mapping.

FIELD AREA

The field area used in this study is located to the west of the oasis town of Gafsa in south-central Tunisia (Fig. 1). Climatically, it is situated on the 150mm isohyet, the boundary between the arid and semi-arid regions of Tunisia. The geological structure of the area, as described by Furon (1963), reflects the juncture between the highly folded and faulted strata of the Atlas mountains and the relatively undeformed Saharan plateau (Fig. 2). The main lithologies are Jurassic and Cretaceous limestones, dolomites, marls and shales, flanked by Mio-Pliocene conglomorates. Extensive Quaternary deposits exist throughout the area.

The geomorphology of this part of Tunisia is very varied and is discussed by Coque (1962) and Coque and Jauzein (1967). Although they are thorough accounts, local detailed studies and geomorphological maps in central and southern Tunisia are quite rare and as a result the area is well suited to demonstrate the applicability of remotely sensed imagery for geomorphological mapping. The landforms in the study area reflect the interactions between the major structural and lithological units, and the many climatic fluctuations which have occured during the Quaternary.

A basin and range topography characterises the area. The mountain in the study area, Djebel ben Youssef, comprises of folded Cretaceous strata which havebeen faulted by the southern Atlas fault. Flanking the mountains are extensive piedmont deposits, which include alluvial fans (Fig. 3) and alluvial plains, often dissected by gullies which drain to larger dry ephemeral channels, known locally as oueds. The alluvial plains and fans often exhibit gypsiferous or calcareous crusts known as croutes which, in places, are extensively broken up

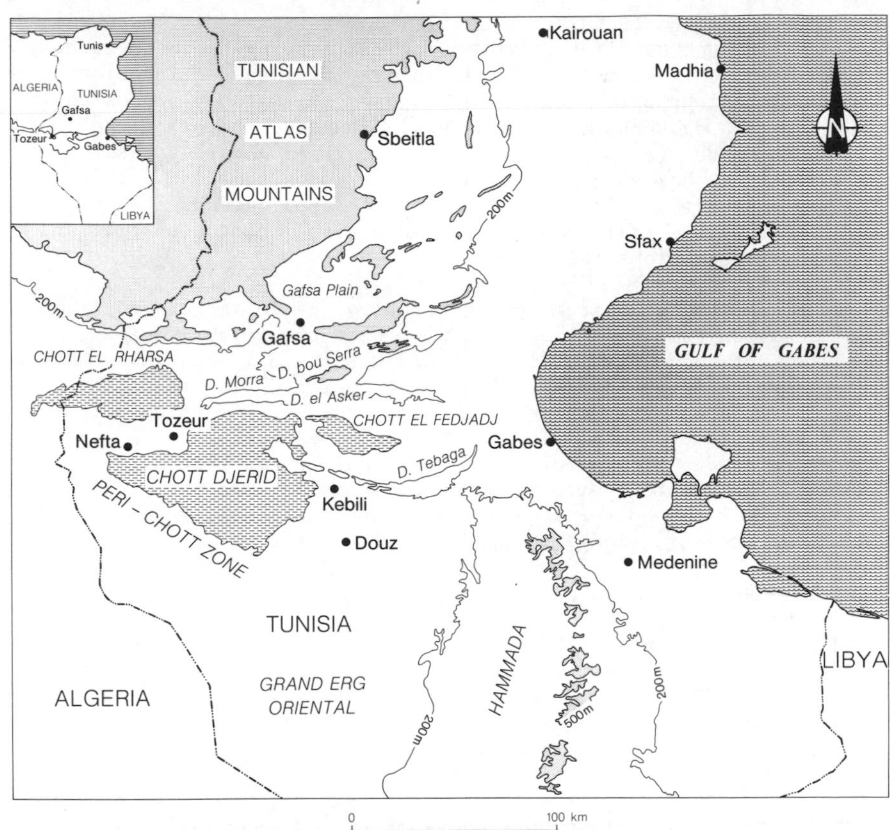

Fig. 1. Map of southern Tunisia.

and covered by Holocene sediments.

At their distal end, the fans grade into saline playa-like landforms known as Chotts. They seasonally flood and have numerous variations in surface morphology which are related to ground water proximity, evaporation regimes, salt solubility and sediment inputs. Scattered around the area are nebkas which are small (less than 10m length) accumulations of aeolian deposits around vegetation. Fuller descriptions of these landforms can be obtained in Cooke and Warren (1973) and Mabbutt (1977).

In attempting to evaluate the potential of TM data for geomorphological mapping, a number of geomorphic phenomena, which are representative of the semi-arid/arid environment, have been selected for study (see Table 1). The effectiveness of the processed imagery in allowing the extraction of the features listed below will be discussed.

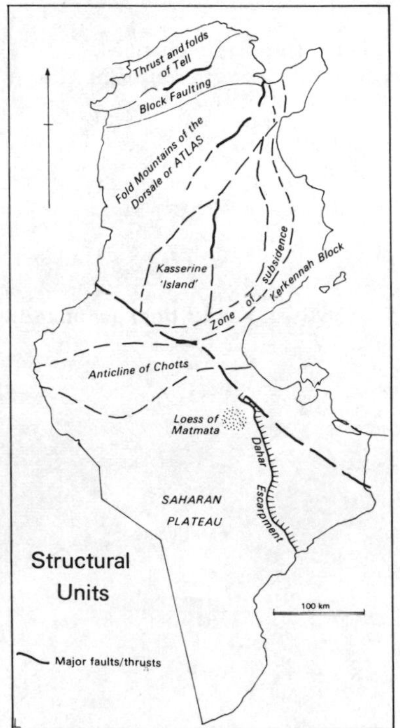

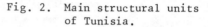

Fig. 2. Main structural units of Tunisia.

TABLE 1. Geomorphological phenomena under investigation

1. AREAS OF MODERATE AND HIGH RELIEF:-

   a) Investigation of outcrop areas and contact between detrital materials.

2. PIEDMONT AREAS:-

   i. ALLUVIAL FAN ENVIRONMENT

      a) The accurate and effective delineation of the alluvial fans.
      b) Discrimination of fans from mountain and playa areas.
      c) Change in sediment characteristics on the fan.
      d) Drainage systems on the fan.
      e) The occurrence of crust (croutes) on alluvial fans and on flanks of cuestas.

   ii. ALLUVIAL PLAIN

      a) Discrimination between croutes and sheet-wash deposits.

b) Extent of channel floodplain and sediment variation within it.
c) Gully networks draining into the main channel.
d) Identification of erosional or depositional pediments.

3. SALINE ENVIRONMENTS:-

a) The accurate delineation of the playa boundary.
b) Surface variation on the surface of the playas.

4. AEOLIAN ACTIVITY:-

a) Identification of areas of blown sand, nebkas and dunes.

Fig. 3. Alluvial fans flanking the southern side of the mountain range in the study areas exhibiting both entrenchment and anastomosing channels. The fans can be seen grading into a fine grained playa. A major fault runs along the mountain front. See Fig. 5 for comparison.

POTENTIAL OF SATELLITE TM DATA FOR GEOMORPHOLOGY

A number of marked differences exist between the first and second generation Landsat satellites. The salient points are summarised in Table 2. The most significant difference is the advent of the Thematic Mapper sensor aboard Landsat 4 and 5. It can be seen that TM differs from MSS in many respects. The major changes include an improvement in spatial resolution (79m to 30m), more spectral bands and narrower bandwidths, better suited for earth science

TABLE 3. Potential geomorphic applications for single TM bands.

TM BANDS	WAVELENGTH (um)	GEOMORPHOLOGICAL APPLICATIONS
1	0.45-0.52 (blue-green)	Studies of sediment laden water:longshore drift, esturine plumes, suspended sediment in lakes and rivers. Identification of sediment source areas. Bathymetry. Surface properties of snow and ice. Soil organic matter.
2	0.52-0.60 (green)	Biogeomorphic indicators-soil erosion. Pedological studies, soil toxicity and disturbed ground. Ratio 2/4 limonitic rock mapping and for redness on desert sand
3	0.63-0.69 (red)	Vegetation cover mapping and identification of cropping practices for erosion studies. Ratio 3/4 geobotanical relationships. Lithological separation (iron rich rocks) and structural studies.
4	0.79-0.90 (near-IR)	Water body delineation (lakes, rivers, wetlands and active ephemeral channels), spring lines and drainage network morphometry. Reconnaissance mapping and geobotanical studies.
5	1.55-1.75 (mid-IR)	Lithological mapping, bedrock/drift separation. Soil moisture mapping. Ratio 4/5 separates hydrous and iron rich rocks, ratio 5/7 for clay mineral differentiation.
6	10.4-12.5 (thermal)	Lithological mapping, geological reconnaissance studies, thermal mapping of sediments. Ground water studies, topographic mapping and extraction of sub-surface anomalies. Bathymetry of lakes and discrimination of silicious rich rock.
7	2.08-2.35 (mid-IR)	Lithological discrimination, metamorphic rocks, hydrous minerals (OH-clay mineral) and carbonates (CO3-calcites etc.) separation. Hydrothermal alteration.

Based on Jones (1984).

applications. The advantages of increased resolution has been described by Jones (1984). The obvious advantages for the geomorphologist are that small elements in the landscape can now be observed.

However, the major emphasis of this study is on the spectral improvements of TM over MSS. Table 3 summarises the potential geomorphological applications for each TM band based on a specific spectral response in that wavelength. The more notable improvements are the shallow water penetration capability of band 1 (blue-green) and the inclusion of the reflective middle infra-red bands (5 and 7) which are very diagnostic for discriminating lithologies, carbonates and clay minerals. Also of interest is the thermal infra-red (band 6) which senses both reflected and emitted energy. This wavelength is very useful in mapping silicious rich rocks but suffers from having a very coarse resolution (NASA, 1984).

TABLE 2. Characteristics of the Landsat satellites

	4,5	1,2,3
ALTITUDE (KM)	705	900
TIME OF OVERPASS	11.15	09.00
REPEAT CYCLE (DAYS)	16	18
SENSORS	TM, MSS	MSS, RBV
	TM	MSS
GROUND RESOLUTION (M)	30	79
SPECTRAL BANDWIDTHS	BAND	BAND (APPROX.) *
0.45 - 0.52	1	
0.52 - 0.60	2	4
0.63 - 0.69	3	5
0.76 - 0.90	4	7
1.55 - 1.75	5	
2.08 - 2.35	7	
10.4 - 12.5	6 (120m PIXEL)	

* No equivalent MSS band 6 in TM.

THEMATIC MAPPER IMAGERY AND SYNOPTIC GEOMORPHOLOGY

Figure 4 shows a TM band 4 image (near IR) of southern Tunisia taken from Landsat 4 in January 1983. It clearly shows intricate geological structures trending NE-SW which are the southern vestiges of the Atlas mountains. Numerous breached anticlines (A) and faults (B) are clearly visible. The darker area in the lower part of the image is one of Tunisia's playa-like features, the Chott Djerid. These saline areas are devoid of vegetation and the change in tone across the surface is due to variation in surface morphology, texture and chemical composition, which in turn are related to the proximity of ground water. The lighter tones suggest the extent of

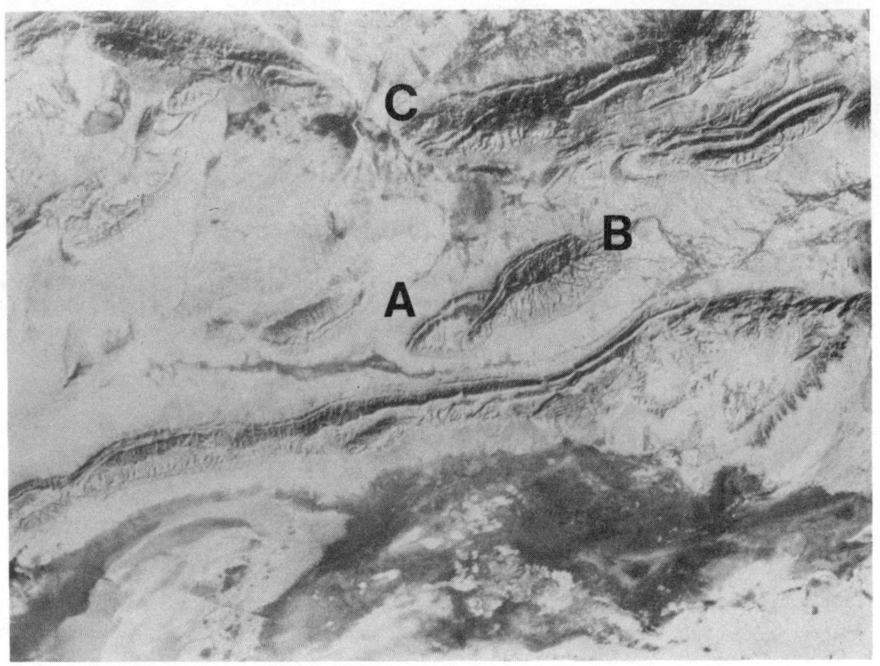

Fig. 4. TM band 4 (near infra-red) scene of southern Tunisia for January 1983. The area depicted is approximately 90 x 90kms. (See text for details).

a saline crust while the darker areas are also crusted but have a higher sediment concentration and as a result cause a decrease in reflected energy. The dark spots in the halide area of the playa are upwellings of fresh artesian water known as aiouns.

Alluvial fans, some over 10kms across, can be seen flanking the mountain ranges while other fluvial features range from gullies to large dry ephemeral channels (oueds). The highly reflective sandy beds of the latter can be clearly distinguished from the adjacent plains in the north of the image. Some of the channels appear dark due to the presence of water (winter image) or associated seasonal vegetation. The dark area in the north (C) is the date plantations of the oasis town of Gafsa.

IMAGE PROCESSING FOR GEOMORPHOLOGY: A CASE STUDY

While the small-scale image provides excellent synoptic information of geomorphological phenomena and general processes operating in the region, full resolution sub-images display greater detail. This is demonstrated by Fig. 5, a 15 x 15kms area to the west of the town of Gafsa (see Fig. 4) which has been thoroughly mapped in the field

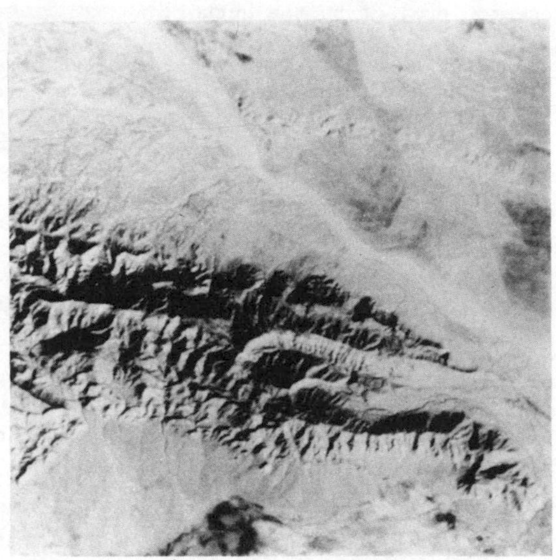

Fig. 5. TM band 4 subscene to the west of Gafsa with a sketch map showing main geomorphological units interpreted from the the above image.

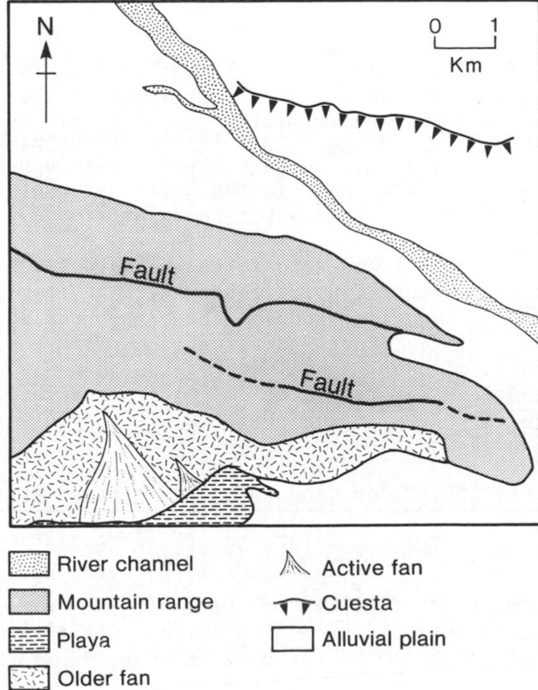

as part of a research project by the author.

The image shows a range of highly folded and faulted strata. Numerous flatirons, hogsbacks and fault scarps are apparent. South of the mountains, unvegetated alluvial fans are found. Fan head entrenchment is clearly visible. This is due not only to the improved spatial resolution of TM but to the detection of lighter gravels in the channel compared to the surrounding fan. This is also shown at the distal end of the fan where anastomosing channels are depositing fresh gravel. The fans merge with the darker tones of a playa. Crossing the image is a large ephemeral channel with a dry sand/gravel bed, into which drain several gully networks.

Not much more detail can be obtained from standard false colour composites. However, much more information can be extracted from the data if computer-assisted digital image processing is applied to the imagery. A basic but powerful image processing technique is the ratioing of two spectral bands. This reduces topographic noise and enhances subtle spectral differences of surface features. However, to effectively use such methods, the relationship between the surface material and its spectral reponse must be known. Hunt and Salisbury (1976) and Kahle (1982) have shown that up to a wavelength of 2.5µm, there is an increase in absorption of reflected energy by clay particles. Thus a ratio of TM band 5 (1.55-1.75µm) and band 7 (2.08-2.35µm) will display the ocurrence of clay particles as light tones. In figure 6a, a 5/7 ratio, the light area in the lower part of the image corresponds with the playa where clay particles are expected. It is interesting to note the low concentration of clays in the river channel.

Another interesting image is produced by ratioing bands 3 and 4. Band 3 coincides with a chlorophyll absorption band while the response of band 4 is controlled by the physical structure of the mesophyll layer of leaves (Townshend, 1984). The resulting ratio image displays vegetation as dark tones (Fig. 6b). Two phenomena stand out in this image. The higher vegetation cover to the north of the mountains and the existence of spring line vegetation along the distal end of the fan.

Multivariate enhancement techniques reduce the variability of multi-band remotely sensed data, and can lead to the extraction of information not apparent on the original imagery. One such technique is principal component analysis (PCA), especially useful if the bands are highly correlated. New images are produced by calculating new uncorrelated principal components which represent more efficiently the variance of the data.

A study of Table 4 shows the eigenvector weights attached to 6 principal component images, derived for the study area, using the 6 reflective TM bands. PC1 is an overall summary of albedo while PC2 (Fig. 7) is the difference between the visible/near-IR and the mid-IR. This explains the brightness of the image, the extraction of areas of current fan deposition and variation within the playa. PC3 has a high band 4 weight which produces an image similar to the 3/4

Fig. 6a. Ratio image of TM bands 5/7.

Fig. 6b. Ratio image of TM bands 3/4. (Both 15 x 15kms.)

ratio displaying vegetation while PC4 is the difference between the two mid-IR bands. Too much noise in the remaining components prevent any useful analysis.

Table 4. Eigenvectors derived from PCA of 6 reflective TM bands for test site.

	SPECTRAL BANDS					
PC	1	2	3	4	5	7
1	-.232	-.197	-.335	-.324	-.685	-.486
2	-.644	-.330	-.395	-.238	.403	.315
3	.372	.126	.040	-.830	-.054	.387
4	-.167	.002	.048	.302	-.601	.718
5	.561	-.194	-.757	.237	.042	.118
6	-.219	.893	-.391	.009	.016	-.017

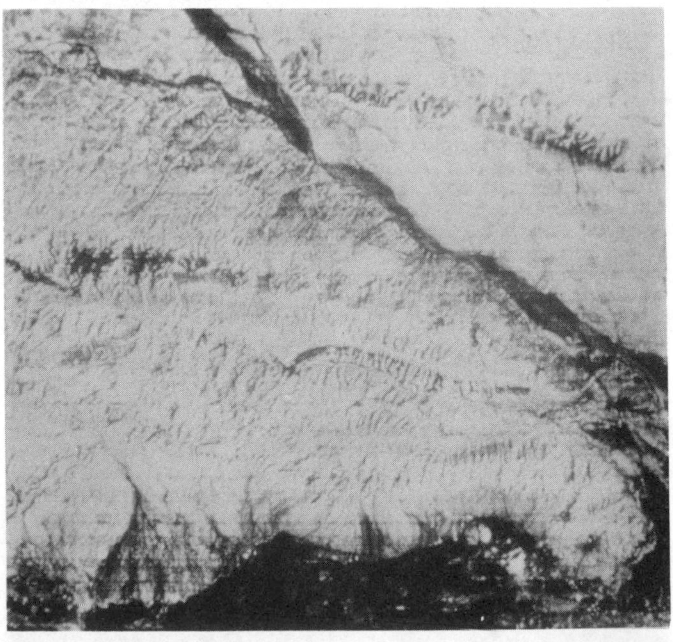

Fig. 7. Second principal component image of the study area derived from 6 reflective TM bands. (15 x 15kms)

Fig. 8a. Unsupervised classification image derived from TM bands 3, 4 and 5 showing 10 classes. (15 x 15kms.)

Fig. 8b. Zoomed lower left quadrat of above image showing detail on active area of alluvial fan.

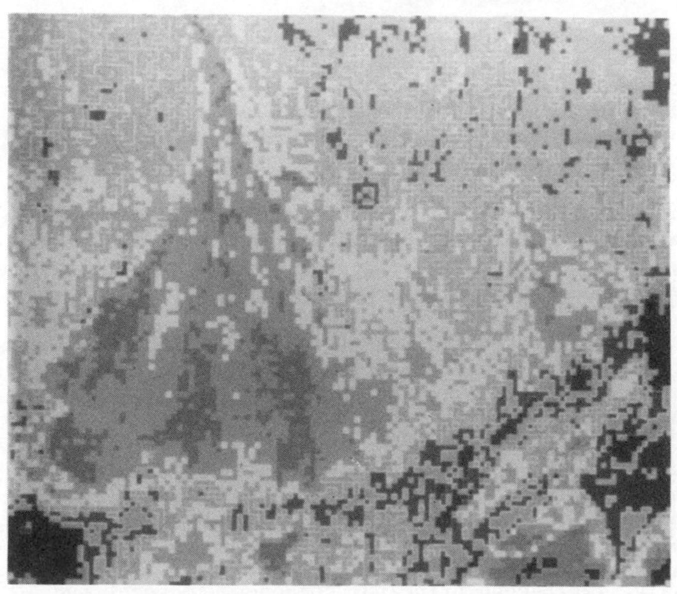

Many remote sensing studies have used supervised classification whereby the image is classified by the user defined training areas where the surface properties are known. This works very well in vegetation studies (Townshend, 1981) but, in geomorphology it is very difficult to obtain unique spectral responses for landforms. Therefore, supervised classification is not very suitable and the potential of unsupervised classification has been tested. This applies a clustering algorithm to the data and requires no previous knowledge of the area. In the algorithm used (the data was clustered according to a maximum likelihood classification) the only main user requirement was to stipulate the number of classes. The resulting image was colour-coded, density sliced, and class statistics were derived allowing the display of individual classes and the proportion of the image contained in each class.

Figure 8a shows the result of an unsupervised classification on a three band image (3,4 & 5) for 10 classes. The results are very encouraging. Some of the features which have been classified are areas of solid geology, variation on the fans, playa, alluvial plain, agriculture and the ephemeral channel. The level of separation is visible in Figure 8b which distinguishes the playas from the fan and areas of contemporaneous deposition. Although there was slight misclassification, overall the technique shows great potential for geomorphological mapping especially if the user wishes to extract main geomorphic units in a new area of interest.

CONCLUSIONS

As a result of this study, the following conclusions have been arrived at:

1. Landsat Thematic Mapper imagery has great potential for geomorphological mapping due to its improved resolution and specific spectral bandwidths compared with MSS.

2. A large and diverse number of geomorphological phenomena are apparent on individual Thematic Mapper bands. However, digital image processing of the data can lead to greater information extraction.

3. Ratios and principal component analysis enhance sedimentological differences in landforms, often at the expense of finer detail. Unsupervised classification corresponds well with actual geomorphological units.

4. No single image processing technique is suitable for all landforms. Best results are obtained when there exists a specific relationship between spectral response and the physical properties of the phenomena under investigation. This relationship can then be exploited by selecting the optimum image processing technique.

5. TM imagery can help our understanding of arid and semi-arid landforms and their constituent surficial materials. With such data, it may be possible to gain new insights into the nature, extent and frequency of the processes that produce them.

ACKNOWLEDGEMENTS

I wish to thank Dr. J. Townshend and Dr. A. Millington for their useful comments during the preparation of this paper, the photographic and cartographic units of the Geography Department, University of Reading. The author is a NERC postgraduate research student GT4/83/GS/87.

REFERENCES

Bailey, G., et al., 1982. Evaluation of image processing of Landsat data for geologic interpretation of the Qaidam Basin, China. International Symposium on Remote Sensing of Environment, Second Thematic Conference, Remote Sensing for Exploration Geology, Fort Worth, Texas. 1982. pp 555-577.

Cooke, R., and Warren, A., 1973. Geomorphology in Deserts. London.

Coque, R., 1962. La Tunisie Presaharienne. Armand Colin, Paris.

Coque, R., and Jauzein, A., 1967. The geomorphology and Quaternary geology of Tunisia, in Guidebook to the Geology and History of Tunisia, (Ed. Martin). Ninth International Conference Petroleum Society of Libya, pp 227-257.

Doornkamp, J., et al., 1980. Geology, Geomorphology and Pedology of Bahrain. Geobooks, Norwich.

Furon, R., 1963. Geologie de l'Afrique. Oliver & Boyd, London.

Hunt, G., and Salisbury, J., 1976. Mid-infrared spectal behaviour of Sedimentary rocks. Environmental Research Paper 543-AFCRL-TR-75-0256.

Jones, A., 1984. The applicability of digital satellite imagery for geomorphological mapping in arid regions. Proc. of tenth International Conference of Remote Sensing Society., Satellite Remote Sensing - Review and Preview. Remote Sensing Society, pp 351-360.

Lillesand, T., and Keifer, R., 1979. Remote Sensing and Image Interpretation. J. Wiley & Sons, New York.

Kahle, A., 1982. Spectral remote sensing of rocks in arid lands. First Thematic Conference: Remote Sensing of Arid and Semi-Arid Lands, Cairo, Egypt, pp 279-291.

Mabbutt, J., 1977. Desert Landforms. MIT Press, Cambridge, Massachusetts.

Mainquet, M., 1984. Space observations of Saharan aeolian dynamics, in Deserts and Arid Lands. (Ed. El-Baz). Martinus Nijhoff, The Hague. pp 59-77.

N.A.S.A., 1984. A prospectus for Thematic Mapper research in the earth sciences. NASA Technical Memorandum 86149.

Sobur, A., et al., 1982. Remote sensing applications in the south-east Sumatra coastal environment. Remote Sensing of Environment,

7,281-303.

Townshend, J., 1981. Image analysis and interpretation for land resources survey, in Terrain Analysis and Remote Sensing. (Ed. Townshend). Allen & Unwin, London. pp 59-108.

Townshend, J., 1984. Agricultural land-cover discrimination using thematic mapper spectral bands. International Journal of Remote Sensing, 5 , 681-698.

Verstappen, H., 1977. Remote Sensing in Geomorphology. Elsevier, Amsterdam.

International Geomorphology 1986 Part II
Edited by V. Gardiner

JOINTING AND AIRPHOTO LINEATIONS IN JURASSIC LIMESTONE FORMATIONS OF AL-ADIRAB AREA, TUWAYQ MOUNTAIN, ADJACENT TO AR-RIYADH, SAUDI ARABIA.

M. R. Mohammad

Department of Geology, Faculty of Science,
Alexandria University, Moharram-Beh, Alexandria, Egypt.

ABSTRACT

Six sets of meso-fractures, mostly joints, are distinguished by field analysis within an area of about 1300 square kilometres, southwest of Ar-riyadh, forming three systems, two of which are generally well developed. System J1 includes two subsystems; J1a, parallel to the general trend of the Dhurma-Nisah graben; and J1b and J1c, a subsystem of two conjugate sets enclosing an acute angle about the graben trend. System J2 consists of two conjugate sets, J2a and J2b, enclosing nearly a right angle dihedral angle about the graben trend. The third system is represented by a weakly developed set, J3, striking normal to the graben trend. The mean trends of fracture sets are mostly poorly correlated to the predominant trends of the airphoto lineations, being almost subparallel. Systems J1a and J3 are interpreted as extensional, J2 as shear and J1b and J1c as transitional. Cliffs and escarpments of Tuwayq Mountain are mostly fracture-controlled.

INTRODUCTION

The project area of about 1300 square kilometres is south-west of Ar-riyadh city, within the interior homocline of the Arabian Peninsula (Fig.1). The homocline consists of strata dipping at rarely more than 1° to the northeast, of Mesozoic-Cenozoic formations cut and disturbed by a regional zone of faulting known as the Central Arabian Arc (Fig.1), which is a system of grabens and troughs (Powers et al., 1966; Beydoun, 1966; Brown, 1972). The arc cuts the southern extremities of the project area with its local segment, the 'Dhurma-Nisah graben', trending in an approximate mean direction of N 74°W.

The project area is part of a regional complex cuesta which overlies the eastern extremities of the Arabian Shield, and consists of a series of scarps facing generally to the west. The biggest is that forming the Tuwayq Mountain, extending for more than 950km and divided by the Dhurma-Nisah graben into: the Northern Tuwayq Mountain trending northwest and the Southern Tuwayq Mountain trending south-southwest (Figs.1 and 2).

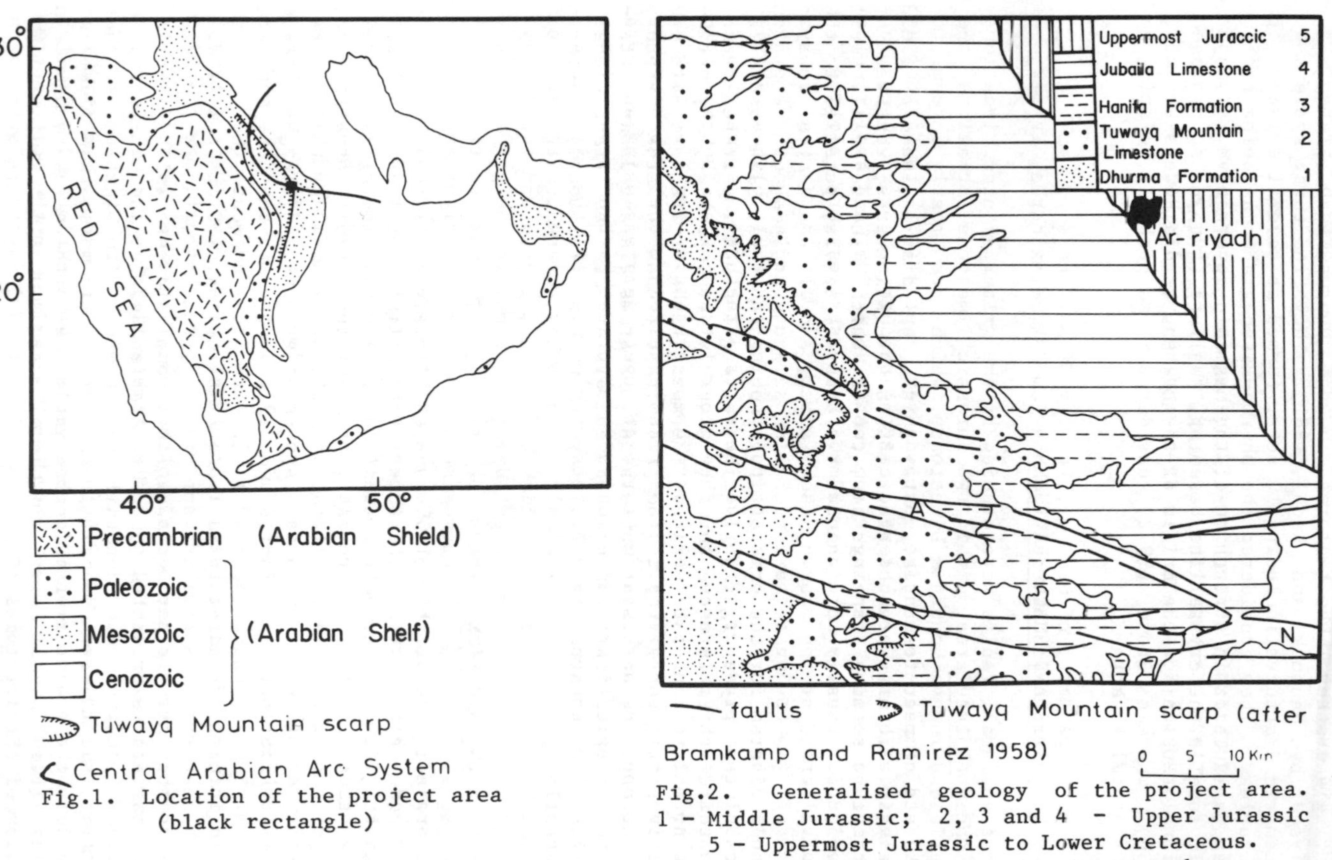

Fig.1. Location of the project area (black rectangle)

Fig.2. Generalised geology of the project area. 1 - Middle Jurassic; 2, 3 and 4 - Upper Jurassic 5 - Uppermost Jurassic to Lower Cretaceous. D - Dhurma graben. A - Awsat graben. N - Nisah graben.

Jointing and other mesofractures in the Dhurma-Nisah graben are analysed by Hancock and Kadhi (1981). They find three systems of systematic fractures including 10 sets, formed at several episodes of fracturing. They considered the Dhurma-Nisah graben as being formed by an early Cenozoic stretching of the Mesozoic cover above a major fault zone in the crystalline basement. Wolfart (1961) also proposed regional tension for the origin of the graben, as did Powers et al., (1966), whereas Brown (1972) considered it as probably relating to left lateral shear.

JOINTS AND SMALL FRACTURES

Joints and small faults were studied by the writer at nine locations in the project area within two Jurassic limestone formations, the Jubaila and Tuwayq Mountain Limestones, which are mostly cream to white and compact to massive, dipping very gently (0.5° to 1°) to the northeast. Joints were observed cutting through the exposures for vertical distances ranging from one to about 20m, and small faults extended for usually more than 20m. Mean strikes are plotted in Fig.3.

A pole diagram for the orientation of joint and fault planes (Fig.4) reveals the presence of five sets of fractures, symmetrically arranged with reference to the general trend of the Dhurma-Nisah graben, and forming two well developed systems. The first, J1, includes two subsystems; J1a strikes parallel to the graben trend; and J1b and J1c, a subsystem of two conjugate sets enclosing an acute dihedral angle (approximate mean 58°) about the graben trend. Indications of extension origin are observed on most of the J1 planes, especially J1a. They are almost vertical, opened, with matching margins, and often filled by films of calcite, gypsum and red clay. This system may thus represent dilation normal to the graben trend, in a NNE-SSW direction. The second system, J2, comprises two conjugate sets, J2a and J2b, enclosing approximately a right angle dihedral angle (approximate mean 100°) about the graben trend. Indications of shear are observed on many fractures of J2. They are straight, tight, steep, and occasionally have smooth surfaces and slickensides. This system is shown in the field to be generally younger and more developed than J1. The subsystem J1b and J1c may be considered as a transitional one developed by failure in shear-extension fracture transition.

Joints forming J2b constitute two maxima close together (Fig.4), giving mean surfaces striking approximately north-south, one of which dips to the east and the second to the west. This may be related to local variations in lithology and/or stress field.

Another (sixth) set of fractures, weakly developed and striking normal to the graben trend, is observed at two locations close to the graben (Fig.3). It does not form a maximum in the pole contour diagram of the fractures (Fig.4). This set (J3) may be related to a third system formed by a separate period of extension normal to the graben trend, as proposed by Hancock and Kadhi (1978).

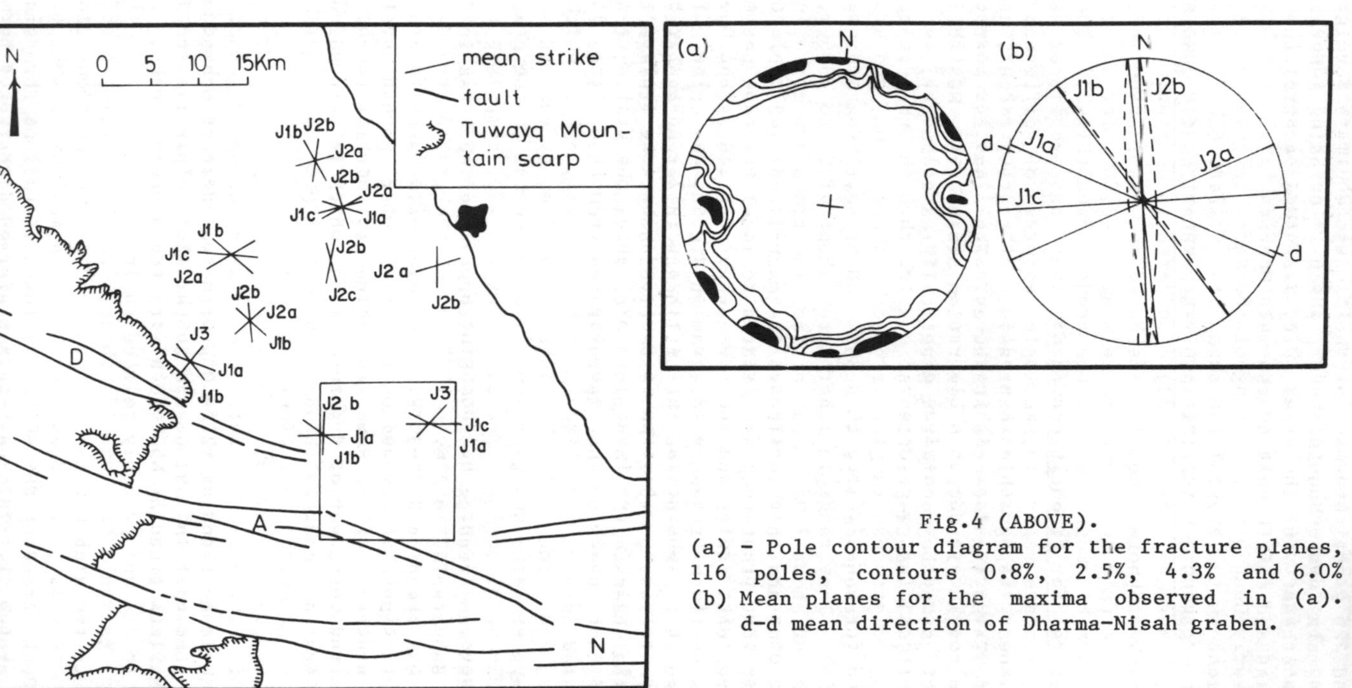

Fig.4 (ABOVE).
(a) - Pole contour diagram for the fracture planes, 116 poles, contours 0.8%, 2.5%, 4.3% and 6.0%
(b) Mean planes for the maxima observed in (a). d-d mean direction of Dharma-Nisah graben.

Fig.3 (LEFT).
Mean strike of fracture sets at the studied locations; the rectangle marks the area studied for lineations.

J1b and J2b constitute most of the limestone cliffs forming the faces of the main Tuwayq Mountain scarp which is trending generally northwest-southeast in the area. J1c, J2a and J3 control the transverse drainage and cliffs cutting the main scarp.

AIRPHOTO LINEATIONS AND THEIR RELATIONSHIP TO JOINT TRENDS

The study of photo lineations is based on examination of a photographic mosaic of approximate scale 1:40,000, for an area of about 210 square kilometres within the project area (Fig.3). The principal parameters for identification of lineations are linear tonal variations, linears separating geologic features of different textures, and linear topographic expression of scarps, ridges, and valleys; included also are detected fractures. The area is mostly dry and barren with virtually no vegetation alignments. Most of the traced lineations may be considered genetically as secondary, as defined by Cloos (1946) and El-Etr (1967).

In the present analysis about 260 linear features are traced (Fig.5a). They range in length from about 120 to 3560m, with an average of about 780m. Most of the lineations are microlinears, less than 2000m, and only 14 linears are macrolinears, 2 to 10km. This may indicate that they are related to local fracture patterns rather than regional faults and shatter zones. The lineations are plotted on an azimuth-frequency diagram in Fig.5b, the upper half of which represents the sum of lengths of linears in percentages (L%) in each azimuthal class of 10°; the respective numbers of linears in the same azimuthal classes (N%) are plotted on the lower half of the diagram. Significant peaks on the diagram range from about 6% to 13%, and may be listed as:

Peaks trending NE-SW

1-Maximum trending N40°-60°E with well defined peak in N50°-50°E, subparallel to J2a.
2-Peak trending N20°-30°E. The relatively large length percent of this peak in comparison to its number percent reflects the presence of a few long linears. It may be related to the transverse set of fractures (J3), which is perpendicular to the graben trend.

Peaks trending NW-SE

3-Maximum trending N20°-40°W with two adjacent peaks. Lineations forming this maximum are subparallel to parallel to J1b.
4-Maximum trending N50°-70°W with two adjacent peaks, subparallel to J1a and probably related to it.

Peaks trending E-W

5-Peak trending N80°-90°E, diffused lineations form smaller peaks around it. This east-west preferred direction fits well with J1c which strikes mostly parallel to it.

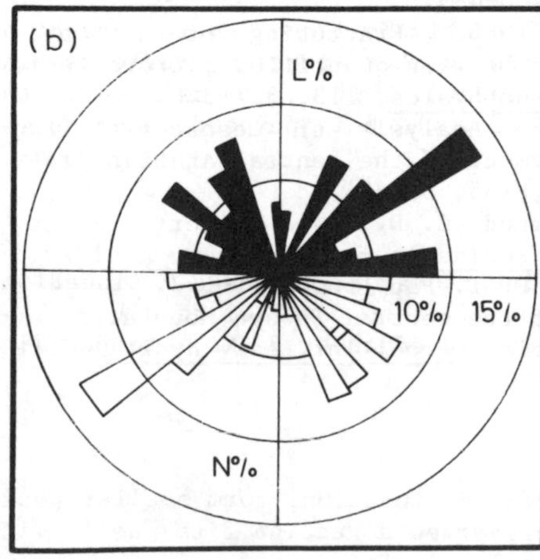

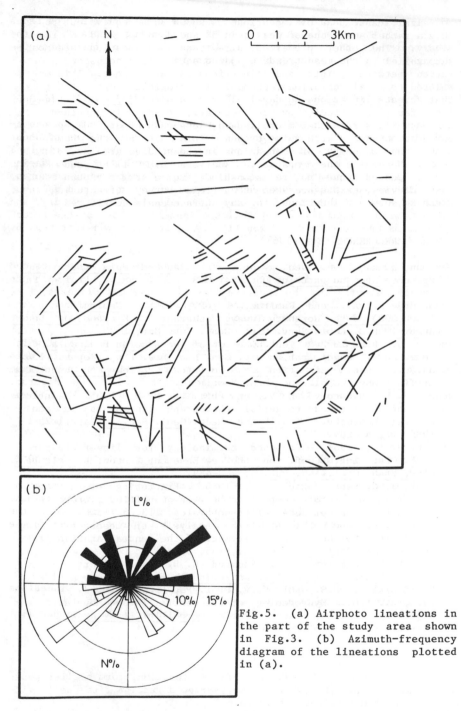

Fig.5. (a) Airphoto lineations in the part of the study area shown in Fig.3. (b) Azimuth-frequency diagram of the lineations plotted in (a).

No significant peak of lineations is shown to be equivalent to J2b, only a rather small one of less than 5% and 3% of L and N respectively. The other preferred directions of lineations are mostly subparallel to the mean trends of joint sets. In conclusion, correlation between joints and lineations is rather poor. This may be related to local variations, probably lithological and/or structural, that create lineations subparallel or even unrelated to jointing and local faulting. Studies by Dean et al., (1985) in south-central Kentucky illustrate that microlinears, less than about 500m, give low correlation values with joint trends. This may be the same here since the majority of lineations in the studied area are microlinears. Meanwhile the comparative analysis by Grillot and Razack (1985) showed that no relationship, from a strictly quantitative point of view, exists between the field survey (fracturing) and airphoto survey (lineations) for any surface scale considered.

REFERENCES

Beydoun, Z. R., 1966. Geology of the Arabian Peninsula, Eastern Aden Protectorate and parts of Dhufar. U.S. Geol. Surv. Prof. Pap. 560-11.

Bramkamp, R. A., and Ramirez, L. F., 1958. Geologic map (at 1:500,000) of the Northern Tuwayq quadrangle, Kingdom of Saudi Arabia. U.S. Geol. Surv. Misc. Geol. Inv. Map, 1-207A.

Brown, G. F., 1972. Tectonic map of the Arabian Peninsula. U.S. Geol. Surv., Saudi Arabian Dir. Gen. Min. Res. Proj. Report, 134.

Cloos, E., 1946. Lineation : a critical review and annotated bibliography. Geol. Soc. Ann. Memoir 18.

Dean, S. L., Blanco, J. M. V., and Phillips, M. W., 1985. Relationship of joint trends to aerial photograph lineaments in south-central Kentucky. Abs.19 Ann. Meet. Geol. Soc. Am. April 25-26, 1985, 17, 5, 284.

El-Etr, H. A., 1967. Proposed terminology for linear features. Proc. 1st Int. Conf. on the New Basement Tectonics, Salt Lake City, Utah, Geol. Ass. Pub. 5, 480-489.

Grillot, J. C., and Razack, M., 1985. Fracturing of a tabular limestone platform: comparison between quantified microtectonics and photogeological data. Tectonophysics, 113, 327-348.

Hancock, P. L., and Kadhi, A., 1978. Analysis of mesoscopic fractures in the Dhurma-Nisah segment of the central Arabian graben system. Jl. Geol. Soc. London., 135, 339-347.

Powers, R. W., Ramirez, L. F., Redmond, C. D., and Elberg, E. L., 1966. Geology of the Arabian Peninsula - Sedimentary geology of Saudi Arabia. U.S. Geol. Surv. Prof. Pap. 560D, D1-147.

Wolfart, R., 1961. Hydrogeology of the central Tuwayq Mountains and adjoining regions, Saudi Arabia. Proc. Int. Ass. Hydrol., 566, 98-112.

International Geomorphology 1986 Part II
Edited by V. Gardiner

COASTAL INSTABILITY AND GEOECODYNAMIC ASSESSMENT IN NICHUPTE LAGOON, CANCUN, YUCATAN PENINSULA, MEXICO

J. F. Cervantes-Borja and M. Meza-Sanchez

Institute of Geography,
UNAM, Mexico.

ABSTRACT

The system of coastal lagoons at Cancun on the northeastern coast of the Yucatan Peninsula is a natural tourist attraction. The morphology of the area has developed from an outline of Pleistocene aeolian dune ridges and subsequent accretion of lime sands in coastal dunes. Holocene processes infilled basins, formed coastal depositional features, and delimited the present system of six lagoonal ponds. At present tidal and current energy is low, salinity is highly variable, and water circulation is sluggish. Three biological realms exist, in lagoonal areas of open circulation and marine salinity; more restricted areas; and in mangrove swamps. There is a relationship between sediment patterns and the distribution of biological realms, and much of the lime mud in the lagoons is the product of local organic activity. The study provides a framework for determining an overall geoecological plan for management of the area.

INTRODUCTION

The purpose of this paper is to illustrate the application of environmental planning, based on a concept of geoecological (geomorphological and ecological) synthesis and evaluation, to resolve geoecodynamic problems in the system of coastal lagoons situated in the northeastern coast of the Yucatan peninsula, bordering the Caribbean Sea (Jordan et al., 1971).

Cancun is an international tourist resort on the Caribbean coast of Mexico. Its attraction depends on aquatic sports in the coastal lagoon system; the clear turquoise-coloured water, and exotic environment, with wildlife and abundant mangrove vegetation, play a fundamental role for the tourist industry. Cancun therefore has to keep its natural beauty in order to guarantee the tourist attraction. However uplift of the Caribbean shelf has caused changes in energy and volume of ebb and flow tidal and coastal currents, and consequently the geomorphological processes tend to modify the lagoon floor, disrupting the stability in aquatic and terrestrial environments and destroying the coastal lagoon system.

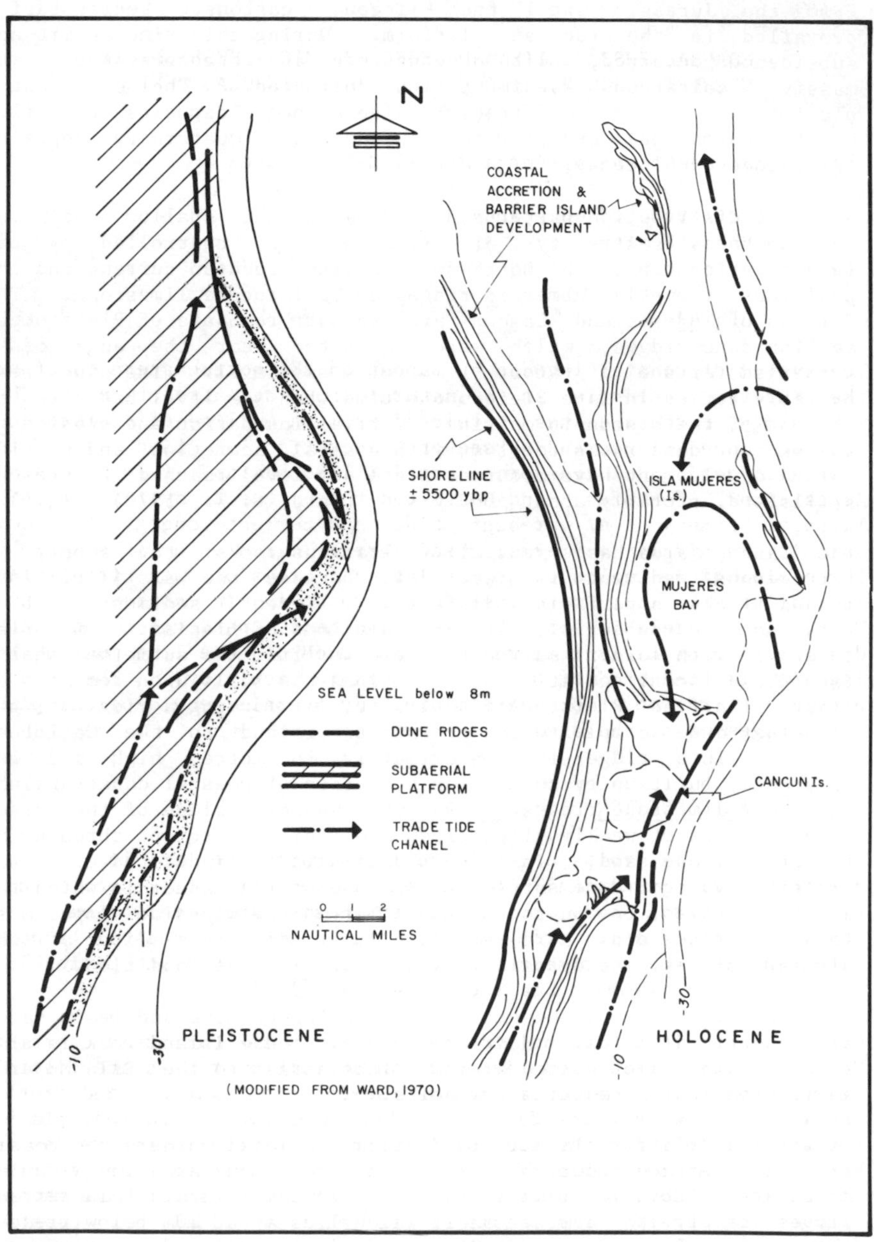

Fig.1. Pleistocene and Holocene morphologies of the area.

GEOMORPHOLOGICAL EVOLUTION

From the Jurassic until the Pliocene carbonate sedimentation prevailed in the Yucatan platform. During this time uplift and subsidence occurred, with regressions and transgressions, and massive calcareous sedimentation dominated. The continental platform is composed of transgressive Pliocene deposits (Carrillo Puerto Formation) and layers of concordant Quaternary deposits (Pleistocene-Holocene).

Sediment distribution patterns on the narrow and shallow shelf of the northeast extremity of Yucatan are controlled by the interrelationship of the northward flowing Yucatan current and the positions of partly submerged ridges of Pleistocene limestone. The islands of Mujeres and Cancun are in part remnants of Pleistocene aeolian dune ridges, which were deposited near the edge of a submerged terrace at about 10m below sea level. The dunes were probably formed in late Pleistocene time when sea level had risen at the end of the Wisconsin melting. Partial inundation and erosion of the ancient dune rocks together with coastal accretion and barrier island development have occurred since sea level reached its present stillstand, probably about 5000 yrs BP (Ward, C. W., 1970) (Fig.1).

The Cancun area has essentially calcareous rocks; the supply of terrigenous sediment is restricted because the karstic platform inland lacks superficial drainage. Recent sediments, both subaqueous and subaerial, are accumulations of bioclastic material together with older sediments reworked from the submerged shelf. Thus the Holocene accretion of lime sands has built upon remnants of older lithified Pleistocene eolianites, forming a ridge of the Holocene dune complex which crosses the island to the Caribbean shore. This ridge is approximately 16 metres high and was apparently built up by continuing accretion of coastal dunes against the older limestone ridges. On the seaward slope of the larger ridge patches of weakly lithified aeolianite outcrop through the veneer of loose sand. In many places slightly indurated crust has preserved younger dune surfaces. Because of the manner in which the dunes are piled, one on top of the other up the windside slope, most leeward surface dips here are shallower than the dips of the windward slopes, the opposite of the "normal" dune condition.

A large percentage of the grains of the Cancun dune and beach sands are superficial ooids. Most of the Pleistocene island rock is also largely composed of coated grains, in contrast to the Isla Mujeres aeolianite, which contains few ooids.

Seaward a platform about 10m below sea level borders the coast, forming a terrace about 3km wide in the Cancun area and widening northward to about 10-15km in the north Contoy island. This terrace slopes steeply to another shelf which lies about 20m below present sea level. The remnants of Pleistocene calcarenite dunes in Cancun and Isla Mujeres were dated by Ward to about 26,360 yrs BP. The dunes were apparently formed when the sea had covered the lower terrace with shallow wave-tossed water. The shelf biota furnished abundant shell debris for the wind to pile up at the shore. The dunes lithified rapidly but were afterwards partly destroyed as the transgression returned. In this epoch the penetration of the principal trans-reef channel between Cancun and Isla Mujeres was achieved, when the formation of Nichupte lagoon begun. Consequently coastal erosion in the south and continental sediments brought by rivers furnished abundant material which was deposited in the low energy environment of the lagoon. This material was mixed and slowly filled up the channel basin till the end of the Wisconsin melting 15,000 to 10,000 yrs BP, when the sea level rose to its present position. At the beginning of the Holocene beaches were formed, which began to accrete over and to seaward of the foundations of the old Pleistocene ridges, so Holocene deposits constitute the bulk of the tombolos and spits of Cancun, Isla Mujeres and Isla Blanca. The Nichupte lagoon was formed by Holocene accretion of lime sands which exist in calcarenite barrier form. Dates from shell material within the lagoon indicate that the lagoon has been formed and filled up by lime muds and peats within the last 5,000 years. Nichupte lagoon was separated into discrete ponds (North Nichupte, Central Nichupte, South Nichupte, La Caleta, Rio Ingles and Bohorques ponds). This configuration was created by tidal and lagoonal currents in the south and central lagoon, which reworked earlier washover fans and spits.

CONTEMPORARY GEOMORPHOLOGICAL PROCESSES

At present the tidal and current energy is low, because of the naturally strong sedimentation rate and artificial obstacles in the entrances and along the Cancun and Nizuc Channels (Fig.2). Consequently there is a major contribution of freshwater and the water in the lagoon has a high variation of salinity, particularly during storms when subterranean drainage flows through lagoon floor sources called *cenotes*. This action, accompanied by wind wave and storm currents, enables important sediment movement, winnowing out mud and piling it into the banks. Thus the mudbanks contain the finest sediment, forming a rough unstable micro-relief on the lagoon floor. During storms this sediment muddies the lagoon water and causes the loss of its magnificient transparency and turquoise colour, both basic characteristics of its tourist attraction.

The vegetation of the marshes and marine grasses (*Thalassia*, sp.) form a bottom filter restricting the sediment movement and making the circulation current sluggish. But the strong variations in both the salinity of the water and the sediment of the bed has harmed the environment of the two communities, and consequently a rapid decrease in their filter function follows.

Thus the water is murky, with suspended organic matter and of variable salinity. Normal marine salinity exists in the open circulation areas but parts of the lagoon have restricted circulation and are metahaline (Rio Ingles 60 to 34 $^{o}/oo$). pH is more or less constant at 8.2 to 8.6. Summer water temperature is from 28°C to 34°C. The thicker sediment attains depths of from

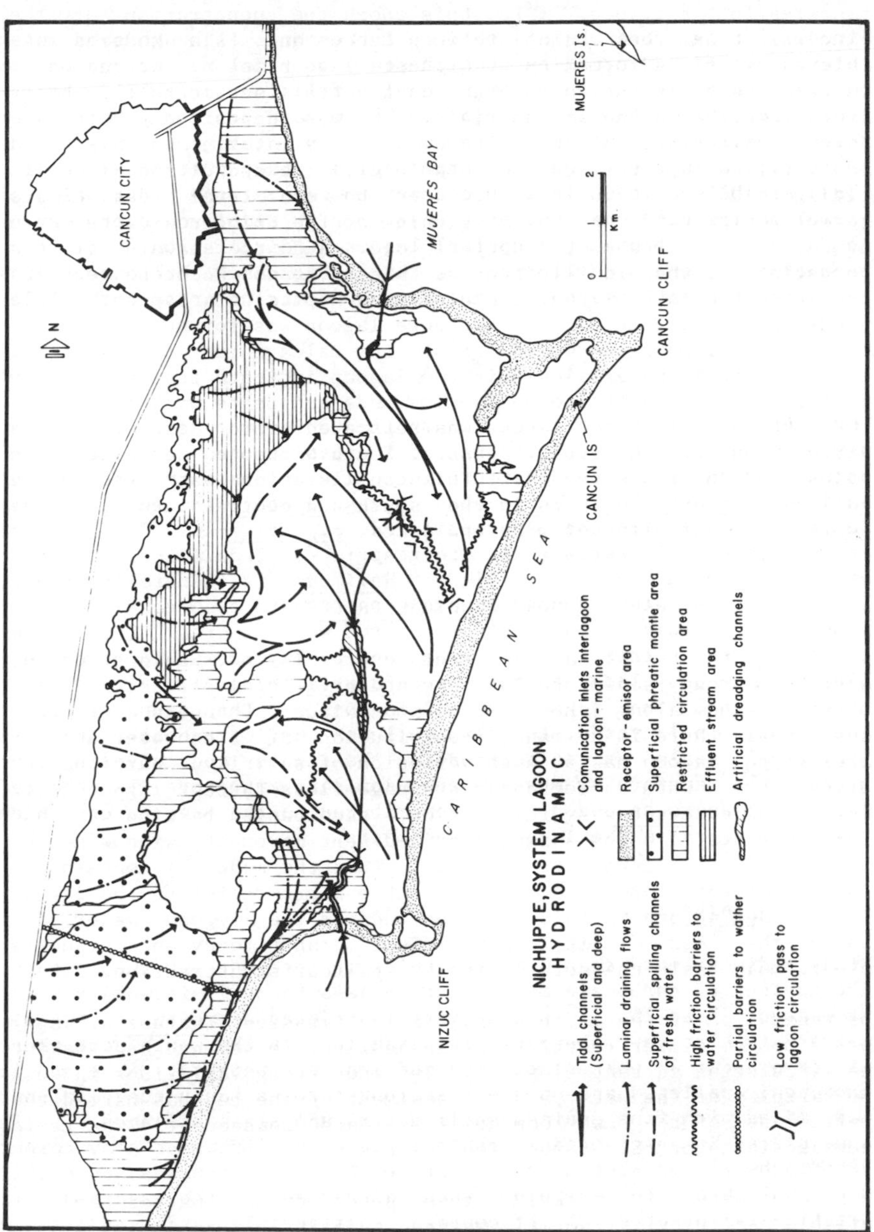

Fig.2. Nichupte lagoon hydrodynamics.

2 to 5 metres in the ponds, and mudbanks are covered with only a few centimetres of water.

Circulation is today sluggish in all the ponds although some water interchange is indicated by the presence of tidal deltas and passes through the outer barriers (high and low friction surfaces), and by linear furrows on the lagoon floor. Storm wash-overs may keep water freshly marine in Cancun south and north near the ends of the principal channels. Our geomorphological study (Cervantes et al., 1984) established that it is necessary to recover the penetration of normal marine currents, not only that of high tidal and storm waves, in order to obtain the optimal lagoonal circulation permitting recovery of the equilibrium of the interphase environment and recovery of healthy marsh, mangrove and marine grass communities (Fig.3).

BIOTA

Three biological realms exist, controlled chiefly by circulation and oxygenation of the water. Lagoonal areas of open circulation and normal marine salinity characteristically contain the green algae *Halimeda*, *Penicillus*, *Rhinocephalus*, and *Udotea*, the turtle grass, *Thalassia*, Miliolid and peneroplid and *Elphidium*-like Foraminifera are abundant. In parts with more restricted circulation, which are often shallow and sediment-filled, *Halimeda* is generally absent. The other green algae appear less sensitive to such conditions but appear very rarely in the most restricted water-bodies. These areas contain more turtle and eel grass and scummy brown and blue-green algae. The dasycladacean alga *Acetabularia* appears in very shallow waters around the edges of all lagoons. The fauna includes gastropods, pelecypods, but few molluscs or ostracods. A third biological realm exists surrounding the lagoon and covering the banks: the mangrove swamps. The dominant mangrove is the red variety. Mangroves undoubtedly contributed to the peat layers which have been cored in the lagoonal sediments.

SEDIMENTS

Muddy sediment more than 6 metres thick is present in the lagoons, the content of carbonate mud (fraction less than 62 microns) varying from about 75 to 25%. In some very restricted areas humic organic matter is also very abundant (for example, in the south part where it forms about 40 percent of the sediment dry weight). The mud is thoroughly pelleted and on the sea floor forms a soupy-to-gelatinous mass mixed with fine grained shell debris and algal filaments. In some places its high organic content gives it a brown or gray colour just inches below the light surface layer. It has a noticeable odour of hydrogen sulphide when disturbed. The sediment is practically unsorted. It contains considerable sand- and gravel-sized shell debris composed of dull unabraded *Halimeda* plates, ostracods, molluscs, and Foraminifera. Below the soupy mass in the upper few inches the mud is more solid, but about 50% of its volume is water-filled void space. This porosity does not vary down to

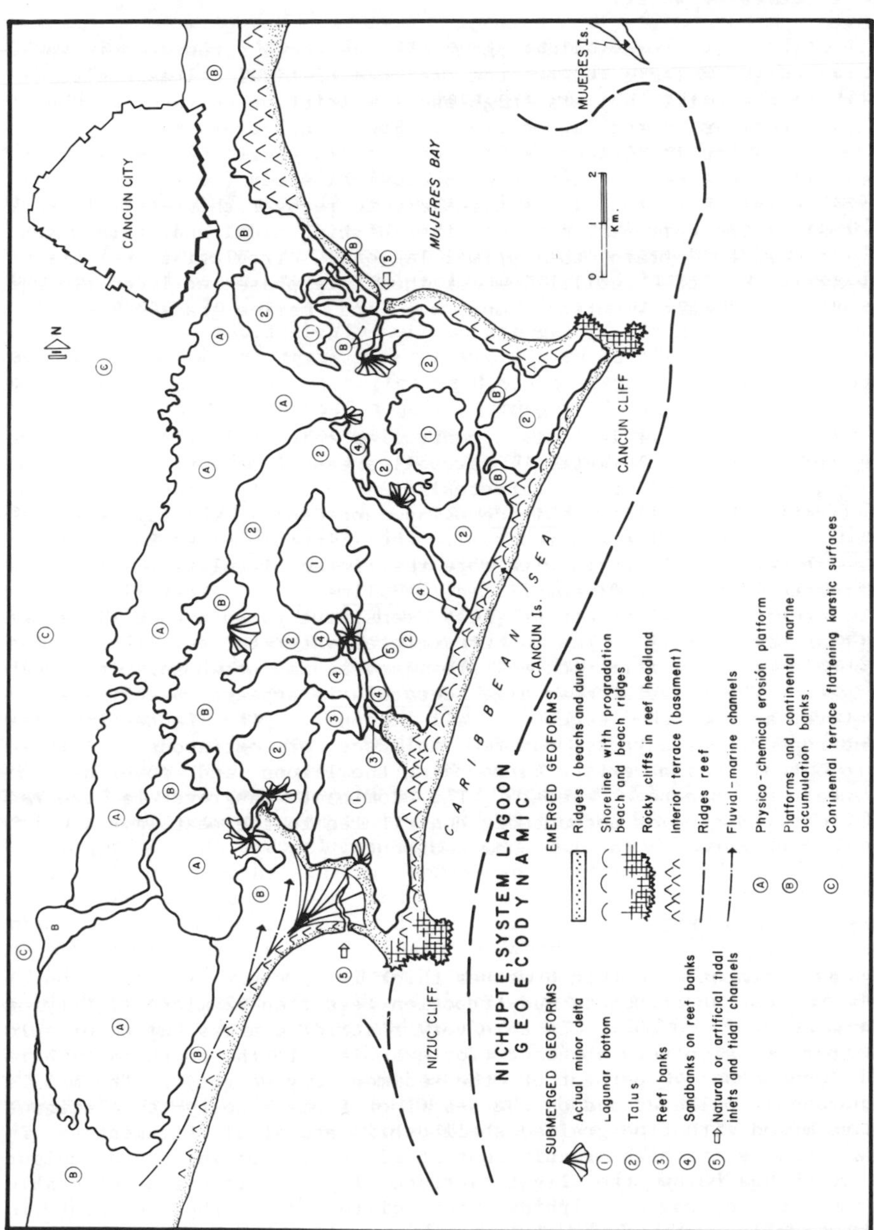

Fig.3. Nichupte lagoon geoecodynamics.

9 or 10 feet, showing that below the top few inches the sediment has not compacted.

Mineralogy is variable but generally different between sediments from areas of open circulation and normal marine salinity and those within the restricted and hypersaline areas. Ward (1970), obtained the following data:
Open circulation areas: 40 to 60% aragonite; 35 to 60% Mg calcite (11 to 12 mol per cent $MgCO_3$); 0 to 11% calcite.
Restricted lagoons: 18 to 46% aragonite; 41 to 82% Mg calcite (up to 16 mol percent $MgCO_3$); 0 to 11% calcite.
The trend from more aragonite in open waters to more Mg calcite in lagoons of restricted circulation may be explained first and most simply by change in the biota. A higher ratio of green algae to Foraminifera exists in lagoons of open circulation. Not only is the content of Mg calcite relatively high in restricted water bodies, but mineralogy changes in the hypersaline ponds. The trend is from a high percentage of aragonite in open lagoons, through a higher amount of Mg calcite in closed lagoons, to the presence of protodolomite in highly restricted hypersaline water.

The mud lime has several origins: breakdown of aragonitic green algae; abrasion and rotting of shell debris of various types and composition; possible biochemical precipitation in areas of high photosynthesis; possible physico-chemical precipitation in relatively highly saline water. Lime mud may be easily transported, and lagoon ponds are really sediment traps for mud produced elsewhere. The beautifully clear water of the Mujeres strait apparently carries a minimum of lime mud sediment. No areas of mud exist on its bottom that could furnish fine grained carbonate sediment which could be stirred into suspension by storms and carried over, through, or around the coastal barriers into the lagoon. The barriers separating the lagoon from the strait or open Caribbean Sea are high (up to 16m) and relatively unbroken. It does not seem probable, therefore, that mud is currently being carried into the lagoon except by minor influx through tidal passes. Possibly periodic storms carry lime mud, as well as coarser material, along the coast and swirl it into the lagoon as the sand spits accrete along the barrier islands. The spits themselves are almost entirely of carbonate sand, most of which is created in the strait and open sea and along beaches by breakdown of skeletal debris. The limited evidence on mineralogical variation cited above suggests that most of the lime mud in the lagoons is a local product derived from breakdown of organisms living there. We know that the accumulation of the muds is comparatively recent and that at present the system in Cancun is more or less closed.

CONCLUSIONS

This study established the methodology and strategies for resolving geoecodynamic problems of the area, proposing a "management landscape," in order to control the changes in the environmental system. Furthermore, landscape evaluation by this method will offer the best decisions on the land use of discrete land units in the

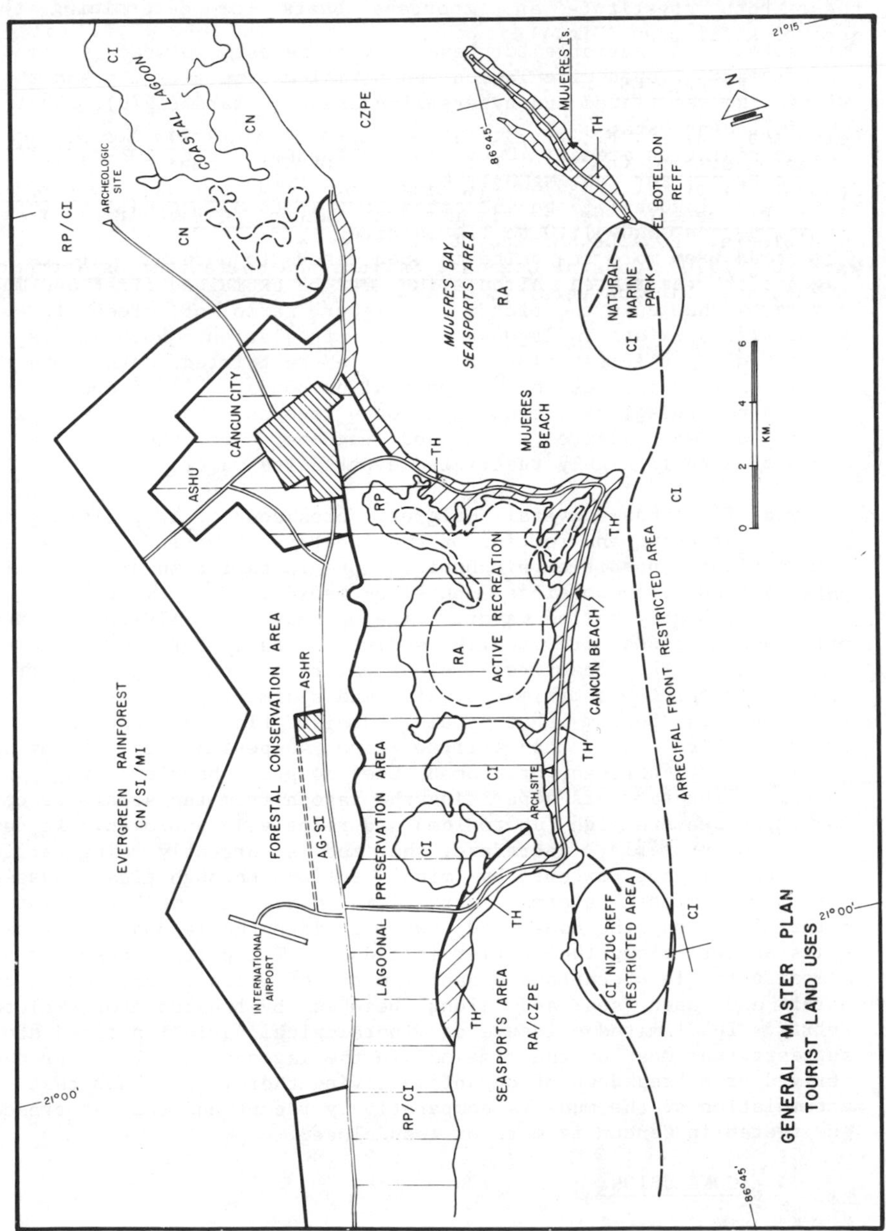

Fig.4. General master plan of tourist land uses.

area, or for optimal location of each category of land use unit; these both constitute an important basis for determining the geoecological plan (Fig.4).

REFERENCES

Cervantes, B. J. F., et al., 1984. Geoecological Plan of Nichupte lagoon System, Cancun Q. R. Mexico. FONATUR. México, D.F.

Jordan, E. et al., 1971. Informe sobre el estudio ecológico de Prospección de la laguna de Cancún, Q.R. Instituto de Biología. FONATUR. México, D.F.

Ward, C., 1970. Coastal Lagoons, Salinas and Eolianites of Northern Yucatan. Rice University, U.S.A.

APPLIED GEOMORPHOLOGY, LAND CONSERVATION AND RESOURCE EVALUATION

International Geomorphology 1986 Part II
Edited by V. Gardiner

APPLIED GEOMORPHOLOGY, LAND CONSERVATION AND RESOURCE EVALUATION

R.P.C. Morgan

Silsoe College, Silsoe, Bedford, U.K.

Although most of the papers in this section make original and interesting contributions, they in no way form a representative collection of the scope of applied geomorphology. There is a bias towards the problems of soil erosion but almost entirely in an agricultural context with no attention being given to urban environments or recreational activity. There are no contributions on coastal erosion, mass movement, arid lands, permafrost or thermokarst. As a demonstration of the dynamic and relevant nature of geomorphology, the papers are disappointing particularly in view of the increasing importance being attached in planning to environmental issues in which the geomorphologist, through studies of landform processes and evolution, ought to have a role. Only four themes are represented here: resource evaluation, soil erosion, man-made landforms and site conservation.

A geomorphological role in resource evaluation has been implicit since terrain analysis was adopted as a standard procedure in resources surveys. This depends on identifying land units which are reasonably uniform internally but significantly different one from another. The units are normally recognized by a combination of geomorphological and soil properties. Although valuable as a technique, there is some concern that the approach provides a rather static view of land resources which has somewhat limited value for predicting environmental impacts. With the recent emphasis in geomorphology on processes and systems it should be possible to modify the technique or replace it with something better. It is unfortunate in this respect that no examples are presented here of the French approach to resource assessment based on geosystems or the Dutch work on landscape ecological mapping. As these newer procedures become more widely used, the role of terrain analysis may be restricted to rapid reconnaissance surveys in which case, according to King in his paper, more attention will need to be paid to the naming of the units. The success or otherwise of terrain analysis depends upon being able to predict surface conditions, particularly soil attributes, in the field. Since this, in turn, depends upon the degree of variability of those attributes within a land unit, it is surprising that more studies have not been made of the variation of terrain properties. Research such as that described by Viville and Ambroise on variability in the hydrological properties of soil is important to judgements that can be made on the value of terrain analysis as well as to the application of any techniques that rely on the recognition of homogeneous land units, for example distributed hydrological and sediment models.

Cooke and Shaw give a broader illustration of the place of geomorphology in resource development using several examples of their work in Botswana. Their studies are impressive because, in addition to describing relatively short-term investigations of terrain classification and soil mapping, they demonstrate the importance of longer term research into landform evolution for predicting the environmental impacts of proposed development projects.

The papers on soil erosion fall into four groups. The first concerns detailed studies of process and is represented by work on stream bank erosion (Haigh and Rydout), surface sealing under raindrop impact (Poesen) and the effect of vegetation on wind speed (Morgan and Finney). Despite their coverage of widely different processes, all three papers bring out the importance of understanding erosion as a series of discrete events, the effectiveness of which depends on the magnitude of the event itself, the antecedent conditions and, in the case of surface sealing, the cumulative effect of the preceding events. The second group, comprising papers by Jeje, Mushala, Marques and Roca, covers measurements made from erosion plots. It is of interest for the data presented for areas of the world where little information was previously available on rates of hillslope erosion as well as for the insights provided on the difficulties of establishing and managing such plots. The third group consists of one paper on erosion modelling. Sharpley, Smith, Williams and Jones describe the use of the EPIC model for assessing the impacts of erosion on soil fertility and therefore on productivity as a result of the removal of phosphates and nitrates.

The fourth group and, in the light of recent research on soil erosion, the most innovative, gives a historical perspective. This is important to the interpretation of present-day measurements of erosion rates in showing whether the erosion is becoming worse or is subject to fluctuations related to the management and intensity of use of the land. The evidence presented from the Central Highlands of Papua New Guinea (Gillieson, Gorecki, Head and Hope) and from the Cevennes (Cosandey, Billard and Muxart) and the Pyrenees (Metaillie) in France strongly supports the last interpretation.

Two examples are provided of man-made landforms. Howard and Huggins relate terracettes to grazing intensity and raise the possibility that studies of terracette morphology may give a guide to optimum stocking rates. Hughes and Sullivan investigated terrace forms in the Eastern Highlands of Papua New Guinea and, after considering several options for their formation, show that they were most likely built in prehistoric times for the growing of taro. Proposals are presented for their conservation as part of the management plan for the Ramu Stage 2 hydro-electric power project. Disturbingly the authors imply that conservation was only acceptable because the terraces were of cultural rather than natural origin.

The case for conserving and managing natural sites of geomorphological significance is put by Gordon in his review of the current situation and policies in the United Kingdom. However, the problem is an international one and it would have been revealing to know what happens in other countries. What is clear from Gordon's paper is that if geomorphologists want key sites preserved they have got to be more positive and publicity-minded in recognising such sites and in justifying the case for their conservation.

There is a need for a more forthright and commercially oriented approach by geomorphologists generally if they are going to play a role in solving the world's environmental problems. Although there are many geomorphologists who feel strongly that their subject should not develop in this way, more and more potential students are concerned about the environment and the way it is managed. The future of geomorphology depends upon attracting these students. The papers presented in this section are not going to do this. They are interesting enough to convince the converted but lack the dynamism to do applied geomorphology justice.

International Geomorphology 1986 Part II
Edited by V. Gardiner

REVIEW OF GEOMORPHIC DESCRIPTION AND CLASSIFICATION IN LAND RESOURCE SURVEYS

R B King

Land Resources Development Centre, Tolworth KT6 7DY, UK

ABSTRACT

Rapid determinations of relevant geomorphic data, whose descriptions are easily and widely understood, are needed for the increasing trend of rapid resource surveys. Methods of land classification for agricultural potential are reviewed, distinguishing in particular the preclassificatory from the objective location specific systems. At the land system level, a simple land unit identification system is suggested consisting of a local name combined with one, or a combination of the following terms: plain, plateau, hills, valleys, escarpment.

In order to determine simple, relevant and widely applicable terminology for land unit descriptions at the land facet level, the most commonly used criteria were counted from a selection of land resource survey reports and compared with a questionnaire sent to colleagues asking them to rank the same criteria according to agricultural importance and ease of recognition. The investigation suggests upland facets should be named and distinguished primarily according to angle of slope, soil type and vertical position in the landscape; whereas lowland land facets should be named and distinguished primarily by their vegetation type and hydrological condition.

INTRODUCTION

Twenty years ago, a concerted international effort to devise a land classification system applicable to a wide range of land resource surveys, recommended adoption of the land system technique (Brink et al., 1966). It was devised by field practitioners with both agricultural and civil engineering backgrounds, and was largely based on the land system method used by the Australian CSIRO (Christian and Stewart, 1953). The classification system was essentially pragmatic; the different hierarchic levels being based on the mapping scale. The lack of rigid definitions to the classifying terminology however provoked considerable criticism, and a certain amount of confusion as to the position of a particular land unit in the hierarchy.

In the late sixties and early seventies, the land system technique was criticised on a number of grounds, particularly by Moss (e.g. 1969, 1981) for not concentrating enough on human, dynamic and ecological processes. Other classification systems were put forward (e.g. Moss, 1969; Thomas, 1969), but they did not appear to be that significantly better than or different from the Brink et al. system to be widely accepted. The Brink et al. system was accepted and adopted by the Land Resources Development Centre (then the Land Resources Division) of the British Overseas Development Administration and British commercial companies undertaking surveys in less developed countries, at a time when there was a perceived need for reconnaissance land resource surveys of large areas of the newly independent countries. Finance was also widely available from aid organisations, and the underlying philosophy at that time was to produce as comprehensive a survey as possible, to avoid the necessity of repeated surveys of the same area for different purposes.

This combination of philosophy and available finance resulted in surveys taking longer than orginally intended, and questions began to be asked whether the surveys were not wasting too much time collecting data which would never be relevant. It was also discovered that more detailed prefeasibility or feasibility surveys still necessitated a repeated survey, because the reconnaissance survey did not provide data in sufficient detail.

These criticisms produced a questioning of the effectiveness of the land system technique; but it is now generally realised that it was not so much the technique that should have been criticised, but that too much effort was put into gathering data of areas unlikely to have much agricultural potential. The current philosophy is to employ the land system technique more to undertake rapid land resource surveys of large areas (e.g. LRDC with Direktorat Bina Program, 1985; and see King, 1982a), and to concentrate the effort more on areas of higher agricultural potential.

The purpose of this paper is to review the methods most commonly employed for providing physical environmental data for mapping agricultural potential at different mapping scales.

GLOBAL AND CONTINENTAL MAPPING

Brink et al. (op.cit.) recommended the 'land zone', describing major climatic type, as the highest order land unit, and most attempts at agricultural potential mapping at global, continental or even national scales, are mainly based on climatic criteria. There does at least seem to be some agreement that climate is the most important criterion at this level of mapping.

The Food and Agriculture Organisation (FAO), which organisation has been most concerned with agricultural potential mapping at these scales, has applied climatic data to produce isohyets of lengths of growing periods which separate 'agroclimatic zones' (FAO, 1978). FAO has gone on to combine this agroclimatic data with other data to assess "potential population supporting capacities of land in the developing world" (Higgins et al., 1982).

NATIONAL MAPPING

Climatic data are also usually the most important criteria for mapping the agricultural potential of countries. However, at this level of mapping, the isohyets are usually not very accurate because of the paucity of meteorological stations in many parts of the less developed world. Extrapolation of climatic data between meteorological stations is best effected using vegetation indicators. In this context, Pratt et al. (1966) devised 'ecoclimatic zones' for Kenya, and the vegetation index mapping from satellite imagery (e.g. Justice et al., 1985) could also be used for this purpose.

Subdivision of a country into its agroclimatic or ecoclimatic zones would appear to be the most important classification level; but national mapping of agricultural potential usually needs more detailed information, more dependent upon geomorphic criteria. Brink et al. (op.cit.) recommended the terms 'land division' for mapping at 1:15 000 000 scale, and 'land province' for mapping at scales of 1:5 000 000 to 1:15 000 000. The land division was defined as "an assemblage of surface forms on a scale expressive of the whole, or a large part of a major structure", whereas the land province was described as "a land unit possessing an assemblage of surface forms and other surface features on a scale expressive of a second order structure or a large lithological association". King (1974) gave the following examples for Zambia and Ethiopia respectively: Bangweulu Craton and Ethiopian Highlands as land divisions; Central Plateau and Rift Valley Lakes Basin as land provinces. In practice, the land division is hardly ever used.

I do not know of any national mapping that has divided the country into land divisions and provinces. A more typical example of a national agricultural potential map legend is indicated in Table 1, from which we can see that an exclusively hierarchic system is not followed; rather the classification is more in the form of a matrix. Thus river floodplains have a different potential depending on their agroclimatic zone. In addition, the geomorphic units are not restricted to a single occurrence, but recur throughout the country. Even the more comprehensive terms 'depositional lowlands' et al., are more a method of grouping geomorphic units for the purpose of the legend, than representative of distinct geographical areas. The map has recognised the phenomenon of recurrence which tends to increase in importance at larger scales.

The map, whose legend is indicated in Table I, was produced using the preclassificatory approach whereby the landscape is first classified according to the attributes that are considered important, after which the landscape is mapped according to this predetermined classification. The Brink et al., system tends to classify the landscape only after it has been mapped. It is also location specific (for scales smaller than or equal to 1:250 000), whereby the landscape is subdivided hierarchically in a spatial

TABLE I An example of a national agricultural potential map legend (after Mitchell and King, in press). 'A' indicates suitability for rainfed agriculture, 'R' for rangeland potential, 'I' for irrigation. '1' indicates highly suitable, '2' moderately suitable, '3' marginal. Where no suitability code is shown, the unit is not suitable for that purpose.

	Key	Land use potential	Land suitability				Area in km^2
			Agroclimtic zones				
			1 Woodland	2 Bushland	3 Semi-desert	4 Desert	
Depositional lowlands							
River floodplains	Lf	1	A3/R1/I1	A3/R1/I1	A3/R1/I1	I1	10 310
Vertisols	Lv	1	A3/R1/I1	-	-	-	6 720
Deltas	Le	1	-	-	-	-	4 660
Major watercourse	Lw	2	A3/R1/I2	-	R3/I3	I3	4 950
Alluvial fans, bajadas	La	2	A3/R1/I2	R2	R3	N	132 290
Alluvial flats, reg	Lr	3	-	-	R3	N	9 680
Sand plain with pans	Lsp	2	R1	R2	-	-	80 490
Sand plain without pans	Ls	2	R1	R2	R3	N	31 210
Dunes: longitudinal	Ldl	2	R1	R2	R3	N	211 740
Dunes: transverse	Ldt	2	-	R2	-	N	6 850
Lowland calcrete	Lc	3	R3	R3	-	-	3 510

TABLE I (continued)

	Key	Land use potential	Land suitability				Area in km^2
			Agroclimtic zones				
			1 Woodland	2 Bushland	3 Semi-desert	4 Desert	
Plateaux							
Little dissected: siliceous	Ps	2	–	R2	R3	N	15 960
calcareous	Pc	2	R2	R2	R3	–	55 720
basaltic	Pb	2	–	R2	–	–	2 990
Much dissected: siliceous	Pds	3	–	–	R3	N	71 340
calcareous	Pdc	3	–	–	R3	N	11 200
basaltic	Pdb	3	–	R2	–	N	5 870
Plains of grit or rock, hamadas	H	3	–	–	R3	N	29 700
Steep slopes	S	3	R2	R3	R3	N	97 640
Wetlands							
Seasonally waterlogged land	Ws	1	R1/I1	R1/I3	–	–	23 170
Periodically waterlogged land, pan	Wp	3	R3	R3	–	–	6 170
Swamp	Ww	3	N	–	–	–	2 060
Area in km^2			206 710	196 350	268 860	152 320	824 230

sense, if not in a taxonomic one. Each subdivision tends to be specific to its higher orders. Thus in Zambia the Chitoshi Plain land system is part of the Chambeshi-Bangweulu Plain land region which is part of the Central Plateau land province. The Chitoshi Plain cannot belong to any other land region, nor can the Chambeshi-Bangweulu Plain land region belong to any other land province.

The preclassificatory recurring unit system is more detailed than the Brink et al. system, and therefore takes longer to map. It is a useful discipline to consider before mapping what are likely to be the useful mapping units. A disadvantage with the preclassificatory approach is that during the course of delineation, unforeseen significant attributes are often discovered, which usually means reviewing the previously delineated land units to incorporate the new attribute - a procedure often producing errors. The more detailed map is useful where the intention is to locate specific areas for development, such as, for example, irrigation schemes.

The location specific systems are better for a general appreciation of the broad development opportunities in different parts of the country. The system provides for more objectivity, in that the development attributes of the Chambeshi-Bangweulu Plain for example can be described without having to follow a predetermined classification.

PROVINCIAL, REGIONAL AND DISTRICT MAPPING SCALES

In the Brink et al. system, the units mapped at the provincial, regional and district level of mapping (i.e. generally at scales of 1:250 000 to 1:500 000 inclusive) are usually the land province, land region and land system. The most important units are the land region and land system, which tend to be employed in fundamentally different ways. A land system was defined by Christian and Stewart (1953) as an "area, or group of areas, throughout which there is a recurring pattern of topography, soils and vegetation". It is classified according to chosen parameters (King, 1970) and contains within it the minimum of variation feasible at that scale of mapping. The lower level land units listed in the first column of Table I e.g. river floodplains, would mostly be land systems. The land region, on the other hand, is "a cluster of land systems, grouped according to considerations of potential development. Generally, the land region tends to be the unit considered for development planning, whereas the land system is a fundamental land unit, around which the project itself is based, particularly with regard to sampling" (King, 1982a).

As indicated in the Christian and Stewart definition, the land system is the highest level in the Brink et al. hierarchy to allow for recurrence. The Brink et al. system recognises the importance of 'land facets' which recur within the land system, but considers them too small to map at this scale. Rather they and their areal extent are included in the land system description. Some land systems are distinguished according to the aerial extent of their land facets.

The procedure for mapping at the provincial, regional or district mapping scales is usually to distinguish the land provinces first, which is normally a major geomorphic feature, such as the East African Rift Valley or the intervening Central Plateau; then to map the land regions usually on the basis of modal angle of slope, altitude and lithology; then to map the land systems within the land regions.

The recurring principle is sometimes applied at the land system level itself. For the land system survey of Indonesia (LRDC/Bina Program, 1985), typical land systems were first delineated. The rest of the landscape was then mapped according to these typical land system descriptions until there was a significant change in the landscape, due to, for example, the influence of a different geomorphic process such as vulcanism. This method of mapping is similar to the preclassificatory system of mapping higher order land units, since after the initial 'objective' land system phase, the rest of the landscape is mapped as far as possible according to a predetermined classification. In practice, the recurring land system method tends to produce more total land system units (including outliers) than the conventional land system method, and the end result is a more detailed map than the conventional land system method.

The most detailed preclassificatory system is the 'parametric approach' which Mabbutt (1968) defined "as the division and classification of land on the basis of selected values". The land system approach assumes an interrelationship between landform, soils and vegetation, which vary in sympathy. The parametric approach avoids this assumption by mapping the significant attributes independently. If there is plenty of time available, this method should be superior to the land system technique; but the time involved in considering each attribute independently is much greater than that involved in the land system approach. The establishment of Geographic Information Systems (GIS) however will necessitate a parametric approach, unless the land system is taken as the basic cell, which has been suggested for less developed countries (Lawrence, pers. comm.,1985).

LAND SYSTEM NOMENCLATURE

At the simplest level, land systems can be described as flat or hilly. (This section also refers to other similar land units mapped at scales of 1:250 000 to 1:1 000 000.) Laymen (i.e. non-geomorphologists) describe landscapes not only according to their basic inherent characteristics, but also according to their relative position with respect to their surrounds. Thus an elevated plain is called a 'plateau', and if the observer is standing on the crest of a hill or a plateau, he describes the 'hills' below him as 'valleys'.

I therefore suggest for ease of comprehension, consideration should be given first to naming land systems using one of these basic names (plain, plateau, hills, valleys) combined with a local name, e.g. Debra Valleys. For some landscapes, description in terms of these units is straightforward; but for others, criteria need to be defined to determine what they should be called. Another complication arises where landscapes consist of a combination of basic units which are not worth separating at the level of mapping required, e.g. an inselberg plain (i.e. plain with hills).

Three basic criteria need to be defined: (1) modal slope to separate plains and plateaux from hills and valleys; (2) relationship between unit and its surrounds, to separate plain from plateau, and hills from valleys; and (3) extent of subordinate unit to justify its inclusion in the name. It should be emphasised that I am not necessarily proposing another classification system (in addition to for example van Lopik and Kolb (1959) and Ollier (1967)), but rather trying to devise a simple nomenclature that gives an immediate simple impression to a non-geomorphologist of the nature of the landscape.

Following the CSIRO slope classification whereby 6° is used to distinguish moderate from gentle slopes (Speight, 1967), I would suggest the distingishing modal slope between plains (and plateaux) from hills and valleys should also be 6°. The distinction between plains and plateaux, and between hills and valleys, should be based upon whether more than 50% of the boundary of the land system is upstanding or depressed with respect to the surrounding land systems. Subordinate units should be included in the nomenclature if they occupy more than 10% of the area of the land system. It is also suggested that the term 'escarpment' should be added to the other four basic units, as it is usually very distinctive, well known and an easily recognisable feature.

The way these criteria are employed to produce the final nomenclature is shown in Figures 1 and 2, but for the majority of situations it will not be necessary to refer to these figures. They are only necessary to resolve borderline situations. Examples of land systems named according to this system (taken from King *et al.*, 1979) are:

Mahata Plain
Mbede Valley
Zimba Hills
Kausinse Plateau
Chizi Escarpment
Kashusha Plain with Hills

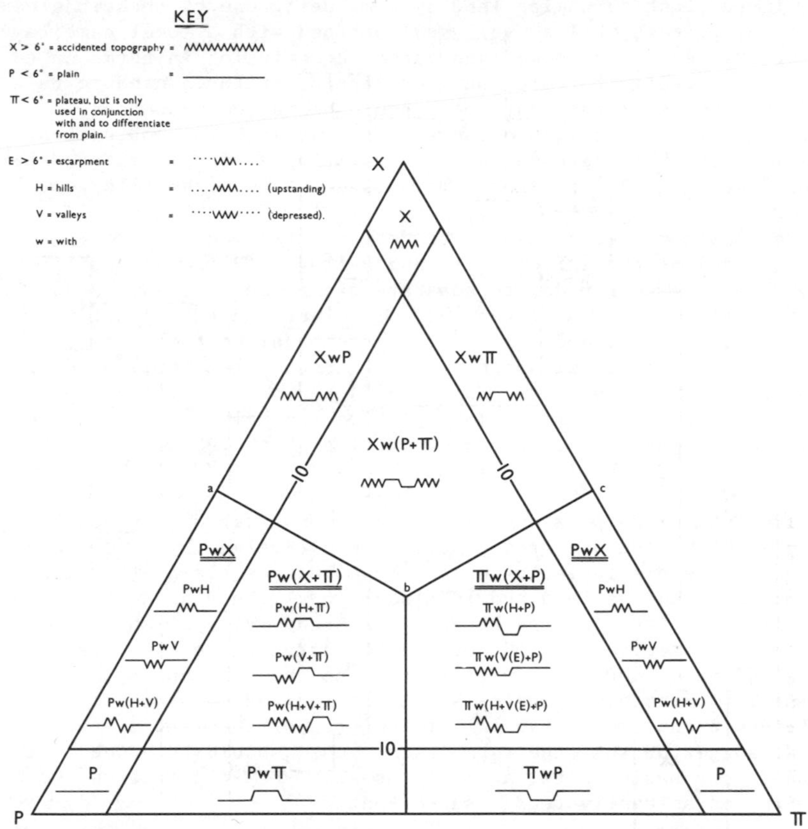

Fig. 1. Ternary classification of landform (after King, 1974)

However users of this system, particularly geomorphologists, may find it too restrictive. For example, why call a floodplain a plain, where the term 'floodplain' is so much more descriptive? This is the classic quandary of classification systems: the conflict between simplicity and comprehensivity.

Nevertheless my intention is to offer a simple land system description which can be used and understood by non-geomorphologists. When the geomorphologist wishes to use a more descriptive nomenclature, he should also provide the simple name, by means of, for example, a 'general morphology' column in the land classification table.

Literature survey assessment. For the purpose of this paper, I examined the actual names given in various land system studies.

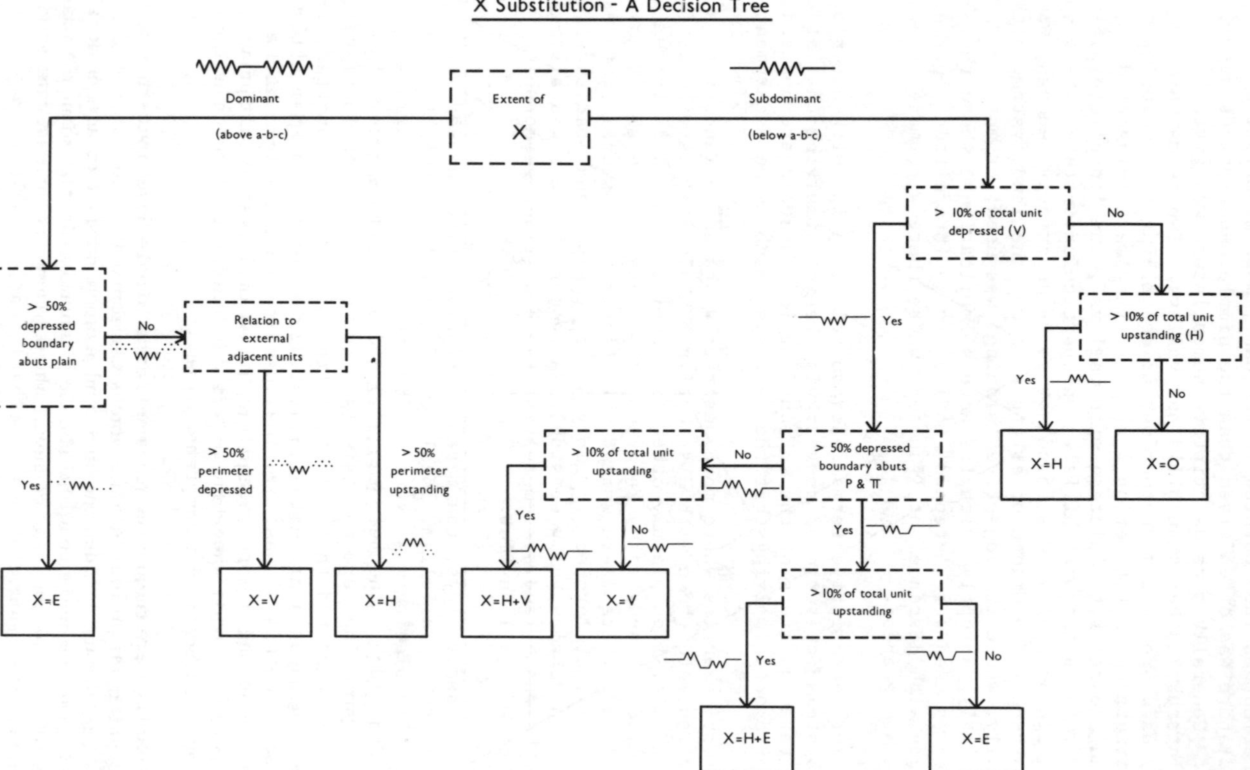

Fig. 2. A decision tree for the substitution of x in Fig. 1 (after King, 1974)

Most studies just give local names, or even just a number, without any indication of the nature of the terrain; but where such information was incorporated in the name, more descriptive nomenclature was often given than the simple descriptions suggested above. Generally more descriptive nomenclature was given to plains and plateaux, rather than hills and valleys. Two terms which were widely used are really another way of describing the inclusion of subordinate units (criterion 3 above). They were 'dissected plateau' to refer to 'plateau with valleys', and 'inselberg plain' to refer to 'plain with hills'. One particular instance where the term plain or plateau was found not to be descriptive enough, was where there was a change of altitude across the land system, either gradually, where the adjective 'sloping' was used; or more importantly, abruptly (with its access implications) where the adjectives used were 'stepped' indicating an intermittent increasing altitude with distance, or 'faulted' where there were more irregular horsts and grabens.

Other more descriptive terms refer to relationships with exterior land systems (criterion 2 above): viz 'basin', implying that at least three quarters of the surrounding land systems are upstanding, and 'piedmont' implying that the land system abuts an escarpment.

Additional examples where more descriptive names to land systems were given were those with distinctive hydrological attributes (e.g. swamp, dambo, floodplain), other distinctive alluvial features (e.g. terrace, delta and bajada which could perhaps be more simply called 'fans'), and where there were distinctive microerosional features (e.g. gullied, 'eroded' to imply sheet erosion, and 'badlands' to imply high density gully erosion). A geomorphic term was also given where the whole land system represented a distinct geomorphic feature, e.g. a lava flow.

For hills and valleys, the more descriptive criteria used were singularity, parallelism, local relief and relationship to exterior units. Singularity was expressed by the use of such terms as gorge, range and ridge. The term range(s) and ridge(s) express parallelism. The term mountain is often used for high local relief; and it is suggested that a local relief of 300 m (Desaunettes, 1977) could be used as a criterion to distinguish mountains from hills. The term 'foothills' was sometimes used to distinguish a hilly land system bordering a more upstanding hilly land system, although this was also often used where there was difficulty in finding a local name (see below).

Other terms were also used to those mentioned above, but they were really only variations of the above terminology. Geological terms were also often included in the land system name; but although of interest to the geomorphologist, they usually do not mean much to the planner; and geological terms should perhaps be restricted to the land system descriptions or classification table.

The local name prefixes were usually villages, sometimes rivers (especially where a village could not be found on the map). Land systems named after large rivers were distinguished by the terms 'upper', 'middle' and 'lower' as necessary, e.g. Middle Katuma Valley, lower Katuma Valley (King et al., 1979). In desperation, the term 'headwaters' was also used. In some locations where no local names could be found, peripheral terms such as 'foothills', 'shore' and 'fringe' were used, e.g. Buhoro Flat, Buhoro Shore, Buhoro Fringe (King,1982b).

Conclusions. Not surprisingly, the literature survey has revealed that many more terms were used to name land systems than the simple nomenclature of plains, plateaux, hills, valleys and escarpments. The number of times the simple terms were used ranged from a quarter to two thirds of the total number of land systems.

Surveys have to be both intelligible to laymen, and yet provide enough information for specialists. The present shortage of development funding has meant surveys have had to be carried out by generalists. The best solution to presenting specialist information to a mixed audience is to provide information at different hierarchic levels. The specialist will want to provide as much information as possible; but as far as land system nomenclature is concerned, it is suggested that he restricts the terminology to those terms discussed in the literature survey above (preceded by a local name) and listed below; but relates it also to the simple nomenclature (underlined in the list below) in the land system descriptions (e.g. under a 'general morphology' heading). Generalists may only want to use the simple nomenclature.

Plains

Plain
Inselberg plain
Sloping plain
Stepped plain
Basin
Piedmont
Swamp
Dambo
Floodplain
Gullied plain
Eroded plain

Plateaux

Plateau
Inselberg plateau
Dissected plateau
Sloping plateau
Faulted plateau
Gullied plateau
Eroded plateau

Hills

Hills
Range(s)
Ridges(s)
Mountain(s)
Foothills

Valleys

Valleys
Gorge

Escarpment

Escarpment

LAND FACET NOMENCLATURE

Reference has already been made to the land facet which was defined by Webster and Beckett (1970) as "a part of the landscape, usually with simple form, on a particular rock or superficial deposit, and with soil and water regime that are either uniform over the whole facet or if not, vary in a simple and consistent way. Each land facet is sufficiently homogenous over its extent to be managed uniformly for all but the most intensive kinds of land use". It is usually the most important land unit for agricultural potential, and it is therefore especially important that there should be recognisable, understandable and accepted terminology at this level, particularly because it is the level at which geomorphology impinges most on other disciplines, especially soil survey. Other disciplines require accepted geomorphic terminology so that they can relate their descriptions to acceptable land units.

Eleven years ago, I attempted a classification of land facets by examining a number of LRDC, CSIRO and MEXE project reports to determine what terminology was used to describe land facets (King, 1974). The investigation showed that land facets were mostly distinguished by means of adjectives, e.g. upper from lower terrace, sandy from silty plain. Land facets were classified by means of a multivariable matrix rather than in an hierarchic fashion. The following were the variables used: angle of slope, slope shape, width or diameter, length, local relief, altitude (absolute and relative), plan shape, relationship with stream order, geomorphic process, vertical position, horizontal position, geology, soils, microrelief, hydrology, age, stability and vegetation. I accordingly devised a land facet description card, which by means of a key, any land facet could be plotted and recorded according to its value with respect to the above variables. The card was never used however because its size was too great a deterrent!

For the purpose of this conference, I decided to re-examine my land facet description card to see if it could be simplified to a more useful form. Firstly, I examined a selection of land resource survey reports to find out which of the descriptive variables (as listed in the paragraph above) were used most often. The results were as follows:

Vertical position	28%
Geomorphic process	20%
Angle of slope	13%
Slope shape	5%
Altitude	5%
Hydrology	5%
Width/diameter	4%
Stream order	4%
Soils	3%
Geology	3%
Age	3%
Horizontal position	2%

Stability	2%
Plan shape	1%
Local relief	1%
Microrelief	1%
Length	0%
Vegetation	0%

Further investigation as to why the first three were preferred, suggested that vertical position was chosen because of its ease of recognition, geomorphic process because of an academic bias on behalf of the authors, and angle of slope because of both its importance and ease of recognition (that is within significant limits). To test these hypotheses, the variable list was circulated among colleagues in the Land Resources Development Centre (who cover a range of environmental sciences), and asked them to list the variables according to agricultural importance, ease of recognition, academic importance and frequency of significance. The last factor was included because some variables are more significant in particular environments. The results of the survey are shown in Table II. If the perceived agricultural importance and ease of recognition criteria are considered in conjunction with the results of the literature survey, the six most important criteria become (in order of importance):

Angle of slope
Vertical position
Vegetation
Hydrology
Altitude
Soils

Although I am not satisfied the respondents fully understood what I meant by academic importance and frequency of significance (some never completed these columns), it would seem that geomorphic process had been principally chosen in the literature survey more because of the academic bias of the author rather than its agricultural (or engineering) importance or ease of recognition. Slope shape scored high on the literature survey largely because of the inclusion of one of my own reports in the survey, where I included the slope shape in the name of nearly every land facet - my own idiosyncratic bias!

TABLE II Land facet criteria ranked according to perceived factors

Criteria	Factors			
	Agricultural importance	Ease of recognition	Academic importance	Frequency of significance
Angle of slope	1	1	4	2
Slope shape	14	5	6	15
Width/diameter	8	6	8	7
Length	10	7	14	12
Local relief	13	3	8	10
Altitude	6	8	12	5
Plan shape	18	12	8	18
Stream order	17	14	3	17
Geomorphic process	12	18	1	13
Vertical position	11	4	5	4
Horizontal position	15	9	12	14
Geology	9	15	2	8
Soils	2	13	8	1
Microrelief	5	11	18	9
Hydrology	3	10	16	2
Age	16	17	7	16
Stability	4	16	15	11
Vegetation	7	2	16	6

Further examination of the use of the six most important criteria suggests angle of slope, soil type and vertical position should be used for upland land facets; whereas hydrological condition, vegetation cover, relative altitude and soil type should be used for lowland land facets.

Upland land facets. Table II demonstrates the significance of angle of slope. It is suggested that defined qualitative terms should be used to suit the project, e.g. as indicated in Table III.

TABLE III An example of angle of slope categories

Category	Slope angle (degrees)
Flat	<0.5
Very gentle	0.5–3
Gentle	3–6
Low moderate	6–10
High moderate	10–15
Moderately steep	15–20
Steep	>20

Because of its importance, the soil type should be indicated, if known, either by texture (e.g. sandy) or according to parent material (e.g. ash)(see Table IV).

Further examination certainly does seem to suggest that vertical position nomenclature is mostly used for ease of recognition of the land facet in the landscape. The most comprehensive determination of vertical position has been by Dalrymple et al. (1968). Their units are however numbered, rather than named. I would suggest the following nomenclature for their system:

1.	Crest or interfluve
2 & 3.	Plateau edge or upper slope
4.	Scarp
5.	Debris slope or midslope
6.	Footslope or lower slope

Units 7, 8 and 9 are lowland facets. Where two alternatives are given, the first usually relates to hillslope elements where a scarp is present. I am here again adopting the principle of progressive complexity. The inexperienced or non-geomorphological land resources surveyor can use these basic units as indicated in Figure 3. The experienced geomorphologist will want to use more descriptive terminology, examples of which are indicated in Table IV.

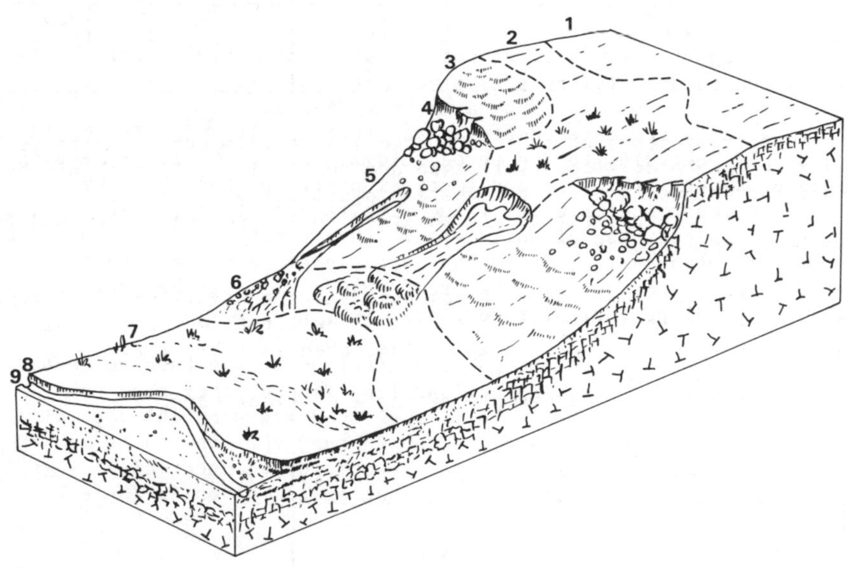

Fig. 3 Diagrammatic representation of land facets classified according to vertical position (after Conacher and Dalrymple, 1981).

TABLE IV Upland land facet classification according to vertical position. Descriptions implying soil type are underlined.

Crest	Scarp	Debris slope Midslope	Footslope Lower slope
Schichtrücken	Cliff	Adventive cone	Alluvial fan
Schichtschwelle	Breakaway	Alluvial cone	Colluvial fan
	Gorge	Ash cone	Cuirasse
	Undercut slope	Avalanche	Gallery laterite
		Barchan	Pediment
		Benched slope	Piedmont
		Block-and-ash cone	Slip-off slope
		Cinder cone	
		Dune	
		Crater	
		Dipslope	
		Dyke ridge	
		Earthflow	
		Inselberg	
		Landslide	
		Seif	
		Vein ridge	

Geomorphologists generally do not accept land unit classifications restricted to one attribute like vertical position, because they rarely cover the significant range of variability in any particular area. What is suggested here, is not that the land facets should be named only according to the three recommended upland attributes of angle of slope, soil type and vertical position; but that the name should at least indicate its value with respect to these attributes (if known in the case of soil type). Thus the term 'cliff' would be sufficient because it implies both steep angle and very shallow soil, if any; whereas 'alluvial fan' should at least be prefaced by an angle of slope indicator, such as 'very gentle'.

Lowland land facets. Although four attributes have been suggested as important for describing lowland land facets, they tend to correlate with each other. It is suggested hydrological condition should be indicated at least to show whether the land facet is normally permanently dry, seasonally waterlogged, periodically flooded, or permanently flooded. Examples are given in Table V.

TABLE V Lowland land facet classification according to hydrological condition. Descriptors implying soil type are underlined.

Dry	Seasonally waterlogged	Periodically flooded	Permanently flooded
Plain Terrace Back<u>reef</u> Beach ridge Lacustrine plain <u>Loess</u> plain Strand plain	Dambo	Floodplain Shore Oveflow channel Backland Beach Bolson Bar Choked valley Delta Floodplain bench Tidal flat Wadi	Swamp Lake Lagoon

It is suggested vegetation cover should be included in the name where it is likely to indicate more information on the hydrological condition. A wooded dambo (with percentage tree cover of 2-20) for example implies waterlogging for less months in the year than for a normal grassland dambo.

Similarly relative altitude should also be used to indicate hydrological condition. In very low lying areas, like the New Guinea lowlands, Mabbutt et al. (1965) for example found the following separations significant: higher from lower backlands, higher from lower tidal flats, upper from lower mangrove zones, higher from lower salt flats, and higher from lower beach ridges.

Further examination of the literature does in fact indicate that vegetation type and relative altitude tend to be used to indicate hydrological condition. The main criteria then that should be used to name lowland land facets should be hydrological condition, whether implied directly, e.g. floodplain, or indirectly by vegetation or relative altitude, and, if known, soil type.

SUMMARY

Classifications tend to be hierarchic or tabular. Hierarchic systems are easier to comprehend but necessitate variable ranking. If a universal system of ranking variables can be justified, then a hierarchic system should be used, but usually this is not the case. Thus although it can be argued that climate should be the first consideration in an agricultural resource survey of the world, below this level different attributes tend to have different significance in different environments. One of the reasons why the land system technique has been popular is because it is based upon an examination of the landscape with no preconceptions, although psychologists will deny that this is possible.

The concern of this paper has mainly been to review and try to simplify land resource classifications and descriptions so that they can be more relevant to surveys with high Reconnaissance Factors (King, 1982a). Simplified terms are suggested as nomenclatures rather than comprehensive descriptors; the terminology decided on the basis of usage and practical significance. It is suggested that the simplified terms should always be included in the reports so that they can be widely understood, while more detailed information can be presented in tabular form. Inexperienced land resource surveyors need only use the simplified terms.

ACKNOWLEDGEMENTS

This paper has been published with the permission of the Director (Mr A J Smyth) of the Land Resources Development Centre of the Overseas Development Administration, United Kingdom Foreign and Commonwealth Office. I would also like to thanks Mr Smyth for reading and criticising the draft. I would also like to thank the scientific staff of the Land Resources Development Centre for taking the trouble to answer my questionnaire on land facets.

REFERENCES

Brink, A. B. C., Mabbutt, J. A., Webster, R., and Beckett, P. H. T., 1966. Report of the working group on land classification and data storage. Rep. MEXE, No. 940, Christchurch.

Christian, C. S., and Stewart, C. A., 1953. General report on survey of Katharine-Darwin region, 1946. Land Research Series CSIRO Australia No.1.

Conacher, A. J., and Dalrymple, J. B., 1981. The catenary sequence of landsurface units and some of its applications. Western Geographer, 5, 82-94.

Dalrymple, J. B., Blong, B. J., and Conacher, A. J., 1968. A hypothetical nine unit landsurface model. Z. Geomorph., 12, 60-76.

Desaunettes, J. R., 1977. Catalogue of landforms for Indonesia: Examples of a physiographic approach to land evaluation for agricultural development, FAO, Rome.

Food and Agriculture Organisation (FAO), 1978. Report on the Agro-ecological Zones Project, Food and Agriculture Organisation of the United Nations, Rome.

Higgins, G. M., Kassam, A. H., Naiken, L., Fischer, G., and Shah, M. M., 1982. Potential population supporting capacities of lands in the developing world, Food and Agriculture Organisation of the United Nations, Rome.

Justice, C. O., Townshend, J. R. G., Holben, B. N., and Tucker, C. J., 1985. Analysis of the phenology of global vegetation using meteorological satellite data. Int. J. remote Sensing, 6, 1271-1318.

King, R. B., 1970. A parametric approach to land system classification. Geoderma, 4, 37-46.

King, R. B., 1974. Suggested methods and terminology for the geomorphological component in land resource surveys. Land Resources Development Centre, Misc. Rep. No. 174, Tolworth.

King, R. B., 1982a. Rapid rural appraisal with Landsat imagery: A Tanzanian experience. Z. Geomorph. N. F., 44, 5-20.

King, R. B., 1982b. A land system map of Mbeya Region, GCP/URT/055/DEN, RIDEP Rep. No. 26, Mbeya.

King, R. B., Rombulow-Pearse, C. W., Kikula, I. S., Ley, G., and Kamasho, J. A., 1979. Land resources of the Rukwa Region: A reconnaissance assessment, BRALUP, Dar es Salaam.

Lopik, J. R. van., and Kolb, C. R., 1959. A technique for preparing desert terrain analogues. US Army Corps Engrs Waterways Exptl Stn., Tech. Rep. No. 3-506.

LRDC/Bina Program., 1985. Regional Physical Planning Programme for Transmigration: Review of Phase I B results, Central Kalimantan, Dept. of Transmigration, Unpubd Rep., Jakarta.

Mabbutt, J. A., 1968. Review of concepts of land classification, in Land evaluation (Ed. Stewart), pp.11-28. Macmillan, Melbourne.

Mabbutt, J. A., Heyligers, P. C., Pullen, R., Scott, R. M., and Speight, J. G., 1965. Land systems of the Port Moresby-Kairuku area, in Lands of the Port Moresby - Kairuku Area, territory of Papua and New Guinea, pp.19-82. CSIRO, Melbourne.

Mitchell, C. W., and King, R. B., in press. Namibia satellite imagery studies for the preparation of a comprehensive economic map: Report to accompany 1:4 000 000 scale map of land use potential, FAO, Rome.

Moss, R. P., 1969. The appraisal of land resources in tropical Africa. Pacif. Viewp. 10, 18-27.

Moss, R. P., 1981. Ecological constraints on agricultural development in Africa. African Research and Documentation, 25, 1-12.

Ollier, C. D., 1967. Landform description without stage names. Australian Geographical Studies, 5, 73-80.

Pratt, D. J., Greenway, P. J., and Gwynne, P. D., 1966. A classification of East African rangeland with an appendix on terminology. *J. appl. Ecol.*, 3, 369-382.

Speight, J. G., 1967. Explanation of land system descriptions, in *Land of Bougainville and Buka islands, territory of Papua and New Guinea*, pp.174-184. CSIRO, Melborne.

Thomas, M. F., 1969. Geomorphology and land classification in tropical Africa, in *Environment and land use in Africa* (Eds. Thomas and Whittington), pp.103-145. Methuen, London.

Webster, R., and Beckett, P. H. T., 1970. Terrain classification and evaluation using air photography: a review of recent work at Oxford. *Photogrammetria*, 26, 51-75.

International Geomorphology 1986 Part II
Edited by V. Gardiner

SPATIAL VARIABILITY OF SOIL HYDRIC PROPERTIES IN THE RINGELBACH CATCHMENT (GRANITE VOSGES, FRANCE)

D. Viville and B. Ambroise

Laboratoire de Géographie Physique, RCP 741 CNRS
3 Rue de l'Argonne, F 67083 Strasbourg Cedex, France.

ABSTRACT

In the small granitic Ringelbach catchment (36 ha), 396 undisturbed soil cores have been sampled at different depths in 15 sites representative of the four morphopedologic units, which have been defined in the basin on the basis of natural criteria. One hundred and ninety four have been sampled in 2m large and 0.75m deep pits, and at the two levels, 0-5 and 15-20cm, in a 100m long transect within the same unit.

Their textural fractions, organic matter content, bulk density, water retention at 10 pF and hydraulic conductivity at saturation have been determined in the laboratory. In these soil units, which are characterized by a coarse texture (more than 70% of sand and gravel), a high organic matter content, an important macro-porosity and a high permeability, water retention is rather low even in the top layers.

Within a unit, the vertical differentiation of these properties in the soil profile is low in the deep layers but important in the top layers, for both the standard deviations (larger on the surface) and the mean values. Within each layer, mean values and variabilities are similar for the transect (unit scale)and for the pit (local scale). The statistical distributions are normal for static properties, lognormal for dynamic ones. No spatial correlation has been found along the transect.

The global comparison of the four units by statistical data analysis (hierarchical ascending classification, stepwise discriminant analysis) has shown that:

- the 4 units are significantly different with respect to texture and retention, even if the differences are mainly textural;

- in each unit, there is a marked vertical differentiation, different from one unit to another, which concerns mainly the retention - well correlated with profiles of organic matter or bulk density;

- the hydraulic conductivity at saturation presents no great differentiation, both laterally and vertically.

These results, which confirm the validity of the naturalist approach used to determine hydrodynamic units in the catchment, and which lead us to analyse the hydric properties of the organic matter, are useful to estimate the representativeness of point measurements.

THE SITUATION

The hydrodynamic properties (hydric retention, hydraulic conductivity) of surface layers and soils characterise the ability of porous environments to retain and transmit water in relation to their hydric or energetic state. These properties control the process of redistribution of precipitation to the soil network (evapotranspiration, surface runoff, throughflow and deep infiltration to the water table). Because of this they play, directly or indirectly, an important role hydrologically (hydrologic balance, origin of floods) and geomorphologically (mass movements, weathering) as well as ecologically and agronomically (primary productivity, irrigation, drainage, fertilisation). The economic implications of knowledge of these characteristics are therefore significant.

These hydric characteristics depend mainly on the structure and texture of the porous environment, thereby on how it was developed and its further evolution. Like textural properties they are in fact very variable, vertically in a profile as well as laterally according to soil type. The spatial variability of the hydrodynamic properties of soils is a constant research subject found in many geomorphological and agronomic studies, but we know from the work of Lobert and Cormary (1964) and especially that of Nielsen et al., (1973) that there has been considerable development, although as yet insufficient. This spatial variability in fact raises the question of the spatial representativeness of localised determination of these properties and, therefore, sampling is carried out in the study zone to estimate their statistical distribution (average, dispersion) with given precision.

As the process of determining these properties is generally long and onerous it is essential to keep the number of samples treated to a minimum. Apart from simple observations of the terrain, it would appear best to proceed with a preliminary division of the area into morpho-pedologic units defined by homogeneous sedimentary bodies, and then to do a relief sampling on these units. It is then necessary to ensure that these units also constitute units from the hydrodynamic viewpoint.

Such a study has been undertaken in the Ringelbach catchment (commune of Soultzeren, Haut-Rhin) which is representative of a large section of the Vosges massif. The testing of a large number of samples in the laboratory has determined the spatial variability of their textural and hydric properties, and elucidated the relationship between variables, to simplify further sampling of this type of environment.

OUTLINE OF THE STUDY AND COLLECTION OF DATA

The Ringelbach Catchment. This small elementary catchment of 36 ha, between 750 and 1000m in altitude, is situated on a vast regional slope of high gradient (20° on average). It is cut along an axis paralleling a fault in two types of Hercynian granites, overlaid on the summits by fragments of Triassic sandstone. The granitic area is overlain by less thick sloping formations (approx. 1m) which are rich in rocks and large particles, with a sandy matrix. Landslides, which have occurred since the end of the last cold period, have transported material to the bottom of the valley. Podzolic (brown earth to podzol) soils have developed within the rocky material, which are rich in organic matter. The anthropic activity is denoted in the landscape by the presence of cultivation terraces and gravels, by pastures on the granitic slopes and by pine forest on the sandstone.

Definition of morpho-pedologic units. Observations of terrain (geological, geomorphological, pedological, vegetation structures) allow definition of four morpho-pedologic units:

- a unit of sandstone and sandstone colluvials associated with podzol, pineforest and heathland (unit GRES).
- a unit on the colluvial formations at the bottom of the valley and the base of the slopes (having hydromorphic characters linked to the presence of a water table), associated with a grassland at Caltha Palustris (unit VALL).
- a unit on coarse grained granite, associated with grassland (unit HUBU).
- a unit of fine-grained acidic granite associated with grassland (unit GEIS).

Sampling of the Units. Three hundred and ninety six samples of undisturbed soil were taken from 14 sites within the study catchment and a site nearby, in order to have several sites per unit and several samples per 5cm level - the depth sampled in each site depended on the presence of a compact area, rocks or water-saturated levels: unit GRES 0-35cm, unit VALL 0-15cm, unit HUBU 0-30cm, unit GEIS 0-75cm. Furthermore, a transect 100m long centred on a pit 2m wide was sampled in unit GEIS at two levels, 0-5 and 15-20cm, to determine the statistical distributions and the spatial variability of the textural and hydrodynamic properties within a single unit.

Treatment of the Samples. The hydric properties of these undisturbed samples were determined in a laboratory using the appropriate technical methods.

- the hydric retention and desorption curve, by measuring the retention (θsat) to saturation, and retentions (θpF) at nine other tensions, up to pF2.7 using the sand and kaolin vat method; up to pF3.5 and pF4.2 using the diaphragm method.
- hydraulic conductivity at saturation by permeameter with constant charge.

For each of the samples the following properties were also determined:

- the apparent density, by heating at 105°C
- the organic matter content, by loss when heated to 850°C
- granulometric portions : (0-2 μ) and fine silts (2-20μ) by the pipette method; 6 classes of sand (between 50 and 2000μ) and larger components (>2000μ) by dry sieving; the larger silts (20-50μ) by the complement to 100%.

RESULTS

In coarser formations (more than 70% sand and stone) with a large organic matter content (10-15% on the surface, 2-3% at depth) and a considerable macroporosity (65-70% on the surface and 45-50% at depth):

- the retention is small and diminishes considerably at depth (for example at pF2.0 the retention comprises between 30-40% at the surface and between 15-30% at depth)
- the hydraulic conductivities at saturation are high (several dozen cm/h).

Variability within the unit. This variability has been the particular subject of study in the GEIS unit, but the results obtained (Ambroise and Viville, 1986) have proved equally valuable for the other units. Certain results, obtained from the coarse granitic formations of unit GEIS, confirm the observations made by other writers, of different environments, for example the reviews of Warrick and Nielsen (1980) or Vauclin (1983):

- the statistical distributions of static properties like the textural fractions, the apparent density, and the water retention are normal, whilst the hydraulic conductivity at saturation, which is a dynamic property, has a log-normal distribution.
- the degree of spatial variability (at the same depth) is small for the water retention, the water content in situ or the apparent density, average for the textural breakdown, considerable for the conductivity.

Other results appear to be related specifically to this type of environment:

- the vertical variability along the profiles is more marked in the organic surface layers than in the deeper, more uniform, mineral layers.
- the local variability within the same layer is greater on the surface than at depth where it is small but not insignificant.
- the analysis of the variograms for the properties being studied has not given rise to any significant spatial correlation along the 100m transect.
- the lateral variability is the same over several hectares (i.e. morpho-pedologic unit) as the local variability monitored over several square metres (i.e. soil pit).

Variability between units. To appreciate the validity of the naturalist division carried out - notably for the hydrodynamic properties - the data collected were treated statistically, using ascending hierarchal classification and discriminant analysis (Viville et al., 1986). This global approach shows:

- the morpho-pedologic units are significantly different in texture as in retention, the differences between the units being textural in relation to lithology.
- in each unit there is a vertical differentiation, varying from one unit to another, which is especially manifest in retention, being correlated to organic matter content or apparent density.
- the hydraulic conductivity at saturation is marginally different within the study catchment, vertically as well as laterally, with little relation to the other variables.

CONCLUSIONS

Therefore, even in a limited environment with little contrast, having very permeable rock structures with little hydric retention, there are significant textural and hydric differences which observations of the terrain and statistical treatment have encompassed and interpreted. The essential factor in spatial variability (vertical, lateral, local) appears to be the organic matter content (Viville, 1985).

In this type of environment the hydric properties and their spatial variability can be evaluated by means of a division into morpho-pedologic units with a preliminary sampling, which is simple to carry out (organic matter or apparent density), giving useful indications for defining a minimal sampling, stratified by unit and by level, allowing surface layers where the greatest dispersions are recorded.

Such an approach, which can provide basic useful data on which to establish spatial hydrologic models, interpretation of data by remote sensing and the management of water and earth resources, is in the process of being applied to grading the whole of the Fecht catchment.

Acknowledgements

We thank the Laboratory of Sedimentology at the Applied Geography Centre in Strasbourg where we carried out granular analyses; also B. Korosec for his assistance in the statistical treatments which were carried out at the CNRS Statistics centre at Strasbourg-Crnenbourg.

This study was carried out within the framework of the PIREN-Eau/ Alsace Programme (National Centre for Scientific Research, Ministry of the Environment, Alsace).

REFERENCES

Ambroise, B., Amiet, Y., and Mercier, J. L.,. 1984. Spatial variability of soil hydrodynamic properties in the Petite Fecht catchment at Soultzeren, France - Preliminary results, in Catchment Experiments in Fluvial Geomorphology, (Eds. Burt and Walling), pp35-53. Geo. Books, Norwich (UK).

Ambroise, B., and Viville, D., (1986). Spatial variability of textural and hydrodynamical properties in a soil unit of the Ringelbach study catchment, Vosges (France), in Geomorphology and Land Management, (Eds. Balteanu and Slaymaker), Z. Geomorph., Suppl. Bd. (In press).

ERA 569 CNRS, 1982. Structure et fonctionnement du milieu naturel en moyenne montagne - Bassins de la Petite Fecht et du Ringelbach (Vosges, France). Rech. Géogr. à Strasbourg 19/20/21, 276 p.

Lobert, A., and Cormary, Y., 1964. Variabilité des mesures de caractéristiques hydrodynamiques. Cahiers ORSTOM, Pédologie, 2, 23-49.

Nielsen, D. R., Biggar, J. W., and Erh, K. T., 1973. Spatial variability of field-measured soil-water properties. Hilgardia, 42, 215-260.

Vauclin, M., 1983. Méthodes d'étude de la variabilité spatiale des propriétés d'un sol. Les Colloques de l'INRA, Ed. INRA Publ., Paris, n.15, 9-43.

Viville, D., 1985. Variabilité spatiale des propriétés physiques et hydriques des sols dans le bassin versant du Ringelbach (Vosges granitiques.) Thèse Doctorat de l'Université Louis Pasteur, Strasbourg, novembre 1985, 151 p.

Viville, D., Ambroise, B., and Korosec, B., 1986. Variabilité spatiale des propriétés texturales et hydrodynamiques des sols dans le bassin versant du Ringelbach (Vosges, France), in Erosion Budgets and their Hydrologic Basis, (Eds. Vogt and Slaymaker), Z. Geomorph., Suppl. Bd. (In press).

Warrick, A. W., and Nielsen, D. R., 1980. Spatial variability of soil physical properties in the field, in Applications of soil physics, (Ed. Hillel), pp319-344. Academic Press, London.

International Geomorphology 1986 Part II
Edited by V. Gardiner

GEOMORPHOLOGY IN DEVELOPMENT: SOME RECENT WORK IN BOTSWANA

H.J. Cooke and P.A. Shaw

Department of Environmental Science
University of Botswana

ABSTRACT

Botswana is a large, thinly-populated country, with a semi-arid climate. The promise of its mineral wealth, surviving wild-life, and the water resources of the north offer excellent opportunities for geomorphologists to contribute in multi-disciplinary teams, to the exploration for and assessment and management of these resources. Four diverse projects are described to illustrate the application of geomorphologically related research during the past decade. These are a Drought Susceptibility Survey; Terrain Classification and Soil Mapping; the application of geomorphological principles to a prospecting problem; and geomorphological and hydrological studies of the lower Okavango Delta. Experience suggests that at present the most valuable contribution to be made by geomorphologists in this type of environment is as part of multi-disciplinary teamwork in broad-based studies.

INTRODUCTION

Botswana is a large (576,000 km^2) country, with a sparse and unevenly distributed population, and dominated physically by the Kalahari thirstland. Much of it is a sand-covered, undulating, monotonous plain. The northern third of the country however, displays a wide variety of landforms variously resulting from the operation of fluvial, lacustrine, and aeolian processes whose temporal succession in the late Pleistocene at least, must have been largely controlled by a climatic regime alternating between drier and wetter episodes, with probably more muted temperature variations also significant. The resulting landscape is one of great scientific interest. Though in many ways hostile to Man, it has nonetheless offered opportunities which human beings have long utilised, as shown by the very widespread occurrence of human artefacts dating from the early Stone Age up to Recent times. To modern Botswana the area is one of much promise, largely because of the huge water resources of the Okavango Delta, the potential for irrigated agriculture, considerable mineral wealth, and the surviving wealth of wildlife (Botswana Society 1976; UNDP 1976;Kalahari Conservation Society 1983)

In the unravelling of the complex pattern of the landscape of this

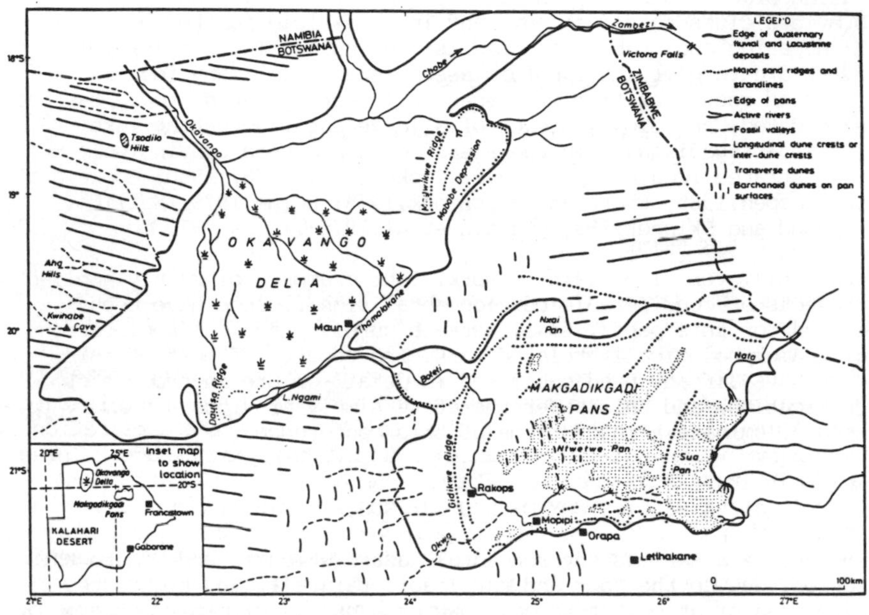

Fig.1 Northern Botswana - major physical features

region, and in evaluating its resources for humanity, the geomorphologist has an important role to play, in both pure and applied fields of research and development. He/she can function most effectively as a member of multi-disciplinary groups which will also ideally include geologists, pedologists, hydrologists, and archaeologists. Ideally they should also themselves have skills in these fields, if their work is to be fruitful and meaningful. This paper describes some examples of constructive work achieved by geomorphologists in collaboration with other specialists in Botswana over the past fifteen years.

As a result of both pure and applied work (usually carried out in tandem) in the area, the broad outlines of the landforms and their relationships in space and time have been studied through detailed fieldwork, often in remote and difficult country, greatly aided by remote sensed imagery and orthodox air photography. Geomorphological mapping on scales ranging ranging from 1:500,000 to 1:25,000 has been carried out. Hypotheses have been put forward linking the evolution of the major geographic features such as the Zambezi, Chobe, Limpopo, and Boteti drainages, the Okavango Delta, and the Makgadikgadi Pans to neo-tectonics and climatic oscillations. The detailed study of a major cave in the region, and its geomorphic setting, has provided correlative evidence of climatic variation, and thus of climate dependent geomorphic processes. Radiometric dating of cave sinters, calcretes, and rarer shells and peats has provided clues to chronology, and in this field further work using U/Th dating and stable isotopes is continuing. The complex polygenetic nature of the landform assemblages, with many fossil features resulting from processes no longer operating today, has been elucidated, and the existence of a large but fluctuating lake (Palaeo-Makgadikgadi) in the Pleistocene has been convincingly demonstrated (Coates et al.,1979; Cooke,1975, 1978, 1980; Cooke and Verstappen,1985; Ebert and Hitchcock,1978; Heine,1979; Mallick, Hapgood and Skinner,1981; Scholz et al.,1975).

From this overall pattern of work, four examples of mainly applied projects of considerable developmental significance have been chosen for this paper, to illustrate purely practical applications of geomorphologically related work. The results of most of this work have already been published in detail or are in process of publication, and this paper therefore simply provides a summary to help illuminate a major theme of this conference.

1. A PILOT STUDY OF DROUGHT SUSCEPTIBILITY

Drought is a serious environmental hazard which is endemic to this region, due to the poor and unreliable rainfall, and high rates of evapotranspiration. It is not however a purely climatic matter, and geomorphic terrain features with related geological, pedological, hydrological and biological characteristics have as great an influence on the medium and large scale distribution of the drought hazard as rainfall and temperature factors alone. A collaborative project involving a multi-disciplinary team from the University of Botswana Department of Environmental Science and the International Institute for Aerial Survey and Earth Science of the Netherlands carried out a pilot Drought Susceptibility Survey in the lower Boteti river region in 1978. This was based on geomorphological mapping of terrain units and their evaluation in terms of the quantity, quality, and seasonal variation of surface water, soil moisture, and groundwater resources. The work involved a number of phases, with a progressive increase in detail. Visual examination of available LANDSAT imagery confirmed patterns of landform relationships already apparent through earlier fieldwork. This was followed by reconnaissance level survey using LANDSAT enlargements at a scale of 1:250,000 as a basis for eventual plotting of geomorphological terrain units from photo-mosaics and detailed study of air-photo stereopairs. Some digital interpretation of LANDSAT data was also used. Groundwork followed with the terrain units identified and detail gathered on landforms, lithology, soils, surface and ground water, and vegetation. Local human adaption to seasonal and recurrent moisture deficiency was also recorded.

Finally an attempt was made to divide the region into Drought Susceptibilty classes (Verstappen and Cooke (ed.),1981)

2. SOIL MAPPING AND LAND QUALITY ASSESSMENT

The agricultural potential of parts of northern Botswana is considerable, based on the availability of water from the Okavango, Chobe, and Boteti rivers. Soil quality is a limiting factor. In the lower Boteti river region between Rakops and Mopipi the local people have for long practised a form of irrigated agriculture based on natural overbank flooding in good years. It became apparent during the Drought Susceptibilty Survey that some of the soils in the region were in fact suitable for crop production. J. Breyer of the University of Botswana Department of Environmental Science carried out a study in 1978-83 to map and evaluate the soils of the region. Her work involved firstly a refinement of the geomorphological terrain classification already carried out. This differentiated between sediment units or associations to which limiting characteristics such as active erosion/deposition, poor or excessive drainage, may be related. This was followed by the identification, classification, and description of the soils related to these units, and assessment with special reference to their soil management properties. Three full colour soil maps on a scale of 1:50,000 were produced (Breyer,1979, 1982, 1983).

3. DIAMOND PROSPECTING

An important kimberlite field was discovered by De Beers geologists in 1967 on the southern fringes of the Makgadikgadi Pans complex, and subsequently two major pipes at Orapa and Letlhakeng have been developed as very important diamond producers. Continued prospecting revealed a spread of heavy minerals westward between Orapa and Mopipi. This stimulated an intensive search for further pipes in this direction, which however proved inconclusive. David Grey, one of the De Beers geologists approached the problem by taking a broader view of the total environment, and using the hypothesis of a large late Quaternary lake in the Makgadikgadi basin, showed that the spread of heavy minerals noted was almost certainly due to the longshore drift of material derived from surface erosion of the Orapa pipe, along the southern shores of the lake between its 920 m and 945 m stillstands. This action produced a complex of spits and bars formed of the coarse gravels in which the heavy minerals were located. An important prospecting problem was thus solved by the application of geomorphological principles, and at the same time much new information was gathered on landform evolution in the region (Grey,1976; Grey and Cooke,1977).

4. OKAVANGO DELTA HYDROLOGY

Lake Ngami, at the southwest extremity of the Okavango Delta system, provides a water source for some 50,000 cattle, and has potential for fisheries and tourism. However, the lake is subject to great fluctuations in level, and has dried up periodically. P.A. Shaw of the University of Botswana department of Environmental Science in collaboration with the Botswana Government's department of Water Affairs, has been studying the behaviour of the lake on different time-scales, namely hydrological modelling from available contemporary data; historical studies covering the past 135 years from documentary evidence; and geomorphological mapping to gain insight into late Quaternary changes. The work has revealed a pattern of remarkable changes in the regime of the lake, particularly on the historical scale, and has indicated a number of difficulties that may arise relating to imminent future development for economic purposes.

The same methodology has been applied to the Mababe depression and the distal sector of the Delta, an area undergoing development of its land and water resources at the present time. Here historical studies have suggested that the mathematical model of the Delta constructed as a planning tool by UNDP/FAO and the department of Water Affairs in the 1970's may be inadequate to cope with the full range of hydrological fluctuations within the Delta. At present a new model is being developed to incorporate palaeohydrological variables, whilst investment by government is being made to collect accurate and up-to-date data for future planning. Shaw's geomorphological work has also greatly extended our knowledge and understanding of the evolution of the Okavango-Makgadikgadi complex of landforms. (Shaw,1983,1984,1985, and in press).

These four examples of work briefly described above, serve to underline the important role that geomorphological research can play in helping to solve practical environmental problems, and in promoting research development, as well as contributing a great deal of data towards a better understanding of the processes and patterns of landform evolution in a varying climatic environment. The examples also show that in developing countries such as Botswana, which have large areas of remote and difficult terrain, and low levels of scientific manpower availability, the most successful strategies involve multi-disciplinary teamwork in broad-based studies combining dedicated fieldwork with correct use of modern technology, especially in the form of remote-sensed imagery and air photography.

ACKNOWLEDGEMENTS

The authors and their colleagues past and present of the University of Botswana Department of Environmental Science wish to express their thanks to the University for continued generous funding, and to De Beers Botswana (Debswana Donations Fund), NUFFIC, and the Royal Society 20th International Geographical Congress Fund, for additional financial assistance. They would also like to express their appreciation to the Botswana Government's Departments of Geological Survey, Surveys and Lands, Agriculture, and Water Affairs for continuing co-operation andassistance. Finally a word of thanks must go to Professor Herman Verstappen and his colleagues of the ITC for effective and friendly collaboration on the Drought Susceptibility Survey.

REFERENCES

Baillieul,T.A., 1979. The Makgadikgadi Pans complex of central Botswana. Bull. Geol.Soc. Amer.,90, 133-136.

Breyer,J.I.E. , 1979. The application of Remote Sensing techniques for geomorphological terrain classification and terrain feature dynamic mapping in northern Botswana. Unpub. M.Sc thesis ITC Enschede 220 pp.

Breyer,J.I.E. , 1982. Reconnaissance geomorphological terrain classification of the lower Boteti region, Botswana ITC Journal Verstappen Jubilee Edition , 317-324.

Breyer,J.I.E. , 1983. Soils in the lower Boteti region, Botswana. Working Paper No.47 NIR University of Botswana, Gaborone.

Botswana Society, 1976. Proceedings of a symposium on the Okavango Delta and its future utilisation. 350 pp. The Botswana Society Gaborone.

Coates,J.N.M et al., 1979. The Kalatraverse one report. Department of Geogical Survey, Lobatse.

Cooke,H.J. , 1975. The palaeoclimatic significance of caves and adjacent landforms in the Kalahari of western Ngamiland, Botswana. Geogr. J.,141, 432-444.

Cooke,H.J. , 1976. The palaeogeography of the middle Kalahari of northern Botswana and adjacent areas. Proc. Symp. on the Okavango Delta , 21-29. The Botswana Society, Gaborone.

Cooke,H.J. , 1980. Landform evolution in the context of climatic change and neo-tectonism in the middle Kalahari of north central Botswana. Trans. Inst. Br. Geogr. , NS 5 , 80-97.

Cooke,H.J. , and H.Th. Verstappen , 1985. The landforms of the western Makgadikgadi basin in northern Botswana, with a consideration of the chronolgy of evolution of Lake PalaeoMakgadikgadi. Zeit. für Geomorphologie , NF 28, 1-19.

Ebert,J.,and Hitchcock,R.K.,1978. Ancient Lake Makgadikgadi, Botswana: mapping, measurement, and palaeoclimatic significance. Palaeoecol. Afr.,10/11 , 47-57.

Greenwood,P.G.,and Carruthers,R.M., 1973. Geophysical Surveys in the Okavango Delta . IGS Geophysics Div. London.

Grey,D.R.C. , 1976 . The prospecting of the Mopipi area. Unpub. Int. Rep AAC Johannesburg.

Grey,D.R.C. ,and H.J. Cooke, 1977. Some problems in the Quaternary evolution of the landforms of northern Botswana. Catena ,4, 123-133.

Grove,A.T. , 1969. Landforms and Climatic change in the Kalahari and Ngamiland. Geogr. J. ,135, 191-212.

Kalahari Conservation Society, 1983. Proceedings of a symposium, Which Way Botswana's Wildlife , 108 pp. KCS Gaborone.

Lancaster,I.N., 1979. Quaternary environments in the arid zone of

southern Africa. Dept. Geogr. Wits. Univ. Occ. paper No.22
Mallick, D.I.J et al.,1981. A geological interpretation of Landsat imagery and air photography of Botswana. Overseas Geol. and Min. Res. 56 pp35
Scholz,C.H.,et al.,1975. Seismicity, tectonics, and seismic hazard of the Okavango Delta. UNDP/FAO Bot./71/506 proj.rep. UNDP Gaborone, Botswana.
Shaw,P.A. , 1983. Fluctuations in the level of Lake Ngami: the historical evidence. Botswana Notes & Records,15 , 81-85.
Shaw,P.A. , 1984. A historical note on the outflows from the Okavango Delta system. Botswana Notes & Records, 16, 127-131.
Shaw,P.A. , 1985. The dessication of lake Ngami: a historical perspective. Geogr. J. , 151 , 318-326.
Shaw,P.A. , 1985. Late Quaternary landforms and environmental change in northwest Botswana: the evidence of lake Ngami and the Mababe depression. Trans. Inst. Br. Geogr., NS 10 , 333-346.
Shaw,P.A. , 1986. The palaeohydrology of the Okavango Delta, Botswana Proc. of VII SASQUA conf. Palaeoecol. Afr. 17.
UNDP/FAO 1976 Investigation of the Okavango Delta as a primary water source for Botswana BOT/71/506 UNDP/FAO and Government of Botswana. Gaborone.
Wright,E.B., 1978. Geological studies in the northern Kalahari Geogr. J. ,144 , 235-250.
Verstappen,H.Th,and Cooke,H.J.,(Editors). A Drought Susceptibilty Pilot Survey in northern Botswana, 237 pp. ITC Enschede. 1981.

International Geomorphology 1986 Part II
Edited by V. Gardiner

EROSION PIN MEASUREMENT IN A DESERT GULLY

Martin J. Haigh and Gary B. Rydout

Oxford Polytechnic, Oxford, England.
and University of Arizona, Tucson,
U.S.A.

ABSTRACT

Gully evolution is controlled by collapse into underlying soil pipes, rapid retreat of near vertical sidewalls, the development of a rilled, rectilinear, lower slope element, and accumulation in the gully basin. Average annual retreat rates on each sidewall morphological unit during the period 1976-1983 were: vertical free face 9.0mm, rectilinear lower slope 1.6mm, and weakly developed upper convexity 5.5mm. This compared with 2.2mm retreat of the neighbouring undisturbed desert surface and 4.4mm deposition in the floor of the gully basin.

INTRODUCTION

This case study concerns the long term measurement of morphological change in an active desert gully. The gully is tributary to the Greene's Canal arroyo. This large trenched channel has developed from a small diversion canal created between 1908 and 1910. It is cut into Quaternary alluvium deposited between the pediment fringed mountains of a typical desert basin: the lower Santa Cruz flats, southwest of Picacho Peak, in southern Arizona. The study focuses on the evolution of the gully sidewall slopes.

SITE DESCRIPTION

The catchment is located in southern Arizona, some 14.5 km west of Interstate Highway I10 at Red Rock and 10 km southwest of Picacho Peak (32°35'5" N 111°29'30" E). It can be found on U.S.G.S. Samaniego Hills 7.5 Minute Series Quadrangle at SW¼, Sec 1, T.10S., R.8E.

The gully lies on the northeast flank of Greene's Canal. This, the only arroyo on the lower Santa Cruz plains, is the product of a failed irrigation scheme. The canal was constructed to divert the waters of the Santa Cruz river towards a shallow reservoir in the desert. The original canal was 1.5 metres deep, 6.1 metres wide, and had a gradient of 0.15 degrees (Fuller 1913:8).

The canal was never intended to contain the floodwaters of the Santa Cruz. Fuller (1913:28-29), the consulting engineer with the project, expected the natural processes of erosion to adjust the channel gradient and alignment to approximate that of the river at the point of diversion. However, the floods of the winter of 1914-1915, called "the worst for generations" in the lower Santa Cruz (Pierce and Kresnan 1984:1), destroyed the project's storage reservoir (Sonnichsen 1974: 2-1) and incised the canal to 3.7 metres, almost twice the depth of the original river channel (Turner *et al.* 1943, Cooke and Reeves 1976:54).

Despite some early attempts at reclamation, the channel continued to divert the flows of the Santa Cruz, to incise, and to become wider. It also began to develop a fringe of tributary gullies, especially on its upslope, southwestern flank. These gullies, in turn, diverted other local watercourses towards the main channel (Smith 1940: 2-7). The U.S.G.S. 15' topographic survey map of 1946 indicates that, in places, gullying had displaced the contours by 500 to 1,000 metres. The photogrammetric survey of 1976 for the 1:24000 (7.5') Samaniego Hills Quadrangle shows this displacement increased to over 1,500 metres. At the time of commencement of this study in 1976, Greene's Canal was 6 metres deep and around 80 metres wide at its point of confluence with the instrumented gully.

Greene's Canal is cut across an old alluvial fan created by the spreadings of the Santa Cruz river. The canal runs to the southwest and above the level of the modern channel (Figure 1). The alluvium ranges in texture from clayey or silty loam up to gravel. Sands and gravel lenses, marking former Santa Cruz channels, are common in well-log records. So are beds of caliche, a matrix of secondary calcium carbonate deposited by percolating waters (Smith 1940:16).

The soils developed on this alluvium are Typic Torrifluvents perhaps mixed with Aridic Cumulic Haplustolls. They are fine textured loams and clay loams of the Gilman-Glenbar-Trix formation. Wet sieve analyses of gully bank samples show 90% of the soil passing the 63μm sieve and 99% the 210μm sieve. The soils are deep, well-drained, and moderately to moderately-slowly permeable (5-15 mm/hr). They are calcareous (pH 7.9-8.4), but there is no cemented layer. The moist bulk density of the soil is quite low (1.35-1.80 gm/cm^3). However, there are clayey compacted layers of higher density dispersed irregularly through the soil profile.

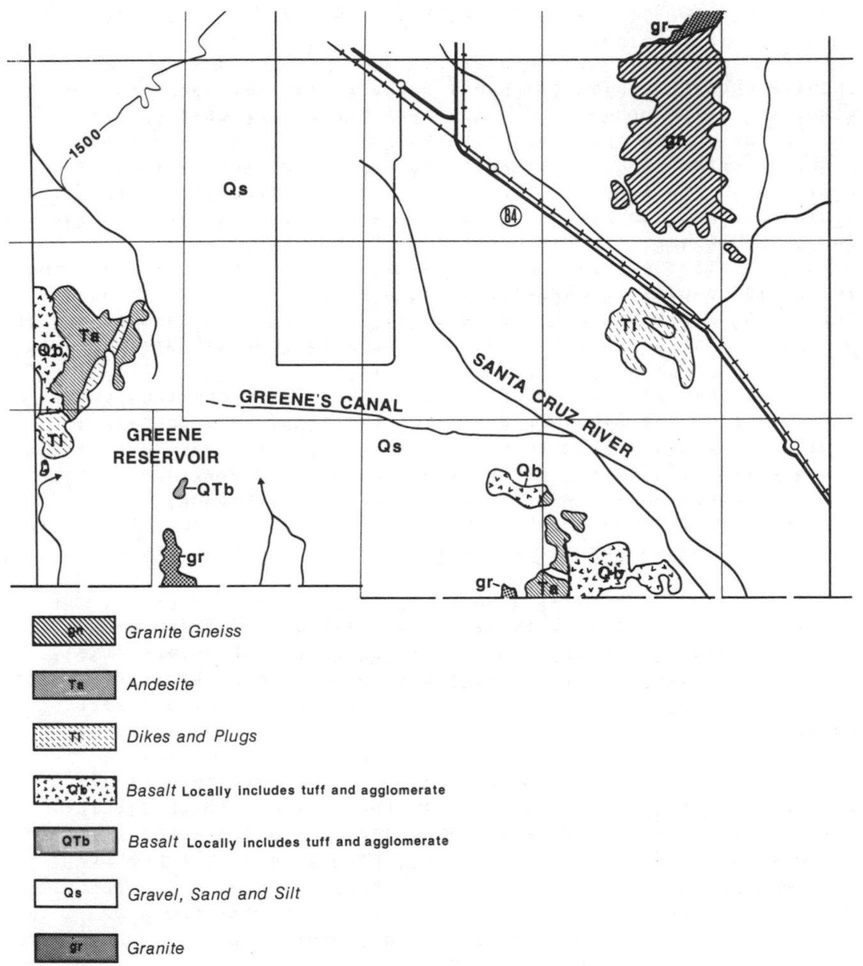

Fig. 1. Greene's Canal (Geology)

The banks of Greene's Canal, into which the instrumented gully is incised, are vegetated with creosote bush (Larrea tridenta). Further inland, some 5 metres beyond the headcut, there are pasture grasses, and some tens of metres beyond this, arable cropland and fallow. Immediately beyond the mouth of the gully in the wall of Greene's Canal, the arroyo floor is colonised by a dense scrub of mesquite (Prosopis sp.) and saltbush (Atriplex polycarpa). The walls and floor of the gully channel, and the flanks of the gully channel between the creosote bushes, supported little vegetation during the study. However, in 1983, grass (30% cover) colonised the floor and northern wall of the channel.

The instrumented gully is one of a series of similar channels extending 10 to 20 metres inland from the arroyo wall on the downslope side of the canal. These gullies are incised against the grain of the land which slopes away gently (0.15°) to the northwest. Consequently, these channels receive almost no surface drainage from the surrounding area since their headcuts are also the water divide.

The gully channel is underlain by a network of large soil pipes. Where they break through into the walls of the main arroyo or into the walls of neighbouring gullies, these triangular soil pipes may be more than two metres from apex to (0.3-0.5 metre wide) base.

The soil pipe network extends well beyond the surface gully area. Collapse hollows with diameters in excess of one metre are common 5 to 10 metres beyond the headcuts of the gullied zone, and smaller zones of subsidence can be found in the fields beyond.

CLIMATE DURING THE STUDY

The Santa Cruz Plains are a desert/semi-desert environment. At nearby Casa Grande there is an 80% probability that the annual rainfall will not exceed 250 mm (Kangieser and Green,1965). At the nearest climatological station (Redrock: 30°49'8" N. 111°36'3" W), rainfall during the 6.7 years of the project was a little higher than average (277 mm/a^{-1} compared to 248 mm/a^{-1}).

During the study, 62% (168 mm/a) of the rainfall came in storms greater than 12.5 mm/dy , which is the critical intensity for application to the gully headcut extension prediction equation of the United States Department of Agriculture (1966). Such events occurred on average 7.65 times each year. The rainfall erosivity factor (R in the Universal Soil Loss Equation) for the area is quite low (75), but the soil is largely unprotected by vegetation and highly erodible (K: 0.32-0.37).

Light frosts were experienced on 11 days/a during the study, but their impact on the soil was very slight. Buol (1964) confirms that even in the winter there is very little natural moisture in the soil and estimates evapotranspiration to be 1100mm/a . So, the main causes of mass movement are shrink/swell, slaking, burrowing and disturbance by ground squirrels, and human trampling.

HUMAN IMPACTS DURING THE STUDY

Running close to, and locally displaced by, the evolving gully network, is an unsurfaced roadway. The gully networks are occasionally used by humans. In the study period, litter, consisting of used cartridge shells and cans, was deposited in and around the gully basin on at least three occasions. It is possible that some soil collapse on the gully walls and the loss of one erosion pin was due to this disturbance.

Beyond the canal-bank track lies irrigated agricultural land. The instrumented gully lies within 20 metres of marginal agricultural land which was cultivated during the earlier years of the project. It is within 950 metres of a major zone of irrigated agriculture which was in production throughout the project.

Irrigation is the key to agriculture in this area. Local farmers operate a cotton/grain rotation which involves the application of water by flood irrigation at the rate of 0.68-0.74 hectare-metres/year during the cotton and about 0.37 hectare-metres/year during the grain rotation. Local studies by Cable (1977) have shown that rainfall events effect remarkably little recharge of soil moisture. Further, the groundwater table lies far below the ground surface. So, it is likely that much of the soil pipe activity in this environment is due to the percolation of irrigation waters.

Irrigation water extraction has one further possible impact on the gully system. Pumpage has created a major zone of groundwater recession and hydraulic subsidence in the Santa Cruz Plains. Compression may have already reduced subterranean pore space by 62,000 hectare-metres and, at its centre, the basin has suffered up to 4 metres of subsidence during the last half century. The site of the instrumented gully may have suffered up to 2 metres of subsidence to 1976. It may have subsided by a further 0.5 metres during the study period and during the process have suffered a slight back tilting. It is unlikely that a differential movement of less than 1mm from mouth to headcut has made much of a geomorphological impact. Nevertheless, Laney et al. (1978) have reported an acceleration of erosion in gullies and drainage ditches whose slope has been positively affected.

MORPHOLOGY OF THE EXPERIMENTAL CATCHMENT

The intention of the project is to monitor ground surface changes in an active desert gully. A particular focus of interest is the evolution of the gully sidewall slopes (cf. Haigh,1985).

The gully selected for study is 15 metres in length from its headcut to its mouth in the wall of Greene's Canal, some 1.7 metres above the arroyo's immediate floor. The channel has an even, gentle south and westerly slope which is broken by two internal headcuts: a degraded 100mm step just 2.2 metres from the main headcut, and an active deep headcut near the mouth of the channel.

Gully Profiles. Gully cross-profiles were surveyed at two metre intervals down the length of the channel from the headcut. The seven profiles were recorded by means of a 150mm slope pantometer. Six complete profiles were surveyed from surface water divide to divide. One profile was abbreviated at the central channel because it coincided with the confluence of a small tributary.

If the new headcut near the gully mouth is ignored, the channel is shown to have a local relief of 0.5 metres at its final headcut and two metres at its mouth. The morphological character of the gully is distinct and quite typical of desert environments (cf. Haigh, 1984). It is depicted in Figure 2, which runs down the channel from its mouth, Profile 1, at the top of the page, to Profile 7 at the headcut, at the foot of the page.

The gully has steep, near vertical sidewalls which abut almost directly onto the undisturbed surface of the Santa Cruz Plains. The sidewalls are best developed in Profiles 1,2,4, and 6. An upper convexity may be almost absent (Profiles 1,4) although in the smaller channel sections, it may be a prominent feature (Profiles 6,7).

However, the development of the upper convexity tends to be minor when compared to that of the rectilinear slope element that joins abruptly with the foot of the vertical sidewall. This element grades from more than 40° to less than 28°, but it tends to preserve a characteristic angle of 34-37°. A diagnostic feature of this slope element is the presence of shallow (10-30mm) rill channels. These features are entirely restricted to this slope zone.

A gently curved (0-4°) slope foot element occupies most of the floor of the gully basin. There may be little (Profiles 2,7) or no evidence (Profiles 3,4,6) of a conventional channel within this basin.

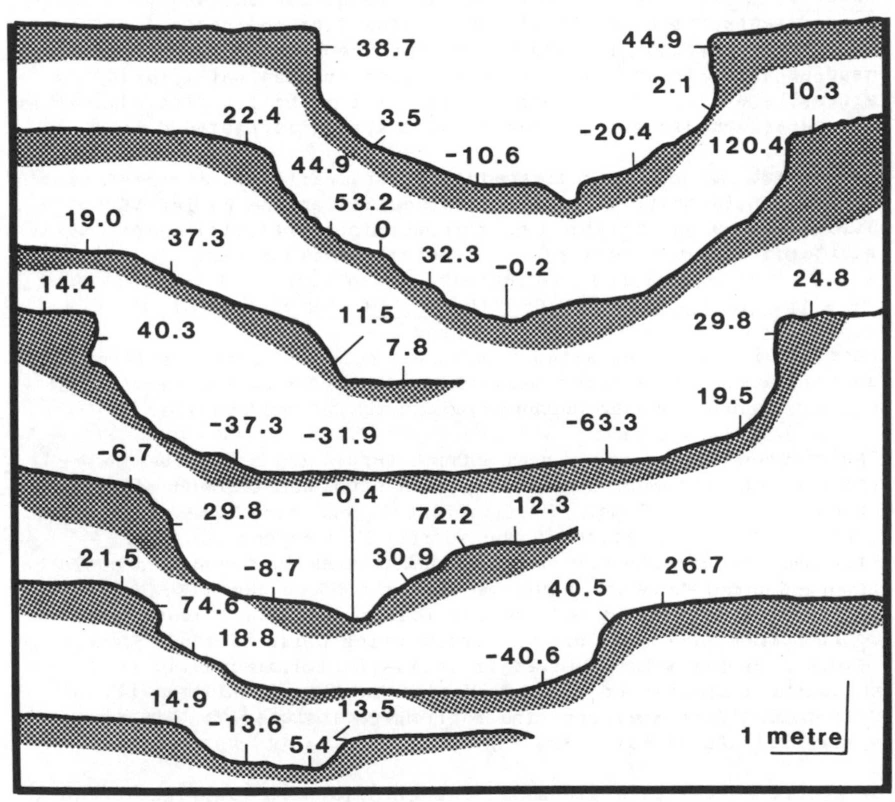

Fig. 2. Morphological changes in a desert gully. Northeast bank, Greene's Canal, Arizona, 1976-1983 (Retreat in millimetres).

In addition to these basic morphological units, several profiles have additional features. These include benches in the sidewalls (Profiles 2,3), which seem to be relics of older gully basin floors, and mounds of debris formed by the collapse of sections of the vertical sidewalls of the basin (Profile 4).

Long Profile. The long profile of the instrumented gully was recorded in a similar fashion by means of a 150mm slope pantometer. Figure 3 depicts the long profile and the location of the measured cross profiles. The long profile preserves a 2° slope through most of its length. However, the upper third of the catchment is steeper and the channel disappears into a short soil pipe section briefly. The lower third of the gully basin also includes a new headcut. This feature developed rapidly during the study. Figure 3 represents its condition in 1981. The feature consists of a shallow foreheadcut incised by concentrated surface wash and a main headcut which feeds directly from a subterranean soil pipe. The sidewalls of the headcut are vertical and unstable. Its channel is irregular because of the repeated deposition of fallen debris.

The desert surface is indicated, but not precisely recorded on Figure 3. In fact, this has an irregular surface relief of 0.10-0.05m with the land being of higher elevation beneath the scattered creosote bushes.

INSTRUMENTATION OF THE EXPERIMENTAL CATCHMENT

Erosion pins were installed along each surveyed cross profile at approximately metre intervals. Care was taken to locate at least one pin in every major morphological unit of the profile.

The erosion pins (Dimensions: 610mm long by 9.5mm diameter) were emplaced on September 20th, 1976. Each pin was allowed an initial exposure of circa 30mm. Changes in this exposure were counted as changes in the elevation of the surrounding ground surface. Provided the erosion pins were not disturbed, such changes could be recorded accurately to about 1mm (Haigh,1977). However, collection of accurate records of erosion pin exposure was not undertaken until October 24th, 1976. The intervening period included several rainfall events which removed the soils disturbed during emplacement and helped the pins to become bedded in. Subsequent data collections were undertaken in March 1977, September 1979, April 1981, April-May 1982, and finally in April 1983.

In addition to their role as reference points for vertical changes in the ground surface, the erosion pins were also used to record changes in the locations of the new headcuts within the gully basin.

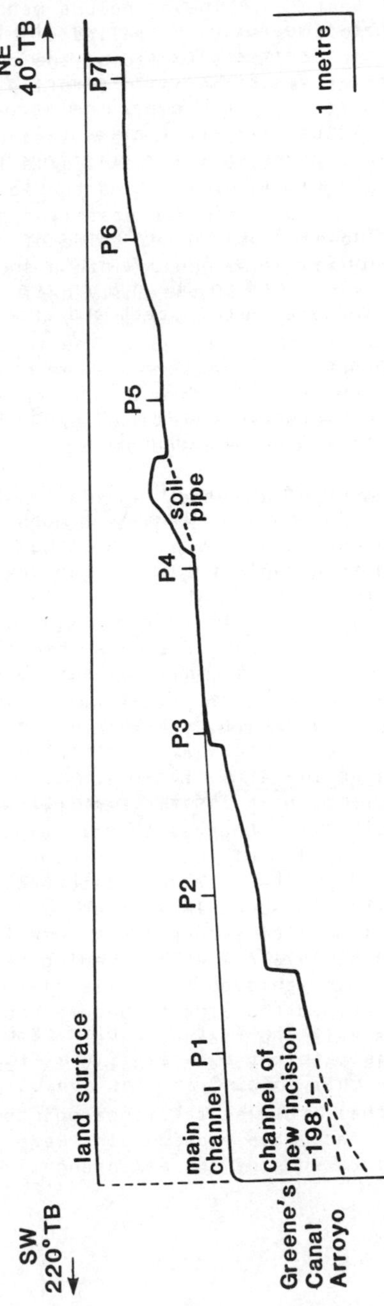

Fig. 3. Long profile of instrumented gully showing new headcut and locations of surveyed cross-profiles.

RESULTS

Total ground retreat measured at each of the surviving erosion pins is recorded adjacent to the erosion pin station on Figure 2. Examination of these results reveals some striking regularities in the retreat rates recorded on comparable morphological units along the different profiles. If these data are averaged across the five major morphological zones, a clear pattern of gully sidewall retreat emerges (Figure 4).

The retreat of the crest sites, undisturbed smooth surfaces surrounding the gully, is slight (14.3mm,S.D. 10.5mm). The variation in these results may be due to the differential shading offered by the creosote bushes scattered across this surface.

Retreat rates on the vertical sidewalls are both high and irregular (59.0mm,S.D. 29.7mm). Most soil leaves these sites by free fall either as individual aggregates, or as clods broken free during shrink-swell or trampling disturbance.

The underdevelopment of the upper convexity is readily explained. Its retreat rate (36.2mm,S.D. 7.9mm) is much lower than that of the free face element below. Consequently, the upper convexity tends to be eliminated more rapidly than it can develop, except in locations where the retreat rate of the sidewall is inhibited.

Free face retreat may inhibit the development of the upper convexity but it promotes the development of the lower slope elements. The essentially rectilinear, transportational slope below the free face undergoes relatively little retreat (10.4mm, S.D. 18.1mm). The high variability of retreat rates recorded on this unit is due to the irregular dumping of fallen debris from above, the differentiation of small parallel finger rill channels, and burial from below.

Retreat rates from the floor of the gully basin are almost all negative. The recorded average (-29.1mm,S.D. 43.2mm) may, however, be something of an underestimation. Three erosion pins were lost by burial during the study and so a result some 5-6mm greater might be indicated.

In sum then, the morphological evolution of the gully was controlled by the parallel retreat of the sidewall free-faces, the infilling of the gully basin, and the evolution of a rectilinear transportational slope linking the two.

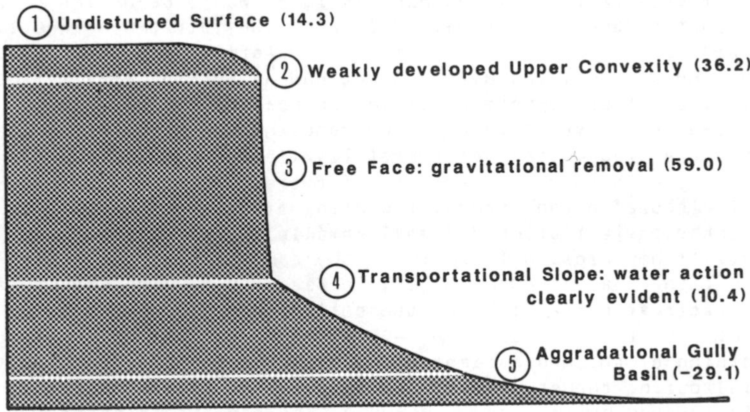

Fig. 4. Pattern of sidewall slope evolution Greene's Canal tributary gully, 1976-1983 (Retreat in millimetres).

Records from each individual data collection are summarized as Table 1. There are large variations in the retreat rates recorded on each morphological zone. However, the pattern of retreat is fairly constant. In four of five records maximum retreat is recorded on the midslope (free face) first, the upper convexity second, the rectilinear lower slope third, and the crest fourth. In all cases ground advance due to deposition is recorded in the gully basin, although as the large deviation scores show, some gully floor sites became involved with new channel incision. Negative retreat scores were also recorded on crest sites, and were responsible for the low free face retreat record in 1981/1982. These results were due, in the main, to expansion of the ground following hydration and on the free face, to the peeling away of sections of the gully wall prior to collapse.

Data in the balance of Table 1 show rainfall and freeze-thaw data for each measurement period. However, there is insufficient evidence to suggest any kind of correlation.

The most dramatic change during the period of the study was the migration of a new headcut through the gully basin. At the start of the project, the headcut was 1.62 metres deep and 0.46 metres wide at the mouth of the gully. By the close, it was 1.08 metres deep and 1.86 metres wide. At Profile 1, it was 0.68 metres deep and 0.32 metres wide in October 1976, and 1.00 metres deep and 1.28 metres wide at the close of the project in April 1983. At the start of the project, the main headcut had a depth of 1.12 metres and was 1.21 metres downstream of the channel pin in Profile 2. Six and two thirds years later, the headcut was just 0.95 metres deep but only 0.58 metres below Profile 2.

During the course of the study, the area of the incision at the mouth of the gully increased from 0.75m² to 2.01m² and that at Profile 1 from 0.22m² to 1.22m². In the same period, the gully basin over the headcut at Profile 1 increased by 0.32m² whilst sidewall retreat (without basal deposition) added 0.36m².

The appearance of the new headcut had relatively little to do with surface erosion, rather, it was due to exposure of a subterranean soil pipe system by collapse. However, the concentration of surface flow caused by the appearance of this cut in the floor of the channel also created a small surface channel headcut (see Figure 3: Profile 3).

TABLE 1. Ground retreat per data collection, and climatic data.

	Erosion pin data					Climatic Data[4]			
Period	Crest	Upper Convexity	Free Face	Rectilinear Lower Slope	Gully Basin	Rainfall (mm)	Days over 12.5mm	Days over 25mm	Frost (Days)
10/76-3/77	-0.7	4.1	4.3	1.2	-2.4	102	96	27	10
3/77-9/79	6.1	6.3	18.6	-0.5	-3.8	726	399	177	17
9/79-4/81	2.1	5.3	9.5	2.6	-9.0	337	176	88	23
4/81-4/82	-0.1	8.4	3.9	1.6	-9.4	247	174	69	15
4/82-4/83	6.9	9.5	22.7	7.1	-4.8	385	265	102	9
(Mean S.D.)[1]	(3.8)	(8.4)	(10.9)	(7.5)	(8.9)	-	-	-	-
10/76-4/83	14.3	36.2	59.0	10.4	-29.1	1797	1110	463	74
(S.D.per pin)[2]	(10.5)	(7.9)	(29.7)	(18.1)	(43.2)	(74)[3]	(56)[3]	(43)[3]	(7)[3]
Annual Averages	2.2	5.5	8.9	1.6	-2.8	25.2	135	66	11

1. Average of deviations about mean of record per morphological unit per record interval (mm)
2. Standard deviation of records per morphological unit for whole study (mm)
3. Standard deviation of annual records about annual average (mm)
4. Climatic data recorded at Eloy (32°34'N 111°31'W) Elev: 486 metres.

Discussion

Surface wash then, plays a minor role in the evolution of this gully system. Soil pipe erosion, pipe exposure by collapse, and the parallel retreat of vertical soil faces are the main erosional agencies responsible for the evolution of this desert gully.

The processes of desert gully exposure have been described by Jones (1968) who worked in Arizona's San Pedro Valley. First, a steep banked arroyo forms in cohesive but erodible sediments which previously lacked a distinct drainage line. This causes an increase in the hydraulic gradient and in the velocity of subsurface flow. These flows pick out lines of weakness in the soil, including desiccation cracks, perhaps developed in response to moisture stresses developed by vegetation. Erosion enlarges these fractures beyond any possibility of closure by swelling. Soil pipes are formed and possibly different layers of soil pipes, as episodic trenching periodically, and abruptly, adjusts base level. The active pipe systems cut backward upslope, especially towards zones of steeper slope and water concentration. However, as the pipes grow, they become more liable to exposure by roof collapse. Gradually, the whole system becomes exposed at the soil surface (cf. Sharma,1980).

Undoubtedly, these same processes are active at Greene's Canal. However here, the process resembles some kind of cut and fill cycle. Soil pipe exposure creates what swiftly become vertical gully sidewalls. These rapidly retreat parallel to themselves, creating a transportational lower slope in their wake, and burying the former pipe channel in sediment. This retreat continues transforming an original high-sided and narrow trench into a wide shallow basin with low, but still near vertical, walls. Surface flows of water tend to diffuse across the floor of the gully basin and there is little incision.

However, ground water, which in this case undoubtedly originates in the main as agricultural irrigation water, still seeps toward the arroyo and excavates soil pipes. The processes of soil pipe expansion continue until, by collapse, the pipe system is again exposed at the soil surface (Figure 5). Harvey (1982) confirms a close correlation between the location of soil pipes and the alignment of surface channels, and attributes this to the properties of the soil at depth.

At some point after exposure, channel burial becomes dominant, due to the rapid retreat of the sidewall free faces and to the redistribution of mobilized debris by relatively unconcentrated wash during rainstorms. These processes give rise to the characteristic morphology of the desert gully. Repetition of the cycle creates the irregularities in the side and long profiles of the flat-bottomed, steep sided, gully basin.

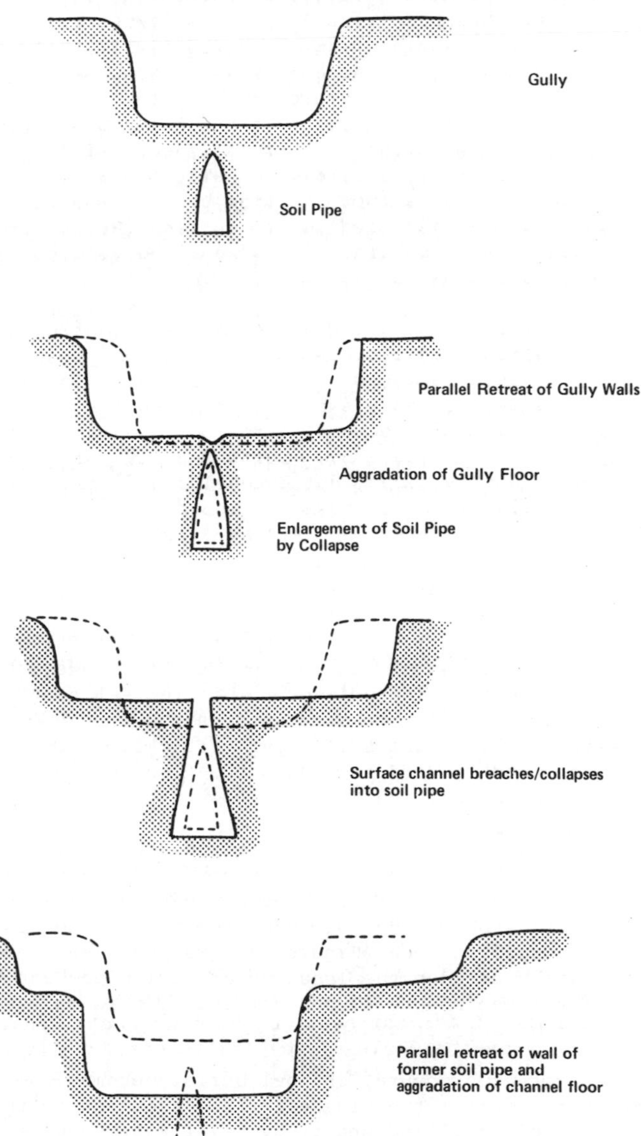

Fig. 5. Evolution of a desert gully

Conclusion

The instrumented tributary gully, evolving on the downslope flank of Greene's Canal, has been created from the exposure, by collapse, of soil pipes carrying water from irrigated fields to the incising arroyo. After exposure, morphogenesis has become dominated by the parallel retreat of the near vertical pipe/gully sidewalls (Retreat: 8.9mm/a), the infilling of the gully basin (Retreat: -2.8mm/a) and the appearance of a rectilinear, rill-scarred, transportational lower slope below the gully sidewall (Retreat: 1.6mm/a). In time, an abrupt junction between the retreating gully sidewalls and undisturbed desert surface (Retreat: 2.2mm/a) may be softened by the appearance of a weakly developed upper convexity (Retreat: 5.5mm/a).

Burial of the original channel does not prevent continued subterranean erosion. Soil pipes continue to form beneath the gully fill, perhaps encouraged by renewed incision in the parent arroyo. Eventually, collapse exposes these newer soil pipes and initiates a new phase of sidewall retreat and channel burial within the older gully basin (Figure 5). Surface gully forms, however, tend to be dominated by slope, not channel processes. The resulting forms closely echo those developed on larger scales in desert landscapes (cf. Carson and Kirkby,1973).

Acknowledgements

Thanks go to the Officers of the Soil Conservation Service, Casa Grande for the soils data, to Dr. R. W. Reeves for introducing us to Greene's Canal, and to Ms. M.V. Harris, Ms. S. Antrobus, and Mr. M. J. Wilkinson for their assistance in fieldwork. Martin Haigh acknowledges the receipt of grants from the University of Chicago and Oxford Polytechnic in support of this project.

Postscript

In October 1983, the area was hit by devastating floods. These were the result of a "30-year intensity storm" which produced a flood stage twice any previously recorded (Saarinen et al,1985). Reich (1984) attributes this phenomenon to upstream reduction in the Santa Cruz basin's capacity for floodwater storage, caused in part by channel trenching. Bank retreat in excess of 20-25 metres affected this part of the northeast bank of Greene's Canal. The erosion carried away the whole of the instrumented gully and neighbouring gully systems, the canal bank road, and parts of the field beyond.

REFERENCES

Buol, S.W., 1964. Calculated actual potential evapotranspiration in Arizona. University of Arizona, Agricultural Experiment Station, Technical Bulletin 162, 48pp

Cable, D.R., 1977. Soil water changes in creosote bush and Bursage during a dry period in southern Arizona. Journal of the Arizona Academy of Science, 12 (1), 15-20

Carson, M.A., and Kirkby, M.J., 1972. Hillslope: Form and Process, pp 338-355, Cambridge Geographical Studies 3, Cambridge University Press

Colorado State Staff, Soil Conservation Service, 1976. Average annual value of factor R. United States Department of Agriculture, Portland, Oregon Map 7-0-23697

Cooke, R.U., and Reeves, R.W., 1976. Arroyos and Environmental Change in the American Southwest, pp 54-55, Oxford Research Studies in Geography/Clarendon Press, Oxford

Fuller, P.E., 1913. Report upon the Santa Cruz Reservoir Project (Collections of the Arizona State Historical Society, Library) 64pp

Haigh, M.J., 1977. Use of erosion pins in the study of hillslope evolution, British Geomorphological Research Group, Technical Bulletin 18, 20-47

Haigh, M.J., 1984. Ravine erosion and reclamation in India. Geoforum, 15 (4), 542-561

Haigh, M.J., 1985. Geomorphic evolution of Oklahoma roadcuts Zeitschrift für Geomorphologie N.F., 29(4), 439-452

Harvey, A.M., 1982. Role of piping in the development of badland and gully systems in southeast Spain (Eds. Bryan, R.B. and A. Yair) Badland Geomorphology and Piping, pp 317-325, Geobooks, Norwich

Jones, N.O., 1968. Development of Piping Erosion (unpublished Ph.D thesis, Department of Geosciences, University of Arizona) 163pp

Kangieser, P.C., and Green, C.R., 1965. Probabilities of precipitation at selected points in Arizona. University of Arizona, Technical Reports on the Meteorology and Climatology of Arid Regions 16, 20pp

Laney, R.L., Raymond, R.H., and Winikka, C.C., 1978. Maps showing water-level declines, land subsidence and earth fissures in South-central Arizona. United States Geological Survey, Water-Resources Investigations 78-83, (Open File Report 2 sheets).

Pierce, H.W., and Kresnan, P.L., 1984. The floods of October 1983. Field Notes from the Arizona Bureau of Geology and Mineral Technology,14 (2), 1-7

Reich, B.M., 1984. Recent changes in a flood series (abstract). Arizona Water Resources Project Information Bulletin, 32,1

Saarinen, T.F., Baker, V.R., Durrenberger, R., and Maddock Jr, T., 1984. Tucson Arizona Flood of October 1983. 105pp (Committee on Natural Disasters, National Research Council, Washington)

Sharma, H.S., 1981. Ravine Erosion in India, pp 49-53, Concept Publishing, New Delhi

Smith, G.E.P., 1940. The ground-water supply of the Eloy District in Pinal County, Arizona. University of Arizona, Agricultural Experiment Station, Technical Bulletin,87, 84pp

Sonnichsen, C.L., 1974. Colonel Greene and the Copper Skyrocket, 325pp, University of Arizona Press, Tucson

Turner, S.R. et al., 1943. Groundwater resources of the Santa Cruz basin, Arizona. United State Geological Survey, Open File Report (in Cooke and Reeves 1976)

United States Soil Conservation Service, 1966. Procedures for determinating rates of land damage and depreciation and volume of sediment production by gully erosion. United State Department of Agriculture, Soil Conservation Service, Technical Release (Geology) 32, 20p

International Geomorphology 1986 Part II
Edited by V. Gardiner

THE ROLE OF SLOPE ANGLE IN SURFACE SEAL FORMATION

J. Poesen

Research Associate, National Fund for Scientific Research, K.U.Leuven, Leuven, Belgium

ABSTRACT

In this paper, two simple surface sealing indices have been used to assess the influence of slope angle on surface seal formation rate: i.e. seal strength, measured with a torvane, and S.I., being the rate at which percolation rate through a sediment layer decreases during a standardized simulated rainfall test. Laboratory and field results clearly indicate that sealing intensity is not only a function of soil (texture) and rainfall properties, but also of surface slope angle, provided that the soil is susceptible to sealing. Surface seal formation becomes less intense as slope steepens essentially because seal development is obstructed by intense rainwash. For aggregated soils other mechanisms add to the explanation of the relation observed. These findings have implications for the measurement of soil erodibility and for runoff modelling on hillslopes.

INTRODUCTION

Surface seal formation, an important process leading to physical degradation of soils (F.A.O., 1979), has been defined as the orientation and packing of dispersed soil particles in the immediate surface layer of the soil, rendering it relatively impermeable to water (Soil Science Society of America, 1984).
Soil surface seals form under the influence of external forces, such as raindrop impact and mechanical compaction, or through slaking and breakdown of soil aggregates during wetting (Moore, 1981). Most research dealing with natural factors of surface seal formation has been focussed on two main factors: either soil characteristics or rainfall properties. However, the role of surface slope angle has not yet been studied in a systematic way. Hence, laboratory experiments and field measurements have been set up in order to detect the role of slope angle in surface seal formation and to elucidate the relevant mechanisms.

THE MEASUREMENT OF SURFACE SEALING INTENSITY

According to the definition of surface sealing, this process leads to an increase of bulk density - and hence to an increase of shear strength (Marshall and Holmes , 1979) - of the top soil and also to a reduction of hydraulic conductivity of the surface layer. Therefore, sealing intensity has been assessed in the laboratory in two different ways. First, shear strength of the 5 mm thick top layer of two loose sediments has been measured in interrill areas with a torvane, equipped with a sensitive vane adaptor, at the end of a simulated rainstorm. This parameter is regarded as an index of seal strength. Secondly, changes of percolation rate through an initially loose 6.5 cm thick sediment layer have been studied during a simulated rainstorm. Since this research aimed at studying structural changes in the sediments' top layer as a function of different surface slopes, for each experiment only the upper 2 cm thick layer was removed 24 h after a simulated rainstorm which lasted long enough to saturate the entire sediment in the plot box. Immediately after this, the top layer was replaced by air-dry loose sediment and a run was made. Typical plots of percolation rate (P) versus rainfall duration (T) for two different sediments are depicted schematically in Fig. 1. From the moment that percolation started, its rate increased rapidly to reach a peak. After that, it decreased more gradual to reach a more or less constant value after a certain time. Attention was given only to the recession phase because the moment at which percolation started and the rising limb of the P-hydrograph were strongly influenced by the pretreatment of the sediment layer at the beginning of each experiment. The recession phase was less influenced by the initial experimental conditions and so reflects more the influence of slope on sealing intensity. The exponential decay of P during the recession phase is entirely attributed to structural changes of the top layer because during this phase the moisture content of the subsurface sediment layer in the tray exceeded field capacity and so the sorptivity of the sediment is negligible. These changes comprised the compaction of the top layer as a result of drop impact stresses and the washing in of the smallest grains into a so-called "washed in" layer or a "filtration pavement" just beneath the surface (Poesen 1981).

From the "percolation rate versus time"-curves, a simple sealing index (S.I.) was deduced: i.e. the rate at which percolation rate through a sediment sample decreases during a simulated rainfall experiment with constant intensity (Fig. 1).

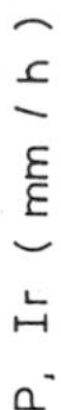
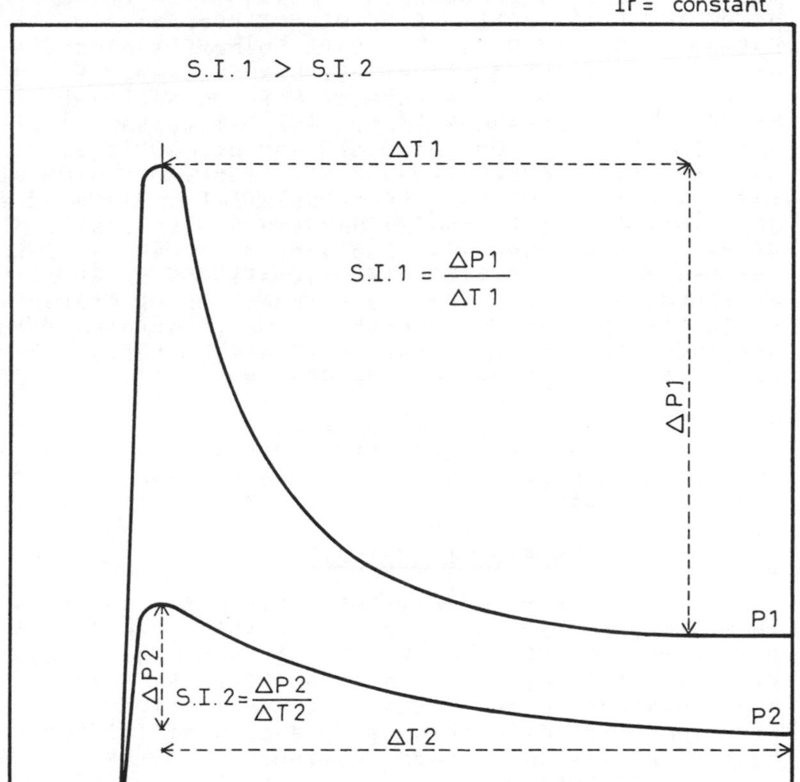

Fig. 1. Definition sketch of sealing index (S.I.) for two different materials (P = percolation rate, Ir = rainfall intensity, T = time, scales are arithmetic).

$$S.I. = \frac{\Delta P}{\Delta T}$$

where

S.I. = sealing index (mm/h/h),

ΔT = time interval (h) between the moment at which P reaches a peak and the moment where P becomes more or less constant,

ΔP = difference in percolation rate (mm/h) corresponding to the above mentioned moments.

For the calculation of S.I., rainfall duration can eventually be replaced by rainfall volume or kinetic rainfall energy.

In the field, sealing intensity has also been assessed by measuring seal strength. In addition, runoff volume produced on 12 small - i.e. with a surface varying between 0.50 and 0.66 m^2 - bare interrill plots, situated on a convex concave slope, has been measured in Huldenberg (central Belgium). Texture of the top soil varies between loam and sandy loam. Initially, the entire experimental plot was ploughed and placed in a conventional seedbed. Rainfall amount, measured with rain gauges having an orifice parallel to the soil surface, did not vary significantly between the interrill plot sites. During the measuring period, moisture content of the top soil was fairly high for all interrill plots, i.e. equal to or superior to field capacity. Therefore, the variation in measured runoff volumes, corresponding to the different interrill plots, is likely to reflect the variation in surface seal development.

The laboratory experimental set-up as well as the procedures used and the field equipment have been described in more detail elsewhere (Poesen,1984; Govers and Poesen,1985).

RESULTS AND DISCUSSIONS

For the laboratory experiments, two different loose sediments, collected in eastern Belgium, were used: a fine sand (89.3 % sand, 2.3 % silt and 8.4 % clay) and a silt (7.7 % sand, 81.5 % silt and 10.8 % clay). These sediments were chosen because of their different granulometric composition and hence their different susceptibility to surface sealing.

TABLE 1. Texture of soils or sediments extremely susceptible to surface sealing

Soil or sediment identification and location	Texture % sand	% silt	% clay
Oakville sand[1] (Indiana, U.S.A.)	94.0	4.0	2.0
Princeton loamy fine sand[1] (Indiana, U.S.A.)	89.0	7.0	4.0
Ferralitic sandy soil[2] (Adiopodoumé, Ivory Coast)	83.6	5.0	11.4
Hamra soil[3] (Sharon plain, Israel)	85.0	2.0	13.0
Loamy sand soil[4] (Jodhpur, India)	84.5	5.6	9.9
Fine sandy sediment[5] (Tongeren, Belgium)	89.3	2.3	8.4

[1]Mannering (1967), [2]Lafforgue and Naah (1976), [3]Morin et al. (1981), [4]Sharma et al. (1983), [5]Poesen (1984)

Fig. 2 shows the influence of sand and silt content for 8 different binary sediment mixtures, obtained after mixing a dune sand (100 % sand) and a calcareous silt (3 % sand, 80 % silt and 17 % clay) in different proportions (Cuyt, 1983), on sealing index. From this figure, it can be seen that a loose sediment consisting of 90 % sand and 10 % silt is the most susceptible to surface sealing. Other field and laboratory results confirm that bare soils or sediments, containing 80 to 94 % sand and 20 to 6 % silt and clay, are extremely susceptible to surface capping (Table 1). Soils and sediments with such a granulometric composition seem to undergo the strongest clogging of the pores in the filtration pavement layer, situated immediately below the surface (Poesen, 1981).

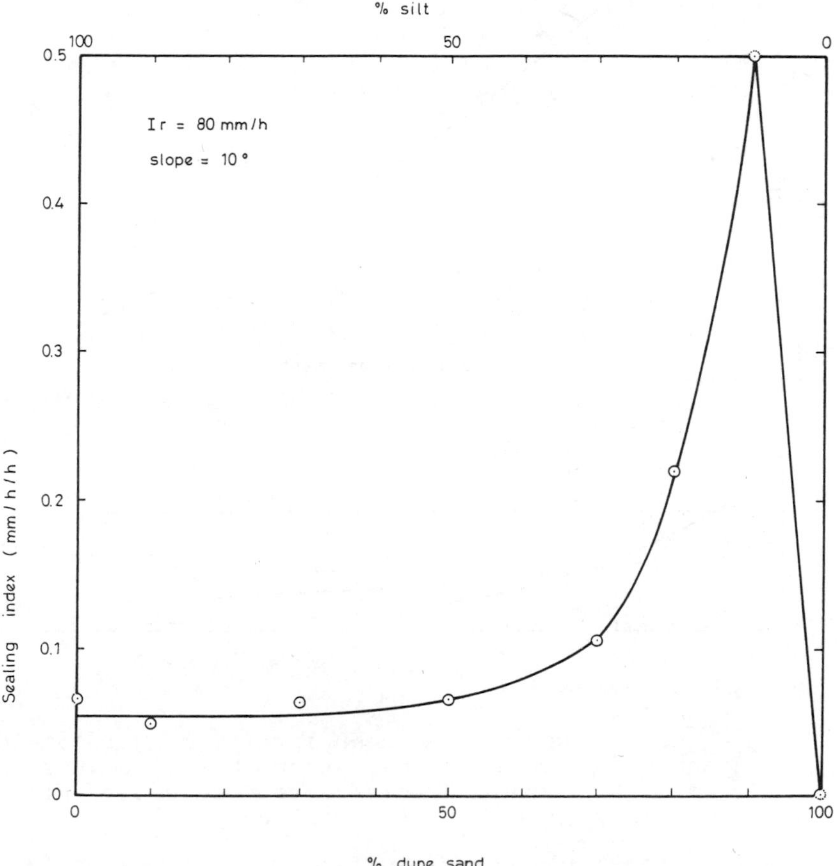

Fig. 2. Sealing index versus sand and silt content of binary sediment mixtures.

Fig. 3 shows the laboratory and the field data on the relation between surface slope angle and mean seal strength of the interrill top layer. As interrill slope steepens, seal strength decreases, suggesting that surface seals are less well developed on steep slopes compared to low slope angles.

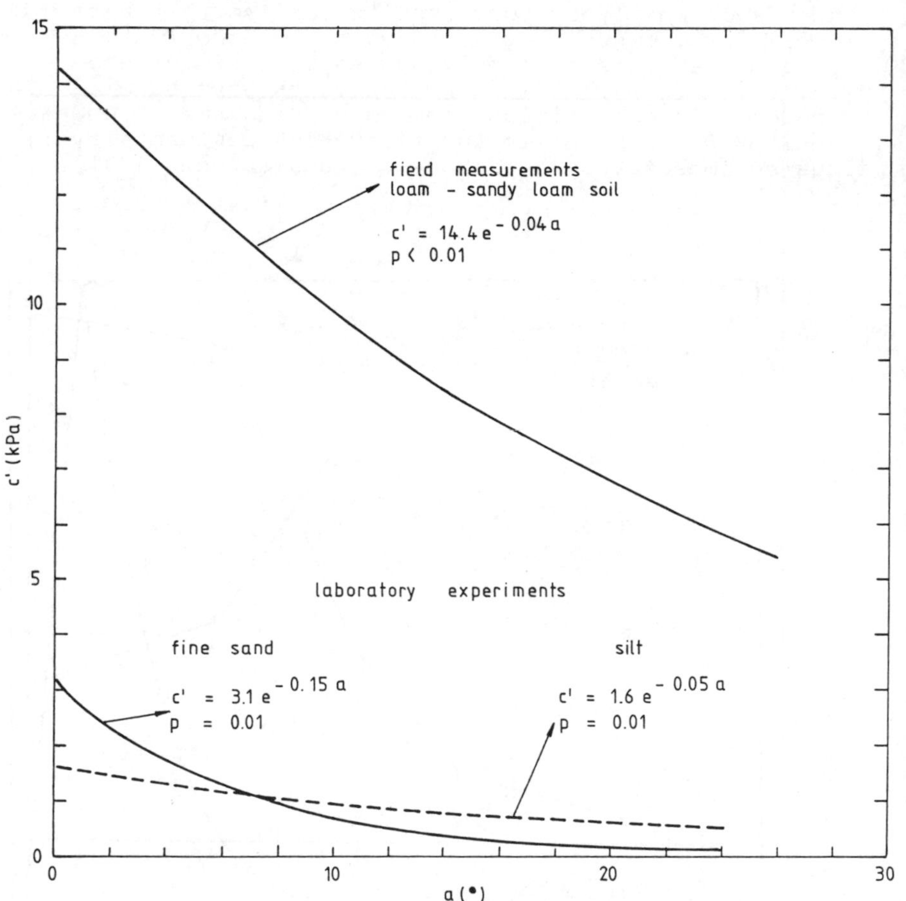

Fig. 3. Relation between interrill surface slope angle (a) and mean seal strength (c') of the 5 mm thick top layer. Water content exceeds field capacity (p = significance level).

In Fig. 4. the influence of surface slope angle on sealing index, deduced from laboratory "percolation rate versus time"-measurements, is shown. From this figure, it can be seen that sealing intensity is inversely

proportional to surface slope angle: i.e. the steeper the slope, the smaller the intensity of surface seal formation. The main explanation for this relation is given by the corresponding increase in interrill and rill erosion. The high erosion rates on the steep slopes prevented the development of a surface seal and through that infiltration remained high. This is confirmed by the significant negative relation found between sealing index and erosion rate:
S.I. =-0.12 (E.R.) + 12.1 with r^2 = 0.97 (Fig. 4).

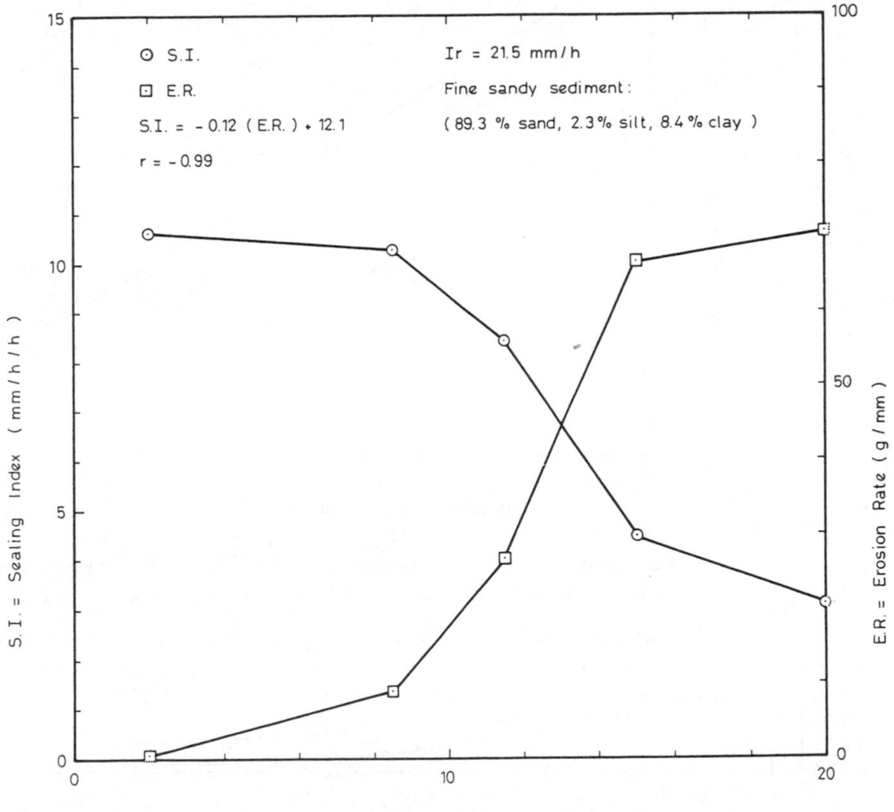

Fig. 4. Influence of surface slope angle on sealing index and erosion rate, measured at the end of a 150 min rainfall experiment.

Finally, Fig. 5. shows the field measurements on the relation between surface slope angle and interrill runoff volume: the steeper the slope, the lower the produced runoff volume. It is likely that some scatter of these field data is due to differences in soil physical characteristics between the plots (Govers and Poesen,1985). Nevertheless, these data corroborate the laboratory findings.

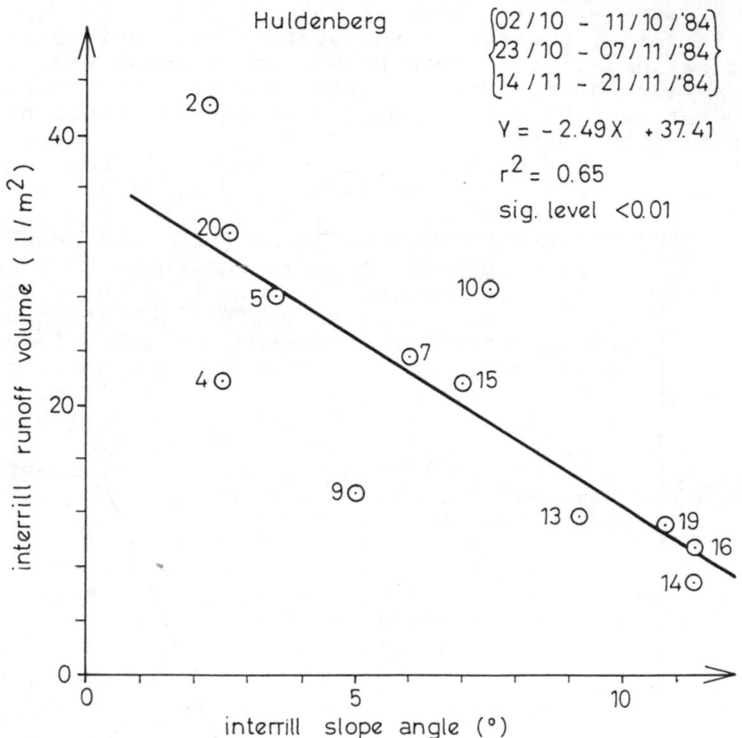

Fig. 5. Relation between surface slope angle and interrill runoff volume, measured in the field.

From the laboratory and field results it is concluded that surface seal formation rate on loose sediments as well as on soils is inversely proportional to slope angle. On steeper slopes the formation of a surface seal is less intense because of several mechanisms. These are:

- the high splash and sheet erosion intensity on interrill areas for the steepest slopes, through which the top layer is eroded constantly (Poesen,1985);
- the increase of rill density and rill depth with increasing slope: hence, sealed interrill surface decreases while exposed sediment surface through which water can infiltrate, increases;
- the smaller amount of drop impacts per surface unit for the steeper slopes when rain falls vertical;
- a decrease of the normal component of drop impact force with increasing slope when rain falls vertical (cosine effect);
- the greater opportunity for the occurrence of a thin

water layer, covering the soil surface, on low slopes through which the compactive force of the impacting raindrop is increased (Poesen,1983).

For soils, characterized by structural units, i.e. aggregates, other mechanisms add to the explanation of the relation observed. Moisture condition of the top soil determines to a large extent aggregate stability when soil water matric potential is high: the closer the water content approaches saturation, the lower the aggregate stability will be (Francis and Cruse,1983). Surface water storage on flatlands will be greater for the same surface roughness found on steeper lands (Stammers and Ayers, 1958). This condition enhances aggregate breakdown more on flat to gentle sloping surfaces than on steeper ones. Low slopes on flatlands also differ from steep slopes because of more difficult internal drainage (Poesen,1984). Soil moisture contents are often higher, and water tables are often high during the cooler seasons (Mutchler and Murphree,1981). As demonstrated by Boekel (1974), these conditions also affect aggregate stability and hence surface sealing. On flatlands, detached micro-aggregates and primary particles from aggregates are not easily evacuated and so, these sediments contribute to the rapid formation of depositional seals (Boiffin,1984). On steep slopes, however, this sediment is more readily removed either by splash transport (Poesen,1985) or by interrill flow and so the development of a depositional seal will be retarded on these slopes.

CONCLUSIONS

The two proposed surface sealing indices, i.e. seal strength and S.I., are reliable and objective measures in order to express surface sealing intensity under different environmental conditions. Consequently, they can be used to evaluate the effectiveness of different techniques, e.g. tilling, mulching, ... in reducing surface sealing intensity.

Laboratory and field data clearly show that sealing intensity is not only a function of soil and rainfall properties, but also of slope angle, provided that the soil is susceptible to sealing. Surface seal formation becomes less intense as slope steepens. Therefore, we recommend to study the susceptibility of different soils to surface sealing in the laboratory on a low surface slope angle.

Seal strength of the top layer determines to a large extent the erodibility of sediments (Poesen,1981). Since seal strength is a function of surface slope angle, it is very likely that relative erodibility classifications of soils with a different sealing susceptibility will be a function of slope angle.

Susceptibility of a given soil surface to rill initiation is inversely proportional to soil shear strength (Savat, 1979; Rauws pers.,comm.). Hence, the negative relation between surface slope angle and seal strength can be an additional explanation for the observed positive relation between slope angle and rillability.

The development of surface seals is very important with respect to the reduction of a soil's infiltrability and to the production of Horton overland flow. As a consequence, our results have implications for runoff modelling on hillslopes (Fig. 6; De Ploey,1985) and for the spatial variability of runoff erosion processes such as rill initiation (Poesen and Govers,1985).

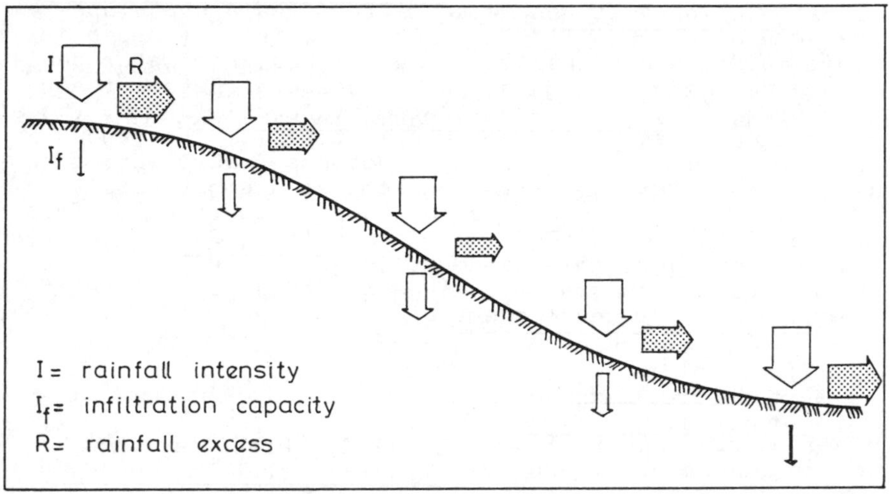

Fig. 6. Sketch to illustrate the influence of local slope angle on infiltration rate for a soil susceptible to surface sealing and on rainfall excess during rainfall. The effect of slope length is not taken into account. Width of arrows is proportional to the intensity of the hydrological processes.

REFERENCES

Boekel, P., 1974. De betekenis van de ontwatering voor de bodemstructuur op zavel- en lichte kleigronden en de financiële consequenties daarvan. Bedrijfsontwikkeling, 5, 875-880.

Boiffin, J., 1984. La dégradation structurale des couches superficielles du sol sous l'action des pluies. Unpubl. Ph. D. Thesis, I.N.R.A. Paris-Grignon, 320 p.

Cuyt, G., 1983. Runoff en erosie op mengsels van zand en loess. Unpubl. M. Sci. Thesis, K.U.Leuven, 220 p.

De Ploey, J., 1985. Experimental data on runoff generation, in Soil Erosion and Conservation (Eds. El-Swaify, Moldenhauer and Lo), pp 528-539. Soil Conservation Society of America, Ankeny.

F.A.O., 1979. A provisional methodology for soil degradation assessment, Rome, 84 p.

Francis, P. and Cruse, R., 1983. Soil water matric potential effects on aggregate stability. Soil Science Society of America Journal, 47, 578-581.

Govers, G. and Poesen, J., 1985. A field-scale study of surface sealing and compaction on loam and sandy loam soils. Part I. Spatial variability of soil surface sealing and crusting, in Assessment of Soil Surface Sealing and Crusting (Eds. De Boodt, Gabriels and Callebaut), Gent, in press.

Lafforgue, A. and Naah, E., 1976. Exemple d'analyse expérimentale des facteurs de ruissellement sous pluies simulées. Cahiers ORSTOM Série Hydrologie, 13, 195-237.

Mannering, J.V., 1967. The relationship of some physical and chemical properties of soils to surface sealing. Unpubl. Ph. D. Thesis, Purdue University, 207 p.

Marshall, T. and Holmes, J., 1979. Soil Physics. Cambridge University Press, 345 p.

Moore, I., 1981. Effect of surface sealing on infiltration. Transactions of the ASAE, 24, 1546-1552, 1561.

Morin, J., Benyamini, Y. and Michaeli, A., 1981. The effect of raindrop impact on the dynamics of soil surface crusting and water movement in the profile. Journal of Hydrology, 52, 321- 335.

Mutchler, C. and Murphree, C., 1981. Prediction of erosion on flatlands, in Soil Conservation. Problems and Prospects (Ed. Morgan), pp. 321- 325, Wiley, Chichester.

Poesen, J., 1981. Rainwash experiments on the erodibility of loose sediments. Earth Surface Processes and Landforms, 6, 285-307.

Poesen, J., 1983. Regenerosiemechanismen en bodemerosiegevoeligheid. Unpubl. Ph. D. Thesis, K. U. Leuven, 368 p.

Poesen, J., 1984. The influence of slope angle on infiltration rate and Hortonian overland flow volume. Zeitschrift für Geomorphologie N.F., Suppl.-Bd. 49, 117-131.

Poesen, J. 1985. An improved splash transport model. Zeitschrift für Geomorphologie, 29, 193-211.

Poesen, J. and Govers, G., 1985. A field-scale study of surface sealing and compaction on loam and sandy loam soils. Part II. Impact of soil surface sealing and compaction on water erosion processes, in Assessment of of Soil Surface sealing and Crusting (Eds. De Boodt, Gabriels and Callebaut), Gent, in press.

Savat, J., 1979. Laboratory experiments on erosion and deposition of loess by laminar sheet flow and turbulent rill flow, in Seminar on Agricultural Soil Erosion in Temperate Non Mediterranean Climate (Eds. Vogt and Vogt), pp.139-143.

Sharma, K., Singh, H., and Pareek, O., 1983. Rainwater infiltration into a bare loamy sand. Hydrological Sciences Journal, 28, 417-424.

Soil Science Society of America, 1984. Glossary of Soil Science Terms. Madison, Wisconsin, 38 p.

Stammers, W. and Ayers, H., 1958. The effect of slope and microtopography on depression storage and surface detention. Publication I.A.H.S., 45, 89-94.

International Geomorphology 1986 Part II
Edited by V. Gardiner

DRAG COEFFICIENTS OF SINGLE CROP ROWS AND THEIR IMPLICATIONS FOR WIND EROSION CONTROL

R.P.C. Morgan and H.J. Finney

Department of Agricultural Engineering, Silsoe College, Silsoe, Bedford MK45 4DT, UK.

ABSTRACT

The effectiveness of single crop rows in reducing wind velocities below the threshold levels for soil particle movement can be evaluated by calculating the frictional drag coefficients of the crop in the lower 0.05 m of the atmosphere from wind speeds measured in the field with cup anemometers. Studies made with winter wheat, spring barley, potatoes, sugar beet, peas, broad beans, oilseed rape, carrots and onions, during their growth period from 0.02 to 0.30 m height, and with 0.15 m high planted straw strips show that the drag coefficients generally decrease with increasing wind speed. However, under conditions of low wind speed variability, drag coefficients increase with increasing wind velocity. When these conditions are combined with newly emergent crops with low densities of biomass, they cause an increase in drag velocity and therefore a greater risk of wind erosion. Onions and sugar beet in particular can enhance erosion risk in this way.

INTRODUCTION

It is generally accepted that a crop cover decreases wind velocity as a result of drag imparted to the air flow and that this reduces wind erosion by decreasing the frequency at which threshold velocities for soil particle movement are exceeded. Previous studies have concentrated on the bulk effect of a plant cover expressed by either a roughness coefficient (z_o) or a bulk drag coefficient (C_D) (Table 1). Neither of these parameters explicitly indicates conditions at or close to the ground surface. This paper examines the effect of different crops on wind speeds close to the soil surface by calculating the frictional drag coefficient (C_d) for the lower 0.05 m of the atmosphere within the crop. The coefficients are determined for single crop rows in order to evaluate the effectiveness of different crops when planted in strips for in-field wind erosion control. Data were obtained for winter wheat, spring barley, potatoes, sugar beet, peas, broad beans, oilseed rape, carrots, onions and planted strawstrips.

TABLE 1. Roughness measures for crops.

1. HEIGHT OF PLANE OF ZERO WIND VELOCITY, SOMETIMES CALLED THE ROUGHNESS LENGTH (z_o)

   From the equation

   $$u(z) = (2.3/k)\ v_* \log (z/z_o) \tag{1}$$

   where u is the mean wind velocity at height, z, k is the von Karman universal constant for turbulent flow, usually taken as = 0.4, and v_* is the drag or shear velocity.

2. BULK DRAG COEFFICIENT (C_D) EXPRESSING DRAG EXERTED ON THE LOWER ATMOSPHERE ABOVE HEIGHT z_o

   Balancing the drag force of the wind profile (τ) against the extraction of momentum by surface friction between the atmosphere the plant elements and the soil surface gives

   $$\tau = \tfrac{1}{2}\ \rho\ C_D\ \bar{u}\ (z)^2 \tag{2}$$

   where is the fluid density and $\bar{u}$ is the mean wind speed at height z.

   Since $\tau = \rho\ v_*^2$,

   $$C_D = \frac{2\ v_*^2}{\bar{u}\ (z)^2} \tag{3}$$

3. DRAG COEFFICIENT (C_d) EXPRESSING THE FRICTIONAL DRAG EXERTED BY THE PLANT ELEMENTS AND THE GROUND DURING THE TRANSFER OF MOMENTUM WITHIN A PLANT CANOPY

   Balancing the drag force of the wind profile (τ) for the ground surface and a plant cover of canopy height (h) with the extraction of momentum due to the frictional surface area of the ground and the plant elements gives

   $$\tau\ (h) = \tfrac{1}{2}\ \rho \int_{z_o}^{z} h\ C_d\ A\ (z)\ u\ (z)^2\ dz \tag{4}$$

   where A is the leaf area per unit volume.

   With $\tau = \rho\ v_*^2$, this gives

   $$C_d\ (z) = \frac{2\ v_*^2}{\int_{z_o}^{z} A(z)\ u(z)^2\ dz} \tag{5}$$

   This equation is applied in this paper to a single layer with z = 0.05 m.

METHODOLOGY

Wind velocities were measured in the field with ten digital cup anemometers (Type HA/6A/151 available from F. Darton & Co. Ltd., but modified at Silsoe College so that all ten could be operated simultaneously from a single control point). Each anemometer comprises three, 35 mm diameter, black acetal cups and gives a digital read out of the average wind velocity over a 10 s period. Velocity recordings were taken 0.1 m in front of a crop row, within the crop row and 0.1 m behind the row. Measurements were made at heights of 0.05 m, 0.10 m (within the crop only), the top of the crop canopy and 1.0 m. Only the data for the 0.05 m height in front of and within the crop are relevant to this paper. Readings were obtained at approximately two-weekly intervals between February and June for two years whilst the crops grew from 0.02 m to 0.30 m in height. With further growth, the wind velocity at 0.05 m was generally reduced to zero.

Between 50 and 60 consecutive 10 s recordings were made for each measurement point for each crop at each period of observation. These were then grouped into wind speed classes at 1 knot (approximately 0.5 m s^{-1}) increments according to the velocity measured at 1 m height in front of the crop. Where three or more recordings for a given wind speed class were obtained, the average velocities were calculated for each measurement position for that wind speed class for that period of observation.

The frictional drag velocity of the air moving through the crop in the lower 0.05 m of the atmosphere was calculated for each wind speed class for each period of observation from the average wind velocity recorded at 0.05 m height within the crop. Calculations were based on equation 1 (Table 1) assuming a zero wind velocity at a height of 0.05 mm ($z_o = K_s/30$ where K_s is the equivalent grain roughness height, approximated by the grain diameter = 1.5 mm for a sandy loam soil). Generally wind speeds within a crop rise rapidly from the ground surface to a value which then remains nearly constant with height until the canopy level is reached (Monteith, 1973). Although equation 1 is normally applied to changes in wind velocity with height above the canopy, data from wind tunnel studies with simulated crops (Table 2) show that it can be used to describe the rapid rise in velocity within a crop close to the ground surface. Where the wind direction was not at right angles to the crop row, the velocity readings for positions below canopy level were first multiplied by cosine θ, where θ is the angle between the wind direction and a line perpendicular to the crop row. Velocity measurements above canopy level were not corrected in this way because cup anemometer records in open ground are generally unaffected by wind direction.

The variability in wind velocity (T) was calculated for each period of observation from consecutive 10 s velocities recorded at 1 m height in front of the crop using the index $T = \sigma/\bar{u}$, where σ is the standard deviation of the velocity readings and $\bar{u}$ is the mean wind velocity.

TABLE 2. Comparison of predicted wind velocity based on the logarithmic law (equation 1) with measured wind velocity for a height of 0.05 m with different crop covers.

Crop Cover	Predicted Wind Velocity (u_p)	Measured Wind Velocity (u_m)	$\frac{u_m}{u_p}$
Potatoes (5 cm tall)	5.67	5.38	0.95
Potatoes (15 cm tall)	7.30	7.38	1.01
Broad Beans (5 cm tall)	5.88	6.04	1.03
Broad Beans (10 cm tall)	6.85	6.38	0.93
Onions (5 cm tall)	6.28	6.12	0.97
Onions (10 cm tall)	5.67	5.90	1.04
Onions (15 cm tall)	5.40	5.72	1.06
Straw Strip (15 cm tall)	7.10	6.56	0.92

Predicted wind velocities (m s^{-1}) are based on a calculated value of z_o = 0.027 mm for the wind tunnel and measured wind velocities at 5 mm height.

The area of the crop facing the wind was measured with a dot planimeter on a photograph taken facing the crop row and against a background of a white board and a 20 mm x 20 mm mesh screen. Leaf area densities (m^2 m^{-3}) were calculated using the projected area of the crop contained in the volume represented by the lower 0.05 m of the atmosphere, a typical 0.10 m length of the crop row and the crop row width. The density of biomass (kg DM m^{-3}) was also determined for the same volume. Values of the frictional drag coefficient (C_d) were then calculated using equation 6 in Table 1. These values combine the effects of the plant cover and soil surface roughness. Differences in the values relate mainly to the plant cover, however, because soil surface roughness was remarkably uniform. Roughness levels expressed as a ratio of the distance between two points over the ground surface to that in a straight line were between 1.03 and 1.05 for all crops apart from the straw strips (1.12) and potatoes (1.15 to 1.24).

RESULTS

The frictional drag coefficients varied from 0.00009 to 3.41 for wind speeds at 0.05 m height in front of the crop ranging from 0.03 to 4.22 m s^{-1}. The wind speeds measured at 1 m height in front of the crop ranged from zero to 9.8 m s^{-1}.

For a wide range of conditions the drag coefficients decrease in value with increasing wind speed in a power function relationship with an exponent approximating -2.0 (Figures 1 and 2; Table 3). The data following this trend fall into two groups according to the density of biomass. The data for the lower biomass densities, < 0.65 kg DM m^{-3}, plot closer to the top right of the graph. This is seen in Figure 1 for onions in comparison with planted straw strips which have biomass densities greater than 21 kg DM m^{-3}. The top right of the graph is occupied by data points which also coincide with drag velocities greater than 0.13 m s^{-1}, a value which is reasonably representative of the threshold for soil particle movement (Savat, 1982). Figure 2 shows that the data for peas and sugar beet plot particularly close to this danger zone. For biomass densities greater than 0.65 kg DM m^{-3}, the data for sugar beet, broad beans and potatoes plot to the right of those for other crops, except that when the biomass for potatoes exceeds 4 kg m^{-3}, the points plot with the others (Table 2).

Under certain circumstances the frictional drag coefficients were found to increase with increasing wind velocity. This occurred when the levels of wind speed variability (T) were less than a threshold value, the magnitude of which was crop specific. For bladed-leaved crops such as onions, winter wheat and planted straw, a T value of about 0.20 appeared to be critical, but for larger-leaved crops like carrots, peas, sugar beet, broad beans and potatoes, the threshold value was about 0.30. As seen from Table 2, however, the threshold levels cannot be defined precisely.

The positive relationship between the drag coefficient and wind speed means that as wind velocity increases, the data plot closer

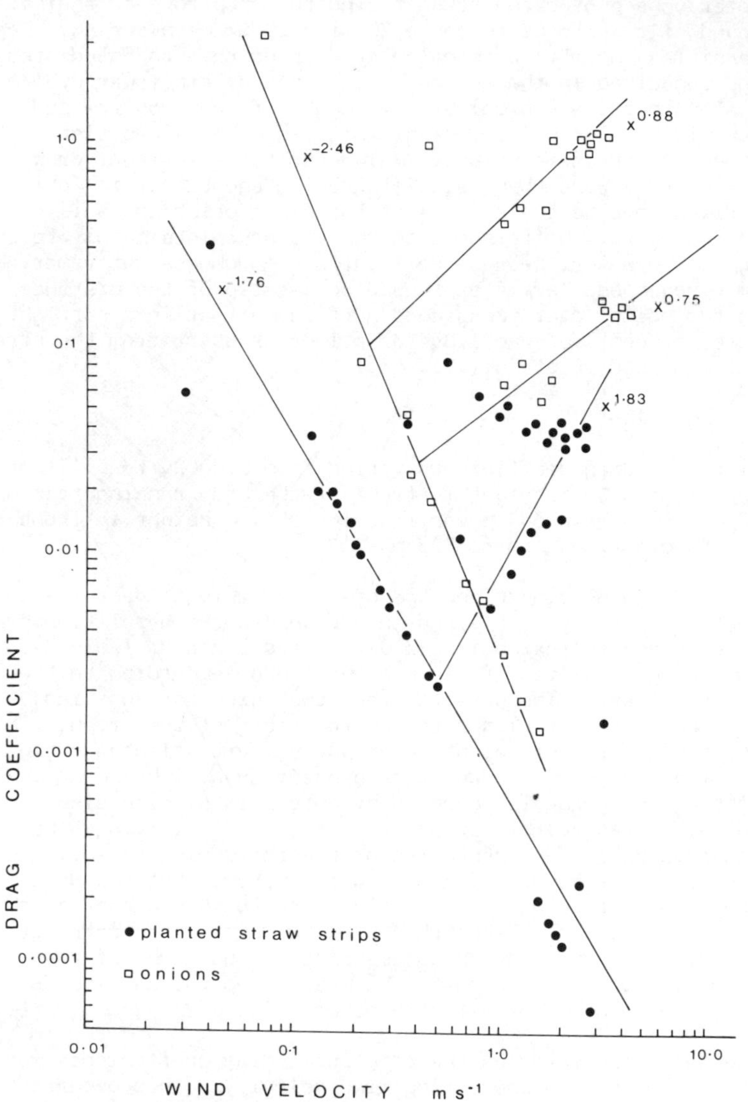

Figure 1. Relationships between drag coefficients (C_d) and wind velocity at 0.05 m height for onions and planted straw strips.

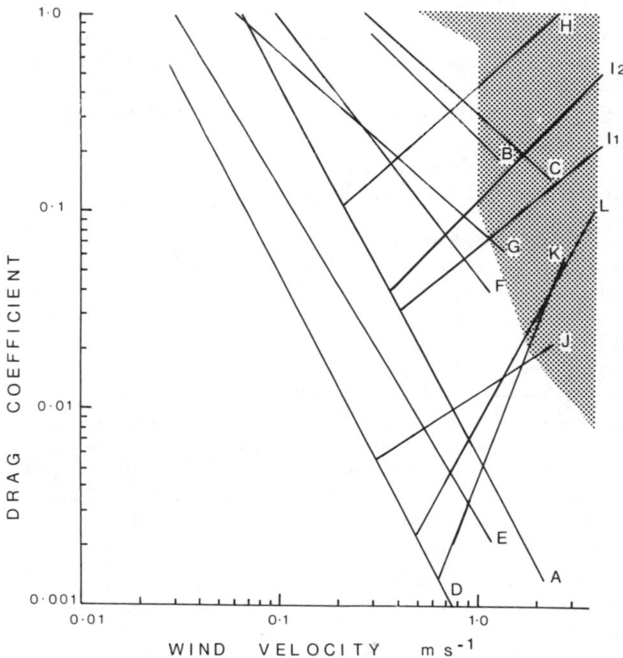

Figure 2. Plots of the regression lines listed in Table 3 showing the relationships between drag coefficients (C_d) and wind velocity at 0.05 m height. Shaded area encloses data points coincident with drag velocities > 0.13 m s^{-1}.

TABLE 3. Relationships between drag coefficients (C_d) and wind velocity (u) at 0.05 m height.

Group	Crop	Conditions	a	b	n	r
A	Winter wheat Onions Potatoes Spring barley Carrots	BM<0.65 T>0.20 L>5 BM<0.65 T>0.20 BM<0.65 BM<0.65 BM<0.65 T>0.30	0.0056	-1.92	36	-0.815
A1	Onions	BM<0.65 T>0.20	0.0030	-2.46	10	-0.991
B	Peas	BM<0.65 T>0.30	0.2366	-0.99	6	-0.946
C	Sugar beet	BM<0.65 T>0.30	0.3290	-0.90	6	-0.904
D	Planted straw Winter wheat Peas Potatoes Spring barley Oilseed rape	BM>22.0 BM>0.65 T>0.18 BM>0.65 T>0.30 BM>4.00 T>0.30 BM>0.65 BM>0.65	0.00059	-1.91	67	-0.976
E	Sugar beet	BM>0.65 T<0.20	0.0028	-1.69	11	-0.917
F	Broad beans Broad beans	BM>0.65 T>0.30 L<25 BM>0.65 L>25	0.0485	-1.28	13	-0.979
G	Potatoes	BM>0.65<4.00 T>0.30	0.0810	-0.87	11	-0.532
H	Onions	T≤0.23 L≤5	0.4020	0.87	10	0.885
I	Onions Winter wheat Carrots Planted straw Sugar beet	T<0.20 L>5<10 T<0.20 L>5<10 T<0.30 L>5<10 T<0.20 BM<22 T<0.30 L≤20	0.0810	0.32	59	0.237
I1	Onions	T<0.20 L>5<10	0.0560	0.75	9	0.893
I2	Sugar beet	T<0.30 L≤10	0.1070	1.02	23	0.443
J	Potatoes Broad beans	T<0.30 T<0.30	0.0117	0.72	13	0.637
K	Peas	T<0.30 L>10<20	0.0043	2.59	9	0.924
L	Planted straw	T<0.20 BM>22	0.0083	1.83	15	0.801

Notes: BM = biomass density (kg DM m^{-3}), T = wind speed variability index (see text), L = leaf width (mm), a and b are the regression coefficient & exponent in the relationship $C_d = a u^b$., n = number of observations in sample, r = correlation coeff.

to the danger zone on the top right of the graph (Figure 2). Location of the data points, however, seems to relate, with the exceptions of sugar beet and planted straw strips, to the width of the plant elements, namely stems and leaves. As the width of these in the lower 0.05 m increases, there is a movement from the top to the bottom of the graph. This is accompanied by a change in the value of the power function exponent from about 0.9 to 2.6. The strength of the correlations between drag coefficients and wind speed is less for the positive relationships than for the negative ones (Table 2). This is a reflection of the greater scatter of the data points, as shown in Figure 1 for onions and planted straw strips.

DISCUSSION

The decrease in drag coefficients with increasing wind speed conforms with the expected relationship. It presumably reflects the greater tendency for streamlining effects as the leaves of the plants point downwind. The values of the exponents for a large range of crops confirm that the drag coefficient varies with the wind velocity to the power of 0.5 (Monteith, 1973).

Instances of increases in drag coefficient with increasing wind velocity have been noted for bulk drag coefficients (C_D) for maize (Wright and Brown, 1967) and rice (Uchijima, 1976). These have been explained by disturbance of the surrounding air through leaf movement which creates a wall effect. This explanation seems likely for the drag coefficients (C_d) calculated here since it is compatible with an association of low levels of wind speed variability which means that the leaves would rarely be still. The results of the study indicate, however, that the wall effect is not transmitted to the atmospheric layers above 0.05 m. A high frictional drag and consequent reduction in windspeed close to the ground with no change in velocity above brings about an increase in the rate of change in velocity with the logarithm of height and therefore in the drag velocity. This explains how high drag coefficients and high windspeeds can be combined with high drag velocities for those data points falling in the top right of the graphs (Figures 1 and 2). It should be noted that if the exceedance of the critical drag velocity results in particle movement the relationships based on equation 1 in this paper may not hold. No soil movement occurred whilst wind velocity was being recorded in this study, probably because the soil was partly compacted by tractor wheelings and also remained moist. Lack of erosion, however, does not nullify the results of the study as indicators of wind erosion potential. From Figure 2 it is apparent that conditions of low variability of wind speed, low biomass and narrow leaf widths provide the greatest risk of wind erosion.

Sugar beet and onions emerge as the two crops most likely to give rise to erosion. Although it has long been recognised that these crops are associated with erosion, this has been explained previously in terms of low plant cover in the early stages of crop growth coincident with the period of the year when winds are strong. This study shows, however, that these crops can actually enhance the risk of wind erosion directly because of their high drag coefficients. The study also demonstrates that planted straw strips are an effective in-field erosion control measure. The data points plot to the bottom and left side of the graphs and even when the drag coefficients increase with wind speed consideration of the regression lines (Table 2) indicates that there would be no risk of erosion until the wind speed at 0.05 m height in front of the strip exceeds 1.7 m s^{-1}.

CONCLUSIONS

The effectiveness of single crop rows in reducing wind speeds below the threshold levels for soil particle movement can be evaluated by calculating the frictional drag coefficients of the crop for the lower 0.05 m of the atmosphere. Although the drag coefficients generally decrease with increasing wind speed,under conditions of low wind speed variability they increase. This results in an increase in drag velocity and, particularly with narrow-leaved crops and low biomass densities, a greater risk of wind erosion. Onions and sugar beet can enhance the likelihood of erosion whereas planted straw strips provide an effective in-field control method.

ACKNOWLEDGEMENTS

The research reported here was supported by a Research Grant (AG 63/170) from the Agriculture and Food Research Council. The authors are grateful to the Directors of the Arthur Rickwood and Gleadthorpe Experimental Husbandry Farms for allowing the wind speed measurements to be made on their land.

REFERENCES

Monteith, J.L., 1973. Principles of environmental physics. Edward Arnold, London.

Savat, J., 1982. Common and uncommon selectivity in the process of fluid transportation: field observations and laboratory experiments on bare surfaces. Catena Supplement,1, 139-160.

Uchijima, Z., 1976. Maize and rice. In Monteith, J.L. (ed), Vegetation and the atmosphere. Volume 2. Academic Press, London, 33-64.

Wright, J.L., and Brown, K.W., 1967. Comparison of momentum and energy balance methods of computing vertical transfer within a crop. Agronomy Journal,59, 427-432.

International Geomorphology 1986 Part II
Edited by V. Gardiner

RUNOFF AND SOIL LOSS FROM EROSION PLOTS IN IFE AREA OF SOUTHWESTERN NIGERIA

L. K. Jeje

Department of Geography, University of Ife,
Ile-Ife, Nigeria.

INTRODUCTION

Accelerated soil erosion is an endemic problem in the humid tropics, especially in the less developed countries that can ill-afford to lose such valuable natural resources, yet little is known of its dynamics. Factors such as high population density and the attendant pressure on the land, the nature of the local geology, the physico-chemical properties of the soil, rainfall amount and the effects of subsurface hydrology are already over-emphasized (Grove, 1952; Ofomata, 1965; 1966; 1978; Floyd, 1965; Ologe, 1973). The dynamic interaction between factors of soil erodibility and rainfall erosivity, with regard to sediment production under different intensities of vegetal cover, has still to be emphasized. This is a failure that has contributed to the significant lack of success in solving erosion problems in most tropical countries. However, experimental studies of soil erosion based on erosion plots and data on rainfall erosivity parameters are now being attempted, for example by Roose (1975) in Ivory Coast; Kowal (1970), Leow and Ologe (1981) and Jeje and Agu (1982) in Nigeria; Hudson and Jackson (1959), Elwell and Stocking (1973; 1976) in Zimbabwe. This study covers the period from 1978 to 1983 and is a further contribution to detailed study of soil erosion processes based on experimental plots in the Ife area of southwestern Nigeria. The study focusses on:

(i) the effects of different intensity of vegetal cover on sediment yield, and
(ii) the influence of some factors of rainfall erosivity on soil loss in order to assess the most important factors of sediment yield with particular regard to vegetal cover and rainfall and runoff parameters.

PARAMETERS STUDIED

Rainfall erosivity involves energy expenditure for breaking down soil cohesiveness, splashing the particles as well as entraining them in overland flow. The most critical property of rainfall erosivity is kinetic energy (Laws and Parsons, 1943; Gunn and Kinzer, 1949; Bubenzer and Jones, 1971; Ellison, 1947; Bisal, 1960; Hudson, 1965; Elwell and Stocking, 1973; Stocking and Elwell, 1973; Lal, 1974; Stocking, 1978), determined by Wischmeier and Smith (1958) as follows:

$$KE = 11.9 + 8.7 \log I \qquad [1]$$

where KE = kinetic energy in Joules m^{-2} mm^{-1}

I = rainfall intensity in mm hr^{-1}

For high intensity tropical rainfall, Hudson (1965) derived the equation:

$$KE = 29.8 - 127.5/I \qquad [2]$$

Several parameters of kinetic energy in relationship to soil loss have been examined by various workers. Thus Wischmeier and Smith (1958; 1962) in the U.S.A., and Stocking and Elwell (1973) in Zimbabwe, found the EI_{30} index to be the best predictor of soil loss from bare surfaces, where E is the storm kinetic energy and I_{30} is the maximum 30-minute rainfall intensity. Hudson (1971), and Ahmad and Breckner (1974), however, found low correlations between this index and soil loss in Zimbabwe and Tobago respectively. Hudson (1971) found the total kinetic energy of rainfall of intensities greater than 25mm hr^{-1} (i.e. KE > 25) to be a more significant predictor of soil loss than the EI_{30} index, implying that 25 mm hr^{-1} represents an intensity threshold at which soil erosion starts. An appraisal of Hudson's work led Stocking and Elwell (1973) to conclude that while the EI_{30} index best predicted soil loss from bare soils, the EI_{15} and EI_5 indices best related to soil loss from surfaces with sparse and dense vegetal covers respectively.

Lal (1976) proposed that the AIm index which is the product of total rainfall amount (A) and peak intensity (Im), best predicts soil loss from Alfisols in south-western Nigeria on standard non-vegetated runoff plots on slopes of 1, 5, 10, and 15% (r-value of 0.81, compared with 0.75 for EI_{30} and 0.80 for KE > 25). Also based on a study from erosion plots, Hadley and McQueen (1961) and Emmett (1970) in the U.S.A. found that runoff more than any other erosivity factor correlated with soil loss on various vegetated surfaces.

All the above indices of rainfall energy, i.e. EI_{30}, EI_{15}, KE >25 and AI_{15}, together with the mean and peak intensities, total rainfall and antecedent moisture condition, are examined in relationship to soil loss in this study. The last parameter is especially important as it has a strong bearing on runoff and erosion (Bridges and Harding, 1971; Barnes and Franklin, 1970). Based on the recommendation of Wischmeier (1959), the EI_{30}, EI_{15}, KE >25 and AI_{15} indices were computed only for rainfalls in excess of 12.5mm. The EI_5 or the AI_5 indices were not computed due to the coarseness of the raingauge charts used, on which only the 15 minute interval can be clearly discerned.

STUDY AREA

Eight erosion plots each measuring 4m x 25m were established in 1978 by the Department of Soil Service in two groups, comprising four plots in the University of Ife farm on first order valley sides of 4% and 8%, and a third group comprising three plots located on a rectilinear part of a valley side with a 10% slope in the basin of Opa river close to Alakowe village, about 16km by road from the University of Ife campus. This was established in 1979 by the Department of Geography, and started to function in 1980. All the sites are underlain by the Egbeda soil series (Rhodic Luvisol or Haplustalfs).

Ife area in southwestern Nigeria is characterized by humid tropical climate (Am of Koppen climatic classification) with a mean annual temperature of 27°C and a mean annual rainfall of 1400mm from March to October, with double peaks in July and September and a short dry spell in August. The onset and withdrawal of the rains are marked by thunderstorms with high rainfall intensity.

STUDY METHOD

Of the 11 plots, the eight on the 4% and 8% slopes and one on the 10% slope were repeatedly cultivated with maize and mulched with maize stovers from the previous season. The thickness of the mulch cover (about 7.5cm) was similar on all the plots. Of the remaining two plots on the 10% slope, one was uncultivated and kept weed free, while the other was located in a degraded secondary forest regrowth characterized by tall emergents up to 25m high, a middle layer 10-15m high with discontinuous and irregular canopy, and a lower layer made up of shrubs and saplings. The floor cover is well open and the litter layer is very sparse. For the purpose of this work, the plots on the 10% slope are designated A, B and C. A was bare, B cultivated and C was located in the forest. The maize population on all the plots was about 8100 stands per hectare during the early rains, and about 6200 per hectare during the late rains.

Each site was equipped with an automatic tilting rain-gauge and a manual-gauge to act as a check on the accuracy of the former. Runoff from the plots was measured in the collecting tanks, after which it was stirred vigorously and samples taken for laboratory analysis to determine suspended and solute loads. The Department of Soil Science averaged and recorded data from each group of erosion plots, thus treating each group as a single unit. However, the data derived from the three plots on the 10% slope were recorded separately, and part of it has already been analysed and published (Jeje and Agu, 1982).

Rainfall parameters such as EI_{30}, EI_{15}, KE >25 and AI_{15} were determined from data derived from the automatic rain-gauges, and analysed following the methods outlined by Morgan (1979). Rainfall amount and peak intensity (for 15 minutes) were obtained directly from the rainfall charts and antecedent precipitation index was determined following Gregory and Walling's (1973) modification of Butler's (1957) method:

$$Pa = Pt.1/t \qquad [3]$$

$$\text{or } Pt.k^t$$

where Pa = antecedent precipitation index (A.P.I)

Pt = precipitation for any given day

t = time elapsed in number of days after the rainfall

k = recession factor which is less than 1.0, but ranging from 0.85 to 0.98. The smaller figure was used in this study.

The amount of sediment load measured from the plots was taken to represent the dependent variable (Y), and the selected rainfall parameters were related to it in a pairwise correlation. A correlation matrix to show the relationship between soil loss and the rainfall parameters was prepared for the plots on the 10% slope.

OBSERVATIONS

Soil Characteristics. Table 1 indicates some of the physical characteristics of the soils in the sites. The upper few centimetres are dark brown to dark greyish brown, loose, clayey fine sand to very clayey fine sand with weak crumb structure to structureless; it has a fairly low concentration of quartz gravel and stones. The soil is relatively deep and well drained. However, with the high relative concentration of silt, the low organic matter content and its poor structure, the top soil down to 50cm appears highly erodible (Richter and Negendank, 1977).

TABLE 1. Physical properties of the ten soils in the erosion plot sites.

	Soil depth in cms.			
	0 - 50	50 - 90	90 - 120	120 - 150
Texture class	Clay-loam	Sandy-clay	Sandy-clay	Sandy-clay loam
Mean % gravel and concretions larger than 2000 μm	13.7	47.8	15.6	18.6
Mean % sand 50 μm to 2000 μm	35.2	50.4	67.8	58.4
Mean % silt 2 μm to 50 μm	32.6	5.4	4.0	14.5
Mean % clay smaller than 2 μm	32.2	44.2	28.2	26.1
Bulk density	1.30	1.24	1.43	1.25
Mean % organic matter	3.47	0.57	-	-

Rainfall and Runoff. Rainfall amount varies from year to year with the highest of 1417mm in 1980 and the lowest of 924mm in 1982. The rainfall peak intensity corresponds fairly well with the rainfall pattern. The highest mean intensity is recorded in May, followed by October - at the onset and withdrawal of the rainy season. During the study period, storms with mean intensities less than 10mm hr^{-1} were dominant (38% of all recorded rainstorms), followed by 10-20mm hr^{-1} mean intensity storms (7%), 20-30mm hr^{-1} (20%), 30-40mm hr^{-1} (20%), 40-50mm hr^{-1} (4%), 50-60mm hr^{-1} (4%), 60-70mm hr^{-1} (3%), 70-80mm hr^{-1} (3%), and above 90mm hr^{-1} (1%).

About 92% of the entire rainfalls generated runoff, but the threshold rainfall amount that generated any runoff varies from month to month and from year to year depending on the frequency and amount of the antecedent rainfall. For instance in 1982, in the relatively dry month of April, the threshold rainfall was 7mm; this declined to 5.4mm in June and 2.5mm in July, rising to 6.5mm in the relatively dry month of August and declining to 2.5mm in October.

TABLE 2. Mean % runoff from the plots.

	1978 - 1983		1980 - 1982		
	4% Slope Plots	8% Slope Plots	10% Slope Plots A	B	C
Mean	6.75	4.04	11.4	7.9	3.2
Standard Deviation	9.43	4.14	10.9	6.5	2.5

The monthly pattern of runoff from the three control plots has been described elsewhere (Jeje and Agu, 1982). As is evident from Table 2, runoff from the plots is generally low. However, the high values of the standard deviation show the high variation in the runoff, which is highest at the beginning of each rainy season declining as leaf coverage increases, but with a slight post-harvest increase.

As expected, runoff was consistently highest from the control bare plot (A), and this was followed by the cultivated plot on the 10% slope (B), and by those on the 4% and 8% slopes respectively. Plot C under the degraded secondary forest produced the lowest runoff. Surprisingly, slope does not seem to exert much influence on runoff volume as the mean runoff value from the 4% slope plots is higher than from the 8% slope plots.

As illustrated in Table 3, based on the data from the three contiguous control plots on the 10% slope, most of the rainfall erosivity indices correlate significantly with runoff except mean intensity and the antecedent precipitation index, which show no clear relationship. Runoff correlates more significantly with both rainfall amount and peak intensity (Table 3). The degree of correlation between runoff and the rainfall parameters varies from plot to plot, being highest in the forest plot with r value of 0.87 for rainfall amount and peak intensity. The r value in the bare plot is 0.83 for each of the

TABLE 3. Correlations between runoff from the plots and some erosivity parameters.

Runoff correlated with	Plot A (bare)	Plot B (Cultivated)	Plot C (under secondary forest)
1. Rainfall amount	0.83*	0.72*	0.87*
2. Mean intensity	0.17	-0.08	0.01
3. Peak Intensity	0.83*	0.73*	0.87*
4. AI_{15}	0.77*	0.68*	0.80*
5. EI_{30}	0.86*	0.56*	0.78*
6. EI_{15}	0.83*	0.56*	0.78*
7. KE >25	0.72*	0.56*	0.62*
8. A.P.I.	0.38	0.14	0.24

* Significant at 0.01 level

parameters while it is 0.72 and 0.73 respectively in the maize plot. AI_{15} has r values of 0.80, 0.77, and 0.68 in plots C, A, and B, while the r values of the EI_{30}, EI_{15} and KE >25 are 0.56 in plot B (cultivated plot). In plot C, the r value is 0.78 for the EI_{30} and the EI_{15} and 0.63 for KE >25.

Soil Loss. The annual values of soil loss declined steadily from the inception of the study (Table 4). The highest mean annual soil loss is recorded from the bare plot, followed by plots on 4% slope, and 8% slope. Soil loss is lowest under the secondary forest.

The mean annual soil loss of 157.80kg ha^{-1} yr^{-1} from the bare plot compares favourably with the 180kg ha^{-1} yr^{-1} (0.18mm) obtained by Leow and Ologe (1981) in the savanna (Zaria) area of northern Nigeria, but it is lower than the 2.3 t ha^{-1} obtained for a similar surface in Zimbabwe by Hudson and Jackson (1959). The mean annual soil losses of 133.9kg ha^{-1}, 128.1kg ha^{-1} and 93.8kg ha^{-1} obtained from the maize plots compare favourably with the 101.6kg ha^{-1} obtained by Lal (1976) from a similar plot on a 15% slope in Ibadan. The mean annual soil loss of 78.9kg ha^{-1} yr^{-1} from the forest plot

TABLE 4. Annual soil loss (kg/ha)

Year	4% slope plots	8% slope plots	10% slope plots A	B	C
1978	196.90	192.70	-	-	-
1979	181.40	185.50	-	-	-
1980	174.75	160.95	232.61	157.35	124.35
1981	135.00	121.75	172.72	107.49	87.54
1982	75.40	71.20	68.00	16.60	25.00
1983	39.70	36.50	-	-	-
Mean	133.86	121.10	157.80	93.80	78.90

is higher than Charreau (1972) obtained from a ferralitic soil in Abidjan in an area with a total rainfall of 2100mm. It is also higher than the 40kg ha^{-1} yr^{-1} obtained in the headwaters of the Gombak river in Malaysia by Douglas (1972).

Interrelationships between soil loss and rainfall parameters. Tables 5 and 6 show interrelationships between soil loss and the selected rainfall parameters. Soil loss from all the plots correlated most significantly with runoff and this accords with the findings of

TABLE 5. Correlation between soil loss and some rainfall erosivity factors on the 8% and 4% slope plots

Erosivity factors	8% Slope plots	4% Slope plots
Rainfall amount	0.49*	0.94*
EI_{15}	0.61*	0.53*
EI_{30}	0.53*	0.91*
AI_{15}	0.55*	0.69*
KE >25	0.54*	0.67*
Mean rainfall intensity	0.21	0.15
Peak intensity	0.65*	0.74*
% runoff	0.74*	0.98*
A.P.I.	0.27	0.38

* r value significant at 0.01 level

TABLE 6. Correlation matrix showing the relationship between soil loss and rainfall erosivity parameters on the 10% slope plots.

	Z_1	Z_2	Z_3	S_1	S_2	S_3	S_4	S_5	S_6	S_7	S_8	S_9	S_{10}	S_{11}
Z_1	1.00	0.64	0.58	0.91	0.70	0.74	0.76	0.08*	0.77	0.70	0.75	0.75	0.71	0.37*
Z_2		1.00	0.57	0.58	0.87	0.56	0.64	0.19*	0.60	0.58	0.41	0.39*	0.49	0.18*
Z_3			1.00	0.57	0.60	0.73	0.18*	0.70	0.70	0.64	0.55	0.58	0.48	-0.01*
S_1				1.00	0.69	0.81	0.83	0.17*	0.83	0.77	0.86	0.83	0.72	0.30
S_2					1.00	0.66	0.72	0.08*	0.73	0.68	0.56	0.56	0.56	0.14*
S_3						1.00	0.87	0.01*	0.87	0.80	0.78	0.78	0.62	0.24*
S_4							1.00	0.02*	0.99	0.93	0.85	0.85	0.68	0.36*
S_5								1.00	0.01*	0.02*	0.03*	0.18*	-0.05*	-0.21*
S_6									1.00	0.93	0.85	0.87	0.67	0.39*
S_7										1.00	0.85	0.80	0.64	0.45
S_8											1.00	0.88	0.67	0.48
S_9												1.00	0.73	0.33*
S_{10}													1.00	0.46
S_{11}														1.00

Emmett (1970) and Hadley and McQueen (1961) in the United States. Correlation is highest for the 4% slope plots with 0.98, and lowest with 0.73 on the forested plot. The EI_{30} and EI_{15} indices also correlate significantly with soil loss from the plots (Hudson, 1971). However, contrary to the findings of Elwell and Stocking (1973b), the EI_{15} index is not highly correlated with soil loss from the maize plots.

With r-value of 0.71, KE > 25 correlates highly with soil loss from the bare plot and this accords with Hudson's (1971) findings in Zimbabwe. As expected, peak rainfall intensity correlates highly with soil loss in all the plots, being the second most significant of the indices tested after percentage runoff on the 8% slope plots, and on the cultivated and forested 10% slope plots. It is however fourth in importance on the 4% slope plots. The AI_{15} index also correlates significantly with soil loss on all the plots (Aina et al., 1977). The lowest correlation values are recorded between soil loss and mean rainfall intensity, the value is even negative for the cultivated plot on the 10% slope. Although the antecedent precipitation index correlates very highly with runoff, it correlates very poorly with soil loss, especially for the maize plots.

DISCUSSION

The main findings in this study are as follows:

- For the duration of the study, the bare plot produced the highest values of runoff, followed by the maize plots on the 10% slope, on the 4% slope and on the 8% slope, and then by the forest covered plot.
- The bare plot produced the highest values of soil loss, followed by the maize plots on the 4% slope, the 8% slope and the forest covered plot. The mean annual values of soil loss from all the plots declined steadily from the inception of the study.
- As sediments are normally transported in suspension, as is to be expected, runoff appears to be the best predictor of soil loss, as it has the highest r values relative to soil loss from all the plots.
- Both the EI_{30} and the EI_{15} indices believed to be the best predictor of soil loss from bare and vegetated surfaces respectively (Stocking and Elwell, 1973) also appear strongly related to soil loss in the study area, most especially on the bare plot. However, the AI_{15} index appears to be more related to soil loss from vegetated surfaces than these indices and this confirms Lal's (1976) finding in the International Institute of Tropical Agriculture (I.I.T.A.).

Key to Table 6 [FACING PAGE] Z_1: Soil loss from plot A; Z_2: Soil loss from plot B; Z_3: soil loss from plot C; S_1: runoff from plot A; S_2: runoff from plot B; S_3: runoff from plot C; S_4: rainfall amount; S_5: mean rainfall intensity; S_6: peak intensity; S_7: A_{15}; S_8: EI_{30}; S_9: EI_{15}; S_{10}: KE > 25; S_{11}: A.P.I. * relationship not significant at 0.1, all others are significant at 0.1.

That both soil loss and runoff were highest in the bare plot was not unexpected as this merely confirms the findings by other workers such as Wischmeier and Smith (1962), Hudson (1971), Stocking and Elwell (1972; 1976), and Lal (1976), amongst others. However, the values of runoff and soil loss recorded in the secondary forest plot appear to be high when compared with the values obtained by Charreau (1972) and Douglas (1972) from similar humid forest environments, or with those obtained by Hewlett and Hibbert (1967), Kirkby (1969), Reinhart et al., (1963), Rothacher (1965) and Whipkey (1969) amongst others who worked in forest areas in humid temperate regions. However, Pierce (1967) and Kessel (1977) observed overland flow and slope wash in the forested temperate region of the United States, tropical Papua New Guinea and Guyana respectively. Such overland flow which could also be observed on the plot towards the end of the rainy season and whenever heavy rains occurred could have accounted for the high values of runoff and soil loss from the forest plot.

The higher value of soil loss on the 4% slope than on the 8% slope is difficult to explain. It may be due to variations in the nature of the soil in the study area, as the soil on the 4% slope may be more erodible than that on the 8% slope. This however, is still subject to confirmation. Mean annual sediment yield declined steadily in all the plots, possibly because the fine soil aggregates were being removed, leading to concentration of the coarser fractions, which are more difficult to entrain except during very high intensity storms. This gravel concentration is now a noticeable feature of the bare plot, and in fact as from 1984, fertilizers have had to be applied on the other plots.

Generally, runoff was low in the maize plots due to mulching by maize stover and the cleared refuse, the high maize density and foliage cover. The mulch reduced runoff and soil loss because of increased surface detention (Borst and Woodburn, 1942). Also mulch can absorb rainwater, transmitting it into the soil at a slow rate, and the maize canopy intercepts rain drops so that drop energy was considerably reduced. However, during the periods when the maize crops were immature and the maize stover of previous season decayed, raindrops can be highly erosive. This may possibly account for the relative importance of the AI_{15} and the KE >25 indices as predictors of soil loss from these plots.

Tables 5 and 6 show that runoff correlates more significantly with soil loss from all the plots than all the other erosivity parameters considered. This shows that while raindrop impact may be very important in soil detachment on many surfaces - woodland, savanna, cropped land, etc., - (e.g. Soyer et al., 1982), it would appear that rain drop impact without a more important transporting agent may be incapable of causing significant downslope sediment movement.

Lal (1976), and Aina et al., (1977) found that the AI_m correlated more significantly than the EI_{30} index with soil loss on bare surfaces. This is not confirmed by this study, from which it appears that the EI_{30} and the EI_{15} indices are more significantly related to soil loss on bare plots while in the maize and forest plots, the

AI_m index is generally more significantly related to soil loss than either the EI_{30} or the EI_{15}.

The KE > 25 index is positively and significantly related to soil loss in all the plots, but it is more important on the bare plot where it has an r value of 0.71, than in all the other plots, where r values are generally less than 0.67. The kinetic energy of rainfall with intensity greater than 25mm hr^{-1} would appear to be more important where rain splash erosion takes place, especially on bare soils, whereas on the other plots its effect is muted by vegetal cover. This confirms Hudson's (1965) results that the KE > 25 index may be a better indicator of soil loss from bare soils.

Rainfall amount which has been recognised as an important parameter in the soil loss equation (Elwell and Stocking, 1973b) also proves to be significantly correlated with soil loss in this study. It is third in importance on the bare plot, and second on the 4% slope plots and on plots B and C on the 10% slope.

The antecedent precipitation index (A.P.I.) correlates poorly with soil loss in all the plots, with r values of 0.27 and 0.38 on the 8% and 4% slope plots, and 0.36, 0.17 and 0.01 on plots A, B, and C respectively.

CONCLUSION

This study represents an attempt to examine the effects of different landuse/vegetal covers on soil loss, and to determine the most important erosivity parameters relative to soil loss from these different surfaces.

As already indicated, the amount of runoff and soil loss was consistently greatest from the bare plot, followed distantly by loss from the maize plots, and both runoff and soil loss were lowest in the forest. In the case of the maize plots, both runoff and soil loss were higher on the 4% than on the 8% slope plots.

Although temporal variations occurred in the production of runoff and sediment from the plots, it is evident that the same volume of rainfall and runoff may not necessarily yield the same amount of sediment from the same plot. It also appears that only very few storms were actually responsible for generating most of the soil loss.

As expected soil loss correlates most significantly with runoff on all the plots. On the bare plot, runoff is followed by peak intensity, rainfall amount, the EI_{30}, EI_{15}, KE >25 and the AI_{15} indices in that order, while on the cultivated plots, runoff is followed by rainfall amount, peak intensity the AI_{15}, EI_{30}, KE >25 and EI_{15} indices. In the forest plot, runoff, rainfall amount, peak intensity and the AI_{15} index correlate significantly with soil loss, while the EI_{30}, EI_{15}, and KE >25 indices correlate least with soil loss.

As runoff correlates so significantly with soil loss from these plots, it is quite obvious that to reduce soil loss from any surface, it is not only important to prevent rain drops from impacting with the ground surface, but an important requirement is to reduce runoff to the barest minimum by enhancing rain water infiltration into the soil. This can be done through ensuring adequate vegetal cover, litter and mulch cover on soil surfaces.

REFERENCES

Aina, P., Lal, R., and Taylor, G. S., 1977. Soil and crop management in relation to soil erosion in the rainforest of Western Nigeria, in Soil erosion: prediction and control. Soil Conservation Society of America S.P., Pub. 21, 75-82.

Ahmad, N., and Breckner, E., 1974. Soil erosion on three Tobago Soils. Trop. Agric., 51, 313-324.

Barnes, D. L., and Franklin, M. J., 1970. Runoff and soil loss on sandveld in Rhodesia. Proc. Grass. Soc. S. Agric., 5, 140-144.

Bisal, F., 1960. The effect of rain-drop size and impact velocity on sand splash. Can. J. Soil Sc., 40, 242-245.

Bridges, E. M., and Harding, D. M., 1971. Micro-erosion processes and factors affecting slope development in the Lower Swansea Valley, in Slopes form and process. (Ed. Brunsden), pp65-69. Inst. Br. Geog. Spec. Publ., 3, London.

Bubenzer, G. D., and Jones, B. A., 1971. Drop size and impact velocity effects on the detachment of soils under simulated rainfall. Trans. American. Soc. Agric. Eng., 14, 625-628.

Butler, S. S., 1957. Engineering hydrology. New Jersey.

Charreau, C., 1972. In Meotuppa, F., 1973. Soil aspects in the practice of shifting cultivation in Africa and the need for a common approach to soil and land resources evaluation, shifting cultivation soil conservation. F.A.O. Soil Bull., 24, Rome.

Douglas, I., 1972. The environment game. Inaugural lecture, University of New England, Armidale.

Elwell, H. A., and Stocking, M. A., 1973a. Rainfall parameters for soil loss estimation in a subtropical climate. J. Agric. Eng. Res., 18, 169-177.

Elwell, H. A., and Stocking, M. A., 1973b. Rainfall parameters to predict surface runoff yields and soil losses from selected field-plot studies. Rhod. J. Agric. Res., 11, 123-129.

Emmett, W. W., 1970. The hydraulics of overland flow on hillslopes. U. S. Geol. Surv. Prof. Paper, 662-A.

Floyd, B., 1965. Soil erosion and deterioration in eastern Nigeria - A geographical appraisal. Nig. Geogr. J., 8, 33-44.

Friese, F., 1936. Das Binneklima von Urwaldern in sub-tropischen Rasitien. Petermans Mitt., 82, 301-307.

Gregory, K. J., and Walling, D. E., 1973. Drainage basin form and process : A geomorphological approach. Arnold, London.

Grove, A. T., 1952. Land use and soil conservation in the Jos Plateau. Geol. Surv. Nig. Bull., 22.

Gunn, R., and Kinzer, G. D., 1949. The terminal velocity of fall for rain droplets. J. Met., 6, 243-248.

Hadley, R. F., and McQueen, I. S., 1961. Hydrologic effects of water spreading in Box Creek Basin, Wyoming. U. S. Geol. Surv. Water Supply Paper, 1532-A.

Hewlett, J. D., and Hibbert, A. R., 1967. Factors affecting the response of small watersheds to precipitation in humid areas, in Procs. of the International Symposium on Forest Hydrology, 1965, Penn. State University (Eds. Sopper and Lull). Pergamon, Oxford.
Hudson, N. W., 1965. The influence of rainfall mechanics on soil erosion. Unpublished M.Sc. Thesis, University of Cape Town.
Hudson, N. W., 1971. Soil conservation. Batsford, London.
Hudson, N. W., and Jackson, C. C., 1959. Results in the measurement of erosion and runoff in Southern Rhodesia. Proceedings, 3rd Annual Inter-African Soils Conference, Dalaba. 575-583.
Jeje, L. K., and Agu, A., 1982. Runoff and soil loss from erosion plots in Ife area of southwestern Nigeria. Geo-Eco-Trop, 6,161-181.
Keltman, M. C., 1969. Some environmental components of shifting cultivation in upland Mindanao. J. Trop. Geogr., 28, 40-56.
Kessel, P. H., 1977. Slope runoff and denudation in the Rupununi savanna. Guyana. J. Trop. Geogr., 44, 33-42.
Kirkby, M. J., 1969. Infiltration throughflow and overland flow, in Water earth and man (Chorley). Methuen, London.
Kohler, M. A., and Linsley, R. K., 1951. Predicting runoff from storm rainfall. U. S. Weather Bur. Res. Paper, 34.
Kowal, J., 1970. The hydrology of a small catchment basin of Samaru, Nigeria: Assessment of soil erosion under varied land management and vegetation cover. Nig. Agric. J., 7, 134-147.
Lal, R., 1974. Soil erosion and shifting agriculture. Soils Bull. F.A.O. Rome, 24, 48-71.
Lal, R., 1976. Soil erosion on alfisols in Western Nigeria, V. Changes in physical properties and response of crops. Geoderma, 16, 419-431.
Laws, J. A., and Parsons, D. A., 1943. The relationship of rain drop size to intensity. Trans. Am. Geophys. Union, 24, 452-460.
Leow, K. S., and Ologe, K. O., 1981. Rates of soil wash under savanna climate, Zaria, Northern Nigeria. Geo-Eco-Trop, 5, 87-98.
Moorkajee, D., 1950. Anti-soil erosion equipment at Araba, West Bengal. Cent. Bd. Irrig. J., 191-193.
Morgan, R. P. C., 1978. Field studies of rain splash erosion. Earth Surface Processes, 3, 295-299.
Morgan, R. P. C., 1979. Soil erosion. Longman, London.
Ofomata, G. E. K., 1965. Factors of soil erosion in the Enugu area of Nigeria. Nig. Geogr. J., 8, 45-49.
Ofomata, G. E. K., 1978. Man as factor of soil erosion in southeastern Nigeria. Geo-Eco-Trop., 2, 243.
Ologe, K. O., 1973. Soil erosion, sediment yield and water resources development. Conference on Environmental Resources Management in Nigeria, University of Ife, July 1973.
Pierce, R. S., 1967. Evidence of overland flow on forest watersheds, in Procs. of the International Symposium on Forest Hydrology, 1965, Penn. State University (Eds. Sopper and Lull). Pergamon, Oxford.

Reinhart, K. G., Eschner, A. R., and Trimble, G. R., 1963. Effect in stream flow of four forest practices in the mountains of West Virginia. U. S. Forest Serv. N. E. Forest Exp. Statn. Res. Paper, N.E.-1.

Richter, G., and Negendank, J. F. W., 1977. Soil erosion processes and their measurement in the German area of the Moselle river. Earth Surface Processes, 2, 261-278.

Roose, E. J., 1975. Erosion et ruissellement en Afrique de l'Ouest. Vingt annees de mesures en parcelles experimentales. O.R.S.T.O.M. Centre d'Adiopodoume, Labo. Pedol., Abidjan.

Ruxton, B. P., 1967. Slope wash under mature primary forest in Northern Papua, in Landform Studies from Australia and New Guinea (Eds. Jennings and Mabbutt), pp85-94. Univ. Press, Cambridge.

Stocking, M. A., 1978. The prediction and estimation of erosion in subtropical Africa: Problems and prospects. Geo-Eco-Trop, 2, 161-174.

Stocking, M. A., and Elwell, H. A., 1976. Rainfall erosivity over Rhodesia. Trans. Inst. Br. Geogr., 1, 231-245.

Soyer, J., Mitt, T., and Aloni, K., 1982. Effects compares de l'erosion pluviale en milieu peri-urbain de region tropicale (Lubumbashi, Shaba, Zaire). Revue Geomorph. Dyn, 31, 71-80.

Tackett, J. L., and Pearson, R. W., 1965. Some characteristics of soil crust formed by simulated rainfall. Soil Sc., 9, 401-413.

Whipkey, R. Z., 1969. Storm runoff from forested catchments by subsurfaces routes. Publ. Intl. Assoc. Sc. Hydrology, 85, 773-779.

Wischmeier, W. H., 1959. A rainfall erosion index for universal soil loss equation. Soil Sc. Soc. Am. Proc., 23, 246-249.

Wischmeier, W. H., and Smith, D. D., 1958. Rainfall energy and its relationship to soil loss. Trans. Am. Geophys. Union, 39, 285-291.

Wischmeier, W. H., and Smith, D. D., 1962. Soil loss estimation as a tool in soil and water management planning. Int. Assoc. Sc. Hydrology. Pub. 59, 148-159.

Zinke, P. J., 1967. Forest interception studies in the U.S., in Procs. of the International Symposium on Forest Hydrology, 1965, Penn. State University, (Eds. Sopper and Lull). Pergamon, Oxford.

International Geomorphology 1986 Part II
Edited by V. Gardiner

VARIATION IN SOIL LOSS ON SMALL SCALE AGRICULTURAL PLOTS IN TANZANIA

Hezekiel M. Mushala

University of Dar es Salaam
P. O. Box 35049
Dar es Salaam, TANZANIA

ABSTRACT

Field results of soil loss investigations on farmland indicate that total rainfall influences splash material displacement, sediment yield and net downslope sediment movement. The contribution of big storms is particularly important in that a few rain events of high intensity increase the volume of both splash material displaced and sediment yield. The influence of wind in net downslope sediment movement is implicit. The influence of crop cover on splash material, net downslope sediment movement and sediment yield is not significant. The major soil losses occur on maize and sorghum fields. The soil loss data seem low in their absolute values but compared to published tolerance values, these data values are large. This implies that rates of erosion are excessive.

INTRODUCTION

In evaluating soil erosion, the development of spectacular gullied landscapes attracts the attention of research workers and obscures raindrop and sheetwash impact under different farming conditions. In many cases the subtle processes of sheetwash and rainsplash are more damaging over the long term compared to the spectacular gullies which tend to stabilize. Sheetwash and rainsplash impact is exacerbated by the absence of groundcover, steep slopes, and inherent soil characteristics that inhibit infiltration or enhance disaggregation. This is especially true under land use systems that do not employ soil conservation measures.

Soil loss remains unnoticed on farmland in many parts of Tanzania. In the Central Plateau erosion has reached alarming proportions, especially in areas where grazing is a major activity. This paper describes the results of estimating soil loss on farmland using sediment traps to determine sheetwash impact, and splashboards to determine splash movement.

STUDY METHODS

Field investigations for the estimation of soil loss on small scale agricultural plots were carried out in Kondoa district, on the Central Plateau. Twenty sample agricultural plots were selected in

Bolisa area along slopes of 6-7 percent. Soil traps, each 0.5m long and 0.11m in diameter, similar to those developed by Gerlach (1967), were installed at the downslope ends of the sample agricultural plots. It was assumed that on a straight slope the catchment area for each trap would be the length of slope times the width of the trap. Slope length was determined by the distance between the trap and the upslope limit of the field. The traps were installed perpendicular to the slope with the lower lip flush with the soil surface. The soil traps were emptied once every week and sediment placed in plastic bags. The bags were then labelled to indicate the date material was collected and the identification number of the plot. After each trap was emptied the cover was closed and the trap inspected to ensure proper installation.

In addition, sixteen splashboards, in the model of those developed by Ellison (1945), were set up perpendicular to the surface and horizontal to the traps, a metre and a half away. The purpose was to measure splash material and indicate the general direction and amount of net sediment movement. The splashboards were supported by wooden anchors for protection against the wind. Splashboards were cleaned and material collected on a weekly basis. For each splashboard, material was collected from each of the upslope and downslope sides and put in separate plastic bags. The bags were then labelled to indicate plot location and whether sediment is derived downslope or upslope.

A standard rain gauge was installed at the control sample plot from which rainfall amount was recorded once every day between November 1984 and April 1985. This was to coincide with the rainy season in Kondoa. A quadrant sighting frame was used to approximate percentage land cover at monthly intervals. The crop canopy was estimated from the shadow cast around midday on clear days. Crop density per square metre, stand quality and status were recorded.

RESULTS

Sediment Transport by Rainsplash. During the study period the average rainfall per week was 25.5mm. An average of 22.4 g/plot-week of splash material moved upslope. For the whole season, the average total upslope splash displacement is 292.3 g per plot, with total rainy season ranges between 504.5 g and 183.6 g. Compared to splash material displacement downslope, the upslope rates are lower. Average splash material displacement downslope per week is 0.6 g higher per plot. The total average splash material downslope is higher by 32.8 g, with a seasonal range between 551.3 g and 214.1 g. The average net downslope sediment movement, which indicates the difference between upslope and downslope splash, is only 4.5 g per week.

Twelve and a half percent of the plots have net downslope sediment movement greater than 100 g. Only 18.75 percent of the plots have net downslope sediment movement higher than 50 g but less than 100 g. Most of the plots (68.75 percent) have net downslope sediment movement below 50 g (Table 1).

TABLE 1. Crop type and sediment yield in Kondoa

Plot I.D. #	Slope Length (m)	Crop Type	Net DSM (g)	Sediment Yield (mt/ha)
1	15	M	0	37.13
13	10	M/S/C	96.7	30.04
2	10	M/S	84.7	29.99
15	10	M/S	0	27.13
5	6	M/P/B	0	26.61
16	8	M/B	0	26.6
9	10	M/S/C	77.2	21.81
3	10	M/P/B	170.7	21.13
12	8	M/S/C	0	19.95
17	10	M/C	n.d.	19.17
6	10	S/P/Pp	0	18.51
4	5	M/B/C	0	17.41
10	8	M/S/C	0	17.37
18	8	S/B	n.d.	16.41
7	10	M/P/Pp	0	15.63
14	5	M/S/C	171.5	12.15
11	5	M/S/Pp	37.1	9.89
20	5	M/S/C	n.d.	9.44
19	5	M/S	n.d.	9.42
8	6	M/S	30	7.08

M = Maize, S = Sorghum, P = Peanuts, C = Cowpeas, Pp = Pigeonpeas, B = Beans

The correlations between total rainfall and splash material upslope and total rainfall and splash material downslope are significant at the 0.05 confidence level. The average correlation coefficient for upslope splash material is 0.83. The established relationship is:

$$U = -2.52 + 0.98R$$

where

U = amount of splash material upslope (in grams), and
R = rainfall amount (in mm.).

The coefficient of determination (r-squared) is 68.9 percent.

The significance of total rainfall in displacing splash material upslope is portrayed by the scatterplot (Figure 1). Over 75 percent of the distribution points are in the lower half, which means less than 25 percent of the rainfall events displaced half the amount of splash material upslope. Thus a few high intensity rainfall events move the majority of the splash material upslope. If local winds would have been downslope in these few events, significant differences in the net effect of rainsplash would result.

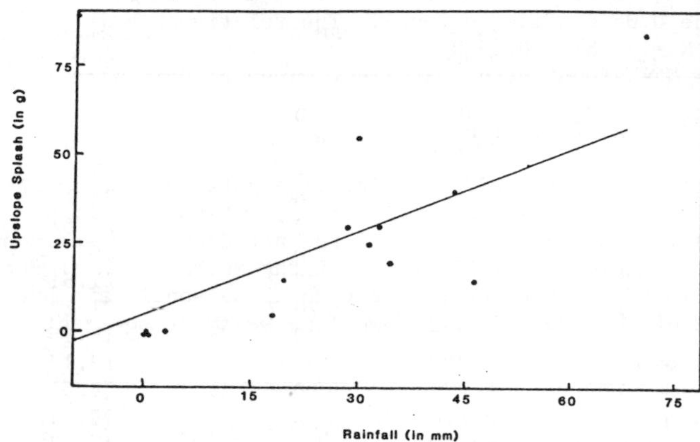

Fig. 1. Relationship between rainfall and upslope splash.

The average correlation coefficient between rainfall amount and splash material downslope is 0.85. The established relationship is:

$$D = -\ 5.26 + 1.11R$$

where D = amount of splash material downslope (in grams), and R = rainfall amount (in mm.).

The coefficient of determination is 72.7 percent.

A trend similar to the scatterplot in upslope splash displacement is observed here (Figure 2). A few rainfall events of high intensity cause over half the amount of displaced splash material downslope.

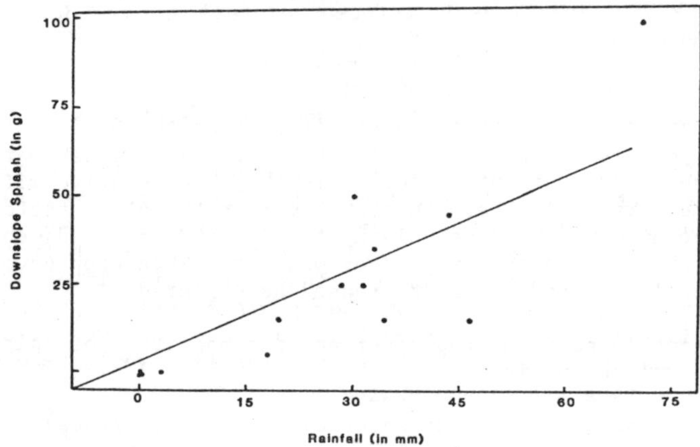

Fig. 2. Relationship between rainfall and downslope splash.

The correlation coefficient between rainfall and net downslope sediment movement is 0.84, the coefficient of determination is 71.3 percent at the 0.05 confidence level. The relationship is:

$$N = -1.87 + 0.252R$$

where N = amount of net sediment movement downslope (in grams), and

R = rainfall amount (in mm.).

The significance of individual rainfall events in displacing splash material is observed here also (Figure 3). Less than 25 percent of the rainfall events cause over half the net downslope sediment movement. For example, rainfall events of the second week of February 1985 contributed 19.8 percent of the total seasonal rain but caused 37 percent of the season's net downslope sediment movement.

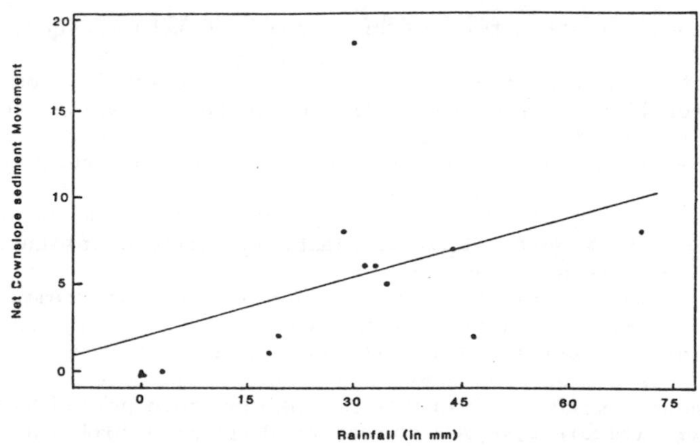

Fig. 3. Relationship between rainfall and net downslope sediment movement.

The correlation coefficients are significant at the 0.05 confidence level and positive between average total rainfall, splash material, trap sediment and sediment yield, while they are negative with crop cover. The correlation between crop cover and the other factors is negative because most of the splash material and trap sediment were collected when percentage crop cover was low. When crop cover increased, soil loss and splash decreased as expected.

The relationship between crop cover and the magnitude of splash material displacement is:

$$U = 18.8 + 0.872R - 0.349C$$

where U = amount of splash material upslope (in grams),

R = rainfall amount (in mm.), and

C = crop cover (in percentage).

The coefficient of determination is 74.6 percent. Thus the additional factor of crop cover increases the explained variation by 6 percent. For downslope splash displacement the relationship is:

$$D = 21 + 0.98R - 0.431C$$

where D = amount of splash material downslope (in grams),
R = rainfall amount (in mm.), and
C = crop cover (in percentage).

The coefficient of determination is 79.7 percent and crop cover has increased the explained variation by 7 percent. For net downslope sediment movement the relationship is:

$$N = 3.74 + 0.224R - 0.092C$$

where N = net downslope sediment movement (in grams),
R = rainfall amount (in mm.), and
C = crop cover (in percentage).

The explained variation here is increased by 6.1 to 77.4 percent.

Although the combined effect of crop cover and rainfall amount accounts for 77 percent of the variation in net downslope sediment movement, the general observation is that plots with high net downslope sediment movement have poor crop cover and poor quality of stands.

Sediment Loss by Runoff. Trap sediment was collected from individual fields to estimate soil loss by surface runoff. The catchment area for the trap was assumed to be the length of the slope times the width of the trap. This is a valid assumption as the slopes were all straight in profile. Most sediment was collected at the peak of the rainy season when crop cover and growth stage were approximately fifty percent. Soil loss was greatest during this period because rainfall amounts were largest (Table 2) while crop cover, a factor expected to offset rainfall impact, was still not dense enough to dampen the impact of the raindrops. Eighty-five percent of the plots had over 70 percent of their sediment yield collected between January and March 1985. Average sediment yield per plot for the study period was 1.37 m tons/ha-week. Although this average figure is modest, the overall sediment yield is very high. Forty percent of the plots have sediment yields of greater than 20 m tons/ha. Another 40 percent of the plots have sediment yields higher than 10 m tons/ha., but less than 20 m tons/ha.; and only 20 percent of the plots have less than 10 m tons/ha. In the study area most sample farm plots are established on soil depths of between 100 to 150cm. If soil depth is the criterion the soil tolerance limit in the area is likely between 0.7 to 0.9 (Morgan 1980), the equivalent of 7 to 9 m tons/ha/yr. Given these figures, 80 percent of the farmplots exceed the established thresholds and only 20 percent of the plots are within these limits. This is a clear indication that the soil loss rates are severe.

TABLE 2. Sediment yield between January and March 1985

Plot	A	B	C
1	37.13	99.6	21.7
2	29.99	93.7	29.5
3	21.13	92.4	33.1
4	17.41	93.8	40.0
5	26.61	87.9	23.2
6	18.51	76.5	14.8
7	15.63	83.1	16.0
8	7.08	77.5	26.7
9	21.81	92.2	34.1
10	17.37	71.8	23.8
11	9.89	86.8	29.9
12	19.95	69.1	28.1
13	30.04	92.6	30.7
14	12.15	81.1	29.6
15	27.13	65.1	21.0
16	26.6	90.1	26.0
17	19.17	56.9	13.5
18	16.41	92.4	16.6
19	9.42	97.7	15.5
20	9.44	78.3	19.7

A = Total Sediment Yield (mT/ha)
B = Sediment Collected in Jan - March (Percentage)
C = Sediment Contributed by highest rainfall (percentage)

A high and significant correlation exists between rainfall amount and sediment yield at the 0.05 confidence level. Its correlation coefficient is 0.86 and its coefficient of determination is 75 percent. Thus 75 percent of the variation in sediment yield is explained by rainfall amount alone. The established relationship is:

$$S = -10.9 + 5.83R$$

where S = sediment yield (in g/sq.m.), and
R = rainfall amount (in mm.).

The combined relationship between total rainfall, crop cover and sediment yield is:

$$S = 83.7 + 5.35R + 1.55C$$

where S = sediment yield (in g/sq.m.),
R = rainfall amount (in mm.).
C = percentage crop cover.

This relationship is significant at 0.05 confidence level. The coefficient of determination is 78 percent.

The scatterplot (Figure 4) shows that a few individual rainfall events cause the majority of the sediment yield. It is indicated that in 80 percent of the cases sediment yield caused between January - March 1985 was contributed by less than 40 percent of the total rainfall (Table 2).

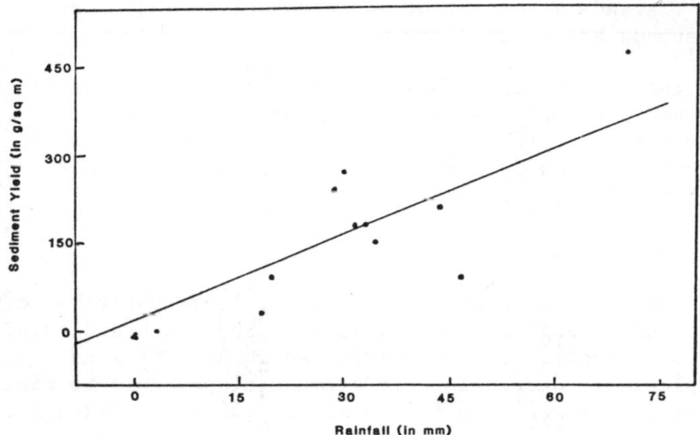

Fig. 4. Relationship between rainfall and sediment yield.

DISCUSSION

Several observations can be made from these results. First, the soil loss data are relatively low for highland areas in East Africa. One possibility is that net downslope sediment movement was influenced by wind in that more splash material was moved upslope than would have been expected. Farmplots located at the margins of the cultivated block show great variations in net downslope sediment movement because they are unshielded against wind. Plots in the central zone are shielded by crops from all sides and therefore the aeolian impact is at its minimum.

Second, most of the soil loss occurs on maize or sorghum plots because of the crop canopy characteristics. Similar observations were made by Lal (1976). For these crops the leaf spread does not cover completely the area around the stem to reduce raindrop impact or runoff detachment, as in other crops. Also, as crops grow and increase their height, they increase the potential volume of water that collects on leaves and stems. As a result stem drop impact increases with crop stage, exacerbating the net raindrop impact.

Third, microorganism activity, visible on twenty five percent of the plots, increases the amount of loose material in the trap catchment area. The additional loose material increases the potential volume of sediment yield and exaggerates the magnitude of erosional processes under normal conditions.

Fourth, the system of land management plays a significant role in creating conditions conducive to soil erosion. Thirty five percent of the plots with high sediment yields were either plowed up and down or had poor crop quality and cover. Over 60 percent of the plots with low sediment yields are of mixed crops, in which the quality of stands is excellent, crop cover is generally very good, and the stands are multi-cropped.

Finally, these data cannot be predictive of erosion conditions in the area because they are collected over a single growing season, which may not be representative. Long term data are required to predict conditions under existing farming conditions due to variability of rainfall, the main controlling factor.

CONCLUSIONS

Processes of rainsplash and sheetwash may seem subtle but contribute great volumes of soil loss on farmland. Sheetwash and raindrop impact can be controlled under improved methods of land management. Contour plowing reduces slope length and slope angle to minimize the impact of surface runoff velocity and volume. Mixed crops can control erosion effectively as long as they provide the crucial groundcover against raindrop impact and surface runoff detachment. Sharma et al. (1976) identify crop combinations which can be adopted for Tanzania conditions to control erosion. Usually better crops are established with the application of fertilizer and manure. All these agronomic strategies can be adopted easily to curtail erosion under existing farming practices. Cut-off drains are suitable mechanisms for the safe disposal of excessive surface runoff that may be incorporated.

ACKNOWLEDGEMENT

Field work for this study was supported by a grant from the Canadian International Development Research Centre, Nairobi, Kenya, with some assistance from the University of Dar es Salaam.

REFERENCES

Ellison, W. D., 1945. Some effects of raindrops and surface flow on soil erosion and infiltration. Trans. Amer. Geophys. Union, 26, 415-429.

Gerlach, T., 1967. Hillslope Troughs for Measuring Sediment Movement. Rev. Geomorph. Dyn., XVII, (4), 173-174.

Lal, R., 1976. No-tillage Effects on Soil Properties under Different Crops in Western Nigeria. Soil Sci. Soc. Amer. Jour., 40, (5), 762-768.

Morgan, R. P. C., 1980. Implications. In Kirkby, M. J. & R. P. Morgan (eds.), Soil Erosion. Chichester, Wiley.

Sharma, S. G., R. M. Gupta, and K. S. Pamwar, 1976. Concept of Crop Protection Factor Evaluation in Soil Erosion. Indian Jour. Agric. Res., 10 (3), 145-152.

International Geomorphology 1986 Part II
Edited by V. Gardiner

SOIL LOSS MEASUREMENTS IN AN AGRICULTURAL AREA OF NORTH-EAST SPAIN

M. A. Marqués and J. Roca

Department of Geomorphology and Tectonics, Faculty of Geology, University of Barcelona, Gran Via 585, Barcelona, 08007 Spain.

ABSTRACT

Soil losses by sheetwash have been measured since November 1982 on experimental plots set up on private agricultural estates 20km west of Barcelona. Climate and runoff data have been collected from meteorological instruments and gathering devices similar to Gerlach troughs. During 1983 total rainfall was 528.5mm and soil losses in this experimental zone were 2.4kg/m^2. The erosion took place at an uneven rate. From 46 rainy days only 11 produced overland flow and three of these were responsible for 92% of the total erosion. For 1984 total rainfall was 628.5mm and soil losses were 0.358kg/m^2; 61 rainy days were recorded but only ten produced overland flow. Only a few trends have been identified between the different parameters.

INTRODUCTION

Most of the soil loss by overland flow in Catalunya, north-east Spain, takes place in the agricultural area. This is due both to the type of crops and to the agricultural methods and techniques. The vegetation is mainly vineyards together with almond trees, olive trees and peach trees, and also cereals. Due to an increase in mechanization, the size of the cereal fields is growing and old terraces are being removed. The direction of the ploughing is essentially determined by the field shape rather than its slope. Between harvest and sowing times, the fields remain bare (ploughed or with hardened soil and stubble) and this period coincides with the main rainfalls. On the other hand, the vineyards and the fruit trees are cultivated in lines in which orientation with respect to slope varies and may be parallel, perpendicular or oblique. Between the stocks or fruit trees, the soil is kept completely bare by means of the plough or weed-killers. These agricultural techniques tend to leave the soil without protection and to increase the rainfall and overland flow aggressivity.

Up to the present day, there are almost no experimental quantitative data on erosion in agriculture. The only ones available are those from Marqués and Roca (1985) related to the same experimental plot that we discuss in the present paper. The only other available data at the moment consist of estimates based on the silting of some reservoirs, which include agricultural as well as forested and uncultivated areas.

On the other hand, during the first term of 1985, the Official Chamber of Commerce, Industry and Navigation of Barcelona accomplished a survey of Economic growth and desertification in Catalunya (Creixement econòmic i desertització a Catalunya, in press). Various estimates of erosion are found in this study based on Fournier's index, the Universal Soil Loss Equation, and qualitative assessment.

A survey of erosion risk was carried out by dividing the area into one kilometre grid squares and categorising the risk in each square according to lithology, vegetation cover and slope. Six different types of risk were established with the highest risks associated with the agricultural areas, which agrees accurately with the field surveys.

The Universal Soil Loss Equation was applied to a set of samples related to every type of erosion risk. The results are only indicative as they were developed from data for the one kilometre grid squares. The values that result range between 0.85 and 171 mt $ha^{-1} yr^{-1}$ from the lowest to the highest risks respectively.

In respect of Fournier's Index, the average rate for Catalunya is 930 mt $km^{-2} yr^{-1}$, which is approximately 0.66mm yr^{-1}. The lowest rates are in the Ter basin (Cuenca del Ter) with 232 mt $km^{-2} yr^{-1}$, and the highest in the Baix Camp with 3,301 mt $km^{-2} yr^{-1}$.

In addition to this regional evaluation, there are some detailed studies, as for example that of Clotet et al., (1983) about the badland areas of the Catalan Prepyrenees, that of Clotet (1984) in a coal mining area and in the high basin of the river Llobregat (suspended sediment load at one sample point). Also, Sala (1981, 1982) and Salvador (1982) have carried out some studies in forested areas.

We consider that the highest erosion rates take place in agricultural soils; the types of agriculture and above all the farming techniques are the main factors controlling the soil loss; and quantitative date are necessary to make the farmers and others aware of these losses.

THE EXPERIMENTAL STUDY

The zone and the erosion trends. The zone under study is located in the Alt Penedes. We chose the agricultural area of Penedes because of the proximity to Barcelona and the economic importance of the wine and fruit production of the region. The selection of the specific place at the village of Masquefa was because of the permission and facilities of the landowners for installation of the experimental plots, as there are no public farms near Barcelona.

The Penedes is a Neogene tectonic depression filled with lutites, sandstones and lense-shaped conglomerates. It lies 100-300m above sea level and is a hilly region with gentle slopes, mainly pediments, which are dissected by a drainage network in the distal part. The soils are mainly Xerorthent (Aquic and Typic) and Xerochrept (Calcixerollic and Typic) (Boixadera, 1983), and they are devoted to vineyards, fruit trees and cereals.

The climate is typically Mediterranean with noticeable maritime influences, gentle temperatures and frost risk for only three months. The lowest absolute monthly average is -3°C and the highest 35°C. The water balance registers a lack of water from June to September and an excess from February to April. The average annual precipitation is 560mm, with maxima in April-May and September-October. The first one reaches 50-70mm per month and the second 70-90mm per month. The rainfall minimum occurs during the months of July and August.

From the erosion point of view, surface water is the main agent and three types of overland flow can be distinguished: gully, rill and sheetwash. The three types are closely related to the landuse and management as well as to the farming techniques. Many of the gully heads are located at the edge of the fields, the land properties, or in relationship with the paths and roads. Drainage channels with an area of about 11 square kilometres where the experimental plot has been installed, have been represented in Fig.1. Man-induced gullies represent a high percentage out of the total.

Rill erosion is closely associated with the distribution and disposition of the crops (parallel, oblique or perpendicular to the slope) and to the ploughing. The rills form mainly where cropping and ploughing occur parallel or oblique to the slope. In the contour worked areas, the decimetric scale rills only form when either this particular area receives a supplementary discharge of water from the upslope fields (for instance overflow of diversion channels), or in the case of unusual rains.

The experimental plots. The survey is being carried out to study the erosion produced by sheetwash. Thus the soil losses measured belong to the lowest category of soil loss because the area is contour-cultivated. This technique reduces surface runoff and generally inhibits rill formation. Nevertheless, rills are formed in some contour-cultivated areas as stated above.

Four experimental plots were installed at the beginning but unfortunately three of the sediment collecting troughs disappeared. Although we were working in private property, it was open to people as there were no fence protections.

The data we present here deal with a plot located in a field of peach trees planted on the contour lines and contour cultivated. The ground remains bare between the trees. No lateral boundaries were placed to delimit the drainage area of the experimental plot, in order to avoid interference with the farming works. We measured the erosion produced by sheetwash with the help of Gerlach type troughs (80 x 30 x 10cm) connected to 30 litre capacity collecting tanks. The total runoff and sediment from the plot is collected and then a runoff/sediment mixture is sampled and analysed.

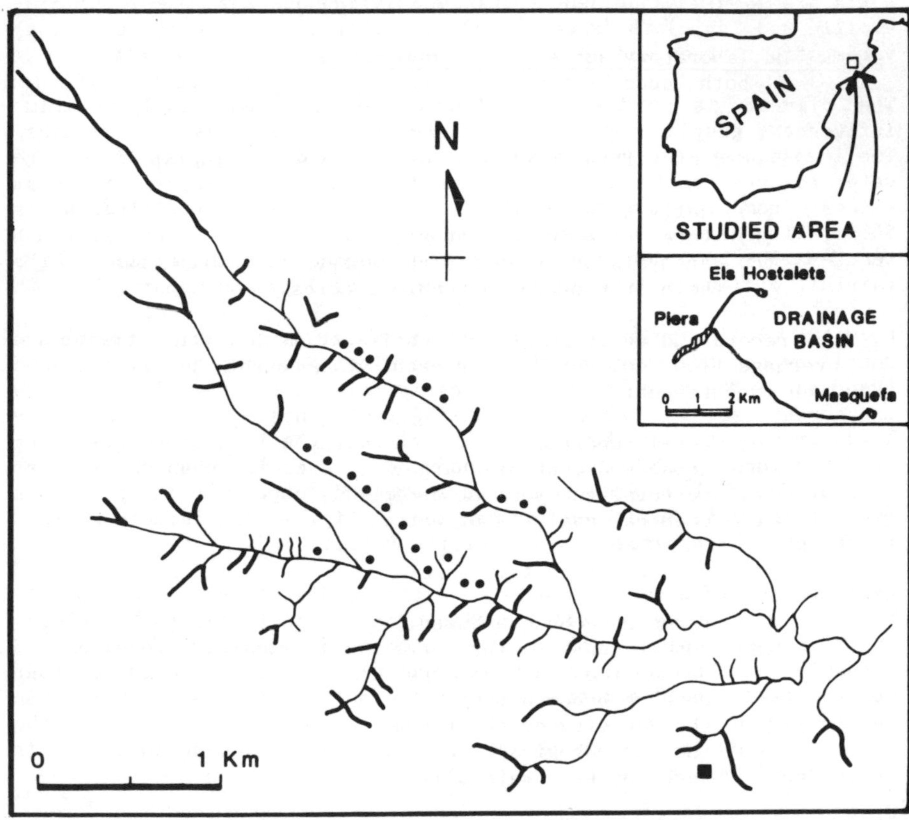

Fig.1. Drainage network. Man-induced gullies are indicated by thick strokes; dots correspond to badland areas. The location of the experimental plot is marked by a square.

The experimental plot has a surface of 6.4 square metres, a slope of 5° and faces north-east. The bedrock is a calcareous siltstone and the soil, according to Boixadera (1983), can be classified as typic xerochrept with an $Ap\text{-}B_2\text{-}2C_1\text{-}3C_2$ profile. The Ap horizon is of ochric type; it is 10cm thick and consists of 27% sand, 47% silt and 26% clay. The pH is 8.2 and the carbonate content approximates 30.6%. It has a granular structure and the percentage of coarse grains is lower than 5%. Some other characteristics are the appearance of organic compounds, the frequency of fauna, the oxidation and the lack of concretions.

The B_2 horizon of Cambic type develops between 10 and 40cm depth. It has a texture of 29% sand, 40% silt and 25% clay. The pH is 8.3 and the carbonate content is 32.5%. The structure is polyedric subangular. Its coarse particle content is less than 5%. It is without apparent organic compounds, is oxidized, and contains some calcareous concretions. The infiltration capacity of the soil was determined with a ring infiltrometer at 0.06mm min^{-1} (Marqués and Roca, 1985).

Field and laboratory survey and measurements. The analysis carried out was both quantitative and qualitative. The quantitative data are:

(1) A record of rainfalls with a Hellman type pluviograph:

- (a) daily total rain,
- (b) total rain in each storm,
- (c) analysis of intensity characteristics with the help of the pluviographic records.

(2) A record and analysis of the runoff-sediment mixture transported by overland flow, collected after each rain event by means of a Gerlach type trough:

- (a) total runoff,
- (b) total sediment discharge,
- (c) textural analysis of sediment,
- (d) chemical analysis of water.

The qualitative data refer to:

- (1) farming techniques used,
- (2) conditions of the vegetation concerning the peach trees as well as the weeds that can develop beneath them,
- (3) the activity of the fauna,
- (4) the morphological results and consequences produced by each rainfall.

RESULTS

As expected considering the Mediterranean character of the area the results show a strong irregularity in both the rainfall and the effects it produces on soil loss. Some rainfalls with very similar pluviometric characteristics produce different results. This apparent disagreement can be explained by:

- (a) different agricultural works carried out,
- (b) preparation of the superficial material by the fauna,
- (c) time gap between the precipitation events.

During the year 1983 (Marqués and Roca, 1985) 46 rain days were registered in the pluviograph located near the experimental plot. Only 11 rain days out of the total were effective in producing erosion. The total rainfall recorded this year was 528.5mm with 384mm in the period September-November. The total soil loss was 2,400 g/m^2, and the runoff was 25.5 l/m^2.

Ninety two per cent of the total erosion took place in only three days:

(a) the 23rd of August 1983 produced 14% of the erosion (331 g/m^2).
(b) the 2nd of September 1983 produced 70% of the erosion (1669.25 g/m^2).
(c) the 10th of November 1983 produced 8% of the erosion (197.82 g/m^2).

The event taking place on the 2nd of September is especially remarkable. Even though we did not succeed in having a pluviograph record, we can estimate that the maximum intensity was about 2.3mm min^{-1} according to eyewitnesses. This event must be considered as an uncommon phenomenon, with a return period of up to 30 years.

In 1984, 61 rain days produced 628.5mm of precipitation. Ten rainfalls were erosive and they gave sediment transport of 358 g/l. Ninety two per cent of the erosion of these effective rainfalls occurred in 4 days of rain:

(a) the 29th of September 1984 produced 40% of the erosion (1729 g/m^2).
(b) the days from the 2nd to the 4th of November caused 20% of the erosion (73.5 g/m^2), in six showers of almost continuous precipitation.
(c) the 5th of November 1984 caused 12.5% of the erosion (44.8 g/m^2).
(d) the 1st of December 1984 produced 12% of the erosion (43 g/m^2).

During 1983 the minimum quantity of rain to produce erosion was 8mm and in 1984 it was 7mm, but it is important to mention that some greater rainfalls did not produce erosion. This is the case, for instance, of a rainfall of 10.5mm with an average intensity of 0.02mm min^{-1} and a maximum intensity of 0.126 mm min^{-1} over 15 minutes. Another rainfall of 8.4mm with an average intensity of 0.018 mm min^{-1} and maximum of 0.05 mm min^{-1} over 45 minutes caused a small erosion of 0.93 g/m^2. These differences are due to the physical condition of the soil during the storm.

RELATIONSHIPS BETWEEN THE MEASURED PARAMETERS

Before discussing this we would like to make some observations. First of all, in one year of observation in an area where the total pluviosity is very reduced and its distribution very irregular, the number of effective rainfalls producing soil loss was very limited. Consequently, many years of observation are required to find out meaningful statistical relationships. This is one of the main problems and may be one of the basic reasons for the few results found in the literature of similar climatic zones.

Second, the period of time between each rainfall can be very different. In some cases it can be more than one month while, in other cases, several rainy days can be consecutive. In response to this fact, the soil conditions vary considerably not only from the moisture and compactness point of view, but also from the supply and preparation of the material by, for example, the fauna and the farming operations.

Finally, as our aim consists of studying the agricultural erosion in a real situation, the experimental plot is cultivated and worked in the same way as the rest of the cropped land, bearing then the same vicissitudes such as theft and trampling.

Some trends have been established between the measured parameters for the period November 1982 to December 1984, even though few data are available. Only 21 rainfalls produced sediment transport; those producing less than 2% (18 g/m^2) out of the annual erosion in 1983 or 1.8% (11 g/m^2) in 1984 have been considered as negligible. Hence the data become quite reduced.

The total effective rainfall versus the maximum rainfall intensity over ten minutes does not indicate a random distribution. Due to the scarce number of significant events, the few available data do not allow us to determine exactly the relationship between both variables, or even to establish whether the exponential relationship fits better than the linear one.

The effective rainfalls that have created a surface runoff equal to or greater than 1.4 l/m^2 have been plotted versus the runoff (Fig.2). The runoff of 1.4 l/m^2 has been taken as a threshold because low runoffs have produced very low quantities of sediment transport (less than 0.9 g/m^2) and these events have not been considered. Moreover, this soil loss may be produced by other processes, such as wind, fauna, tillage, since low quantities of sediment have sometimes been trapped in the troughs in dry periods. The distribution tends to represent a linear relation, with a correlation coefficient of 0.97. Two events for which the relationship between the runoff and the total rainfall is very low have not been plotted. In one case, the soil offered a vegetation cover that as well as helping the infiltration, also operated as a barrier to raindrop impact. In the other case, due to the period of the year, evaporation occurred between the event and the collection of the sample. With the available data low correlation appears between the effective rainfall and the soil loss.

The plot between the total rainfall of each effective event versus the particle concentration or density of overland flow (g l^{-1}) seems to present a random distribution, although two groups can be distinguished (Fig.3). One of them has a distribution with the peculiarity that the density of the flow is close to 10 g l^{-1} independently of the quantity of rain. In the other one the density increases with the rain. The two distributions may be related to different processes of transport. The first process shows a low density flow only, caused by the washing down of fine particles. The second process shows a high density flow and it may even carry gravel size particles. In this case the material is seen as being moved down the hillslope in bulk as a micro-scale mass movement by shear failure.

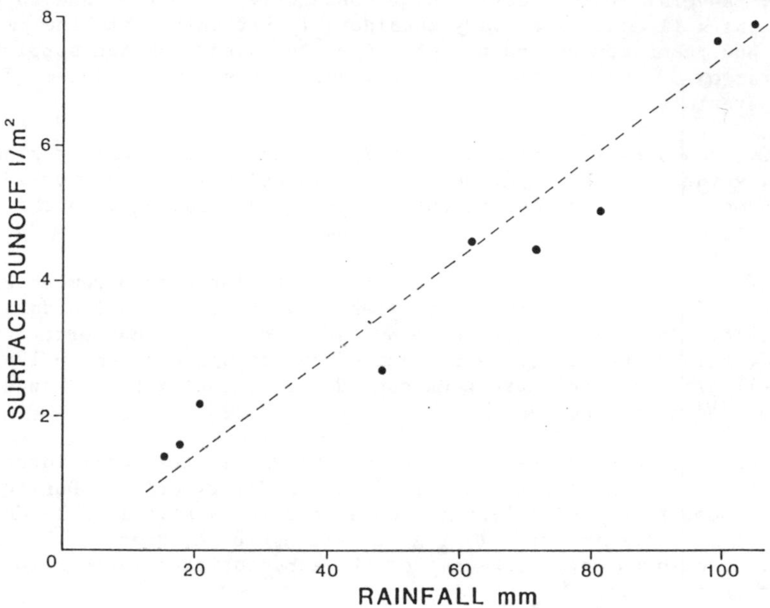

Fig.2. Relationship between surface runoff and rainfall.

The shear failure mainly results from an increase of pore water pressure and from the micromorphological changes of the soil surface derived from the washing down of fine particles. The genesis of these two types of processes can be explained by the conditions of the soil and availability of the material. Dense flows are usually generated by rainfall following a long period of drought or when the soil has been prepared by the plough or biologic activity (mainly ant hills).

The relationship between the density of the overland flow and the soil loss (Fig.4) shows a distribution that tends to be an exponential curve (correlation coefficient 0.88, standard error 0.36) more than to a straight line (correlation coefficient 0.85, standard error 2.9). A similarity appears when comparing these results with the ones of the rainfall versus density of overland flow. A group of points has a soil loss magnitude more or less constant and with a rate approaching 10 g/m^2. In the other group however, the soil loss increases with the density.

The soil loss versus surface runoff plot shows a linear distribution (Fig.5), though some points are anomalous. For instance the 2.7 l/ m^2 runoff with 11.4 g/m^2 soil loss is the response to a denser vegetal cover. The results from the 2nd September 1983 storm are not included because their range (7.5 l/m^2 and 1660g/m^2) is off the standard values of the zone.

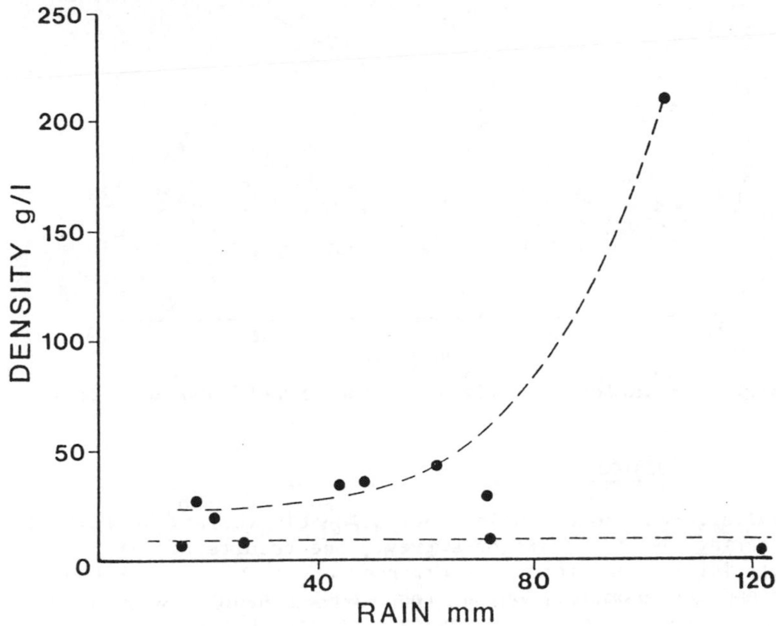

Fig.3. Relationship between particle density of the overland flow and rainfall.

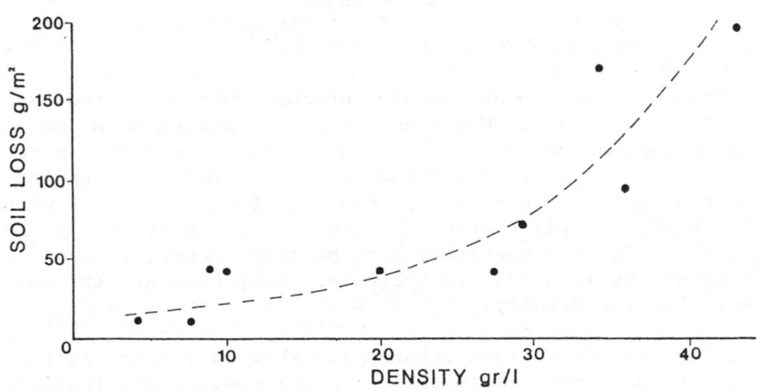

Fig.4. Relationship between soil loss and particle density of the overland flow.

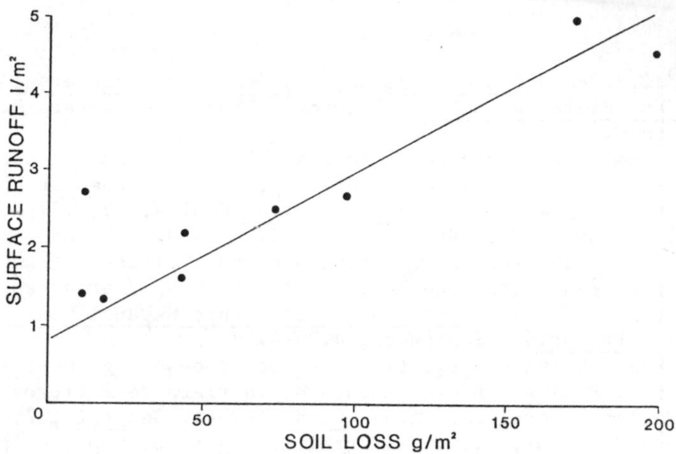

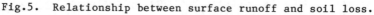

Fig.5. Relationship between surface runoff and soil loss.

CONCLUSIONS

The Mediterranean zone studied had noticeable variations in soil loss rates during the two years of survey. The results confirm the uneven characteristics of the Mediterranean areas in terms of the erosive agents and the geomorphological responses. In such a context long periods of recording are necessary in order to obtain representative data.

Acknowledgements

The project has been partially funded by CIRIT (1984) and University of Barcelona (1985) research grants. The authors would like to thank the landowners of Can Massana and Can Bonastre for the permission to install the experimental plots and to Mr. J. Monsó and Mr. J. Casanova for their assistance in field work.

REFERENCES

Boixadera, J., 1983. Proyecto de un área modelo de conservación de suelos en Piera-Masquefa. Proyecto final de carrera Etsia de Lleida (ined.).

Clotet, N., 1984. La conca de la Baells (Alt Llobregat) : processos geomorfològics actuals responsables del subministrament sòlids i balanç previ de sediments. Act. Geol. Hispanica, 19, 177-191.

Clotet, N., Gallart, F., and Calvet, J., 1983. Estudio de la dinámica de un sector de badlands en Valcebre (Prepirineo Catalán): puesta a punto de la metodología y primeras aproximaciones cuantitativas. Act. 2a Reunión Grupo Espanol Geol. Amb. y Ord. del Territorio. Santander, 4.20-4.38.

Sala, M., 1982. Datos cuantitativos de los procesos geomorfológicos fluviales actuales en la cuenca de la riera de Fuirosos (Montnegre, macizo litoral catalán). Cuad. Invest. Geograf., 7, 53-70.

Sala, M., and Salvador, F., 1982. Preliminary report on the process measurements on the Catalan Mediterranean slopes. Studia Geomorph. Carpatho-Balcanica, 15, 65-70.

Marqués, M. A., and Roca, J., 1985. Resultados preliminares de las pérdidas de suelo en zona agrícola del Penedes (Cataluña, Espana). Cuad. Inv. Geograf., x, 149-157.

SUMMARY

[illegible]

[illegible] the modified universal soil loss equation (MUSLE) [illegible] [illegible] changes in soil fertility and productivity as a result of 30 years [illegible] 620 Mg/ha/yr average [illegible] soil loss) were obtained with [illegible]

INTRODUCTION

Changes in soil chemical [illegible] and [illegible] and land use and management [illegible]. Consequently, the [illegible] of [illegible] (30 years of [illegible]) [illegible] soil and erosion of soil and associated nutrients as a result of different land management, is important from agronomic [illegible] and environmental standpoints.

[illegible] and Adams (1976, 1959) and [illegible] (1973) suggested that the phosphorus content of the parent material of a soil ultimately controlled the phosphorus (P), nitrogen (N), carbon (C) and sulfur contents of the soil. This was developed further by [illegible] (1976) who, [illegible] examining the P content of several [illegible] and [illegible] suggested that phosphorus availability [illegible] would control nitrogen fixation and that this, in turn, would ultimately control organic matter accumulation. [illegible] key element in pedogenesis, [illegible] significance [illegible]

International Geomorphology 1986 Part II
Edited by V. Gardiner

SIMULATION OF SOIL FORMATION AND DEGRADATION PROCESSES

A. N. Sharpley, S. J. Smith, J. R. Williams and C. A. Jones

USDA- Agricultural Research Service, Durant, Oklahoma and Temple, Texas, U.S.A.

ABSTRACT

A model simulating soil N, P, and C cycling was formulated and incorporated into the Erosion-Productivity Impact Calculator model (EPIC), previously developed to assess the effect of erosion on soil productivity. Crop uptake and transformations among several inorganic and organic pools of N, P, and C are simulated. Accurate simulations of long-term changes (more than 50 years) in plant available soil nutrients, organic P and N, and fertilizer requirements and recommendations for maize and wheat were obtained for a range of soils in the continental U.S. To predict the movement of soil in runoff, the modified Universal Soil Loss Equation (MUSLE) was employed for several grassed and cropped watersheds in 3 Major Land Resource Areas of the U.S. MUSLE and measured soil losses were similar over study periods of 3 to 5 years. Annual total P and N losses, predicted using the logarithmic relationship between enrichment ratio (nutrient content of eroded soil/source soil) and soil loss, were similar to measured values. Reasonable estimates of changes in soil fertility and productivity as a result of 50 years of erosion (20 Mg/ha/yr average soil loss) were obtained with EPIC.

INTRODUCTION

The change in soil chemistry through nutrient cycling and erosion due to land use and management may be a slow process. Consequently, the accurate prediction of long-term nutrient cycling (50 years or more) in the soil and erosion of soil and associated nutrients, as a result of different land management, is important from geomorphologic and environmental standpoints.

Walker and Adams (1958, 1959) and Smeck (1973) suggested that the phosphorus content of the parent material of a soil ultimately controlled the phosphorus (P), nitrogen (N), carbon (C), and sulfur (S) contents of that soil. This was developed further by Walker and Syers (1976), who, after examining the P content of several chrono-, topo-, and climosequences, suggested that phosphorus availability to organisms would control nitrogen fixation and that this, in turn, would ultimately control organic matter accumulation. As P is probably the key element in pedogenesis, because of its ecological significance (Walker, 1965) and its recently acknowledged role in accelerated eutrophication (Schindler, 1977; Vollenweider, 1980), slightly more emphasis will be placed on P than the other nutrients in the following discussion.

Over a long time (50-200 years) nutrient cycling and erosion can have a major effect on soil formation and degradation and consequently, soil fertility and productivity. Since soil P, N, and C are normally concentrated in the top layers of the soil, they are particularly susceptible to removal by erosion. Over the same period, however, P, N, and C fertility can also be affected by fertilizer application, crop removal, leaching, and changes in the organic matter content of the soil. Consequently, a simple model simulating soil P, N, and C cycling was developed for incorporation into the Erosion-Productivity Impact Calculator (EPIC), which is composed of physically based components for simulating erosion, plant growth and related processes, and economic components for such assessments as the cost of erosion and determining optimum management strategies (Williams et al., 1983; 1984a). The processes involved are simulated realistically using readily available inputs. Since erosion can be a relatively slow process, EPIC is capable of simulating hundreds of years if necessary. The model is generally applicable, computationally efficient, and capable of computing the effects of management changes on outputs.

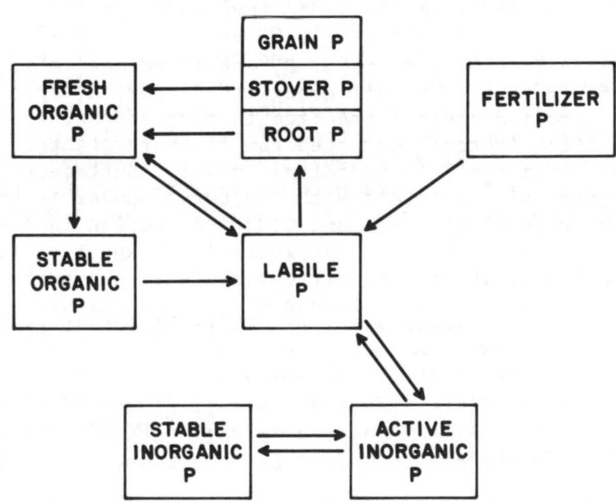

Fig. 1. Pools and flows of phosphorus in the EPIC model

For P, the model simulates uptake and transformations between several inorganic and organic pools in up to 10 soil layers of variable thickness (Fig. 1) (Jones et al., 1984 a). Fertilizer P is added to the labile inorganic P pool which rapidly achieves equilibrium with active inorganic P. The relative sizes of the labile and active inorganic P pools are soil specific and are based on soil classification, texture, and chemical properties (Sharpley et al., 1984). Movement of inorganic P from the active to the stable pools simulates slow adsorption of inorganic P. Crop P uptake from a soil layer is sensitive to crop P demand and the amounts of labile P, soil water, and roots in the layer. Stover and root P are added to the fresh organic P pool upon their death and/or incorporation into the soil. Decomposition of fresh and stable organic matter may result in net immobilization of labile P or net mineralization of organic P.

Soil organic matter and N transformations closely follow P flows and are based on Seligman and Van Keulen's (1981) PAPRAN model. They are divided into fresh pools, consisting of decomposing crop residue and microbial biomass N and stable pools consisting of organic matter humus and humus N. The nutrient models are sensitive to soil chemical and physical properties, nutrient requirements of crops, tillage practice, fertilizer rate, soil temperature, and soil water content.

Prediction of soil erosion is based on the Modified Universal Soil Loss Equation (MUSLE) (Williams, 1975). In this equation the rainfall energy factor of USLE (Wischmeier and Smith, 1960) is replaced by a runoff energy factor ($11.8\ (Qq_p)^{0.56}$), resulting in an improvement in prediction, as sediment yield variation is generally more dependent upon runoff than rainfall. The MUSLE is:

$$Y = 11.8\ (Qq_p)^{0.56}\ KMESL \qquad [1]$$

where Y is sediment yield (Mg); Q, runoff volume (m^3); q_p, peak runoff rate (m^3/sec); K, soil erodility factor; M, crop management factor; E, erosion control practice factor; and SL, slope length-gradient factor. Computed runoff energy factors were determined using the EPIC hydrologic model. Surface runoff was computed from daily rainfall with the curve number technique (SCS, 1972). The removal of P, N, and C during erosion is a selective process with respect to the preferential transport of fine material (less than 2 µm), of higher nutrient content than surface soil (Sharpley, 1980). Nutrient losses are, thus, calculated from the total P, N, and C content of surface soil and respective enrichment ratios (Menzel, 1980; Sharpley, 1980, 1985):

$$\text{Ln PER} = 2.48 - 0.27\ \text{Ln Soil Loss} \qquad [2]$$
$$\text{Ln NER} = 2.00 - 0.20\ \text{Ln Soil Loss} \qquad [3]$$
$$\text{Ln CER} = 2.00 - 0.20\ \text{Ln Soil Loss} \qquad [4]$$

where the units of soil loss are kg/ha.

This paper presents an overview of the modeling approaches used to simulate soil nutrient cycling and erosion as it impacts on soil forming processes and soil fertility and productivity. Simulations are compared with measured field data where available and in the case of 200 year simulations, estimates of changes in the content of soil P forms are presented in an attempt to simulate soil weathering processes.

METHODS

Published data were used to test simulations of changes in P, N, and C content of soil over approximately 50 years. Initial soil properties and crop management information were obtained from published data and from personal communications. When soil pedon data were not available, data from published and unpublished U.S. Soil Conservation Service (SCS) pedon descriptions were substituted. Weather sequences at the site of experiments were simulated (Nicks and Harp, 1980; Richardson, 1981, 1982).

The predictions of soil and associated nutrient loss in runoff were compared to measured data from 23 watersheds in the Southern Plains area of the U.S. The watersheds encompassed a range of sizes (1 to 122 ha), soils (Mollisols, Vertisols, and Inceptisols), slopes (1 to 9%), grasses (native and introduced), crops (wheat, oats, sorghum, and cotton), fertilizer P (0-40 kgP/ha/yr) and N (0-134 kgN/ha/yr) application rates, and study periods (3-5 yr). More detailed information on management of the watersheds and runoff collection and analysis is given by Sharpley et al. (1985) and Smith et al.(1983).

RESULTS AND DISCUSSION

Nutrient Cycling. Measured and predicted long-term (50 yr) changes in surface soil organic P, total N, and organic C at several locations in the Great Plains of the U.S. are summarized in Table 1. Measured and simulated values after the period of cultivation were not significantly different at the 0.10 probability level (as determined by analysis of variance for paired data). This suggests that EPIC produces reasonable estimates of the long-term (50 yrs) effects of cultivation on surface soil P, N, and C content in the Great Plains. Sensitivity analyses have shown that the slight overestimation of mean topsoil organic P, total N, and organic C may be due in part to the fact that soil erosion (by water and wind) was kept minimal during the simulations.

Although cultivation affects subsoil nutrient cycling to a lesser degree than surface soil, little field information is available (Sharpley and Smith, 1983; Walker et al., 1959). The nutrient model predicted a smaller change in organic P, total N, and organic C content in the subsoil than in surface soil (Fig. 2). Fargo silty clay loam cropped with small grains and alfalfa for 70 years and Houston Black clay cropped with grain sorghum for 60 years are presented as examples.

It is apparent, therefore, that EPIC can simulate changes in organic P, total N, and organic C in a soil profile, occurring over approximately 50 years, reasonably well in comparison with field data. An accurate simulation of plant available P over similar time periods has also been shown (Jones et al., 1984b). During pedogenesis, however, the relative amounts and forms of nutrients, particularly P, can change (Walker and Syers, 1976). In general, mineralized organic P accumulates in the stable P pool. The amounts of labile, active, stable, and organic P were simulated over a period

TABLE 1. Measured and simulated changes in soil surface organic P, total N and organic C in the Great Plains. Measured data from Haas et al. (1957 and 1961).

Location	Duration of study	Rotation	Organic P			Total N			Organic C		
			Virgin	Cultivated Meas.	Cultivated Sim.	Virgin	Cultivated Meas.	Cultivated Sim.	Virgin	Cultivated Meas.	Cultivated Sim.
	Years		(mg/kg)						(g/kg)		
Havre, MT	31	Sp. Wheat-fallow	157	102	108	1510	900	1135	17.5	8.3	12.4
Moccasin, MT	39	"	308	183	169	3000	2050	1787	32.4	21.9	19.6
Dickinson, ND	41	"	292	148	174	2930	1490	1957	36.4	15.1	21.5
Mandan, ND	31	"	139	132	97	1600	1160	1172	24.2	17.0	12.8
Sheridan, WY	30	"	120	93	86	1590	1210	1149	16.6	11.9	12.6
Laramie, WY	34	"	142	91	96	1220	820	900	13.3	7.8	9.9
Akron, CO	39	"	115	82	81	1340	800	911	14.2	7.7	10.0
Colby, KS	31	W. Wheat-fallow	158	61	92	1650	1050	952	18.3	10.1	10.4
Hays, KS	30	W. Wheat	174	97	108	2200	1220	1360	24.7	12.1	14.9
Lawton, OK	28	"	128	71	73	1540	740	904	17.3	8.0	9.9
Dalhart, TX	29	Maize	84	39	53	670	420	444	7.2	4.4	4.9
Big Spring, TX	41	W. Wheat	55	30	29	600	410	328	6.7	4.0	3.6
Mean	34		156	94	97	1654	1023	1083	19.1	10.7	11.9

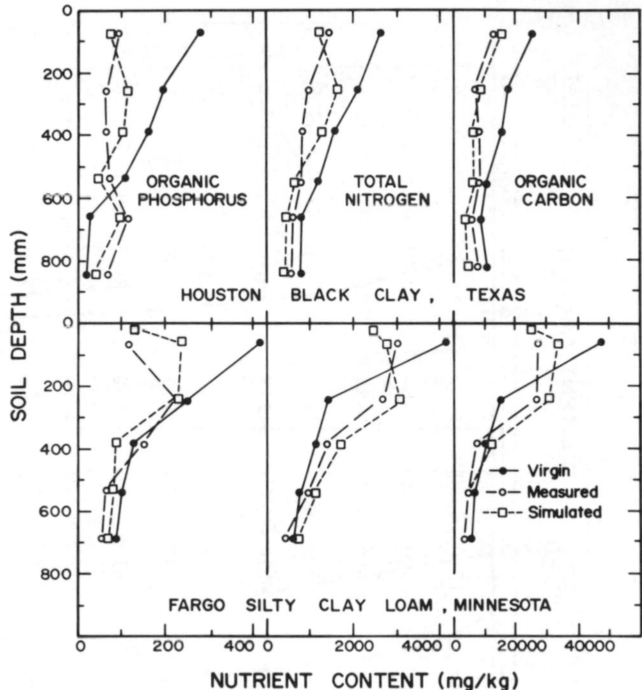

Fig. 2. Measured and simulated changes in the nutrient content of Houston Black and Fargo soil profiles after 60 and 70 years cultivation,respectively. Measured data from Sharpley and Smith (1983).

of 200 years for a Venezuelan soil (Colabosa clay) under grass and subject to temperate and tropical climatic conditions (Fig. 3). For the tropical climate, annual maximum and minimum temperatures were 31.6 and 21.4 C, respectively, and annual rainfall was 1404 mm averaged for the 200 year simulation. For the temperate climate, annual maximum and minimum temperatures were 11.7 and -1.3 C, respectively, and annual rainfall was 377 mm averaged for the 200 years. In order that weathering could be the dominant process affecting the changes in P content, fertilizer P input and erosion (by water and wind) were set to zero. Although plant material was recycled back into the soil organic matter pool, a net mineralization of organic P and increase in stable inorganic P occurred during the 200 year simulation (Fig. 3). The rate of transformation of P from the organic to stable pool, however, was greater under tropical than temperate climates, due to more favourable soil moisture and temperature conditions for both inorganic and microbial soil reactions in the former climate (Fig. 3). The changes in amounts of organic, active, and stable P were consistent with pedogenic studies summarized by Walker and Syers (1976). Consequently, the model may provide accurate simulations of the effect of weathering on the distribution of soil P forms.

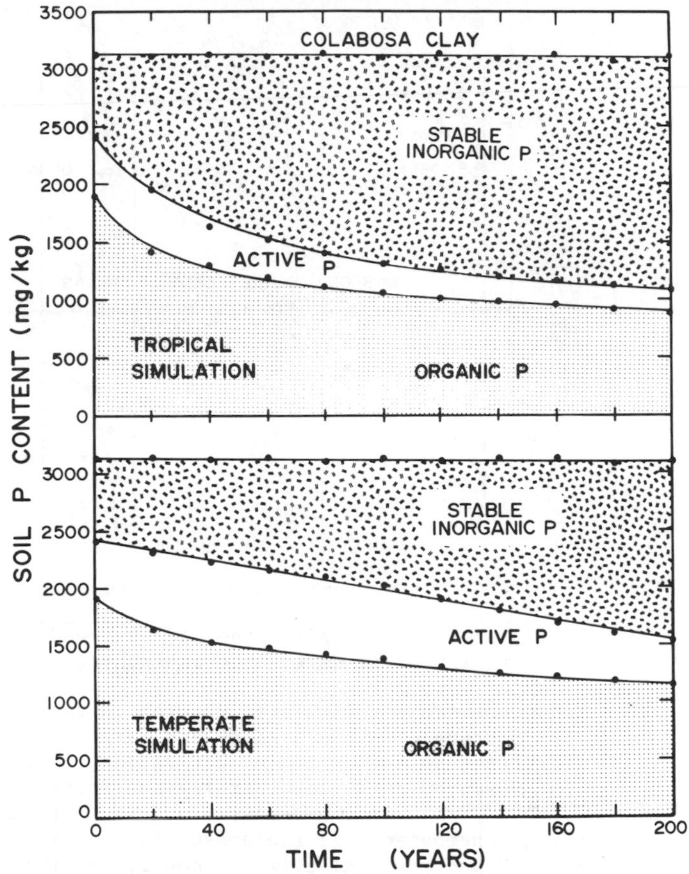

Fig. 3. Simulated changes in the forms of phosphorus in Colabosa clay under tropical and temperate climates over a 200-year period.

Erosion. The mean annual sediment yield predicted by MUSLE and actual measured values from the watersheds are compared in Figure 4. The MUSLE predictions were obtained using computed runoff energy factors (Qq_p, Eq. [1]), as measured factors are not usually available. The mean annual sediment yields were predicted over a range in yields from less than 100 kg/ha from the grasslands to nearly 4000 kg/ha from the croplands (Fig. 4). Predictions were good as slope (0.87) and intercept values (0.76) of the measured-predicted regression were close to unity (Fig. 4). As the computed runoff energy term predicted satisfactory sediment yields, MUSLE will have application beyond just the calibrated watershed situation.
Mean annual losses of P and N associated with sediment were also closely predicted, with regression slope and intercept values close to 1 and 0, respectively (Fig. 5). A similar comparison for C losses

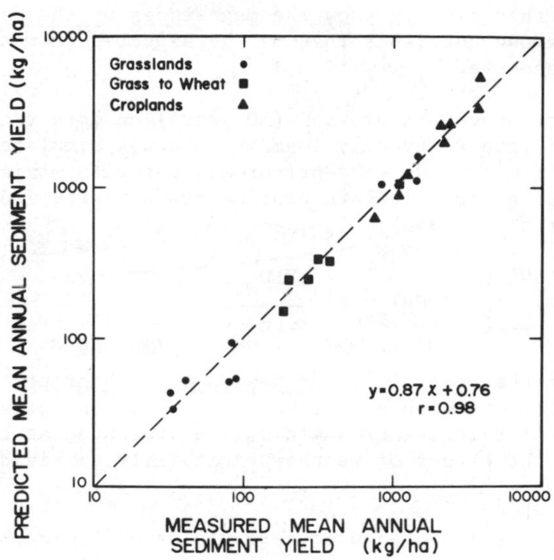

Fig. 4. Comparison of predicted and measured mean annual sediment yields for the watersheds.

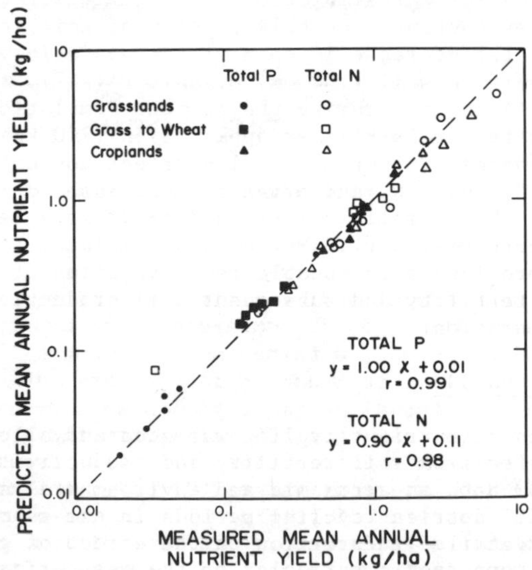

Fig. 5. Comparison of predicted and measured mean annual nutrient yields for the watersheds.

was precluded by a lack of measured data at the present time. Although the nutrient losses are small from a soil fertility standpoint, they clearly show the magnitudes of the differences within and among nutrients that may be expected from a wide range in watershed land-use.

The effect of long-term erosion (50 years) on corn yield and fertility of Tama silty clay loam in Iowa was simulated using EPIC. Two 50-year simulations were performed, one with maximum and one with minimum erosion, to calculate erosion productivity index (EPI) values (Williams et al., 1984b);

$$EPI_a = \frac{YLD_{ea}}{YLD_{oa}} \qquad [5]$$

where YLD is the crop yield for year a and subscripts e and o refer to the simulations with and without erosion, respectively. Calculation of relative crop yields are presented as EPI does not account for the effect of weather variability on yields from year to year.

Figure 6 depicts a decrease in corn yields of 6.4 Mg/ha during 50 years of cultivation without conservation practices on a Tama silty clay loam, receiving 50 kgN and 10 kgP/ha/yr. The average annual erosion rate was 20 Mg/ha/yr, which is approximately twice the acceptable erosion rate and is the average annual loss from U.S. croplands (Duttweiler and Nicholsen, 1983). As the Tama soil is a deep loess (1550 mm profile depth), erosion of the surface 91 mm of soil during the 50-year simulation, had little effect on texture, available water content, and bulk density of soil in the root zone. Consequently, the decrease in corn yields was mainly a result of the decrease in surface soil (150 mm) organic P (22%), total N (51%), and organic C (19%) content during the 50-year simulation (Fig. 6). With recommended rates of fertilizer application (200 kgN and 50 kgP/ha/yr), changes in surface soil nutrient content were minor (Fig. 6). Organic P, total N, and organic C decreased only 7, 12, and 4%, respectively, while labile P increased 66%. As a result little decrease in corn yields (1.3 Mg/ha) was simulated (Fig. 6). The EPIC model, thus provided a reasonable representation of the change in surface soil fertility and subsequent soil productivity as a result of long-term erosion.

CONCLUSIONS

The simulation of nutrient cycling was good and allowed estimation of management effects on soil fertility and productivity, which will be important from both an agronomic and environmental standpoint. Simulations of soil nutrient cycling periods in excess of 200 years may provide more detailed information on the effect of given climatic, soil, and/or topographic variables on the weathering process in a soil. The model EPIC, used to simulate soil nutrient cycling and erosion can, thus, have application to a study of climatic and land management effects on soil formation and degradation processes.

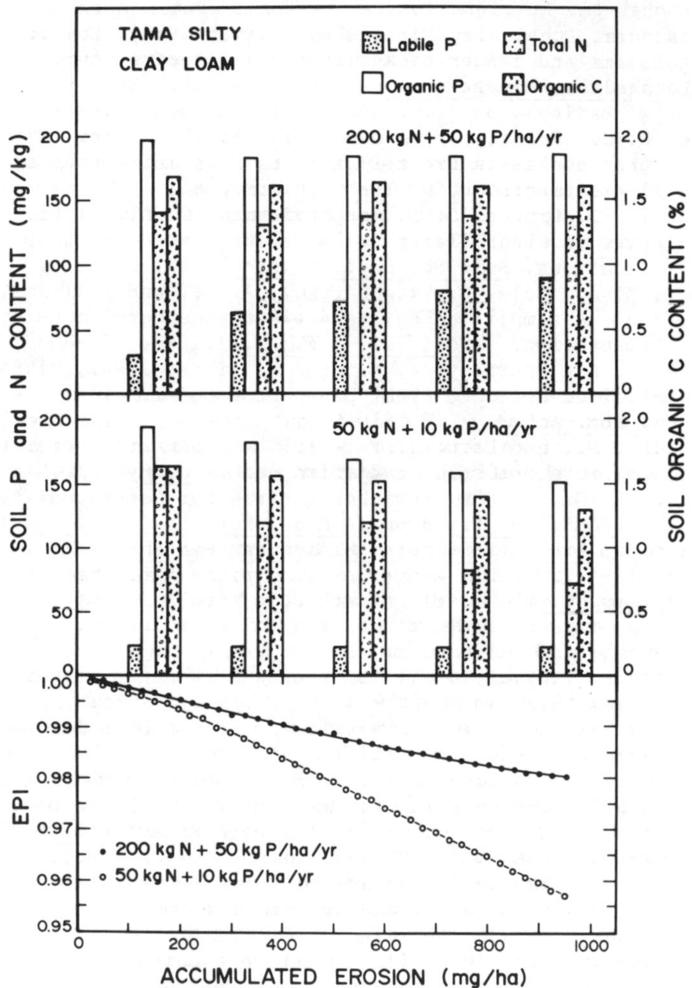

Fig. 6. Simulated changes in relative corn yield (EPI) and soil phosphorus, nitrogen, and carbon content of Tama silty clay loam under recommended (200 kgN and 50 kgP/ha/yr) and low (50 kgN and 10 kgP/ha/yr) fertilizer applications, as a result of 50 years of erosion.

ACKNOWLEDGEMENTS

The authors wish to acknowledge the help of Wes Fuchs, USDA-SCS, Temple, Texas, U.S. and Orlando Guenni, Ministerio De Agricultura Y Cria, FONAIAP-CENIAP, Maracay, Venezuela for providing soil information.

REFERENCES

Duttweiler, D. W., and Nicholsen, H. P., 1983. Environmental problems and issues of agricultural nonpoint source pollution, in Agriculture Management and Water Quality. (Eds. F. W. Schaller and G. W. Bailey), pp 3-16. Iowa State University Press, Ames, Iowa.

Haas, H. J., Evans, G. E., and Miles, E. F., 1956. Nitrogen and carbon changes in Great Plains soils as influenced by cropping and soil treatments. USDA Tech. Bull. 1164.

Haas, H. J., Grunes, D. L., and Reichman, G. A., 1961. Phosphorus changes in Great Plains soils as influenced by cropping and manure applications. Soil Sci Soc. Am. Proc., 25, 214-218.

Jones, C. A., Cole, C. V., Sharpley, A. N., and Williams, J.R., 1984a. A simplified soil and plant phosphorus model: I. Documentation. Soil Sci. Soc. Am. J., 48, 800-805.

Jones, C. A., Sharpley, A. N., and Williams, J. R., 1984b. A simplified soil and plant phosphorus model. III. Testing. Soil Sci. Soc. Am. J., 48, 810-813.

Nicks, A. D., and Harp, J. F., 1980. Stochastic generation of temperature and solar radiation data. J. Hydrol., 48, 1-17.

Menzel, R. G., 1980. Enrichment ratio for water quality modeling, in CREAMS: A Field Scale Model for Chemicals, Runoff, and Erosion from Agricultural Management Systems. (Ed., W. Knisel), pp 486-492. USDA, Conser. Res. Rep. No. 26. Washington, DC.

Richardson, C. W., 1981. Stochastic simulation of daily precipitation, temperature, and solar radiation. Water Resour. Res., 17, 182-190.

Richardson, C. W., 1982. Dependence structures of daily temperature and solar radiation. Trans. ASAE., 25, 735-739.

Schindler, D. W., 1977. Evolution of phosphorus limitation in lakes. Science, 195, 260-262.

Seligman, N. G., and Van Keulen, H., 1981. PAPRAN: A simulation model of annual pasture production limited by rainfall and nitrogen, in Simulation of Nitrogen Behavior of Soil-Plant Systems. (Eds., M. J. Frissel and J. A. Van Veen). pp 192-221. Proc. of a Workshop, Wageningen, January 28 - February 1, 1980.

Sharpley, A. N., 1980. The enrichment of soil phosphorus in runoff sediments. J. Environ. Qual., 9, 521-526.

Sharpley, A. N., 1985. The selective erosion of plant nutrients in runoff. Soil Sci. Soc. Am. J., 49, 1527-1534.

Sharpley, A. N., and Smith, S. J., 1983. Distribution of phosphorus forms in virgin and cultivated soil and potential erosion losses. Soil Sci. Soc. Am. J., 47, 581-586.

Sharpley, A. N., Jones, C. A., Gray, C., and Cole, C. V., 1984. A simplified soil and plant phosphorus model. II. Prediction of labile, organic, and sorbed P. Soil Sci. Soc. Am. J., 48, 805-809.

Sharpley, A. N., Smith, S. J., Berg, W. A., and Williams, J. R., 1985. Nutrient runoff losses as predicted by annual and monthly soil sampling. J. Environ. Qual., 14, 354-360.

Smeck, N. W., 1973. Phosphorus: an indicator of pedogenetic weathering processes. Soil Sci., 115, 199-206.

Smith, S. J., Menzel, R. G., Rhoades, E. D., Williams, J. R., and Eck, H. V., 1983. Nutrient and sediment discharge from Southern

Plains grasslands. J. Range Mgt. ,36, 435-439.

Soil Conservation Service, 1972. National Engineering Handbook, Section IV, Hydrology, U.S. Govt. Print. Off., Washington, D.C.

Vollenweider, R. A., 1980. The loading concept as a basis for controlling eutrophication: Philosophy and preliminary results of the OECD program on eutrophication. Prog. in Water Tech., 12, 5-38.

Walker, T. W., 1965. The significance of phosphorus in pedogenesis, in Experimental Biology. pp 295-316. Williams, Clowes and Sons Ltd., London and Beciles.

Walker, T. W., and Adams, A. F. R., 1958. Studies on soil organic matter: I. Influence of phosphorus content of parent materials on accumulation of carbon, nitrogen, sulfur and organic phosphorus in grassland soils. Soil Sci., 85, 307-318.

Walker, T. W., and Adams, A. F. R., 1959. Studies on soil organic matter: 2. Influence of increased leaching on various stages of weathering on levels of carbon, nitrogen, sulfur and organic and total phosphorus. Soil Sci., 87, 1-10.

Walker, T. W., and Syers, J. K., 1976. The fate of phosphorus during pedogenesis. Geoderma., 15, 1-19.

Walker, T. W., Thapa, B. K., and Adams, A. F. R., 1959. Studies on soil organic matter: 3. Accumulation of carbon, nitrogen, sulfur, organic and total P in improved grassland soils. Soil Sci., 87,135-140.

Williams, J. R., 1975. Sediment yield prediction with universal equation runoff energy factor, in Present and Prospective Technology for Predicting Sediment Yield and Sources. pp. 244-252. USDA, ARS-S-40.

Williams, J. R., Dyke, P. T., and Jones, C. A., 1983. EPIC-a model for assessing the effects of erosion on soil productivity, in Analysis of ecological systems: State-of-the-art in ecological modeling. (Eds., W. K. Lauenroth, G. V. Skogerboe, and M. Flug). pp 553-572. Elsevier Scientific Publ. Co., Amsterdam.

Williams, J. R., Jones, C. A., and Dyke, P. T., 1984a. A modeling approach to determining the relationship between erosion and soil productivity. Trans. ASAE, 27, 129-144.

Williams, J. R., Putman, J. W., and Dyke, P. T., 1984b. Assessing the effect of soil erosion on productivity with EPIC, in Proc. National Symp. on Erosion and Soil Productivity. (Chm., D. K. McCool). New Orleans, LA. December 1984. Publ. by Am. Soc. Agric. Eng., St. Joseph, MI. No. 8-85.

Wischmeier, W. H., and Smith, D. D., 1960. Predicting rainfall erosion losses. USDA Sci. Ed. Adm. Handbook 537, Washington, DC.

International Geomorphology 1986 Part II
Edited by V. Gardiner

SOIL EROSION AND AGRICULTURAL HISTORY IN THE CENTRAL HIGHLANDS OF PAPUA NEW GUINEA

D.Gillieson, P.Gorecki, J.Head and G.Hope.

Department of Geography, Australian Defence Force Academy, University of New South Wales; Department of Prehistory, Research School of Pacific Studies, Australian National University; Radiocarbon Laboratory, Australian National University; Department of Geography, The Faculties, Australian National University, Canberra.

ABSTRACT

Evidence from rockshelter and swamp deposits suggests that intensive dryland swidden horticulture and swampland cultivation commenced at least 9000BP in the Highlands of Papua New Guinea (1000 - 2500m altitude). Several phases of intensification occurred, culminating with the introduction of sweet potato (*Ipomoea batatas*) some 300 - 400 years ago, and accompanied by rapid population increase. The sediment record suggests that in the period 32000 to 9000BP, human impact was minimal and slopewash processes predominated under primary forest. After 9000BP, rates of erosion increased dramatically and were accompanied by changes in secondary forest structure. The last few hundred years has seen a dramatic increase in erosion rates, with widespread soil stripping on hillslopes. Much of this appears to be the result of soil nutrient depletion. In the Highland fringe areas (300 - 1000m altitude), swamp margin cultivation systems were established by at least 3500BP, and have resulted in an eight fold increase in swamp sedimentation rates. Sediment analysis shows that increases in the proportions of secondary forest pollen taxa are paralleled by peaks in magnetic susceptibility and chemical parameters suggesting widespread clearing and burning of forest. Extensive anthropogenic grasslands may be a consequence of soil nutrient depletion.

INTRODUCTION

From the air the Highland valleys of New Guinea look like a green patchwork carpet in which the traces of present and past human activity are clearly visible. These cultural landscapes occupy roughly one quarter of the 800,000km of the island of New Guinea. In many regions there is evidence of severe soil erosion and slope stripping. This may be a consequence of recent intensification of subsistence agriculture and the innovation of cash cropping. It can be demonstrated, however, that the intensification of prehistoric agricultural systems led to deforestation and accelerated rates of hillslope erosion. Perhaps contemporary problems are an inheritance of earlier agricultural regimes, and the contemporary environment is still adjusting to that impact.

Throughout the Highlands traces of past agricultural systems can be seen in the form of infilled ditches, garden boundaries and terraced slopes. Much of the evidence for prehistoric agriculture in New Guinea comes from the Kuk swamp site in the Wahgi valley (Figure 1). There Golson and his co-workers (Golson,1982; Harris and Hughes,1978; Blong,1982; Gorecki,1985) have employed the record from swamp sediments and associated prehistoric drains, gardens and dwellings to reconstruct prehistoric people - land relationships over the Holocene. In the Wahgi, a change from organic clay to inorganic grey clay deposition at c.9000BP is accompanied by a major increase in the rate of sedimentation (Golson and Hughes, 1980). This change coincides with archaeological evidence for wetland cultivation systems.

Unfortunately rates of erosion after 6000 BP in the Wahgi cannot be inferred, for two reasons: firstly open water disposal channels in the swamps carried away most hillslope erosion products, and secondly the deposits that did accumulate were severely disturbed from time to time by tillage and raised bed gardening. The record of the last few millenia is important; within this period major changes in agricultural practice and intensity occurred.

At one point in the Kuk Swamp sequence there is a marked change in sediment type from clays to soil aggregates. This has been interpreted by Golson (1982) as representing the advent of soil tillage as an adaptation to grassland cultivation, and is dated by radiocarbon and tephrachronology to circa 2500 BP. This replacement of forest by grassland results in a drastic biomass loss, so burning of the grassland is insufficient to restore the nutrient balance of the soil. Progressive deleterious changes in soil structure may therefore lead to erosion. Restoration of structure by planting of tree species such as _Casuarina_, _Trema_ and _Dodonaea_ may redress these changes: it is significant that marked rises in _Casuarina_ pollen occur at about 1200 BP in sequences from three Highlands provinces (Enga, Western Highlands, Simbu). This change in the pollen record parallels archaeological evidence for increasing intensification and complexity of prehistoric garden systems. Thus it is likely that these changes caused increases in erosion rates, though these cannot be estimated from the Kuk swamp deposits.

Around three hundred years ago a distinctive volcanic ash fall, named Tibito (Blong,1982) mantled much of the Central Highlands. It is widespread in soil, lake and swamp sections and forms an important marker horizon. Tibito ash marks a series of changes in Highlands land use. Firstly, the rates of sedimentation in lake sediments increase dramatically after that time (Oldfield _et al._, 1980) and their analysis suggests both increased inorganic erosion in the

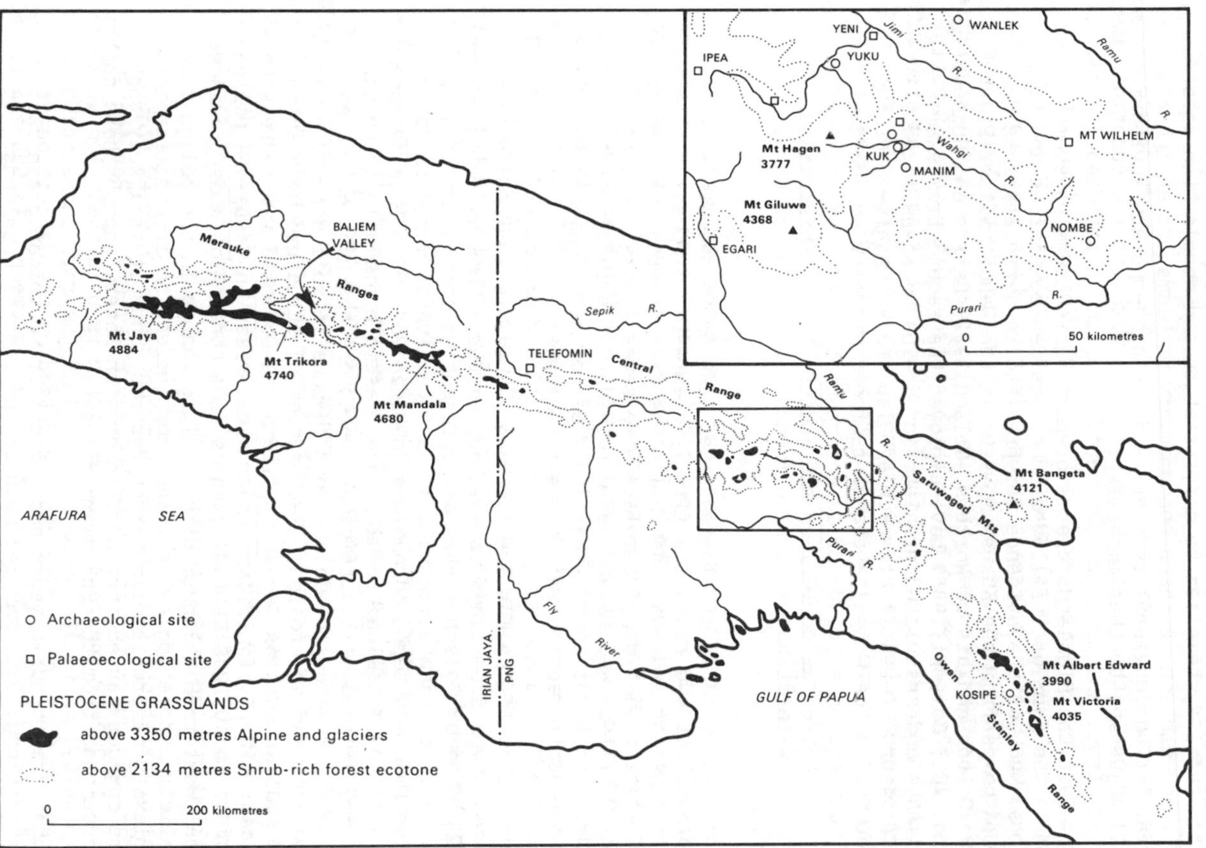

Fig. 1. Palaeoecology and archaeology sites in the New Guinea Highlands. The Jimi valley drains the Sepik - Wahgi divide and the Bismarck Range.

catchment and attendant forest clearance. At higher altitude lake sites in Enga and Simbu, similar forest clearance is indicated, with intensification in the last three hundred years (Walker and Flenley, 1979: 339; Corlett, 1984).

These environmental changes herald the arrival of sweet potato cultivation in the region (Golson,1982). With short maturation time, partial frost and drought resistance, and wide altitudinal range, the sweet potato permitted expansion of population and attendant intensification of social customs such as ceremonial exchange. Perhaps the most significant human effect has been pressure on land resources. That pressure has caused groups such as the Melpa, who live in the Wahgi, to move into lower altitude areas in the Highlands fringe. In this paper we explore the historical dimensions of soil erosion in the Highlands and the lower Highland fringe , an area into which present populations are expanding. Evidence is drawn from both swamp deposits and hillslope deposits accumulating in caves.

ROCKSHELTER DEPOSITS AND PREHISTORIC AGRICULTURE IN THE HIGHLANDS

Records of accumulation rates of lake and swamp sediments are remarkably consistent throughout the Highlands (Table 1). These rates increase through time with a dramatic change in the last five centuries. But what of the surrounding hillslopes? Hughes (1981) has suggested that the colluvial deposits at Manim rockshelter (Wurup Valley, 1600m altitude) reflect changes in the intensity of land use. The rubbly basal sediment accumulated since 10,000 BP at an average rate of 350 mm/1000 yrs. This rate declined around 6000 BP to 170 mm/1000 yrs, then increased dramatically around 2400 BP to 8000 mm/1000 yrs. This change parallels the Kuk sequence in the adjoining Wahgi Valley.

A more detailed study of rockshelter deposits has been carried out by Gillieson (1983 and in press). Two limestone rockshelters in the Elimbari land system of Eastern Simbu yielded a record of slopewash accumulation for the last 30,000 years. Nombe rockshelter is at 1720 m altitude and has been infilled by fans of slopewash sediment, along with cultural deposits. It is surrounded by sweet potato gardens on the steep (30 to 40 degree) hillslope. Lemouru rockshelter at 2450m is close to the altitudinal limit of gardens, and has been similarly infilled by slopewash deposits.

Rates of hillslope sediment accumulation can be inferred from a depth- age curve for the Nombe sediments (Figure 2). Between 30,000 BP and 16,000 BP the rate of accumulation is relatively low at 25 to 35 mm/1000y. From 16,000 BP to 10,000 BP the rate increases to an average 72 mm/1000y. This rate is raised by the deposition of the thick stratum of air-fall tephra (Ep ash) between c.14,500 and 11,000 BP (Golson, *pers. comm*; Blong, 1982). Between 10,000 BP and 6000 BP the rate declines slightly to an average of 64mm/1000y. This more truly reflects increased rates of hillslope sediment accumulation. A dramatic change occurs in the few hundred years between 6380±90 BP (ANU 3075) and 5850±180 BP (ANU 3074). The rate of accumulation

TABLE 1. Rates of accumulation of lake and swamp deposits, PNG.

Site	Rates of Accumulation	Interpretation
Kuk Swamp 1550m alt. (Golson, 1982)	50000-9000 BP 40mm/1000y 9000-6000 BP 100mm/1000y 6000 BP- present NO RECORD	Slopewash, some burning Forest clearance Increased drainage, sediments flushed
Telefomin 1500m alt. (Hope, 1983)	12000-8000 BP 80-88mm/1000y 4000 BP- present 120mm/1000y	Greater fire frequency, replacement of forest by grassland
Lake Egari 1800m alt. (Oldfield, 1980)	500 BP 0.44 $g/cm^2/y \times 10^{-2}$ 200 BP 0.21 " " " 100 BP 0.61 " " " 40 BP 0.90 " " "	Postdates introduction of sweet potato, reflects increased human activity
Lake Ipea 2500m alt. (Oldfield, 1980)	500 BP 0.21 $g/cm^2\ y \times 10^{-2}$ 200 BP 0.26 " " " 100 BP 0.77 " " " 40 BP 1.92 " " "	Accelerated sediment input to lake

changes to between 53 and 166mm/1000y. Since c.6000 BP the rate of sediment accumulation is difficult to determine. A band of cemented sediment forms a level 20cm above the present ground surface and is very similar to the uppermost deposits. At least 20cm of deposit has been lost due to the depredations of people and pigs, indicating reduced rates of sediment accumulation in the last five millennia.

At Lemouru the rate of accumulation was high (200 to 240mm/1000y) between c.1800BP and 400-500BP (Figure 3). This is higher than the maximum rate estimated for older deposits at Nombe. The rate increases after 400-500 BP to a rate of 1300-1800mm/1000y. This rate can be compared with that gained from Lakeba (Fiji) by Hughes *et al.* (1979). Here demonstrably severe prehistoric land degradation resulted in rates of swamp infilling of between 1500 and 2700mm/1000y. The severity of hillslope erosion in the Elimbari area indicated by the rockshelter data is further evidenced by exposed limestone pinnacles with sub-soil solution notches between one and two metres above present ground level.

The data from the two Elimbari rockshelter sites studied suggest that rates of hillslope erosion have increased dramatically from low

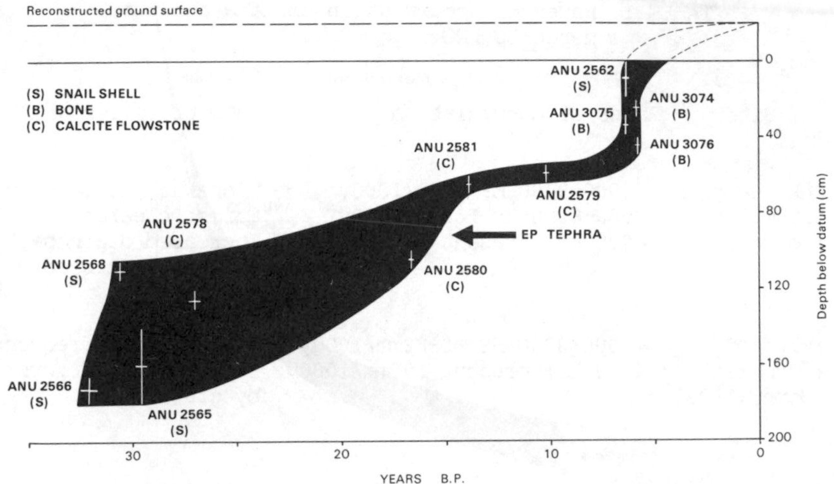

Fig. 2. Depth - age curve for deposits in Nombe rockshelter, Chimbu region, PNG. Error bars indicate depth range of dated samples and one standard deviation on radiometric ages. The likely age correction on both flowstone and snail shell dates is less than 1000 years.

values of 25-35mm/1000y in the Pleistocene to high values of 1300-1800mm/1000y in the last few hundred years. This increase accords well with the predicted consequences of intensification of horticulture in the area, especially since the introduction of sweet potato. Repeated clearing and burning of the same area at short intervals may interrupt the vegetation succession, leading to a mosaic of degraded secondary forest, ferns and grass. Associated with this successional change is a decline in the rates of replacement of soil nutrients, as recycling is interrupted.

A major methodological problem with the above approach is the determination of the relationship between rates of accumulation in rockshelters and the areal loss of soil on the contributing slopes. Contemporary process data is lacking. At Lemouru, there appears to be a great degree of sediment focussing down a shallow gully system into the rockshelter: its trap efficiency may therefore be quite high. But estimation of that efficiency will be fraught with difficulty.

In mountain terrain where enclosed basins (lakes, swamps) are few, reconstruction of historic erosion rates is difficult. If care is exercised in study site selection, rockshelters may offer a useful interpretable record of hillslope soil erosion. A better record is obtainable from small swamps whose contributing catchment can be well defined. Rates of sedimentation can be expressed in volume terms and converted to areal soil loss rates in the catchment.

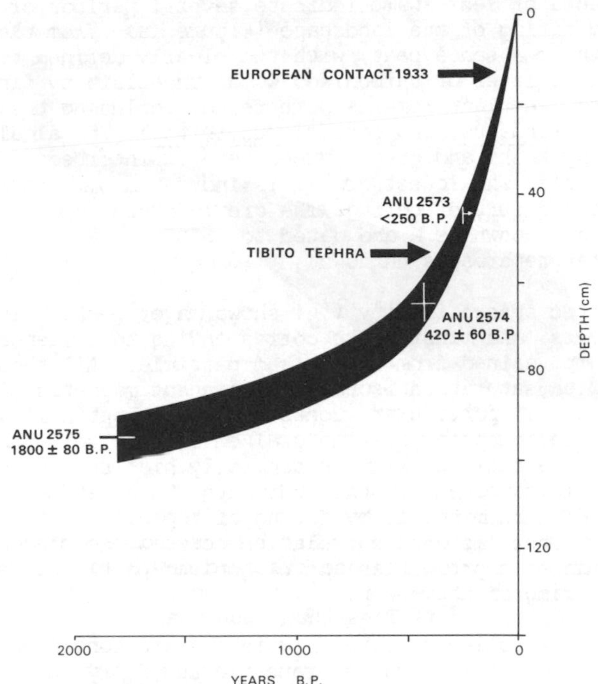

Fig. 3. Depth - age curve for deposits in Lemouru rockshelter, Simbu province, PNG. Error bars indicate depth range of dated samples and one standard deviation on radiometric ages.

SWAMP DEPOSITS AND PREHISTORIC AGRICULTURE

North of Mount Hagen, the Jimi River slices through the backbone of the Highlands on its way to the Sepik plains. Along the Jimi are extensive valley floor grasslands, sparsely populated today. These _Themeda_ grasslands indicate poor soil nutrient status. Modern drainage ditches in the Yeni swamp, near the Jimi - Lai confluence at 550m altitude, revealed heavily patinated stone tools and associated prehistoric garden systems (Gillieson, Gorecki and Hope, 1985). A multidisciplinary project has therefore investigated the environmental history of the lower Jimi and the Sepik - Wahgi divide which flanks it. Attention has been focussed on grassland swamps and rockshelters within the altitude range 350 to 2200m. The area therefore spans both the Highlands proper and the Highland fringe, an important resource zone into which past and present people have moved. Certainly the agricultural history of valleys such as the Wahgi cannot be fully understood without knowledge of the lower, overflow areas such as the Jimi.

The sediments of Yeni swamp indicate several periods of vegetation change and firing of the landscape (Figure 4a). From the surface to 100cm depth is a sedge peat, with two clearly defined tephra layers. At 100cm is a layer of carbonised wood, underlain by fine algal mud in which edible plant remains were found, including the sago palm *Metroxylon*, *Pandanus*, and gourds (*Cucurbitae*). The algal mud is underlain by khaki and grey - green tephras interbedded with organic clayey gravels. The lowest gravel is indurated with ferricrete and this prevented further coring. The grey - green tephra may correlate with a tephra known as R and dated to 5500BP at Kuk. The correlation of the other tephras is unknown.

Mass specific susceptibility (χ) shows major peaks coincident with tephra layers and minor peaks corresponding to evidence of burning in the swamp, gained from carbonised particles. All these peaks are enhanced with saturation isothermal remanent magnetisation (SIRM) measurements. In particular, zones of low susceptibility and higher SIRM indicate inwashing of fine grained organic soil from the catchment. The ratio SIRM/ χ is generally high suggesting a secondary ferrimagnetic component in the sediments. This may be due to the formation of such minerals by firing of topsoil, or to inwashing of fine clays. There is good correlation between the presence of broad peaks in magnetic properties and carbonised particles, suggesting periodic firing of the swamp.

Changes in the pollen spectra (Figure 4b) are consistent with these conclusions. Conifers such as *Araucaria* and *Dacrydium* are present throughout, as are angiosperm trees such as *Nothofagus*, *Elaeocarpus* and Sapindaceae. Many trees in the area are insect pollinated, wind dispersion being poor. Thus the pollen record is mainly representative of local plants. A significant decrease in the sedge - grass ratio above 110 to 120cm reflects infilling of the swamp and locally drier conditions; this change is accompanied by an increase in monolete spores, ferns and fern allies, which may also reflect disturbance. Regrowth plants also show increased proportions above this depth zone, which coincides with evidence of burning and the carbonised plant remains. In particular, the proportions of secondary growth colonisers such as *Macaranga*, *Mallotus*, *Casuarina*, *Plantago* and legumes all increase at this point. Similar disturbance indicators are present at the base of the sequence, though some accumulation on the gravel layer may have occurred.

These major changes in the nature of sedimentation and the pollen record can be related to the evidence for prehistoric agriculture. Gravels associated with the earliest disturbance phase fill in the ditches of island bed gardens on the swamp margin. The later disturbance, around 1000BP, relates to mound and hollow features which are infilled with Olgaboli tephra (c. 1200BP) and resemble dryland gardens. Thus two phases of disturbance and a major change in the agricultural system are indicated.

Personal observations of Yeni and other upland swamps in Papua New Guinea suggest that the surface *Sphagnum* and fringing *Cyperus* sedge fen act as very efficient fine sediment traps. Rapid settling and

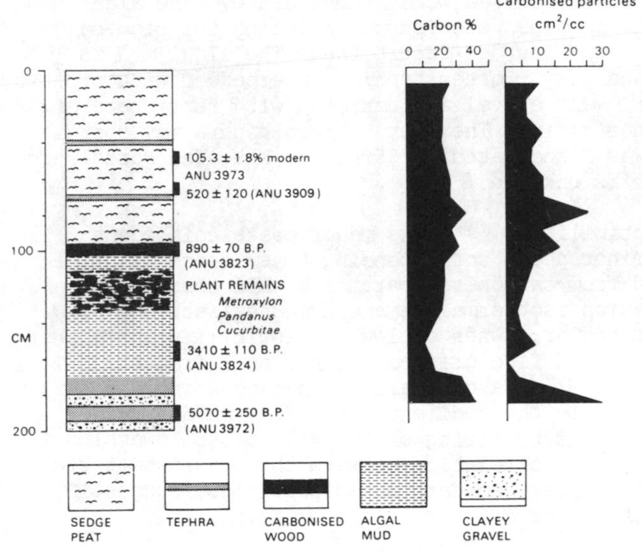

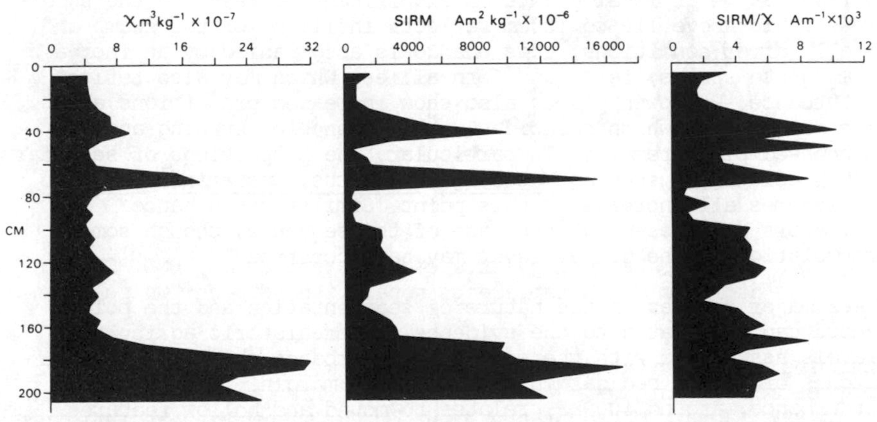

Fig. 4a. Stratigraphy, chronology and sediment properties of site MSI, Yeni swamp, PNG. = mass specific susceptibility, SIRM = saturation isothermal remanent magnetisation.

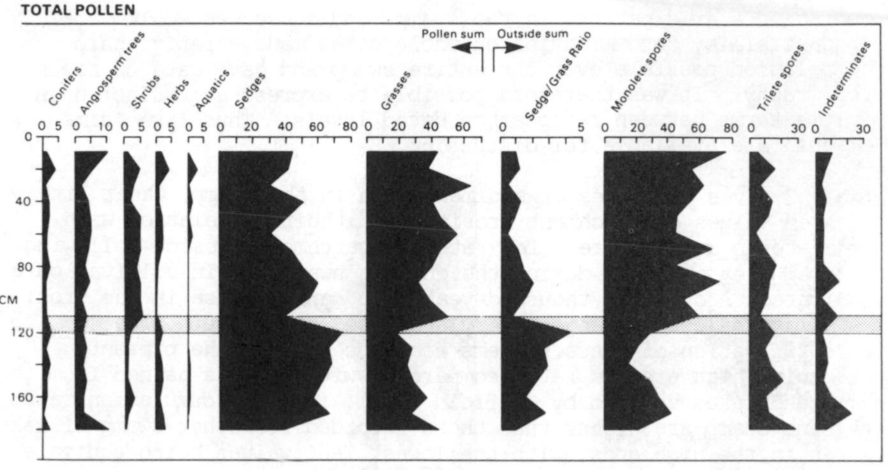

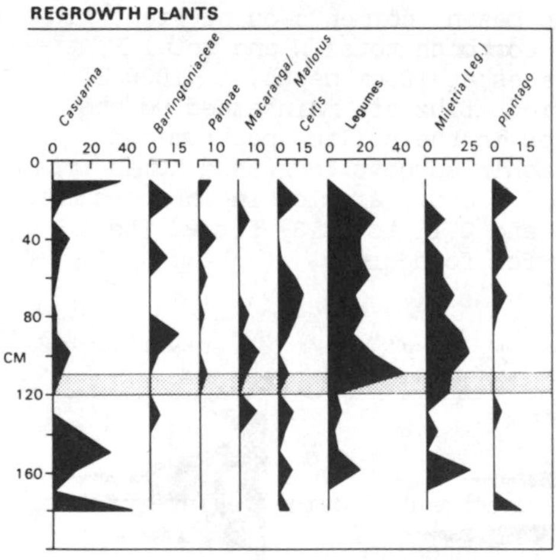

Fig. 4b. Palynology of site MSI, Yeni Swamp, PNG. Shaded bar at 110cm depth represents separation between two pollen zones.

trapping of suspended load seems to occur. Estimates of reservoir trap efficiency in Papua New Guinea (Ramu, 92%) and Tanzania (range 92 to 99% : Rapp, 1975) suggest that little suspended sediment passes through deeper reservoirs. We therefore assume that the trap efficiency of Yeni Swamp is at least 90%, possibly more. The relationship between swamp sedimentation and catchment erosion can therefore be defined. Organic productivity in the swamp is very low, as a dysmictic status is indicated by species such as Nepenthes and Gleichenia. Biogenic silica (phytoliths, diatoms) does not form a major proportion of the sediments.

Coring on a grid pattern in the swamp, and subsequent bulk susceptibility determination on whole cores made stratigraphic correlation possible over the entire swamp and gave data on basin topography. It was therefore possible to express accumulation in volume terms between radiocarbon dated levels. Thus five 'time slices' are available for discussion .

Table 2 gives the rates of sedimentation in the swamp, their likely related values of catchment erosion, and their correlation with prehistoric agriculture. Low rates of catchment erosion following initial clearance and during the ensuing swamp margin cultivation are indicated. Around one thousand years ago, an increase in the erosion rate may reflect a change to dryland based agriculture. The intensification of these systems has proceeded to the present, producing high erosion rates comparable with results gained from the Highland lakes studied by Oldfield et al. (1980).Accumulation rates at Yeni Swamp are higher than those recorded for other swamp or lake sites in the Highlands, with the lowest Yeni values being equivalent to the highest at Kuk and Lake Pipiak.

Given the dysmictic nature of the basin, comments on nutrient flux in the catchment can also be made. For both total N and total P, there is a dramatic increase in flux rates at 100cm depth, c. 1000 BP (Figure 5). High levels of nutrient flux are maintained to the present. Coupled with the changes in the pollen spectrum and magnetic properties, these data again suggest increased human impact on the catchment. Today both soil N and P are low in the catchment, ranging from 0.16 to 0.36% for N and 0.11 to 0.18% for P. These values are likely to be limiting for food crops.

Table 2. Estimated Prehistoric Erosion Rates, Yeni Swamp, PNG.

Time Span	Mean depth sediment (mm)	Volume (m^3)	Catchment Erosion tonnes/km^2/y	Interpretation
5000 -3500BP	430	688	11.5	Initial forest clearance
3500 -1000BP	520	3458	8.5	Swamp margin irrigation
1000 -500BP	250	2280	20	Further clearance with dryland horticulture
500 -200BP	450	7820	61	Intensification of horticulture
200BP - present	300	7500	60	Further use and European impact

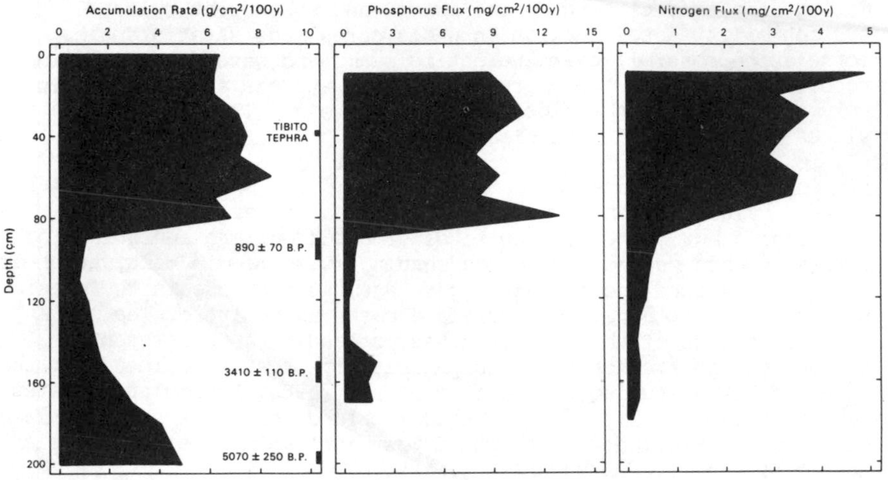

Fig. 5. Rates of accumulation and nutrient flux for N and P by depth for site MSI, Yeni swamp, PNG.

Despite the evidence of human impact on the Yeni catchment, no people lived on the swamp margins at the time of European contact in 1933. Rather the sparse settlement was concentrated in several small villages in the foothill zone. This depopulation of the Yeni grasslands and the adjacent, extensive Ruti grasslands may be in part due to soil nutrient depletion. Another factor to be invoked here is a prehistoric epidemic which depopulated the Wahgi and possibly the Jimi at least one generation prior to European contact in 1933.

Located high on the Sepik - Wahgi divide at 1950m, Nurenk swamp is close to the present upper limit of gardens in the area. The swamp catchment is not cultivated today, but a mosaic of secondary forest attests to past land clearance. The swamp is the result of infilling of a valley by ten metres of peat and mud. It has formed on a buried soil dated at 4000BP. A depth - age curve for Nurenk (Figure 6) suggests rapid initial sedimentation (2800mm/1000y), a reduced rate of 1330mm/1000y between 3700 and 2100BP, and a great increase in the rate of accumulation in the last thousand years to 4500mm/1000y. Olgaboli tephra is a useful temporal marker for the start of that increase. The rate of accumulation has increased further in the last few hundred years, since the fall of Tibito tephra, and this may reflect intensification of land use in the catchment. Archaeological evidence from nearby rockshelters and a stone quarry site gives a minimum age for human occupation as 1500BP. Non - directional magnetic properties of the Nurenk sediments show clearly defined peaks coincident with the tephras, and other small peaks reflecting periodic catchment instability. This may relate to phases of forest clearance for individual gardens. The complete record of Nurenk will be published elsewhere.

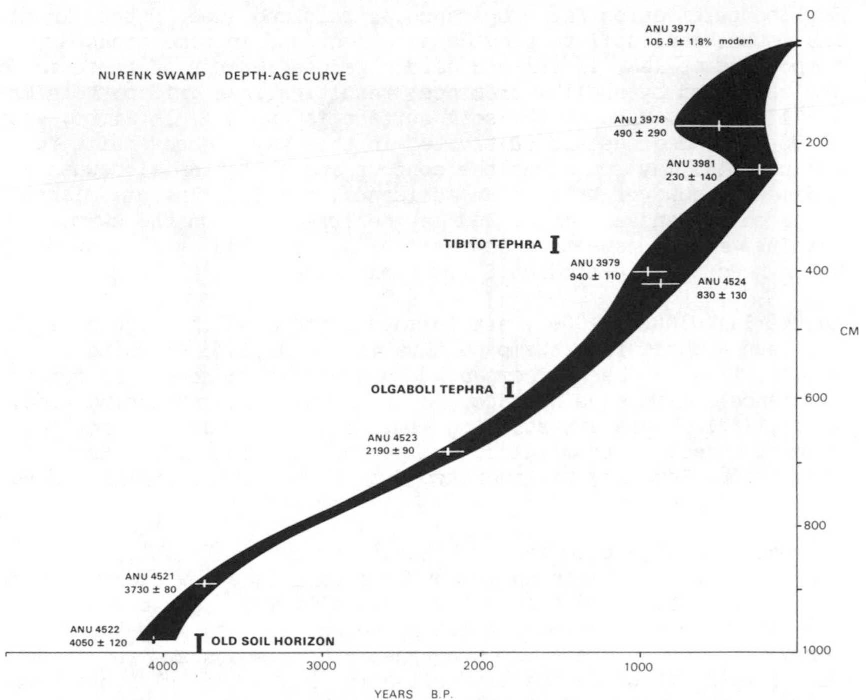

Fig. 6. Depth - age curve for site MSPC, Nurenk swamp, PNG. Error bars indicate depth range of dates samples and one standard deviation on radiometric ages.

RATES OF SOIL EROSION AND AGRICULTURAL SYSTEMS

The initial forest clearance phases at Yeni and Kuk are associated with archaeological evidence for irrigated, island bed agriculture on the swamp margins and in the swamps. In this system water filled trenches separate low mounds upon which crops are grown. Water tolerant species are grown in the trench sides. In the Baliem valley of Irian Jaya, this form of swamp cultivation is still practised. Here extensive island beds are separated by deep ditches, from which organic mud is frequently plastered on to the garden beds. Blue - green algae thrive in the eutrophic ditch water, and play a crucial role in nitrogen fixation. Algal nitrogen fixation can only be effective in neutral to alkaline pH, and where there is sufficient available phosphorus (Sanchez, 1976). Taro will thrive under these conditions, whereas sweet potato and yams will not. In time, the infilling of ditches and the decomposition of organic matter produce acidic conditions and inhibit nitrogen fixation. At that point the ditches are re - dug or the garden allowed to lie fallow.

Dryland cultivation for crops such as taro and sweet potato involves the mounding of soil to provide aeration, and in some areas dry composting of leaf litter and garden refuse. Groups of these mounds are separated by shallow trenches, resulting in a grid pattern of small gardens. Much of the soil surface is exposed. In Simbu, slopes of up to 40 degrees are cultivated in this way. Fences built to exclude pigs may run along the contour and thus trap slopewash sediment , however this is an accidental benefit. In some places terraced gardens are made, and may be irrigated. On the swamp margins, mounded sweet potato gardens are drained by shallow ditches which need to be periodically re - excavated.

In the last 500 to 1000 years in many regions of the Highlands there has been a shift from swamp cultivation to dryland cropping for sweet potato. This has been accompanied by dramatic increases in forest clearance and in erosion rates, a trend which is continuing today (Hope,1980). There are striking similarities in the evidence for this among the records from lakes, swamps and rockshelters. Many of today's problems may be inherited from former overcropping and soil nutrient depletion.

Two points emerge from the prehistoric evidence. Firstly, the rehabilitation of these soils may be accomplished by the improvement of soil structure and nutrient status following organic mud plastering on garden beds, mulching techniques and longer fallow periods. Secondly there is much under - utilised swampland in New Guinea which might become productive if island bed cultivation were to be reintroduced. Preliminary moves in this direction have been taken in Vanuatu, and in the Kuk swamp itself.

Acknowledgements: This work was funded by the Australian Research Grants Scheme and the Faculties Research Fund, Australian National University. Permission was granted by the Institute of Papua New Guinea Studies and the Provincial Governments of the Simbu and Western Highlands. We are grateful to Neil Ryan and Martin Gunther of Mount Hagen and Don Matheson of Ruti Station for hospitality. Laboratory assistance was provided by Jon Luly, Jim Tulip and Julia McIntyre. We are grateful to Eileen McEwan and Paul Ballard for their help in producing the text.

REFERENCES

Blong, R.J., 1982. The time of darkness. Australian National University Press, Canberra.

Burton, J., 1985. Quarrying in a tribal society. World Archaeology, 16, 234-247.

Corlett, R.T., 1984. Human impact on the subalpine vegetation of Mount Wilhelm, Papua New Guinea . J.Ecol. 72,841-854.

Gillieson, D., and Mountain, M.J., 1983. The environmental history of Nombe Rockshelter, in the Highlands of Papua New Guinea. Archaeology in Oceania, 18, 53-62.

Gillieson, D., Oldfield, F., and Krawiecki, A., in press. Records of Prehistoric Soil Erosion from Rockshelter Sites in Papua New Guinea. In C. Pain (ed.), Geoecology of the Papua New Guinea Highlands, Mountain Research and Development.

Gillieson, D., Gorecki, P.P., and Hope, G.S., 1985. Prehistoric Agricultural Systems in a Lowland Swamp, Papua New Guinea. Archaeology in Oceania, 20, 32 - 37

Golson, J., 1982. The Ipomoean Revolution revisited: society and the sweet potato in the upper Waghi valley. In A. Strathern (ed.). Inequality in the New Guinea Highlands. pp. 102-136. Cambridge University Press.

Golson, J., and Hughes, P.J., 1980. The appearance of plant and animal domestication in New Guinea. Journal de la Societe des Oceanistes, 36, 294-303.

Gorecki, P., 1985. The conquest of a new "wet and dry" territory: its mechanism and its archaeological consequence. In I.S. Farrington (ed), Prehistoric Intensive Agriculture in the Tropics, BAR International Series 232, pp. 321-345.

Harris, E., and Hughes, P.J., 1978. An early agricultural system at Mugumamp ridge, Western Highlands Province, Papua New Guinea. Mankind, 11, 437-444.

Hope, G.S., 1980. Tropical Mountain Forest in Retrospect and Prospect. In J.N. Jennings and G.J. Linge (eds.), Of Time and Place, Canberra, ANU Press, pp. 153-172.

Hope, G.S., 1983. The vegetational changes of the last 20,000 years at Telefomin, Papua New Guinea. Singapore Journal of Tropical Geography, 4, 26-33.

Hughes, P.J., 1981. Prehistoric Human Induced Soil Erosion: Examples from Melanesia. Paper presented to Conference on Prehistoric Intensification of Agriculture in the Tropics. Canberra August, 1981.

Hughes, P.J., Hope, G.S., and Latham, M., 1979. Prehistoric man-induced degradation of the Lakeba Landscape : evidence from two inland swamps. UNESCO/UNEPA Fiji Island Reports No. 5 (Canberra, ANU for UNESCO).

Oldfield, F., Appleby, P.G., and Thompson, R., 1980. Palaeoecological studies of lakes in the highlands of Papua New Guinea. Journal of Ecology, 68, 457-77.

Rapp, A., 1975. Soil erosion and sedimentation in Tanzania and Lesotho. Ambio, 47, 154-163.

Sanchez, P., 1976. *Properties and Management of Soils in the Tropics*. New York.

Walker, D., and Flenley, J.R., 1979. Late Quaternary vegetational history of the Enga Province of upland Papua New Guinea . *Phil. Trans. Roy. Soc. Lond B.*, 286, 265-344.

International Geomorphology 1986 Part II
Edited by V. Gardiner

PRESENT DAY EVOLUTION OF GULLIES FORMED IN HISTORICAL TIMES IN THE MONTAGNE DU LINGAS, SOUTHERN CEVENNES, FRANCE

C. Cosandey, A. Billard and T. Muxart

Laboratoire de Géographie Physique "Pierre BIROT", CNRS, Meudon 92190, France

ABSTRACT

Overgrazing in the fragile environment of the granitic Montagne du Lingas, has led to gullies up to 100m long and 10m deep cut into the landscape. As the intensity of grazing is now greatly reduced, it is important to know whether this type of erosion still constitutes a serious land management problem. To establish the severity of erosion at present, six ravines were instrumented with measurement strikes, along 2 to 4 cross-sections in each case, on steep and gentle slopes and in enclosures and open pasture. Three years results show the present evolution, enlargement and flattening of the gullies. Although some material is still being removed, the gullies are being infilled. Present-day grazing seems to be having little effect on erosion, but many small mammals have colonised the gullies, their burrowing augmenting the material released from the gully walls by rain, wind and frost.

INTRODUCTION

Situated at the extreme southeast of the Massif Central, close to Mont Aigoual, the Montagne du Lingas covers an area of about 90km^2, lying between 800 and 1445m a.s.l. Dissected by a network of small streams, the region presents a series of rounded crests, linked by convexo-concave slopes to often marshy valley floors. The Lingas is made up essentially of the porphyritic granite batholith of Saint-Guiral, intruded at the close of the Hercynian orogeny.

The deep weathering of the rock, locally to 10m or more, appears to be linked to a complex tectonic history and also contributes to the area's vulnerability to erosion. Furthermore, as the Lingas is at the meeting place of Mediterranean and Oceanic air masses, it receives a mean annual precipitation of 2242mm (Saint-Joiyny, 1985) with some of the highest storm rainfall intensities in mainland France, for example 130mm h^{-1} in 6 minutes on 6 October 1961 and a total of 548mm in 36 hours from 6-8 November 1982. These heavy downpours occur most frequently in autumn, but also in winter and sometimes in Spring. The strong winds which accompany these storms add to the impact of raindrops on the soil.

The relatively elevated situation means that nocturnal frost is likely between September and May, 140 days of frost with 85 diurnal freeze-thaw cycles occurring at 1567m on Mt Aigoual in 1983. Human occupation of this fragile environment has introduced instability through woodland clearance and overgrazing leading to the discontinuous vegetation cover and deep gully erosion, still characteristic of the Lingas.

At present, the human pressure on the landscape is decreasing, with rural depopulation since the mid-nineteenth century and, since 1945, a decrease in the transhumant herds coming to the Lingas for the summer from the garrigue around Montpellier and the lowland parts of the Cevennes.

In view of these relatively recent changes, it was necessary to find out whether the reduced intensity of grazing was still creating a real erosion hazard or whether the denuded areas were recovering. To gain an answer to this question, we investigated present-day processes in some gullies in a particularly severely eroded part of the area: the Airette valley.

FIELD INSTRUMENTATION

The 30ha Airette valley lies between 1200 and 1292m in the northeast of Lingas. Its infilling can be reconstructed from the peat on the valley floor (Muxart et al., 1985). Calluna and broom occupy the more degraded slopes while grasses cover the convex summits and the valley floor. The slopes are dissected by many gullies which are linked by runoff channels to small debris fans leading to the valley floor. Little vegetation protects these bare erosional and depositional features.

Six gullies on the 8% to 16% western slope of the valley were investigated, their west-east orientation providing strong contrasts in exposure to wind and sunshine on their north and south facing slopes. One of the gullies was enclosed for protection, while the remaining gullies were grazed by transhumant sheep from 15 June to 15 September. In the investigated gullies, several cross-sections were monitored using 60cm long, 12mm diameter concrete stakes, driven 35cm into the ground, generally a metre apart. A painted mark on each stake, exactly 20cm above the ground at the time of the installation, served as a reference for future measurements. Remeasurement on every stake in all cross-sectons each season of the distance between this reference mark and the ground surface enabled the evolution of the gullies to be established.

The field measurement design (Fig. 2) comprises:

a) in the 8% slope part of the valley, the protected enclosed gully (R1) and two open, free access gullies (R2 and R3) with a total of 7 cross-sections and 76 measuring stakes;
b) in the steeper zone with coalescing gullies, one protected gully (R41) and two free-access gullies (R42 and R43) comprising 7 cross-sections and 77 measuring stakes.

ANALYSIS OF THREE YEAR'S OBSERVATIONS

The first year's results are not easy to interpret. A succession of resurveys from the installation date in June 1982 to October 1983 for gully R41 (Fig. 3) shows no clear pattern, although infilling appears to exceed erosion. However, after three years, seasonal cycles become apparent, with a trend towards infilling in the frost-free summer period from June to September, and erosion in the winter, frost-prone, October-May period (Fig. 4). Overall, the three years saw a net infilling, both in the grazed, steep gullies (Fig. 5) and the protected, grazing-free steep gully (Fig. 6).

To eliminate the local peculiarities of individual gullies, two typical profiles for the gullies on steep slopes and those on the gentler 8° slope showing the July 1982 - July 1985 changes were drawn (Figs. 7 and 8). The following trends emerge:
a) an accumulation zone at valley floor with the talweg being systematically offset towards the south-facing slope;
b) a mid-slope erosion zone, regardless of exposure;
c) an accumulation zone upslope of the gullies.

From these observations a general model of gully development in the Airette valley could be established.

INTERPRETATION: PRESENT-DAY GULLY EROSIONAL PROCESSES

Four zones of morphological developments, each dependent on the other, can be distinguished in a gully cross-profile (Fig. 9):

a) The upslope sector corresponding to the upper edge of the gully wall. Undercutting beneath the roots of calluna tussocks leads to eventual instability and the fall of the tussock and a considerable amount of soil to the valley floor. These fallen tussocks are either disintegrated to add to the material eroded, or become re-established and thus contribute to the colonisation of the gully floor by pioneer vegetation. It appears that gully enlargement is regulated by the number of freeze-thaw cycles producing needle-ice action and to loosening soil particles in the arc layer at the base of the calluna roots. Furthermore, a large population(of field mice) is concentrated on the gully wall at the calluna root lead and thus by their burrowing probably contributing to the cutting of the roots and detachment of soil;
b) Slightly further down and gully side is an accumulation zone made up of the material released from the upper part of the gully wall;
c) Below the accumulation zone is a zone where erosion is dominant;
d) Finally, at the floor of the gully is a general area of infilling, some of whose material is carried along the gully to the debris fans where the gullies open on to the main valley floor.

Figures 7 and 8 show the contrasts in the changes between the north and south facing sides of the gullies. The talweg becomes displaced towards the south facing slope which thus tends to steepen, leading to gully asymmetry and preferential evacuation of debris from the south facing slope. The exposure of this slope to the sun increases the number of diurnal freeze-thaw cycles, as well as to the strongest winds coming from the east south east, so augmenting the kinetic energy of raindrop impact.

The overall trend is for infilling of the gullies, albeit irregularly because of frequent displacement of the talweg and many scour episodes during heavy rains.

In all instances, the evolution of the cross-profiles is clear in zones a (talweg infilling), c (slope erosion) and b (upslope accumulation). However zone a is less clearly distinguished as the slope evolve here by the random, irregular fall of blocks of calluna. Such events are difficult to record at individual measurement sites over a short period.

Overall, Figure 9 shows the trend towards gully elimination by both infilling and lowering of the slopes which slows runoff, reduces the erosive power of gully flows and favours plant succession. In some cases the so-called erosion zone (c) can have years of no erosion or even some accumulation. Nonetheless, this zone retains less debris than the valley floor and thus tends to add to the process of sedimentation and gully flattening.

When the low intensity of grazing is taken into account (300 sheep in 100ha covering the Airette valley according to Albaladejo, 1982), the animals do not inhibit the recovery of the gullied areas. The example (Fig.4) of a grazed gully area shows erosion in the winter, but infilling during the summer period when the sheep are present. Our measurements also indicate no general difference between enclosed, protected gullies and those open to grazing. On approximately the same slope there is no more erosion in the protected gully (R41 Fig.5) than in the unprotected gullies (Fig.6).

CONCLUSION

The three year's observations have established the present morphological evolution of the gullies by widening and flattening, which explains the apparent contradiction between continuing removal of material from the gullies and their gradual elimination. The situation is no longer one of chronic erosion, but more one of slow evolution where although some sediment is still being removed, infilling is more important than erosion. The importance of aspect is shown by the asymmetry of deposition in the gully floors and the greater freeze-thaw slopes, as well as their greater exposure of strong winds. Finally, the present intensity of grazing is insufficient to affect the trend in recovery of this degraded landscape.

REFERENCES

Albaladejo, C., 1982. Comportement alimentaire et utilisation du territoire pastoral par des ovins et caprins (Cévennes Gardoises). Mémoire de D.E.A. Publications CNRS P.I.R.E.N., et I.N.A., Paris-Grignon.

Saint-Joigny, J. Etudes et échos climatiques. Un climat de moyenne montagne méditerranéenne: le cas du Massif de l'Aigoual (Cévennes). La Météorologie, 6.

Muxart, T., Billard, A., Cohen, J., Cosandey, C., Denefle, M., Fleury, A., and Guerrini, M. C., 1985. Dynamique physique récente des versants sur les hautes terres cévenoles (Esoériy-Lingas), en relation avec l'occupation humains. Revue Géographique des Pyrénées et du Sud-Ouest.

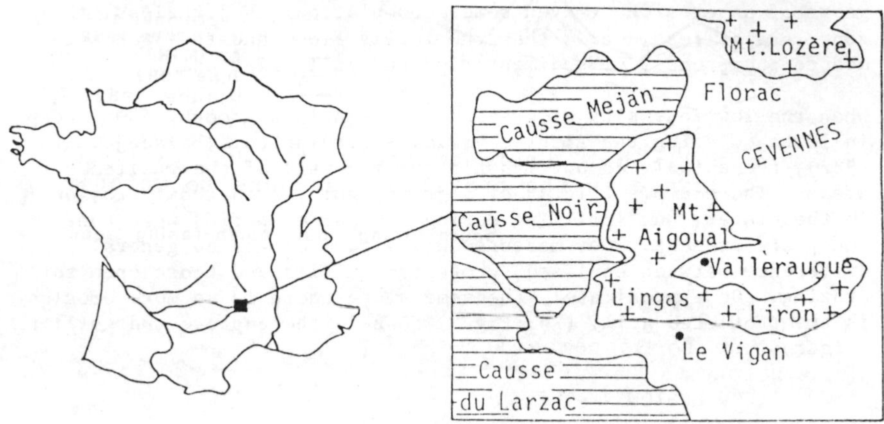

Fig. 1. Location of the Massif du Lingas

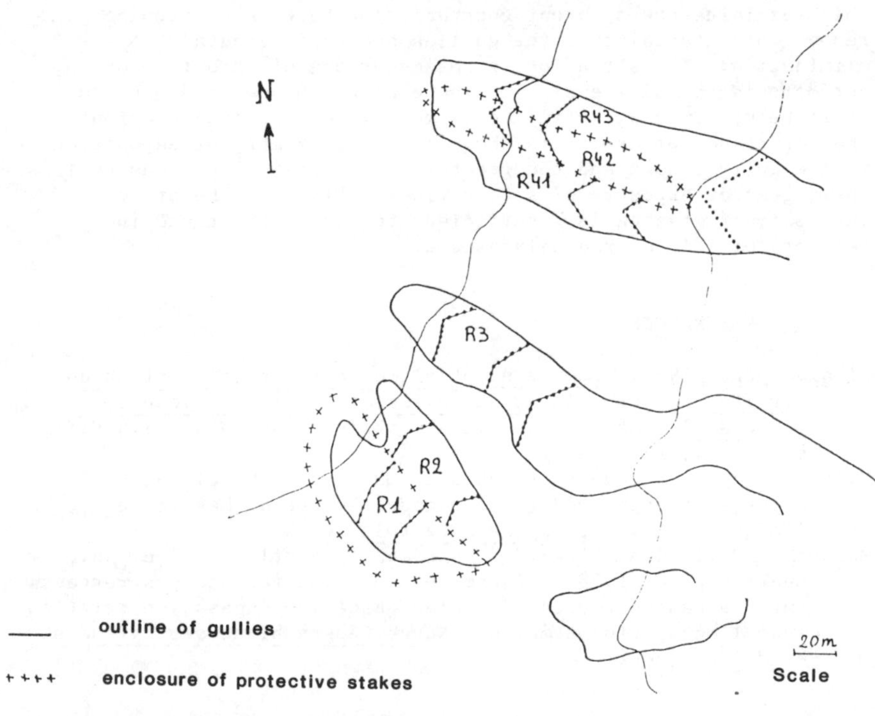

Fig. 2. Location of the gullies studied in the valley of the Airette

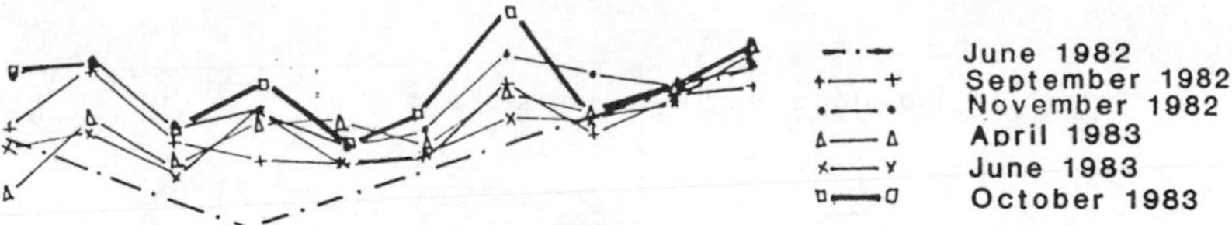

Fig. 3. Gully R41, Upstream transect.
Change in the profiles from June to October 1983

south-facing slope north-facing slope

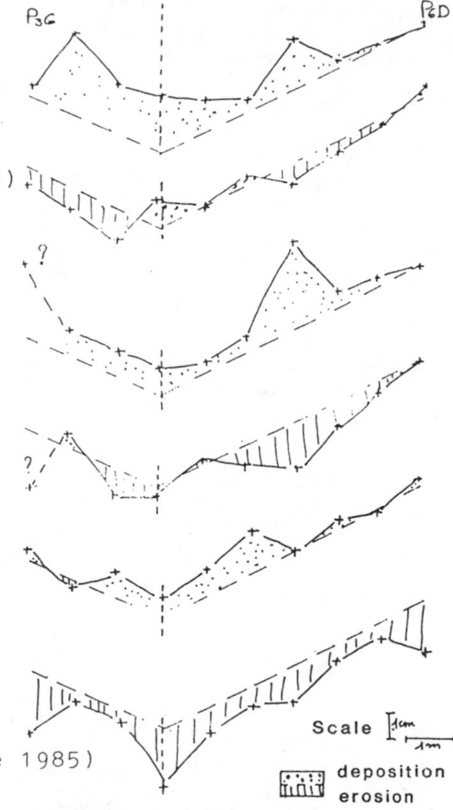

Summer 1982
(from June to September)
Depositing

Winter 1982-3
(from September to June 1983)
Erosion

Summer 1983
(from June to October)
Depositing

Winter 1983-4
(to October)
Erosion

Summer 1984
(from May to September)
Depositing

Winter 1984-5
(from September 1984 to June 1985)
Erosion

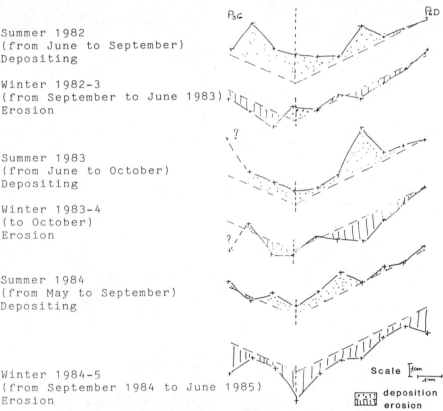

Fig. 4. Gully R41, Upstream transect

Evolution of profiles and seasonal changes from June 1982-June 1985.
(each profile represents the change in the slope since the preceding measurement, represented by - - - : it therefore shows the change during the season considered. The change resulting during the entire period is given in Fig. 5.

The dates bounding each period are not exactly the same each year, depending on the dates of fieldwork.

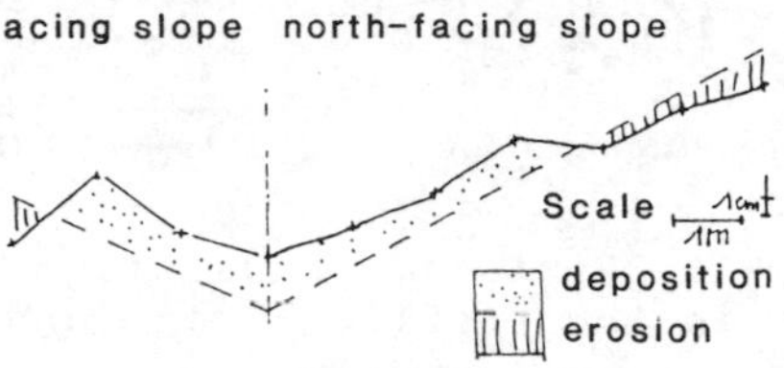

south-facing slope
north-facing slope
Scale
1cm
1m
deposition
erosion

Fig. 5. Gully R41, Upstream transect.
Change in the profile from June 1982 to June 1985.

south-facing slope
north-facing slope
Same legend as Fig. 5

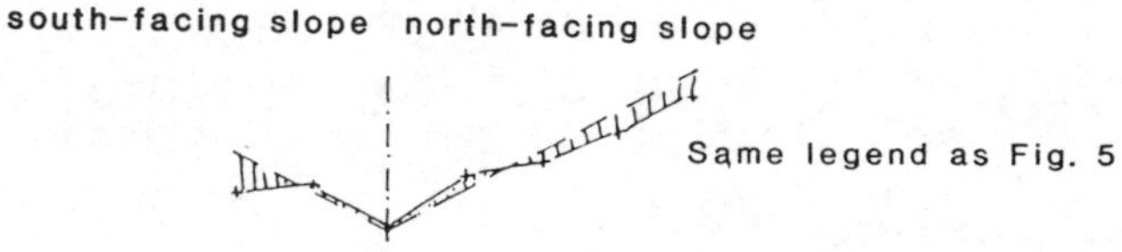

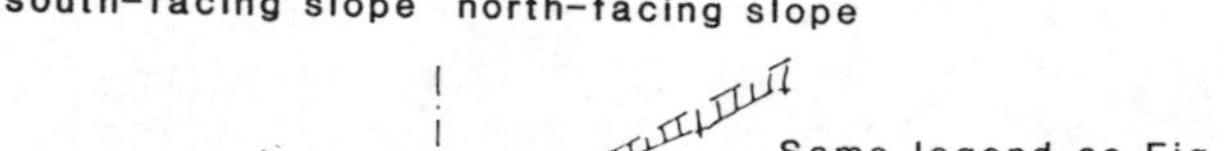

Fig. 6. Gully R42, Upstream transect.
Change in the profile from June 1982 to June 1985.

south-facing slope
north-facing slope
deposition
erosion
initial talweg
shifted talweg
1cm
1m
Scale

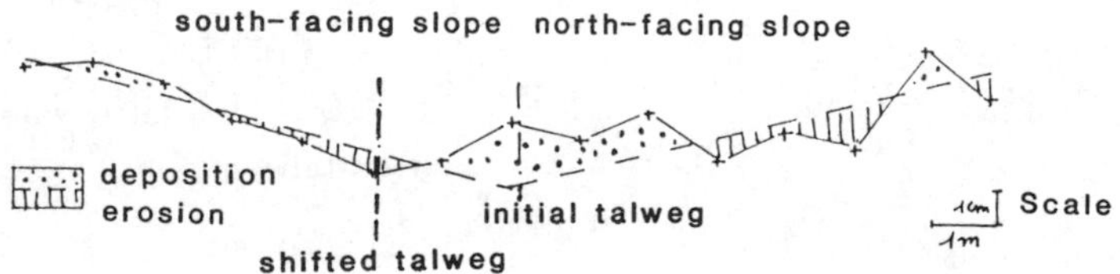

Fig. 7.
Mean change in the gully profiles with gentle gradients (R1, R2, R3) for the period June 1982 to June 1985.

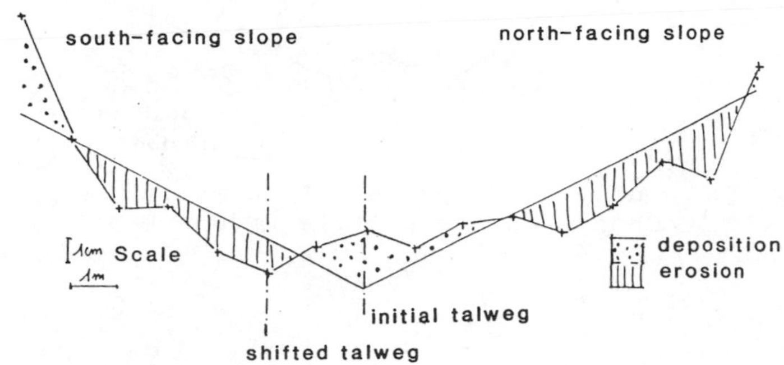

Fig. 8.
Mean change in the gully profiles with high gradients (R41, R42, R43) for the period June 1982 to June 1985.

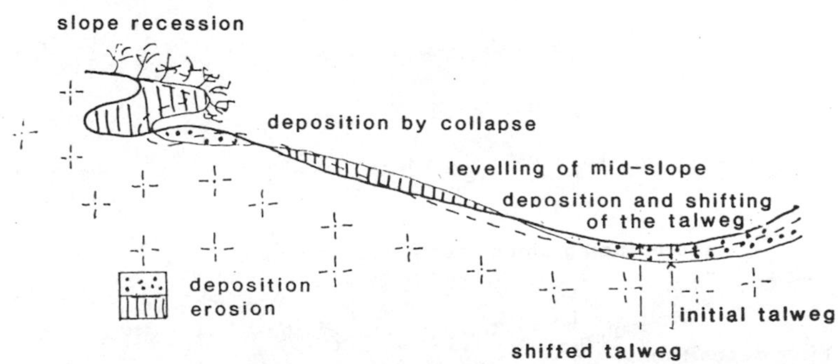

Fig. 9. Proposed scheme for slope evolution of erosional gullies.

International Geomorphology 1986 Part II
Edited by V. Gardiner

THE DEGRADATION OF THE PYRENEES IN THE NINETEENTH CENTURY - AN EROSION CRISIS?

J-P. Metailie

CIMA - UA 366 CNRS, Toulouse, France.

ABSTRACT

In France the first policy about reafforestation and restoration of the mountains appeared during the nineteenth century. It was based on observations of erosion in the Alps and the great periodic floods. The events on which were based the theories about the anthropic degradation of the slopes are analysed. In the Pyrenees the "erosion crisis" was complex and the true anthropic erosion limited to particular areas, namely the winter pastures on fragile substrata. It seems that the great catastrophic events were linked to exceptional meteorological phenomena, particularly frequent in the nineteenth century. The anthropic influence appears to have been over-estimated by the scientists of the time.

INTRODUCTION

The first French development policy linked to a study of the natural environment dates back to the years 1850-1860. It was based on observation of morphological evolution, first in the Alps and then in the Pyrenees, and on the concept of degradation of the mountains, a precursor of the current concept of natural risks. Surrel's book "Etude sur les torrents des Hautes-Alpes" appeared in 1841. This work set the precedent for the later analyses of erosion in the mountains and the ways of combatting it. In the years to come it was the hub of a movement in which engineers, scientists and foresters were to play an important part. The restoration of degraded mountains was presented as indispensable to the modernisation of the French economy and a new planning policy put into action by the reafforestation laws of 1860 and 1882 was to result from this movement. No longer was it merely a question of the extension of cultivated areas or of rehabilitation, but an entirely new policy which aimed at a complete transformation of the mountain regions. The engineers of the period considered that the traditional sylvo-pastoral practices were completely misguided and that they resulted in over-exploitation, degradation and erosion. The great periodic floods of the Rhone, the Loire and the Garonne (in particular those of 1855, 1860 and 1875) fuelled their arguments and the mountain-dwellers were made the culprits of their own misfortunes and of the devastation of the plains. There was thus, for the first time, a direct link between:

- geographic analysis (empirical geomorphological and socio-economic analysis undertaken by Civil Engineers and the Forest Service),
- widely-held theories (which were propagated by the press),
- and a development policy (which failed due to the stubborn resistance of the mountain peasantry).

In the light of this, and with the benefit of a century's hindsight, it is worth asking a few questions about the facts which fuelled the theories about reafforestation and restoration of the mountains. What were the causes, anthropic or natural, of erosion in the Pyrenees? Was there an "erosion crisis" in the nineteenth century, and what was the extent of the degradation? Were contemporary theories as applied to the Pyrenees justified?

Research for this paper was based mainly on archives, and in particular on photographic archives which provide an extremely precise picture of erosion in the last century, the effects of which are now largely hidden by a century's vegetation growth. Two photographic examples illustrating occurrences which are quite exceptional in their extent are provided below.

REALITY AND LIMITS OF ANTHROPIC EROSION

From the very beginning of research engineers and foresters established a correlation between geological conditions and erosion : the fragility of morainic deposits, marls and schists was obvious. But the processes of erosion were less well-defined than they are now. In the middle of the nineteenth century analytical methods were primitive and historical perception of morphological evolution almost non-existent. Two schools of thought reigned. The first held that erosion was the inevitable consequence of deforestation and overgrazing, these phenomena being extremely recent (eighteenth century). The second was that the pastoral communities themselves were the sole cause of the degradation of the slopes and of the flooding.

It cannot be denied that between the sixteenth and nineteenth centuries there was extremely intense exploitation of the French mountains, almost to the point of overloading. The density of population in certain valleys around 1840 was sometimes as high as that of the plains, at 30 people per square kilometre. That is not to say, however, that there is necessarily a direct link between intensity of exploitation and erosion. By increasing the drainage systems, terraces, dykes and so on the traditional grading of the slopes may in fact help decrease the risks of erosion. Furthermore, the overgrazing theory, which is still widely held today (cf. Iranian examples in Hourcade, 1984) often far from corresponds with the facts.

Photographs from the Pyrenees throw light on what the mountain landscape was like in the last century. The area was without doubt "overstressed", with cultivation going on even in situations of extremely uncertain profitability, forests reduced to the minimum, and almost over-loaded pastures. Yet in spite of this the signs of disequilibrium are rare and limited to a few particular areas.

The degraded areas were in fact the middle pastures between 800 and 1400m, which were used from autumn to spring. Given the fodder shortage in the mountains in the nineteenth century these pastures were particularly precious as they were near to the villages, with good exposure, and rarely under snow - which permitted regular use even in the middle of the winter. They were vital in early spring when fodder stocks were exhausted and the new grass was not yet through in the higher pastures, and they were protected by a very precise set of rules. These rules however, were constantly broken. Given the increases in population and the size of the flocks they had to be broken for the mountain communities to be able to survive and thus overloading was inevitable. These pastures are generally situated on fragile substrata : stabilized screes, the steep slopes of glacial troughs covered in morainic deposits, and unstable glacial terraces. The trampling of hooves was quickly destructive, especially in winter and spring when freeze-thaw cycles and daily temperature variations of 10-20°C are common due to the powerful regional föhn. Completely saturated and destabilised by the thaw, the ground was easily damaged by the livestock. The resulting small landslips were exacerbated by spring avalanches or by storms.

Plates la and lb and Fig.l provide an example of this type of anthropic erosion. The slope shown is the middle pasture area of the village of Montauban-de-Luchon (Pique valley, Haute-Garonne). Morainic deposits cover the whole of the very steep slope (20-25°), from the bottom of the valley to the summit, which is a kame terrace. The deposit is not consolidated and quite fine though there are some coarse elements including drift boulders of several square metres in size. It lies on Devonian schists which since the end of glaciation have been grooved by the deep gorges (10-20m) of two mountain streams. The village is built on one of the alluvial fans and throughout the nineteenth century was choked by gravel each time there was a flood.

It is difficult to put a date on the erosion shown by the 1888 photograph as there is no information in the archives. It seems recent, perhaps dating from the seventeenth or eighteenth centuries, but its localisation and layout show clearly that it is the result of very ancient and varied exploitation of the whole of the slope. The stripes correspond to the daily movement of the livestock and the rectilinear marks are caused by the unloading of wood which was either thrown or dragged down. It is clear that exploitation of the forest area played a considerable part in the degradation of the pastoral area here, and that it was also one of the causes of erosion under the forest below the summit line.

The tree-trunks were brought down in spring at the very moment when the ground, already destabilised by climatic forces, underwent the highest pastoral pressure. This concentration of uses during critical periods is the reason behind the degradation of the slope. Moreover, here as in the other communities, the archives bear witness to the breaking of the regulations. For example, irrigation above the village was forbidden due to the danger of landslips but the rules

Plate 1a. Montauban-de-Luchon, 1888.
Archives of Office National des Forêts, Toulouse.

Plate 1b. Montauban-de-Luchon, 1984.

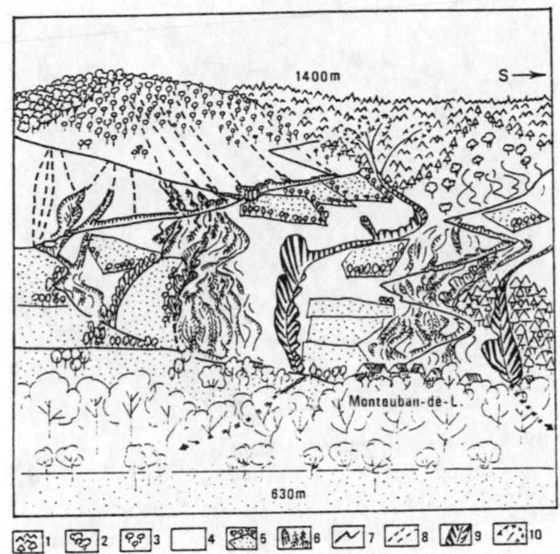

Fig.1. Montauban-de-Luchon (Haute-Garonne) 1888.

1) Abies pectinata woodland
2) Fagus sylvatica woodland
3) Betula alba woodland
4) Calluna vulgaris-Brachypodium pinnatum moorland
5) Cultivations and meadows
6) Hoof-trampled ravines
7) Main paths
8) Trunk dragging ravines
9) Ancient torrential gorges
10) Landslips in the forest

Plate 2a. Ravine of the Gravel, 1889.
Archives of Office National des Forêts, Toulouse.

Plate 2b. Ravine of the Gravel, 1984.

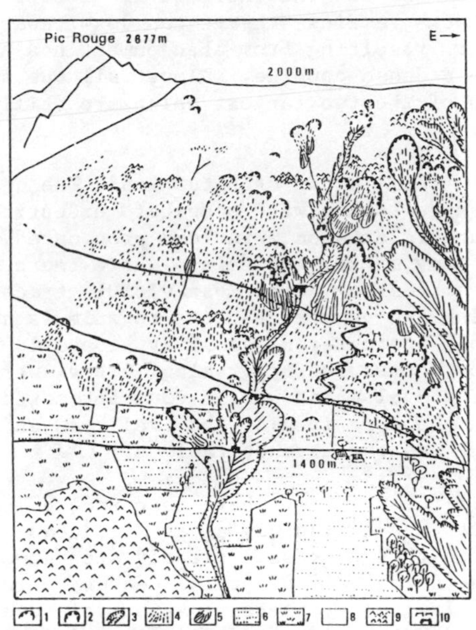

Fig.2. Ravine of the Gravel (Vicdessos, Ariège) 1889.

1) Superficial landslips
2) Deep landslips
3) Ravines
4) Gravel covering the slope
5) Scree
6) Meadow and cultivation
7) Abandoned cultivation
8) Moorland (Calluna vulgaris, Genista pilosa, Pteridium aqilinum
9) Ordovician schist outcrops
10) Restoration works of the Forest Service (terraces and dams)

were ignored and illegal irrigation provoked three disastrous landslips in 1860 [1].

The causes of erosion have since disappeared. Nowadays the logs are brought down by the road and the increase in fodder supplies means that what livestock remains there can do without these pastures. Spectacular regrowth resulting from abandonment has hidden the signs of erosion beneath a dense coppice. Almost all the scars have healed with the exception of the two largest which are still there, hidden beneath the trees.

How frequent was this kind of situation? The degradation of the slopes at Montauban-de-Luchon was somewhat exceptional. Generally the traces of anthropic erosion were less pronounced when spread over a larger area and they marked the points where there was the greatest concentration of livestock. The most visible traces of erosion are in limey areas which are very attractive to animals and have fragile soils. On the old photographs the limey slopes look bare, stripped by the trampling of hooves and by grazing. They were often strewn with a layer of gravel which was in constant motion because of the animals. This was the case in all the valleys and in particular those of the Ariège and the Gave de Pau.

For a century all these sectors have been undergoing recolonisation by vegetation, sometimes as rapidly as at Montauban-de-Luchon. There were two stages to this process. Firstly spontaneous stabilisation and regrowth in the first half of this century, in spite of the persistence of relatively extensive pastoral use, and then rapid regrowth from the 1950s with the acceleration of rural exodus.

ANTHROPIC DEGRADATION AND ADVERSE WEATHER CONDITIONS

A well-known fact : inordinate weather conditions aggravate the effects of over-stressing. It is clear from the erosion at Montauban-de-Luchon that the presence of a fragile sub-stratum and localised over-stressing (pastoral or otherwise), together with a violent meteorological event, can be disastrous. There is no shortage of examples of this in the course of the nineteenth century.

The erosion of the south slope of the Artigue valley in the Vicdessos area (upper Ariège) is a good example of this (see Plates 2a and 2b and Fig.2). The whole of the slope, which is between 1500 and 2000m in altitude, is covered in thin weathered outcrops of schist with the morainic terrace reaching as high as 1450m (the cultivated area on Plate 2a). It lies above extremely friable Ordovician slates which have subvertical dips and are easily cracked by frost. In the 1889 photograph the schist can be clearly seen in the bare patches of the slope.

The ravine in the centre is that of the Gravel and the one on the right is that of the Moulinas, which is bigger but only partly visible on the edge of the photograph. The photograph dates from the beginning of the restoration work in 1884. The access paths to the sites stripe the slope which is strewn with avalanche terraces. The vegetation is acidophilous moorland, sparse, stunted _Calluna vulgaris_, _Genista pilosa_ and _Pteridium aquilinum_.

How the gullies developed is well-known. The Vicdessos valley had been occupied since ancient times (the first clearings, attested by palynology, date from 4000 B.P.) and very quickly deforested. There was both intense pastoral exploitation (with 30,000 sheep and goats in 1852) and many forges which destroyed the forests for charcoal production (the most important iron mine of the Pyrenees was in the valley). In the nineteenth century over-population reached such a point that some of the meadows had to be turned over to food production. This decrease in available grazing land accompanied by an increase in the size of the flocks forced the breeders to move to winter pastures in the plains, but also to over-load the middle pastures around the barns. Moreover, at that time, most of the summer hamlets had become permanent dwellings and this put further pressure on the surrounding area. Here we have a classic example of a traditional agro-pastoral system firmly anchored in its economic logic and practices, incapable of increasing its productivity and forced to overload its resources.

Moreover, from 21st to 24th June 1875 there were exceptionally heavy rains averaging 200-300mm over the whole of the Pyrenees. Such heavy precipitation caused considerable damage everywhere, which was exacerbated by the fact that June 1875 had already been a particularly wet month and the ground was already saturated. The effects were felt for years afterwards and were still quite visible fourteen years later (see Plate 2a). We can see that the outcrops of schist, already destabilised by animals, were the sites of small localised landslips which covered the slope in gravel. The larger breaks of slope were the site of actual rockfalls. The concentration of transport and landslips in the slide furrows, which were probably already bare, formed mudslips which dragged down the morainic deposits and then attacked the easily-weathered rock.

This type of situation was frequent in the nineteenth century in the Vicdessos area because it suffered all the disadvantages : deforestation, over-grazing and fragile substrata. Each time there was a landslip it was triggered off by a climatic incident : a snowless winter with deep ground frost, late snows followed by heavy spring rains or huge autumn storms after a period of drought. There were so many catastrophes towards the end of the century, in particular enormous avalanches, that the hamlets in the upper valley were abandoned. The nineteenth century events are now historical evidence of soil erosion but the erosion of the gullies is much earlier. The great Pyreneean gullies and ravines seem very ancient. They were probably formed in periglacial or immediately post-glacial times, as they experienced a succession of alternating equilibrium states and crisis periods.

An unknown fact : the relationship between climate and catastrophe in the Pyrenees during the nineteenth century. As has been shown, the coincidence of economic pressure and violent meteorological incidents had terrible consequences. However it was not the scattered damage in the Pyrenees which fuelled the arguments about degradation and restocking, but the serious floods in the valley and particularly in the plains. The whole of the nineteenth century is marked by these disasters, notably in the years 1850-1860, but the biggest between 1875 and 1900.

In 1875 there were two sets of floods in the south of France: those of 21st-24th June and 9th-13th September. The Pyrenees were spared the second but the Languedoc area suffered both, with rainfall more than 500mm in the Cevennes. The June rains brought about the greatest rise in the level of the Garonne ever recorded to this day (8.47m in Toulouse): the whole valley was ravaged and more than 500 people were killed. In the mountains the slopes and the river beds were destabilised for many years to come, and afterwards the slightest storm provoked disproportionate damage.

We know more about the events in 1897 because of the flood warning system installed after 1875. Once again there were two catastrophic incidents, 2nd-4th July and 5th-7th October, with rains touching the west and east of the Pyrenees respectively. The central Pyrenees were hit both times. A great many valleys were devastated regardless of whether they were completely deforested like the Bastan (Hautes-Pyrenees), or thickly forested like the Barousse and Campan valleys in the Hautes-Pyrenees, or the Pique and Garonne valleys in the Haute-Garonne, and also regardless of the dominant form of exploitation. The elements which determined the extent of flooding and erosion were the duration of the rainy front and the accentuation of the thermals by certain relief patterns.

This double rainy event, at the beginning and end of the summer, was a common occurrence at that time. Other examples took place in 1900 and 1901. The frequency of this phenomenon is surprising. In the central Pyrenees there were 37 recorded cases of rain followed by flooding erosion or landslides between 1810 and 1913 (and records are incomplete for the period before 1875).

Interpretation of this climatic picture is difficult due to the lack of data covering the whole of the nineteenth century. The only usable statistics for the mountains are those of the Pic-du-Midi observatory where records began in 1873. Until now these have only been looked at as averages, which does not provide any new information about the pattern of the European climate last century, but a more detailed analysis is under way. Research on scattered data in archives, newspapers, travelogues and limited series of meteorological observations is also being undertaken in order to have a clearer picture of the disasters and the types of weather associated with them.

At this stage in research we can only put forward a very general hypothesis. The meteorological situations which bring about disasters in the Pyrenees are classic and well known. There is a persistent depression over the Atlantic or the Mediterranean, generally associated with a strong anticyclone in the area. Then cold air descending towards the South brings about violent thermal contrasts with warm air masses rising towards the north. This was the case in 1875 and 1897 when extremely violent winds occurred, blowing in different directions in different valleys, with the rainy front coming and going across the whole width of the mountain chain. The frequency of strong anticyclones would appear to be a feature of the climate at the end of the nineteenth century. They were responsible for several years of drought (1872, 1873 and 1874 for example) and also for cold winters (with the freezing of the Garonne in 1882). The blocking of depressions in winter also brought heavy snowfall, which caused devastating avalanches. It would thus seem that catastrophic weather was particularly frequent in the Pyrenees towards the end of last century, but this still remains to be proved [2].

It is clear that the observers of that period could not fully understand these circumstances and that they systematically underestimated the importance of the meteorological phenomena which were taking place. For them, the dogma of recent deforestation remained the only explanation for flooding and erosion.

CONCLUSION

It is undeniable that there was some degree of anthropic erosion in the Pyrenees but it was limited to particular areas. Anthropic erosion was only one of the elements of the crisis whose complex origins were misunderstood by contemporary observers. It was complex because of the extreme heterogeneity of the impact of the erosion, with badly eroded areas next to completely stable ones. In assigning to the forest an almost magical role (both from an hydrological and a social point of view), the engineers of the time, entrenched as they were in scientific theories and contemptuous of the mountain communities, developed an analysis of the situation that was overgeneralised and too simplistic.

SOURCES

[1] County archives of Haute-Garonne (353C-40, 6 P 50), Common archives of Montauban-de-Luchon.

[2] County archives of Haute-Garonne (S 89, S 90, 10 M 18); Service Navigation Midi-Garonne; Société Météorologique de France: Nouvelles Météorologiques, 1875, 2° partie: faits météorologiques (Nov.1874-Nov.1875); Belgrand E. : "Résumé du tome 3 du Bulletin Météorologique Spécial de l'Association Scientifique de France et des observations qui serviront à publier le tome 4" (1876).

REFERENCES

Bucher, A., 1980. Analyses de données météorologiques de la région de Midi-Pyrénées et d'Aquitaine. Thesis, Pau.

Hourcade, B., 1985. Pastoralisme et protection du tapis végétal dans l'Alborz central (Iran). Acta Biologica Montana, 4, 567-572.

Jacob, C., Casteras, M., 1928. La morphologie des vallées luchonnaises. Bull. Soc. Hist. Nat. Toulouse, 67, 17-54.

Kalaora, B., 1980. Fôret et société au XIXe siècle: la sève de Marianne. INRA; Laboratoire d'Economie et de Sociologie Rurales.

Lamb, H. H., 1985. Climate history and the modern world. Methuen, London.

Metailie, J-P., 1985. Cent ans de paysage pyrénéen. Exposition 23 panneaux; catalogue. Université Toulouse-Le Mirail.

Taillefer, F., 1939. Le Vicdessos. Etude géographique. Revue Géogr. Pyr. et Sud-ouest., 10, 161-268.

Tricart, J., 1974. Phénomènes démesurés et régime permanent dans les bassins montagnards (Queyras et Ubaye, Alpes françaises). Revue Géomorpho. Dyn., 3, 99-114.

Trutat, E., 1898. Inondations dans les Pyrénées centrales. Bull. Soc. Géographie de Toulouse, 17, 177-214.

International Geomorphology 1986 Part II
Edited by V. Gardiner

DIMENSIONS OF GRAZING-STEP TERRACETTES AND THEIR SIGNIFICANCE

Jeffrey K. Howard and Charles G. Higgins

California Department of Water Resources
Sacramento, California, 95814 USA

Department of Geology
University of California,
Davis, California
95616 USA

ABSTRACT

Statistical analyses of the dimensions of grazing steps show that such terracettes have a mean spacing that is dependent on the size of the animals using the pasture, the slope angle of the hillside, and the density of stocking. Significantly greater spacing is associated with larger types of livestock, steeper slopes, obstacles on the slopes, and sparse stocking. The observed spacing of steps on slopes grazed intensively by sheep, cattle, and horses corresponds closely with that predicted by mathematical modeling of livestock grazing behavior.

Such analyses not only provide partial confirmation that the steps are of biogenic rather than of purely physical origin, but also yield data with which to evaluate the origins of other types of terracettes, such as "catsteps" or "frost steps", which may begin as grazing steps but are later modified by other processes.

More information is needed about such modifications and the effects that grazing steps have on slopes. In many climates grazing steps seem to inhibit surface erosion, but under other conditions they promote shallow slope failures or, where converted to other types of terracettes, may be accompanied by severe erosion. There are some indications that grazing-step development leads to formation of rectilinear midslopes on concave-convex hillsides. Other indications suggest that step spacing becomes more regular and approaches a minimum predicted by the grazing model as grazing pressures increase, thereby providing a possible guide to optimum stocking practices.

INTRODUCTION

Grazing-step terracettes, like those illustrated in Figure 1, are common microrelief features of hillslopes in pasturelands throughout the world. The origin of these and other kinds of terracettes has been debated ever since Charles Darwin described some in England and speculated that they might be formed by worms (Darwin, 1881). This long-standing and lively controversy has been reviewed by Vincent and Clarke (1976).

Figure 1 (above). Rangeland terracettes of the grazing-step type, with bare flat treads and sloping grassy risers, on a 34-1/2° rectilinear hillslope in the central California Coast Ranges at Briones Regional Park, near Martinez. June, 1981.

Figure 2 (right). The CalTrans steps at transect A-A', as initially formed in 6 weeks by 60 sheep in 1977 (photo A), and then subsequently modified by continued grazing by small bands of sheep for brief intervals in succeeding years and by a deep-seated arcuate slump in January 1978 (photos B, C, D).

All photographs were taken in early morning: A) July 1977; B) January 1979, after the slump and further modification by grazing; C) November 1979, after more grazing; D) September 1980, after still more grazing. Note anastomozing pattern of steps, with interconnecting ramps, characteristic of steps grazed by sheep, in contrast with more uniform, parallel steps grazed by cattle, as shown in Figures 1 and 3. A profile of the slope shown in photo A is reproduced in Figure 12.

Note that one or two major trails that serve mainly as transportation pathways can be recognized just above the fenceposts in all four photographs. Usage of other steps varied during the study interval, with new ones appearing, others being modified or shifting slightly in position, and still others disappearing altogether. Such changes caused minor variations in histograms of vertical spacing (V) between steps measured between 1977 and 1984, as shown in Figure 14. Despite these changes, the overall appearance of the slope has changed little since the steps were first formed.

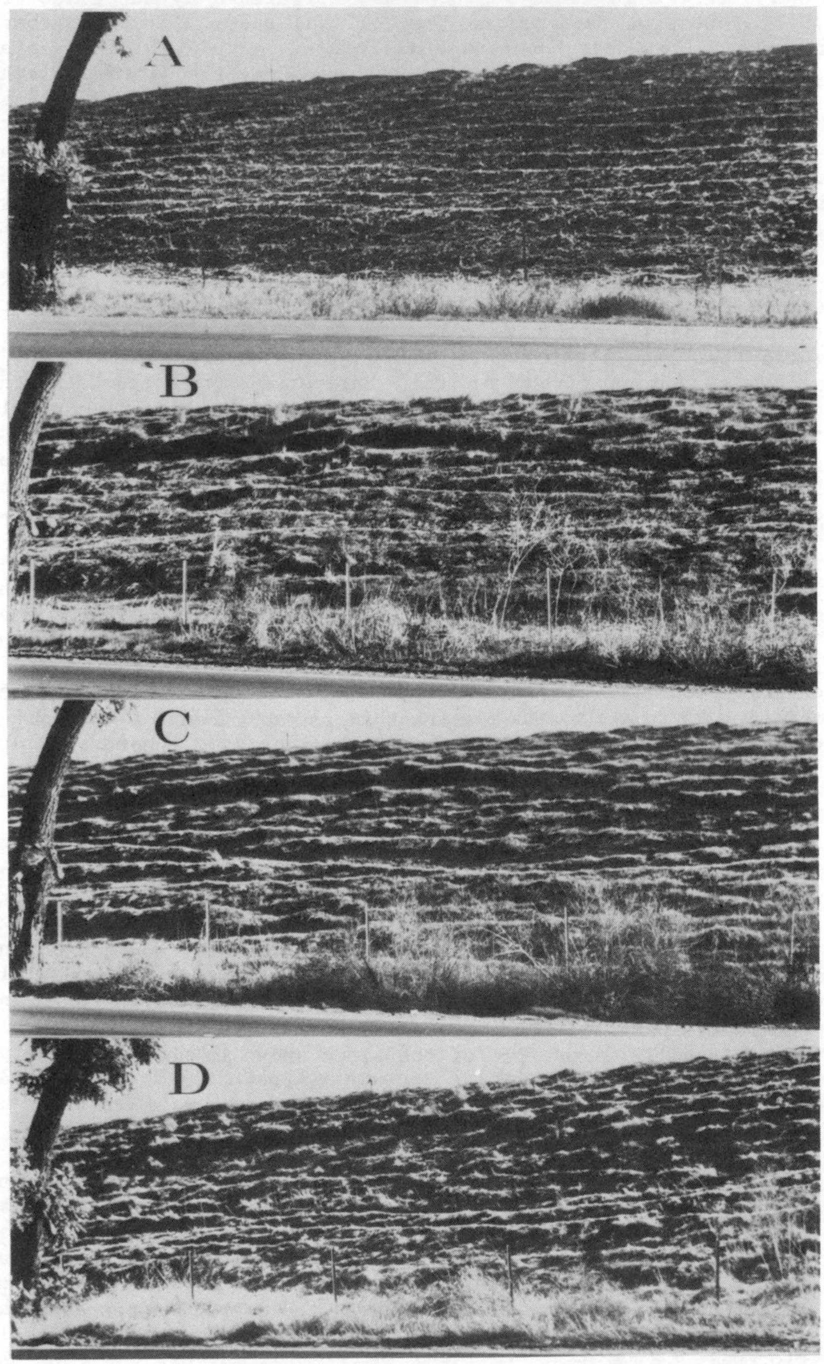
A
B
C
D

Our own interest in these features was stimulated by our discovery in 1977 of what we have called the CalTrans steps, in the southern Sacramento Valley near Woodland, California. At the CalTrans site, an initially smooth highway embankment was converted into a flight of well-defined steps by a flock of 60 sheep in only 6 weeks (Higgins, 1982). The resulting stepped slope is pictured in Figure 2A. These observations seemed to confirm that terracettes are of animal origin. However, careful study of the literature disclosed that there are several forms of terracettes with differing morphologies and thus possibly differing origins (Higgins, 1979, 1982).

The steps formed by the sheep at the CalTrans site (Fig. 2), like those shown in Figures 1 and 3, have sloping vegetated risers with nearly level bare treads. We call this type "grazing steps" to distinguish them from other types, such as the "catsteps" of the U.S. Great Plains (Brice, 1958) or the "delapsive terracettes" of the northern Caucasus (Chemekova and Chemekov, 1975), which have nearly vertical bare risers with level vegetated treads, or the "frost steps" or "needle-ice steps" of alpine regions, which Demangeot (1951) and others describe as being characterized by overhanging vegetated risers and bare, pebble-veneered treads. However, we found that reports of measurements of all types of terracettes seem to indicate that many have similar dimensions, suggesting that they may be related in some way. We hypothesized that all or many of the other types may originate as grazing steps and then are transformed by other processes such as slopewash erosion or needle-ice action. To investigate these possible relationships, we first set about to demonstrate that the grazing-step variety of terracette is indeed formed by grazing animals. To do this, Howard (1982) developed a dimensional model of ideal step development based on observed grazing behavior of domestic livestock.

GRAZING BEHAVIOR

The first element incorporated into the model was the behavior of grazing animals on a sloping pasture. Livestock tend either to follow one another along a major track or trail ("trailing behavior") or to graze slowly along individual steps ("grazing behavior"), as pictured in Figure 3. Sheep change steps fairly often, and so develop anastomozing pathways with level steps and sloping interconnecting ramps (Fig. 2). Cattle and horses rarely change steps, especially when the soil is damp, plastic, and slippery--the conditions under which steps are apparently formed or modified (Higgins, 1982)--so that their stepped pathways tend to be continuous and parallel for long distances (Figs. 1, 3).

Sheep, cattle, and horses use grazing steps by standing on the treads while grazing the risers (Fig. 3). The maximum distance an animal can reach upslope from its stance depends on the length of its neck and the slope angle. After the animal has eaten the accessible vegetation it will move forward along a horizontal route and continue to graze the riser. Rarely will it reach outboard of the path to graze. As a result, grass on the outer edge of a recently grazed step is generally tall and untrimmed, whereas the grass on

Figure 3. Grazing steps, with the cattle that presumably formed them, on a hillslope near Paskenta, Tehama County, northern California. Winter, 1982. Note grazing behavior, in which each animal occupies a specific step or track.

the inner edge and riser is trimmed in a smooth curve that reflects the swing of the animal's neck and head. This aspect of the steps is clearly shown in Figure 4. Such variation of grass height has long been noted in the terracette literature. Thomas (1959) attributed the differences to manuring on the pathway, but close inspection of the grasses and observations of the behavior of the animals seem to confirm that the differences owe instead to grazing.

In this way a grazing animal proceeds across a hillside. Other livestock that follow the lead animal choose ungrazed steps and graze along them until, if enough animals are present, the entire slope becomes grazed. This produces the observed pattern of increasing grass height above the tread of each step.

This behavior appears to be prompted by both efficiency and caution. Sloping California grassland soils typically contain enough clay to exhibit plasticity and become slippery and deformable when damp. At such times it is hazardous to traverse the slope. The least hazardous course across a steep slope or one with poor footing is generally a near-horizontal route. Patton (1971) notes that cattle generally avoid walking directly up or down slopes greater than 30 percent (17 degrees). Beside a concern for safety, livestock may simply prefer more horizontal travel because it involves the least expenditure of energy. Heavier animals work disproportionately more than lighter ones when moving, and appear to prefer

Figure 4. Detail of slope profile in Briones Regional Park (Fig. 1), showing variations in grass height across a step. Note untrimmed grass outboard of tread, but trimmed swath on inboard riser. This pattern owes to the animals' grazing behavior, as depicted in Figure 5.

more horizontal routes (May, 1981). This could explain why sheep cross from step to step more than cattle.

Commonly cited arguments against the development of terracettes by livestock have been based on the assumption that the steps would be formed by trailing behavior. It would then follow that livestock could not create such steps because steps often lead to impassable barriers such as fences. However, while grazing, an animal may proceed along a step right to such barriers, so that such arguments are invalid if grazing action is the responsible process.

Observed livestock grazing behavior seems well suited to the form of grazing steps. Whether the one causes the other can be addressed statistically. Howard (1982) has compared the measured spacing between grazing steps from pastures grazed exclusively by sheep, cattle, and horses, and has found that for similar slopes the spacing is greater for larger animals. Table 1 presents a summary of the variation of V, the vertical spacing or height, of steps with slope angle and animal type. In the table the data are divided into five slope groups and three animal types. Within each division is shown the number of observations (n), mean vertical spacing ($\bar{V}$) and standard deviation (s). T-test results show that when identical slope groups are compared, the mean vertical spacing of steps grazed by sheep is significantly smaller than for those grazed by cattle.

TABLE 1. Variation of average vertical spacing with slope angles (V vs Θ) for grazing steps used by horses, cattle, and sheep, based on measurements at seventeen sites in central California. Sample characteristics of various slope groups indicated by: n = sample size, $\bar{v}$ = mean vertical spacing (cm), s = standard deviation (cm).

	20 - 24°	24 - 28°	28 - 32°	32 - 36°	36 - 40°
HORSES	n = 0	n = 2 $\bar{v}$ = 83 s = 24	n = 16 $\bar{v}$ = 101 s = 28	n = 43 $\bar{v}$ = 115 s = 28	n = 18 $\bar{v}$ = 128 s = 18
CATTLE	n = 65 $\bar{v}$ = 65 s = 21	n = 96 $\bar{v}$ = 77 s = 24	n = 80 $\bar{v}$ = 81 s = 20	n = 130 $\bar{v}$ = 96 s = 24	n = 58 $\bar{v}$ = 112 s = 23
SHEEP	n = 10 $\bar{v}$ = 26 s = 6	n = 24 $\bar{v}$ = 37 s = 11	n = 20 $\bar{v}$ = 48 s = 12	n = 2 $\bar{v}$ = 42 s = 13	n = 0

Likewise, mean heights of cattle-grazed steps are significantly smaller than for steps grazed by horses (Howard, 1982). Furthermore, we and others find that terracettes of the grazing-step type are generally restricted to slopes greater than 15 to 18 degrees, which agrees with the slope limit to cattle travel mentioned by Patton (1971).

THE GRAZING MODEL

The grazing behavior described above was modeled geometrically to test the hypothesis that grazing behavior causes terracettes. After the model was developed, a mathematical formula was derived from it in order to compute the step spacing that should result for any variation in slope angle, animal size, and grass height. These hypothetical values were then compared with measured spacing of steps grazed by different animals on a variety of slopes. If grazing behavior is indeed the mechanism of step formation, then the hypothetical and measured values should agree.

Figure 5 is a cartoon of the geometrical grazing model showing the parameters that may influence step spacing--θ, Ha, Hg, and Wt--and the parameters that describe the step spacing--SD, V, and H. In the model, the swing of an animal's head as it grazes up the riser is approximated as a circular arc having its center at the animal's shoulders and a radius equal to the animal's shoulder height, Ha. According to the hypothesis, the slope distance, SD, that the animal can graze uphill determines the minimum separation between two steps.

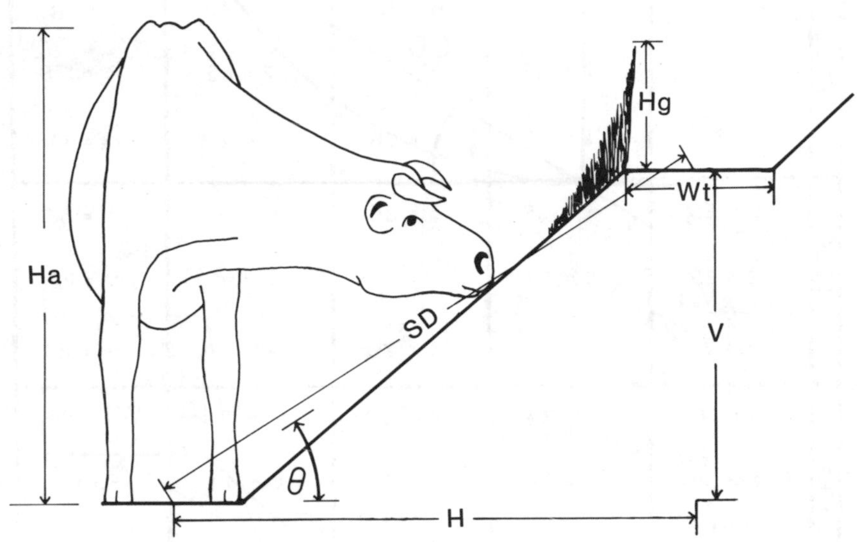

Figure 5. Cartoon of grazing model. Note parameters used: θ = macroslope angle; SD = slope distance between successive steps; V = vertical spacing, or height between successive steps; H = horizontal distance between steps; Wt = width of tread; Ha = height of animal; Hg = height of grass. From Howard (1982)

Figure 6A shows a simple geometric approximation of the model in which the step spacing, here indicated as SD, is a function of only the animal height, Ha, and the slope angle, θ. As shown, SD is the length of the chord formed between the two points where the grazing arc and slope line intersect. Given that the slope line forms an angle, θ, with respect to the horizontal, then SD is found by

$$SD = 2\ Ha\ \sin\theta \qquad \text{(equation 1)}$$

This simple model's drawbacks are that it does not account for two other variables--the initial grass height, Hg, and the width of the animal's stance, which can be approximated as the trail width, Wt. Consequently, SD tends to be underestimated.

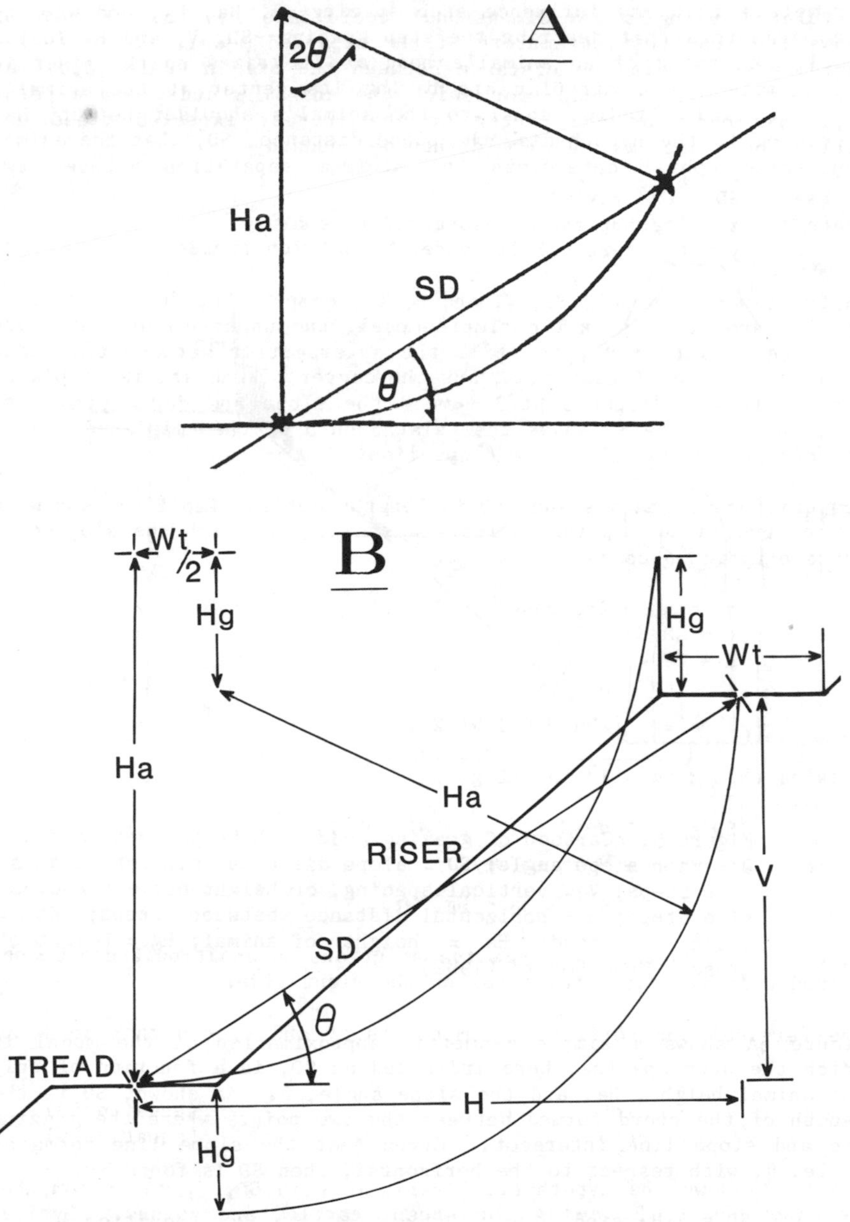

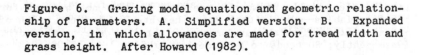

Figure 6. Grazing model equation and geometric relationship of parameters. A. Simplified version. B. Expanded version, in which allowances are made for tread width and grass height. After Howard (1982).

An expanded model that incorporates these variables can be formulated using a two-dimensional coordinate system. Figure 6B shows the geometric parameters of the expanded model. In Cartesian coordinates, SD is the distance between the origin (point 0,0), or the center of the tread upon which the animal stands, and a point (x,y) representing the center of the nearest expected tread uphill. The slope distance would then be found by

$$SD = (x^2 + y^2)^{1/2}$$

where x = the horizontal distance, H, and
y = the vertical distance, V, between treads.

In the expanded model, SD, V, and H, are computed as functions of θ, Ha, Hg, and Wt. As in the simple model, the upper end of the slope distance, point (x,y), is still the intersection between the slope line and an arc of radius Ha, but the center of the arc is displaced horizontally a distance Wt/2 toward the slope and downward a distance Hg. Figure 6B shows the relationship of the displaced arc to the step positions along the slope line.

Computation of the distance SD is lengthy but is simplified somewhat by conversion to a polar coordinate system, in which the slope-distance equation becomes

$$r^2 = R^2 + 2rc\cos(\theta - a) - c^2$$

where
$r = SD$
$R = Ha$
$c = (Ha - Hg)^2 + (Wt/2)^2$
$a = \tan^{-1}[(Ha - Hg)\ Wt/2].$

Solving this quadratic for SD gives

$$SD = A^{1/2}\cos B \pm [Ha^2 - A(\sin^2 B)]^{1/2} \qquad \text{(equation 2)}$$

where
$A = (Ha - Hg)^2 + (Wt/2)^2$
$B = \theta - \tan^{-1}[2(Ha - Hg)/Wt].$

It can be shown that non-negative solutions to equation 2 can be obtained only by adding the terms on the right side.

The vertical and horizontal components, V and H, can then be found by

$$V = SD\sin\theta \qquad \text{(equation 3)}$$
$$H = SD\cos\theta \qquad \text{(equation 4)}$$

Figure 7 shows the hypothetical variations in SD, V, and H computed by equations 2, 3, and 4 for sheep, cattle, and horses. Typical mature heights of these livestock types were determined to be 65 cm, 120 cm, and 150 cm respectively, and were used for values of Ha. Hg was set at 35 cm, and Wt was set at 25 cm, 38 cm, and 40 cm, respectively, corresponding to the means of typical trail widths observed for these animals. The trends of the hypothetical values of V shown in Figure 7 are nearly linear and exhibit the greatest variation with θ. For this reason, the observed vertical spacings of grazing steps were chosen for linear-regression analysis and for comparisons with the predicted, hypothetical trends.

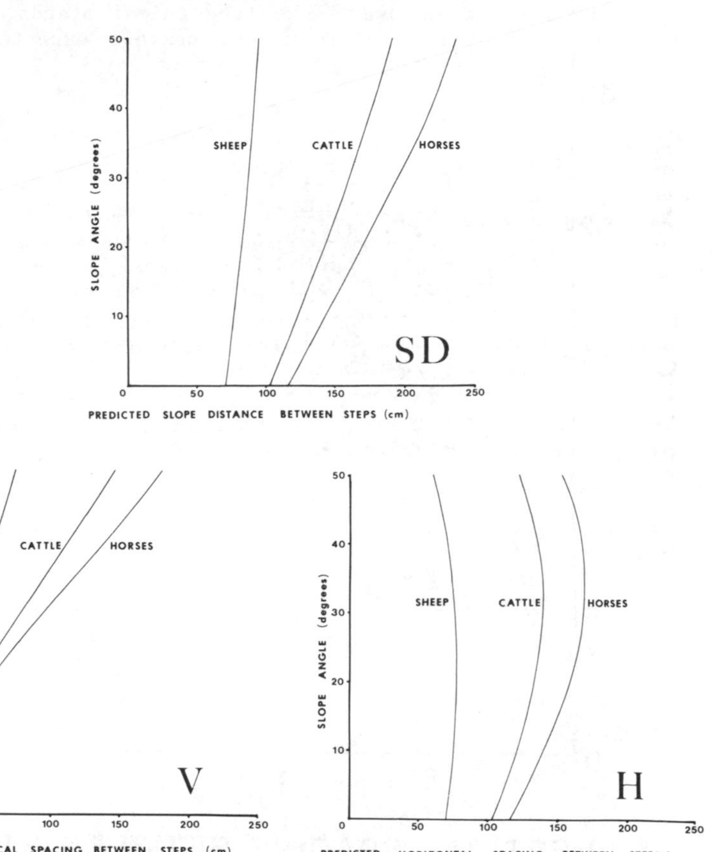

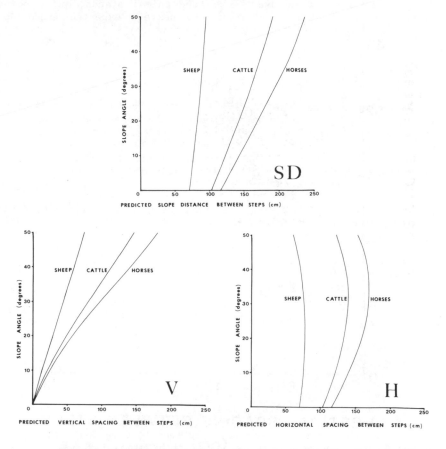

Figure 7. Variations of SD, V, and H with slope angle θ, as predicted by the grazing model equation for sheep, cattle, and horses.

Figure 8 shows the vertical spacing of steps grazed by cattle, from data measured or compiled by us, largely from the Coast Ranges of central California. The data points plot about a solid line of predicted values that is derived from the same model parameters used for cattle in Figure 7V. Thirty-eight percent of the points are within the range of predicted values--between the dashed lines--that result when reasonable variations of animal and grass heights are allowed. Figure 8 includes a suite of mean values of V reported by Robert (1963) for steps grazed by cattle in eastern Belgium. These are plotted as open circles, and fall close to the predicted values.

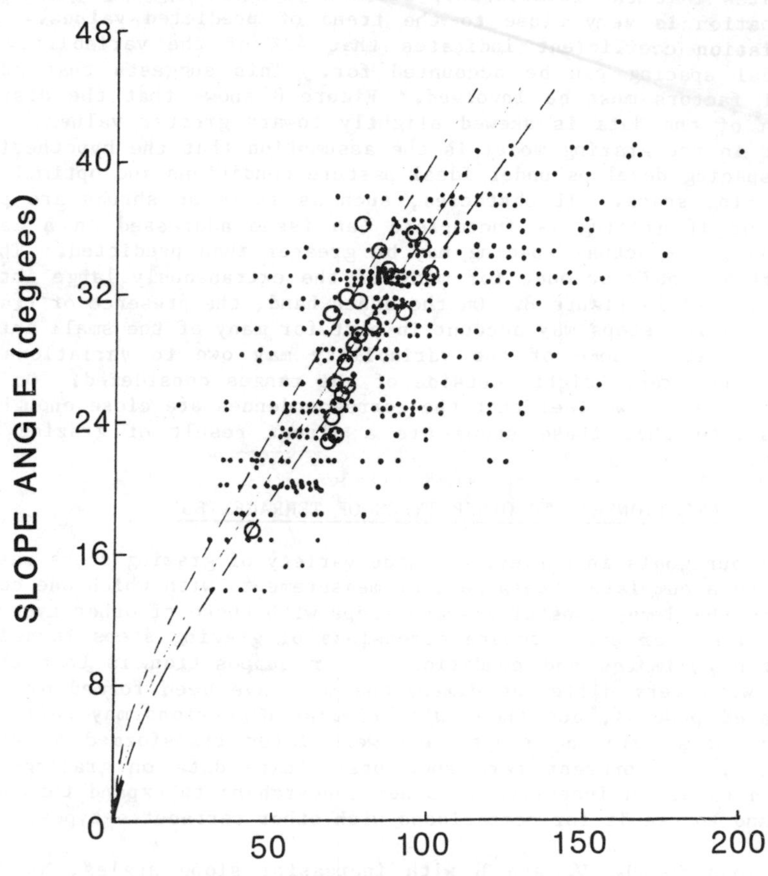

Figure 8. V vs θ for grazing steps found on slopes grazed exclusively by cattle. Dots represent measurements in California combined with some by Barbara O. Bach (personal communication) in Montana. Circles represent measurements by Pierre Robert (1963, Diag. 6) in eastern Belgium. Lines of values predicted by the grazing model equation are superimposed; the dashed lines enclose values expected for the range of slopes where grass height varies from 20 to 50 cm and animal height varies from 110 to 130 cm. After Howard (1982).

Linear regression and correlation analysis of the California data indicates that the relationship between vertical spacing and slope inclination is very close to the trend of predicted values. The correlation coefficient indicates that 42% of the variability of vertical spacing can be accounted for. This suggests that additional factors must be involved. Figure 8 shows that the distribution of the data is skewed slightly toward greater values. Implicit in the grazing model is the assumption that the hypothetical step spacing develops under ideal pasture conditions and optimal use of grazing space. If obstacles, such as rocks or shrubs are present, or if grazing is incomplete (an issue addressed in a later section), the actual spacing may be greater than predicted. These factors probably account for many of the extraneously large intervals plotted in Figure 8. On the other hand, the presence of trails among grazing steps may account in part for many of the small intervals. Lastly, some of the variability may owe to variations in animal and grass heights outside of the ranges considered. On the whole, however, we feel that the correspondences are close enough to demonstrate that these steps are a direct result of grazing behavior.

RELATIONSHIP TO OTHER TYPES OF TERRACETTES

One of our goals in surveying a wide variety of grazing steps was to begin to accumulate a data base of measurements with which one could compare the dimensions of grazing steps with those of other types of terracettes, or even compare dimensions of grazing steps formed in different climates and conditions. Our supposition is that those steps with very different dimensions may have been formed by some unrelated process, but those with similar dimensions may well have originated as grazing steps, and were later transformed by other proceses. We present here some preliminary data on grazing-step dimensions as an incentive to other researchers to expand this data base and to use it for comparisons with other terracette types.

Variations in SD, V, and H with increasing slope angles, as predicted by the grazing model, are depicted in Figure 7. V is seen to have the greatest dependence on θ, H has the least variability, and SD is intermediate. However, SD, the slope distance, is actually the easiest of the parameters to measure; one simply stretches a tape measure down a hillside from a step near the top to one near the bottom. The total distance divided by the total number of step intervals gives the mean SD. The macroslope angle can be measured with a clinometer. If mean values for V and H are wanted, they can be obtained from SD and θ by equations 3 and 4.

Most of the measurements presented in this paper concern the dimension V, the vertical interval, or height, between steps, for steps formed by cattle. In a later publication we expect to present the results of a much larger sample of field measurements of steps formed by sheep and cattle.

GEOMORPHIC IMPLICATIONS

In much of the literature, terracettes have been treated as minor, even trivial features of the landscape. However, they may affect the landscape significantly in several ways. Some of these are discussed here.

The role of grazing steps in surface erosion. Livestock trails may initiate accelerated erosion in the form of gullying because they can intercept and channel runoff. However, trails, which are generally inclined, are different from grazing steps, which are generally horizontal. Grazing steps are analogous to soil-conservation methods of contour plowing and terracing that are used to inhibit rilling and gullying. Howard (1982) notes that rills or gullies rarely occur on slopes with grazing steps in California's Mediterranean climate. It is difficult to find adjoining grazed and nongrazed slopes in California for studies of comparative erosion, but useful analogs for such studies occur along some highways, where artificial steps of about the same dimensions as those formed by cattle have been created by highway engineers. Experiments conducted by the California Department of Transportation (CalTrans) have shown that these steps not only retain loose rock debris, which is their primary function, but also significantly decrease surface wash and rilling in erosible materials (Wagner et al., 1979). Results of one such study are shown in Figure 9.

Figure 9. Experimental cut on Interstate-280 near Palo Alto, California. Upper left part of the slope was artificially stepped; the remainder was left with normal grading. Note severe rilling on non-stepped areas and absence of rills on stepped portions. Photograph August 1974, courtesy of Tom Hoover, Transportation Laboratory, California Department of Transportation.

Stepped highway cuts and observations of grazing steps suggest that such steps tend to reduce surface runoff and erosion in climates like that of California. In other climates, however, especially where needle-ice action is effective, development of grazing steps may help to accelerate surface erosion (Higgins, 1982). These detrimental effects need further investigation.

The role of grazing steps in mass movement. Even where surface erosion is lacking, relatively small surficial soil failures are commonly associated with grazing-step terracettes on steeper slopes, as shown in Figures 10 and 11. CalTrans engineers have found that in

Figure 10. Shallow slope failure on stepped hillslope in the Dunnigan Hills north of Woodland, California. Overall relief is 25 to 30 m. Note absence of rills and other evidence of surface runoff on both this slope and the one pictured in Figure 11, as well as on those shown in Figures 1, 2, and 3.

Figure 11. Shallow slope failures on stepped hillslope near Fairfield, California. These developed in Spring 1982, after above-normal rainfall. Overall relief about 110 m; oak trees on slope are about 10 m high.

some semi-consolidated materials, their stepped roadcuts are more susceptible to shallow mass movements than are nonstepped cuts (Tom Hoover, personal communication, August, 1985). Other observers have noted an association between slope failures and grazing steps. Dalyrymple et al. (1968) and During and Radcliffe (1962) associate terracettes with slopes dominated by mass movement erosion, and Saul (1973) has noted that grazed grassland slopes in the central California Coast Ranges are more susceptible to failure than are adjacent non-grazed slopes. Grazing steps themselves have generally not been identified as influential factors in mass movement, although other factors, such as vegetation conversion and soil desiccation cracking that are associated with them, have been. In fact, terracettes are more often cited as resulting from slope instability rather than as causing it.

The absence of surface wash and rilling, noted above, would logically be associated with increased water infiltration into the slope. Evidence indicates that steps do act to intercept runoff and/or provide more effective slope surfaces on which precipitation can infiltrate. Sears (1951), During and Radcliffe (1962), and Radcliffe (1968), among others, have noted that the treads of grazing steps retain more moisture than do the risers. Indeed, water can be seen standing on some steps shortly after heavy rains.

It could be argued that the treads of grazing steps may become compacted to a degree that inhibits water infiltration, which would counteract the effect of greater interception of precipitation. Many studies have been made of the impact of livestock grazing on soil compaction, and the results have been inconclusive (Heady, 1975). However, a review of these studies indicates that where compaction does occur it is typically associated with severe stocking rates on pastures with slopes less than 10 degrees.

In New Zealand rangeland, under conditions very similar to those of the California Coast Ranges, Radcliffe (1968) measured bulk densities of soil on stepped slopes and found no significant differences between treads and risers. The reason that compaction may not occur on stepped rangeland slopes may be related to the moisture condition of the soil at the time that treading occurs. Livestock hoof pressures are approximately 0.65 kg/cm^2 for sheep and 1.7 kg/cm^2 for cattle and horses (Heady, 1975). While these are sufficient to deform plastic soils when wet, they are much less than the dry strength of the clayey soils typical of the California Coast Ranges. Furthermore, significant soil compaction cannot occur under saturation or near-saturation conditions because the soil interstices are supported by porewater. Under these conditions, soils on grazing-step treads are simply remolded by the kneading action of hoof pressure. As these soils dry, the moisture conditions may pass rapidly through the range under which significant compaction can occur. Because livestock graze widely over a hillside, their use of the steps during this brief range of moisture conditions is generally unlikely to result in any significant compaction.

If grazing steps increase the efficiency by which a slope intercepts precipitation, then the duration of the rainfall event required to cause failure of that slope is decreased. Shallow landsliding like that shown in Figures 10 and 11 would then be associated with shorter, more frequent storms, and the incidence of failures would be accelerated. Still greater infiltration could lead to increased area of instability or to deeper-seated mass movements, as seen in Figure 2B.

The role of grazing steps in hillslope modification. In California Coast Range pasturelands, profiles of hillslopes exhibiting grazing steps are almost invariably rectilinear (Howard, 1982) and susceptible to shallow slope failures. In their nine-unit landsurface model, Dalyrymple *et al.* (1968) associate terracettes with the rectilinear "transportational midslope" component that has typical slope angles from 26 to 35 degrees.

Even where slope profiles appear to be smoothly curved, they may have pronounced rectilinear segments, as Clark (1965) found in the English chalk downs. The hillslopes shown here in Figure 1 at first appear to be simply concave-convex, but measured profiles demonstrate that they consist mainly of rectilinear midslopes with macroslopes ranging from 31 to 34-1/2 degrees. Virtually none of the careful studies that have been made of rectilinear hillslope profiles (see references in Clark , 1965) have considered the possible role of grazing steps in their development. However almost all of the photographs reproduced with these studies show recognizable grazing steps and/or the animals that form them.

Eyles (1971) reported that accelerated shallow mass movement is associated with expansion of rectilinear midslopes on hillslopes that formerly were concave-convex in New Zealand rangeland. He did not identify grazing steps as a cause of the mass movement, but instead speculated that it had been the result of vegetation conversion in combination with heavy rainfall events. However, at least part of the cause may have been grazing-step development, because photographs accompanying Eyles' report show that steps are abundant there.

The rectilinear midslope profiles in New Zealand appear to have been formed at the expense of convex summits. Eyles identified the mechanism of failure as debris sliding on slopes between 24 and 42 degrees. The slide planes were shallow, averaging 0.6 m, and rectilinear, within 2 degrees of the original slopes. From this, Eyles inferred that the slopes were backwearing by parallel retreat.

Theoretically, on a slope with grazing steps, if a general condition exists whereby the rate of soil removal is proportional to the slope angle, it can be shown by mathematical modeling that the slope will retreat in parallel fashion, eventually evolving a straight profile (Scheidegger, 1961). This agrees with Eyles' observations in New Zealand. Carson and Kirkby (1972) note that a straight midslope, which they call a "main slope", is produced primarily by weathering-limited processes, particularly mass movement, and may evolve by parallel retreat, slope decline, or slope shortening. They further state that parallel retreat could result in a slope profile that contains two or more straight segments, in which the upper, steeper unit retreats by landsliding and is eventually consumed by the lower, gentler slope. Such conditions are also evidenced in Eyles' study, and appear to be the case in our own studies, where the angles of the many slopes we have surveyed have a strong bimodal distribution centered about 25 and 34 degrees (Howard, 1982). These angles appear to correspond with colluvial footslopes and weathering-limited erosional slopes respectively. This tends to confirm that parallel retreat has been occurring, because if slope decline had been operating, a more uniform distribution of slope angles would be expected.

The existence of significant convex and concave segments above and below rectilinear midslopes in the California Coast Ranges suggests that slope modification and parallel retreat may have been initiated there only recently, perhaps with the introduction of grazing about 160 years ago. This, in turn, leads to the speculation that similar slope modification and development of rectilinear midslopes elsewhere may owe chiefly to effects of grazing rather than to the purely physical processes that traditionally have been cited in the literature.

Summary. Theory, analogy, and circumstantial evidence all indicate that grazing steps may indeed increase the frequency of shallow slope failure, while at the same time they may decrease the impact of erosion by surface wash. Over time, affected slopes may evolve a predominant rectilinear profile composed of one or several different segments, often at the expense of the existing topography. Given the sheer numbers of grazing steps that have been formed on the world's landscapes, further research into their effects should be encouraged.

RANGE-MANAGEMENT IMPLICATIONS

Perhaps the most useful application of grazing-step study and measurement is in determination of the carrying capacity of sloping pasturelands. Paul Krupin's unpublished studies of grazing steps in west-central Idaho for the Bureau of Land Management showed that livestock can and do graze much steeper slopes than had previously been thought. He measured steps developed primarily by cattle on slopes as steep as 41 degrees (Krupin, 1982). The Bureau's leasing arrangements had formerly been based on the assumption that slopes steeper than 50% (26°34') were not subject to grazing. As a result of Krupin's studies, steeper slopes were later apportioned in a different manner (Paul Krupin, personal communication, August 1985).

Krupin was also first to perceive that spacing of steps depends in part on the density and history of stocking and grazing. In an unpublished paper delivered at the 1982 annual meeting of the Society for Range Management, in Calgary, Alberta, he observed that denser, more optimum stocking is associated with closer spacing. Our own studies tend to confirm this conclusion.

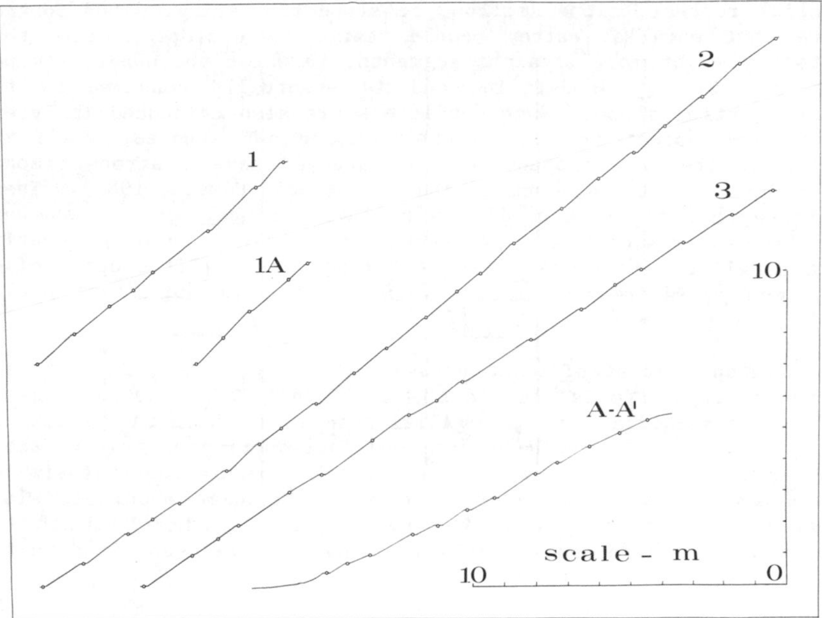

Figure 12. Measured profiles (1 through 3) of stepped slopes on West Butte of Sutter Buttes, Sacramento Valley, California, compared with profile A-A' on CalTrans slope. Dots indicate centers of treads. All slopes grazed by sheep, with relatively low-density stocking on West Butte and denser stocking on CalTrans. Note resulting closer spacing of steps on CalTrans slope. The CalTrans slope is illustrated in Figure 2.

The West Butte profiles were surveyed 18 May 1985. Macroslopes: 1 = 39°; 1A = 42° (this is one of the steepest stepped slopes we have surveyed); 2 = 32° for steps 1 through 6, 39° for the remainder; 3 = 32°. All are rectilinear segments of concave-convex hillsides, except profile 2, which lies on a debris fan or talus cone. The CalTrans profile was surveyed 23 August 1977, the year it was formed (see Fig. 2A). Macroslope angle = 23°-28°.

Figure 12 shows profiles of four slopes on West Butte of the Sutter Buttes, near Yuba City in the Sacramento Valley, California. Open dots show centers of treads. Slopes 1, 1A, and 2 face south; slope 3 faces southwest. The terrain is grassy rangeland, lightly stocked with sheep. Figure 13 shows histograms of the heights, or vertical spacing, (V), between steps for these four flights. There is a wide range of heights, from about 45 cm to about 140 cm. Mean heights are shown by small black arrows. Large arrows indicate the heights expected for these slopes from the grazing model; the width of the base of the arrows shows the variability expectable for the par-

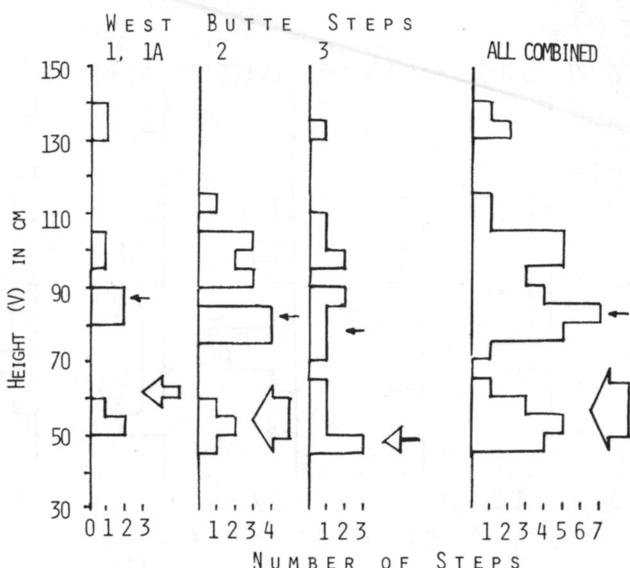

Figure 13. Histograms of vertical spacing between steps for four flights on West Butte (Fig. 12). Small arrows indicate mean heights for each group--87.7 cm for flights 1 and 1A, 82.5 for 2, 78.3 for 3, and 82.3 for all combined. Macroslopes of 1 and 1A range from 39° to 42°, of 2 from 32° to 39°, of 3 are 32°.

Base widths of the wide arrows show the ranges of heights predicted for the slopes by the grazing model, assuming 40 cm grass height--60 to 63.5 cm for flights 1,1A; 49 to 60 for 2; 49 for 3; and 49 to 63.5 overall. Note increases in means as macroslopes increase, but also large discrepancies between the means and expected values, which we attribute to understocking and underdevelopment of the pasture.

ticular slope angles. Note that although both large and small arrows show progressively smaller vertical spacing on gentler slopes, the means of the measured heights are 25 to 30 cm greater than the expected values. On the steepest slopes, 1 and 1A, wide and irregular spacing of steps may owe in part to scattered outcroppings of sandstone bedrock. Similarly, on slope 2, a debris fan or talus apron, scattered small shrubs may have prevented optimal spacing of the steps. However, the lack of obstacles of any kind on slope 3 and the consistently large spacing of steps on all four slopes suggest that the departures from the expected values owe to some other factor. The responsible factor is probably the lack of intense grazing, because field observations showed that at all transects much of the tall grass between the steps had not been trimmed.

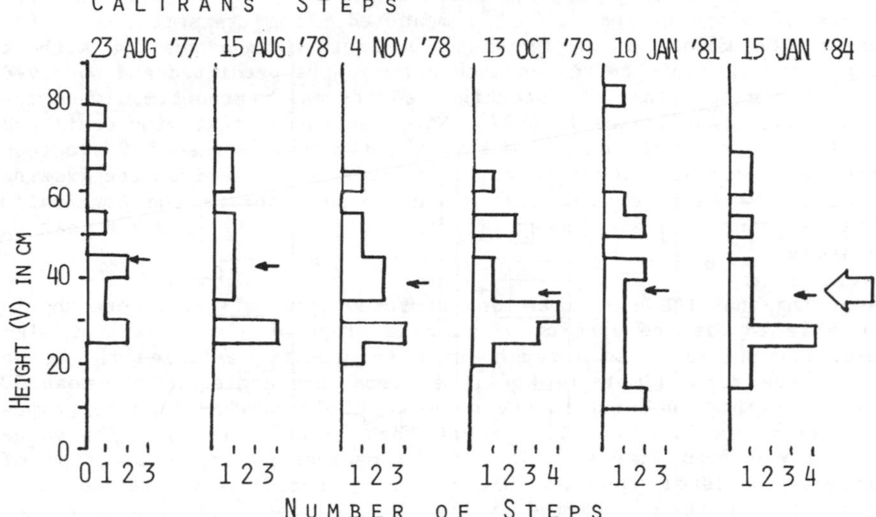

Figure 14. Histograms of vertical spacing between steps along CalTrans transect A-A' (Figs. 2, 12), as surveyed on six separate occasions over a 6-1/2 year interval. Small arrows indicate mean heights, decreasing from 44.4 cm in 1977 to 37.3 in 1984. Base width of wide arrow shows range of heights--35 to 41 cm--predicted by the grazing model for this slope of 23° to 28°. Note close correspondence of means with expected values in 1977, when the steps were first formed, with still closer correspondence as the steps developed over time.

Figure 12 also shows the first of several profiles surveyed along transect A-A' of the CalTrans slope. This slope faces southwest and has been grazed exclusively by sheep. In 1977 when this profile was surveyed, shortly after the steps were developed, stocking pressure was relatively great because there was little fodder on the summit and east slope of the embankment. Also, there were no obstacles of any kind on the slope. Note the relatively close and regular spacing of steps. This contrasts sharply with the spacing on the West Butte flights.

Figure 14 shows that the heights (V) of the CalTrans steps, which range from less than 20 to more than 80 cm, are indeed less than those of the West Butte steps, which range between 45 and 140 cm (Fig. 13). Figure 14 also shows a more significant difference between the CalTrans and West Butte steps: the mean values determined for each of the six different surveys of CalTrans transect A-A', indicated by small black arrows, are shown to correspond closely with the values expected from the grazing model. The latter are indicated by the large open arrow. The width of the arrow's base shows the range of heights predicted by the grazing model for the range of slope angles (23°-28°) measured along transect A-A'. Because the model was designed to predict minimum spacing without overcrowding, this correspondence between the predicted and observed values suggests that the stocking density was near optimum when the slope was first grazed in 1977. Note, moreover, that with continued periodic grazing in successive years, the mean value of V progressively approaches closer to the mean value expected from the grazing model. We believe that this tends to corroborate the conclusion that spacing of steps depends in part on grazing density and grazing history.

Assuming that these observations represent general cases, one should be able to use the spacing of grazing steps on slopes without significant obstacles as a rough guide to stocking efficiency. To do this, one would simply measure the macroslope angle and the mean SD by the method outlined in the section, "Relationship to other types of terracettes". This figure can then be compared with the value determined from Figure 7-SD for the macroslope angle and type of animals pastured there. Where the measured value is markedly greater than that predicted by the grazing model, as at West Butte, the grazing density must be less than optimum. That is, it must be less than what could be supported by the pasture. This method of analysis only identifies understocking of pasturelands. Overstocking would not be recognizable from step spacing, which should be the same as for optimal grazing, but which might be accompanied by excessive wear, exposure of bare soil, and consequent erosion on the treads and lower parts of the risers.

CONCLUSIONS

Far from being a trivial academic pursuit, the study and measurement of grazing-step terracettes may, in time, yield considerable dividends in advancing our understanding of hillslope processes and evolution. The role of grazing steps in reducing surface wash and soil erosion in many climates may be valuable in range management and conservation, although it must be recognized that steps may aid mass movement under some conditions. In many places grazing steps may be an important recent factor causing parallel retreat as evidenced by rectilinear midslopes in terrain that was initially formed by other processes. Finally, as grazing-step spacing seems to owe at least in part to stocking density and grazing history, such spacing may provide a useful tool for more efficient management of hillside pasturelands.

REFERENCES

Brice, J. C., 1958. Origin of steps on loess-mantled slopes. U. S. Geological Survey, Bulletin 1071-C, p. 69-85.

Carson, M. A., and M. J. Kirkby, 1972. Hillslope form and process. Cambridge University Press, 475 p.

Chemekova, T. Yu., and Yu. F. Chemekov, 1975. Hillslope terracettes. Soviet Geography: Review and Translation, v. 16, p. 609-615. Translated from: Izvestiya Vsesoyuznogo Geograficheskogo Obshchestva, v. 4 (1974), p. 323-328.

Clark , M. J., 1965. The form of chalk slopes. Southampton Research Series in Geography, v. 2, p. 3-32.

Dalrymple, J. B., R. J. Blong, and A. J. Conacher, 1968. An hypothetical nine-unit landsurface model. Zeitschrift für Geomorphologie, N.F., v. 12, p. 60-76.

Darwin, Charles, 1881. The formation of vegetable mould through the action of worms, with observations on their habits. London: John Murray. 313 p.

Demangeot, J., 1951. Observation sur les "sols en gradins" de l'Apennin Central. Revue de Géomorphologie Dynamique, v. 2, p. 110-119.

During, C., and J. E. Radcliffe, 1962. Observations on the effect of grazing animals on steepland soils. International Society of Soil Science, Transactions of the joint meeting of Commissions IV and V, p. 685-690.

Eyles, R. J., 1971. Mass movement in Tangoio conservation reserve, northern Hawkes Bay. Earth Science Journal, v. 5, p. 79-91.

Heady, H. F., 1975. Rangeland management. New York: McGraw-Hill, 460 p.

Higgins, C. G., 1979. "Terracettes": A classification. Geological Society of America, Abstracts with Programs, v. 11, p. 443-444.

Higgins, C. G., 1982. Grazing-step terracettes and their significance. Zeitschrift für Geomorphologie, N. F., v. 26, p. 459-472.

Howard, J. K., 1982. The origin and significance of "grazing-step" terracettes. M. S. thesis, Department of Geology, University of California, Davis. 137 p.

Krupin, P. J., 1982. Erosion and morphology of grazed, steep mountain slopes. Society for Range Management, 35th Annual Meeting, Abstracts, p. 32.

May, R. M., 1981. Mammal tracks on mountain slopes. Nature, v. 292, p. 672-673.

Patton, W. W., 1971. An analysis of cattle grazing on steep slopes. M. S. thesis, Brigham Young University, Utah. 42 p.

Radcliffe, J. E., 1968. Soil conditions on tracked hillside pastures. New Zealand Journal of Agricultural Research, v. 11, p. 359-370.

Robert, Pierre, 1963. Etude de géomorphologie de la région de Nessovauz. Mémoire de Licence en Sciences Géographiques, Université de Liège. 179 p. (Parts of this study were published in 1964 as: Quelques problèmes géomorphologiques dans la région de Nessonvaux: étude des terrassettes. Societé Géologique de Belgique, Annales, v. 87, Bulletin 9, p. B273-B293).

Saul, R. B., 1973. Geology and slope stability of the southwest quarter Walnut Creek quadrangle, Contra Costa County, California. California Division of Mines and Geology, Map Sheet 16.

Scheidegger, A. E., 1961. Mathematical models of slope development. Geological Society of America, Bulletin, v. 72, p. 37-50.

Sears, P. D., 1951. The technique of pasture measurement. New Zealand Journal of Science and Technology, v. 33A, p. 1-30.

Thomas, A. S., 1959. Sheep paths. Journal of the British Grassland Society, v. 14, p. 157-164.

Vincent, P. J., and J. V. Clarke, 1976. The terracette enigma - a review. Biuletyn Peryglacjalny, v. 25, p. 65-77.

Wagner, M., J. Egan, G. F. Warn, and R. B. Howell, 1979. Erosion measurements on a smooth and stepped highway slope. California Department of Transportation, Interim Report CA-TL-79-15, 49 p.

ACKNOWLEDGEMENTS

We are indebted to the East Bay Regional Park District, the East Bay Municipal Utility District, and many ranchers and landowners for permission to survey stepped slopes on their property; to Cathy Martin, Gary Pedroni, and Lori Summa for occasional assistance in the field; to Paul Schneman and Dennis Ojakangas of the Computer Center, University of California, Davis, for consultation on statistical analyses; and to Mary Graziose, Ellen Guttadauro, and Susan Raubach of the Department of Geology for manuscript preparation. Some financial assistance for these studies was provided by a Penrose grant from the Geological Society of America. Additional funding was furnished by the Committee on Research, University of California, Davis.

International Geomorphology 1986 Part II
Edited by V. Gardiner

THE GEOMORPHIC SETTING OF PREHISTORIC GARDEN TERRACES IN THE EASTERN HIGHLANDS OF PAPUA NEW GUINEA

M. Sullivan and P. Hughes

University of Papua New Guinea, P.O. Box 320,
University, Papua New Guinea

ABSTRACT

Prehistoric agricultural terracing involving either irrigation or the control of natural runoff is known from several parts of the Pacific, but has not been previously recorded from Papua New Guinea. Horizontal slope forms in the Ramu River valley near Yonki, Eastern Highlands were previously recorded as lake bed features, bench-like slumps or landforms produced by differential erosion. Preliminary geoarchaeological investigations have indicated that they are in fact cut garden terraces, used to control water on slopes. Their use almost certainly predated the entry of sweet potato into the country about 500 years ago, and they were probably used for growing taro, perhaps under irrigation.

These terraces, the physical manifestation of prehistoric engineering works, lie in an area which is the site of a current major engineering project, the Ramu Stage 2 hydroelectric power scheme. The timely recognition of these terraces as cultural rather than natural features allowed the inclusion in the environmental assessment of the project of a strategy for their management.

INTRODUCTION

The terraces in the upper Ramu River Valley at the eastern edge of the Papua New Guinea Highlands, are a dominant feature of the landscape. From Yonki Dome (Figure 1) there is a panoramic view over the Ramu River and its tributary valleys, and above the level of entrenchment, on the upper valley side slopes and the footslopes of the ranges, is a dramatic band of horizontally terraced land, which from a distance appears to consist of regularly spaced continuous terraces (Plate 1). So striking and extensive are the terrace formations that on the one hand it is difficult to explain why they are so rarely mentioned in the literature on what is a relatively well known area, while on the other it is possible to understand why, when they have been mentioned, a natural origin for them has been assumed.

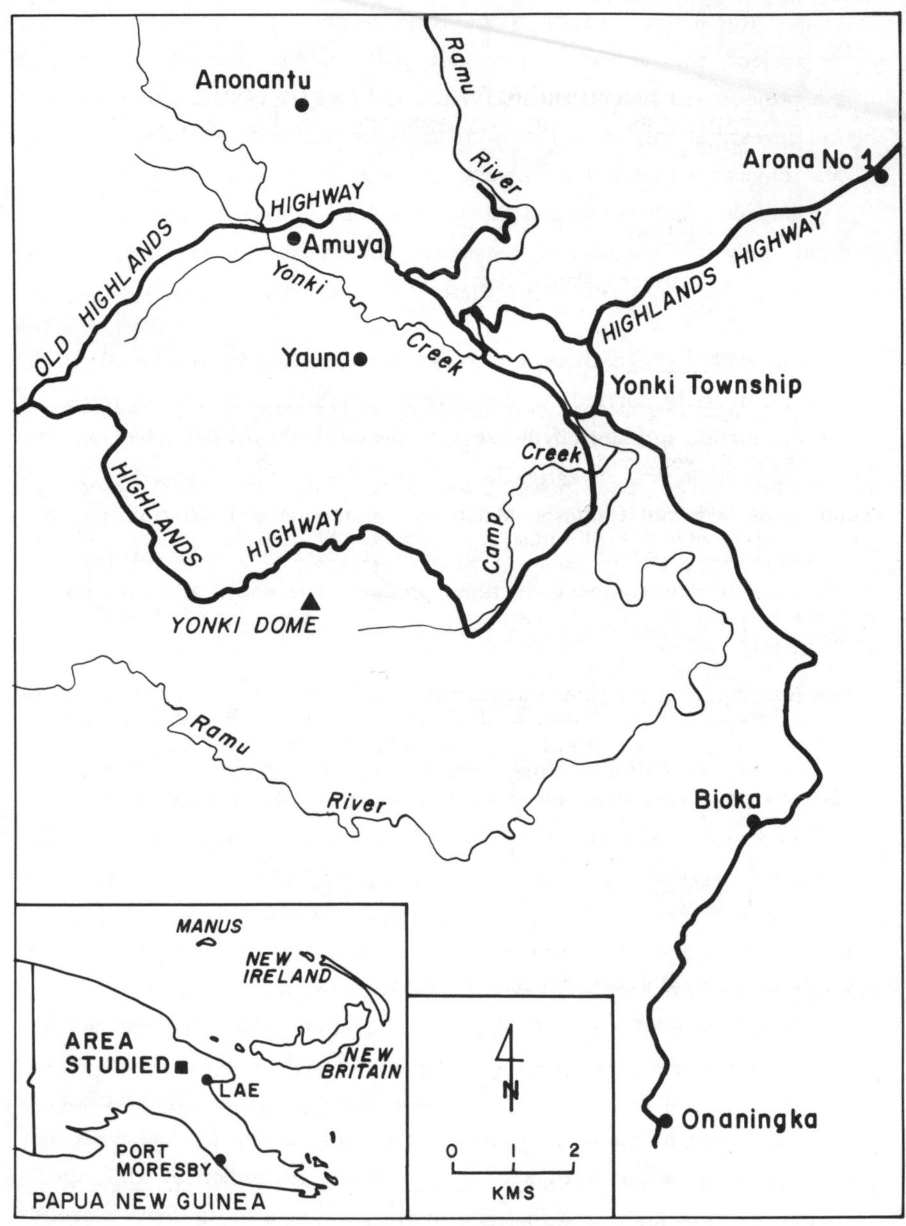

Fig. 1. The Arona or Yonki area of the upper Ramu River Valley, Eastern Highlands Province, Papua New Guinea.

The upper Ramu River valley is the focus of the Ramu hydroelectric power project, the largest such project in Papua New Guinea. Stage 1 of the project was completed in 1972, and Stage 2, with attendant environmental investigations, is currently in progress (Cameron McNamara Kramer, 1985). Environmental impact assessment work commissioned by Elcom in 1971-2 included anthropological and archaeological assessments, a biophysical study and engineering feasibility studies (see Sullivan, Hughes and Golson in press, for a review of these investigations).

In 1984 and 1985 the authors and Professor Jack Golson, of the Australian National University carried out preliminary investigations of the terraces (Sullivan, Hughes and Golson, in press). In addition Hughes undertook specific archaeological investigations in relation to the proposed Stage 2 Ramu power scheme (Cameron McNamara Kramer, 1985). In this paper emphasis is placed on the geomorphic aspects of these investigations, and on the effects that the Stage 2 Ramu power scheme will have on the terraces, rather than on their archaeological significance.

ENVIRONMENTAL SETTING

The Arona Valley, situated in the humid equatorial tropics at an altitude of about 1300m above sea level, has an equable climate with a mean annual maximum temperature of about 26 degrees C and a mean minimum of about 15 degrees C (Heyligers and McAlpine, 1971). Mean annual rainfall is about 2000mm with a pronounced dry season peaking in June-July.

The Arona Valley through which the upper Ramu River flows is a broad, gently sloping basin which rises to hilly to mountainous terrain on all sides. The river drains this basin through the Ramu Gorge which has cut into low hills to the north. In the Pleistocene this basin was filled with fluvial sediments of the Kainantu Formation, consisting largely of conglomerate and reworked tephra, capped with *in situ* tephra (Rogerson *et al.*, 1982). The present drainage system has dissected these sediments down to solid Tertiary bedrock. The terrain in which the terraces occur was mapped as the Abiera Land System by Haantjens *et al.* (1970), the landforms of which are shown schematically in Figure 2. The benches of Unit 1 illustrated in the land system block diagram (Figure 2) are the same features as the

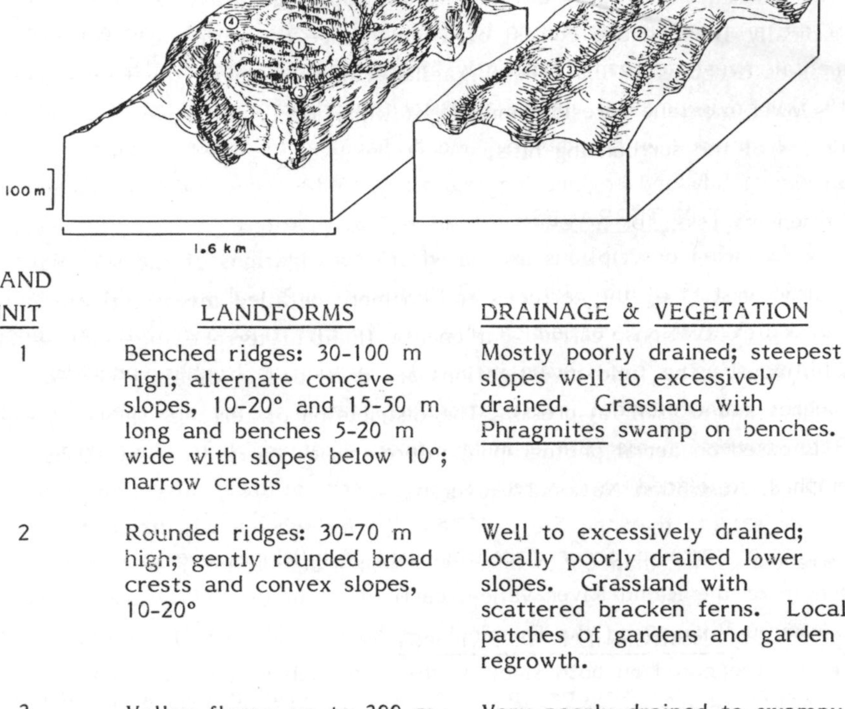

LAND UNIT	LANDFORMS	DRAINAGE & VEGETATION
1	Benched ridges: 30-100 m high; alternate concave slopes, 10-20° and 15-50 m long and benches 5-20 m wide with slopes below 10°; narrow crests	Mostly poorly drained; steepest slopes well to excessively drained. Grassland with Phragmites swamp on benches.
2	Rounded ridges: 30-70 m high; gently rounded broad crests and convex slopes, 10-20°	Well to excessively drained; locally poorly drained lower slopes. Grassland with scattered bracken ferns. Local patches of gardens and garden regrowth.
3	Valley floors: up to 300 m wide, level, hummocky, without drainage channels	Very poorly drained to swampy. Phragmites swamp; some grassland.
4	Summit flats: slopes up to 5°; undissected; up to 600 m wide	Well to poorly drained, but subject to rapid drying of topsoil. Grassland and garden regrowth.

Fig. 2. The Abiera Land System, with a brief description of the landforms, drainage and vegetation of the land units. After Haantjens et al. (1970).

terraces described in this study.

The landscape is largely covered by anthropogenic grasslands which are typically 1m high and consist of a wide range of trailing and sub-erect grasses (Robbins,1970). These grasslands have replaced the former cover of lower montane forest, remnants of which occur on the crests and upper slopes of the surrounding hills.

THE TERRACES

The following descriptions are based on investigations of the surficial characteristics of the terraces and on more detailed investigations, including excavations, carried out on one flight of terraces near the village of Amuya. The field investigations are described in detail in Sullivan, Hughes and Golson (in press). The distribution of terraces shown in Figure 3 is based on aerial photographic interpretation carried out by Chris Ballard, Australian National University.

Flights of terraces occur discontinuously on hillslopes over a considerable length of the Ramu River valley, particularly in the 10km zone between Anonantu Village and the Onaningka-Bioka road (Figure 3). Terraced hillslopes were noted on both sides of the Ramu River, but they are most extensive on the western side, in the catchments of Yonki and Camp (Arona) Creeks (Figure 1). The terraces, which appear to be confined to the sediments of the Kainantu Formation, cover an area of about 16 km^2.

Terraces were noted on a range of slopes between 8° and 30° with an apparent concentration on slopes around 15° to 25°. Old slump scars appear to have been favoured terrace locations. In the poorly consolidated sedimentary rocks rotational slumps are common, and the steep backwalls and contained deposits of disturbed sediments were ideal for terrace construction.

The terraces are generally horizontal platforms with steep backwalls. The walls are commonly 3 to 4m in height with slopes from vertical to 45°. Occasional terraces were recorded with backwalls as high as 8m. The terrace platforms or steps are generally from 10 to 15m wide and run along the hillslopes for highly variable distances between 20 and 200m.

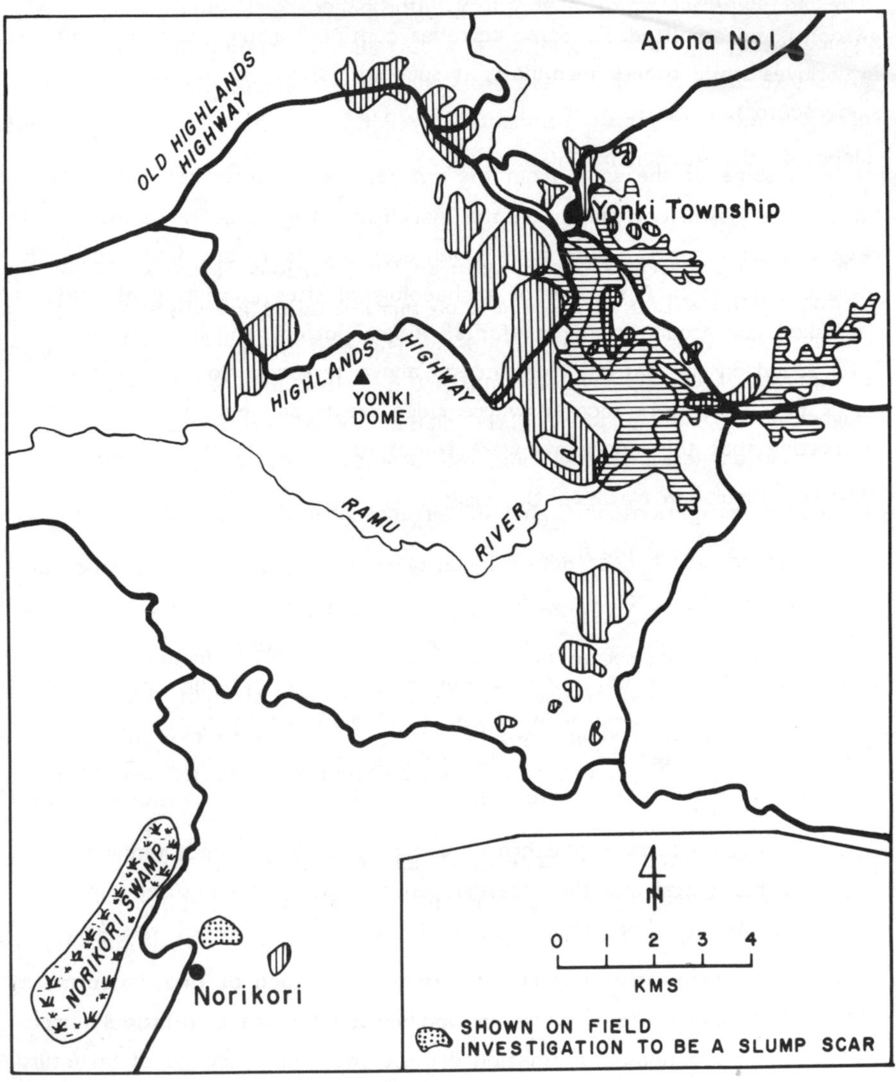

Fig. 3. The area around Yonki in which terraced hillslopes occur. Vertical shading indicates the areas in which field investigations confirmed the presence of terraces identified from airphotos. The area which will be inundated by the Ramu Stage 2 dam is horizontally shaded.

Flights of three to eight terraces down a slope are common, and adjacent slopes of similar gradient and length may display different terrace frequencies and amplitudes. Some terraces continue around spurs or rounded interfluves while others terminate at such interfluves or within concave slope sectors.

At least some of the spoil from the cut terraces appears to have been piled up as downslope ridges on the margins of flights of terraces. These ridges, or emergents, would have given easy access to any level within a terrace sequence. A number of archaeological sites consisting of scatters of flaked and ground stone artefacts, broken pottery and hearths have been found on the natural ridge crests above terraced slopes. In most cases the sites were located where ridge crests and emergents intersected, suggesting that the emergents were routes of access to the terraces.

Detailed investigations of flights of terraces near Amuya and Yonki revealed the following features:

* At Amuya a set of seven clearly defined terraces was found on a slope sector about 300m across and with an average slope of 20°. Most of the terraces stopped at a straight ridge or emergent running down the slope but two terraces continued around this emergent onto an adjacent slope sector with three terraces.
* Eleven terrace platforms were accurately levelled; ten were horizontal and the other sloped at 2%. The measured dip of Kainantu Formation beds in this part of the valley was 4%.
* A narrow trench was excavated across each of two platforms at Amuya. The terraces proved to be formed in relatively uniform Kainantu Formation deposits with no evidence of structural or lithological control over the development of the junction between the platform and the backwall. The junction was marked by sharply angled cuts which were remarkably like the artificial cuts characteristic of prehistoric and contemporary dug garden features in similar tephra-rich deposits throughout the Highlands.

* A distinctly cut drain-like feature along the 6.5m wide platform of one of the Amuya terraces was intersected in the excavation. This feature, which runs along the centre of the platform parallel to the backwall, appears to have been an integral part of the terrace.

* A trench through the Yonki flight of terraces similarly revealed contemporaneous dug features on the platforms. As at Amuya there was no lithological control over the terrace forms.

EXPLAINING THE TERRACES

If the terraces are not the result of human activity there must be a natural explanation for their occurrence. There has been no previous suggestion published that the terraces have other than a natural explanation, therefore it is appropriate to consider the explanations suggested previously for their origin.

Several such suggestions have been advanced, but it seems likely that these have all been made without detailed geomorphic field evidence. The suggestions, all of which sought to explain the strong horizontal element in the structures, their superficially continuous and regularly spaced layering, and their limited areal occurence within the valley landscape included:

* bench-like slumps (Heyligers and McAlpine , 1971),
* Pleistocene (or thereabouts) lake beds (Dow and Plane , 1965; repeated by Tingey and Grainger , 1976),
* differentially eroded more and less resistent bedrock layers (Rogerson _et al._ , 1982),

but each of these can be discounted for a number of reasons.

Bench-Like Slumps. These imply mass movement or slope failure on a major scale. Not only would a sequence of parallel extensive vertical slumps be geomorphologically unlikely, the local evidence indicates that this did not occur. Slumping can be seen to have occurred over large areas of these slopes, but this takes the form of rotational slumps with steep backwalls and lobate deposits. Some of these slump scars have terraces within them while others transgress the terraces. The best indication that the terraces are not mass movement features is that some continue around interfluves onto transportational, not accumulating slope elements. In addition there are no slump backwalls, no accumulated colluvial deposits and no displaced geological strata. Stepped landforms with vertical walls are known from previously glaciated or periglaciated areas in Papua New Guinea above about 4000m, but these are very poorly developed (Löffler, 1977). The Arona valley is at a very much lower altitude, 1300m, and none of the necessary landform elements or climatic processes for such solifluction terracettes is or was present in the Arona Valley.

A sequence of terrace-like debris flows was investigated at the foot of Yonki Dome. These terraces had a superficial resemblance to the horizontal terraces under consideration but were composed of poorly sorted gravels, had distinctly domed surfaces, lobate fronts, and were not continuous over several metres. These mass moved deposits provided a clear contrast to the terrace sequences.

Pleistocene Lake Beds. The suggested origin as Pleistocene lake beds could explain the horizontal bedding of the sedimentary layers, but not the stepped slope profile. If it were conceivable that variations in putative lake levels could have caused variations in the resistance to erosion of the sedimentary beds, the factor of differential slope erosion might apply. However subsequent geological investigations have demonstrated that the sediments are terrestrial, not lacustrine, in origin (Rogerson *et al.*, 1982).

Differential Erosion. Such erosion of horizontally lying beds of varying resistance could give rise to a landscape superficially similar to the terraced slopes of the Arona Valley. Closer inspection of this terraced landscape however, reveals two sets of features which negate that explanation. These are: offset terraces on adjacent slope elements, and several occurrences of terrace frequency changes across a slope, indicating that there is no bedrock control of the terraces. Additionally it can be noted that the terrace steps are widest in gully lines, and that occasional terraces continue around interfluves crossing transportational slope elements, both features inconsistent with a suggested origin as erosional features.

Hence even at a broad scale, after careful observation it is apparent that the terracing in the Yonki area is unlikely to be a natural phenomenon within the landscape.

THE OCCURRENCE OF TERRACES WIDER AFIELD

There are airphoto and ground indications that similar terraces occur widely on Kainantu Formation sediments to the west of Yonki. Apart from the Yonki series, the most definite is a group situated on the western slopes of Yonki Dome, on both sides of the Highlands Highway (Figure 3): ground observation suggests that these terraces are also likely to be artificial. Two apparent occurrences of terraces near Norikori Swamp were field checked, and one in fact proved to be an extensive complex of natural rotational slumps with irregular, lobate surfaces. The airphoto impression of linear stepping is more apparent than real, and it is clear from this example that field checking is necessary before the artificial nature of such features can be accepted.

Poorly preserved terrace-like features that may be the butts of terraces similar to those described at Yonki have also been noted on hillslopes near Henganofi, 50km west of Yonki (Sullivan, Hughes and Golson, in press).

GARDEN TERRACES

Closer geomorphic investigation has shown that these flights of terraces have attributes that cannot be adequately accounted for by natural causes but which are entirely consistent with a human origin. Limited archaeological excavations support this contention and suggest further that the terraces were used for gardening (see Sullivan, Hughes and Golson, in press).

Agricultural terracing is a well-known and widely distributed feature of the traditional cultural landscape of the South Pacific, and includes non-irrigated as well as irrigated systems (see for example Spriggs, 1982). The primary role of the Yonki terraces is interpreted as having similarly been agricultural, aimed at relieving the soil moisture stress to which crops in this region would have been subject, especially during the dry season.

It is most likely that the principal crop was the water-dependent taro (Colocasia esculenta) which was planted in the relatively moist soils of the horizontal, water retaining terrace platforms. This soil moisture may have been derived largely from local surface runoff and subsurface seepage. However the discovery of what is interpreted as a ditch cut into the excavated terrace raises the possibility of irrigated cultivation of taro, particularly given the frequent association of this with terracing in the Pacific Islands.

That it would be technically feasible to divert water along canals from streams in the upper catchments of local creeks has been demonstrated by recent gold mining activities along Yonki Creek. Water races for gold sluicing in the middle and lower part of the creek have extended for several kilometres from the upper catchment and across terraced slopes.

Certainly the present day gardens with their staple crop of sweet potato (Ipomomea batatas) which was indirectly introduced from South America into island Southeast Asia about 500 years ago bear no resemblance to these terrace systems. It is likely that not only do the terraces have nothing to do with sweet potato cultivation, they may well predate it.

THE IMPACT OF THE RAMU STAGE 2 HYDRO POWER SCHEME

It was fortuitous that the recognition of these features as garden terraces rather than natural landforms occurred immediately prior to the time further environmental investigations were being planned in relation to the proposed Ramu stage 2 hydro power scheme. When appraised of the cultural significance of these terraces the Papua New Guinea Electricity Commission made provision that in the project environmental study a detailed investigation of the nature, distribution and archaeological significance of the terraces be carried out, along with an assessment of the likely impact that the development project would have and of ways in which such impact might be mitigated. The findings of these investigations, carried out by Hughes, are described in Cameron McNamara Kramer (1985).

Briefly, the proposed hydro power scheme would affect not more than 10% of the terraces in the Arona Valley in the vicinity of Yonki (Figure 3). The major impact will be in the reservoir where about 6% of the terraces will be inundated. Between 0.5% and 4% of the terraces will be affected by quarrying of earth and rock fill for the dam wall; the degree of impact will depend on where the quarries are located. The re-routing of the Highlands Highway will affect less than 0.5% of the terraces and the dam wall and associated structures will have negligible effects.

Nothing can be done to minimise the impact on the terraces which will be inundated. Elsewhere, measures to minimise the impact of construction work have been proposed, especially in relation to the re-routing of the Highlands Highway and the siting and operation of the quarries.

For those terraces which will inevitably be affected, a recording programme involving airphoto and field mapping of the features will be implemented. Limited excavations involving the use of both heavy machinery and more traditional archaeological methods such as trowelling will be undertaken. These investigations will be integrated into the teaching and research programmes of both the University of Papua New Guinea and the Australian National University.

CONCLUSIONS

Earlier explanations of these terraces as natural features were apparently based on very limited geomorphic investigation and without the possibility of a human origin being considered. More detailed geoarchaeological investigation has led to their reinterpretation as agricultural terraces of some antiquity, possibly associated with water control and the cultivation of taro. This reassessment has been a salutary lesson about the danger of looking at Papua New Guinea landscape history with predetermined expectations.

The form and operation of these garden terraces show that the prehistoric engineers manipulated the unusual combination of geomorphic and geological settings to great advantage. In particular they were aware, as were other prehistoric and contemporary agriculturalists in the Highlands, of the character of ash-rich soils that allowed the easy digging of ditches and terraces which, once formed, remained geomorphically stable during their period of use and commonly for considerable periods after their abandonment. This stability may explain the long term survival of the terraces in this landscape.

ACKNOWLEDGEMENTS

In carrying out these investigations we have received considerable support from staff of the Papua New Guinea Electricity Commission both in the field at Yonki and at their head office in Port Moresby. The detailed investigations at Amuya were carried out with the permission of, and assistance from, the local village people.

The University of Papua New Guinea and the Australian National University, through Professor Jack Golson, have provided financial and material support for this project.

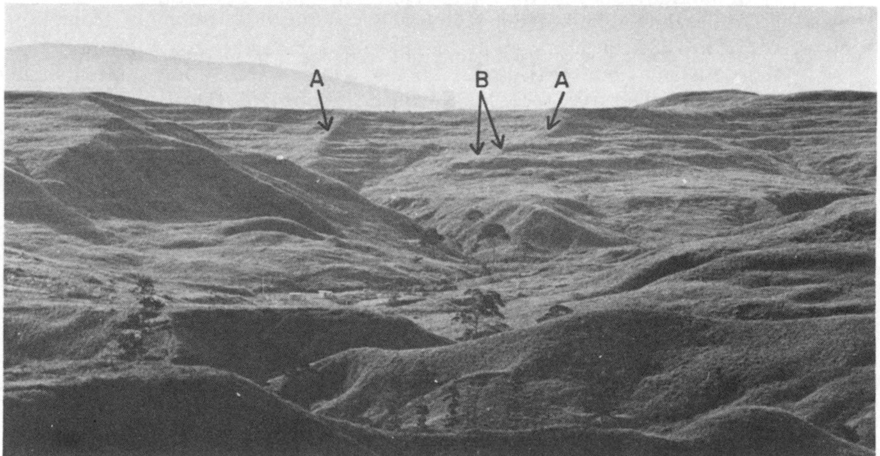

Plate 1. Terraced hillslopes with emergent interfluves (A) apparently built up from the dumping of spoil. Occasional terraces (B) continue around such interfluves.

REFERENCES

Cameron McNamara Kramer Pty Ltd, 1985 Yonki Dam project environ mental assessment: physical environment, Papua New Guinea Electricity Commission, Boroko.

Dow, D.B. G,and Plane, M.D., 1965. The geology of the Kainantu goldfields, Report No. 76, Bureau of Mineral Resources, Canberra, Australia.

Haantjens, H.A, Reiner, E.,and Robbins, R.G., 1970. Land systems of the Goroka - Mount Hagen area. In Lands of the Goroka - Mount Hagen area, Territory of Papua New Guinea, Land Research Series No. 27, Part IV, CSIRO, Melbourne.

Heyligers, P.C.,and McAlpine, J.R., 1971. An ecological reconnaissance of the upper Ramu River catchment, Technical Memorandum No. 71/12, Division of Land Use Research, Commonwealth Scientific and Industrial Research Organisation, Canberra, Australia.

Löffler, E., 1977. Geomorphology of Papua New Guinea. Australian National University Press, Canberra, Australia.

Robbins, R.G.,1970. Vegetation of the Goroka - Mount Hagen area. In Lands of the Goroka - Mount Hagen area, Territory of Papua New Guinea, Land Research Series No. 27, Part VII, CSIRO, Melbourne.

Rogerson, R., Williamson, A., Francis, G.,and Sandy, M.J., 1982. Geology and mineralisation of the Kainantu area. Report No. 82/23, Geological Survey of Papua New Guinea, Konedobu.

Spriggs, M.J.T., 1982. Irrigation in Melanesia: formative adaption and intensification. In May, R.J. and Nelson, H. (eds.). Melanesia: beyond diversity, Vol 1., Research School of Pacific Studies, Australian National University, Canberra , Australia.

Sullivan, M.E., Hughes, P.J.,and Golson, J., in press. Pre-historic engineers of the Arona Valley. Science in New Guinea.

Tingey, R.J.,and Grainger, D.J., 1976. Markam, Papua New Guinea - 1:25,000 Geological Series, Explanatory Notes SB/55-14, Bureau of Mineral Resources, Canberra, Australia.

International Geomorphology 1986 Part II
Edited by V. Gardiner

CONSERVATION OF GEOMORPHOLOGICAL SITES IN BRITAIN

J. E. Gordon

Nature Conservancy Council,
Pearl House, Bartholomew Street,
Newbury, Berkshire, RG14 5LS, UK

ABSTRACT

The Nature Conservancy Council is responsible for the identification, assessment and conservation of key sites of national and international importance for earth science research and education in Britain. To redress previous deficiencies in site coverage, the Geological Conservation Review has been instigated to provide a national overview of earth science sites and a sounder, scientific foundation for their conservation. Work is currently in progress to identify systematically a national network of key sites for geomorphology and Quaternary studies. Key sites will be accorded protection under the legal and planning frameworks within which conservation in Britain functions. Specific examples illustrate the operation of geomorphological conservation in Britain, including the 1981 Wildlife and Countryside Act which places greater emphasis on the principle of compensation payments to landowners for conservation constraints placed upon them. To meet growing threats to key fieldwork localities, geomorphologists have a significant role to play in stimulating greater public awareness of their subject and in providing scientific expertise for site assessment, site management and evaluation of the impact of proposed developments or management changes.

INTRODUCTION

For its size, Great Britain has a remarkably diverse range of geomorphological features spanning the major constituent branches of the discipline. Many of these features are internationally renowned as classic textbook examples, but equally there are many, perhaps lesser known but no less significant localities, that have played a fundamental role in understanding form - process - material relationships or in the development of concepts and theory. Concomitant with the substantial growth of the discipline over the last few decades and the increasing demand for field facilities for research and education, there has been a parallel growth in threats to field sites from diverse sources eg. urban and industrial developments, sand and gravel extraction, landfill, recreational developments and coastal defences. In this context, earth science conservation may be viewed as having two fundamental objectives: firstly to ensure the continued existence of a national network of field facilities for research and education representing the major variations in the geology and geomorphology of Britain; secondly to ensure that due consideration is given to geological and geomorphological features in environmental assessments.

A need to identify and conserve important geomorphological features has been recognised by Government in Britain since the late 1940 s when the original Nature Conservancy was established. This has been continued in the statutory duties of its successor, the Nature Conservancy Council (NCC), which is the official body established by Act of Parliament in 1973 to be responsible for the conservation of flora, fauna and geological and geomorphological features throughout Great Britain. It is financed by the Department of the Environment but is free to express independent views. It has four principal statutory functions - to establish, maintain and manage National Nature Reserves; to advise Government on nature conservation policies and on how other policies may affect nature conservation; to provide advice and dissemination of knowledge on nature conservation; and to undertake necessary research. As part of these functions, the NCC has a statutory responsibility to protect earth science features and this has been discharged through the Geology and Physiography Section (planning casework) and the Geological Conservation Review Unit (site assessment and selection). Key sites are conserved through a network of National Nature Reserves and Sites of Special Scientific Interest and are also protected under the provisions of the Wildlife and Countryside Act (1981).

SITE ASSESSMENT AND SELECTION

Historical Perspectives.

Early developments in earth science conservation took place during the late 19th. century when special efforts were made by local authorities to preserve individual features of exceptional significance eg. Fossil Grove in Glasgow and the so-called Agassiz Rock at Blackford Hill in Edinburgh. However, it was not until the 1940 s that a firm statutory basis was established at a national level. In 1941 the Society for the Promotion of Nature Reserves convened a conference to consider the place of nature conservation in post-war reconstruction. One outcome was the establishment of a Nature Reserves Investigation Committee and later a sub-committee to consider the requirements of earth science. The Geological Sub-committee comprised six geologists who consulted widely with local advisors. Their report (Anon, 1945) identified 390 localities in England and Wales that should be protected as Geological Reserves. In 1948 Professor J. G. C. Anderson listed some 60 sites in Scotland. These lists provided the basis for earth science conservation undertaken by the Nature Conservancy during the years following its formation in 1949. However, it was widely recognised that the site lists were incomplete and further sites were considered on an _ad hoc_ basis following recommendations made by individual geologists. Such proposals were scrutinised in consultation with recognised experts before conservation measures were undertaken. By the early 1960 s it was apparent that this procedure was unsatisfactory in that a generally acceptable and balanced national network of conserved sites could not be constructed merely by lumping together occasional or random proposals. A programme of progressive site revision was therefore formulated with the ultimate goal of achieving systematic coverage of all aspects of geological and geomorphological interests in Britain in full consultation with all appropriate experts. Key aspects of this programme involved the first development of site selection criteria, a consensus approach to site selection and the use of outside voluntary expertise. This rolling programme of site coverage revision was superseded by a more radical approach following a decision by the Nature Conservancy Council in 1974 to proceed with a Geological Conservation Review, an earth science equivalent of the _Nature Conservation Review_ (Ratcliffe, 1977).

Geological Conservation Review.

Work commenced on the Geological Conservation Review (GCR) in 1977 with the aim of identifying and documenting all localities in Britain of national or international importance for earth science education and research. This is still in progress and has involved a complete reappraisal of both existing and potential sites. In due course, a comprehensive series of volumes comprising individual site assessments will be published. These will contain a brief description and review of the scientific interest of each site followed by a thorough assessment and evaluation of its importance, placing it in the context of related sites and setting out reasons for its selection. This document will provide the scientific basis for practical earth science conservation in the future. It is not intended to be a "once and for all" statement, and both the science content of the database and the site coverage will need to be continually revised and updated to follow developments in the various disciplines of the earth sciences.

Site assessment and selection is undertaken using a variety of approaches according to the nature of the particular field and the availability of in-house expertise. In some cases the GCR employs on a temporary basis young post-doctoral workers with appropriate expertise; in others established scientists or consortia work part-time on the project. The established procedure involves a review of all relevant literature since many potentially important sites may not have been studied recently, or in the case of some Quaternary sites lain buried for years. Draft lists of sites are prepared and circulated among appropriate experts to solicit as wide appraisal as possible. Sites are then examined in the field and sometimes excavated where necessary before final selection is made. A full account of site selection methodology and criteria is outwith the scope of this paper, but the overall objective is to produce a national network of inter-related sites representing the major regional variations in Quaternary environments and geomorphology in Britain. It is anticipated that the GCR will describe a total of 2700 individual earth science interests. Of these, the current working estimate is that around 600 or 22% will be listed for geomorphology or Quaternary studies. Because of areal overlaps of interests the total number of Sites of Special Scientific Interest that will be notified will reduce to around 1500 for the whole spectrum of earth sciences. Many of these will also overlap with biological SSSIs of which there are 3166 currently scheduled. The finished product will not only provide a sound basis for earth science conservation, but will also be a unique summary of the remarkably rich variety of earth science interests in Britain.

GCR sites of interest to geomorphologists cover a variety of features which differ greatly in character and scale, ranging for example from the glaciated Cuillin Hills of Skye to the famous Pleistocene sections in East Anglia; from the classic coastal features such as Chesil Beach to the sandstone cliffs of Orkney. A full complement of sites will represent the more important lithostratigraphic and biostratigraphic localities for interpreting and reconstructing the Quaternary history of Britain. Site selection is approaching completion for coastal geomorphology (84 sites), caves (48 sites), karst (40 sites), mass movement features (27 sites), Quaternary of Scotland (ca. 100 sites), Quaternary of Wales (ca. 70 sites). Work on the Quaternary of England is well in progress (ca. 250 sites) and that on fluvial geomorphology is at an early stage.

SITE PROTECTION

National Nature Reserves and Sites of Special Scientific Interest.

Site protection is organised through a system of National Nature Reserves and a national network of Sites of Special Scientific Interest.

A National Nature Reserve is an area managed for the conservation of wildlife or features of geological or geomorphological interest. Reserves are either owned or leased by the NCC or managed under a formal agreement with the owner. At present there are over 200 National Nature Reserves covering more than 150,000 hectares. Five of these are reserves primarily for geomorphological or Quaternary interests: the Parallel Roads of Glen Roy, the Swanscombe Skull Site in Kent, the active coastal land-slips of Axmouth-Lyme Regis in Devon, the sarsen stones of Fyfield Down near Marlborough in Wiltshire and the important cave system at Ogof Ffynnon Ddu, in Powys. Other reserves also include geomorphological features of the highest importance within their total range of interests: for example the Cairngorm Mountains of Scotland with an exceptional assemblage of glacial and periglacial landforms, Braunton Burrows and Oxwich with sand dune systems and the island of Rhum in the Inner Hebrides with glacial, periglacial and other interests. The

NCC encourages geomorphological research on National Nature Reserves both for its own value and for the light it sheds on biological problems.

Nearly all earth science sites are protected as Sites of Special Scientific Interest (SSSIs). These are areas of outstanding biological, geological or geomorphological interest that the NCC is obliged by statute to notify to local planning authorities under Section 28 of the Wildlife and Countryside Act 1981. Many other bodies are also kept informed about SSSIs in the interests of sensible land-use planning. SSSIs form a national series and in the earth sciences are intended to provide a representative cross-section of the geology and geomorphology of Britain. The site network is regularly revised and modified to keep abreast of the most recent developments in geological and geomorphological research. For the first time this revision will be systematically completed at a national level through the Geological Conservation Review. The NCC does not own SSSIs and access to them remains completely in the control of the landowner; neither NCC staff nor the general public have automatic rights of access. However, NCC has endeavoured to promote co-operation between landowners and scientists to ensure the continued existence and maintenance of key sites, and access to them for study.

Conservation and Planning Procedure.

Planning authorities are required to consult NCC for advice over planning applications affecting SSSIs. When a proposal affecting a geological or geomorphological SSSI is received, the appropriate NCC Regional Office informs the Geology and Physiography Section. After seeking expert advice through the GCR liaison network the Section advises the planning authority, through the NCC Regional Office, how the scientific interest can best be maintained. If an acceptable compromise between the differing viewpoints cannot be reached, the Nature Conservancy Council presses the case to the highest level, including Public Inquiries called by the Department of the Environment, the Scottish Office or the Welsh Office. At such inquiries the widest possible support from the scientific community is vital to the successful presentation of the case in defence of a site.

The Wildlife and Countryside Act 1981 introduced new provisions to protect SSSIs against developments or threats not covered by planning legislation; for example relating to agricultural or forestry operations. Under the Act, NCC are now required to notify all SSSI landowners and occupiers of those activities that could damage the interest of a site. Before undertaking any of the notified activities, the owner or occupier of an SSSI is required to obtain NCC's approval and a three-month consultation period is provided by the legislation. NCC may enter management agreements and pay compensation to landowners for restrictions placed upon them that are necessary for conservation purposes. Alternatively, management agreements may include positive elements under which a landowner agrees to carry out work to enhance the interest of an SSSI. If agreement cannot be reached, the Secretary of State has powers under Section 29 of the Act to extend the consultation period. Very occasionally, the NCC may exercise its powers to purchase compulsorily land which in the national interest should be managed as a Nature Reserve.

CONSERVATION IN PRACTICE

As outlined above, the vast majority of protected earth science sites are SSSIs and are not owned by NCC. Site safeguard therefore operates mostly through local government planning procedures and occasionally through management agreements with landowners. In the latter case the Wildlife and Countryside Act 1981 has strengthened NCC's hand in negotiating management agreements through the provision of Government funding to compensate landowners for financial disadvantages resulting from site management constraints placed upon them. In the former case, it should be borne in mind that while site designation gives NCC the statutory right to be consulted over planning applications affecting SSSI s, nature conservation is only one of a number of considerations which planning authorities are required to take into account in reaching planning decisions or developing planning strategies. Where developments do not conflict with conservation of scientific interest, it has been the approach of NCC to attempt to find a suitable agreement whereby both development and conservation can proceed together. Where this has not proved possible, or where the scale of development has been such as to effectively destroy the interest, then NCC has normally lodged a planning objection. To do so as effectively as possible, it is vital that NCC can demonstrate widespread support from the earth science community. Individuals can help and do help through provision of specialist scientific information, through considered professional assessments on the likely impact of developments, by writing letters of support which can be used at planning inquiries and by offering to appear as expert witnesses at such inquiries. Given that the geomorphological interest may be only one of a number of competing land use demands, geomorphologists can also contribute at a more general level through promoting greater public interest in their subject and public awareness of its relevance.

In reviewing the work of NCC over the last few years certain types of development threat have been recurrent:

Mineral Extraction.

Mineral extraction can have both positive and negative effects on geomorphology. On the one hand it can provide important new exposures, often on a scale unobtainable under normal scientific investigations. On the other, quarrying may substantially or totally destroy landforms of particular interest. Very often NCC is placed in the difficult position of weighing up these conflicting effects and assessing how much, if any, quarrying is acceptable on scientific grounds or at what point extension of quarrying becomes unacceptable. Where classic landforms are involved or if morphology is of prime importance, NCC may support total protection if this is in line with current scientific thinking; for example in the case of the braided esker at Kildrummie near Inverness, NCC has strongly opposed quarrying but has adopted a more flexible attitude over some other esker systems where morphology is arguably less important. However, in the case of cave and karst sites, scientific benefits rarely arise from quarrying.

Site Restoration.

Many GCR sites are working quarries or gravel pits. However, it is not normally the aim of NCC to interfere with normal working procedures but rather to safeguard interests in exposures remaining when extraction ceases. Typically, quarries and gravel pits are prime sites for rubbish disposal and landfill, or if they can be flooded, for recreational use. NCC has a role to play in ensuring that important scientific interests are given due consideration in restoration plans, providing that these interests are conservable in practice. A case example is Wolston Sand Pit, the type locality for the Wolstonian Stage of the British Pleistocene which has been threatened with total infilling. Following two planning appeals by the developers, the Secretary of State for the Environment has directed that planning restrictions on tipping should be retained in order to conserve a key face in the former sandpit. In the case of working quarries it is important that NCC are made aware of the scientific interests at an early stage in order to expedite appropriate consultations with both quarry operators and planning authorities.

Coastal Defences.

The coastline of Britain is remarkable for its diversity of cliff, sand and shingle formations. Frequently also the coastline provides good sections in Pleistocene deposits, often continuous over long distances in areas where inland exposures are rare. Given high population densities and development pressures in coastal areas, together with the inherently dynamic nature of soft coasts, it is not surprising that coastal protection is frequently a substantial threat to geomorphological interests, particularly since this normally involves major civil engineering solutions to problems of coastal erosion. These typically obliterate the interests in the coastal exposures or alter the natural balance of geomorphological form-process systems. As in the case of all SSSIs, NCC is able to provide advice and recommendations on the impact of coastal defences on geomorphological interests but these are not always heeded when risks to people and property are involved, for example in the case of recent work at Chesil Beach (Carr, 1983a,b). The position also arises where NCC needs to appraise modified engineering solutions designed to attempt some compatibility between conservation and coast protection, for example in the case of West Runton. This is clearly a difficult field in which to operate given the complexity of coastal process systems and it is one that requires greater consideration.

Site Management Changes and Activities Not Requiring Planning Permission.
A range of site management changes and activities not requiring planning permission may have significantly damaging effects on geomorphological sites, for example tipping of farm waste in gravel pits, afforestation, ploughing and drainage improvements. The 1981 Wildlife and Countryside Act has now brought such activities within the site safeguard system. For example, one particular GCR site has been selected as an important example of linear subsidences associated with the Cheshire salt karst. A significant part of the interest includes small slip scars and subsidence features typically less than 0.5m high. Before notification the landowner had ploughed part of the site and wished to plough the remainder, effectively destroying the small scale features of interest. However, under the provisions of the 1981 Act, NCC is now seeking a management agreement with the landowner to leave particular areas unploughed, in return for which he will receive appropriate compensation for loss of income.

These examples provide only a flavour of geomorphological conservation in practice; many more examples are described in the newsletter, Earth Science Conservation, issued free to all interested individuals and organisations. Typically, large geomorphological sites are prone to a multiplicity of pressures from substantial national interests; for example the shingle foreland at Dungeness is currently under threat from nuclear power station developments, expansion of Ministry of Defence training areas, sand and gravel extraction and water abstraction. Carr (1983a,b) reports a similar position at Chesil Beach. Given the economic, social and political pressures on land use in Britain to-day, it is difficult to envisage any relaxation in development threats to many geomorphological sites.

CONCLUSIONS

1. The Geological Conservation Review and its continued updating will provide for the first time a substantial scientific database for conservation of earth science features. It should help to set individual sites into a national context and establish the scientific reasoning for their conservation. Further, it will help to identify localities where further research or detailed survey work is necessary.

2. Conservation of geomorphological sites in Britain operates principally through a system of SSSIs. These do not have absolute protection but their designation allows NCC a statutory right to advise and make recommendations on their conservation to planning authorities and at planning inquiries. The 1981 Wildlife and

Countryside Act provides additional site protection from activities not covered by planning legislation and places a greater emphasis than in the past on the principle of compensation payments to landowners for conservation constraints placed upon them.

3. To help safeguard sites and to counter the multiplicity of threats to important fieldwork and research localities, geomorphologists have a significant role to play - firstly in providing scientific expertise for assessing the significance of sites and evaluating the impact of proposed developments or site management changes, and secondly in stimulating greater public understanding of the relevance and interest of their discipline both in its pure and applied aspects.

REFERENCES

Anon., 1945. National Geological Reserves in England and Wales. Report by the Geological Reserves Sub-committee of the Nature Reserves Investigation Committee. Conference on Nature Reserves in Post-war Reconstruction, Memorandum No. 5, The Society for the Promotion of Nature Reserves, British Museum (Natural History), London.

Carr, A. P., 1983a. Scientific conservation as a minority interest. Earth Science Conservation, 20, 3-10.

Carr, A. P., 1983b. Chesil Beach: environmental, economic and sociological pressures. Geographical Journal, 149, 53-62.

Ratcliffe, D. A., 1977. A Nature Conservation Review, Cambridge University Press, Cambridge.

WEATHERING

WEATHERING

International Geomorphology 1986 Part II
Edited by V. Gardiner

WEATHERING

V. Gardiner

Department of Geography, University of Leicester,
Leicester, U.K.

The topic of weathering was not initially included in the Conference programme as a separate symposium, but responses to the First Circular demonstrated that this was a mistake; at the Conference a small but stimulating session on weathering was held. Weathering is obviously, in some senses, the most basic of all geomorphological processes, and without it there would be no basis of material upon which other processes could operate. Consideration of weathering was therefore evident in many of the papers presented in other sessions, and published in these Proceedings. For example, some papers in the Basin Sediment Systems and Long-Term Landform Development symposia include much on weathering, and some studies of weathering under arid regimes are included amongst the papers on Desert landforms.

The papers included in this section provide a representative cross-section of the major directions in which research on weathering has proceeded in recent years. That by Blair focuses on the landforms produced by weathering processes, and exemplifies a perhaps traditional, but still very effective approach to the subject. The paper by Douglas et al., is typical of a rather more recent trend, in which laboratory experiments are used to shed light on the mechanisms operating during weathering. A third major research direction, in which microscopic and geotechnical techniques are used in an attempt to elucidate the ways in which weathering processes are operating in the field, is illustrated by the paper of Robinson and Williams. Finally, a comprehensive study by Smith reminds us that weathering, whatever the details of the processes involved, operates on the landscape within the contexts of particular rock types and changing morpho-genetic regimes. The fuller integration of these approaches, in which laboratory simulations, detailed laboratory analyses and field examination of landforms are linked together, represents a most challenging goal for further research endeavours on weathering.

International Geomorphology 1986 Part II
Edited by V. Gardiner

DEVELOPMENT OF NATURAL SANDSTONE ARCHES IN SOUTH-EASTERN UTAH

Robert W. Blair, Jr.

Department of Geology, Fort Lewis College
Durango, Colorado 81301 U.S.A.

ABSTRACT

The region around Arches National Park and Canyonlands National Park in south-eastern Utah displays the highest concentration of natural sandstone arches in the United States and perhaps the world. Rock fins are a prerequisite to arch formation and develop either from closely spaced joints or from canyon wall retreat from opposing sides of adjacent drainage courses. Specific criteria important in arch formation include 1) the horizontal attitude of vertically jointed sandstone formations, 2) the presence of less competent horizons, 3) the systematic release of stressed rock via exfoliation slabs, 4) the continued weakening of surface materials by hydration and wet-dry cycles, and 5) the erosion of weathered products by wind, running water, and gravity. The arches have probably all evolved within the Quaternary Period.

INTRODUCTION

More than 300 natural sandstone arches are found in semi-arid south-eastern Utah, U.S.A., which may represent the highest concentration of this type of landform found anywhere in the world. Most of the arch openings are concentrated within two prominent sandstone formations, the Cedar Mesa Sandstone Member of the Cutler Formation (Permian) and the Entrada Sandstone (Jurassic). This preliminary study is concerned with the factors responsible for the development of these sandstone arches.

According to Cleland (1910) a natural arch is not the same as a natural bridge and, therefore, should be treated separately. Bridges span present or former water courses while arches do not. Bridges result from the corrasion and hydraulic action of streams or possibly waves, while arches result from weathering and less intense erosional processes.

PREREQUISITES TO ARCH FORMATION

The most important prerequisite to the formation of sandstone arches is the development of a thin wall of rock, or rock fin. Two types of fin development are recognized, those which evolved from closely-spaced jointing and those which result from headward erosion and valley wall retreat from adjacent drainage courses. The best examples of joint-controlled fins are found in the Devil's Garden section of Arches National Park, Utah (Fig. 1), where hundreds of joints striking N40°W run parallel to the axis of the Salt Valley anticline. These are bending moment joints associated with the development and collapse of the Salt Valley Anticline. The anticline has been evolving since Late Pennsylvanian time from the slow upward migration of a ridge of salt and has been synchronous with the deposition of Late Paleozoic and Mesozoic rocks (Baars, 1972). During the Quaternary and possibly Late Tertiary, solution and removal of the salt along the fold axis has resulted in the partial collapse of the fold limbs towards the axis, to produce a topographic valley. Because of the brittle nature of the Entrada Sandstone, the limb folding event has resulted in the development of open joints narrowing toward the base and rock fins 3 to 6 m wide, with a relief of 6 to 40 m (Blair et al., 1975). Some fins run for hundred of metres without a break (Fig 1).

Many of the joints have separated to produce long narrow passages several metres across. Sand and silt along with spalled rock fragments have filled some of the joint passages. In others, weathering and stream channel erosion has resulted in the widening of the joint passages; these are important in that they allow the walls to be exposed to rain, wind, and sun.

Rock fins are also found as sandstone ridges which separate parallel drainage courses or two opposing valley headwalls. The fin development accompanies the later stages of entrenchment and valley widening of these stream courses. Salt Canyon in Canyonlands National Park exhibits this second type of fin development.

DEVELOPMENT AND GEOMETRY OF ARCH OPENINGS

Many factors interact to produce a resultant arch. These are discussed under the headings, Lithology, Structure, Climate, Weathering and Erosion, and Time.

Lithology. A natural arch will form only in a rock competent enough to support a roof (Blair et al., 1975); therefore, it is not surprising that the most stable arches develop in sandstones. There are a number of sandstone formations in south-eastern Utah competent enough to maintain an arch, but, curiously, nearly all the known arches are confined to either the Entrada Sandstone or Cedar Mesa Sandstone. Arches have been noted in the Jurassic Navajo Sandstone and Triassic Windgate Sandstone, but they are exceedingly rare.

None of these sandstone formations are truly homogeneous; thus, there are horizons within them which are less competent. It is observed that initial breakthroughs most often occur along horizontal re-entrants composed of calcareous mudstones or shale partings (Fig. 2).The re-entrants or hollows develop at the same level on opposing sides of the fin, and as they grow, eventually intersect

Fig. 1. Joints and associated rock fins in the Entrada Formation in the Devil's Garden section of Arches National Park, Utah.

Fig. 2. North Window Arch in Arches National Park. Note the influence of lithology (statigraphic contact) and the tensional release fracture (upper right of opening).

to form a breakthrough or an incipient arch. In Arches National Park, the arches found in the Windows section develop at the contact between the Dewey Bridge Member and the Slick Rock Member in the Entrada Sandstone (Lohman, 1975). Some arches, such as those in the Devil's Garden section of Arches National Park, show no obvious lithologic contact.

A qualitative observation of the degree of effervescence of dilute hydrochloric acid applied to the sandstone suggests that the per cent of carbonate cement varies vertically within the Slick Rock Member of the Entrada Sandstone (Blair et al., 1975). The richer carbonate horizons frequently reveal themselves by leaving a string of small holes or hollows parallel to the bedding. Although not confirmed by petrographic studies, variations in the per cent of clay cement may also lead to those less competent horizons.

Structure. All of the arches observed in Canyonlands and Arches National Park are confined to sandstone layers which are horizontal or gently dipping, such as fold limbs and, therefore, are exposed over a broad region. This partly accounts for the wide distribution of arches in this part of Utah. Vertical joints crosscutting the fins are rare. Where they do occur, arches usually do not form; although in Horse Canyon in Canyonlands National Park, openings have developed along lower portions of joints and still have allowed a small roof (marred by the joint) to exist.

In addition to the master joints which control rock fin formation, tensional joints were found which closely mirror the existing surface morphology (Fig. 2). These joints are influenced by the weight of the supporting rock column and the release of residual or locked-in stresses from the time of lithification. These unloading joints are the direct response to the internal distribution of stresses within the rock. Dr. Elsaged Ahmed Eissa (Hoek and Brown, 1980) has modelled the distribution of the principal and minor stresses and the stress trajectories of various tunnel openings within an idealized homogeneous rock (Figs. 3, 4 and 5). These diagrams tend to support the observations of tensional joints forming in the regions along the upper walls and ceilings where tensional stresses overcome low compressive stresses. Expansion occurs along these joints normal to the rock surface as defined by stress trajectories within the rock (lefthand of figures) and causes exfoliation of rock slabs from the surface.

The removal of these arcuate shell-shaped slabs both shape and enlarge alcove and arch openings (Hunt, 1953; Lohman, 1975; Gregory, 1938). The arch design is the most stable geometric form an opening can take because it distributes most efficiently the release of internal stresses (Blair et al., 1975).

Depending on the distribution of stresses, variations in the arch geometry will ensue. Compare Figs. 3, 4 and 5. In Fig. 3, where the horizontal or lateral compressive stress is twice the vertical load, potential tensional failure will occur at the sites where compressive stress values are least, which is at the side walls.

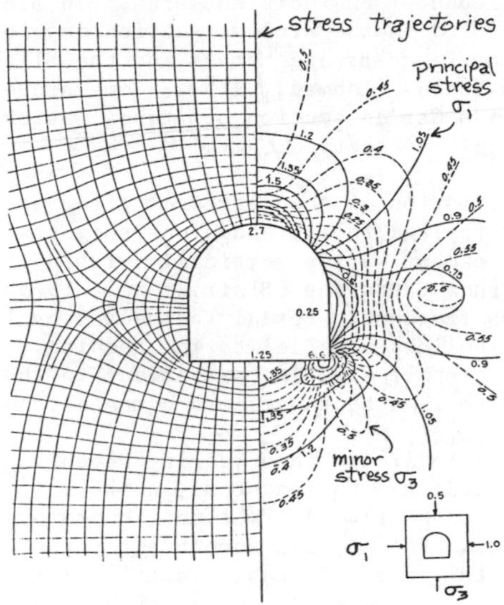

Fig. 3. Shows the stress trajectories of an arch-shaped tunnel where the horizontal stress is twice the vertical stress. Modified from Hoek and Brown (1980).

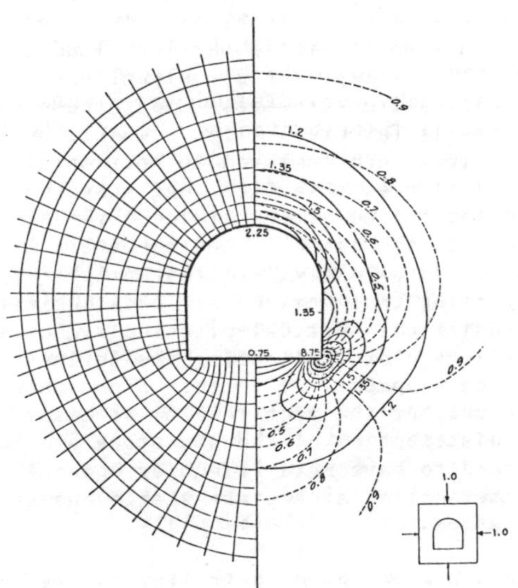

Fig. 4. Shows the stress trajectory distribution around a tunnel where the horizontal stress equals the vertical stress. Modified from Hoek and Brown (1980).

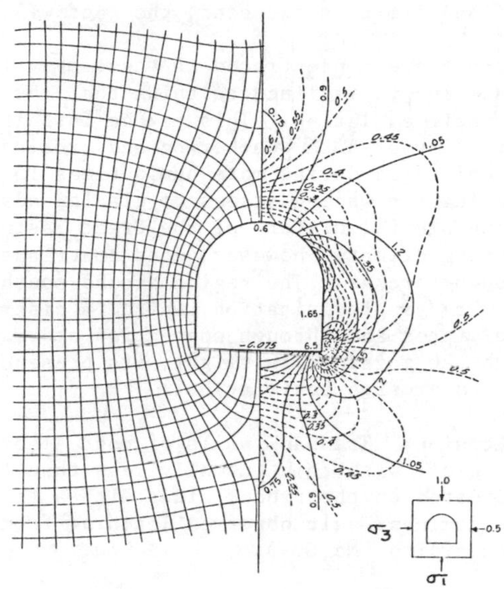

Fig. 5. Shows the stress trajectories around a tunnel when the vertical stress is twice the horizontal stress. Modified from Hoek and Brown (1980).

The stress trajectories indicate that if lateral expansion takes place, a more oblong geometry will result. The most extreme oblong arches, such as the Landscape Arch in Arches National Park, have thin roofs and support little vertical load compared to horizontal load. Fig. 4 represents a situation of equal vertical and horizontal compressive stresses. Under these conditions, a more circular shape is favoured. Fig. 5 models an opening subjected to a large vertical load compared to horizontal compressive stress. In this instance, a vertical egg shape would be most stable, as indicated by the stress trajectory pattern. Flat-lying oblong arches tend to be common in south-eastern Utah while true circular arches are rare. Very few true upright egg-shaped arches were found, suggesting that horizontal loads dominate arch formation. Several additional vertical-oriented arches are known but their shapes are irregular, indicating the influence of other factors in their development. A uniform arch shape is rarely achieved because neither the arch-forming processes nor the internal stress distribution of the sandstone are homogeneous. Smaller arches tend to have more irregular openings than larger arches because the smaller arches are stable enough not to necessitate the arch shape.

The larger arches, however, have their limits. As an arch grows larger laterally, the surface area under the arch will increase by a length factor squared compared to the overlying volume or mass which increases by a length factor cubed. This puts a constraint on the upper size limit of the arch opening. If the upper range of surface area to load limit is exceeded, the roof will collapse.

Climate. The arches are concentrated in the semi-arid southwestern desert and one is inclined to think that the arid climate is required for their evolution. This may be fortuitous because the rate and intensity of both weathering and erosion processes were probably accelerated during more humid times in the past. This occurred during the waning stages of glacial advances as suggested from the pluvial history of the Great Basin (Morrison, 1965). The existing climate, however, does contribute to the ongoing evolution of arches. The region around south-eastern Utah averages about 20 cm/yr precipitation, which is distributed as snow and sleet in the winter and through convection storms during late summer. Although the moisture is slight, its presence is enough to fuel weathering and erosion activity.

Weathering and Erosion. The back walls of most alcoves (pre-arch stage) are flaking off with thin (usually less than 1 cm thick) chips which are coated on their back sides with a thin layer of white salt. The dominant salt observed leaching from sandstones and shales is thenardite (Na_2SO_4).

Most likely a combination of hydration, wet-dry cycles, and salt weathering operate simultaneously to loosen the surface veneer. At the back of many alcoves, there is evidence of water having dribbled out of the rock, especially just above a shale or horizon less permeable than the overlying sandstone. Some even support active springs. The presence of this moisture accelerates weathering and, thus, alcove development. Freeze and thaw also operates but its specific impact is not known. The removal of surface materials is by gravity, wind, and water. Wind is important in arch formation only as an agent which removes loose grains from the surface. No sandblasting effects were observed.

Time. The rate of arch formation can be rapid or slow depending upon the influence of many variables. If exfoliation fractures develop, the opening may form rapidly from the collapse of slabs as occurred with the Skyline Arch in Arches National Park in 1940. Most arches continue to enlarge slowly as grains of sand are gently loosened and removed. These arches exhibit smooth, rounded surfaces in contrast to arches with angular edges associated with fracture release (Blair et al., 1975).

No criteria were observed for assigning absolute ages to the arches. Desert varnish has been found on the exterior of many of the canyon walls but has not been dated in this area of the Colorado Plateau.

CONCLUSION

The evolution of an arch is dynamic and complex. It involves the delicate interplay between rock, structure, process, and time. The magnitude of input from each of these factors determines the style of arch formed. One significant realization from this study is that each arch is a separate entity and that the factors controlling the inception and growth differ from one arch to another. Hence, there exists a wide variety of arch shapes and sizes.

REFERENCES

Baars, D.L., 1972. Redrock Country, Doubleday, Natural History Press, New York.

Blair, Jr., R.W., 1975. Origin and classification of natural arches in south-eastern Utah, in Four Corners Geol. Soc. Guidebook, 8th Field Conf., Canyonlands. 81-86.

Cleland, H.G., 1910. North American natural bridges, with a discussion of their origins. Geol. Soc. American Bull., 21, 313-338.

Gregory, H.E., 1938. The San Juan Country. U.S. Geol. Survey Prof. Paper 188, 105-106.

Hoek, E. and Brown, E.T., 1980. Underground Excavations in Rock, The Inst. of Mining and Metallurgy, London.

Hunt, C.B., 1953. Geology and geography of the Henry Mountains region, Utah. U.S. Geol. Survey Prof. Paper. 288, 171-172.

Lohman, S.W., 1975. The Geologic Story of Arches National Park, U.S. Geol. Survey Bull.

Morrison, R.B., 1965. Quaternary geology of the Great Basin, in The Quaternary of the United States, (Eds. Wright, H.E. and Frey, D.G.), pp. 265-284. Princeton Univ. Press, Princeton, N.J.

International Geomorphology 1986 Part II
Edited by V. Gardiner

THE USE OF STRAIN GAUGES FOR EXPERIMENTAL FROST WEATHERING STUDIES

G.R. Douglas, J.P. McGreevy and W.B. Whalley

Mentaskolinn við Hamrahlið, Reykjavik, Iceland
Public Record Office, Belfast. UK
Department of Geography, Queen's University, Belfast, UK

ABSTRACT

Frost damage to rock specimens subjected to experimental freezing and thawing is usually evaluated in terms of weight loss measurements but this is of limited use in assessing frost damage. It is uninformative as regards weakening effects which can occur without weight loss. The use of other, more sensitive, indicators of frost damage offers promise. Foil strain gauges have been used in laboratory experiments to examine temporal patterns of strain development across water-filled cracks in rocks during freeze-thaw cycles. A basalt and a stylolite seam in chalk have been used for tests reported here. The presence of smectite clays in these rocks provided an opportunity to investigate the relative importance of clay-related breakdown (hydration) and freezing-induced damage. Results indicate that the technique should prove useful in frost weathering studies as it shows strains developing in critical zones, e.g. at right angles to the stylolite seam, before damage is visible or weight loss occurs. Adequate precautions need to be taken with specimen/strain gauge preparation.

INTRODUCTION

In the numerous experimental weathering studies which have been conducted by geomorphologists (see McGreevy (1981) for review), frost damage has usually been assessed by measurements of weight loss from test specimens. Although it may enable relative frost susceptibilities to be established for different rock types, this criterion is of restricted value, being uninformative with regard to weakening effects which occur without weight loss (microcracking for example) and it is not particularly suggestive as to the possible nature of mechanisms which cause rock breakdown. The use of other, more sensitive, indicators of frost damage offers promise in these respects.

Dilatometric methods have commonly been used by building researchers to monitor strains which occur in rocks during freezing (Thomas, 1938; Cady, 1969; Mellor, 1970; Davison and Sereda, 1978; Litvan, 1978). Viewed simply, these methods record patterns and amounts of length change experienced by a rock specimen subjected to freezing and thawing. Because length changes are usually integrated over the entire specimen, they provide indications of "bulk" linear strain and are not always the most appropriate means of evaluating frost damage effects since it is often likely that "local" strain, much higher than that experienced in the whole specimen, will be significant in causing frost damage. Such local strain could occur in association with (i) concentrations of water within rock, and (ii) inhomogeneities in rock. The most obvious example of the first case would be water-filled cracks, particularly in crystalline rocks. In more porous sedimentary rocks, preferential saturation of particular areas of rock, due either to pore structure variations or the tendency for surface layers of rock to experience relatively high levels of saturation, would also be expected to promote local strain during freezing. Inhomogeneities can be either physical (i.e. discontinuities) or mineralogical or perhaps both. In addition to favouring accumulation of water, discontinuities such as cracks will contribute to local strain by acting as stress concentrators (Whalley et al., 1982; Hallet, 1983; Walder and Hallet, 1985), whilst mineralogical concentrations, particularly if they contain swelling clay minerals (Douglas, 1981; McGreevy, 1982), can be preferred locations for frost damage since strains induced by freezing water can be complemented by those related to clay mineral swelling (cf. Cady, 1969).

Conditions such as those just described may well be common. It is thus desirable to employ methods which would enable monitoring of any local strains in rocks subjected to freezing. In this paper we report on the use of strain gauges for this purpose. The basic principles of strain gauge measurements are described and results are presented of two case studies in which strain gauges have been used to examine freezing strains in association with water-filled cracks in basalt and a stylolite seam in chalk. Both rocks are from Co. Antrim, Northern Ireland.

STRAIN GAUGE MEASUREMENTS - BASIC PRINCIPLES

At its simplest, a strain gauge is a length of wire which changes its resistance when strained. If each end of the wire is bonded on to the material under study the strain in the object between the anchored ends (along the active axis) is determined by measuring the change in resistance with suitable circuitry. The sensitivity of the gauge is maximized by increasing the overall length of wire deformed by folding it in a zig-zag pattern. This configuration also minimizes effects at right angles to the measurement direction (transverse sensitivity along the passive axis).

Gauges are backed with a variety of materials but all have two main purposes: they protect the gauge from damage during handling and mounting and they transmit the strain from the test object to the gauge. The fixing of a strain gauge to a test specimen by use of various adhesives is straightforward but requires care if reliable results are to be obtained. Numerous texts are available which provide full details of methods and precautions (e.g.Dove and Adams, 1964; Vaughan, 1975; see also manufacturers' literature).

Although strain gauges are inexpensive and, in principle, easy to use, some problems can be encountered. Assuming that the gauge has been properly mounted, these fall mainly within two categories: temperature effects and moisture effects.

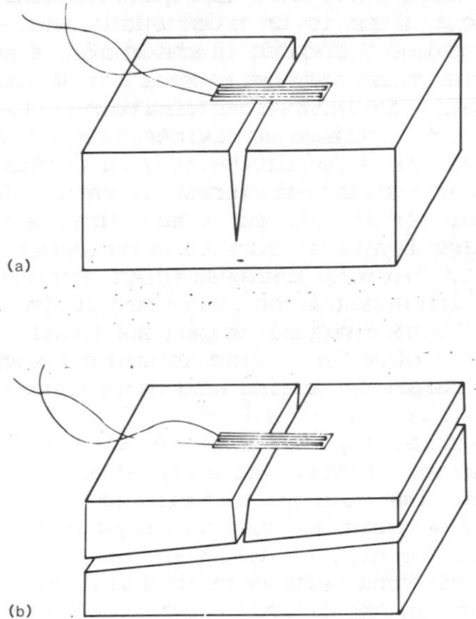

Fig.1. Schematic diagram to show types of crack examined in the basalt experiment; a. single wedge, b. completely open crack.

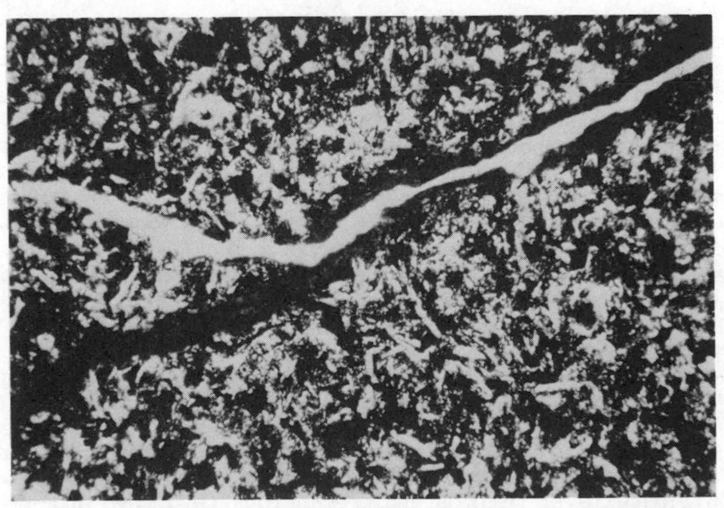

Fig.2. Micrograph of typical basalt microcrack, picture width = 3 mm.

Ambient temperature variations can affect strain gauge readings in two ways. First, a change in temperature causes a change in the resistance of the gauge. Second, because of differential thermal expansion between the gauge and the material to which it is fixed, an apparent strain will be induced in the gauge during temperature variation. There are various methods whereby temperature-induced measurement errors can be eliminated - principally by using thermally compensated gauges and so-called 'dummy' gauges. The latter is a gauge having the same resistance and sensitivity as the active strain gauge on the test specimen. It is mounted on another specimen of the same material, placed close to the test piece and is subjected to the same temperature conditions but is unstrained. The dummy then forms one of the arms of a Wheatstone bridge such that any temperature-induced resistance changes in the active gauge occur equally in the dummy gauge which causes the effects to be cancelled out (Vaughan, 1975).

Direct contact with water or water vapour affects strain gauges in three ways. Absorption of moisture by both the backing material and the adhesive used for mounting can change the electrical resistance between the gauge and the ground potential and thus affect the output-resistance readings. Moisture also decreases the strength and rigidity of the bond between gauge and specimen and thereby causes a reduction in the transmission of strain to the gauge. Lastly, absorption may induce backing material expansion which is not that of the actual specimen. Various methods have been devised for moisture-proofing strain gauges, usually involving complete encapsulation in sealing compounds or hard plastic or metal covers which are themselves sealed (Garfield and McLain, 1978). In porous materials (such as concrete or rock) there is a possibility of moisture interference from the underside of the gauge. To counter this, Dove and Adams (1964) have suggested that the simple precaution should be taken of placing a thin strip of aluminium foil between the gauge and the test specimen. This idea has recently been endorsed by Kohlbeck and Scheidegger (1984).

A typical resistance of a gauge is 120 ohms and the excitation voltage is of the order of a few volts. A variety of measuring instruments is now available for making strain measurements, either manually operated or as part of a continuous data logging system. A half-bridge arrangement probably provides the most useful configuration in rock deformation measurements. Electrical errors can arise from interference picked up by the bridge network but this is usually evident in the experiment and localized screening will generally cure the problem.

FREEZING STRAINS IN WATER-FILLED CRACKS IN BASALT

Douglas (1972) first employed strain gauges to examine patterns of opening and closing of water-filled cracks in basalts subjected to freezing and thawing. The study comprised one element in a broader investigation of weathering in relation to rockfall activity from a basalt freeface (Douglas, 1972; 1980). Two experiments were conducted to measure strain across single wedge (Fig. 1a) and completely open (Fig. 1b) cracks in polished surfaces. Fig. 2 shows a typical basalt microcrack from a basalt flow in Co. Antrim.

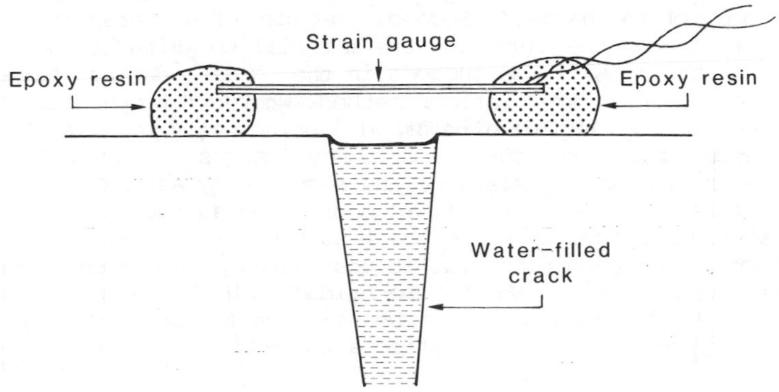

Fig. 3. Attachment of strain gauge over a crack in basalt

Methods. Strain gauges [TML(PR-10)] were given a light waterproof coating of epoxy resin and attached to the polished block surfaces as shown in Fig. 3. As noted above, it is normal practice to bond strain gauges completely on to the surface under investigation. However, the likelihood of ice extrusion from the cracks and its possible effects on strain readings dictated the course adopted here. Because the height of gauges above the specimen surfaces will influence the magnitude of the strain determined, attention is focussed mainly on patterns as opposed to absolute strains. Following attachment of the strain gauges, the samples were coated with wax on all but the polished surfaces and subjected to vacuum saturation during which the polished surfaces were only just covered by water. This procedure was followed to ensure that the effective "pore" space of the samples was completely saturated, while the wax coating prevented water escaping. Surplus water was removed after saturation, particular attention being paid to the areas around the strain gauges. Dummy gauges were mounted on dry specimens to provide temperature compensation.

The prepared samples were subjected to rapid, short-term freezing and thawing in a freezing cabinet. Air temperatures in the cabinet were recorded using a thermometer. Actual rock/water temperatures were not monitored; it was assumed that because of the small amounts of water within the cracks, the water temperature would not differ greatly from that of air (see however, McGreevy and Whalley, 1982). The strains were measured directly with a manually-controlled Wheatstone bridge (Peekel strain bridge) and the readings in microstrains recorded.

Results from Wedge-fracture. Strain readings were commenced soon after the water had been added at a temperature of +18 °C. Strain across the crack (Fig. 4a) initially showed a gradual rise to 25 microstrains after which, when temperature had dropped to between -2 and -3°C, there was a sudden sharp increase to 250 microstrains. As temperature was lowered to -10°C, strain increased further, but only

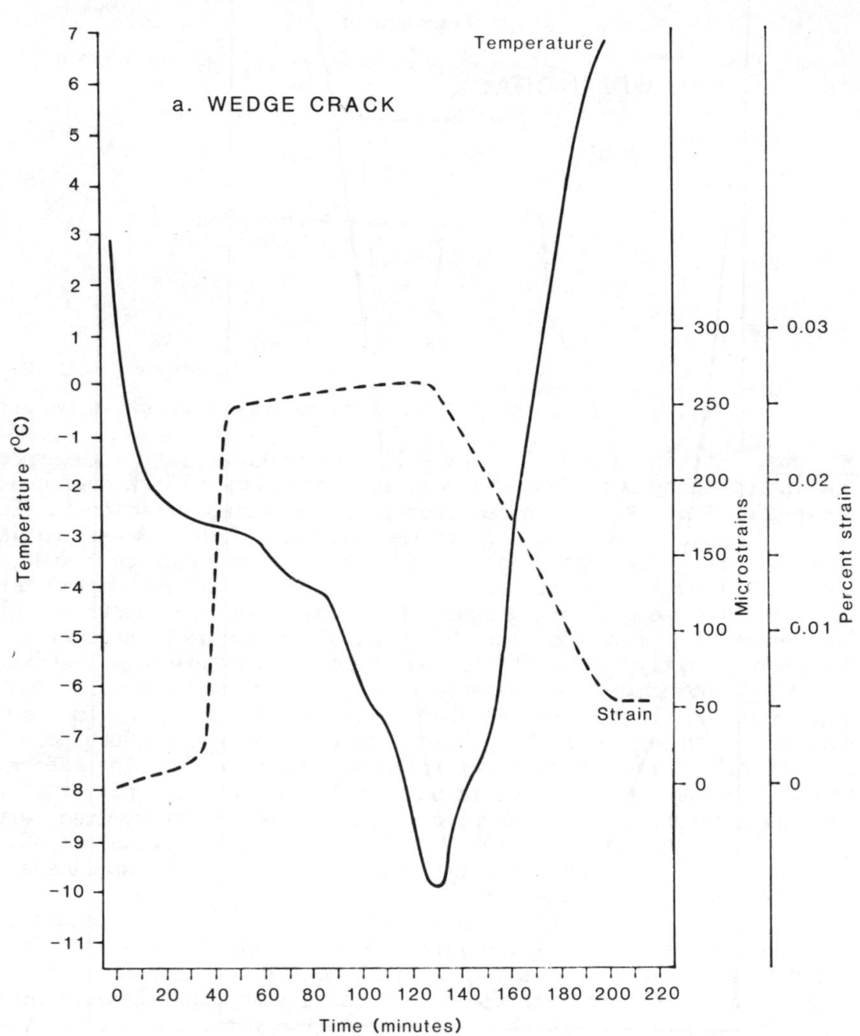

Fig.4a. Strain versus temperature curve for a single wedge fracture.

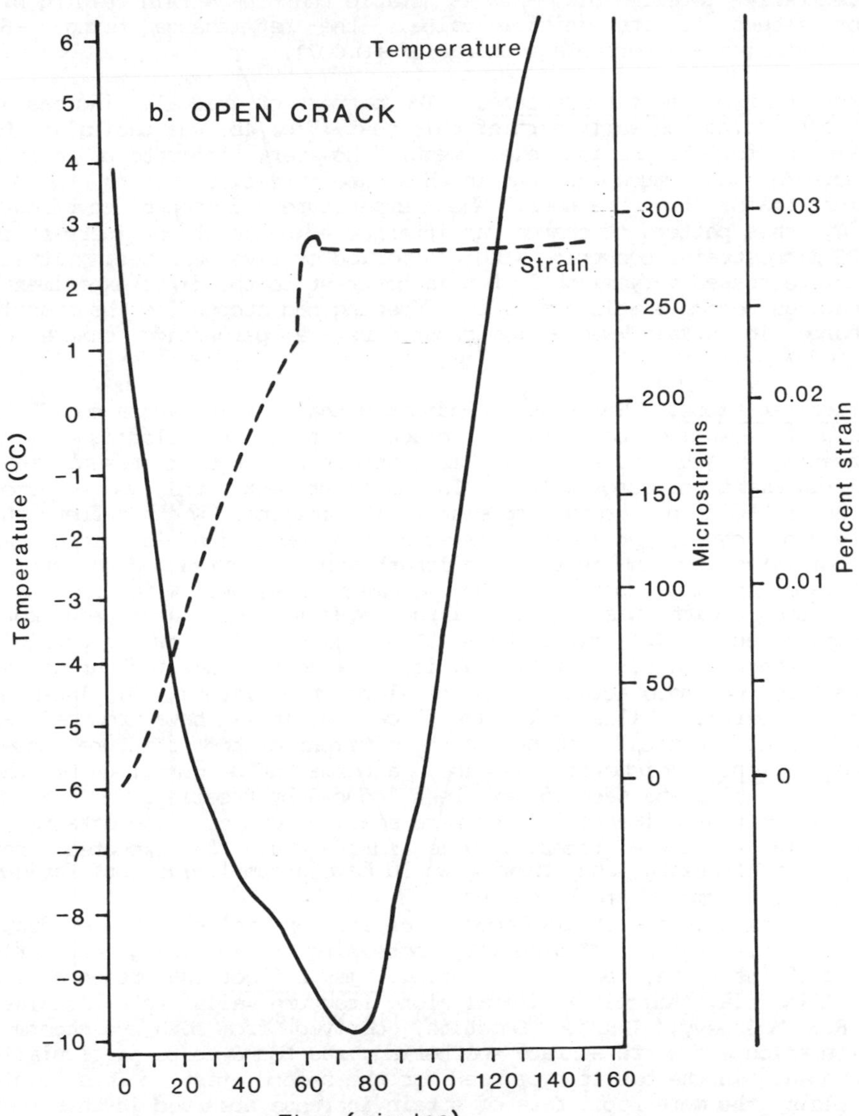

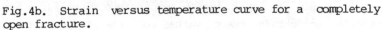

Fig.4b. Strain versus temperature curve for a completely open fracture.

slightly. During thawing, strain decreased relatively rapidly and eventually levelled off. It is notable that the strain reading did not return to its initial value, the net change being +68 microstrains - a percentage change of +0.007%.

Results from an open fracture. The pattern of gradually increasing strain during the early part of this test (Fig. 4b) was basically the same as for the previous experiment. However, the rate of strain increase was much greater in this case, readings rising to 230 microstrains after one hour. When temperature had dropped to below -8°C, this pattern of change was interrupted by an abrupt increase to 280 microstrains occurring within a period of five minutes. Strain then decreased very slightly but in contrast to the first experiment, readings became stable even after freezing had stopped. The overall change in strain was +278 microstrains - a percentage change of +0.028%

Interpretations. The results indicated that strain gauges could be used to examine patterns of crack opening and closing during freezing. Net increases in microstrain readings represent real displacement of crack walls. In the first test, the wedge-shaped fracture did not return to exactly its original width after the freezing cycle, presumably because the expansion in width was accompanied by an increase in length and a corresponding strain release on its surfaces. The movement observed with the open fracture, which was part of a larger system, would have been more complicated. It might have taken place in two, possibly simultaneous, ways. From Fig. 5, it is suggested that the horizontal fracture may have acted as a plane along which lateral displacement of the vertical walls could take place; or it may have produced an additional vertical moment in the movement of the fracture under observation. Whichever the case, a permanent deformation of the fracture walls was seen to have been induced by freezing.

In both experiments strains were small. However, the data refer to only a single freezing cycle and it would be expected that repeated freezing and thawing would have a cumulative or fatigue effect in terms of crack opening.

Several phases of displacement of fracture walls were evident. First, the period of gradually increasing strain during the early part of the tests, prior to freezing, may reflect the presence of a swelling clay (smectite) mineral along fracture walls (cf. Douglas, 1981; McGreevy, 1982). Smectite, derived from both hydrothermal alteration and earth surface weathering, was found to be particularly abundant in the basalt type used for the second test. This could explain the more rapid rate of strain increase observed in this test (Fig. 4b). Secondly, the ensuing sudden rise in strain can most likely be attributed to expansion induced by ice formation causing a moving apart of the fracture walls. In these tests it was not possible to obtain exact values for the temperatures at which freezing was initiated within the fractures. It seems certain, however, that freezing did not begin until temperatures had dropped to below -2°C. Thirdly, in the first test, the further increase in strain with negative temperatures may be due to continuing clay mineral swelling activity or, possibly, the freezing of additional

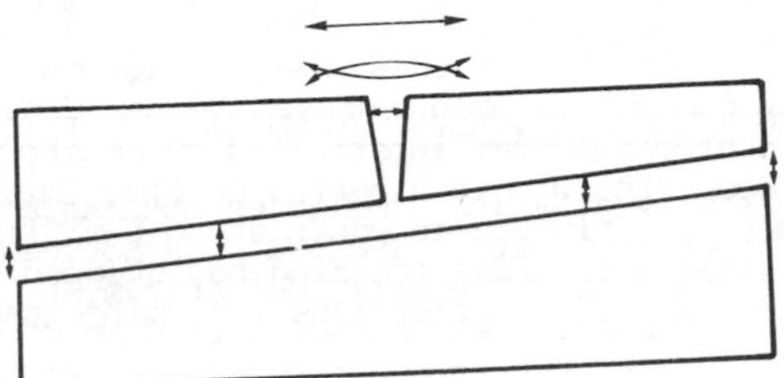

Fig.5. Possible freezing-induced movements in an open fracture system.

Fig.6. Experimental breakdown of large and small chalk blocks along stylolite seams.

amounts of water within the fracture. The final phase of decreasing strain observed in the first test is related to the thawing of water within the fracture and an associated strain relief on the fracture walls.

The experiment indicated the usefulness of strain gauges in the examination of freezing-induced deformation of rocks. The same basic method was therefore used by McGreevy (1982) in a laboratory study of frost weathering of a chalk.

FREEZING STRAINS DEVELOPED IN CHALK STYLOLITES

In a series of frost weathering simulation studies conducted on a variety of rock types (McGreevy, 1982) it was noticed that for one rock type, a hard chalk from Glenarm, Co. Antrim, N. Ireland, daily freeze-thaw cycling caused preferential breakdown of samples along stylolite seams. This occurred irrespective of the sizes of specimens used (Figure 6) - the importance of specimen size in frost weathering experiments has been stressed elsewhere (McGreevy and Whalley, 1982).

The characteristics and origin of stylolites have been discussed in detail elsewhere (Park and Schot, 1968; Guzzetta, 1984), but require some brief consideration here. Stylolite seams are pressure solution phenomena formed in consolidated rocks, particularly limestones. The seams are basically contact surfaces marked by the mutual interpenetration of two bodies of rock. The stylolite surface itself is indicated by a thin deposit of insoluble residue which in carbonate rocks is usually clay. Although this residue is often a minor constituent of the rock, the process of stylolite formation (pressure solution) causes it to concentrate in relatively high amounts along the seams. One further characteristic which is of some relevance in the present context, is the reduction of porosity in the vicinity of stylolites due to the reprecipitation of dissolved rock material (cf. Sprunt and Nur, 1977).

In the chalk under study, the maximum amplitude of the stylolite seams is approx. 1 cm, with values usually lying between 0.1 and 0.5 cm. In thin section (Fig. 7) it can be seen that the insoluble residue present in the seams (identified as consisting primarily of smectite, with smaller amounts of illite/mica) is more abundant at the "crests" and "valleys" of the stylolites than along the vertical walls, a characteristic also reported by Park and Schot (1968).

The fact that there is lower porosity around stylolites reduced the likelihood that preferential frost weathering was due to accumulation of relatively large amounts of water in their vicinity. The presence of smectite along the seams prompted the alternative hypothesis that disruption along stylolites reflected a dual mechanism i.e. stylolite opening due to ice formation but supplemented by swelling of smectite on contact with water (cf. Cady, 1969; Douglas, 1972; Konishchev, 1978). It was thought that strain gauges might profitably be used to monitor strain across a stylolite seam during freezing and enable the relative importance of ice- and water (swelling)-induced strains to be assessed.

Methods. In contrast to the method described above for the basalt, strain gauges were cemented on to the actual surfaces under study.

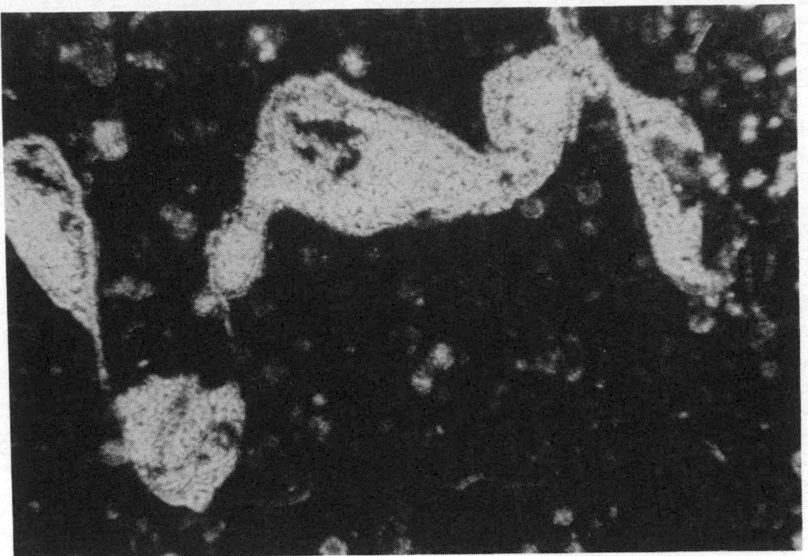

Fig.7. Thin section micrograph of stylolites, picture width = 2 mm.

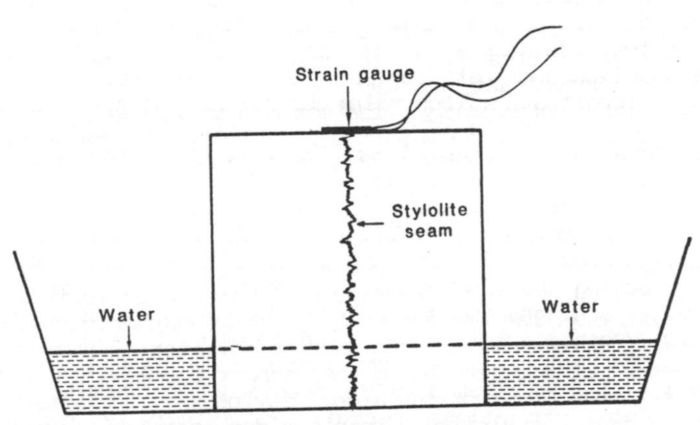

Fig.8. Chalk cube standing in water, showing stylolite traversing the cube and strain gauge on the upper surface.

Because of the chalk's low porosity (= 8.27%) and extremely fine pores (median pore throat radius = 0.065 µm) it was believed that ice extrusion would not occur to any great extent and would not adversely affect strain measurements.

Two 5 cm cubes of chalk were cut so that each contained a single stylolite seam. The surfaces on to which active and dummy strain gauges were to be mounted were traversed by the stylolites (Fig. 8). The relative abundance of smectite at the crests and valleys of the stylolites (see Fig. 7) would ensure that swelling during wetting would be greatest normal to the stylolite seam (i.e. left to right in Figure 8). Lateral movements, parallel to a seam, would be expected to be minimal due to the thinness of the stylolite walls. Thus, by fixing a strain gauge normal to a seam, the maximum disruptive effects of clay swelling could be assessed. Following polishing and thorough cleaning of test surfaces, two small pieces of aluminium foil were cemented across the stylolite seams and on to these were mounted the strain gauges (R.S. Components, type 11). Thin films of quick-set epoxy resin were used as adhesive for both operations. The strain gauges were then completely covered with a moistureproof coating of silicone rubber compound.

The cube on to which the active gauge had been attached was placed in a small plastic basin filled with water to a depth of 1 cm. The strain gauge rested on the uppermost face of the cube and was not in direct contact with the water (Fig. 8). After 48 hours, during which the cube absorbed water by capillarity, the basin containing the specimen was placed on a bed of expanded polystyrene in a freezing cabinet. The cube bearing the dummy gauge was not soaked and was placed on the polystyrene beside the basin containing the wetted specimen. A J-type thermocouple was taped on to the surface of the wetted cube to monitor rock temperature. The cubes were subjected to daily freeze-thaw cycling during which rock temperature and electrical resistance changes were recorded every twenty minutes using a multi-channel data logger. Resistances were converted to equivalent microstrain values using manufacturers' specifications.

Results. Data are presented here for a ten day period. Recording problems restricted observations but the results proved to be of considerable interest.

Rock surface temperatures and associated changes in strain across the stylolite seam at different stages in the experiment are shown in Fig. 9. During the first three freeze-thaw cycles (Fig. 9a) strain showed only a slight increase, with the rate of increase becoming noticeably greater during the third cycle. In contrast to the results depicted in Fig. 4, the strain curve showed no variations which could be attributed to the freezing and thawing of water. However, such variations became increasingly evident as the experiment progressed. Figure 9b shows strain patterns during cycles 4, 5 and 6. It is seen that short, sharp increases in strain began to occur when temperature dropped below 0°C, usually at a temperature of -1.9 °C. Between these distinct steps in the curve, strain continued to increase through both the freezing and thawing periods of each cycle. During the thaw phase of cycle 6, strain decreased to produce a pattern of change similar to that shown in Fig. 4. Towards the end of the experiment (Fig. 9c), a definite pattern of strain

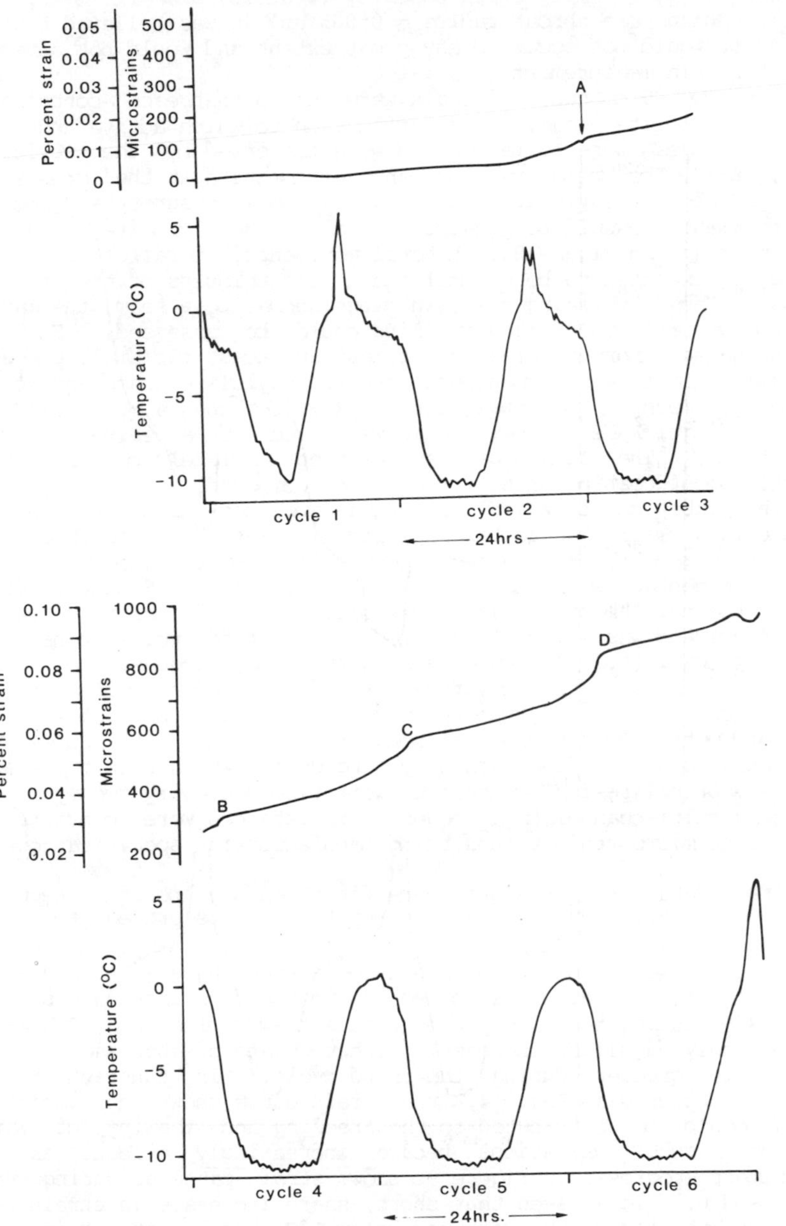

Fig.9. Strain versus temperature curves for stylolitic chalk. 9a (top) cycles 1 - 3; 9b (bottom) cycles 4 - 6.

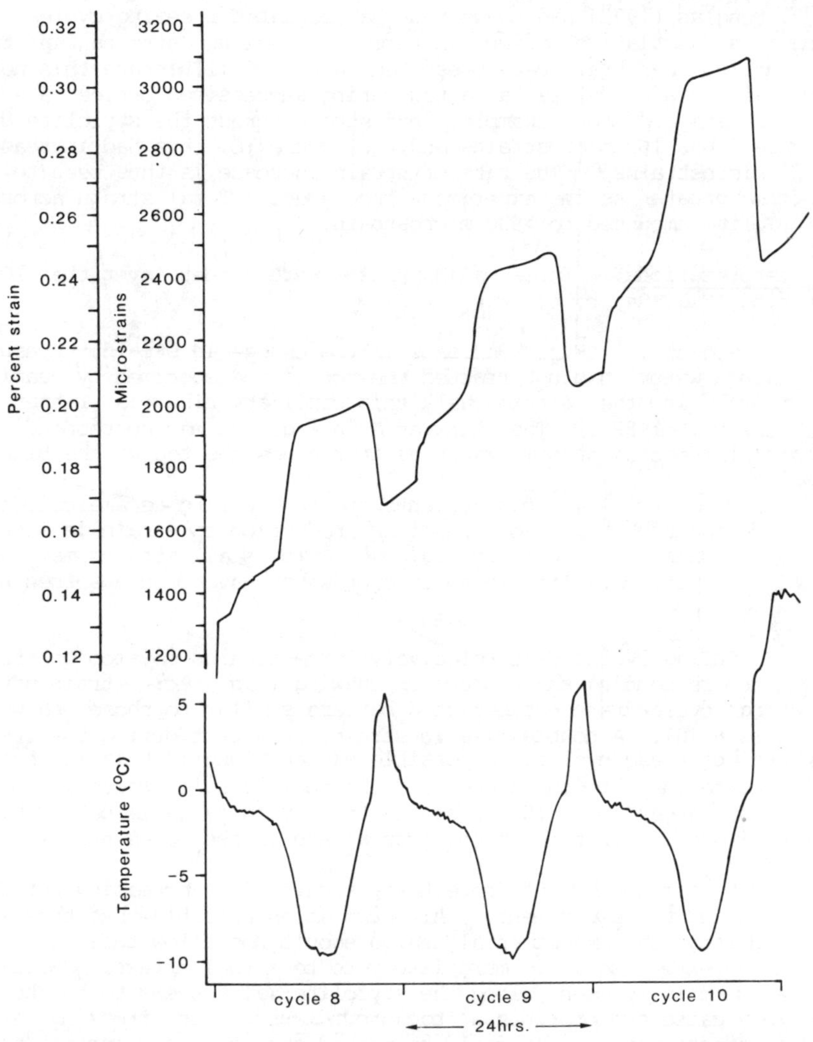

Fig.9c. Strain versus temperature curves for stylolitic chalk, cycles 8 - 10.

variation had emerged. Three periods of relatively slow strain increases were punctuated by two abrupt increases, the first occurring at a temperature of -0.9°C and the second, and main, increase at around -2.0°C. A rapid and quite substantial strain decrease occurred during thawing the decline usually beginning in the -2.0 to -2.5°C temperature range.

Douglas (1972) had suggested that repeated freezing cycles would have a cumulative effect in terms of freezing strains in rocks. Visually, the strain data presented in Fig. 9 illustrate this notion. Actual strain values attained during successive cycles are also illustrative. For example, net strain across the stylolite during cycle 6 was 162 microstrains and, by cycle 10, this had increased to 356 microstrains. The rate of strain increase is thus seen to have become greater as the experiment proceeded. Total strain across the stylolite amounted to >900 microstrain.

Interpretations. Three main phases were evident over the 10 day period (Fig. 9a, b, c).

1. Cycles 1-3 (Fig. 9a) show little change in strain, presumably because water had not reached the top of the specimen by capillary action. In the Antrim chalk this capillary rise is rather slow (McGreevy, 1982). The kink at A in Fig. 9a may correspond to the initial freezing of some water as it reaches the top of the block.

2. During cycles 4-6 dispacements due to freezing became evident (B, C and D on Fig.9b). No significant reduction in strain is seen upon thawing until cycle 6 (E on Fig, 9b). The small strains may suggest that, as yet, only limited amounts of water have been absorbed by the chalk.

3. During cycles 7-10 relatively large strains develop on freezing with corresponding sharp drops on thawing (Fig. 9c; strain patterns during cycle 7 are not depicted but are similar to those shown for cycles 8-10). A double rise in strain is evident during the freezing phase of these cycles. A possible explanation for this is that the stylolite has begun to open and the rock to fail (which is usually evident after 10 to 15 cycles) so the strain gauge is detecting the freezing of water both in the 'crack' and in the pores next to it.

It is worthy of note that strain is increasing gradually throughout the experiment. Although it is possible that this is due to drift, the experimental design should not allow this to occur to such an extent so it is more likely to be a real effect. Swelling of the smectite present along the stylolite walls seems to be the most likely cause and provides a 'background' upon which freezing strains are superimposed. It could be argued that an added contribution is the freezing of further amounts of water within the finer pores near to the stylolite. Countering this, however, are the following considerations. First, the gradual increases occur during both freezing and thawing phases which would suggest that freezing is not the cause (this also rules out temperature drift as a possible factor). Second, previous workers (e.g. Thomas, 1938) have shown in their length-change measurements on rocks subjected to freezing that when contained water freezes there is a sudden sharp length increase. Thereafter, even when negative temperatures persist, contraction usually occurs which is believed to reflect strain relief due to expulsion of water and extrusion of ice from the specimen.

GENERAL DISCUSSION AND CONCLUSIONS

Our preliminary results suggest that, provided certain precautions are taken, particularly with regard to moisture-proofing, strain gauges offer a useful means for assessing frost weathering effects in micro-cracked rocks. Their sensitivity allows detection of freezing-induced displacements long before rupture actually occurs. Thus there is no need for recourse to relatively crude indicators of damage such as weight loss.

Strain gauges are available in a wide range of sizes and configurations and can be adapted to suit most experimental needs. Those which we have used are the most basic type Refinements in experimental design could be achieved by using more specialized types, especially the 'crack-opening' displacement gauges which bridge or arch over cracks and would allow precise evalauation of crack wall movements during freezing.

The experiments described are laboratory-based and were conducted under conditions somewhat removed from those likely to be experienced naturally. Although acquisition of field data on rock temperature and moisture conditions should ultimately provide a basis for more realistic laboratory tests, it may be more useful to experiment with strain gauges actually in the field (Cochran, 1980; Kohlbeck and Scheidegger, 1984).

In continuing our laboratory experiments we aim to compare freezing strains developed in water-filled cracks with those in uncracked areas of the same specimens. It is also hoped to monitor freezing strains over longer time periods and study patterns leading up to complete failure of specimens.

ACKNOWLEDGEMENTS

We thank the laboratory staff at Queen's University Geography Dept. for their help, in particular Mr A. Edwards, and Ms M. Pringle, Ms S. McWilliams, Mr T. Molloy for diagrams, typing and photography.

REFERENCES

Cady, P.D., 1969. Mechanisms of frost action in concrete aggregates. Journal of Materials, 4, 294-311.

Davison, J.I. and Sereda, P.J., 1978. Measurement of linear expansion in bricks due to freezing. Journal of Testing and Evaluation, 6, 144- 147.

Douglas, G.R., 1972. Processes of weathering and some properties of the Tertiary basalts of County Antrim, Northern Ireland. Unpublished PhD, Thesis, Queen's University of Belfast, 293pp.

Douglas, G.R., 1980. Magnitude-frequency study of rockfalls in Co.

Antrim, N. Ireland. Earth Surface Processes, 5, 123-129.

Douglas, G.R., 1981. The development of bonded discontinuities in basalt and their significance to freeface weathering. Jökull, 31, 1-9.

Dove, R.C. and Adams, P.H., 1964. Experimental Stress Analysis and Motion Measurement. Merrill, Columbus.

Garfield, D.E. and McLain, B.G., 1978. Waterproofing strain gauges for low ambient temperatures. U.S. Army Cold Regions Reseach and Engineering Laboratory, Special Report, 78-15, 20pp.

Guzzetta, G., 1984. Kinematics of stylolite formation and physics of the pressure solution process. Tectonophysics, 101, 383-394.

Hallet, B., 1983. The breakdown of rock due to freezing: a theoretical model. in Proceedings of the Fourth International Conference on Permafrost, pp 433-438, National Academy Press, Washington, D.C.

Kohlbeck, K and Scheidegger, A.E., 1984. Application of strain gages on rock and concrete. in International Symposium on Field Measurements in Geomechanics, (Ed. Kovari), Vol.1, 197-207, A.A.Balkema, Rotterdam.

Konishchev, V.N., 1978. Frost weathering. in USSR Contribution, Proceedings of the Second International Conference on Permafrost, pp 176-181, National Academy of Sciences, Washington, D.C.

Litvan, G.G., 1978. Adsorptive systems at temperatures below the freezing point of the absorptive. Advances in Colloid and Interface Science, 9, 253-302.

McGreevy, J.P., 1981. Some perspectives on frost shattering. Progress in Physical Geography, 5, 56-75.

McGreevy, J.P., 1982. Some field and laboratory investigations of rock weathering, with particular reference to frost shattering and salt weathering. Unpublished PhD thesis, Queen's University of Belfast, 486pp.

McGreevy, J.P. and Whalley, W.B., 1982. The geomorphic significance of rock temperature variations in cold environments: a discussion. Arctic and Alpine Research, 14, 157-162.

Mellor, M., 1970. Phase composition of pore water in cold rocks. U.S. Army Cold Regions Research and Engineering Laboratory, Research Report, 292, 61pp.

Park, W.C. and Schot, E.H., 1968, Stylolites: their nature and origin, Journal of Sedimentary Petrology, 33, 175-191.

Sprunt, E.S and Nur, A., 1977. Reduction of porosity by pressure solution: experimental verification. Geology, 4, 463-466.

Thomas, W.N., 1938. Experiments on the freezing of certain building materials. Department of Scientific and Industrial Research (Great Britain), Building Research Station, Technical Paper, 17, 146pp.

Vaughan, J., 1975. Application of B & K equipment to strain measurement. Bruel and Kjaer, Copenhagen, 125pp.

Walder, J. and Hallet, B., 1985. A theoretical model of the fracture of rock during freezing. Geological Society of America Bulletin, 96, 336-346.

Whalley, W.B., Douglas, G.R. and McGreevy, J.P., 1982. Crack propagation and associated weathering in igneous rocks. Zeitschrift für Geomorphologie, 26, 33-54.

International Geomorphology 1986 Part II
Edited by V. Gardiner

SURFACE CRUSTING OF SANDSTONES IN SOUTHERN ENGLAND AND NORTHERN FRANCE

D.A. Robinson and R.B.G. Williams

Geography Laboratory, University of Sussex

ABSTRACT

The exposed surfaces of many sandstones in temperate regions develop hard crusts or rinds within their surface layers. Detailed studies of crusts developed in the Ardingly Sandstone of south-east England and the Fontainebleau Sandstone of the Paris Basin, France, indicate that, although the crusts are frequently coloured by iron and carbon compounds, their dominant component is secondary silica which is deposited on the surfaces of quartz grains and in the pores. The surface permeability of crusted sandstone has been measured in the field by means of small infiltration cylinders glued onto the surfaces of the rock. In the case of the Ardingly Sandstone, the presence of a crust can reduce infiltration through the rock surface to less than 5% of the rate through the underlying rock, and in Fontainebleau Sandstone reduce the rate to less than 20%. The surface hardness of fresh rock surfaces and crusted rock surfaces was measured by means of a Schmidt Hammer. Crusted rock does not give consistently higher readings. This is probably because of a sub-surface layer of soft, incoherent sandstone that frequently lies immediately beneath the crust.
The crusts described in this paper are believed to exert a profound control on the morphology of natural outcrops of the sandstones and upon the origin and distribution of micro-weathering features that they exhibit. Preliminary study of other soft porous sandstones in Britain suggests that silica crusting is a common phenomenon.

INTRODUCTION

Weathering is frequently portrayed as having a purely destructive effect on rocks, leading to their eventual disintegration and decomposition. However, this is an oversimplification because, for limited periods at least, weathering can have a markedly protective role. Desert varnish, for example, which develops as an external coating on many rocks in arid regions (Potter and Rossman, 1977; Dorn and Oberlander, 1982), strengthens and waterproofs rock surfaces, thus helping to prevent further weathering. Also protective, but less well studied, are crusts that develop within the surface layers of sandstones in humid temperate regions. These are invariably darker than the fresh rock owing to enrichment with iron and carbon compounds, and perhaps compounds of manganese (Ollier, 1984), but there have been no detailed investigations as to whether these colouring substances are also responsible for giving the crusts protective properties. In this paper we report on the character and protective properties of the crusts that develop on two sandstones: the Ardingly Sandstone of the central Weald of south-east England, and the Fontainebleau Sandstone of the Paris Basin, France. In both cases, we suggest that it is not the colouring substances but silica that hardens and waterproofs the crusts.

THE ARDINGLY SANDSTONE

The Ardingly Sandstone is a Hastings Beds sandstone of Valanginian (lower Cretaceous) age. Over most of its outcrop it is a massively jointed, fine grained sandstone that contains less than 5% non-quartz material. When freshly quarried, it is soft and friable with little resistance to abrasion and a low crushing strength. Grains of sand can be detached by gentle rubbing with the fingers. The colour of the unweathered rock is white to buff-yellow, or occasionally orange-brown. It has been suggested that limonite is the principal cementing agent between the quartz grains, and that the colour of the sandstone varies according to how much limonite is present (Milner, 1923; Dines *et al.*, 1969). However, the quantities of iron present are always very low. We have obtained values that vary from 0.02% to 0.30% as Fe_2O_3, which agree with the very limited data published previously (Sabine, 1969). Examination of the unweathered sandstone under the scanning electron microscope (Photo 1) does not support the idea that iron oxide cements the grains together. There is very little cement either filling the pores or cementing the quartz grains at their points of contact. Limonite may be the colouring agent on or within the grains, but it appears to play little or no part in strengthening the rock. The limited strength possessed by the sandstone appears to be due to the partial interlocking of the grains. The grain surfaces show slight solution and reprecipitation and many upturned plates, but the grains are very loosely packed together leaving abundant pore spaces in between.

On steep valley sides the Ardingly Sandstone outcrops as discontinuous lines of cliffs up to 15 m high (Robinson and Williams, 1976; 1981). Surface crusts or weathering rinds have developed on both the tops and faces of the cliffs. Typically, the crust is only a millimetre or so thick, and it tends to be best developed on the faces of the cliffs, where it is brownish red to black in colour. It is generally thinner and more greyish on the cliff tops. In many places it has probably developed over a period of several thousand years, but it is not a fossil feature for it continues to form at the present time. Many inscriptions carved through the crust only a century or so ago have completely crusted over, as have the faces of quarries that were still in use fifty to one hundred years ago. Crusting can develop even on surfaces not directly exposed to the weather. Near Uckfield, for example, a crust has developed on the roof and sides of a tunnel excavated in the late eighteenth or early nineteenth century (Mantell, 1833).

Our analyses of the crust confirm that it is frequently enriched with iron, but the quantity remains low, with the proportion of Fe_2O_3 varying between 0.02% and 0.90%. Occasionally, the content of iron in the crust is actually lower than in the interior of the rock. Viewed with the scanning electron microscope the quartz grains in the crust are seen to be much better cemented than in the unweathered rock, and the pore spaces are largely infilled (Photo 2). However, the surface appearance of the cement suggests that it is composed of silica rather than iron, and this has been confirmed by energy dispersive X-ray analyses (EDAX) of samples from several outcrops. The silica is deposited as a smooth coat on the surface of the grains, as small irregular "blobs" on the coating, and as "bridge-work" across the pores.

The dark colour of the crust appears to be largely due to deposits of carbon. The colour survives heating to around 100°C. in concentrated hydrochloric acid but is lost when dry specimens are heated to 800°C.

THE FONTAINEBLEAU SANDSTONE

The Fontainebleau Sandstone is a very pure quartzitic sandstone of Oligocene age that outcrops extensively in the woods and heaths surrounding the town of Fontainebleau, some sixty kilometres south of Paris (Loiseau, 1970; Robinson and Williams, 1982). The sandstone occurs as an indurated cap rock, up to 5 m in maximum thickness, on the summits of a series of long parallel ridges aligned approximately northwest - southeast. The ridges are believed to have developed as sand dunes along the southern margins of a shallow arm of the Atlantic which extended across northern France in mid-Tertiary times (Aliman, 1936; Pomeral and Feugueur, 1968). The cap-rock is thought to have formed by the evaporation of silica-rich waters that were drawn up through the dunes and evaporated at the summits leaving the silica behind as a coating on the sand grains. The sandstone on many ridge tops has no soil cover and forms bare rock pavements, or "platières", that are criss-crossed by widened joints. The edges of the platières are often terminated by low cliffs which overlook the sides of the ridges that are covered by jumbled masses of boulders, aptly called "chaos" by French geomorphologists. Many of the boulders appear to be joint blocks that have broken away from the platières. In some places, scattered boulders of sandstone also lie on the surface of the platières. Many of the boulders of sandstone within the chaos and on the platières are weathered into strange mushroom and animal-like shapes, and their surfaces are crazed by extensive patterns of polygonal cracks.

Superficially, Fontainebleau Sandstone appears very similar to Ardingly Sandstone. When freshly exposed, it is brilliant white or pale buff, but it weathers on the surface to a red or more usually greyish - black colour. The unweathered sandstone consists of quartz grains with crystal overgrowths displaying pyramidal truncations that sparkle in strong sunlight. Scanning electron microscope studies reveal that the overgrowths fit closely together and are often interpenetrated (Photos 3 and 4). There is virtually no cement between the grains, and it is their tightness of fit that gives the rock its strength. Different beds vary greatly in hardness and porosity, depending on the amount of void space. Some samples of fresh sandstone are so soft and weak that it is impossible to view coherent samples either in thin section or beneath the electron microscope. On the other hand, the cap rock of the platières tends to be particularly hard with very low porosity.

The quartz grains have extensive overgrowths that closely interlock. In contrast, the boulders in the chaos and on top of the platières are more variable in character with many having a hard, thin, outer crust protecting softer, more porous rock beneath. Examination of the crust under the scanning electron microscope reveals that the sutures between the sand grains, and many of the pore spaces, are plugged by nodular and filamentous growths of secondary silica (Photos 5 and 6). Very little iron is present and it is evidently the deposits of secondary silica that give the crust increased strength.

WATERPROOFING PROPERTIES

The waterproofing properties of the silica crusts on the Ardingly and Fontainebleau sandstones have been measured by means of a series of infiltration experiments. Small, perspex infiltration cylinders, 32 mm in diameter and 75 mm high, were temporarily glued on to dry surfaces of the sandstones with Loctite. The cylinders were filled with water to a depth of 50 mm, and the head of water in the cylinders was maintained, for a period of normally two hours, by adding measured quantities of water to replace the water lost through downward percolation into the rock. In a few instances where the rates of infiltration were very rapid the measurement period was reduced. At each site at least five replicate cylinders were used.

TABLE 1. Infiltration rates (ml cm^{-1} hr^{-1})

Site type	Location	Readings	Mean	St.Dev.
Ardingly Sandstone				
Crusted cliff tops	Harrison's Rocks	5	1.56	1.707
	High Rocks	5	1.08	0.798
	Chiddingly Wood	5	4.70	8.332
Crust removed	Harrison's Rocks	5	44.36	15.870
	High Rocks	5	24.06	12.188
Bedding planes	Philpots Quarry	5	18.78	17.851
Quarried blocks	Philpots Quarry	5	21.30	16.976
Fontainebleau Sandstone				
Crusted boulders	Larchant	5	0.58	0.228
Crust removed	Larchant	5	3.16	2.066
Fresh Quarry Stone	Larchant	5	81.52	71.123

Measurements were made on the crusted surfaces of rock outcrops, and on surfaces from which the crust was deliberately removed. They were also made on the bedding planes of rock recently exposed in quarries, and in the laboratory on fractured faces of blocks removed from quarries.

Some of the results are shown in Table 1. High Rocks and Harrison's Rocks are near Tunbridge Wells. Philpots Quarry is at West Hoathly 25 km west southwest of Tunbridge Wells. The cliffs of Ardingly Sandstone in Chiddingly Wood lie within 200 m of Philpots Quarry. The measurements on natural outcrops at Larchant were made at Rocher de Dame Jouanne about 2.5 km from the quarry.

Although the infiltration rates vary at each sample site, the table clearly shows that the rates are greatly reduced by the development of a crust. In the case of the Ardingly Sandstone the waterproofing is especially dramatic because the crust is very thin and the layer immediately beneath is much more permeable, in fact even more permeable than the unweathered rock. The crust on the Fontainebleau Sandstone tends to be thicker, and the subcrustal layer, although on average nearly five times more permeable than the crust, is many times less permeable than the fresh rock.

STRENGTH CHARACTERISTICS

A Type L Schmidt Hammer has been used to measure the strength of the crusts. Originally designed for testing the *in situ* strength of concrete (Proceq, 1977), the Schmidt Hammer has been shown to be a useful estimator of the relative strength of natural rock surfaces provided certain precautions are observed in selecting the sites and surfaces for testing (Day and Goudie, 1977; Williams and Robinson, 1983). The results (Tables 2 and 3) are much more complex than for the infiltration experiments. No clear relationship appears to exist between crusting and increased surface strength for the two sandstones tested. Indeed, the highest values were recorded on samples of fresh Fontainebleau Sandstone in Larchant quarry, the surfaces of which were very poorly cemented. The sand grains could be rubbed away with the lightest touch of the fingers and the impact of the hammer dislodged sufficient grains to leave holes several millimetres deep. Similarly, the highest values on the Ardingly Sandstone were obtained on freshly exposed bedding planes at Hook Quarry, which lies 1 km southeast of Philpots. The surfaces of these bedding planes were coated in iron and silica deposited by percolating groundwater, but even uncoated surfaces gave readings as high as those recorded on many crusted surfaces elsewhere in the Weald (Table 2).

The lowest readings were recorded on crusted surfaces of Fontainebleau Sandstone that displayed polygonal cracking. Some of these surfaces could be seen to bounce under the impact of the hammer, and average values obtained were little more than a third of those for neighbouring uncracked surfaces. Polygonal cracking is much less common and less well developed on Ardingly Sandstone outcrops (Robinson and Williams, 1982), but measurements in Chiddingly Wood suggest that, where cracking is present, the surface of the sandstone is weaker than usual.

TABLE 2. Schmidt Hammer readings - Ardingly Sandstone

Site Type	Location	Readings	Mean	St.Dev.
Crusted cliff tops	Sheffield Forest	15	31.07	2.344
	Harrison's Rocks	15	31.14	2.489
	High Rocks	60	35.72	2.315
	Chiddingly Wood	40	33.75	2.329
Crusted cliff tops with polygons	Chiddingly Wood	40	27.05	3.456
Crusted cliff faces	Sheffield Forest	45	33.57	3.728
	Harrison's Rocks	30	32.11	4.139
	High Rocks	60	36.08	3.651
Sub-crustal surfaces	Harrison's Rocks	30	23.05	2.698
Freshly exposed bedding planes	Hook Quarry	40	38.93	3.68
Fresh vertical faces	Hook Quarry	60	33.60	3.595

The low readings recorded on polygonised surfaces are not thought to result primarily from the presence of the cracks, but from the presence of a soft layer of incoherent rock that lies immediately beneath the crust separating it from more solid sandstone beneath. At Fontainebleau, some boulders give a sharp ring if struck with a metal rod, but others sound dull and hollow. The latter have suffered subsurface weathering, and often display fine polygonal cracking. The subsurface layer can be seen by chipping off the overlying crust, and it is occasionally exposed where the crust has been breached by weathering. It is extremely incoherent and its weakness is very clearly demonstrated by the very low hammer values shown in Table 3.

A similar layer of weakened rock underlies the crust that develops on Ardingly Sandstone. In places this subsurface layer is so incoherent that it runs out freely when the crust is removed (Robinson and Williams 1976; 1981). However, the layer tends to be only a few millimetres in thickness and has a much less dramatic affect on the Schmidt Hammer readings (Table 2).

The Schmidt Hammer used in this study would seem to be of limited use for estimating the strength properties of surface crusts and a field based abrasion test would probably be more appropriate.

TABLE 3 Schimdt Hammer readings - Fontainebleau Sandstone

Site Type	Location	Readings	Mean	St.Dev.
Crusted surfaces without polygons (platières)	Apremont	50	32.28	9.017
	Cuvier Châtillon	50	41.44	6.606
	Franchard	50	39.94	8.897
	Larchant(1)	50	35.48	9.342
	Larchant(2)	50	35.50	10.017
Crusted surfaces with polygons	Apremont	50	13.94	3.248
	Cuvier Châtillon	50	13.86	4.518
	Franchard	50	9.40	3.143
	Larchant(1)	45	9.58	2.981
	Larchant(2)	50	14.48	6.665
Crust removed	Larchant(1)	16	13.20	3.902
Crust disintegrating	Larchant	25	7.76	1.877
	Franchard	25	8.92	2.886
Fresh sandstone	Larchant Quarry	40	33.05	3.602
	Larchant Quarry	10	56.90	2.849

DISCUSSION

The exact source and cause of mobilisation of the silica that is deposited in the crusts of the two sandstones remains uncertain. Crystalline quartz has a very low solubility in water at ambient temperatures if the pH is below 9. Solubilities of 6 - 14 mg/l are normally quoted (Yariv and Cross, 1979). However, natural quartz particles tend to have amorphous or disturbed layers around their surfaces that are more soluble, and solubilities of 100-180 mg/l have been recorded. The solubility of quartz also increases in the presence of certain natural organic acids (Krauskopf, 1979; Seiver and Scott, 1963).

Paraguassu (1972) demonstrated, under laboratory conditions, that the passage of pure water through a sandstone can dissolve and carry sufficient silica to an evaporating surface to form a crust in only a few months. The crust was reddened by iron oxides, but cementation was by silica. In a similar experiment, Whalley (1978) also successfully silicified the surface of a sample of silica sand, and analysis of the cement that formed led him to conclude that silification was by amorphous silica.

The development of crusts on outcrops of the two sandstones is presumably the result of pore water movement. In the absence of a well developed crust, rainfall can infiltrate into the rocks and dissolve silica, perhaps aided by the presence of organic acids. This silica is subsequently left behind in the surface of the rocks when water drips away or evaporates from exposed surfaces. Crust development must depend upon an adequate supply and circulation of water within the rock outcrops. Gravestones of quarried Ardingly Sandstone show little crusting because there is insufficient percolating water to supply more than a small amount of silica. Quarried blocks in the walls of buildings tend to remain relatively dry, and also therefore develop little or no crust.

On natural outcrops, as the crust develops it will increasingly tend to seal the surface of the sandstone and reduce the infiltration of water. On isolated upstanding outcrops this will greatly reduce the supply of water, and crust formation will slow down or even cease. In some instances the crust begins to crumble away, allowing the supply of water to increase and crust formation to recommence. The occurrence and thickness of a crust on such outcrops is therefore the result of a circulation of water largely controlled by the strength and permeability of the crust itself.

Where an outcrop passes back into a hillside, much of the pore water in the sandstone may enter by percolation through overlying soil or strata. In the absence of a crust, water entering the sandstone will be drawn towards the exposed surfaces where evaporation occurs. The development of a crust will not cut off the supply of much of the water, but it will greatly reduce the evaporation rate as the pores become filled with deposits of silica. Water will have increased difficulty escaping, except as seepages along bedding planes, and crust formation will slow down or cease entirely. Weathering and erosion of the outer surface gradually destroys the crust and this will allow water evaporation and crust formation to recommence. Thus, here also, the crust exists in a dynamic equilibrium with a circulation of water that is largely controlled by the crust itself.

The weakening of the sandstone beneath the surface crust is presumably caused by the solution of silica. Once dissolved the silica migrates to the surface where it is deposited when the water evaporates. As long as water is able to penetrate the crust, solution will be very active in the subsurface layer because of the high frequency of wetting and drying cycles. Why subsurface weakening is not always present remains unknown. One possibility still being investigated is that marked subsurface weakening is a feature of outcrops where crustal formation is primarily the result of water penetration through the exposed rock surface while crusts without subsurface weakening are the result of ground water evaporation.

CONCLUSIONS

Many sandstones in Western Europe form surface rinds or crusts on exposure to the weather. The Ardingly and Fontainebleau Sandstones described in this paper are particularly good examples of relatively porous sandstones that develop surface crusts. Deposition of silica in the crusts weatherproofs the rock by reducing the ingress of water, and strengthens the rock by increasing the cementation and interlocking between grains. Preliminary examination of other sandstones, such as the Ashdown Sandstone of south-east England, suggests that crusts with properties similar to those described in this paper are characteristic of many of the softer more porous sandstones of the region. Harder, less porous sandstones, such as the Millstone Grit, Pennant and Old Red Sandstones, also develop crusts, but these crusts may be of a rather different character because such rocks never allow the easy entry and circulation of large volumes of water. On outcrops such as the Millstone Grit of the Pennines, for example, some of the crusts may result from little more than rain-borne pollutants collecting on the surface and penetrating the pores. Rather than strengthening or hardening the rock, such crusts may be extremely acidic and actually destructive.

Crusts on the soft sandstones described in this paper exercise a profound control on the morphology of the outcrops. The curious weathering forms found on many outcrops of the Ardingly and Fontainebleau Sandstones are, in part, due to uneven crust formation and differential weathering. Many micro-weathering features, such as honeycombing and polygonal cracking, are best developed in crusted sandstone, and their origin would seem to be intimately bound up with crust formation and destruction. A fuller understanding of crusting processes, and of the character of the crusts, appears to be an essential first step towards any satisfactory explanation of the origin and distribution of such micro-weathering features.

ACKNOWLEDGEMENTS

The authors would like to thank the Royal Society and the University of Sussex for financial assistance towards the cost of field work, Dr. M. Ford-Smith, School of Molecular Sciences, University of Sussex for many helpful discussions, and Dr. T. Browne, Computing Centre, University of Sussex for help in producing the camera-ready text.

REFERENCES

Alimen, H., 1936. Etude sur le Stampien du Bassin de Paris, Memoir S.G.F., 31.

Day, M.J. and Goudie, A.S., 1977. Field assessment of rock hardness using the Schmidt Hammer, British Geomorphological Research Group Technical Bulletin, 18, 19-29.

Dines, H.G. et al., 1969. Geology of the Countryside around Sevenoaks and Tonbridge, Memoir Geological Survey Great Britain, H.M.S.O., London.

Dorn, R.I. and Oberlander, T.M., 1982. Rock Varnish, Progress in Physical Geography, 6,3, 317-367.

Krauskopf, K.B., 1979. Introduction to Geochemistry, 132-35. McGraw Hill, New York.

Loiseau, J., 1970. Le Massif de Fontainebleau, Tome 1, 92-138. Vigot Frères, Paris.

Mantell, G., 1833. The Geology of South-East England. Longmans, London.

Milner, M.B., 1923. Notes on the Geology and Structure of the area around Tonbridge Wells, Proceedings Geologists' Association, London, 47-55.

Ollier, C.D., 1984. Weathering, (2nd Edition). Oliver and Boyd, Edinburgh.

Paraguassu, A.B.,1972. Experimental silicification of sandstone,

Bulletin Geological Society America, 83, 2853-2858.
Pomeral, Ch. and Feugueur, R.L., 1968. Guide Geologique du Bassin de Paris (Ile de France Pays de Bray), Masson editeur.
Potter, R.M. and Rossman, G.R., 1977. Desert varnish: the importance of clay minerals, Science, 196, (4297), 1446-1448.
Proceq, S.A., 1977. Operating instructions: concrete test hammer, Zurich, Switzerland.
Robinson, D.A. and Williams, R.B.G., 1976. Aspects of the geomorphology of the sandstone cliffs of the Central Weald, Proceedings Geologists' Association, London, 87, 93-100.
Robinson, D.A. and Williams, R.B.G., 1981. Sandstone cliffs on the High Weald landscape, Geographical Magazine, 53,(9), 587-592.
Robinson, D.A. and Williams, R.B.G., 1982. Sandstone sculptures in the Fontainebleau Woods, Geographical Magazine, 54,(10), 572-579.
Sabine, P.A., 1969. Geochemistry of Sedimentary Rocks, Institute of Geological Sciences, Report 69/1, 62. H.M.S.O., London.
Seiver, R. and Scott, R.A., 1963. Organic Geochemistry of Silica, in Organic Geochemistry (Ed. Breger, I.A.), 579-95. Pergamon, Oxford.
Whalley, W.B., 1978. Scanning electron microscope examination of a laboratory-simulated silcrete, in Scanning Electron Microscopy in the Study of Sediments (Ed. Whalley, W.B.) , 399-405, Geo-Abstracts, Norwich.
Williams, R.B.G. and Robinson, D.A., 1983. The effect of surface texture on the determination of the surface hardness of rock using the Schmidt Hammer, Earth Surface Processes and Landforms, 8, 289 -92.
Yariv, S. and Cross, H., 1979. Geochemistry of Colloid Systems, 247-86, Springer-Verlag, Berlin Heidelberg.

Scanning Electron Micrographs of samples of Ardingly and Fontainebleau Sandstones. The white scale bars near the bottom of each photograph are all 100 microns in length except in Photo 6 where the bar is only 10 microns in length.

Photo. 1 Typical fresh Ardingly Sandstone showing loosely packed quartz grains with some crystal overgrowths but little or no cement between the grains.

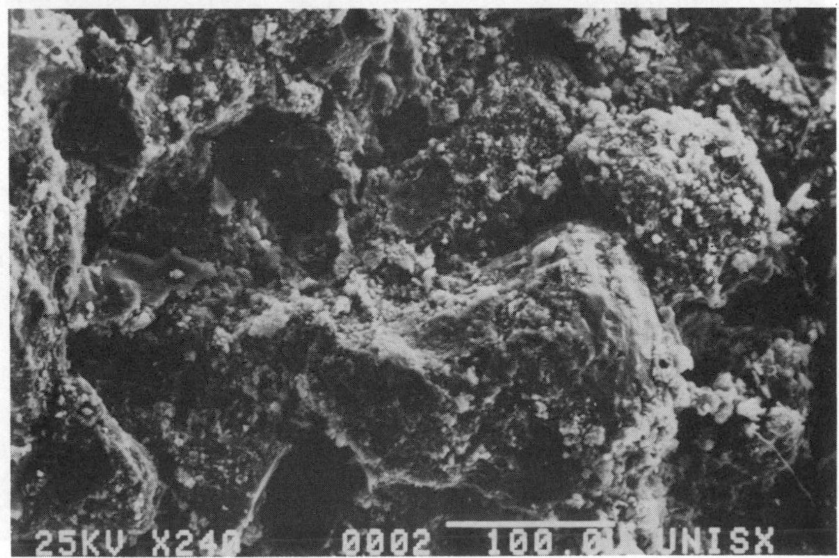

Photo. 2 Exterior view of crusted Ardingly Sandstone showing secondary deposits of silica both on the surfaces of quartz grains and blocking the pores between the grains.

Photo. 3 Fresh Fontainebleau Sandstone showing typical interlocking quartz grains with well developed pyramidal overgrowths.

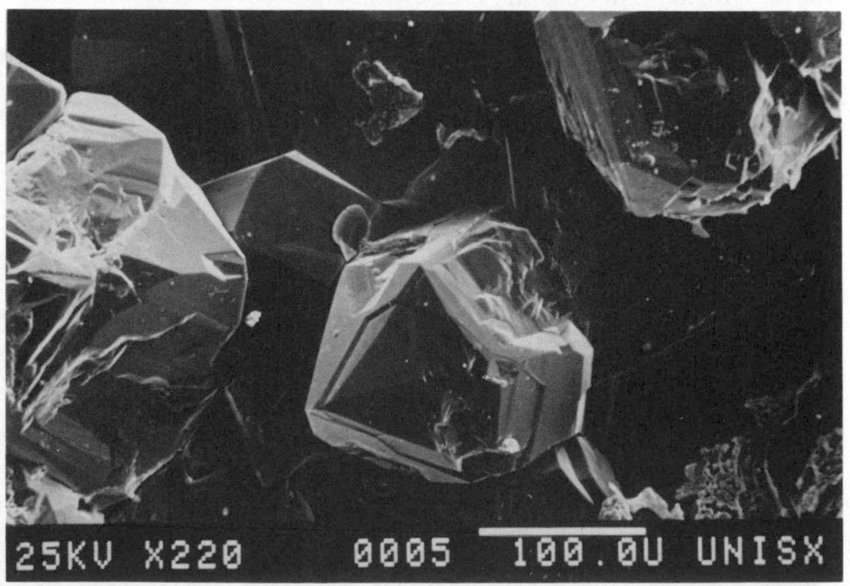

Photo. 4 Detail of pyramidal crystal overgrowth and interpenetration of quartz in Fontainebleau Sandstone.

Photo. 5 An oblique section through a crust on Fontainebleau Sandstone showing secondary deposits of nodular and filamentous silica on and between the quartz grains. The exterior of the crust is towards the bottom of the photograph.

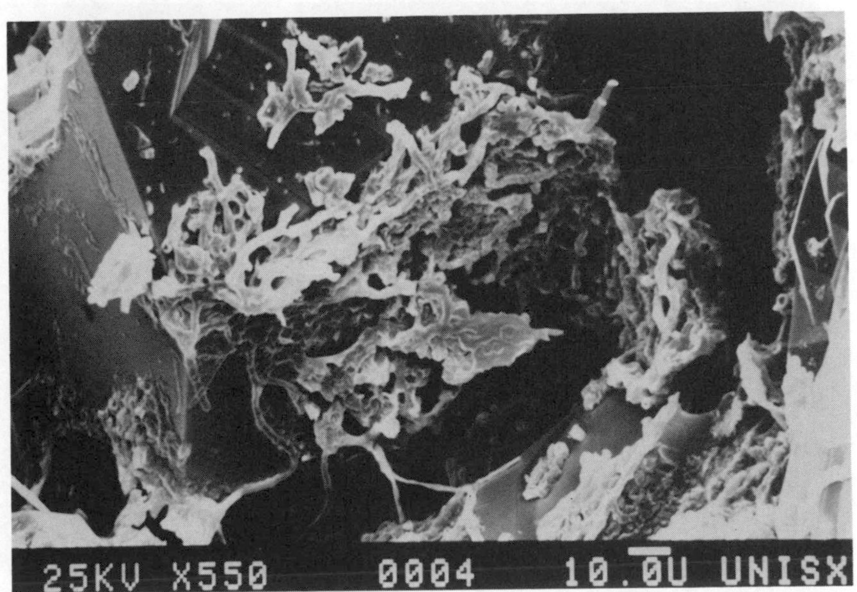

Photo. 6 Detail of a crust in Fontainebleau Sandstone showing the anastomosing growth form of secondary deposits of silica on quartz surfaces and across the pores.

International Geomorphology 1986 Part II
Edited by V. Gardiner

AN INTEGRATED APPROACH TO THE WEATHERING OF LIMESTONE IN AN ARID AREA AND ITS ROLE IN LANDSCAPE EVOLUTION: A CASE STUDY FROM SOUTHEAST MOROCCO

Bernard J. Smith

Department of Geography, Queen's University,
Belfast BT7 1NN, Northern Ireland, U.K.

ABSTRACT

A range of Macro, Meso and Micro-scale weathering features of karstic and mechanical origin are described from the Cretaceous limestones of the Hamada de Meski. Karstic features include escarpment valleys (resulting initially from spring-sapping), large and small sediment filled solution hollows, infilled cave systems, and shallow caves that are being modified by active cavernous weathering. All these features appear to be relict from former periods of greater available moisture. Some of the larger features may date back as far as the late Tertiary, whereas the smaller cave systems were active during Quaternary pluvial periods. Active solution appears to be restricted to the pitting of limestone surfaces. Mechanical weathering is mainly in the form of widespread cavernous weathering, which creates both sidewall and basal tafoni.

Weathering has been instrumental in effecting landscape development through scarp retreat, especially by selective backwearing along lines of subterranean drainage. It appears, however, that development occurred primarily under previous, moister conditions, and that only limited modification of the landscape has taken place since the mid-late Quaternary (Tennsiftien).

INTRODUCTION

Textbooks tell us that rock weathering in hot deserts is a major influence upon landscape development, especially through the production of debris to be removed by subsequent aeolian and fluvial processes. In many desert environments this removal of rock waste keeps pace with preparatory weathering (Mabbutt, 1977) and landforms exhibit a well developed adjustment to underlying geology. Because of this influence upon sediment supply, and hence upon the rate of erosion, it could be argued that such landscapes are in some ways 'weathering controlled'.

Despite this accepted significance of weathering, few studies have attempted a concerted evaluation of the contributions made by, and interactions between, the full range of weathering mechanisms operating in an arid landscape system. Instead, there has been an increasing tendency to isolate individual weathering phenomena, study them in great detail and seek understanding through segregation rather than integration. Such segregation is, however, at variance with the ultimate aim of the geomorphologist, which must be to explain the nature and origins of the physical landscape (Douglas, 1980 and Mark, 1980). Detailed weathering studies must therefore be viewed as a means to an end. Eventually they must be placed in a wider temporal and spatial context in the hope that they will help us to understand the landscapes within which they occur.

It is thus the aim of this study to exemplify the full range of weathering mechanisms and phenomena that occur within a single landscape unit. The unit chosen is part of the Hamada de Meski in arid southeast Morocco, which in many respects is typical of the numerous limestone plateaux found across the northwest Sahara. In turn, the role of these phenomena in the long-term evolution of the landscape will be suggested, especially in the context of present-day aridity and a Quaternary climatic history of fluctuating arid and humid conditions.

FIELD AREA

Location: The 'Cretaceous' Hamada de Meski lies immediately south of the High Atlas Mountains in the Moroccan Pre-Sahara, and consists of a gently inclined (2-3°) structually controlled plateau dipping approximately north-northeast (NNE). It lies on the northern flank of a large, unroofed dome structure which is centred on the Tafilalt Oasis (Fig. 1a) and was once overlain by Miocene and Oligocene deposits as well as the Cretaceous. The unroofing of the dome and hence the exposure of the Hamada de Meski was accomplished during the late Tertiary, mainly by a number of large perennial rivers which drain south from the High Atlas (Smith, 1977a). In the field area the principal river is the Oued Ziz, which begins to incise the Hamada near to the town of Er Rachidia and flows through an increasingly deep gorge (Fig. 2) until it breaks through the southern escarpment into the Tafilalt Oasis. The main fieldwork has been centred on the area to the east and northeast of the engorged middle Ziz valley and is shown in detail in Fig. 1b.

Climate and Climatic Change: The climate of the area is arid with spring and autumn rains, a hot summer and a cool winter (as defined by Meigs (1953)). There is, however, a strong north-south variation in mean annual rainfall, from 140mm at Er Rachidia to 70mm at Erfoud. Possibly because of the altitude of the area (Er Rachidia, alt. 1060m), there are marked seasonal differences in air temperature. The hottest month at Er Rachidia is July (31.6^{o}C) and the coolest is January (9.2^{o}C). During the winter months overnight temperatures commonly drop below zero and Er Rachidia has an average of 26 frost nights per year. Most rainfall is brought by moist Atlantic winds which are deflected by the western slopes of the Atlas Mountains (Reading University, 1969). These form depressions that can result in heavy rainfall followed by instability and showers, especially over the northern and western part of the Mountains. To a certain extent these showers penetrate over the mountains to the Er Rachidia

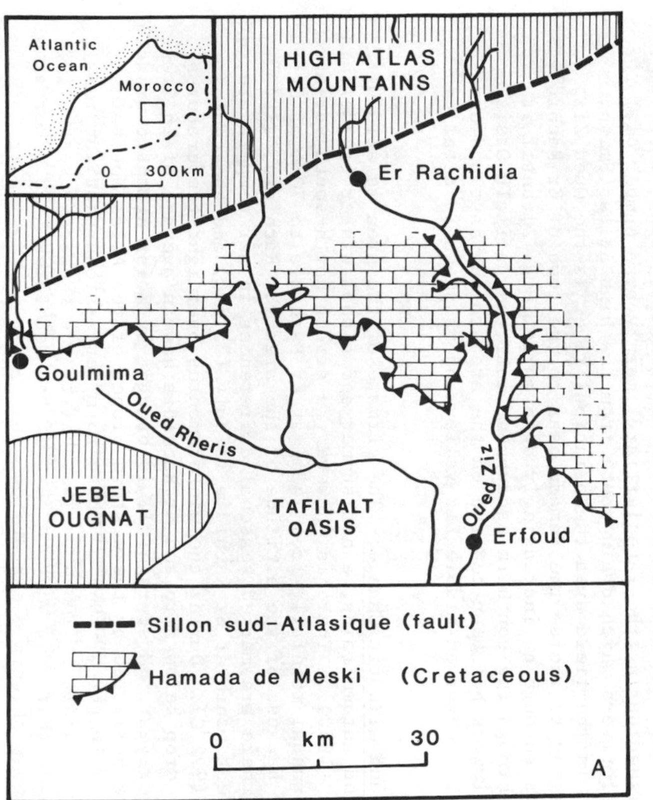

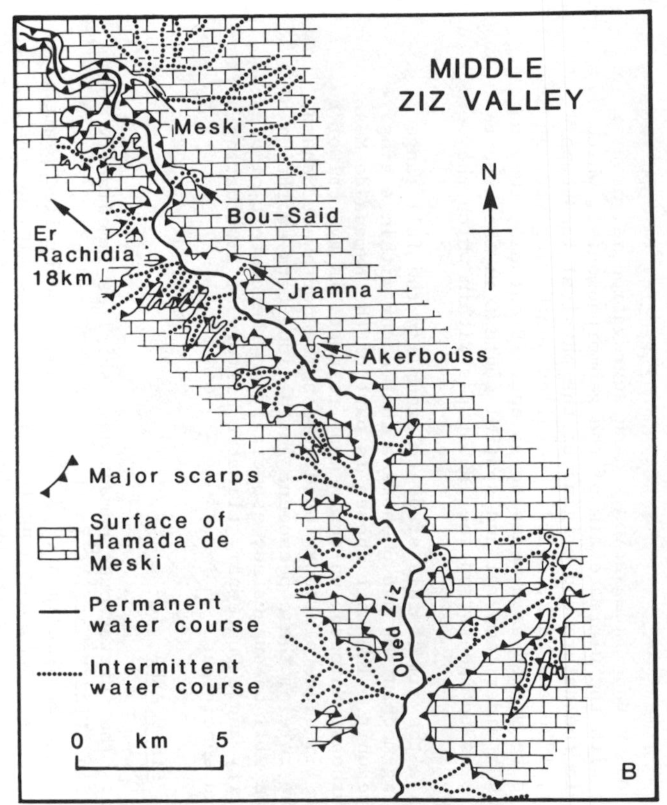

Fig. 1. Location Maps

Fig. 2. General view of the middle Ziz valley and Hamada de Meski, showing the flat plateau surface and free faces and debris slopes of the major escarpments.

Fig. 3. South-facing valley side of Wadi Jramna with intermittent spring (S), rockfall scars and debris and undermining of massive limestones.

region and give a characteristic light rain.

Present-day aridity is not, however, representative of climate during the Quaternary and the region has experienced a number of so-called 'pluvial' periods when climate was evidently moister, either through increased precipitation or decreased evapotranspiration. These pluvials and associated interpluvials are shown in Table 1, which is an interpretation of the work by Choubert (1961) and 1965). There are two points that should be

TABLE 1. Quaternary chronologies for Morocco in which pluvial periods are equated with glacial episodes in higher latitudes (Choubert 1961 and 1965).

Interpluvial Periods (Coastal Evidence)	Pluvial Periods (Continental Evidence)
	Actuel
	Rharbien
Mellahien	
	Soltanien
Ouljien	
	Pre-Soltanien*
Rabatien	
	Tennsiftien
Anfatien	
	Amirien
Maarifien	
	Saletien
Messaoudien	
	Regreguien
	Moulouyen
Moghrebien	

* May be second phase of Tennsiftien

noted about this chronology. First, the key to the sequence has been the interpretations of coastal evidence of a fluctuating Quaternary sea-level, in which high sea-levels have been equated with interglacial periods in high latitudes. Pluvial periods have then been interdigitated between the high sea-levels on the assumption that pluvials equate with high latitude glacial periods (Smith, 1977b). Second, the chronology is given in the knowledge that the assumed correlation between pluvials and glacials has been called into question. One would expect, for example, that increased continentality and reduced evaporation associated with glacials should encourage aridity in low latitudes (see Goudie (1983) and Warren (1985) for recent reviews). At present, however, no attempt has been made to reassess the Moroccan Quaternary, and whilst the traditional trans-Mediterranean correlations may be questioned, the pluvial sequence itself has withstood considerable local fieldwork. It is therefore considered to provide a suitable relative framework within which the geomorphological history can be fitted.

Geology and Topography: The Hamada is underlain by Turonian and Cenomanian marine limestone which consists of alternating bands of massively bedded, well-jointed micrite and biomicrite, and thin-bedded, flaggy, marly limestone. Below these are interbedded red and green marls and sands of Infra-Cenomanian age. At the northern end of the field area, the Ziz valley escarpment is cut entirely within the limestone. Further south, escarpment height increases and sands and marls are exposed at the base.

The Hamada can be subdivided into three principal landforms:

1. An undulating plateau surface covered in the north by a thin veneer of Quaternary deposits. The underlying limestone only crops out extensively near the escarpment edge and elsewhere is covered by a stone pavement of limestone cobbles and gravel. There is no marked incision into the Hamada, although there is an extensive network of shallow, intermittent streams and in the north of the area are several large, shallow depressions into which local drainage concentrates.

2. The Ziz valley escarpment; which comprises complex free faces cut in limestone standing above high angle debris slopes (Fig. 2). Free faces vary in height from approximately 10m in the north to 30m plus in the south. Detail is controlled by alternating massive and thin-bedded bands which often results in a stepped form. They retreat principally through undermining and collapse either as rockfalls of joint guided blocks (on minor escarpments) or slab failure of large rock columns (major escarpments). Debris slopes likewise increase in height and extent towards the south, where they are underlain largely by the Infra-Cenomanian sands and marls. Major debris slopes consist of jumbled rockfall debris immediately below the free face, above a complex of ancient slump sequences which grade downwards into smooth, concave debris mantled slopes.

3. Last, at intervals along the eastern valley side of the Ziz the escarpment is breached by small steep wadis (e.g. Meski, Bou Said, Jramna and Akerbôus - Fig. 1b). These are effectively dry, except at Meski where there is a permanent spring at the base of the escarpment. Groundwater for this derives from the High Atlas Mountains from where it flows southwards under piezometric head against regional geological dip (information from the Division de Résource en Eau, Er Rachidia). Similar evidence of intermittent groundwater seepage is seen near the mouth of the Wadi Jramna. Further south, no evidence of contemporary spring sapping has been seen and it is assumed that regional groundwater does not reach beyond Jramna under present conditions.

Except at Meski, the valleys are of limited extent. This is because Meski has tapped surface drainage from a considerable area underlain by Quaternary deposits, whereas the other valleys are cut into much higher escarpments, and are backed by the NNE sloping hamada surface which tends to direct any surface runoff away from the valleys.

Consequently, surface catchments are restricted to their incised sections and a narrow zone immediately behind the hamada edge.

WEATHERING FEATURES AND MECHANISMS

It is possible to group weathering phenomena on the hamada into a number of categories based upon size.

MACRO-SCALE FEATURES

Escarpment Valleys: These are normally a stepped V-shape in cross profile and exhibit pronounced geological control on form. Massive limestone produces small free faces and the thin-bedded limestones give short, steep debris slopes. Valley-side erosion is primarily by weathering and erosion of the thin-bedded bands, often as a series of small caverns, which undermines the massive limestones and leads to collapse. Evidence of this is seen in the form of small rock-falls and arcuate rockfall scars (Fig. 3). Over the long-term, however, slope debris has not accumulated, debris slopes carry only a thin mantle of loose material and the narrow valley floors are virtually debris free. Transport of material on the slopes and basal removal can obviously be delayed, but over the timespan of slope development removal is effectively unimpeded. Even at the present-time it appears that the limited catchments available periodically generate sufficient runoff to flush-out the comminuted products of slope weathering and erosion. It is this effectively unimpeded removal that has allowed the structural control of slope form to manifest itself. In turn, structural control and paucity of debris implies that long-term transport rates exceed those of weathering and debris supply, and that the latter controls the rate of slope development.

While occasional streamflow is capable of removing weathered debris, there is little evidence that it has been or is responsible for lineal erosion of the valleys. Channels are barely discernible or non-existent and, like valley-sides, there is marked structural control of valley long profiles. Moreover, in the absence of any initial catchment, overland flow cannot have been responsible for either the initiation of the valleys or their early development against the geological dip and surface slope of the hamada. Instead, it seems probable that they originated through a process of spring-sapping and collapse (fed by regional groundwater) at selected points along the main escarpment. In support of this contention there are a number of similarities in both the location and form of the valleys that can be used to suggest a possible developmental sequence (Fig. 4).

Because of its direction of flow, regional groundwater favours spring-sapping and scarp recession in an approximately NNE direction. This recession is, however, 'opposed' by the dip of the strata which produces structurally strong south-facing slopes. Any nascent valleys on these slopes are therefore restricted to isolated rock-falls or slumps (Fig. 4, No. 2). As an apparent compromise between the influences of groundwater source and geological structure, the favoured locations for valley initiation appear to have been on west-facing escarpments on the outside of meanders in the Ziz valley (Fig. 4, No. 2). Again, these valleys probably originated as rockfalls. However, once embayments in the escarpment had formed they became natural foci for local surface and sub-surface runoff, while continuing to tap regional groundwater. In response to this, valleys developed eastwards, principally by headward retreat (Fig. 4, No. 3). As valleys developed, two new slopes were created; a south-facing valley-side which is prone to spring sapping but resistant to collapse (e.g. in the Wadi Jramna, Fig. 3), and a north-facing structually weaker slope. The latter slope is more prone to collapse and also commands a considerable area of dip-slope drainage. In response to these factors a northwards flowing tributary may develop (e.g. at Bou Said and Jramna), which can eventually outgrow the initial valley (Fig. 4, No. 4). It is not possible

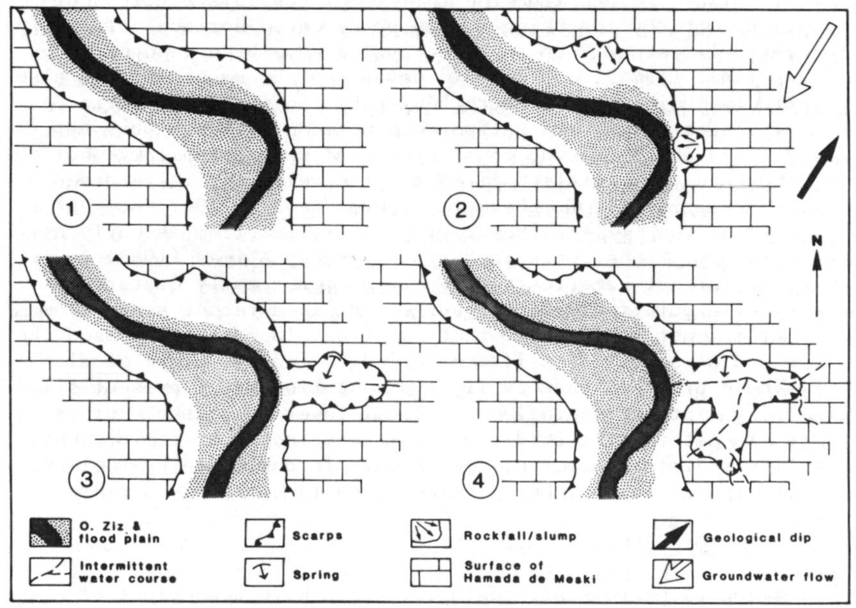

Fig. 4. Diagrams illustrating a possible evolutionary sequence for escarpment valleys.

to determine whether the valleys post-date the incision of the Ziz or whether they developed contemporaneously with that incision. It is clear, however, that development has been chiefly through headward recession driven by spring-sapping rather than downwearing. Apart from the previously mentioned seepage in the Wadi Jramna, no evidence of active groundwater flow has been observed. In view of this it is concluded that under present-day climatic conditions (both locally and in the High Atlas Mountains) there is little if any headward extension of the valleys - except where they can tap a sufficient quantity of dip-slope drainage (e.g. Meski).

Large, Sediment Filled Solution Hollows: The dominant karstic landforms of the hamada surface are large, circular to elliptical depressions which resemble the dolines of more humid regions. Characteristically these hollows (local name 'daya') may be over 100m across but are generally only up to 5-6m deep (Clark *et al.*, 1974). Normally they are areas of internal drainage, but in the two examples from the Hamada de Meski described by Mitchell and Willimott (1974) both have outflow channels. The outer margins of the dayas comprise a stony hamada surface, but the centres contain a sedimentary infill (sand-sized and finer) derived from a combination of surface wash and aeolian deposition. In the present climate they occasionally contain ponded water after rainstorms and there is some evidence of active leaching (clay translocation) within the surface horizons of the infill (Clark *et al.*, 1977). There is also evidence from the infills of possible cyclic sedimentation, with cycles being marked by cemented horizons of brecciated limestone. Up to three cycles have been recognised by Clarke *et al*. (1974), not only in dayas near Er Rachidia, but also in southern Tunisia and near to Béni Abbès in western Algeria. Similar phases of calcretisation were noted near Béni Abbès by Conrad (1959 and 1969) who considered that while such large hollows are undoubtedly of karstic origin their development is retarded by present-day aridity. Furthermore, Conrad *et al*. (1967) suggest that most of the dayas in that region are relict from as early as the Late Tertiary. Other work in North Africa by Alimen (1965) would place the largest karstic features as at least early Quaternary in age while Quinif (1983) has suggested, in a recent study from Hammamet in eastern Algeria, that karst development had ceased there by the end of a cold Tennsiftien period. Finally, Clark *et al*. (1974) have proposed that cyclic sedimentation may represent similar cyclical variations in inflow and local lowering of the depressions related to alternating arid and pluvial periods. At the present time, the sediment budget of the hollows is dominated by aeolian addition and deflation, of which addition is currently dominant.

MESO-SCALE FEATURES

In addition to macro-scale features, which have dimensions of hundreds of metres, there are a range of surface and subsurface features, the size of which can be expressed in metres, or at most tens of metres. These include:

Minor, Infilled Solution Hollows and Sinkholes: These are normally restricted to within 20-30m of the escarpment edge. In common with the larger dayas they are relatively rare and it is more usual to see a complex of solution widened joints near the hamada edge than a single concentrated sink. A typical infilled hollow is of the order of 5-10m across, but excavation has shown them to be only of the order of 0.5m deep. The infill is characteristically an undifferentiated, reddish brown fine sandy loam containing a few fragments of weathered limestone. Beneath the fill is a smooth, rock-cut surface criss-crossed by rounded, solution widened joints infilled with the same sandy loam.

The few sinkholes that have been observed show similar signs of current inactivity. The holes may be between 0.5-1.0m across, and the deepest hole observed reached down approximately 1.5m before it deteriorated into solution widened joints. There is little evidence of smoothing or rounding of the limestone that one would normally associate with an active swallet. In contrast the holes are angular, often irregular and the limestone surrounding them carries a cover of active and inactive micro-solutional features together with loose debris. The presence of this debris, the infill and other features suggests, in conjunction with the absence of marked channels draining into the hollows, that both phenomena are relict. They may even have originated beneath a former sediment or soil cover.

Karstic Caves and Pipes: An obvious extension of solution hollows and sinkholes are systems of subterranean solution pipes and caves. The latter are relatively common features of both the Ziz escarpment and tributary valleys. Two forms can be recognized. First, narrow (normally <1m x 1m) but deep (often >5m) caves which show signs of former solution by flowing water. Second, much more common 'solution recesses'. These are shallower (normally <2-3m) but much more open (up to 2m high and 10m across). They show no evidence of solution by concentrated flowing water (e.g. widened joints), but may have formed in association with slow, widespread seepage of moisture from the recess backwall. Seepage would not only cause direct erosion by solution but would encourage mechanical weathering through flaking and granular disintegration. In addition, this seepage has been responsible for the patchy deposition of flowstone in some recesses and the cementation of mechanically produced debris on some recess floors. No recent redeposition of limestone was seen in any of the recesses studied and many of the recess interiors carry an oxidised layer of red ferroan calcite up to 0.5mm thick. This, and the absence of active mechanical disintegration, has been taken to indicate the relict character of most deeper recesses (Smith, 1978). The only 'intermittently active' example of a solution recess is on the south facing valley side of the Wadi Jramna (Fig. 3). In this example, there is no clearly defined subterranean passage and, in common with many recesses, seepage appears to concentrate at a lithological boundary. Often this is where well jointed, massive limestone overlies flaggy, and perhaps less permeable, thin-bedded limestone.

More difficult to study, are the subterranean passages and pipes that feed into karstic caves. None of the caves are large enough to permit exploration and pipes are generally visible only in mass movement scars. The commonest pipes are small (normally <15 cm diameter) densely packed systems which give a vermiform appearance to a zone 2-3m below the hamada edge. This minor piping is not universal and its occurrence may be a factor of some, as of yet unidentified, geological control. Much larger cave systems are sometimes seen however, and an opportunity to study these is provided by a deep road cutting through the hamada approximately 2km southeast of Wadi Akerboûss. The karstic features exposed range from part of a large horizontal cave system in massive limestone (1.5m high and 8m long) to vertical pipes up to 5m high and 1m diameter. The base of the largest pipe opens out into an elliptical chamber approximately 2m wide which may once have been a scour pool. In addition, these are filled with a reddish brown clay infill. Similar fills have been excavated from the larger cavities, but residual amounts remain, often trapped in the honeycombed sides of the caves.

When examined in thin section the clay is seen to contain numerous pedo-relicts of sub-angular, blocky soil fragments up to 2mm across. These are often surrounded by oriented clay cutans, and appear similar to pedo- and litho-relicts described from cave infills in Israel by Goldberg (1979). It is presumed that they indicate an origin for the fill as a one-time soil cover on the hamada surface. This cover was then eroded and elements of it were washed into the cave systems. There is no evidence whatsoever of such a clay rich soil cover on the present hamada surface.

It has been suggested by Marker (1972) that the above sequence of events indicates soil formation during a relatively humid period, and erosion during the transition towards dryer conditions. The presence of a soil cover, combined with greater available water and a more extensive plant cover, might, as pointed out for example by Schmid (1963), have been able to supply the CO_2 required for development of cave systems. Similar waters rich in dissolved CO_2 could also have been responsible for the network of smaller pipes and solution widened joints in the massively bedded limestones. The age of the infill, and hence the date of effective fossilisation of the cave system is not known with any certainty. It has, however, been proposed by Butzer (1961) that during the Mindel/Riss interglacial, and to a more limited extent during the Riss/Wurm, the margins of the Mediterranean Basin were characterised by moist tropical conditions, and the formation of Terra Rossa soils. He also suggests equivalent soil development in the northern Sahara. This suggestion would conform with that made by Conrad (1969) to the effect that the Mindel/Riss climate of the northwest Sahara was non-arid , but is contrary to traditionally held views of 'interglacial aridity' embodied in the chronologies for southeast Morocco. This may perhaps lend further support to the need for a reinterpretation of these chronologies.

Tafoni: In addition to karstic caves there are numerous smaller caverns that are actively developing, principally through mechanical weathering. These can be grouped under the general name of 'tafoni' and consist of two types: basal and sidewall (Whilhelmy, 1958). Basal Tafoni occur at the junctions between free face and debris slope. It has been suggested (Dragovich, 1969) that they originate through accelerated weathering resulting from a concentration of direct precipitation (dew and possibly frost) near ground level. Basal tafoni are normally <1m high, 1-2m wide and 1-2m deep, and as shown in Fig. 5 they sometimes have a frontal 'lip' or 'hood' behind which they open out. It has been proposed that these hoods are

Fig. 5. Basal tafoni in massively-bedded limestone.

instrumental in tafoni formation, in that they represent hardened outer skins which are breached to allow excavation of weakened sub-surface zones (e.g. Wilhelmy, 1964 and Winkler, 1979). However, although this mechanism may explain some tafoni, it does not account for Tafoni occurring on rocks which lack outer rinds (e.g. Bradley et al., 1978). An alternative, and possibly more universal, explanation of tafoni growth is that once hollows are initiated (perhaps by exploitation of structually or lithologically weaker areas), they create a microenvironment in which weathering is further enhanced (Smith, 1978). Weathering enhancement is most probably a response to reduced diurnal temperature regimes and increased relative humidity within the tafoni (Dragovich, 1967 and 1981). In arid environments this favours direct precipitation and moisture absorption (especially near ground level), while still permitting its eventual evaporation. Repeated wetting/drying and heating/cooling will enhance a variety of weathering mechanisms including chemical decay (e.g. Conca and Rossman, 1985). However, in the non-crystalline, non-granular limestone, breakdown is primarily accomplished by flaking of cavern roofs and walls. Flaking may be further helped by salt crystallisation or hydration behind the flakes, but as of yet no evidence of salt weathering has been encountered in these tafoni. Debris from weathering accumulates on the tafoni floors, where it is further comminuted and eventually removed, most probably by aeolian deflation.

Tafoni development does not continue indefinitely and weathering is perhaps curtailed once they reach a size whereby moisture is no longer regularly precipitated and/or diurnal temperature regimes no longer permit thorough evaporation. This is illustrated by the manner in which shallow karstic recesses are being modified by cavernous weathering mechanisms, while deeper caves and recesses remain untouched (Smith 1978).

Sidewall tafoni are not restricted to cliff foot zones, but apparently originate through the initial exploitation of zones of weakness in the rock mass - wherever these occur. An example of such a zone could be where the massively bedded limestones are locally more fossiliferous. Alternatively a frequent location for sidewall tafoni is in the structurally weaker thin-bedded limestones.

Of all meso-scale weathering phenomena, the cavernous weathering of rock surfaces is by far the most widespread and consistently active under present-day conditions. Tafoni show no preferential location in terms of aspect or rock type, but are universally engaged in the localised backwearing of cliff faces.

MICRO-SCALE FEATURES

Considerable areas of the hamada surface, especially near the escarpment edge, comprise a poorly developed form of arid 'limestone pavement' (Fig. 6A). Clints and grykes are clearly visible, but there has been little joint widening. Most joint-bounded blocks retain sharp edges and joints contain infills of angular limestone fragments and fines derived from slopewash and aeolian addition. Upper surfaces of blocks, and to a lesser extent vertical faces of cliffs and large debris, exhibit a wide range of surface weathering features, many of which are apparently of solutional origin. A survey of the most prominent of these features is given below.

Rillenkarren (Solution Flutes): First, it must be emphasised that these are rare, and when found they are restricted to surfaces inclined between approximately 5-10°. In form they resemble the classic features described by Bögli (1960), and have a semi-circular cross section with sharp intervening crests (Fig. 6B). Most are linear features but some bifurcate and contain minature channels within the valleys formed by the karren. In both instances they are clearly the product of solution by water flowing over the surface of the limestone.

In a survey conducted between the Wadis Jramna and Akerboûss, Kerr (1983) measured rillenkarren with an average width of 27mm, depth of 12mm and length of only 55mm. This short average length may, as Dunkerly (1979) suggests, be a function of the relatively gentle slopes on which they occur or, as Bögli (1960) contends, it may reflect an inverse relationship between rillenkarren length and rainfall intensity. A contributing factor to thier shortness must, however, be the way in which many karren are traversed by fractures in the limestone blocks.

Similar rillenkarren have been described from the Namib Desert by Sweeting and Lancaster (1982). They deduced that karren were

Fig. 6. A. Poorly developed limestone pavement. B. Rillenkarren. C. Small-scale solution pitting. D. Localised desert varnish.

forming at the present time, primarily through surface solution by slow runoff of possibly saline moisture precipitated directly by advective fog. Such a consistent source of moisture is not available on the Hamada de Meski. Instead, rillenkarren are in some instances being destroyed by other features such as small solution pits. Likewise, the fracturing of flute covered blocks is more suggestive of flute destruction than active development.

Small Solution Pits: These are by far the most common of all micro-solutional forms in the area, and are estimated to cover some 75% of all bare limestone surfaces (Fig. 6C). A survey of solution pits near Wadi Akerboûss by Kerr (1983) identified differences in pit size between vertical surfaces (average diameter 14.5mm and depth 10.9mm) and horizontal surfaces (average diameter 9.4mm and depth 7.2mm). As of yet, no explanation has been found for these differences.

These pits have previously been termed 'rainpits' (e.g. Jennings, 1971), but it is clear from their occurrence on vertical surfaces that their origin has little to do with either the direct action of falling rain or persistence of trapped rainfall. A more recent interpretation is that based upon work in arid areas of Israel where similar pits are prevalent on limestone surfaces. Danin and Garty (1983) and Danin *et al*. (1983) have proposed that preferential weathering of pits may be caused by the activity of endolithic cyanobacteria which inhabit the upper 0.1mm of the weathered rock. These bacteria can survive in deserts such as the Negev at sites with less than 100mm of annual rainfall. This is possible because they can absorb moisture directly, especially from dew, but remain dormant when moisture is unavailable. The Negev has an exceptionally high dewfall (e.g. 190 days/annum with nightly dewfall >0.02mm at Avdat (Danin and Garty, 1983)), but even so, this microbiotic explanation is worthy of further study on the Hamada de Meski, especially as it provides a viable explanation of the perceived freshness of the pitting.

Solution pits will often coalesce to produce pan-like depressions beneath the original rock surface. This pattern of weathering is particularly prevalent on case-hardened limestone and leaves patches of case-hardening as miniature mesas and buttes. The case-hardening is itself quite common and may also have a partly biological explanation. Traditionally (e.g. Merrill, 1898) such crusts have been interpreted as resulting from the evaporation of moisture which has brought dissolved limestone to the surface from deeper within the rock. Recently, however, interest has been expressed by several workers (see the review by Viles (1984 p. 525)) in the possible role of microbial solution and precipitation of calcium carbonate and the formation of case-hardening. Certainly, case-hardening does produce a stable surface on which colonies of lichen occur. However, the fact that these surfaces are being destroyed by active pit development, implies either that case-hardening forms very slowly at present or is a relict feature. It is not possible to say which of these is correct, and clearly more work is required on the weathering environments, mechanisms responsible and the rates at which case-hardening forms.

Kamenitzas (Solution Pans): Occasionally on horizontal or near horizontal surfaces it is possible to see pans up to 50cm across and 5-10 cm deep. These have steep margins (which invariably overhang) and sometimes contain a thin layer of dried organic matter and limestone debris on a flat floor. Such pans originate through dissolution of limestone by water trapped in them during and after rainfall. Dissolution is greatest around the pan margins, whilst their downward extension is inhibited by accumulated debris and organic matter (Allen, 1982, p.232). Dissolution will only occur if there is solvent motion and Allen (1982) has suggested three possible mechanisms for generating this. First, convection due to cooling by surface evaporation, second, Taylor-Gortler instability flow of dense, solute rich water down the pan sides or third, through wind generated waves. Irrespective of the detailed mechanisms that operate, there seems little doubt that pans are filled with water after rainstorms and continue to develop, even if infrequently, at the present time.

Other Weathering Forms: As well as the four weathering forms already described there are innumerable individual features, which often reflect peculiarities in the underlying rock. These include, for example, irregularly shaped, irregularly spaced holes and tubes.

These may have formed as pipes and cavities within the bedrock, but they have been subsequently exposed and consequently partially destroyed by surface solution. Other forms include very smooth surfaces, perhaps related to solution by occasional sheetflow, and isolated patches of desert varnish on partly silicified limestone (Fig. 6D). However, the overall occurrence of any one feature is very restricted, and beyond demonstrating the role of solution in shaping rock surfaces their significance is likewise considered to be limited.

DISCUSSION: WEATHERING AND LANDSCAPE EVOLUTION

The principal weathering mechanisms and phenomena that have been described are summarised and placed in a landscape perspective in Figure 7. Within this landscape, weathering serves two specific roles: the comminution of rock to a size that can be entrained and transported by other processes, and the direct removal of material in solution. Through these agencies, weathering creates zones and lines of weakness that are either exploited by erosion processes or lead directly to landscape adjustments such as rockfalls. In addition, through its particular emphasis on the enlargement and extension of joint systems, weathering is important as a means of weakening the rock mass and controlling the production of large rock debris.

In terms of effecting landscape evolution it is not necessarily the variety of weathering phenomena that is important, but their spatial and temporal concentration. In this context, one of the first things to be noted about the present landscape, is that weathering appears to have had little influence upon its overall initiation. This was principally accomplished through erosion and incision by the O. Ziz, and was begun at least as early as the Villafranchian. We know this because of a late-Villafranchian conglomerate that crops out in the Tafilalt Oasis near Erfoud at a height below that of the Hamada de Meski (Margat, 1954). Since deposition of that conglomerate, the oasis has been lowered a further 140m, together presumably with the bulk of the incision by the O. Ziz through the Hamada de Meski. Joly (1962) is of the opinion that most erosion was accomplished during the Amirien and Early Saletien pluvial periods (see Table 1), and that the material removed was deposited beyong the Tafilalt Oasis as a series of vast alluvial fans. This contention is supported by his interpretation of the uppermost slumps in the Ziz valley (as well as talus flat-irons below the Hamada and the major alluvial surfaces around Er Rachidia) as Tennsiftien in age (Joly 1962 and 1965). Since then, there has been sporadic incision into, and only limited reworking of these deposits.

Once the landscape was initiated, weathering, in its broadest sense, has had a major influence upon the subsequent pattern of development. This development has been primarily by scarp retreat around the margins of the hamada. Despite the presence of large, shallow solution hollows there is little evidence of Quaternary lowering of the landscape, and Joly (1962) considers the hamada surface to be Moulouyen in age. Scarp retreat has been concentrated however, both

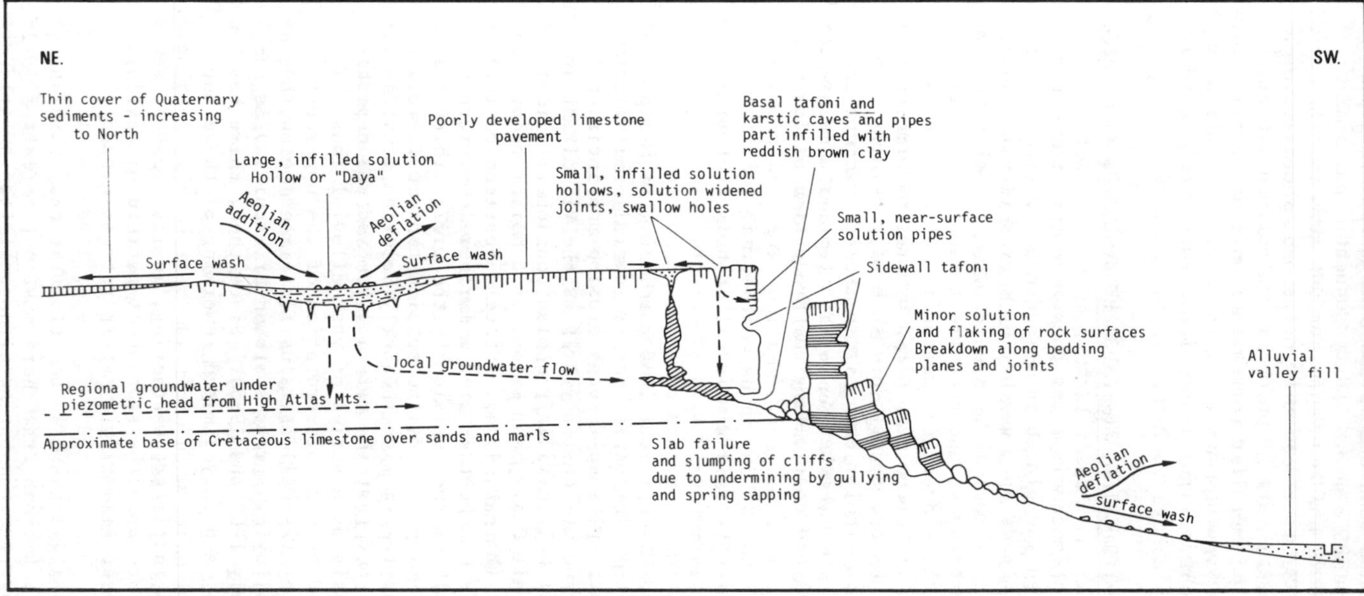

Fig. 7. Composite diagram illustrating the range of weathering features and geomorphological processes found on the Hamada de Meski.

in space and time, and has been accomplished by undermining and episodic collapse at specific points along the major escarpments. Between these points there has been only limited retreat. Undermining appears to have been largely the product of spring-sapping, and retreat has been influenced initially by the pattern of subterranean drainage fed by regional groundwater from the north. This retreat has led to the formation of a number of short, steep valleys along the main escarpment, whose further development has sometimes been enhanced when they manage to intercept local runoff.

Within these valleys, valley-side erosion has been achieved by widespread spring-sapping and collapse, largely by locally derived groundwater. The location of caves and seepage recesses has largely been determined by local geological factors. This, when combined with apparently unimpeded basal removal of debris from the slopes, has produced stepped valley-sides the form of which is dictated by geological structure.

There appear to be strong links, therefore, between the patterns of scarp retreat and karstic weathering of the hamada. However, the bulk of the evidence assembled points to the current inactivity of most macro- and meso-scale karstic phenomena. Regional groundwater no longer reaches most of the escarpment valleys, locally fed cave systems are infilled with eroded soil, shallow caves and recesses are undergoing modification by cavernous weathering, and large solution hollows are infilled and may be relict from as long ago as the Tertiary or early Quaternary.

Instead, the major contribution of active weathering in the present environment is through the extensive cavernous weathering of free faces and larger debris. This appears to be very efficient at producing rock fragments and fines, and if sufficient tafoni are concentrated on a cliff face they will gradually coalesce and cause it to wear back. Cavernous weathering does not, though, appear to be responsible for the undermining and collapse of large rock faces. This is demonstrated by the failure of cavernous weathering mechanisms to operate within the deeper karstic caves and recesses. On the hamada surface there is abundant evidence of active weathering. The principal weathering mechanism is, however, one of small-scale pitting, which is unlikely to make a major contribution to landscape development.

The overall impression, therefore, is of a landscape unit that contains, contrary to popular impressions of hot deserts, a wide range of karstic features. Karstic processes have played a major role in influencing landscape development, but that role was played out largely at times in the past, when water was more readily available. Under present conditions of aridity landforms are undergoing superficial modification by weathering, predominantly mechanical, but the landscape appears essentially relict.

Finally, it must be stressed that, as the title states, this has been a case study. Conclusions drawn here must not be extrapolated to other desert limestone areas without similar detailed field study. It is not, for example, the aim of the study to suggest that all karstic phenomena in all hot deserts are relict. Indeed, there are numerous observations of apparently active karst phenomenon in deserts such as the Negev (Yaalon - personal communication), and on a variety of lithologies (see Jennings (1983) for a recent review). Ultimately, active solution may depend on parameters such as rainfall duration, frequency, intensity and the size of individual storms, rather than the often meaningless values of mean annual rainfall that are normally used to describe aridity.

The main aim of the study has, however, been to illustrate an approach in which detailed weathering studies are no longer satisfactorily concluded until they have been placed within the broader spatial and temporal context of landscape evolution.

ACKNOWLEDGEMENTS

The writer is indebted to Colin Mitchell for his help in discussion and in the field and to Jim McGreevy for his comments on the original manuscript. Financial assistance for the work was supplied initially by the Natural Environment Research Council and latterly by Queen's University, Belfast. Thanks are also due to the technical and secretarial staff of the Geography Department, Queen's University (Maura Pringle, Gill Alexander, Trevor Molloy, Sharon Wright and Suzanne McWilliams), for preparing the diagrams and photographs and typing the paper.

REFERENCES

Allen, J.R.L., 1982. Sedimentary Structures, Vol. II, Elsevier, Amsterdam, 664p.

Alimen, H., 1965. The Quaternary era in the northwest Sahara. Geol. Soc. Amer. Spec. Pap., 84, 273-291.

Bögli, A., 1960. Kalklosung und Karrenbildung. Zeit. für Geom., 2, 4-21.

Bradley, W.C., Hutton, J.T. and Twidale, C.R., 1978. Role of salts in development of granitic tafoni, South Australia. J. Geol., 86, 647-654.

Butzer, K.W., 1961. Palaeoclimatic implications of Pleistocene stratigraphy in the Mediterranean area. Ann. N.Y. Acad. Sci., 95, 449-456.

Choubert, G., 1961. Quaternaire du Maroc. Biuletyn Periglac. 10, 9-29.

Choubert, G., 1965. Essai de correlation des formations continentales et marine du Pleistocene au Maroc. Notes, Serv. Géol. Maroc., t.15, 35-55.

Clark, D.M., Mitchell, C.W. and Varley, J.A., 1974. Geomorphic evolution of sediment-filled solution hollows in some arid regions. Zeit. für Geom., Suppl. 20, 130-139.

Conrad, G., 1959. Observations preliminaires sur la sedimentation dans les daias de la Hamada du Guir. C.R. Soc. Geol. Fr., t.1 (7), 156-162.

Conrad, G., 1969. L'évolution continentale post-Hercynienne du Sahara Algerien. Centre Nat. Réch. Sci., Paris, 527p.

Conrad, G., Gèze, B. and Paloc, Y., 1967. Phénomènes karstiques et pseudo-karstiques du Sahara. Rev. Géogr. Phys. Géol. Dyn., 9, 357-369.
Danin, A. and Garty, J., 1983. Distribution of cyanobacteria and lichens on hillsides of the Negev Highlands and their impact on biogenic weathering. Zeit für Geom., 27, 423-444.
Danin, A., Gerson, R., Marton, K. and Garty, J., 1982. Patterns of limestone and dolomite weathering by lichens and blue-green algae and their palaeoclimatic significance. Palaeogeog. Palaeoclim. Palaeoecol. 37, 221-233.
Douglas, I., 1980. Climatic geomorphology, present-day processes and landform evolution, problems of interpretation. Zeit für Geom., Suppl. 36, 27-47.
Dragovich, D., 1967. Flaking, a weathering process operating on cavernous rock surfaces. Bull. Geol. Soc. Amer., 78, 801-804.
Dragovich, D., 1969. The origin of cavernous surfaces (tafoni) in granite rocks of Southern Australia. Zeit für Geom., 13, 163-
Dragovich, D., 1981. Cavern microclimates in relation to preservation of rock art. Stud. Conservation, 26, 143-149.
Dunkerly, D.L., 1979. The morphology and development of rillenkarren. Zeit. für Geom., 23, 332-348.
Goldberg, P., 1979. Micromorphology of sediments from Hayonim cave Israel. Catena, 6, 167-182.
Goudie, A.S., 1983. Environmental Change 2nd Edition, O.U.P., Oxford, 258 p.
Jennings, J.N., 1971. Karst. M.I.T. Press, Cambridge, Mass., 252p.
Jennings, J.N., 1981. Morphoclimatic control: a tale of piss and wind or a case of the baby out with the bathwater. Proc. 8th Int. Cong. Speleol., 1, 367-368.
Jennings, J.N., 1983. The disregarded karst of the arid and semi-arid domain. Karstologia, 1, 61-73.
Joly, F., 1962. Études sur le relief du sud-est Marocain. Trav. Inst. Scient. Cherif., 10, 578 p.
Joly, F., 1965. Remarques sur l'emboîtment des formes Quaternaires continentales dans le sud-est Marocain. Notes. Mém. Serv. Géol. Maroc., t.25, 71-78.
Kerr, A., 1983. Rock temperature measurements from southeast Morocco, and their implications for weathering and weathering forms. Unpubl. Undergrad. Thesis, Queen's University, Belfast, Geography Dept., 136p.
Mabbutt, J.A., 1977. Desert Landforms. M.I.T. Press, Cambridge, Mass. 340p.
Mark, D.M., 1980. On scales of investigation in geomorphology. Can. Geogr., 24, 81-82.
Marker, M.E., 1972. Karst landform analysis as evidence for climatic change in the Transvaal, South Africa. S. Afr. Geogr. J. 54, 152-162.
Meigs, P., 1953. World distribution of arid and semi-arid homoclimates. Reviews of research on arid zone hydrology, UNESCO, Paris, 203-209.
Merrill, G.P., 1898. Desert Varnish. U.S. Geol. Surv. Bull., 150, 389-391.
Mitchell, C.W. and Willimott, S., 1974. Dayas of the Moroccan Sahara and other arid regions. Geog. J., 140, 441-453.

Quinif, Y. , 1983 . La reculée et le réseau karstique de Bou Akouss (hammamet, Algerie de l'Est) Geomorphologie et aspects evolutifs. Rév. Belge de Géogr., 107 , 89-111.
Reading University , 1969 . Expedition to Moroccan Sahara, Department of Geography, Reading University, 60p.
Schmid, E. , 1963 . Cave sediments in Prehistory. In Brothwell, D. and Higgs, E. (ed.) Science in Archaeology, a survey of progress and research. Thames and Hudson, London, 720p.
Smith, B.J. , 1977a . An introduction to the landscape evolution of the Tafilalt Oasis region of southeast Morocco. Reading University Geographical Papers, 53 , 22-30
Smith, B.J. , 1977b . A review of climatic geomorphology and Quaternary climatic change in the northwest Sahara. Reading University Geographical Papers, 53 , 1-21.
Smith, B.J. , 1978 . The origin and geomorphic implications of cliff foot recesses and tafoni on limestone hamadas in the northwest Sahara. Zeit. für Geom., 22 , 21-43.
Sweeting, M.M., and Lancaster, N., 1982 . Solutional and wind erosion forms on limestone in the central Namib Desert. Zeit. für Geom., 26 , 197-207.
Viles, H.A. , 1984 . Biokarst: review and prospect. Prog. Phys. Geogr., 8 , 523-542.
Warren, A. , 1985 . Arid Geomorphology. Prog. Phys. Geogr. 9, 434-441.
Wilhelmy, H., 1958 . Klimamorphologie der Massengesteine, Braunschweig, 283p.
Wilhelmy, H., 1964 . Cavernous rock surfaces, (tafoni) in semi-arid and arid climates. Pak. Geogr. Rev., 19 , 9-13.
Winkler, E.M. , 1979 . Role of salts in development of granitic tafoni, South Australia. A discussion. J. Geol., 87 , 119-120.

GLACIAL GEOMORPHOLOGY

International Geomorphology 1986 Part II
Edited by V. Gardiner

GLACIAL GEOMORPHOLOGY

David Sugden

Department of Geography,
University of Aberdeen,
Aberdeen,
Scotland.

The one-day session on Glacial Geomorphology included 13 papers and 15 posters. The presentations came from thirteen countries, namely the USA, Canada, China, Australia and nine European countries (Belgium, Denmark, Finland, France, Italy, Poland, Spain, Sweden and the UK). The day was subdivided into themes on glacial processes, glacial landforms (erosional and depositional) and models of deglaciation.

The papers in this section of the symposium proceedings represent a selection of papers and posters presented in Manchester. M. Lassila (Sweden) draws attention to previously unnoticed large trough forms associated with ice flow around hills. Two Polish papers by E. Drozdowski on surge moraines and by S. Kozarski and L. Kasprzak on facies analysis highlight the enormous benefits of testing glaciological theory against the landform and sediment record. They contribute to our understanding of the dynamics and basal thermal regime of the southern margin of the North European ice sheet during the Pleistocene. H. Zilliacus (Finland) reviews the origin of a long-puzzling landform, the De Geer moraine, while R.P. Bourman (Australia) adds insight into the nature and geomorphic effects of Late Palaeozoic glaciers in Australia. J. Gordon *et al* study glacier fluctuations in the plateau environment of northern Norway and assess the implications for models of deglaciation.

Some excellent papers on glaciological processes presented at the conference have been published elsewhere. They involve the effects of glacier surging by M. Sharp (UK) and R. Anderson (USA), the origin of basal ice beneath Alpine glaciers by R. Lorrain and R. Souchez (Belgium), nivation in Sweden by A. Rapp (Sweden) and push moraines in Iceland by J. Kruger (Denmark).

We are all very grateful for the enthusiasm of all participants at the session, to the referees of the papers and to W. Theakstone who did so much to organize the session.

International Geomorphology 1986 Part II
Edited by V. Gardiner

GLACIAL TROUGHS CAUSED BY STEEP ROCK FORMS

M. Lassila

Department of Geography, University of Umeå,
S-901 87 Umeå, Sweden.

ABSTRACT

Glacial processes are more intensely studied than glacial forms. Some depressions created close to protruding rocks are described in order to demonstrate the lack of information linking process and form. The above landforms, virtually ignored before, have been studied by means of air-photo interpretation, field measurements and soundings, predominantly in northern Sweden.

Strictly proximal troughs are found in front of square-shaped hills or other glacially abraded, positive landforms, where the inclination of the proximal slopes exceeds 30-35°. In front of more streamlined hills proximal furrows without overdeepening are formed. Lateral troughs also exist.

The troughs mentioned above seem to be quite frequent in granitic, hilly terrain characterized as a knock-and-lochan topography (Linton, 1963). In the Caledonian mountain region of Scandinavia however, as well as in Greenland, there are no well-defined troughs, possibly due to a lower velocity of the basal ice and/or shifting ice divides in that zone. It is suggested that further investigations of the glacial sculpturing of both positive and negative landforms will lead to a more definite estimation of various ice movement directions, their succession, strength and duration.

INTRODUCTION

The last decades have been characterized by a considerable amount of achievement as regards the theories concerning glacier sliding and other related problems. Many authors have made valuable contributions to these fields of research (Boulton, 1964, Kamb, 1964, Lliboutry, 1968, Nye, 1952, 1973, Weertman, 1957, 1964). However, the links between process and form are not yet clear. Sugden and John (1976, p 170) state that 'one of the serious gaps seems to be the lack of accurate observations on form. Without this, theoreticians are deprived of one source of highly significant information with which to constrain their theories'.

This gap is illustrated by the case of ice flowing around obstacles of various sizes, a problem that has been discussed for a long time. The morphological results of the process have been only vaguely pointed out, however. The present paper is an attempt to reduce the lack of information linking process and form. Another gap is disclosed by the fact that although the positive landforms of glacial origin have been defined and described in detail, the negative landforms have not. Therefore, some hitherto virtually ignored negative glacial landforms created close to steep obstacles have been described here. The aim has also been to discuss the parameters that have affected the formation of these landforms. These facts, plus knowledge of the regional distribution of the forms, might establish their relation to various ice movements and possibly also contribute to the determination of their age.

PREVIOUS WORK

Glacial troughs associated with steep rock forms were first observed and described by Hall (1815) in Scotland. The 'hollow valleys' close to crag and tail ridges were created by high floods, he stated. A correct glacial origin was proposed for similar forms on the stoss side of rock drumlins in New Zealand (Park, 1910). The troughs were first noted in a textbook by Charlesworth (1957). Heidenreich (1964) briefly mentioned 'horseshoe-shaped depressions' in front of drumlins in Canada. Recently 'crescentic furrows or troughs' were mentioned by Aario (1977a) in Finland and treated in greater detail by Markgren and Lassila (1980) in Sweden. According to Lassila (in press) 'proximal troughs' seem to indicate the ice movement of the continental ice sheet at its maximum.

TERMINOLOGICAL REMARKS

Most authors using the term 'glacial troughs' refer to valleys or fjords created by ice flowing in channels (Linton, 1963, Sugden, 1968, Sugden and John,1976). A landscape of selective linear erosion is thus formed. In contrast, glacial erosion over the entire land surface produces a landscape of areal scouring, called a 'knock and lochan' topography (Linton, 1963). The irregular depressions restricted and determined by joints, faults and dykes are mostly referred to as 'glacial basins'. Troughs in the sense meant by the present paper, were first regarded as an unnamed negative counterpart to crag and tail ridges (Charlesworth, 1957).

Landform terminology should be well-defined genetically and also be descriptive, if possible. In this case, however, no adequate term has gained footing for the forms in question. The words proximal and lateral together with troughs indicate depressions that are the result of ice flowing around obstacles and might thus be applicable. A 'proximal trough' should therefore be an overdeepened depression on the stoss side of a positive rock form, while a 'lateral trough' extends along a flank (Fig. 1 a-b). Sometimes troughs extend both along the proximal and lateral sides of a rock knob. A 'crescentic trough' is thus formed (Fig. 1c). In the same position, slight erosion without overdeepening creates a 'crescentic furrow', with an even bottom (Fig. 1d). Glacially moulded depressionssituated between two rock drumlins exist too, but are not as yet described and therefore have no definite term associated with them.

ICE FLOW AROUND OBSTRUCTIONS ON THE GLACIER BED

Flowing ice passes over and around an obstacle in two ways. The stress against the stoss side leads to pressure melting, followed by transport of meltwater to the lee side, where it refreezes. This process does not work when the ice is permanently cold-based, or when the bumps on the glacier bed are bigger than about one metre (Weertman, 1957). The process is important only when the latent heat released on the lee side can be transferred back through the bump and assist in further melting on the stoss side. For the formation of troughs, pressure melting and regelation play a subordinate role.

Of greater importance for the formation of glacial troughs is the plastic flow of ice, the so-called enhanced basal creep (Weertman, 1957). The stress induced by the obstacle leads to a multiplied increase in strain rate. The velocity of the ice is proportional to the strain rate times the distance from the obstacle. The stress concentrations are extended over a distance from a cubical obstruction of the order of the obstacle size (Weertman, 1979, cf. also Carol, 1947). The sliding velocity of the ice increases with the obstacle size provided the roughness of the bed is not changed (Weertman, 1979).

Troughs close to steep rocks are formed by that plastic flow. The ice is thus diverted around the obstruction so that debris-rich, erosive, basal ice flows along the lateral flanks, while ice depleted in debris flows over the summit (Boulton, 1974, 1979). Evidences for this process include observations from under recent glaciers and the rather good preservation of old striations on the upper parts of roche moutonnées, which is a feature of large parts of northern Sweden.

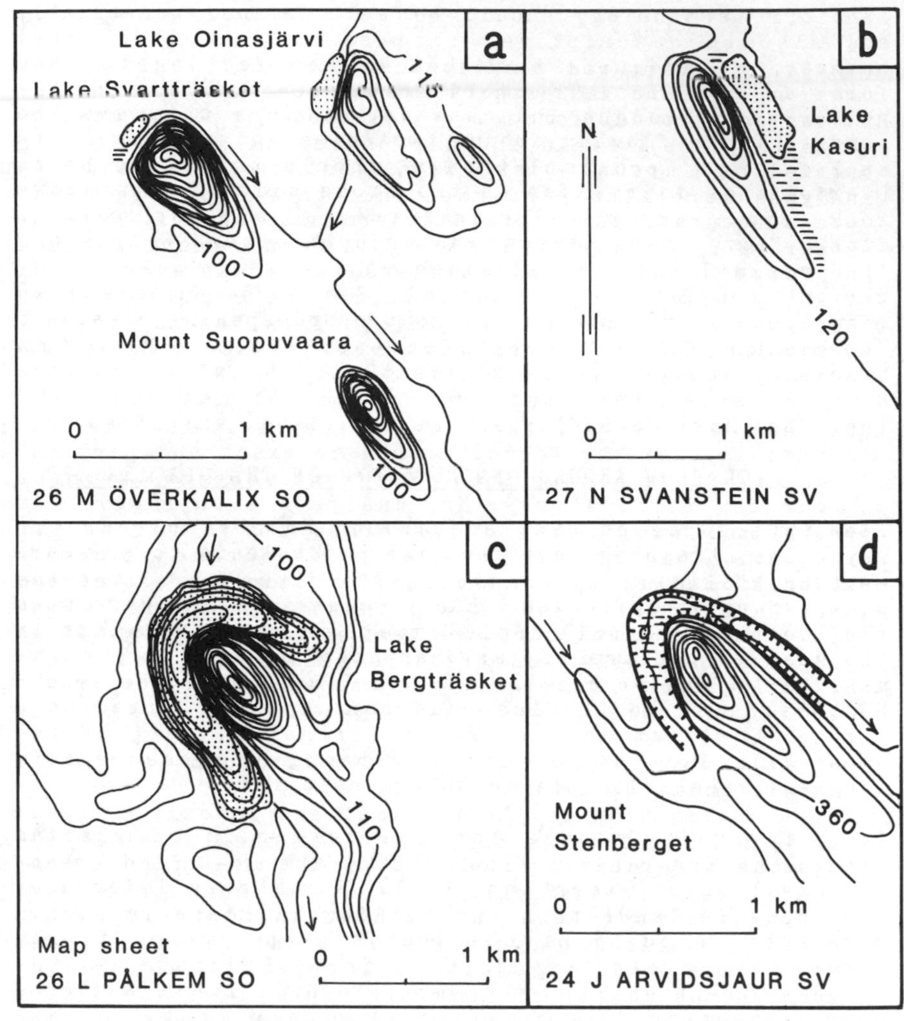

Fig. 1. Various trough forms in N. Sweden and Finland. Ice movement direction was approximately from NW. The contour interval is 5 m. a/ Lake Svartträsket and Lake Oinasjärvi are proximal troughs in front of steep mountains. No trough is formed close to Mount Suopuvaara due to its streamlined form and gentler slopes. b/ Lake Kasuri and adjoining peat lands form a lateral trough close to the steeper eastern slope. c/ A steep-sided mountain is almost circumscribed by a crescentic trough, Lake Bergträsket. The water depth is 12.5 m. d/ In front of the steep but streamlined landform, Mount Stenberget, a crescentic furrow is found.

RESULTS

Glacially sculptured positive landforms of various sizes and compositions in northern Sweden were surveyed on air-photos and typical forms were selected for field measurements (Fig. 2b). Field and air-photo analyses were also carried out around Disko Bay, western Greenland. Map analyses were also performed, but the topographic maps were inaccurate for profile determinations (cf. Rose and Letzer 1975, Goudie, 1981, p 62). A requirement was that the forms should be solitarily situated in even or only slightly undulating ground, where the continental ice sheet was characterized by uniform flow, as far as could be judged. For the most part troughs accompanied the positive landforms.

The main difference between measurements in the field and on maps is that the former method is exact and permits a division of the slopes into different vertical segments of varying inclinations. Map analyses always give more generalized profiles and therefore lower values. The inclination figures below refer to measurements on rock bodies of medium size, which differ from perfect streamlined landforms in such a way that the proximal slopes form plane inclined surfaces facing the ice flow. If the landforms show blunt or more streamlined proximal slopes, the inclination figures must be raised before a furrow or a trough is formed.

The field work shows that positive landforms almost always are built up by till and have no visible rock core if the inclination of the proximal and lateral slopes is less than 25°. This is only valid for the lowest 20-30 m above the base plane. In higher levels the bedrock crops out at lower inclinations. Till-covered summits of large drumlins are therefore rarely seen in northern Sweden. Slope inclinations between 25-30° mean that these areas do not have any lodgement till cover at all, only a discontinuous veneer of flow till in some places. Close to rock drumlins or drumlinized mountains steeper than 30-35°, furrows or troughs are found (Fig. 3). Smoothly rounded landforms seem to be accompanied by proximal furrows extended along the lateral flanks, whereas more square-shaped forms have resulted in deep troughs.

Unfortunately, there are very few crescentic furrows in Sweden, although this landform seems to be more frequent in the northeastern part of Finland (Aario 1977b). Its parameters have been studied very little by the present author. On the other hand, proximal and lateral troughs are quite common in Sweden. Soundings have been made in about 80 water-filled and 40 sediment-filled basins predominantly in northern Sweden and on the island of Gotland in the Baltic (Fig. 2). There appears to be a clear correlation between the increasing height, breadth, and steepness of rock drumlins, and the increasing depth of the troughs (Fig. 3). A difference between the two types

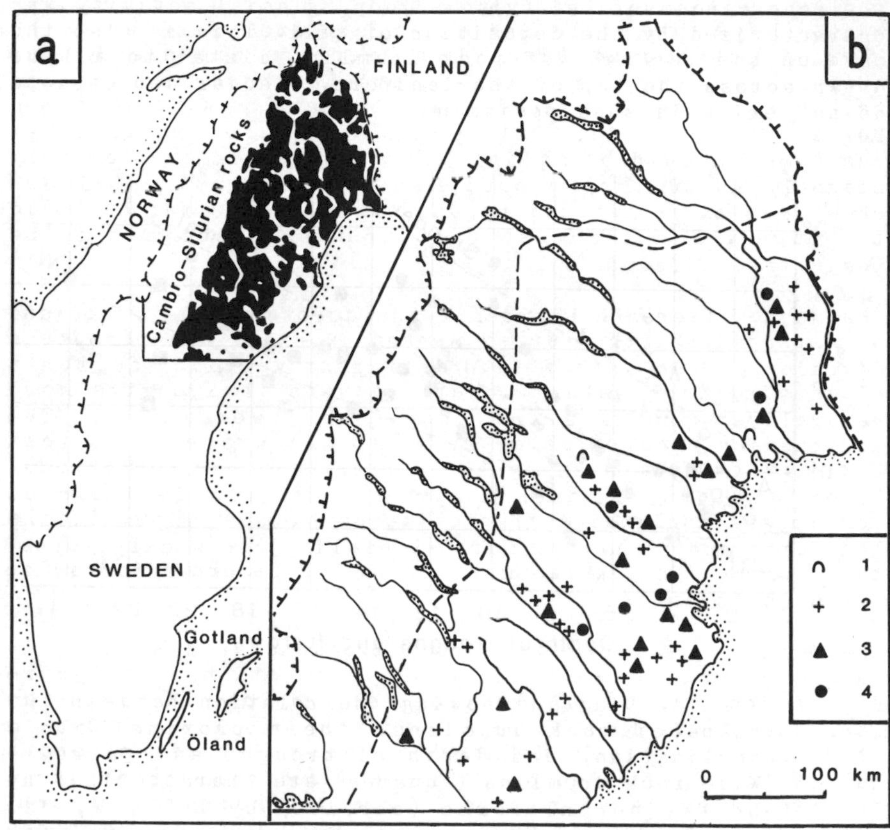

Fig. 2. The location of well-sculptured trough forms in N. Sweden. A single symbol may represent several troughs. 1/ Crescentic troughs, 2/ proximal troughs, 3/ lateral troughs, 4/ crescentic furrows. In the beginning the ice divide was situated in the Caledonian mountains close to the Norwegian border, but moved eastwards to the dashed line, according to striations (b). No field evidence for more easterly divides exist. A broad zone close to the mountains shows long, deep, and narrow lakes, indicating linear erosion by valley glaciers. Black colour indicates granitic bedrock. The formation of troughs seems to be controlled by tectonic conditions rather than petrographic.

of troughs is that the proximal type is almost twice as deep as the lateral one. When a contour line is drawn around both the positive landform and the negative basin a perfect lemniscate loop is formed (Fig. 1b). Outside the zone circumscribed by this loop, glacial activity was characterized by the deposition of till. This is also the case on the lee side of rock drumlins distal to a line drawn across the top of the landform. Glacial depressions do not exist in such positions.

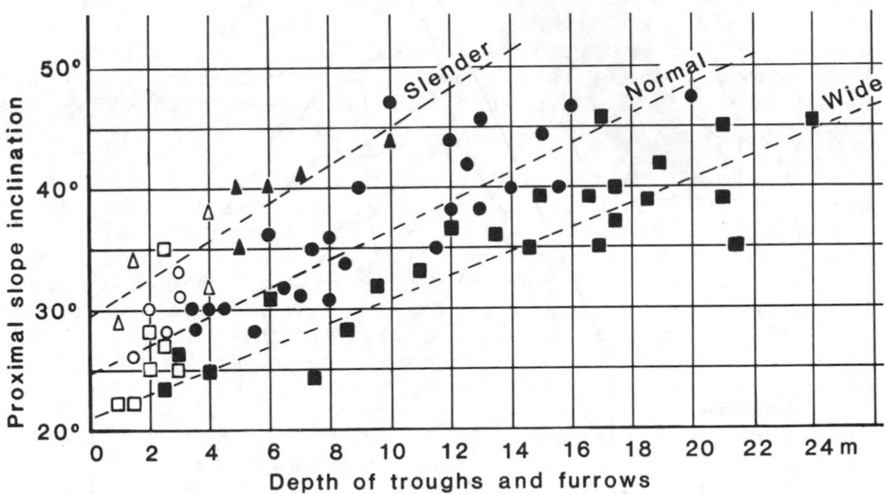

Fig. 3. Diagram showing the relation between the shape of rock drumlins, their proximal slope inclination, and depth of troughs and furrows. Wide rock drumlins (squares) are characterized by L<2·W, normal forms (circles) by L=2-4·W, and slender forms (triangles) by L>4·W. Filled symbols represent troughs, open symbols furrows.

According to the soundings, the bottom of the troughs is always rather smoothly abraded. The greatest depths are found close to the steepest parts of the positive landforms (Lassila, in press). Whether the troughs are overdeepened basins in the bedrock or not has not yet been resolved satisfactorily. The easily erodible limestones of Cambro-Silurian age in Gotland, southern Sweden, show well-developed overdeepened basins in front of protruding hills of harder limestones (Lassila, in press). Such basins were also noticed in front of extremely smallscale features in limestone on the island of Öland, southern Sweden (Svensson and Frisén, 1964). Similar minor troughs in Precambrian rocks exist too, as can be attested by personal observations in photos and in the field. It has not been definitely established if there is overdeepening in large-scale troughs, because borings or seismic investigations are necessary in such cases.

For the most part, glacial troughs are cut in bedrock on the stoss side of steep hills and in till on the others (Lassila, in press). Till and sometimes small quantities of glaciofluvial or even littoral sediments are found on the bottom of these forms. Manual soundings are therefore complicated to carry out and assess. Provided the depths of the troughs are more than about 10 metres the postglacial sedimentation of gyttja, detritus and silt has been rather uniform (Lassila, in press). Today's sediment surface therefore reflects the glacially moulded surface below. Small ponds that remain today in basins that have been almost filled-up are thus situated right above the deepest parts of the original troughs.

The regional distribution of the landforms discussed above depends on several factors, i.e. the occurrence of steep rock forms, the distance from the ice divide, and the flow characteristics of the ice. In even or only slightly undulating ground there are no well-developed troughs, whereas in hilly terrain outside the ice divide zone the landforms are frequently found (Fig.2b). The bedrock in this area mostly consists of granites (51 %), while gneisses, porphyry and other volcanic rocks cover 37 % and sedimentary derived rocks only 12 % (Fig. 2a). Therefore, the petrographic conditions for the creation of troughs are favourable. The most important controlling factor seems to be the tectonic features, above all vertical jointing and sheeting structures dipping downglacier. Evidently, in areas where the ice flow was uniform or extending, the landforms seem to be more frequent and well-sculptured.

DISCUSSION AND CONCLUSIONS

A meaningful discussion of the formation of glacial troughs is possible only if a sufficient number of almost perfectly moulded rock drumlins, situated in even terrain, and of varying size and steepness, are studied. In spite of certain difficulties in finding these landforms accompanied by troughs and in establishing their glacial form, the survey has thus far attained some promising results. This can be demonstrated for instance by the good correlation between slope inclination and the depth of the troughs (Fig. 3). Furthermore, it has been possible to predict not only the position but also the approximate figure for the maximum depth of troughs in northern Sweden without previous field investigations.

The glacial moulding of the positive landforms and troughs belonging to these eminences is controlled by several parameters, of which only a few, connected to the shape of the rock bed and the flow characteristics of the ice, are of special interest in this study. Low, streamlined bedforms cause only a slight resistance to the overflowing ice, a fact demonstrated by the conformal sequences of till beds in drumlins and other moraines. Another evidence is the slight erosion of the bedrock surface in level ground. An increased inclination of the up-glacier slopes results in a gradual transition from depositional to erosional conditions, giving rise to bare bedrock forms and also erosion outside the obstacles, i.e. formation of troughs. According to the creep theory (Weertman, 1957, 1979) the velocity of the overflowing ice increases with the strain rate, which is multiplied close to square-shaped and large obstacles on the glacier bed. There is also a general increase in the flow velocity of an ice sheet in down-glacier slopes and in higher levels above the ground. Since depositional and erosional processes show good correlation with form parameters even in the case of rather small obstacles, the flow velocity of a warm-based, possibly also a cold-melting ice sheet seems to be the most important factor controlling the sculpturing of the substratum. The importance of the temperature gradient between troughs and summits for the deposition of till, proposed by Nobles & Weertman (1971), may have a certain significance concerning large bedrock forms only. In summary, under a warm-based uniformly flowing ice sheet or where pressure melting occurs, topography strongly affects the depositional and erosional processes at the interface.

The observations above also support the idea that the ice will mould streamlined bodies, both positive and negative, giving the least resistance to the overflowing ice (cf. Tarr, 1894, Chorley, 1959, Smalley and Unwin, 1968, Boulton, 1974). This is fully achieved only where the initial form was suitable and the flow direction of the ice was uniform most of the time. Another idea is that proximal slopes of moraines steeper than 25° should be deposited either by an ice characterized by compressive flow or in dead ice environments. There are also great similarities between flowing ice, running water and wind, as regards flow patterns and resulting forms (Mattsson, 1976, Boulton, 1982), which may have a certain significance when discussing glacial sculpturing.

The negative landforms mentioned above, close to steep rocks, have been found neither in the Caledonian mountains in Sweden, where the ice divide was situated during most of the Weichselian glaciation, nor in the vicinity of the inland ice sheet in Greenland (Fig.2b). The morphological results of the ice divide moving eastward outside the Caledonian mountains during the maximum and late stage of the glaciation have been neglected by most scientists and the present state of knowledge is therefore weak (Rudberg, 1954). Provided the ice flow was sufficiently vigorous and long-termed, formation of troughs of this type could be expected. However, the zone seems devoid of these landforms, which possibly indicates only slight erosion in proximal positions, or perhaps that the tectonic/petrographic properties of slightly inclined strata and soft bedrock were unfavourable for creating steep forms. The succeeding ice flow, coming from an opposite direction than that of the previous phase, might also have destroyed the troughs by depositing till in lee-side positions. No definite evidence for either opinion has been found as yet, however.

Glacial troughs of the types discussed above always seem to exist in close connection with drumlin fields (cf. Aario and Forsström, 1979). The ice movements creating these troughs may be somewhat difficult to determine, since the landforms represent the morphological sum of several glaciations, with the present sculpture dating from the last Weichselian phase. A survey of troughs in various parts of Sweden has revealed a good accordance between the ice movements forming these landforms, adjacent drumlin fields, and a roughly striated bedrock (Lassila, in press). Later ice movements were rather insignificant, producing new striations but only a slight retouching of the previous morphology. Further investigations of the glacial sculpturing of both positive and negative landforms of different size, combined with striae and fabric analyses, will lead to a more definite estimation of various ice movement directions, their succession, strength, and duration.

ACKNOWLEDGEMENTS

Grateful thanks are due to Prof. Lennart Strömqvist and Mr Hans Ivarsson for their valuable comments on the manuscript and to Mrs Jessie Karlén for correcting the English.

REFERENCES

Aario, R., 1977a. Classification and terminology of moraine landforms in Finland. Boreas, 6, 87-100.

Aario, R., 1977b. Associations of flutings, drumlins, hummocks and transverse ridges. GeoJournal, 1, 6, 65-72.

Aario, R., and Forsström L., 1979. Glacial stratigraphy of Koillismaa and north Kainuu, Finland. Fennia, 157, 2, 1-49.

Boulton, G. S., 1964. Processes and patterns of glacial erosion. In Coates, D.R., (Ed.) Glacial geology, State Univ. of NY, Binghamton. pp 41-87.

Boulton, G. S., 1979. Processes of glacier erosion on different substrata, J. Glaciol., 23, 15-37.

Boulton, G. S., 1982. Subglacial processes and the development of glacial bedforms. Research in glacial, glacio-fluvial and glacio-lacustrine systems, Geo Books, Norwich, pp1-31.

Carol, H., 1947. The formation of roche moutonnées. J. Glaciol., 1, 57:59.

Charlesworth, J. K., 1957. The Quaternary Era, E. Arnold, London, 2 vols, 1700 pp.

Chorley, R. J., 1959. The shape of drumlins, J. Glaciol., 3, 339-344.

Goudie, A., (Ed.) 1981. Geomorphological Techniques, George Allen & Unwin, 395 pp.

Hall, J., 1815. On the revolutions of the earths surface, Tr. Royal Soc. Edinb., Vol 7, 171-199.

Heidenreich, C., 1964. Some observations of the shape of drumlins. Can. Geogr., VIII, 101-107.

Kamb, B., 1964. Glacier geophysics, Science 146, 353-365.

Lassila, M., in press. Ice movements and proximal troughs in Gotland, S. Sweden, Zeitschrift für Gemorphologie.

Linton, D. L., 1963. The forms of glacial erosion, Trans. Inst. Br. Geogr., 33, 1-28.

Lliboutry, L., 1968. General theory of subglacial cavitation and sliding of temperate glaciers, J. Glaciol., 7, 21-58.

Markgren, M., and Lassila, M., 1980. Geomorfologi i valda delar av Vindelälvsområdet, GERUM, A 26, Umeå, 63 pp. (With an English summary).

Nobles, L.H., and Weertman, J., 1971. Influence of irregularities of the bed of an ice sheet on deposition rate of till, In Goldthwait, R.P. (Ed.) Till, a symposium, Ohio State Univ. Press, 117-126.

Nye, J. F., 1952. The mechanics of glacier flow, J. Glaciol., 2, 82-91.

Nye, J. F., 1973. The motion of ice past obstacles. In Whalley, E., Jones, S.J., and Gold, L.W., (Eds.), Physics and chemistry of ice, R. Soc. Can., Ottawa, 387-395.

Park, J., 1910. The geology of New Zealand, Whitcombe and Tombs, Melbourne and London, 488pp.

Rose, J. and Letzer, J. M., 1975. Drumlin measurements: a test of the reliability of data derived from 1:25 000 scale topographic maps. Geol. Mag., 361-371.

Rudberg, S., 1954. Västerbottens berggrundsmorfologi. Geographica 25, 457 pp.

Smalley, I.J., and Unwin, D.J., 1968. The formation and shape of drumlins and their distribution and orientation in the drumlin fields. J. Glaciol., 13, 255-264.

Sugden, D. E., 1968. Landscapes of glacial erosion in Greenland and their relationship to ice, topographic and bedrock conditions. Polar Geomorphology, 4, 177-193.

Sugden, D. E., and John, B. S., 1976. Glaciers and landscape, 376 pp. E. Arnold.

Svensson, H., and Frisén, R., 1964. Hällmorfologi och isrörelser inom ett Alvarområde vid Degerhamn. Svensk geografisk årsbok, 40, 19-30.
Tarr, R. S., 1894. The origin of drumlins. Amer. Geologist, 13, 393-407.
Weertman, J., 1957. On the sliding of glaciers. J. Glaciol., 3, 33-38.
Weertman, J., 1964. The theory of glacier sliding. J. Glaciol., 5, 287-303.
Weertman, J., 1979. The unsolved general glacier sliding problem. J. Glaciol., 23, 97-111.

International Geomorphology 1986 Part II
Edited by V. Gardiner

SURGE MORAINES

Eugeniusz Drozdowski

Polish Academy of Sciences,
Institute of Geography and Spatial Planning,
Department of Lowland Geomorphology and Hydrology,
87-100 Toruń, Kopernika 19, Poland.

ABSTRACT

End moraines and hummocky moraines occurring in the northern part of the lower Vistula region, south of the Baltic Sea, were formed by ice lobes which surged across the temporary Gdańsk ice-dammed lake. The end moraines consist of closely-spaced arcs of push ridges, frequently containing material derived from the floor of the former proglacial lake situated up-glacier from them. Hummocky moraines, situated behind the end moraines, are generally twofold in their structure and origin. Pushed and extruded masses of lake sediments commonly form their cores, indicating the initial stage of their formation, whereas superimposed glacio-aqueous sediments and meltout tills redeposited by sediment gravity flows relate to the subsequent stage associated with meltout of stagnant ice. The suggested surge origin for the described former end moraines and hummocky moraines seems to be justified by the occurrence of comparable features in the forefields of the modern Svalbard glaciers, Sefströmbreen and Aavatsmarkbreen, which are known to have surged across fjord or bay sediments.

INTRODUCTION

It has long been recognised in areas of Pleistocene glaciation that arcuate ridges of push moraines are associated with hummocky moraines. However, the origin of these two types of landform is generally interpreted in different ways, without attempts to consider in more detail the possible glacio-dynamic interrelations and the effects of drainage conditions beyond the ice margin on the geomorphological activity of the glacier.

Field studies recently carried out in the northern part of the lower Vistula region, northern Poland (Fig.1), submerged during the retreat of the Scandinavian ice sheet, point out that the processes involved in the formation of these landforms may have been associated with surging lobes of the retreating ice sheet. The primary objective of this paper is to present evidence for the genesis of these features, with an emphasis on the less known subglacial hummocky moraines. Finally, an attempt is made to generalise the described characteristics of the landforms in an interpretative genetic model.

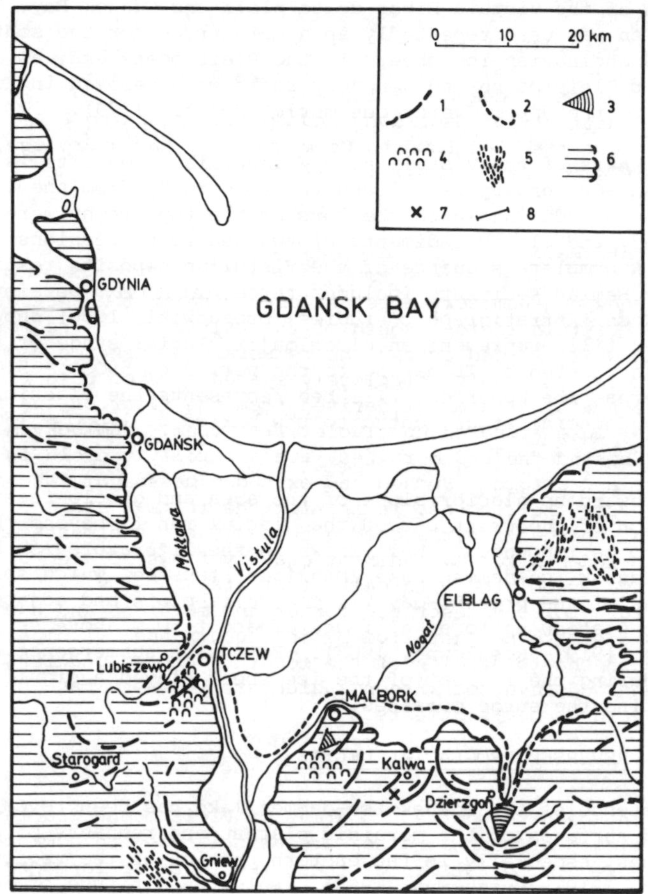

FIG.1. Map of the northern part of the lower Vistula region showing positions of the retreating ice margin and drumlin swarms after Roszkówna (1963), and location of features and geological cross-section discussed in text.

1. end moraines of the Cashubian-Warmian Substage (a subordinate stage of the Pomeranian Stage); 2. possible line of the retreating ice margin; 3. reversed proglacial deltas; 4. surge hummocky moraines; 5. drumlin swarms; 6. morainic plateau; 7. large thrust raft of Senonian marl; 8. cross-section through the field of surge hummocky moraines near Tczew.

GEOLOGY OF THE AREA

The landforms under consideration are located on a morainic plateau, close to the southern border of the tectonically controlled depression of the Vistula River delta-plain and Gdańsk Bay. This depression was used repeatedly as a main track for the southward-moving Scandinavian ice sheets in the Pleistocene and, as a result, the older Pleistocene and Cenozoic rocks were removed in its axial part up to the Upper Cretaceous marls (Mojski, 1979).

The thickness of the Vistulian (= Weichselian) deposits is dependent on the sub-Quaternary relief and the topography from the last Eemian Interglacial, during which the lows of the topography were lined with sandy and clayey sediments of two sea transgressions (Makowska, 1979). A complete sequence of the Vistulian deposits resting on the boundary Eemian sediments includes three laterally persistent till beds. Recent stratigraphic studies (Drozdowski, 1986) show that the two lower tills represent an older major glacial stage of the Vistulian related by TL dating to the period c. 59000 - 51000 years ago, whereas the uppermost till bed represents the classical Late Vistulian glacial stage, dated to the period c. 17000 - 15000 years ago.

The northward up-glacier slope of the area and glacio-isostatic depression of the crust caused the glacial runoff beyond the ice margin to be impounded. Well-known in the literature is the so-called Gdańsk ice-dammed lake (Danziger Staussee) which spread over the areas of the present-day Vistula delta-plain and adjacent morainic plateau up to an altitude of about 50 m above sea level (Sonntag, 1919; Roszkówna, 1963). Extensive submergence of the region during the retreat of the ice sheet played a significant role in creating the surge moraines.

REVERSED PROGLACIAL DELTAS

The existence of the Gdańsk ice-dammed lake has been revealed by the terraced surfaces of the morainic plateau and remnants of displaced shorelines, especially in the eastern portion of the region (Sonntag, 1919; Roszkówna, 1963). Below an altitude of 50 m previously unrecognised glaciofluvial deltas were also preserved, delineating accurately the extension of the lake. The largest delta is in the central part of the Dzierzgoń lobe about 30 m above sea level (Fig.2). At present, it is dissected by the Dzierzgoń stream entering the Vistula delta-plain (Fig.1). Another small but exceptionally well-exposed glaciofluvial delta occurs south-east of Malbork. The top facies and part of the foreset facies of this delta were truncated by flow till moving down the delta slope from the adjacent vanishing ice mass (Figs. 3 and 4). Since the glaciofluvial deltas do not mark the glacier margin, but the lake shoreline beyond the margin, it is proposed to call them 'reversed proglacial deltas'. The top of their foreset facies may be regarded as a good indication of the water level in the former lake.

FIG.2. Sandy-gravelly deposits of a reversed proglacial delta at Dzierzgoń. Note the characteristic foreset facies and capping topset facies in the upper background.

FIG.3. Reversed proglacial delta truncated and buried by flow till, located southeast of Malbork. The top surface of the foreset facies indicates the water level during deglaciation of the area.

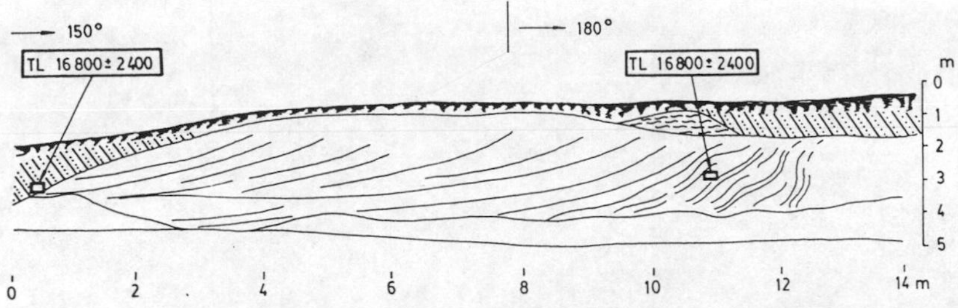

FIG.4. Main morphological and structural features of the delta southeast of Malbork with thermoluminescence age estimates of the sediment.

END MORAINES

The main morphological features of the morainic plateau south of the Vistula delta-plain are more or less continuous arcs of end moraines belonging to the Pomeranian Stage of ice sheet retreat on Polish territory. Their spatial distribution reflects distinct lobation of the ice sheet. Based on the map elaborated by Roszkówna (1963), three ice lobes may be distinguished (Fig.1) - the central lobe along the Vistula valley called the Vistula lobe; the Dzierzgoń lobe in the east, and a relatively small but well-developed lobe in the west, near Tczew, called recently by Sylwestrzak (1984) the Lubiszewo lobe. The majority of the end moraines are push moraines, in which thrusted local material and rafts of older Pleistocene and sub-Pleistocene deposits derived from the floor of the former proglacial lake may be found (Wolff, 1914; Roszkówna, 1955; Galon, 1961).

One of the numerous examples reflecting strong erosional and glaciotectonic activity of the ice sheet is a hill included in an end moraine arc of the Vistula lobe at Kalwa (Fig.1). It is almost entirely built of a huge massive raft of Senonian marl, quarried for a century for soil fertilisation and road construction. As a consequence a large crater-like pit appears now in the landscape (Fig.5). Further to the north there are end-moraine forms composed predominantly of clay-rich till with extensive inclusions of reddish-brown clays (Jentzsch, 1896) transported by ice from the floor of the Gdańsk ice-dammed lake depression situated up-glacier from the end moraines.

FIG.5. Exploited thrust raft of the Senonian marl in a hill at Kalwa.

HUMMOCKY MORAINES

Distribution and morphology. Hummocky moraines occur in the region in two groups or fields on the surface of the morainic plateau near Tczew and Malbork formerly occupied by the Vistula and Lubiszewo lobes respectively. The western field near Tczew will be discussed here in more detail.

The hummocky landforms near Tczew were firstly described by Sonntag (1919) who regarded them as drumlins. Later, Roszkówna (1961) reinterpreted these landforms as push-moraines, based mainly on the impressive glaciotectonic disturbances of the sediment. She concluded correctly that the deforming ice sheet moved from northwest to southeast. However, she did not consider the upper sedimentary element of the landforms and its irregular distribution.

The hummocky moraine field forms a zone elongated in a meridional direction and is superimposed on a pre-existing topographic rise composed of older Vistulian deposits which pre-dates the last ice advance in the area (Fig.6). The landforms do not differ in their outer appearance from typical ice-distintegration moraines resulting from the collapse of supraglacial sediments due to meltout of the underlying ice. They consist of randomly distributed, irregular hummocks and ridges of various sizes interspersed with numerous undrained depressions (Fig.7). However, their internal structure differs considerably from the simple ice-disintegration moraines.

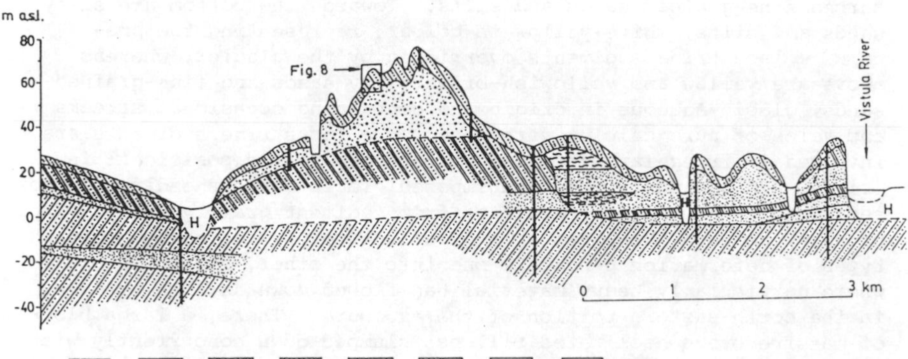

FIG.6. Geological cross-section through the field of surge hummocky moraines near Tczew.

1. pre-Vistulian glaciogenic deposits; 2. older Middle -Vistulian till; 3. younger Middle-Vistulian till; 4. Late-Vistulian glacial and glacio-aqueous deposits; 5. lacustrine silty sands with admixture of coarser-grained particles; 6. fine-grained sand; 7. Clayey silt; 8. clays; 9. Holocene deposits.

FIG.7. Surge hummocky moraine landscape near Tczew.

Internal structure. The internal structure of the hummocky moraines is revealed in a large sand pit dissecting a 15 m high hummock southwest of Tczew. As is shown in Fig.8, the hummock is built of disturbed fine-grained sands and silts. Towards the bottom are silty sands and silts, white-yellow in colour, representing the pre-existing lacustrine sediments overridden by the glacier, whereas above are yellow and yellowish-brown silty sands and fine-grained sands, glacio-aqueous in origin and containing occasional streaks and veins of pure calcium carbonate. The lacustrine sediments are involved in irregular extrusion structures, post-depositional in origin. In contrast, the superimposed glacio-aqueous sediments show sedimentary deformation resulting from sediment gravity flows. The boundary between these two sets of structures is not sharp; both types of deformation penetrate one into the other, except in places where particularly heavy material has flowed down as, for example, in the north-eastern portion of the exposure. There, a large block of massive overconsolidated till has slumped down concurrently with the flow of water-saturated sandy silts, forcing the development of an extensive flow fold (compare Figs. 8 and 9). The laminae around the till block, reflecting former flow lines of the sediment, create a kind of crescent flow mark converging at the tapering end of the block.

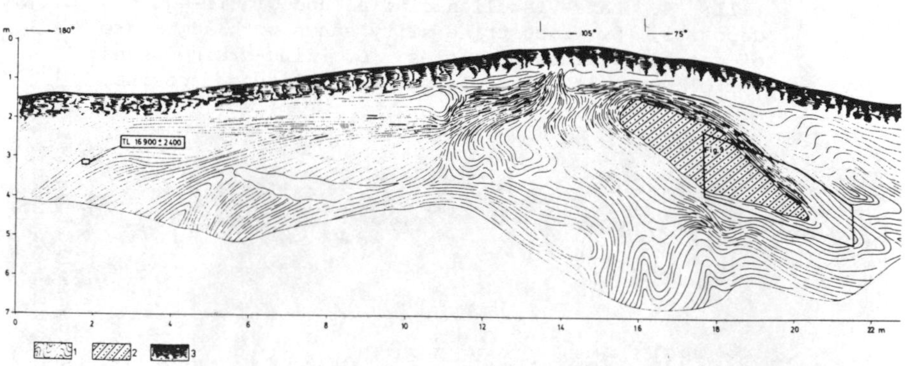

FIG.8. Internal structure of a hummock situated southwest of Tczew (see Fig.6 for geomorphological and stratigraphic situation).

1. Lacustrine sandy-silty sediments below and glacio-aqueous sediments above, disturbed as a result of extrusion and sediment gravity flows; 2. till raft; 3. soil.

FIG.9. Raft of overconsolidated till which slumped into a sub-ice cavity concurrently with flows of surrounding silty sands. In the circle, purse 11 cm long.

The deformation structures combined with the geomorphological evidence imply that the till block was entrained together with the silty sands in the basal portion of the ice sheet, perhaps in a bottom crevasse, and later during stagnation and meltout of the ice sheet was released and slumped down into a sub-ice cavity concurrently with the sandy silts. This implies that the initial formation of the hummock is related to sub-glacial processes.

Another example of distinct deformation structures occurs in an exposure in an adjacent hummock 12-13 m high. On the right side of Fig.10, which shows a general view of the exposure, the crest of a glacio-tectonically sheared fold can be seen, involving silty-sandy lacustrine sediments. The axial plane of this fold dips 80-84° toward the northwest, pointing to the direction of the applied glacier shear stresses. Further to the left of the photograph, a distinct flow fold occurs, resulting from viscous-plastic flow of extremely heterogenous material, including silty clay, sand, till, gravel, and even boulders several decimetres in diameter (Fig.11). On account of the varying plasticity of the material, particular layers are deformed in a somewhat different way. The measured dip of the fold plane is 8° toward the north-northeast, but the original dip as the sediment gravity flow proceeded was probably higher.

FIG.10. Top part of a next hummock southwest of Tczew. Crest of an overturned fold sheared by ice push can be seen on the right. Higher to the left a flow fold of glaciogenic deposits occurs. Ice movement from the left to the right.

FIG.11. Lithology and structure of the flow fold. Note the presence of boulders, till, and pure calcium carbonate concentrations. The scale on the left is 1 m long.

The genesis of the material involved in the flow fold in this exposure is less clear. It seems likely that the bulk of the material could have been entrained and carried forward by the advancing ice sheet and afterwards deposited as the ice became stagnant. However, such an interpretation is not conclusive since the formation of the constituents of the flowed material might have been derived from the down-wasting ice sheet above.

DISCUSSION ON THE FORMATION OF FORMER SURGE MORAINES

The formation of the hummocky moraines in the sub-glacial environment suggests the existence of bottom crevasses which possibly appeared when the Lubiszewo lobe moved across the Gdańsk ice-dammed lake. This interpretation seems to be reasonable. However, there is no independent evidence for a rapid readvance of the ice sheet. The only geochronometric data are the TL-age estimates of the sediments, which are too inaccurate to reconstruct ice velocities. Instead, they permit chronostratigraphic correlations that, for example, the sediments in the hummocky moraines near Tczew are correlative with the deltaic deposits in the reversed proglacial delta southeast of Malbork. Therefore, evidence accounting for a surge is, as yet, of less conclusive, indirect nature. Based on the field evidence, it may be summarised under the following points:

1. lobation of the former glacier margin indicated by series of arcuate push-moraine ridges, reflecting great shear stresses applied by the advancing glacier snout;

2. location of the hummocky moraines behind the push-moraine ridges, suggesting the existence of a close glaciodynamic interrelationship between these two types of landform;

3. deformation of local sediments in the hummocky-moraine cores in response to glacier shear stresses and the weight of ice and sediments, suggesting crevassing of the basal portion of the ice sheet;

4. superimposition of stagnant-ice deposits on the hummocky moraine cores.

To comment on these points, it should be noted that the bottom crevasses as well as the moving forward margin of the ice lobe may have been capable of producing glaciotectonic structures of various scales from simple folds and faults in the pre-existing sediments to large-scale folds and thrust features. Moreover, the stagnant-ice sedimentary element of the hummocky moraines is elsewhere distributed in irregular patches, being of limited thickness (usually less than 2 m) and lithofacially variable. Most frequently, it is represented by meltout till with some sorting and stratification, implying high water content during deposition. In this respect the moraines differ considerably from the surge moraines in the terminal zone of Klutlan Glacier, Yukon Territory, Canada (Driscoll, 1980; Wright, 1980) which are characterised by a thick mantle of supraglacial debris. This difference probably stems from the fact that the described moraines were formed by a glacier that surged across a large water body. Under such conditions, extending flow of ice may be expected in the surging glacier as a whole rather than compressive flow, and this does not permit a thick till mantle to be produced.

Worthy of note is the conclusion that surging occurred here in areas composed at the surface of slightly permeable sediments, such as fine-grained tills and clayey deep-water lake sediments, which may have prevented subglacial escape of water, thereby favouring high subglacial water pressures. These factors are considered by Clayton *et al.* (1985) to be the primary cause of surging in the Prairie region of Canada and the United States.

The possible stages of formation of the hummocky moraines described are depicted in a schematic drawing in Fig.12. It shows grounded ice calving into the lake, for it is assumed that the maximum depth of the Gdańsk ice-dammed lake (the altitudinal difference between the sub-Holocene surface in the Vistula delta-plain and the maximum level of the lake) was insufficient for an active ice sheet to be afloat. Thus, deformation of subglacial water-saturated sediments might have readily occurred, particularly in the vicinity of submerged topographic protuberances transverse to the ice movement direction, as in the case depicted (compare Fig.6).

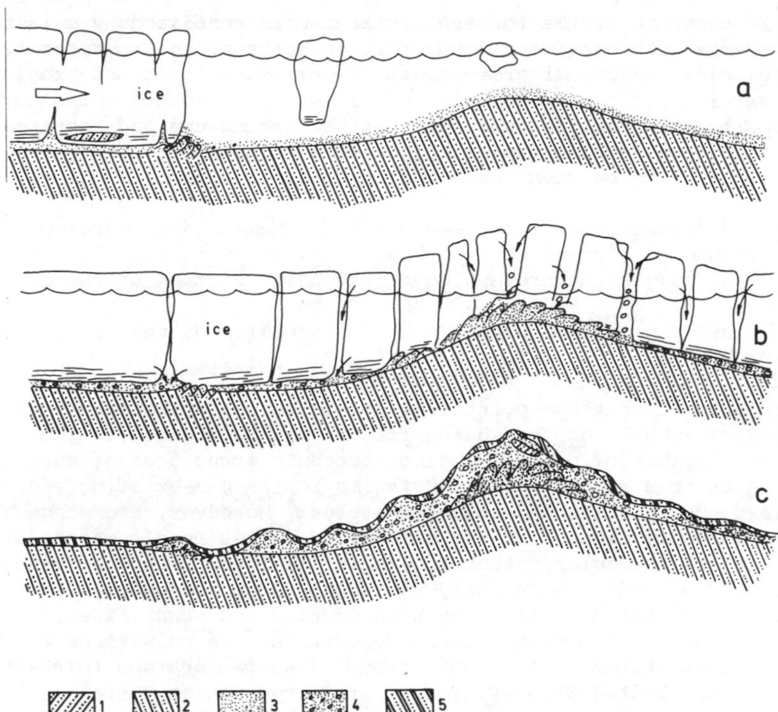

FIG.12. Main sequential stages of formation of the surge hummocky moraines.

1. thrust raft from the glacier bed; 2. glacier bed composed predominantly of slightly permeable deposits; 3. clayey and sandy-silty deposits of the proglacial lake; 4. disturbed lacustrine sandy-silty sediments mixed locally with coarser-grained fractions; 5. meltout tills and flow tills.

Taking into account the evidence, it is proposed to call the landforms described by specific genetic terms, adopting here the more general term 'surge moraines' used by Driscoll (1980) and Wright (1980), viz. surge end moraines and surge hummocky moraines. The surge hummocky moraines differ from the hummocky collapse moraines resulting from lateral redeposition of supraglacial material as the underlying ice melted, known from modern glaciers (Clayton, 1964; Eyles, 1983) and areas glaciated in the Pleistocene (Hoppe, 1952; Gravenor and Kupsch, 1954). They are rather similar to those which are interpreted as having originated by squeezing-up of subglacial material into crevasses and basal irregularities of stagnant ice (Hoppe, 1952; Stalker, 1960), except that they have not been related to surging ice lobes.

MODERN SVALBARD ANALOGUES

In order to understand the environmental conditions and factors forming the discussed landforms, it seems useful to refer to modern glaciers which are known to surge periodically across proglacial water bodies. Several such glaciers may be found in Svalbard, perhaps the most comparable being Sefströmbreen and Aavatsmarkbreen in central and northwest Spitsbergen (Fig.13).

FIG.13. Schematic map of Svalbard showing location of glaciers discussed in text.

Sefströmbreen surged at the end of the nineteenth century into Ekmanfjord and onto Cora Island, where its margin was observed in 1908 by Lamplugh (1911). The terminal zone of the glacier was in that time intensively crevassed and considerable volumes of shell-bearing mud from the fjord was squeezed up into bottom crevasses. The glacier subsequently retreated, leaving on Cora Island "a complex reticulate pattern of steep ridges which reflect the original pattern of crevasses" (Boulton, 1976). The maximum extension of the former surge marks a series of pushed ridges beyond the subglacially-formed hummocky moraines. Similar series of arcuate pushed ridges, composed of fjord sediments, may be observed at the present glacier margin (Clapperton, 1975).

Aavatsmarkbreen terminates in Hornbaekbukta with an ice cliff over 4200 m long and up to 50 m high (Fig.14), but partly also on land on both sides of the bay. The maximum surge of this glacier as identified from field and cartographic data by Niewiarowski (1982) occured at the end of the nineteenth century. Marginal moraines delineating a glacier readvance from the mid-1920s are of particular interest here. They consist of ice-free sharp-crested ridges and cones, with thrusted till slices and shell-bearing sediments which, quoting Niewiarowski, "could have been formed only in result of the glacier readvance and pushing of the sediment incorporated in the glacier bottom". Evidence of a strong compressive flow of ice, which could characterise a quiescence stage before the next surge, has been reported from the present stagnant margin (Boulton, 1970; Drozdowski, 1977, 1985).

FIG.14. Ice cliff of Aavatsmarkbreen covered with coarse supraglacial debris as seen from its southern margin in August, 1982.

Thus, the observations reported appear to account convincingly for the incorporation of marine sediments by glaciers surging across a fjord or bay. Although the mechanism of sediment entrainment at the glacier bottom is not explained by Niewiarowski, it is likely to be closely connected with bottom crevasses opened during the surge as in the case of Sefströmbreen.

It should be added here that it is presumed the Spitsbergen glaciers referred to were grounded on the fjord or bay floor when they surged across them, because their thicknesses do not approach the critical value for moving ice to be afloat. Pietrucień (1983) showed that in 1975 the depth of Hornbaekbukta near the ice cliff was 35-40 m, less than the thickness of ice above sea level. This obviously precludes floating of the glacier.

This interpretation is in full agreement with new data from Svalbard glaciers concerning the well-known Bråsvellbreen, which surged offshore at the southern margin of Austfonna-Sörfonna ice cap in Nordaustlandet in 1936-1938 (Schytt, 1969). Recent airborne radio echo measurements reported by Drewry and Liestøl (1985) indicate that the Bråsvellbreen lobe is not floating, though its average thickness is 250 m, and the fall of the ice/bedrock surface is from over 200 m above sea level to 500-1000 m below at the ice cap margin.

SUMMARY

Surging glaciers, in particular those surging across proglacial lakes or marine bays, produce characteristic landform assemblages in which the major landform types are surge end moraines and surge hummocky moraines. The characteristics of these landforms may be summarised in the following way:

> Surge end moraines marking the outermost positions of the surging glaciers are represented by series of closely spaced push ridges. If the glacier readvances across a proglacial lake or marine bay, they may be built of lacustrine or marine deposits as a result of frontal ice push and squeezing-up of the sediments into bottom crevasses.
>
> Surge hummocky moraines occur behind push end-moraine ridges. Their initial formation is indicated by cores of local sediments deformed as a result of thrusting and squeezing-up of the material into bottom crevasses opened during the surge. Sediments mantling the landforms relate to the subsequent and final stage of their formation associated with meltout of debris and thrust rafts from the downwasting ice sheet.
>
> The existence of former proglacial lakes, which promoted surges of the ice sheet, may be indicated, among others, by reverse proglacial deltas. They can provide convincing evidence for the water level altitudes of the former lakes.

REFERENCES

Boulton, G., 1970. On the origin and transport of englacial debris in Svalbard glaciers. J. Glaciol., 9, 213-229.

Boulton, G., 1976. Glaci-tectonic structures and landforms. Manuscript of paper presented at the symposium 'Till/Sweden-76', p.8.

Clapperton, C.M., 1975. The debris content of surging glaciers in Svalbard and Iceland. J. Glaciol., 14, 395-406.

Clayton, L., 1964. Karst topography on stagnant glaciers. J. Glaciol., 5, 107-112.

Clayton, L., Teller, J.T., and Attig J.W., 1985. Surging of the southwestern part of the Laurentide Ice Sheet. Boreas, 14, 235-241.

Drewry, D.J., and Liestøl, O., 1985. Glaciological investigations of surging ice caps in Nordaustlandet, Svalbard, 1983. Polar Rec., 22, 357-378.

Driscoll, Jr., F.G., 1980. Formation of the Neoglacial surge moraines of the Klutlan Glacier, Yukon Territory, Canada. Quat. Res., 14, 19-30.

Drozdowski, E., 1977. Ablation till and related indicatory forms at the margins of Vestspitsbergen glaciers. Boreas, 6, 107-114.

Drozdowski, E., 1985. On the effects of bedrock protuberances upon the depositional and relief-forming processes in different marginal environments of Spitsbergen glaciers. Palaeogeogr., Palaeoclimatol., Palaeoecol., 51, 397-413.

Drozdowski, E., 1986. Stratygrafia i geneza osadów zlodowacenia Vistulian w północnej cześci dolnego Powiśla : Stratigraphy and origin of the Vistulian deposits in the northern part of the lower Vistula region. Prace Geograficzne IGiPZ PAN 146 (in press).

Eyles, N., 1983. Modern Icelandic glaciers as depositional models for 'hummocky moraine' in the Scottish Highlands. In Tills and related sediments. Evenson, E.B., Schluchter, C.H. and Rabassa, J. (eds.), Balkema, 47-59.

Galon, R., 1961. General Quaternary problems of North Poland. In Guide-book of excursion 'From the Baltic to the Tatras', part I, North Poland, 9-53.

Gravenor, C.P., and Kupsch, W.O., 1959. Ice-disintegration features in western Canada. J. Geol. 67, 48-69.

Hoppe, G., 1952. Hummocky moraine regions, with special reference to the interior of Norbotten. Geogr. Ann., 57, 1-71.

Jentzsch, A., 1896. Das Interglacial bei Marienburg und Dirschau. Jahrb. Königl. Preuss. geol. Landesanstalt und Bergakademie, 16, 165-208.

Lamplugh, G.M., 1911. On the shelly moraine of the Sefström glacier and other Spitsbergen phenomena illustrative of British glacial conditions. Proc. Yorksh. Geol. Soc., 17, 216-241.

Makowska, A., 1979. Stratigraphy of Vistulian Glaciation deposits in the lower Vistula valley at the background of marine and continental key series of Eemian Interglacial. In Symposium on Vistulian stratigraphy, Poland 1979, Guide-book of excursion, 51-63.

Mojski, J.E.,1979. Zarys stratygrafii plejstocenu i budowy jego podloza w regionie gdańskim : Outline of the stratigraphy of the Pleistocene and structure of its basement in the Gdansk region. Biul. Instytutu Geologicznego,317, 5-50.

Niewiarowski, W.,1982. Morphology of the forefield of Aavatsmark glacier (Oscar II Land, NW Spitsbergen) and phases of its formation. Acta. Univ. Nicolai Copernici, Geografia,16, 15-43.

Pietrucień, C.,1983. Influence of ablation of the Aavatsmark glacier on thermic and salinity conditions in the Hornbaekbukta in summers 1975 and 1977. Acta Univ. Nicolai Copernici, Geografia,18, 147-158.

Roszkówna, L.,1955. Moreny czolowe zachodniego Pojezierza Mazurskiego : End moraines of the western Mazurian Lake County. Studua Soc. Sci. Torunensis, Sectio C 2, 24-134.

Roszkówna, L.,1961. End moraines near Tczew. In Guide-book of excursion 'From the Baltic to the Tatras', part I, North Poland, 86-87.

Roszkówna, L.,1963. L'influence de la surface sousquaternaire sur le developement du relief de la Pomeranie. In Report of the VIth Intern. Congress on Quaternary, Warsaw 1961, 315-326.

Schytt, V.,1969. Some comments on glacier surges in eastern Svalbard. Can. J. Earth Sci.,6, 867-873.

Sonntag, P.,1919. Geologie von Westpreussen, p.240, Berlin.

Stalker, A.M.S.,1960. Ice-pressed drift forms and associated deposits in Alberta. Bull. Geol. Surv. Canada,57, p.38.

Sylwestrzak, J.,1984. Zagadnienie recesji zachodniego skrzydla lobu Wisly na Pojezierzu Kociewskim : The question of recession of western part of the Vistula lobe in the Kociewskie Lake District. Kwartalnik Geologiczny,28, 367-386.

Wolff, W., 1914. Die geologische Entwicklung Westpreussens. Schriften der Naturforschenden Gesellschaft in Danzig. Neue Folge,13, 3/4, 59-105.

Wright, Jr., H.E.,1980. Surge moraines of the Klutlan Glacier, Yukon Territory, Canada : origin, wastage, vegetation succession, lake development and application in the Late-Glacial of Minnesota. Quat. Res.,14, 2-18.

ACKNOWLEDGEMENT

I am very grateful to Jane Calder, Department of Geography, University of Aberdeen, for preparing the camera-ready copy.

International Geomorphology 1986 Part II
Edited by V. Gardiner

FACIES ANALYSIS AND DEPOSITIONAL MODELS OF VISTULIAN ICE-MARGINAL FEATURES IN NORTHWESTERN POLAND

Stefan Kozarski and Leszek Kasprzak

Quaternary Research Institute,
Adam Mickiewicz University,
Poznań, Poland.

ABSTRACT

Detailed facies analysis of relict ice-marginal features is a powerful way of developing depositional models which vary according to differing ice-front dynamics. The procedure is time-consuming but an efficient and promising means of testing theoretical models of former ice sheets. The methodology is worked out in relation to two former ice sheet margins in Poland, the Poznań Phase and the Chodziez Subphase. It is shown that the ice sheet was warm-based and sliding during till deposition. Only under steady-state conditions during the Poznań Phase was there a wedge of cold-based ice at the margin which encouraged compressive ice flow. These latter conditions were accompanied by ice wedge formation in the proximal parts of the outwash plain and allochthonous flow till fans originating on the ice.

INTRODUCTION

Highly complicated glaciological and depositional processes which operated at the margins of Pleistocene ice sheets resulted in the creation of forms and sediments of differing nature which are now used for the reconstruction of the extent of those vast continental ice bodies. The objective of reconstruction is not only to establish the position of the ice sheet margin, but also to reveal its dynamics (Kasprzak and Kozarski, 1984; Boulton *et al*, 1985). Knowledge about the latter is needed to constrain simple process-form models (Boulton, 1972a; Clayton and Moran, 1974), to construct and test theoretical models of Pleistocene ice sheets (Boulton *et al*, 1977; Boulton and Jones, 1979; Andrews, 1982), to develop kinetostratigraphic schemes (Berthelsen, 1978), and to reconstruct general thermal regimes which may have prevailed in the marginal zones of stationary or slowly retreating ice sheets (Kozarski, in press; Ruszczyńska-Szenajch, 1982).

Studies of last ice sheet marginal features in northwest Poland have been limited by (a) lack of detailed studies and (b) the tendency to interpret these features in terms of over-simplified models. Three such models have been proposed to explain the origin of Vistulian ice-marginal forms and deposits in northwestern Poland (Table 1 - see end of article). The 'alpine model' was adopted by mapping geologists of the Königliche Preussische Geologische Landesanstalt (Berendt *et al*, 1898) during the last years of the 19th. and the first decade of the 20th. century. The concept of 'push moraine' (Stauchmoräne) was preferred subsequently and resulted from studies in Spitsbergen (Gripp, 1929). Most recently the 'dead-ice form model' has gained preference (Barkowski, 1967, 1969).

Detailed studies carried out in northwestern Poland in the last decade (Kasprzak, 1985; Kasprzak and Kozarski, 1984; Kozarski, 1978, 1981a) have shown that the above models cannot be considered universal since particular zones reveal great variations in deposits and forms as well as in modes of their occurrence. It is this fact, together with progress in the knowledge of glacial sedimentary environments, that encouraged the present authors to develop detailed studies. Their purpose is to seek new geologic and geomorphic facts and to re-interpret marginal features on a basis which will permit the construction of new, more realistic depositional models. Complex facies analysis has been used in these studies as the most efficient research tool (Kasprzak and Kozarski, 1984). The studies have been carried out in only a few case study areas as yet (Fig.1). Two of them will be presented and discussed briefly in this paper as examples showing different ice-margin dynamics, namely the Poznań Phase (Pz) and the Chodziez Subphase (Ch).

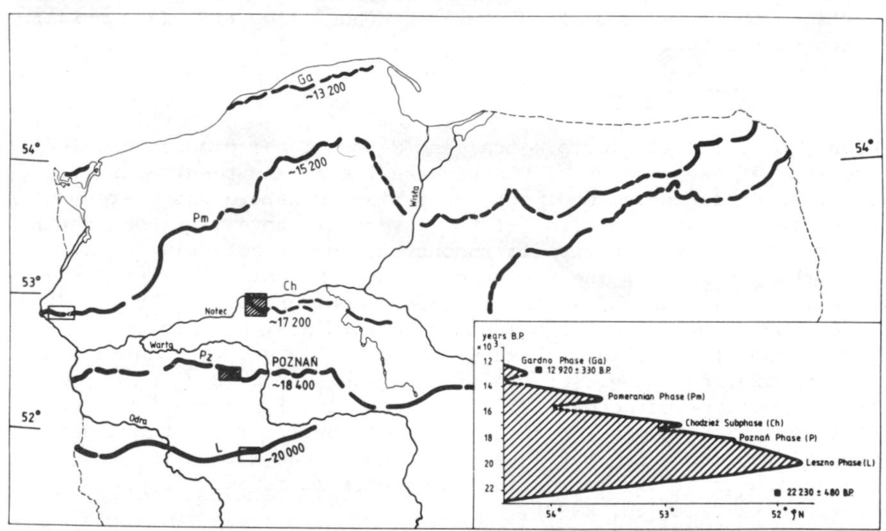

FIG.1. Location of case study areas and major Vistulian ice sheet position in northwestern Poland with estimated radiocarbon age BP (after Kozarski, in press). Hatching shows the case studies discussed in this paper. Inset : time-distance diagram of the last ice-sheet in northwestern Poland (Source : S. Kozarski, in press).

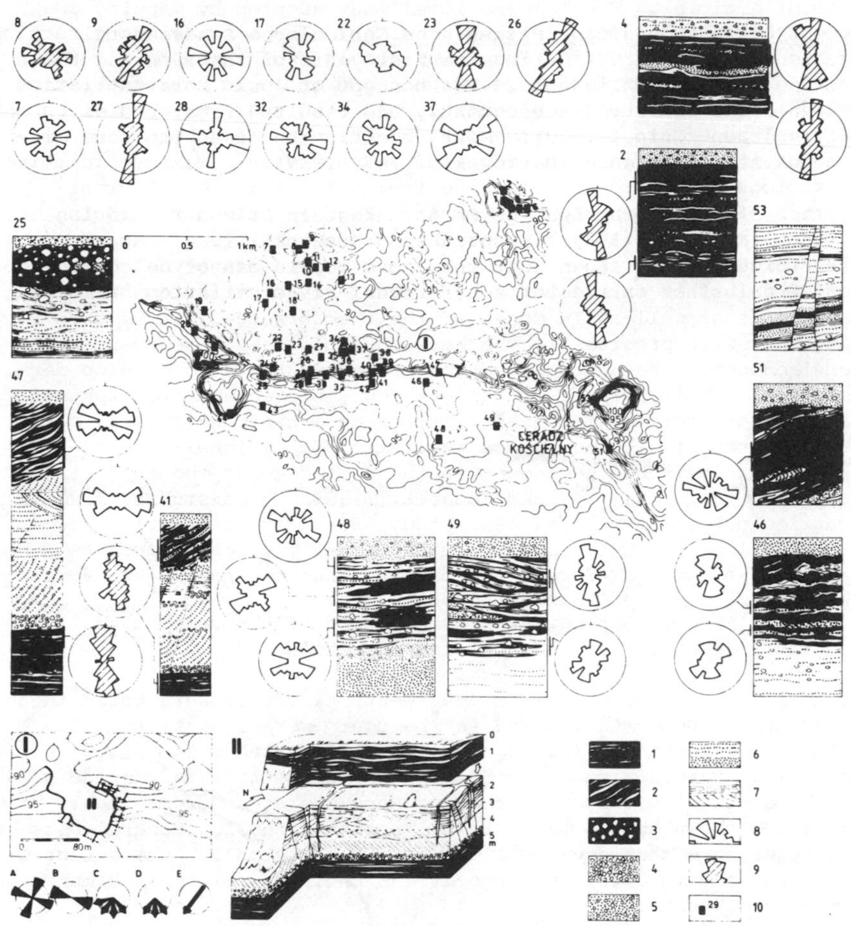

FIG.2. Sedimentary structure and texture of deposits in the marginal zone of the Poznań Phase near Ceradz Kościelny.

1. lodgement till with horizontal sandy streaks; 2. allochthonous flow till; 3. flow till rich in block material; 4. gravel; 5. sand with pebbles; 6. sand; 7. silt; 8. fabric diagrams of flow till; 9. fabric diagrams of lodgement till; 10. excavated pit.

I Structural and directional elements: A - orientation of ice-wedge casts; B - orientation of fault-planes; C - glaciofluvial flow directions; D - aggregated orientation of pebbles in allochthonous flow tills; E - aggregated orientation of pebbles in lodgement till.

A CASE STUDY OF STEADY-STATE ICE-MARGINAL FEATURES OF THE POZNAŃ PHASE

Geomorphic and geologic setting of sites. The marginal zone section in the vicinity of Ceradz Kościelny consists of three main terrain units displaying different geomorphic and geologic characteristics (Fig.2). One of them is a centrally located scarp which divides the marginal zone into two parts, i.e. northern and southern, that represent its hinterland and foreland respectively. Maximum relative elevations within the scarp range from a few to 10 m. A cross-section of the scarp shows asymmetry (Fig.3): the north-facing slopes are considerably steeper than those inclined southward. The area which is referred to as hinterland represents a vast depression bordered further in the north by irregularly distributed hummocks.

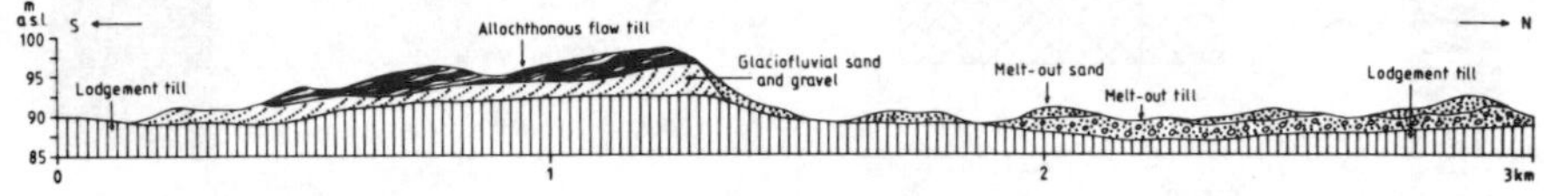

FIG.3. Geological cross-section of the scarp zone (Poznań Phase) near Ceradz Kościelny.

Fifty-six pits were excavated and twenty-two holes were bored in the study area. Deposits exposed in the scarp were investigated in detail in a large pit (Fig.2, II). The excavated pits (Fig.2, Nos. 1-53) were 2.5 to 3.5 m deep. The sediments were sampled down to the depth of 4 m by the use of a hand-operated corer. Apart from the sites shown on the map (Fig.2), about 40 shallow control pits and holes were made down to 1 m in order to establish the extent of sediments and to obtain a three-dimensional view of the sediments as lithofacies units.

Lithofacies units related to landforms. Lodgement till occurs in the study area as a continuous bed approximately 7 m thick (Fig.3) (Assman and Dammer, 1916). As can be seen in a schematic geologic section, there is a distinct depression in the till. The latter is situated at about 93 m a.s.l. at the scarp and slopes down north of the scarp to about 87 m a.s.l. It is again about 90 m a.s.l. at the northern fringe of the study area.

The particle-size distribution of the lodgement till reveals a generally accepted diagnostic characteristic (Kozarski and Szupryczyński, 1973; Boulton and Paul, 1976), i.e. higher clay contents of lodgement till which are 26 per cent; single samples contain from 12 to 70 per cent of clay particles (Fig.4). The orientation of clasts was studied using Krüger's procedure (1970). The preferred orientation of elongated clasts ($\geqslant$ 3 cm) is parallel to a north-south or northeast-southwest direction (Fig.2, Nos. 9, 23, 24, 26, 27, 41, 47). This orientation is constant in the vertical profile. The structure of lodgement till is hardly detectable. Distinct sedimentary structures occurring as horizontally 10 cm thick sand streaks between lodgement till subunits were found only in two pits (Fig.2, Nos. 2, 4).

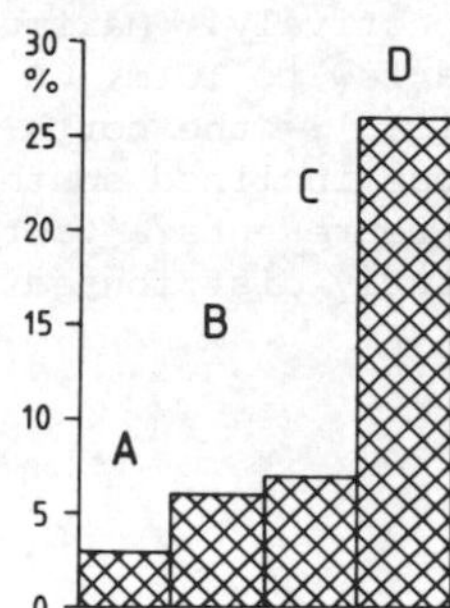

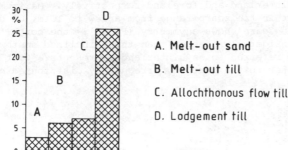

FIG.4. Clay content of till deposits and melt-out sand in the scarp zone.

Allochthonous flow tills are composed of sediments varying in particle size. Their main characteristic is quasi-rhythmic bedding (Fig.5). It is due to streaks of clay-rich material containing pebbles and to sandy or sandy-stony intercalations, i.e. diagnostic features described in the literature (Boulton, 1968, 1976; Marcussen, 1975; Krüger and Marcussen, 1976). The average clay contents calculated for 14 samples are 7 per cent (Fig.4). However, allochthonous flow tills may be distinguished as clayey (Fig.2, No.51) or clayey-sandy-stony (Fig.2, Nos. 41, 47, 48, 49) tills. Such variability may be associated with a varying contribution of meltwater to the process of till deposition. Allochthonous flow tills have only been detected within the scarp and proximal side of the outwash plain where they cover glaciofluvial sediments (Fig.3) and form flat fans which attain lengths of up to 900 m (Fig.6). The thickness of allochthonous flow tills ranges from 1 to 3 m on the average at the scarp and sporadically it is 5 m.

Melt-out tills and sands have a varied structure. Clay lenses present in sandy tills, well-sorted sand and gravel lenses and clusters of chaotically arranged gravel-stone material have been found. Fluid structures sometimes preserved in the inclusions are indicative of minor lateral displacement occurring throughout the deposition of melt-out sands and tills. The study of the orientation of the long axes of clasts (Fig.2, Nos. 7, 16, 17, 22, 28, 32, 34, 37) and elongated sand bodies points to a distinct relationship of flow reflecting the local slope angle of the terrain surface.

FIG.5. Allochthonous flow till over the proximal part of the outwash fans.

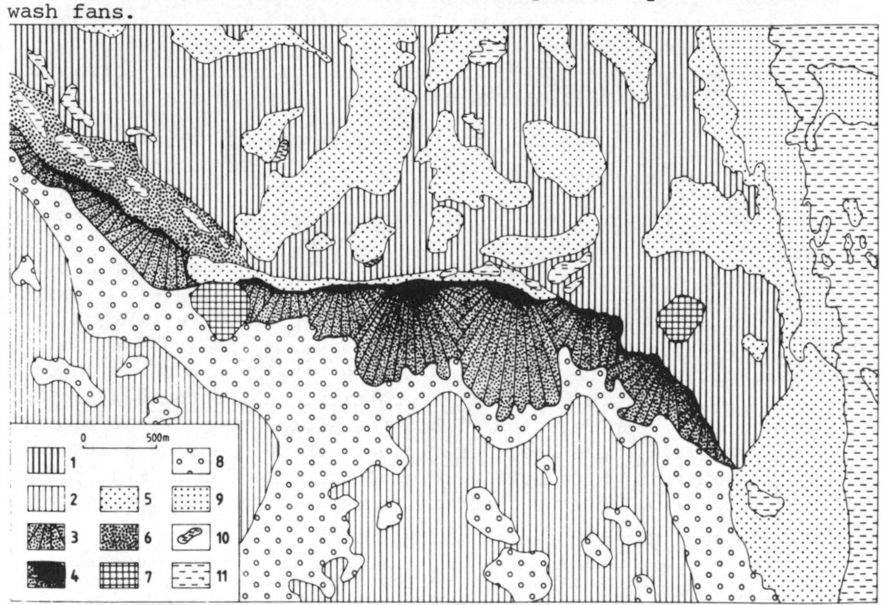

FIG.6. Main lithofacies distribution in the marginal zone of the Poznań Phase near Ceradz Kościelny.

1. lodgement till in the hinterland; 2. lodgement till in the foreland; 3. allochthonous flow till fans; 4. concentration of block material in the scarp; 5. melt-out tills and sands; 6. melt-out sandy-stony deposits; 7. glaciolacustrine silts and clays of kame plateaux; 8. glaciofluvial sands and gravels; 9. sandy-silty deposits of kame terraces; 10. varved deposits; 11. organic deposits.

The above properties permit a comparison to be made between the lithofacies under investigation and facies No.2 of supraglacial deposits reported from Icelandic and Alpine glaciers by Eyles (1979). Melt-out sands and tills 1 to 4 m thick lie north of the scarp where they build up small hummocks (Figs. 2, 3).

Melt-out sandy-stony deposits are structureless. They are much coarser than melt-out sands. Occasionally irregular clayey streaks are present in them. They are indicative of multiple redeposition. A specific characteristic of this lithofacies is its association with shallow closed depressions elongated parallel to the scarp (Fig.6). Low rims are the geomorphic expression of the deposits under consideration. Many pebbles are included in the sandy-clayey matrix on the south side, i.e. at the scarp, whereas fine sands and silts similar to glaciolacustrine sediments prevail on the north side.

Glaciofluvial sands and gravel and their structure are well displayed in a gravel-pit located in the scarp (Fig.2, No.11). Small- and large-scale cross-stratification, horizontal stratification and small-scale trough stratification have been discovered in the glaciofluvial series. Upward fining of sediments is indicative of a decrease in stream power. The study of cross-stratification provided the basis for reconstructing the southward direction of streamflow (Fig.2, No.1, C). Deformation structures found in the glaciofluvial series are of different origin. They include (1) syngenetic and epigenetic ice-wedge casts which are a few tens of centimetres to over three metres long (Fig.7), (2) gravity faults parallel to the scarp (Fig.2, No.1, B). Displacements towards the scarp are indicative of the presence of a dead-ice wedge beneath the glaciofluvial series, as can be inferred from the analysis of throw values. Close to the scarp glaciofluvial sediments form outwash fans which become sandur tracks in the foreland (Fig.6). The glaciofluvial sediments separate lodgement till from the allochthonous flow till series in the scarp alone. Their maximum thickness is about 5 m (Fig.2, No.11).

Glaciolacustrine silts and clays with horizontal stratification are common in the study area. The stratification remains rhythmic in some pits, with alternating sandy silts and clays. Thick silty-clayey series have been found in two hummocks lying in the scarp zone (Fig.6). The hummocks have distinct flat-topped surfaces and steep slopes. Gravity faults which deform the silty-clayey lacustrine series were detected in the marginal zone of the eastern hummock (Fig.2, No.53).

<u>Local lithostratigraphy, palaeoenvironmental implications and depositional model</u>. Lodgement till is the earliest sedimentary unit in the portion of the Poznań Phase marginal zone in the vicinity of Ceradz Kościelny (Fig.8). Clast orientation suggests it was laid down by an ice sheet moving from a north-east direction as far as the ice front along the Poznań Phase line (Fig.2). Flat outwash fans (glaciofluvial sand and gravel lithofacies) and the overlying

FIG.7. A syngenetic ice-wedge cast in glaciofluvial deposits.

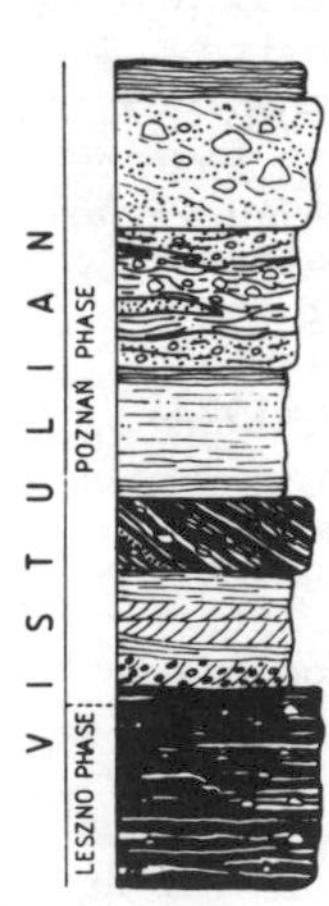

SUMMARY SEQUENCE	GEOLOGICAL AND GEOMORPHOLOGICAL RECORD		ICE-FRONT DYNAMIC	− | +
clay sand and pebbles	varved clays sandy and pebbles deposits of ice-cored moraines	ice-cored moraine rims		
sand till sand	melt-out till and sand	ice-disintegration hummocks and flat surfaces	retreat	
clay sand silt clay	glaciolacustrine clay, sand and silt	kames		
till silfy sand sand gravel	allochthonous flow till glaciofluvial sand and gravel	ice-marginal fans	steady state	
till	basal till	undulated moraine surfaces	advance	

FIG.8. A depositional model for the ice-marginal features of the Poznań Phase.

ablation fans (allochthonous flow till lithofacies) were deposited during steady-state conditions associated with a still stand of the active ice-sheet front. From the absence of subglacial channels and the occurrence of faults generated by dead-ice melting (Fig.2, No.11), it can be inferred that outwash deposits were laid down by supra-glacial waters. The presence of syngenetic ice-wedge casts in glaciofluvial deposits suggests severe climatic conditions and perma-frost aggradation at the time of deposition. The cessation of intensive glaciofluvial activity resulted in the deposition of allochthonous flow tills onto the proximal parts of outwash fans.

Subsequently, the lowermost portion of the ice-front produced ice-cored ridges and rims (melt-out sandy-stony deposit lithofacies) in the north-west extension of the marginal zone and kames (glacio-lacustrine silts and clays) close to the margin in the eastern portion of the marginal zone. Finally, dead ice lying in the hinter-land melted away and ablation melt-out tills and sands accumulated as a discontinuous cover.

The research results are contradictory to Bartkowski's statements (1967) since (1) not all the forms present in the marginal zone of the Poznań Phase are kames and/or crevasse infillings, and the kame-like forms exhibit no pattern of parallel fissures, and since (2) during the Poznań Phase the ice-sheet front first remained active (steady-state) and then became stagnant. The presence of syngenetic ice-wedge casts in the outwash and the palaeoclimatic implications of this fact cast doubt on a recent interpretation by Ruszczyńska-Szenajch (1982) that glaciofluvial sediments are indicative of the amelioration of climate.

The standstill position of the ice-sheet front along the Poznań Phase line resulted from a dynamic equilibrium between ablation and accumulation. High proportions of sediments of supraglacial and glaciofluvial origin have been found in the marginal deposits. The great thickness and extent of supraglacial sediments at the ice-sheet front implies compressive ice flow. The compressive flow was generated by the presence of cold ice at very extreme margin of the ice-sheet which adhered to a permafrost wedge in the substratum. Thus, the view proposed here is that the incorporation of debris derived from the bedrock and its transport into a supraglacial position may have occurred in a way described by Boulton (1972, Fig.5c).

Detailed analysis of the distribution of lithofacies associated with landforms also permits construction of a scheme of local litho-stratigraphy and a summary sequence. At the present stage of research the summary sequence should be regarded as a depositional model (Fig.8). It points to important palaeoenvironmental impli-cations and identifies different conditions of ice-front dynamics.

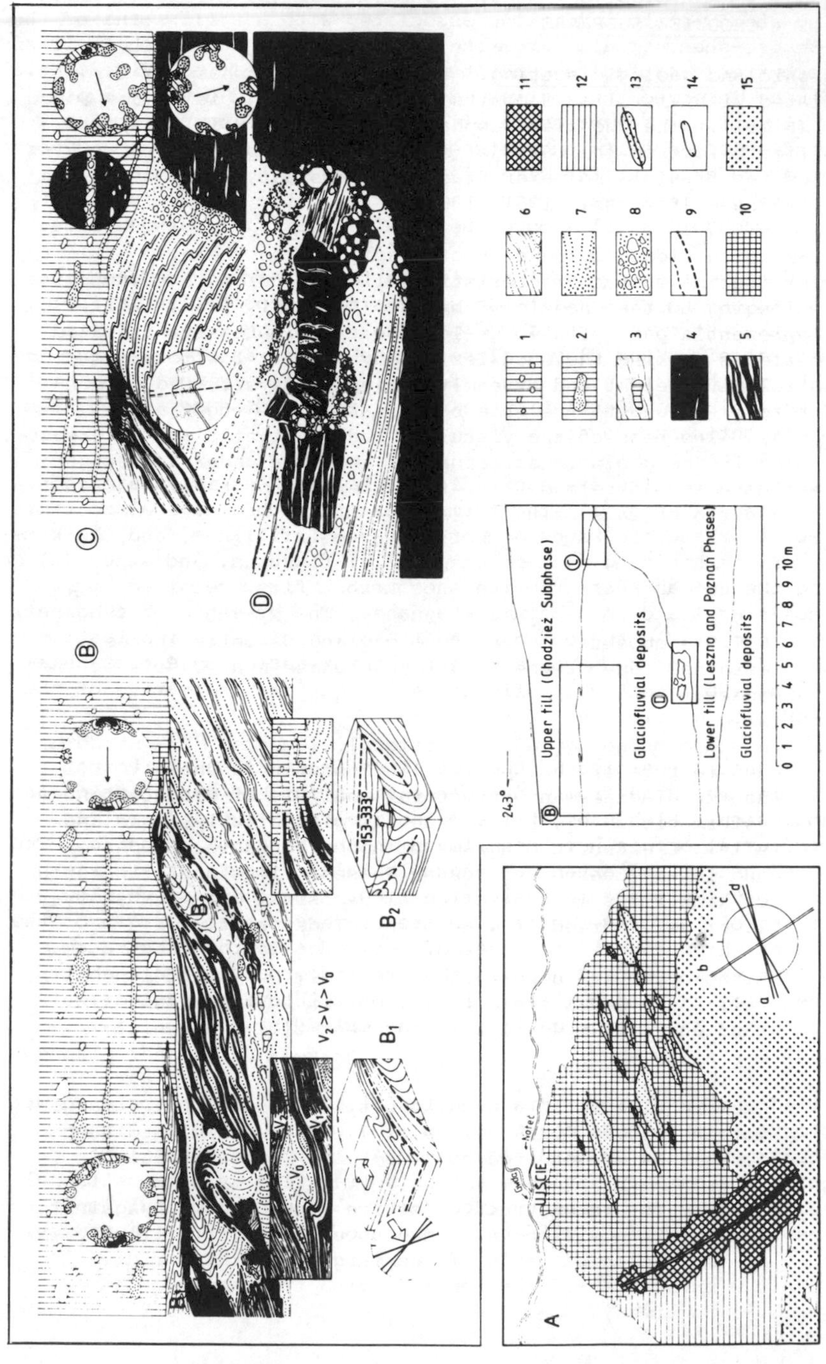
A
UJŚCIE
hotel
Gwda
a
b
c
d
B
C
D
$V_2 > V_1 > V_0$
B_1
B_2
153-333°
243°
Upper till (Chodzież Subphase)
Glaciofluvial deposits
Lower till (Leszno and Poznań Phases)
Glaciofluvial deposits
0 1 2 3 4 5 6 7 8 9 10 m
1
2
3
4
5
6
7
8
9
10
11
12
13
14
15

A CASE STUDY OF RE-ADVANCE ICE-MARGINAL FEATURES OF THE CHODZIEZ SUBPHASE

Geomorphic and geologic setting. The Chodziez Subphase is largely associated with Vistulian deposits occurring at Ujście on the Noteć river (Fig.1). The Chodziez Subphase (Kozarski, 1981b) has been classified before on the basis of morphostratigraphical criteria as the Chodziez Stadial (Kolmarer Stadium, Woldstedt, 1932) or as the Chodziez Phase (Kozarski, 1961, 1962).

Thrust-moraine ridges lying east of Ujście in the vicinity of Chodziez are the most characteristic relief features in the marginal zone belonging to the Chodziez Subphase (Kozarski, 1959, 1961). They represent a part of a 60 km-long belt of impressive isolated thrust ridges marking the Chodziez re-advance. A low morainic ridge 5 m in relative height and streamlined landforms situated perpendicular to it belong to the same belt in the case study area (Fig.9, A). Two beds of the Vistulian lodgement till are of greatest importance in the geologic structure of the area under investigation. However, note should be made that the upper bed of the Vistulian till lies only north of this belt.

FIG.9 (opposite). Geomorphology and deposits in the marginal zone of the Chodziez Subphase near Ujście on the Noteć.

A - Geomorphology and directional elements. B - Structure and texture of the upper till (Chodziez Subphase). Till fabric diagrams and structural analysis of glacitectonic deformations (B_1, B_2) in subtill glaciofluvial sand and silt. C - Fabric diagrams of upper (Chodziez Subphase) and lower (Leszno and Poznań Phases) tills and orientation of kink-folds in glaciofluvial deposits. D - Lumps of lower till buried in glaciofluvial deposits.

Lithology: 1 - upper lodgement till; 2 - sand lenses in till; 3 - small-scale till lumps in subtill glaciofluvial sand and silt; 4 - lower lodgement till; 5 - silt; 6 - silty sand; 7 - sand; 8 - gravel; 9 - shear planes.

Geomorphology: 10 - undulated till surface; 11 - marginal moraine ridge; 12 - till plain; 13 - streamlined landforms; 14 - elongated hollows; 15 - outwash plain.

Diagram for reconstruction of ice flow direction against ice margin position (A): a - upper lodgement till fabric; b - morphological axis of the marginal moraine ridge; c - main stress direction according to glaciotectonic deformations; d - streamlined landforms.

The profile of deposits in Ujście on the Noteć river which provides a detailed record of depositional and geomorphic events during the Upper Pleni-Vistulian is of crucial importance in establishing a lithostratigraphical base for the reconstruction of ice-sheet margin dynamics during the Chodziez Subphase. Sediments belonging to the Upper Pleni-Vistulian are on the average 10 m thick in 300 m-long exposures. The upper 3 m of the profile contain a sand-gravel series which is underlain by two tills 3 and 3.5 m in mean thickness respectively, and the intervening discontinuous layer of glacio-fluvial sediments is about 3 m thick. The two till beds, together with sands and gravel containing numerous small-scale glaciotectonic deformations have been studied in detail (Fig.9). Features distinguishing the till beds, including the sequence of sedimentary units in the profile and their differing colour, have been supplemented by data concerning grain-size distribution, carbonate content and by the so-called directional elements (Berthelsen, 1978) which comprise the orientation of clasts and deformation structures.

Lithofacies. The grain-size distribution and carbonate content of the tills are shown in Fig.9 (inset C and 10). The samples for laboratory investigations were taken at 15 cm intervals. A graph showing the percentage size distribution in the vertical plane reveals significant deflections of individual curves except for fractions coarser than 0.71 mm, 0.175-0.125 and 0.125-0.09 mm at the contact between the two tills. The contact is best expressed by the curves of clay, silt and sand ranges (Fig.10). A decrease in the percentage of sands by over 20 per cent and a concurrent increase in the percentage of silts and clays by nearly 20 per cent and about 3 per cent respectively occurs in the upper dark-brown till. Both tills also differ in carbonate content. The mean contents of the lower and upper tills are 6.4 and 8.3 per cent respectively. A graph of carbonate content in the vertical profile shows a sharp boundary at the contact between both tills, as has been the case with the grain-size distribution (Fig.10). The causes of existing differences in the grain-size distribution and carbonates content are not discussed in this study. Descriptions have been given in order to demonstrate the presence of two distinct till beds.

The orientation of clasts with distinct long axes was systematically studied at different depths and over large exposed surfaces of till. The orientation of 50 clasts was studied at each test site. Special measurements were also made in the contact zone between the two tills, in a 10 cm-thick basal layer of the upper dark-brown till, and in a 10 cm-thick top layer of the lower light-brown till. An important difference occurs in the orientation of the long axes of clasts in the two tills (Fig.9, B and C). The prevailing orientation is 22° for the lower till (300 clasts) and 78° for the upper till (450 clasts). From the above distribution of clast orientations, it follows that the lodgement till beds under investigation were laid down by two ice-sheet advances differing remarkably in direction.

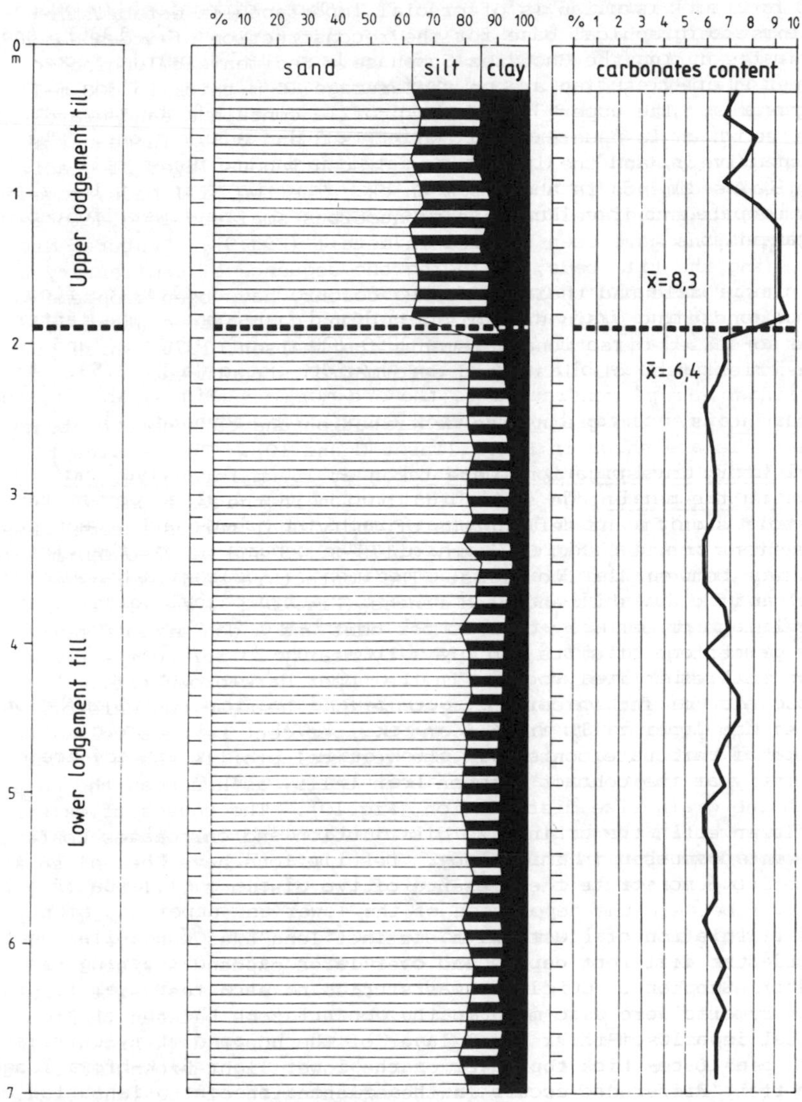

FIG.10. Grain-size distribution and carbonate content of Vistulian lodgement tills at Ujście on the Noteć.

Deformation structures genetically related to the upper till bed occur in sand gravel deposits which locally separate the two till beds from each other (Fig.9, B, C, D). Three-dimensionally prepared and exposed drag-fold, kink structures (Van Loon *et al*, 1985), and boudinage or torpedo structures (Berthelsen, 1978) permit reconstruction of the main stress direction responsible for their formation (Fig.9, B_1 and B_2). Horizontal alignment of the main stress axis indicates a glaciotectonic origin of the deformations. The orientation is similar to that of detritus in the upper till (Fig.9, Aa, c). It proves a quite different direction of ice sheet flow compared to that during the deposition of the lower light-brown lodgement till.

The clasts and deformation structures suggest a similar ice flow direction to that indicated by streamlined landforms. The latter occur as parallel morainic ridges 2250-125 m long (axis a) and 500-75 m wide (axis b), with a length/width ratio (a:b) 3.53. As the ridges are perpendicular to the end moraine (Fig.9, Ab, d), they support arguments for the Chodziez Subphase re-advance.

<u>Local lithostratigraphy, palaeoenvironmental implications and depositional model.</u> The analytical techniques applied permit the construction of a local lithostratigraphical scheme and a sequence of events for the Chodziez Subphase. The separation of the Vistulian deposits from earlier Pleistocene sediments (Kozarski and Nowaczyk, 1985) and the identification of kinetostratigraphical units related to relief features are of principal importance in this respect. They permit recognition of a key horizon, the bed of lower light-brown till which lies at a depth of 8-10 m in the north, but is exposed at the surface in the south beyond the thrust ridges belonging to the Chodziez Subphase (Kozarski, 1962). Till fabric analysis indicates that it is a bed deposited during the maximum advance of the last ice sheet from a north direction (Fig.9, C).

The lower till is overlain by a discontinuous bed of glaciofluvial sediments containing buried lower till lumps (Fig.9, D). It is a fact of importance since it is indicative of the occurrence of a time interval between the deposition of the lower and upper lodgement tills. This interval was linked to the recession of the ice sheet outside the area into which the ice sheet re-advanced during the Chodziez Subphase. Evidence of its transgressive character is provided by glaciotectonic deformation occurring at the top of glaciofluvial deposits (Fig.9, B_1, B_2 and C) and the zone of mylonitization which penetrates into the bottom part of the upper dark-brown lodgement till (Fig.9, B), as well as the quite different orientation of clasts. Streamlined landforms and thrust-moraine ridges occurring in the vicinity of Chodziez are a geomorphological expression of the direction of the ice sheet advance, which parallels the directions reflected in till fabric and glaciotectonic deformations (Kozarski, 1959, 1961).

Therefore, a re-advance model in the form of summary sequence, Fig.11, is proposed for the Chodziez Subphase. It is based on certain facts: (1) the occurrence of two lodgement tills, a lower one which belongs to the maximum advance and an upper one which provides evidence of a new regional ice-front advance from a quite different direction, (2) the deposition of the lower till followed by a melt-water cutting episode represented by glaciofluvial deposits entrenched in the lower till; the glaciofluvial deposits contain buried lower till lumps which originated from lateral erosion in a meltwater channel, (3) the occurrence of a mylonityzation zone in the top parts of glaciofluvial deposits and at the base of the upper till manifested by small-scale glaciotectonic deformations and sand-till interdigitation features, (4) the appearance of streamlined landforms closely related to ice-thrust ridges which occur in the southernmost part of the upper till cover.

The upper till produced by the Chodziez re-advance is covered by a discontinuous mantle of sands and gravels and delicately laminated sands and silts (Fig.11). This cover which is 3 m thick reveals the presence of gravity faults and records the change of depositional processes which operated after the re-advance during ice-sheet disintegration and the formation of partially buried dead-ice blocks. The lithofacies analysis shows that the dead-ice phase contributed only to the ice-marginal features and was not dominant, as assumed by Bartkowski (1967, 1969).

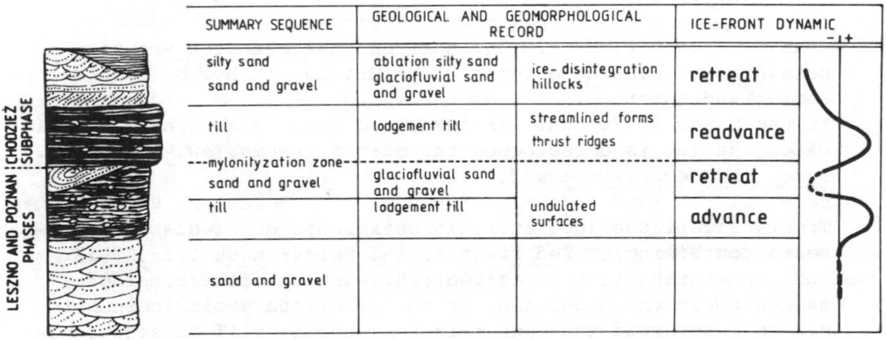

	SUMMARY SEQUENCE	GEOLOGICAL AND GEOMORPHOLOGICAL RECORD		ICE-FRONT DYNAMIC − +
CHODZIEŻ SUBPHASE	silty sand sand and gravel	ablation silty sand glaciofluvial sand and gravel	ice-disintegration hillocks	retreat
	till	lodgement till	streamlined forms thrust ridges	readvance
LESZNO AND POZNAN PHASES	---mylonityzation zone--- sand and gravel	glaciofluvial sand and gravel		retreat
	till	lodgement till	undulated surfaces	advance
	sand and gravel			

FIG.11. A depositional model for the ice-marginal features of the Chodziez Subphase.

CONCLUSIONS

The two case studies which illustrate the research procedure and analysis of Vistulian ice-marginal features call for expansion of such a type of detailed investigation in order to gain a better basis for the construction of realistic depositional models reflecting ice-front dynamics. The procedure is time-consuming but an efficient and promising means of testing theoretical models of the last ice sheet.

The former models of Vistulian ice-marginal features in northwestern Poland are oversimplified and they incorrectly postulate a single general genetic mode which has been extended to all end moraines and end-moraine-like landforms in the area. The evidence presented here shows important differences in ice-front dynamics between different major ice sheet positions. It must be stressed that they do not agree with the strongly exaggerated 'dead-ice form model' treated by Bartkowski (1967, 1969) as universal for all ice-marginal features.

It can be assumed from lodgement till studies, which reveal a well-pronounced orientation of clast a-axes, that the ice sheet was warm-based and sliding during till deposition. Only in a steady-state position at the extreme margin may the occurrence of a cold-based ice which adhered to a permafrost wedge and generated compressive ice flow be assumed. The direct evidence of permafrost occurrence during the Poznań Phase are syn- and epigenetic ice-wedge casts in the proximal parts of the outwash plain, while evidence of compressive flow is provided by the overlying allochthonous flow till fans.

ACKNOWLEDGEMENT

I am very grateful to Jane Calder, Department of Geography, University of Aberdeen, for preparing the camera-ready copy.

REFERENCES

Assmann, P., Dammer, B., 1916. Geologische Karte von Preussen und benachbarten Bundesstaaten. Blatt Gr.Gay, 1:25,000. Kgl. Preuss Geol. Landesanst.

Bartkowski, T., 1967. Sur les formes de la zone marginale dans la Plaine de la Grande Pologne (in Polish). Pozn.Tow.Przyj.Nauk., Prace Kom.Geogr.-Geol., 7, 1-260.

Bartkowski, T., 1967. Zonaler Eisabbau-ein 'normaler' Eisabbau des Tieflandsgebietes (erlautert an Beispielen aus West- und Nordpolen (in Polish). Bad.fizjogr. nad Polska zach., 23, 7-33.

Berendt, G., Keilhack, K., Schröder, H., Wahnschaffe, F., 1898. Neuere Forschungen auf dem Gebiet der Glacialgeologie in Norddeutschland, erläutert an einigen Beispielen.III Aufschüttungsformen des Inlandeises. H.S.a, Endmoränen.Jb.Preuss.Geol.Landesanst., 18, 88-103.

Berthelsen, A., 1978. The methodology of kineto-stratigraphy as applied to glacial geology. Bull.geol.Soc.Denmark, 27, Spec. Issue, 25-38.

Boulton, G.S., 1968. Flow tills and related deposits on some Vestspitsbergen glaciers. J.Glaciol., 7 (51), 391-412.

Boulton, G.S., 1972a. Modern Arctic glaciers as depositional models for former ice sheets. Jl.geol.Soc.Lond., 128, 361-393.

Boulton, G.S., 1972b. The role of thermal regime in glacial sedimentation. Inst.British Geogr., Special Publ. 4, 1-19.

Boulton, G.S., 1976. A genetic classification of tills and criteria for distinguishing tills of different origin. Zesz.Nauk.UAM, Geografia, 12, 65-80.

Boulton, G.S., Jones, A.S., 1979. Stability of temperate ice caps and ice sheets resting on beds of deformable sediments. J. Glaciol., 24, 29-43.

Boulton, G.S., Jones, A.S., Clayton, K.M., Kenning, M.J., 1977. A British ice sheet model and patterns of glacial erosion and deposition in Britain. In Shotton, F.W. (ed.) British Quaternary Studies. Oxford Univ.Press, Oxford, 231-246.

Boulton, G.S., Paul, M.A., 1976. The influence of genetic processes on some geotechnical properties of glacial tills. Q.Journ.Engng. Geol., 9, 159-194.

Boulton, G.S., Smith, G.D., Jones, A.S., Newsome, J., 1985. Glacial geology and glaciology of the last mid-latitude ice sheets. J.geol.Soc., London, 142, 447-474.

Clayton, L., Moran, S.R., 1974. A glacial process-form model. In Coates (ed.) Glacial Geomorphology, State Univ., Binghamton, New York, 88-119.

Eyles, N., 1979. Facies of supraglacial sedimentation in Icelandic and Alpine temperate glaciers. Can.Journ.Earth Sci., 16, 1314-1361.

Gripp, K., 1929. Glaziologische und geologische Ergebnisse der Hamburgischen Spitzbergenexpedition 1927. Naturwissenschaftlicher Verein. In Hamburg, Abhandl. aus dem Gebiet der naturwissenschaften, 22, 146-249.

Kasprzak, L., 1985. The origin of glacio-tectonic deformations within the push-moraine near Leszno (in Polish). Bad.fizjogr. nad Polska zych., 35, 63-82.

Kasprzak, L., Kozarski, S., 1984. Facies analysis of marginal zone deposits produced by the Poznań Phase of the last glaciation in Middle Great Poland (in Polish). Geografia, 29, 1-54.

Kozarski, S., 1959. On the origin of the Chodziez end moraine (in Polish). Bad.fizjogr.nad Polska zach., 5, 45-72.

Kozarski, S., 1961. Terminal push-moraine of the Chodziez phase, in the Great Poland Lowland. Guide-Book of Excursion A, VI Congress INQUA, Poland 1961, 24-25.

Kozarski, S., 1962. Recession of last ice sheet from northern part of Gniezno Pleistocene Plateau and formation of the ice-marginal valley of the rivers Noteć-Warta (in Polish). Pozn.Tow.Przyj. Nauk,Prace Kom.Geogr.-Geol., 2, 1-154.

Kozarski, S., 1978. Lithologie und Genese der Endmoränen im Gebiet der skandinavischen Vereisung. Schriftenr.geol.Wiss., 8, 179-200.

Kozarski, S., 1981a. Ablation end moraines in western Pomerania, NW Poland. Geogr.Ann., 63, 169-174.

Kozarski, S., 1981b. Vistulian stratigraphy and chronology of the Great Poland Lowland (in Polish). Geografia, 6, 1-44.

Kozarski, S. (in press). Ablation deposits of the last glaciation in west central Poland : identification and palaeogeographical implications (in Polish). Acta Univ.Wratislaviensis.

Kozarski, S. (in press). Timescales and the rhythm of the Vistulian geomorphic events in the Polish Lowland (in Polish). Czas.Geogr.

Kozarski, S., Nowaczyk, B., 1985. Stratigraphy of Pleistocene deposits at Ujście on the River Noteć (Introductory report) (in Polish). Pozn.Tow.Przyj.Nauk,Wydz.Mat.-Przyr., Sprawozdania, 101, 49-51.

Kozarski, S., Szupryczyński, J., 1973. Glacial forms and deposits in the Sidujökull deglaciation area. Geogr.Polonica, 26, 255-311.

Krüger, J., 1970. Till fabric in relation to direction of ice movement. A study from the Fakse Banke, Denmark. Geogr.Tidsskrift, 69, 133-170.

Krüger, J., Marcussen, I.B., 1976. Lodgement till and flow till : a discussion. Boreas, 5, 61-64.

Marcussen, I., 1975. Distinguishing between lodgement till and flow till in Weichselian deposits. Boreas, 4, 113-123.

Ruszczyńska-Szenajch, H., 1982. Depositional processes of Pleistocene lowland end moraines and their possible relation to climatic conditions. Boreas, 11, 249-260.

Van Loon, A.J., Brodzikowski, K., Gotowala, R., 1985. Kink structures in unconsolidated fine-grained sediments. Sedimentary Geology, 41, 283-300.

Woldstedt, P., 1932. Über Randlagen der letzten Vereisung in Ostdeutschland und Polen und über die Herausbildung des Netze-Warthe the Urstromtales. Jb.d.Preuss.Geol.Landesanst, 52, 59-67.

TABLE 1. Various models of genetic interpretation of ice-marginal features in NW Poland

Deposits and structures	Processes	Preferred model
Tills rich in block material	Dumping	'Alpine'
Clays, tills, sands and gravels Glaciotectonic structures	Shearing, thrusting, folding	Thrust ridges and push moraine
Sands, silt and gravels Distinct bedding and gravitational deformations (mainly faults)	Deposition in meltwater on and in glacier ice	Stagnant and dead ice forms

International Geomorphology 1986 Part II
Edited by V. Gardiner

THE DE GEER MORAINES IN FINLAND

H. Zilliacus

Department of Geography, University of Helsinki,
Hallituskatu 11-13, 00100 Helsinki, Finland.

ABSTRACT

The study is a review of De Geer moraines in Finland. Descriptions are given of their morphology and material, including granulometry, petrography, morphometry, structure and three-dimensional orientations. In addition, the distribution of their main occurrences is shown, particularly taking into consideration sub- and supra-aquatic areas and the formation of large ice lobes during deglaciation. Finally, the distances between successive ridges in groups of De Geer moraines from all parts of Finland were estimated and compared with clay varve data on the annual rate of the ice recession permitting an evaluation of the annual moraine problem. The De Geer moraines in Finland are low and ridge-like forms consisting mainly of till. Some differences between the material in separate parts of the ridges were found, and sometimes till is influenced by glaciofluvial drainage. A relationship between De Geer moraines and eskers, especially in interlobate complex areas, is also obvious from spatial evidence. The main ice movement is not usually reflected in the fabrics; instead, oblique orientations and large dipping inclinations are found. The average distance between successive ridges is about 100m for nearly all areas of Finland. This constant figure points to a regular deposition process under circumstances including glaciofluvial drainage and deep proglacial waters in topographical basins. Also typical is a location near the margins of ice lobes, where there probably have been cracks and crevasses. The De Geer moraines in Finland are probably submarginally deposited till ridges and thus not true end moraines. The depositional sites of the ridges and the material seem to be influenced by abundant melt waters squeezing up till into subglacial cavities. The spacing of the moraine ridges does not conform with the inferred rate of deglaciation in Finland based on clay varves, and therefore, the De Geer moraines lack a geochronological meaning.

INTRODUCTION

Small moraine ridges called De Geer moraines were originally described in Sweden by De Geer (1889). He called them 'annual moraines', because he thought that they were formed annually each winter at the ice front during deglaciation. Hitherto it has not been possible to show an exact geochronological correlation between

Fig.1. De Geer moraine in Mäntsälä, southern Finland. The height of the ridge, viewed from its proximal side, is c. 2.5m. The lowest parts of the moraine are obscured by clay deposits laid down in a subaquatic proglacial environment.

the deposition of these moraines and the annual retreat of the ice margin - as manifested for example by clay varves (Hoppe, 1948; Elson, 1953; Strömberg, 1965; 1971; Zilliacus, 1982). In recent decades, the so-called 'annual moraines' have therefore been given other names such as De Geer moraines (Hoppe, 1959; Prest, 1968; Flint, 1971; Sugden and John, 1976) or washboard moraines, as they were first called in North America (Mawdsley, 1936; Norman, 1938). Today more is known about the distribution, morphology, material and structure of the De Geer moraines, but their genesis and geochronological significance are still not clear.

DIMENSIONS AND MORPHOLOGY

The De Geer moraine ridges in Finland are typically about 200m long, 2-3m high and 10-15m wide (Fig.1). The length of the ridges, however, varies from some tens of metres up to about 2km (Sauramo, 1929; Zilliacus, 1976; 1981; Aartolahti, 1977; Tikkanen, 1981). The height rarely exceeds 5m, but parts may be obscured by clay deposits (Mansikkaniemi, 1973). The width of the ridge seldom exceeds 25m.

The cross-section is often symmetrical with both sides inclining between 15-30°. It is common to find asymmetrical ridges with a somewhat steeper distal side, too. A steeper proximal side, however, is clearly exceptional (Zilliacus, 1976; 1981). The crest of the ridge is usually fairly sharp and ridge-like, but sometimes the whole form is smoothed out over large areas. The levelled cross-section may be original, but it is more likely to result from littoral evening out and redeposition during emergence from the sea.

DISTRIBUTION OF DE GEER MORAINE AREAS

The map in Fig.2 shows the distribution of major De Geer moraine areas in Finland. It is based mainly on existing studies and geomorphological maps (Aartolahti, 1972; Fogelberg and Seppälä, 1979), but also includes information from field observations and investigations by the author during the years 1974 to 1984 (Zilliacus, 1976; 1981; 1982; 1984). The sites of the largest moraine areas were superimposed on a recently revised map (Eronen and Haila, 1981) showing sub-and supra-aquatic areas during final deglaciation in Finland.

It can be inferred from Fig.2 that the De Geer moraines were deposited in circumstances characterized by an ice front standing in water. They occur in low-lying areas, where the proglacial water depth varied from about 20m to almost 270m during deglaciation (Hyyppä, 1966). For instance the ridges in the largest De Geer moraine area on the islands of the Vaasa archipelago (Fig.1) and on the mainland around the town of Vaasa in western Finland were deposited in water some 260-270m deep (Zilliacus, 1981). The moraine ridges in the Vaasa archipelago appear as narrow and stony shoals, islets and headlands (Aartolahti, 1982; Jaatinen, 1982). Two other major concentrations of De Geer moraine areas occur in south-western and southern Finland. Small occurrences not recorded in Fig.2 are situated in southern and western Finland.

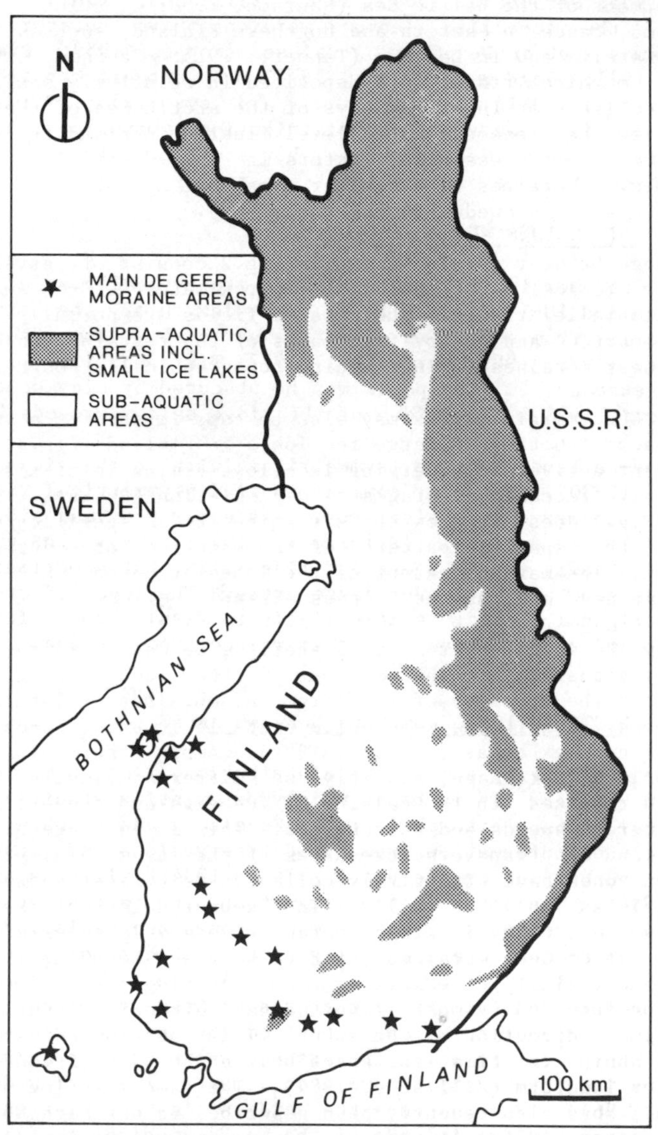

Fig.2. The main De Geer moraine areas in Finland.
The distribution is superimposed on a map showing sub- and supra-aquatic areas in Finland (Eronen and Haila, 1981; simplified). All moraine areas are situated in terrain once submerged by the Baltic.

The deposition of the large De Geer moraine areas in Finland took place both in the sea and in lakes during different stages in the development of the Baltic Sea (Sauramo, 1958; Ignatius et al., 1980). In some cases in eastern and northern Finland, such as Ilomantsi in North Karelia and in Lapland (Tanner, 1915; Sauramo, 1929; Ström, 1980), moraine ridges were deposited in relatively small and temporary ice lakes or in narrow bays of the Arctic Sea of that time. This indicates that seasonal water level fluctuations and the saltiness of the water are not essential factors in the deposition of De Geer moraines. Moraines of a similar type have also been discovered in a mountain area in Sweden (Borgström, 1979).

The large De Geer moraine areas in Fig.2 consist of several smaller moraine fields each including approximately 10-60 successive and almost parallel ridges. The moraine fields are normally well delimited, narrow and occupy flat parts of the terrain with a total lack of De Geer moraines in intervening topographically rougher areas.

The De Geer moraine areas may also be superimposed on a map showing the reconstruction of large ice lobes in Finland (Fig.3). The lobes represent active and diverging ice-flows coming into existence during successive recessional stages of deglaciation (Punkari, 1980). Therefore, Fig.3 does not show the situation at a specific moment, but rather the spatial pattern of the main ice lobes during different periods. The maximum extent of the lobes is often marked by large end moraines, such as the Salpausselkäs.

It may be inferred from Fig.3 that the De Geer moraines are predominantly situated near the margins of the ice lobes. They often appear in the vicinity of the large end moraines or laterally between the lobes. Conditions were quite unstable in these so-called interlobate complex areas (Punkari, 1984). Among other things, intensive glaciofluvial drainage, possibly on different levels in the ice sheet, resulted in the calving of icebergs and converging ice-flows of different speeds and thicknesses. This in turn resulted in the formation of transverse crevasses at the ice margin. These deglaciation zones have effectively collected melt water and glaciofluvial material as well as till. The geomorphology of the interlobate complex areas today is often characterized by large eskers, kames and groups of De Geer moraines (cf.Fogelberg and Seppälä, 1979). In the immediate vicinity of eskers, the moraine ridges sometimes bend from a transverse to almost longitudinal direction, thus adopting the approximate direction of the esker. This pattern indicates a close relationship in time and space between the deposition of these two types of landform (Zilliacus, 1984). The same bending of De Geer moraines has also recently been described in northern Norway (Sollid and Carlsson, 1984). In more central parts of the ice lobes such as south-western Finland (Fig.3), successions of De Geer moraines are encountered in connection with eskers, too.

The De Geer moraines were deposited during practically all temporal stages of deglaciation in Finland, representing a period of approximately 3,000 years from 12,000-9,000 B.P. (Sauramo, 1958; Hyvärinen, 1975; Ignatius et al., 1980). Only in certain parts of the country, however, did the retreat of the ice margin during this time result in the formation of large De Geer moraine areas.

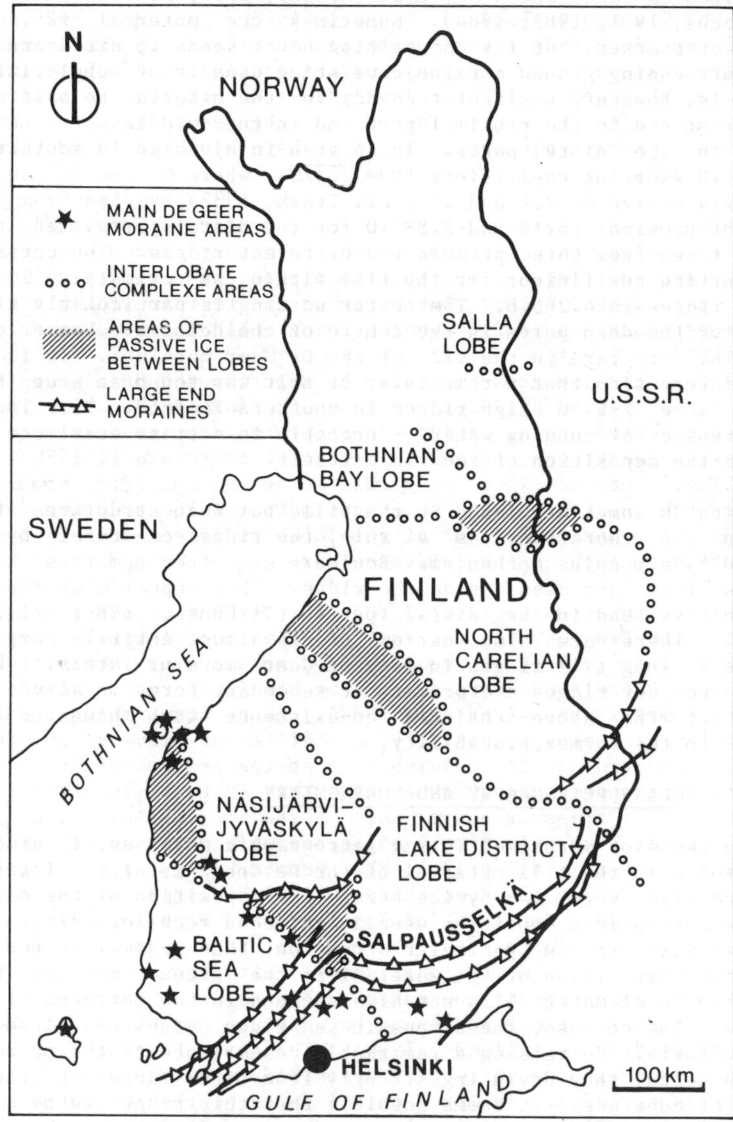

Fig.3. The distribution of De Geer moraine areas in Finland, superimposed on a map showing the reconstruction of large metachronous ice lobes during deglaciation (lobes after Punkari, 1980). The moraines are predominantly situated in the vicinity of interlobate complex areas.

COMPOSITION AND STRUCTURE OF THE MATERIAL

The material in the De Geer moraines in Finland may normally be described as sandy and stony till (Mölder, 1954; Mansikkaniemi, 1973; Zilliacus, 1981; 1982; 1984). Sometimes the material is somewhat finer or coarser, but its composition never seems to differ much from the surrounding ground moraine consisting usually of subglacial till. There is, however, a slight tendency for the material to be finer and better-sorted in the proximal part and in the middle of the ridge than in the distal parts. In an area in Hyvinkää in southern Finland, the sorting coefficient ($Sd=\sqrt{Q_3/Q_1}$, where Q_3 and Q_1 represent the grain-size at 75% and 25%; cf. Trask, 1932), varies from 3.2-6.0 for the proximal parts and 3.8-8.0 for the distal parts. The samples were taken from three pits in two different ridges. The corresponding sorting coefficient for the till within approximately 20m from these ridges is 4.2-5.8. The better sorting is particularly significant for the deep parts in the centre of the deposit. Lenses of fine material are rare in the till of the De Geer moraines. It is therefore interesting that a thin layer of silt was found at a depth of 3m under some 2-2.5m high ridges in southern Finland. This indicates the presence of running water - probably in a transversal crevasse - before the deposition of the ridge itself.

Layering is sometimes found in the till, but such structures are not common and therefore, as a rule, the ridges can rarely have been shaped by a pushing mechanism. Boulders are often numerous both on the surface and deep inside the ridge. They appear most frequently on the crest and in the distal parts (Tikkanen, 1981; Zilliacus, 1981). There are also narrow ridges almost entirely composed of boulders along till ridges in many De Geer moraine areas. Whether these boulder ridges are primary or secondary forms is as yet uncertain, but the above-mentioned co-existence with unwashed ridges points to the former probability.

CLAST PETROGRAPHY AND MORPHOMETRY

On the basis of morphometric and petrographic analyses, it seems that the stones in the till material of the De Geer moraine ridges were not carried very far by ice before the deposition of the material. The petrographic composition usually reflects rock occurrences some 2-10km away in an up-glacier direction, thus resembling the corresponding composition of the material in the ground moraine nearby. Strictly local material is not always found in the petrography of the stones. In one case, the stones in a ridge situated close to a glaciofluvial delta show a remarkable resemblance to the petrography of the delta, thus deviating strongly from other ridges in the same group of moraines. It seems possible that this ridge has been influenced by glaciofluvial material being carried by melt water to the delta. This assumption is strengthened by relatively high roundness values.

The stones in the De Geer moraines in Finland are generally subangular with roundness values corresponding to those of stones in the surrounding areas (Mansikkaniemi, 1973; Zilliacus, 1981; 1984). In many ridges, the material seems to be less rounded in the proximal parts than in other parts of the ridge. The average roundness values (2r/L x 1000; cf.Cailleux, 1961) for the three pits mentioned above are 146-172 for proximal parts and 166-222 for the rest of the ridge. Other observations of clast roundness from De Geer moraines in Finland suggest greater homogeneity (Mansikkaniemi, 1973; Zilliacus, 1981). The roundness of till samples from ground moraine at a depth of 0.5m in the Hyvinkää area varies from 181 to 203.

CLAST ORIENTATION

The orientation of elongated clasts in De Geer moraines has been studied in only a few parts of Finland (Ignatius, 1949; Mansikkaniemi, 1973; Zilliacus, 1984). Only rarely have three-dimensional analyses been carried out and the results from these investigations are far from clear. There is often a distinct peak in the fabric, but its orientation may vary from a direction parallel with the long axis of the ridge (Mansikkaniemi, 1973), normal to the long axis of the ridge, or in a direction corresponding to the assumed ice-flow in the area. It seems obvious, therefore, that the fabric of De Geer moraines does not reflect ice movement trends as clearly as the adjacent till material of the ground moraine. Nor do the fabrics show definite signs of pushing by the ice. Local factors seem to have been important in the depositional process. Likewise, there is no clear difference in the orientation of the clasts from the proximal and distal parts of the ridge, although the peak normal to the long axis of the deposit is usually somewhat stronger distally than proximally. The fabrics vary more in the central parts of the ridges. Orientations of the large boulders on the surface of De Geer moraines are often hard to measure, but in one case, on a curved ridge in southern Finland, 52 clearly visible boulders (>0.5m) were measured showing an orientation consistently at right angles to the long axis of the ridge. Most boulders were situated in distal parts of the deposit.

The stones in the De Geer moraines which have been studied in Finland usually dip a little less than the surface above, but dip in the same direction as the ridge itself. Thus, the stones in the proximal parts of the ridge dip up-glacier, while stones in the distal parts dip down-glacier (Fig.4). In the central parts of the ridge, both directions of dips can be found. Thus, in the middle of one ridge in the same swarm of forms as the one in Fig.4, the ratio of up/down-glacier dips is 26%:62% at a depth of 0.5m and 54%:36% at a depth of 1m. Some clasts have no dip or dip in the direction of the crestline of the ridge. In the till underneath the ridge and in tills of the ground moraine some 15m south of the ridge in Fig.4 most stones dip gently in an up-glacier direction. In northern Norway, fairly similar inclinations and an up-glacier dip are reported at nearly all De Geer moraine sites analysed (Sollid and Carlsson, 1984). This observation, then, is in contradiction with observations made from ridges in southern Finland. Inclinations of greater than 45° are excluded from the three-dimensional analyses in order to make the horizontal measurements more reliable (cf.Krüger, 1970). These stones usually make up 5-15% of all stones in a sample pit.

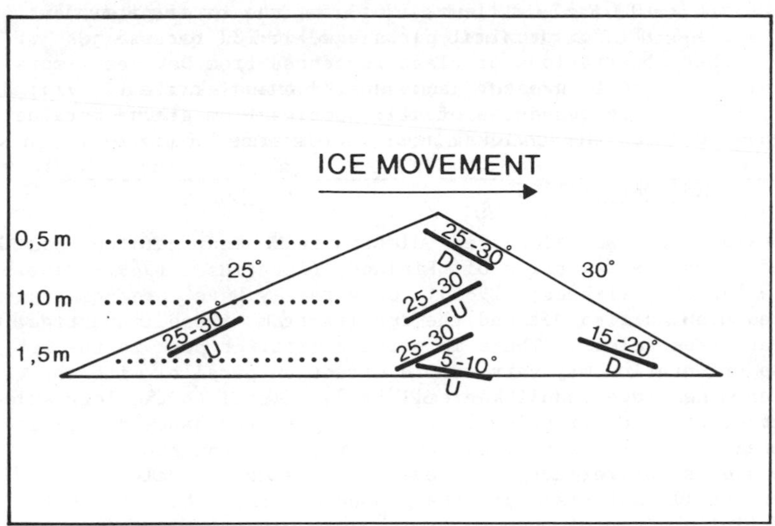

Fig.4. Cross-section of a De Geer moraine from Hyvinkää, southern Finland, indicating the dips of stones in different parts of the ridge. Modal dip of classes are shown. The height of the ridge is approximately 2m. U = more then 50% of the stones in the sample dip in an up-glacier direction, D = more than 50% dip in a down-glacier direction. Each sample comprises 50 stones and inclinations greater than 45° are excluded.

THE GEOCHRONOLOGICAL PROBLEM

Distances between successive ridges in the De Geer moraine areas were estimated in order to correlate these values with ages derived from varved clays for the rate of ice recession in Finland (Sauramo, 1918; 1929; Niemelä, 1971). This distance, measured in the assumed direction of the ice recession from the crest of one De Geer moraine to the next in a series of ridges, has been called the 'interdistance' (Zilliacus, 1981). In order to throw some light on the geochronological significance of De Geer moraines, a comparison can be made between the average interdistance for a group of moraines and the values given by the varve chronology for the annual retreat of the ice margin in the same area. This has been done for many areas, particularly in Sweden, but the results point mostly to a rejection of 'the annual moraine theory' (Hoppe, 1948; Strömberg, 1965; 1971). Only a few studies (Järnefors, 1956; Möller, 1962) support the old theory. In Finland geochronological investigations of this kind were very rare until recently (Ignatius, 1949; Zilliacus, 1976; 1981; 1984). The De Geer moraines are mainly used in reconstructions of the deglaciation pattern (Aartolahti, 1972; Tikkanen, 1981); the studies have not focused on the geochronological problem or on the deposition of De Geer moraines.

TABLE 1. Interdistances from five De Geer moraine areas in Finland, and clay varve data on the annual speed of deglaciation (extrapolated if necessary)

Location	Number of inter-distances	Longest inter-distance (m)	Shortest inter-distance (m)	Average inter-distance (m)	Annual deglaciation speed (m/year)
Björkön, W.Finland	63	210	20	100	500
Karvia, W.Central Finland	18	200	60	120	340
Vehmaa, S.W. Finland	41	320	30	120	580
Ridasjärvi, S.Finland	22	140	40	100	70
Elimäki, S.E. Finland	15	160	40	80	60

The interdistances were studied in more than 50 swarms of ridges in the main De Geer moraine areas in Finland. The mean value for this distance, in the groups which were studied, varies from 60 to nearly 200m. A common value for many moraine fields is approximately 100m. This is also a typical value for many moraine fields in Sweden, Norway and in North America (Hoppe, 1948; Elson, 1957; Løken and Leahy, 1964; Strömberg, 1965; Sollid et al., 1973; Sollid and Carlsson, 1984), although longer distances are sometimes mentioned, especially in North America (Ignatius, 1958; Elson, 1968; Sugden and John, 1976). It is interesting to note that this distance is very near the annual velocity of the ice recession given for southern Finland by the clay varve chronology (Table 1 ; cf.Sauramo, 1918; Niemelä, 1971), but in central Finland, the clay varve chronology indicates a more rapid deglaciation in the order of 200 up to 500m/year. For the majority of the De Geer moraine areas, this would then imply the formation of several moraine ridges each year which, according to the old theory (De Geer, 1912; Möller, 1962) would mean a larger one during the winter and several smaller ones during the summer. It is, however, hard to find such a morphological periodicity in the moraine fields in Finland; indeed, it has been possible only in one locality in western Finland (Zilliacus, 1981). Thus - and since the material and the structure of successive ridges are fairly similar - it is impossible to distinguish between winter and summer moraines in Finland.

CONCLUSIONS

Studies of the De Geer moraine ridges in Finland seem to emphasize the hydrological circumstances in the marginal zone of the ice sheet during the deglaciation process. As a result of hydrostatic pressure in depressions inside the ice margin standing in deep proglacial water, melt water found its way transversally into basal cracks, enlarging them and determining the depositional sites and distances between the moraine ridges. The material in the De Geer moraines is slightly influenced by glaciofluvial drainage, but on the whole, the ridges are built up of subglacial till squeezed up in cracks from underneath the ice (cf.Price, 1973). Boulders and finer fractions may have accumulated on top of the ridge along with its deposition, assuming there to have been abundant melt water and cracks in the basal parts of the ice. According to this mechanism of deposition, the bottom of the ice did not move actively during or after the deposition of the De Geer moraines.

The De Geer moraines in Finland can hardly be given a precise geochronological meaning. The distance between successive ridges in a group of moraines, is approximately 100m in all areas in Finland. Only accidentally does this value coincide with clay varve data on the annual retreat of the ice margin. The De Geer moraines were not deposited annually, but rather as a result of a regular accumulation process inside the ice margin. The evidence seems to point to a rejection of the genesis of these moraines as true end moraines. Instead, their deposition is more likely to be connected with the formation of basal cracks and the melting of an ice sheet in the presence of both submarginal and proglacial waters.

Acknowledgements

I particularly wish to hank Professor Toive Aartolahti, Head of the Department of Geography at the University of Helsinki, for supporting me in my work since the beginning in 1973. I would also like to thank Mrs. Iris Lampila from the same department for drawing the figures and Mrs. Joan Nordlund, Helsinki, for checking the language of the text. The work has been carried out mainly with the financial help of the Research Council for the Natural Sciences, part of the Academy of Finland.

REFERENCES

Aartolahti, T., 1972. On deglaciation in southern and western Finland. Fennia, 114,1-84.

Aartolahti, T., 1977. Suomen geomorfologia (The geomorphology of Finland). Helsingin yliopiston maantieteen laitoksen opetusmonisteita, 12, 1-150.

Aartolahti, T., 1982. Suomen luonnonmaisemien kehitys (The development of natural landscapes in Finland). Terra, 94, 33-42.

Borgström, I., 1979. De Geer moraines in a Swedish mountain area? Geografiska Annaler, 61A, 35-42.

Cailleux, A., 1961. Application à la géographie des méthodes d'étude des sables et des galets. Curso de altos estudos geográficos, 2, Universidade do Brasil.

De Geer, G., 1889. Om en serie ändmoränvallar i trakten mellan Spånga och Sundbyberg. Geologiska Föreningens i Stockholm Förhandlingar, 11, 395-397.

De Geer, G., 1912. A geochronology of the last 12,000 years. Compte Rendu du XI:e Cong. Géol. Int., Stockholm 1910, 241-253.

Elson, J. A., 1953. Periodicity of deglaciation in North America since the Late Wisconsin maximum. Part II, Late Wisconsin recession. Geografiska Annaler, 35, 95-104.

Elson, J. A., 1957. Origin of wash-board moraines. Bulletin of the Geological Society of America, 68, 1721.

Elson, J. A., 1968. Washboard moraines and other minor moraine types, in Encyclopaedia of Geomorphology, (Ed. Fairbridge), pp1213-1219.

Eronen, M., and Haila, H., 1981. The highest shore line of the Baltic in Finland. Striae, 14, 157-158.

Flint, R. F., 1971. Glacial and Quaternary Geology. Wiley & Sons, New York.

Fogelberg, P., and Seppälä, M., 1979. General Geomorphological map 1:1,000,000. Atlas of Finland. Landforms, geomorphology. Folio 122.

Hoppe, G., 1948. Isrecessionen från Norrbottens kustland i belysning av de glaciala formelementen. Geographica, 20, 1-112.

Hoppe, G., 1959. Glacial morphology and inland recession in northern Sweden. Geografiska Annaler, 41A, 193-212.

Hyvärinen, H., 1975. Myöhäisjääkauden Fennoskandia - käsityksiä ennen ja nyt (Late-glacial palaeogeography of Fennoscandia). Terra, 87, 155-166.

Hyyppä, E., 1966. The Late-Quaternary land uplift in the Baltic sphere and the relation diagrams of the raised and tilted shore levels. Annales Academiae Scientiarum Fennicae A III, 90, 153-168.

Ignatius, H., 1949. Vuosimoreeneista (On annual moraines). Unpublished M.Sc. Thesis, University of Helsinki.

Ignatius, H., 1958. On the Late-Wisconsin deglaciation in Eastern Canada. Acta geographica, 16, 1-34.

Ignatius, H., Korpela, K., and Kujansuu, R., 1980. The deglaciation of Finland after 10,000 B.P. Boreas, 9, 217-228.

Jaatinen, S., 1982. Maamme saaristomaisemien luonne ja muuttuminen (Structure and change of the Finnish archipelagos). Terra, 94, 43-52.

Järnefors, B., 1956. Isrecessionen inom Uppsalaområdet. Geologiska Föreningens i Stockholm Förhandlingar, 78, 310-315.

Krüger, J., 1970. Till fabric in relation to direction of ice movement. Geografisk Tidsskrift, 69, 133-170.

Loken, O., and Leahy, E. J., 1964. Small moraines in southeastern Ontario. Canadian Geographer, 8, 10-21.

Mansikkaniemi, H., 1973. De Geer -moreenin rakenteesta Turussa (The structure of a De Geer moraine, Turku). Geologi, 25, 97-102.

Mawdsley, J. B., 1936. Wash-board moraines of the Opawica-Chibougamau area, Quebec. Transactions of the Royal Society of Canada. Ser 3, Sec IV, 30, 9-12.

Mölder, K., 1954. Maalajikartan selitys. Lehti B3 Vassa (Explanation to the map of Quaternary deposits. Sheet B3 Vaasa). Geological Map of Finland 1:100,000, Geological Survey of Finland.

Möller, H., 1962. Annuelle och interannuella moräner. Geologiska Föreningens i Stockholm Förhandlingar, 84, 134-143.

Niemelä, J., 1971. Die Quartäre Stratigraphie von Tonablagerungen und der Rückzug des Inlandeises zwischen Helsinki und Hämeenlinna in Südfinnland. Bulletin de la Commision géologique de Finlande, 253, 1-79.

Norman, G. W. H., 1938. The last Pleistocene ice-front in the Chibougamau district, Quebec. Transactions of the Royal Society of Canada. Ser 3, Sec IV, 32, 69-86.

Prest, V. K., 1968. Nomenclature of moraines and ice-flow features as applied to the glacial map of Canada. Geological Survey of Canada Papers, 67-57.

Price, R. J., 1973. Glacial and fluvioglacial Landforms. Oliver & Boyd, Edinburgh.

Punkari, M., 1980. The ice lobes of the Scandinavian ice sheet during the deglaciation in Finland. Boreas, 9, 307-310.

Punkari, M., 1984. The relations between glacial dynamics and tills in the Eastern Part of the Baltic Shield. Striae, 20, 49-54.

Sauramo, M., 1918. Geochronologische Studien über die spätglaziale Zeit in Südfinnland. Bulletin de la Commission géologique de Finlande, 50, 1-44.

Sauramo, M., 1929. The Quaternary geology of Finland. Bulletin de la Commission géologique de Finlande, 86, 1-110.

Sauramo, M., 1958. Die Geschichte der Ostsee. Annales Academiae Scientiarum Fennicae A III, 51, 1-522.

Sollid, J. L., and Carlsson, A. B., 1984. De Geer moraines and eskers in Pasvik, North Norway. Striae, 20, 55-61.

Sollid, J. L., Andersen, S., Hamre, N., Kjeldsen, O., Salvigsen, O., Sturod, S., Tveitå, T., and Wilhelmsen, A., 1973. Deglaciation of Finnmark, North Norway. Norsk Geografisk Tidsskrift, 27, 233-325.

Ström, O., 1980. Drumliinit ja vakoumat Muotkatunturien alueella Suomen Pohjois-Lapissa (Drumlins and flutings in the Muotkatunturit area of northern Finnish Lapland). Turun yliopiston maantieteen laitoksen julkaisuja, 91, 1-31.

Strömberg, B., 1965. Mappings and geochronological investigations in some moraine areas of south-central Sweden. Geografiska Annaler, 47A, 73-82.

Strömberg, B., 1971. Isrecessionen i området kring Alands hav. Forskningsrapport, Stockholms universitet, Naturgeografiska Institutionen, 10, 1-156.

Sugden, D. E., and John, B. S., 1976. Glaciers and landscape. Edward Arnold, London.

Tanner, V., 1915. Studier öfver Kvartärsystemet i Fennoskandias nordligaste delar, III. Bulletin de la Commission géologique de Finlande, 38, 1-815.

Tikkanen, M., 1981. Georelief, its origin and development in the coastal area between Pori and Uusikaupunki, south-western Finland. Fennia, 159, 253-333.

Trask, P. D., 1932. Origin and environments of source sediments of petroleum. Natural Research Council, Report of the Commission of Sedimentology, 67-76.

Zilliacus, H., 1976. De Geer -moräner och isrecessionen i södra Finlands östra delar (De Geer moraines and the ice recession in the eastern parts of southern Finland). Terra, 88, 176-184.
Zilliacus, H., 1981. De Geer -moränerna på Replot och Björkön i Vasa skärgård (The De Geer moraines on the islands of Replot and Björkön in the Vasa archipelago, western Finland). Terra, 93, 12-24.
Zilliacus, H., 1982. De Geer -moränerna i Finland. Unpublished Lic.Phil. Thesis, University of Helsinki.
Zilliacus, H., 1984. On the moraine ridges east of Pohjankangas in northern Satakunta, western Finland. Geologiska Föreningens i Stockholm Förhandlingar, 106, 335-345.

International Geomorphology 1986 Part II
Edited by V. Gardiner

A REVIEW OF CONTROVERSIAL ISSUES RELATED TO THE LATE PALAEOZOIC GLACIATION OF SOUTHERN SOUTH AUSTRALIA

R. P. Bourman

School of Applied Science, South Australia
College of Advanced Education, Underdale Site,
Holbrook Road, Underdale, South Australia. 5032.

ABSTRACT

Controversial issues related to the Late Palaeozoic glaciation that affected southern South Australia include considerations of the nature of the evidence used to demonstrate glaciation, the identification of genuine glacigene sediments, the interpretation of current landforms as relics of the glaciation, the age of the glaciation, the direction of ice movements and the nature of the glaciation, including the possibility of multiple phases.

During the Late Palaeozoic a wet-based ice mass of continental proportions transgressed southern South Australia in a general northwesterly direction. Superimposed upon this movement were numerous local variations in ice flow. The precise age of the glaciation has yet to be firmly established and it may straddle the Carboniferous/Permian boundary. Contrary to interpretations from elsewhere in Australia there is as yet no definitive evidence supporting multiple glaciation in southern South Australia. All evidence for multiple glaciation presented so far can be accounted for by the progressive build up, transgression, and decay of a single ice mass. However, this interpretation may be modified in the light of new evidence. Some glacial landforms, preserved by burial and only recently exhumed, persist in the modern landscape, whereas some other landforms ascribed to glacial origins may be non-glacial landforms of post-Permian age.

INTRODUCTION

The passage of glacier ice over South Australia has long been known, with the first discovery of glacial action in Australia being made in 1859 in the Inman Valley, south of Adelaide, by A. R. C. Selwyn, a Victorian Government Geologist working under contract to the South Australian Government (Selwyn, 1860). Selwyn discovered a large area of strongly grooved and striated Cambrian metasedimentary bedrock, now known as 'Selwyn Rock', in the channel of the Inman River, which had also exposed a section of "drift composed of fragments of all sizes, irregularly imbedded through the clay." Interest in the imprints of glacial action in South Australia was reawakened by the discovery of additional areas of striated bedrock at Hallett Cove, immediately south of Adelaide (Tate, 1879). David and Howchin (1897) found further evidence of glaciation when they sought the original

'Selwyn Rock', and in the ensuing 30 years, Howchin published prolifically on the glaciation, demonstrating its occurrence on Fleurieu Peninsula, Kangaroo Island and Yorke Peninsula.

Many controversies have surrounded the glaciation in South Australia, and this paper examines those related to the original evidence for the glaciation, the recognition of glacigene sediments, the possible preservation of glacial landforms in the modern landscape, the age of the glaciation, the nature of the glaciation, the direction of ice movement and the number of glacial episodes.

EVIDENCE FOR THE GLACIATION

The report of evidence of glacial action at Hallett Cove by Tate (1879) was greeted with some scepticism and led to the first controversy in Australian geomorphology (Scott, 1977). In particular, Scoular (1879 and 1885) preferred to explain the striated bedrock in terms of water erosion and sand driven by the north wind. He deduced that glacial climatic conditions could not have prevailed near sea level in South Australia during the Pleistocene, to which period Tate (1879 and 1885) assigned the glaciation. In hindsight both workers were partly correct: the evidence for glaciation south of Adelaide is incontestable, although the glaciation has been shown to be of Late Palaeozoic and not Pleistocene age.

Indisputable evidence of an extensive glaciation in southern South Australia has now been provided by reports of numerous striated rock surfaces, the widespread occurrence of erratics from both local and distal sources and a wide range of distinctive glacigene sediments including lodgement and ablation tills, ice-contact fluvio-glacial and outwash fan deposits, and glacio-lacustrine and glacio-marine beds with associated dropstones (Fig.1). Those in the north of the State dominantly occur in sub-surface positions and are of particular interest because some are associated with commercial gas and oil reserves and coal.

It is now generally accepted that at the time of glaciation Australia together with the other southern hemisphere continents and India formed a super-continent, Gondwana, which was situated in high latitudes (Crowell and Frakes, 1975) and southern South Australia was at approximately 70°S (Ludbrook, 1980).

GLACIAL SEDIMENTS AND LANDFORMS

The early publication of discoveries of glacial action in South Australia was followed by a spate of unproven reports of glacial phenomena in the state. Some of the reports were undoubtedly influenced by the view of Tate (1879 and 1885) that the glaciation was of Pleistocene age. Examples include: <u>roches moutonnées</u> in the Mount Lofty Ranges at Crafers and Kaiserstuhl; glaciated pebbles from the Torrens Gorge and moraine near the gorge of the Torrens River (Tate, 1879); Quaternary alluvium at Ardrossan (Tepper, 1881); smoothed bedrock surfaces on Caroona Hill (Cleland, 1887); striated rock surfaces on Cambrian limestone at Curramulka (Pritchard, 1891); and

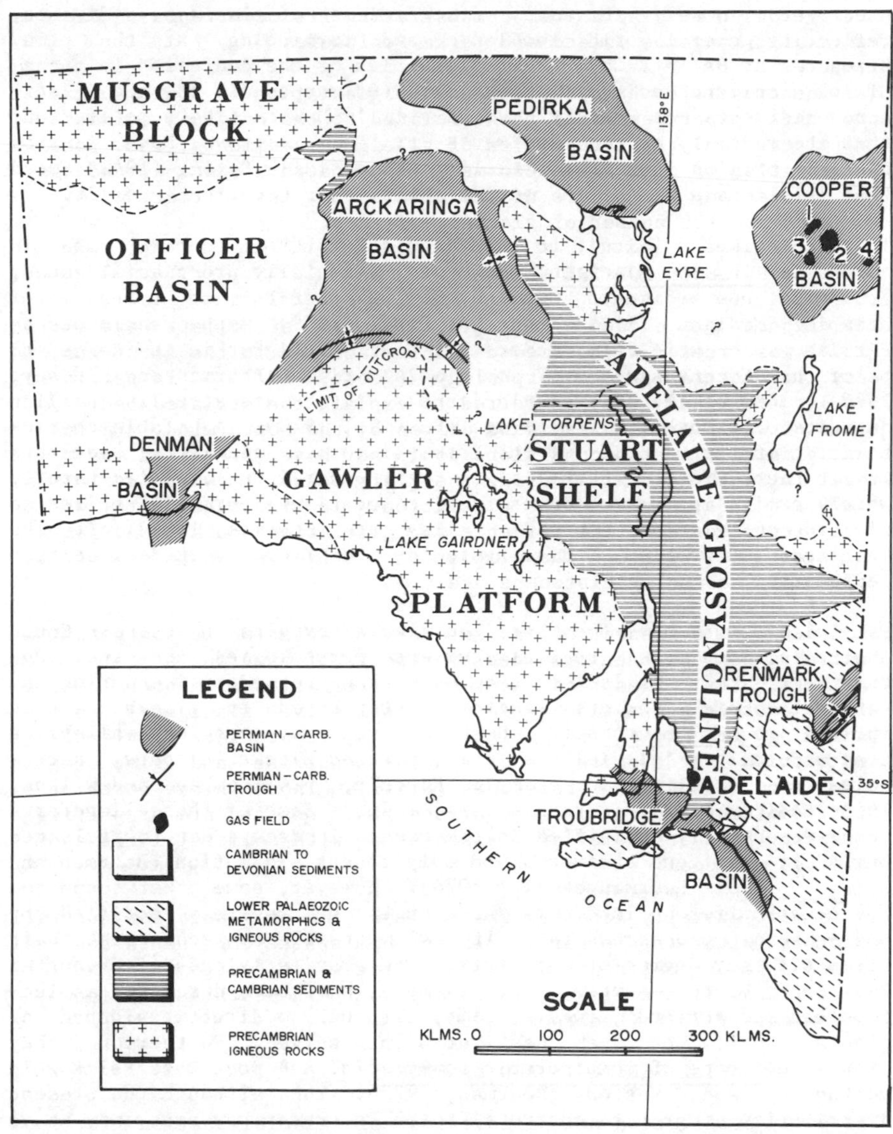

Fig.1. Basins of Late Palaeozoic sedimentation and distribution of basement rocks in South Australia (after Wopfner, 1970).

angular and rounded stones in sand deposits, smooth limestone surfaces and rounded hills near Mount Gambier (Priestley, 1892). All of the above were erroneously attributed to ice action.

In South Australia terms such as 'drift', 'till' and 'moraine' were used very loosely in early descriptions of glacigene sediments, reflecting general world knowledge and terminology at the time. Exposures at Hallett Cove provide examples of the confusion in recognising glacigene sediments. Segnit (1940) mapped all of the glacigene sediments at Hallett Cove as 'till', Mawson (1940) pointed out that the majority of the sediments at Hallett Cove have fluvio-glacial origins and that 'till' is rare, while Sprigg (1942) noted that 'till' is abundant and mapped three major occurrences of it.

It is sometimes difficult to distinguish both in outcrop and in boreholes *in situ* glacigene sediments, especially pro-glacial sands, from glacigene sediments reworked during post-Permian times. For example, Madigan (1925) and Crawford (1959) mapped sediments as Permian, which were later demonstrated on stratigraphic evidence to be of post-Eocene age and of non-glacial origin (Bourman and Lindsay, 1973). Some of the alleged glacigene sediments are actually aeolian deposits of probable Pliocene age (Bourman, 1973). A notable characteristic of the majority of the Late Palaeozoic glacigene sediments in Southern South Australia is their general un-lithified nature, which permits the use of investigative techniques normally applied to Pleistocene glacial deposits (Alley and Bourman, 1984). It also heightens the problem of distinguishing them from reworked glacigene sediments.

Many landforms have been ascribed glacial origins in southern South Australia, some merely on their current morphologies and knowledge that the areas including them were transgressed by ice during the Late Palaeozoic. Various landforms attributed to glacial action include *roches moutonnées*, U-shaped valleys, hanging valleys, a crag-and-tail, a 'glacial-platform', perched blocks, cirques, kettle holes and drumlins (for example, Howchin, 1926; Campana and Wilson, 1955; Madigan, 1925; Bourman et al., 1976). Some of these landforms occur essentially unmodified in the modern landscape due to prolonged burial by glacigene sediments and only recent exhumation (Bourman and Milnes, 1976; Bourman et al., 1976). However, some other suspected Permo-Carboniferous landforms have been so severely modified by post-glacial processes that it is impossible to reconstruct their forms with any degree of certainty. The granite islands of Encounter Bay near Victor Harbor, for example, have been described as ice-smoothed and abraded (Howchin, 1926) although no direct evidence of glacial action on them has yet been observed. Furthermore, they display evidence of considerable non-glacial and post Late Palaeozoic weathering and erosion (Bourman, 1973). Thus although the present distribution of granite outcrops in the Encounter Bay area may have been broadly determined by glacial erosion during the Late Palaeozoic (Milnes and Bourman, 1972), any traces of glacial action on the islands have been obliterated. Some other landforms that have been ascribed glacial origins can be shown to have developed via non-glacial processes in post-Permian times. For example, Howchin (1926) attributed hummocky ground (Fig.2) to preserved Late Palaeozoic kettle holes whereas it has actually resulted from recent gilgai formation (Bourman, 1969).

Fig.2. Hummocky ground reflecting gilgai development at the head of Ducknest Creek, Fleurieu Peninsula, South Australia.

AGE OF THE GLACIATION

Selwyn (1860) made no assertion concerning the age of the glaciation, but Brown (1892) associated the striated Selwyn Rock with the adjoining 'drift', which he mapped as Tertiary. Tate (1979) correlated the glaciated bedrock at Hallett Cove with the Pleistocene glaciations of the Northern Hemisphere, but excavations later revealed that the glacigene sediments are overlain by fossiliferous marine sandstone, then thought to be of Miocene age (Tate et al., 1895) but now known to be Pliocene. Howchin (1895) correlated the glacial deposits at Hallett Cove with fossiliferous glacigenic sediments of Bacchus Marsh and thus assigned them a Permo-Carboniferous age, but Tate (1895) remained unconvinced about this correlation because he regarded the striated surfaces and the sediments at Hallett Cove as much fresher than those at Bacchus Marsh.

The possibility of a Cretaceous age for the glaciation, or for a Cretaceous glaciation additional to that of the Late Palaeozoic in South Australia has also been raised. Segnit (1940) suggested an Early Cretaceous age for the glacial beds at Hallett Cove but this was refuted by Sprigg (1942). Support for a Cretaceous age was given by Crespin (1954), but the fossils on which this suggestion was based actually occurred in Eocene sediments (Ludbrook, 1969). David (1950) discussed the possibility of a Cretaceous glaciation in South Australia, but Parkin (1956) and Flint et al., (1980) have demonstrated that the critical deposits were probably Late Palaeozoic glacigene sediments reworked by the Cretaceous seas.

Prior to the work of Ludbrook (1957; 1967) no diagnostic fossils had been found in the glacigene sediments. Brackish water, arenaceous foraminifera of presumed Early Permian age found in glacio-marine sediments in the Troubridge Basin on Yorke Peninsula and Fleurieu Peninsula were used to fix the age of the glaciation, particularly as Ludbrook (1967) considered the fossils to come from near the base of the sedimentary sequence. However, Alley and Bourman (1984) showed that the marine fossiliferous unit represents only the final deglaciation stage of the glacial sequence at Cape Jervis.

Exacerbating the problem of determining the precise age of the Late Palaeozoic glaciation in South Australia are the difficulties of defining the Permo/Carboniferous boundary in Australia and correlating it with international chronostratigraphic sections. Some biostratigraphers have assigned the glaciation a Permian age (e.g. Ludbrook, 1957 and 1967; Foster, 1974), although Foster noted that a Late Carboniferous age was not impossible. Using palynological criteria established by Balme (1980), Cooper (1981 and 1983) suggested a revision of the Permo/Carboniferous boundary in South Australia and placed the early part of the glaciation in the Late Carboniferous. This suggestion was criticised on both palynological (Foster, 1983) and faunal grounds (Archbold, 1982), and the timing of the Permian/Carboniferous boundary in Australia remains controversial.

There is no doubt that some of the glacigene sediments in southern South Australia are of Early Permian age. C. B. Foster and J. B. Waterhouse (personal communication) have shown that sediments in Western Australia palynostratigraphically correlative with those at Waterloo Bay, southern Yorke Peninsula (Foster, 1974) are of Permian age. The palynological assemblages belong to the Granulatisporites confluens Zone, which, in Western Australia occurs with marine fauna (brachiopods, crinoids, bivalves, gastropods, bryozoans and foraminiferida) dated as Late Asselian to Sakmarian. These dates were derived from pro-glacial sediments.

However, the resolution of the age of the glaciation will require that biostratigraphic dates be obtained from facies clearly associated with distinctive glacigenic sediments, including basal lodgement tills, which seldom yields diverse or well preserved fossils. Recently, however, Early Permian pollen were recovered from two basal lodgement till samples at Hallett Cove and from a borehole in the Murray Basin (N. F. Alley - personal communication), which favours the restriction of the glaciation to the Permian. Further work on the critical lodgement till facies will help to resolve the age of the glaciation in South Australia, which will also depend on the resolution of the location of the Australian and international Carboniferous/Permian boundaries.

NATURE OF THE GLACIATION

Many views concerning the nature of the glaciation have been presented. Howchin (1926) envisaged a terrestrial glaciation of continental proportions after he reconstructed former ice movements from evidence in Tasmania, Victoria and South Australia and concluded that ice had spread from a central point to the southwest of Tasmania.

On the other hand, Campana and Wilson (1955) maintained that over-deepening of valleys and the presence of glacial bars and alleged cirques in the area are indicative of a mountain or valley type glaciation. The amount of valley deepening is considerable as revealed by boreholes at Myponga and Back Valley (Fig.3). At the latter site the glaciated bedrock floor is 240m below sea level, while glacial sediments nearby (not shown on section) occur up to 300m above sea level. There is no evidence of significant post-Permian tectonic dislocation to account for this distribution and the surviving glacial relief is at least 540m. If the highest bedrock occurrence, 10km away at Clarke Hill is used, and the evidence suggests that the entire area was submerged by ice, sub-glacial relief was at least 678m. Glacial deepening and irregular bedrock topography, however, are not restricted to areas affected by valley glaciations. Moreover, the recognition of cirque-like forms as providing evidence for a mountain type glaciation is questionable, particularly in view of the difficulties in the reconstruction of the glacial relief in the light of the evidence of extensive erosion in post-glacial times. The direction of glacial movement is relevant to the nature of the glaciation, for if the site of the present Mount Lofty Ranges had been the source of the ice-mass, then there should be evidence of glacial movement radiating from this area, but the evidence indicates general movement of ice from the southeast (Fig.4) overriding the remnants of a dissected fold mountain range.

Glaessner (1962) commented that the areas of Late Palaeozoic glaciation could have resulted from individually glaciated highlands rather than being related to parts of a circumpolar ice-cap. Wopfner (1970) favoured this interpretation and suggested that because of the dominance of local clasts in the erratic assemblages, the glaciers may have been localised plateau-types developed on up-faulted highlands. He argued that abundant local erratics in the Troubridge Basin of southern South Australia (Horwitz, 1960) indicated a similar style of glaciation in this area. It has been shown, however, that local lithologies dominate lodgement tills regardless of the type of glacier (Dreimanis, 1976). More importantly, many of the erratics of the Troubridge Basin are exotic; some have no known sources in Australia and may have been derived from Antarctica (Harris, 1971), which is considered to have been contiguous with Australia in the Late Palaeozoic (Crowell and Frakes, 1975). Consequently, in southern South Australia at least, the ice-mass was of large extent, being more than 540m thick with ice flow of at least 500km from the southeast to the northwest as indicated by the presence of diagnostic erratics (Harris, 1971; Milnes and Bourman, 1972; Alley and Bourman, 1984) so that a continental ice-mass must be envisaged. This is not to deny that in favoured localities local plateau-type glaciers may have existed in the north of the State. However, Flint et al., (1980) suggested that fossiliferous Lower Devonian boulders in Cretaceous sediments of the Great Australian Basin had been transported from the Cobar region of New South Wales by Late Palaeozoic ice of continental dimensions as suggested by Crowell and Frakes (1975) and

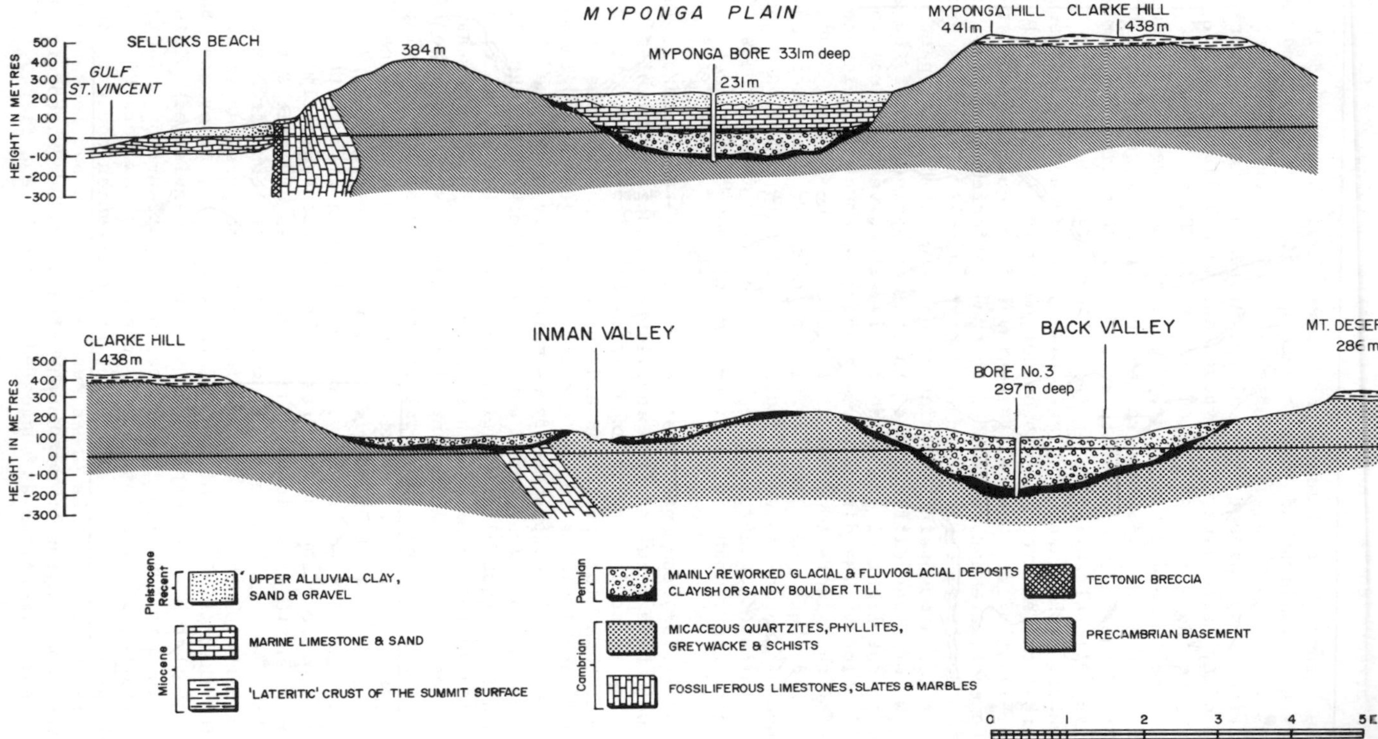

Fig.3. Late Palaeozoic glacial valleys on Fleurieu Peninsula (after Campana et al., 1953)

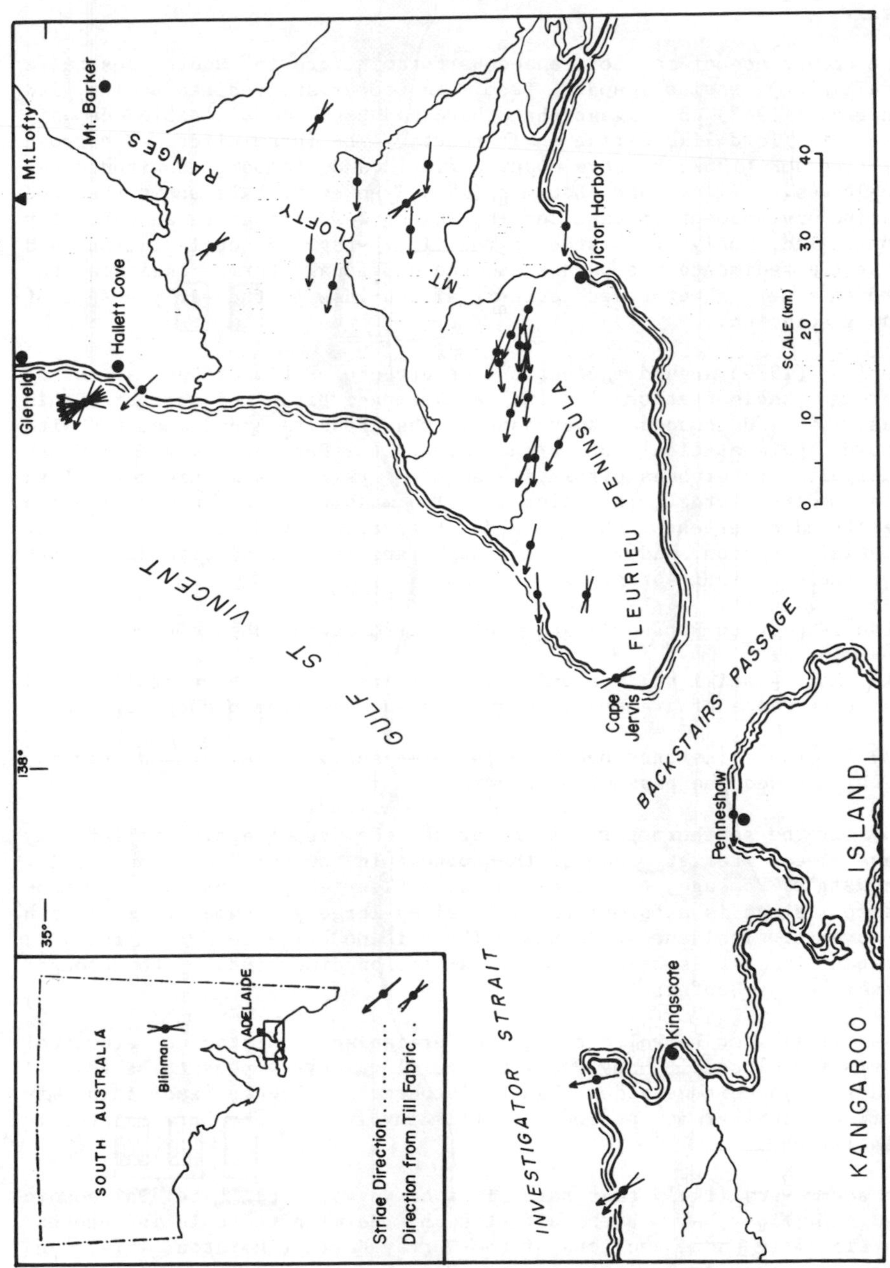

Fig.4. Late Palaeozoic ice movements across southern South Australia.

were subsequently reworked by the Cretaceous seas. Hence, continental style glaciation may have occurred in the north of the State as well.

The occurrence of marine arenaceous foraminifera in South Australia sedimentary basins, apart from the Cooper and Pedirka Basins, led Ludbrook (1967) to suggest that there had been considerable deposition in fiord-like marine environments. The foraminifera were shown by Ludbrook (1967) to have modern fiord and iceberg environmental analogues. Alley and Bourman (1984), however, have shown that the marine environment of Cape Jervis, which was probably eustatically controlled, only persisted in the final stages of deglaciation, and that the sedimentary sequence and the erosional forms indicate the presence of a wet-based terrestrial ice-mass in the early stages of the glaciation.

Wopfner (1970) argued against the occurrence of fiords as sites of Permian sedimentation in the Arckaringa Basin of northern South Australia. He conceded that some of the deposits are compatible with fiord sedimentation but argued that the Permian troughs in South Australia are grabens displaying subdued relief and greater width than modern fiords. The point was also made that pre-Permian in situ weathered corestones, which could not have survived the process of glacial erosion, underlie the troughs and provide additional support for their tectonic origin.

With respect to Wopfner's arguments the following should be noted:

(1) Many glacial valleys and fiords are tectonically controlled and the occurrence of grabens does not prevent ice from occupying them.

(2) If the corestones had been in permanently frozen ground they may have survived the passage of glaciers.

(3) In the southern part of the State the topographic relief may have been greater than in the north. In southern South Australia, Backstairs Passage, the strait between Fleurieu Peninsula and Kangaroo Island is a former glacial valley largely exhumed from beneath a cover of glacigene sediments. There is no evidence for it being a graben and it is of similar cross-section dimensions to the troughs described by Wopfner (1970).

(4) It may not be valid to relate Permian and Pleistocene erosional forms too closely. Many Permian glacial valleys appear to be several orders of magnitude wider than Pleistocene analogues and this may reflect the great period of time involved in the Late Palaeozoic glaciation.

It seems very likely that some sites of very thick Late Palaeozoic sedimentation have been affected by Permian tectonic influences, particularly under sections of the Murray Basin (Thornton, 1974) and also in the Arckaringa Basin. However, the possible existence of former fiords or fiord-like forms cannot be dismissed.

At its maximum extent, the glaciation is best represented as a large-scale terrestrial ice-mass covering much of southern South Australia, and in places over-riding a topography of considerable relief. Insofar as the extensive post-Permian erosion of the Mount Lofty Ranges allows, the ice-mass at its maximum may have been at least 1000m thick with the topographic relief of the ranges being possibly 2000m. However, both during the onset of glacial conditions and during deglaciation, discrete ice tongues and plateau and cirque glaciers may have existed, giving rise to forms akin to those of valley glaciers. In the final deglacial stages, particularly in coastal lowland environments, fine-grained sediments may have been deposited in eustatically-controlled marine to brackish environments with restricted access to the open sea. Such a model could account for the very poor marine fauna of the southern South Australian Permian, quite unlike the more diverse shelly macro-faunas recorded elsewhere in eastern and western Australia.

Fig.5. Large erratics of metasandstone and granite 3km northwest of Victor Harbor, derived from source areas 10km to the east

DIRECTION OF GLACIAL MOVEMENT

Evidence for the direction of glacial movement has been derived from striated rock surfaces, erosional forms such as roches moutonnées, the provenance of erratics (Fig.5) and fabrics of lodgement tills.

At present there are 25 areas of glacially striated rock surfaces that have been found on Fleurieu Peninsula, Kangaroo Island and at Hallett Cove. These display grooves, striae (including needle striae), polished surfaces, various forms of friction cracks, 'p' forms and other micro-erosional features. Care needs to be taken to distinguish genuine glacial striae from similar markings resulting from tectonic activity and mass movements. Roches moutonnées in the area do not provide sound evidence for accurate ice movements as the morphology of some of them is dominated by bedrock structures.

Some erratics in southern South Australia are particularly useful as indicators of ice movement, but many of the larger erratics appear to be ice-rafted. Consequently they may not be as critical as indicators of glacial movement as erratics contained within lodgement tills or flow tills. Erratics related to some flow tills formed during glacier decay by sliding from the ice may have been relocated only relatively short distances from their original positions in the ice so that they can provide a general indication of ice movement.

Ice-transported Devonian micro-fossils have been recognised in Late Palaeozoic glacigene sediments on Yorke Peninsula in South Australia. These have possible source regions in the Grampians of Victoria or in Antarctica (Harris and McGowran, 1971), but because of the fragile nature of some of the spores, these authors postulated a more local, but now eroded, source in the area of the modern Mount Lofty Ranges. Erratics similar to the Encounter Bay Granites (Milnes and Bourman, 1972) occur on Yorke Peninsula (Crawford, 1965) and distinctive erratics on Fleurieu Peninsula have possible sources in the South East of South Australia (Alley and Bourman, 1984) and Dergholm in Victoria (Milnes and Bourman, 1972), 500km to the southeast. Where neither striated surfaces nor indicator erratics occur, fabrics of basal lodgement tills have been utilised to throw light on the direction of ice flow (Alley and Bourman, 1984). Using a combination of these various criteria some reasonably useful reconstructions of ice flow have been made (Fig.4).

Many earlier workers including Crowell and Frakes (1971a and 1971b) considered that the direction of ice movement across southern South Australia was generally from the south to the north and even to the northeast (Crowell and Frakes, 1975). However, distinctive erratics derived from igneous rocks in the Murray Basin and from Western Victoria suggest that the ice passed across southern South Australia from the southeast to the northwest. Superimposed upon this general direction are various deviations as revealed by striae and till fabrics. For example, in the Mount Lofty Ranges, during the glacial maximum, it appears that ice was funnelled in a more westerly direction through the Inman Valley across Fleurieu Peninsula (Milnes and Bourman, 1972; Bourman and Milnes, 1976; Bourman et al., 1976) and through Backstairs Passage between Kangaroo Island and the mainland. On the western side of the ranges the ice converged with another ice stream moving in a more northerly direction along the site of the modern Gulf St. Vincent and together they continued northwesterly across the gulf to Yorke Peninsula. These ice-flow directions probably reflect variations in movement beneath a single ice-cap, rather than the passage of separate ice-lobes.

In some places ice flow was affected by bedrock topography such as at Hallett Cove where a pre-existing bedrock valley caused the ice flow to vary from 280° to 15° (Sprigg, 1942; Milnes and Bourman, 1972). However, on central Fleurieu Peninsula ice passed transversely across a pronounced north-south trending structural valley developed along the contact of Archaean and Proterozoic bedrock. In detail, much variability in striae direction is revealed on small stoss and lee features. At Blinman in the Flinders Ranges an isolated outlier of basal glacigene sediments indicates a general northwesterly flow of ice (Morton et al., 1984).

The ice flow directions discussed above probably relate only to the stage of maximum glaciation, and during the advance and build up of the ice mass smaller tongues of ice may have followed pre-existing depressions.

MULTIPLE GLACIATION: EVIDENCE FROM SOUTHERN SOUTH AUSTRALIA

Various workers have proposed multiple phases of glaciation in southern South Australia, but to date no crucial evidence suggesting that any more than one glaciation occurred has come to light. In this paper, 'multi glaciation' refers to major glaciations separated by non-glacial periods, and does not include localised glaciers formed immediately prior to the advance of and during the decay of the major ice mass.

Fig.6. Crossing striae on striated bedrock surface, Hallett Cove, South Australia.

Crowell and Frakes (1971a and 1971b) considered that crossing striae at Hallett Cove (Fig.6) indicated the readvance of a second ice-mass across a previously striated surface. However, the several sets of striae do not bear consistent cross-cutting relationships and can be adequately explained by alteration in ice movement during one glacial advance by erosion of the underlying bedrock or by variations in stress distributions within the ice. The variation in stress can be attributed to alterations in the ice-overburden.

Maud (personal communication) suggested that lithified fluvioglacial sediments, squeezed sediments and drumlin-like features represent evidence of multiple glaciation, but all of these features are equally explicable in terms of a fluctuating ice-mass during a single glaciation.

Bowen (1959) interpreted deposits at Cape Jervis and Hallett Cove as containing multiple tills that indicate two separate glacial advances at the former locality and three at the latter. Recent work at Cape Jervis (Alley and Bourman, 1984) demonstrated that there is only one genuine lodgement till at the base of the sequence and that other till-like deposits represent flow tills derived by sliding from the ice surface and from icebergs during the stagnation and decay of the ice-mass. The section at Cape Jervis appears to reveal evidence of the deteriorating climatic conditions prior to the glaciation, the existence of a temperate terrestrial ice-mass and successive environments, which reflect the progressive decay of that ice-mass. Work at Hallett Cove reveals a similar sequence, with only one lodgement till being present, but with at least nine diamictons (flow tills) that superficially resemble lodgement tills. From the north of the state, Wopfner (1964) described a sequence of events in the Lower Permian of glaciation-marine incursion-freshwater deposition in a swamp environment, which is also compatible with a single major glaciation.

Nowhere in southern South Australia have unquestionable interglacial sediments been reported to separate genuine glacial deposits, nor has more than one complete glacial-deglacial sequence been described. It is possible that with successive glaciations, large parts of the older glacial materials may be eroded so that two different lodgement tills may occur without intervening beds typical of deglaciation being preserved. Thus, the possibility of multiple glaciation in Southern Australia remains, but on present evidence there appears to have been no more than one glacial episode related to the Late Palaeozoic glaciation preserved and observed in southern South Australia.

CONCLUSIONS

The investigation of the Later Palaeozoic glaciation in southern South Australia has produced considerable controversy. The following summarises the current state of these controversies.

(1) The former presence of an ice-mass in southern South Australia is unquestionable.

(2) The age of the glaciation is undoubtedly Late Palaeozoic and some of its related sediments are of Early Permian age, but the possibility of whether the glaciation was initiated in the Late Carboniferous remains unresolved.

(3) The glaciation was of continental style, but forms commonly associated with mountain type glaciations may have developed during deglaciation when discrete ice-masses may have existed, although the evidence for these forms is dubious. In suitable environments isolated highland plateau glaciers may have existed at the same time.

(4) The preserved landforms and glacigene sediments suggest temperate (wet-based) ice during the glacial advance. During the final deglacial stage glacio-marine sediments may have accumulated in a brackish environment, which would account for the paucity of marine macro-faunas in the glacigene sediments.

(5) There is no critical evidence suggesting multiple glaciation in southern South Australia. Moreover, all of the observed phenomena can be accounted for by the passage and progressive decay of a single ice-mass.

(6) It may be impossible in some cases to distinguish glacigene from reworked glacigene sediments and to identify resurrected Late Palaeozoic glacial landforms.

(7) The direction of ice movement across southern South Australia was generally from the southeast to the northwest, but superimposed on this direction were more westerly and northerly trends, which reflect topographic influences and variations in the stress field of the ice.

Acknowledgements

I am grateful to Dr. N. F. Alley, Mr. G. Crawford, Dr. C. B. Foster, Dr. N. H. Ludbrook, Dr. A. R. Milnes, Prof. H. Wopfner and an unknown referee for constructive comments on an early draft of this paper, and to Mr. M. Tscharke for drafting the figures.

REFERENCES

Alley, N. F., and Bourman, R. P., 1984. Sedimentology and origin of Late Palaeozoic glacigene deposits at Cape Jervis, South Australia. Transactions Royal Society South Australia, 108, 63-75.

Archbold, N. W., 1982. Correlation of the Early Permian faunas of Gondwana: implications for the Gondwana Carboniferous-Permian boundary. Journal Geological Society Australia, 299, 276-76.

Balme, B. E., 1980. Palynology and the Carboniferous-Permian boundary in Australia and other Gondwana continents. Palynology, 4, 143-155.

Bourman, R. P., 1969. Landform Studies near Victor Harbour. Unpublished B.A. (Hons) Thesis, University of Adelaide, pp145.

Bourman, R. P., 1973. Geomorphic evolution of Southeastern Fleurieu Peninsula. Unpublished M.A. Thesis, University of Adelaide, pp126.

Bourman, R. P., and Lindsay, J. M., 1973. Implications of fossiliferous Eocene marine sediments underlying part of the Waitpinga drainage basin, Fleurieu Peninsula, South Australia. Search, 4,77

Bourman, R. P., and Milnes, A. R., 1976. Exhumed roche moutonnée. Australian Geographer, 13, 214-216.

Bourman, R. P., Maud, R. R., and Milnes, A. R., 1976. Late Palaeozoic glacial features near Mount Compass, South Australia. Search, 7, 488-490.

Bowen, R. L., 1959. Late Palaeozoic glaciation of eastern Australia. Unpublished Ph.D. Thesis, Melbourne University, pp265.

Brown, H. Y. L., 1892. Geological report upon a shale deposit in the Hundreds of Encounter Bay and Yankalilla. Parliamentary Papers South Australia. 23, 13-14.

Campana, B., Wilson, R. B., and Whittle, A.G.W., 1953. Geology of the Jervis and Yankalilla Military Sheets. Geological Survey of South Australia. Report of Investigation No.3, 1-26.

Campana, B., and Wilson, R. B., 1955. Tillites and related glacial topography of South Australia. Eclogae Geologicae Helvetiae, 48, 1-30.

Cleland, W. L., 1887. Caroona Hill (Lake Gilles). Transactions, Royal Society South Australia, 10, 74-77.

Cooper, B. J., 1981. Carboniferous and Permian sediments in South Australia and their correlation. Quarterly Geological Notes, Geological Survey South Australia, 79, 2-6.

Cooper, B. J., 1983. Carboniferous and Permian sediments in South Australia and their correlation. A reply to C. B. Foster. Quarterly Geological Notes, Geological Survey South Australia, 87, 10-15.

Crawford, A. R., 1959. Encounter Geology Sheet 1:63360. Geological Survey, South Australia.

Crawford, A. R., 1965. The geology of Yorke Peninsula. Bulletin Geological Survey South Australia, 39, 1-96.

Crespin, I., 1954. Stratigraphy and micropalaeontology of the marine Tertiary rocks between Adelaide and Aldinga, South Australia. Rep. Bureau of Mineral Resources, Geology and Geophysics 12, pp65 and plates 1-7.

Crowell, J. C., and Frakes, L. A., 1971a. Late Palaeozoic glaciation of Australia. Journal, Geological Society Australia, 17, 115-155.

Crowell, J. C., and Frakes, L. A., 1971b. Late Palaeozoic glaciation; Part IV Australia. Bulletin, Geological Society America, 82, 2515-2540.

Crowell, J. C., and Frakes, L. A., 1975. The Late Palaeozoic Glaciation, in Gondwana Geology, (Ed. Campbell), pp313-331. University Press, Canberra.

David, T. W. E., 1950. The Geology of the Commonwealth of Australia. Vol.I, pp746. Edward Arnold, London.

David, T. W. E., and Howchin, W., 1897. Notes on the glacial features of Inman Valley, Yankalilla and Cape Jervis districts. Transactions, Royal Society South Australia, 21, 61-67.

Dreimanis, A., 1976. Tills: Their origin and properties, in Glacial Till, (Ed. Legget) pp11-49. Special Publication, Royal Society of Canada, 12.

Flint, R. B., Ambrose, G. J., and Campbell, K. S. W., 1980. Fossiliferous Lower Devonian boulders in Cretaceous sediments of the Great Australian Basin. Transactions, Royal Society South Australia, 104, 57-66.

Foster, C. B., 1974. Stratigraphy and palynology of the Permian at Waterloo Bay, Yorke Peninsula, South Australia. Transactions, Royal Society South Australia, 98, 29-42.

Foster, C. B., 1983. Carboniferous and Permian sediments in South Australia and their correlation by B. J. Cooper: a discussion. Quarterly Geological Notes, Geological Survey of South Australia, 87, 5-10.

Glaessner, M. F., 1962. Isolation and communication in the geological history of the Australian fauna, in The Evolution of Living Organisms (Ed. Leeper) pp242-249. Symposium Royal Society Victoria.

Harris, R. F., 1971. The geology of Permian sediments and erratics, Troubridge Basin, South Australia. Unpublished B.Sc. (Hons) Thesis, University of Adelaide, pp33.

Harris, W. K., and McGowran, B., 1971. Permian and reworked Devonian microfossils from the Troubridge Basin. Quarterly Geological Notes, Geological Survey of South Australia, 40, 5-11.

Horwitz, R. C., 1960. Geologie de la region de Mt. Compass (feuille Milang) Australia Meridionale. Eclogae Geologicae Helvetiae, 53, 211-263.

Howchin, W., 1895. New facts bearing on the glacial features at Hallet Cove. Transactions, Royal Society of South Australia, 19, 61-69.

Howchin, W., 1926. Geology of the Victor Harbour, Inman Valley and Yankalilla districts with reference to the great Inman Valley glacier of Permo-Carboniferous age. Transactions, Royal Society South Australia, 50, 89-116.

Ludbrook, N. H., 1957. Permian foraminifera in South Australia. Australian Journal of Science, 19, 161.

Ludbrook, N. H., 1967. Permian deposits of South Australia and their fauna. Transactions, Royal Society South Australia, 91, 65-75.

Ludbrook, N. H., 1969. The Permian Period, in Handbook of South Australian Geology, (Ed. Parkin) pp117-129. Geological Survey South Australia.

Ludbrook, N. H., 1980. Geology and Mineral Resources of South Australia. Department of Mines and Energy, South Australia, pp230.

Madigan, C. T., 1925. The geology of the Fleurieu Peninsula, Part I: The coast from Sellick Hill to Victor Harbour. Transactions, Royal Society South Australia, 49, 198-212.

Mawson, D., 1940. Tillite and other rocks from Hallett Cove, South Australia. Transactions, Royal Society South Australia, 64, 362.

Milnes, A. R., and Bourman, R. P., 1972. A Late Palaeozoic glaciated granite surface at Port Elliott, South Australia. Transactions, Royal Society South Australia, 95, 149-155.

Morton, J. G. G., Alley, N. F., Hill, A., and Griffiths, M., 1984. Possible Late Palaeozoic glacigene sediments near Blinman, Flinders Ranges. Quarterly Geological Notes, Geological Survey South Australia, 91, 9-15.

Parkin, L. W., 1956. Notes on the younger glacial remnants of northern South Australia. Transactions, Royal Society South Australia, 79, 148-151.

Priestley, P. H., 1892. Notes on glacial phenomena about Mount Gambia. Transactions, Royal Society South Australia, 15, 123-124.

Pritchard, P. H., 1892. On the Cambrian rocks at Curramulka. Transactions, Royal Society South Australia, 15, 179-182.

Scott, H. J., 1977. The development of Landform Studies in Australia, pp282. Bellbird Publishing Services, Artarmon, N.S.W.

Scoular, G., 1897. The Geology of the Hundred of Munno Para. Transactions, Royal Society South Australia, 2, 60-69.

Scoular, G., 1885. Past climatic changes with special reference to the occurrence of a glacial epoch in Australia. Transactions, Royal Society South Australia, 8, 6-48.

Segnit, R. W., 1940. Geology of Hallett Cove with special reference to the distribution and age of the younger till. Transactions, Royal Society of South Australia, 64, 3-44.

Selwyn, A. R. C., 1860. Geological notes of a journey in South Australia from Cape Jervis to Mount Serle. Parliamentary Papers Adelaide, 20, 1-4.

Sprigg, R. C., 1942. Geology of the Eden-Moana fault block. Transactions, Royal Society South Australia, 66, 185-214.

Tate, R., 1879. Leading physical features of South Australia. Transactions, Royal Society South Australia, 2, 64.

Tate, R., 1885. Post-Miocene climate in South Australia. Transactions, Royal Society South Australia, 8, 49-59.

Tate, R., 1895. The anniversary address of the president. Transactions, Royal Society South Australia, 19, 266-277.

Tate, R., Howchin, W., and David, T. W. E., 1895. Report on the research committee on glacial action in Australasia. Australasian Association Advancement of Science, 6, 315-320.

Tepper, J. G. D., 1881. Geological and physical history of Hundred of Cunningham. Transactions, Royal Society South Australia, 4, 61-70.

Thornton, R. C. N., 1974. Hydrocarbon potential of Western Murray Basin and Infrabasins. Report of Investigations, Geological Survey of South Australia, 41, pp43.

Townsend, I. J., and Ludbrook, N. H., 1975. Revision of Permian and Devonian nomenclature of four formations in and below the Arckaringa Basin. Quarterly Geological Notes, Geological Survey of South Australia, 54, 2-5.

Wopfner, H., 1964. Permian-Jurassic History of the Western Great Artesian Basin. Transactions, Royal Society South Australia, 88, 117-128.

Wopfner, H., 1970. Permian palaeogeography and depositional environment of the Arckaringa Basin, South Australia, in Second Gondwana Symposium, South Africa, pp273-291.

International Geomorphology 1986 Part II
Edited by V. Gardiner

GLACIERS OF THE SOUTHERN LYNGEN PENINSULA, NORWAY

J. E. Gordon, W. B. Whalley,
A. F. Gellatly and R. I. Ferguson

Respectively:
Geological Conservation Unit, Nature Conservancy Council, Pearl House, Bartholomew Street, Newbury, Berkshire, U.K.
Department of Geography, Queen's University, Belfast, U.K.
Dept. of Geography, Sheffield University, Sheffield, U.K.
Dept. of Environmental Sciences, University of Stirling, Stirling, U.K.

ABSTRACT

Several glaciers in the southern part of the Lyngen Peninsula are described with reference to their recent recession and supposed mass balance changes. Summit plateau glaciers above c.1600m appear to be little changed in recent years but those below this altitude show very substantial ice loss. The valley glaciers have receded by as much as 0.5km in the last 30 years. They are mostly fed by avalanches from the plateaux and thus respond to this ice supply. Hence the usual ideas of accumulation area ratios and glaciation level for the valley glaciers need to be modified. This has implications for modelling Pleistocene glaciers. The highest plateau glaciers currently appear to have almost stable mass balances and beds frozen to the underlying blockfields.

INTRODUCTION

The southern part of the Lyngen peninsula (70°N, 20°E; 1:50,000 topographic maps, AMS 1533 I, 1633 IV) of maritime North Norway is a deeply dissected mountain region composed of gabbroic rocks (Randall, 1971). In the terminology of Sugden and John (1976), an alpine type of landscape on the western side gives way eastwards to one of selective linear erosion where plateau surface remnants survive between glacial troughs. In the central part of the peninsula there are small ice caps up to several square kilometres in extent (Figs.1 and 2). These appear to be cold-based on the plateau remnants above c.1600m, the highest of which is Jiek'kevarri at c.1833m (Whalley et al., 1981). To varying degrees, avalanches from the ice caps nourish a series of valley and corrie glaciers (Østrem et al., 1973) which, apart from their margins where the winter cold wave may not be completely eliminated in summer, are probably temperate. The highest plateau ice caps appear to have been relatively stable in size over the last three decades, but some of the lower valley glaciers which descend to altitudes around 400m a.s.l. have receded by 0.5km. Farther east on the lower plateaux (1400-1500m), small thin ice patches which lie close to or below the present glaciation threshold have wasted appreciably in recent years.

In terms of glacier modelling and developing realistic modern analogues for Pleistocene mountain glacier environments, the Lyngen peninsula is instructive in several respects: first, in demonstrating topographic controls on glacier nourishment and particularly the relationships between plateau ice caps and valley glaciers; secondly, in demonstrating that cold-based plateau ice caps may be a substantial element in regional glacier cover but leave little significant geomorphological impact on the landscape; thirdly, by illustrating relationships of periglacial features to cold-based glaciers; and fourthly, in demonstrating the variations of glacier response to short-term variations in climate.

GLACIERS OF THE LYNGSDALEN AREA

Balgesvarri. Balgesvarri is a sloping plateau remnant at an altitude of c.1500-1625m, surrounded on all sides by precipitous cliffs. Most of its surface is covered by a small ice cap c.0.6 square kilometres in area (Figs.3 and 4). Several thin ice tongues draining from the ice cap plunge over the cliff edge (Whalley et al., 1981), feeding glaciers of various sizes to north and south of the plateau. A narrow ice-free "promenade" up to 100m wide extends discontinuously between the ice cap and the plateau edge. The largest ice-free area is a 0.25km long promontory extending beyond the northern edge of the ice cap. Much of the ice-free ground is covered by blockfields, frequently sorted into stone circles and lobes (Whalley et al., 1981).

There is now no snow/ice at the spot from which Mrs. Aubrey Le Blond took a photograph (Fig.4) probably in 1898 (Le Blond, 1908), but changes of thickness of the main plateau ice in the last hundred years appear to have been slight. This is in marked contrast to valley glacier changes marked by recessional moraine sequences, and to the estimated volume loss of Strupbreen (north Lyngen), over a similar period (Whalley, 1973). Some recession of the Balgesvarri ice which has reduced the discharge from the plateau to the glaciers below can be seen from examination of aerial photographs taken in 1952 and 1979, but it has not been possible to estimate fluctuations of elevation or volume of ice on the plateau itself. The Balgesvarri ice cap was surveyed in 1979 and again in 1984. During this period there has been little significant change in the position of the ice edge, but overall the ice cap has thickened by up to 2m in the summit region. A trench dug into the ice edge in 1979 showed that the margin of the glacier was frozen to its bed, that sorted stone circles extended beneath the glacier from the ice-free promenade, and that no debris was entrained in the basal layers of the glacier (Whalley et al., 1981). Preliminary results of stable isotope analyses of the basal ice support an apparent absence of melting and refreezing. Examination of lichen diameters on the plateau and its blockfield shows some recent ice recession, generally restricted to the south-western (and lowest altitude) portion arrowed in Fig.3 .

Recent snowpit stratigraphy (Fig.5) indicates no apparent significant deterioration in the state of health of the ice cap. In 1984, summer surfaces were clearly identified at depths of 0.98, 2.04, 2.76 and

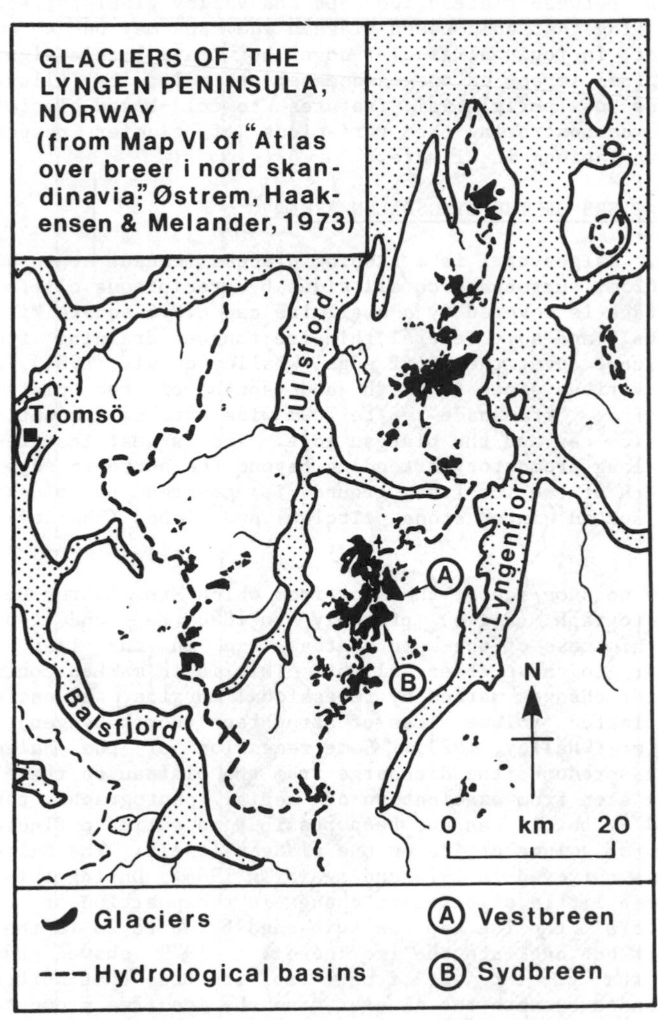

Fig.1. Location of the Lyngen Peninsula.

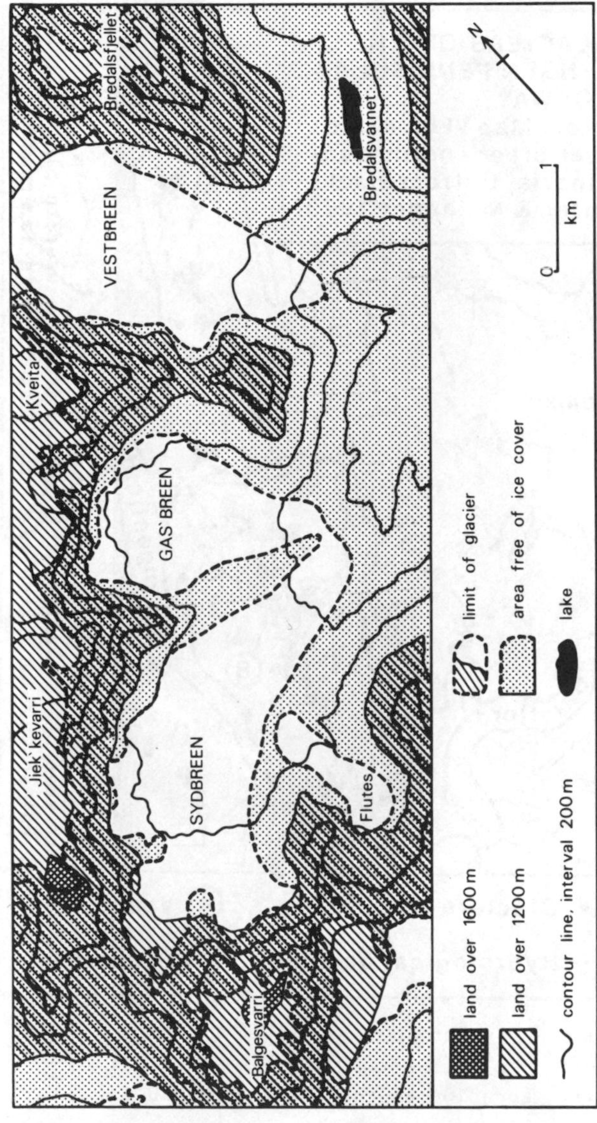

Fig.2. The main glaciers and mountains referred to in the text.

Fig.3. View of the northern edge of the Balgesvarri ice cap from the western plateau, July 1984. The zone which has shown most recession in the last 30 years is arrowed.

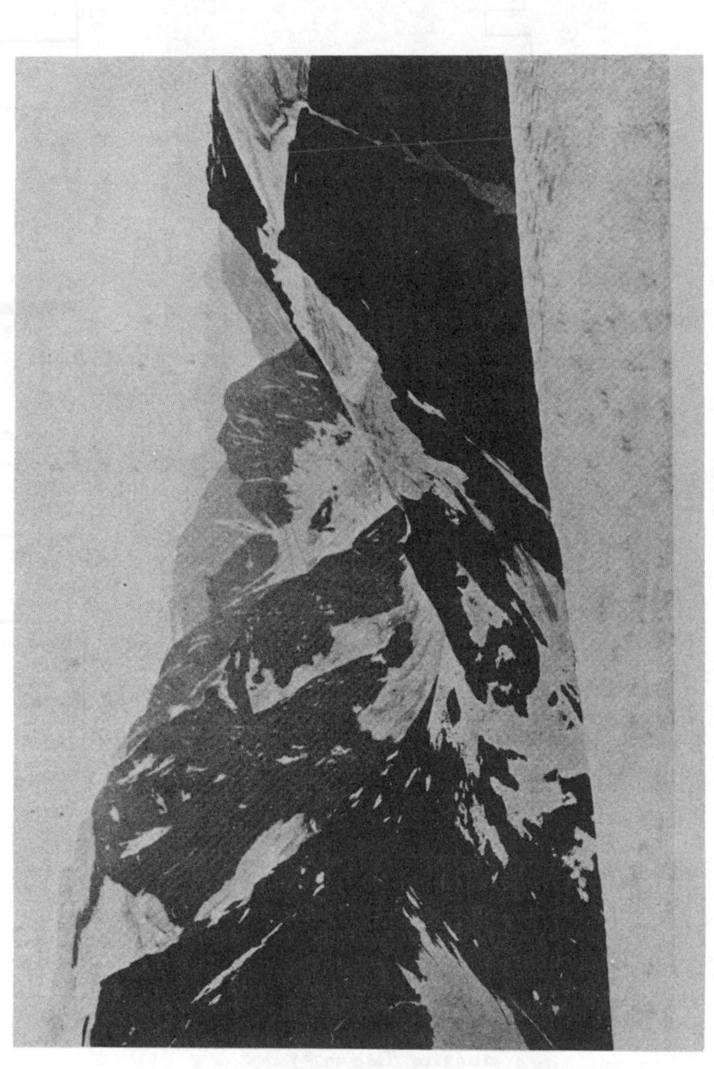

Fig.4. A similar view to Figure 1 taken by Mrs A. Le Blond in the summer of (probably) 1898.

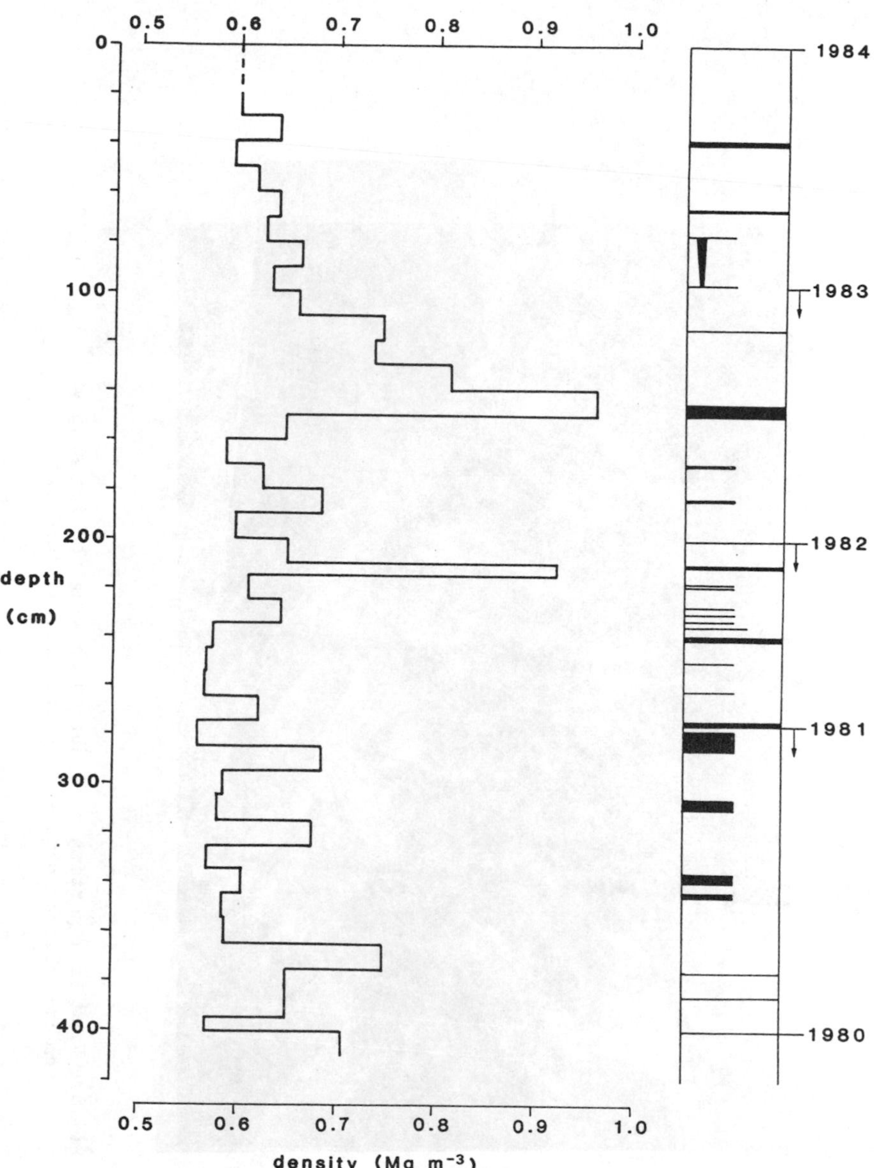

Fig.5. Snowpit stratigraphy from the Balgesvarri summit, August 1984. Main ice layers, veins and summer surfaces are indicated. Discontinuous ice layers are represented on the left hand side of the column.

Fig.6. View of a portion of Sydbreen showing the ice supply from the plateau of Jiek'kevarri (left), July 1984.

3.95m, appearing as slightly darker, dirtier layers 10-15cm thick in the snowpack. The top 2cm of the summer surfaces were less hard than adjacent firn layers. Thin ice layers existed immediately above the top and also at or towards the bottom of each summer surface. Interpretation of annual layers in a 1979 snowpit, where darker or dirty horizons were not encountered, is more tentative. Water equivalents estimated for individual balance years are given in Table 1. However, it should be noted that these figures do not necessarily represent the true net accumulation for each year because of the percolation of surface melt and its refreezing at depth in the snowpack.

TABLE 1. Water equivalents for summit snow pits on Balgesvarri (dug 1979 and 1984)

Year	Water equivalent (cm)
1983-84	60.0
1982-83	75
1981-82	44.3
1980-81	68.2
1979-80	indeterminate
1978-79	53.8
1977-78	61.5
1976-77	63.7

Sydbreen. Sydbreen (Figs.1 and 2) is a 4.5km long valley glacier occupying a deep trough, bounded to the north and northeast by 900m high rock walls that rise to the Jiek'kevarri plateau, and to the south and southeast by similar cliffs rising to the Balgesvarri plateau and tributary ridges. To the west, an arm of the glacier extends to c.1200m on the low col linking the two plateaux, while to the south a tributary valley occupied by an unnamed glacier (informally called 'Flutes glacier') breaks the continuity of the encircling trough walls. Sydbreen comprises three main septa fed by three separate accumulation areas. The northernmost and largest of these septa, which forms the main bulk of the glacier and the entire snout area, is sustained primarily by ice avalanches which calve from the summit icecap on Jiek'kevarri (Fig.6). The middle septum, fed from a corrie basin accumulation area, feathers out along the southern margin of the glacier. The southernmost septum is nourished largely by ice avalanches from the plateau icecap on Balgesvarri. Fluctuations of the icefront of Sydbreen are therefore closely linked to glaciological and climatic conditions on the glacier itself, and especially to those on the adjacent plateau icecaps, Jiek'kevarri in particular.

Former ice margin positions have been determined from air photographs taken in 1953, 1978 and 1979. These, together with a ground survey completed in 1984, indicate that Sydbreen receded by up to 580m between 1953 and 1984.

Exploration of ice caves along the southern margin of Sydbreen showed that the ice there is not frozen to the underlying bedrock. However, excavations along the front ice edge revealed isolated fluted moraine frozen beneath the glacier, a phenomenon much more strikingly developed on the forefield of the southern tributary glacier of Sydbreen ('Flutes glacier'). Here, preliminary results of stable isotope analyses suggest that basal ice melts and refreezes. The valley glaciers therefore appear to be temperate, with the possible exception of their front margins.

A series of low moraine ridges is well developed in front of the Sydbreen snout. They are 1-2m wide and up to 1m high and in section appear to consist entirely of subglacial till which overlies a relatively more consolidated till unit. In 1979 the ridges were interpreted as annual moraines, a view confirmed in 1984 by the formation of a further five ridges during the intervening period.

The spacing of the annual moraine ridges provides a measure of net annual recession over the last eleven years (Table 2). Those formed during the three winters 1981-82, 1982-83 and 1983-84 are notably more closely spaced than those formed in the previous years, suggesting a significant change in short-term glacier dynamics. Bearing in mind the dangers of over-simplistic correlations between glacier behaviour and weather patterns, there is nevertheless a striking correspondence of decreased glacier retreat with relatively lower mean temperatures during 1981, 1982 and 1983.

<u>The eastern plateaux : Bredalsfjell, Rundfjell and Daltind</u>. Ice on Bredalsfjell is thinning rapidly and may soon cease to exist. The 1952 aerial photographs show a continuous ice cover over the plateau. This subsequently decayed into the thin northern and southern icefields now visible. Fig.7 shows the current state of the northern section. From drilling in 1979 we estimate that the maximum thickness of the northern field is no more than 3m. No significant geomorphological evidence of this wastage, for example meltwater channels or moraines, is visible on the surface. Bredalsfjell still supports small glaciers on all but its southern flank (Fig.6), but they are no longer supplied from the summit ice.

The other eastern plateaux below c.1500m (Rundfjell and Daltind) support even less perennial ice than does Bredalsfjell, but lichen growth zones on blockfields (Fig.8) show that, as on Bredalsfjell, perennial ice was probably extensive until the rapid decay of the last one hundred years or so. All three plateau ice fields once contributed mass to small glaciers on their flanks, as do those of Balgesvarri and Jiek'kevarri today. The supply of ice to Vestbreen (Figs.1 and 2) now comes totally from the Jiek'kevarri side.

DISCUSSION

During the last five years there has been little substantial change in the extent of the plateau ice cap on Balgesvarri. Over the same period, however, the Bredalsfjell icefields have wasted appreciably and the front of the valley glacier Sydbreen has receded by some 41m.

Fig.7. Bredalsfjell, northern ice patch, July 1984.

Fig.8. View from the Daltind summit blockfield northwards to Bredalsfjell (B), Rundfjell (R), Jiek'kvarri south summit (J), and Vestbreen (V).

TABLE 2. Spacing of annual moraine ridges in front of Sydbreen.

Ridge	Distance from datum (m)	Net annual recession (m)
Winter 1973-74	-29.0	-
Winter 1974-75	-14.0	15.0
Winter 1975-76	0.0	14.0
Winter 1976-77	12.3	12.3
Winter 1977-78	24.9	12.6
Winter 1978-79	39.1	14.2
Winter 1979-80	54.3	15.2
Winter 1980-81	73.1	18.8
Winter 1981-82	82.2	9.1
Winter 1982-83	90.8	8.6
Winter 1983-84	96.7	5.9
Winter 1984-85	110.4	13.7
Icefront 27 July 1985	116.1	5.7

A similar pattern appears to have characterised the last three decades. These observations emphasise both the different response of different types of glacier in the same area to recent and contemporary climatic conditions and the existence of significant glacio-climatic thresholds in Lyngen between the altitude of Bredalsfjell (c.1500m) and Balgesvarri (1625m).

Nourishment of many valley and corrie glaciers in the southern Lyngen peninsula is by avalanching of ice and firn from the plateaux rather than by accumulation in extensive, contiguous accumulation areas. Under present day conditions the plateau glaciers are probably vital for the maintenance of the valley and corrie glaciers. The known limits of Neoglacial glaciers suggest it is likely that the plateaux have always been important during the existence of mountain glaciers in Lyngen. Thus, the conventional ideas of equilibrium line, equilibrium line altitude and accumulation area ratio (AAR) do not apply here. If the ice supply from Jiek'kevarri fails, then the valley glaciers will decay very rapidly. Ice on the Bredalsfjell plateau is no longer supplying the three glaciers which once descended from the plateau. At present, only small quantities of ice are contained in gullies and patches on the slopes of the mountain.

In areas of Pleistocene glaciation where mountains consist of plateau areas, the supply of ice from mountain tops to valley bottoms, as seen in Lyngen today, may have been an important control of glacier behaviour during deglaciation. The Lyngen model may be more representative of conditions during deglaciation than some conventional ideas which envisage valley-contained glacier systems. Accordingly, it is not possible to use calculated equilibrium line altitudes to reconstruct the valley ice masses: the dynamics of the valley and corrie glaciers are controlled by the accumulation on the highest summits. Unfortunately, virtually nothing is known about past accumulation and ablation on these summit accumulation areas.

It is possible that the Lyngen model also applies to other Norwegian mountain glacier systems, such as those at Svartisen and Okstindan to the south and Oksfordjøkul to the north of Lyngen. The summit plateau glaciers of Seilandsjøkul and Nordmannsjøkul, some 120km northeast of Lyngen, were reported to be much smaller in 1956 than in 1895-96 (Ford, 1958). These caps lie at 690-981m and 720-1075m respectively; in Lyngen, mountains at this altitude are unlikely to support any perennial ice. Deglaciation of some highland areas in the British Isles where plateau areas would have contributed ice to valleys could be related to the model also.

Østrem (1964) and Østrem et al., (1973) published maps of mean glacier elevation and glaciation levels in North Norway, based on aerial photography. However, interpretation of glaciation (glacierization) levels is not easy because of their time-transgressive nature. The source used by Østrem (1964) was the 1952/3 aerial photography; use of the 1978-79 photography probably would give different results because of the marked recession or disappearance of the lowest plateau ice. There are problems of evaluating the controls of glaciation levels and, indeed, the existence of the glaciers themselves; Flint (1971), Østrem (1972), Miller et al., (1975) and Porter (1975) favour slightly different influences of such controls as summer and winter mass balances, topographic effects and windward-leeward precipitation gradients. Climatic gradients in the Lyngen area cannot be evaluated due to lack of data. A further control may be a threshold altitude related to topography. The complex dynamics of rapidly-ablating valley glaciers fed by avalanches from plateaux, clearly seen in Lyngsdalen, makes it difficult to apply traditional accumulation area ratio models (for example, Sissons and Sutherland, 1976), where the assumption is of equilibrium conditions of mass balance producing static snout (terminal moraine) positions.

The rather low snow accumulation on Balgesvarri (about one metre per year), apparently in equilibrium with, or slightly in excess of, ablation, allows basal ice to be below the ice pressure melting point. All the investigated plateau ice masses show negligible subglacial debris entrainment. As such cold-based plateau ice caps probably leave little if any geomorphological evidence of their former existence, it is difficult to identify such conditions clearly in glaciated areas.

Periglacially-formed sorted stone circles are emerging undeformed from under the ice, indicating that the presence of periglacial features may not indicate the absence of glacier ice. This again suggests caution in the interpretation of glacier limits in glaciated plateau areas.

CONCLUSIONS

Studies of present glacier limits indicate the recent dynamic behaviour of a plateau and valley glacier system and the complex response of the latter to the former. Precipitation and altitude are probably important in maintaining the near neutral net mass balances of the higher plateau ice masses (1500m). However, the recession of the glacier snouts is governed primarily by the summer temperature; this asymmetry of response is accentuated by the large vertical height range of these glacier systems. Long-term climatic data need to be used to assess glacier response to changing precipitation and temperature effects. It is not yet possible to say how longer-term (more than 100 years) changes of climate have affected the limits of both plateau and valley glaciers in Lyngen. The lack of geomorphological evidence of the presence of plateau ice (despite its importance for the nourishment of the valley glaciers) suggests that care must be taken in the interpretation of ice limits in glaciated highland plateau areas.

Acknowledgements

We thank the members, helpers and sponsors of the 1979 and 1984 British Schools Exploring Society expeditions to Lyngen as well as friends and helpers in Furuflaten, Tromso and Lyngseidet. We gratefully acknowledge financial assistance from the University of Stirling, University College, Dublin, the Queen's University of Belfast and the Royal Society. We also thank the referees whose comments greatly improved the paper.

REFERENCES

Flint, R. F., 1971. Glacial and Quaternary Geology. Wiley, New York.

Ford, D. C., 1958. Seilandsjøkulen and Nordmannsjøkulen, Finnmark, Norway. Journal of Glaciology, 3, 249-252.

Le Blond, A., 1908. Mountaineering in the Land of the Midnight Sun. T. Fisher Unwin, London and Leipzig.

Miller, G. H., Bradley, R. S., and Andrews, J. T., 1975. The glaciation limit and lowest equilibrium line altitude in the high Canadian arctic: maps and climatic interpretation. Arctic and Alpine Research, 7, 155-168.

Østrem, G., 1964. Ice-cored moraines in Scandinavia. Geografiska Annaler, 48A, 126-138.

Østrem, G., 1972. Height of the glaciation level in northern British Columbia and south eastern Alaska. Geografiska Annaler, 54A, 76-84.

Østrem, G., Haakensen, N., and Melander, O., 1973. Atlas over breer i Nord-Skandinavia. Norges Vassdrags - og Elektrisitetsvesen. Meddelelse fra Hydrologisk Avdeling, Nr 22.

Porter, S. C., 1975. Glaciation limit in New Zealand's Southern Alps. Arctic and Alpine Research, 7, 33-37.

Randall, B. A. O., 1971. The igneous rocks of the Lyngen Peninsula, Troms, Norway. Norges Geologiske Undersøkelse, 269, 143-146.

Sissons, J. B., and Sutherland, D. G., 1976. Climatic inferences from former glaciers in the south-east Grampian Highlands, Scotland. Journal of Glaciology, 17, 325-346.

Sugden, D. E., and John, B. S., 1976. Glaciers and Landscape. Edward Arnold, London.

Whalley, W. B., 1973. A note on the fluctuation of the level and size of Strupvatnet, Lyngen, Tromso, and the interpretation of ice loss on Strupbreen. Norsk Geografisk Tidsskrift, 27, 39-45.

Whalley, W. B., Gordon, J. E., and Thompson, D. L., 1981. Periglacial features on the margins of a receding plateau ice cap, Lyngen, North Norway. Journal of Glaciology, 27, 492-496.

QUATERNARY GEOMORPHOLOGY

International Geomorphology 1986 Part II
Edited by V. Gardiner

QUATERNARY GEOMORPHOLOGY

INTRODUCTION

Peter Worsley

Department of Geography, University of Nottingham,
Nottingham, UK

The term Quaternary geomorphology is a somewhat enigmatic one since it has not, as yet, developed an independent identity of its own. Consequently, it is not surprising to discover that the eight papers included here are rather diverse in nature and the cynic will be excused if he or she thinks that we have created a convenient refuge for studies which do not fit readily into other categories. Nevertheless, each does have a concern with events in time and this is their linking characteristic.

It is worthwhile to recall the definition of Quaternary. If we follow the American Geological Institite we find that it is "the younger of the two geological periods or systems in the Cenozoic era .... it comprises all geologic time and deposits from the end of the Tertiary until <u>and including the present</u> " - the emphasis being due to the present writer. Perhaps we might suggest that an improvement would be the addition of the word landforms to highlight the importance of morphological factors in recent earth history. Unlike all the other elements of time occurring since the origin of the earth, the Quaternary is unique in that it progressively becomes longer by the day and present day processes are one of its prime concerns. Accordingly it is difficult to conceive of a viable geomorphology without any linkage with the Quaternary but once the time domain becomes prominent we find ourselves using stratigraphical methodology and soon become engrossed with litho and biostratigraphy. In effect we enter the field of Quaternary geology <u>per se</u>. Indeed, recently Clayton (1985) has suggested that British Quaternary geomorphologists should either recognise that their research is Quaternary geology and change their departmental affiliations from geography to geology or adopt a different kind of research activity more in keeping with the geographical heartland. Yet the paradox is that all the papers included here emanate from Departments of Geography.

The situation is not, of course, unique to geomorphology and we find the same issue in association with other disciplines e.g. the botanists become labelled palaeo-ecologists when they undertake studies of plants in relation to former environments yet they too have to master stratigraphic principles. Over 50 years before the notion of an international grouping of geomorphologists was seriously mooted, the International Union for Quaternary Research or

INQUA was launched in Copenhagen (1928) in order to bring together those workers from diverse disciplines who wished to have a forum within which their common interest in the changing face of the earth during the last two million years or so could be stimulated. Geomorphologists have been prominent in the activities of INQUA and at least one has served as President. However, it is clear that the vigorous growth of geomorphology is such that it is now urging upon becoming a discipline independent of its traditional guardians - geography and geology. It is to be hoped that the relatively minor component which Quaternary geomorphology forms in these proceedings will not be the pattern for the future. Geomorphology has a vital contribution to make to the elucidation of Quaternary events and the growing current awareness of the value of lessons to be learnt from past geomorphological activity, highlights its value to predictions of future changes.

REFERENCE

Clayton, K.M., 1985. The state of Geography. Transactions of the Institute of British Geographers. New Series, 10, 5-16.

International Geomorphology 1986 Part II
Edited by V. Gardiner

PALAEO-ENVIRONMENT OF THE LAST GLACIAL (DALI) STAGE IN NORTH CHINA

Sun Jianzhong and Li Xingguo

Xian Geological College, Xian, China
Institute of Vertebrate Palaeontology and Palaeoanthropology, Academia Sinica, Beijing, China

ABSTRACT:

Limited glacial ice was present in the Taibaishan and Baitoushan Mountains. In the northern area of North China, periglacial environments were dominant during the last glacial (Dali) stage, while cool climatic conditions prevailed in the south. The indicators of the cold climate include: permafrost, periglacial structures and cold faunas and floras. On the basis of pollen analysis and data from ^{14}C and U-series dating, the glacial stage can be subdivided into three substages, Early Dali (70,000-53,000 yr B.P.) with a desert or steppe environment (mean annual temperature about 10^0 lower than present); Middle Dali (53,000-23,000 yr B.P.), including the Liufangtun Interstadial (53,000-36,000 yr B.P.), Ashihe Stadial (36,000-32,000 yr B.P.) and Shangetun Interstadial (32,000-23,000 yr B.P.) (mean annual temperatures 4^0C respectively lower than present) and Late Dali consisting of Beizhuangcun Stadial (23,000-13,000 yr B.P.) and Fengzhuang Interstadial (13,000-11,500 yr B.P.). The Beizhuangcun Stadial, the coldest in the Dali Stage, (mean annual temperature 12^0C lower than present). The southern limit of permafrost has oscillated between 38^0-49^0N latitude, and the coastline has moved across the continental shelf.

EVIDENCE OF COLD CLIMATES

1. Glaciation. In the Taibaishan mountains Shaanxi Province, (elevation between 3,490-2650^m) cirques and five U-shaped valleys are found, and two terminal moraines developed at the elevation 2850-3100 m (Qi Shuhua et al. 1980). In the Baitoushan mountains, Jilin Province (elevation between 2,400-2,200 m) there are two groups of cirques leading to U-shaped valleys (Sun Jianzhong, 1982). They are assigned to the Taibai Glacial stage and Baitoushan Glacial Stage respectively, and are correlated with the Dali Glacial Stage.

2. Permafrost. The permafrost of North China is contiguous with that of North Eurasia. The continuous permafrost is 100 m thick and located at the northern end of the Great Xinganling mountains. The southern limit of discontinuous permafrost lies at latitude 49^0N (Fig. 1) and approximately coincides with the present mean annual 0^0C isotherm. In addition, in the Baitoushan and

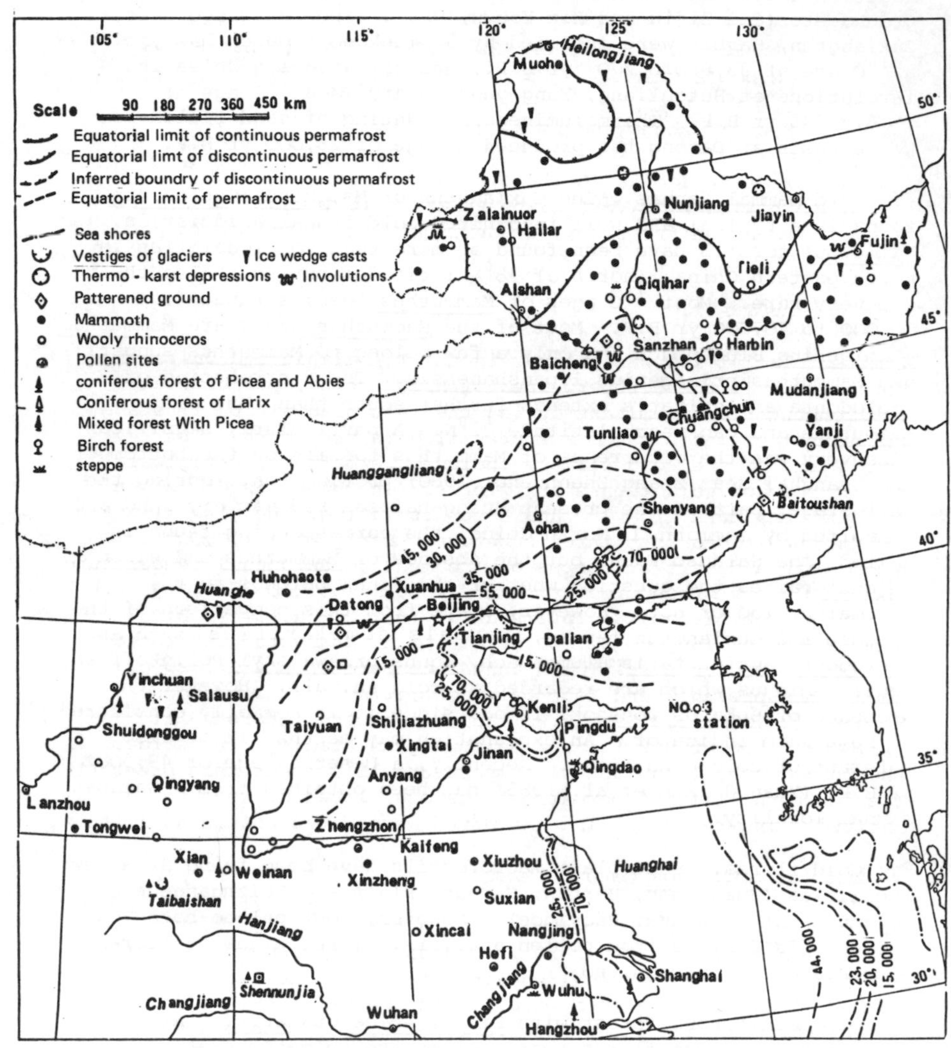

Fig. 1. Palaeo-environmental map of the last glacial stage in North China.

Huanggangliang Mountains, scattered areas of permafrost are known.

3. Fossil periglacial phenomena. Ice wedge casts and involutions are common in north eastern China and Inner Mongolia (Sun Jianzhong, 1983), at Datong, Shaanxi Province (Yang Jingehun et al., 1983) and

at Yangyuan, Hobei Province (Wu Zirong and Gao, Fuqing 1982). In addition there are thermokarst features, blockfields, patterned ground etc. (Cui Zijiu and Xie Youyu, 1984). The ice wedge casts at Guxiangtun, Harbin were infilled by a black soil which has yielded a ^{14}C age of 33,660±3,270 yr B.P., and calcareous nodules in involutions at Hutouliang, Yang-yuan county has a ^{14}C age of 27,675±745 yr B.P. Thermoluminescence dating of sand filled ice wedge casts at Datong has produced an age of 26,000 yr B.P.

4. Cold mammal faunas. The Guxiangtun or *Mammuthus-Coelodonta* fauna is a typical and well recognized cold faunal association. *Mammuthus* fossils have been found at more than 200 localities in north eastern China, (north of 38^0N latitude) but to the south, they are very rare. Most ^{14}C ages of *Mammuthus* fossils range from 10,000 to 40,000 yr B.P. Most of the *Mammuthus* finds are *Mammuthus primigenius* Blumenbach and only a few belong to *Mammuthus sungari* and Mammuthus *primigenius lipanshanensis*. The distribution of *Coelodonta antiquitatis* extends further south than that of *Mammuthus* and can reach latitude 32^0N. A particularly significant discovery was the occurrence of *Mammuthus* fossils on the bottom of the Huanghai Sea (Zhang Zhenhueng, 1980) proving that, during the Dali Stage, part of the present Huang-hai sea bed was dry land and inhabited by mammoth. In North China, the corresponding fauna is called the Salausu Fauna but the absence of *Mammuthus* and *Bison exiguus* raises the question whether the fauna belongs to the glacial period or not. However, apart from this, comparison of the Salausu and Guxiangtun faunas, reveals similarities such as *Coelodonta antiquitatis* Blumenbach, *Equus przewalskyi* Poliakoff and *Equus hemionus* which are regarded as cold animals. However the presence of *Bubalus wonsjoki* is odd since it is normally considered to be a warm indicator. An explanation may be that it is also tolerant of cold conditions. Recently, a U-series age of 49,500±200 yr B.P. (Yuan Shixun et al., 1983) has been obtained from a Salausu faunal locality.

5. Cold floras. Some plant macrofossils have been found at a few localities (Huangshan, Harbin; Sanjang Plain and Beizhuangeun, Weinan County, Shaanxi Province). However, most palaeo-botanical evidence is derived from pollen analysis. Three kinds of floral assemblages are recognised:

a) Coniferous forest with *Picea*, *Abies* or *Larix*. At present, these floras characterise the cold-temperate zone and the high mountains. When such floras are found in deposits of the southern low plains it is regarded as clear evidence of cold climatic conditions during a glacial stage.

b) Birch forest or forest-steppe. These have ecological requirements similar to that of the coniferous forest, but are somewhat warmer and dryer. Nevertheless they are regarded as cool climate indicators.

c) Steppe flora with Chenopodiaceae, Gramineae or Compositae. Species determination using pollen analysis is currently not possible and hence climatic interpretation is problematic. By using other methods, we consider that steppe vegetation in the last glacial stage marks a colder and drier climate than that indicated by the coniferous forest of *Picea* and *Abies*. For example, a steppe vegetation with *Artemisia* was present in Fenzhuang, Beijing and Beizhuangeun during 23,000-13,000 yr B.P. (Kong Zhaochen and Du Naiqiu,1980, Institute of Botany et al.,1966) when the climate was at its most severe in the last glacial stage.

EVOLUTION OF VEGETATION AND THE SUBDIVISION OF THE LAST GLACIAL STAGE

Currently much material from North China has been acquired enabling the pattern of vegetation evolution during the last glacial stage to be recognised and affording the basis for a climatic subdivision of the stage. There are three typical pollen diagrams: Huangshan, Harbin; Zhoujiayoufang, Yushu County, Jilin Province and Jiangnanwopo, Yongji County, Jilin Province (Sun Jianzhong, 1982). A common characteristic of these is that, in the middle part of the profiles,there is a zone rich in pollen indicative of luxuriant tree growth. In contrast the upper and lower parts of these profiles are either poor in pollen or the pollen consists of herbs and a few trees. This three-fold contrast provides a basis for pollen diagram subdivision with zones: I.II.III in ascending order. In the Huang-shan and Zhoujiayou-fang profiles, (see Fig. 2) the middle of zone II, shows a short interval when tree frequency was reduced and herbs increased, permitting the recognition of three subzones: II_1, II_2, III_3. On the evidence from pollen diagrams at Fenzhuang (Kong Zhaochen and Du Naiqiu, 1980) and the Baizhuangcun section (Institute of Botany et al.,1966) zone III can be split into two subzones: III_1 and III_2.

Zone I (70,000-53,000 yr B.P.) at Huang-shan, *Artemisia* was dominant, with a forest-steppe landscape. At Jiang-nan-wopo, there is poor pollen suggesting a cold interval. Until recently (Sun Jianzhong et al.,1980) the age limitation on the ^{14}C method of 40,000 years forced us to use long distance correlations with the early Wisconsin of North America (Dreimanis and Karrow,1972). However, new U-Series dates are now available from the laboratory at the Department of Archaeology, Beijing University. At the bottom of the profile at Zhoujia-youfang, a ^{230}Th age has been determined as 59,800±3,300 yr B.P. whilst a ^{231}Pa age on the same sample is $63{,}100^{+9{,}200}_{-7{,}600}$ yr B.P. These two ages are in very close agreement and suggests that the sample is from a closed geological system without uranium migration.

Zone II (53,000-23,000 yr B.P.).

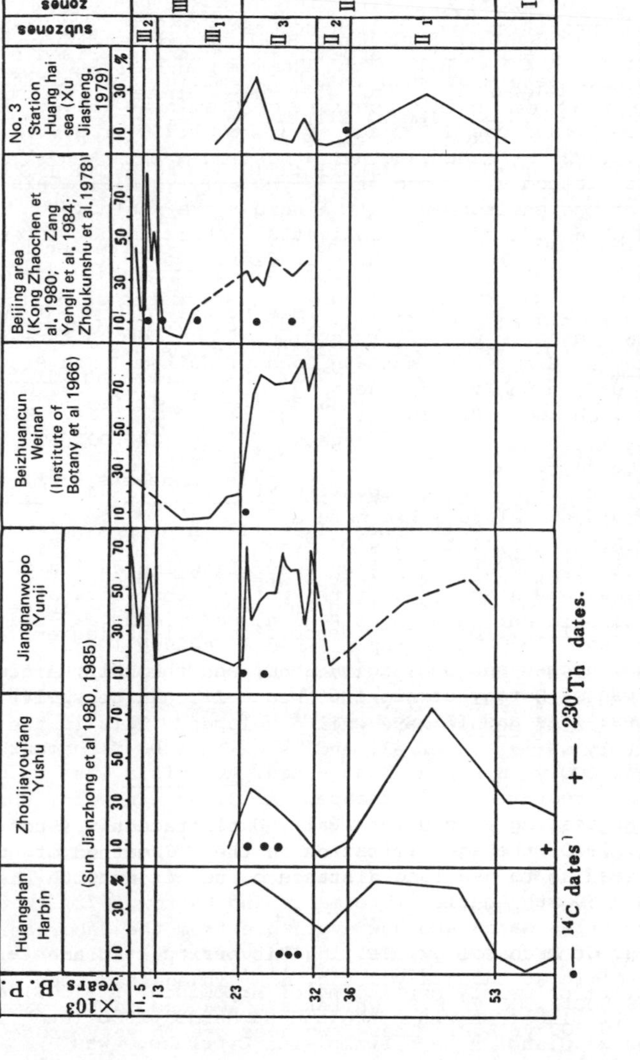

Fig. 2. Correlation of the tree pollen curves of the last glacial stage in North China.

Subzone II_1 (53,000-36,000 yr B.P.). Two tree pollen peaks appear simultaneously in the Huangshan and Zhoujiayou-fang profiles. Here tree pollen reaches 75% and consists mainly of Betula with some Picea, Abies and Pinus. Pure coniferous forest of Larix (its pollen frequency reaches 42%) with a ^{14}C age of 36,300±1500 yr B.P. is known from a site in the Sanjang plain (Kong Zhaochen and Du Naigiu,1984). This is the warmest interval currently known in the last glacial stage, and is called the Liufangtun Interstadial.

Subzone II_2 (36,000-32,000 yr B.P.). In the Huangshan and Zhoujia-youfang profiles, tree pollen reduces rapidly, and herb pollen increases. The landscape appears to be characterised by forest-steppe with Betula or steppe with Artemisia. At Zalainor, the steppe vegetation mainly consists of Gramineae (51.6%) and Artemisia (31.3%), and has a ^{14}C age of 33,760±1.700 yr B.P., (Li Xingguo et al.,1982). At Guxiangtun, a horizon of black soil was found, and has yielded a ^{14}C age of 33.660±32.70 yr B.P. This represents a steppe environment with a herb flora including Artemisia and Chenopodiaceae and indicates a short colder interval in the Middle Dali, called the Ashihe Stadial.

Subzone II_3 (32,000-23,000 yr B.P.). This is the most prominent phase characterised by a wide distribution of coniferous forest with Picea and Abies at Huangshan; Wuchang County, Heilongjiang Province (Liu Zerong, 1979, unpublished); Beijing Hotel (C^{14} age-29,300±13 50 yr B.P.; Kenli, Shandong Province (C^{14} age 24, 400±1,100 yr B.P. Zhou Kunshu et al.,1978) and Beizhuangeun (C^{14} age 23,100±850 yr B.P. Institute of Botany et al., 1966). In the past few years, a coniferous forest pollen assemblage with Pinus, Picea and Abies was found at the Salausu palaeolithic site and has a ^{14}C age of 27,900±600 yr B.P. Involutions were also found in the same strata (Zhou Kunshu et al.,1982b). A pollen assemblage indicative of mixed forest with Picea was also found at the palaeolithic site of Shuidonggou, Ningxia Hui Autonomous Region, and yields a ^{14}C age of 26,230±800 yr B.P. (Zhou Kushu,unpublished). Meanwhile, at Zhoujiayoufang; Jiang-nan-wopo; Guxiangtun; Dehui County, Jilin Province and Sanjiang Plain (Kong Zhaochen and Du Naigiu,1984) widespread forests of Betula were well developed. This is the second relatively warmer interval, and is called the Shanggentun Interstadial.

Zone III (23,000-11,500 yr B.P.). When the vegetation was poor only herbs were present and this corresponds to the coldest environmental conditions. This is the late Dali substage, and it can be subdivided into two subzones:

Subzone III_1 (23,000-13,000 yr B.P.). This period is characterized by steppe vegetation mainly consisting of Artemisia (95-100%) e.g. at Fenzhuang, Beijing (Kong Zhaochen and Du Naiqui,1980) and Beizhuangcun. At Jiangnan-wopo, swamp with Cyperaceae was surrounded by steppe with trees. Recently, a pollen assemblage of steppe with Artemisia and Chenopodiaceae has been found at Jiazhouwan Bay, Qingdao, Shandong Province and has yielded a ^{14}C age of 18,700±200 yr B.P. (Wang Yong Ji and Li Shan Wei,1983). Even in the south at Wuhu, Anhui Province; in the interval between

20,600 and 12300 yr B.P. a steppe pollen assemblage was found (Zhang Shuwei, quoted from Pu Quingyu, 1984). This interval of climatic deterioration is called the Beizhuangcun Stadial.

Subzone III_2 (13,000-11,500 yr B.P.). Coniferous forest with *Picea* and *Abies* re-appeared at Fenzhuang during the interval between 13,100 and 11850 yr B.P. and is called the Fenzhuang Interstadial. It is a transitional period from the glacial stage to the post glacial.

RECONSTRUCTION OF THE PALAEOTEMPERATURE CURVE

The curves of tree pollen in the pollen diagrams basically represent the palaeotemperature changes during the last glacial stage. During the intervals of 32,000-23,000 yr B.P. and 53,000-36,000 yr B.P. there are two peaks of tree pollen, showing two relatively warmer events. In contrast during the intervals of 23,000-11,500; 36,000-32,000 and 70,000-53,000 yr B.P. there are three phases of herb pollen dominance reflecting three cold and dry periods.

Estimates of the palaeotemperature values during the interval between 32,000-23,000 yr B.P. can be derived by comparing the elevation of the localities where the fossil vegetation is found, with that of the nearest mountains where the same vegetation is currently present. Taking a lapse rate temperature gradient of 0.5°C/100 m, we may derive the values given in the following table:

TABLE 1. Calculation of temperature reduction

Elevation of the fossil coniferous forest localities	Huangshan Harbin 150 m	Beijing 50-100 m	Beizhuangcun Weinan, Shaanxi Province 490 m
Elevation of the modern coniferous forest on the nearest mountains	Baitoushan 1400-1800 m	Xiaowutai 1600-2000 m	Caolianling 2100-2300 m
Difference of temperature °C	6°	7.5°	8°

Hence we may conclude that mean annual temperatures were lowered by 7°C on average.

In the interval between 53,000-36,000 yr B.P. *Betula* forest was developed on the Sonliao Plain at an elevation of 200 m. *Betula* is present in the modern mixed forest in the Changbeishan region at an elevation of 1,000-1,400 m. Hence, the temperature at that time should be some 4°C lower than that of the present.

The estimation of the reduced temperature value for the interval between 23,000-13,000 yr B.P., that is the coldest interval of the last glacial stage, is both difficult and controversial. This is the same problem as the estimation of the maximum southward extension of permafrost in the last glacial stage and similarly there is dispute over its limit. For example, Sun Jianzhoug et al. (1980) suggested that the reduced annual temperature was about 12°C; Yang Huairen and Xu Xin (1980) estimated 7-12°C; Pu Zingyu, (1984) 14°C, Cui Zijiou and Xie Youyu (1984) 11°C. Currently a consensus is being attempted using two additional kinds of evidence. First, ice wedge casts are known at Aohan, Inner Mongolia and Datong, where the present day mean annual temperature is 6.5°C. According to Washburn (1979) ice wedge growth requires at least a mean annual temperature of -5°C. Thus the mean annual temperatures of these two localities were 11.5°C lower than that of present. After allowing for the neotectonic uplift in these regions, present temperature requires a small adjustment upwards, consequently, a value of 12°C is adopted for the total temperature reduction. Second, most localities with *Mammuthus* fossils are situated to the north of Tianjing-Dalian. Mean annual temperature at Tianjing is 12.2°C and if we assume that the area colonised by *Mammuthus* corresponded to the area of permafrost, then the former mean annual temperature of Tianjing should be 0°C giving a minimum reduced temperature of 12°C.

In the interval between 70,000-53,000 yr B.P. the climate would not be so cold as that in the interval of 23,000-13,000 yr B.P. and the estimate temperature reduction is 10°C and that for 36,000-32,000 yr B.P. is 8°C. Thus the reconstructed temperature curve of the last glacial stage in this region was derived (see Fig. 3).

TRANSGRESSION, REGRESSION AND ALLIED SEA LEVEL CHANGES

With the development of the oceanographical research, the pattern of sea level changes on the maritime plain and continental shelf during the last glacial stage is becoming known. It consists of three regressions separated by two transgressions. In order of decreasing age they are: Early Chengshantou regression → Haizhouwan Bay transgression → Late Chengshantou regression → Guannan transgression → Haiyangdao regression (Xu Jiasheng et al., 1981). Evidence from Foraminifora, Ostracoda and palynology show that the transgressions correspond to the warm peaks of the palaeotemperature curve, while regressions occurred during the cold peaks (Xu Jiasheng et al., 1981, Xu Jinsheng, 1982; Wang Jingai and Wang Pinxian, 1980). At the time of maximum cold about 15,000 yr B.P. the greatest shoreline regression shows a sea level some 150-160 m below present (Zhu Yongqi et al., 1979). Then the coastline was located on the outer margin of the continental shelf some 600 km from the present position. The whole continental shelf was emergent with Japan, Taiwan and other islands directly connected with the

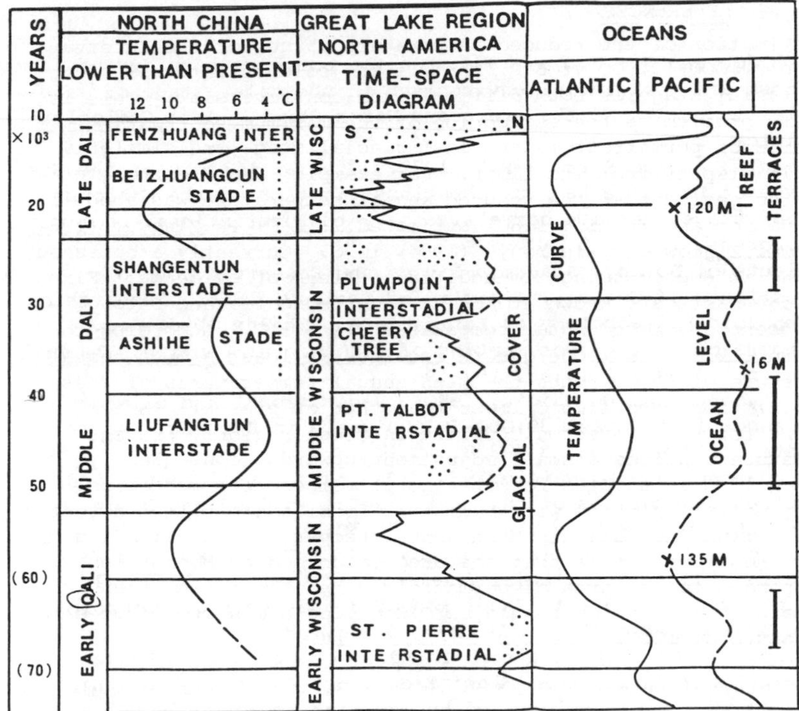

Fig. 3. Correlation of the temperature curves of the last glacial stage in North China with that of different areas (the latters are quoted from A. Dreimanis and P.F. Karrow 1972, the dots are ^{14}C and U-series dates)

continent by landbridges, providing passages for the migration of animals and plants, and for human cultural exchange. In contrast the transgressions associated with relatively warmer interstadials might extend 200 km into the maritime plain. Sychronism between sea level change and climate change supports the theory of "Glacial control" that is when the climate became colder, the volume of glacier ice on the continents increased and oceanic volume decreased. Sea level accordingly dropped and coastlines retreated onto the continental shelf. Whereas, when the climate improved the reverse occurred.

ACKNOWLEDGEMENTS

The authors would like to express their gratitude to Mr Zhou Kunshu, Mr Kong Zhaochen, Mr An Zhisheng, Mr. Yang Jingchun and Mr. Cui

Zhijou, for their kind help in the work presented in this paper.

REFERENCES

Cui Zhijou and Xie Youyu, 1984. On the southern boundary of permafrost and periglacial environment in the taler stage of late Pleistocene in Northeastern and North China. Acta Geologica Sinica, 2, 165-175.

Dreimanis, A. and Karrow, P.E.,1972. Glacial history of the Great Lake-St.Lawrence Region, the classification of the Wisconsin (an) stage, and its correlations, 24th International Geological Congress, 12, 5-15.

Institute of Botany, Academia Sinica and Research Group of Institute of Geology, Ministry of Geology, 1966. Research on Cenozoic palaeobotany at Lantian area, Shaanxi Province. Papers of on-the-spot meeting of Cenozoic at Lantian, Shaanxi, Science Press, Beijing, 157-196.

Kong Zhaochen and Du Naiqiu, 1980. Vegetational and climatic changes in the past 30,000-10,000 years in Beijing. Acta Botanica Sinica, 4, 330-338.

Kong Zhaochen and Du Naigiu,1984. The macrofossil plants and pollen assemblages of the Last Glacial in Sanjiang Plain. Scientia Geographica Sinica, 1, 76-80.

Li Xingguo, Liu Guang lian, Xu Guoying, Li Feng chao, and Wang Fulin, 1982. Periglacial phenomena at east opencut coal mine of Zhalainuor in Inner Mongolia and their geochronology. Journal of Glaciology and Cryopedology, 3, 65-72.

Pu Qingyu, 1984. The southern boundary of permafrost horizon in East China during Dali Glaciation Epoch and change of this boundary. Hydrogeology and Engineering Geology, 4,49-51.

Qi Shuhua, Gan Zhimao and Xi Zhende, 1980. Vestiges of ancient glaciation and problem of glacial stages at Taibei mountain. Selected papers of scientific and technical results. Geographic session, Shaanxi pedogogic college, China, 8-1 - 8-18.

Sun Jianzhong, 1981. Subdivision of the Quaternary periglacial stages at SongLiao plain. Scientia Geographica Sinica, 2, 163-170.

Sun Jianzhong, 1982. Classification of the Quaternary glacial stages in Jilin Province. Acta Geologica Sinica, 2, 174-186.

Sun Jianzhong, 1983. Stratigraphy of the Dali glacial stage in Northeastern China. Journal of Stratigraphy, 1, 1-11.

Sun Jianzhong, Wang Shuying, Wang Yuzhuo, Zhou Yajie, Lin Zerong, Zhang Qingyun and Chen Shuhan,1980. Palaeoenvironment of the Last glacial stage in Northeastern China. Jilin Geology, 4,38-60.

Wang Jingtai and Wang Pinxian, 1980. The relationship between sea level variation and climatical changes in eastern China since later Pleistocene . Acta Geographica Sinica, 4, 299-311.

Wang Yong Ji and Li Shan Wei, 1983. Palaeovegetation and palaeoclimate in the past 20,000 years in Jiaozhou Bay, Qingdao District. Acta Botanica Sinica, 4, 385-392.

Washburn,A.L.,1979. Geocryology - A survey of periglacial processes and environments, Edward Arnold, London p 406.

Wu Zirong and Gao Fuqing, 1982. Analysis of mechanism of thawing-freezing deformation of the upper Nihewan Group and discussion on the time of its occurrence. Papers on Prehistoric Earthquakes and Quaternary Geology, Shaanxi Scientific and Technical Publishing House, Xian, 143-148.

Xu Jiasheng, Gao Jianxi, and Xie Fuyan, 1981. Huanghai (Yellow Sea) in the last glacial stage. Scientia Sinica, 5, 605-613.

Xu Jiasheng, 1982. Spore-pollen assemblages of the peaty samples from the Huanghai sea and their signification in palaeogeography, Selected papers from the first Symposium of the Palynological Society of China. Science Press, Beijing, 22-31.

Yang Huairen and Xu Xin, 1980. Quaternary environmental changes in Eastern China. Acta of Nanjing University, 1, 121-144.

Yang Jingchun, Li Shude, Sun Jianzhong and Yang Yixun, 1983. Fossil ice wedges and Late Pleistocene environments in Datong Basin, Shanxi Province. Scientia Geographica Sinica, 4, 339-344.

Yuan Shixun, Chen Tiemei and Gao Shijun, 1983. Uranium series dating of "Ordos man" and "Sjara-osso-gol culture". Acta Anthropologica, 1, 90-94.

Zhang Yingli, Huang Xinggen, Jiaozhenxing, Sun Xiouping, and Zhao Xitao, 1984. Stratigraphy of the last stage of late Pleistocene at Beijing Plain. Journal of Stratigraphy, 1, 56-61.

Zhang Zhenhueng, 1980. New discovery of fossils of Mammuthus and Coelodonta in the Northern Huanghai. Quaternaria Sinica, 1, 96-97.

Zhou Kunshu, Yan Fuhua, Lang Xioulong and Yie Yongying, 1978. Pollen analysis of the late Quaternary in Beijing Plain and its significance. Scientia Geologica Sinica, 1, 58-63.

Zhou Kunshu and Li Xinggua, 1982b. The division of the periglacial period of the valley of the Salawusu river in Inner Mongolia and its significance. Papers of Prehistoric Earthquakes and Quaternary geology. Scientific and Technical Publishing House of Shaanxi Province, Xian, 149-153.

Zhou Kunshu, Liang Xiulong and Liu Ruiling, 1982a. A preliminary discussion on palaeoenvironments of North China since late Pleistocene. Selected papers from the First Symposium of the Palynological Society of China, Science Press, Beijing, 6-12.

Zhu Yongqi, Li Cheng Yi, Zheng Chengkai and Li Bogen, 1979. On late Pleistocene lowest sea level on the continental shelf of the East Sea. Ke xie Tong Bao, 7, 317-320.

International Geomorphology 1986 Part II
Edited by V. Gardiner

ESTABLISHMENT OF SOIL COVER ON TILLS OF VARIABLE TEXTURE AND IMPLICATIONS FOR INTERPRETING PALEOSOLS - A DISCUSSION

Anne F. Gellatly

Department of Geography,
The University of Sheffield, UK

ABSTRACT

Soil development will proceed more quickly on fine, gravelly till than very coarse bouldery deposits. The delay in the establishment of a soil cover can be readily detected within recent deposits where, with surface age-determination, anomalies in the depositional landform sequence can be identified. On older surfaces, however, detection of the original surface conditions is difficult but important since the soil cover may not yield an accurate age-estimate of the underlying surface. Coarse-textured tills are characterised by a dominant lichen colonisation stage and these slowly decomposing micro-residuals can be radiocarbon dated. This organic fraction provides approximate dates pertaining to surface stabilisation of the underlying substrate.

INTRODUCTION

Relative soil age is recognised as a useful dating tool by geomorphologists. A sequence of soils sharing a similar history throughout the duration of their formation and which developed under similar environmental conditions of climate, parent material and topographic situation may be examined within the concept of a chronosequence. Various workers have attempted to describe the progressive development of soil profiles in which time alone is assumed to account for the main changes which have taken place e.g. Mokma et al.,1973; Ross et al.,1977; Jacobsen & Birks,1980; Fitze,1980; Heikinen & Fogelburg,1980; Birkeland,1982, 1984. However, field descriptions of soil profiles are generally limited by what can be recorded at one point in time. In a critical review of the application of the concept of soil chronosequences, Stevens & Walker (1970) stressed the difficulties of chosing a series of strictly 'monogenetic' soil profiles for the examination of soil change over time. In a changing depositional environment, such as that associated with a fluctuating glacier margin, the deposition of new material must, to some extent, affect the rate of surface weathering processes. Stevens & Walker (1970, p. 345) suggested that "all sequence studies have failed in one way or another to 'control' the soil-forming factors other than time, and all occur in different regions with features peculiar to each individual sequence".

On a closely-spaced sequence of soils on young geomorphic surfaces, time may be more appropriately substituted for space; but the distinction is less well defined in older chronosequences. Moraines near presently active glaciers have been described as: 'excellent natural laboratories within which to chart the course of soil formation' (Gerrard,1981, p. 143).

Another fundamental assumption is that due to the time factor involved, more weathered and leached soil profiles are to be expected on older deposits. Similarly, it is often assumed that the age of that geomorphic surface increases with distance from the ice margin. However there are exceptions to this general rule as demonstrated by Ugolini (1966), Østrem (1965) and Sigafoos & Hendricks (1961).

The influence of parent material texture and subsequent, post-depositional modifications are discussed in relation to both the establishment of soil chronosequences and interpretation of C14-dated paleosols in New Zealand. A model is presented which highlights the need to examine the evidence of individual soil profile histories (Figure 1). Examples are given from two types of chronosequence.

INTERPRETING POST-INCISIVE SOIL CHRONOSEQUENCES

Post-incisive chronosequence are described by Vreeken (1975, p. 380) as "an array of soils that began forming at different moments in the past and that are either still exposed or that were simultaneously buried". Little is known about the similarity or otherwise in soil development between members of any one sequence. Furthermore studies of soil chronosequence rely on a number of basic assumptions c.f. (Vreeken,1975). The older the age-sequence, the more difficult it is to determine the original conditions at the commence of soil development. Thus, with time, coarse-textured surfaces will weather and break down to produce a soil cover, albeit at a slower rate than fine-textured material (Figure 1). Ultimately, the original parent material texture will be obscured. Chemical and physical analysis of the soil cover may give a rough approximation as to the duration of soil development but may greatly underestimate the actual age of the deposit. Similarly, subsequent, post-depositional modifications may influence the rate of soil development to varying extents (Figure 1). In the Classen Valley, New Zealand, an apparent reversal in the development of a sequence of soils was detected on late-Holocene moraines. Independent dating control highlighted the anomaly in soil development. The Classen Glacier (43° 30'S) descends to an altitude of 1006m, and terminates in a 60m high ice cliff at the northern end of a glacial lake, approximately 1km in length

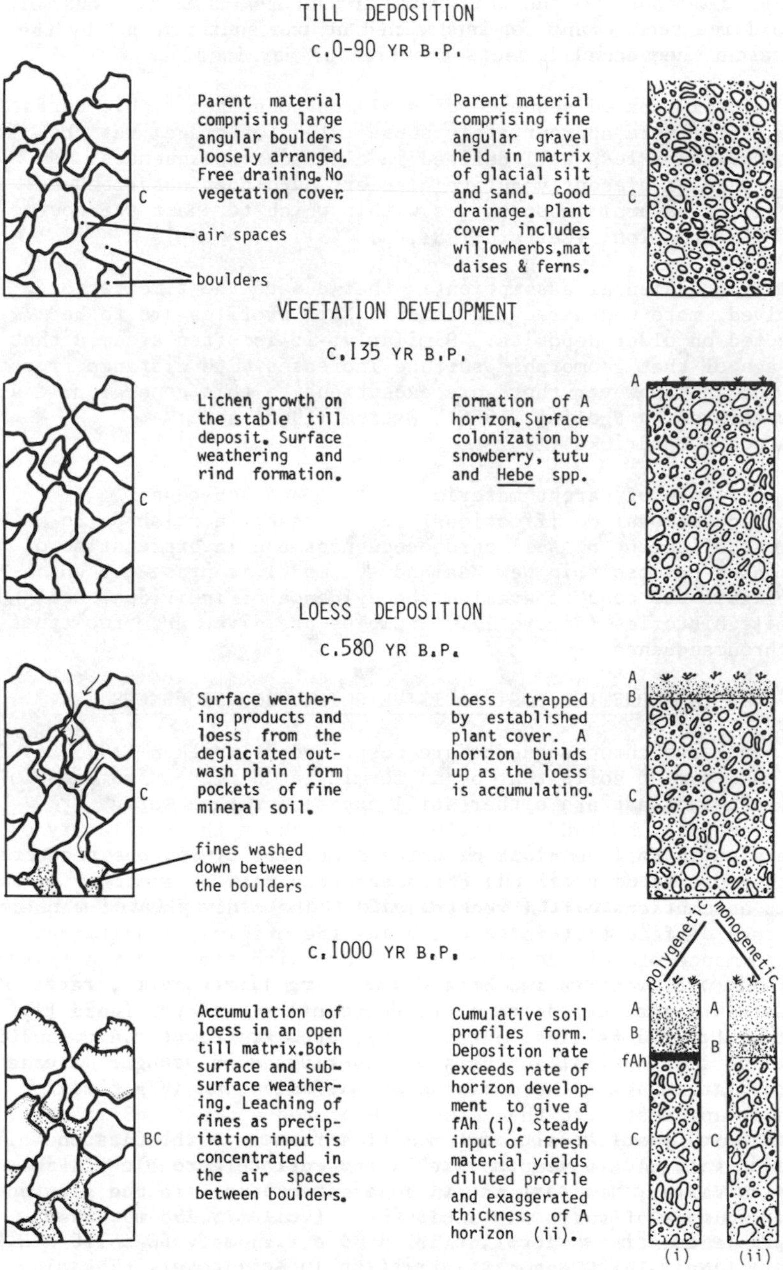

TILL DEPOSITION

c.0-90 YR B.P.

Parent material comprising large angular boulders loosely arranged. Free draining. No vegetation cover.

C

air spaces

boulders

Parent material comprising fine angular gravel held in matrix of glacial silt and sand. Good drainage. Plant cover includes willowherbs, mat daises & ferns.

C

VEGETATION DEVELOPMENT

c.135 YR B.P.

Lichen growth on the stable till deposit. Surface weathering and rind formation.

C

Formation of A horizon. Surface colonization by snowberry, tutu and Hebe spp.

A

C

LOESS DEPOSITION

c.580 YR B.P.

Surface weathering products and loess from the deglaciated outwash plain form pockets of fine mineral soil.

C

fines washed down between the boulders

Loess trapped by established plant cover. A horizon builds up as the loess is accumulating.

A

B

C

polygenetic

monogenetic

c.1000 YR B.P.

Accumulation of loess in an open till matrix. Both surface and subsurface weathering. Leaching of fines as precipitation input is concentrated in the air spaces between boulders.

BC

Cumulative soil profiles form. Deposition rate exceeds rate of horizon development to give a fAh (i). Steady input of fresh material yields diluted profile and exaggerated thickness of A horizon (ii).

A

B

fAh

C

(i)

A

B

C

(ii)

Fig. 1. Model of soil development on recent till deposits of variable texture. Rates of change are estimated from measurements of surface rock weathering rind thickness.

(Fig. 2). The lake began to form in the 1940's in the wake of rapid ice recession. It is drained at the southern end by the Classen River which bisects a series of moraines.

Fig. 2. Aerial photograph of the Classen Glacier terminus (March 1965) showing Holocene moraine sequence. Recent till deposits overlap onto the older vegetated moraines. (N.Z. Aerial Mapping Ltd.)

The glacial sequence has been dated using lichenometry, rates of weathering rind development and documentary evidence (Gellatly, 1982, 1984a, 1984b). The oldest surfaces are dated c.4200 ± 1090 and c.3350 ± 870 a B.P. They are overlapped by younger material deposited during glacier advances this millenium (Fig. 3).

Soil development on the sequence of moraines in the Classen Valley was investigated using an established surface chronology (Fig. 3). The material comprising the till is fine-medium grained greywacke that is described in detail elsewhere (Gellatly, 1984a). Two separate moraine surfaces, c.135 ± 35 a B.P. and 580 ± 150 a B.P., respectively, show a marked variation in soil cover. The older deposit consists almost entirely of large boulders and has virtually no soil cover, whilst the younger deposit comprises chiefly of fine to medium sized clasts (c.0.3m) and exhibits early

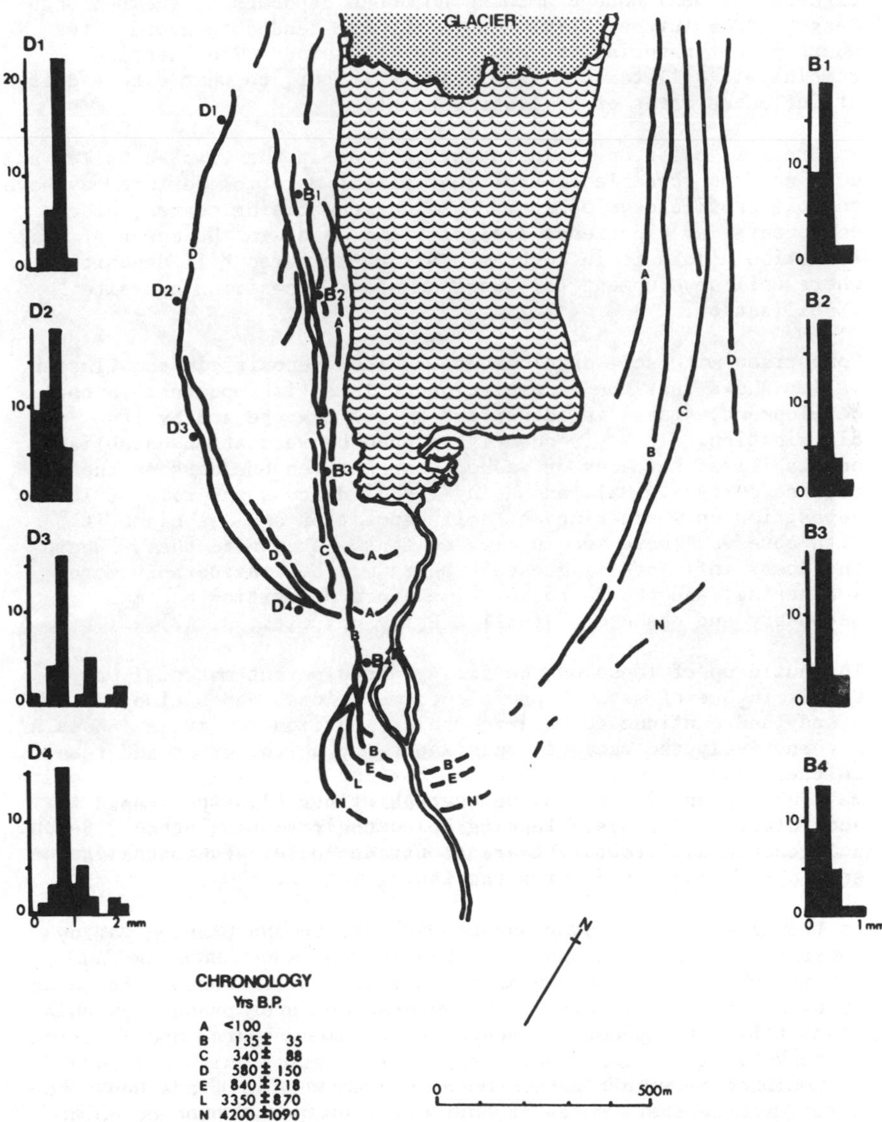

Fig. 3. Location of moraines at the Classen Glacier terminus, rock weathering-rind histograms and estimated surface ages.

stages of soil profile development as well as loess accumulation. The absence of fines between blocks in moraines was noted by Birkeland (1982) when examining Holocene deposits in the Ben Ohau Range, (30km further West). Workers have tended to avoid sites where parent material is very coarse textured. The question remains, if till texture is a primary factor, to what extent does it influence rates of soil development?

Figure 1 is based on field investigations in the Classen Valley and outlines the possible control that parent material texture may have on soil profile development. Moraines comprising coarse, blocky components may experience a significant delay in the onset of soil formation. This is in contrast to fine-textured till deposits where soil development follows soon after deposition and site stabilisation.

Comparison with soil development on older deposits in the Classen Valley shows that the accumulation of loess is important in soil development. Loess is identified by its texture and by its distribution. It is frequently trapped by vegetation established on stabilised surfaces or washed down between boulders on the moraine surface. Hall and Heiny (1983) discuss the role of loess deposition on weathering and soil deposition and associate it with observed increases in clay content. They note that; "Where the loess influence is greatest there may be considerably more weathering, and the factor of time since deposition becomes secondary and obscured," (Hall & Heiny op. cit., p. 47).

The build up of loess on the fine-grained parent material may result in one of several profile forms: loess deposition may be steady and continue to enlarge the 'A' horizon to give a 'Loess A'. Alternatively the rate of loess input may be too great and result in the eventual burial of the 'A' horizon (fAh). A new 'A' horizon may develop and it too may be eventually buried by the rapid accumulation of loess. This leads to the formation of a polygenetic soil (c.f. Figure 1) which should not be used in soil chronosequence studies (Stevens & Walker,1970).

On the fine-grained loamy parent material in the Classen Valley loess quickly helps to build up the 'A' horizon. However the blocky moraine deposited nearly 450 years earlier shows little or no evidence of soil cover. The vegetation cover present on this older till after about 580 years is extremely sparse and restricted to pockets of loess and weathered rock material which may have accumulated between clasts. Once such pockets of fines begin to accumulate however, rates of soil development may proceed quite rapidly as leaching processes concentrate between the boulders (Howell & Harris,1978).

Boulton & Dent (1974) describe post-depositional textural changes in a much younger glacial sequence from Iceland. They stress the importance of wind in redistributing silt and fine sand leaving a stony lag surface. In addition, they recorded a progressive downwashing of fines by percolating water which show a tendency to accumulate on the upper surface of the compact till.

The important point to note is that it takes over 500 years and often more than 1,000 years for the bouldery till matrix to reach a similar point of "inception" as that described for soils initiated on fine-grained parent material. Furthermore, whilst the discrepancy may be obvious in a young sequence there is no information available to suggest at what stage in chronosequence development the effect of such a 'delayed start' to soil development becomes imperceptible, given the diminishing resolution of techniques for age-determination. That is to say, when examining a surface deposited during the early and mid-Holocene period what indication is there that these soils began to accumulate close to the time of deposition or at some substantially later date? Whilst loess deposition may be accelerating the apparent rate of weathering on fine-medium textured material, assessments of the duration of soil formation on coarse-grained till may be underestimating the length of time since the surface was stabilised.

INTERPRETING VERTICAL SOIL CHRONOSEQUENCE

Vreeken (1975) recognises time-transgressive chronosequences whereby soils began forming and were buried at different moments in time. Within the Tasman Valley, New Zealand, a number of time-transgressive chronosequences have been described from actively eroding superimposed lateral moraines. The sequences have been documented and dated by C14 (Burrows,1980; Gellatly *et al*.,1985). A knowledge of the limiting assumptions used to interpret time-transgressive chronosequences has led to a careful re-examination of the significance of C14 dated paleosols.

In analysing organic material buried within superimposed till sequences different organic fractions were dated according to the nature of the material sampled (Geyh et al.,1985). Where possible dates were obtained from in situ wood as this yielded the closest approximation to the actual date of burial. In the absence of contamination, the humic acid fraction was also shown to give a good approximation to the date of superimposition. A third organic fraction, the micro-residuals, was found to closely approximate the initiation of soil development and moraine stabilisation. Micro-residuals are comprised of slowly decomposing lichen material. The significant age differences between the different organic fractions highlights the impact that variations in parent material texture has, not only on contemporaneous soil development but also on the interpretation of paleosols (Geyh *et al*.,1985, Fig. 3).

CONCLUSION

Two important points arise from a consideration of the nature of parent material texture when analysing the rate of soil development and the interpretation of paleosols in Holocene glacial deposits in New Zealand. Observations suggest that variations in soil profile development may reflect differences in texture of the original parent material. The anomalous relationship between surface age and soil development should be readily detected in young sequences and thus in chronosequence interpretations. With the apparent convergence in trends, time will tend to reduce the influence of parent material on soil formation. The distinction, which is clearly apparent on young till deposits, may be masked by progressive changes in post-depositional surface modifications. Ages assigned to phases of soil profile formation may be underestimates if the original parent material was markedly blocky in texture. Alternatively, the excessive accumulation of loess may accelerate rates of soil development. Clearly interpretations should be undertaken with care. Detailed examination of the composition of unweathered and weathering parent material is required as is the need to examine soil development at all sites and not just those free of surface and subsurface boulders. The nature of original parent material is clearly an important primary complicating factor in studies of soil development and the model of soil formation presented in this paper suggests possible influences of two very different inherited textures on patterns of surface weathering on recent till deposits.

A second consideration of the importance of parent material texture arises from examination of radiocarbon dated micro-residuals present in time-transgressive paleosol chronosequences in the Tasman Valley. Detailed interpretation of separate dateable organic fractions in buried soil material suggest that slowly decomposing lichen material characterising the early stages of post-depositional surface modification (weathering and soil development), can provide close approximations for the date of surface stabilisation. The presence of sufficient organic material for dating the 'micro-residual' fraction is interpreted as reflecting the blocky texture of the original parent material which in turn inhibits soil profile development.

ACKNOWLEDGEMENTS

This research was funded by a New Zealand Commonwealth Scholarship. I wish to thank Phil Tonkin, Lincoln College, Canterbury for many shared discussions and Brian Whalley and John Gordon for their comments.

REFERENCES

Birkeland, P.W., 1982. Subdivision of Holocene glacial deposits, Ben Ohau, New Zealand, using relative-dating methods. Geological Society of America Bulletin, 93, 433-449.

Birkeland, P.W., 1984. Holocene soil chronofunctions, Southern Alps, New Zealand. Geoderma, 34, 115-134.

Boulton, G.S., & Dent, D.L., 1974. The nature and rates of postdepositional changes in recently deposited till from south-east Iceland. Geografiska Annaler, 56A, 121-134.

Burrows, C.J., 1980. Radiocarbon dates for post-Otiran glacial activity in the Mount Cook region, New Zealand. New Zealand Journal Geology and Geophysics, 23, 239-248.

Fitze, P., 1980. Zur Bodenentwicklung auf Moränen in den Alpen. Geographica Helvetica, 3, 97-106.

Gellatly, A.F., 1982. The use of lichenometry as a relative age-dating method with specific reference to Mt. Cook National Park, N.Z. New Zealand Journal of Botany, 20, 343-353.

Gellatly, A.F., 1983. Revised dates for 2 recent moraines of the Mueller Glacier, Mt. Cook National Park, N.Z. (Note). New Zealand Journal of Geology and Geophysics, 26, 311-315.

Gellatly, A.F., 1984a. The use of rock weathering-rind thickness to re-date moraines in Mt. Cook National Park, N.Z. Arctic and Alpine Research, 16, 2, 225-232.

Gellatly, A.F., 1984b. Historical records of glacier fluctuations in Mt. Cook National Park, New Zealand - A century of change. The Geographical Journal, 151, 1, 86-99.

Gerrard, A.J., 1981. Soils and Landforms. Allen & Unwin, London, 219 p.

Geyh, M.F., Röthlisberger, F. & Gellatly, A.F., 1985. Reliability tests of C14 dates from paleosols in glacier environments. Zeits. Gletscherkunde und Glazial-geologie, 21, 275-281.

Hall, R.D., & Heiny, J.S., 1983. Glacial and post glacial physical stratigraphy and chronology, North Willow Creek and Cataract Creek drainage basins, Eastern Tobacco Root Range, Southwestern Montana, U.S.A. Arctic and Alpine Research, 15, 19-52.

Heikinen, O., & Fogelburg, P., 1980. Bodenentwicklung im Hochgebirge: Ein Beispiel vom Vorfeld des Steingletscher in der Schweiz. Geographica Helvetica, 3, 107-112.

Howell, J.D., & Harris, S.A., 1978. Soil-forming factors in the Rocky Mountains of southwestern Alberta, Canada. Arctic and Alpine Research, 10, 313-324.

Jacobsen, G.L., & Birks, H.J., 1980. Soil development on recent end moraines Yukon Territory. Quaternary Research, 14, 87-100.

Mokma, D.L., Jackson, M.L., Syers, J.K. & Stevens, P.R., 1973. Mineralogy of a chronosequence of soils from greywacke and micaschist alluvium, Westland, New Zealand. New Zealand Journal of Science, 16, 769-797.

Østrem, G., 1965. Problems of dating ice-cored moraines. Geografiska Annaler, 47, 1-38.

Ross, W.C., Mew, G.,& Searle, P.L., 1977. Soil sequences on two terrace systems in the North Westland area, New Zealand. New Zealand Journal of Science, 20, 231-244.

Sigafoos, R.S.,& Hendricks, E.L., 1961. Botanical evidence of the modern history of the Nisqually Glacier, Washington. U.S. Geological Survey Professional Paper, 387-A, 20 p.

Stevens, P.R.,& Walker, T.W., 1970. The chronosequence concept and soil formation. The Quarterly Review of Biology, 45, 333-350.

Ugolini, F.C., 1966. Soils. Soil Development and Ecological Succession in a Deglaciated area of Muir Inlet. in Mirsky, A. (ed.). Institute of Polar Studies, Ohio State University Research Report, 20, 29-56.

Vreeken, W.J., 1975. Principal kinds of chronosequence and their significance in soil history. Journal Soil Science, 26, 378-394.

International Geomorphology 1986 Part II
Edited by V. Gardiner

QUATERNARY PROBLEMS IN THE GERMAN ALPINE FORELAND

Konrad Rögner

University of Paderborn,
Warburger Str. 100, D-4790 Paderborn, West Germany

ABSTRACT

In the southern part of the Iller-Lech-Platte (i.e. the area between the rivers Iller and Lech in the German alpine foreland) there is evidence of five different Pleistocene piedmont glaciations. In addition to this there are at least two fluvioglacial gravel terraces which cannot be connected with terminal moraines.

The piedmont glaciations are named after Penck: Würm, Riss, Mindel, Günz and Eberl (Donau). There is little justification at present for the age of such names as Haslach, Roth or Paar.

INTRODUCTION

The Iller-Lech-Platte - the area lying between the rivers Iller in the west and Lech in the east, the Danube in the north and the border of the Alps in the south (Fig. 1) - was the focus of research by the German geographer, geomorphologist and Quaternary geologist Albrecht Penck (1858-1945). After he proved in 1882 the existence of three different piedmont glaciations ("Unter Glacialschotter", i.e. lower glacial gravel; "Oberer Glacialschotter", i.e. upper glacial gravel; "Diluviale Nagelfluh", i.e. diluvial cemented gravel) in the alpine foreland, Penck in 1898 documented the presence of the four 'classic' glaciations (Würm, Riss, Mindel, Günz) using geological and geomorphological evidence. Penck was the first scientist to combine geological and geomorphological results in Quaternary research and this enabled him to establish his Glacial Series ("Glaziale Serie"). The Glacial-Series consists of a combination in time and space of glaciofluvial outwash ("Schotter"), terminal moraine ("Endmoräine") and glacial tongue basin ("Zungenbecken") and are characterised by their geomorphological and sedimentological unity.

In the three volumes of "Die Alpen im Eiszeitalter" published between 1901-1909, Penck and his friend and co-author Brückner named

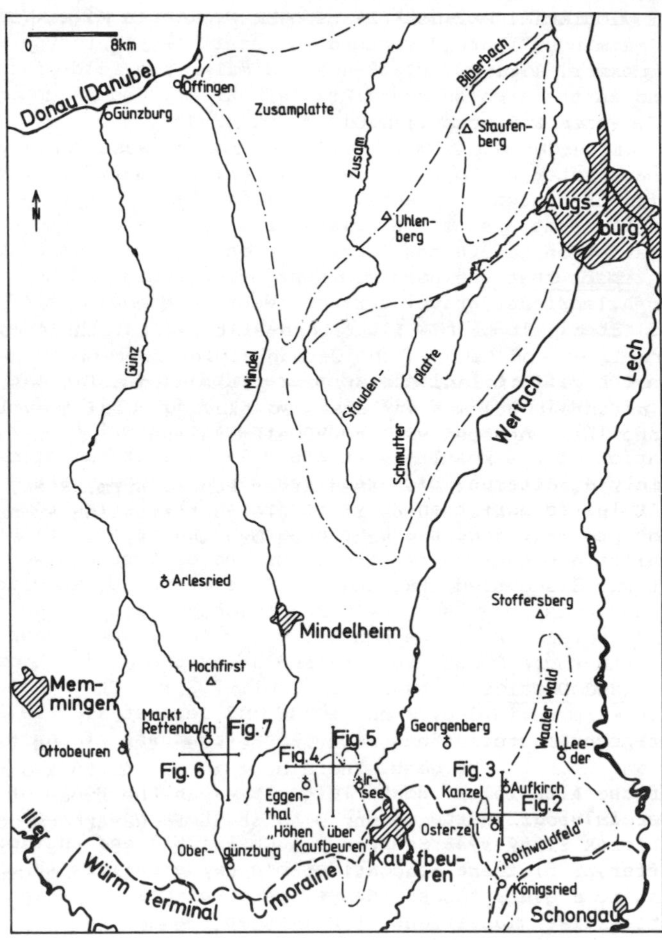

Fig. 1. Location map with the positions of figures 2-7.

the four piedmont glaciations - Würm ("Niederterrasse"), Riss ("Hochterrasse"), Mindel ("Unterer Deckenschotter") and Günz ("Oberer Deckenschotter") after small rivers in the alpine foreland. The four fluvioglacial outwash plains near Memmingen (i.e. "Die vier Felder von Memmingen") are described in detail as stratotype regions (Penck & Brückner, 1901-09, p. 28-40). Similar fourfold divisions can be found in the area surrounding Kaufbeuren ("Gegend unterhalb Kaufbeuren"; Penck & Brückner, 1901-09, p. 40-45).

About 30 years later B. Eberl, a catholic priest, dared to question the world famous concept of Penck. Eberl, working with the same methods and the same system ("Glacial Series"), was able to establish an older glaciation before the Günz: the Donau-glaciation. It was during this epoch that the Hochfeld west of Augsburg or the Plattenberg-Arlesrieder gravel terrace (west of Mindelheim) accumulated (Eberl, 1930, map 1).

In addition the different glaciations are subdivided into two or three stadials (W I, W II, W III; R I, R II; M I, M II; G I, G II; D I, D II, D III). Another well known aspect of Eberl's work was the correlation of his results with the Milankovitch radiation curves, mainly an attempt which received support from workers outside Germany. It lasted more than 20 years during which time some of Eberl's concepts were accepted by the German Quaternary scientists. However, whilst a Donau-age for the older gravel terraces was admitted, it was discovered that the moraines of the Donau glaciation either did not exist or else they were of younger age. Even the "Ottobeurer" gravels, Pliocene according to Eberl, were shown to be younger. Furthermore doubts were raised about the division of Mindel, Günz and Donau into stadials (Troll, 1931). The results of Eberl's work were revised by Sinn (1972) and Jerz et al. (1975) and their stratigraphic conclusions were closer to those of Penck.

In other areas of the Iller-Lech-Platte, e.g. in the southeastern part (Rögner, 1979) or in the middle part (Rögner, 1980), west of the Iller (Eichler & Sinn, 1975; Schreiner & Ebel, 1981) and east of the Lech (Schaefer, 1975) there is considerable evidence for the existence of more than four glaciations. Habbe (personal communications, 1985) has documented five glaciations at Memmingen too.

For a better understanding concerning the dating of the Pleistocene gravel terraces in the German alpine foreland and the division of the same gravel terraces it is important to remember that the different terraces accumulated as steplike terrace flights. The 'en bloc' uplifting of the whole alpine foreland after the sedimentation of the Miocene "Obere Susswassermolasse" (OSM) caused the rivers to entrench their channels hence the oldest accumulations were raised the most and today they form the highest elevations. The youngest sediments fill the valley bottoms.

Penck called this steplike occurrence of the Pleistocene gravel terraces the "Suebian type" (Schwäbischer Typ) of fluvioglacial outwash plain landscape. The opposite is the "Bavarian type"

(Bayerischer Typ) with normal geological aggradation represented only in the area surrounding Munich ("Münchner Schiefe Ebene; Münchner Schotterebene").

The concept of the Glacial Series has not been modified by later findings; it has remained until recently the foundation of each Quaternary morphological and stratigraphicical study.

REAPPRAISAL

1. Penck's "Kanzelschotter" (Günz) and "Rothwaldfeld" (Mindel); Fig. 2 and 3. Kanzel and Rothwaldfeld are described by Penck as the Günz- and Mindel-Glacial Series of the area near Kaufbeuren ("Gegend unterhalb Kaufbeuren"). Studies carried out during 1974 and 1977 revealed that this view is too simple. The youngest sediment in this area, situated in the foreland of the eastern lobe of the Pleistocene Lech glacier, is the Würm gravel in the valley bottom of the Hühnerbach valley (sometimes called 'cold valley', "Kaltes Tal") partly covered by Holocene sediments (Fig. 2). During the Riss a small glacier lobe penetrated into the Mindel erosion-channel (Fig. 2). Buried soils of R/M interval age are found at some points below the Riss sediments. The buried soil of

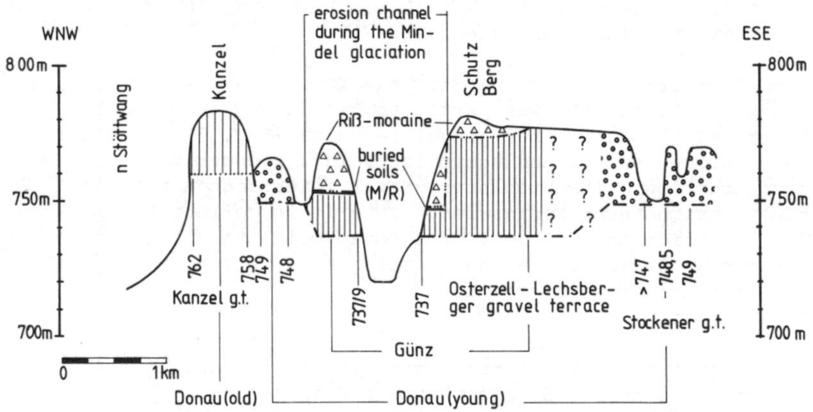

Fig. 2. Cross profile of the gravel terraces near Osterzell. The numbers indicate the elevation (in m) of the boundary between the Miocene and the Pleistocene sediments.

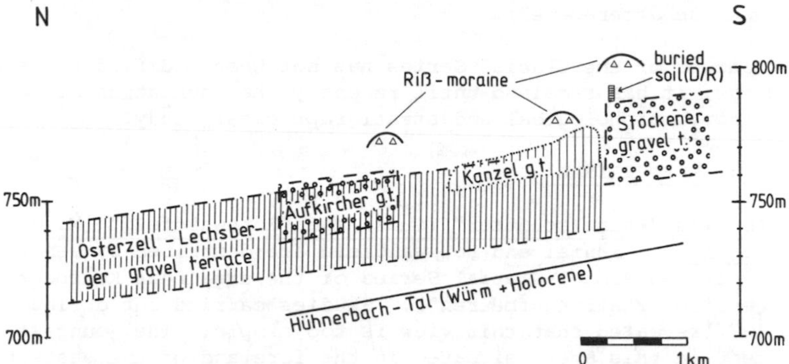

Fig. 3. Longitudinal profile of the different gravel terraces near Osterzell (i.e. Kanzel, Hühnerbach valley, western part of the Rothwaldfeld).

Stocken (Fig. 3) is an older one developed above a Donau gravel terrace and its genesis occurred during the D/G to the M/R interval. During the Mindel the glaciers advanced as far as the region just south of Stocken. The erosion channel mentioned above was formed by proglacial meltwater. Gravel terraces of Mindel age can be found in the west (Georgenberg), the east (Leeder), the north (Waaler Wald) but they are missing near Osterzell.

The broad Osterzell - Lechsberger gravel terrace was formed during the Günz glaciation and was partly eroded during the Mindel. The Stockner and the Aufkircher gravel terraces (Fig. 3) are older than those mentioned so far and accumulated in a younger phase of the Donau glaciation. The Stockener gravel intertongues with the terminal moraine near Königsried. Thus it is correct to speak of a Donau glaciation and not only of a Donau cold stage. The oldest sediment near Osterzell is the Kanzel gravel, which was deposited during an older phase of the Donau.

Concerning conclusions to be discussed later it is important to note that Penck's Günz gravel terrace consists of two and if the Stoffersberg is included (also Penck's Günz) of three gravel terraces of different age. The "Rothwaldfeld" east of the Hühnerbach valley (Mindel according to Penck) consists of three different cemented ("Nagelfluh") gravel terraces (younger Donau, Günz, Mindel) and of two unconsolidated sediments from the Riss and Würm.

2. Penck's "Höhen über Kaufbeuren" (Günz); Fig. 4 and 5. The gravel terraces of the Kanzel and the Stoffersberg cover a relatively small area and are isolated. Therefore Penck drew attention to the "Höhen über Kaufbeuren" (1901-09, p. 44) in other words the region

of Irsee. The following cross profile (Fig. 4) is an excellent example of the Suebian type of a terrace flight.

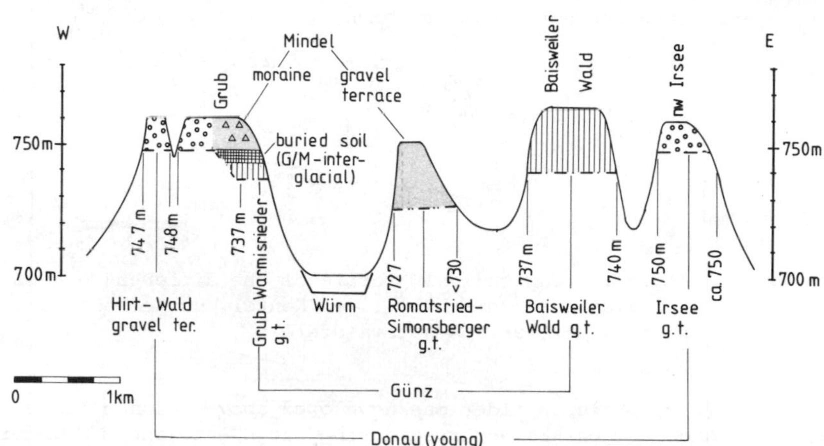

Fig. 4. Cross profile near Irsee (i.e. "Höhen über Kaufbeuren").

The youngest sediment is the Würm gravel in the Frisenrieder erosion channel (partly covered by Holocene peat). The huge gravel terrace of Eggenthal (Fig. 5) accumulated during the Riss. The Romatsrieder gravel belongs to the sediments of the Mindel glaciation and is dated as Mindel (near the Mindel confluence with the Danube river) from buried soils in loess sediments (Leger, Löscher & Puissegur,1972). The Romatsrieder gravel terrace is situated along the Mindel river and is well represented near Mindelheim. Therefore the region of the Mindel river should be considered as the stratotype region of the Mindel glaciation. This is necessary because the 'locus typicus' of the Mindel glaciation ("Grönenbacher Feld") is older than the Mindel drift (Löscher,1976).

The Baisweiler Wald gravel terrace is Günz in age and contains five to ten times more crystalline pebbles of the central Alps than younger sediments. The Lech river today does not extend from such source regions in the Alps. The gravel terrace of Irsee is the oldest and intertongues with moraine south of Irsee near Bickenried. Hence in this area evidence of five glaciations can also be found whereas Penck described the whole area as one single Günz gravel terrace. Just as at the Kanzel or the Rothwaldfeld there are three different terrace aggradations within one gravel terrace. Their age spans from the Mindel to the Donau glaciation.

West of the Friesenrieder erosion channel there are also at least three different middle and early Pleistocene gravel terraces. At Grub a buried soil separates the overlying Mindel moraine from the underlying Günz gravel; the latter is clearly younger than the Hirt-Wald gravel terrace (Donau). The longitudinal profile (Fig. 5) makes clear that the gravel terraces mentioned above extend long distances and are not local phenomena.

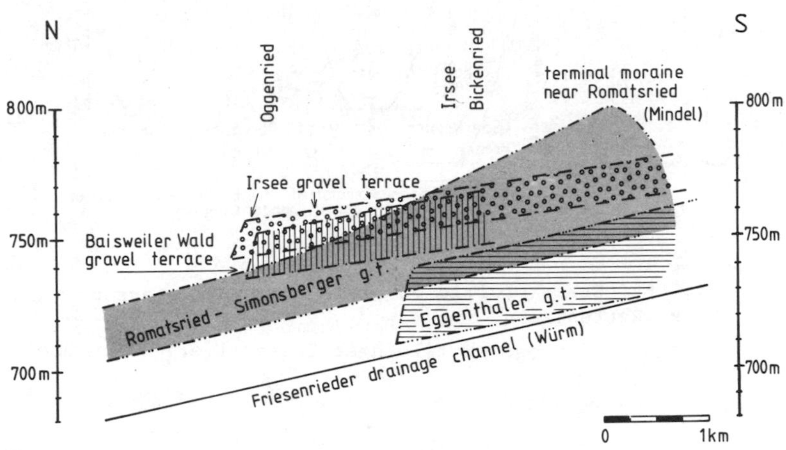

Fig. 5. Longitudinal profile of the terrace flight near Irsee.

<u>3. "Gravel terraces near Markt Rettenbach"; Fig. 6 and 7.</u> This area, not mentioned in the work of A. Penck, figures largely in the research of B. Eberl. The results presented below indicate that the conditions described above are not isolated examples but are typical for the whole Iller-Lech-Platte. At Markt Rettenbach the valley bottom of the eastern Günz creek was formed by a Würm late-glacial to Holocene level of erosion. The terrace on which Markt Rettenbach is built consists of Würm fluvioglacial outwash. The age of the terrace in the Auerbach valley is Riss. It is difficult to date the Unterburger terrace, since its base level is clearly below that of the other Mindel sediments east of it in the upper Mindel valley. Field studies carried out in 1985 have proved that the terraces of Unterburg and that of the upper Mindel valley converge on the same level in the Mindelheim area.

The Eheim/Rempholzer gravel terrace is an accumulation of Günz age and the younger Donau glaciation is represented by the Speckreuer

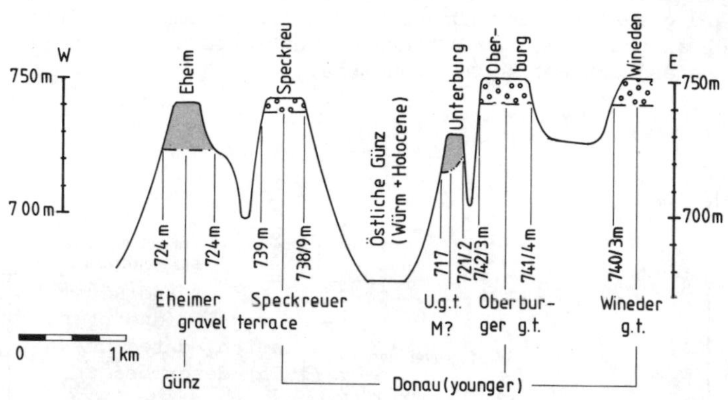

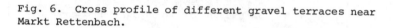

Fig. 6. Cross profile of different gravel terraces near Markt Rettenbach.

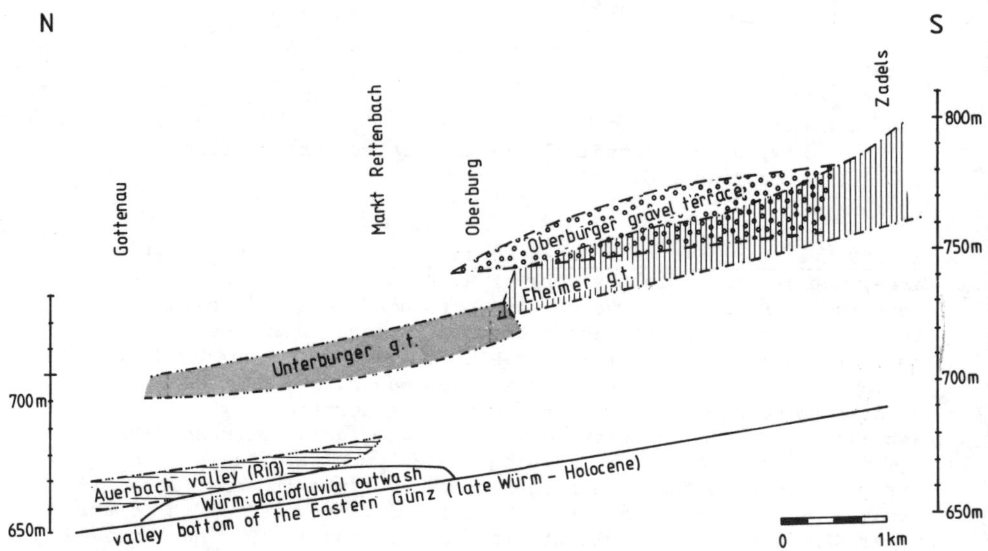

Fig. 7. Longitudinal profile in the area of the eastern Günz and the Auerbach valley.

and the Oberburger terraces. The small relict of glaciofluvial outwash at the top of the Hochfirst is at least of Biberian age.

Detailed reasons for the age determination of the different gravel terraces is outside the scope of this paper. They are available in Rögner (1979) covering the region near Osterzell ("Kanzel, Rothwaldfeld, Stoffersberg"); Rögner (1980), for the region west of Kaufbeuren ("Höhen über Kaufbeuren"); and Rögner (1986), for the region near Markt Rettenbach. These results are integrated for the first time in the present paper.

DISCUSSION AND COMPARISON WITH OTHER WORK

In the introduction it was implied that west of the Iller and east of the Lech there are also five Glacial Series. Eichler & Sinn (1975) named one glaciation ROTH (between Mindel and Riss); Schreiner & Ebel (1981) found in the same region (and partly using the same buried soils!) the HASLACH glaciation, dated between Mindel and Günz. East of the Lech, Schaefer (1975) described the PAAR glaciation (between Mindel and Riss).

It seems clear that the results in the area near Osterzell, in the Rothwaldfeld or near Kaufbeuren are not singular phenomena. Problems arise because west of the Iller (Eichler & Sinn,1975; Schreiner & Ebel,1981) and east of the Lech (Schaefer,1975) these workers have introduced new terms for glaciations. Both the author and M. Löscher, who has mapped the whole northern part of the Iller-Lech-Platte, could not find evidence supporting the definition of 'new' glaciations.

Eichler & Sinn (1975), Schaefer (1975) and Schreiner & Ebel (1981) argue that most of the 'upper cover debris' (i.e. "Obere Deckenschotter") should be dated as Günz according to Penck. It is agreed that there are gaps in the arrangement of Penck's stratigraphical scheme. However, our own results show that Penck's gravel terraces of Günz (Kanzel and Stoffersberg; Höhen über Kaufbeuren) and Mindel age (Rothwaldfeld) consist of various different gravel terraces. The insertion of a single 'new' glaciation thus could not solve the problem, one would have to include all the 'new' glaciations in a stratigraphical scheme. Therefore the author follows Eberl (1930) in naming the glaciations from Würm to Donau. Older gravel terraces which do not interfinger with moraines aggraded during the older Donau and the Biber cold epochs.

In West Germany there is a general consensus of opinion concerning the dating of the 'classic' glaciations; the Würm, Riss, Mindel and Günz ice advances are considered to be younger than the Matuyama palaeomagnetic epoch (Jerz & Grottenthaler,1981). Unfortunately, there are only a few places in the German alpine foreland where palaeomagnetical studies of older sediments can be carried out. The most important locality is the Uhlenberg west of

Augsburg (Scheuenpflug,1979). Above the Zusamplatten gravel a loam can be dated either from the Jaramillo event or from an earlier age. In any case it accumulated during a normal polarity event during the reverse Matuyama epoch and hence the gravel beneath the loam must therefore be older (Brunnacker et al.,1976).

A fossil peat layer ("Schieferkohle") above the loam shows a pollen association of at least the Waalian in age (Schedler,1979) and the associated molluscs are dated at least from the D/G-interval (Dehm,1979). The gravel terrace (i.e. the Zusamplatte; c.f. Fig. 7 in Penck & Brückner, 1901-09, p. 51) underneath these sediments (Penck's Günz) is, according to this evidence, surely older (from older Donau).

In this paper it is assumed, as Eberl (1930) did, that the fifth picdmont glaciation belongs to the Donau period and that this took place before the Brunhes epoch. The 'classic' ice advances (Würm, Riss, Mindel, Günz) would therefore be younger than 630 000 yr B.P. and this would generally coincide with Penck's estimation concerning the duration of the ice age (600 000 years). The longer part of the Pleistocene antedates the four famous classic piedmont glaciations (Würm, Riss, Mindel, Günz)when ice advanced as far as the alpine foreland.

REFERENCES

Brunnacker, K., Boenigk, W., Koci, A., & W. Tillmanns, 1976. Die Matuyama/Brunshes-Grenze am Rhein und an der Donau, Neues Jahrbuch für Geologie und Paläontologie, Abhandlungen, 151, 358-378.

Dehm, R., 1979. Artenliste der altpleistozänen Molluskenfauna vom Uhlenberg bei Dinkelscherben, Geologica Bavaria, 80, 123-125.

Eberl, B., 1930. Die Eiszeitfolge im nördlichen Alpenvorlande, Filzer, Augsburg, p. 427.

Eichler, H., & P. Sinn, 1975. Zur Definition des Begriffes "Mindel" im schwäbischen Alpenvorland, Neus Jahrbuch für Geologie und Paläontologie, Monatshefte, 1975, 705-718.

Jerz, H., & W. Grottenthaler, 1981, Glazialer und fluvioglazialer Bereich, Erläuterungen zur Geologischen Karte von Bayern (1:500 000), 135-141.

Jerz, H., Stephan, W., Streit, R., & H. Weinig, 1975. Zur Geologie des Iller-Mindel-Gebietes, Geologica Bavaria, 74, 99-130.

Leger, M., Löscher, M., & J.-J. Puissegur, 1972. Les terrasses de la valle de la Mindel en aval de Jettingen, Bulletin de l'Association francaise pour l'étude de Quaternaire, 2, 135-151.

Löscher, M., 1976. Die präwürmzeitlichen Schotterablagerungen in der nördlicher Iller-Lechplatte, Heidelberger Geographische Arbeiten, 45, p. 157.

Penck, A., 1882. Die Vergletscherung der Deutschen Alpen, Barth, Leipzig, p. 483.

Penck, A., 1899. Die vierte Eiszeit im Bereich der Alpen, Vorträge des Vereins zur Verbreitung naturwissenschaftlicher Kenntnisse, Wien, 39, 1-20.

Penck, A.,& E. Brückner, 1901-09. Die Alpen im Eiszeitalter, Tauchnitz, Leipzig, p. 1199.

Rögner, K.J., 1979. Die glaziale und fluvioglaziale Dynamik im östlichen Gletschervorland, Heidelberger Geographische Arbeiten, 49, 67-138.

Rögner, K.J., 1980. Die pleistozänen Schotter und Moränen zwischen oberem Mindel - und Wertachtal (Bayerisch-Schwaben), Eiszeitalter und Gegenwart, 30, 125-144.

Rögner, K.J.,1986. Die quartären Ablagerungen beiderseits des östlichen Günztales zwischen den Marktorten Rettenbach und Ronsberg, Jahresberichte und Mitteilungen des Oberrheinischen Geologischen Vereins, in press.

Schaefer, I., 1975. Die Altmoränen des diluvialen Isar-Loisachgletschers, Mitteilungen der Geographischen Gesellschaft München, 60, 115-153.

Schedler, J., 1979. Neue pollenanalytische Untersuchungen am Schieferkohlenvorkommen des Uhlenberges bei Dinkelscherben, Geologica Bavaria, 80, 165-182.

Scheuenpflug, L., 1979. Der Uhlenberg in der östlichen Iller-Lech-Platte, Geologica Bavaria, 80, 159-164.

Schreiner, A.,& R. Ebel, 1981. Quartärgeologische Untersuchungen in der Umgebung von Interglazialvorkommen im östlichen Rheingletschergebiet, Geologishes Jahrbuch, A. 59, p. 64.

Troll, C., 1931. Die Eiszeitfolge im nördlichen Alpenvorland, Mitteilungen der Geographischen Gesellschaft München, 24, 215-226.

International Geomorphology 1986 Part II
Edited by V. Gardiner

A NOTE ON THE QUATERNARY PALAEOGEOGRAPHY OF THE SONG-LIAO PLAIN, NORTHEAST CHINA

Qiu Shanwen[1] Xia Yumei[1] Li Fenghua[1] Sui Xiulan[1] and Li Shupei

Changchun Institute of Geography, Academia Sinica, Jilin, China

The First Hydrogeological Team of the Geological Bureau of Jilin Province China[2]

ABSTRACT

Previously it was considered that the Song-Liao Plain was a former megalake basin infilled with Middle Pleistocene bedded clays. New palynological and palaeomagnetic data suggest that the lower part of the clays are of Lower Pleistocene age, whilst the upper part are Middle Pleistocene. The lake basin was already in existence during the Early Pleistocene.

INTRODUCTION

The Song-Liao Plain is one of the three great plains of China and is situated in the centre of the northeastern part of the country. The study of its palaeogeographical evolution in the Quaternary is not only of theoretical significance to the study of natural environmental change but also of practical importance in the study of the ground water-bearing strata, in the estimation of the water resources as well as in the development of industry and agriculture.

The Song-Liao Plain was considered in the past to be a megalake basin formed in the Middle Pleistocene. Its extent roughly coincides with a long ellipse drawn between Qiqihar - Shuangliao - Zhaoyuan - Lindian and Qiqihar. The Plain is some 50,000 km^2 in area, and is underlain by widespread lacustrine clays between 30-70 m thick. These clays cap a regional aquifer.

New studies of the palaeogeography of the Plain in the Quaternary have used micropalaeontological, magnetic stratigraphical and isotope techniques supplemented by data from over 200 boreholes. Amongst the many boreholes, that of the Ling-zi well, located in Qianan county at the centre of the Plain is particularly instructive. This hole, 112.5 m deep, reveals four lithostratigraphical units:- 1. 0-20 m, loess-like silty sands; 2. 20-75 m, grey and yellow green clays interbedded with silt layers containing lacustrine mollusca; 3. 75-80 m, grey sands and gravels, and 4. 80-112.5 m, sandstones and conglomerates. The traditional stratigraphical division assigns the four lithological units respectively to the Upper Pleistocene, Middle Pleistocene (Daqinggou Group), Lower Pleistocene (Baitushan

Group) and the upper Tertiary system. Palaeomagnetic measurements show that the Brunhes-Matuyama epoch boundary lies at a depth of 45 m and the start of the Olduvai event within the Matuyama is at about 80 m.

PALYNOLOGY

Pollen has been abstracted from the core between 25 and 75 m. Six pollen assemblage zones can be recognised and in order of decreasing age these are:-

I. An assemblage of *Betula*, *Artemisia*, and grass dating from about the time of the Olduvai event.

II An assemblage of *Picea*, *Salix*, *Polygonum* and grass extending from before and after the Jaramillo event. The difference between assemblages I and II lie in that the latter has a higher percentage of tree pollen. Among them *Picea* accounts for 23%, at the first peak of the *Picea* curve, whereas *Salix* forms 9.6% and *Phragmites* 3.8%. It is thought that lakes and bogs were present in the early Pleistocene, when the associated climate was mild and moist. This warm period is regarded as the last phase of the early Pleistocene.

III An assemblage including *Ephedra*, *Tamarix chinensis* and Chenopodiaceae lies at a depth of about 45 m, among which Chenopodiaceae makes up to 33.6%, mainly of *Suaeda* and *Eurotia*. This is interpreted as representing a dry steppe with abundant saline plants. The climate appears to have been extremely dry and cold with permafrost, the latter probably causing the soil to become salinized. This cold period forms the first stage of the Middle Pleistocene, and is located just below the Bruhues-Matuyama polarity reversal.

IV An assemblage of *Pinus*, *Betula*, *Ulmus* and *Chrysanthemum* appears between 30-43 m. *Betula* is the most abundant broadleaf tree and makes up to 36.6% of the tree pollen followed by *Ulmus*, *Quercus*, *Juglans* and *Tilia*. There is a small amount of pollen from aquatic plants together with spores of *Zygnema*, and also a large amount of *Pediastrum*. The vegetation cover is either steppe with *Betula* or a low density broad-leaf forest. The climate appears to have been mild and wet and both lakes and bogs were present.

V An assemblage of *Pinus*, *Picea*, Chenopodiaceae and grass occurring between 25-30 m with Gymnosperm spores accounting for 65% of the n.a.p. and with *Pinus* at 40% and Picea 38% of the aboreal pollen. A dense coniferous forest with *Pinus* and *Picea* appears to have expanded under a cold and wet climatic regime.

VI An assemblage of *Pinus*, *Betula*, *Salix* and grass occurs between 20-25 m, and is similar to assemblage IV, but the climate was mild and wet. This warm period marks the upper limit of the Middle Pleistocene.

According to the palaeobotanical evidence presented above it is clear that the lacustrine clays of the northern part of the Song-Liao Plain span a substantial part of the Quaternary. In the Ling-zi borehole between 20 and 45 m Middle Pleistocene sediments are present and these correspond roughly to the period 800,000-200,000 yr B.P. The Middle Pleistocene is characterised by climatic alternations consisting of dry - cold, mild - wet and cold - wet, phases. The lakes and bogs expanded and shrank in sympathy with the prevailing climate. We propose to designate the 25 m thick sequence of lacustrine clays which accumulated in this interval, the Dabusu Group.

The sequence extending from 45-75 m may approximate to an absolute age of 800,000 to 1.87 M yr B.P. and hence belongs to the Lower Pleistocene. During this period the climate alternated from mild and cool to mild and wet and a 30 m thick lacustrine succession was deposited. The underlying lacustrine clays formerly attributed to the Baitushan Group of the Lower Pleistocene appear to have an upper age limit of 1. 87 M yr B.P. According to the view point of whether the lower limit of the Quaternary should be 1.7 - 1.8 M yr or 2.4 M yr B.P., then this underlying lacustrine clay may belong to the beginning of the Pleistocene or the end of the upper Pliocene.

CONCLUSION

From the above, it is clear that the present day Song-Liao Plain was occupied by a lake basin not only in the Middle Pleistocene, but also significantly farther back in time to at least the Early Pleistocene. Hence the lowest lacustrine clays are possibly pre-Pleistocene in age. In the later stage of the Early Pleistocene, the lake basin of the Song-Liao Plain was very extensive. The rivers around it including the East Liaohe, West Liaohe, Taoerhe, Nenjiang, and the second Songhuangjiang, all flowed into the lake basin forming a centripetal drainage network with an outlet in the northeast of it. Comparison and analysis of the terrace deposits of the Song-Liao Plain and the Sanjiang Plain reveal a similar pattern of development and suggests that the Songhuajiang river before the Middle Pleistocene flowed from the lake basin northeastward to the Sanjiang Plain.

International Geomorphology 1986 Part II
Edited by V. Gardiner

LATE QUATERNARY SUB FOSSIL WOOD BEDS IN THE ZHUJIANG DELTA DEPOSITS

Huang Zhenguo, Li Pingri, Zhang Zhogying and Li Konghong

Guangzhou Institute of Geography, China

ABSTRACT

Three horizons of buried sub-fossil wood in deltaic sediments are described. Each wood bed includes in situ material and is directly overlain by a marine transgressive horizon. All are younger than 30 ka B.P.

INTRODUCTION

Buried sub fossil wood is extensively distributed in the Zhujiang Delta (Pearl River Delta) but its preservation is variable. Two opinions have been expressed about the distribution and number of beds containing the wood. In one opinion, there is only one horizon of sub fossil wood in the Quaternary deposits and it lies 1-5 m beneath the plain surface whereas in the other, there are three beds, an upper at 2-4 m, a middle at 9-13 m and a lower at about 20 m deep. During the study of over 1100 drill hole cores collected in the area, wood was found, in 158. Some 16 wood samples have been dated by ^{14}C dating. These samples appear to be representative of the whole delta and since they are from different depths, they probably represent different aged deposits. According to their stratigraphic position and comparison with many other wood layers, we support the notion that they can be divided into 3 basic beds: upper, middle and lower. The depositional facies reflected by these beds is interpreted as a transition from terrestrial to marine conditions.

STRATIGRAPHY

Table 1 describes the depths and stratigraphic positions of the 16 samples which were ^{14}C dated. These samples can be classified into 3 groups according to their radiocarbon ages: (1) about 20,000-30,000 yr. B.P.; (2) about 6000 yr. B.P.; (3) about 2000 yr. B.P. When compared with three *Ostrea* shell bed dates (i.e. about 20,000, 5000, 2000 yr. B.P. respectively) in this area, an interesting relationship can be found with the ages of the wood beds. The latter are a little older than the ages of the related *Ostrea* shell beds. This fact suggests that the three wood beds were later covered by a marine facies with *Ostrea* shells, and it demonstrates that the wood beds represent transitions from terrestrial to marine environments. In Table 1, the stratigraphic positions of the first 9 samples (about 20,000-30,000 yr. B.P.) are

Sample No	Burial Depth	Lithology of wood containing bed	Lithology of overlying bed	C^{14} age (yr B.P.)
1	15.9	Gravelly sand	Clayey silt	37,000±1480
2	9.3	Clayey silt	Silty clay	36,170±2700
3	2.7	Fine sandy clay		35,000±2800
4	10.9	Sandy gravel	Silty-fine sand	33,000±3000
5	22.5	Silty clay	Silty clay with _Ostrea_	30,440±2300
6	25.6	Silty-fine sand	Clayey silt	28,240±2220
7	15.9	Sand	Clay	25,140±500
8	20.3	Gravelly sand	Silty clay	24,400±1950
9	33.0	Coarse sand	Silty-fine sand	23,170±980
10	3.95	Silty clay	Muddy sand with shells	6510±170
11	7.0	Clayey silt	Muddy sand with marine diatoms	6300±330
12	12.9	Gravelly sand	Muddy sand	6150±160
13	4.7	Sandy gravel	Muddy sand with brackish diatom	5940±300
14	4.0	Silty-fine sand	Muddy silt	2350±110
15	2.8	Silty mud		2050±100
16	1.5	Silty-fine sand	Mud	2270±110

TABLE 1. C^{14} Dating of buried subfossil wood in the ZhujiangDelta

in or just above basal gravels, which are overlain by silty clays with _Ostrea_ or dark grey clay. Marine sedimentary features are more obvious in the top layers of the next sample group (about 6000 yr. B.P.). In some cases, the wood lies in or just above the weathered variegated clays which are overlain by sediments with brackish fossil diatoms or marine shells. In other cases, the wood is found between gravelly sands below and muddy silts with brackish fossil diatoms above. As for the remaining samples (about 2000 yr. B.P.), differences of lithological character between the beds above and below are not clear, but when the wood bed is compared with the sediments below, a transition from marine facies (lower) to fluvial facies (wood bed) can be seen.

As mentioned above, all the 3 main wood beds reflect a transition in sedimentary facies - from terrestrial to marine. The dates from the wood beds approximate to the time when they were submerged, i.e. when a transgression took place. Thus the fact that the wood dates are little older than those from the Ostrea shell beds is consistent with this interpretation.

Table 1, also gives the depths of the 3 wood beds: the upper is 1.5-4.0 m; the middle is 5-7 m (on the margin of the plain) or 12.9 m (in the centre of the plain); and the lower is over 10 m (on the margin of the plain) or 20-30 m (on the centre or the south of the plain). The depth of burial of the wood is related in general to its age, but there is not a simple rule which decrees that the deeper wood is older and vice versa.

The lower woods beds were found in 37 places. They are mainly distributed where the lowland Quaternary is over 25 m thick. It seems to suggest that the lowlands were broad valleys at that time. The inclined landform of the former valley was revealed by the fact that the wood depth increased from the upper reaches to the lower reaches. The trees appear to have grown in the valley bottom where favourable conditions of water and heat prevailed. Then they were later submerged by sea water and covered by marine deposits.

The middle group of beds were found in 46 places, with a distribution similar to that of the lower beds. However the area covered is much greater, and it stretches from the upper reaches in the delta down to the river mouths. The greater distribution of the middle beds shows that the delta had grown towards the sea, with an expansion of the terrestrial environment.

The upper group of wood deposits were found in numerous places - at least 153. Most are concentrated in the northern part of the delta but some occur in the central or southern parts. The environment which they reflect is again terrestrial. A feature of their locations i.e. mainly in the north, shows that the degree of marine regression in this period was not so large as that in the previous two periods.

PALAEOENVIRONMENT

The tree species in the upper beds are mangroves around the former river mouths. Acidic zones are associated with the buried mangroves. Water pines (Glyptostrobus pensilis) are the principal species in other areas with a concentration at the confluence area of the Xijiang and Beijiang rivers. In recent years a great number of buried trees have been dug out from below the plain surface up to several metres in depth. The local people call them "Underground Forests". Some of the trees still retain good xylems, they also stand upright and the roots spread out. Most of the trunks and branches lie horizontally and they cross one another. Generally, trunks are some 10 m in length although the longest ones can be up to 18 m long, 0.3-0.4 m in diameter, while the biggest ones can be

1.5 m. The mass of wood per ha is 300-3000 tons and the wood with good xylems is used as raw material for the cork industry. The geomorphic contexts of the areas where the wood is concentrated are waterlogged lowlands and hollows. The depth below the present day surface and the top of the wood bed ranges on average between 0.5-2 m. Bed thickness is 2-3 m and usually they cover areas of hectares to dozens of hectares.

The change in sedimentary environment can be deduced from the character of the upper wood beds. The water pines apparently died because of climatic deterioration. This was followed by an amelioration and allied rise of sea level in conjunction with tectonic submergence so the dead water pines were buried by sedimentation. In the outer part of the delta to the north west ^{14}C dating of the wood of a depth of 2 m, has yielded an age of 2940±110 yr. B.P. Taken together with the other 3 wood samples from the upper bed (see Table 1) all these samples are bracketed by the period 2000-3000 yr. B.P. According to Zhu Kezheng, there was a cold period at 3000 B.P. and a cool period at 2200 B.P. The 4 samples mentioned above lie within these periods of cold or cooler climates. There then followed a warm period between 1850-1250 B.P. when sea level rose, in conjunction with land subsidence and as a consequence the water pines were buried. Evidence for tectonic subsidence is as follows: (1) Water pines have a preference for river bank environments, but they are now found several metres underneath the surface; (2) Below the wood beds are flood deposits, whereas above there are swamp deposits consisting of grey-black muds with yellow-white clays associated with the present alluvial deposits; (3) The wood is generally concentrated beneath lowlands where the Quaternary sediments are 20-25 m to 30-35 m thick. Some lowland surfaces are below sea level, with a minimum altitude of -1.7 m.

It has been claimed by Wu Zhengyi and others that the natural forests of water pines grew in Guangdong Province at least 1000-2000 years ago on swamp soils or in hill foot hollows. Later they disappeared gradually. We hold that the natural forests of water pines in the Zhujiang Delta thrived before the cold period of 3000 years ago, during a warm period which lasted for some 2000 years. These forests of Water pines died between 2000-3000 years ago and were buried by later deposits. In the last 2000 years a fall of air temperature has been the general trend, with the result that Water pine forests are unable to regenerate. Hence natural forests of Water pines are absent in the Zhujiang Delta today.

Trees species in the middle and lower wood beds cannot at present be determined although the cause of death and preservation was similar to that of the upper bed. The middle bed dates from the relatively warm Atlantic period, but the forest died out in the subsequent cold sub Boreal period. The lower bed corresponds to the middle Würm when interstadial conditions witnessed a climatic warming.

CONCLUSION

As mentioned above, the stratigraphic positions of these 3 beds of sub fossil wood show successively three periods of change from terrestrial to marine environments and concurrently a climatic amelioration. All those beds are extensively distributed and although the species of trees were different, the processes of burial are the same. The relation of the wood beds with other sedimentary facies indicates repeated depositional cycles in the late Quaternary of this area.

International Geomorphology 1986 Part II
Edited by V. Gardiner

QUATERNARY TRANSGRESSIONS, EUSTATIC CHANGES AND MOVEMENTS OF SHORELINES IN NORTH AND EAST CHINA

Yang Huai-jen and Chen Xi-qing

Department of Geography, Nanjing University, China

ABSTRACT

Hundreds of cores have been taken from the coastal plains of North and East China and studies of biostratigraphy, magnetostratigraphy and radiometric age determination reveal a close relation between climatic and sea-level changes. This paper deals with the complicated relationship between sea-level changes, buried marine horizons, neotectonics and evolution of the coastal plain. Foraminiferal faunas and sedimentary facies show eight to ten marine horizons at different depths in the Great Plain of North China. In the coastal plain of North China and the Gulf of Puhai six marine horizons have developed since the late Pleistocene and can be correlated with eustatic high stands of sea levels. In the early Pleistocene major neotectonic movements caused the sea to transgress far inland along fault bounded features. The time intervals between two successive late Pleistocene marine horizons are much shorter, possibly due to the increasing amplitude of sea level fluctuations linked to the Quaternary glacial cycles. Since the coastline of China spans different geotectonic units, the influence of tectonic movements varies greatly. Specific attention is paid to the regional Quaternary sea level flucuations and dynamic stratigraphy. The factors that have controlled the development of cheniers and its implication in reconstructing the history of sea level changes are discussed together with a consideration of the rate of coastal progradation and the impact of sea-level fluctuation on the human activity since the mid-Holocene.

CHARACTERISTICS OF LATE QUATERNARY MARINE SEDIMENTS

The coastal regions of the Gulf of Puhai, the Yellow Sea (North Jiangsu Province), the Yangtze Delta and Fujian Province are typical of coastal China, but each has its own tectonic background and development model.

In recent years, there has been considerable discussion on the chronology and distribution of marine sediments of late Quaternary age. It has been agreed that three transgressive units are widely developed along the western coastal area of the Puhai in the late Quaternary. Several local names have been given to the same marine sediments by different authors according to either the borehole location or characteristic foraminifera. Here we shall deal with it in a simplified way, as follows:

marine unit I : 4000-8000 yr B.P., at depths of -5 to -15 m

marine unit II : 30,00-40,000 yr B.P., at depths of -20 to -40 m

marine unit III: 70,000-110,000 yr B.P., at depths of -60 to -80 m

In different locations, the variation in depth value may be between ±10 m to ±20 m. These successive marine strata are synchronous with high sea-levels and can be correlated with the deep-sea isotopic stages 1, 3 and 5a-5c. The core from Nanpai is typical of this area (Fig. 1).

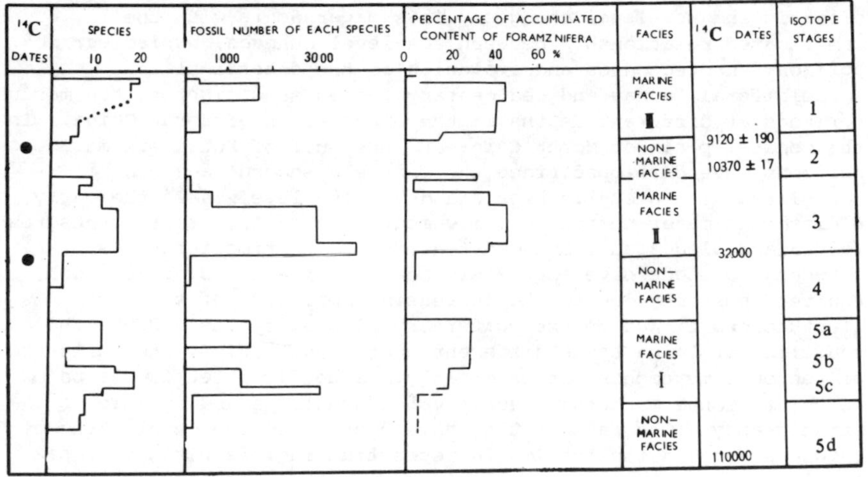

Fig. 1. Correlation between changes of foraminifera content in the Nanpahe Core, Hebei and isotope stages

Figure 2 shows the extents of the three transgressions which roughly coincide with one another. Since the mid-Holocene, four cheniers running nearly parallel to the present shoreline have developed in the western coastal area of the Puhai (Zhao Xitao et al., 1980). Their ages are 5235, 3330, 1080 and younger than 800 yr B.P. respectively. Numerous boreholes in the Subei-Yangtze delta coastal area have also revealed three transgressive units (Zhao Songling and Chin Yun-Shan, 1982) and appear to correspond with the three separate marine units in the Puhai western coastal

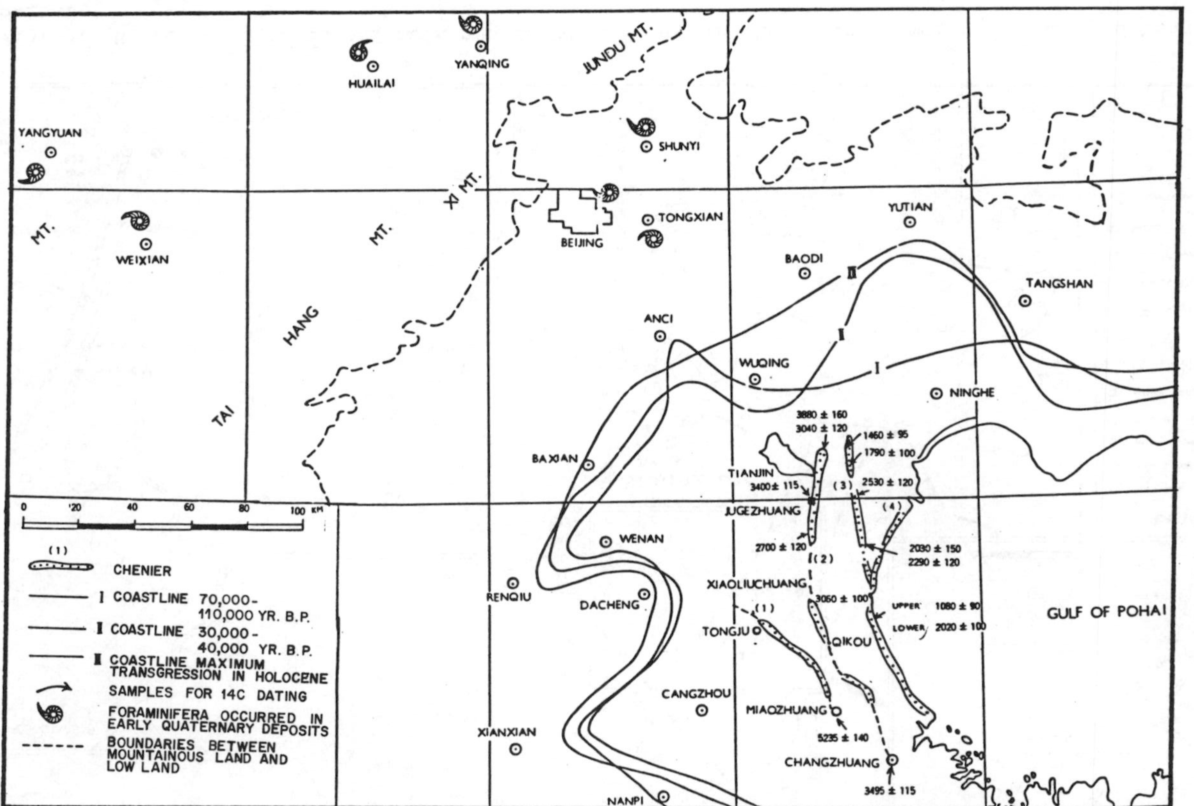

Fig. 2. Quaternary transgressions in the North China Plain, where early Quaternary foraminifera and Holocene chenier ridges were found.

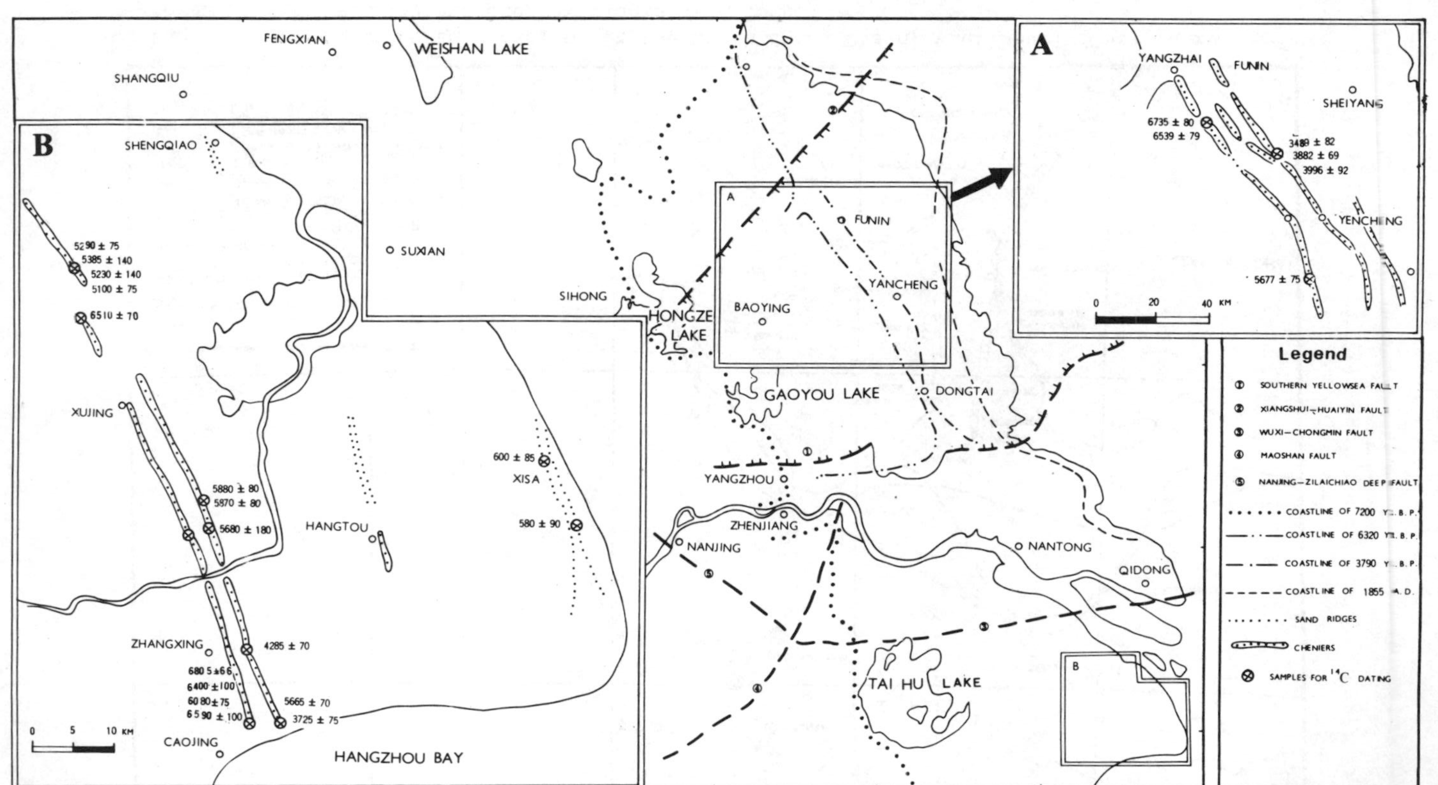

Fig. 3. Holocene cheniers and main neo-tectonic lines in the plain of North Jiangsu and Yangtze delta.

area. Unlike those of the Gulf of Puhai, the Subei-Yangtze delta area faces an open sea and, due to neotectonic movements it has a more complicated history of coastal development. There are also four cheniers and sand ridges of different ages (Fig. 3). Problems concerned with the development of the cheniers will be discussed later.

In the Fujian coastal area of southeastern China only two marine units were produced in the late Quaternary and they may correlate with marine units II and III in North China. The older one is known as "old red sand" for it has undergone strong weathering in the Holocene after being elevated above sea level by neotectonic movements. At present the so-called "old red sand" is limited in its distribution to depositional terraces 10 to 20 m above sea-level, while the younger marine sediments are restricted to fault-basins and are usually found at depths of 15 to 30 m in boreholes.

LATE QUATERNARY NEOTECTONICS

The Puhai western coastal area is located in a region of continuous subsidence due to a series of horsts and grabens now lying beneath the land surface. The Gulf of Puhai is also confined to this tectonic unit and Figure 4 shows that it is surrounded by four tectonic uplift blocks. Tectonic movements can be traced back to the Jurassic but it was not until the late Tertiary that the basin began to subside and act as a sediment sink. Since the late

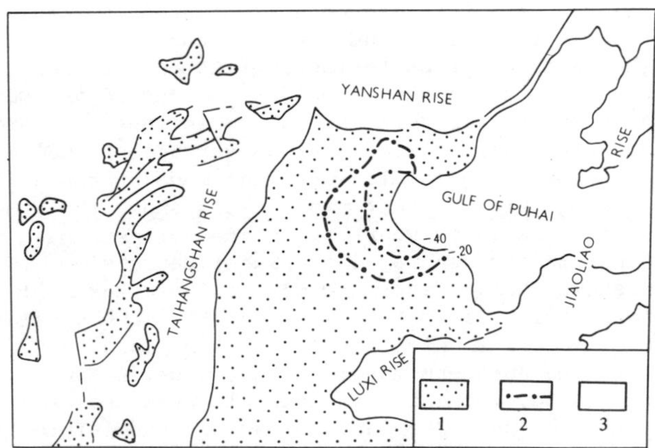

Fig. 4. Tectonic map and rates of vertical deformation in recent years of North China Plain

1. Grabens and faulted basins, have been active since the late Tertiary
2. Deformation isolines from 1952 through 1972 (in mm) (from seismic surveying, 1977)
3. Rising blocks

Tertiary, about 1000 m of fluvio-lacustrine deposits have accumulated but in the centre of the Puhai Basin up to 5000 m of sediments have been deposited since the Tertiary.

Geodetic levelling from 1953 to 1972, indicates that the Puhai Basin is still sinking with the maximum rate in the centre. Based on the gravity data, Liu Yuan-lung et al. (1978) have made a calculation which suggests that the Moho discontinuity is located at the depth of 29 km in the centre of the Puhai and it rapidly increases to 34 km in the west side of the basin. This pattern is consistent with the pattern of vertical deformation shown in Figure 4. It is inferred that a hot spot underlies the lithosphere of the basin and as a result, the crust has been gradually thinning and sinking, inducing a series of grabens. By comparing Figure 2 with Figure 4, it can be seen that the maximum marine transgressive limits since the late-Pleistocene have reached the same position as the 20 mm vertical deformational isoline. This suggests that the distribution of the marine sediments was controlled by neotectonic movements and associated landform evolution.

There are two fault lines bordering the Subei coastal plain on the north and south (Fig. 3). Tectonically, the plain is a part of Subei-South Yellow Sea Depression and is bounded on the north by the Lusu Rise. To the south of the Subei Plain lies the Yangtze Delta which has been sinking since the beginning of the Quaternary at a rate smaller than that of Subei Plain Coast. Apart from subsidence, differential tilting is another pronounced feature in the coastal plains. This can be demonstrated by the inclination of the Pleistocene formations and loess terraces along the Yangtze River. The "Xiashu" loess terrace, of Late-Quaternary age, is 30 m high around the Maoshan area (around Jurong and Danyang County of Jiangsu Province), but it gradually declines eastwards so that in the estuary area, it becomes a buried terrace some 30 m below sea-level. Repeated geodetic levelling made by the Seismological Bureau of Jiangsu in the last two decades shows that the greatest rate of subsidence (4 mm/year) is in the northern part of Yangtze delta and not in the centre of the Subei Depression. Such a migration shows that a wave-like movement has taken place in the Subei-Yangtze Delta Coastal region and this may be related to migration of material under the crust. The Yangtze River estuary was close to the northern margin of the Subei Basin in the early stage of delta development but later, it progressively migrated to its present location. Two factors may be responsible, one is the migration of the depressional centre mentioned above; the other might be the effect of the Coriolis force.

The Fujian coastal area experienced major faulting in the Mesozoic and intermittent uplift since the Cenozoic era has resulted in several step planation surfaces. Based upon tectonic pattern and intensity, the tectonic history can be divided into three stages since the Neogene. From Miocene to Early Pleistocene it was characterised by differential block-fault movement. During the Mid-Pleistocene the tectonic movement became relatively weak and

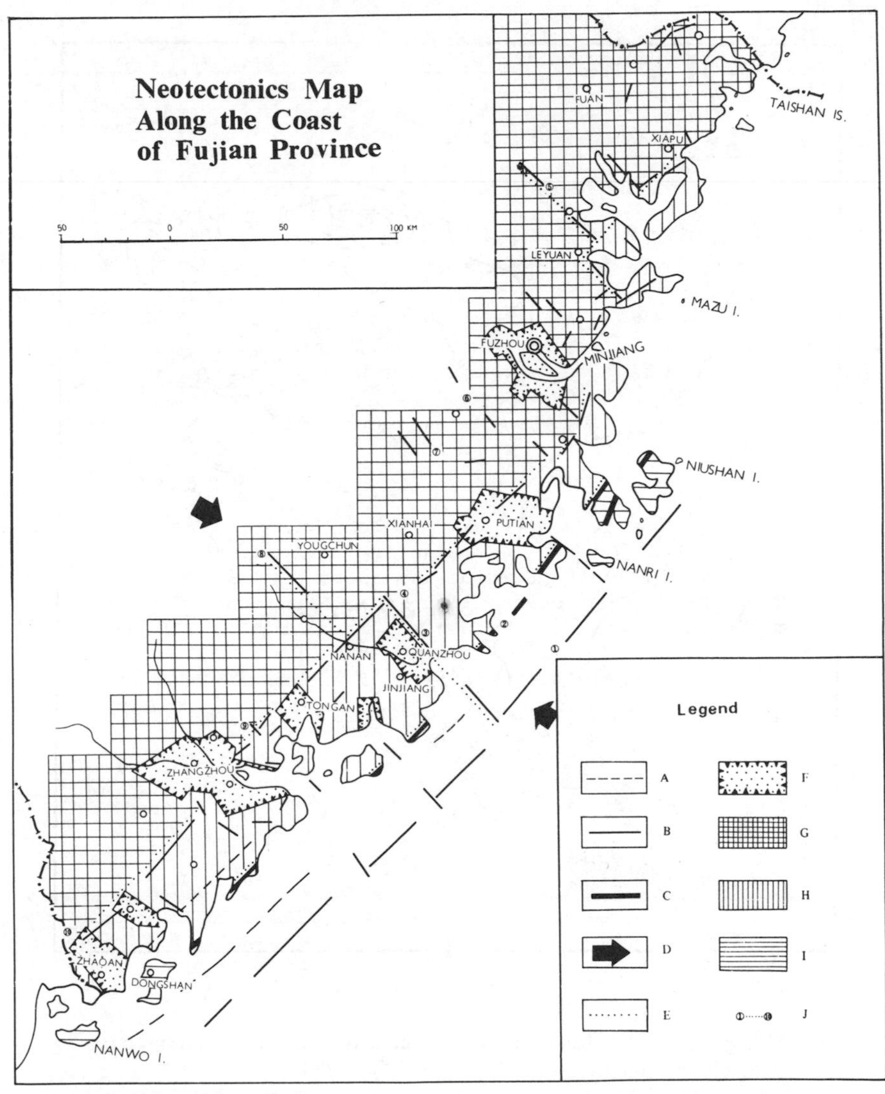

A. LESS ACTIVE FAULT **SINCE LATE PLEISTOCENE**

B. QUATERNARY ACTIVE FAULT

C. MORE ACTIVE FAULT SINCE LATE

D. MAIN COMPRESSIVE STRESS PLEISTOCENE

E. BOUNDARIES OF TECTONIC REGIONS

F. FAULTED DOWN SINKING BASINS SINCE LATE PLEISTOCENE

G. SECULARLY RISING REGIONS IN QUATERNARY

H. RELATIVELY STABLE REGIONS SINCE LATE PLEISTOCENE

I. RISING REGIONS SINCE LATE PLEISTOCENE

J. FAULTS ALONG THE SOUTHEASTERN COAST OF FUJIAN PROVINCE

Fig. 5. Neotectonic map along the coast of Fujian Province, southeastern China

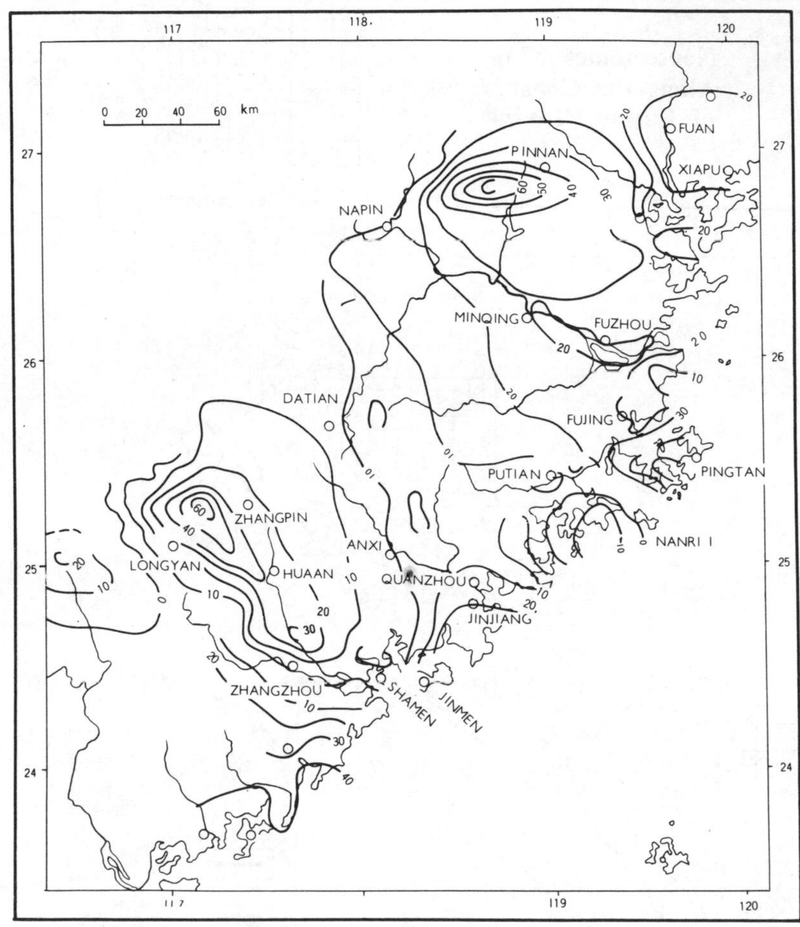

Fig. 6. Vertical deformation map (in mm) along the coast of Fujian Province, (1965-1974)

planation dominated. Since the Late-Pleistocene, the differential block-faulting tectonics has strengthened again; NE-NNE, NW and E-W trended faults were reactivated creating a remarkable network of block-faulted structures(Fig. 5). Among these block-faulted structures the most active ones are the faulted-basins. The oldest deposits in these fault basins are mostly of Late-Pleistocene age. The differential uplift and subsidence during the late-Quaternary has formed coastal cliffs, terraces and drowned coasts.

DEPOSITIONAL ENVIRONMENT OF THE CONTINENTAL SHELF

Sea-level change has been the major control on the recent continental shelf environments of North and East China. Using data from 15 cores in the Yellow Sea, Xu Jiasheng et al. (1981) proposed that three marine units interbedded with the continental ones had developed since Würm I. The lower marine unit, for example, at a depth of 3-4 m overlies lower non marine sediments in Core H_{29} and Core H_{72}. According to palaeomagnetic measurements these marine sediments were deposited during the interval 69,000-60,000 yr B.P. and abundant foraminiferal and ostracod evidence indicates that the former sea-level was 50 m lower than present. The middle and upper marine units under the Yellow Sea corresponded to the marine units II and III in the Puhai coastal plain. The location of the lower continental sediments indicate that sea-level around 72,000 yr B.P. was 81 m lower than that of present. The middle continental sediments represented by a peat layer some 78 m below the present sea-level, was formed about 42,000 yr B.P. The upper continental sediments were deposited during the Würm II.

Recently, the BC-1 core in the Puhai has disclosed a complete record of transgression events since the Last Interglacial (Fig. 7). The chronology, based on sedimentation rate extrapolation, indicates that the horizon assigned to 125,000 yr B.P. was located in the middle of marine unit VI. Consequently, there have been 6 episodes of marine transgression in the Gulf of Puhai during the last 125,000 years. It is worth noting that the marine units III and VI in BC-1 core did not exist in the western coastal plain of the Puhai. This problem will be discussed later.

The latest marine regression from the shelf started about 26,000 yr B.P., and is documented by the age of a sample taken from the bottom of marine unit II in BC-1 core. Peat deposits with radiocarbon dates have shown that the coastline had retreated to a position of -110 m at 23,00 yr B.P. when the climate of continental China, was cold and dry.

Recently, Yang and Xie (1984) have suggested that the maximum lowering of sea-level was 120-130 m at 18,000 yr B.P. After 15,000 yr B.P. the sea-level began to rise rapidly, and by 12,000 yr B.P. it had reached -56 to -58 m i.e. a rise of 90 m in 3000 years. Here the hydroisostatic adjustment after the full glacial is not taken into consideration. Radiocarbon dating of peat deposits in the shelf area also reveals that sea-level was 59 m lower than that of present.

COASTAL PROGRADATION AND SEA-LEVEL FLUCTUATION IN EAST CHINA

The coast of Fujian in southeastern China is mainly a tectonic coast with a zigzag outline. Depositional processes have played only a subsidiary role in coastal development, nevertheless attention will now be focused on the progradation rates in the

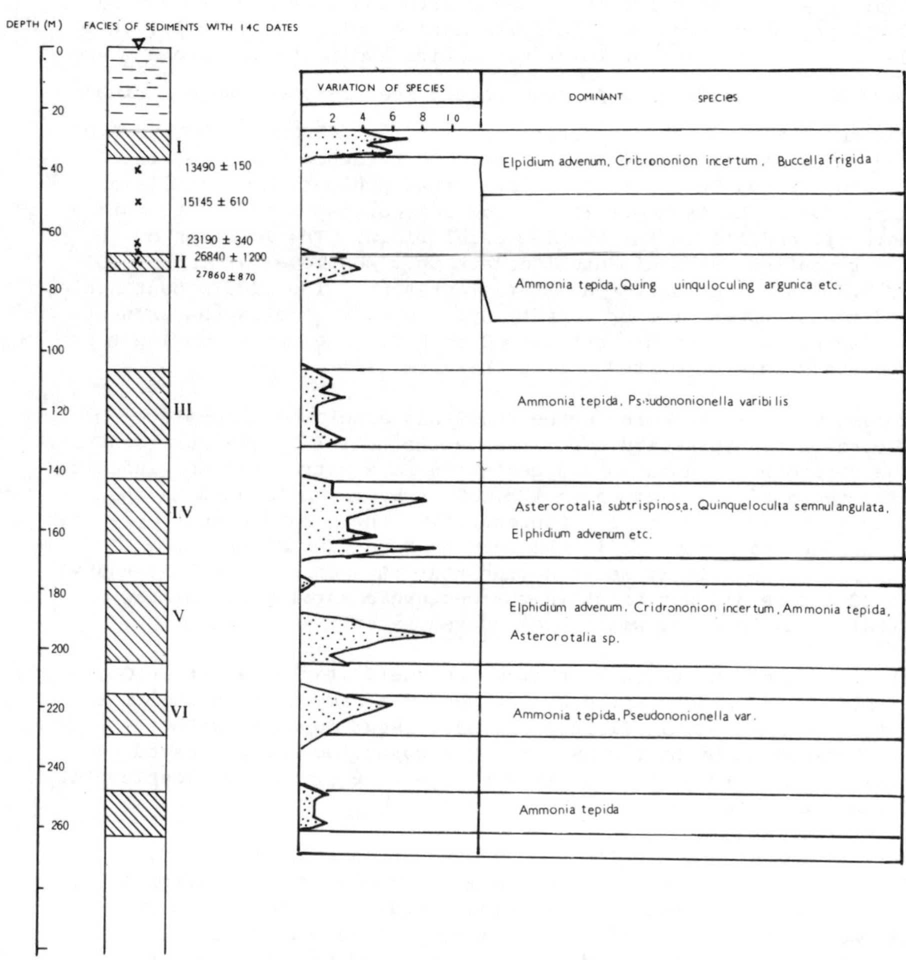

Fig. 7. Variation and distribution of species of foraminifera in Core BC-1 in the Gulf of Puhai (After Huang Qingfu and Cang Shuxi, 1985)

depositional plains of east China since the mid-Holocene.

Table 1 shows the rates of progradation in three coastal segments based on the occurrence of cheniers with different ages. Rates increase rapidly since the mid-Holocene when the rising sea level become relatively stable. This accelerated accretion is initiated by the following factors: (1) The vigorous uplift of the Qinghai-Tibetan Plateau and the high mountains in West China, which has created a sharp topographical contrast between eastern and western China and, in consequence, increased the erosional and transportational capacity of the rivers; (2) Changing atmospheric circulation patterns over continental China caused by the uplift has led to a deterioration in the climate of West China with associated intensification of mechanical weathering processes, increased rates of denudation and acceleration of loess deposition; (3) Increasing human activity, such as the destruction of natural vegetation and allied anthropo-geomorphological effects, which yield

NAME OF COAST	TIME SPAN	RANGE OF COASTLINE ADVANCE	HORIZONTAL DISTANCE (KM)	RATES OF COASTAL PROGRADATION (M/YR)
COAST OF NORTH JIANGSU	6317–1128 YR. A. D.	XIGANG–DONGGANG	4–15	0.7–2.7
	1128–1425 YR. A. D.	DONGGANG–XINGANG	12	36
	1425–1983 YR. A. D.	XINGANG–PRESENT COAST	33	65
DELTA COAST OF THE SOUTH BANK OF YANGTZE R.	6380–5410 YR. B. P.	SHAGANG–ZHUGANG	1.5–4	1.5–4.1
	5410–1500 YR. B. P.	ZHUGANG–SHENGQIAO–HONGTOU	20	5.1
	1500–580 YR. B. P.	SHENGQIAO–HONGTOU–DONGSHA	25	27.2
WESTERN COAST OF THE PUHAI BAY	5235–3330 YR. B. P.	SHELL RIDGE (1) – (2)	5–12	2.6–6.3
	3330–1080 YR. B. P.	SHELL RIDGE (2) – (3)	10–19	4.4–8.4
	1080 YR. B. P–PRESENT	NORTHWEST COAST		0–20

TABLE 1. Rates of coastal progradation in the coastal plain of East China since the mid-Holocene

high sediment supplies for the coastal progradation.

As shown in Figure 2 the North China Plain was invaded by the sea during three high sea-level periods. The aggradation of the lowlands was achieved by alluvial fans and river delta growth but not until 2000 yr B.P. did the swampy land scattered with many lakes disappear.

The accretion of Subei coastal plain was also completed in the last 2000 years when the Yellow River has repeatedly captured the lower course of the Huai River and built a large submarine delta. The sediments transported by the palaeo-Yellow River and many other small rivers such as the Huai, Shu, Qi and the Guan had played an essential role in the coastal accretion. After the Yellow River returned to its old course in 1855, the Subei Coast has been undergoing adjustment.

The depositional plains of North and East China are very sensitive to sea-level changes because of their low elevation and gradient. For instance, most of the area of the plain west of the Gulf of Puhai is lower than 2.5 m in elevation with a gradient of 1/20,000 to 1/30,000. As a result, the coastline will advance or retreat by 20 km in response to only one metre of sea-level fluctuation. Studies on numerous cultural relics and stratigraphy indicates that three successive major environmental changes were caused by sea-level fluctuations around the Taihu Lake during the mid-Holocene. In consequence, the human habitants could either change their lowest dwelling site level or make a long distance migration. The vertical amplitude attained was 2.7 m (Fig. 8) whilst the horizontal distance was up to 100 km.

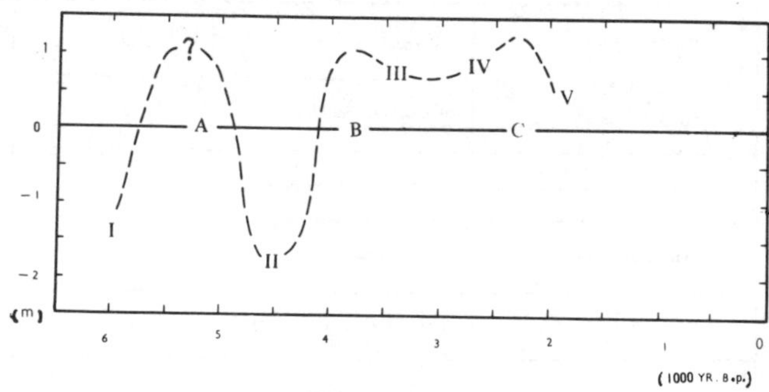

Fig. 8. Changes of elevation of the lowest level of the dwelling sites in ancient Chinese history in the plains of Jiangsu, Zhejiang and Yangtze delta (A.B.C. indicate cultural discontinuities or hiatuses)

In the coastal plain west of the Gulf of Puhai, two episodes of environmental changes are directly connected with the sea-level fluctuations (Han Jiagu, 1984). The fall of sea level after the mid-Holocene made the coastline retreat to the position of the second chenier east of Tianjin (Fig. 3). Nevertheless, the Tianjin Plain had still not been effectively cultivated before 2425 B.P. because of low elevation and sea flooding during spring tides but after that date cultivation began and lasted to the West Han dynasty (about 2000 yr B.P.) when the sea water once again began invading this region. This is well illustrated by the abrupt disappearance of then existing cultural remains which were buried under later marine deposits in Ninghe County. The coastal plains of North and East China have been especially vulnerable to the fluctuations of sea level and disastrous floods caused by rising of sea level as recorded in ancient Chinese history.

THE FREQUENCY OF TRANSGRESSIONS AND REGIONAL DIFFERENCES

Recent work suggests that there were nine transgressive events with associated marine sediments during the past 2 million years. For instance, there were 8 layers of foraminifera-bearing strata in Haixing 7-16 core (Fig. 9). In different regions the number of marine layers varied considerably. In Suzhou 827 core, which lies to the east of Taihu Lake, 7 layers of foraminifera-bearing strata have been identified in the upper 70 m of the core. Therefore, caution must be used in correlating these marine sediments especially when the palaeomagnetic time scale is utilized. Data from many boreholes in coastal plains of North China have shown that the time interval between two successive transgression events in the last 100,000 years took only 20,000-30,000 years. Thus regional stratigraphic correlation of the early Quaternary is very difficult, not to mention the tectonic impact that may increase the variability of marine strata in different regions.

Some scientists have attempted to construct a unified transgression model in east China. The philosophy underlying this is that the marine strata at the same depth in different areas would be formed simultaneously. This assumption obviously neglects regional tectonic effects on the occurrence of transgressive events and the variability of sedimentation such that some transgressions may be only local events not related to high sea-levels. As already noted topographical differentiation and tectonic movement might also influence both where and when the marine sedimentation occurred.

Chronostratigraphical work has shown that the time intervals between two successive transgressive events in North China were much shorter in the late Quaternary than in the early Quaternary. The following two aspects may be responsible. Firstly, the amplitude of continental ice-sheet change as reflected by the variation of oxygen isotope ratios in deep-sea cores (Fig. 10) is greater after 900,000-700,000 yr B.P. than before. In response, the amplitude of sea-level change is also greater after 700,000-900,000 yr B.P. Secondly, the broad low-lying coastal plain of North and East China

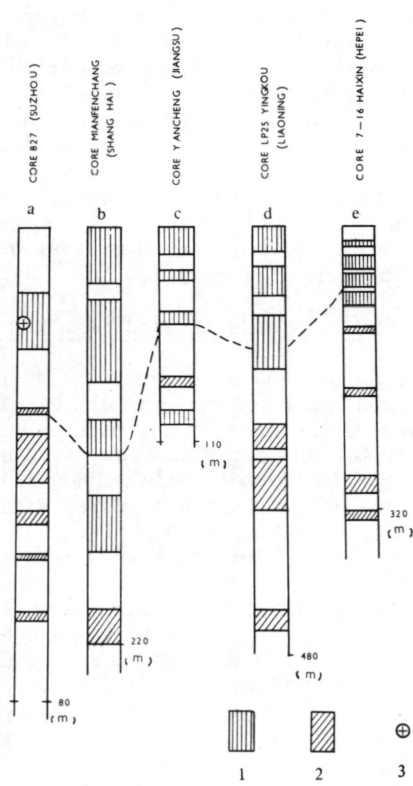

Fig. 9. Correlation and distribution of marine deposits in type cores from East China.

1. marine sediments with abundance of fossils
2. marine sediments with a small amount of fossils
3. 33090 370 Yrs B.P. (radiocarbon date)

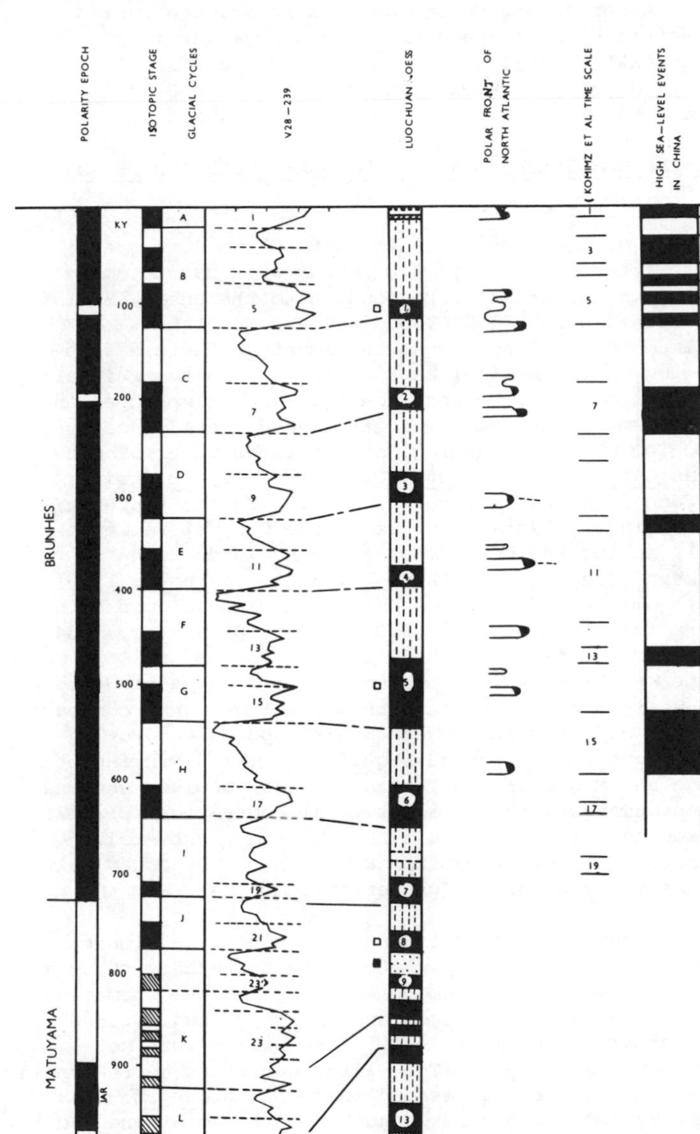

Fig. 10. Correlation of sea-level changes in North China with the loess sequence of Luochuan and isotope stages of V28-239, showing their palaeoclimatic significance.

has influenced extensive marine sedimentation. Since such a landform had not been formed in the early Quaternary, sea water was not able to advance and retreat over such a wide area as that in the late Quaternary. After a long period of subsidence, the rapid uplift of the western high plateaus provided a huge volume of sediments for the east coastal area, the North China plain in particular witnessed massive sedimentation in the late Quaternary and thereby became highly susceptible to sea-level changes.

MARINE TRANSGRESSIONS DURING HIGH SEA-LEVELS AND TECTONIC ACTIVITY

There are two kinds of transgressions in North China during the Quaternary: those after 1,500,00 yr B.P. which are restricted to the North China Plain and those corresponding to the Puhai, Yellow Sea and Peijing events before 1,500,000 yr B.P. The latter extended far inland into the northwest intermontane fault-basins such as the Yanqing Basin, Huailai Basin (Fig. 2) and Fen-wei valley. All of these are components of a long Cenozoic graben which extended as far as Xian, Shaanxi Province. The marine sediments discovered at a depth of 80-100 m in the Yanqing core contains such foraminiferal species as *Cibicides refulgens*, *Cibicides* sp., *Trifarina angulosa* (Chen Fangji and Huang Xinggen, 1980). In the Yuncheng Basin (Shaanxi Province) which is a part of the Fen-wei valley (Fen-wei graben), marine sediments containing *Evolutononoin shanxiense* has been found at a depth of 77-83 m (Jia Fuhai, 1959). According to palaeomagnetic measurements, the marine sediments in Yanqing core were formed about 1,550,000 yr B.P. (Qian Fang, 1984). Furthermore, Upper Pliocene marine sediments at depths of 1000-2000 m were found in the Wei River valley, which occupies the southwestern part of the Fen-wei graben. Foraminifera assemblages were composed of warm water types and occurred together with spores and pollen including *Gingo*, *Magnolia*, *Carya* and *Liquidambar* (Wang Pinxian et al., 1982), *Gingo* and *Magnolia* are Tertiary relic species, whereas *Carya*, *Liquidamber*, amongst others, are typical subtropical plants. This floral assemblage also reflects a Late Tertiary warm-moist climate. Penetration of sea water such a distance inland can only have been possible via a subsiding graben system.

Another noticeable feature is that transgressive event - high sea-level relationship was much weaker in the early Quaternary than in the late Quaternary. Neo-tectonic movements causing the formation of a deep graben system explain the occurrence of transgressions in the early Quaternary. The microfossil assemblages were not necessarily of a warm water type. The marine sediments corresponding to the so-called "Peijing transgression" that took place about 2.4 million years ago contains a large number of foraminifera and calcareous nannofossils which lived in a cold open sea water. For instance, *Coccolithus pelagicus* Wallich is now the dominant species in sub-arctic ocean and hence it is clear that not all the transtression events in the late Cenozoic could be ascribed to interglacial climates. Meanwhile, Wang and He (1982) think that the climate was similar to that currently prevailing in Kuril Islands or even in Bering Strait.

Because the regional tectonic movement could either increase or decrease the impact marine transgression, we believe that the age of marine sediments in different tectonic units might vary considerably. Therefore, close attention should be paid on the regional differentiation rather than the regional coincidence.

THE CHENIER PLAINS OF THE EAST COAST

We have already mentioned the rates of coastal progradation suggested by the radiocarbon dates from cheniers formed from the mid-Holocene to the present in the coastal plain of China. Many scientists have taken a special interest in the study of cheniers because of their significance in sea-level changes but some problems remain. For instance, does the oldest chenier (6300 yr B.P.) represent the coastline during a mid-Holocene maximum sea-level? The maximum extent of mid-Holocene marine transgression as revealed by numerous boreholes was far landward of the oldest chenier. Further, the ages of cheniers in the coastal plain of East China are different from each other, which indicates that, besides the changes in sea-level, many other factors had controlled the development of cheniers. These include the amount of sediments provided by rivers, the wave strength, and the coastal slope. The two oldest cheniers in the Subei (northern Jiangsu) Plain and the Yangtze Delta Plain are of the same age, but the younger cheniers in these plains were formed at different times. This phenomenon suggests that, after the formation of the oldest cheniers, the amount of sediments transported to these coastal areas by the Yangtze River increased. The estuary of the Yangtze River has migrated from north to south. The oldest chenier in the coast plain of Puhai formed approximately 5200 years ago, can be correlated with a younger one in the Yangtze Delta Plain. The information, so far as we can judge suggests that the mean high water level had maintained itself at the base of the chenier for some time if the later uplift or subsidence of land surfaces were not taken into consideration.

MARINE SEDIMENTS, NEOTECTONICS AND SEA-LEVEL CHANGES DURING THE LAST INTERGLACIAL

During the last full-interglacial (oxygen isotope substage 5e), sea-level reached a maximum around 125,000 yr B.P. However, no marine sediments corresponding to this episode have been discovered in the coastal area of the Puhai, whereas the marine sediments corresponding to substages 5c and 5a are widely distributed. This very interesting abnormality might be due to stratigraphic error, or might be caused by uplift of the coastal region. A possible explanation may be found in Figure 11 which shows six successive episodes of high sea-level representing six transgressive events in the BC-1 core taken from the middle of the Puhai. The dotted line in Fig. 12 indicates the water depth reflected by the microfossils. After separating the influence of subsidence, each of the I-V high

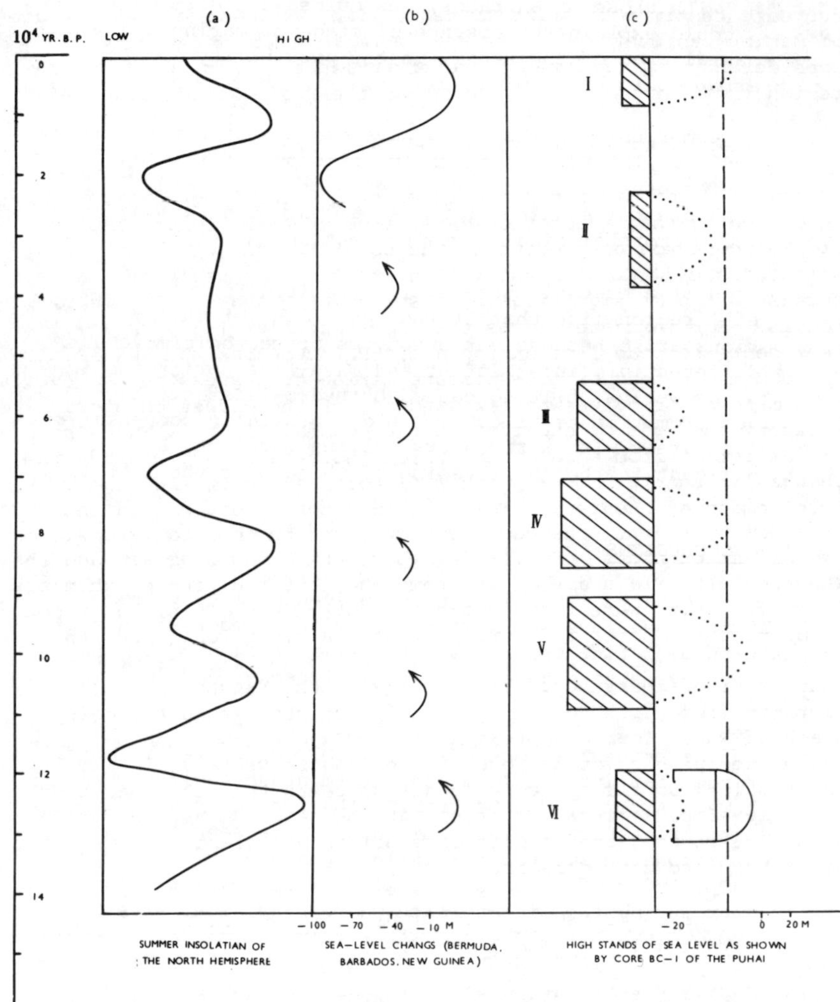

Fig. 11. Correlation of high sea levels from Core BC-1, Puhai with summer insolation curve of the Northern Hemisphere and high sea levels of other regions.
(a) From Broecker, 1966
(b) Generalized sea level curve
(c) Depth of sea water

sea-levels coincide very well with those of Barbados, New Guinea and Bermuda. If the depositional base of marine formation VI were raised 20 m, all the six high stands could be satisfactorily correlated with those of Barbados, New Guinea and Bermuda. Neotectonics could explain the absence of stage 5e marine sediments in the coastal area of the Puhai, if the possibility of stratigraphic error could be eliminated.

Extensive geomorphological studies have demonstrated that intermittent and oscillatory crustal movement is one of the most remarkable features of the coastal region of China. The intermittent crustal sinking of the Puhai may be determined by the movement of geomorphological waves as mentioned before.

Construction of a eustatic sea-level curve is difficult, especially in a coastal region like that of the east coast of China, complications arise because various types of tectonic movements, such as differential, intermittent and wave-like uplift subsidence movements, as well as tilting caused by hydro-isostasy, have taken place from time to time. The hydro-isostatic compensation is weaker in areas with a rigid crystalline basement, but it is very significant in the shelf region of China over the last 15,000 years.

CONCLUSIONS

1. Late Quaternary. The transgressive events were sychronous with the high world sea-levels. Nevertheless, the number and distribution pattern of marine sediments were still influenced by the neotectonic movements and geomorphological background in a given coastal region.

2. The marine transgressions before 1,500,000 yr B.P. in North China extended far into the western intermontane faulted basins and rift valleys. After 1,500,000 yr B.P. their extents were limited to the North China Plain. The time interval between two successive marine formations became shorter in late-Quaternary. This implies that the amplitudes of climatic change and sea-level fluctuations were greater during the mid and late Quaternary, i.e. after 900,000-700,000 yrs B.P. The sea water temperature reflected by the microfossils in early and late Quaternary marine sediments also varies greatly. Tectonic movement rather than eustasy determined the occurrence of transgressive events in the early Quaternary, hence, the advent of transgressions was not necessarily synchronous with high sea-levels, the sea was able to invade far inland through linear tectonic features. Thus it is necessary to distinguish these two types of transgressive events and related marine sediments.

3. Since the long coastline of China transverses several geotectonic units, the frequency and the age of marine sediments between different units cannot be easily correlated. Attention must be paid to the reconstruction of the regional

palaeoenvironment, coastal evolution and interaction model.

4. It has been recognized that around 125,000 yr B.P. world sea-level was 4-8 m higher than the present. However, no marine sediments of this age have been found in the coastal plain of North China. Two possible causes have been proposed, one, stratigraphic error, and two, intermittent and oscillating neotectonic movements.

5. The highly complicated tectonic movements hinder the compilation of a sea-level curve that could be applied to each segment of the long Chinese coastline. In addition, geoidal changes along the east coast during the Quaternary must have increased the complexity of sea level changes.

6. The sea water repeatedly moved back and forth over the broad continental shelf zone as the sea-level fluctuated during the Late Quaternary.

7. The coastal plain of east China is so sensitive to sea-level fluctuations that even small oscillations in the Holocene forced changes in the habitation sites of early settlers.

8. The rate of coastal progradation in North and East China has increased rapidly since the mid-Holocene.

9. Apart from the changes of sea level, the development of cheniers was also controlled by several other factors. Therefore, the age of cheniers in different coastal plains varies greatly.

REFERENCES

Broecker, W.S., 1966. Absolute dating and the astronomical theory of glaciation. Science, 151, 299-304.

Chen Fangji, and Huang Xinggen, 1980. The latest discovery and its significance of early Quaternary marine fossil assemblages in Yanging Basin. Quaternaria Sinica, 5, 97-98 (in Chinese).

Han Jiagu, 1984. The impact of geographical changes on the history of Tianjin, Paper submitted to the Symposium of Environmental Changes (Chongqing, Sichuan).

Huang Qingfu, and Cang Shuxi, 1985. The Late Quaternary palaeoenvironmental history of Bohai on assemblage of faunas. Marine Geology and Quaternary Geology, 5, 27-38 (in Chinese).

Jin Fuhai, 1959. A preliminary understanding on San Meng System in San Meng Gorge reservoir of Yellow River. Proceedings of Conference on Quaternary Geology in San Meng Gorge Area, Science Press, Peking (in Chinese).

Liu Yuan-lung, Wang Qian-shen, and Zhao Jian-hua, 1978. A preliminary study based on gravity data of the crustal structure of the Peking-Tientsin area and its neighbouring regions. Acta Geophysica Sinica, 21, 9-17 (in Chinese).

Qian Fang, 1984. A preliminary study on Magnetostratigraphy of Quaternary transgressive series on China Continental. Marine Geology and Quaternary Geology 4, 89-100 (in Chinese).

Wang Naiwen, and He Xixian, 1982. The discovery and its significance of calcareous nannofossils in Peking Plain region. Kexue Tongbao, 27(13).

Wang Pinxian, Wang Naiwen, and Bai Jinsung, 1982. Discovery of the Cenozoic foraminifera in the Fen-Wei Basin and its significance. Geological Review, 28, 94-100 (in Chinese).

Xu Jiasheng, Gao Jianxi, and Xie Fuyuan, 1981. The Yellow Sea in the latest ice age - the acquisition and study of some new data about palaeogeography of the Yellow Sea. Scientia Sinica, 5, 605-613, (in Chinese).

Yang Huai-jen (Yang Huairen), and Xie Zhiren, 1984. Sea-level changes in east China over the past 20,000 years. In: Whyte, R.O. (Ed.), The Evolution of the East Asian Environment. Volume I, Geology and Palaeoclimatology, pp 288-308. Centre of Asian Studies, University of Hong Kong.

Zhao Songling, and Chin Yun-shan, 1982. Transgressions and sea-level changes in the Eastern Coastal Region of China in the last 300,000 years. In Quaternary Geology and Environment of China, pp 147-154. China Ocean Press, Peking, (in Chinese).

Zhao Xitao, Zhang Jingwen, Jiao Wenqing, and Li Guoying, 1980. The cheniers in the western coastal area of Bohai Bay, Kexue Tongbao, 6, 279-281 (in Chinese).

International Geomorphology 1986 Part II
Edited by V. Gardiner

A RE-EXAMINATION OF THE DIMLINGTON STADIAL GLACIGENIC SEQUENCE IN HOLDERNESS

Clive T. Foster

Department of Geography
Liverpool University Liverpool L69 3BX

ABSTRACT

Sedimentary and geotechnical properties of the Dimlington Stadial tills (Skipsea, Withernsea Tills) of Holderness were identified from three boreholes. Both sedimentary units were found to possess characteristics typical of diamicts deposited by lodgement beneath a wet based sliding ice sheet. No evidence could be found for the downwasting of a complexly stratified ice sheet as previously proposed for the sequences. A depositional model is proposed which explains the multiple till stratigraphy, in terms of a lateral shift in the east coast ice flow units carrying distinct mixes of lithologies along individual flow lines within a single ice sheet.

INTRODUCTION

Glacial sequences are often regarded as one of the most complicated and variable suite of sediments in the geological record. Attempts have been made to rationalize the complexity of these deposits through the application of depositional theories which link process, landform and sedimentology into a single unifying model or landsystem (Boulton and Paul, 1976; Eyles, 1983). Of particular importance to the glacial deposits of Holderness is the recognition of "subglacial" and "supraglacial" landsystems which have proved useful in the interpretation of diamict sequences deposited by lowland ice sheets across areas of sedimentary strata (Boulton et al., 1977).

The subglacial model can be applied to warm based glaciers that transport a thin layer of debris at the glacier sole, maintained by high rates of basal melt. Deposition occurs when the tractive force imposed by the moving glacier is inadequate to maintain in motion the basal debris against the frictional resistance offered by the bed (Boulton 1975). The balance between lodgement and erosion is dependent on a critical balance between a number of inter-related variables e.g. water pressure at the ice base, nature of transported load, and the permeability of the subglacial bed (Boulton and Paul, 1979). The control of several critical lodgement parameters leads to the slow accretion of a massively bedded, overconsolidated, poorly sorted diamict which can possess remarkably uniform textural and mineralogical properties (Luttenegger, Kemmis and Hallberg, 1983). Particularly diagnostic of this system is the development of an anisotropic meso-fabric of blade shaped clasts, orientated with the long axis parallel to ice flow and localised clast "smudges" caused by the shearing out of soft incompetent lithologies (Eyles, 1983). Other significant components of a till sheet deposited in this manner include subglacial channel fill sequences which result from active basal drainage beneath melting ice. Such forms can be recognized as distinct from the proglacial sediment association by a marked lateral discontinuity, forming "stacked" sequences with direct contact with true basal lodgement tills (Figure 1). Channel cross-sections can show evidence of hydro-static flow and generally fall within a smaller size range (1.5-12m) than fully developed proglacial outwash sequences (Eyles and Sladen, 1981).

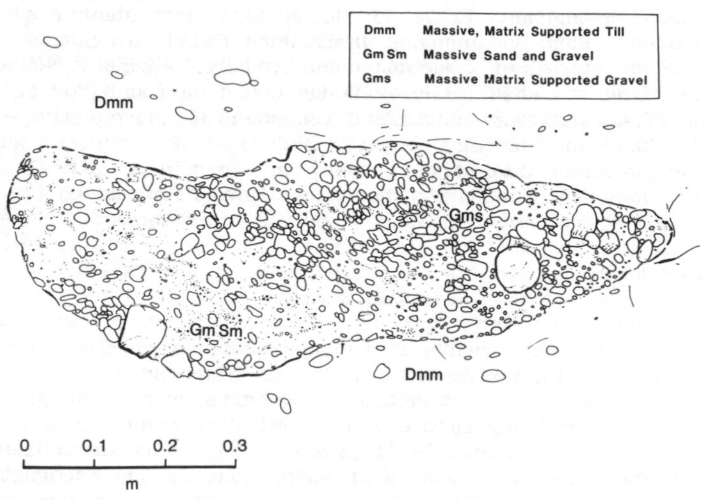

Figure 1. Cross sectional detail of a "cut and fill" subglacial channel sediment in the Withernsea Till. 150m north of Aldbrough (TA 256398).

By way of contrast the supraglacial landsystem describes the sediment association produced by the decay of ice transporting significant quantities of supraglacial or englacial debris. In the case of lowland continental ice sheets, this is usually related to compressive flow or the development of a cold ice margin leading to refreezing at the sole and the thrusting of basal debris into an englacial position (Boulton and Paul, 1976).

Eventually ablation causes debris to be released and deposited by processes of mass wasting as mobile flow tills (Boulton, 1970; Lawson, 1982). Slow melting of buried ice beneath flow tills releases more debris in situ (meltout till) which can be more stable and retain its englacial structure. Diamict sequences deposited in this manner characteristically deform as a result of sediment collapse or settling during the melt of underlying or buried ice masses, and such features include faulting, folding and slump structures (Shaw, 1972). On final ice melt an irregular hummocky kamiform relief is produced with a complex vertical profile formed of inter-stratified reworked diamicts and melt water deposits (Paul, 1983).

REGIONAL SYNTHESIS

The current depositional model for the Devensian glacial succession in Holderness outlined by Catt and Penny (1966) and Madgett and Catt (1978) after the original suggestion by Carruthers (1953), proposes that the Withernsea and Skipsea till sheets were deposited contemporaneously by a stratified ice sheet containing distinct basal and englacial debris that moved into the Holderness embayment during the Late Devensian. Both till units have been shown to overlie organic rich silt lenses at Dimlington Cliff (TA 386224 - 408184) which have yielded radio-carbon dates of 18,500 ± 400 BP (I-3372) and 18240 ± 250 BP (Birm-108) (Penny, Coope and Catt, 1969). The oldest date yet obtained from material above either till is 13,045 ± 270 BP at The Bog, Roos (TA274288), (Catt, 1977). The decay of the ice sheet during the latter part of the period bracketed by these two dates, is thought to have resulted in the superimposition of two till sheets without the development weathering, contortion or incorporation across the junction, indicative of two separate ice advances (Madgett and Catt, 1978). This sequence defines the Dimlington Stadial of Rose (1985).

Analysis of the till petrography has shown that the diamicts possess specific mineralogical, sedimentological and lithological properties, the result of different zones of provenance (Madgett, 1975) and possibly reflecting transportation at discrete levels within the stratified ice sheet. This situation is believed to have occurred when "Withernsea" ice flowing east from the Lake District through the Stainmore Gap moved into contact with the outcrop of Mercia Mudstone (argillaceous facies of the former Keuper, Triassic) in the lower Tees valley, before overriding a pre-existing coastal ice stream carrying the Skipsea Till as basal debris (Madgett and Catt, 1978). Although the exact nature of the deglaciation is as yet undetermined, Catt (1977) suggests that both ice streams stagnated soon after moving into Holderness and features such as the Killingholme and Hogsthorpe ridges represent irregular deposition of flow till and melt-out till deposited during the slow decay of dead ice rather than the terminal moraine complex of an active ice front (Straw,1979).

AIMS

The Withernsea-Skipsea diamict sequence has been re-examined within the context of recent work on subglacial and supraglacial sedimentation in order to determine whether the sequence was deposited in a sub or supraglacial setting. This has been done using new borehole information and field work in the Cowden - Aldbrough area (TA 254404-258373).

METHODS AND SAMPLING

Three dry boreholes nominated CS 1 - 3 were sunk during the summer of 1982 at Cowden, Humberside, each to a depth of 10m (Figure 2). Drilling and testing of the cores recovered from CS1 and CS2 took place during a three week on-site programme. CS3 was sampled a month later during a separate five day drilling schedule. All three boreholes were located in a 5m radius. In total 47 samples were produced with a mean length of half a metre. Core recovery was good, averaging 73%.

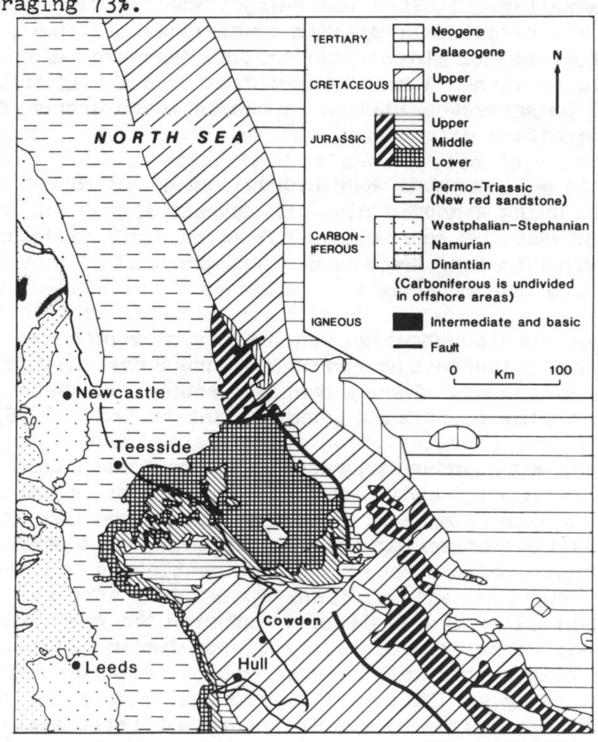

Figure 2. Regional geology and site location.

The problems of obtaining reasonably undisturbed samples of till have been reported by a number of workers (McKinlay, Tomlinson and Anderson,1974). Disturbance leading to significant re-moulding, can have an important effect on the measured properties of a sediment causing a reduction in undrained strength, a decrease in preconsolidation pressure and a reduction in the coefficient of consolidation. The following procedures were adopted to minimise this crucial factor.

a) Thin walled tubes were used throughout the sampling programme. These were construced from 1.5 mm gauge metal with an internal diameter of 98 mm which provided a cutting edge ratio of 3.5%. Standard U100 tubes with removable cutting shoes have an edge ratio of 25%.

b) Using a file, the cutting face was sharpened to a knife edge. The thinning of the leading edge caused some problems due to buckling in contact with flints, but this effect was minimal since the majority of the clasts encountered were formed from medium hard lithologies which were cleanly cut with no dragging.

c) All the tubes were cleaned and lightly greased as an aid to both sampling and extrusion. Frictional drag between the core and the internal wall created a characteristic warping of the sample at the margins. However, the degree of disturbance was limited to within 1 mm of the circumference. Greasing also helped to prevent oxidation in sample tubes stored for long periods before extrusion.

d) The thin walled tubes were hydraulically jacked into the profile using a mobile rig. The straight drive was at a constant rate of approximately 1m/min. This technique has been shown to produce less disturbance than either driven or cored sampling methods.

The release of in-situ stresses was minimised by on-site testing in a mobile triaxial laboratory supplied by the British Building Research Establishment. Twenty samples were placed under confining stress equal to the calculated total overburden within 24 hours of sampling, the average delay being approximately three hours. All of the tests followed standard undrained unconsolidated test procedures. Pore pressure was monitored throughout by pressure transducers to allow complete effective stress analysis. One dimensional consolidation was carried out using an oedometer with a 76 x 19 mm sample size.

A 500 gm sub-sample was wet sieved to provide the size distribution of the 4Øto - 5Ø fraction. Complete dispersion was ensured by pre-treatment in dilute sodium hexametaphosphate solution followed by ultrasonic agitation. The particle size characteristics of the 4Ø - 9Øfraction was determined by the Andreasen's pipette apparatus (BS 3406).

Bulk index properties were calculated from the mass of the geotechnical specimens, trimmed to a known volume. Additional till samples were kindly supplied (for comparative purposes) by the British Geological Survey from offshore vibro-core returns. Large scale structural detail of the tills was gained through field work undertaken on the Cowden-Aldbrough cliff section (TA 254404 - 258373), which exposes the Skipsea-Withernsea boundary at 1-2m OD. The same feature was recorded in all three boreholes at an average depth of 5.2m below ground level.

RESULTS

Diamict lithofacies and fabric.
The boreholes prove two diamicts differentiated in the field largely on the basis of colour, the Withernsea Till being dark greyish brown (10 YR 4/2) and the Skipsea Till dark brown (7.5 YR 3/2). This interpretation was supported by the adjacent coastal section. There was little visual evidence in the upper profile of significant reworking or dissaggregation of the Withernsea Till, both units are massively bedded, matrix dominated tills, the groundmass being stiff and cohesive. Apart from the superficial effects of desiccation in the upper 1-2m and stress relief features developed parallel to the free face, both tills display a distinct lack of fissuring or structural discontinuity. Fissuring was not recorded from depth in any of the boreholes, a fact noted by previous workers on the Holderness tills who have had access to good quality borehole returns (Marsland, Prince, and Love,1982). A weakly developed micro-foliation created by the parallel alignment of silt grains, as described by Sitler and Chapman (1955) for basal tills in Pennsylvania and Ohio and Penny and Catt (1969) for the Holderness tills, was observed in the lower Skipsea Till at a depth of 6-7m.

The existence of crushed clast "smudges" was noted from both till units, being particularly noticeable in the Skipsea Till where shattered chalk clasts provide a strong visual contrast against the darker matrix (Figure 3). Similar structures could be seen in the Withernsea Till commonly developed in Triassic and Liassic lithologies.

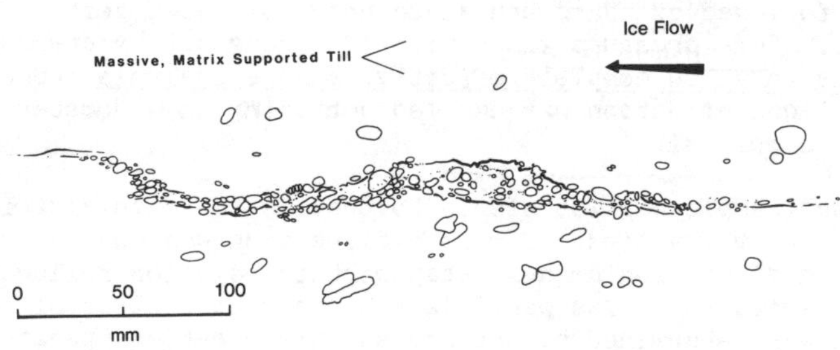

Figure 3. Detail of a clast "smudge" created by the distintegration and shearing out of a chalk cobble at the glacier base. These features form without any visible fracture development in the adjacent till. Skipsea Till, Cowden Cliff (TA 253404).

Although no measurement was taken of clast meso-fabric, strongly orientated fabrics have been previously recorded from both till units throughout the whole length of the Holderness coastline. (Penny & Catt, 1969). This is unlike those fabrics reported for melt-out units (Lawson,1982).

Particle Size Distribution.

The particle size analysis of the borehole samples supports the visual evidence, as to the lack of any obvious reworking or disaggregation of the Withernsea Till. Both units possess a narrow range of matrix texture, a uniformity characteristic of basal lodgement tills (McGown and Derbyshire, 1977). Further treatment of the data shows that a classification of clay 27-44%, silt 34-48%, sand 18-31% accounts for 83% of all samples analysed, including offshore material (Figure 4). Within this classification no clear distinction could be made between the Withernsea and Skipsea Tills on the basis of silt content as proposed by Madgett and Catt (1978).

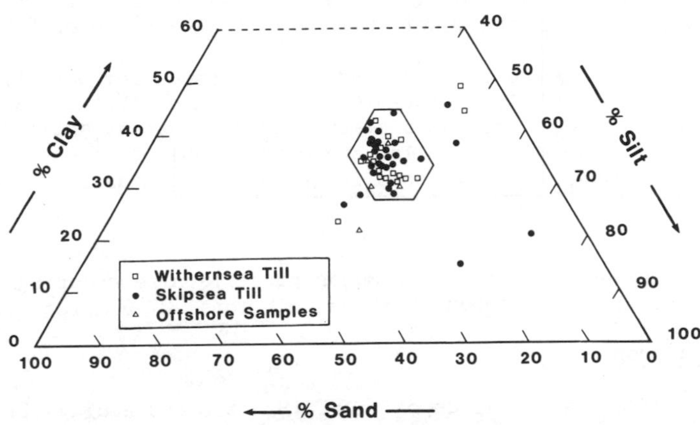

Figure 4. Textural envelopes for the Holderness tills as sampled in boreholes at Cowden and offshore returns. Total 56 samples.

The range of matrix texture and complex interbedding with fluvioglacial material as predicted by the supraglacial model was not observed in the vertical section, where the Withernsea Till maintains the general sand-silt-clay ratios which characterize the Skipsea Till (Figure 5). The Withernsea Till does not possess higher degrees of sorting, both units being extremely poorly sorted (sorting coefficient 4.2 - 5.46, Folk and Ward,1957).

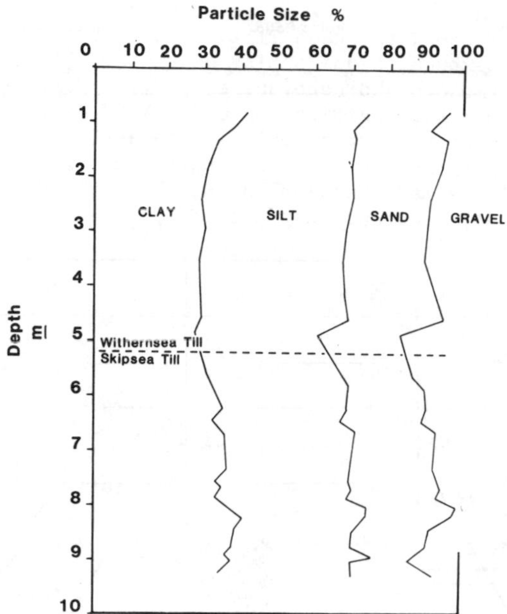

Figure 5. Textural variation in major particle size groupings with depth. Borehole CS1, Cowden, Humberside. Total 27 samples.

Bulk Index Properties.

In studies of supraglacial deposition, (Boulton and Paul,1976; Paul, 1983) it is suggested that differences in density exist between supraglacial materials and basal tills, with the latter being significantly more dense than supraglacially deposited materials. Both bulk density and voids ratio provide a suitable index of compaction. Voids ratio is frequently used in geotechnical analysis as a volume change coefficient during stages of consolidation and can be used to relate the in-situ conditions to past conditions of effective stress.

A summary of density measurements for the till sequence sampled in boreholes CS1-3 is shown in Figure 6. Both the Withernsea and Skipsea Tills fall within the range of densities typically associated with basal tills (2.13 - 2.3 Mg/m^3) as reported by Eyles and Sladen (1981) and Radhakvishna and Klym (1973). Differences between the average bulk density of the two units can be attributed to soil development in the upper profile, since there was no evidence for a change in bulk density across the colour boundary. The mean value for voids ratio, 0.438, reflects the dense nature of

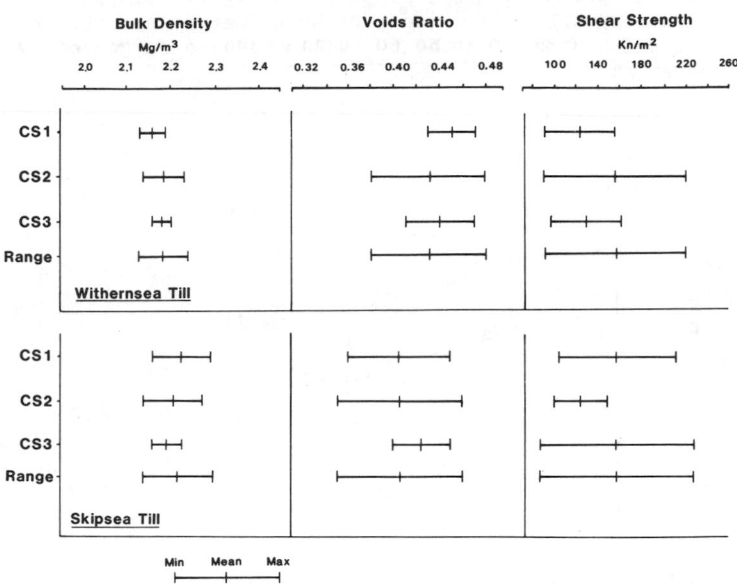

Figure 6. Variation of basic soil parameters for the Skipsea and Withernsea Tills. Boreholes CS1-3, Cowden, Humberside. Total 43 tests.

the material, the data showing consistent reduction with depth in line with the corresponding increase in confining stress. The high densities and reduced range of data values across all three boreholes suggest a uniform process of subglacial deposition rather than the increased scatter of values and reduced densities characteristic of supraglacial processes.

Stress History and Consolidation.

Owing to the disaggregation and fluvial reworking suffered by debris in the supraglacial environment most workers recognize that significant differences in consolidation exist between basal and supraglacial material (Boulton and Paul,1977; Marcussen,1975). Assuming fully drained conditions, subglacial till can experience normal loads in excess of 2000 kN/m² (Sladen and Wrigley,1983) producing the typically overconsolidated condition in exhumed sections. Owing to the lack of any significant depositional load, supraglacial material commonly displays lower pre-consolidation loads, of order 150-250 kN/m², related to post-depositional drying (Paul, 1983).

Results of oedometer tests show that both tills possess similar consolidation characteristics, with pre-consolidation pressures between 300-425 kN/m² indicating moderately overconsolidated soil (Figure 7). At all depths the till displayed a highly dilatant response in shear typical of a dense overconsolidated deposit. A similarity in stress history for the two tills is also suggested by the range of shear strengths recorded.

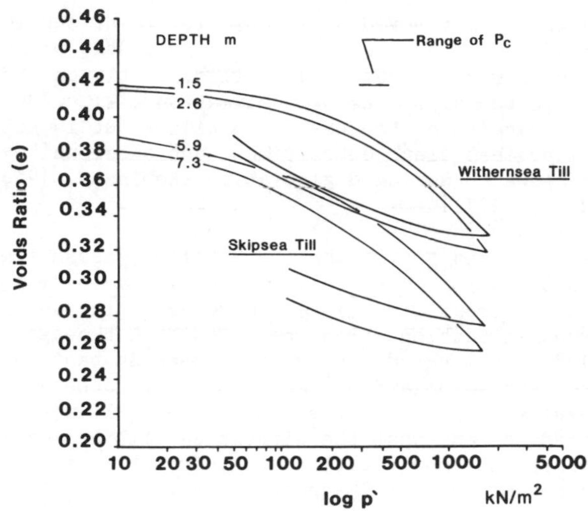

Figure 7. Consolidation curves for the Withernsea and Skipsea Tills. Samples recovered from borehole CS1, Cowden, Humberside.

The low values of pre-consolidation load displayed by material which otherwise shows all the sedimentary features of basal lodgement tills has been reported by a number of workers, (Dremanis, 1974) and has been accommodated in models of subglacial deposition (Boulton, 1975).

DISCUSSION

A complexly stratified ice sheet model involving basal melt-out, clearly forms an inadequate explanation for the deposition of the Holderness tills, taking into account the range of sedimentary and geotechnical evidence presented so far. There is no evidence for the melt-out debris from a stagnant ice mass. If melt-out had occurred the sediment may have retained its subglacial characteristics, but the material would inherit a new set of geotechnical properties and a rather different stratigraphy and geometry related to the rapid melt at the underlying ice sheet and subsequent resedimentation.

The suggestion that the melt of the intervening ice sheet might not have produced sufficient water to rework the upper till (in discussion, Catt and Penny (1966) p.419) is improbable, considering that the terminus of the lower advance is generally accepted to be marked by the Hunstanton moraine, 100 km further south and that, during its passage through Holderness, the same ice stream was responsible for the deposition of 15-20 m of lodgement till (Skipsea Till).

Similarly, the possibility that the englacial fabric of the till could have been wholly preserved by intense levels of sublimation has been dismissed by Eyles, Sladen and Gilroy, (1982) as an unrealistic proposal for the east coast tills, partly on the grounds that the process is too slow, (2m of sublimation till c.7000 years - Shaw, 1980). The margins of the Devensian till sheet in Holderness and Lincolnshire possess clear depositional and erosional evidence for the passage of water during deglaciation (Robinson, 1968; Dingle, 1970; Catt, 1977).

The geotechnical analysis proved that both tills possess similar strength and consolidation characteristics which can be interpreted in terms of related stress histories. The wide variations in strength and preconsolidation stress reported by Lutenegger, Kemmis and Hallberg (1983) as marking the junction between basal and supraglacial sediment associations were simply not observed at Cowden. The results support the view that both tills were deposited beneath warm based ice and that the Withernsea Till has experienced little post-depositional modification apart from the development of a Flandrian weathering profile (Madgett, 1974).

The distinct till lithology and petrography (Madgett and Catt, 1978) can be explained by a change in the mix of lithologies deposited at any one point due to a lateral shift in ice flow or by the alteration of lithologies carried along a single flow line. This is essentially the process of "unconformable facies superimposition" as proposed by Eyles, Sladen and Gilroy (1982) for the basal tills of Northumberland. The ice streams in question can be defined as a Pennine - Lake District flow moving through the Stainmore and Tees Gaps providing the initial advance which deposited the Skipsea Till, only to be displaced in the coastal regions by a vigorous ice stream moving out from the eastern Southern Uplands and Cheviots via the Tweed lowlands and down the east coast, a route which provides the Triassic/Chalk mix typical of the Withernsea Till. The area of the Tees lowlands is still seen as critical, not as an area of overriding, but as a region of confluence of active ice streams.

Acknowledgements

I am grateful to Edward Derbyshire, Nicholas Eyles and Mike Love for their critical assessment of this paper. My thanks also to A. Marsland and the British Building Research Establishment, Geotechnical Division who kindly allowed the use of equipment, on-site facilities and drilling crew.

Martin Edge provided assistance during the testing programme at Keele University and in Holderness. This work was supported a Research Studentship awarded by the Natural Environment Research Council.

REFERENCES

Boulton, G.S., 1970. On the deposition of subglacial & meltout tills on the margin of certain Svalbard glaciers. Journal of Glaciology, 9, 231-45.

Boulton, G.S., 1975. Processes and patterns of subglacial sedimentation: a theoretical approach. In A.E. Wright & Moseley F. (eds.) Ice ages: ancient & modern; 7-42.

Boulton, G.S., 1982. Subglacial processes and the development of glacial bedforms, Proceedings of the 6th Guelph Symposium on Geomorphology 1980, 1-31.

Boulton, G.S., Jones, A.S., Clayton, K.M., Kenning, M.J., 1977. A British ice sheet model and patterns of erosion and deposition in Britain. In Shotton, F. (ed.) British Quaternary Studies Recent Advances, 231-216, Clarendon Press, Oxford.

Boulton, G.S.,& Paul, M.A., 1976. The influence of genetic processes on some geotechnical properties of glacial tills, Quarterly Journal of Engineering Geology, 9, 159-94.

Carruthers, R.G., 1953. Glacial Drifts and the Undermelt Theory, H. Hill, Newcastle upon Tyne, vi + 38pp.

Catt, J.A., 1977. Yorkshire & Lincolnshire. International Union for Quaternary Research, 10th Congress, Geoabstracts, Norwich, 56pp.

Catt, J.A.,& Penny, L.E., 1966. The Pleistocene deposits of Holderness East Yorkshire, Proceedings of the Yorkshire Geological Society, 375-420.

Dingle, R.V., 1980. Quaternary sediments & erosional features of the North Yorkshire coast, Western North Sea. Marine Geology, 9, 17-22.

Dremanis, A., 1976. Tills: their origin and properties In Legget, R.F. (ed.) Glacial Till. Special Publication Royal Society Canada, Ottawa, 11-49.

Eyles, N. , 1983. Glacial geology : a landsystem approach. In Eyles, N. (ed.) Glacial Geology. Pergamon Press, Oxford.

Eyles, N.,& Sladen, J.A., 1981. Stratigraphy & geotechnical properties of weathered lodgement till in Northumberland, England. Quarterly Journal of Engineering Geology, 14, 129-141.

Eyles, N., Sladen, J.,& Gilroy, S. , 1982. A depositional model for stratigraphic complexes & facies superimposition in lodgement tills. Boreas, 11, 317-333.

Folk, R.L.,& Ward, W.C. , 1957. Brazos river bar : A study in the significance of grain size parameters. Journal of Sedimentary Petrology, 27, 3-26.

Lawson, D.E., 1982. Mobilization, movement and deposition of active subaerial sediment flows, Matanuska Glacier, Alaska, Journal of Geology., 90, 279-300.

Lutenegger, A., Kemmis, T.,& Hallberg, G. , 1983. Origin & properties of glacial till & diamictons. Special Publication on Geological Environment and Soil Properties. ASCE Convention, 310-331.

Madgett, P. , 1975. Re-interpretation on Devensian Till stratigraphy in eastern England. Nature, 253, 105-107.

Madgett, P.A.,& Catt, J.A. , 1978. Petrography, stratigraphy & weathering of late Pleistocene tills in East Yorkshire, Lincolnshire & North Norfolk. Proceedings of Yorkshire Geological Society, 42, 55-108.

Marcussen, I. , 1975. Distinguishing between lodgement till and flow till in Weichselian Deposits. Boreas, 4, 113-123.

Marsland, A., Prince, A.,& Love, M.A. , 1982. The role of fabric studies in the evaluation of the engineering parameters of offshore deposits. Proceedings 3rd International Conference on the Behaviour of Offshore Structures. 1, pp. 181-202.

McGown, A.,& Derbyshire, E. , 1977. Genetic influences on the properties of tills. Quarterly Journal of Engineering Geology, 10, 389-410.

McKinlay, M.J., Tomlinson,& Anderson, W.F., 1974. Observations on the undrained strength of a glacial till. Geotechnique, 24, 4, 503-516.

Paul, M.A. , 1983. The Supraglacial Landsystem. In Eyles, N. (ed.) Glacial Geology 71-90, Pergamon Press, Oxford.

Penny, L.F.,& Catt. , 1969. Stone orientation & other structural features of tills in East Yorkshire. Geological Magazine,104, 344-360.

Penny, L.F., Coope, G.,& Catt, J. , 1969. Age & insect fauna of the Dimlington silts East Yorkshire. Nature,224, 65-67.

Radhakrishna, H.S.,& Klym, T.W. , 1973. Geotechnical properties of a very dense glacial till. Canadian Geotechnical Journal, 11, 396-408.

Robinson, A. , 1968. The submerged glacial landscape off the Lincolnshire coast. Trans. of the Institute of British Geographers,44, 119-132.

Rose, J.,1985. The Dimlington Stadial/Dimlington Chrono-zone: a proposal for naming the main glacial episode of the Late Devensian in Britain. Boreas,14, 225-230.

Shaw, J. , 1972. Sedimentation in the ice contact environment, with examples from Shropshire (England), Sedimentology,18, 23-62.

Shaw, J. , 1980. Application of present day processes to the interpretation of ancient tills. In Stankowski W. (ed), Tills and Glaciogene Deposits, 49-56 Geografia 20, Poznan.

Sitler, R.F.,& Chapman, C.A. , 1955. Microfabrics of till from Ohio and Pennsylvania, Journal Sedimentary Petrology, Vol. 25, No. 4, 262-269.

Sladen, J.,& Wrigley, W. , 1983. Geotechnical properties of Lodgement till-a review. In Eyles, N. (ed.) Glacial Geology,184-212 Pergamon Press, Oxford.

Straw, A., 1979. Eastern England, Methuen Press, London. 139pp.

International Geomorphology 1986 Part II
Edited by V. Gardiner

GLACIAL GEOMORPHOLOGY, QUATERNARY GLACIAL SEQUENCE AND PALAEOCLIMATIC INFERENCES IN THE ECUADORIAN ANDES

Chalmers M. Clapperton

Department of Geography,
University of Aberdeen,
Aberdeen, AB9 2UF, Scotland, UK.

ABSTRACT

Detailed field mapping and air photo interpretation in parts of the eastern, western and southern cordilleras of Ecuador have identified landforms of glacial erosion and deposition. There are three morphologically fresh glacial moraine stages. On volcanic massifs rising above 5,000 m these stages commonly terminate at 4,250 m, 3,900 m and 3,500 m. On surrounding glaciated plateaux where ground rises above 4,000 m, but is lower than 4,700 m, the two higher moraine stages are usually missing. Deeply weathered till lies stratigraphically below the lowest moraine stage and extends to 2,750 m altitude. Radiocarbon dating of associated palaeosols and peats suggests the following ages for the glacial stages: the weathered till is older than 45,000 BP, the outer moraine stage is younger than 35,000 BP, the middle stage formed between 10,000 BP and 12,000 BP, and the inner stage is younger than 6,000 BP. Rough calculation of ELAs for the Neoglacial, Late-glacial and Full-glacial stages using a modified median altitude method suggests lowering of approximately 290 m, 570 m and 860 m for the western cordillera and 300 m, 585 m and 1,040 m for the eastern cordillera. Mean annual temperature lowering in the Ecuadorian Andes during these stages was of the order of 1.8°C, 3.9°C and 7.4°C respectively.

INTRODUCTION

Glacial features in the Ecuadorian Andes have been described in a number of papers (e.g. Wolf, 1892; Reiss and Stübel, 1892-1898; Meyer, 1907; Sievers, 1908; Kennerly and Bromley, 1971; Hastenrath, 1981), but the sequence of four glacial periods separated by interglaciations proposed by W. Sauer in 1950 has remained unchallenged as the basic model for Quaternary events for 35 years. Recently, the detailed mapping of glacial landforms in parts of the Ecuadorian Andes and analyses of sediments exposed in contiguous sedimentary basins have indicated the need for a complete reinterpretation of Sauer's work (Clapperton and McEwan, 1985, Clapperton and Vera, 1986).

Sauer's Quaternary sequence was based mainly on his interpretation of sediments exposed in three stream-cut sections 5-10 km east of Quito at Puente Rio Chiche, Quebrada Guarangupugru and Guangopolo (Fig.1). Whereas Sauer (1950, 1965, 1971) concluded that the sediments are of glacial, glacio-fluvial, glacio-lacustrine and 'pluvio-glacial' origin, recent work demonstrates that they are entirely of volcanic, volcano-loessic, laharic, fluvial, coluvial and pedogenic origin (Clapperton and Vera,1986). Fundamental errors made by Sauer include the interpretation of recent laharic boulder deposits from Cotopaxi volcano as moraines of his 'third glaciation'. Sauer's sequence also depends heavily on his interpretation of an environmental marker deposit, the loess-like sediment known in Ecuador as 'cangahua'. He considered this to be a distinctive interglacial deposit marking warm and dry conditions. This sediment is now believed to have accumulated almost entirely during the last glaciation (Hall,1979; Clapperton and Vera,1986). This paper presents geomorphological and sedimentary evidence from sites in the eastern, western and southern cordilleras of Ecuador suggesting that the Quaternary glacial sequence in these mountains cannot be traced earlier than the penultimate glaciation.

THE ECUADORIAN ANDES : RELIEF, CLIMATE, PRESENT GLACIERS

The *Cordillera de los Andes* in Ecuador forms a N-S barrier of mountainous terrain 100-120 km wide rising abruptly to over 4,000 m above lowlands of the Guayas and Amazon basins (Fig.1). From the Colombian border to Alausi in southern Ecuador the landscape is characterised by two parallel mountain ranges forming dissected plateau-like surfaces at 3,500-4,500 m altitude. These are surmounted by impressive strato-volcanoes which commonly exceed 5,000 m in height. Chimborazo (6,310 m) is the highest. Between the main ranges lies a series of intermontane basins known as the 'inter-andean depression'. Varying in width from 20-40 km, the basins are zones of relative subsidence lying at altitudes of 1,600-2,000 m. Some central volcanoes have developed within the interandean depression at major fracture zones and separate one basin from another, e.g. Mojanda, Rumiñahui, Cotopaxi and Igualata volcanoes lie between the basins of Guayllabamba, Latacunga, Ambato and Riobamba (Fig.1). South of Alausi a fundamental tectonic boundary (separating plate segments), represented by a broad zone of NE-SW trending fractures, marks the southern limit of andesitic strato-volcanoes, the interandean depression and the two parallel mountain ranges. It also separates the zone of accreted oceanic crust to the north from continental crust to the south (Hall and Wood,1985). In the tectonically distinct southern part of Ecuador mountain ridges and plateaux rise above 4,000 m in a few places to maximum elevations of around 4,600 m, but since the ground does not exceed 5,000 m, there are no permanent glaciers. The latter occur only on the twelve highest central volcanoes, all of which rise above 4,800 m, and are described by Hastenrath (1981) who produced an inventory of most glaciers in Ecuador.

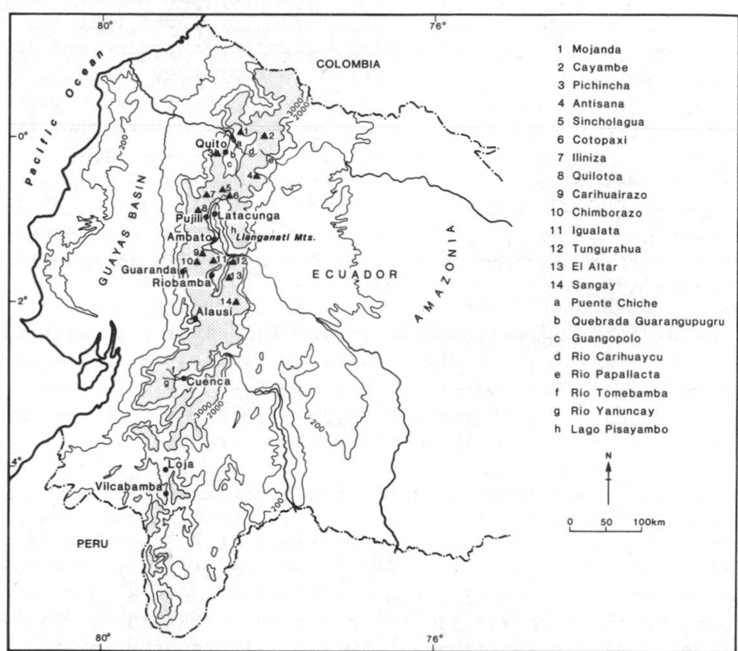

Fig.1. Location map showing the main cordilleras, the interandean depression, major central volcanoes and places mentioned in the text.

The climate of the Ecuadorian Andes is influenced mainly by latitude, altitude and the persistence of warm humid air masses derived from tropical trade wind systems over the Atlantic Ocean. The latter move generally westwards to the Andes over the Amazon basin. Upper air-flow patterns over the mountains indicate the dominance of easterly winds from the western Atlantic area at all seasons (Hastenrath,1981, p.11). Throughout most of the cordilleras precipitation maxima occur during March/April and October/November. The main precipitation minimum occurs in July/August, with a secondary one in January/February (Hastenrath, 1981). Absolute values for climatic parameters in Ecuador are extremely crude because of the lack of climatic recording stations, but the generalised map of precipitation in the Atlas of Ecuador (Delavaud, 1982) indicates mean annual totals of 1,300-2,000 mm along the crest of the eastern cordillera, 800-1,300 mm in the western and southern cordilleras and less than 800 mm in the interandean basins. A recording station at 3,560 m on Cotopaxi volcano measured a mean annual precipitation of 1,215.3 mm for the period 1964-1978

(Pourrut, 1983, p.29). Since the station is more than 1,000 m lower than the glacier-clad upper slopes of the mountain, the latter possibly receive much more than 1,215 mm precipitation per annum. The difference in precipitation totals between the eastern and western cordilleras is due to the predominantly easterly source of moist air masses. Orographic precipitation is induced by the eastern cordillera, but a distinct rain-shadow effect is felt in the interandean basins and on the western cordillera. Mean annual temperature in equatorial latitudes is very much dependent on altitude and seasonal contrasts are small. Within Ecuador the mean annual temperature at 2,500 m is c. 13°C (Pourrut, 1983) and according to Hastenrath (1981, p.16) the annual mean elevation of the 0°C isothermal surface is "somewhat below 4,900 m, with an annual range of the order of 100 m".

There are no published mass balance data for Ecuadorian glaciers and the only glaciological information on tropical Andean glaciers comes from Peru. Measurements on the Quelccaya ice cap located in the eastern cordillera of southern Peru places the Equilibrium Line Altitude (ELA) at 5,250 m and show that the glacier is temperate throughout its depth of 120 m (Thompson, 1980). The ice cap is situated on a gently undulating plateau and terminates along much of its margin in spectacular vertical ice cliffs. Above 5,400 m there is no ablation and although some melting takes place in the lower-lying marginal zone of the ice cap, most of the annual ice loss seems to be through calving at the fringing cliffs (Hastenrath, 1978, p.97). Since many parts of Ecuadorian ice fields are drained by glacier tongues descending steep slopes to warmer altitudes below the probable equilibrium line, most ablation may be by melting and evaporation. Prominent ice cliffs are present in places, however, such as on the northern and western sides of Chimborazo. The terminal altitude of Ecuadorian glaciers has been roughly estimated by Hastenrath (1981) on the basis of field observations and interpretation of vertical air photographs. This has been corrected in a few places during the present study. Table 2 shows present and former ELA values calculated for glaciers on the eastern and western sides of six mountains, three in the eastern cordillera and three in the western cordillera. Values for some plateau areas are also indicated. Calculations used a modified version of the median altitude method for reasons explained later in the text. The results indicate pronounced asymmetry in glacier extent (Figs. 2 and 7), and show consistently lower glaciation limits and ELA values on the eastern sides of individual mountains and in the eastern cordillera. These features reflect the dominant easterly source of precipitation. Local anomalies exist because of a pronounced 'snow-fence' effect caused by local topography. For example, on El Altar the glacier on the western side of the mountain is the lowest in Ecuador. A few decades ago the glacier terminated at 4,100 m, and even today large masses of stagnant debris-covered ice remain in the rock basin at this altitude. It is evident that avalanching massifs and an east-west transect from Pujili to Quilotoa; in the southern cordillera, the Tombebamba and Yanuncay valleys and plateau surface west of Cuenca. In addition, stereoscopic study of vertical air photographs of Cayambe, Antisana, Cotopaxi, Sincholagua and Laguna Cubillin, together with some field reconnaissance, has permitted identification of the spatial pattern of most glacial landforms in these areas (Fig.1). Since a reinterpretation of the sedimentary evidence used by Sauer forms the basis of a separate publication (Clapperton and Vera,1986), it is not discussed here.

GLACIAL GEOMORPHOLOGY

Landforms of Glacial Erosion. In terrain so tectonically and geomorphologically dynamic as the Andes, the evidence of previous glaciation is sometimes more obvious from erosional landforms than from sediments which are more easily removed by denudation. Thus features of glacial erosion may often help establish the minimal extent of former ice cover. Erosional landforms are also large enough to remain obvious despite a cover of subsequent volcanic sediment. The most prominent landforms of glacial erosion in the Ecuadorian Andes are those of areal scouring (knock-and-lochan topography), glacial valleys (troughs) and cirques.

Landscapes of areal scouring are widespread above an altitude of 3,800 m where the rugged outlines of glacially roughened terrain contrast strikingly with the smoother profiles of lower-lying land. Typically, the ice-scoured plateau summits consist of abraded, mammilated bedrock knolls in some places and craggy 'stoss-and-lee' forms in others, all frequently interspersed with bedrock basins filled with water or peat bogs (Fig.3). Relief amplitude is commonly 50-200 m, but may exceed this in higher areas such as the Llanganati mountains. Glacially polished and striated bedrock is present in many places, particularly good examples occurring in the Papallacta Valley (eastern cordillera) and on the plateau west of Cuenca (southern cordillera). Large blocks of bedrock glacially quarried from the knolls also litter such landscapes and it is clear that mechanisms of bedrock fracture propagation, of debris entrainment and of basal abrasion have operated at the base of ice caps and ice fields formerly covering these areas. Such glaciers were probably temperate-based like the present sole-surviving tropical ice cap at Quelccaya in Peru. The ice-scoured terrain is obvious on large-scale topographical maps (1:25,000, 1:50,000), on vertical air photographs and on satellite imagery and, with some field checking, it can be mapped for large areas. Ice-scoured bedrock is clearly more widespread in the eastern cordillera which contains a greater area of plateau above 3,800 m than the other cordillera and experiences much higher precipitation, factors which would have resulted in larger and probably more dynamic ice masses in glacial times. West of Cuenca the extent of ice-scoured terrain indicates the former presence of an ice cap almost 2,500 km^2 in from massive snow cornices built up along the crest of the 1,000 m high back wall of the large caldera-like cirque basin nourished glacier ice at an unusually low altitude.

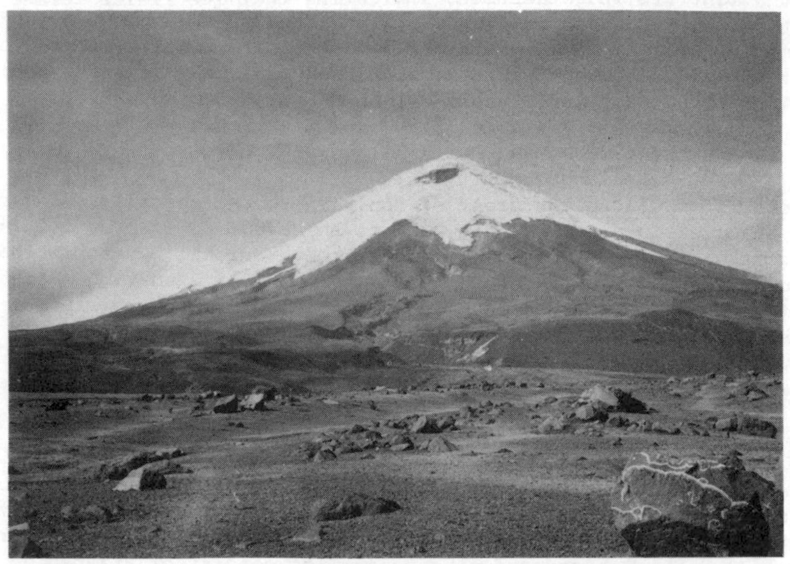

FIG.2. Cotopaxi volcano viewed from the northeast. Glaciers descend to lower altitudes on the east and south-east slopes because most annual precipitation comes to these sides of the mountain.

FIELD SITES

Fieldwork for this study focused on four main localities - the Sangolqui basin east of Quito and selected parts of the eastern, western and southern cordilleras. The study had two main aims - to test the model sequence of Sauer by reinvestigating sedimentary evidence at the sections he described and to map in detail the morphological evidence of glaciation in key localities of the main cordilleras, thereby identifying former glacier limits which would permit a reconstruction of former ELAs and an assessment of past climatic change. The sites studied in detail include the following: in the eastern cordillera, the El Altar massif, the Lago Pisayambo area and a west-east transect across the Carihuaycu and Papallacta valleys; in the western cordillera, the Carihuairazo and Chimborazo area, located only 3°S of the equator and 35-70 km from the tropical Pacific ocean. Since relief amplitude of the scoured terrain is commonly 200 m, the volume of the former ice cap was probably at least 500 km^3. Ice-scoured features also occur on the flanks of some large central volcanoes, indicating that ice fields formerly covered the mountains much more completely than now. The distribution of such terrain on Cayambe, El Altar and the Chimborazo-Carihuairazo massif clearly demonstrates that their former ice covers merged with surrounding plateau ice caps. The direction of former ice movement in some places can be established from striations and the alignment of roches moutonnées (Fig.3). In the Carihuaycu-Papallacta area these show that the ice shed was situated more or less along the crest of the eastern cordillera.

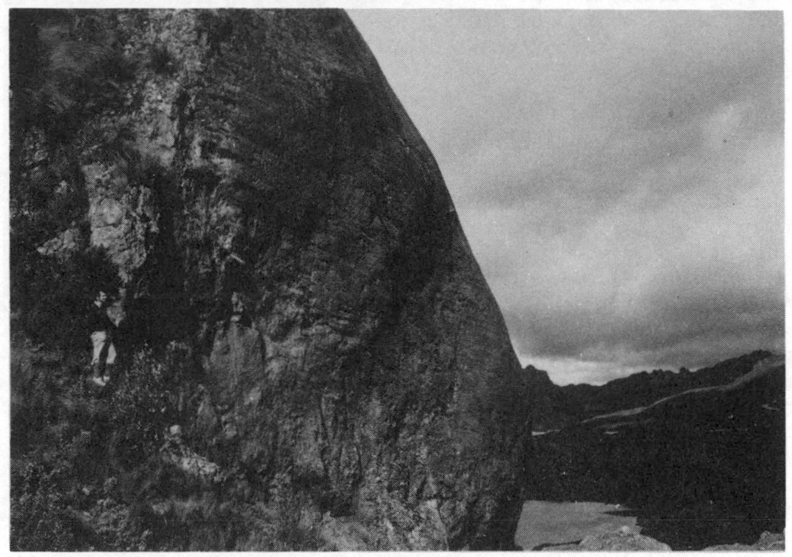

FIG.3. Landscape of areal scouring typical of many plateau areas in the eastern and southern cordilleras. This scene is on the plateau west of Cuenca at an altitude of c.4,000 m.

Glacial troughs with well-developed catenary cross profiles are produced by the linear concentration of glacial erosion (Sugden and John, 1976). Such features are prominent in parts of the Ecuadorian Andes, particularly in the eastern cordillera and on the flanks of some central volcanoes. In the former, troughs that are commonly 8-10 km long and 200 m deep lead from the high plateau between Cayambe and Antisana volcanoes and from the Jaramillo-Achiriqui massif in the Llanganati mountains. Trough features are either very short or poorly developed in the two other cordilleras where lower altitudes and less precipitation probably meant that there was less ice discharging down outlet valleys. The degree of development of glacial troughs on central volcanic massifs partly reflects the age of the volcanic complex. For example, troughs are absent from very active volcanoes such as Cotopaxi, Tungurahua and Sangay despite the fact that they are ice covered and rise above 5,000 m. Short, shallow troughs exist on volcanoes that are infrequently active, dormant or recently extinct, such as Pichincha, Antisana, Cayambe and Chimborazo. Troughs are well developed on the older, long-extinct volcanoes such as Carihuairazo, El Altar and Sincholagua. In parts of the eastern cordillera glacial troughs provide some indication of the former ice extent where limits are not clearly defined by moraines. For example, the upper Papallacta valley is an obvious glacial trough extending towards Amazonia. At an altitude of c. 2,800 m the valley cross profile becomes V-shaped and sinuous marking the point at which glacial erosion ended. There are no moraines, but glacial till is plastered onto the trough sides to the same altitudinal limit. Similar examples occur in the cordillera west of Cuenca in southern Ecuador.

Cirques are present in various stages of development throughout the Ecuadorian Andes. Mature forms consisting of classic semi-circular hollows are confined to the higher parts of the eastern cordillera, the flanks of extinct volcanoes such as Cayambe and El Altar and narrow mountain ridges in southern Ecuador just high enough to have supported local glaciers. Most cirques and cirque-like valley heads in the high plateau areas appear to have been buried beneath continuous ice-fields during maximal phases of glaciation and, like most cirques in Scotland, their bounding interfluves have been scoured by over-riding ice. This implies that the cirque forms *per se* were developed during periods of lesser ice extent. The altitude of the basin floors of such features can therefore not be used in reconstructing the former snowlines of glacial maxima (c.f. Hastenrath, 1971; Nogami, 1976). Exceptions are the cirques located on narrow mountain ridges in southern Ecuador where crests rising to 3,600-3,800 m have been affected only by cirque glaciation because interfluves are too narrow to have supported ice caps. For example, east of the Loja-Vilcabamba road part of the Cordillera Oriental is a N-S ridge averaging 2 km in width and extending for 20 km. It contains over 40 cirques. The mean cirque threshold altitude is 3,200 m and the mean elevation of the ridge crest (i.e. the 'snowfence') is 3,450 m (data from the 1:50,000 topographical maps 'Rio Sabanilla' and 'Vilcabamba'). Assuming a former accumulation area extending to 80 per cent of the basin, the data indicate a former ELA for the area of c. 3,250 m. The present ELA is probably close to 4,900 m. Cirques also occur on almost all of the extinct central volcanoes rising above 4,000 m, but are absent from presently active volcanoes. It is clear that the longer a volcano has been inactive or extinct, the more scope has there been for cirque development, so that the largest features have formed on El Altar, Carihuairazo and Sincholagua. The most spectacular form of cirque development in Ecuador has resulted from the enlargement of hollows initially created by explosive volcanism and such features are present on the long-extinct volcanoes of Corazon, Atacazo, El Altar, Rumiñahui and Pasochoa. These mountains contain horse shoe-shaped basins up to 3 km in diameter and over 1,000 m deep. All of them open out to the west and north-west. Although cirque-like basins can be created by magmatic inflation causing collapse of a volcano's superstructure, followed by a rock-slide avalanche and laterally-directed eruptions of pyroclastic flows (as at Mount St. Helens in 1980), it seems improbable that all such activity in Ecuador would have occurred on the west-northwest sectors of the volcanoes. A more plausible explanation of the exceptionally large cirques is that original explosion craters, similar to that on the presently active Quilotoa volcano (3 km diameter and 350 m deep) were breached on their western sides during periods of glacial expansion. The build-up of more ice on the eastern rim of a crater as a result of a 'snowfence' effect with easterly-derived precipitation would have caused greater discharge of ice westwards, ultimately breaching the crater to form an apparently oversized cirque. The so-called caldera on the west side of El Altar is a magnificent example of such a feature. It leads into a short but equally impressive glacial trough in the upper valley of the Rio Collanes (Fig.4).

Landforms of Glacial Deposition. Glacial deposits are widespread in the Ecuadorian Andes above an altitude of 3,600 m and consist primarily of conspicuous moraine systems, patches of till and scattered erratic boulders. Sauer (1950) mentioned the presence of what he believed were glacial deposits as low as 1,800 m in the Cuenca area and 'pronounced moraines' at 800 m near Piñas in south-west Ecuador. Glacial deposits could not be found at such low altitudes during the present study, but sedimentary features resulting from landslides and mudflows, superficially resembling glacial landforms, were commonly encountered on the Pacific slopes of the southern and western cordilleras. It is possible that these were mistakenly interpreted by Sauer as glacial features.

Till sheets are generally thin in Ecuador. On glaciated plateaux they commonly occur as discontinuous patches in the lee of rock knolls and are seldom more than 2-3 m deep, but till sheets may be 15-20 m thick in some of the major outlet valleys. In such localities the till is characteristically well compacted and contains abundant faceted and striated clasts. The matrix normally has a pronounced clay-silt content which, together with the degree of compaction and multi-lithological composition, helps distinguish till from lahar deposits.

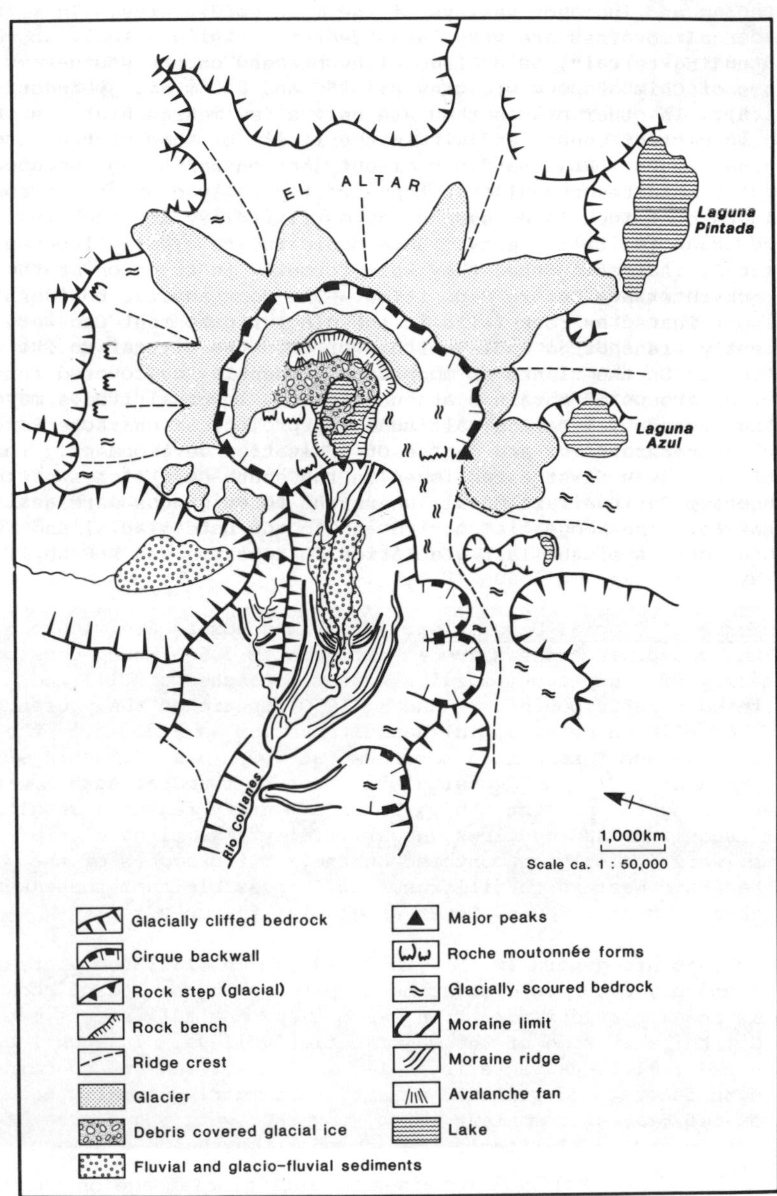

Fig.4. Map of glacial features in the El Altar massif.

The most conspicuous landforms of glacial deposition are the lateral-terminal moraine systems located on many of the extinct central volcanoes and in major valleys of the main cordilleras. In places individual moraines are very large features rising steeply above surrounding terrain, as at Lago Pisayambo and on the south-east flanks of Chimborazo where they are 250 and 270 m high respectively (Fig.5). In other places they are only a few metres high, as on the plateau west of Cuenca and within the limits of some of the larger moraine ridges. Road- and stream-cut sections exposing the upper 3-8 m of moraines reveal that they are typically composed of coarse boulders contained in a loose sandy matrix. They are much less compacted than the till sheets. Some boulders are clearly faceted and striated, indicating that they were formerly in traction at the ice-bedrock interface before deposition. The more angular boulders may have originated as rock-falls in the glacier catchment and were subsequently transported wholly within or upon the glacier to its terminus. On the flanks of mountains presently ice-covered three distinct groups of moraines at successively lower altitudes may be distinguished by location, altitude, morphological characteristics, depth of tephra cover and degree of vegetation development. This distinction was first established on the south-east slopes of the Chimborazo-Carihuairazo massif where the three groups were assigned to the Holocene Neoglacial period and to the Late-glacial and Full-glacial stages of the last glaciation (Clapperton and McEwan,1985).

FIG.5. Multiple moraines of Full glacial age on the north-east slopes of Chimborazo; Neoglacial superposed moraines over 100 m in height are contiguous with the large glaciers.

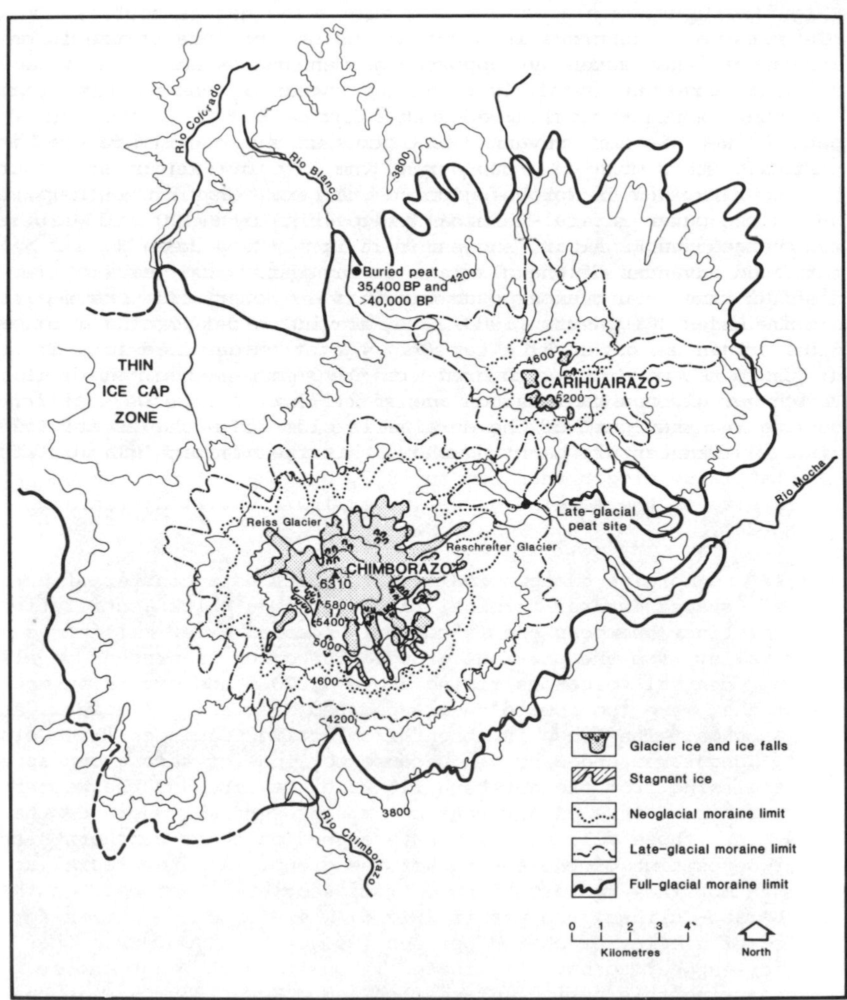

Fig.6. Map of the three moraine groups on the Chimborazo-Carihuairazo massif.

The three groups have now been mapped throughout the entire massif and confirmed on other central volcanic massifs examined in detail (Figs. 4 and 6). Each moraine group is normally composed of 3-4 ridges clustered closely together within a horizontal distance of 500-2,000 m, suggesting that the glaciers may have fluctuated in position close to the limits of each group over a period of time.

The first moraine group is located close to existing glaciers or to catchments that recently supported permanent ice fields. It lacks a tephra cover and normally is either barren or bears a thin carpet of vegetation. Historical documents indicate that glaciers in Ecuador were in more advanced positions than now and close to the limits of this moraine group during the last few centuries (Hastenrath,1981). Morphological characteristics of the group, in which the outer lateral-terminal ridges rise over 100 m in height on Chimborazo and El Altar, suggest that it may have been formed by repeated advances of the glaciers during the ~5,000 years of Neoglacial time. Such superposed moraines are known from other parts of the Andes (Clapperton,1983). Some evidence supporting a composite origin exists at El Altar (Fig.4). A boulder bed forming part of the outer moraine is covered with a 10 cm peat bed, the base of which has given a radiocarbon age of 2,170 ± 50 BP (SRR - 2587), indicating that part of the moraine is older than the Little Ice Age. A 19th century painting shows that the adjacent glacier was substantially bigger than now in 1872, but that it terminated within the limits of the moraine group (Meyer,1907, vol.2; Hastenrath,1971, photo 24).

The second moraine group normally terminates as a cluster of three closely-spaced lateral-terminal ridges or as a thick accumulation of indistinct hummocks and is commonly located a few kilometres down-valley from the previous group. The group is present on all extinct central volcanoes rising above 4,200 m and occurs on mountains that were too low to have supported Neoglacial glaciers, such as Pasochoa. The group is absent from active volcanoes like Cotopaxi, Tungurahua and Sangay. Because moraines of this group appear to be missing from the northern flanks of Cayambe and the western and northern flanks of Antisana but are present elsewhere on the two mountains, those volcanoes may have eruped on their northern sides after deposition of the second moraine group. A Late-glacial age for the moraines was established from associated peat beds on the Chimborazo-Carihuairazo massif (Fig.6) where evidence exists for a period of glacier expansion between 10,000 and 12,000 BP. The glaciers appear to have fluctuated in position at least two or three times during this period (Clapperton and McEwan,1985). Peat beds in basins contiguous with the equivalent moraine system on El Altar began accumulating after 10,000 BP. This is indicated by a radiocarbon age of 9310 ± 60 BP (SRR-2585) for the lowest peat exposed in banks of the Rio Collanes (the true base of the sedimentary sequence is below river level and could not be uncovered). Although the moraine group is clearly developed on almost all extinct central volcanoes, it is absent from most parts of the main cordilleran plateaux, other than where catchments rise above 4,600 m (e.g. in the Llanganati mountains and south of Laguna Cubillin). In southern Ecuador cirques cut into the western slopes of narrow ridges aligned N-S contain well-developed moraines of the second group and emphasise the importance of the 'snowfence' effect with easterly-derived precipitation.

The third group of moraines is characterised in many places by a massive lateral-terminal ridge 100-200 m high, within which smaller moraines are normally present. Such features are best developed where glacier tongues several kilometres long extended from plateau ice fields and from the flanks of ice-covered central volcanoes. Evidence that some of these large moraines may have been constructed by more than one glacial advance is present at Lago Pisayambo where a large quarry excavated from the inner side of the eastern lateral moraine exposes 8-10 m of till overlying a 50-120 cm band of silt, peat and soil which, in turn, had developed on top of 90 m of till. The buried organic layer extends along the quarry face for approximately 60 m before tapering out. It may represent the site of a shallow hollow in the former surface of the original moraine before it was overridden by a subsequent readvance of the glacier. Similar evidence is present on the northern slopes of Carihuairazo where road cuts expose little-weathered till overlying thick beds of peat and fluvial or glacio-fluvial sediments. At one site (Fig.6) the upper peat layer gave a radiocarbon age of 35,440 ± 680/630 BP (SRR-2583), but the lower peat was beyond the range of ^{14}C dating, with an age in excess of 40,000 BP (SRR-2584). The limit of the third moraine group, generally between 3,600-3,800 m, is not always clearly defined by terminal moraines. In localities where glacier tongues descended from plateau catchments into steep narrow valleys terminal moraines were either not deposited or have not survived subsequent denudation, particularly in valleys descending to the Amazon basin. Large moraines are also absent from valleys cut into the western cordillera where low precipitation appears to have inhibited the development of extensive outlet glacier systems during the last glaciation, but limits of the glacial stage may be deduced from the extent of fresh till plastered against the valley sides.

Moraines have not been found beyond the limits of the third group, but an older glacial stage is indicated by deposits of weathered till stratigraphically below the third moraine group and which extend to lower altitudes. Such deposits are present to 2,750 m in valleys on the southwest side of the southern cordillera and to similar altitudes on the eastern side of the eastern cordillera. Up to 4 m of till, weathered throughout its depth, is exposed in road-cut sections on the northern slopes of Carihuairazo. Some clasts of coarse-grained andesite are chemically decomposed to soft rock, but others, usually of fine-grained rock, remain relatively intact and have only a 4 mm weathering rind. This till crops out below 8-10 m of weathered tephras in places and also appears to lie stratigraphically below the peat and gravel beds which are older than 40,000 BP. Till with similar characteristics is exposed at 2,800 m in the Rio Tomebamba valley west of Cuenca. A buried soil horizon lies above 20 m of this till and in turn is overlain by 1-2 m of soliflucted weathered till. The age of the soil is in excess of 40,000 BP (SR-2588). The deeply weathered nature of these deposits suggests that they may date from the penultimate glaciation, but a precise age remains to be established, possibly through fission-track dating of associated tephras. The limits of the weathered till roughly coincide with those of glacial troughs and together they represent the absolute limits of Quaternary glaciation in the Ecuadorian Andes.

QUATERNARY GLACIAL SEQUENCE

The sequence of events indicated by glacial landforms, glacial stratigraphy and radiometric dating in the Ecuadorian Andes is shown in Table 1. No other glacial deposits were found during this study and it is firmly believed that glaciation has never extended below c. 2,750 m altitude. There are two main implications arising from this sequence. One is that late Quaternary glacial fluctuations in Ecuador appear to have been more or less synchronous with those identified in adjacent parts of the tropical Andes (e.g. Thouret and Van der Hammen,1981; Van der Hammen *et al.*,1981; Clapperton,1972, 1981, 1983). Glaciers were most extensive during the penultimate glaciation and the maximum advance of the last glaciation may have been between 36,000 and 45,000 BP (Van der Hammen *et al.*,1981). Although more precise dating is required before this conclusion can be confirmed for Ecuador, it seems likely that the complex of moraines comprising the third group described in the previous section were deposited between 13,000 and 45,000 BP. The smaller inner moraines may be equivalent to readvances identified elsewhere in the Andes between 20,000 and 19,000 BP and between c. 16,000 and 14,500 BP (Clapperton,1983). The sharp climatic reversal indicated by the Late-glacial moraines in Ecuador may have followed a period of warmth peaking at c. 13,000 BP in other parts of the Andes (Mercer,1976; 1983; Van der Hammen *et al.*,1981). The cold period endured for c. 2,000 yrs., implying that tropical regions experienced a significant climatic reversal at a time similar to the Upper Dryas of north-west Europe, and suggests that the Late-glacial temperature drop may have been felt globally (Clapperton,1985). Neoglacial advances in Ecuador are not well dated, but at least one occurred shortly before c. 2,000 BP and others are well-known from the last 500 yr. (Hastenrath,1981). In view of the size and apparent composite nature of some Neoglacial moraines in Ecuador, it is possible that glaciers advanced earlier in the Neoglacial. They did so in other parts of the tropical Andes and in Patagonia between c. 4,700 and 4,200 (Thouret and Van der Hammen,1981; Mercer,1976;

TABLE 1. Late Quaternary glacial sequence, mean lower limits and ages, where known for the Ecuadorian Andes.

Glacial stage	Characteristics	Lower Limit (mean value)	Age and duration
Neoglacial	Superposed and multiple moraines	4,250	c. 5,000 - 100 BP
Late-glacial	Multiple lateral-terminal moraines : commonly 3-4	3,900	12,000 - 10,000 BP
Full-glacial	Multiple lateral-terminal moraines : commonly 3-4	3,500	< 35,000 - > 12,000 BP
Penultimate glacial (?)	Weathered till : erosional landforms	2,800	isotope stage 6 ?

Clapperton,1983).

A second implication is that glacial deposits older than the penultimate glaciation do not exist in Ecuador because either they have been buried by volcanic and other sediments in subsiding basins or they have been removed by erosion. An alternative explanation is that there was no earlier glaciation. Since other parts of the Andes are known to have been glaciated at various times during the Pliocene and Pleistocene (Clapperton,1979, 1983), the Ecuadorian mountains would have remained ice-free during early cold periods only if they had not been high enough to reach the lowered snowlines.

The Ecuadorian cordilleras are part of a physiographic province tectonically distinct from the adjacent Andes of Peru and northern Colombia because of the segmented nature of the subducting Nazca plate and overriding crust (Hall and Wood,1985). It is therefore possible that differential uplift has occurred within the northern Andes and that the Ecuadorian region became elevated above the glaciation threshold only within the last c. 200,000 - 300,000 yr. A major factor affecting uplift of the Ecuadorian sector has been subduction of the Carnegie Ridge, a massive topographical 'high' on the Nazca plate, and this may have commenced only 2-3 M yr. ago (Pennington,1981; Pilger,1983). The original altitude of the Ecuadorian crust and the rate of uplift are unknown, but parts of the mountains appear to be still rising relatively rapidly. For example, sedimentary evidence from the Quito area suggests that the Ilumbisi horst on the eastern side of the city has risen c. 150 m during the last c. 50,000 yr. (R. Vera pers. comm.) and Late-glacial moraines south of Lago Pisayambo are vertically displaced by 8 m. An assumed constant rate of regional uplift during the last few hundred thousand years of 150 m per 50,000 yr. would infer that the Ecuadorian Andes were 600 m lower c. 200,000 yr. ago. For the moment such assumptions are too speculative to support conclusions about the lack of glacial evidence prior to the penultimate glaciation, but the tectonic variable is likely to be important in interpreting variations in the pattern of older glaciations throughout the Andes.

Many of the large central volcanoes have been active during the last 200,000 yr. and some have had a much longer history of evolution. For example, Cotopaxi and Chimborazo have undergone at least two stages of growth and destruction before the development of their present edifices (Hall,1977). Thus even if surrounding plateaux were unable to support glaciers during early cold periods, any volcano rising above 5,000 m would have done so, but evidence of such glaciation will only be uncovered if tillite is present in the stratigraphy of the volcanic basement.

GLACIATION LIMITS AND RECONSTRUCTED ELAs

Altitudinal limits of the different stages of glaciation were established by field observation and measurement (aneroid barometer) at four sites in each of the three cordilleras. These were cross-checked where possible by mapping from vertical air photographs and by plotting moraine positions onto contoured topographical maps (1:25,000, 1:50,000). Such maps are not available for El Altar. Data for other sites were obtained from air photographs, topographical maps and Hastenrath (1981). Mean values were derived for the limits of present glaciers and limits of the three moraine groups for the western and eastern sides of the western and eastern cordilleras. These are plotted schematically in Fig.7. Although the data are approximate, they do illustrate significant differences in the altitude of glaciation limits on either side of each cordillera and between the two cordilleras. It is clear that the distribution of former glaciers was asymmetrical, similar to that of present-day glaciers, indicating that the main source of precipitation was also from the east. Thus most glaciers reached lower altitudes on the eastern/south-eastern sides of individual mountains, and on the eastern sides of the main cordilleras. The lowest limits of Quaternary glaciation were probably on the eastern side of the eastern cordillera. The diagram also shows that the gradients between glaciation limits on opposite sides of the eastern cordillera were much steeper than those on opposite sides of the western cordillera. This presumably results from the predominant easterly source of precipitation and a sharper rain shadow effect.

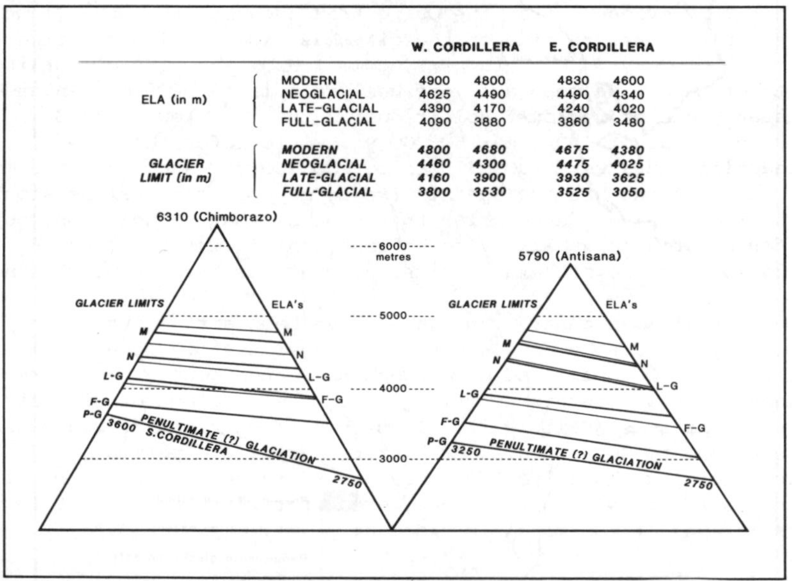

Fig.7. Schematic diagram depicting glaciation limits and calculated ELAs for the eastern and western cordilleras represented for illustration by Antisana and Chimborazo respectively. Limits of weathered till (penultimate glaciation?) in the S. Cordillera are shown on the left diagram.

Mean values for the altitudinal limits of the last glaciation were used to reconstruct the broad outlines of areas covered by permanent snow and ice (Fig.8). Allowance was made for the fact that lower limits were achieved in major valleys draining from the cordillera, thus slightly higher values were assumed for the plateau margins.

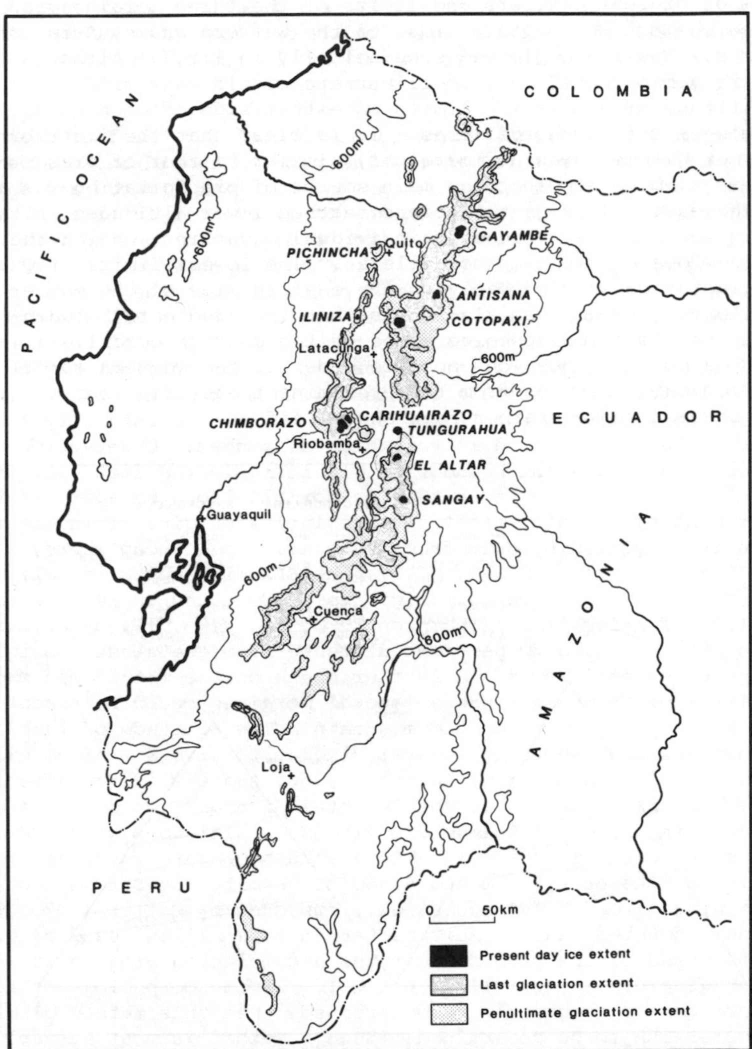

Fig.8. Extent of glaciation in the Ecuadorian Andes during maximum stages of the last and penultimate glaciations (isotope 5c and 6 respectively).

A similar reconstruction was made for the period of maximum (penultimate) glaciation on the basis of the lowest limits of glacial erosion and weathered till. The resulting map clearly shows much more ice in the eastern cordillera, reflecting the presence of widespread surfaces above 3,600 m and high precipitation. The map also emphasises the much smaller scale of the last glaciation compared with the maximal extent at an earlier period. This compares favourably with evidence from Colombia, Peru and Bolivia where the weathered till of an earlier glaciation extends a few kilometres beyond the fresh moraine systems marking limits of the last glaciation (Clapperton,1983).

Equilibrium line altitudes were reconstructed for the four glacial periods discussed in this paper. The results should be considered only as crude approximations because of the lack of accurate data on present-day glaciers, the coarse (40 m) contour interval on topographic maps and the moderate to severe distortion on the vertical air photographs. Calculation of the ELAs was based on the median altitude method which takes the mid point between the glacier's origin and terminus as an approximate value for the ELA (Meierding, 1982). The method's assumption that the ELA will be 50 per cent of the altitudinal distance from the origin to the terminus may be reasonable for high latitude glaciers with marked seasonal variation in snowline altitude, but it appears to be less satisfactory for tropical glaciers located in very high altitudes. Observation of the apparent ELA on present Ecuadorian glaciers, in the field and from air photographs, suggests that it is normally located approximately 80 per cent of the distance from the glacier origin. This observation is supported by data from the Quelccaya ice cap where, according to Thompson (1980, p.72), the ELA lies approximately at 5,250 m. Taking the ice cap origin as 5,600 m and an average value of 5,150 m for its limit (Thompson,1980, Fig.3, p.71), calculation of the ELA, assuming 80 per cent distance from the glacier origin, gives a value of 5,240 m. A further check on the method was made at localities on Chimborazo where lateral moraines could be traced to their true upper limits on the mountain. The altitude of lateral moraines has been shown in several studies to give a minimum elevation of former ELAs (e.g. Anderson,1968; Andrews,1975; Larsen *et al.* 1984). Lateral moraines at the Reschreiter and Reiss glaciers on the east and west sides of Chimborazo extend to altitudes of 4,400 m and 4,600 m respectively (Fig.6). These values compare favourably with calculated ELAs of 4,360 m and 4,560 m. A more accurate method of determining former ELAs successfully employed by Porter (1970, 1977), Sissons and Sutherland (1976) and Larsen *et al.* (1984) used calculations based on the premise that the accumulation area ratio (AAR) of most glaciers is normally 0.6 ± 0.1. If the morphology of former glaciers can be accurately reconstructed, then this method permits the former ELA to be determined. Such a method is most suitable where topography results in well defined glacier basins in which accumulation and ablation zones can be precisely delimited. It also relies on the existence of detailed contour maps. This situation does not exist on most Ecuadorian mountains which supported either plateau ice caps/ice fields with ill-defined margins and accumulation basins or amorphous carapaces of ice extending down the flanks of central volcanoes. Such considerations suggested that the modified median altitude method was more suitable for the data presently available in Ecuador. Table 2 illustrates the values derived for the sites studied and indicates the amount of ELA lowering during the four glacial stages.

The approximate drop in mean annual temperature was calculated from the values obtained for the amount of ELA lowering, assuming an atmospheric thermal gradient of 0.65°C per 100 m. Because of the important influence of a rain shadow effect in the Ecuadorian Andes, calculations involved only the eastern slopes of the eastern cordillera, anticipating that precipitation changes may have been less in that area compared to other parts of the mountains. Results suggest temperature lowering of 1.8°C, 3.9°C, 7.4°C and 10°C for the Neoglacial, Late-glacial, Full-glacial and Penultimate glacial periods respectively. An analysis of Pleistocene snowlines in the western United States by Porter *et al.* (1983, p.103) drew attention to a flaw in calculations which assume that annual accumulation at the ELA of a modern glacier is similar to that which occurred at the lower ELAs of former glacial periods. They pointed out that in some parts of the mountains precipitation increases with altitude by as much as 1 m per 1,000 m at the present day and that if this effect is ignored, the inferred Pleistocene temperatures derived from the standard lapse rate could be too high by several degrees centigrade. It is difficult to judge how important this effect would have been in the tropical mountains of Ecuador where above an altitude of 3,000 m precipitation also appears to increase with elevation. The lack of climatic stations precludes an accurate assessment of this, however. Another factor which is perhaps appropriate only to the tropical Andes is that precipitation actually increases with decreasing altitude on the eastern flanks of the eastern cordillera because of the influence of convectional systems in Amazonia. Whatever the interplay between these two precipitation regimes may have been during glacial times, it appears that the overall amount was probably lower than now. Biological studies have suggested that during parts of the glacial period the Amazonian rain forest was fragmented into small patches by greatly increased aridity (Vuilleumier, 1971; Van der Hammen, 1974; Tricart, 1974; Colinvaux, 1979). If precipitation totals were also reduced over the Ecuadorian cordilleras, then it is possible that temperature depression for the glacial stages may have been a few degrees greater than that calculated, at least during the last and penultimate glaciations.

One implication of the drop in temperature in equatorial latitudes and the presence of extensive ice caps over the Ecuadorian cordilleras is that very cold and dry conditions would have prevailed

TABLE 2. Calculated ELA values for sites in the Eastern, Western and Southern Cordilleras of Ecuador.

Period	Present		Neoglacial		Late-glacial		Full-glacial	
	West	East	West	East	West	East	West	East
EASTERN CORDILLERA								
CAYAMBE								
Glacier Origin	5600	5600	5600	5600	5600	5600	5600	5600
Glacier Limit	4625	4350	4350	3900	3700	3500	3300	3000
ELA	4820	4600	4600	4240	4080	3920	3760	3520
ELA Lowering	-	-	220	360	740	680	1040	1080
ANTISANA								
Glacier Origin	5600	5600	5600	5600	5600	5600	5600	5600
Glacier Limit	4800	4400	4600	4150	4400	3750	3600	3150
ELA	4960	4640	4800	4440	4640	4120	4000	3640
ELA Lowering	-	-	160	200	320	520	960	1000
EL ALTAR								
Glacier Origin	5200	5200	5200	5200	5200	5200	5200	5200
Glacier Limit	4600	4400	3800	-	3700	-	3600	-
ELA	4720	4560	4080	-	4000	-	3920	-
ELA Lowering	-	-	640	-	720	-	800	-

Period	Present		Neoglacial		Late-glacial		Full-glacial	
	West	East	West	East	West	East	West	East
EASTERN CORDILLERA (Continued)								
CARIHUAYCU/PAPALLACTA								
Glacier Origin							4400	4400
Glacier Limit							3600	3000
ELA	4890*	4620*					3760	3280
ELA Lowering							1130	1340
LAGO PISAYAMBO								
Glacier Origin							4200	
Glacier Limit							3600	
ELA	4800**						3720	
ELA Lowering							1080	

* Average of Cayambe and Antisana
** Estimated value

Period	Present		Neoglacial		Late-glacial		Full-glacial	
	West	East	West	East	West	East	West	East
WESTERN CORDILLERA								
ILINIZA								
Glacier Origin	5100	5100	5100	5100	5100	5100	5100	5100
Glacier Limit	4800	4800	4400	4300	4000	3900	3700	3500
ELA	4860	4860	4540	4460	4220	4140	3980	3820
ELA Lowering	-	-	320	400	640	720	880	1040

Continued over .....

Period	Present		Neoglacial		Late-glacial		Full-glacial	
	West	East	West	East	West	East	West	East
WESTERN CORDILLERA (Continued)								
CARIHUAIRAZO								
Glacier Origin	4900	4900	4900	4900	4900	4900	4900	4900
Glacier Limit	4650	4550	4400	4200	4100	3800	3900	3600
ELA	4700	4620	4500	4340	4260	4020	4100	3860
ELA Lowering	-	-	200	360	440	600	600	760
CHIMBORAZO								
Glacier Origin	5800	5800	5800	5800	5800	5800	5800	5800
Glacier Limit	5000	4700	4600	4400	4400	4000	3800	3500
ELA	5160	4920	4840	4680	4680	4360	4200	3960
ELA Lowering			320	240	480	560	960	960
PUJILI/QUILOTOA								
Glacier Origin							4200	4200
Glacier Limit							4000	3800
ELA	4900*						4040	3880
ELA Lowering							860	1010

* Average of Iliniza,Carihuairazo and Chimborazo (west sides)

Period	Present		Neoglacial		Late-glacial		Full-glacial	
	NW	SE	NW	SE	NW	SE	NW	SE
SOUTHERN CORDILLERA								
Cuenca Plateau								
Glacier Origin							4300	4300
Glacier Limit	5000*	4900					3800	3200
ELA							3900	3420
ELA Lowering							1100	1480

* Estimated values

in the intermontane basins. The present mean annual temperature of the Quito area is 13°C. This would have dropped to around 6°C or less during the last glaciation. In addition to the absolute temperature, the enhanced precipitation shadow effect induced by the large ice cap over the eastern cordillera and the probable frequent development of strong glacial and katabatic winds would have created extremely dry and cold conditions for the biosphere in the inter-andean depression. Such conditions are partly indicated by the widespread accumulation of the volcanic-loessal sediment known as cangahua, in which have been found the remains of large grazing animals and the fossilised nests of dung beetles. The latter presently live only in arid parts of South America and are no longer found in Ecuador.

ACKNOWLEDGEMENTS

Grants supporting the project in Ecuador were generously made by the Carnegie Trust for the Universities of Scotland, the Royal Society, Aberdeen University and the Royal Scottish Geographic Society.

Many people helped in Ecuador, but particular thanks are due to Dr. Nelson Gomez, Roy Ryder and Juan Hidalgo of CEPEIGE, Colin McEwan and Ing° Ramon Vera. Dr. Douglas D. Harkness undertook all the radiocarbon analyses.

REFERENCES

Andersen, B.G., 1968. Glacial geology of Troms, North Norway. Norges Geologiske Undersøkelse, 265, 160 pp.

Andrews, J.I., 1975. Glacial Systems - An Approach to Glaciers and their Environments. Duxbury Press, Massachussets.

Clapperton, C.M., 1972. The Pleistocene moraine stages of west-Central Peru. Journal of Glaciology, 62, 255-263.

Clapperton, C.M., 1979. Glaciation in Bolivia before 3.27 M yr. Nature, 277, 375-377.

Clapperton, C.M., 1981. Quaternary glaciations in the Cordillera Blanca, Peru, and the Cordillera Real, Bolivia. Memoria del Primer Seminario Sobre el Cuaternario de Colombia, Bógota, Agósto 25 al 29 de 1980. Revista Centro Interamericano de Fotointerpretacion C.I.A.F. (Bógota), 6, 93-111.

Clapperton, C.M., 1983. The Glaciation of the Andes. Quaternary Science Reviews,2, 83-155.

Clapperton, C.M., 1985. Significance of a Late-Glacial readvance in the Ecuadorian Andes. In Rabassa, J. (ed.) Quaternary of South America and Antarctic Peninsula, 3. (In press)

Clapperton, C.M. and McEwan C., 1985. Late Quaternary moraines in the Chimborazo area, Ecuador. Arctic and Alpine Research, 17, 135-142.

Clapperton, C.M. and Vera, R., 1986. The Quaternary glacial sequence in Ecuador. A reinterpretation of the work of W. Sauer. Journal of Quaternary Science. (In press)

Colinvaux, P., 1979. The Ice-Age Amazon. Nature, 278, 399-400.

Delavaud, A.C., 1982. Atlas del Mundo : Ecuador : Banco Central del Ecuador, Quito.

Hall, M., 1977. El volcanismo en el Ecuador. Instituto Panamericano de Geografía e Historía, Quito.

Hall, M. and Wood, C.A., 1985. Volcano-tectonic segmentation of the Northern Andes. Geology,13, 203-207.

Hastenrath, S., 1971. On the Pleistocene snow-line depression in the arid regions of the South American Andes. Journal of Glaciology,10, 255-267.

Hastenrath, S., 1978. Heat budget measurements on the Quelccaya Ice Cap, Peruvian Andes. Journal of Glaciology,20, 85-97.

Hastenrath, S., 1981. The Glaciation of the Ecuadorian Andes. Balkema, Rotterdam.

Heusser, C.J., 1974. Vegetation and climate of the southern Chilean Lake District during and since the last interglaciation. Quaternary Research, 4, 290-315.

Heusser, C.J., 1984. Late-glacial-Holocene climate of the Lake District of Chile. Quaternary Research,22, 77-90.

Kennerley, J.B. and Bromley, R.J., 1971. Geology and geomorphology of the Llanganati Mountains, Ecuador. Instituto Ecuatoriano de Ciencias Naturales (Quito), 73, 3-16.

Larsen, E., Eide, F., Longva, O. and Mangerud, J., 1984. Allerød-Younger Dryas climatic inferences from cirque glaciers and vegetational development in the Nordfjord area, Western Norway. Arctic and Alpine Research, 16, 137-160.

Meierding, T.C., 1982. Late Pleistocene glacial equilibrium-line in the Colorado Front Range : A comparison of methods. Quaternary Research, 18, 289-310.

Mercer, J.H., 1976. Glacial history of southernmost South America. Quaternary Research, 6, 125-166.

Mercer, J.H., 1983. Cainozoic glaciation in the Southern Hemisphere. Annual Review of Earth and Planetary Science, 11, 99-132.

Meyer, H., 1907. In den Hochanden von Ecuador, 2 vols. Dietrich Reimer-Ernst Vohsen, Berlin. Vol.1 522 pp. Vol.2 picture atlas.

Nogami, M., 1976. Altitude of the modern snowline and Pleistocene snowline in the Andes. Tokyo Metropolitan University Geographical Reports, 11, 71-86.

Pennington, W., 1981. Subduction of the eastern Panama basin and Seismotectonics of north western South America. Journal of Geophysical Research, 86, 10753-10770.

Pilger, R., 1983. Kinematics of the South American subduction zone from global plate reconstructions. In Cabre, R. (ed.) Geodynamics of the eastern Pacific region, Caribbean and Scotia Area. American Geophysical Union Geodynamics Series, 11, 113-125.

Porter, S.C., 1970. Quaternary glacial record in Swat Kahistan, West Pakistan. Geological Society of America Bulletin,81, 1421-1466.

Porter, S.C., 1977. Present and past glaciation threshold in the Cascade Range, Washington, U.S.A. : topographic and climatic controls and palaeoclimatic implications. Journal of Glaciology, 18, 101-116.

Porter, S.C., Pierce, K.L. and Hamilton, T.D., 1983. Late Wisconsin Mountain Glaciation in the Western United States. In Porter, S.C. (ed.) Late-Quaternary Environments of the United States, Ch.4, 71-111. University of Minnesota Press.

Pourrut, P., 1983. Los climas del Ecuador, fundamentos explicativos. Centro Ecuatoriano de Investigacion Geografica, Documentos de Investigacion 4, 9-44.

Reiss, W. and Stübel, A., 1892-98. Reisen in Südamerika. Das Hochgebirge der Republik Ecuador. Petrographische Untersuchungen, Vol.1, West-Cordillere; Vol.2, Ost-Cordillere. Asher & Co., Berlin.

Sauer, W., 1950. Contribuciones para el conocimiento del cuaternario en El Ecuador. Anales de la Universidad Central, Quito, 77 (328), 327-364.

Sauer, W., 1965. Geología del Ecuador. Editorial de Ministerio de Educación, Quito.

Sauer, W., 1971. Geologie von Ecuador. (Beitrage zur Regionalen Geologie der Erde, Vol.11). Gebrueder Borntraeger, Berlin-Stuttgart.

Sievers, W., 1908. Vergletscherung der Cordillera des tropischen Südamerika. Zeitschrift für Gletscherkunde, 2, 271-284.

Sissons, J.B., and Sutherland, D.G., 1976. Climatic inferences from former glaciers in the south-east Grampian Highlands, Scotland. Journal of Glaciology, 17, 325-46.

Sugden, D.E., and John, B.S., 1976. Glaciers and Landscape. Arnold, London.

Thompson, L.G., 1980. Glaciological investigations of the tropical Quelccaya ice cap, Peru. Journal of Glaciology, 25, 69-84.

Tricart, J., 1974. Existence de périodes sèches du Quaternaire en Amazonie et dans les regions voisines. Revue Geomorphologique dynamique, 23, 145-158.

Thouret, J.C., and Van der Hammen, T., 1981. Una secuencia Holocenica y Tardiglacial en la Cordillera Central de Colombia. In Memoria del Primer Seminario sobre el Cuaternario de Cólombia, Bógota, Agósto 25 al 29 de 1980. Revista Centro Interamericano de Fotointerpretacion, C.I.A.F. (Bógota), 6, 609-634.

Van der Hammen, T., Barelds, J., De Jong, H., and De Veer, A.A., 1981. Glacial sequence and environmental history in the Sierra Nevada del Cocuy (Colombia). Palaeogeography, Palaeoclimatology, Palaeoecology, 32, 247-340.

Vuilleumier, B.S., 1971. Pleistocene changes in the Fauna and Flora of South America. Science, 173, 771-780.

Wolf, T., 1892. Geografía y geología del Ecuador. Brockhaus, Leipzig.

QUATERNARY OF LOW LATITUDES

International Geomorphology 1986 Part II
Edited by V. Gardiner

QUATERNARY OF THE LOW LATITUDES : INTRODUCTION

V. Gardiner

Department of Geography, University of Leicester,
Leicester, LE1 7RH, U.K.

Quaternary geomorphology of low latitude areas was the subject of a joint symposium with the INQUA Palaeoclimate Commission. Six papers were presented orally; five considered aspects of the Quaternary in South Africa, Mexico, Brazil, Venezuela and the Andes, and one considered parts of India, Africa and Australia. Papers presented as posters were concerned with Malaysia, Namibia and South Africa. The papers included in these Proceedings give an adequate representation of the symposium as a whole, and illustrate the great range of questions which remain to be answered in this field.

Adamson et al., attempt to link Quaternary alluvial histories of three widely separated basins to the Southern Oscillation. Such wide-ranging correlation must necessarily be founded upon detailed work in more localised areas. This is exemplified by Reynhardt, who relates colluviation to climatic fluctuations, on the basis of detailed examination of deposits in part of South Africa. One major difficulty in elucidating Quaternary landscape evolution in lower latitudes is that there is often only very limited evidence on which to base conclusions. This is exemplified by Marker's paper, in which she draws attention to the extremely localised high-altitude scree tongue deposits of South Africa. A second difficulty is that in some parts of the low latitudes there are relatively few workers active in this field. Vivas highlights this problem for Venezuela, and describes how even the most basic Quaternary chronology is not yet firmly established, despite the major significance of Quaternary events for almost half of the country.

International Geomorphology 1986 Part II
Edited by V. Gardiner

COMPLEX LATE QUATERNARY ALLUVIAL HISTORY IN THE NILE, MURRAY-DARLING, AND GANGES BASINS: THREE RIVER SYSTEMS PRESENTLY LINKED TO THE SOUTHERN OSCILLATION

D. Adamson
School of Biological Sciences and Quaternary Research Unit, Macquarie University, North Ryde, N.S.W., 2113, Australia

M.A.J. Williams and J.T. Baxter
Geography Department, Monash University, Clayton, Victoria, 3168, Australia

ABSTRACT

In river basins in Africa, India and Australia, very high and very low river discharges have been remarkably synchronized with the Southern Oscillation over at least the last 100 years. These river systems are linked by common responses to present climatic fluctuations. In each of these river basins behaviour in the past was also linked. In particular an upward fining alluvial sequence characterises the transition from the full glacial late Pleistocene to the Holocene. This general sedimentary sequence may be obscured by aeolian inputs, by varying degrees of aggradation or incision, and by the strong influence of inherited floodplain features. Detailed examples of each are presented: the importance of dust in producing unusually fine fluvial sediments during the late Pleistocene in India; aggradation on low-angle floodplains in the late Pleistocene, and incision in the Holocene are compared for the Blue Nile and the Lachlan; the importance of inherited features of floodplains is examined for the Lachlan fan.

UPWARD FINING ALLUVIAL DEPOSITION

Pleistocene to Holocene upward fining alluvial sequences are common on the large low-angle fans and floodplains of rivers in the low to middle latitudes. In Africa, India and Australia many examples exist of Holocene clays overlying coarser or more mixed sediments of late Pleistocene age. We have observed such sequences in each area, and they have been well described in Australia. In the Nile basin of Sudan they occur on the Bahr el Arab floodplains of Southern Darfur (Williams et al.,1980b), and on the Gezira alluvial fan between the Blue and White Niles (Williams and Adamson, 1974; Adamson et al.,1980; Williams and Adamson,1980; Williams et al.,1980a). In India, upward fining sequences occur in the valleys of the Son and Belan rivers, southern tributaries of the Ganges (Williams and Royce,1982; Williams and Clarke,1984). In Australia they occur on the vast alluvial fans and floodplains of the central and eastern inland as studied by many workers including

Bowler (1978), Taylor (1978), Veevers and Rundle (1979), Bowler and Wasson (1984), and by Rust (1981) who placed the transition from sandy to muddy sedimentation during the mid Holocene.

The late Pleistocene to Holocene upward fining sequence on low-angle floodplains is probably a phenomenon which is circum-global for middle to low latitudes and which can best be explained by climatic changes from glacial to interglacial working via rainfall, vegetation cover, erosion, and stream behaviour. A purpose of this paper is to describe some examples of how this broad generalisation can be obscured by aeolian inputs, by variable aggradation and incision, and by inherited attributes of the floodplains.

A second purpose is to show that the three areas, although so widely separated, contain rivers which today show a remarkable degree of synchroneity in discharge. Extreme high and extreme low discharge tend to occur at the same time. Extremes in river discharge reflect extreme fluctuations in rainfall which in turn are linked to large swings of the Southern Oscillation. This circum-global oscillation of air pressure and rainfall reflects one of the main present controls of world weather. The behaviour of the Southern Oscillation in the past is now a crucial question for palaeoclimatology. At this stage however the remarkable synchroneity of river behaviour shown in this paper is ample additional justification for considering together these three trans-equatorial river basins.

PRESENT RIVER DISCHARGE AND THE SOUTHERN OSCILLATION

The major rivers in northeastern Africa, in central and northern peninsular India, and in inland eastern Australia have flooded or failed together during the last hundred years or so. The Nile and its tributaries, the Krishna and Godavari, and the Darling are highly seasonal rivers which depend on tropical rain, itself linked to fluctuations in the Southern Oscillation. In eastern Australia, in years of extreme fluctuation, flow of the Lachlan, Murrumbidgee and Murray rivers may be linked in.

Since early in this century, a large area of the globe centred on India-Australia has been claimed to form a coherent climatic zone (Lockyer and Lockyer, 1904; Walker, 1924) and in 1925 Bliss extended the analysis to show that Nile discharge was linked to the same zone. These dramatic and adventurous claims have been a long time gaining credibility but the coherence of this huge area is now firmly established (Barnett, 1984). Analysis of river discharges has not, however, progressed beyond the pioneering work of Bliss.

Figure 1 shows years of particularly high or particularly low river discharge since 1877, together with fluctuations of an index of the Southern Oscillation about its mean value. Most of the years of very high (+) or very low (-) river flow across all basins are notable for the extreme fluctuation of the index (SOI) in the same direction. The more positive the index the lower the atmospheric pressure from Africa across India to Australia. Heavy precipitation results which is integrated by the rivers into high discharges and memorable floods. A strongly negative SOI indicates the reverse, leading to drought and below average river discharges from the Nile to the Darling. In the case of the Darling, the most variable of the rivers under consideration, it may become reduced to a chain of waterholes for months at a time.

Focussing on certain extreme years or short periods illustrates the tendency of the rivers to flood or fail together. In 1877 there was a great drought in the Darling basin, vast famine and drought in India and later in China, the lowest annual Nile flood recorded in Egypt since gauging recommenced in 1825, and the lowest SOI between 1851 and 1974. Other notable years of drought and low discharge throughout this vast inter-tropical zone included 1902 when the Nile recorded a new record low, when the Darling ceased to flow in its middle reaches for 11 months and when extremely low discharges also prevailed in the Lachlan, Murrumbidgee and Murray. In that year, drought and low river flow also occurred in India. The period between 1912 and 1914 was also one when the Nile again recorded its lowest high season discharge since 1825, the Darling again ceased to flow and low discharges were recorded in the other rivers of the Murray-Darling basin. The Indian rivers were also low. In the drought of 1918 the Darling ceased to flow for 10 consecutive months but the reasonably normal discharge of the Main Nile in Egypt that year concealed very low discharges of the Blue Nile and Atbara - the two major Ethiopian tributaries - because their low flows were balanced by delayed discharge from the White Nile floods of the two previous remarkably wet years. The period around 1940 was also notable for low SOI, low rainfall and low river discharge in the three regions.

In this broad meridional zone of the world, major floods also tend to occur synchronously. Again the Darling is notable, this time for floods in its middle reaches which last for many months because of water storage on its flood plain and slow return of only some of it to the main channel. In the 1892-4 period severe flooding occurred in 15 months, and in 1954-6 for over 10 months. Flood lags on the White Nile such as in 1916-7 occur because of huge spillage into the swamps and floodplains of southern Sudan.

Fig. 1 simplifies the picture by concentrating on years of extreme climate when the rivers and the SOI were fluctuating together. There are only few years when the synchroneity of extreme behaviour is not so neat. For example the SOI was roughly normal in 1887 but river discharges were high. Again, in 1944-5 the SOI was not very different from normal but river discharges were low throughout the region. Occasionally one region stands separate from the other two, as in 1961 when the Darling was low but the SOI, the Indian rivers and the Nile were high. In years of near average SOI, factors more local to each region or external to the Southern Oscillation allow the rivers to react independently of each other, but only rarely in an extreme manner. It seems to be the circum-global meridional extremes reflected in extreme SOI values which pull these rivers into synchroneity.

The present synchroneity raises the intriguing possibility that in the past, under different world climates from those of to-day, river behaviour may also have been linked across this region.

COMPLEXITIES IN THE RECORD OF FLUVIAL SEDIMENTATION

Dust. The removal and deposition of dust has occurred unevenly during the climatic fluctuations of the late Quaternary, and to different extents in each of the three basins. For example wind blown dust remobilised by fluvial erosion is a major component of the late Pleistocene sediments of the Son and Belan river terraces in north central India, immediately north of the Krishna and Godavari basins (Williams and Clarke, 1984). By contrast wind deposited dust forms an important, extensive but more diffuse component in the sediments of the Murray-Darling basin (Butler, 1958). The differences between northern India and eastern inland Australia reflect the enormous sediment production from the collision zone of India and Asia, in contrast to the trivial sediment production from the low altitude tectonically inactive margin of eastern Australia. The great Australian floodplains are all sediment starved compared with those of northern India, where sediment dumped on the Ganges floodplain could have been blown over the Son and Belan catchments during the late Pleistocene. The contribution of dust to the Blue and White Nile floodplains has not been assessed. Khartoum, at the toe of the Blue Nile Gezira fan, at present receives a notable but unmeasured annual addition of dust to judge from roof-top accumulations.

Terraces in the valleys of the Son and Belan Rivers illustrate how aeolian dust deposited over the catchment area and transported by the rivers can form a large proportion of the floodplain deposits in late Pleistocene times (Williams and Clarke, 1984). Away from the main channels, the bulk of the floodplain was, with minor interruptions, built from water transported loess during the period from about 26,000 until about 10,000 years B.P. These fine silt-rich deposits (up to 69% silt) were being laid down at a period when rivers in the other two regions were laying down coarser deposits (Williams and Adamson, 1980; Bowler and Wasson, 1984). In the Son and Belan, the aeolian input accumulated in the floodplain therefore modifies and obscures the generalisation of an upward fining alluvial sequence from the late Pleistocene full glacial period to generally moister Holocene times.

Fig. 2 shows where volcanic ash has been deposited on the Gezira floodplain by the Blue Nile, or more probably its minor tributary the Rahad River, flowing from Ethiopia. As with the Indian loess described above, the airfall deposit was shifted onto the fan by water flow - but in this case a thick slurry of ash choked the rivers for one or two wet seasons after the fall. Two types of deposit are present in the Gezira (Adamson, Williams and Gillespie, 1980): broad spreads on the surface of the fan near its distal end, and channel-fill deposits buried beneath Holocene and terminal Pleistocene clays but now exposed by incision of the Blue Nile. We have found an ash-choked palaeochannel in a terrace of the Son

River in India. This late Pleistocene ash, the first discovered in this region, resembles in its chemical composition ash from Toba volcano in Sumatra (work in progress).

Although wind blown and water redeposited ash makes a negligible quantitative contribution to sedimentation on floodplains in the Nile basin and in India, these deposits are useful chronostratigraphic markers on low-angle floodplains and terraces. For example, the up-fan burial of Pleistocene ash beneath early to mid Holocene clays shows that the uneroded surface of the low-angle alluvial fan of the Blue Nile is time-transgressive.

Incision and Aggradation. The three river basins show different amounts of incision during the Holocene: the Ganges tributaries are now deeply incised into their late Pleistocene floodplains which are left as wide terraces 20 to 30 m above river level; the Blue Nile is incised 10 to 15 m into its now abandoned late Pleistocene and early Holocene alluvial fan, the Gezira; the present day rivers of the Murray-Darling basin are incised a mere 1 to 2 m into their floodplains and the floodplains and older channels still carry flood water.

The switch from aggradation to incision on low-angle deposits under fluvial and erosional regimes reacting to climatic change was used by Fairbridge (1962, 1963) and Adamson et al. (1980) to explain the complex depositional history of the Main Nile in Egypt during the late Quaternary. Along the length of the river there was discontinuous behaviour, with erosion and channel incision in the steep headwaters during maximum glacial times entraining a heavy mixed sediment load which was deposited on the low-angle sections of the basin. Late Pleistocene incision in the headwaters was associated with aggradation on the distal low-angle floodplains. The Holocene, with minor reversals well illustrated by the Indian rivers (Williams and Royce,1982; Williams and Clarke,1984), has been the reverse of the late Pleistocene regime - sediment has accumulated in the steeper uplands under climatic conditions more favourable for soil stability whereas the rivers in their distal reaches have incised their floodplains and valley fills.

Fig. 2 shows the alluvial fans of the Blue Nile and the Lachlan rivers, fans with similar areas and slopes (10^{-4} to 10^{-5}). The differences between these two fans are instructive. About three quarters of the Gezira is covered by a skin of clay 2 to 4 m thick derived from the basaltic highlands of Ethiopia and spread during the terminal Pleistocene and early Holocene. The few palaeochannels on the surface of the plain are major distributaries functional into the Holocene. In the upper and middle sections of the fan older palaeochannels are buried by the most recent clays. Deposition on the fan was rapid until incision commenced. By contrast, large areas of the Lachlan fan are still covered by high floods and aggradation still occurs. Generations of Pleistocene palaeochannels are still well exposed on the surface of the fan, many still carrying flood waters. Deposition on the Lachlan fan in the Pleistocene was slow and it is probably even slower today. Similarly, incision by the rivers has been extremely slight during the Holocene.

These great contrasts between the Blue Nile and Lachlan fans are accounted for by the great difference in discharge between the two rivers. Average annual discharge for the Blue Nile is about 51,000 million m^3 while that for the Lachlan is about 970 million m^3, a discharge equivalent to that of the Rahad River, one of the minor tributaries of the Blue Nile. The Lachlan fan is analogous to a terminal flood-out because the through-flowing Lachlan is a minor stream unable to entrench its own deposits to any marked degree. History is therefore long preserved on the surface of such fans but quickly buried on the Gezira to be revealed mainly by incision.

<u>Inherited Landforms on Low-Angle Alluvial Fans.</u> As emphasised earlier, the Australian rivers are relatively starved of water and sediment. Some consequences of such a river flowing out onto a large low-angle fan are illustrated by the Lachlan River. They include:

1. Preservation on the surface of the fan of a long time sequence of palaeo-fluvial, -lacustrine, and -aeolian features.

2. Channel(s) of the modern river can be distinguished from palaeochannels by their morphology, their lack of numerous relict features, and by their carrying the non-flood flows.

3. Flood flows may reoccupy palaeochannels and lakes formed under different climatic and/or discharge regimes. Palaeochannels may even readjust to their new role as floodways.

4. Minor topographic irregularities and sedimentary differences developed on the fan over long periods exert profound influences during and after the establishment of a new river channel - floodplain inheritance controls both location and behaviour of the new channel.

5. A major change of drainage pattern across the fan, which may or may not have a climatic basis, will alter the environment over large areas - by beheading river channels, lakes and swamps, and by altering the pattern of ground water recharge.

6. The presence of a suite of relict and active features allows a comparison of present processes and their features with palaeo-features.

The floodplain of the Lachlan has been mapped at a scale of 1:500,000 mainly from airphoto interpretation (Butler et al.,1973). We have begun more detailed work. Fig. 3 shows the modern Lachlan and a few of the many distributaries which radiate across its low-angle fan. The present river occupies a narrow sinuous gutter usually with muddy bed and banks of high stability. Along its course there are relatively few meander cut-offs and no nested sets of scroll bars. The channel is remarkably devoid of relict features. Mud-lined channels are known to be stable (Riley and Taylor,1978; Woodyer,

1978; Jackson,1981) and a muddy section of the Murray is known to have kept a stable but meandering channel for over 8,000 years (Bowler,1978).

On the Lachlan fan, Marrowie (or Box) Creek has a channel similar to that of the present Lachlan. They form two of the only three distributaries with channels exiting the fan. The third is Willandra Billabong Creek. The Lachlan is the only one which regularly discharges water from the fan now. Willandra Creek follows an extraordinary course westwards beyond the fan onto a relict Tertiary surface covered by sand dunes where it turns south through the huge relict Willandra Lakes, ultimately connecting with the Murray River. The Willandra has not flowed through the lakes since the late Pleistocene, but its channel across the fan still acts as a flood distributary. The morphologies of the Lachlan and of the Willandra are compared in Fig. 4. Willandra Creek has abundant large meander cut-offs, nested meander scrolls, and substantial lengths of abandoned channels with their own large-scale relict features. Its morphology and abundant large relict features suggest an antiquity far greater than that for the present Lachlan. The parallel to other present rivers and palaeochannels in the Murray-Darling system such as the Goulburn and Murrumbidgee is striking (Butler et al.,1973; Bowler, 1978).

Our interpretation is that the Lachlan is a young channel formed de novo across the fan, probably in response to slow progressive aggradation of the northern distributaries including Willandra Creek. Stream flow shifted southwards to occupy a slightly steeper gradient. The present Marrowie Creek became active at about the same time but the more southerly channel of the present Lachlan dominated and carried the non-flood flows. In flood all the easily distinguishable distributaries carry water.

The date of establishment of the present Lachlan channel across the fan is not known. Dates from rivers in the southern parts of the basin suggest it is probably younger than last glacial maximum times in the late Pleistocene (Bowler,1978). It may have become established within the last 14,000 years. The modern Lachlan channel diverted all regular flow and some flood flow from Willandra Creek, thereby contributing to the Holocene failure of flow into the huge and archaeologically important Willandra Lakes (Williams, Adamson and Baxter, in press).

At about the same time as drainage across the Lachlan fan shifted to its southern margin, the Murrumbidgee probably shifted towards the northern margin of its fan, abandoning aggraded channels to the south. These two rivers moved towards each other and joined in a low lying area of extensive present-day swamps (Fig. 3). Tectonic movement has occurred on the plain in the Pleistocene resulting in diversion of the Murray River around the Cadell Fault (Butler et al., 1973). The swampy zone at the junction of the Murrumbidgee and Lachlan rivers lies on a possible extension of well recognized north-south fault in Victoria and also corresponds with the north-south oriented western boundary of the Lachlan fan and the whole

Riverine Plain. The swampy Lachlan-Murrumbidgee junction zone may reflect minor tectonic movement and/or compaction of floodplain sediments and/or differential aggradation of both fans.

A consequence of the de novo formation of a river channel across an ancient slowly-accreting fan is that the channel will be established and will then develop with characteristics predetermined by the surface of the fan. That is, floodplain inheritance will determine future fluvial characteristics. In forming a new channel, water will initially find its route from low point to low point across the fan, probably from a new distributary take-off near its apex. Minor variations of topography of a few centimetres to less than one metre and variations in sediment both influence the location and character of the new channel. Figs. 4.2a and 4.2b show the importance of floodplain inheritance for two almost adjacent segments of the present Lachlan. The upstream segment (Fig. 4.2a) is centred on Booligal, a river crossing point, where the channel is less sinuous and is not flanked by floodouts and swamps. The downstream segment (Fig. 4.2b) is flanked by an extensive reticulate network of flood channels similar to those found on the floors of some ephemeral lakes (Fig. 4.1). Similar differences occur elsewhere with sandier sediments marking less sinuous swamp-free sections. Fig. 3 shows that the Lachlan connects a series of large flood-out areas, presumably reflecting the lower-lying southern flank of the fan away from the previous generation of aggraded channels. It is tempting to suggest that the modern Lachlan established its channel by joining these low points starting at the apex and working down the fan. However, processes of de novo channel formation on huge low-angle alluvial fans are unknown.

Lake formation on low-angle floodplains is also poorly understood apart from the importance of ground water and salinity (Bowler and Wasson,1984). Fig. 4.1 shows a small kidney-shaped lake with flood inflow from the east and a wider zone of outflow. The lake has a minor lunette on its eastern shore. The inflow has not breached the lunette, rather the lunette has been built by wind on both sides of the inflow channel at the same time as the lake deepened and its shores became well defined. Incipient lake formation may be occurring at the flood off-take from the Willandra palaeochannel. Minor variations in topography and in sediment properties inherited from earlier phases of fan development are probably also important in determining the loci of lake initiation.

CONCLUSIONS

Several broad generalisations have been drawn about river and floodplain characteristics in the intertropical zone, and some complexities have been illustrated which obscure but do not necessarily invalidate the generalisations.

It is no longer surprising that rivers in widely separated but climatically coherent parts of the world, such as from the northeastern quadrant of Africa across the Indian subcontinent to eastern Australia, flood and fail in concert. However this simple fact has not been noticed previously. It is now obvious (Barnett, 1984) that climate, upon which the rivers depend, has a global scale of meridional patterns as well as the long recognised zonal patterns. Correlations of Nile discharge with atmospheric properties at Darwin and the Aleutian Islands (Bliss,1925) no longer seem perverse, as they did for most of the 50 years since the analysis was published. Atmospheric circulation is now seen to be linked around the equator as well as from equator to both poles - it was always intuitively obvious. The synchroneity of extreme river discharges in Africa, India and Australia and their linkage to the Southern Oscillation becomes obscured if values for all years are treated equally. The synchroneity weakens because of local or wide ranging effects other than the Southern Oscillation which can then operate.

A second generalisation is that on low-angle floodplains finer sediments have been laid down in the Holocene than was the case during the last glacial maximum of the late Pleistocene. This generalisation will probably survive the complex factors which tend in certain places to obscure it. Aeolian dust may dilute or even swamp the sediment eroding from within a basin but this does not invalidate the generalisation. Similarly, on a time-transgressive alluvial fan, particularly one relatively starved of water and sediment, great care is needed lest one look at the evidence through the wrong window in time.

Rivers can be both incising and aggrading at the one time in different parts of their course, depending on broad climatic factors and position within the basin. River behaviour is discontinuous from headwaters to the distal low-angle fans. The degree of aggradation and incision is also exceptionally variable even in rivers which behave synchronously within the one climatic zone. The generalisation of Holocene incision and late Pleistocene aggradation during full glacial time in low-angle inland alluvial fans in the intertropical zone will, however, probably survive.

Finally, more attention should be paid to the effect of inherited features which occur on the surface of stable old alluvial floodplains which accrete and degrade slowly. The inheritance affects subsequent fluvial and lacustrine behaviour in ways that have been little noticed.

Acknowledgements.

The Australian Research Grants Scheme, Macquarie University Research Grants, and Monash University are thanked for their generous and far-sighted support.

REFERENCES

Adamson, D.A., Gasse, F., Street, F.A., and Williams, M.A.J., 1980. Late Quaternary history of the Nile. Nature, 287, 50-55.

Adamson, D.A., Williams, M.A.J., and Gillespie, R., 1980. Palaeogeography of the Gezira and of the lower Blue and White Nile valleys, in Land Between Two Niles (Eds. Williams and Adamson), pp.165-219. A.A. Balkema, Rotterdam.

Barnett, T.P., 1984. Interaction of the monsoon and Pacific trade wind system at interannual time scales. Part III: A partial anatomy of the Southern Oscillation. Monthly Weather Review, 112, 2388-2400.

Bliss, E.W., 1925. The Nile flood and world weather. Memoirs of the Royal Meteorological Society, 1, 79-85.

Bowler, J.M., 1978. Quaternary climate and tectonics in the evolution of the Riverine Plain, southeastern Australia, in Landform Evolution in Australasia, (Eds. Davies and Williams), pp.70-112. Australian National University Press, Canberra.

Bowler, J.M., and Wasson, R.J., 1984. Glacial age environments of inland Australia, in Late Cainozoic Palaeoclimates of the Southern Hemisphere (Ed. Vogel), pp.183-208. A.A. Balkema, Rotterdam.

Butler, B.E., 1958. Depositional Systems of the Riverine Plain of South-Eastern Australia in Relation to Soils. Soil Publication No. 10, Melbourne, Commonwealth Scientific and Industrial Research Organisation, 35 p.

Butler, B.E., Blackburn, G., Bowler, J.M., Lawrence, C.R., Newell, J.W., and Pels, S., 1973. A Geomorphic Map of the Riverine Plain of South-Eastern Australia, Australian National University Press, Canberra, 39 p.

Fairbridge, R.W., 1962. New radiocarbon dates of Nile sediments. Nature, 196, 108-110.

Fairbridge, R.W., 1963. Nile sedimentation above Wadi Halfa during the last 20,000 years. Kush, XI, 96-107.

Jackson, R.G., 1981. Sedimentology of muddy fine-grained channel deposits in meandering streams of the American middle west. Journal of Sedimentary Petrology, 51, 1169-1192.

Lockyer, N., and Lockyer, W.J.S., 1904. Behaviour of the short-period atmospheric pressure variation over the earth's surface. Proceedings of the Royal Society of London Series A, 73, 457-470.

Riley, S.J., and Taylor, G., 1978. The geomorphology of the upper Darling River system with special reference to the present fluvial system. Proceedings of the Royal Society of Victoria, 90, 89-102.

Rust, B.R., 1981. Sedimentation in an arid-zone anastomosing fluvial system: Cooper Creek, central Australia. Journal of Sedimentary Petrology, 51, 745-755.

Taylor, G., 1978. A brief Cainozoic history of the upper Darling basin. Proceedings of the Royal Society of Victoria, 90, 53-59.

Veevers, J.J., and Rundle, A.S., 1979. Channel country fluvial sands and associated facies of central-eastern Australia: modern analogues of Mesozoic desert sands of South America. Palaeogeography, Palaeoclimatology, Palaeoecology, 26, 1-16.

Walker, G.T., 1924. Correlation in seasonal variations of weather. IX. A further study of world weather. Memoirs, India Meteorological Department, 24, 275-332.
Williams, M.A.J., and Adamson, D.A., 1973. The physiography of the central Sudan. The Geographical Journal, 139, 498-508.
Williams, M.A.J., and Adamson, D.A., 1980. Late Quaternary depositional history of the Blue and White Nile rivers in central Sudan, in The Sahara and the Nile (Eds. Williams and Faure), pp.281-304. A.A. Balkema, Rotterdam.
Williams, M.A.J., Adamson, D.A., and Abdulla, H.H., 1980a. Landforms and soils of the Gezira: a Quaternary legacy of the Blue and White Nile rivers, in A Land Between Two Niles (Eds. Williams and Adamson), pp.111-142. A.A. Balkema, Rotterdam.
Williams, M.A.J., Adamson, D.A., Williams, F.M., Morton, W.H., and Parry, D.E., 1980b. Jebel Marra volcano: a link between the Nile valley, the Sahara, and central Africa, in The Sahara and the Nile (Eds. Williams and Faure), pp.305-337. A.A. Balkema, Rotterdam.
Williams, M.A.J., and Royce, K., 1982. Quaternary geology of the middle Son valley, north central India: implications for prehistoric archaeology. Palaeogeography, Palaeoclimatology, Palaeoecology, 38, 139-162.
Williams, M.A.J., and Clarke, M.F., 1984. Late Quaternary environments in north-central India. Nature, 308, 633-635.
Williams, M.A.J., Adamson, D.A., and Baxter, J.T., in press. Late Quaternary environments in the Nile and Darling basins. Australian Geographical Studies.
Woodyer, K.D., 1978. Sediment regime of the Darling river. Proceedings of the Royal Society of Victoria, 90, 139-147.
Wright, P.B., 1975. An index of the Southern Oscillation, Report No. 4, Climatic Research Unit, University of East Anglia, Norwich.

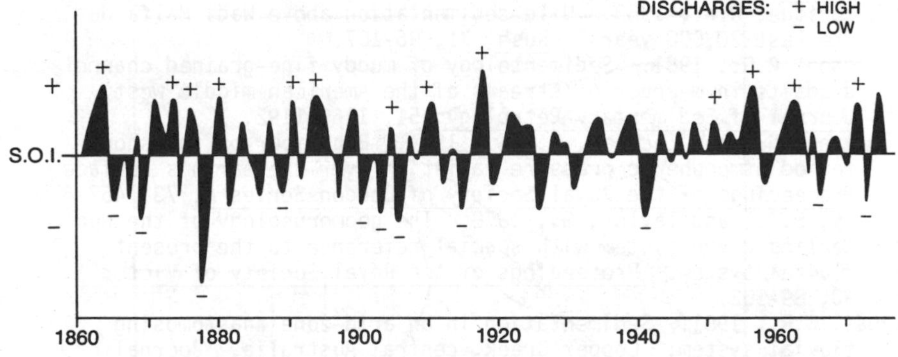

Fig. 1. Fluctuations of the Southern Oscillation Index (SOI) of Wright (1975) matched against years when discharges of the Nile, Darling and Krishna Rivers were all either high (+) or low (-). Because of the absence of discharge data for the Krishna before 1900 rainfall data is used to extend the record. Detailed data sources from Williams and Adamson (in press).

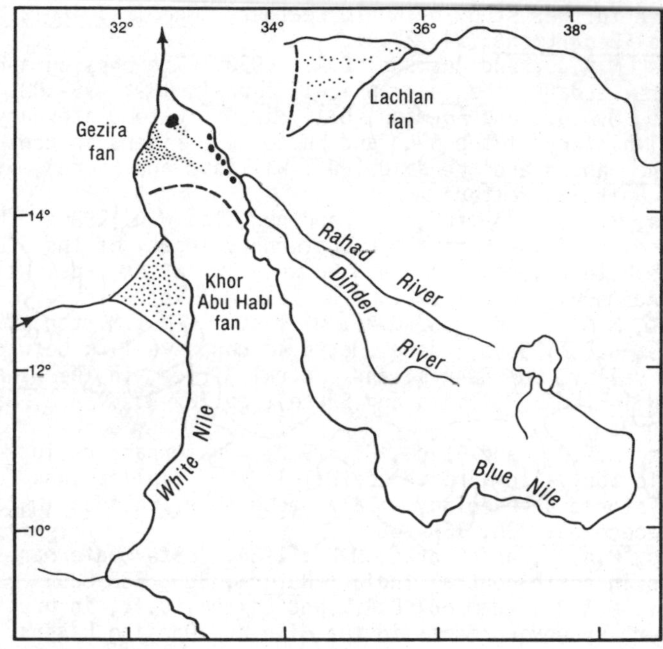

Fig. 2. Gezira (Blue Nile) and Lachlan alluvial fans compared. Stippling represents sandy deposits at the surface of the fans. Black areas represent volcanic ash.

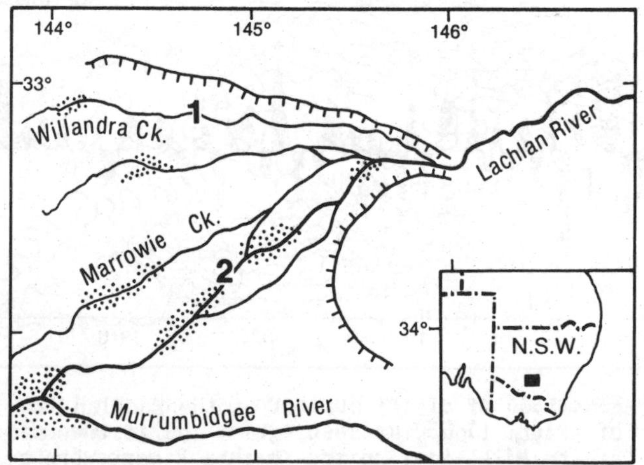

Fig. 3. Lachlan fan. Stippling represents areas of dense reticulate flood channels and swamps. Numbered sites shown in detail on Figure 4.

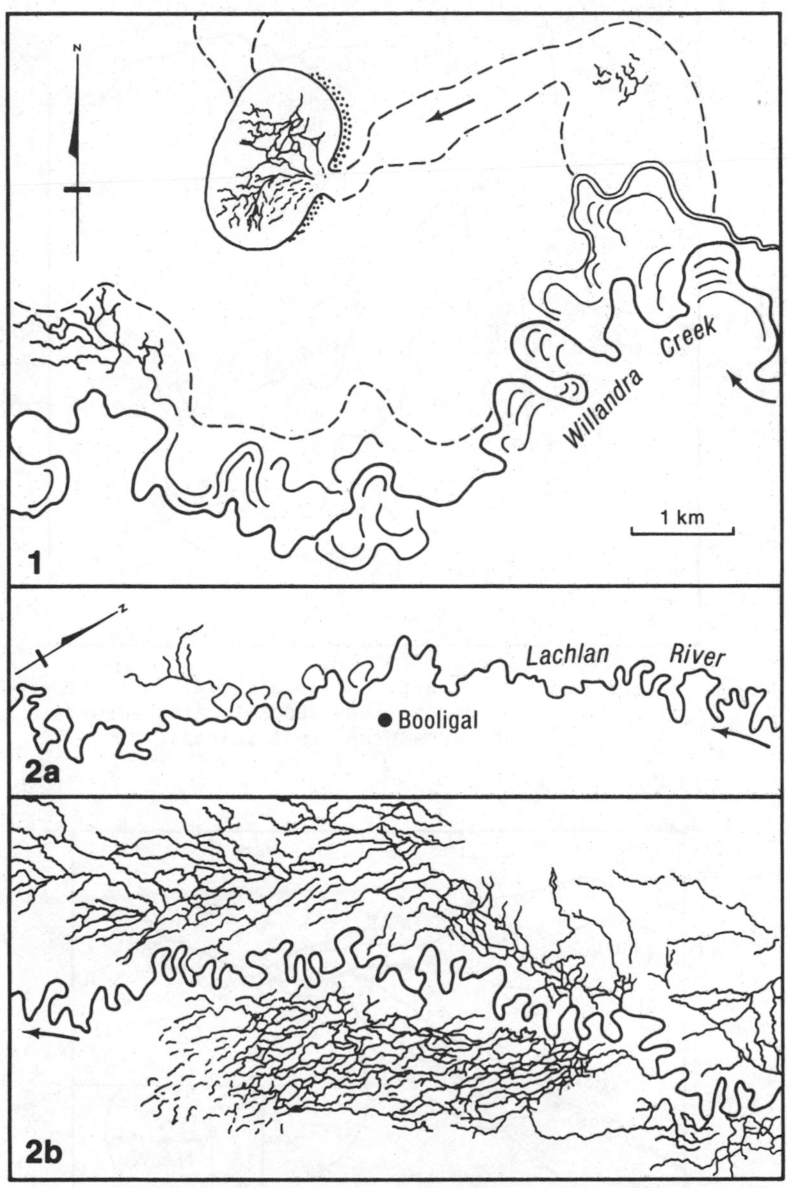

Fig. 4. Morphology of distributaries on the Lachlan fan.
4.1 (site 1 on Fig. 3). Palaeo and active features flanking Willandra Creek. Dashed line is limit of frequent flooding. Stippling is lunette.
4.2a and b (site 2 on Fig. 3). The Lachlan River near Booligal. Segment 2b is about 5 km downstream of 2a.

International Geomorphology 1986 Part II
Edited by V. Gardiner

COLLUVIAL DEPOSITS AND CLIMATIC FLUCTUATIONS IN THE PRETORIA VICINITY, SOUTH AFRICA

J. Reynhardt

Department of Geography, Vista University, Mamelodi Campus, Private Bag X03, Mamelodi 0100, South Africa.

ABSTRACT

Using field work and laboratory analysis it was found that three separate colluvial layers could be distinguished in the deposits of the Pretoria area. Two gravel layers, Colluvium I and Colluvium II, could be distinguished on the northern slope of the Daspoortrand. It was possible to subdivide Colluvium II into three phases. The lowest continues as a gravel layer from the knickpoint along the pediment on the pre-colluvium landscape. The upper two phases of Colluvium II change gradually into a thick sediment deposit on the lower part of the pediment. Colluvium I occurs only in the knickpoint of the Daspoortrand and a thick sand layer, Colluvium III, was found on the pediment of the ridge. It was also possible to subdivide this layer into three phases. Further analyses reveal that the deposits were the result of fluvial action occurring under varying energy conditions of the Quaternary Period. Snow and frost action may also have played a part in the weathering and erosion of the material.

The phases of Colluvium III may be tentatively associated with the wetter periods of the Holocene and Colluvium II may be tentatively correlated with the last pluvial period, which corresponds to the Würm Glacial period of Europe. Colluvium I may be associated with the penultimate pluvial which is regarded as contemporaneous with the Riss Glacial period.

INTRODUCTION

The longitudinal valley west of Pretoria upon which this research was based occurs on the Transvaal Sequence. Of the three units in the Transvaal Sequence only the Pretoria Group is encountered in the research area. Two east-west trending quartzite ridges, the Magaliesberg and Daspoortrand, form the northern and southern physiographic boundaries of the longitudinal valley, which occurs on shale. On the eastern side of the area the Moddersppruit drains from west to east into the Apies River; on the western side the Swartspruit drains from east to west into the Crocodile River (Fig.1). The study area is situated on the semiarid landscape of the Highveld surrounding Pretoria.

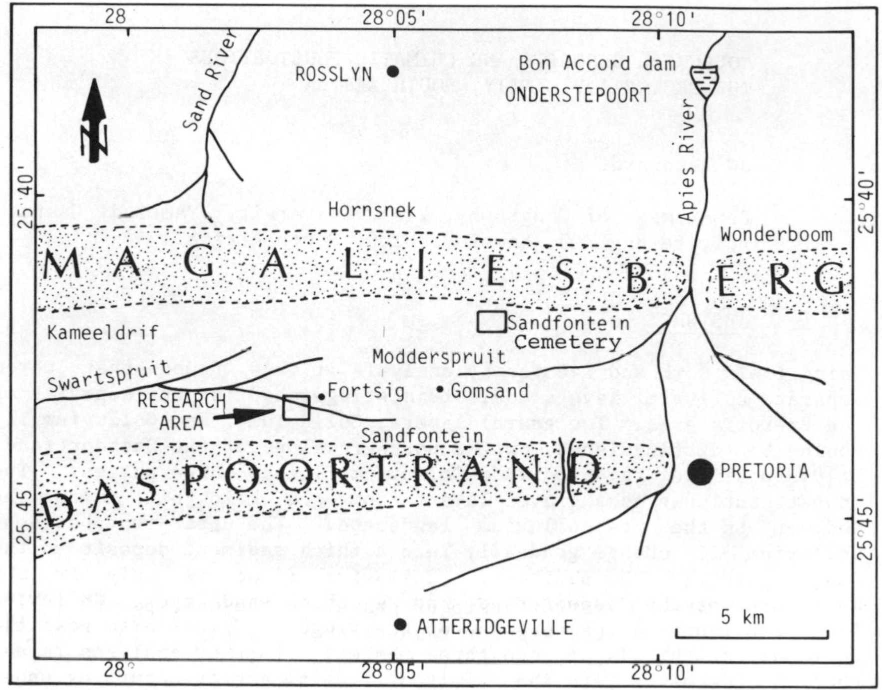

Fig.1. Locality of study area.

The colluvial material in the study area can be classified into two groups according to granular composition: deposits consisting mainly of gravel (particles >2mm) and those consisting mainly of fine sediment (particles <2mm). Apart from a narrow strip of gravel (Plate 1) in the knickpoint along the whole length of the Daspoortrand, fine sediment occurs upon the pediment from the knickpoint up to the southern tributary of the Swartspruit.

THICKNESS OF DEBRIS DEPOSITS

The thickness of the gravel deposits in the knickpoint of the northern slope of the Daspoortrand could be measured only in one stripped area at the northern entrance to the Daspoortrand tunnel (Reynhardt, 1978). Fine sediment deposits in the study area were examined by means of a sediment auger (Thompson type) which can drill to a depth of 9.2 metres. This special auger was essential since excavations in the study area were not always deep enough to reveal the underlying rock.

According to Reynhardt (1978), the gravel deposits in the Daspoortrand knickpoint attain a thickness of about 6 metres and pinch out on the northern side of the rectilinear slope to expose

Plate 1. Narrow strip of gravel in the knickpoint of the Daspoortrand northern slope.

Plate 2. Iron aggregates from the infilled valley of the Swartspruit.

the quartzite of the Daspoortrand. When analysing the thickness of the fine sediments at different points along the pediment one finds that the depth of the sediments, measured down to the weathered shale, varies considerably from one point to the next. There is a sudden increase in depth on the middle of the pediment from about 4.8 metres to 6.5 metres and on the upper lower part from about 4.3 metres to a valley of 10.5 metres. This infilled valley, in which the Swartspruit once flowed, continues for 80 metres whereafter the depth decreases suddenly to 3.0 metres. The fine sediment deposits become thinner for the next 150 metres until at the bottom of the pediment bedrock outcrops (Fig.2).

COLLUVIAL LAYERS AND DEPOSITIONAL PHASES

Field observations of gravel deposits. According to Reynhardt (1978) the gravel deposits in the northern knickpoint of the Daspoortrand show four distinct layers of various fragment sizes (Fig.2). The first layer is about 1.0 metre thick and pinches out against the dipping Daspoort quartzite. This layer consists of a loamy yellowish red (5YR 4/6) sand with 30.0% volume of subrounded (c.31% rho (Folk,1965)), poorly orientated quartzite fragments with a mean long axis length of 85mm. The transition to the next layer is gradual.

Layer two is about 2.2 metres thick and also pinches out against the rectilinear slope. The sediment matrix is a soft coarse sandy loam with a yellowish red colour (5YR 4/6). Fresh, subrounded (c.30% rho) poorly orientated quartzite fragments with a mean long axis length of 117mm make up 45% of the volume of the deposits. The transition to the next layer is moderately clear. Layer three is appreciably thinner (0.4 metre) than layer two and consists of a yellowish red (5YR 4/6) soft, coarse, sandy clay loam with fresh, subangular (c.24% rho), poorly orientated quartzite fragments. The mean long axis length of the fragments is 152mm. Towards the bottom of the layer infrequent hard iron and manganese concretions occur. The transition to the underlying layer is abrupt. Layer four, just above the weathered shale, is thin (0.8m) and contains a moderate number (c.15% volume) of well rounded (c.55% rho), poorly orientated quartzite fragments with a mean long axis length of 85mm. The yellowish red (5YR 4/6) matrix consists of soft apedal clay. Infrequent, hard iron and manganese concretions occur in the layer and there is an abrupt transition to the weathered shale bedrock.

Field observations of sand deposits. On the upper part of the pediment a reddish brown (5YR 4/4) medium, sandy loam to sandy clay soft loam fine sediment exists containing fine particles of partly weathered shale. Below this reddish brown zone of about 0.5 metre, a dark red (10R 3/6) medium sand to sandy loam forms an outcrop on the middle of the pediment. This soft dark red zone of about 6.0 metres contains infrequent, subangular quartzitic and quartz gravel. In the lower part of this zone there is a moderate amount of yellow mottling which increases lower down. At the bottom of the zone there are infrequent, smooth, round iron and manganese concretions (Fig.3).

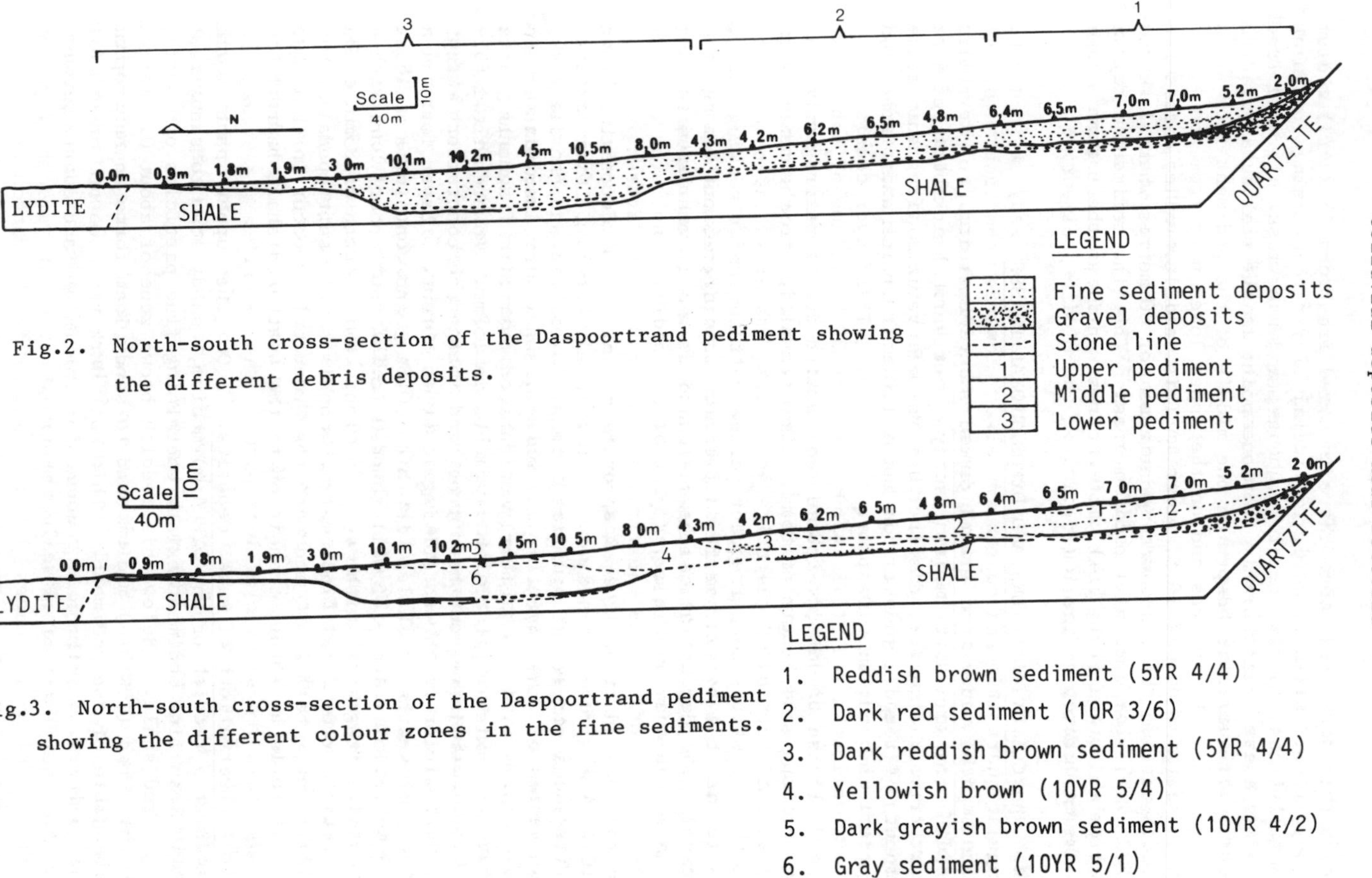

Fig.2. North-south cross-section of the Daspoortrand pediment showing the different debris deposits.

Fig.3. North-south cross-section of the Daspoortrand pediment showing the different colour zones in the fine sediments.

Below this dark red zone there is a reddish brown (5YR 4/3) medium sandy loam, slightly hard, colluvial layer containing frequent quartzitic and quartz gravel in amongst a dense mass of round iron and manganese concretions. This gravel consists of subangular to subrounded quartzite fragments.

Three sediment colour zones occur one on top of the other across the middle part of the Daspoortrand pediment. The top one is dark red (10R 3/6) followed by a reddish brown (5YR 4/4), and then by a yellowish brown (10YR 5/4) gravelly zone. The reddish brown zone forms an outcrop on the lower part of the middle pediment.

The upper dark red loamy medium sand contains very infrequent, angular quartzitic and quartz gravel. Towards the bottom of the zone smooth, round iron and manganese concretions occur. The middle reddish brown soft medium sandy loam contains slightly more quartzitic and quartz gravel than the overlying zone. A moderate amount of round, smooth iron and manganese concretions and iron aggregates occur in this zone.

The yellowish brown gravel base zone consists of a medium sandy loam to medium sandy clay loam, slightly hard, fine sediment with infrequent subrounded quartzite and quartz gravel at the base of it. At the top of the zone frequent yellow mottling and streaking occur, while at the bottom one finds frequent iron aggregates among the gravel. No compaction can be discerned in the gravel deposits, as is also the case on the upper part of the pediment.

On the lower part of the Daspoortrand only a dark greyish brown (10YR 4/2) zone of sediment is found, manifesting clear textural differences across the profiles from top to bottom. Over the first two metres of the profile the sediment is a soft, medium sand to loam medium sand containing very infrequent quartzitic and quartz gravel. In this part of the profile infrequent yellowish red (5YR 5/6) mottling and iron aggregates occur (Fig.3). An abrupt transition exists between the upper dark greyish brown zone and a lower thick grey (10YR 5/1) zone. This grey sticky zone of about 8.0 metres consists of medium loam to medium sandy clay loam with a moderate amount of gravel. Three iron aggregate (Plate 2) accretions were found in the profile, the first at the top of the zone, the second 2 metres lower and the third 3 metres from the top of this zone. Brown (7.5YR 5/2) mottling occurs throughout the sandy loam to sandy clay loam part of the profile, whereas red (2.5YR 4/6) mottling is confined mainly to the middle and lower parts of the bed. Directly beneath the red mottling zone a yellowish brown (10YR 5/6) mottling occurs.

At the base of this fine sediment profile gravel laminae occur with subangular to well-rounded fragments (Plate 3). Transitions over the pediment between colour zones are gradual whereas those between the gravel and weathered shale are abrupt.

Plate 3. Well-rounded pebbles from the gravel lamina on the floor of the infilled valley of the Swartspruit.

Plate 4. Middle Stone Age implements on the ancient landscape of Colluvium II.

Determination of colluvial layers and depositional phases by statistical analysis. In order to determine the number of colluvial layers, number of depositional phases per layer and horizontal correlations between layers and depositional environments, statistical analysis (Davis, 1973) consisting of grouping analysis, discriminant analysis and principal components analysis were used. In these analyses the variables were the percentage of fine sand, mean grain size (Ø), sorting, skewness, kurtosis, percentage of heavy minerals, percentage of tourmaline, percentage of topaz, percentage of rutile and anatase, percentage of garnet, percentage of leucoxene, the ratio of zircon to epidote plus pyroxene, pH and resistance. These analyses established that the sediment on the Daspoortrand pediment consists of three colluvial layers, namely I, II and III (Fig.4). The two youngest colluvial layers, II and III, can be divided into three depositional phases. Morphometric measurements and statistical analysis established that Colluvium I is confined to the knickpoint, while Colluvium II phase I occurs all over the pediment in the form of a gravel lamina. The possibility also exists that Colluvium II phases II and III undergo a facies change from gravel to fine sediment from the Daspoortrand knickpoint across the pediment. Unlike Colluvium I and II, the depositional material of Colluvium III is confined to fine sediments on the pediment of this ridge only.

In all instances the contact between the various colluvial layers is sharp and rectilinear, giving the impression that the deposition of each colluvial layer was followed by a period of erosion. At various exposures on the Daspoortrand pediment in the vicinity of Sandfontein near Fortsig station, the sediment deposits of Colluvium II on the lower pediment have been stripped over sizeable areas. On the lower pediment this exposed surface consists mainly of hard ferricrete. The microrelief of the exposed Colluvium II surface shows that the ferricrete constituted part of an ancient landscape which had been subjected to some measure of erosion. The microrelief of this landscape is a distinctive feature with clear signs of streamlets cut down into the ferricrete bed and subsequently filled up by younger deposits where streamlets appear to have flowed around ferricrete islands.

A great number of implements and rocks from the Middle Stone Age were found on this exposed erosion surface (Plate 4). The implements were manufactured mainly from quartzite and lydite. The majority occur on Colluvium II and a smaller number in the sediment deposits of Colluvium III. It was not established whether these implements also occur in Colluvium I since this layer is not sufficiently exposed anywhere in the Swartspruit valley head. The concentration of implements on the microrelief of Colluvium II suggests that Middle Stone Age people lived on this landscape. The quartzite implements appeared quite fresh, but the lydite ones were weathered.

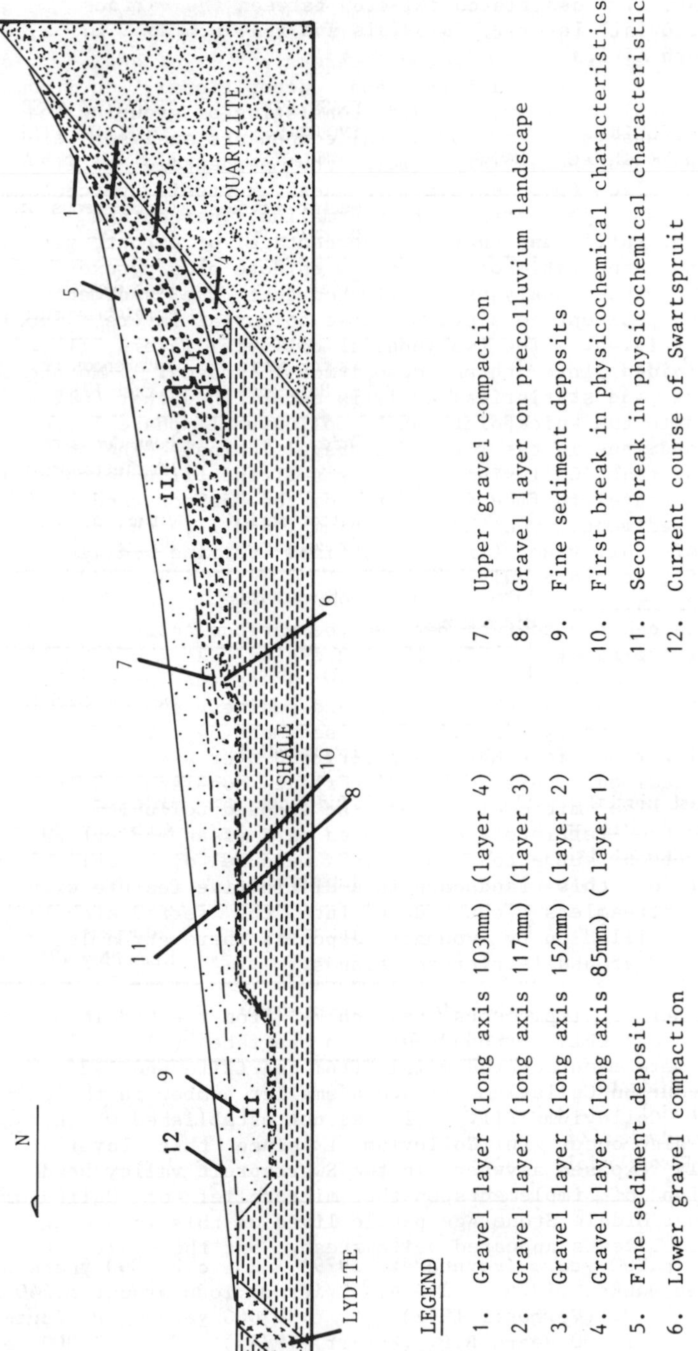

Fig.4. Schematic representation of the succession and continuation of the different debris layers on the Daspoortrand pediment.

TABLE 1. Possible correlation between the various colluvial layers, pluvials and research done in South Africa.

COLLUVIAL LAYERS	ASSOCIATED WETTER PERIODS	PHASES	PHASE AND INTERPHASE-CORRELATES	TENTATIVE PHASE CORRELATES WITH RESEARCH IN S.A.
		III	Wetter period of Holocene	Carbonaceous layer Sandfontein [1]
			Drier	Peat deposits Morelettaspruit [3]
III	Holocene	II	Wetter period of Holocene	Inferences of Butzer [7]
			Drier	Peat deposits Morelettaspruit [4]
		I	Wetter period of Holocene	Inferences of Butzer [6]
		Lithological discontinuity	Interpluvial	
		III	IIIrd stage	Florisbad - wet and cold [2]
			Interstadial	
II	Last pluvial (equivalent of Würm glacial)	II	IInd stage	Calc-pans Kimberley [5]
			Interstadial	
		I	Ist stage	Calc-pans Kimberley [8]
		Lithological discontinuity	Interpluvial	
I	Penultimate pluvial (equivalent of Riss glacial)	?	?	?

LEGEND 1. c.200 years (Reynhardt, 1979) 2. c.25 000 years B.P. (Van Zinderen Bakker, 1976) 3 & 4. Drier periods around c.440 and c.5 220 years B.P. (Verhoef, 1972) 5. c.30 000 years B.P. (Butzer, 1974b) 6. c.8 000 years B.P. (Butzer, 1974a) 7. c.3 000 years B.P. (Butzer, 1974a) 8. c.115 000 years B.P. (Butzer, 1974a)

DISCUSSIONS AND CONCLUSIONS

It is noticeable that the different colour zones do not correspond with the colluvial layers in the study area. Colour zones in this area are the result of the oxidation status of iron in the sediments and recent hydromorphic conditions (Reynhardt, 1979).

One can tentatively infer that the three colluvial layers distinguished in the study area represent three climatological periods associated with wetter and colder conditions than those prevailing at present (Table 1). If so, Colluvia I and II would be associated with periods of more intense weathering and fluviatile action, and Colluvium III was deposited under less intense weathering and transportation conditions. The various phases can be associated with climatic fluctuations during the deposition of the colluvial layers. The chronostratigraphy of South Africa is still extremely sketchy and tentative. This makes it very difficult to demonstrate any correlation between the colluvial layers and phases and South African stratigraphy. Since Colluvium III is the youngest of the various layers its phases can be correlated with the wet conditions during the Holocene. Phase III of Colluvium III corresponds with the carbonaceous layer observed in the vicinity of the Sandfontein cemetery near the study area. The age of this layer is about 200 years B.P. (Reynhardt, 1979).

Phases II and I of Colluvium III may correspond respectively with the wetter periods of about 8 000 and 3 000 years B.P. which existed in Southern Africa (Butzer, 1974a). Drier periods between the phases of Colluvium III may be correlated respectively with two drier periods of about 5 220 and 440 years B.P. in the Morelettaspruit (Pretoria) (Verhoef, 1972).

Quite possibly Colluvium II can be correlated with the last Pluvial (equivalent of the Würm glacial), in which case the various phases of this layer will correspond with the stadials of the last Pluvial. Phase III of Colluvium II could possibly be correlated with the wet, cold conditions prevailing at Florisbad around 25 000 B.P. (Van Zinderen Bakker, 1976), while phases II and I of Colluvium II may be compared to wet periods around 30 000 B.P. and around 111 500 B.P. in the Kimberley area (Butzer, 1974b). It was impossible to demonstrate any correlation between Colluvium I and research done in South Africa. It could possibly be correlated with the penultimate Pluvial (equivalent of the Riss glacial). Table 1 shows possible correlations between the colluvial layers and their phases, the various pluvials and other research on South Africa.

REFERENCES

Butzer, K. W., 1974a. Reflections on the stability of Holocene environmental Zonation South Africa. South African Archaelogical Society, Goodwin Series No.2. Progress in later Cenozoic studies in South Africa, 37-38.

Butzer, K. W., 1974b. Geo-Archaeological Interpretation of Acheulian Calc-Pan sites at Doornlaagte and Rooidam (Kimberley, South Africa). Journal of Archaeological Science, 1, 1-25.

Davis, J. C., 1973. Statistics and data analysis in Geology. Wiley, New York.

Folk, R., 1965. Petrology of sedimentary rocks. Hemphills, Austin, Texas.

Reynhardt, J. H., 1978. 'n Ondersoek van die puinafsettings in die noordelike knakpunt van die Daspoortrand naby Pretoria. South African Geographer, 6, 157-165.

Reynhardt, J. H., 1979. Die kolluviale afsettings in die Moot, wes van Pretoria en die verband daarvan met hangontwikkeling. Unpublished Ph.D. Thesis, University of South Africa, Pretoria.

Van Zinderen Bakker, E. M., 1976. The evolution of late Quaternary paleoclimates of South Africa. Palaeoecology of Africa, 9, 160-202.

Verhoef, P., 1972. A stream deposit with interstratified peat, at Pretoria. Palaeoecology of Africa, 6, 147-148.

International Geomorphology 1986 Part II
Edited by V. Gardiner

PLEISTOCENE EVIDENCE FROM THE EASTERN CAPE SOUTH AFRICA : THE AMATOLA SCREES TONGUES

M. E. Marker

University of Fort Hare,
Private Bag X1314, Alice, Ciskei, South Africa.

ABSTRACT

Evidence for a Pleistocene southern African snow-line gradient is scanty (Hastenrath, 1972). Various Pleistocene periglacial forms have been identified in Lesotho and along the Natal-Transkei Escarpment between latitudes 28°00'S and 30°00'S and above 2800m altitude. Large semi-circular hollows are associated with remnants of the 3000m surface. Within some of the hollows, rock tongue deposits interdigitated with organic soils are preserved. At lower altitudes true nivation cirques and incipient protalus ramparts have been identified. Along the southern Cape coastal zone, 34°00'S, cave habitation sequences include eboulis sec attributed to rock spalling caused by frost (Butzer, 1973). Recent work along the Eastern Cape Escarpment zone, latitude 32°30'S, has identified scree boulder tongues composed of sub angular dolerite boulders. These scree tongues head at altitudes exceeding 1600m and extend downslope over an altitude range of up to 130m. The screes are localised on south facing slopes but not all potential sites carry screes. Stream gullies have dissected the original screes into distinct parts so the scree tongues antedate the formation of the gullies. They also overlie remnants of deeply weathered soils. The existence of these scree tongues is attributed to Pleistocene periglacial conditions. Their altitudes indicate a possible snowline gradient in agreement with the postulated 9°C temperature depression (Butzer, 1973).

INTRODUCTION

An acceptable meridional Pleistocene snowline gradient for Africa remains to be constructed. It may be attempted either via current glacial and snowline levels or from geomorphic evidence for former nival action. A combination of these approaches has been applied to South America and Australasia (Hastenrath, 1971; 1972; Peterson, 1968) and has highlighted the difficulties facing the construction in Africa. Southern Africa extends to only 34°30'S, and reaches a maximum elevation of 3482m in Thabana Ntlenyana, the highest point above the Lesotho-Natal Great Escarpment. Southern Africa is an arid subcontinent with no permanent snowline. Aridity was possibly also accentuated during Pleistocene cold phases, inhibiting the development of icefields and glaciers. Marked local variations in precipitation accentuate the difficulty of successful projection.

Over the past 20 years or so, there have been numerous reports of landforms attributed to former periglacial activity (Alexandre, 1962; Dyer and Marker, 1979; Ellenberger, 1960; Harper, 1969; Hastenrath and Wilkinson, 1972; Marker and Whittington, 1971; Sparrow, 1964; 1967a; 1967b; 1971). Most of these studies have concentrated on the high country of Lesotho and its adjacent areas along the Great Escarpment where maximum elevations are reached. In a summary of the evidence, Butzer (1973) suggested that the landforms can be separated into three sets: large semi-circular hollows or cirques above 2900m which have a preferred northern aspect, secondly oversteepened south-facing slopes and isolated nivation niches with associated 'head' and talus deposits, and thirdly frost effects and ice planing above 3300m. More recently frost spalls have been recognised within the

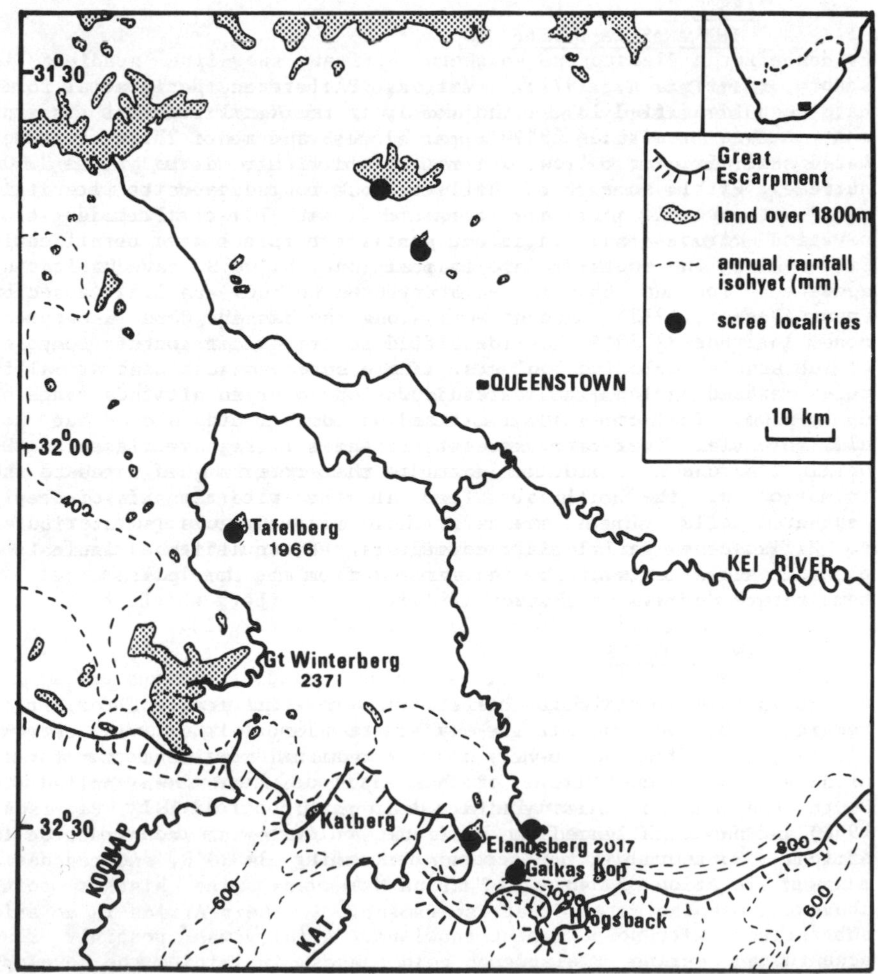

Fig.1. Scree tongue distribution along the eastern Great Escarpment.

southern Cape archaeological record from sites close to sea level and attributed to Pleistocene frost action (Butzer, 1972; 1984). As Butzer (1973) points out, more than one category of landforms is involved and probably more than one period of formation. The various phenomena are not geographically juxtaposed nor latitudinally continuous nor do all occur under similar moisture regimes.

As another contribution to the evidence for former periglacial activity in southern Africa, attention has been focussed on the eastern Cape Great Escarpment between latitudes 31°00'S and 32°30'S which there reaches altitudes exceeding 1800m. The locality is geographically intermediate between the Lesotho-Natal high country and the southern Cape coastal belt and thus its investigation also serves a bridging function.

THE AMATOLA SCREES

The Eastern Cape Great Escarpment, known there as the Amatola Mountains, rises steeply and continuously to mountain residuals exceeding 1800m altitude (Fig.1). The upper slopes of some of these mountains support unvegetated scree tongues which are locally conspicuous elements of the landscape. The screes are believed to be relict Pleistocene cold phase phenomena and it was felt that detailed study might elucidate their origin and contribute to a better understanding of temperature depression in the past.

The scree tongue distribution was mapped from air photos on a scale of approximately 1:15,000 onto 1:50,000 topographic map sheets to elucidate the salient variables of distribution and dimensions. The scree tongues are markedly localised. They are very frequent on Elandsberg which has a plateau summit rising to 2017m, on Gaika's Kop reaching 1962m, and on a mesa behind Gaika's Kop reaching a mere 1800m altitude. Isolated examples also occur on Tafelberg, 1966m, some 60km to the north west and in the vicinity of Queenstown (Fig.1). The screes are related to residuals rising to over 1800m altitude. However not all residuals of the requisite altitude support scree tongues. They are absent from the Hogsback and Katberg, both narrow ridges, but also from Groot Winterberg which has extensive plateaux about 2000m and rises to 2371m (Fig.1).

For all the Amatola screes, variables including maximum and minimum altitude, length of tongue and aspect were extracted. The greatest concentration of scree tongues exists on Elandsberg, the next highest is on Gaika's Kop. Elsewhere only occasional scree tongues occur (Table 1). The average maximum altitude where the screes head is above 1700m and is clearly at least in part controlled by the altitude of the cliff base from which the boulders were derived. Minimum altitude is variable, no screes descending below 1550m altitude. Length of scree tongue is a function of volume of scree and slope steepness.

Most scree tongues are small, 70% having less than 100m fall. Tongues extending over 150m are confined to the slopes of Elandsberg or to Tafelberg. Scree tongues are not only spatially and

TABLE 1. Amatola scree tongue variables.

Place	No	Mt. Alt (m)	Av. Rainfall (mm)	Dom Aspect	Altitude (m)				Fall (m)	
					Max $\bar{X}$	Min $\bar{X}$	Max	Min	$\bar{X}$	Max
Tafelberg	3	1966	480	SSW	1800	1660	1820	1660	140	180
Elandberg	26	2017	700	S	1757	1673	1859	1600	77	229
Gaika's Kop	12	1810	700	SSW	1771	1720	1810	1676	51	92
Rockford	6	1707	600	WSW	1684	1582	1707	1554	102	153

TABLE 2. Amatola scree tongues : relationship between number, size and aspect.

FALL (m)	E	ENE	NE	NNE	N	NNW	NW	W	WSW	SW	SSW	S	SSE	SE	ESE	NO	%
0-49					2			2	1	1	2	8	4	2		22	46.8
40-99						2			1	1	3	2	2			11	23.4
100-149									1	1	4	2				8	17.0
150-199									1		2	1				4	8.5
200-249												2				2	4.3
Number					2	2		2	4	3	11	11	6	2		47	100
%					4.3	4.3		4.3	8.5	6.4	23.4	31.9	12.8	4.3			100

altitudinally localised but are also strongly constrained by aspect (Fig.2, Table 2). Over 75% of all tongues have a preferred southern aspect. All large scree tongues, exceeding 100m fall, face south. This marked preference indicates an origin related to frost and frost cycles.

The scree tongues are formed of Karoo System dolerite boulders derived from the cliffs immediately above them and little or no fines are trapped between the boulders. The screes are essentially stable at the present time. Most of the boulders are encrusted with lichens and only immediately above semi-permanent subscree streams are the boulders loose and tilted as though undermined by the present streams. Unstable scree occupies a very minor percentage of the total scree area.

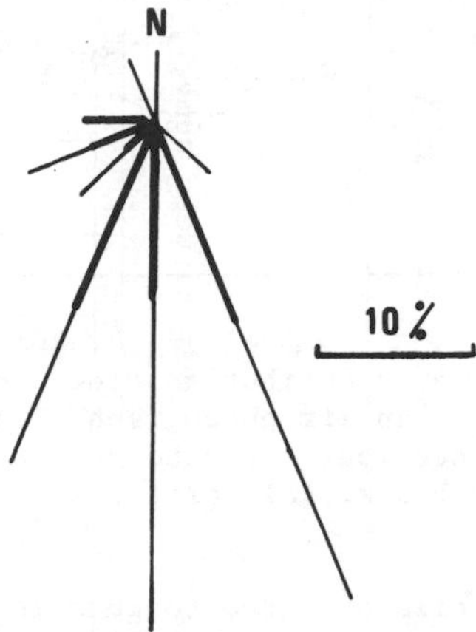

Fig.2. Scree tongue orientation (thick lines represent tongues exceeding 100m fall, thin lines represent all smaller tongues).

The scree tongues usually have a surrounding fringe of woody vegetation, brambles and creepers that may form an impenetrable thicket. The width of the thicket is variable. On occasion median lines of woody vegetation also occur.

The south-facing slopes of Elandsberg exemplify many of the characteristic features of these scree tongues (Fig.3). The density of scree tongues and scree patches is high there. The screes rest on steep grassed bedrock slopes with a veneer of rounded boulders. However sections exposed in gullies indicate former deep weathering preceding emplacement of the screes. Broad grassed swales scar the hillside and appear to have been incised into this weathered material. Some now separate parts of formerly larger scree tongues. Individual scree tongues never abut directly onto the cliff above them. They commence at varying distances from the cliff foot ranging from 30m minimum to 200m maximum. This cliff foot "flat" slopes at an angle of 10 to 12 degrees.

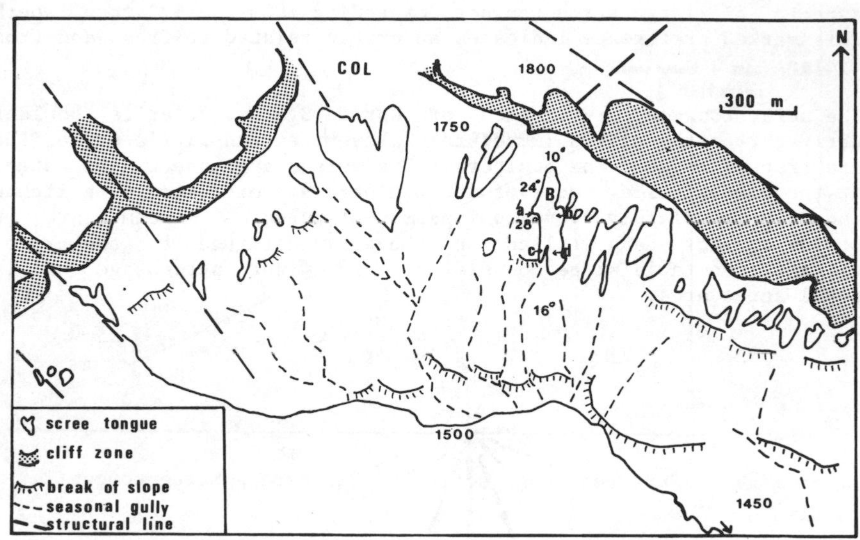

Fig.3. Scree tongue distribution on the southern slopes of Elandsberg, drawn from an air photograph. B = surveyed scree, a-b central cross section, c-d tongue sections above and below protalus block, (all altitudes in metres).

As the frequency and size of scree tongues is so much greater on the southern slopes of Elandsberg than elsewhere, one Elandsberg scree tongue was selected for detailed study. The biggest but most disturbed scree tongue is heart-shaped so another less disturbed tongue (B) with 128m fall in altitude, having the typical vegetation fringe, was chosen for analysis (Fig.3). The long profile of this scree tongue was surveyed (Fig.4). It heads some 90m from the subvertical well-jointed dolerite headwall, at a break of slope between the 10 to 12 degree cliff foot flat and the mountain slope proper. The boundary is abrupt, the scree starting as a clear line. The tongue falls 128m from an altitude of about 1770m. The gradient is only 18 degrees for the first 25m but steepens through 24 degrees in the middle section to 28 degrees on the lower tongue (Fig.4).

Cross sections were measured for the central portion and some 40m above the terminal snout, both above and below a large block with 4m faces (64 m^3) which appeared to have affected scree flow patterns resulting in streaming (Fig.5). These cross sections show that the screes are raised above the grass slope on which they rest by at least 1 to 2m but elsewhere they may stand up to 3m proud. Overall the tongue width varies little. It has a 40m width at the start, expands to a maximum of 75m in the central portion but shrinks to 37m immediately before it terminates.

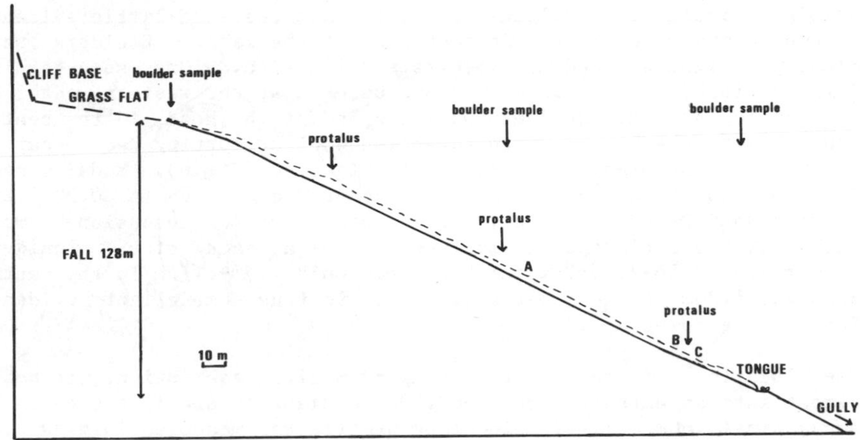

Fig.4. Surveyed long section of Elandsberg scree tongue B.

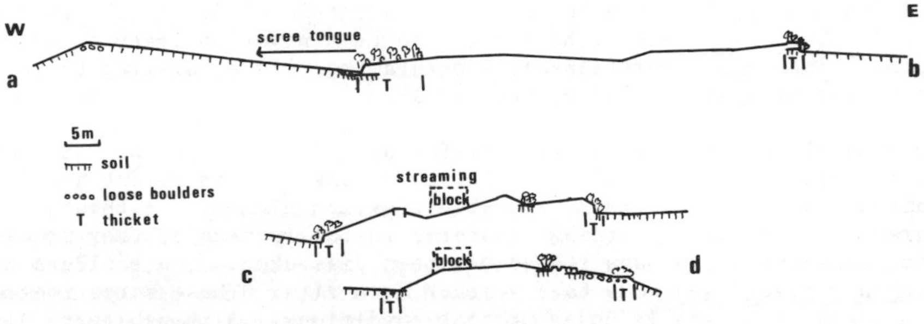

Fig.5. Surveyed cross sections of Elandsberg scree tongue. B (a-b central section on 28° slope, c-d tongue sections above and below protalus zone), (see Fig.3 for locations).

The tongue surface, even ignoring local irregularities due to variation in boulder size, is uneven. Some hollows are clearly caused by streaming effects as exemplified by the pronounced hollows surveyed above and downslope of the large boulder (Fig.5). Elsewhere very large blocks overlie or impede the scree. These occur at specific locations in the profile indicated as protalus blocks on the long profile (Fig.6). There are signs that the precise location of the scree tongues on the slope is related to initial slope irregularities. The tongues appear to be preferentially situated in former gullies and to reach their maximum thickness overlying former hollows (Fig.5). Depths exceeding 3.50m were measured without encountering the basal slope.

Throughout its length, the tongue is composed of angular to subangular dolerite boulders. These are settled in position and dominantly stable. Almost all are lichen encrusted and there is little evidence of contemporary movement. In fact some of the larger boulders have clearly weathered and disintegrated in situ. One such tabular jointed block, 3.5m x 1.75m, was encountered at the western edge of section c and d. Boulders were sampled at the head, in the centre and close to the lower end to establish whether sorting had occurred downslope. Abnormal size blocks were ignored (Fig.6). Modal length is 0.50-0.75m at the head and in the centre, declining to 0.50-0.25m at the end. Breadth however, shows a modal decrease downslope. More marked is the difference in maximum lengths at each site. Boulders can reach 2.50-2.75m at the head, but only 1.50-1.75m in the centre and 1.00-1.25m at the lower end. There is thus some slight evidence for sorting downslope.

The location of this scree tongue merits description. It heads almost into an amphitheatre some 180m in diameter and is backed by a subvertical 60m cliff. The long profile was measured in midwinter 3 days after a snowfall. The upper third of the tongue remained in shadow even at midday. Snow depth at the base of the mountain slope at 1400m altitude was 5cm, at the tongue snout it was 10cm and at the head about 25cm. This amphitheatre has features akin to the snow nivation hollows described from Wales (Watson, 1966) and from the Caledon valley, latitude 28°30'S at an altitude of 2012-2115m (Nicol, 1973). The Elandsberg niche appears to be a nivation feature accentuated into an oversteepened southfacing slope, similar to those described by Sparrow (1967b) from Natal.

The Amatola scree tongues are preferentially located on southern steep slopes below cliffs. In some cases the incipient amphitheatres, in which they head, resemble nivation niches. Although the screes are relatively fresh landforms in the landscape, they predate some dissection and have themselves been dissected. The boulders are lichen covered and some have weathered in situ. These scree tongues are essentially stable under present conditions. Although texturally they resemble the isolated rock rivers of eastern Lesotho, they show no evidence of derivation as a lag by fines removal. They are also strongly localised. Altitude and southern aspects are clearly major controls. Further it is likely that not all dolerite sills, that form the cliffs, are equally susceptible to frost action. The massive caprock of Groot Winterberg for instance has resisted disintegration into scree.

When the scree distribution is plotted in relation to present day mean annual precipitation, scree frequency seems positively associated with moisture regimes (Fig.1). Mist and cloud incidence is undoubtedly higher in the east along the Amatola escarpment in the region where screes are more frequent. Up to 20% extra moisture is precipitated from mist and cloud along the Escarpment although 2km inland this drops to only 10%.

The Amatola nivation niches associated with the larger scree tongues are very similar to those described from the Caledon Valley, latitude

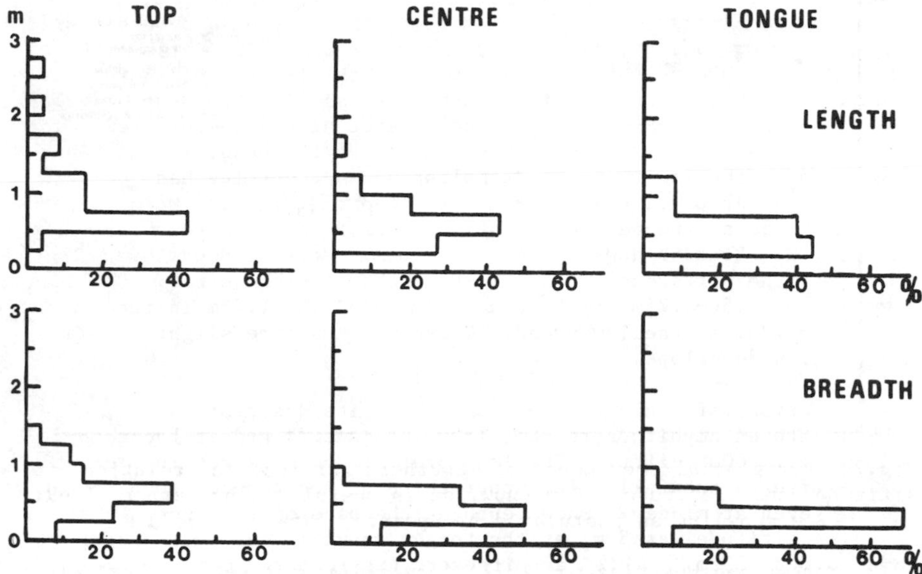

Fig.6. Boulder sample histograms

28°30'S (Nicol, 1973). Since protalus blocks emplaced by sliding across snow patches are features of the Caledon nivation niches, it seems probable that the large boulders associated with specific levels of the Elandsberg scree, are also protalus ramparts. Deposition downslope of a snow patch banked against a shaded cliff would also explain the absence of scree from the cliff foot flat.

Together the evidence is sufficient to accept that the Amatola scree tongues are Pleistocene frost derived landforms. Situated between 32°00'S and 32°30'S and heading at altitude from 1700m and 1850m, they provide a further point on the meridional curve.

GENERAL DISCUSSION

Scree tongues, such as those described, are not prominent elements of the Lesotho high country. Boulder streams are occasionally encountered in valleys above 2600m altitude and mixed gravel and boulder tongues emanate from some nivation cirques (Marker and Whittington, 1971). Field reconnaissance of the intermediate area lying between 31°00'S and 32°00'S shows that scree tongues are rare and localised there too. Clearly altitude, southerly aspect and moisture regime are not the sole controls for scree development.

When scree tongue altitudes are plotted in relation to other periglacial evidence with a view to the construction of a meridional curve, certain aspects become clearer (Fig.7). The supposition put forward by Butzer (1973) that various distinct sets of phenomena are involved is supported. The large, preferentially north-facing nivation cirques of Lesotho described by Marker and others (Dyer and Marker,

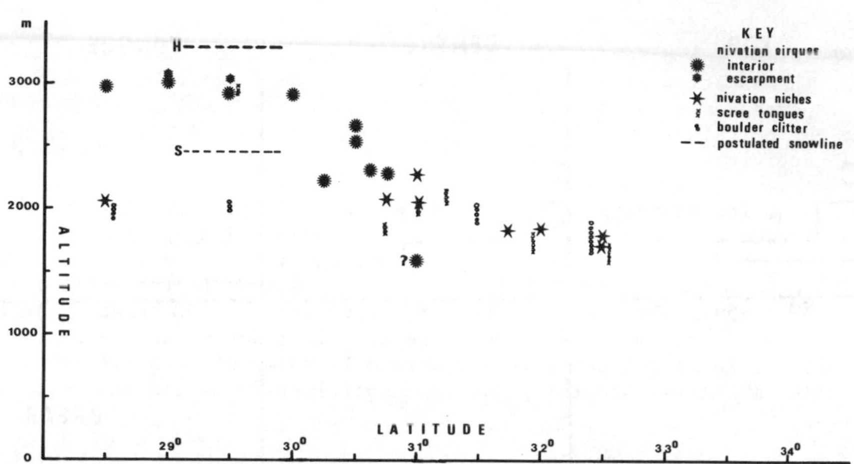

Fig.7. Periglacial phenomena of Southern Africa in relation to altitude and latitude (H = snowline as postulated by Harper, 1969; S = snowline as postulated by Sparrow, 1964 and 1967).

1979; Marker and Whittington, 1971) form a distinct set. They also extend further south as suggested by Sparrow (1967b) (Fig.7). The south-facing but morphologically similar landforms of the high country north of Queenstown, may belong to a different set. The floor altitudes of these cirques decrease with higher latitude and demonstrate a meridional gradient. It also seems likely that the first and most intense phase of periglacial activity associated by Harper (1969) with a snowline at 3300m is contemporaneous with these nivation cirques. This association would add credence to the belief that the nivation cirques are actually the product of arid glaciation rather than solely the product of nivation. The truncation of many of these landforms by recession of the Great Escarpment on the Natal-Lesotho border (Dyer and Marker, 1979) and by fluvial rejuvenation along the southern highlands bordering Lesotho, confirms their antiquity.

The boulder clitter of summits and slopes above 1500m to 1600m together with the scree tongues form a further set. The scree tongues are altitudinally as well as spatially related to the localised nivation niches with their associated head deposits and to oversteepened south-facing slopes. This set occurs at lower altitude and the data available at present demonstrate no perceptible meridional gradient (Fig.7).

Nivation cirques and nivation niches are quite different landforms. The former are large, semi-circular hollows that are predominantly north-facing. The latter are small but steep notches not exceeding 200m diameter, restricted to south-facing slopes. Except where most strongly developed they are difficult to distinguish unequivocally from the more planar oversteepened south-facing slopes. Although it appears that all large scree tongues emanate from such nivation niches, the nivation niches of the Caledon valley are the source of unsorted head deposits rather than of scree. Nivation niches and, by association, the scree tongues are younger elements of the landscape than the nivation cirques. The position of the scree tongues within the landscape also suggests that they are relatively young elements. They may pertain to Harper's (1969) younger and less severe period of nival activity.

The change in landscape topography to smooth rolling terrain occurs at altitudes of 1700m to 1800m at 31°00'S, above 1600m at 32°00'S and above 1500m at 32°30'S again along a meridional gradient. Butzer (1973) in confirming Sparrow's (1969b) observations in the Witteberg district immediately south of Lesotho, attributed the change to a change in landform process and suggested nival action as the cause. However the rolling country is predominantly rock-cut with little or no soil mantle so smoothing due to a solifluction mantle is not the cause.

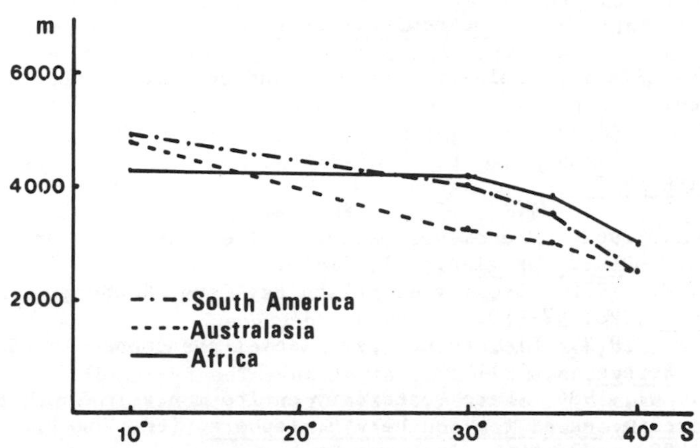

Fig.8. Annual freesing level depression for southern continents (after Hastenrath, 1972).

The evidence discussed is sufficient to link the scree tongues to nivation niches and to attribute them both to frost action. Their distribution is, however too localised to enable them to be utilised for the construction of a meridional gradient. They merely add to the body of evidence available. Hastenrath (1971, 1972) approached the problem of meridional gradients from the present altitude of the 0°C isotherm. When the annual 0°C altitudes for the three southern continents are superimposed, it is apparent that at the present day, at least, the southern African isotherm lies above that of Australasia and South America (Fig.8). Although the annual 0°C isotherm can only act as an indication of nival levels, since moisture is also involved, it does suggest the level from which Pleistocene depression occurred. Africa is an arid sub-continent and although up to 9°C temperature depression has been accepted, it is likely that effects of the Pleistocene meridional gradient will only be perceptible at high altitudes.

CONCLUSION

The scree tongues of the Amatola Mountains of the eastern Cape Great Escarpment are one type of periglacial landform dating from the Pleistocene Period. They belong to a set of phenomena that includes oversteepened south-facing slopes and occasional nivation niches and, in some places, typical head deposits. This set of phenomena is distinct from and younger than the nivation cirques of the Lesotho high country. The detailed analysis of the scree tongues has clarified relationships between the various categories of landforms and confirmed the existence of at least two sets. Indications of a meridional gradient can be perceived but its precise alignment is not yet established.

Acknowledgements

Funding from CSIR for fieldwork expenses and conference attendance is acknowledged.

REFERENCES

Alexandre, J., 1962. Phenomenes periglaciaires dans le Drakensberg due Natal. Biul. Periglac., 11, 11-13.

Butzer, K. W., 1972. Geology of Nelson Bay Cave, Robberg, S.A. S.A. Arch. Bull., 28, 97-110.

Butzer, K. W., 1973. Pleistocene periglacial phenomena in Southern Africa. Boreas, 2, 1-11.

Butzer, K. W., 1984. Late Quaternary environments in South Africa. in Palaeoclimates of the Southern Hemisphere, (Ed. Vogel). Proc. SASQUA 1983, pp235-264. Balkema, Rotterdam.

Dyer, R. J. G., and Marker, M. E., 1979. On some aspects of Lesotho hollows. Zeit. F. Geomorph., 23, 256-270.

Ellenberger, P., 1960. Le Quaternaire au Basutoland. Bull. Soc. Prehist. Francaise, 57, 439-475.

Harper, G., 1969. Periglacial evidence in Southern Africa during the Pleistocene epoch. Palaeoecol. Africa, 4, 71-101.

Hastenrath, S., 1971. On snowline depression in the arid regions of the South American Andes. J. Glaciol., 10, 255-267.

Hastenrath, S., 1972. A note on recent and Pleistocene altitudinal zonation in Southern Africa. S.A.J. Sc., 68, 96-102.

Hastenrath, S., and Wilkinson, J., 1972. A contribution to the periglacial morphology of Lesotho, Southern Africa. Biul. Periglac., 22, 157-167.

Marker, M. E., and Whittington, G. W., 1971. Observations on some valley forms and deposits in the Sani Pass area of Lesotho. S.A. Geog. J., 52, 96-99.

Nicol, I. G., 1973. Landforms in the Little Caledon valley, O.F.S., _S.A. Geog. J._, 55, 56-68.

Peterson, J. A., 1968. Cirque morphology and Pleistocene ice formation conditions in Southern Australia. _Aust. Geog. Studies_, 6, 67-83.

Sparrow, G. W. A., 1964. Pleistocene periglacial landforms in the Southern Hemisphere. _S. A. J. Sc._, 60, 143-147.

Sparrow, G. W. A., 1967a. Pleistocene periglacial topography in Southern Africa. _J. Glaciol._, 6, 551-559.

Sparrow, G. W. A., 1967b. Southern African cirques and aretes. _S.A. Geog._, 1, 9-11.

Sparrow, G. W. A., 1971. Some Pleistocene studies in Southern Africa. _S. A. Geog._, 3, 809-814.

Watson, E., 1966. Two nivation cirques near Aberystwyth, Wales. _Biul. Periglac._, 15, 135-140.

International Geomorphology 1986 Part II
Edited by V. Gardiner

SOME PROBLEMS OF THE VENEZUELAN QUATERNARY PERIOD

L. Vivas

Universidad de Los Andes,
Merida, Venezuela

ABSTRACT

The Venezuelan Quaternary has been little studied. The Tricartian nomenclature, which roughly corresponded to the Alpine model, has been replaced by another which does not assume any correspondence with other models. The Andean cold period called the Merida Glaciation at the end of the Pleistocene is thought to correspond with the Wurm; this can be divided into two stages. Four stages of fluvial terrace formation can be distinguished in the Andes, but debate exists concerning their ages. The greatest advances have been made in examination of marine and fluviomarine sediments on the northern coast, where correlations have been made with stages from Villafranchian to Holocene, although largely on the basis of relative evidence.

Plio-Pleistocene and Pleistocene-Holocene boundaries can be identified in depositional sequences in Venezuela, and some absolute dates have been ascribed, particularly to the latter. The Quaternary was directly responsible for about 45% of Venezuela's relief, and the elucidation of Quaternary chronologies is therefore of considerable academic and practical value.

INTRODUCTION

The Venezuelan Quaternary Period has been relatively little studied by comparison with that of many countries with a greater scientific tradition and observations of the particular characteristics which define it are few. This is a problem in itself and, at the same time, is the fundamental explanation for the existence of other more specific problems, along with others derived from the difficulties and limitations which still exist concerning scientific knowledge of the Quaternary Period on a worldwide level.

The antecedents of pioneer observations of the Venezuelan Quaternary go back to the second half of the past century and the first decade of the present one, and above all include works of the German naturalist W. Sievers in 1886 and the Venezuelan A. Jahn in 1911. Apart from these, and a few other scarce exceptions, the study of the Quaternary in Venezuela began very late and it is even said that it did not begin until the 1950's, fundamentally with the work begun by Professor Jose Royo Gomez. He was the first to write, in 1956, a systematic synthesis entitled "The Quaternary in Venezuela", which was published in the first edition of the "Venezuelan Stratigraphical Dictionary". Even today there are very few renowned scholars who are engaged in the systematic study of this field, and those who do so do it more as a personal scientific interest than as a response to the policies adopted by public or private Venezuelan organizations.

It could be said that until now there has not existed in Venezuela any planning or institutional systematization related to the study of the Quaternary as a geological period of special interest in the earth sciences. This, of course, represents a major obstacle when refined notions of this period in Venezuela are needed. It is appropriate to add that in the last two decades much progress has been made concerning our knowledge of the Quaternary in Venezuela. Geologists, geomorphologists and other professionals in the earth sciences employed by universities, autonomous institutes and various ministries have worked on teaching, research, the compilation of material, and publication of work concerning diverse topics related to the Venezuelan Quaternary Period. This effort has awakened a great academic and professional interest in the Period under discussion. Two of the many valuable works in this area are the extensive work done by C. Schubert since 1968 and the laudable work of C. Gonzalez de Juana et al. (1980). Institutions such as the Schools of Geology of the Central University of Venezuela and the Eastern University, as well as the Schools of Geography in the Central University of Venezuela and the University of the Andes, have played an important role in the teaching and investigation of the Quaternary in Venezuela. The Division of Geology of the Ministry of Energy and Mines has also contributed with special programmes.

Until now the efforts made have been timid and few. Because of the great academic and practical value of studies of the Quaternary it is to be hoped that a greater emphasis will be made in the study of the Quaternary in Venezuela by government and private institutions.

NOMENCLATURE AND STRATIGRAPHY

Apart from the two major and universally accepted divisions of the Quaternary, the Pleistocene and Holocene Epochs, there has not yet been developed in Venezuela a more detailed and uniform subdivision of the Period which is in general use. However, some attempts have been made, though they are partial and incomplete. This is the case for those of Tricart (1963) and Tricart and Millies-Lacroix (1962), who for the first time applied to Venezuela a nomenclature created by Tricart to subdivide Quaternary sequences in France, as well as in other countries. That subdivision was applied by these authors to alluvial sediments of the Andean region and its piedmonts, and is as follows:

$$t_O$$
$$t_I$$
$$t_{II}$$
$$t_{III}$$
$$t_{IV}$$
$$t_V$$

This nomenclature, which Tricart and his disciples later extended to other regions of Venezuela, has only a relative connotation and does not assume either dates or precise correlations with other better-known nomenclatures. In spite of this, upon reading the works of Tricart, one is left with the conviction that he tried to correlate his nomenclature in an approximate way with the Alpine model, roughly equating glaciations as Günz (t_{IV}), Mindel (t_{III}), Riss (t_{II}) and Würm (t_I), where t_O is equivalent to the Holocene, and if necessary regarding t_V as an equivalent of the Villafranchian. Nevertheless, such an equivalence is not necessarily valid, as Tricart himself has recently admitted.

In the last few years the above subdivision has been replaced, principally by geomorphologists, by another which although similar to the previous one does not assume any correlation, either directly or indirectly, with other models which are dated absolutely. In a rough way it does correspond to wide subdivisions of the Pleistocene. This nomenclature is as follows:

Q_0 .... corresponding to the Holocene

Q_1 .... corresponding to the upper Pleistocene

Q_2 .... corresponding to the middle upper Pleistocene

Q_3 .... corresponding to the lower to middle Pleistocene

Q_4 .... corresponding to the lower or ancient Pleistocene.

The letter Q is internationally accepted as indicative of the Quaternary. Until another more precise one is devised this scheme of subdivision is the most convenient for locating Quaternary events of various natures on a relative geological time scale.

C. Schubert (1980) has made an important contribution towards the future establishment of a more complete and precise nomenclature of the Venezuelan Quaternary when, basing his ideas on lithological, morphological, palaeontological and radiocarbon criteria, he individualized the Venezuelan-Andean cold period by denoting it the Merida Glaciation. As we shall see later, there exists quite exact information as to the absolute age of this cold period in the Venezuelan Andean central region. This is of great significance in the search for a national or regional stratigraphic model.

As for the stratigraphic problems themselves, as noted by Gonzalez de Juana et al. (1980), in Venezuela there is not really adequate knowledge concerning the stratigraphy of Pleistocene sediments. These authors admit that the model which could be applied directly to this country is "that related to the definition proposed by The International Geological Congress held in London in 1948" (page 678). As a consequence, the limiting palaeoclimatic model for a given area of the country would be that which is established between a glacial and interglacial period. This has already been applied to Venezuela, more as a morphogenetic model than as a stratigraphic one, by J. Tricart and other authors. The same was done by Van der Hammen in 1974 for Northern South America and by Bigarella et al. in 1969 for some regions of Brazil. In this sense, we believe that as a morphogenetic model it is reasonably valid for all of Venezuela, but a great deal of care must be taken when applying it as a stratigraphic model, especially in those geographic areas furthest from the Andean cordillera. In any case, as we understand it, the model of the International Geological Congress of 1948 to which Gonzalez de Juana et al. referred has only an indicative value, and is too relative to be applied with reasonable margins of security to the Quaternary stratigraphy of Venezuela.

The knowledge we have of Venezuelan Quaternary stratigraphy comes in part from studies of moraine deposits in the Andes and in part from studies of alluvial sediments in the Andes and other areas. In addition to this, there exists information from other types of deposits, such as lacustrine, fluvial, marine, aeolian deposits and so on, and especially information about marine sediments deposited in deep water and on the continental shelf.

At first, it was believed that there were possibly four cold periods in the Venezuelan Andes during the Pleistocene. Until now, though, it has been possible to identify only two types of glacial moraine deposits: a possible one which authors such as J. Tricart (1960) correlate with the Riss glaciation and another quite certain one, which several authors consider to be representative of the last Andean cold period. Although this latter suffered oscillations of glacial advances and recessions, it is in general believed to coincide with the last worldwide glaciation, the Würm (Wisconsin or Devensian).

This Andean cold period at the end of the Pleistocene was recognized by Sievers (1888), Jahn (1925), Royo y Gomez (1959), Cardenas (1962), Tricart (1966) and Schubert (1980), who called it the Merida Glaciation. According to Schubert it is distinguished by two principal moraines in the Sierra de Santo Domingo and the Paramo de La Culata: the lower level between 2,600 and 2,700m and the upper level between 3,000 and 3,500m Schubert indicates that the radiocarbon dating of samples collected by him and by Giegengack and Grauch (1975) suggests one age of more than 10,000 years B.P. for the upper moraine deposit, and possibly even more than 13,000 years B.P. This would support a Late Wisconsin Age for the Merida Glaciation. It is believed that the lower moraine deposit probably corresponds to the principal Wisconsin glacial advance and that the upper level corresponds to the last glacial advance, the Late Wisconsin. In any case, for Schubert the two moraine levels being discussed can be interpreted as two stages within the Merida Glaciation.

These conclusions advanced by Schubert are practically the only ones concerning the Venezuelan glacial stratigraphy which are based upon absolute dates; this stratigraphy is in itself very incomplete as it only covers the late Pleistocene. We need to cover the rest of this period in order to have a Quaternary stratigraphic column for the Venezuelan Andean region that will serve as a reference for correlation of the rest of the national territory.

In the case of fluvial terraces Sievers (1888) correlated the Andean ones with the Pleistocene glaciations. Similarly Tricart and Millies-Lacroix (1962) distinguished four different levels in the region, named t_{IV}, t_{III}, t_{II} and t_{I}, where the first is possibly of the Plio-Villafranchian age. Shagam (1972), also referring to the Andean fluvial sequences, distinguished two levels: one that is at least in part previous to the oldest glaciation and the other, without doubt, of Pleistocene origin.

Numerous other geologists and geomorphologists have established at least four levels of deposits of fluvial terraces in the Andes and its piedmonts, virtually always maintaining the criterion of relative dating. However Schubert (1980), using radiometric dating, suggests that level three of the sediments of the alluvial terrace of the Esnujaque Formation in the Motatan River valley could be 40,000 years old, corresponding to the Early Middle Wisconsin, and levels two and one could be Late Middle Wisconsin and Holocene respectively. However, as Schubert himself admits, there are insufficient isotopic dates and precise palaeontological analyses to be able to establish a valid chronostratigraphy that, in its turn, could be correlated with other regions of the country.

The case of the Andean region and its piedmonts can, in general terms, be extended to cover all of the national territory. Thus, in relation to Quaternary stratigraphy, where fluvial sequences are concerned, we are still in the phase of relative studies and very little has been done in relation to absolute dating. That signifies a fundamental obstacle to the establishment of an acceptable national stratigraphic model. Although it must be admitted that in the last few years great advances have been made in Venezuela concerning the correlation of Quaternary fluvial sediments, as we shall see below, they are based only on relative dating criteria.

Equally, we must acknowledge that in Quaternary stratigraphy in Venezuela the greatest advances have been made in the study of marine and fluviomarine sediments on the northern coasts of the country, with the reservation, of course, that our present knowledge is still insufficient upon which to base an accurate stratigraphic model. For a review of some of the relevant studies see Oshsenius (1980) and Gonzalez de Juana et al. (1980). From these we find that it is possible to establish correlations based on absolute datings in conjunction with palaeo-faunal analysis of sediments of the marine terraces along the Caribbean coast. The study of these marine sediments permits correlation of the Cabo Blanco sequences (Abyssinian Formation), which are assigned an age of 300,000 years by Weisbord (1965), with those of La Tortuga Island and Juan Griego in Margarita Island. They also point out that the lower level of the marine terraces on the island La Blanquilla, in Cubagua and Araya, was assigned radiometric ages of 130,000 ± 7,000, 135,000 ± 15,000 and 125,000 ± 7,000 years respectively. For the second level of La Blanquilla an age of 325,000 ± 7,000 years was obtained.

Upon studying some of the shore deposits of the northern coast of Venezuela, in particular, the peninsulas of Araya and Paraguana, the Island of Margarita and the region of Puerto Cumarebo, Danielo (1976) identified six marine levels between 0-80m, which he correlated with diverse stages of the European Quaternary, from Villafranchian to the Holocene. It should be clarified that this correlation and the stratigraphic sequences of Danielo are purely relative and based solely upon the altitude of the deposits.

THE PLIO-PLEISTOCENE AND PLEISTOCENE-HOLOCENE BOUNDARIES

If the demarcation of these boundaries is difficult and full of controversy in countries with a long tradition and considerable progress in the study of the Quaternary , it is much more difficult in Venezuela, where research into this geological period has just begun. The demarcation of these limits is still very incomplete, but in spite of this some relevant comments can be made.

In the Cordillera of Merida and its piedmonts there exists the possibility that the Plio-Pleistocene boundary is marked by a fundamental attenuation of the Andean Tertiary epeirogenic paroxysm in conjunction with the beginning of a deterioration of the climate. This change of climate, in its turn, signifies the beginning of the first Andean cold period which, although no conclusive evidence has been found, would be contemporary with the first glaciation, well recognized as affecting the world climate and in particular that of the Northern Hemisphere. Field evidence for this boundary is the contact between the upper section of the Betijoque-Isnotú Formation and the coarse alluvial sediments which normally rest discordantly upon it, that is, the Carvajal Formation. The spatial expression most characteristic of this evidence is frequently found along the northern piedmonts of the Merida Cordillera. Field evidence is also given by disconformities existing between the older sediments of the Rio Yuca-Parángula Formations and the coarse alluvial sediments of the Guanapa Formation found in the southern piedmonts of the Cordillera de Merida. In the Central and Eastern Plains of the country the Plio-Pleistocene limit can be marked by the discordant contact between the finer sediments of Pliocene geological formations such as the Sacacual Group (Las Piedras Formation) and the coarse alluvial sediments of the Mesa Formation (Zinck and Urriola, 1970). Similar expressions of lithostratigraphic contacts can also frequently be found in the valleys of the Caribbean Cordillera and in the plains of the Maracaibo lake near Maracaibo city. It must be noted that none of the cases mentioned have been dated by absolute techniques and no study has been made of diagnostic fossils that would enable the limits under consideration to be geochronologically dated.

On the other hand there are several examples of marine sediments which are very useful for marking the Venezuelan Plio-Quaternary contact. Such is the case of the Cabo Blanco area which contains an interesting succession of Pliocene and Pleistocene sediments. For instance, the upper Pliocene-Playa Grande Formation is rich in the foraminifera Globorotalia tosaensis and it is discordantly covered by the Mare Formation, which is considered to be Pleistocene. The Mare Formation, is then covered by the Middle Pleistocene Abyssinian Formation, which is assigned an age of 300,000 years, thus placing it in the Sangamon interglacial between the cold periods of the USA stratigraphic model.

A similar case to that of the Cabo Blanco area is found in the Cumana region, in the Peninsula de Araya and Tortuga Island. In these areas, judging from the foraminifera Globorotalia tosaensis and Globorotalia truncatulinoides contained in the marine sediments, the Cumana Formation represents the Plio/Pleistocene transition. Therefore, it seems that the Cumana Formation probably represents the Plio/Pleistocene limit of the marine section corresponding to the eastern part of the country, be it the terminal Pliocene or the basal Pleistocene.

In Venezuela, as in most of the world, it is less difficult to establish the Pleistocene/Holocene limit. Thus in the Andean region there are varied evidences of a much colder period than the present, when important parts of the land situated above 3,000m suffered a palaeoclimatic action which produced a relatively extensive area of glacial and periglacial activity. The radiocarbon and pollen analyses that have been done indicate that the Andean cold period ended between approximately 13,000 and 10,000 years B.P. (Schubert, 1980; Salgado Labouriau et al., 1977). This date would establish the Pleistocene/Holocene boundary for the central Andes of Venezuela, corresponding to the accepted world limit between these Quaternary epochs.

For other areas of Venezuela, a date of around 11,000 years B.P. is considered to mark the limit under discussion, as can be deduced from the studies of the Valencia basin of Peeters (1970) and Schubert (1980). Platt-Bradburg et al. (1981) found a good correlation between the aridity of the lowlands of that area and the glacial period of the highlands of Venezuela. They also found good correlations between the increase in sea level and the existence of a more humid climate in the areas of low elevation. Developing this line of thought, they point out that the dry conditions of the Valencia lake existing between 13,000 and approximately 10,500 years B.P. correlate with the glacial conditions of elevations of 3,000m in the Andes of Venezuela and Colombia and with the low sea levels and arid conditions in the tropical regions of Colombia, Maracaibo Lake, British Guiana and Panama. In addition, they indicate that logical correlations could also be made with surveys of the Caribbean sea floor since they indicate low sea temperatures and arid continental conditions. Finally, they state that the humidity increase in the Valencia basin began more or less 10,000 years B.P., which corresponds to an increase in sea level. As a result, all of these studies show that the Pleistocene/Holocene transition was characterized by a transition from a dry epoch corresponding to the late Pleistocene to a much rainier epoch corresponding to the Holocene. This assertion seems to coincide with the approximate dates of 11,000 - 12,000 years B.P. given to some palaeosoils found in the sand dunes of the central Venezuelan Plains, which would mark the end of the Pleistocene and the beginning of the Holocene.

As a final note, it must be pointed out that on Venezuela's Caribbean coasts fluctuations of sea level give a good indication of the Pleistocene/Holocene boundary. The Holocene transgression left evidence whose dating indicates that the increase in sea level in Venezuela coincides with the worldwide records established around 6,000 and 8,000 years B.P. for similar Holocene transgressions in other parts of the world.

THE QUATERNARY AND VENEZUELAN RELIEF

In order to illustrate the importance of the Quaternary Period, not only for the role it plays in modelling the Venezuelan relief, but also because of the large areas of Quaternary origin in the country, an attempt is made in Table 1 to classify Venezuelan territory in areal units according to their fundamental geological origins.

Table 1. Geological origins of Venezuelan territory.

Geological origins	Areas	Total km^2	Surface area %
PRECAMBRIAN SHIELDS	Guayana Shield Arco de El Baul	419,193	45.73
MOUNTAIN SYSTEMS AND SECONDARY-TERTIARY ISLANDS	Central and Eastern Caribbean Cordilleras Falcon-Lara Relief or Coriano System Caribbean Islands	43,208	4.71
TERTIARY ANDEAN MOUNTAIN SYSTEM	Cordillera of Perija Cordillera of Merida	42,018	4.58
CONTINENTAL QUATERNARY, COASTAL, INTERMONTANE & GUAYANA DEPRESSIONS	Western, Southern and Central Eastern Plains Orinoco Delta System Lake of Maracaibo Plains Intermountain and Guayana Depressions	412,347	44.98
		916,766	100.00

The Quaternary Period is responsible for the sculpting of a vast proportion of the national territory. This area is conservatively estimated at no less than 45% of the whole, without including other areas in which the Quaternary also plays a role of great importance in the final configuration of the relief. For example, in the Andes the land above 2,800m was in a large measure retouched by the action of ice and the cold of the Pleistocene, as witnessed by clearly glacial and periglacial palaeoforms of accumulation and erosion. We must also add to the regions and areas already mentioned of Quaternary origin other areas which are not strictly Quaternary in origin but which were affected by diverse morphogenetic mechanisms during the Pleistocene epoch.

In summary, the majority of Venezuela has been affected in one way or another by the sculpting processes which operated during the Quaternary. That is why the study of the Quaternary in Venezuela is of academic and practical interest, as it signifies the development of a better understanding which would lead to the means by which our physiographical reality might be better utilized.

REFERENCES

Bigarella, J. J., Mousinho, M. R., and Da Silva, J. X., 1969. Processes and environments of the Brazilian Quaternary, in The Periglacial Environment (Ed. Pewe), pp 417-487. McGill-Queen's University Press.

Cardenas, A. L., 1962. El glaciarismo Pleistoceno en las Cabeceras del Chama. Rev. Geog., 3.

Danielo, A., 1976. Formes et dépôts littoraux de la côte septentrionale de Vénézuéla. Annales de Géographie, 85, 68-97.

Giegengack, R., and Grauch, R. Boconó Fault, Venezuelan Andes. Reimpreso de Science.

Gonzalez de Juana, C., et al., 1980. Geología de Venezuela y de sus Cuencas Petrolíferas. Volumes one and two. Foninves, Caracas.

Jahn, A., 1911. Informe de la Comisión Cientifica Exploradora del Occidente de Venezuela. Rev. Tecnica del M.C.P., vol. 1.

Jahn, A., 1925. Observaciones Glaciológicas en los Andes Venezolanos. Cultura Venezolana No.64.

Ochsenius, C., 1980. Cuaternario en Venezuela. Cuadernos falconianos. UNEFM, Coro, Venezuela.

Peeters, L., 1970. Origen y Evolucion de la Cuenca del Lago de Valencia, Venezuela. Instituto para Conservacion del Lago de Valencia, Caracas.

Platt Bradbury, J., Leyden, B., Salgado-Labouriau, M., Lewis, W. M. jr., Schubert, C., Binford, M. W., Frey, D. G., Whitehead, D. R., and Weibezahn, F. H., 1981. Late Quaternary environmental history of Lake Valencia, Venezuela. Science, 214, 1299-1305.

Royo y Gómez, J., 1956. Cuaternario en Venezuela. Léxico Estratigráfico de Venezuela. Ministerio de Minas e Hidrocarburos, Caracas.

Salgado Labouriau, M. L., Schubert, C., and Valastro, S. jr., 1977. Paleoecologic analysis of a Late Quaternary Terace from Mucubají, Venezuelan Andes. Journal of Biogeography, 4, 313-325.

Schubert, C., 1980. Aspectos Geológicos de los Andes Venezolanos. Historia Breve Síntesis. El Cuaternario y Bibliografía. En estudios ecologicos en Los Paranos Andinos. (Ed. Monasterio), University of the Andes, Merida, Venezuela.

Schubert, C., and Valastro, S. jr., 1977. Evidencias de Levantamiento Reciente de la Costa Norte-Central (Cordillera de la Costa), Venezuela. Acta Científica Venezolana, 28, 363-372.

Shagam, A., 1972. Geología de Los Andes Centrales de Venezuela. Congreso Geologico de Venezuela, IV. Memoria Boletin Geolog. Publicación especial 5.

Sievers, W., 1888. Die Kordillicre von Mérida Nebst Bemerkungen über das Karibische Gebirge. Geog. Abhandl., 3, 238.

Tricart, J., 1963. Estudio integral de la Cuenca del Chama. ULA, Faculty de Ciencias Forestales, IGCRNR, part 1. (Lagumillas de Urao - geomorfologia).

Tricart, J., 1966. Geomorfología del Area de Mecuchíes. Revista Geográfica, 7, 31-42.

Tricart, J., and Millies-Lacroix, A., 1962. Les Terrasses Quaternaires des Andes Vénézueliénnes. Soc. Geol. de France Bull., 7, 201-208.

Van der Hammen, T., 1974. The Pleistocene changes of vegetation and climate in tropical South America. Journal of Biogeography, 1, 3-26.

Weisbord, N. E., 1965. Notes on the geology of the Cabo Blanco area, Venezuela. Bull. Amer. Paleont., 38, 5-25.

Zinck, A., and Urriola, P., 1970. Orígen y evolución de la Formación Mesa. Un Enfoque Edafológico. Ministerio de Obras Públicas, Barcelona.

LONG-TERM LANDFORM DEVELOPMENT

International Geomorphology 1986 Part II
Edited by V. Gardiner

LONG-TERM LANDFORM DEVELOPMENT: EDITORIAL INTRODUCTION

M.A. Summerfield and M.F. Thomas

Department of Geography,
University of Edinburgh, Edinburgh, U.K.
and
Department of Environmental Science,
University of Stirling,
Stirling, U.K.

INTRODUCTION

The six sessions on long-term landform development at the First International Conference on Geomorphology, although revealing a broad diversity of current interests, nonetheless indicated a focus around three major themes: the role of climate and climatic change and the identification and geomorphological significance of exhumed landscapes; the importance of inherited features in contemporary landscapes and the rate of landscape change through time; and finally the operation of tectonic and structural controls on long-term landform development. A general assessment of some of these issues is presented in the following article (Thomas and Summerfield, this volume) but this introductory paper serves to highlight some of the major points raised in the verbal presentations given in the sessions on long-term landform development and to provide a context for the submitted papers published here.

THE ROLE OF CLIMATE AND CLIMATIC CHANGE

In Europe at least, interest in the ´geomorphological era´ beginning in the Late Cretaceous and encompassing ´tropicoid paleo-earth´ which persisted from the Late Mesozoic until towards the close of the Paleogene (Budel, 1968, 1982) has not diminished during more than twenty years of research. Gradually the evidence for ancient saprolites and duricrusts, bauxitic and lateritic sedimentation, and the formation and preservation of erosional landforms, such as tors, inselbergs and pediments, has accumulated. The palaeoclimatic significance of these landforms and deposits was stressed by Linton (1955) and successively by Godard (1965), Dury (1971), Thomas (1978), Battiau-Queney (1984) and Hall (1985).

In the papers given on this theme during the sessions on long-term landform development three trends can be detected: the elaboration of detailed evidence for palaeoclimates and climatic change from widely separated localities in Europe - from north-west Spain to north-east Scotland and Scandinavia; the extension of interest in palaeoclimates as far back as the Late Carboniferous arising from a need to understand the exhumed landscapes of ancient cratonic surfaces; and finally, the development of interest in the morphogenetic significance of palaeoclimates and climatic change throughout this long period of time.

Geological evidence for tropical climates during the Paleogene is not new, but acceptance of possible humid tropical conditions in high latitudes during the Late Cretaceous and Paleocene has not gone undisputed. However, the weight of evidence is now in favour of warm, moist conditions at the start of the Tertiary in the North Atlantic realm (Nilsen and Kerr, 1978), and this is corroborated by research in Spain (Molina et al., this volume), central Europe (Juhasz, 1985), the British Isles (Battiau-Queney, this volume; Hall, this volume; Smith and McAlister, this volume), and from Scandinavia (Lidmar-Bergstrom, 1982, 1985; Miskovsky, 1985). In addition, the recognition that the ´geomorphological era´ needs to be extended to pre-Cretaceous time has arisen from an increasing interest in exhumed sub-Cretaceous and earlier landscapes (Bloom and Strecker, 1985: Gorelov, 1985; Juhasz and Kertesz, 1985; Lidmar-Bergstrom, 1985), in the reworking of Early Mesozoic (or even Paleozoic) sediments (Gorelov, 1985; Juhasz, 1985), and in the persistence of ancient forms in present-day landscapes (Battiau-Queney, this volume; Starkel, this volume). Increasing recognition now needs to be given to the probable humid-arid oscillations within the warmer climates of the Mesozoic and Early Cenozoic. This not only has relevance for the Paleogene, but more particularly for the Late Carboniferous-Permian-Triassic transition (Juhasz, 1985; Miskovsky, 1985).

Interest in the morphogenetic significance of palaeoclimatic reconstructions is expressed in several of the studies presented here, from the generation of individual landforms that persist in present-day landscapes (Battiau-Queney, this volume), to the consideration of entire landscapes or landform complexes (Hall, this volume) and the inferences derived from such studies for the nature of landscape modification. In this last context the overwhelming opinion is for landscape lowering associated with the formation and removal of deep weathering profiles, strikingly and significantly illustrated by the study of pediments from the Mojave desert (Dohrenwend et al.) but widely considered within the European context (Battiau-Queney, this volume; Hall, this volume; Juhasz, 1985; Lidmar-Bergstrom, 1985; Miskovsky, 1985; Molina *et al.*, this volume).

Within these arguments the occurrence and role of pediments remain ambiguous. Molina *et al.* (this volume) refer to inselberg-pediment landscapes under savanna conditions in Spain, but favour the lowering of weathering mantles on old landsurfaces. In none of the papers presented is the case for the backwearing of hillslopes and escarpments with the formation of pediments presented. It is clearly important to investigate these and other issues relating to models of landform development in the light of new evidence concerning both palaeoclimates and tectonics.

LANDFORM INHERITANCE AND RATES OF LANDFORM CHANGE

The investigation of present-day landscapes can be approached in two distinct ways: the functional approach is concerned with the relationship between contemporary processes and forms whereas the historical approach is founded on the identification and interpretation of a developmental sequence of landform change through time. As Starkel (this volume) points out, the former approach fails to emphasize the possible importance of inherited forms while the latter may not take sufficient account of recent landscape modification. Starkel´s analysis of the significance of inherited landforms in the Polish Carpathians shows that the relative importance of inherited forms depends to a large extent on the spatial scale being considered with the largest forms being related primarily to Miocene tectonics and the smallest (present floodplains and channels) being in equilibrium with current hydrological conditions.

The importance of landscape inheritance is also highlighted by Battiau-Queney (this volume) who argues that detailed field surveys have begun to reveal the extent of the survival of Tertiary landforms in the British Isles, even in regions subject to Pleistocene glaciation. Such landforms include remnants of tropical deep weathering profiles, palaeokarst, pediments, tors and inselbergs. A major contributory factor to the preservation of such relict landforms is the crustal stability of some regions of the British Isles; consideration of both the macroscale tectonic setting and the prevailing tropical to sub-tropical climates of at least the Early Tertiary are regarded by Battiau-Queney as a vital prerequisite for a full understanding of the complexity of the present-day British landscape.

Detailed assessments of the relative importance of

contemporary geomorphological processes vis-a-vis the significance of inherited forms are frequently constrained by the paucity of data concerning rates of landscape change. This renders the association of specific landforms with dateable strata (such as lava flows) especially valuable as is illustrated by Dohrenwend et al. (this volume) who relate the positions of modern pediment surfaces to pediment remnants occurring below lava flows of various ages and thereby provide quantitative estimates of long-term erosion rates. Similarly in eastern Australia Bishop et al. (1985a, 1985b) have been able to relate the present valley system of the Lachlan River to its Early Miocene predecessor preserved and dated by extensive basalt lava flows. Rates of channel incision in this summit region of the East Australian Highlands are seen to be low (around 8 mm/ka) in comparison with modern rates measured from similar environments.

TECTONIC AND STRUCTURAL CONTROLS

Active plate margins have been an important focus of research within global tectonics but few geomorphological studies have specifically addressed the relationship between landform evolution and the complex variety of structures developed within these tectonically active regions. The importance of strike-slip (transform) and oblique movements, in addition to the more purely orthogonal motions at convergent boundaries is illustrated by Schubert (1985) in the context of the southern Caribbean plate boundary which extends some 1300 km from Trinidad westwards to the Venezuelan Andes. In his study of the numerous pull-apart basins along this boundary zone Schubert demonstrates, on the basis of sedimentological evidence and radiocarbon dating that these structures are of only Quaternary age.

Recently geomorphologists have begun to evaluate the significance of the tectonic development of passive (rifted and sheared) margins. This has arisen from an increasing appreciation that the vertical movements and lithospheric flexure which characterise such margins, together with their off-shore sedimentary record interpreted on the basis of both boreholes and seismic stratigraphy, is crucial to an understanding of the relationship between base level changes and the denudational response of the continents (Summerfield, 1986; Thomas and Summerfield, this volume). The geomorphological evolution of passive margins must be seen in the context of four primary factors (Summerfield, 1985a; 1985b); the timing and spatial extent of the rift-related uplift along the margin, the nature of the thermally-driven post-rift subsidence, the isostatic response of the margin to denudational unloading (largely through escarpment retreat?), and the flexure of the margin associated with sediment loading off-shore.

Lageat (1985) presents a detailed study of the geomorphological development of an area of south-eastern Africa and at this regional scale demonstrates that structural and lithological factors, together with morphoclimatic factors, must be considered in addition to the broader tectonic setting (Lageat and Robb, 1984). This point was also emphasized in the informal workshop on passive margins organised at the conference by Yvonne Battiau-Queney. The continuing significance of tectonic activity along some passive margins is demonstrated by Zeese (1985) who documents evidence of Neogene uplift on the basis of dislocated and tilted etchplains in northeast Nigeria. A similar emphasis on the differentiation of small regions on the basis of their lithological and structural characteristics and detailed tectonic evolution is to be found in Vergnolle´s study of the Tertiary geomorphological evolution of the north-west Iberian Peninsula (Vergnolle, this volume). This regional investigation also illustrates the way in which morphological and sedimentological evidence needs to be combined if a comprehensive landscape history is to be constructed.

REFERENCES

Battiau-Queney, Y. 1984. The pre-glacial evolution of Wales. Earth Surface Processes and Landforms, 9, 229-252.

Battiau-Queney, Y. this volume. Tertiary inheritance in the present landscape of the British Isles.

Bishop, P., Young, R.W., and McDougall, I., 1985a. Stream profile change and longterm landscape evolution: Early Miocene and modern rivers of the East Australian Highland crest, central New South Wales, Australia. Journal of Geology, 93, 455-474.

Bishop, P., Young, R.W., and McDougall, I., 1985b. Stream profile change and longterm landscape evolution early Miocene and modern rivers of the East Australian Highland Crest, Central New South Wales Australia, in Abstracts of Papers for the First International Conference on Geomorphology (Ed. Spencer), p. 41. University of Manchester, Manchester.

Bloom, A.L., and Strecker, M. R., 1985. Exhumed Gondwana erosion surfaces, Sierras Pampeanas, Northwest Argentina, in Abstracts of Papers for the First International Conference on Geomorphology (Ed. Spencer), p. 42. University of Manchester, Manchester.

Budel, J., 1968. Geomorphology, Principles, in Encyclopaedia of Geomorphology (Ed. Fairbridge), pp 416-422. Reinhold, New York.

Budel, J., 1982. Climatic Geomorphology (Translated by L. Fischer and D. Busche) Princeton University Press,

Princeton.

Dohrenwend, J. C., Wells, S.G., McFadden, L.D., and Turrin, B.D., this volume. Pediment dome evolution in the eastern Mojave Desert, California.

Dury, G.H., 1971. Relict deep weathering and duricrusting in relation to the palaeoenvironments of middle latitudes. Geographical Journal, 137, 511-522.

Dury, G.H., 1972. A partial definition of the term "pediment" with field test in the humid climate areas of southern England. Transactions of the Institute of British Geographers, 57, 139-152.

Godard, A., 1965. Recherches de Geomorphologies en Ecosse de Nordouest. Belles Lettres, CNRS, Paris.

Gorelov, S.K., 1985. Paleogeomorphology - A new perspective trend in the development of geomorphological studies, in Abstracts of Papers for the First International Conference on Geomorphology (Ed. Spencer), p. 220. University of Manchester, Manchester.

Hall, A.M., 1985. Cenozoic weathering covers in Buchan, Scotland, and their significance. Nature, 315, 392-395.

Hall, A.M., this volume. Cenozoic weathering covers in Buchan, Scotland.

Juhasz, A., 1985. Paleogeomorphologic significance of the redeposited dolomite formation in the Transdanubian Mountains, in Abstracts of Papers for the First International Conference on Geomorphology (Ed. Spencer), p. 303. University of Manchester, Manchester.

Juhasz, A., and Kertesz, A., 1985. Buried Mesozoic forms in the Transdanubian Mountains, in Abstracts of Papers for the First International Conference on Geomorphology (Ed. Spencer), p. 304. University of Manchester, Manchester.

Lageat, Y., 1985. L´analyse des formes structurales, methode d´approche de l´evolution morphologique-regionale: L´example du Bushveld Sud-Africain, in Abstracts of Papers for the First International Conference on Geomorphology (Ed. Spencer), p. 345. University of Manchester, Manchester.

Lageat, Y., and Robb, L.J., 1984. The relationships between structural landforms, erosional surfaces and the geology of the Archaean Granite Basement in the Barberton region, eastern Transvaal. Transactions of the Geological Society of South Africa, 87, 141-159.

Lidmar-Bergstrom, K., 1982. Pre-Quaternary geomorphological evolution in southern Fennoscandia. Sveriges Geologiska Undersokning Ser. C., No. 785, 202pp.

Lidmar-Bergstrom, K., 1985. Exhumed Mesozoic landforms in south Sweden, in Abstracts of Papers for the First International Geomorphology Conference (Ed. Spencer), p. 363. University of Manchester, Manchester.

Schubert, C., 1985. Basin formation along the southern Caribbean Plate boundary, in Abstracts of Papers for the First International Geomorphology Conference (Ed.

Spencer), p. 532. University of Manchester, Manchester.

Smith, B.J., and McAlister, J.J., this volume. Tertiary weathering environments and products in northeast Ireland.

Starkel, L., this volume. The role of inherited forms in the present-day relief of the Polish Carpathians.

Summerfield, M.A., 1985a. Plate tectonics and landscape development of the African continent, in _Tectonic Geomorphology_ (Eds. Morisawa and Hack), pp 27-51.

Summerfield, M.A., 1985b. Tectonics and landscape evolution along passive continental margins, in _Abstracts of Papers for the First International Geomorphology Conference_ (Ed. Spencer), p. 577. University of Manchester, Manchester.

Summerfield, M.A., 1986. Tectonic geomorphology: macroscale perspectives. _Progress in Physical Geography_, 10, 227-238.

Thomas, M.F., and Summerfield, M.A., this volume. Long-term landform development: key themes and research problems.

Vergnolle, C., this volume. Tertiary geomorphological evolution of the marginal bulge of the North-West of the Iberian Peninsula and lithostratigraphy of the grabens of the North-East of Galicia (Spain).

Zeese, R., 1985. Neogene tectonics and landforms in Northeast Nigeria, in _Abstracts of Papers for the First International Geomorphology Conference_ (Ed. Spencer), p. 679. University of Manchester, Manchester.

International Geomorphology 1986 Part II
Edited by V. Gardiner

LONG-TERM LANDFORM DEVELOPMENT: KEY THEMES AND RESEARCH PROBLEMS

M.F. Thomas and M.A. Summerfield

Department of Environmental Science,
University of Stirling, Stirling, U.K. and
Department of Geography, University of Edinburgh,Edinburgh.

ABSTRACT

Emphasis in geomorphology on short-term surface process studies needs to be balanced by enquiry into exogenetic and endogenetic forces in the longer term if the speculative status of most theory concerning long-term landform evolution is to be advanced. Analysis of large-scale tectonics, particularly of passive continental margins, reveals sequences of rifting, domal uplift, and subsidence, and the response of denudation in terms of surface alteration and reduction is revealed partly in the development and removal of weathered mantles and in geologically determined differential denudation patterns, but also in off-shore sedimentation, the record of which is now augmented by new seismic data. Rates of epeirogenic uplift have been matched in many instances by the simultaneous lowering of the weathering front and the landsurface by dynamic etchplanation creating acyclic landsurfaces over less resistant rocks, while relief increases where weathered mantles are protected by duricrusts or where outcrops of resistant rock emerge, and if faulting of the basement intervenes. Applications of these ideas are discussed in relation to etching under both palaeotropical and temperate conditions in European high latitudes, and also for the Atlantic slope of part of the West African Craton in Sierra Leone.

SURFACE PROCESS STUDIES AND HISTORICAL GEOMORPHOLOGY

Fashions in geomorphology have greatly influenced the progress of the subject, conditioning the entry of postgraduates to the discipline, and dictating the pattern of research funding. But fashions only clothe the body of the subject; they do not necessarily reveal its outlines in true proportion. The concentration of research on surface processes during the 1960s was understandable, partly because it was new and challenged the conventional view of the subject, couched then in a 'Davisian' evolutionary/historical frame of reference. But twenty years on in the 1980s, a single-minded concentration on short-term surface process measurements to the exclusion of other endogenetic processes and longer timescales of enquiry is misplaced. The value and limitations of 'process geomorphology' have long been evident, for instance within the framework of temporal and spatial scales examined by Schumm and Lichty (1965). Similarly, appraisal of theories of long-term landform evolution has not been absent from the literature (Adams, 1975; Melhorn and Flemal, 1975; Brunsden and Thornes, 1979; Ollier, 1979). Yet many studies have been retrospective, reinterpreting or evaluating existing theories more often than advancing new evidence and fresh concepts to challenge prevailing wisdom.

Perhaps the greatest problems in geomorphology concern the linkages between studies undertaken on different temporal and spatial scales, and the relative importance of exogenetic and endogenetic processes within these spectra (Summerfield, 1981). Sedimentology has benefited greatly from an understanding of the evidence for shortlived depositional events contained within strata collectively spanning perhaps 10^6 years and recording the subsidence of major crustal units. On the other hand *in situ* weathering profiles appear to evolve continuously until truncated by increases in surface erosion rates; while slope profiles may similarly become palimpsests with fresh soils, deposits and declivities superimposed on old macroforms (Starkel, in these Proceedings), or may develop into facetted mosaics of forms and deposits of different ages.

Understanding the long-term denudation of landscapes remains speculative, despite attempts to find bridges between theories and the evidence which supports them. The existence of planation surfaces is asserted by a host of writers, yet few attempt any serious explanation of their development. Our apparent inability to handle this problem may result in part from a reluctance to challenge or dispense with existing theories, but more likely it is a result of periodic excursions into what has become unfamiliar territory for most geomorphologists, and a corresponding lack of new field evidence to support our notions of long term change.

It is therefore not only timely but also imperative that geomorphology is increasingly studied within the context of the new global tectonics; a plea made long ago by Lester King (1962) before plate tectonic theory had gained wide acceptance in geology. It is also important to stress that King was able to advance a comprehensive theory for the formation of planation surfaces even if it does not now command universal acceptance, and it is this linkage between tectonics and denudation that must form the foundation of what Ollier (1979) has called 'evolutionary geomorphology'.

Understanding mechanism and system amongst earth surface processes is valuable in itself, as well as for applied studies concerning floods and sedimentation. But, alone, it is not sufficient for either theory or application in a broader geomorphological context. Earth resources, whether seen as soil parent materials, or as economic minerals from aggregate and sand to gold and diamonds, result from the detailed patterning of the surface by successive generations of formative processes many of which are endogenetic in origin. Similarly, surface stability and instability frequently respond to the presence of inherited forms and deposits reacting within a spectrum not only of precipitation events of varying intensity and duration, but also of inputs of energy from tectonism. These inputs often go unrecognised and are likely to occur at varying rates and over long time periods. Such considerations cannot be ignored lightly, however preoccupied geomorphologists may have become with the capture of flow and system at a point in time and place.

A neglect of geology by geomorphologists has always been unwise, and has become a hallmark of much research into surface processes. Ostensibly this is because the 'past' of a dynamic system is either irrelevant or unknowable, and inherited materials within the system can be accepted as state variables without enquiry into their origins. But this view is limiting to the scope of research enquiry and for the inferences which can be made from field observations and measurements.

LONG-TERM LANDFORM DEVELOPMENT

It is important for geomorphology that 'the state of the art' of historical landform analysis be reviewed and assessed. It has long been apparent that earlier preoccupation with erosional slopes and landscapes led to a neglect of much of the evidence for landscape evolution which was locked up in weathering mantles and in sedimentary covers. This neglect has now been redressed by many studies and a new understanding of weathering mantles, but the use of correlative sedimentary deposits has been less convincing, because of the spatial separation of source area from deposit and the ambiguities which this imposes on interpretations of sediments the provenance of which is often not closely defined.

It is often the aim of such reconstructions to relate sedimentary formations and their unconformities to erosional events in the long term, and particularly to the formation of 'erosion surfaces'. It is perplexing that after a century of argument and observation of the continents, no generally accepted mechanism for planation has been forthcoming. Although pediplanation as advanced by King is accepted by many, it remains a largely untested theory, as often refuted as it is defended. Planation as a cyclic or evolutionary progression of changing form is seen as essentially Davisian (Melhorn and Flemal, 1975), and is opposed by many on the grounds that the landscape is an open system tending always towards equilibrium. But the place of the concept of 'dynamic equilibrium' (Hack, 1960) in the context of long-term landform studies has proved elusive, and different kinds of equilibrium have been advanced.

Büdel (1982) argued that Penck's (1924) concept of the *Primarrumpf* should be employed to imply the maintenance of landsurfaces of low relief during prolonged, slow uplift, by the continuous lowering of the 'Doppelten Einebnungsflachen', as a form of dynamic etchplanation. Hack's (1960) view of the simultaneous lowering of channels and divides, ideally without changing the form of the landsurface during substantial lowering, has many similarities with this concept, and also incorporates a basis for differential erosion between bedrocks of varying resistance to weathering. Büdel (1982) was also able to extend his theory to warped and faulted landsurfaces within which 'etchplain stairways' might be formed, as the lower portions of the warped surface etch back into the uplift until a marked escarpment is formed (see his Fig. 49). The age relationships between the etchplains above and below the escarpment are thus defined by modification of a single warped landsurface. This view is quite different from that of King (1957), who saw scarp formation, hillslope retreat and the extension of pediments to form pediplains as universal combinations of process and form, in which a younger surface would ultimately completely replace an older one.

Studies which have sought to test essentially Davisian concepts of a cycle of erosion or planation, involving the eradication of orogenic relief (Schumm, 1975, Ahnert, 1970), leave many of these conflicts unanswered, though they attempt to specify the nature of tectonism (orogeny and epeirogeny) more carefully. In particular, the generation of escarpments and the ways in which they subsequently evolve have been subjects surrounded by controversy. But where a broad uplift or cymatogen is developed it seems possible that Büdel, Hack and King would all allow for the steepening of the flanking slope to form both an escarpment and a surface of low relief, in equilibrium with both rock mechanics (King, 1977) and the resistance to weathering of the constituent rock formations. The extent and importance of scarp retreat as a subsequent mechanism of relief modification remains uncertain.

It is clear that past neglect of the tectonic framework and dynamics of landscapes has led to a serious loss of information, and an inevitable continuation of a body of untested and unchallenged theory. It is refreshing now to see that studies of stable landsurfaces beneath which weathering may have penetrated deeply are being complemented by enquiries into mobile terrains containing complex sedimentary formations. Equally, the realistion that some quite 'old' terrains have been tectonically disrupted and disturbed during the later Cenozoic is important for the interpretation of evidence and for our perceptions of landscape change through time.

New perceptions of landscape change represented in the papers published in these Proceedings avoid the unproductive extrapolation of short-term measurements to longer time periods. Instead they concentrate upon the evidence of dating from lava flows (Dohrenwend et al.), of tectonic history and sedimentation (Vergnolle), from relict weathering mantles (Molina et al.; Smith and McAlister), and from the whole complex of palaeoforms and deposits (Battiau-Queney; Hall; Starkel). Starkel's view of the incorporation of inherited forms within the contemporary morphogenetic system articulates a common understanding and experience that is too seldom the focus of comment. Yet this attempt to create a qualitative model for the landscape is vital to its comprehension within studies of present-day processes or for applied purposes.

In an attempt to focus present and future debate on some of these issues we offer here views based on our own experience concerning two major issues; the influence of large-scale tectonics on landscape evolution, involving both the generation of new erosional relief and formation of sedimentary sequences, and the character and style of surface alteration and reduction.

LARGE-SCALE TECTONICS AND LANDSCAPE EVOLUTION

In contrast to the emphasis over the past two decades on investigations concerned with the nature and rate of contemporary surface processes, geomorphologists as a whole have been remarkably reluctant to consider the operation of endogenetic mechanisms in general, and tectonic processes in particular. This neglect is unfortunate for at least two reasons: first, no comprehensive model of long-term landscape development can ignore the tectonic regime within which landforms evolve since this controls the spatial and temporal variations in rates of uplift; secondly, geomorphological evidence, especially in the form of uplift and warped planation surfaces, could potentially provide a valuable empirical basis for evaluating the wide range of geophysical models currently being proposed to explain epeirogenic uplift.

The most significant contrast between tectonic regimes is that between convergent plate margins on the one hand, and passive margins and continental interiors on the other. The relationship between convergent margin tectonics and long-term landform development has been graphically demonstrated by Adams (1985) in his study of oblique plate convergence along the Alpine Fault in New Zealand, but there have been few other studies of this kind. Similarly the relationship between tectonic mechanisms generating uplift in regions remote from active plate boundaries and landscape evolution has barely begun to be addressed. One potentially fruitful line of investigation is the role of hot spots in generating domal uplifts which may migrate across a continent as a result of plate motion with respect to more or less stationary sub-lithospheric thermal anomalies (Summerfield, 1985a). This idea has been applied to the interpretation of patterns of erosion and deposition in the sedimentary record (Crough, 1979) but has yet to be evaluated with respect to the likely effect on large-scale drainage patterns. Complementary to such an application of tectonic models to geomorphological problems is the use of landform evidence in evaluating competing geophysical models; this is well illustrated by the current debate over the mechanism responsible for the uplift of the Lesotho Highlands in southern Africa where both a hot spot and a phase change model have been applied (Hartnady, 1985; Smith, 1982).

A primary focus of geomorphological interest in investigations of long-term landform development is the continental margin since uplift and subsistence in this zone affects the base level to which all continental drainage systems are ultimately related. Most continental margins are of the passive type, either rifted (divergent orthogonal to oblique motion) or sheared (divergent oblique to transform motion), and the outlets of the great majority of the world's major drainage systems are located along such margins. The tectonic regime of passive margins and their evolution from their initiation through continental rifting is therefore a subject of considerable geomorphological significance (Ollier, 1985; Summerfield, 1985b, 1985c, 1986).

Passive margins undergo two phases of development: a rifting phase characterised by thinning of the lithosphere with probable associated uplift, and a later subsidence phase where the margin subsides as it cools and is subject to sediment loading along the newly-formed continental edge. The timing and magnitude of uplift to be expected in association with continental rifting and the development of new passive margins depends significantly on the nature of the rifting mechanism involved.

Two major types of rifting sequence have been suggested (Keen, 1985). In *active rifting* volcanism and uplift are considered to precede continental rupture and the rifting event is driven by a convective upwelling of the asthenosphere. Geophysical modelling suggests that a 200 km diameter convective upwelling could produce a domal uplift some 1000 km across with a maximum elevation of around 1000 m. By contrast *passive rifting* is driven by extensional stress in the lithosphere and the resulting lithospheric stretching leads to both a thinning of the lithosphere and the rise of hot asthenosphere towards the surface. In this model volcanism is considered to follow rifting and whether uplift occurs will depend on the balance between the degree of lithospheric thinning and thermal isostatic effects. Estimates of the amount of uplift to be expected vary considerably but most models suggest fairly modest figures of up to 500 m or so. Some nascent passive margins, which apparently experienced rifting prior to uplift, such as those bordering the Red Sea and Gulf of Suez, exhibit rift flank elevations in excess of 1000 m and this has led to the suggestion that the thermally related uplift associated with lithospheric extension may be augmented by the effects of small-scale mantle convection induced by rifting (Steckler, 1985).

As a passive margin ages, subsidence begins to predominate over uplift, at least along its oceanward flank. Subsidence is driven by both lithospheric cooling and the isostatic loading of the margin by the growing wedge of sediment accumulating offshore. This sediment loading in turn induces rotation and flexure along the margin which may promote further uplift inland. With time sediment loading becomes progressively more important with respect to thermal subsidence while cooling leads to an increase in the flexural rigidity of the lithosphere and consequently a broadening of the zone over which flexure occurs (Summerfield, 1985b).

A further contribution to the rotation of the margin is provided by denudational unloading. As erosion proceeds inland from a newly-formed passive margin either through escarpment retreat or the growth of etchplains, a broad zone, perhaps several hundreds of kilometres across, is likely to experience net uplift through denudational unloading because the rigidity of the lithosphere augments the area affected by the isostatic adjustment. Although suggested in the 1950s as an explanation for the escarpment retreat model, periodic uplift and the generation of successive erosion cycles on the African continent (King, 1955; Pugh, 1955) it has not been subject to a recent evaluation in the context of current thinking about passive margin tectonics. A further consequence of isostatic adjustments to lithospheric unloading through escarpment retreat or etchplanation is that the escarpment rim can be elevated to a height well above that achieved during an initial thermal uplift since the magnitude of isostatic uplift in this zone will far exceed the minimal plateau summit erosion.

A final contribution to continental margin morphology which may be important locally is the constructional topography arising from the accumulation of rift-related volcanics and intrusives. Thicknesses of Karoo basalts in southern Africa exceed 1000 m and appear to have extended well inland of the continental margin zone. The thickest remnants of these vast lava flows are to be found in the Lesotho Highlands which, of course, form the highest relief in southern Africa.

In summary six factors relevant to landform development along passive margins can be identified (Fig. 1). An important future direction of geomorphological research will be the quantification of these effects both through the application of existing geophysical models and by reference to evidence for landsurface uplift and flexure. Already a major revision of the landscape history of southern Africa is underway (Partridge and Maud, in prep. cited in Patridge (1985)) and there is currently an active debate between geomorphologists and geophysicists in research on the uplift history of the East Australian Highlands (Summerfield, 1986).

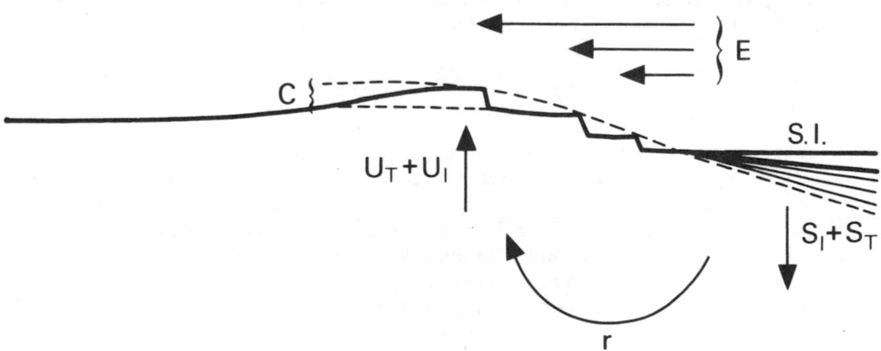

Fig. 1. Six major tectonic factors controlling the long-term morphological evolution of passive margins. 1) U_T = thermally-driven uplift; 2) U_I = isostatic uplift associated with denudational unloading; 3) S_T = thermally-driven subsidence; 4) S_I = isostatic subsidence associated with sediment loading; 5) r = rotation of margin associated with U_I and S_I; 6) E = escarpment retreat (?) related to episodic rejuvenation and/or structurally-controlled; C = constructional topography generated by rift-related volcanics and intrusives.

An example of the relevance of studies of long-term landform development to the geophysical modelling of the tectonic evolution of passive margins is provided by the role of sub-aerial erosion. In some of the early models of the development of nascent passive margins it was assumed that sub-aerial erosion of the uplifted rift flanks would contribute significantly to lithospheric thinning (Sleep, 1971). Although it is now generally agreed that the significance of this effect has been over-estimated, sub-aerial erosion is still considered by most geologists and geophysicists to be a possibly critical factor in the tectonic and sedimentological evolution of passive margins (Pitman and Golovchenko, 1983; Watts and Thorne, 1984). Unfortunately, the somewhat naive assumptions made in the estimation of denudation rates in such studies of passive margin tectonics, with rates frequently being assumed to vary as a simple function of mean elevation, cast considerable doubt on the rates of lithospheric thinning proposed. Although mean elevation may be highly correlated with local relief (which has been shown to be a good predictor of regional erosion rates (Ahnert, 1970)) in areas of mountainous terrain, this is not likely to be the case along most passive margins where an elevated interior plateau with minimal local relief is separated from a fringing coastal plain by a major escarpment. Whereas erosion rates in the escarpment zone are no doubt high, those inland have been demonstrated to be very low (Bishop et al., 1985a, 1985b). Clearly further research is required to establish long-term erosion rates in relation to a range of evolutionary landscape models which incorporate climatic effects and realistic assumptions about rates of continental margin uplift.

STRATIGRAPHIC EVIDENCE FOR LONG-TERM SEA LEVEL CHANGE AND PATTERNS OF CONTINENTAL DENUDATION

The often sparse evidence of long-term landform development preserved on the continents heightens the importance of the record of sub-aerial denudation provided by the sedimentary sequences of continental margins. Geomorphologists have become familiar with the utilisation of lake and channel sediments to document changes in sedimentation rates over relatively short time spans and areas of modest extent but have yet to appreciate fully the possibilities of such a technique applied at vastly greater temporal and spatial scales. There are of course considerable problems involved in relating the sedimentary record of continental margins to the tempo of denudation within their hinterlands; particular difficulties include the estimation of the location and areal extent of the source region and the redistribution of sediment by near-shore and deep ocean currents once it has left the sub-aerial zone. Nevertheless, applied over regions of appropriately large size and over long periods of time (millions of years), this technique can provide good first order estimates of temporal changes in denudation rates and a valuable basis for the comparison of what are frequently seen to be the anthropogenically inflated erosion rates calculated from present day sediment yield data. In the Natal region of southern Africa, for instance, contemporary erosion rates appear to exceed the average for the past 100 Ma or so by a factor of between 12 and 22 (Martin, 1984; Murgatroyd, 1979).

Data on off-shore sedimentary sequences have grown rapidly over the past two decades or so both from the work of the Deep Sea Drilling Project and boreholes drilled as a part of oil exploration programmes. Access to the latter source of information is restricted by confidentiality requirements but many generalised records have been published. During the past ten years seismic stratigraphy has provided an increasingly important means of acquiring information on the sedimentary sequences of continental margins. Far larger areas can be covered at a much lower cost than with isolated boreholes although the latter are of course required to establish a chronology for the seismic profiles obtained. Data of this kind are potentially of great value in attempts to correlate temporal patterns of off-shore deposition along continental margins with records of long-term continental erosion (Summerfield, 1985b, 1986).

The original interpretation of passive margin seismic stratigraphy presented by Vail and his co-workers was based on an inferred global correlation of seismic sequences and postulated eustatic sea level changes characterised by geologically instantaneous falls and more gradual rises (Vail et al., 1977). Although this model has subsequently been revised, its key element - the significant role attributed to eustatic sea level falls in generating supposedly world-wide unconformities on passive margins - has been retained, and indeed, further supported in recent interpretations of the temporal correlation between depositional hiatuses in ocean cores and the major unconformities identified from seismic stratigraphy (Schlee, 1984). Apart from the lack, so far, of a rigorous statistical evaluation of these suggested relationships, several significant problems in the 'eustatic interpretation' of seismic sequences have been raised. In addition to the now acknowledged effect of passive margin subsidence (Pitman, 1978; Pitman and Golovchenko, 1983) and its probable inter-regional synchrony following a phase of continental break-up (Watts, 1982) a number of further problems remain (Thorne and Watts, 1984). For instance, outside periods of significant glaciation, relative rates of eustatic sea level change are too slow to generate unconformities except along old, slowly subsiding margins. Moreover, the limitations of seismic and biostratigraphical resolution mean that it is unlikely that hiatuses with a duration of <4 Ma can be identified on thermally subsiding margins.

Although improvements in the resolution and interpretation of seismic data may overcome these difficulties there remains a further problem which both proponents and critics of seismic stratigraphy have hitherto largely ignored but which is fundamental to the understanding of the temporal and spatial patterns of deposition and erosion along passive margins. This is the assumption that a eustatic sea level fall, if of sufficient rate, duration and magnitude, will invariably give rise to an unconformity and that, more generally, eustatically generated changes in base level are related in a simple manner to rates of sub-aerial fluvial erosion and hence the depositional sequences recorded in marine sediments. Variations in the rate of sediment supply to passive margins have been acknowledged as a factor affecting regressions and transgressions, but they have not been explicitly linked to base level changes. As outlined by Summerfield (1985b) changes in gradient across the shelf zone over which the shoreline migrates during sea level change will have an important influence on the balance of erosion and deposition along continental margins. Rejuvenation of fluvial systems inland during a period of eustatic sea level fall is likely to be confined to rivers where channel gradient over the newly exposed shelf is greater than that in the original coastal zone. Along extensive sections of passive continental margins, however, offshore gradients are less than those of channel gradients immediately inland.

The importance of variations in coastal morphology along continental margins in influencing patterns of sedimentation and the occurrence of unconformities is now being appreciated but has yet to receive sufficient emphasis in eustatic interpretations of seismic stratigraphy. Hiatuses may occur, for instance, during periods of high sea level through the trapping of sediment in estuaries and other near-shore environments (Loutit and Kennett, 1981). A further factor concerns the relationship between off-shore deposition and uplift inland. During a high sea level phase increasing shelf sedimentation and resulting onlap will promote isostatic subsidence on the shelf which may, depending on its extent and duration, and the flexural rigidity of the lithosphere along the passive margin, generate uplift inland through rotation about the land-shelf hinge line (Chappell, 1983). Depending on the coastal profile gradients across the margin this is likely to cause increased river incision inland and accelerated deposition off-shore.

In spite of the emphasis on non-eustatic mechanisms of base level change in a number of recent overviews of continental denudation chronology (King, 1976; Melhorn and Edgar, 1975) major eustatic sea level changes have clearly occurred over the past 150 Ma both as a consequence of changes in ocean basin volume and, especially during the Late Cenozoic, as a result of changes in the volume of ocean water associated with glacial-interglacial cycles. The chronology, rate and magnitude of the latter are now well known from ocean core oxygen isotope data but estimates of the progressive post-Late Cretaceous sea level fall generated by a decreasing ocean ridge volume associated with reduced rates of sea-floor spreading are less certain. A recent detailed evaluation, however, indicates a probable global sea level at 80 Ma B.P. some 230 m above present sea level (Kominz, 1984). Little attempt has been made to evaluate the impact on long-term landscape evolution of the progressive, but irregular, fall in global sea level since the Late Cretaceous and to differentiate the effects of such eustatically generated base level changes from the regional consequences of continental margin tectonics.

SURFACE ALTERATION AND REDUCTION

The twin processes of weathering and erosion have been the traditional concern of historical geomorphology for nearly a century, during which the approach has swung from the anthropogenic 'cycle of erosion' (Davis, 1899) to the open system in 'dynamic equilibrium' (Hack, 1960). Both concepts implicitly support the view that 'the present is the key to the past' in the earth sciences. But the understanding that earth environments and continental configurations have evolved through geological time, creating unique landscapes containing inherited and superimposed features (Ruxton, 1968, Starkel, in these Proceedings) limits the usefulness of this dictum.

Ruxton (1968) has emphasised the problems of 'multicomplexity of process and inheritance' in creating 'disorder' in natural landscapes, while Brunsden and Thornes (1979) have more recently approached similar questions in terms of 'landscape sensitivity and change', focussing on the establishment and disturbance of equilibria in landscapes over long time periods. The work of Hack (1960, 1965, 1975) also reveals an interest in the long term effects of the maintenance or departure from equilibrium in the landscape and focusses on questions related to both the formation of weathering mantles and the development of planate surfaces. Debates about planation surfaces have often become unproductive and circular in argument, yet plains exist and associated with them are many of the weathering mantles that receive so much discussion.

One source of debate is the role of weathering itself in planation. Some argue that it promotes the development of the ultimate plain (Büdel, 1982, Twidale, 1981), but evidence from the passive margins suggests that, during the last 100 Ma at least, rock decay has assisted in the differentiation of terrain according to the geochemical properties of rocks (Thomas, 1980; Kroonenberg and Melitz, 1983). The behaviour of weathering systems (and therefore of weathering profiles) during geological time periods thus becomes a critical question in the development of models for long term landform evolution.

The survival of strongly differentiated profiles is indicative of long maintained landsurface stability at the site, but even quite deep profiles may develop beneath landsurfaces that are not immune from surface erosion. These are commonly undifferentiated and less advanced in loss of replaceable ions than those of the former kind. Other profiles were developed below stable surfaces but have become truncated. Some way of bringing these observations together is required and it can be suggested that profile behaviour in the long term may conform to one of the following:

1. *Profile differentiation* beneath stable landsurfaces and where groundwater levels are maintained at a constant level over long periods.

2. *Profile deepening*, beneath stable surfaces often protected by forest or duricrusts, and when deep water circulation is promoted by relief development.

3. *Profile lowering* in dynamic equilibrium as landsurface & weathering front are simultaneously lowered during slow ground-surface reduction.

4. *Profile thinning* during gradual lowering (2) wherever resistant rock formations are encountered at depth.

5. *Profile truncation* due to the onset of tectonic or bioclimatic rhexistasie, leading to erosion/sedimentation phases.

6. *Profile renewal* following any of 3, 4, 5 above.

These conditions (1-6) may be related to the concept of etching, etchsurface formation and modification (Thomas, 1974, 85; Büdel, 1982; Fairbridge and Finkl, 1980). They also respond not only to external disturbance or input to the morphogenetic system but also to geochemical controls within the bedrock. In this way a geochemically controlled differential erosion takes place and may contain elements in dynamic equilibrium.

BIOCLIMATIC ENVIRONMENT AND LANDFORM DEVELOPMENT

It is also necessary to take into account the effects of changing bioclimatic environments through geologic time, a topic not always well covered by recent studies of weathering mantles in Europe. Hall (in these Proceedings and 1985) is one of few authors who have considered carefully the possible contribution of temperate Mio-Pliocene and early Pleistocene climates to gruss (arènes) formation in northern latitudes, although Bakker and Levelt (1964) considered the problem more than twenty years ago and Smith and McAlister (in these Proceedings) support the view that arenisation can result from weathering under temperate climatic conditions. In Spain, Molina and others (in these Proceedings) write of mineralogical 'homogenisation' accompanying the lowering of the weathered mantle with the relief, and implying the adjustment of surface mineralogy to external environmental conditions which have remained near-tropical, semi-arid throughout much of Spain since the Palaeocene; the profiles responding more obviously to tectonic upheaval than to environmental changes. In higher latitudes, in Britain, Ireland and Scandinavia for example, strong environmental gradients (with latitude and with altitude) are also reflected in sharp climatic oscillations over time. Such changes accompanied by the effects of Alpine tectonics, including widespread volcanism and faulting, make palaeogeomorphological reconstruction more complex.

Nevertheless, when Dury (1971) called for the wider recognition of old weathering crusts in Britain, he was recognising the weight of geological evidence (Montford, 1970) and converging with well established interests in continental Europe. In the intervening 15 years much of the evidence has been further elaborated, but less attention has been paid until recently to the implications for long-term landform development. In a review of the British experience following the London Conference of the BGRG in 1976, Thomas (1978) called attention to the need for the application of models for tropical landform development to formerly sub-tropical or tropical landscapes in high latitudes. Such models include the formation of 'Doppelten Einebsnungflächen' (Büdel, 1957) and the development of etchplains (Thomas, 1968, 74) which have been described from southern Sweden by Lidmar-Bergström (1982) and from northeast Scotland by Hall (1985, and in these Proceedings).

The formation of pediments and glacis also has to be considered. In the work of L.C. King (1957), the pediment is seen as a surface extended by backwearing of hillslopes, but according to Hack (1960) pediments may be an expression of differential erosion and the establishment of different equilibria on adjacent and contrasting lithologies, while glacis are dominantly surfaces of accumulation or mantle truncation in the French literature. Dury (1972) again drew attention to the occurrence of pediments in the British landscape but avoided the question of their long-term development. Yet broad piedmont surfaces which are neither of glacial nor obviously of fluvial origin occur widely in the British Isles, and not always in the sedimentary scarplands, where special explanations were forthcoming many decades ago.

There remains a need to evaluate the available models for landform development in the context of higher latitude landscapes, and to reconcile the observation of widespread plains with the obvious features of differential erosion. The extent to which it may be justifiable to match the fragmentary evidence from Palaeogene Europe to models of landform development based on studies of the present-day tropical cratons studied in Africa and other continents has yet to be tested further.

RELIEF DEVELOPMENT ON PASSIVE CONTINENTAL MARGINS

The common histories of passive margins involving rifting and uplift, followed by subsidence hinged close to the depositional edge of the cratonic basement, appear consistent with the generation of escarpments and the elevation of base-levelled plains. These may in turn be transformed into strongly differentiated terrains controlled by rock mineralogy and lithology (cf. Thomas, 1980, and Kroonenberg and Melitz, 1983), involving the duricrusting of weathering profiles and the further advance of weathering fronts into susceptible rocks during landsurface lowering, creating a variety of etchsurfaces. According to Kroonenberg and Melitz (1983) the appearance of steps in the landscape of Surinam can be attributed to lithological contrasts differentiated by deep weathering under humid-tropical conditions, but later stripped of saprolite and emphasised by surface processes under semi-arid, savanna conditions, a model which fits closely to an interpretation of the relief of Sierra Leone (Thomas and Thorp, 1985).

The applicability of such a model to north west Europe depends not only on the alternation of humid and semi-arid episodes within the early Cenozoic, together with an appreciation of the effects of climatic cooling and glaciation in the Pleistocene, but also on the rate and amount of uplift and the way in which pulses of erosional energy have passed through the landscape (Brunsden and Thornes, 1979) entraining slope mantles and river deposits.

Over susceptible rocks, slow degradation of the landscape can take place by a process of dynamic etchplanation (Büdel, 1982; Thomas and Thorp, 1985), providing that uplift is sufficiently slow. The case of Buchan in NE Scotland cited by Hall (in these Proceedings) is important in that the eastern coastal plains form an etchplain, which is diversified into differentially weathered and scoured basins which increase in relief in a westerly direction towards the warped and uplifted main divide between Atlantic and North Sea drainage. To the west of this divide, the Scottish mountains exhibit the effects of stronger tectonic deformation and increased differential denudation. Initially, the imprint of any etchplanation in the Neogene appears merely superficial, a decoration on the macroforms of the landscape; this view is taken by King (1967, 82) concerning the relief within the, so-called, Natal Monocline, South Africa. Yet Linton (1951) long ago pointed to the occurrence of granite basins in the Scottish Highlands, and the existence of broad granite surfaces of low relief such as Rannoch Moor recall the granite etchsurfaces of the tropics, despite their history of uplift, dissection and multiple glaciation (Thomas, 1978). However, the metamorphic rims of such basins are nowhere deeply weathered today, and we should perhaps recognise that the major part of the Cenozoic was a period of volcanism, tectonism and climatic cooling, leading to a widespread dismantling of the older tropical landscapes of the Palaeogene. Indeed Lidmar-Bergström (1982) has placed this phase earlier, in the middle Mesozoic. Nevertheless, when the depth of erosion below ancient levels in both eastern and western Scotland (Stewart, 1972; Hall these Proceedings) has been calculated it has been found to be small except along lines of strong linear erosion.

Where the temperate landscapes differ markedly from those of tropical cratons, is in the absence of deep profiles beneath duricrusted hills, profiles which exceed 100 m depth in parts of Africa and South America. Such duricrusts (cuirasses), if they were formed at all were probably a product of the Mesozoic or earliest Palaeogene. Crusts of this age are found widely in Africa, but they appear to have been largely destroyed in higher latitudes, except where buried by younger sediments. It is, however, possible that they were never formed to the extent or the depth of tropical examples. Rock formations are also different, perhaps in important respects. Perhaps this observation supports the view that few fragments of Palaeogene and earlier landscapes survive in the Atlantic margins of Europe. The saprolites and their protective crusts were possibly too easily destroyed by the erosion accompanying Cenozoic tectonism. A type of silcrete duricrust is found, however, as flaggy 'sarsen stones' (Summerfield, 1979) overlying the Chalk in southeast England, and the buried 30 m thick weathering profile recorded by Nilsen and Kerr (1978) from the basalts of the Iceland-Faroe Ridge is also testament to such, largely vanished landscapes. The Eocene and Oligocene sedimentation in southern England is further evidence of these events. It is therefore all the more important to the argument developed here that the cooler temperate climates of the Neogene were apparently able to sustain a modified form of etchplanation, seen today in the widespread survival throughout Europe of the arènes or gruss derived from granites.

The palaeoenvironmental significance of surviving Tertiary (and older) deposits in present-day landscapes should therefore be seen as the starting point for argument concerning the development of the landsurface, and not the end product of historical reconstruction. Klein (1974), in reference to the relief development of the Armoricam Massif in France, listed, i) tectonic reactivation, ii) acyclic evolution, and iii) morphogenetic sequence as the three concepts guiding his interpretation. This 'acyclic' view of landscape has been prevalent among process geomorphologists for twenty years, yet its embodiment in the ideas of Hack (1960, 1975) has not often led to its application to long-term landform development, possibly because it does not afford an easy model for the dating of events. Thomas (1980) and Kroonenberg and Melitz (1983) have all attempted to employ such ideas, if only tentatively in accounts from tropical cratonic landscapes. There seems little doubt that different parts of a landscape behave in contrasting ways, recalling Crickmay's 'hypothesis of unequal activity' (1975). If this contrasting behaviour can be tied to an understanding of tectonics and lithology, and in turn to the behaviour of weathering profiles over the different rocks, then perhaps new concepts for the development of landscapes in the long term can be developed.

Melhorn and Edgar (1975), on the other hand, have argued for episodic continent-wide erosion surface development, during at least four intervals of crustal quiescence, each lasting c.25 Ma during the Mesozoic and Cenozoic, and similar to the constructs of King (1950, 67). But neither the existence of the surfaces nor the theoretical requirement for such a development has been established, and a combination of etchplanation and geochemically controlled differential etching and denudation during pulses of uplift and accelerated stream erosion, satisfy the requirements for vertical lowering and incision of the landsurface. Within such a framework escarpments may be generated along fault lines or by the development of etchscarps in the flanks of domal uplifts (Büdel, 1982), and pediments may form by lateral extension, in unconsolidated materials, in quasi-horizontal rock formations, and at the disconformities between sedimentary covers and basement rocks. Pediment or glacis surfaces will also form during differential lowering of less resistant rocks and without significant backwearing (Hack, 1960). According to Finkl and Fairbridge (1981), this 'cratonic regime' is characterised by long periods of etchplanation interrupted by short phases of pediplanation, associated with pulses of uplift. In areas and at times of orogenic relief development, Ahnert (1970) has shown that much longer periods of time would be required to form planation surfaces of minimal relief.

TECTONICS AND LANDFORM DEVELOPMENT: THE WEST AFRICAN CONTINENTAL MARGIN IN SIERRA LEONE

In Sierra Leone and adjacent parts of the West African Craton, the basement has been subject to a domal uplift (Leo Uplift), accompanied by faulting (McFarlane, 1981), possibly associated with igneous events in the Triassic and Cretaceous periods, the latter a kimberlite event which may have been associated with post plate motion crustal flexuring (Haggerty, 1982). Although there are possible planation surface remnants to be found in this area, agreement on correlations according to altitude and age has not been reached (Thomas, 1980), and the regional significance of some altitudinally correlated remnants must remain in doubt. Two supposedly extensive 'surfaces' occur: a Coastal Plain Surface, widely developed across less resistant rocks, and a Main Plateau Surface that covers the northeastern half of the country and is internally so complex that its recognition must be in doubt. The metamorphic grain of the basement is followed by the Atlantic flowing drainage in the east of the country and here there is a clear interdigitation of different relief levels, etching out the lithological differences which are exploited by the rivers (Figures 2, and 3). To the west, rivers break across the duricrusted schist belts that form a marked escarpment which is continued in broken form where granitoid rocks outcrop.

The role of the linear drainage in carrying waves of erosion inland is obvious, but it is only really effective where it can exploit differential weathering patterns. In fact the detail of the cratonic landscape is everywhere controlled either by outcrop pattern or by structural lineaments, including some quite major faults. Near the axis of the uplifted dome, faulted blocks and increased altitude and dissection give rise to striking inselberg landscapes in the Loma Mountains and the Tingi Hills. Any former surfaces of early Cretaceous or Jurassic age have undoubtedly become fragmented if not entirely destroyed, while possible Palaeocene duricrusts which preserve summits at levels between 600 and 800m asl on the flanks of the uplift, may also be submerged beneath coastal plain (Bullom Group) sediments below sea level near the coast. In fact, although there is abundant evidence for local 'planation' (P.K. Hall, 1969), the relief can be interpreted in terms of differentiation from a weathered landsurface (cf Mabbutt, 1965 on central Australia) preserved by duricrusts on the schist belts, and which has been subject to domal uplift and faulting (Figures 2, and 3) and subsequently to dissection and geochemically controlled differential erosion to form a mosaic of local landscapes.

These relationships are accompanied by the observation that much of the country is being lowered over susceptible rocks, by dynamic etchplanation (Thomas and Thorp, 1985), and that pediment forms are of limited extent, insofar as this can be determined. The weathering profiles beneath the duricrusts are strongly differentiated and continue to deepen, while many over the granitoid rocks are possibly being lowered and renewed in dynamic equilibrium, although thinning from below can take place, as resistant kernels of massive rock are intersected by the eroding land surface, emerging as whalebacks and inselbergs. There is also widespread evidence for major semi-arid episodes which have stripped the weathered mantle widely from granite hills and other formations, and formed glacis across the lower zones of weathering profiles within the drainage basins.

The understanding of this terrain thus requires details of tectonic uplift and subsidence, including confirmation of faulting and its

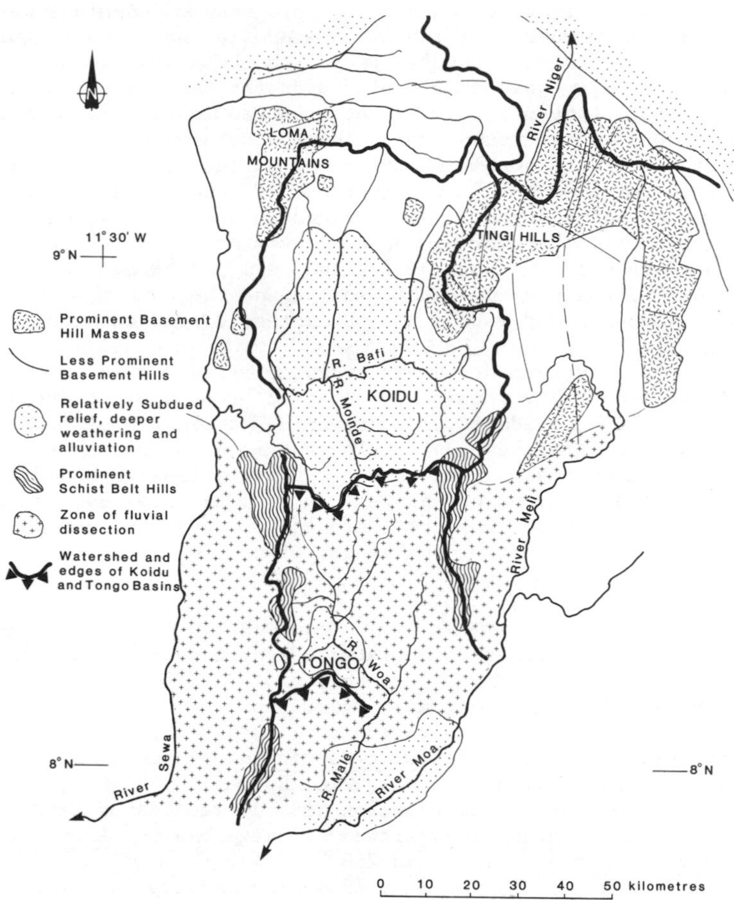

Fig.2. North eastern Sierra Leone, aspects of Geomorphology

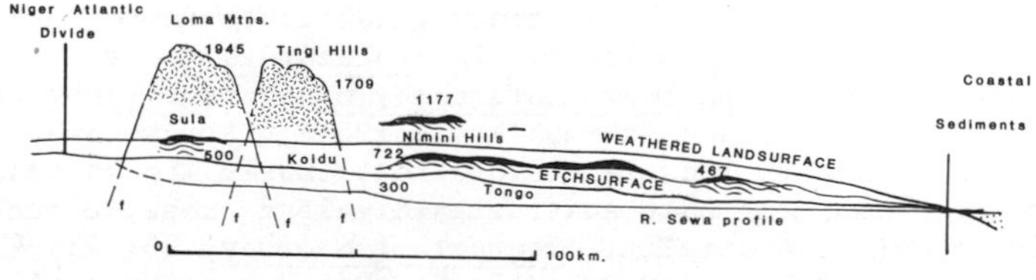

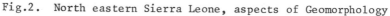

Fig.3. Schematic composite N-S relief profile and interpretation

relations to present relief. It needs a full appreciation of the geochemical control exercised over the progress of chemical weathering, and of the manner in which surface erosion patterns in turn reflect such controls. Finally, there remains a need to provide dates for any major episodes of planation, if these have indeed occurred. Such time periods will be measured in 10^7 yr., but there is also a need to understand more clearly how erosional/depositional forms such as glacis become superimposed on the etchplains within time periods of 10^5 or 10^4 yr. In this way landform evolution in the long term can be constructed and refined as new information is supplied from field research. Some might wish to substitute 'pediplain' or even 'peneplain' for the term 'etchplain' used here. This is not ultimately important, so long as the manner of landscape lowering is understood within the parameters discussed here.

Such models of development remain oversimplified and in part unsubstantiated; they require elaboration or possibly contradiction in the face of new evidence or concepts. The wider applicability of such theories of landscape development is also important to determine if geomorphology is to benefit fully from an understanding of the new global tectonics, as well as from the progressive refinement of its traditional concerns with the processes of denudation whether within the short or the longer timescales.

REFERENCES

Adams, G.F. 1975. Planation Surfaces - Peneplains, Pediplains, and Etchplains. Benchmark Papers in Geology 22, Dowden, Hutchinson and Ross, Stroudsburg, 476pp.

Adams, J. 1985. Large-scale tectonic geomorphology of the Southern Alps, New Zealand, in Tectonic Geomorphology (Eds. Morisawa and Hack), pp 105-128. Allen and Unwin, Boston.

Ahnert, F. 1970. Functional relationships between denudation, relief, and uplift in large mid-latitude drainage basins. American Journal of Science, 268, 243-263.

Bakker, J.P. and Levelt, Th. W.M. 1964. An enquiry into the problems of a polyclimatic development of peneplains and pediments (etchplains) in Europe during the Senonian and Tertiary Period. Publication Service Cartographie Géologie, Luxembourg, 14, 27-75.

Battiau-Queney, Y. (in press). Tertiary inheritance in the present landscape of the British Isles. In Proceedings First International Geomorphology Conference, Manchester 1985, Wiley.

Bishop, P., Young, R.W. and McDougall, I. 1985a. Stream profile change and longterm landscape evolution. Early Miocene and modern rivers of the East Australian Highland crest, central New South Wales, Australia. Journal of Geology, 93, 455-474.

Bishop, P., Young, R.W. and McDougall, I. 1985b. Stream profile change and longterm landscape evolution early Miocene and modern rivers of the East Australian Highland Crest, Central New South Wales, Australia, in Abstracts of Papers for the First International Conference on Geomorphology (Ed. Spencer), p.41 University of Manchester, Manchester.

Brunsden, D. and Thornes, J.B. 1979. Landscape sensitivity and change. Transactions Institute British Geographers, NS 4, 463-84.
Büdel, J. 1957. Die "Doppelten Einebnungsflächen" in den feuchten Tröpen. Zeitschrift für Geomorphologie, NF 1, 201-88.
Büdel, J. 1982. Climatic Geomorphology (Engl. trans. L. Fischer and D. Busche), Princeton Univ. Press, Princeton, 443pp.
Chappell, J. 1983. Aspects of sea levels, tectonics, and isostasy since the Cretaceous, in Mega-Geomorphology (Eds. Gardner and Scoging), pp 56-72. Clarendon Press, Oxford.
Crickmay, C.H. 1975. The hypothesis of unequal activity. In Melhorn and Flemal (Eds.), 1975, 103-110.
Crough, S.T. 1979. Hotspot épeirogeny. Tectonophysics, 61, 321-333.
Davis, W.M. 1899. The geographical cycle. Geographical Journal, 14, 481-504.
Dohrenwend, J.C., Wells, S.G., McFadden, L.D., and Turrin, B.D. (in press). Pediment dome evolution in the eastern Mojave Desert, California. In Proceedings First International Geomorphology Conference, Manchester 1985, Wiley.
Dury, G.H. 1971. Relict deep weathering and duricrusting in relation to the palaeoenvironments of middle latitudes. Geographical Journal, 137, 511-522.
Dury, G.H. 1972. A partial definition of the term 'pediment' with field tests in the humid climate areas of southern England. Transaction Institute British Geographers, 57, 139-52.
Fairbridge, R.W. and Finkl, C.W. Jr. 1980. Cratonic erosional unconformities and peneplains. Journal of Geology, 88, 69-86.
Hack, J.T. 1960. Interpretation of erosional topography in humid temperate regions. American Journal Science, 258A, 80-97.
Hack, J.T. 1965. Geomorphology of the Shenandoah Valley, Virginia and west Virginia, and origin of the residual ore deposits. United States Geological Survey, Professional Paper 484, 84pp.
Hack, J.T. 1975. Dynamic equilibrium and landscape evolution. In Melhorn and Flemal (1975).
Haggerty, S.E. 1982. Kimberlites in Western Liberia.: an overview of the geological setting in a plate tectonic framework. Journal Geophysical Research, 87, 10, 811-10, 826.
Hall, A.M. 1985. Cenozoic weathering covers in Buchan, Scotland and their significance. Nature, 315, 392-395.
Hall, A.M. (in press) Weathering and relief development in Buchan, Scotland. In Proceedings First International Geomorphology Conference, Manchester 1985. Wiley.
Hall, P.K. 1969. The diamond fields of Sierra Leone. Bulletin Geological Survey, Sierra Leone, 5, 133pp.
Hartnady, C.J.H. 1985. Uplift, faulting, seismicity, thermal spring and possible incipient volcanic activity in the Lesotho-Natal region, SE Africa: The Quathlamba hotspot hypothesis. Tectonics, 4, 371-377.
Keen, C.E. 1985. The dynamics of rifting: deformation of the lithosphere by active and passive driving forces. Geophysical Journal of the Royal Astronomical Society, 80, 95-120.
King, L.C. 1950. The study of the world's plainlands. Quarterly Journal of the Geological Society, London, 106, 101-31.

King, L.C. 1955. Pediplanation and isostasy; an example from South Africa. Quarterly Journal of the Geological Society,London, 111, 353-359.

King, L.C. 1957. The uniformitarian nature of hillslopes. Transactions Edinburgh Geological Society, 17, 81-102.

King, L.C. 1962. The Morphology of the Earth (2nd edtn., 1967) Oliver and Boyd, Edinburgh.

King, L.C. 1976. Planation remnants upon high lands. Zeitschrift für Geomorphologie, 20, 133-148.

King, L.C. 1975. Bornhardt landforms and what they teach. Zeitschrift für Geomorphologie,NF 19, 299-318.

King, L.C. 1982. The Natal Monocline. Durban, S. Africa.

Klein, Cl. 1974. Tectogenèse et morphogenèse Armoricaines et péri-Armoricaines. Revue de Géographie Physique et de Géologie Dynamique,16, 87-100.

Kominz, M.A. 1984. Oceanic ridge volumes and sea-level change - an error analysis. American Association of Petroleum Geologists Memoir, 36, 109-127.

Kroonenberg, S.B. and Melitz, J.P. 1983. Summit levels, bedrock control and the etchplain concept in the basement of Surinam. Geologie en Mijnbouw,62, 389-399.

Lidmar-Bergström, K. 1982. Pre-Quaternary Geomorphological Evolution in Southern Fennoscandia. Sveriges Geologiska Undersökning, Serie C. No. 785, 1-202. Uppsala.

Linton, D.L. 1959. Morphological contrasts between eastern and western Scotland. In Miller R. and Watson J.W. (Eds.) Geographical essays in Memory of Alan G. Ogilvie, Edinburgh, 16-45.

Loutit, T.S. and Kennett, J.P. 1981. New Zealand and Australian Cenozoic sedimentary cycles and global sea-level changes. American Association of Petroleum Geologists Bulletin, 65, 1586-1601.

Mabbutt, J.A. 1965. The weathered landsurface of central Australia. Zeitschrift für Geomorphologie,NF 9, 82-114.

McFarlane, A., Crow, M.J., Arthurs, J.W., Wilkinson, A.F. and Aucott, J.W. 1981. The Geology and Mineral Resources of Northern Sierra Leone. Institute of Geological Sciences Overseas Memoir 7, London, H.M.S.O.

Martin, A.K. 1984. Plate tectonic status and sedimentary basin in-fill of the Natal Valley (S.W. Indian Ocean). Joint Geological Survey/University of Cape Town Marine Geology Programme Bulletin, 14.

Melhorn, W.N. and Edgar, D.E. 1975. The case for episodic continental-scale erosion surfaces: a tentative geodynamic model. In Melhorn, W.N. and Flemal, R.C. 1975.

Melhorn, W.N. and Flemal, R.C. (Editors) 1975. Theories of Landform Development. George Allen and Unwin, London, pp 306.

Molina, E., Blanco, J.A. Pellitero, E., and Cantano, M. (in press) Weathering processes and morphological evolution of the Spanish Hercynian massif. In Proceedings First International Geomorphology Conference, Manchester 1985, Wiley.

Montford, H.M. 1970. The terrestrial environment during Upper Cretaceous and Tertiary times. Proceedings of the Geologists Association,81, 181-203.

Murgatroyd, A.L. 1979. Geologically normal and accelerated rates of erosion in Natal. South African Journal of Science, 75, 395-396.

Ollier, C.D. 1979. Evolutionary geomorphology of Australia and New Guinea. Transactions Institute of British Geographers, NS 4, 516-39.

Ollier, C.D. (Ed.) 1985. Morphotectonics of passive continental margins. Zeitschrift für Geomorphologie, Supplementband, 54.

Partridge, T.C. 1985. The palaeoclimatic significance of Cainozoic terrestrial stratigraphic and tectonic evidence from Southern Africa: a review. South African Journal of Science, 81, 245-247.

Pitman, W.C. 1978. Relationship between eustacy and stratigraphic sequences of passive margins. Geological Society of America Bulletin, 89, 1389-1403.

Pitman, W.C. and Golovchenko, X. 1983. The effect of sea-level change on the shelfedge and slope of passive margins. Society of Economic Palaeontologists and Mineralogists Special Publication, 33, 41-58.

Pugh, J.C. 1955. Isostatic readjustment and the theory of pediplanation. Quarterly Journal of the Geological Society London, 111, 361-369.

Ruxton, B.P. 1968. Order and disorder in landform. In Stewart G.A. (Editor) Land Evaluation, Macmillan, Melbourne, 29-39.

Schlee, J.S. (Ed.) 1984. Interregional Unconformities and Hydrocarbon Accumulation Association of American Petroleum Geologists Memoir 36.

Schumm, S.A. 1975. Episodic erosion: a modification of the geomorphic cycle In Melhorn W.N. and Flemal, R.C. (Editors) 1975, 69-85.

Schumm, S.A. and Lichty, R.W. 1965. Time, space and causality in geomorphology. American Journal of Science, 263, 110-119.

Sleep, N.H. 1971. Thermal effects of the formation of Atlantic continental margins by continental break up. Geophysical Journal of the Royal Astronomical Society, 24, 325-350.

Smith, A.G. 1982. Late Cenozoic uplift of stable continents in a reference frame fixed to South America. Nature, 296, 400-404.

Starkel, L. (in press) The role of the inherited forms in the present-day relief of the Polish Carpathians. In Proceedings First International Geomorphology Conference, Manchester 1985, Wiley.

Steckler, M.S. 1985. Uplift and extension at the Gulf of Suez: indications of induced mantle convection. Nature, 317, 135-139.

Stewart, A.D. 1972. Precambrian landscapes in north-west Scotland. Geological Journal, 8, 111-24.

Summerfield, M.A. 1979. Origin and palaeoenvironmental significance of sarsens. Nature, 281, 137-39.

Summerfield, M.A. 1981. Macroscale geomorphology. Area, 13, 3-8.

Summerfield, M.A. 1985a. Tectonic background to long-term landform development in tropical Africa, in Environmental Change and Tropical Geomorphology (Eds. Douglas and Spencer), pp 281-294. Allen and Unwin, London.

Summerfield, M.A. 1985b. Plate tectonics and landscape development on the African continent, in Tectonic Geomorphology (Eds. Morisawa and Hack), pp 27-51. Allen and Unwin, Boston.

Summerfield, M.A. 1985c. Tectonics and landscape evolution along passive continental margins, in Abstracts of Papers for the First International Geomorphology Conference (Ed. Spencer), p.577. University of Manchester, Manchester.

Summerfield, M.A. 1986. Tectonic geomorphology: macroscale perspectives. Progress in Physical Geography, 10, 227-238.

Thomas, M.F. 1968. Etchplain. In Fairbridge R.W. (Editor) The Encyclopaedia of Geomorphology. Reinhold, New York, 331-333.

Thomas, M.F. 1974. Tropical Geomorphology. Macmillan, London 332 pp.

Thomas, M.F. 1978. Denudation in the tropics and the interpretation of the tropical legacy in higher latitudes - a view of the British experience. In Embleton, C. (Editor). Geomorphology - Present Problems Future Prospects. Oxford University Press, Oxford, 185-202.

Thomas, M.F. 1980. Timescales of landform development on tropical shields - a study from Sierra Leone. In Cullingford, R.A., Davidson, D.A., and Lewin, J. (Editors). Timescales in Geomorphology, Wiley, London, 333-54.

Thomas, M.F. and Thorp, M.B. 1985. Environmental change and episodic etchplanation in the humid tropics of Sierra Leone: the Koidu etchplain. In Douglas, I. and Spencer, T. (Editors). Environmental Change and Tropical Geomorphology. George Allen and Unwin, London, 239-67.

Thorne, J. and Watts, A.B. 1984. Seismic reflectors and unconformities at passive continental margins. Nature, 311, 365-368.

Twidale, C.R. 1981. Pediments, peneplains and ultiplains. Revue de Géomorphologie Dynamique, 32, 1-35.

Vail, P.R., Mitchum, R.M. and Thompson, S. 1977. Seismic stratigraphy and global change of sea-level, Part 4: Global cycles of relative changes of sea-level. American Association of Petroleum Petrologists Memoir, 26, 83-98.

Vergnolle, C. (in press) Tertiary geomorphological evolution of the marginal bulge of the north west of the Iberian peninsula and lithostratigraphy of the grabens of the north-east of Galicia (Spain). In Proceedings First International Geomorphology Conference, Manchester 1985.

Watts, A.B. 1982. Tectonic subsidence, flexure and global changes of sea level. Nature, 297, 469-474.

Watts, A.B. and Thorne, J. 1984. Tectonics, global changes in sea level and their relationship to stratigraphic sequences at the US Atlantic continental margin. Marine and Petroleum Geology, 1, 319-339.

International Geomorphology 1986 Part II
Edited by V. Gardiner

WEATHERING PROCESSES AND MORPHOLOGICAL EVOLUTION OF THE SPANISH HERCYNIAN MASSIF

E. Molina, J. A. Blanco, E. Pellitero and M. Cantano

Departamento de Geomorfología, Facultad de Ciencias, University of Salamanca, Spain.

ABSTRACT

The Spanish Hercynian Massif is located on the western side of the Iberian Peninsula and its present morphology is the outcome of the movement of blocks caused by the Alpine Orogeny, and a sequence of weathering and flattening processes which has been taking place from Mesozoic times until the present. The sequence and nature of these processes are studied here, for which we have chosen five weathering profiles. Profile 1 is located on the western side of Spain, and shows the lower part of a weathered and kaolinized granite, fossilized by a fluvial series about 58 million years old. Both the fluvial series and the uppermost part of the weathered granite are silicified by C T opal. Profile 2 is located in the volcanic region of Campo de Calatrave (central Spain) in the piedmonts of the inselbergs of quartzites and shales. The profile displays a ferruginous cuirasse which is fossilized by the calcareous continental deposits of Upper Miocene-Pliocene age. Profiles 3, 4 and 5 are found at three different points of the North piedmont of Montes de Toledo (central Spain). Profile 3 is a weathered granite with a progressive enrichment in smectites towards the top. Its age is also pre-Upper Miocene. Profiles 4 and 5 present different types of calcareous crusts, their ages being Pliocene and lower Pleistocene respectively.

In summary we can say that over the Spanish Hercynian Massif there have been found the remains of an old kaolinitic weathering mantle more than 58 million years old, and the remains of a smectitic mantle of pre-Upper Miocene age; during the Upper Miocene-Lower Pliocene transition the development of calcareous crusts was the dominant process.

INTRODUCTION

According to Sole Sabaris (1952) the oldest structural unit of the Iberian Peninsula is the Hercynian Massif, also called by Hernandez Pacheco (1932) the Hesperian Massif, which is formed by the geological materials affected by the Hercynian Orogeny. At the end of this Orogeny a series of faults developed, which fractures it into a set of blocks. These blocks were later unevenly moved by the Alpine Orogeny throughout the Tertiary.

The westernmost end of the Hesperian Massif has remained uncovered from the end of the Hercynian Orogeny, such that its present relief is a combination of two superimposed mechanisms (Sole Sabaris, 1952; 1958):

1. the effects of the Alpine Orogeny on a rigid and fractured basement
2. the temporal succession of the different cycles of erosion during the Mesozoic and the Tertiary

In the most recent works on this topic the existence of ancient weathering mantles over this basement is pointed out, and these appear fossilized by the Mesozoic (Virgili et al., 1974) or Tertiary (Bustillo et al., 1980) series; their morphological consequences have been analyzed by Molina et al., (1980) and Garcia Abbad et al., (1980).

The aim of the present work is to establish a first approximation to the temporal succession of these weathering mantles on the Hesperian Massif during the Tertiary and Quaternary, indicating their morphological implications. To do so, we chose five well defined weathering profiles within the general geological context. Their presentation and study is conducted as a function of the antiquity of the processes distinguished in them. Their age was determined according to stratigraphic, structural and morphological criteria.

GEOLOGICAL SITUATION OF THE PROFILES STUDIES

The profiles chosen for study are located on the Hercynian basement or in the covering formations which fossilize it (Fig.1). The relief of this Hercynian basement in the North-West part of the Castilian Plateau, close to the Portuguese border, is characterized by the existence of the remains of an old peneplain of polygenetic origin, developed at an altitude of 750-850m. To the East, this peneplain is progressively fossilized by the Palaeocene series of the Duero basin. Profile 1 (Fig.1, No.1) is located in the province of Salamanca and is composed of a weathered granite basement over which the Palaeocene base lies unconformably. Its top is dated at about 58 million years (Blanco et al., 1982) and after this date a very important tectonic phase has been identified which gave rise to the first definition of elevated and lowered blocks in these zones (Cantano et al., in press).

Profile 2 (Fig.1, No.2) is located in the volcanic region of Campo de Calatrava in the province of Ciudad Real, where the Hercynian basement progressively sinks to the east under the marls and limestones of the Neogene. A common finding here is the presence of relics of ferruginous <u>cuirasses</u> (ferricretes) at 600-650m, which are always located in piedmont situations around the massifs of quartzites and shales, which are not found above 1000m. In many sites these <u>cuirasses</u> are fossilized by the Neogene materials (Molina, 1975; Redondo et al., 1980).

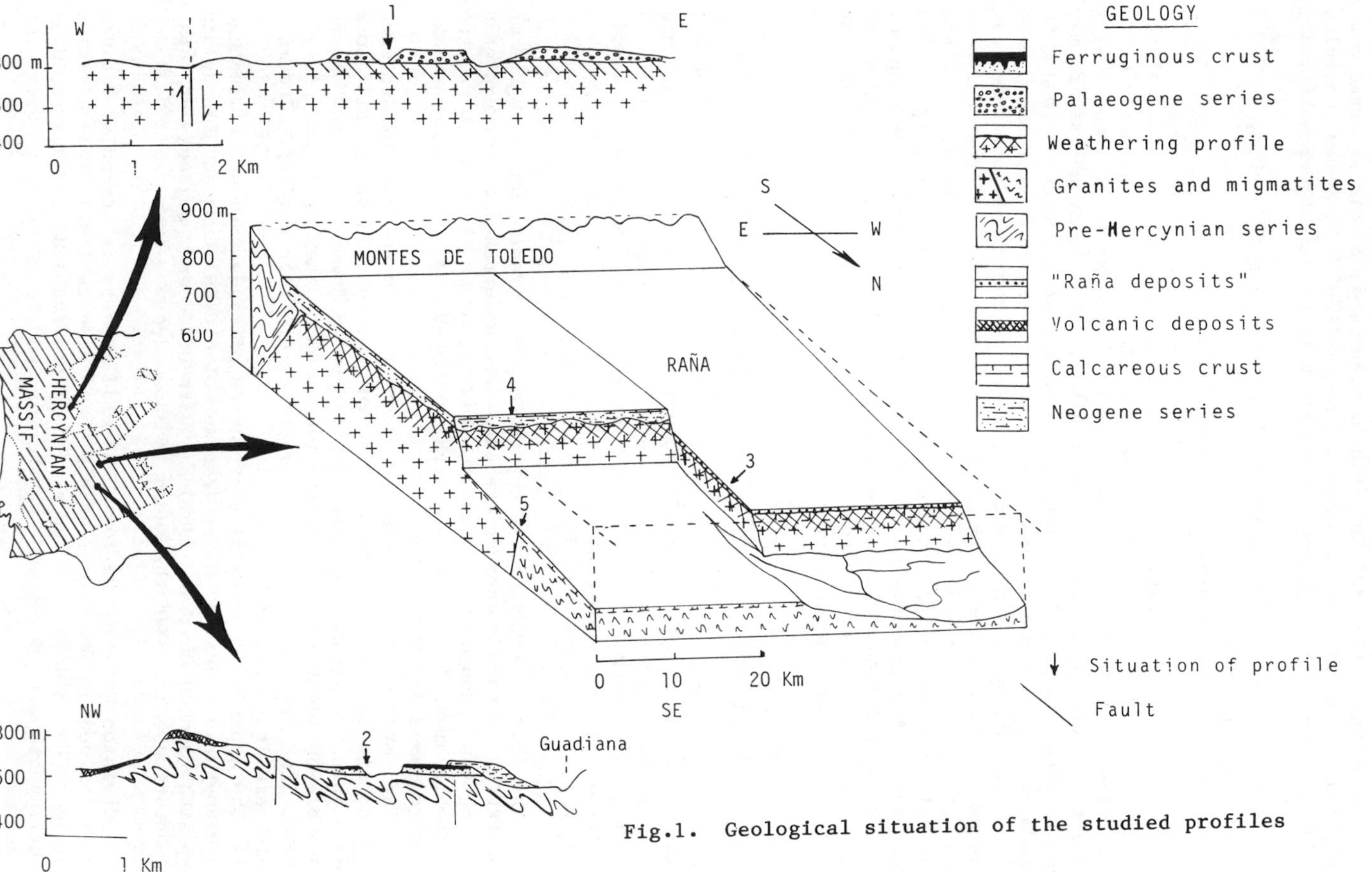

Fig.1. Geological situation of the studied profiles

The remaining three profiles (Fig.1, Nos. 3, 4, 5) are situated over the "Plataforma externa de Montes de Toledo" of Perez Gonzalez (1982), which is a polygenetic piedmont within which are identified:

- the Upper Piedmont surface of Pliocene age (700-800m)
- the Raña surface encased below
- the Quaternary drag surface (630-660m).

Underneath the Raña (Fig.1, No.3) and the carbonated materials of the Upper Piedmont (Fig.1, No.4) is found a relic weathering mantle which has already been studied by Vaudour (1979) and Molina (1980). Owing to its morphology and stratigraphic situation we believe that it may be related to profile 2 of Campo de Calatrava. Indeed, both are in piedmont positions, both are later than the Alpine tectonic phase mentioned above and both are fossilized by the Mio-Pliocene series. Its age is thus imprecise though it could cover most of the Palaeogene to the lower Miocene. The tectonic phases of the Pliocene and the Plio-Pleistocene (Aguirre et al., 1976; Perez Gonzalez, 1979) originated the dismantling of the Tertiary cover and the development of the Upper Piedmont and the Rañas. Accordingly, the ages of profiles 4 and 5 are, respectively, the upper-middle Pliocene and the lowest Quaternary sensu stricto.

PROFILE 1: NEIGHBOURHOOD OF SAN PELAYO DE GUARENA, SALAMANCA

Situation and description. This profile is to the North-west of the city of Salamanca (41°07'N; 5°51'W) over a granite basement fractured into blocks dipping to the South-East. The profile is situated in a scarp whose summit is at about 850m and about 45m above the floodplains of the present rivers. Three parts can be distinguished (Fig.2).

(1) The lower part (friable saprolite) is a granite whose degree of weathering increases towards the top. The occurrence of core-stones at the base changes to a complete arenization at the top. Here, the alterable minerals can be seen with the naked eye. The mean thickness is about 30m.
(2) The middle part (cemented saprolite) is composed of arenized granite similar to the previous case but here with a strong siliceous cementation. Alterable minerals disappear towards the top. It is possible to identify the fracture planes which the opal has fossilized, and also the conserved structures of block sliding. The main thickness is about 10m.
(3) The upper part contains the remains of sandstones and Palaeocene conglomerates forming a succession of fluvial bars and floodplain deposits (Alonso Gavilán, 1981) all with different degrees of opal cementation. The mean thickness is about 2-4m.

The division between the lower and middle parts is irregular because it is defined by the degree of siliceous cementation. The division between the middle and upper part is erosive and involves the dismantling of part of the old weathering mantle developed over the granite.

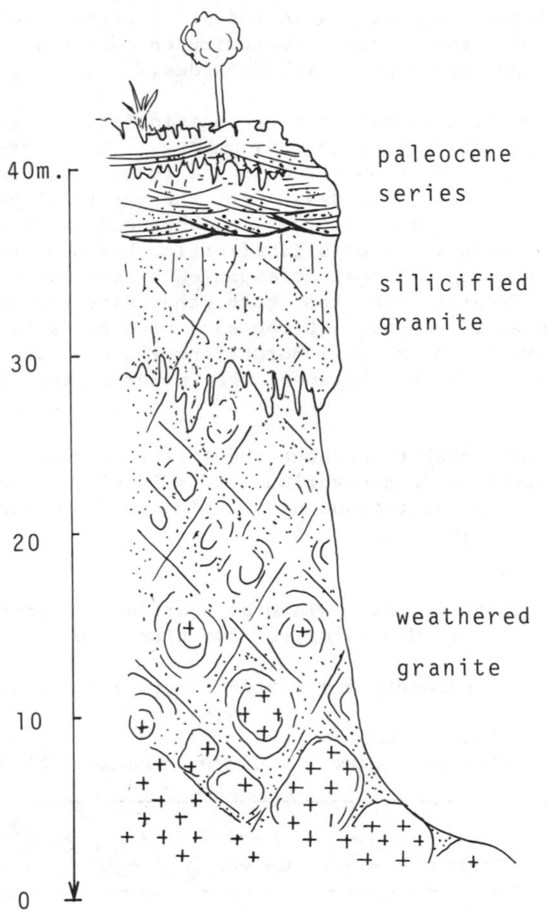

Fig.2. Profile of San Pelayo de Guarena (Salamanca)
(explanation in text)

Micromorphological study. This was only carried out on the lower and middle parts of profile 1.

The Lower part : two different levels are distinguished:

(1) Level of corestones of granite with different generations of quartz, feldspar and mica. Noticeable sericitization and albitization of plagioclases with a slight chloritization of biotite. Trans-mineral microfissures occur, less than 1mm wide, with borders of reaction between the plagioclases and the material filling the fissures.

(2) Arenized granite level. Increase in fissuring with strong argilization of the plagioclases. Quartz, k-feldspar and muscovite all show slight corrosion on margins. Biotites indicate marked segregation of oxihydroxides.

The Middle part: disappearance of plagioclases. Their places are occupied by kaolinite-like clay masses. The progressive disappearance of biotites is accompanied by the formation of clay aggregates surrounded by oxihydroxides. In the uppermost part only corroded quartz and remains of muscovites encompassed within a ferruginous clay matrix exist. Later silicification has occurred with the filling of fissures by C T opal accompanied by some smectites and oxihydroxides. Towards the top both the size and number of fissures filled by opal increases. At the same time there is a loss of resolution between the mineral border and the opal which is interpreted as a result of reaction between the mineral and the contributing solutions.

Bearing in mind that throughout the silicified set there is conservation of structures (conservation of volume) we can define these mineralogical substitutions as an epigenesis of granite saprolite by opal.

TABLE 1. San Pelayo de Guarena (Salamanca). Mineralogy of fraction < 2μ.

+ + + + = very dominant + + + = dominant + + = frequent + = scarce

Part of the profile	Approx. depth (m)	Illites Micas	Kaolinite	Smectite	Chlorite	C.T. Opal	Oxihydroxide Fe
Silicified	7	++	+++	++		+++	+
Saprolite	15	++	++++	+		+	+
Non-	25	++	++++	+			
Silicified	30	++	+++	+++			
Saprolite	40	++	+	++++	+		

<u>Study of clay fraction</u>. Samples for x-ray diffraction were taken from the lower and middle parts. The results obtained are shown in Table 1. The most outstanding points of Table 1 are the kaolinite/smectite ratio throughout the profile and the appearance of C T opal at the top. The kaolinites increase with the degree of weathering of primary minerals. By contrast the smectites exhibit firstly a maximum at the base of the profile, which may correspond to badly drained zones or to very slow weathering processes (Fritz, 1975 ; Leprun, 1979), and secondly an increase towards the top which we interpret as neoformation from the kaolinite and the solutions rich in silica, as shown by the fact that these smectites are aluminous.

PROFILE 2: SIERRA DE MEDIAS LUNAS, CIUDAD REAL

Situation and description. The profile chosen is situated to the West-South-West of Ciudad Real, close to the road from Corral to Alcolea de Calatrava (38°56'N; 4°05'W). It is composed of the remains of a ferricrete at about 640m developed over alluvial/colluvial detrital materials resting on a basement of folded shales and quartzites. The profile appears in a gentle scarp about 20m high, in which the geological features are a lower part of slightly altered Hercynian basement and an upper part of colluvial deposits with a brown reddish clay matrix. Towards the top the set becomes richer in oxihydroxides until a ferruginous cuirasse can be observed. The whole thickness of the colluvial deposits is about 12m; the ferruginisation affects the upper 4m.

In the ferruginous level it is possible to distinguish the following horizons, described from the base to the top (Fig.3):

(1) Mottled horizon in which can be found domains with a loss of clay and enrichment of oxihydroxides forming nodules and grey and/or greenish toned clay domains. The limit between the two is sharp and irregular. The mean thickness of this horizon is about 0.5m.
(2) Transition horizon towards the upper ferruginous "carapace". The ferruginous nodules become very abundant and welded together leaving the clay domains isolated. The mean thickness of this horizon is about 0.2m.
(3) Ferruginous carapace in which the walls of the nodules increase in thickness acquiring a metallic texture with an intense dark violet colour. Most of the clay domains are ferruginized and those which are not have been eroded giving rise to a runiform relief. The mean thickness of the horizon is 0.6m.
(4) Cuirasse horizon where the dark violet colour is dominant. It is possible to distinguish a lower part in which the nuclei of the old nodules, more sandy and brittle, can be observed, and an upper part with a tendency to exhibit an alveolar, tabular or laminar structure according to zones. The mean thickness of the horizon is about 2m.

The original sedimentary structures can be identified down to the lower part of the carapace above which they become indistinct and finally disappear.

Micromorphological study. This was carried out for all the horizons and in summary the following features are of interest:

(1) In the mottled horizon two domains are distinguished:

(a) Domains with a clay matrix. Skeletal grains of coarse sand include quartz, quartzite, some tourmaline, feldspar, zircon and andalucite. The related distribution is intertextic with an insepic plasmic fabric (Brewer, 1964). There are features of strong hydromorphism.

(b) Ferruginous domains. Dark masses of oxihydroxides exhibit strong reduction of content in the skeletal grains with corrosion in their borders and a porphiroskelic related distribution (Brewer, 1964).

In both domains a later contribution of red clays and carbonates filling the fissures is observed (see Fig.3, Micromorphology).

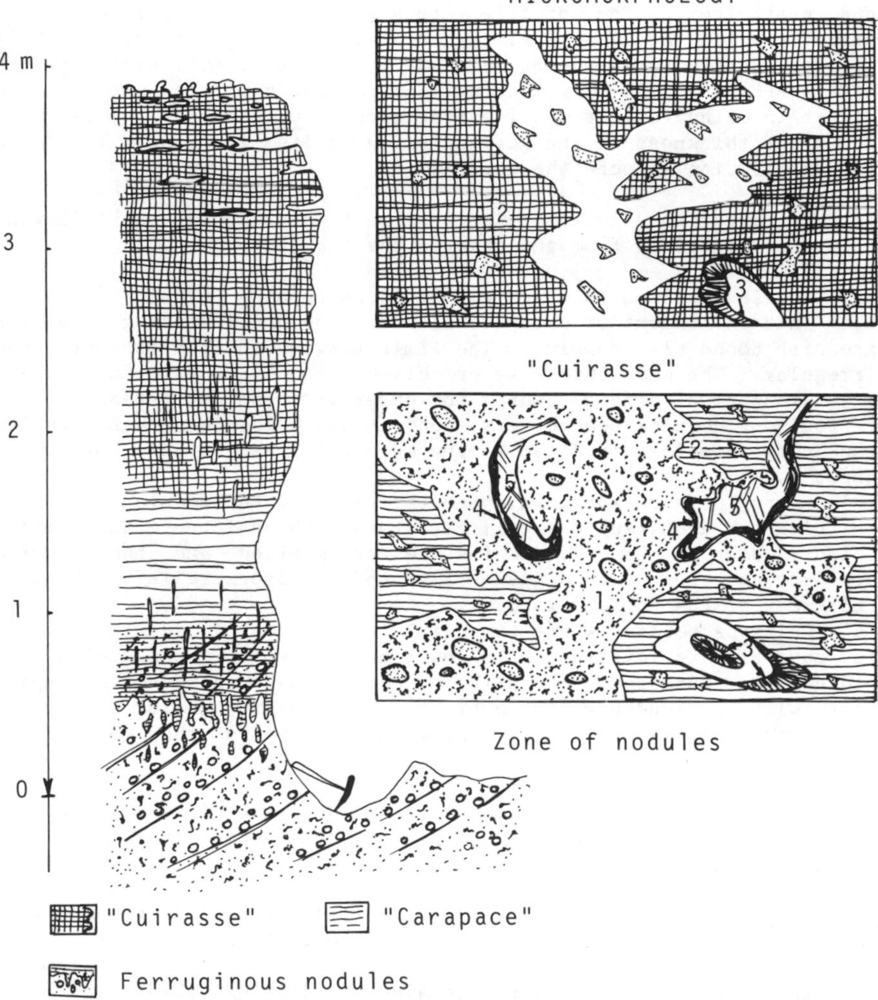

Fig.3. Profile of Sierra de Medias Lunas (Ciudad Real) (explanation in text). Micromorphology : 1 - original deposits; 2 - ferruginous plasma; 3 - goethitic concentration; 4 - illuviation clays; 5 - needles of carbonate.

(2) In the ferruginous carapace the clay plasma has disappeared and is replaced by oxihydroxides. The hollows are filled in many cases by radial goethite and there begin to appear areas with a granular or agglomeroplasmic related distribution pattern (Brewer, 1964) (Fig.3, Micromorphology).
(3) Several samples have been studied in the cuirasse at different levels and in them an increase of soil features can be observed towards the top, with conservation of zones with radial goethite and an increase in areas which have lost ferruginous plasma (Fig.3, Micromorphology).

As a summary of the micromorphological study two observations are of interest. Firstly, towards the top, the original clay plasma is substituted by ferruginous plasma which is accompanied by reduction of skeletal grains. Second, in the cuirasse it is possible to note clear indications of losses of ferruginous plasma.

Mineralogical study. First, a study was carried out of the clay fraction of the different horizons, after which an analysis was conducted of the total rock in each of them and of the different concentration zones of the oxihydroxides. These are manifested in the colour tones which range from red-violet (10 R 3/2) in most hardened zones, through brilliant red (2.5 YR 4/8) to yellow (10 YR 4/8) in the friable zones within the nodules. The analyses were performed on samples taken from the mottled horizon, from the carapace and from the upper cuirasse. The results are in Tables 2 and 3. From these data the following are deduced:

(1) Toward the upper part of the profile a decrease is observed in inherited clays (illite-mica) and an increase in kaolinite. Smectite and the interstratified clay minerals remain concentrated in lower horizons.

TABLE 2. Sierra de Medias Lunas (Ciudad Real)
Mineralogy of fraction $< 2\mu$ (no oxihydroxides)

+ + + + = very dominant + + + = dominant + + = frequent
+ = scarce . = traces

Part of the profile	Illite Micas	Kaolinite	Smectites	Inter-stratified
"Cuirasse" (inf.)	+ +	+ + +		.
"Carapace"	+ +	+ + +		.
Transition Horizon	+ +	+ + +	.	.
Mottled Horizon	+ + +	+ +	+	+ + +

TABLE 3. Sierra de Medias Lunas (Ciudad Real). Mineralogy of oxihydroxides distribution.

+ + + = dominant + + = frequent + = scarce . = traces

Part of the profile	Part analysed	Quartz	Goethite	Hematites	Clays
"Cuirasse"	Total rock	+ +	+ + +		.
	Yellowish domains	+ + +	+	+	+
"Carapace"	Total rock	+ + +	+ +	+	.
Mottled Horizon	Total rock	+ + +	+ +	+	.
	Core of nodule	+ + +	+	+	+

(2) Toward the top an enrichment in oxihydroxides takes place together with a reduction in quartz content.
(3) In the friable zones inside the nodules and in the cuirasse a relative decrease in goethite can be seen. This is interpreted as being because goethite is more mobile than hematite in dissolution processes (Nahon, 1976; Nahon et al., 1977).
All these findings suggest that the cuirasse represents an accumulation of oxihydroxides by epigenesis of the primary minerals, which are progressively reduced towards the top of the profile.

PROFILE 3: VICINITY OF NAVAHERMOSA, TOLEDO

Situation and description. The profile studied is close to km 44 of the Toledo to Navahermosa road (39°39'N; 4°25'W) and is developed over the granite which forms the nucleus of the antiform situated between the river Tajo and the Montes de Toledo (Aparicio, 1971). In these zones the Quaternary fluvial system is entrenched some 100m below the Raña surface which at this point is situated at about 730m. The profile appears on the hillside of one of these valleys and within it three major parts can be distinguished, from bottom to top (Fig.4):

(1) The lower part from 635m (thalweg) up to 700m exhibits granite cores formed by the dismantling of an ancient weathering mantle. Towards the top the granite cores show a transition into spherical scales of brown tones surrounding a less altered nucleus.

(2) The middle part between 700m and 728m exhibits arenized granite which appears homogeneous and friable and in which it is easy to distinguish aplite and pegmatite dykes. There is conservation of structures and a progressive increase of weathering towards the top.

Biotite is transformed into areas rich in oxihydroxides with reddish brown tones. Plagioclase is transformed into powdery masses of whitish tones. Quartz and muscovite are partly conserved. At the top it is possible to note fissures filled with carbonates and clays of greenish tones with slickenside planes. The uppermost part has been truncated by erosion.

(3) The top of the profile is formed by the Raña deposits some 2m in thickness. Its base cuts off the weathered granite by an erosion surface which descends from Montes de Toledo toward the North with a mean slope of 0.004-0.006. The Raña surface is slightly concave and over it are developed gley luvisols and planosols (Monturiol, 1984).

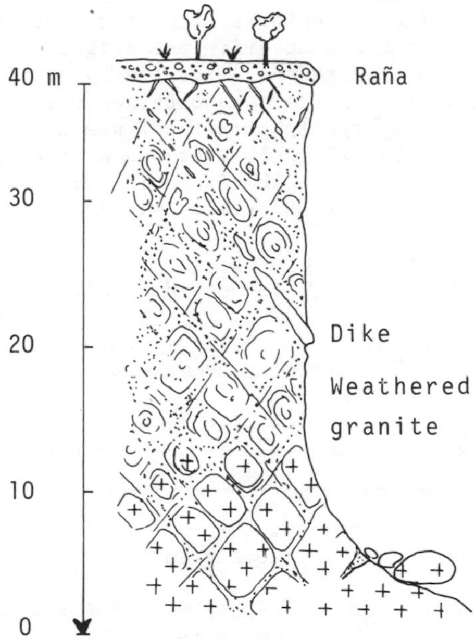

Fig.4. Profile of Navahermosa (Toledo) (explanation in text)

Micromorphological study. This was only carried out on six samples of the weathered granite taken at depths from 645m (compact granite) to 725m (completely arenized). The summary of the micromorphological data is:

(1) The lower part is a granite composed of quartz, plagioclase feldspar, microcline, biotite and muscovite with apatite, zircon and cordierite as accessory minerals. Hydrothermal alteration processes are evidenced by chloritization of biotite, the presence of acicular rutile, formation of pinnite from cordierite and strong seritization of plagioclase. Fissuring increases towards the top and the fissures are partially filled with brown illuviated clays. At the top of the lower part biotites are defringed and there is important illuviation of reddish brown clays.

(2) The defringed biotite of the middle part has released oxihydroxides which mask this mineral towards the top. Plagioclase feldspar is progressively substituted by a mass of white clays. Microcline is less altered and quartz has strong corrosion pits. Inter- and intra-mineral hollows are increasingly filled by a new supply of red clays. In thin sections we did not find any carbonates though their presence is clear in the top of the profile.

Study of the clay fraction. This was carried out with samples taken from 700m, owing to the low degree of weathering of the granite below this level. However in samples taken from levels lower than 700m, x-ray diffraction studies were carried out on the total rock. These analyses revealed the existence of well crystallized kaolinite and an abundance of illite and chlorite and traces of smectite. From 705m upwards samples were taken at different heights and from these the $< 2\mu$ fraction was separated. Its x-ray diffraction led to the result shown in Table 4. The data obtained point to the identification of a first supergenic process of weathering with kaolinite as the characteristic clay mineral, a second weathering process superimposed over the first with the formation of a smectitic mantle, and truncation of the smectitic mantle by the *Raña*.

TABLE 4. Navahermosa (Toledo). Mineralogy of fraction $< 2\mu$.

+ + + + = very dominant + + + = dominant + + = frequent + = scarce . = traces

Part of the profile	Depth (m)	Illites Micas	Kaolinite	Smectites	Chlorite	Inter-stratified
Middle Part	3	+	+ weathered	+ + + +		+
	4	+	+ weathered	+ + + +	+	.
	5	+	+ weathered	+ + +	+	
	10	+	+ + weathered	+ + +	+	
Top of lower part	25	+	+ +	+ +	+	

This smectite mantle is extended throughout the Upper Piedmont of Montes de Toledo towards the North and is truncated by the *Rañas* and by the Pliocene formations at different heights as a function of the age of cover which overlies it. Vaudour (1979) suggests that this weathering mantle would have a Vindobonian age. However, towards the

East it is fossilized by formations attributed to the middle Miocene (near Mora de Toledo) such that an earlier age should not be discounted.

PROFILE 4: VICINITY OF SONSECA, TOLEDO

Situation and description. The profile is situated at km 99 of the Madrid-Toledo-Ciudad Real road, between the villages of Sonseca and Orgaz (39°40'N; 3°56'W). Its morphological situation corresponds to the remains of the Upper Piedmont of Perez Gonzalez (1982) at about 775m. Below the present surface there are Pliocene materials which fossilize the smectitic mantle. These have been studied by Molina et al., (1978) and Perez Gonzalez (1982). From the base to the top the profile exhibits the following characteristics (Fig.5):

(1) Arenized granite with visible weathering features more than 10m deep.
(2) A grooved erosion scar which at many points is crossed by carbonate grids filling cracks of varying thickness. In the cracks red clays associated with carbonate appear.
(3) Above the scar, a conglomerate of quartz, quartzite, granite, some slate and remains of weathered granite in a clay and/or carbonate matrix, ordered in two domains. The first is clay domains with colours from 2.5 YR at the base to 10 R towards the top, this clay mass being in turn crossed by a fine carbonate grid; secondly carbonated domains at the base of the conglomerate exhibit irregular isolated shapes of several centimetres diameter. Towards the top they become joined giving rise to a columnar-like structure. Within the domains there is a strong reduction of detrital materials and the carbonate has a powdery aspect. Sometimes these domains are crossed by fissures filled with red clays. The size of the pebbles of the base (maximum diameter about 30cm) rapidly decreases towards the top.

As a whole the maximum visible height of the profile is about 3m and for the purpose of this study eight zones have been distinguished (Fig.5):

zone A weathered granite
zone B conglomerate with clay matrix
zone C conglomerate of the lower carbonate domain
zone D calcarous crust type *dalle*
zone E microconglomerate and sand with clay matrix
zone F upper carbonate domain
zone G carbonated and leafy calcareous crust
zone H worked soil with fragments of crust.

Micromorphological study. Zone A: the commonest minerals are corroded quartz, weathered k-feldspar, plagioclase epigenized by carbonate and extremely altered biotite and muscovite with altered borders.

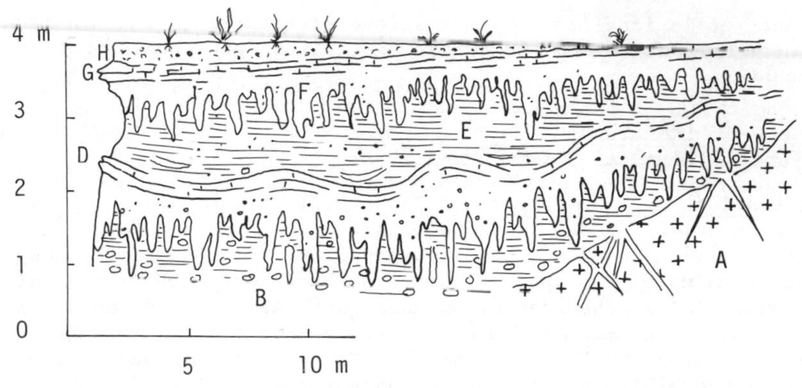

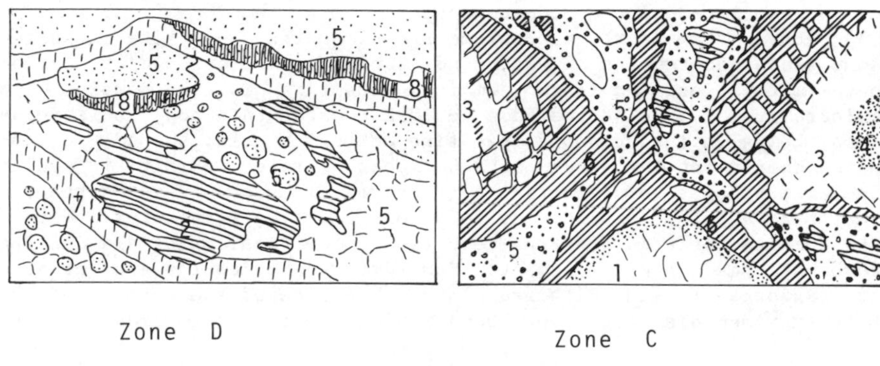

Fig.5. Profile of Sonseca (Toledo). A - H, zones explained in text. Micromorphology: 1 - quartzite; 2 - old argilliplasma; 3 - old crystalliplasma: 4 - epigenized mineral; 5 - new crystalliplasma of different generations; 6 - illuviation clays; 7 - carbonates of phreatic cementations; 8 - carbonate of vadose cementations.

Zones B and C: quartzite, quartz and fragments of granite are encompassed in red argilliplasma and/or carbonated crystalliplasma. The argilliplasma exhibits abundant papules of red clay. The crystalliplasma shows irregular crystals in mosaic form which towards their edges tend to exhibit rhomboidal forms about 80-100µ across the major axis (Fig.5, Micromorphology). There is remobilization of argilliplasma and formation of argilans whose clays serve as a tool for disaggregation of crystalliplasma. New contributions of carbonates in various stages occur, the latter producing the needles which fill the cracks.

Zone D: important biological activity is evident with areas having a perlitic structure. Features of fracturing by drying episodes occur, and later in-filling by carbonates in alternating vadose and phreatic conditions (Fig.5, Micromorphology).
Zone E: cracks are filled with carbonates in alternating vadose/phreatic stages, the last being the genesis of needles.
Zones F and G: micritic crystalliplasma occurs with strong epigenesis of primary minerals. In the upper crust it is possible to note the multiple repetition of a lower level with relative abundance of detrital material, ooides and nodules with a thickness less than 1mm, and a massive upper level with very little detrital material and microstructures of coussinets-type of algal origin (Vogt, 1984). The thickness of this level may be about 1mm.
The recrystallization processes are not important and always appear in vadose conditions. Needles filling hollows are very common.

The micro and macro morphological data indicate:

(1) The epigenetic processes in the Pliocene materials have been very intense and they affect even the gravel fraction.
(2) In the lower part of the profile (zones B and C) there is no vadose cementation and the processes of disaggregation of carbonate into rhombohedral crystals are very important.
(3) In the crust in dalle (zone D) vadose and phreatic conditions have been repeated several times.
(4) In the upper part (zones F and G) only vadose cementation can be observed.

TABLE 5. Sonseca (Toledo). Mineralogy of fraction < 2µ.

+ + + = dominant + + = frequent + = scarce . = traces

Part of the profile	Depth (m)	Illites Micas	Kaolinite	Smectite	Inter-stratified	Paly-gorskite
Zone G	0.5	+ +	+	.	+	.
Zone F	0.7	+ +	+	.	+	.
Zone E	1.0	+ +	+	.	.	
Zone C	1.5	+ +	+	.	+	
Zone A, filling of crack	2.7	+	+	+ + +		
Zone A, weathered granite	3.0	+ +	+	+ + +		
Zone A, weathered granite	6.0	+ +	+	+ + +		

These aspects suggest that the genesis of both crusts is independent and that each of them corresponds to different water movement, fundamentally horizontal in the case of the lower crust and vertically in that of the upper crust.

Study of the clay fraction. The samples taken comprise two from the weathered granite, one from the clay filling a crack in the granite and four from the Pliocene covering. Prior to separation of the $< 2\mu$ fraction the samples were treated with 0.1 acetic acid and the data obtained are shown in Table 5. As an interpretation of these data it is possible to confirm the generalized existence of the previously mentioned smectitic mantle. Moreover we suppose that the kaolinite and the illite-micas of the Pliocene covering have been inherited from the weathering mantle, during erosion. Some smectites may be included in the interstratified clays.

PROFILE 5: BURGUILLOS DE TOLEDO

Situation and description. This profile is situated to the south of Toledo (39°47'N; 4°00'W) over the remains of an ancient Quaternary erosion surface at some 660m, below the Raña and the Pliocene surfaces. The river Tajo is about 200m lower than this surface. The profile has been studied in a diagonal sense along the scarp of the road joining the villages of Cobisa and Burguillos. From base to the top the features are as follows (Fig.6):

(1) The lower part contains migmatites altered to about 0.5m without any important arenization. Carbonates fill fissures in a grid-like fashion. The thickness of the fissures increases toward the top.

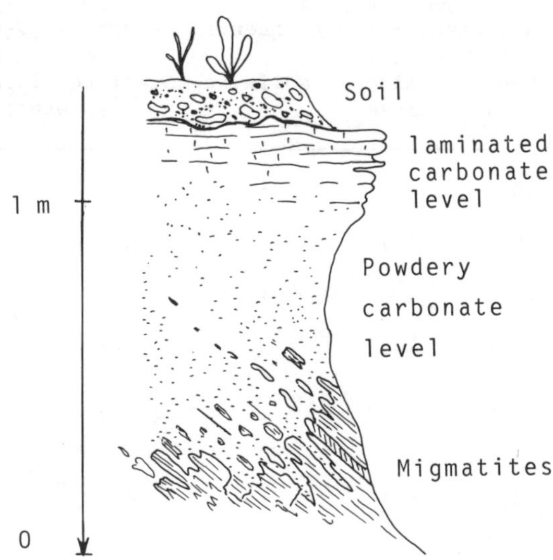

Fig.6. Profile of Burguillos de Toledo (explanation in text).

(2) The middle part has a lower limit which is very irregular, with massive and friable carbonate encompassing extremely altered remains of the basal migmatites. The quartz dykes conserve directions. Towards the top the massive structure of the carbonate begins to be substituted by a laminar structure with definition in plates some centimetres thick and of greater consistency. Mean thickness of the set is about 0.3-0.5m. Above this zone exists an erosion scar.

(3) Worked soil containing pebbles and gravels of quartz, migmatite, and fragments of laminar crust, all encompassed in a carbonate mass. The mean thickness is about 0.4m.

Micromorphological study. This was only carried out on the lower and middle parts, from which several samples were taken. The outstanding features are;

(1) The lower part shows migmatites, quartz, orthoclase, plagioclase and biotite as principal minerals and garnet, cordierite, sillimanite and zircon as secondary ones. Ancient hydrothermal weathering is manifested by chloritization of biotite and seritization of feldspar. Inter- and intra-mineral fissures are filled with recrystallized carbonate under vadose conditions, repeated several times. Strong carbonated epigenesis of the silicates has occurred on the borders of the fissures, with a final stage of carbonate contribution forming needles filling hollows.

(2) The middle part shows silicates at the base encompassed in a micritic crystalliplasma with a certain clay content. This crystalliplasma has recrystallized in many places giving rise to a separation of clay and carbonate and the formation of microsparitic crystals of 10-40μ. Toward the top the degree of recrystallization decreases whereas nodules increase owing to biological activity with the appearance of organic remains (roots). Porosity also increases with the appearance of joint planes (Brewer, 1964) corresponding to the genesis of laminar structures. Later supply of carbonate fills hollows.

TABLE 6. Burguillos de Toledo. Mineralogy of fraction $< 2\mu$.

+ + + = dominant + + = frequent + = scarce . = traces

Part of the profile	Depth (m)	Illites Micas	Kaolinite	Smectites	Chlorite	Palygorskite
Middle Part	0.4	?	.	+ +		+
	0.6	.	?	+	+	
	0.8	+ +	+ +	+ + +	.	
Lower part	1.0	+ +	+	+ + +	.	

A first interpretation of these data suggests a progressive loss of material from the upper to lower parts, mainly by migration of carbonate. This migration took place in several stages, the needles being the actual visible stage. According to Ruellan (1984) these needles are progressively incorporated into the carbonate mass by multiple dissolution and recrystallization processes.

Study of the clay fraction. The sampling for this was carried out in the lower and middle parts of the profile. Prior to the separation of the <2u fraction the samples were treated with 0.1 N acetic acid. The results of x-ray diffraction are shown in Table 6. The profile features a lower level in which there is a smectitic domain and another of illite-mica with presence of kaolinite, and an upper level where there is reduction of the previous clays with appearance of palygorskite. The limit between these two levels corresponds to the start of the appearance of the strong biological activity detected in the micromorphological study.

CONCLUSIONS

Since the 1970s collaboration between soil scientists and geologists has led to a new research line in geomorphology, according to which there is a direct link between relief forms and the geochemical evolution of landscapes. The main idea is that the development of weathering profiles of tropical and arid regions is accompanied by an evolution of the relief consisting fundamentally of:

(1) A mineralogical homogeneization of the topographic surface by epigenetic processes giving rise to clay mantles, silcretes, calcretes or ferricretes.

(2) A progressive and more or less homogeneous sinking of these weathering mantles on the old reliefs.

The mineralogical homogeneization implies a substitution of most of the primary minerals by new ones whose nature is in accordance with the exterior conditions. In the case of tropical humid conditions the epigenesis gives rise to the kaolinites and oxihydroxides. In dry/humid tropical conditions the new stable minerals are oxihydroxides and smectites, depending on the drainage regime. In semi-arid and Mediterranean conditions the carbonated epigenesis is dominant (Nahon et al., 1977; Ruellan, 1984). The result of the mineralogical homogeneization of the uppermost part of the profile is that the erosive processes take place on the same lithology, which is practically independent of the nature of the substratum.

The micromorphological study has shown that there is a parallelism in the evolution of silcretes, ferricretes and calcretes. The mechanism consists of a progressive loss of material in the upper part of the profile and its accumulation in the lower part where the epigenesis is taking place. This is accompanied by a progressive destruction of inherited structures in the uppermost parts which, however, are still conserved in the lower parts.

The mobility of clay, silica, oxihydroxide and carbonate has been produced by water. In some cases the water has moved vertically, such as profiles 2 and 5, indicating vadose conditions. In other cases (the lower part of profile 4) the water movement has been mainly horizontal, where the phreatic micromorphological features are dominant.

Reactions between the surfaces of minerals and the solutions filling the fissures have been the most common process of weathering, which explains the abundance of resistant minerals with corroded surfaces.

Accordingly, the regional evolution of the relief under tropical and semi-arid conditions looks like a consequence of the relationship between the rate of sinking of the weathering mantles and that of the drainage patterns in the landscape. The first process is conditioned by the nature and resistance of the rocks to be weathered at the "weathering front" beneath the surface, which tends to progress downwards. The second is conditioned by tectonic, isostatic and climatic changes, and controls the dynamics of the regional groundwater flows.

The existence of extensive surfaces of almost perfect plains, developed on different lithologies, is a consequence of a more rapid sinking of the weathering mantles than of the drainage pattern. These testify to the existence of geological periods with relative tectonic stability.

In this work we have found two argillic mantles directly over the Spanish Hercynian basement - a kaolinite mantle of Mesozoic age (?), and another intra-Tertiary mantle of smectitic nature. The first is associated with an ancient flattening, which is today unevenly removed and is conserved in certain particular areas. The second is associated with the development of inselbergs and pediments which form part of the piedmont of the present mountain systems. With this latter we associate the development of the cuirasse of the Campo de Calatrava, according to stratigraphic and morphological data. There is furthermore a stage of silcrete formation of Palaeocene age located between the previous argillic mantles, Moreover, calcretes were developed over a large part of the Tertiary, their maximum development taking place during the upper Miocene and middle Pliocene. However, according to the available data, the silicification and carbonation were possibly repeated several times during the Tertiary.

REFERENCES

Aguirre, E., Diaz Molina, M., and Pérez González, A., 1976. Datos paleomastológicos y fases tectónicas en el Neógeno de la Meseta sur Española. Trabajes sobre el Neógeno/Cuaternario, 5, 23-36.

Alonso Gavilán, G., 1981. Estratigrafía y sedimentología del Paleógeno del borde suroccindental de la Cuenca del Duero (provincia de Salamanca). Doctoral Thesis. Faculty of Sciences, University of Salamanca, (2 vols).

Aparicio, A., 1971. Estudio geológico del Macizo cristalino de Toledo. Estudios Geológicos, 27, 368-414.

Blanco, J. A., Corrochano, A., Montigny, R., and Thuizat, R., 1982. Sur l'age du debut de la sedimentation dans le bassin tertiaire du Duero (Espagne). Atribution au Paléocène par datation isotopique des alunites de l'unité inferieure. C. R. Acad. Sc., 295, 559-62.

Brewer, R., 1964. Fabric and mineral analysis of soils. Wiley, London.

Bustillo, M. A., and Martin Serrano, A., 1980. Caracterizacion y significado de las rocas silíceas y ferruginosas del Paleoceno de Zamora. Tecniterrae, 244, 14-29.

Cantano, M., and Molina, E., (in press). Approximacion a la evolución morfólogica de la Fosa de Ciudad Rodrigo. Salamanca. Bol. Real. Soc. Española Hist. Nat.

Fritz, B., 1975. Etude termodynamique et simulation des alterations entre mineraux et solutions. Application á la geochimie des alterations et des eaux continentales. Sci. Geol. Memoires, 41.

Garcia Abbad, F. J., and Martin Serrano, A., 1980. Precisiones sobre la génesis y cronología de los relieves apalachianos del Macizo Hespérico (Meseta Central Española). Central Geológicos, 36, 391-401.

Hernández Pacheco, E., 1932. Síntesis fisiográfica y geológica de España. Trabajos del Museo Nac.Cienc.Nat. (Serie Geol.), 38, 584.

Leprun, J. C., 1979. Les cuirasses ferrugineuses des pays cristallins de l'Afrique occidentale sèche. Genèse, tranformations, degradation. Sci. Geol. Memoires, 58.

Molina, E., 1975. Estudio del Terciario Superior y del Cuaternario del Campo de Calatrava (Cuidad Real). Trabajos sobre el Neógeno/Cuaternario, 3, 106.

Molina, E., and Alexandre, T., 1978. Estudio de las acumulaciones de carbonato y procesos de alteración desarrollados bajo la superficie pliocena en la Cuenca del Tajo, in Proceedings of 5th Internat. Working Meet. on Soil Micromorph. Granada, Spain, I, 501-521.

Molina, E., 1980. Alteración relicta y morfogénesis del Macizo cristalino de Toledo. Studia Geológica Salmanticensia, 16, 15-25.

Molina E., and Blanco, J. A., 1980. Quelques précisions sur l'alteration du Massif Hercynien Espagnol. C. R. Acad. Sc., 290, 1293-1296.

Monturiol, F., 1984. Estudio agrobiológico de la Provincia de Toledo 1, suelos. Inst. Prov. de Investigaciones y Estudios Toledanos, Disputacion Provincial de Toledo, 20-146.

Nahon, D., 1976. Cuirasses ferrugineuses et encroutements calcaires au Senegal occidental et en Mauritanie. Systemes evolutifs: geochimie, estructures, relais et coexistence. Sci. Geol. Memoires, 44.

Nahon, D., and Millot, G., 1977. Geochimie de la surface et formes du relief. V. Enfoncement geochimique des cuirasses ferrugineuses par épigénie du manteau d'alteration des roches meres greseuses. Influence sur le paysage. Sci. Geol. Bull., 30, 275-282.

Pérez González, A., 1979. El límite Plio-Pleistoceno en la submeseta meridional en base a los datos geomorfológicos y estratigráficos. Trabajos sobre el Neógeno/Cuaternario, 9, 23-36.

Pérez González, A., 1982. Neógeno y Cuaternario de la llanura manchega y sus relaciones con la Cuenca del Tajo. Doctoral Thesis. Ediciones de la Universidad Complutense. Madrid, 179.

Redondo, E. and Molina, E., 1980. Bosquejo morfológico de la Cuenca del rio Bullaque (Cuidad Real). Bol. Geol. y Minero, 91, 32-40.

Ruellan, A., 1984. Les sols calcaires. Les principaux trabaux des pedologues Français. Livre Juvilaire du Cinquentenaire. Ed. par A.F.E.S., 111-121.

Solé Sabarís, L., 1952. Geografía de España y Portugal. I Geografía Física. Montaner and Simons S.A., Barcelona.

Solé Sabarís, L., 1958. Observaciones sobre la edad de la penillanura fundamental de la Meseta española en el sector de Zamora. Breviora Geol. Astúrica II, 1 and 2, 3-8.

Vaudour, J., 1979. La region de Madrid. Alteration, sols et paléosols. Editions Ophris, France.

Virgili, C., Paquet, H., and Millot, G., 1974. Alteration du soubasement de la couverture Permo-Triasique en Espagne. Bull. Groupe Français Argiles, 2, 277-285.

Vogt, Th., 1984. Croutes calcaires : Types et génese. Exemples d'Afrique du Nord et de France mediterraneenne. Universite Louis Pasteur, Inst. de Geographie, Strasbourg.

International Geomorphology 1986 Part II
Edited by V. Gardiner

TERTIARY INHERITANCE IN THE PRESENT LANDSCAPE OF THE BRITISH ISLES (examples from Wales, the Mendip Hills and South-West Ireland).

Yvonne BATTIAU-QUENEY

Universite de Lille 1,
59655 Villeneuve d'Ascq Cedex, France.

ABSTRACT

With advances in techniques of field survey, evidence of Cenozoic morphogenesis in the British Isles has become widespread, even in regions which were glaciated during the Pleistocene. Deep weathering profiles and palaeosols have been observed *in situ* on various rock types in Wales, Shropshire and the Mendip Hills. They imply a hot or warm temperate climate and are considered to be preglacial. Carboniferous Limestone exhibits palaeo-karstic landforms subsequently buried with deposits derived from tropical soils. Gibbsite has been found at the top of an *in situ* weathering profile in South-West Ireland.
Relict landforms are related to deep weathering. They are tors, inselbergs of St David's Land and Anglesey, pediments in South Pembrokeshire and the Mendip Hills, swallow-holes and closed basins on Millstone Grit outcrops. The concept of morphotectonic equilibrium is important in explaining the preservation of relict landforms.

INTRODUCTION

Twenty years ago, the idea that widespread Tertiary weathering products and associated landforms could survive in the British Highlands would not have been easily accepted. However more and more evidence has been reported of the survival of such features (Battiau-Queney, 1978, 1981, 1984; Mitchell, 1980; Isaac, 1981, 1983; Hall, 1985). In this paper, several examples of weathering profiles are described from Wales, South Ireland and the Mendip Hills and arguments in support of the widespread preglacial remnants in the present landscape are advanced. The geomorphological and tectonic implications of this survival are discussed.

WIDESPREAD REMNANTS OF DEEP WEATHERING PROFILES.

These weathering products occur *in situ* or have been transported into karstic pockets developed in Carboniferous Limestone.
With regard to the former, nearly every rock type has been affected, but the best preserved remnants have been observed on igneous rocks (dolerite, andesite, granophyre, rhyolite) and on sandstones, especially in Wales and the Mendip Hills.

	% quartz	CO_3Ca %	Montmorillonite (1)	Illite/mica (1)	Chlorite (1)	Kaolinite (1)
72-17	14	13	5	10	0	34
71-57	9	27	5	8	0	30
71-58	20	45	0	12	0	20
72-16	4	72	0	0	0	10

	SiO_2	Al_2O_3	Fe_2O_5	FeO	CaO	MgO	Na_2O	K_2O	TiO_2	MnO	P_2O_5	H_2O	ign. loss	Total
72-17	54,72	13,05	5,32	0,34	7,95	1,15	0,20	2,45	0,78	0,01	0,08	2,52	11,06	99,63
71-57	49,96	8,08	4,72	0,31	15,34	0,87	0,20	1,55	0,56	0,03	0,07	1,79	15,87	99,35
71-58	46,58	3,52	2,09	0,33	23,63	0,60	0,25	0,90	0,33	0,03	0,04	0,87	20,47	99,64
72-16	19,76	3,00	2,16	0,35	40,04	0,72	0,15	0,55	0,20	0,03	0,04	0,82	32,43	99,90

Fig.1

(legend on the following page)

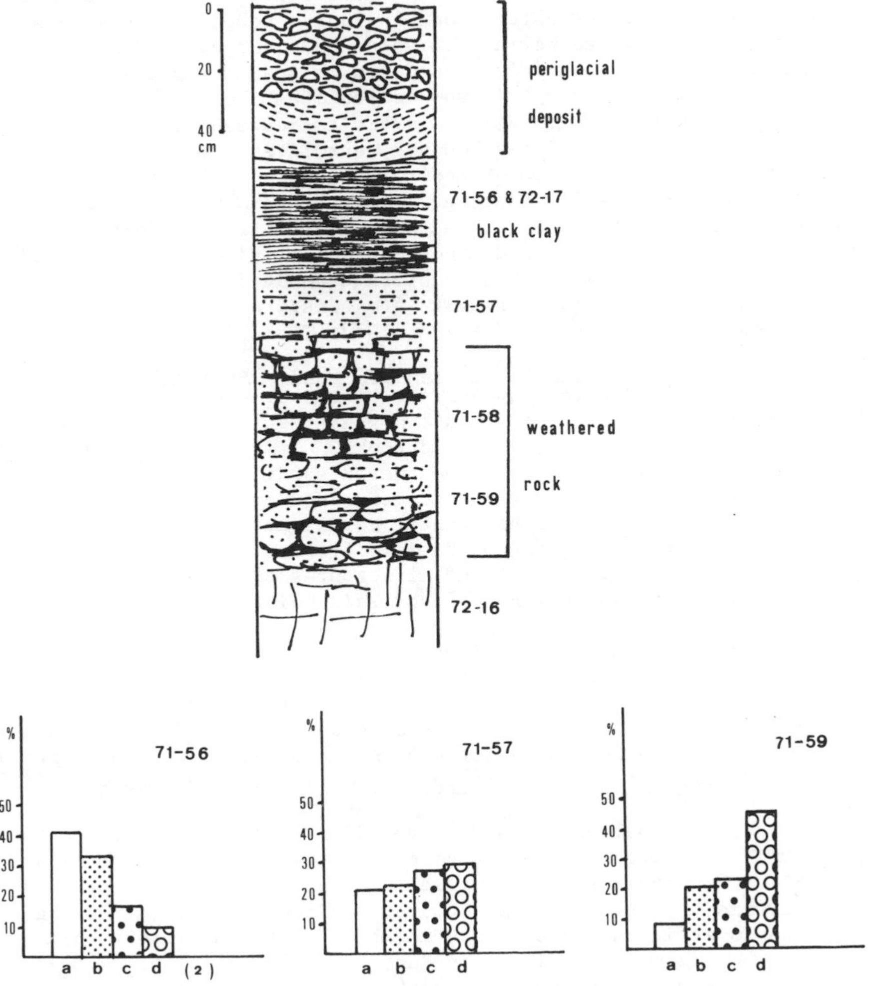

Fig.1: MINERALOGICAL DATA,TEXTURE AND CHEMICAL ANALYSIS OF DIFFERENT HORIZONS OF THE WEATHERING PROFILE OF THE DOL GOCH QUARRY (SHROPSHIRE,SJ 278248)

(1) numbers indicate the height of the main diffraction peaks.

(2) a: particles less than 2μm - b: particles from 2 to 20μm - c: particles from 20 to 50μm - d: particles from 50μm to 2mm.

Saprolites several metres deep are charaterized by the preservation of the structure of the parent rock including bedding joints,veins and fissures.During weathering,the volume has not changed despite a significant loss of material in solution.This explains why the rock is no longer cohesive except in some corestones.The weathered rock is generally sandy with 50 to 70% of particles between 50μm and 2mm in size.On shales and siltstones,however,saprolites are much finer (less than 25% sand particles) and thinner (generally less than 1 metre). Clay minerals are mostly inherited from parent rocks but,where present,plagioclase feldspars are decomposed.A little kaolinite is often present on andesites and sandstones.Montmorillonite has been found at the base level of a weathering profile derived from dolerite on Titterstone Clee Hill.These results are consistent with studies made in the tropics (cf: Birot and Muxart,1977; Thomas,1979).

Everywhere,the basal surface of weathering is quite irregular and nowhere parallel with the landsurface.On the Clee Hills,the columnar joint pattern of dolerite explains why sub-vertical columns of hard rock stand up between deeply rotted rock.On Namurian sandstones,hard beds of fresh rock may lie over weathered sandy beds because of the complex circulation of underground waters.Nowhere has it be possible to see a horizontal or sub-horizontal basal surface of weathering.The marked irregularities of the weathering front are fundamental to an understanding of various superficial features such as dolines,closed basins,tors and hummocky landsurfaces.This point will be developed below.

Although deep weathering profiles are widespread in Wales and the Mendip Hills,it is uncommon to find upper structureless horizons. However,a more complete profile has been exceptionally preserved on Titterstone Clee Hill,in the active Dhustone quarry (SO 592763),on Westphalian Carboniferous sandstones.Chemical,X-ray and granulometric analyses reveal a true ferrallitic palaeosol (Battiau-Queney,1984, fig.5).

Another type of palaesol was discovered in a disused quarry to the south-west of Oswestry in Shropshire.Here,the Dol Goch quarry (SJ 278248) works impure siliceous limestone.This rock bears a profile characterized by an upper black clayey horizon.Chemical,mineralogical and granulometric analyses have been made for the different horizons (fig.1).They show that this is in fact a true tropical black soil (vertisol) formed in a low-lying area at the base of slopes where shales and sandstones outcrop.A little montmorillonite is present in association with kaolinite.It can be explained by a bad drainage. Today this palaeosol has been buried under periglacial colluvium.

On limestone outcrops,*in situ* weathering profiles are generally absent but a great number of quarries exhibit remarkable fossilized karstic landforms; these have been reported elsewhere (Battiau-Queney,1973,1978,1986).They were buried by a loose unsorted material which resulted in the destruction of deep tropical soils developed on nearby shales and sandstones.This buried palaeokarst provides supplementary proof of the preservation of tropical landforms in these regions.

An interesting example of a relict weathering profile has been recently discovered in south-west Ireland (Kerry) in a quarry near the road from Castleisland to Tralee (Q 956112).The Ballyegan quarry works Carboniferous Limestone.It is localized in a flat plain lying at an elevation of less than 40m; this plain is bordered to the south west by the Slieve Mish Mountains reaching a height of more than 750m and to the north by a plateau with shales and sandstones lying at 300m.The Ballyegan quarry has been visited several times since 1981. Large vertical pockets are filled with a non-carbonate material.One of them was analysed in 1981 and 1982.The filling material appears to be Namurian shales and siltstones trapped in limestone.These rocks have been deeply weathered in situ and a pedogenetic profile been formed where the percentage of gibbsite increases upwards.At the uppermost level it is as abundant as kaolinite.It seems improbable that this type of weathering could be post or interglacial (Battiau-Queney and Saucerotte,1985).The preservation of what seems to be an old palaeosol at the surface of a low limestone depression proves that the limestone platforms of south Ireland have been lowered very slowly over the past several millions years and throws doubt on the idea of rapidly lowered limestone surfaces in that region.On the contrary,it is in accordance with chalk relicts at Ballydeenlea, near Killarney at an elevation of 120-220m (Walsh,1966).

Widespread occurences of such weathering products which are no longer in equilibrium with the present environment but inherited from an earlier different morphogenetic regime indicates a new interpretation of long-term landform development.In this approach,if we take the example of hills standing above a lower platform,our attention will not focus on slope debris and Holocene and Devensian processes but on the origins and development of the hills themselves.This requires an approach with a long temporal scale and it raises the problem of the survival of palaeoforms.

TORS AND INSELBERGS IN RELATION TO DEEP-WEATHERING.

Tors exist on various rocks in south and east Wales.The more spectacular instances are found in Dyfed,on the Treffgarne and Dudwell Mountains and in the Mynydd Prescelly.The best area to study them is the Great Treffgarne Mountain which reaches a height of 162m.Two tors have been formed in rhyolite on the east slopes at an elevation of 110-125m (Poll Carn and Maiden Castle).Downwards to the south-east, a large disused quarry working Silurian andesites exhibits a rock weathered through several metres at the summit of the west front (SM 959240).The Cleddau River flows just below in a narrow gorge at 25m.A badly sorted and weathered alluvial deposit truncates and caps the weathered rock.It was laid down by a palaeo-Cleddau in a larger and higher valley than the present one.The Cleddau gorge has been incised transversally through the Treffgarne axis which is a very old west-south-west trending anticline.

This axis was active a very long time.Slow and prolonged uplift explains why weak mudstones outcrop at the summit of the Great Treffgarne Mountain.Widespread remnants of deep weathering occur in that region.The depth of the weathering front is very irregular and varies on the different rock types.In response to slow uplift,subaerial denudation may keep pace with the rate of chemical weathering; but if subaerial erosional processes become more efficient,the weathering front gets shallower especially where uplift is more rapid.In some places,the weathering front meets the land surface and tors might subsequently develop if rock structure is favourable.

The Treffgarne case suggests that tors need several conditions to develop:

- deep weathering with sharp lateral discontinuities in the thickness of saprolites.
- slow and prolonged uplift so that subaerial erosion is locally faster than chemical denudation (a general stripping of the weathering mantle does not lead to the formation of tors but to an irregular surface with knolls and hollows).
- a more or less orthogonal joint pattern leading to the exhumation of massive corestones,as it is observed in the massive Precambrian rhyolites of Treffgarne or in some facies of Millstone Grit.

It is not necessary to postulate two successive stages (a phase of weathering followed by a phase of stripping) as was proposed by Linton (1955).This problem has been clearly reviewed by Thomas (1979, p.186).

The limits of the Pleistocene glaciations passed to the north of the Treffgarne axis (Battiau-Queney,1981) and tors were slightly degraded by frost processes during glacial periods: large blocks of rhyolite lie on gently inclined slopes at the foot of Plumstone Rocks and Maiden Castle.It is well known that tors exist also in glaciated areas (for example in Finland),consequently they cannot be used as an indicator of the limit of glaciations.In Wales,tors exist in glaciated areas,for example on Blorenge Plateau,near Abergavenny.Here, tors built in quartzitic Millstone Grit have been moulded by ice into roches moutonnées.

If tors are not a good indicator of past climatic changes,on the contrary they are an excellent indicator of a morphotectonic equilibrium between the rates of superficial denudation and deep chemical weathering leading to the local exhumation of the weathering front and corestones.A sharp change of climate (cooler or drier) is not required to accelerate subaerial denudation: a persistent local uplift offers the best conditions.In the case of Treffgarne Mountain a prolonged upheaval is confirmed by the deepening of the transverse Cleddau Valley which is antecedent.

Once formed,tors can survive a long time because their bare and steep sides dry rapidly after rain (cf.Thomas,1979,p.187).A thin snow cover is probably more efficient but this point has not been seriously studied.

On the eastern border of Wales,on Sweeney Mountain (SJ 277258),Millstone Grit tors can be seen,still rooted in a sandy saprolite.Corestones are controlled by a widely spaced joint pattern.It is interesto note that Sweeney Mountain is situated on the uplifted edge of an hinge separating the stable Triassic plain of Shropshire and the Palaeozoic outcrops of the Welsh Massif.Because of the long preservation of tors,it is not easy to give them an age.In the case of Sweeney Mountain,their development must be related to the activity of the structural hinge of the Welsh eastern border.Several facts suggest a Neogene activity of this structural accident when climate was still wet and hot enough to produce the black "tropical" soil of the nearby Dol Goch quarry.

The landscape of St David's Land is characterized by several prominent hills standing abruptly above a low and relatively flat platform.It has long been interpreted as a Pliocene marine platform with old sea-stacks (cf.George,1970).It is a spectacular example of where new field data has prompted a completely different interpretation. In that area,hills are always formed in rocks particularly resistant to chemical weathering.The base of their steep slopes exactly follows lithological contacts.One of the best examples is the dolerite hill of Clegyr Boia (SM 738251).Thick saprolites are widespread on the lower platform.In Anglesey,a similar arrangement is frequently observed: For instance the hard and massive meta-quartzite of Mynydd Bodafon stands above deeply weathered Devonian sandstones seen in a small quarry at the foot of the south-east slope (SH 470852).It is evident that these residual hills have resulted from differential weathering and have emerged through downwearing of surrounding waste products and the progressive erosion of more resistant rocks: they are true inselbergs.It is impossible to give them an age.Some of them are probably very old and related to a long period of subaerial evolution in hot climates on stable crustal blocks.Once formed these residual hills tend to survive as has been demonstrated in the tropics (cf.Thomas,1979).

POSSIBLE RELICT PEDIMENTS IN NON-GLACIATED AREAS.

The Castlemartin-Bosherston platform lies at a height of 30-48m at the foot of sandstone hills which reach 70m to the north (fig.2).It truncates not only folded Carboniferous Limestone strata but also deeply weathered Millstone Grit (in the axis of Busslaughter Bay, SR 940943) and the supposed Oligocene clays of Flimston.The platform carries widespread alluvial quartz and quartzitic gravels mostly derived from the Old Red Sandstone.Gravels are subangular or subrounded excepted for some quartz derived from a Devonian conglomeratic facies.They are poorly sorted and belong to an alluvial cover deposited by widespread floods on an undissected surface.A break of slope exists between the relatively flat platform and the steeper sandstone slopes to the north.At Bullslaughter Bay and Flimston the alluvial sheet has truncated underlying weathering profiles.From these convergent field observations I consider that the Castlemartin-Bosherston platform is a pediment.

Other remnants of what could be pediments have been found in the Mendip Hills.An interesting site is Moon's Hill,near Shepton Mallet (ST 663461) where a huge quarry works Silurian andesites.Rock is deeply weathered through more than 20m.A transitional zone between solid and rotted rock exhibits spheroidal or dense orthogonal fissuration.Above,10 to 20m of saprolite with good preservation of the original rock structure are capped with 80 to 90cm of quartz,Old Red Sandstone and andesite fragments set in a sandy loamy matrix. This material is non-stratified and poorly sorted.Rounded quartz gravel comes from conglomeratic facies within the Old Red Sandstone. I think that this is a depositional formation derived from the stripping of regoliths developed on the Old Red Sandstone and other formations which outcrop on hillslopes just above (fig.3).The same type

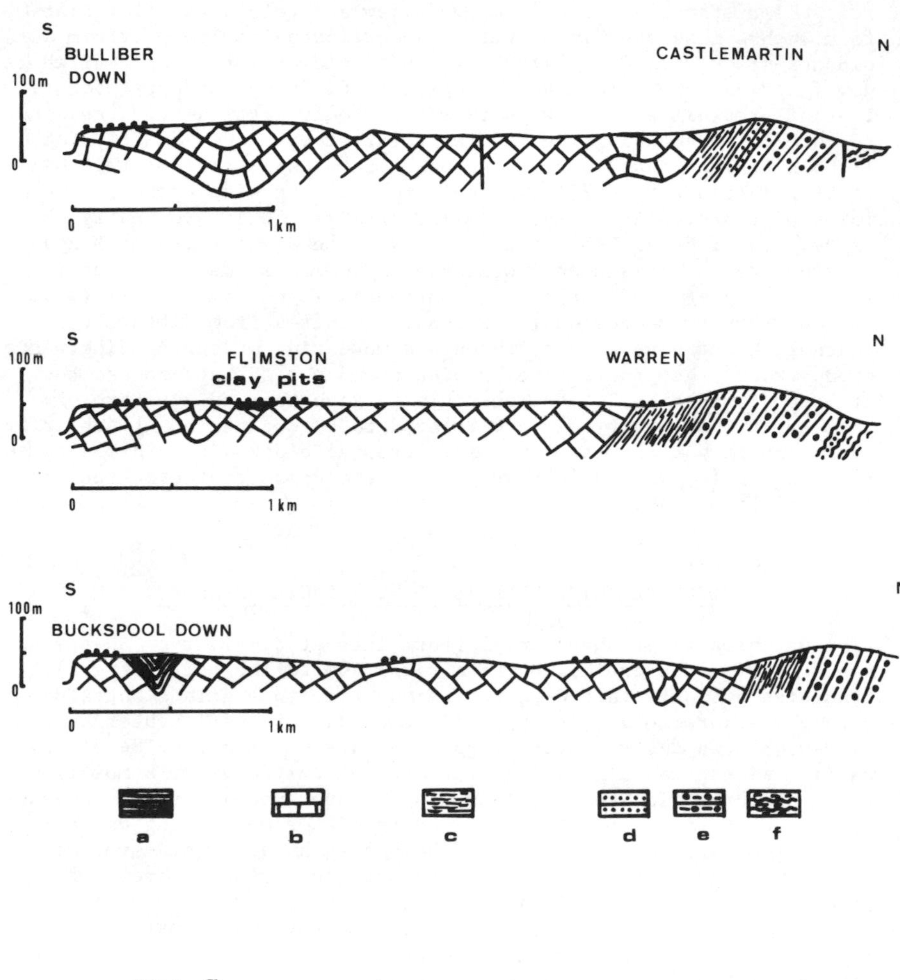

Fig.2: GEOLOGICAL SECTIONS THROUGH THE CASTLEMARTIN-BOSHERSTON PLATFORM (SOUTH DYFED).
a: Millstone Grit - b: Carboniferous Limestone - c: Lower Limestone Shales- d and e: Old Red Sandstone - f:Silurian and Ordovician
g: quartz gravels.

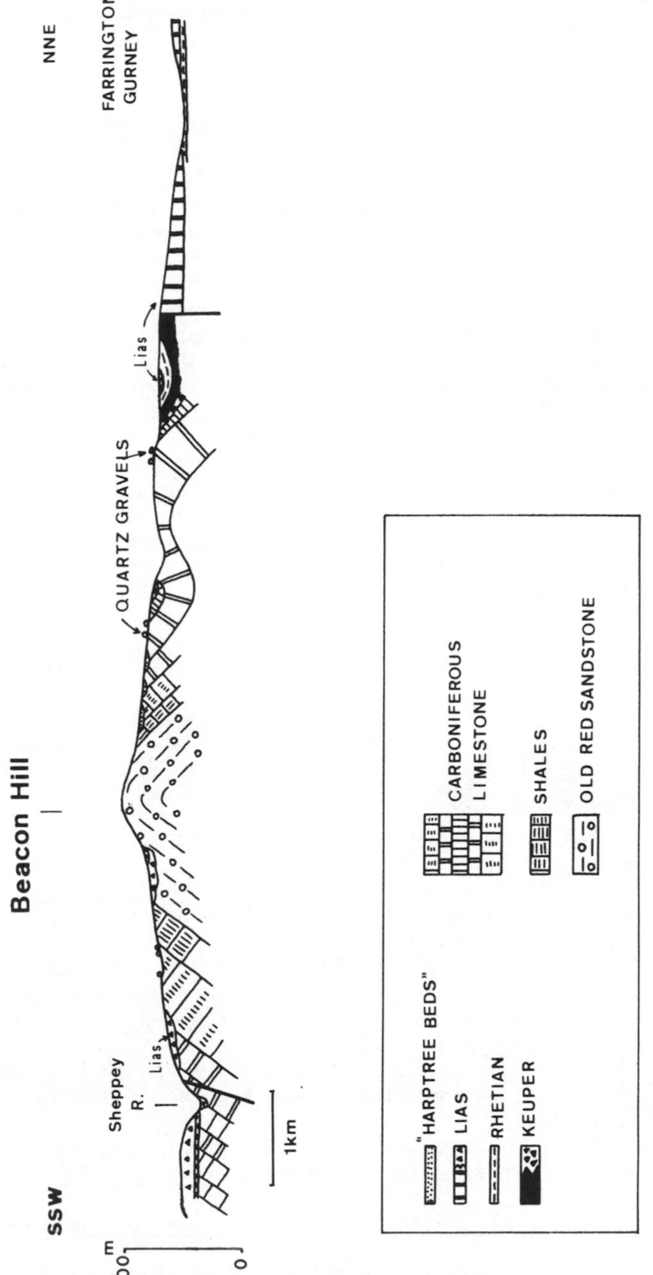

Fig.3: GEOLOGICAL SECTION THROUGH EASTERN MENDIPS WITH LOCALIZATION OF QUARTZ AND QUARTZITIC ALLUVIAL DEPOSITS.

of material is widespread in the Mendip Hills, exclusively on interfluves. I suggest it is a pediment cover laid down after a long phase of weathering and before the deepening of valleys in response to a rapid uplift of the Mendips (Battiau-Queney, 1978, p.403-413).

CONCLUSION.

In our temperate regions, landforms must be considered in the context of a long time-scale if we want to understand the complexity of present landscape. However, most geomorphologists have focussed their attention on the effects of periglacial or glacial morphogenesis and neglected the development of macroforms. In many European regions, for instance in Wales, Scotland, South-West England and Ireland, glacial periods have lasted a very short time compared to the 60 millions years or more of duration of subaerial evolution with hot or warm temperate climates. At that macroscale level of investigation, crustal (and even lithospheric) behaviour has to be taken into account. The remarkable crustal stability of Anglesey since at least Triassic times has been crucial to the exceptional durability of some inselbergs, as Mynydd Bodafon. It appears more and more probable that landforms can survive a very long time through varying climates when crustal properties are appropriate. In the British Isles, I think that a new approach of landscape development at the macroscale is required to elucidate the effects of endogenetic processes combined with successive morphogenetic processes.

REFERENCES.

Battiau-Queney, Y., 1973. Mise en évidence d'un karst tropical fossile au Pays-de-Galles. Norois, 77, 136-140.

Battiau-Queney, Y., 1978. Contribution à l'étude géomorphologique du Massif Gallois (G.B.), Thèse Lettres, Université de Brest, 797p., published in 1980, Honore Champion, Paris.

Battiau-Queney, Y., 1981. Les effets géomorphologiques des glaciations quaternaires au Pays-de-Galles. Rev. Geom. Dynam., 30 (2), 63-73.

Battiau-Queney, Y., 1984. The pre-glacial evolution of Wales. Earth Surface Processes and Landforms, 9, 229-252.

Battiau-Queney, Y., 1986. Buried paleokarstic features in South Wales, in New Directions in karst (Eds. Sweeting M. and Paterson K.), chap. 29, Geobooks, Norwich, in press.

Battiau-Queney, Y., and Saucerotte, M., 1985. Paléosol préglaciaire de la carrière de Ballyegan (Kerry, Irlande). Hommes et Terres du Nord, 3, 234-237.

Birot, P. and Muxart, T., 1977. L'altération météorique des roches, 279p, Université Paris IV, Paris.

George, T.N., 1970. British Regional Geology: South Wales, 3rd ed. H.M.S.O. London.

Hall,A.M.,1985.Cenozoic weathering covers in Buchan,Scotland,and their significance.Nature,315 (6018),392-395.

Isaac,K.P.,1981.Tertiary weathering profiles in the plateau deposits of East Devon.Proc.Geol.Assoc.,92 (3),159-168.

Isaac,K.P.,1983.Tertiary lateritic weathering in Devon,England and the Palaeogene continental environment of South West England.Proc.Geol. Assoc.,94 (2),105-114.

Linton,D.L.,1955.The problem of tors.Geogr.Jl.,121,470-487.

Mitchell,G.F.,1980.The search for Tertiary Ireland.Jl.Earth Sci.,3, 13-33.

Thomas,M.F.,1979.Tropical Geomorphology,2nd ed.332p..Macmillan Press Ltd,London.

Walsh,P.T.,1966.Cretaceous outliers in South West Ireland and their implications for Cretaceous palaeogeography.Quart.Jl.Geol.Soc.Lond., 122,63-84.

International Geomorphology 1986 Part II
Edited by V. Gardiner

WEATHERING AND RELIEF DEVELOPMENT IN BUCHAN, SCOTLAND.

A. M. Hall

Fettes College,
Edinburgh. EH4 1QX, Scotland, U.K.

ABSTRACT

The main stages in the geomorphic evolution of Buchan, north-east Scotland, are discussed. Outliers of Old Red Sandstone resting on deeply-denuded Caledonian basement demonstrate that erosion was already close to present levels by the Late Palaeozoic. Lowering of relief continued throughout much of the Mesozoic and was interrupted only by minor phases of uplift and by marine transgressions in the Late Jurassic and Late Cretaceous. During the Cenozoic prolonged weathering under warm to temperate humid environments caused the surface accumulation of siliceous residues, the formation of at least two different generations of weathering covers and the development of a subdued relief showing subtle litho-structural control. The high ground of central Buchan forms residual relief of Miocene or older age which retains a mantle of highly-kaolinitic saprolites and fluviatile flint and quartzite gravels. Below this lies the extensive Buchan Surface, underlain by thick grusses and finally elaborated under the humid temperate environments of the Pliocene and early Pleistocene. Glacial erosion has exploited pre-existing weathering patterns but generally failed to modify earlier relief significantly. The exceptionally long morphogenic history of Buchan reflects post-Palaeozoic tectonic stability related to the position of the region between the buoyant Grampian massif and the subsiding North Sea and Moray Firth basins.

INTRODUCTION

Buchan is the area of low relief which lies north of the River Ythan and east of the River Deveron in north-east Scotland (Figs.1 and 2). The region offers many opportunities for the study of long-term landscape evolution. Despite repeated glaciation, deep pre-glacial weathering and associated gravel deposits survive widely (FitzPatrick, 1963) and allow identification of the main stages of development of the present relief (Hall, 1983; 1985). In addition, offshore sediments in the Moray Firth and North Sea basins provide evidence of a longer history of morphogenesis stretching back to the Devonian. This paper aims to establish the main stages in the geomorphic evolution of Buchan.

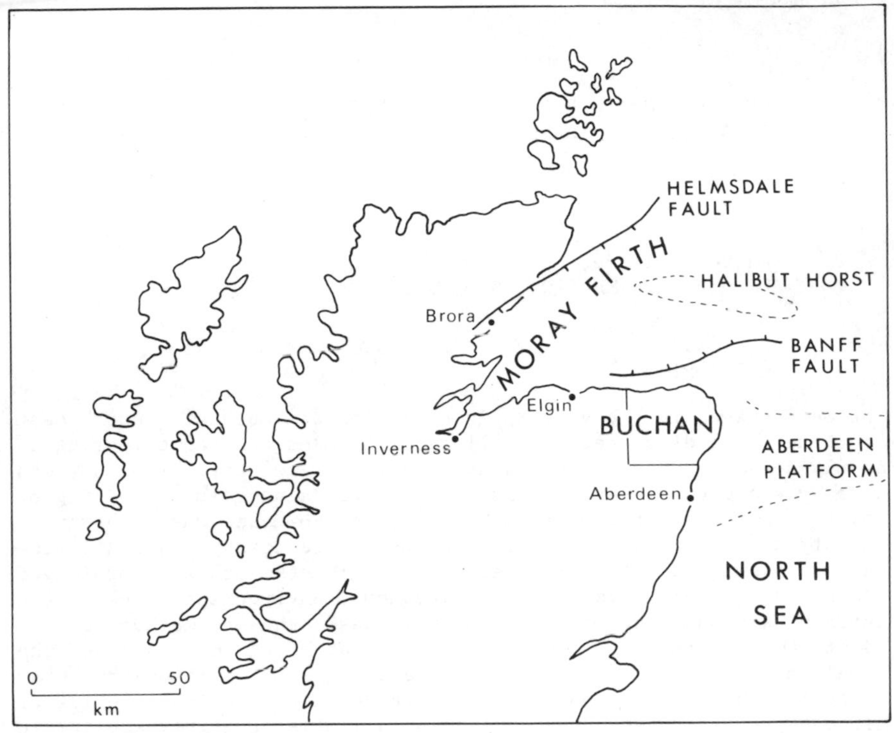

Fig.1. Location

GEOLOGY AND PRE-NEOGENE GEOMORPHIC EVENTS

Metasediments of Dalradian and, possibly, pre-Cambrian age form the oldest and most extensive rocks in the Buchan area (Fig.3). During the Caledonian orogeny, sediments of varied character were metamorphosed to different degrees to give a complex suite of migmatic gneises, psammitic schists and pelites which are now exposed in a shallow syncline across Buchan. At the climax of metamorphism at c.500 Ma numerous, large Younger Basic masses were emplaced at depths of c.15km (Droop and Charnley, 1985). A final phase of post-orogenic magmatism took place at c.400 Ma with widespread intrusion of the Newer Granite masses.

Metamorphic mineral assemblages imply original overburden thicknesses in the Buchan area of c.10km (Ashcroft et al., 1984). Yet by 380 Ma erosion was already close to present levels (Hudson, 1985). At various localities in north-east Scotland, Lower and Middle Devonian Old Red Sandstone sediments, mainly sandstone and conglomerates, rest directly on an irregular erosion surface cut across Dalradian metasediments and across unroofed and deeply-denuded Caledonian intrusions, including the Insch and Belhelvie mafic intrusions and the Peterhead and Aberdeen Newer Granites. The present exposure of Bennachie granite remains close to its Devonian level, for the granite was unroofed and contributed pebbles to the Turriff Old Red Sandstone basin (Mackie, 1923) yet retains roof rocks on Cairn William (Fig.4).

The Old Red Sandstone may have originally covered Buchan entirely (Mykura, 1983, Fig.8.13). This suggestion receives support from the minor contribution from the unroofed Maud basic and the Strichen granite masses to the adjacent Turriff basin (Archer, 1978) and from the recent discovery of a small outlier of red sandstone, of Old Red aspect, resting on Peterhead granite at Moss of Cruden. Post-Devonian erosion has been largely confined to the removal of a thick cover of Old Red Sandstone, of which c.1km remains in the Turriff basin (Ashcroft and Wilson, 1976), together with a thin slice of basement.

The Old Red Sandstone is the youngest sedimentary rock known in Buchan. Post-Devonian palaeogeographic reconstructions generally show Buchan as a persistent positive area (for example, Ziegler, 1981), and post-Palaeozoic erosion of this gently rising area has produced the present annular outcrop of late Palaeozoic sediments around Buchan (Fig.2). However, thick accumulations of post-Devonian sediments occur offshore and to the west around Lossiemouth, and evidence from these rocks suggests episodic transgression into the Buchan area.

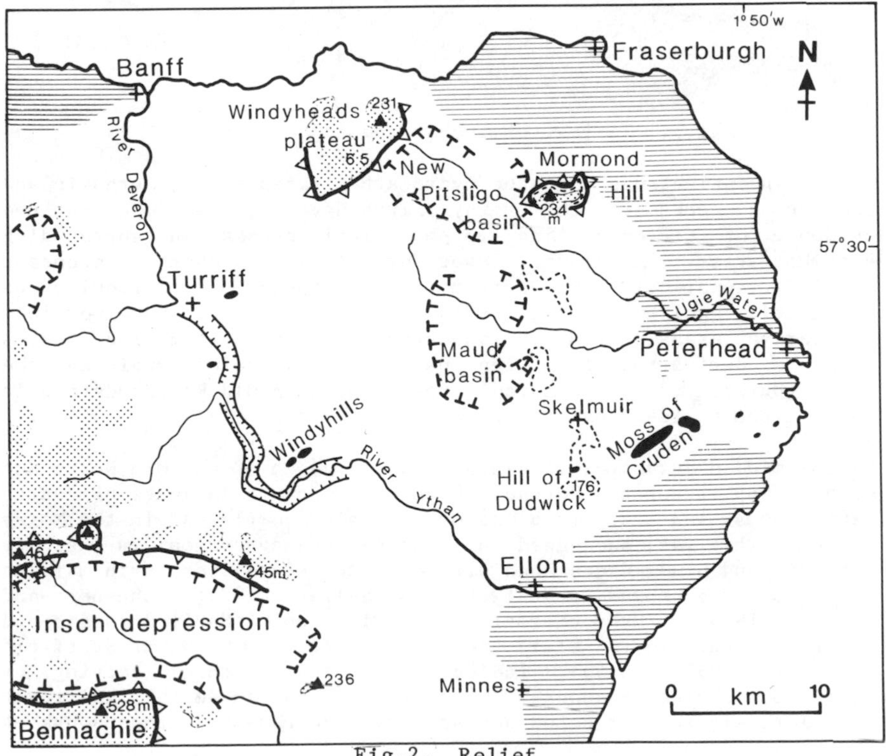

Fig.2. Relief

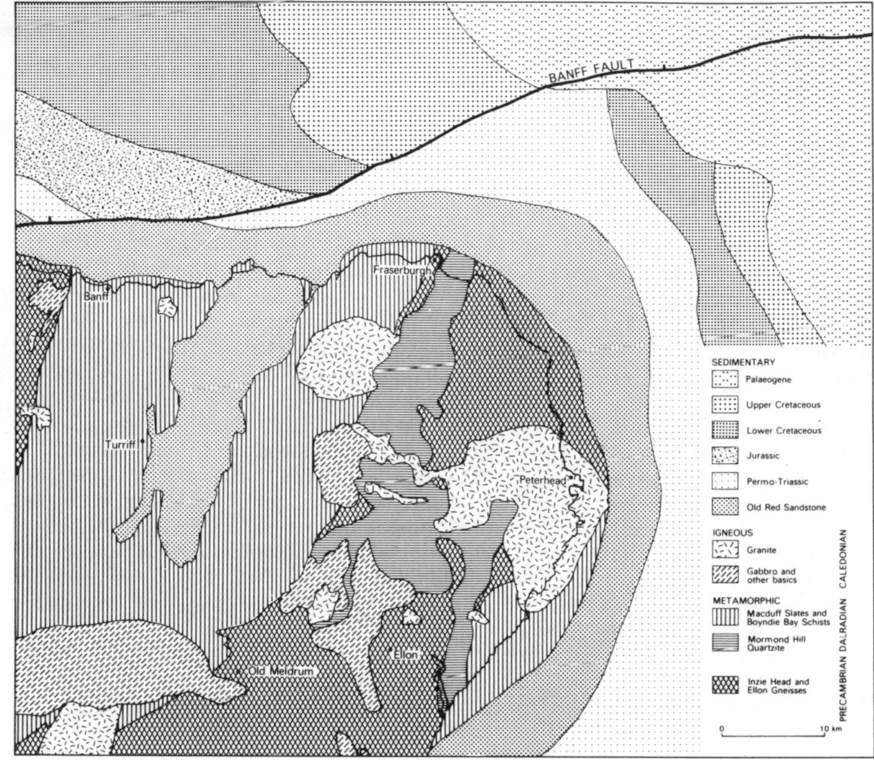

Fig.3. Geology of Buchan and adjacent areas.

Recent investigations of the hydrocarbon potential of Jurassic and Cretaceous rocks in the Moray Firth have provided important new evidence of tectonic and morphological events in surrounding basement areas. At Brora, lower Jurassic shales contain abundant kaolinite apparently derived from the stripping of Carboniferous kaolinitic regoliths (Hurst, 1985a). At Lossiemouth, the mineralogy of Early Jurassic fluvial sandstones and shallow marine shales suggests derivation from igneous and metamorphic terrain to the south (Hurst, 1985b), and indicates erosion of Old Red Sandstone in Moray close to its present outcrop.

In the middle Jurassic, volcanism and emergence of the Mid North Sea High and uplift of the Northern Highlands brought influxes of coarse terrigenous sediment into major deltaic complexes in the Moray Firth. This was succeeded by marine transgression in the Late Jurassic and Lower Cretaceous with overstep of earlier basin margins north of the Moray First (Neves and Selley, 1975; Chesher and Lawson, 1983). Hydrothermal mineralisation of Early Jurassic and older sediments in the Elgin area, dated to 140 ± 60 Ma at Stotfield (Moorbath 1962, p.55), indicates a former thick capping of impervious shales (Peacock et al., 1968) and suggests significant Late Jurassic transgression across levelled basement areas south of the Moray Firth.

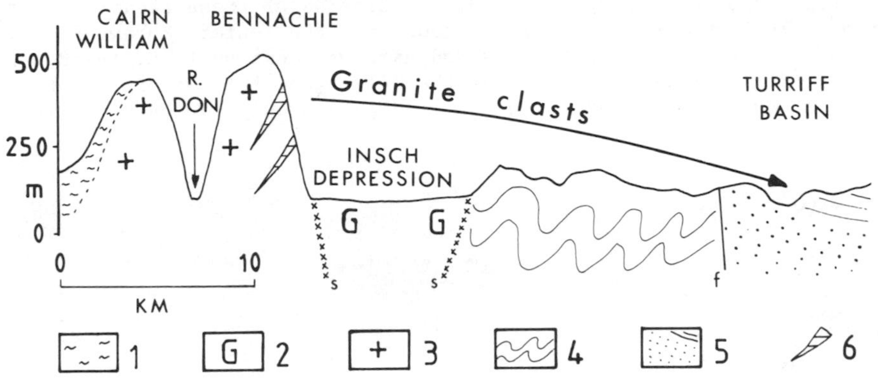

Fig.4. Devonian unroofing and minor post-Devonian erosion of the Bennachie granite mass.
1. Cordierite gneiss. 2. Basic igneous, mainly gabbro. 3. Bennachie leucogranite. 4. Andalusite schist. 5. Old Red sandstone and conglomerate. 6. Zones of hydrothermal alteration. s. Shear zones. f. fault.

The Bennachie granite was emplaced at c. 415 Ma. By c. 385 Ma the granite had been unroofed and was supplying clasts to the Lower and Middle Old Red Sandstone basin around Turriff. Today erosion remains in the upper zone of the granite, as shown by the preservation of roof rocks and by the presence of near-surface hydrothermal effects. The present elevation of the Bennachie granite probably reflects Tertiary uplift.

Tectonic activity was renewed at the Jurassic-Cretaceous boundary and a Palaeozoic massif, the Halibut Horst, emerged in the centre of the Moray Firth basin. Rapid erosion of poorly-consolidated Carboniferous Calciferous Sandstones shed sands to the south but there is no evidence of significant uplift of the Buchan area at this time (Anderton et al., 1979).

The Late Cretaceous marine transgression extended deep into the Scottish Highlands. Chalk occurs today in the eastern Moray Firth and on the Aberdeen Platform east of Buchan (Fig.5), but probably originally extended towards the Great Glen and northwards across parts of Caithness (Hancock, 1975). In Buchan, the presence in Tertiary gravels of large volumes of Cretaceous flint and of occasional silicified Greensand clasts demonstrates transgression across this area. As reworked flints now occur at elevations of up to 150m in central Buchan and, as the Chalk was originally deposited in water depths of 100 to 600m (Hancock, 1975), it is clear that much of north-east Scotland was covered by the Late Cretaceous sea.

In the early Tertiary, magmatic activity in western Scotland was accompanied by uplift of the Scottish Highlands and of the Orkney-Shetland Platform (Ziegler, 1981). Erosion of these areas caused rapid build-up of submarine fans in the outer Moray Firth. Downfaulting of Late Cretaceous and earlier rocks north of the Banff fault (Fig.5) took place during the Palaeogene but the preservation of Cretaceous residues in Buchan indicates only modest vertical movement in this area.

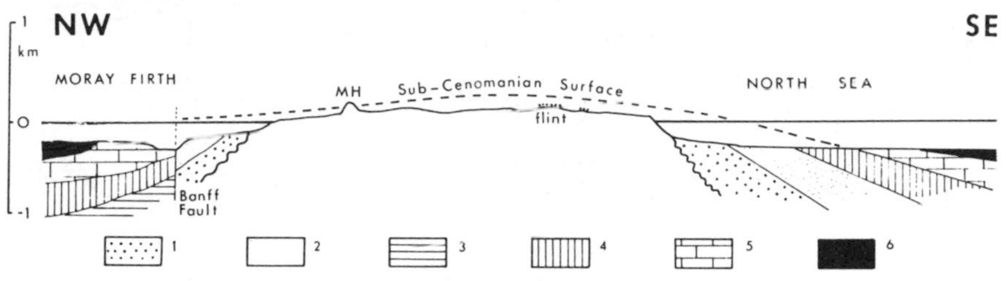

Fig.5. Schematic geological cross-section across Buchan. 1. Old Red Sandstone. 2. Permo-Triassic sandstones. 3. Jurassic shales. 4. Lower Cretaceous sandstones and shales. 5. Upper Cretaceous chalk. 6. Palaeogene lignitic sands. MH Mormond Hill.

The pre-Cenozoic geomorphic history of Buchan is summarised in Table 1. The area has not experienced major uplift since the Devonian and several lines of evidence indicate that post-Palaeozoic erosion of basement has been modest, namely:

(a) the widespread survival of Devonian outliers and evidence of minor post-Devonian erosion of Caledonian intrusions (cf. Watson, 1985).
(b) the survival of Permian weathering effects in the Old Red Sandstone. Haematite growth was reactivated by the deep penetration of oxygen during this period of extreme aridity and the Old Red Sandstone retains a Permian palaeomagnetic signature (Tarling et al., 1976).
(c) the widespread retention of residues from Cretaceous cover rocks.

These indicators of modest long-term erosion suggest that the subdued relief of Buchan is an inherited feature. Levelling of the area was probably first achieved in the Triassic. Buchan contributed little to the thick Mesozoic accumulations in the Moray Firth basin, and lowering of relief during this period was modest.

This long-term stability reflects the distinctive character and position of the Buchan 'craton', partly decoupled from the Grampian Caledonides to the west and south by major shear belts (Ashcroft et al., 1984) and bounded to the north and east by the Moray Firth and North Sea basins, areas of persistent late Palaeozoic and Mesozoic subsidence.

TABLE 1. Summary of the main geomorphic events in the shaping of Buchan

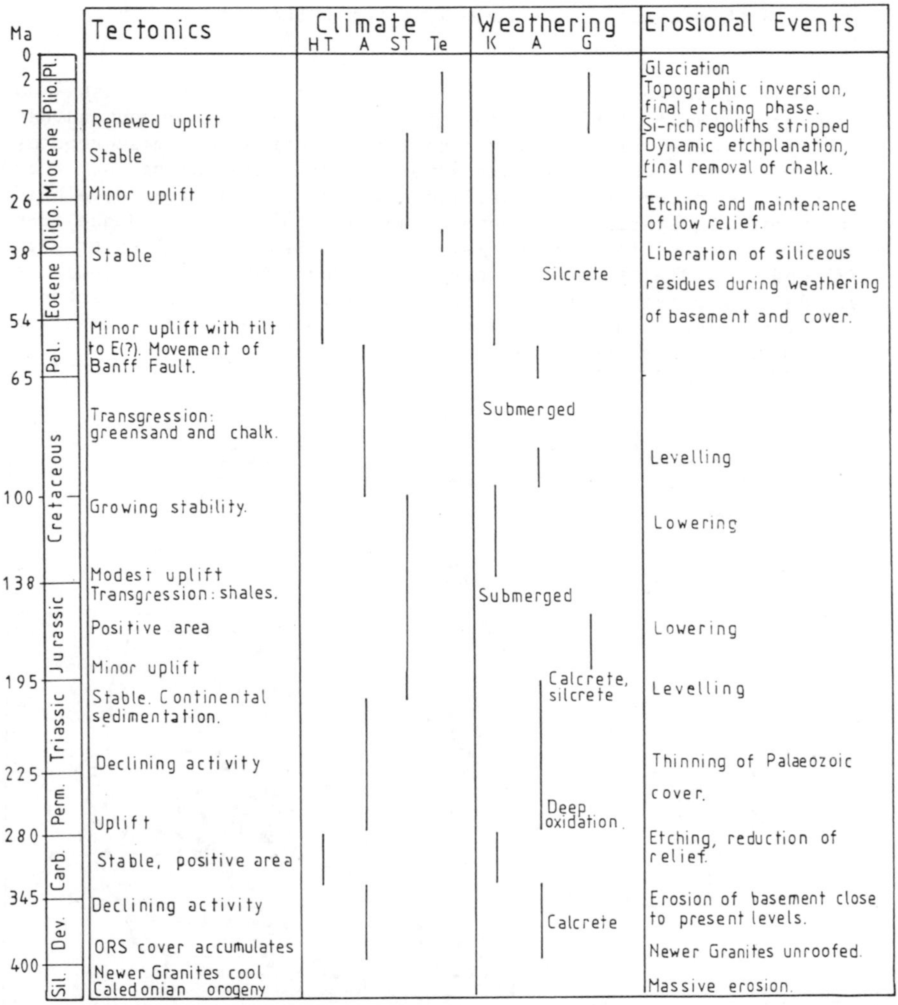

Ma	Period	Tectonics	Climate (HT A ST Te)	Weathering (K A G)	Erosional Events
0	Pl.				Glaciation
2	Plio.	Renewed uplift			Topographic inversion, final etching phase. Si-rich regoliths stripped
7	Miocene	Stable Minor uplift			Dynamic etchplanation, final removal of chalk. Etching and maintenance of low relief.
26	Oligo.				
38	Eocene	Stable		Silcrete	Liberation of siliceous residues during weathering of basement and cover.
54	Pal.	Minor uplift with tilt to E(?). Movement of Banff Fault.			
65	Cretaceous	Transgression: greensand and chalk.		Submerged	Levelling
100		Growing stability.			Lowering
138	Jurassic	Modest uplift Transgression: shales. Positive area Minor uplift		Submerged	Lowering
195	Triassic	Stable. Continental sedimentation. Declining activity		Calcrete, silcrete	Levelling Thinning of Palaeozoic cover.
225	Perm.	Uplift		Deep oxidation.	
280	Carb.	Stable, positive area			Etching, reduction of relief.
345	Dev.	Declining activity ORS cover accumulates		Calcrete	Erosion of basement close to present levels. Newer Granites unroofed.
400	Sil.	Newer Granites cool Caledonian orogeny			Massive erosion.

CENOZOIC WEATHERING COVERS AND THEIR CHARACTERISTICS

Deep weathering covers of Cenozoic age are widely developed in Buchan (Fig.6). All rock types are affected to various degrees and depths of weathering commonly reach depths of several tens of metres, with maximum depths exceeding 50m (FitzPatrick, 1963; Hall, 1985).

Two types of weathering can be identified in Buchan: gruss and clayey gruss (Hall, 1985). Gruss is by far the most common and has low clay contents, modest fines contents, varied and immature clay mineral assemblages, which are strongly influenced by parent rock mineralogy, and low soluble base losses. These saprolites developed under humid temperature environments and mineralogical evidence from Neogene sediments in the central North Sea basin (Karllson et al., 1979) and from Pleistocene sediments in Buchan (Hall, 1984) indicate that deep grusses began to form in the late Miocene, thickened during the Pliocene and had taken on their present characteristics by the middle Pleistocene (Hall, 1985).

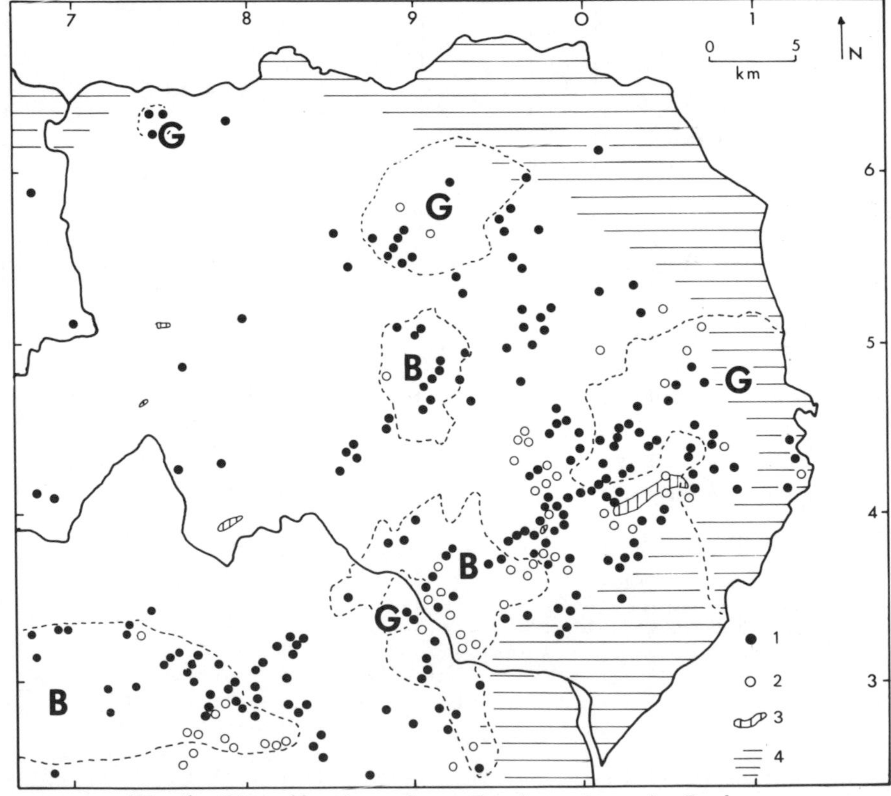

Fig.6. Distribution of weathering sites in Buchan. 1. Weathered rock in exposures. 2. Weathered rock in boreholes. 3. Buchan Gravels. 4. Coastal drift plain. G. Granite. B. Basic igneous

Clayey grusses are virtually confined to central Buchan. These saprolites have higher clay and fines contents, mature clay mineral assemblages dominated by kaolinite and illite, and are often rubefied and show etching of quartz. These characteristics indicate weathering under relatively warm and humid environments, probably before the final establishment of temperate conditions in the North Sea area in the Pliocene (Buchardt, 1978). Clayey grusses appear to pre-date deposition of the Buchan Gravels, which contain high proportions of detrital kaolinite (Hall, 1982) and locally rest on kaolinitic saprolites, but, unfortunately, the precise age of the gravels is unknown. However, the preservation of residual K-feldspar and even, occasionally, biotite, in these saprolites makes a Neogene, rather than a Palaeogene or Cretaceous, age seem most likely.

DEEP WEATHERING PATTERNS

The primary control over thickness and continuity of weathering covers in Buchan has been the variable intensity of glacial erosion. In general, saprolites have been removed from coastal districts by vigorous, warm-based ice streams whilst deep saprolites have survived inland under protective, cold-based masses (Fig.6; Hall, in press). Within individual zones of glacial erosion, however, weathering patterns which reflect geological and topographic controls are apparent.

The incidence and depth of weathering varies with rock type. Biotite-rich granites, gabbros and gneisses provide relatively few fresh outcrops and are associated with the greatest depths of weathering recorded in boreholes. Weathering is least developed on fine-grained and quartzitic schists. Rapid lateral variations in lithology, however, can obscure these relationships, as where deep pockets of weathering are developed in less resistant schists interbanded with quartzites. Fracturation adds further complications, with, for example, troughs of deep weathering aligned along fracture belts crossing lithological boundaries.

Differential weathering of the complex basement structures has created distinctive weathering patterns. Alteration has picked out zones of low rock resistance to give basins and troughs of weathering which often have topographic expression as basins and valleys (Fig.7).

The distribution of weathering types, however, is apparently independent of geology. Clayey grusses developed on varied metamorphic rocks occupy a belt of high ground in central Buchan (Fig.1). At Hill of Dudwick and Moss of Cruden clayey grusses are juxtaposed with or overlain by flint gravels of the Buchan Ridge Formation (McMillan and Merritt, 1981). These kaolinitic saprolites and associated fluviatile gravels define areas of residual relief standing above the Buchan Surface with its thick mantle of grusses.

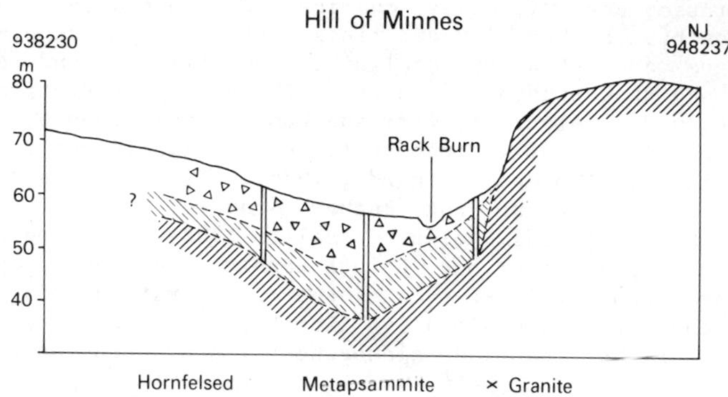

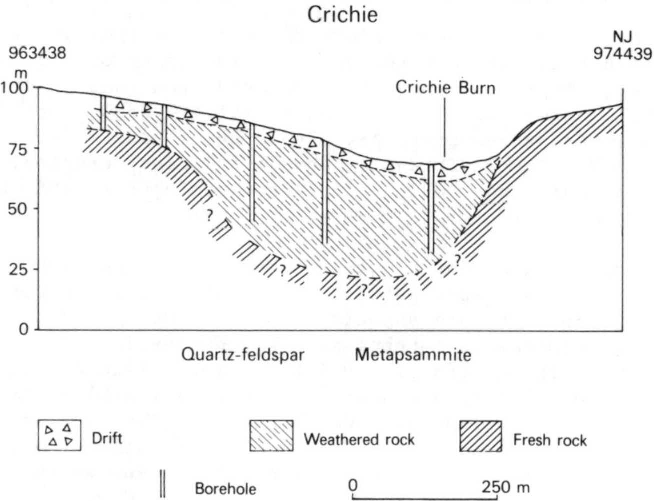

Fig.7. Linear zones of deep weathering and their exploitation by meltwater.

WEATHERING, GEOLOGY AND MORPHOLOGY

The relief of Buchan is dominated by a single erosion surface, the Buchan Surface, developed across complex geology. This feature is inherited from earlier Mesozoic lowlands but there is no firm evidence for survival of Mesozoic levels in the present relief. In situ Mesozoic rocks are unknown in Buchan and the kaolinitic clayey grusses are most likely to be the remnants of Neogene, rather than Cretaceous, weathering covers. Nonetheless, the cavernous flints at Moss of Cruden indicate that lowering of the sub-Cretaceous surface has been modest and allows speculation that the prominent quartzite inselberg of Mormond Hill, with its pockets of deep kaolinisation, is in outline a Mesozoic relic.

The relative relief of the Buchan Surface, together with its upper residual elements seldom exceeds 60m. In detail, however, this subdued terrain resolves into a tiered landscape showing pervasive litho-structural control. An upper relief of isolated, low hills and broad interfluves developed on quartzitic and pelitic schists and with pockets of clayey gruss passes downslope into extensive levelled areas incorporating open, saucer-like basins developed on deep grusses. Set into this middle relief are forms of negative resistance located on zones of weakness. These include the large, shallow basins of Maud and New Pitsligo and broad valleys such as that of the South Ugie Water which follows a septum of biotite granite through the quartzite belt of central Buchan.

Two important anomalies exist in this general accordance between topographic position and relative rock resistance. Firstly, certain drainage divides in central Buchan are developed in rocks of low resistance, as at Skelmuir Hill where local diorites are weathered to depths of over 20m. These ridges appear to owe their present elevation to a former protective cover of flint gravels, evidenced by localised concentrations of flint in drift. Secondly, the Windyheads plateau, south-east of Gardenstown, is formed from a fault-bounded block of Old Red Sandstone, sediments which elsewhere in north-east Scotland tend to form low ground. The Windyheads plateau is interpreted as the easternmost representative of a series of low horsts developed in the Moray Firth border zone by differential tectonic movement in the Tertiary (Hall, 1983).

GEOMORPHIC EVOLUTION

In Buchan the maintenance of low relief during long-term surface lowering and the close links between landforms, relative rock resistance and the incidence of weathering in the meso-scale relief indicate that the present terrain has evolved by a form of etchplanation. Climatic conditions suitable for the formation of deep saprolites prevailed in Scotland throughout much of the Mesozoic and Tertiary. In Buchan, climates favouring profound etching of differential rock resistance were combined with relative tectonic stability so that rates of weathering may have exceeded rates of erosion for long periods.

Early events in the shaping of Buchan, such as Cretaceous transgression, can be inferred from the geological record but cannot yet be related to present topographic levels. The oldest surface known in Buchan, apart from small areas of exhumed sub-Devonian relief, is the high ground of central Buchan with its residual mantle of clayey grusses and flint gravels (Fig.8). This terrain preserves large volumes of siliceous material liberated during intense weathering and gradual surface lowering in the Palaeogene. This material includes quartzite recycled from Old Red sandstone conglomerates, flint and chert released from Cretaceous sediments, and vein quartz derived from alteration of crystalline rocks. This material now constitutes the gravels of the Buchan Ridge Formation, deposits which point to an important phase of uplift which led to mobilisation of siliceous and kaolinitic regoliths and concentration

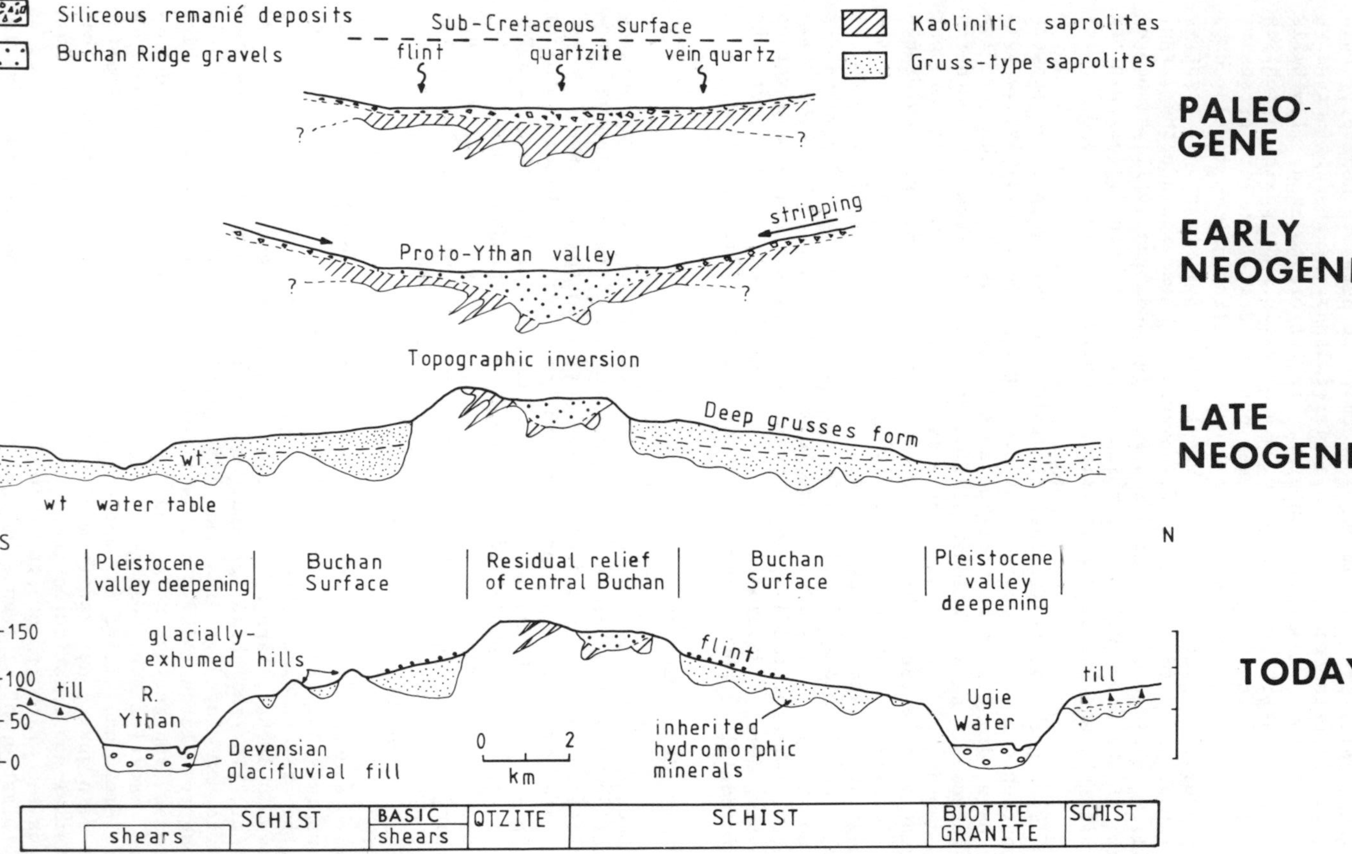

Fig.8. Summary of Tertiary relief development in central Buchan

of siliceous clasts along a proto-Ythan valley. The clayey grusses found beneath the flint gravels may be remnants of such Palaeogene regoliths but interpretation is complicated by the fact that the gravels are themselves intensely weathered (Koppi and FitzPatrick, 1980), with kaolinisation of clasts extending down to 25m at Moss of Cruden (McMillan and Merritt, 1980).

These gravels now occupy summit sites at up to 150m OD in central Buchan and reflect topographic inversion in response to post-depositional uplift. This uplift eventually resulted in stripping of clayey grusses from wide areas. Their replacement by geochemically-immature but thick grusses indicates that uplift continued into the humid temperate Pliocene. Continued etching of bedrock variability is demonstrated by the influence of geology on gruss weathering patterns and by the landforms of differential weathering and erosion which comprise the meso-scale relief of the Buchan Surface.

Pleistocene processes have merely modified a pre-existing land surface. In zones of relatively intense glacial erosion, along valleys and near the coast, basal surfaces of weathering have been exposed and scoured. Inland, however, the widespread survival of deep pre-glacial regoliths demonstrates that successive ice sheets have failed to lower the pre-glacial landsurface significantly.

The geomorphic evolution of Buchan is similar in many aspects to that of other lowland crystalline terrains in western Europe, including Brittany (Estouelle-Choux, 1963), Dyfed (Battiau-Queney, 1984) and Skane (Lidmar-Bergström, 1982). These areas have a common climatic history, preserve Mesozoic and Tertiary weathering covers and sediments and experienced neither major uplift in the Tertiary nor deep erosion in the Pleistocene. Low relief has been maintained by etchplanation since the Mesozoic due to relative tectonic stability which stems from the similar structural positions of these regions at the edges of basement blocks and adjacent to sedimentary basins.

REFERENCES

Anderton, R., Bridges, P. H., Leeder, M. R., and Sellwood, B. W., 1979. A Dynamic Stratigraphy of the British Isles. Allen and Unwin, London.

Archer, R., 1978. The Old Red Sandstone outliers of Gamrie and Rhynie, Aberdeenshire. Unpub. Ph.D. Thesis, Univ. of Newcastle.

Ashcroft, W. A., and Wilson, C. D. V., 1976. A geophysical survey of the Turriff basin of the Old Red Sandstone, Aberdeenshire. J. Geol. Soc., London, 132, 27-43.

Ashcroft, W. A., Kneller, B. C., Leslie, A. G., and Munro, M., 1984. Major shear zones and autochthonous Dalradian in the north-east Scottish Caledonides. Nature, 310, 760-762.

Battiau-Queney, Y., 1984. The pre-glacial evolution of Wales. Earth Surface Processes and Landforms, 9, 229-252.

Buchardt, B., 1978. Oxygen isotope palaeotemperatures from the Tertiary period in the North Sea area. Nature, 275, 121-123.

Chesher, J. A., and Lawson, D., 1983. Geology of the Moray Firth. Rep. Inst. Geol. Sci., 83/5.

Droop, G. T. R., and Charnley, N., 1985. Comparative geobarometry of pelitic hornfelses associated with the newer gabbros: a preliminary study. J. Geol.Soc. London, 142, 53-62.

Estouelle-Choux, J., 1973. Contribution a l'étude des argiles du Massif Amoricain. Thesis, Univ. of Rennes.

FitzPatrick, E. A., 1963. Deeply weathered rock in north-east Scotland, its occurrence, age and contribution to the soils. J. Soil Sci., 14, 33-42.

Hall, A. M., 1982. The 'Pliocene' Gravels of Buchan: A Reappraisal: Discussion. Scott. J. Geol., 18, 336-338.

Hall, A. M., 1983. Weathering and landform evolution in north-east Scotland. Unpub. Ph.D. Thesis, Univ. of St. Andrews.

Hall, A. M., 1984. Buchan Field Guide. Quaternary Research Association: Cambridge.

Hall, A. M., 1985. Cenozoic weathering covers in Buchan, Scotland, and their significance. Nature, 315, 392-395.

Hall, A. M., in press. Deep weathering patterns in north-east Scotland and their geomorphological significance. Zeit. für Geom.

Hancock, J. M., 1975. The petrology of the Chalk. Proc. Geol. Assoc., 86, 499-536.

Hancock, J. M., 1984. Cretaceous, in Introduction to the Petroleum Geology of the North Sea, (Ed. Glennie), pp133-150. Blackwell, Oxford.

Hinxman, L. W., 1896. The geology of Western Aberdeenshire, Banffshire and parts of Elgin and Inverness. Mem. Geol. Surv. Scotland, Sheet 75.

Hudson, N. F. C., 1985. Conditions of Dalradian metamorphism in the Buchan area, north-east Scotland. J. Geol. Soc. London, 142, 63-76.

Hurst, A., 1985a. The implications of clay mineralogy to palaeoclimate and provenance during the Jurassic in north-east Scotland. Scott. J. Geol., 21, 143-160.

Hurst, A., 1985b. Mineralogy and diagenesis of Lower Jurassic sediments of the Lossiemouth borehole, north-east Scotland. Proc. Yorks. Geol. Soc., 45, 189-197.

Karllson, W., Vollset, J., Bjorlykke, K., and Jorgensen, P., 1979. Changes in the mineralogical composition of Tertiary sediments from North Sea wells. Proc. 6th. Int. Clay Conf., 27, 281-289.

Koppi, A. J., and FitzPatrick, E. A., 1980. Weathering in Tertiary gravels in north-east Scotland. J. Soil Sci., 31, 525-532.

Lidmar-Bergström. K., 1982. Pre-Quaternary geomorphological evolution in southern Fennoscandia. Sver. Geol. Unders. Ser. C. No. 785.

Linsley, P. N., Potter, H. C., McNab, C., and Racher, D., 1980. The Beatrice Field, inner Moray Firth, U.K. North Sea. Am. Ass. Petrol. Geol. Mem., 30, 117-119.

Lovell, J. P. B., 1984. Cenozoic, in Introduction to the Petroleum Geology of the North Sea, (Ed. Glennie), pp151-169. Blackwell, Oxford.

Mackie, W., 1923. The principles that regulate the distribution of particles of heavy minerals in sedimentary rocks, as illustrated by the sandstones of the north-east of Scotland. Trans. Edin. Geol. Soc., 11, 138-164.

McMillan, A. A., and Merritt, J., 1980. A reappraisal of the Tertiary deposits of Buchan, Grampian region. Rep. Inst. Geol. Sci. No. 80/1.

Moorbath, S., 1962. Lead isotope abundance studies on mineral occurrences in the British Isles and their geological significance. Phil. Trans. Roy. Soc. Ser. A., 254, 295-360.

Mykura, W., 1983. Old Red Sandstone, in Geology of Scotland, (Ed. Craig), pp205-251. Scottish Academic Press, Edinburgh.

Neves, R., and Selley, R. C., 1975. A review of the Jurassic rocks of North-east Scotland, in Proceedings of the Jurassic of the northern North Sea Symposium. (Eds. Finstad and Selley), JNNSS 5, pp1-29. Norwegian Petroleum Society, Stavanger.

Peacock, J. D., Berridge, N. G., Harris, A. L., and May, F., 1968. The geology of the Elgin district. Mem. Geol. Surv., Sheet 95.

Tarland, D. H., Donovan, R. N., Abou-Deeb, J. M., and El-Batrouk, S. I., 1976. Palaeomagnetic dating of haematite genesis in Orcadian Basin sediments. Scott. J. Geol., 12, 125-134.

Watson, J. V., 1985. Scotland as an Atlantic-North Sea divide. J. Geol. Soc. London, 142.

Wilson, J. S. G., and Hinxman, L. W., 1890. The geology of central Aberdeenshire. Mem. Geol. Surv. Scotland, Sheet 76.

Ziegler, P. A., 1981. Evolution of sedimentary basins of North-West Europe, in Petroleum Geology of the Continental Shelf of North-West Europe, (Eds. Illing and Hobson), pp3-39. Heyden, London.

International Geomorphology 1986 Part II
Edited by V. Gardiner

TERTIARY WEATHERING ENVIRONMENTS AND PRODUCTS IN NORTHEAST IRELAND

B.J. Smith and J.J. McAlister

Department of Geography, Queen's University,
Belfast, BT7 1NN, Northern Ireland

INTRODUCTION AND BACKGROUND

As long ago as 1954, Proudfoot felt confident enough to assert in his paper on erosion surfaces in the Mourne Mountains that: "The major features of the physical landscape in Ireland are the result of erosion forces acting during Tertiary times upon Tertiary and older rocks." (p. 26). Outside of Northern Ireland however, there is an almost total absence of Permian, Mesozoic and Tertiary strata. This, as Davies (1970) pointed out, has made it virtually impossible to reconstruct accurately Ireland's history between the Carboniferous and the Pleistocene. "Yet it was during that interval that the physique of modern Ireland was blocked out" (p. 1). Possibly because of the incomplete geological record, and despite recognition of the significance of Tertiary erosion and deposition, mainstream Irish geology and geomorphology has been characterized by efforts to explain the Irish landscape principally in terms of the effects of Quaternary glaciations. It is only recently that this pre-eminence of the Quaternary has again been called into question, most notably in papers by Mitchell (1980 and 1985). In these he acknowledges his previous convictions about the efficacy of glacial processes, but now notes that he has: "begun to see what I believe are Tertiary features only lightly rounded or veneered by the action of frost and ice" (1980, p. 13). Under this stimulus, interest has been revived in assessing the extent to which the present landscape of Ireland is relict from the Tertiary and, in particular, improving our understanding of the nature of Tertiary geomorphic environments.

The Tertiary environment in Ireland has been reviewed on several occasions, recently by Davies and Stephens (1978) and earlier by Davies (1970). In both reviews it was pointed out that interpretations of Irish landscape evolution have traditionally centred on subaerial denudation of a raised erosion surface. Initially, Jukes (1862) considered that the surface originated at the end of the Carboniferous. Later, Linton's (1951) 'Sub-Cenomanian Surface' was extended across the Irish Sea, and it was proposed that the landscape was initiated by Tertiary dissection of this surface from an initial altitude in excess of 1500m. However, the description of low altitude (75m) deposits of Early Tertiary age from Co. Tipperary by Watts (1957), initiated a reappraisal of denudation history in which Tertiary earth movements now played a major role (eg. Whittow, 1958 and Davies, 1960). Thus it is now accepted that the key to understanding the Irish landscape lies in the idea that during the Tertiary some areas were uplifted and subjected to vigorous denudation while others subsided and were largely protected from erosion (Davies and Stephens, 1978). It is in these latter areas that some of the few deposits of Cretaceous and Tertiary age in southern Ireland have been preserved (Walsh, 1960 and 1965 and Mitchell, 1981). The area in which the clearest evidence for this Tertiary diastrophism exists is in the northeast, where, in addition to evidence of mid-Tertiary faulting and warping, there are extensive igneous outcrops of early Tertiary (Palaeocene) age. These latter include the intrusive complexes of the Mourne Mountains, Carlingford and Slieve Gullion and the Plateau basalts which underlie most of County Antrim (Fig. 1). Beneath the basalts, and protected by them, is an early Tertiary landscape on top of the most complete sequence of Permian and Mesozoic strata in Ireland. It is the presence, in particular, of the thick (up to 150m) Cretaceous chalk that provides one of the more cogent arguments for a former Cretaceous cover over much of the island. If this were so, the implications are that the cover was removed during the Tertiary, and George (1967) has suggested that in the area around Newry considerable stripping occurred even during post-Cretaceous, pre-basalt times. This is shown by the presence of basalt remnants without underlying chalk near, for example, Carlingford. In addition, the Tertiary has seen the removal in places of more than 850m of country rock (mainly Silurian shales and grits) to expose the Mournes Massif in Co. Down.

There is, however, debate as to how and when the bulk of this erosion was accomplished. In his paper on landform and structure in Ulster, George (1967) considers that the major elements of the landscape originated in the second half of the Tertiary (the Neogene Period), and developed from an emergent Mid-Tertiary, stepped wave-cut plateau. This surface would have truncated Mid-Tertiary, Mesozoic and Hercynian structures and rocks, but was not flat and would have contained depressions reflecting a mid-Tertiary synclinal pattern. This surface, he suggested, can be seen today in the accordant summits not only of northeast Ireland but of the British Isles as a whole. Opposed to this viewpoint, is the more recent suggestion of Reffay (1972) which lays emphasis on initial planation by subaerial processes followed by continued fluvial erosion and corrosion throughout the Tertiary, together with considerable tectonic distortion. Certainly, given, as we shall see, the evidence for Tertiary subaerial weathering in the northeast of Ireland, and the absence of depositional evidence for marine inundation, the latter interpretation appears more convincing. Although, as Mitchell (1980) points out, it would be dangerous to accept completely Reffay's invocation of excessive tectonic displacement in which; "she is prepared to move blocks up and down, just as elevators in a sky-scraper can be moved up and down at will" (p. 4).

In all the models of Irish landscape evolution, it is apparent that the northeast corner has seminal significance, in providing almost the sole, unequivocal indication as to the nature of late Mesozoic

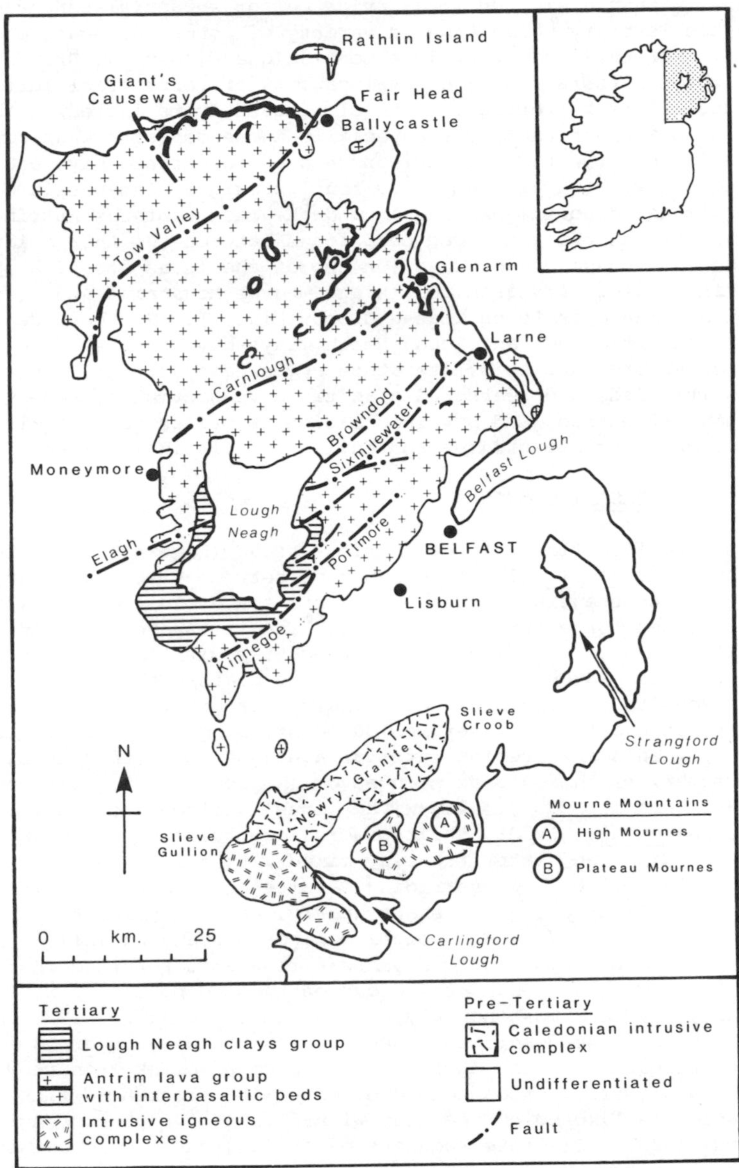

Figure 1: Location map showing the Tertiary geology of northeast Ireland.

and Tertiary tectonics and denudation. In particular, the northeast provides, within a limited area, evidence for subaerial conditions during the Tertiary which takes a variety of forms and spans most of the era. For this reason it is almost unique within the British Isles, which suffers from a general paucity of terrestrial Tertiary deposits. It is therefore the aim of this study to introduce this evidence to a wider audience; especially those deposits that can be used to indicate weathering environments and how these have changed. The evidence we will examine takes four forms: 1. Sub-Basaltic palaeosols and deposits exposed around the margin of the Antrim Plateau. 2. Interbasaltic red beds found between the Antrim lavas. 3. Oligocene 'Lough Neagh Clays' on top of the lavas and 4. Deep weathering of Tertiary granites in the Mourne Mountains and Caledonian granodiorite on Slieve Croob (Fig. 1). Of these deposits, the Interbasaltic beds and Lough Neagh Clays have been studied in detail elsewhere, and this literature reviewed. However, the Sub-Basaltic Beds and weathered granites have previously received only scant attention, and new information is presented on their composition, possible origins and environmental significance.

SUB-BASALTIC BEDS

The basaltic lava flows of the Antrim Plateau (Figure 1) have been dated by Evans et al. (1973) to a period between 65-62 m.y. ago, although circumstantial evidence suggests that volcanic activity may have continued for a further 2-3 m.y.. The lavas are predominantly tholeiitic basalts and fine-grained trachytes with locally occurring rhyolite lavas. Around the margins of the plateau the lavas are usually seen to overlie Cretaceous chalk, but between the two strata there is often a thin, clayey residue containing reddened flints (Plate 1). This intervening material averages 2m in thickness, though sometimes there is as much as 6m (Wilson, 1972). Although they have been mapped as 'clay-with-flints' (Wilson and Robbie, 1966 and Manning et al., 1970), their precise composition is variable and includes: dark-red, lateritic clay mixed with flints which has been partially 'baked' by the overriding basalts, grey marl containing thin beds of lignite (eg. Belshaws quarry, north of Lisburn; Irish Grid reference J231671), and solution pipes containing infills of banded sands and clays (eg. Moneymore; Irish Grid reference H964933). Very little work has been carried out on these deposits apart from some early studies, most notably by Cleland (1928, 1929, 1930 and 1938). In these he describes a section in Knockadona quarry, 5 km north of Lisburn in which approximately 30cm of black lignite was found. This overlay calcareous grey clay with fragmented flints and chalk which he likened to the clay-with-flints of southern England (Cleland, 1928). Sizeable remnants of 'tree trunks' were obtained from the lignite, but they had been so badly burnt by the basalt that he was unable to identify them. From further north, Cleland (1929 and 1930) described a very different profile in Magheramorne quarry, 5 km southeast of Larne. This consisted of 1.5m of calcareous clay and flints overlain by 60cm of reddish brown clay, (containing only trace amounts of calcium carbonate) and 90cm of stratified chalk. He interpreted this sequence as the result of deposition by running water, and suggested that the source of the

Plate 1: Sub-Basaltic Bed, Spy-Window Quarry, Glenarm showing dark red clay derived from solution of underlying Cretaceous chalk and shattered flint nodules.

Plate 2: Boundary between Tertiary Basalts and Cretaceous chalk, Spy-Window Quarry, Glenarm. Note: solution hollows in the chalk infilled with red clay and flints. Dashed line shows the base of the basalt.

reddish brown clay was nearby outcrops of Triassic marl, which now lie directly beneath the basalt. Finally, in 1938 Cleland described a number of profiles around the margins of the Antrim plateau in which reddened flint gravels of variable thickness separated the chalk and basalt, filling-in many of the hollows that once existed in 'the old landscape.' The red character of these flints and associated clays was seen as a product of iron-staining rather than baking by the basalts, and Lamont (1946) used Cleland's observations to invoke a tropical climate at the time of their deposition.

Solution hollows similar to those that Cleland described can be seen in a large quarry on the east Antrim coast south of the village of Glenarm,Irish Grid reference (D333146). Here they are infilled predominantly with dark red clay and basalt and can exceed 10m in depth (Plate 2). In addition to the solution hollows, sections can also be seen in the quarry which show distinctive down profile variations indicative of in-situ weathering of the chalk and the origin of the clay as the insoluble residue of this weathering. A typical profile of this type is described in Table 1.

It is generally assumed that the Sub-Basaltic Beds represent an early Palaeocene 'soil' that developed over a generally flat and regular landscape on newly emerged Cretaceous strata. The clays would appear to be predominantly the insoluble residue from solution of the limestone. However, there is some evidence of local redistribution of overlying materials combined with the lowering of the bedrock. This pattern is similar to that envisaged for the 'clay-with-flints' of the English Chalk (Thomas, 1978), although a singular advantage of the Antrim deposits is that a definitive limiting date can be placed on their development.

Both Wilson (1972) and Reffay (1972) have used the general thinness of the soil, its redness and apparent slowness of formation to suggest that it formed under warm, arid conditions. However, where in-situ profiles are retained, for example at Glenarm, they exhibit many of the characteristics of Terra Rossa soils associated with present-day Mediterranean environments (Fitzpatrick, 1971). These include a strong blocky structure, low calcium carbonate content and high iron content (for survey of analytical data see Bardossy (1982) ch. 6). Although Terra Rossa soils are widespread around the Mediterranean, the general consensus is that they are relict from former climates and show: "a high degree of maturity established during a long period under a climate which provided an abundance of moisture, but also seasonal drying intense enough for the dehydration of iron oxides" (Limbrey (1975), quoted in Crabtree et al. 1978, p. 17). These conditions have not prevailed in the Mediterranean during the Holocene (Crabtree et al.,1978), and Butzer (1961) has proposed that Terra Rossa soils formed under tropical interglacial conditions. Moreover, when clay-rich solution hollow infills are found in present-day arid regions they are, like Mediterranean Terra Rossa soils, generally considered to be relict from former pluvial periods (eg. Smith, 1986).

TABLE 1. Sub-Basaltic Weathering Profile, Spy-Window Quarry, Glenarm

1. Plateau Basalt - dated elsewhere at between 65-62 m.y. B.P. This has a characteristically very sharp, smooth lower boundary and contains thin (20-30 cm) red beds which may represent nascent palaeosols developed between individual lava flows.

2. 20-30 cm of dark red (7.5 R 2.5/4), indurated, baked clay with a very strong, angular blocky structure. Ped faces are stained by iron and manganese coatings. $CaCO_3$ - 2.75%, Fe_2O_3 - 14.30%.

3. 100-150 cm of partially baked clay (10 R 3/4) also characterised by angular blocky structure, but unlike the baked clay peds can be crumbled by hand. Occasional granules of unweathered chalk occur, but no flints were found. Clay minerals are predominantly kaolinite with trace amounts of sepiolite. $CaCO_3$- 2.55%, Fe_2O_3 - 21.45%.

4. 50-75 cm of mixed red clay (7.5 R 3/6) and flints with occasional pebbles of chalk. No bedding is seen and flints appear to float in a clay matrix. Flint surfaces are stained by dendritic growths of manganese. (Plate 1). $CaCO_3$ - 1.27%, Fe_2O_3 - 23.59%.

5. 200-300 cm of chalk containing flint nodules and veins of clay penetrating vertically into the rock mass. Clay percentage is highest near the top of the horizon and decreases with depth accompanied by a colour change from red (7.5 R 3/6) to greyish brown (5 YR 4/8). Clay minerals are predominantly kaolinite, although trace amounts of palygorskite have been recorded. $CaCO_3$ (of clay) 1.50 - 0.82%, Fe_2O_3 (of clay) 20.02 - 5.72%.

6. Unweathered chalk with flint nodules.

Given, therefore, the pedological similarities between the Sub-Basaltic Beds and Terra Rossa soils, the substantial quantities of vegetation debris in some profiles, and extensive solution of the chalk, their formation under conditions of aridity seems unlikely. Instead, a seasonally humid tropical to sub-tropical palaeoenvironment would seem to be a more realistic possibility.

INTERBASALTIC BEDS

The Antrim basalts have traditionally been divided into three major components (Patterson, 1952 and 1955). The lower basalts, which are the thickest and most extensive (533m maximum), the middle basalts (136m maximum) and the upper basalts, which occur now as a series of isolated residuals (Davies and Stephens, 1978). Each of the major phases comprises numerous individual flows, and during intervening periods the surfaces of the flows were weathered to varying degrees to give dark red weathering profiles. Often these profiles may be no more than a few cm thick, but the intervals between the major units, especially after the lower basalts, were marked by long periods of quiescence. During these periods considerable erosion occurred, sizeable valleys were cut in underlying basalts and there may have been some tectonic tilting of the previous flows. In addition to which, two major Interbasaltic red beds were able to develop under conditions of subaerial weathering. The best known exposure of the lower bed is at the Giant's Causeway on the north Antrim coast (Plates 3 and 4), but outcrops of the upper bed are common over much of Antrim (Fig. 1) and the lower bed has also been observed in boreholes west of Lough Neagh (Wilson, 1972 and Fowler and Robbie, 1961). The thickness of the beds is variable, but can be up to 30m and they are extensive enough to have been mined commercially for iron ore and bauxite. Composition is likewise variable, and as well as in-situ palaeosols there are related sediments of water borne detritus including sand and pebble beds (George, 1967). Montford (1970) has reviewed available observations and produced a composite profile, which it must be stressed is not everywhere complete:-

Composite Interbasaltic profile (Montford, 1970)

6. Upper basalt.
5. Lignite up to 10 cm or coal up to 2m.
4. Ferruginous laterite up to 30 cm.
3. Lithomarge (ferruginous kaolinitic clay) usually 10-5m but up to 30m.
2. Spheroids of basalt weathering to Lithomarge.
1. Lower basalt.

The lignite most probably derives from the vegetation that was growing on the soils when it was overridden, but it is thought that the coal, which often contains the remains of tree trunks, represents local swamps or marshes. Both lignite and coal are, however, commonly absent from the profiles.

On the basis of mineralogy, Eyles *et al.* (1952) recognized three types of deposit:-

1. Highly ferruginous red bauxites, with iron oxides usually >20-30%,but silica normally only 10-5%. These form from the basalts.

2. Siliceous grey bauxites, with iron oxides usually <10% and often <5%, but silica is often 20% and can be >40%. These usually form from rhyolitic debris.

3. Bauxites of mixed origin, these are of variable colour and are intermediate between 1 and 2 in composition.

All of these profiles comprise a complex mineralogy which Cole *et al.* (1912) and Eyles *et al.* (1952) recognised as predominantly kaolinite, meta-halloysite and gibbsite with haematite and goethite at some sites. More recently, McAlister and McGreal (1983) and

Plate 3: Giant's Causeway, Co. Antrim the Interbasaltic Bed (arrowed) between the Middle and Lower Basalts of the Antrim Lava Series.

Plate 4: Base of lower Interbasaltic Bed, Giant's Causeway showing incipient corestone development.

McAlister *et al.* (1984) have added lepidocrocite, cristobalite and possibly chlorite to the list using differential thermal techniques. The complexity was seen by Eyles *et al.* (1952) to derive from the operation of two processes:-

1. *Kaolinisation*, whereby rock is altered to a 'lithomarge' in which complex silicate minerals are broken down into their oxides, and the silica and aluminium thus freed combine with water to form kaolinite or halloysite. Considerable amounts of silica are lost in the process leading to a relative enrichment of aluminium, and the oxides of ferric iron and titanium to form an integral part of the lithomarge.

2. *Lateritization*, primarily a desilicification of the lithomarge through the decomposition of hydrated aluminium silicates and the removal in solution of the silica. The residual aluminium then absorbs water to form gibbsite and the result is a mixture of this and the iron and titanium oxides. There may, however, be internal segregation into an upper iron and titanium rich layer and a lower gibbsite rich bauxite.

Contemporary conditions under which these processes occur are the humid tropics, and as long ago as 1908, Cole proposed an *in situ* origin for the Interbasaltics under 'tropical (warm) conditions.' Montford (1970) is of the opinion that the Interbasaltic horizons represent ferrallitic soils formed under conditions in which:

1. Soil temperatures were 24-27 °C (annual average)

2. Rainfall was well distributed over the year and would have averaged 3000-5000mm *per annum*.

3. Soils carried little humus, and organic matter decayed rapidly.

4. Terrain was rolling with good drainage initially, but this deteriorated locally as surface laterite formed.

Montford believes that alternating wet and dry seasons are not required to explain the lateritization, but this is contrary to observations made by Reffay and Ricq de Bourd (1970) at the Giant's Causeway. In their paper they consider that the lower Interbasaltic horizon (lower/middle basalt) formed as a ferruginous soil under seasonally humid (savanna) conditions, whereas only the upper horizon (middle/upper basalt) represents a ferrallitic soil of perennially humid (rainforest) origin. However, many soils of the present-day savannas may be inherited from former rainforest conditions, and the distinction between ferruginous and ferrallitic soils may be as much a function of the duration of chemical weathering as any differences in type of weathering (Carter and Pendleton, 1956). Moreover, there is increasing evidence that weathering processes in the present tropical zone are neither exclusive or distinctive (Thomas, 1974, p. 13). Thus it would be very difficult to identify definitively the precise tropical conditions under which the Interbasaltic horizons may have formed. Indeed, it is perhaps worth noting that despite pedological evidence for humid tropical conditions, lignite from the interbasaltics, while displaying a 'broadly Palaeocene' flora, also contains plant remains characteristic of sub-tropical or warm temperate forested conditions (Wilson, 1972). These include spores of Pinus (Watts, 1970), cones of Pinus Plutonis, fragments of Cupressus (Cupressites) Machenrii and Castanca Ungeri - which is now represented by the Spanish chestnut of southern Europe (Hartley, 1938).

LOUGH NEAGH CLAYS

Indirect evidence of the Mid-Tertiary weathering environment can be obtained from the extensive deposits which partly infill the Lough Neagh basin (Fig. 1). These underly much of the present Lough to a maximum thickness of some 350m at its southwest corner, where they overlie approximately 7m of weathered basalt (lithomarge) on top of the Upper Basalts of the Antrim plateau lavas (Wright, 1924). These deposits, known as the 'Lough Neagh Clays', are of late Oligocene (Chattian) in age (Wilkinson et al., 1980). They consist predominantly of pale coloured, kaolinite and titanium rich clays but these are intercalated with sands, ironstone nodules, silicified wood and extensive lignite deposits laid down under estuarine conditions in a predominantly shallow lacustrine environment (Fowler and Robbie, 1961).

It is thought that the clays may have derived predominantly from the erosion of a post-basaltic lithomarge that formed on the surrounding lavas. However, the overall composition of the beds varies widely across the basin and this is thought to reflect their deposition by numerous rivers draining different sources but mainly from the west (George, 1967). They therefore contain, for example, pebbles of Triassic sandstones, Old Red Sandstone and Cretaceous flints as well as basalt and weathered basalt debris washed in from eroded soils of the surrounding lavas. The great thickness was able to accumulate because of continuing subsidence of the basin produced by a combination of synclinal folding and the reactivation of block faulting (Fig. 1) along earlier (Caledonoid) structural lines (George, 1967). In places this down faulting was greater than 300m.

Studies of the flora associated with the Lough Neagh Clays (eg. Watts (1962 and 1970) and Wilkinson et al., (1980)) suggest an environment of Taxodium swamps surrounded by ferns, palms and other tropical and sub-tropical plants. There are also substantial quantities of conifer pollen - mostly Sequoia Couttsia (Watts, 1970, Wright, 1924) - perhaps derived from upland forests at some distance or on hummocks within the lake and swamps. The indications are therefore of a low-lying, warm, frost-free stable climate (Wilkinson et al., 1980) dominated by fluviatile erosion and freshwater sedimentation. Within the beds there is little botanical variability, which might suggest either a long period of stability or rapid deposition (Watts, 1970).

DEEP WEATHERING OF GRANITE AND GRANODIORITE

At a number of localities on the igneous complexes of Slieve Croob and the Mourne Mountains (Fig. 1), it is possible to see sections in which the underlying granite and granodiorite (respectively) have been broken down into a friable, partially disaggregated grus. On Slieve Croob this disaggregation has been observed to a depth of at least 10m. Again, little has been written on these weathering profiles, apart from a note by Proudfoot (1958) in which he likened those on Slieve Croob to fossil rotlehm soils. In an attempt to remedy this situation, characteristic profiles from both areas are described below.

Deeply Weathered Granodiorite: Legananny Road Quarry (J284442).

The abandoned quarry just off the Legananny road, is the deepest of several quarries on Slieve Croob and nearby Slievenaboley, in which sections cut in deeply weathered granodiorite (Old Red Sandstone age) are exposed. These weathered profiles appear to be located in isolated pockets, (where they were perhaps protected from stripping by overriding ice during the Quaternary) and are adjacent to glaciated surfaces cut in fresh rock. The sections were first studied in detail by Proudfoot (1958), who described a section to the west of Slievenaboley as the B-horizon of a 'relict Rotlehm' soil. This would have formed under 'a very much warmer climate than the present' (presumably moist tropical) and most likely during the Tertiary.

In the Legananny Road Quarry the exposure (Table 2) comprises at least 10m of weathered granodiorite capped in places by up to 2m of slopewash/head material comprising pebbles and cobbles of country rock set in a reddish brown sandy matrix (Plate 5). The weathered profile itself is pale brownish/grey in colour (10YR 6/2) and two weathering zones can be identified.

Although the quarry face is sub-divided by iron-stained joints, no evidence of spheroidal weathering and corestone development is seen, and the lower weathering zone is evenly disaggregated throughout. There is, however, evidence of block production from the granodiorite in the form of rounded and sub-rounded boulders within the deposits mantling nearby hillslopes. Perhaps, as the base of the weathered zone is not seen in the quarry, these boulders might originate at a lower, and earlier stage of weathering as joint-bounded blocks of unweathered rock. Similar blocks have been observed at the base of weathered granite profiles of the Mourne Mountains where rounded boulders also mantle many of the slopes.

TABLE 2. Weathered Profile in Granodiorite, Slieve Croob

Weathering Zone I (0-4 m)

The upper weathering zone is up to 4m thick in the main quarry face and consists of a friable grus that can be broken down by hand. The structure of the granodiorite can still be clearly seen and the deposit is shown to be in situ by a number of veins of unweathered, fine grained rock that cut through it. The grus is a mixture of sand and gravel (average 54% >2mm and 46% <2m) caused by the breakdown of the rock along crystal boundaries. Proudfoot (1958) identified chemical weathering of crystal margins to form dense red to reddish brown deposits. However, the clay content of the <2mm size fraction is very low, and decreases with depth from 4% (<2μm) at the top to 2% at c.2m and 0% at the base where the grus is composed of 100% sand and gravel sized material. Clay minerals (determined by XRD and DTA) are predominantly a kaolinite/halloysite mix with minor amounts of chlorite and muscovite and trace amounts only of gibbsite. Thin sections (Proudfoot, 1958) reveal a spongy mineral fabric with occasional dark flecks and concretions within it, but in general weatherable minerals such as feldspars and mica are clearly discernible to the naked eye. Weathering has occurred primarily through disaggregation rather than decomposition.

Weathering Zone II (-4 - -10m)

The lower weathering zone consists of at least 6m of in situ, partially disaggregated granodiorite. It is friable in the hand on exposed faces but within about 5cm it becomes much more compacted and can only be sampled using a hammer. Most of the quarry face retains the outward appearance and colour of the bedrock but it is criss-crossed by a series of find joints. Along the joints iron minerals have been oxidised to give a dark red colour and there has been partial decomposition to give a certain amount of clay (96.8% sand, 1.2% silt and 2.0% clay in the <2mm fraction). For the main rock mass, however, particle size analysis (hydrometer method) revealed no measurable amounts of either clay or silt, and the loosened material divides into crystals and aggregates of approximately 50% >2mm and 50% sand <2mm.

Mineralogical analysis (XRD and DTA) of the clay sized fraction from the joints, and what clays there are in the rock mass, revealed predominantly a kaolinite/halloysite mix, with minor amounts of chlorite and muscovite, but no gibbsite. Again, weatherable minerals such as feldspars and mica are clearly visible within the granular debris, though the former may be cloudy.

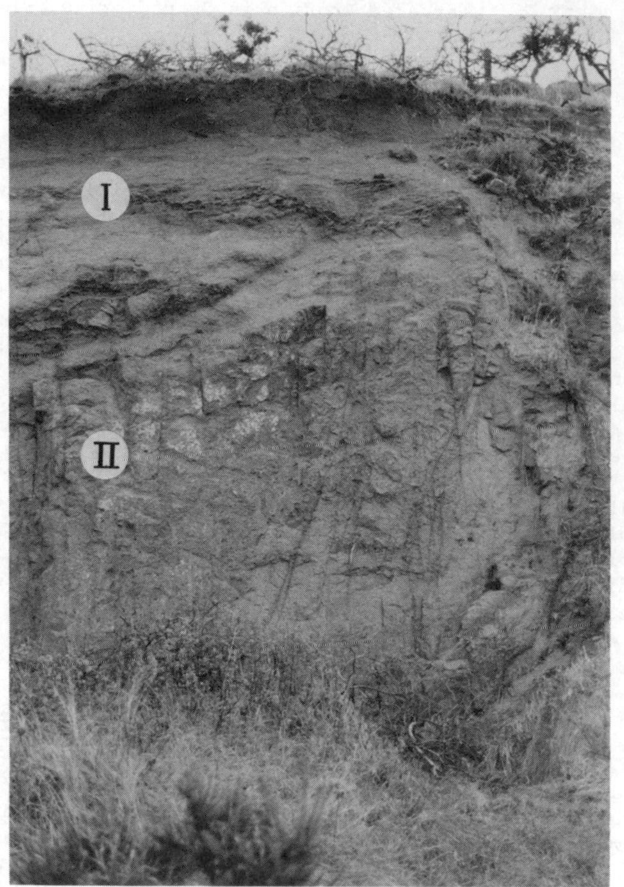

Plate 5: Section in deeply weathered granodiorite, Legananny Road Quarry, Slieve Croob, illustrating weathering zones described in Table 2. Height of section c. 10m.

Deeply Weathered Granite, Northwest Mourne Mountains. Within, and in close proximity to the Rocky River catchment in the northwest Mournes, numerous sections in deeply decayed granite (G4 and G5 of Emeleus (1955)) have been exposed along stream channels and in small quarries excavated for road fill (Plate 6). The products of granite weathering are, however, not restricted to these sections and they occur in the form of a grus of sand and gravel on all slopes and at all altitudes. Thus, although weathering profiles may have been truncated or removed from certain areas by subaerial erosion, the implications are that granite weathering has been very extensive across this part of the Mournes. Because of subsequent erosion and deposition the weathering profiles in each exposure vary

Plate 6: Section in weathered granite below Castle Bog, Mourne Mountains. Note overlying peat and head of Midlandian age.

considerably, but a composite profile is summarised in Table 3. This consists of different combinations of soil and till or periglacial head of Midlandian age (Last Glaciation) similar to that described from Dartmoor by Green and Eden (1973). These overly a sandy, structureless, but in situ grus which readily crumbles in the hand. Beneath this, and sometimes separated by a sharp boundary, is more coherent, partly decomposed material that retains the general appearance of the parent rock, but disaggregates under the impact of a hammer. Neither of these horizons has been observed to contain corestones but beneath them is a zone of angular, locked joint blocks that could have provided the numerous granite boulders which mantle many of the slopes.

The nature of the weathering is 'arenaceous' (Millot, 1970), in that it is characterised by granular disintegration with some limited alteration of mica, but little alteration of feldspars, and no observed dissolution of quartz grains. Production of clay sized material during weathering is limited and the fine fraction (<2mm) contains no more than 6% clay (<2μm) and normally less than 2%, whereas the sand sized fraction (2mm-20μm) is rarely <90%. The degree of disaggregation decreases with depth, as is shown by a comparison of the percentages of material >2mm and <2mm in the loose debris. This shows approximately 50-55% <2mm in the upper, disaggregated zone (zone I) compared to only 20-30% in the lower, partially disaggregated zone (zone II).

PROFILE CHARACTERISTICS	Thickness ranges observed	Size distribution of loose fraction (%)		Size distribution of fine fraction <2mm (%)			Composition of <2µm size fraction (XRD & DTA)		
		>2mm	<2mm	Sand 2mm-20µm	Silt 20-2µm	Clay <2µm	Strong Peak	Minor Peak	Trace
Peat OR Coarse Podzolic Soil	<75cm								
Granite rich head over bedded grus OR Sandy till with pebbles/boulders of granite and Palaeozoics	0-2m								
Weathering Zone I Structureless, disaggregated granite with decomposition of micas and partial decomposition of feldspars. Retains a high percentage of cloudy but otherwise unweathered feldspars. Dark red iron-staining along joints, together with clay accumulations. Contains veins of quartz and unweathered microgranite with some displacement downslope, but in general weathered material remains *in situ*. No corestones	0->2m	45 46 52	55 54 48	94.0 89.6 94.8	4.0 6.4 5.2	2.0 4.0 0.0	Kaolinite Halloysite	Muscovite Chlorite	 Gibbsite
Weathering Zone II Partially decomposed but coherent granite retaining the structure of the parent rock. Decomposition is restricted to grain boundaries and primary joints are still visible. Feldspar grains, in particular remain largely unaltered, though there is iron-staining between grains and along joints No corestones observed	0->2m	73 82	27 18	96 88.5	1.5 5.7	2.5 5.8	Kaolinite Halloysite	Muscovite Chlorite	
Joint bounded granite blocks, no visible alteration of micas and feldspars but iron-staining along joint faces.	0->1.5m								
Unweathered granite with incipient joints									

Table 3: Composite table, illustrating weathering profiles found on granite in the northwestern (plateau) Mourne Mountains. Note: all horizons are not necessarily found at the same exposure, and chemical and particle size data is representative of horizons and not drawn from the same profile.

This 'trend' of decreasing weathering with depth is apparently contradicted by data on clay content, as this increases with depth in some sections. However, this could reflect increased lateral elluviation of clay from the more open, disaggregated upper zone. Such a contention is supported by the most complete of the profiles below Castle Bog (J231242). Here there is a consistent increase in calcium content down-profile which could reflect increased weathering and removal of calcium through hydrolysis of feldspars from the upper weathering zone. Care must be taken, however, in interpreting these analyses as there is in fact very little consistent down-profile variation. In many instances there is considerably more variability within the weathering zones related in particular to the occurrence of joints along which weathering and elluviation are concentrated. This appears to be a common feature of arenaceous weathering profiles (Calvo et al., 1983) and is reflected for example, in the uneven distribution of iron-staining in many profiles. This, in turn reflects the mobility of iron under the often water-logged conditions that prevail across much of the area. Overall iron content in weathering zones I and II is normally between 1.7 to 1.9%.

Origins and Age of Deep Weathering. The observed assemblages and amounts of clay minerals, the high proportions of weatherable minerals and the physical characteristics of the profiles are similar to those found in, for example, the weathered granites of Dartmoor. Traditionally, the latter were interpreted as products of sub-aerial weathering under humid tropical conditions (see Linton (1955) and the discussion of his paper), during the Mid- to Late-Tertiary with little alteration during the Pleistocene (Brunsden, 1964). Views are, however, changing as to the origins of arenaceous, kaolinite-rich weathering profiles.

First, with regard to texture, both Eden and Green (1971) and Green and Eden (1971) drew attention to the disparity between the low clay contents of Dartmoor profiles (2-7%) and those of weathered granite in the present-day humid and sub-humid tropics (30-50+%, e.g. Bakker, (1960) and Verheye (1974 and 1979)). Brunsden (1964) considered the coarseness to be a product of excessive elluviation, whereas Eden and Green (1971) assign it to limited chemical weathering. In support of their contention they cite descriptions of similar profiles in southern Europe, which are seen as products of meso-humid, sub-tropical climates, probably during the Late Pliocene or Early Pleistocene.

Second, with regard to clay mineral assemblages, the most contentious issue has been the palaeoclimatic significance of gibbsite found in weathered profiles. This mineral is often thought of as indicating prolonged, intensive weathering of silicate minerals under humid tropical conditions. Although, given sufficient time, weathering of granite under any constant climate should lead to gibbsite production (Tardy et al., 1973). More recently, work in Galicia on humid temperate saprolites (Vazquez, 1981) has shown that gibbsite can also form in the initial weathering stages of various aluminosilicates, especially plagioclase. The only requirements appear to be open, well-drained, unsaturated local environments. Emphasis upon local environmental factors was also stressed by Eden and Green (1971) when they too suggested that gibbsite could be an initial product of weathering, a proposition which is supported by hydrological studies in the same area where the present-day formation of gibbsite in well-drained regoliths has been reported (Ternan and Williams, 1979 and Williams et al., 1984).

Similar observations have been made regarding the origins of kaolinite and halloysite - the principal clay minerals in the Mournes and Slieve Croob profiles. Torrent and Benayas (1977) have, for example, identified halloysite as an intermediate product in the transformation of primary minerals (e.g. feldspars) to gibbsite in granite from central Spain. The same authors found kaolinite only in modern soils, thus implying formation under current, warm temperate conditions. Work in southern New South Wales (Dixon and Young, 1981) has also suggested that kaolinite can form from granite under temperate conditions in arenaceous, corestone-free profiles, a view which is shared by Ternan and Williams (1979) for regoliths on Dartmoor, and is similar to that proposed in Basham (1974). In the latter, Basham contended that kaolinite could be produced by intense leaching of gabbros in Aberdeenshire under acid conditions beneath a peat cover - similar conditions to those observed over much of the Mourne Mountains.

Reports of deep weathering profiles on granitic and other rocks are not restricted to the unglaciated areas so far mentioned. Increasing evidence is, for example, available for their occurrence in many areas of Scotland, where depths of over 60m have been recorded in boreholes. Although Scottish profiles were described as long ago as the late nineteenth century, attention was first drawn to their widespread distribution in a paper by Fitzpatrick (1963). In this he suggested a tropical origin and pre-Pliocene age. He also noted that weathering products had been widely incorporated into later glacial deposits and soils, a view that has been reiterated, especially with respect to gibbsite inheritance, by Wilson (1967, 1969 and 1985), although it is also suggested that some of the clay may be of interglacial rather than pre-glacial age (Wilson, 1973 and Wilson and Brown, 1976.

Following on from Fitzpatrick's work, a series of detailed studies by Hall (1983, 1984, and 1985) have identified two types of granite weathering profile in Buchan (N.E. Scotland). First, there is the so-called 'clayey grus', characterised by the advanced alteration of primary minerals to mature kaolinite/illite suites, and produced under sub-tropical conditions prior to the mid-Miocene. Second, there are the much more common 'grus profiles'. These exhibit characteristics of 'arenaceous weathering', with granular disintegration, abundant little-altered primary minerals and heterogeneous and immature clay mineral assemblages. Hall believes that these profiles formed in a humid temperate environment during the Pliocene to Early Pleistocene. He dismisses a totally

Interglacial origin, and uses their preservation to indicate tectonic stability and either the local absence or ineffectiveness of glacial erosion.

The 'grus profiles', like those described by Dixon and Young (1981), are markedly similar to those found on both Slieve Croob and the Mournes, especially in terms of low clay contents, paucity of corestones, clay mineral species, and preservation of weatherable minerals. In turn this would seem to imply a similar origin through humid temperate, subaerial weathering. Under these conditions weathering is predominantly by granular disintegration, in which grain boundary microcracks are widened with little alteration of primary minerals. Microcracks may originate due to pressure release as the granite mass is exposed leading to differential expansion of crystals (Pye, 1985). Thus disintegration can occur without chemical weathering and expansion of minerals such as biotite. Instead, microcracks can be exploited by a number of mechanisms. These could include: stress corrosion of crack tips, swelling pressures exerted by ordered water in fine capillaries, or possibly even freeze-thaw (see Pye, (1985) for a summary). The fact that these mechanisms exploit rock weaknesses of this order of magnitude, rather than major joints which normally define corestones, may explain the frequent absence of corestones from arenaceous profiles. In addition, the presence or absence of suitable microcracks in a granite may determine its susceptibility to 'arenaceous weathering' under humid temperate conditions.

The age of the Mournes and Slieve Croob profiles cannot be determined with any great precision, but certain limits can be placed upon their formation. Slieve Croob profiles are, for example, overlain by head of Midlandian age and therefore pre-date that period. However, they generally occur in small basins where they would have been protected from erosion not only during the Midlandian, but during preceding glacial periods. There is the possibility, therefore, that deep weathering may have been initiated prior to the Pleistocene, and that what we see today are truncated remnants of once much deeper profiles. Furthermore, it has been suggested that the upper weathering zone seen in Legananny Quarry may have been affected by a second phase of weathering (Le Coeur, personal communication) which could have occurred under interglacial conditions.

The shallow profiles in the northwest Mournes pose different questions to those on Slieve Croob. Like those on Slieve Croob they can be seen to be overlain by glacial and periglacial deposits of Midlandian age. However, they frequently occur not in protected pockets but on steep exposed hillsides. Examples of such profiles can be seen below Castle Bog (J231242) at altitudes between 350m and 400m. These lie above the suggested Midlandian ice limit in the area (c.265m - Hannon (1974)), but well below that for the Munsterian (penultimate) Glaciation which overran and scoured this part of the Mournes (Synge and Stephens, 1960 and Hannon, 1974). It seems unlikely that exposed granite grus would have survived this erosion. In turn this suggests that the currently observed weathering profiles may post-date the Munsterian, and that they formed predominantly during the last Interglacial (Gortian). Such an interpretation does not preclude earlier phases of deep weathering, the material from which no longer remains in situ but has been incorporated into subsequent glacial and fluvio-glacial deposits.

Certainly, previous suggestions of granite weathering during the last Interglacial are not unknown, even from Scotland (e.g. Synge, 1956). Moreover, in view of the free-draining character of the profiles, acid soils and high rainfall (1500mm/annum) it seems possible, given the earlier remarks by Basham (1974), that weathering and production of minerals such as kaolinite may be continuing at the present time. Contemporary deep weathering of granite has indeed recently been hinted at elsewhere in Ireland, for example at Carnsore in Co. Wexford (Mitchell, 1985). Such an interpretation, which lays emphasis upon relatively recent granite weathering, would concur locally with the need to unroof the Mourne granite complex of overlying sediments, and possibly basalts, before extensive weathering of the granites could commence. Such unroofing must, of course, post-date granite emplacement which Evans et al. (1973) has placed at between 54.9 m.y. and 56.1 m.y. ago. However, remnants of the former Palaeozoic cover can still be seen, for example in the Silurian strata beneath Deers Meadow in the central Mournes. This suggests that exposure of the granite complex may have been a relatively 'late' feature of the Tertiary, especially considering the undoubted amount of erosion that has taken place in post-Tertiary times under glacial and periglacial conditions. Similarly, the occurrence of deeply weathered granite at low altitudes on the northwest margins of the Mourne Mountains (for example, in the Rocky River catchment; J242269) might also suggest a relatively young age for the weathering. Unfortunately, however, it is impossible to advance beyond this speculation in the absence of any associated mineral or organic deposits that can be either dated, or used to infer the vegetational environment at the time when the granites were weathered.

SUMMARY AND CONCLUSIONS

The principal aim of this paper has been to present a review of the evidence available in the northeast of Ireland that can be used to reconstruct terrestrial conditions during the Tertiary. The bulk of this evidence consists of weathered material and can therefore be used to indicate the nature of weathering environments.

The earliest evidence described is a range of soils and fluvial deposits, which once covered an Early Palaeocene landscape cut in newly uplifted Cretaceous chalk. These deposits are available to us because of their burial beneath Early Tertiary lava flows which were laid down some 65 m.y. ago. Although the overall composition of the Sub-Basaltic beds is highly variable, in situ weathering profiles exhibit many characteristics of Terra Rossa soils. A seasonally humid, sub-tropical climate is therefore suggested for their formation.

Within the overlying basalts of County Antrim two major Interbasaltic beds occur which formed as deep weathering profiles between major volcanic episodes. On the basis of their mineralogy and morphology these beds have been interpreted as the products of humid tropical weathering, although there are suggestions that the lower Interbasaltic beds may have formed under seasonally humid conditions. There is however, an unresolved disparity between this interpretation and evidence provided by vegetation remains. The latter include, for example, spores and cones of Pinus which might indicate a climate more akin to that of present-day southern Europe.

On top of the basalt lie outcrops of heterogeneous lacustrine deposits of Oligocene age which occupy a structural depression known as the Lough Neagh Basin. These 'Lough Neagh Clays' contain floral remains which indicate, at the time of deposition, a warm, frost-free environment with a vegetation cover dominated by conifers.

Lastly, to the south of the Antrim basalts outcrops of Caledonian granodiorite (Slieve Croob) and Tertiary granite (Mourne Mountains) carry a number of arenaceous weathering profiles. Traditionally these have been represented as the products of humid tropical weathering at some time during the Tertiary. However, a review of recent literature on 'arenaceous weathering' of granites suggests that this pattern of disaggregation is more typical of humid temperate than humid tropical conditions. It is accepted that some of the deeper profiles retained in pockets on Slieve Croob may be of Late Tertiary or possibly Early Pleistocene in age. However, shallower profiles which occur on exposed slopes in the Mourne Mountains are seen as the possible products of weathering during the last Interglacial period. If this is the case, these particular deposits can no longer be used as evidence of Tertiary environmental conditions. Nonetheless there remains in northeastern Ireland, a record of Tertiary environmental conditions which is unique within the British Isles both for the timespan it covers and for the range of evidence it offers.

ACKNOWLEDGEMENTS

The writers would like to thank Sharon Wright and Suzanne McWilliams for typing the original manuscript of this paper, Gill Alexander for drawing Figure 1 and Trevor Molloy for preparing the photographs. Also, we would like to thank Dr J. McGreevy for his valuable comments on the draft version of this paper.

REFERENCES

Bakker, J.P., 1967. Some observations in connection with recent Dutch investigations about granite weathering and slope development in different climates and climate changes. Zeitschrift für Geomorphologie, suppl., 1, 69-92.

Bardossy, G., 1982. Karst bauxites: Bauxite deposits on carbonate rocks, Elsevier, Amsterdam, 441p.

Basham, I.R., 1974. Mineralogical changes associated with deep weathering of gabbro in Aberdeenshire. Clay Minerals, 10, 189-202.

Brunsden, D., 1964. The origin of decomposed granite in Dartmoor. In: Dartmoor Essays, (ed. I.G. Simmons) Devonshire Association, Torquay, pp. 97-116.

Butzer, K.W., 1961. Palaeoclimatic implications of Pleistocene stratigraphy in the Mediterranean area. Annals, New York Academy of Science, 95, 449-456.

Calvo, R.M., Garcia-Rodeja, E., and Macias, F. 1983. Mineralogical variability in weathering microsystems of a granitic outcrop of Galicia, Spain. Catena, 10, 225-236.

Carter, G.F., and Pendleton, T.L., 1956. The humid soil: process and time. Geographical Review, 46, 488-507.

Cleland, A. McI., 1928. An old land surface in a chalk quarry. Irish Naturalists' Journal, 2, 37-40.

Cleland, A. McI., 1929. "Flour" of flint, Irish Naturalists' Journal, 2, 204-207.

Cleland, A. McI., 1930. On some recent rock exposures at Magheramorne quarries, County Antrim. Irish Naturalists' Journal, 7, 5-8.

Cleland, A. McI., 1938. Red Flints. Irish Naturalists' Journal, 7, 5-8.

Cole, G.A.J., 1908. The red zone in the basaltic series of County Antrim. Geological Magazine, 5, 341-344.

Cole, G.A.J., et al., 1912. The interbasaltic rocks of north-east Ireland, Memoirs of the Geological Survey of Ireland, Dublin, 129p.

Crabtree, K., Cuerda, J., Osmaston, H.A., and Rose, J. (ed.), 1978. The Quaternary of Mallorca. Quaternary Research Association, Guide to Field Meeting, December 1978, 114 p.

Davies, G.L., 1960. The age and origin of the Leinster mountain chain: a study of the evolution of south-eastern Ireland from the Upper Palaeozoic to the later Tertiary. Proceedings of the Royal Irish Academy, 61B, 79-107.

Davies, G.L., 1970. The Enigma of the Irish Tertiary, in Irish Geographical Studies, (ed. N. Stephens and R.E. Glasscock), Queen's University, Belfast, pp 1-16.

Davies, G.L., and Stephens, N., 1978. The geomorphology of the British Isles: Ireland. Methuen, London, 250 p.

Dixon, J.C., and Young, R.W., 1981. Character and origin of deep arrenaceous weathering mantles on the Bega Batholith, southeastern Australia. Catena, 8, 97-109. See also comment by Ollier (1983) Catena, 10, 57-59 and reply (1983) Catena, 10, 439-440.

Eden, M.J., and Green, C.P., 1971. Some aspects of granite weathering and tor formation in Dartmoor, England. Geografiska Annaler, 53A, 92-99.

Emeleus, C.H., 1955. The granites of the western Mourne Mountains, Co. Down. Scientific Proceedings of the Royal Dublin Society, 27 (N.S.), 33-50.

Evans, A.L., Fitch, F.J., and Miller, J.A., 1973. Potassium-argon age determinations on some British Tertiary igneous rocks. Journal, Geological Society of London, 129, 419-443.

Eyles, V.A., Bannister, F.A., Brindley, G.W., and Goodyear, J., 1952. The composition and origin of the Antrim laterites and bauxites. Memoirs of the Geological Survey, Belfast, H.M.S.O., 85p.

Fitzpatrick, E.A., 1963. Deeply weathered rock in Scotland, its occurrence, age and contribution to the soils. Journal of Soil Science, 14, 33-43.

Fitzpatrick, E.A., 1971. Pedology: A systematic approach to soil science. Oliver and Boyd, London, 306p.

Fowler, A., and Robbie, J.A., 1961. The geology of the country around Dungannon. Memoir of the Geological Survey, Belfast, H.M.S.O., 274p.

George, T.N., 1967. Landform and structure in Ulster. Scottish Journal of Geology, 3, 414-448.

Green, C.P., and Eden, M.J., 1971. Gibbsite in the weathered Dartmoor granite. Geoderma, 6, 315-317.

Green, C.P., and Eden, M.J., 1973. Slope deposits on the weathered Dartmoor granite, England. Zeitschrift für Geomorphologie Suppl., 18, 26-37.

Hall, A.M., 1983. Weathering and landform evolution in northeast Scotland. Unpublished Ph.D. thesis, University of St. Andrews.

Hall, A.M., 1984. Central Buchan: The Buchan gravels, deep weathering and the inland series. In Buchan, Quaternary Research Association, Field Guide, September 1984. (ed. A.M. Hall) pp. 27-45.

Hall, A.M., 1985. Cenozoic weathering covers in Buchan, Scotland and their significance. Nature, 315, 392-395.

Hannon, M.A., 1974. The Late Pleistocene geomorphology of the Mourne Mountains and adjacent lowlands, Unpublished M.A. Thesis, Queen's University of Belfast.

Hartley, J.J., 1938. On plant remains from the Interbasaltic Beds of Portrush. Irish Naturalists' Journal, 7, 46-48.

Jukes, J.B., 1862. On the mode of formation of some of the river valleys in the south of Ireland. Quarterly Journal Geological Society of London, 18, 378-403.

Lamont, A., 1946. Red Flints. Irish Naturalists' Journal, 8, 398-399.

Linton, D.L., 1951. Problems of Scottish scenery. Scottish Geographical Magazine, 67, 65-85.

Linton, D.L., 1955. The problem of tors. Geographical Journal, 121, 470-487.

Manning, P.I., Robbie, J.A., and Wilson, H.E., 1970. Geology of Belfast and the Lagan valley. Memoir of the Geological Survey, Belfast, H.M.S.O., 242p.

McAlister, J.J., and McGreal, W.S., 1983. An investigation of deep weathering products from a fossil laterite horizon in central Antrim, N. Ireland. Proceedings, II International seminar on lateritisation processes, Sao Paula, July 4-12, 1982. pp 345-357.

McAlister, J.J., McGreal, W.S., and Whalley, W.B., 1984. The application of X-Ray and thermoanalytical techniques to the mineralogical analysis of an interbasaltic horizon. Microchemical Journal, 29, 267-274.

Millot, G., 1970. Geology of clays, Springer, New York, 429p.

Mitchell, G.F., 1980. The search for Tertiary Ireland. Journal of Earth Science (Dublin), 3,13-33.

Mitchell, G.F., 1981. Other Tertiary events. The Geology of Ireland, (ed. C.H. Holland) Scottish Academic Press, Edinburgh, p. 231-234.

Mitchell, G.F., 1985. The preglacial landscape. In The Quaternary History of Ireland, (ed. K.J. Edwards and W.P. Warren), Academic Press, London, pp. 17-37.

Montford, H.M., 1970. The terrestrial environment during Upper Cretaceous and Tertiary times. Proceedings of the Geologists Association, 81, 181-204.

Patterson, E.M., 1952. A petrochemical study of the Tertiary lavas of north-east Ireland. Geochimica Cosmochimica Acta, 2, 283-299.

Patterson, E.M., 1955. The Tertiary lava succession in the northern part of the Antrim plateau. Proceedings of the Royal Irish Academy, 57B, 79-112.

Proudfoot, V.B., 1954. Erosion surfaces in the Mourne Mountains. Irish Geography, 3, 26-35.

Proudfoot, V.B., 1958. Relict rotlehm in Northern Ireland. Nature, 189, 1287.

Pye, K., 1985. Granular disintegration of gneiss and migmatites. Catena, 12, 191-199.

Reffay, A., 1972. Les Montagnes de l'Irlande septentrionale: contribution a la géographie physique de la montagnes atlantique, Grenoble, 615p.

Reffay, A. and ricq de Bourd, M. 1970. Contribution a l'étude des palaeosols interbasaltiques à la Chaussée des Géants (compte d'Antrim, Irelande du Nord). Révue de Géographie Alpine, 58, 301-338.

Smith, B.J., 1986. An integrated approach to the weathering of limestone in an arid area and its role in landscape evolution: A case study from southeast Morocco, Proceedings of 1st International Geomorphological Conference Manchester, September 1985. In Press.

Synge, F.M., 1956. The glaciation of north-east Scotland. Scottish Geographical Magazine, 92, 129-143.

Synge, F.M., and Stephens, N., 1960. The Quaternary period in Ireland - an assessment. Irish Geography, 4, 121-130.

Tardy, Y., Bocquier, G., Paquet, H., and Millot, G., 1973. Formation of clay from granite and its distribution in relation to climate and topography. Geoderma, 10, 271-284.

Ternan, J.L., and Williams, A.G., 1979. Hydrological pathways and granite weathering on Dartmoor. In Geographical Approaches to Fluvial Hydrology, (ed. A.F. Pitty), GeoBooks, Norwich, pp. 5-30.

Thomas, M.F., 1974. Tropical Geomorphology, Macmillan, London, 332p.

Thomas, M.F., 1978. Denudation in the tropics and the interpretation of the tropical legacy in higher latitudes - A view of the British experience. In Geomorphology: present problems and future prospects, (ed. C. Embleton, D. Brunsden and D.K.C. Jones), O.U.P., Oxford, pp. 185-202.

Torrent, J., and Benayas, J., 1977. Origin of gibbsite in a weathering profile from granite in west-central Spain. Geoderma, 19, 37-49.

Vazquez, F.M., 1981. Formation of gibbsite in soils and saprolites of temperate humid zones. Clay Minerals, 16, 43-52.

Verheye, W., 1974. Nature and evolution of soils developed on the granite complex in the subhumid tropics (Ivory Coast) I. Morphology and classification. Pedologie, 24, 266-282.

Verheye, W., 1979. Le profil d'alteration pedo-géologique sur granodiorites en Côte d'Ivoire centrale. Révue de Geomorphologie Dynamique, 28, 49-60.

Walsh, P.T., 1965. Possible Tertiary outliers from Gweestin valley, Co. Kerry. Irish Naturalists' Journal, 15, 100-104.

Walsh, P.T., 1966. Cretaceous outliers in south-west Ireland and their implications for Cretaceous palaeogeography. Quarterly Journal Geological Society of London, 122, 63-84.

Watts, W.A., 1957. A Tertiary deposit in County Tipperary. Scientific Proceedings of the Royal Dublin Society, N.S., 27, 309-311.

Watts, W.A., 1962. Early Tertiary pollen deposits in Ireland. Nature, 193, p. 600.

Watts, W.A., 1970. Tertiary and interglacial floras in Ireland, In Irish Geographical Studies, (ed. N. Stephens and R.E. Glasscock) Queen's University, Belfast, pp. 17-33.

Whittow, J.B., 1958. The structure of the southern Irish Sea area. Advancement of Science, 14, 381-385.

Wilkinson, G.C., Bazley, R.A.B., and Boutter, M.C., 1980. The geology and palynology of the Oligocene Lough Neagh clays, Northern Ireland, Proceedings of the Geologists Association, London, 137, 65-75.

Williams, A.G., Ternan, J.L., and Kent, M., 1984. Hydrochemical characteristics of a Dartmoor hillslope. In Catchment experiments in fluvial geomorphology, (ed. T.P. Burt and D.W. Walling), Geo Books, Norwich, pp. 379-398.

Wilson, H.E., 1972. Regional Geology of Northern Ireland, Geological Survey of Northern Ireland, Belfast, H.M.S.O., 115 pp.

Wilson, H.E., and Robbie, J.A., 1966. Geology of the country around Ballycastle, Memoir of the Geological Survey, Belfast, H.M.S.O., 370p.

Wilson, M.J., 1967. The clay mineralogy of some soils derived from a biotite-rich quartz-gabbro in the Strathdon area, Aberdeenshire. Clay Minerals, 7, 91-100.

Wilson, M.J., 1969. A gibbsitic soil derived from the weathering of an ultrabasic rock on the island of Rhum. Scottish Journal of Geology, 5, 81-89.

Wilson, M.J., 1973. Clay minerals in soils derived from lower Old Red Sandstone till: effects of inheritance and pedogenesis. Journal of Soil Science, 24, 26-41.

Wilson, M.J., 1985. The mineralogy and weathering history of Scottish soils. In Geomorphology of Soils, (ed. K.S. Richards, R.R. Arnett and S. Ellis), Allen and Unwin, London, pp. 233-244.

Wilson, M.J., and Brown, C.J., 1976. The pedogenesis of some gibbsitic soils from the Southern Uplands of Scotland. Journal of Soil Science, 27, 513-522.

Wright, W.B., 1924. Age and origin of the Lough Neagh Clays. Quarterly Journal Geological Society of London, 80, 468-488.

International Geomorphology 1986 Part II
Edited by V. Gardiner

THE ROLE OF THE INHERITED FORMS IN THE PRESENT-DAY RELIEF OF THE POLISH CARPATHIANS

L. Starkel

Department of Geomorphology and Hydrology,
Institute of Geography, Polish Academy of Science,
Cracow, Poland.

ABSTRACT

The existing slope-valley floor system incorporates the forms of different ages. These inherited forms play various roles depending on the vertical movements, climatic changes, and the resistance of rocks. In the case of the Polish Carpathians the megaforms originated at the end of the synorogenic phase in the Miocene. The macroforms on the flysch rocks of Pliocene-Quaternary age have been transformed by periglacial processes and the primary features of the mature relief of older cycles can only be reconstructed. The present-day shape of slopes has been inherited from the last cold stage, changed only locally by landslides during the Holocene. The existing floodplains and channels are fully adapted to the present-day hydrologic regime.

PRESENT LANDSCAPE AND ITS PAST

In the present landscape there exist side by side elements connected very closely with contemporaneous climate, such as hydrology and - to some extent - vegetation, and elements inherited from the previous morphogenesis, for example geological structures, many relief forms and soils.

In modern geomorphology we can distinguish two different approaches; I will call one of them functional/geophysical, and the other historical/palaeogeographical. The former concerns the relief as a complex of forms identified by different morphometric parameters (length, height, gradient, etc.) which undergo transformation by different physical and chemical processes. The rates of those processes and the tendency for change in the primary form profile are at the centre of investigations.

The other approach is historical or palaeogeographical. The present-day relief is considered as a complex of forms of different ages. While making geomorphological maps we identify forms of different genesis and age. In studying the relations between forms and lithology and the tectonics of the substratum, between forms and weathering profiles, forms and correlative deposits, we try to decipher the evolution of relief (Klimaszewski, 1963).

The first approach does not emphasize the incorporation of inherited forms into the existing geomorphic slope-valley floor system. The second one does not take enough care to reconstruct secondary changes, because the relief rooted in the Neogene (for example, the Carpathians) has not been developed under a protective umbrella. Therefore taking different indicators into consideration we can usually only reconstruct the relief of ancient epochs.

Continuous transformation and incorporation of inherited forms into new climatogenetic systems is the way of relief evolution (Starkel, 1976). These different approaches are reflected in Fig.1. In this paper I will try to present the role of inherited forms in the present-day geomorphic systems exemplified in the Polish Carpathians, built to a great extent of sandstone and shale flysch rocks of medium and low resistance.

PRESENT-DAY MECHANISM OF PROCESSES

The northern slope of the Carpathians lies in the temperate forest zone. Due to differences in elevation, between 200-1700m a.s.l. in the Flysch Carpathians and up to 2500m in the High Tatra, the vertical zonation is well expressed, and the mean annual temperatures vary between +8° and -2°C. The precipitation varies from 700mm in the foothills to 1800mm near the upper timber line, which rises to 1450-1550m a.s.l. Among the processes characteristic of the natural conditions in the forest belts the leading role is played by the subsurface runoff, causing leaching and piping. Among the gravitational processes it is the fall, creep and flow which are typical of the cryonival belt, and at lower elevations under favourable conditions, landslides are widely developed (Kotarba and Starkel, 1972). The fluvial sediment load is therefore dominated by the dissolved load during normal years (Froehlich, 1982).

After deforestation the character and intensity of processes changed. On cultivated fields the dominant role is played by slope wash which rises rapidly when rainfall intensity exceeds 1mm min , and the total precipitation is about 10mm (Gil, 1976). Shallow slides and creeping are common on steeper, poorly drained soils (Starkel, 1960). In areas with strong winds deflation may transfer up to 20,000 t km^{-2}, which exceeds the effect of other processes (Gerlach, 1976).

Most of the material transported on slopes and in river channels is removed during extreme events (Starkel, 1976). Local heavy downpours with a high intensity up to 3-4mm min^{-1} at the beginning, and of variable duration, cause slope wash, piping and earth slumps. During continuous rains with a total amount exceeding 400-500mm in 2-3 days there also occur earth and debris flows, shallow landslides, linear erosion and piping, as well as the transformation of valley bottoms. The total sediment load during such a year with floods can be ten and more times greater than during a year without floods (Fig.2). In the flysch Carpathians we may also distinguish other types of extreme events. These are rainy seasons with a recurrence interval of 20-50 years, when summer and autumn rains of low intensity and long

Fig.1. Three different approaches in geomorphological research (functional, historic and integrated).

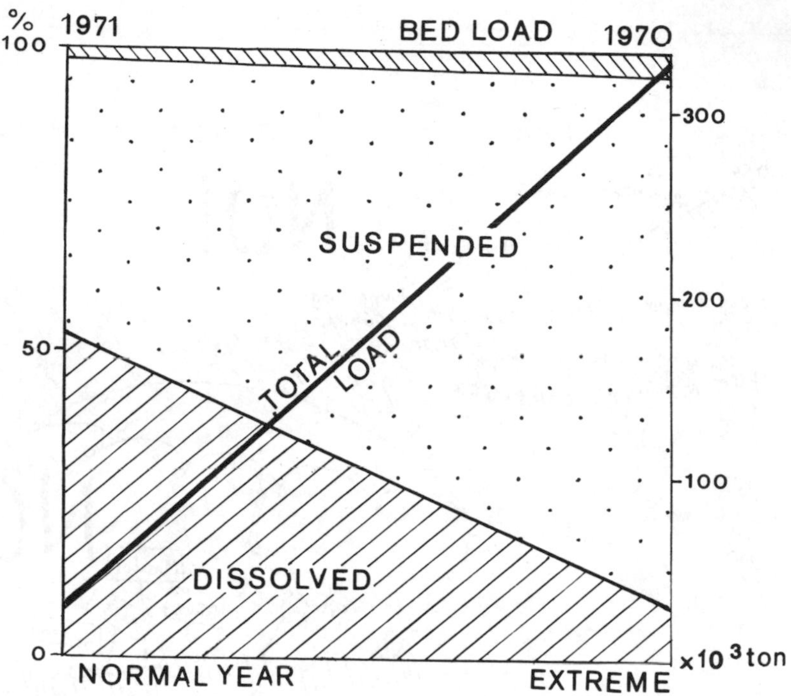

Fig.2. Percentage of dissolved, suspended and bed load in the Kamienica flysch catchment basin. Curve of total load showing the difference between normal and extreme years (after Froehlich, 1975).

duration cause a general reactivation of landslides on slopes (Gil and Starkel, 1979).

In general, the floodplains of the Carpathian rivers are under continuous transformation by floods excluding reaches controlled by water reservoirs and regulation of channels, but, on the contrary, the slopes undergo a total transformation only locally, preserving their mature profile inherited from the Pleistocene.

STAGES OF RELIEF EVOLUTION

Considering the differential preservation of inherited forms and the quantity of data (amount of information) increasing when we are closer to the present day, I have used a logarithmic scale for presentation of the relief evolution of the Polish Carpathians (Fig.3). In the relief history of the flysch zone we may distinguish two phases: synorogenic and a postorogenic one.

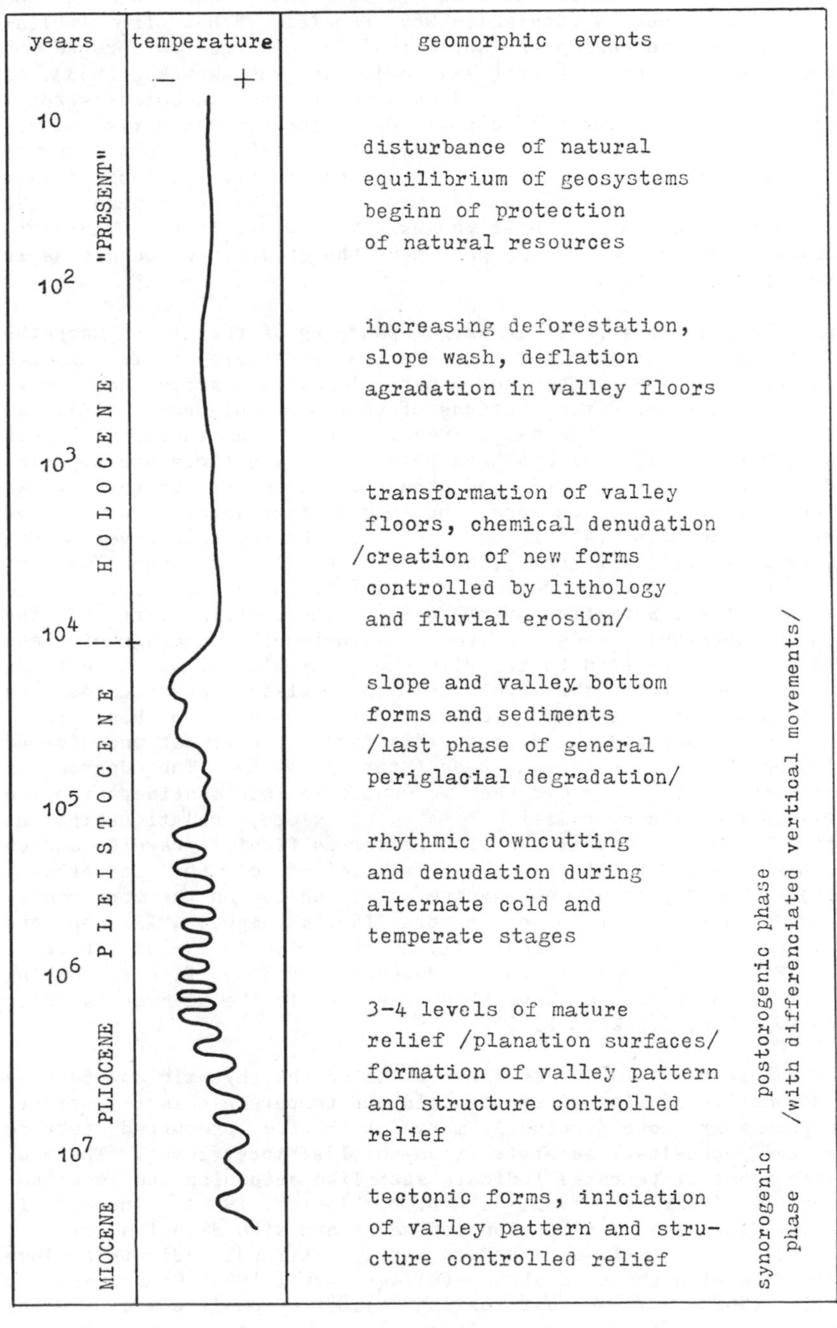

Fig.3. Main events in the relief evolution of the Polish Flysch Carpathians (logarithmic scale used)

The synorogenic phase may be equated to the period from the late Oligocene up to the upper Miocene (Sarmatian), when folding and overthrusting occurred (Starkel, 1969; Henkiel, 1978). The simultaneous erosional truncation of anticlinal structures is proved by re-deposited Oligocene beds (Jasionowicz and Szymakowska, 1963), as well as by the great hypsometric diapazone of the Badenian strata, preserved both in an interfluve position on the truncated flysch, and in deep valleys (Starkel, 1957; Ksiazkiewicz, 1965). This synorogenic phase was also the period of formation of large tectonic forms: mountain blocks and depressions (Birkenmajer, 1978), even when these were reactivated during later phases (Zuchiewicz, 1984). Therefore this synorogenic phase was the period of the creation of megaforms in the Polish Carpathians.

During the following phase of large upwarping of the Flysch Carpathians between the Late Miocene and early Quaternary there existed conditions for the creation of mature planate landscapes under favourable climates and during periods of tectonic quiescence (Klimaszewski, 1934, 1965; Starkel, 1965). Three destructional levels younger than the main valleys have been formed and these are superimposed on the older truncated and structurally controlled relief. We distinguish, in the Flysch zone, the intramontane level of a relative relief of 230-300m (450-1000m a.s.l.), the foothill level rising 120-200m above valley bottoms (360-800m a.s.l.), and the "valley" level with a relative relief of 40-100m (300-700m a.s.l.). The occurrence of the Sarmatian overthrusts in the Eastern part of the Polish Carpathians leads to the conclusion that the age of these levels should be shifted to the Pliocene (Starkel, 1975). As a rule, the older and higher the level, the more resistant are the rocks on which it has been preserved. The Quaternary denudation has transformed these surfaces and most distinctly the lowest one, formed usually on the least resistant beds (Starkel, 1965). The degree of transformation is so great that we must take into consideration the details of interfluve relief, breaks of slope, relationships to structure and rare localities with preserved fluvial gravels, and on this basis reconstruct the Pliocene relief of the Carpathians (Fig.4). The degree of downwearing is variable. On the more resistant sandstones it is recorded in tors 10-15m high, while on the shales forming the lowermost level they rise to 30-50m (Starkel, 1965). Fragments of older, not rejuvenated valley floors, dating even from the Pliocene, have been preserved in the headwaters (Klimaszewski, 1960; Starkel, 1965).

During the last million years there followed the rhythmic downcutting and denudation during alternate cold and temperate climatic stages. These phases are more distinctly marked by better preserved terrace forms and deposits, separate from the flat interfluves. The erosional benches of terraces indicate step-like deepening and continuous uplift (Starkel, 1965, 1969b; Zuchiewicz, 1984). The alluvia interfingering with solifluction sediments and with deposits of the local Tatra, as well as Scandinavian glaciations, indicate a close relationship with the cold stages (Klimaszewski, 1948; Dziewański and Starkel, 1962). These alternating climatic conditions caused the lowering of interfluves and slopes and the excavation of more

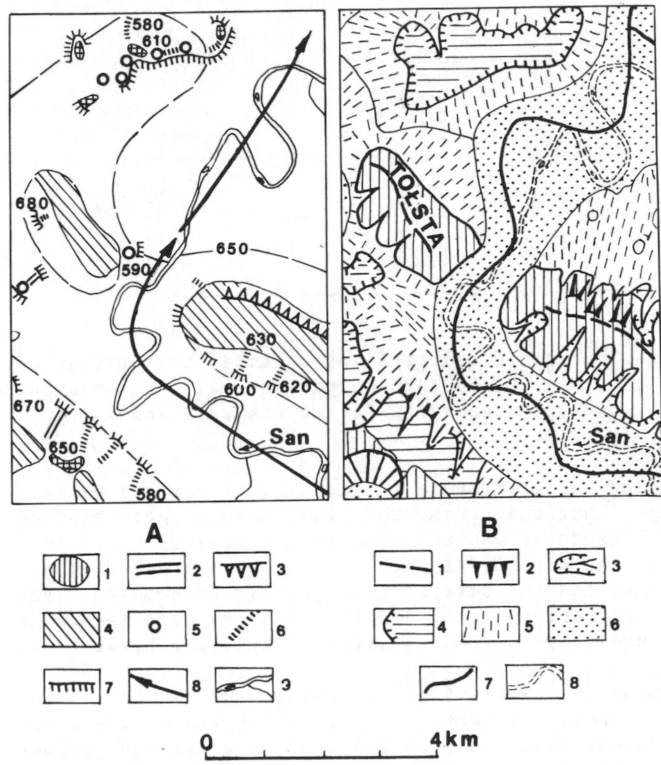

Fig.4. Examples of mapping of interfluve relief and reconstruction of planation levels in upper San river basin (Starkel 1965)

A. Elements of reconstruction of the foothill level: 1. flattenings in 200m level 2. wide rounded humps in the level of flattenings 3. structural steps 4. structure controlled ridges rising above the level 5. rounded hills (lowered in relation to flattenings) 6. humps lowered (with levelled axis) 7. edges separating the 200m relief from younger slopes. 8. direction of river outflow 9. present day river 10. elevation of level above sea level.

B. Palaeomorphologic map of the 100m valley-side level: 1. structure controlled ridges 2. denudation escarpments 3. valleys dissecting the slopes of ridges 4. monadnocks and larger fragments of higher level relief 5. inclined surfaces of glacis-pediments 6. valley floor (locally with gravels) 7. direction of river outflow 8. present-day river channel.

resistant beds due to selective denudation. Part of these materials buried the Quaternary terraces, even forming the accumulational glacis. Due to the step-like rejuvenation of valleys, their tributary valleys have preserved older features and the longitudinal profile has until now preserved a step-like character (Fig.5).

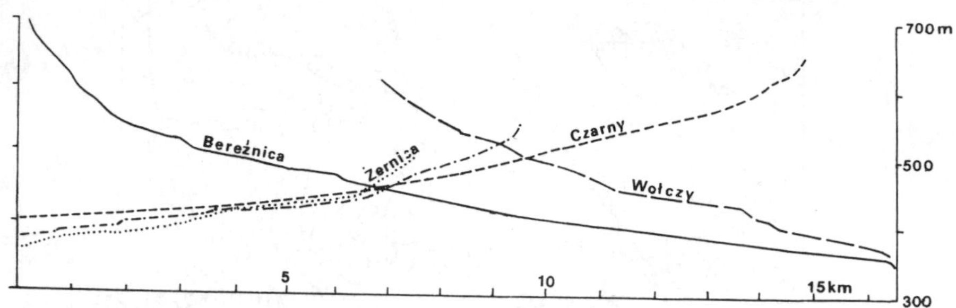

Fig.5. Longitudinal profiles of small tributaries of the San river in the Eastern Carpathians with un-rejuvenated upper segments (after Starkel 1965).

An especially important role in the present-day relief is played by the forms inherited from the last cold stage: they are very well preserved because it was the most recent period of periglacial morphogenesis. This period is reflected in the block fields, cryoplanation terraces and the tors of structurally controlled ridges, as well as in loamy solifluction debris and silty deluvial covers building lower concave slope segments (Klimaszewski, 1971b; Starkel, 1969c).

The thickness of solifluction, deluvial and fluvial deposits indicates the scale of degradation, which was of the order of 10m on shaly-sandstone beds. At the scale of the Flysch Carpathians the last cold stage mesoforms occupy 90% of the area. This last cold phase has given a new shape to all older forms. In the Tatra Mountains this was also the last phase of glacial transformation of pre-Quaternary valleys into U-shaped ones (Klimaszewski, 1960).

The Holocene phase, due to the changes in the vegetation and hydrological regime (retreat of permafrost, deep infiltration and rise in precipitation), caused a general increase in slope stability and the leaching of inherited periglacial covers. The underloaded streams started to incise their channels which locally caused the dissection of slopes by piping, landsliding and gullying (Starkel, 1960). Most recently a very distinct rise in the intensity of processes was caused by the extension of farming and road building.

GENERAL TENDENCIES OF EVOLUTION

In the relief evolution of the gently uplifted Carpathians during rhythmic climatic changes we can distinguish the following general tendencies:

(a) The rejuvenation of valley floors progressively upstream is reflected in the preservation of undissected hanging valleys in headwater areas (e.g. Fig.5), which at higher elevations were transformed into glacial corries and troughs. The change of sediment load

caused by climatic oscillations under conditions of tectonic uplift is reflected in the sequence of cut and fill terraces (Fig.6c).
(b) The rejuvenation of slopes caused by undermining and downcutting, leading to the excavation of structurally-controlled relief (Fig.6a,b), and the close correlation of alternate cold phases with areal solifluction, and temperate ones, with soil formation and localised landslides.
(c) General downwearing (Fig.6d) tending to change the step-like polycyclic relief into convex-concave slopes with aggradation on the lower steps of Quaternary terraces (Fig.6c). This long-term trend on less resistant Flysch rocks is characterised by the incorporation of all older inherited elements into one system.

PRESENT-DAY TENDENCIES

The present-day geomorphic system of the temperate climatic zone also shows some tendencies of relief transformation. These are as follows (Fig.7):

(a) Slope stability. Most slopes inherited from the last cold stage are so gentle that the threshold value for mass wasting and erosional processes cannot be reached. Therefore soil formation and leaching up to 4-5m deep is a dominant process under natural conditions.

(b) Continuation. After deforestation on the cultivated fields intensive slope wash and some creep and deflation continue the formation of inherited periglacial convex-concave profiles.

(c) Rejuvenation. On steeper slopes usually undercut by rivers there follows the rejuvenation by landslides or piping. This is mainly by the removal of periglacial slope deposits and the exposure of the bedrock (eroded only in exceptional cases).

(d) Accumulation of debris slopes. In the Alpine belt of the Tatra mountains, the steep valley sides undergo transformation by downwearing of the upper part and burying of the lower one by talus debris (Klimaszewski, 1971a; Kotarba, 1984).

(e) Erosion or aggradation in the valley bottoms. The natural trend towards down-cutting in the forest belt has been replaced in the deforested areas by aggradation due to overloading of rivers by suspended and bed load. Current changes in land use may cause a return to downcutting, especially downstream of water reservoirs.

FINAL REMARKS

The existing relief is a complicated system which operates by the incorporation of past geomorphic elements of various ages. Among them the mega-forms controlled by tectonics and lithology (for example, mountain groups and basins) are inherited from the late stage of the synorogenic phase and the early stages of the post-orogenic one; they also reflect younger neotectonic, differential movements. The macroforms, mainly the ridges of pre-Quaternary

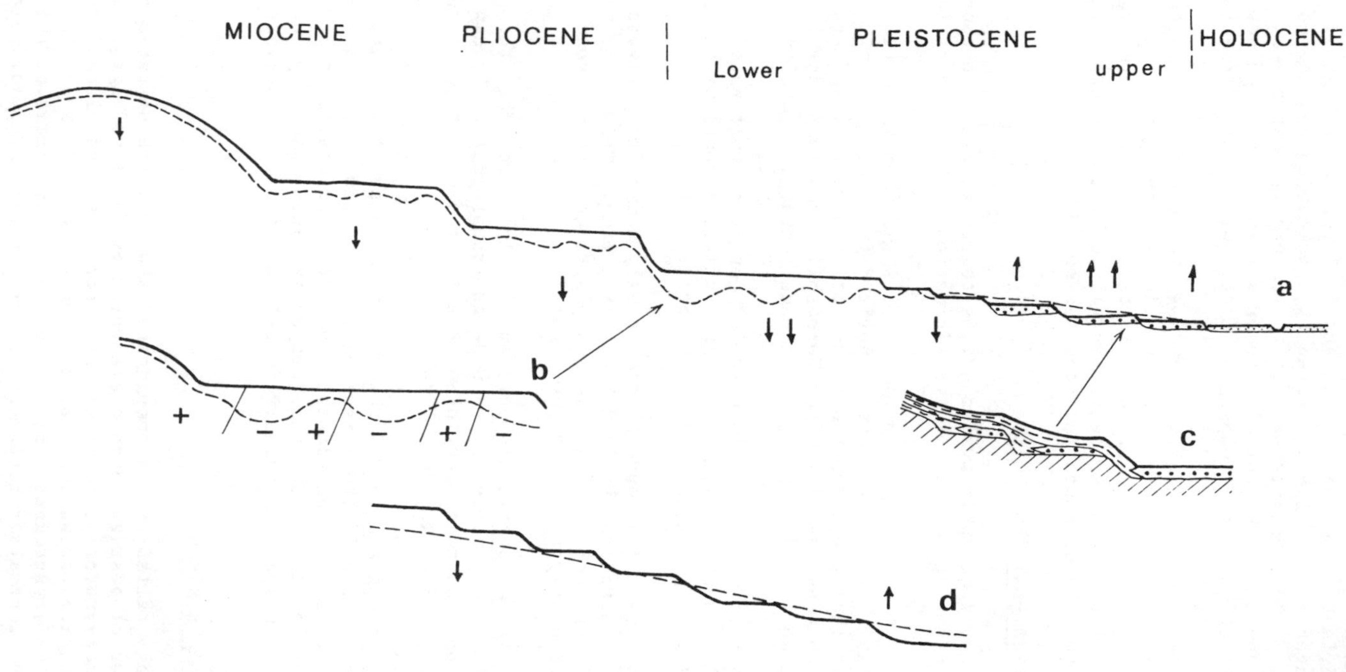

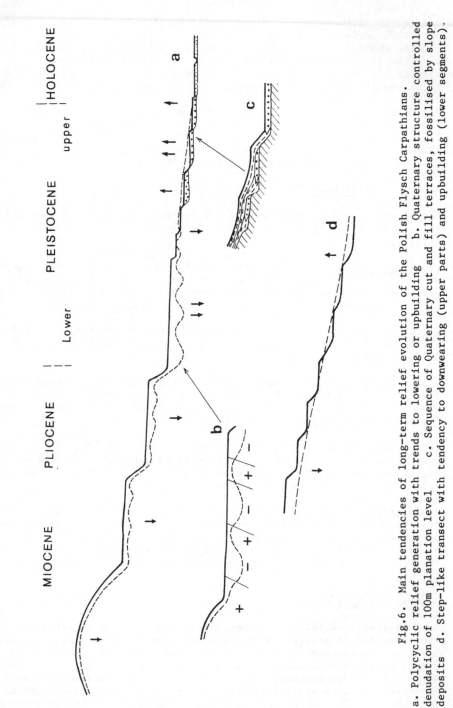

Fig.6. Main tendencies of long-term relief evolution of the Polish Flysch Carpathians.
a. Polycyclic relief generation with trends to lowering or upbuilding b. Quaternary structure controlled denudation of 100m planation level c. Sequence of Quaternary cut and fill terraces, fossilised by slope deposits d. Step-like transect with tendency to downwearing (upper parts) and upbuilding (lower segments).

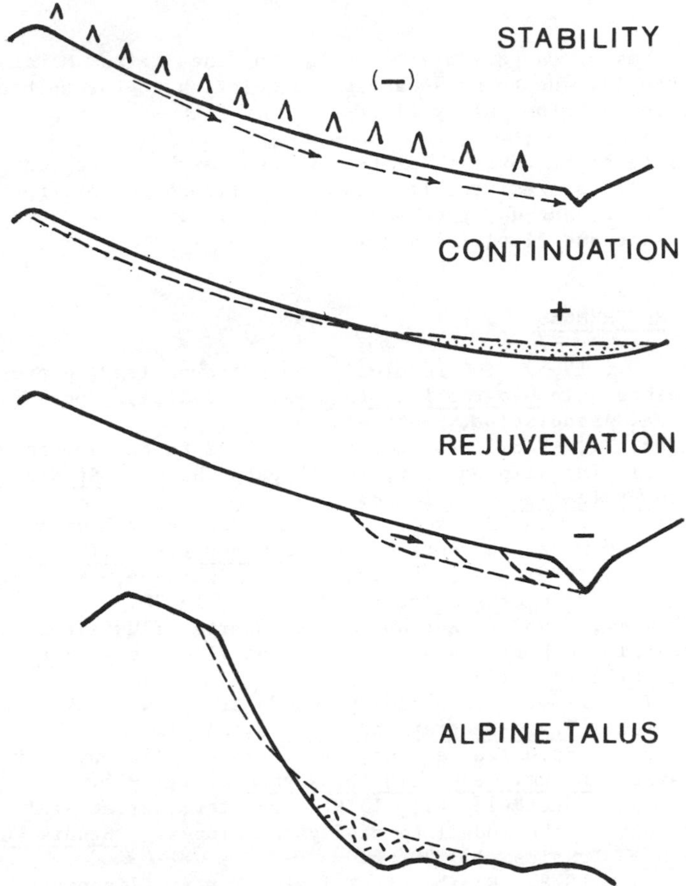

Fig.7. Holocene tendencies of relief evolution in the Carpathians a. stability (only leaching under forest) b. stability (slope wash after deforestation and upbuilding of valley floor) c. rejuvenation (downcutting and retreat of slopes) d. alpine slopes inherited from cold phase transformed by talus.

age, later excavated or deepened, now play the role of upper slope segments.

The Quaternary terraces, forming the middle slope segments, were mainly denuded or fossilised by slope deposits. Long slopes, resulting from the transformation of valley sides under periglacial conditions, are still preserved, and only in some regions have they been reworked by Holocene landslides and gullies. In contrast the flood plains and river channels have been adjusted to the existing hydrologic regime, although the main part of the alluvia comes from the reworking of unconsolidated periglacial deposits filling the valley bottoms.

The mesoforms of the cryonival belts in the Tatra Mountains are slowly changing due to mass wasting in relation to the unchanged base level of the hanging valley floors.

The coexistence of forms of different ages within the contemporary morphogenetic system results in the obliteration of their primary features by present-day processes, which incorporate them into a unified slope and fluvial system.

REFERENCES

Brunsden, D., 1985. The revolution in geomorphology: a prospect for the future, in Geographical futures. (Ed.King), pp30-35. Geographical Association, Sheffield.

Birkenmajer, K., 1978. Neogene to early Pleistocene subsidence close to the Pieniny Klippen Belt, Polish Carpathians. Studia Geomorph. Carpatho-Balcanica, 12, 17-28.

Dziewański, J., and Starkel, L., 1962. Dolina Sanu miedzy Solina a Zwierzyniem w czwartorzedzie. Prace Geograficzne IGPAN, 36, 1-86.

Froehlich, W., 1975. Dynamika transportu fluwialnego Kamienicy Nawojowskiej. Prace Geograficzne IGPAN, 114, 1-122.

Froehlich, W., 1982. Mechanizm transportu fluwialnego i dostawy zwietrzelin do koryta w górskiej zlewni fliszowej. Prace Geograficzne IGiPZ PAN, 143, 1-133.

Gerlach, T., 1976. Współczesny rozwój stoków w Polskich Karpatach Fliszowych. Prace Geograficzne IGPAN, 122, 116.

Gil, E., 1976. Splukiwanie gleby na stokach fliszowych w rejonie Szymbarku. Dokum. Geogr. IGPAN, Warszawa, 1, 1-65.

Gil, E., and Starkel, L., 1979. Long-term extreme rainfalls and their role in the modelling of flysch slopes. Studia Geomorph. Carpatho-Balcanica, 13, 207-220.

Henkiel, A., 1978. Rzezba strukturalna Karpat fliszowych. Annales UMCS Lublin, 32/33, 2, sec.B.

Jasionowicz J., and Szymakowska, F., 1963. Próba wyjaśnienia genezy platów magurskich w okolicy Jasla oraz plata podślaskiego z okolicy Wielopola Skrzyńskiego. Rocznik Pol. Tow. Geolog., Kraków, 33, 3, 363-385.

Klimaszewski, M., 1934. Z morfogenezy Polskich Karpat Zachodnich. Wiad. Geogr., 12, Kraków, 30-44.

Klimaszewski, M., 1948. Polskie Karpaty Zachodnie w okresie dyluwialnym. Prace Wrocl. Tow. Nauk. Ser.B., 7, 1-236.

Klimaszewski M., 1960. On the influence of the pre-glacial relief on the extension and development of glaciation and deglaciation of mountainous region. Przegl. Geogr., 32, Suppl.

Klimaszewski, M., 1965. Views on the geomorphological evolution of the Polish West Carpathians in Tertiary times. Geomorph. Problems of Carpathians, vol. 1, Bratislava, 91-126.

Klimaszewski, M., 1971a. A contribution to the theory of rock-face development. Studia Geomorph. Carpatho-Balcanica, 5, 139-151.

Klimaszewski, M., 1971b. The effect of solifluction processes on the development of mountain slopes in the Beskidy (Flysch Carpathians). Folia Quaternaria, 38. Kraków.

Kotarba, A., 1984. Elevational differentiation of slope geomorphic processes in the Polish Tatra Mts. Studia Geomorph. Carpatho-Balcanica, 18, 117-133.

Kotarba, A., and Starkel, L., 1972. Holocene Morphogenetic Altitudinal Zones in the Carpathians. Studia Geomorph. Carpatho-Balcanica, 6, 21-36.

Ksiazkiewicz, M., 1965. Trzeciorzed mlodszy Podkarpacia i Karpat, in Ksiazkiewicz, Samsonowicz and Rühle, Zarys geologii Polski, Wyd. Geolog. Warszawa.

Saunders, I., and Young, A., 1983. Rates of surface processes on slopes slope retreat and denudation. Earth Surface Processes and Landforms, 8, 473-501.

Selby, M. J., 1982. Hillslope materials and processes, Oxford Univ. Press, Oxford.

Starkel, L., 1957. Rozwój morfologiczny progu Pogórza Karpackiego miedzy Debica a Trzciana. Prace Geograficzne, IGPAN, nr.11, Warszawa, 1-152.

Starkel, L., 1960. Rozwój rzezby Karpat fliszowych w holocenie, Prace Geograficzne, IGPAN, 22, 1-239.

Starkel, L., 1965. Rozwój rzezby polskiej cześci Karpat Wschodnich. Prace Geograficzne, IGPAN, 50.

Starkel, L., 1969a. The age of the stages of development of the relief of the Polish Carpathians in the light of the most recent geological investigations. Studia Geomorph. Carpatho-Balcanica., 3, 33-44.

Starkel, L., 1969b. Climatic or tectonic adaptation of the relief of young mountains in the Quaternary. Geogr. Polonica, 17, 209-229.

Starkel, L., 1969c. L'evolution des Versants des Carpates a flysch au Quaternaire. Biuletyn Peryglacjalny, 18, 349-379.

Starkel, L., 1975. Communique on sujet de l'etat actuel des recherches sur le development des surface d'aplanissement dans les Carpates Polonaises (in Russian). Studia Geomorph. Carpatho-Balcanica., 9, 75-81.

Starkel, L., 1976. The role of extreme (Catastrophic) meteorological events in the contemporaneous evolution of slopes, in Geomorphology and Climate, (Ed. Derbyshire), pp203-246. Wiley, London.

Starkel, L., 1978. The features of the past and of the future in the present-day relief of the Polish Flysch Carpathians. Beiträge zur Quartär-und Landschaftforschung. Festschrift zum 60. Geburtstag von J. Fink, Hirt, Wien, 585-600.

Zuchiewicz, W., 1984. Neotectonics of the Polish Carpathians: facts and doubts. Studia Geomorph. Carpatho-Balcanica., 17, 29-44.

International Geomorphology 1986 Part II
Edited by V. Gardiner

PEDIMENT DOME EVOLUTION IN THE EASTERN MOJAVE DESERT, CALIFORNIA

John C. Dohrenwend[1], Stephen G. Wells[2], Leslie D. McFadden[2], and Brent D. Turrin[1]

[1]United States Geological Survey
Menlo Park, California 94025

[2]Department of Geology
University of New Mexico
Albuquerque, New Mexico 87131

ABSTRACT

Relations between pediment surfaces and potassium-argon-dated lava flows of the late Cenozoic Cima volcanic field record a four-million-year history of pediment dome evolution in the eastern Mojave Desert of California. Collectively, the more than 60 basaltic lava flows of the Cima field cover an area of approximately 150 km^2 on the crests and flanks of several active and inactive pediment domes. Progressively younger flows spread across and partly bury progressively younger and lower pediment surfaces. These pediment surfaces cut across Cretaceous granitic rocks and Cenozoic arkosic gravels that are pervasively weathered to depths of at least 40 m. Comparison of the relative positions of lava-flow-covered pediment remnants and modern pediment surfaces indicates that (1) downwasting has been the dominant mode of pediment modification, and (2) although flow emplacement has caused some local perturbations, the general form of these pediment domes has changed little since at least early Pliocene time. Downwasting has been greatest in crestal areas and has progressively decreased downslope. Mid-flank areas have remained in a state of approximate erosional equilibrium, and lower flanks have probably aggraded to some extent. Erosion rates on upper flanks have ranged between 1.2 and 2.8 cm/10^3yr, and with an apparent lack of sensitivity to climatic change, downwasting has proceeded more or less continuously at an average rate of approximately 2.0 cm/10^3yr.

INTRODUCTION

Pediments are one of the least understood and most controversial of all arid landforms. "Disagreement often begins with the problem of definition ....and continues throughout all subsequent phases of inquiry" (Cooke and Warren, 1973, p. 188). Lack of reliable long-term time control in most pediment-related research is one of the primary reasons for this limited understanding. This paper uses detailed radiometric dating of late Cenozoic lava flows in the Cima

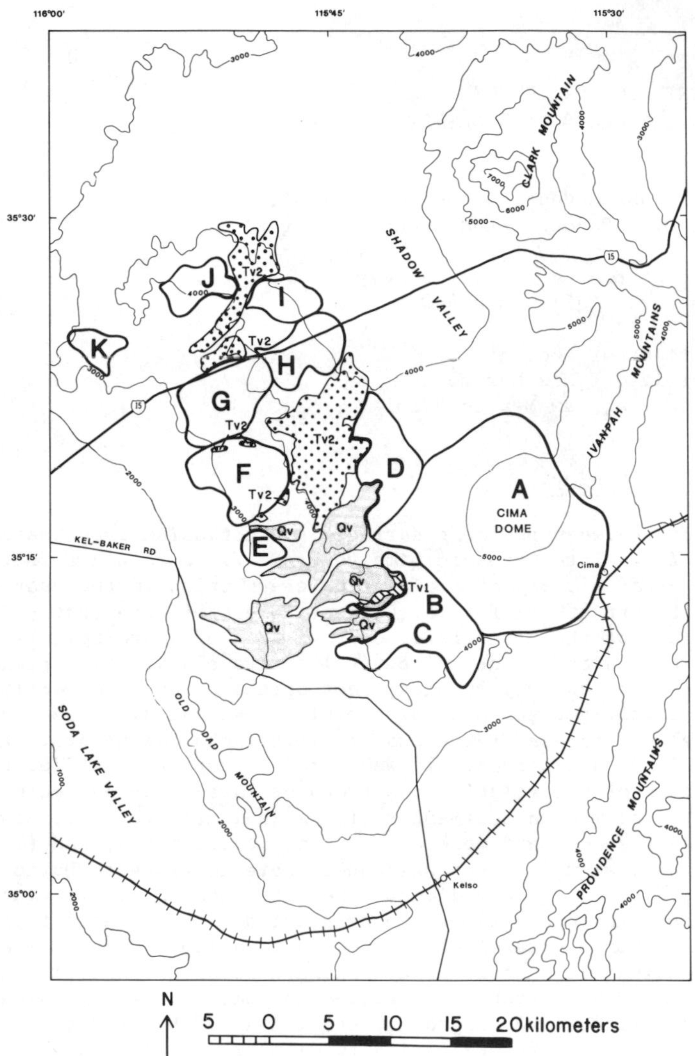

Fig 1. Generalized map of the Cima volcanic field showing the locations of major pediment domes: Tv_1 - late Miocene volcanic flows; Tv_2 - latest Miocene and early Pliocene volcanic flows; Qv - Quaternary volcanic flows. Letters mark the locations of pediment domes: A = Cima dome; B = Cimacita dome; C = Marl Mountain dome; D = Cow Cove dome; E = Indian Springs dome; F = Granite Springs dome; G = Halloran Wash dome; H = Yucca Grove dome; I = Solomons Knob dome; J = Squaw Mountain dome; K = Turquoise Mountain dome. Contour interval is 1000 ft.

volcanic field of southeastern California and extensive exposures of remnant pediment surfaces buried by these flows to document four million years of pediment evolution. Topographic positions of remnant and modern pediment surfaces are compared and variations in pediment erosion rates are discussed.

PEDIMENTS OF THE CIMA AREA

The largest and best developed pediment domes in the eastern Mojave Desert are located in the area immediately surrounding the Cima volcanic field. These landforms have been the subject of several lengthy and detailed geomorphic analyses (Davis, 1933; Sharp, 1957; Warnke, 1969; and Oberlander,1974). Davis (1933) described many of the pediment domes of the Mojave Desert region and proposed two general mechanisms of pediment formation, one involving backwasting of bounding scarps on upfaulted terrain of low relief and the other involving downwasting of upwarped masses of low relief. Davis speculated that Cima Dome, the largest of these pediment domes, was probably formed by the first of these two mechanisms,namely backwasting of an upfaulted terrain. Sharp (1957) marshaled geomorphic, geologic, and geophysical evidence to explain Cima Dome and adjacent granitic domes as the product of a modified version of the second of Davis' two dome-forming mechanisms, namely complex upwarping of an ancient erosion surface with subsequent modification by stream erosion and regrading. Warnke (1969) argued that pediment evolution in the Halloran Hills on the northwest margin of the Cima field could be described in terms of Davisian stages of geomorphic evolution and that downcutting followed by lateral corrasion and backwasting were the most important processes of pedimentation in that area. Warnke also concluded that the combined presence of a suitable rock type (quartz monzonite or its sedimentary derivatives) and a temporary local baselevel is the principal determinant of pediment formation in the Halloran Hills. Oberlander (1974) challenged Warnke's premise that the granitic pediments of the Mojave were the "consequence of continuous arid morphogenesis" by developing a model of landscape inheritance wherein the present granitic domes are interpreted as relict forms produced by erosional stripping of late Tertiary deep-weathering profiles developed during periods of greater effective moisture. The present paper is compatible with Oberlander's general concept of landscape inheritance. It documents a Pliocene and Pleistocene history of essentially continuous pediment downwasting across late Tertiary pediments underlain by deeply weathered granitic rock and poorly consolidated arkosic sediments.

In the area of the Cima volcanic field, eleven pediment domes have been continuously evolving over the past several million years, and the remnants of at least three other pediment systems have been partly buried and preserved by flows of the Cima field (Fig. 1). These pediment domes are broad, 5 to 16 km across; low, 0.1 to 0.4 km high; and gently sloping, 1.4 to 4.5^{o} slopes. They are more conical than domelike in overall form with nearly straight slopes

Fig. 2. Granite Springs pediment dome and the 100- to 200-m-high erosional scarp along the west flank of the Pliocene lava flows (on the skyline). These flows cap an ancient pediment dome that truncates quartz monzonite and terrigenous sedimentary deposits. The Granite Springs pediment dome, also cut across quartz monzonite, extends from the foreground to the erosional scarp. Note the irregular patchwork of dissected and undissected areas on the flank of this dome. Aerial view east in the general vicinity of topographic profiles A and B (Figs. 4 and 5).

Fig. 3. Quaternary cones and flows of the southern part of the Cima volcanic field in the general vicinity of topographic profiles C through G (Figs. 4, 6, and 7). Aerial view east towards the crest of the Cimacita pediment dome; the New York Mountains form the skyline beyond.

that generally vary by less than $\pm$ 0.5^{o} along any radial profile (Sharp, 1957). Although occasionally interrupted by prominent irregularities such as inselbergs and atypically deep drainageways, dome form is generally smooth and regular; local relief is typically less than 5 meters. In some areas, a more complex morphology occurs where shallow drainageways locally incise the pediment surface into irregular patchworks of dissected and undissected topography (Fig. 2). Undissected areas of temporary aggradation and transport are mostly flat with anastomosing drainageways and low indistinct interfluves, whereas dissected areas are scored by well-defined subparallel drainageways (3 to 10 m deep) that form shallow regularly spaced valleys separated by rounded interfluves.

Lithology has significantly influenced the distribution and development of pediments in the Cima area (Davis, 1933; Sharp, 1957; Warnke, 1969). Most pediment surfaces cut indiscriminantly across Cretaceous plutonic rocks (primarily the Teutonia Quartz Monzonite) or Tertiary terrigenous clastic rocks (Hewett, 1956; Sharp, 1957); however, inselbergs of Precambrian metamorphic rocks, including biotite-hornblende gneiss and marble, stand as much as 250 m above the pediment surfaces. The Teutonia Quartz Monzonite is a light-gray holocrystalline rock composed of orthoclase crystals, 20 to 75 mm in diameter, in a groundmass of quartz and feldspar grains, 5 to 15 mm in diameter, with sparse plates of biotite (Hewett, 1956). This rock weathers rapidly into its constituent mineral grains, and low-energy seismic measurements (K. D. Mahrer, unpublished report) indicate pervasive subsurface weathering to depths of at least 40 m in the area of the Cima volcanic field. However, the resulting grus appears to be relatively resistant to further decomposition; indeed, the entire piedmont surface surrounding the Cima field is dominated by this material. The Tertiary clastic rocks are poorly to moderately indurated and variable in composition and texture. Although fluvially deposited grus constitutes the bulk of many exposures, clast-supported monolithologic quartz monzonite boulder conglomerates and intercalated beds of exotic gravel, sand, silt, and mudstone are also present. The exact nature of the contact between the Tertiary deposits and the underlying plutonic rocks is not well understood; however, it is characterized by substantial irregularity and high relief and may include both tectonic and depositional elements.

RELATIONS BETWEEN LAVA FLOWS AND PEDIMENT SURFACES

The majority of the vents and flows that constitute the Cima volcanic field have been superimposed across several pediment domes. Volcanic activity within the Cima field has been sufficiently continuous to produce flows of widely ranging late Cenozoic age. However, these flows are small and spatially separated and have only partly buried these pediments. Field relations indicate that the pediment surfaces buried by these flows were either active or relatively stable just before flow emplacement. Many flows rest directly on weathered bedrock surfaces; others locally bury soils developed in thin alluvial veneers. This

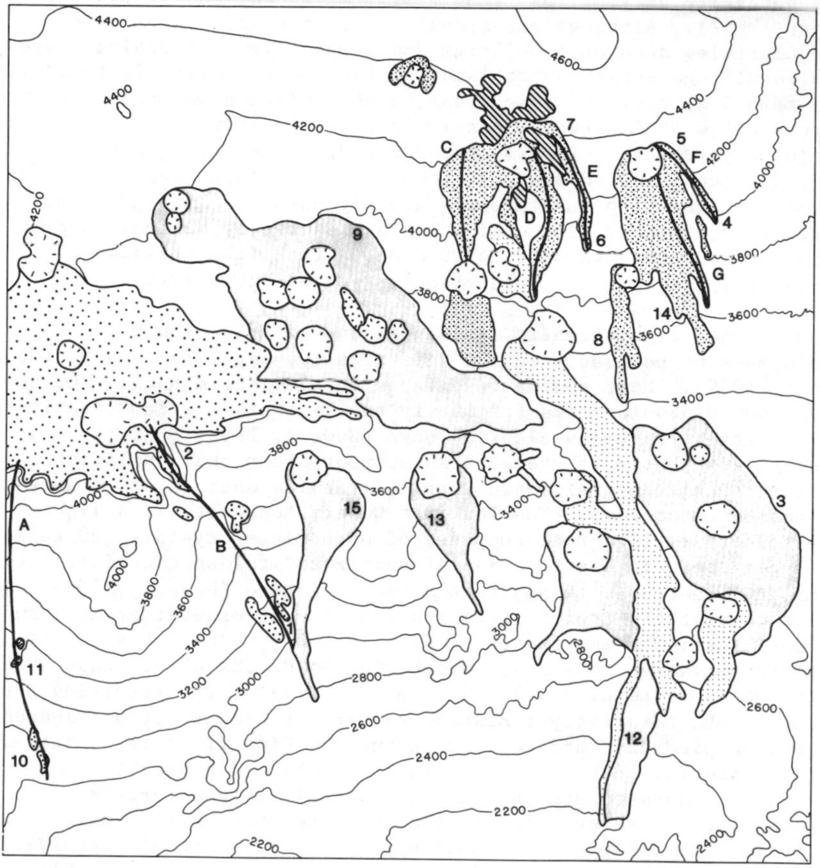

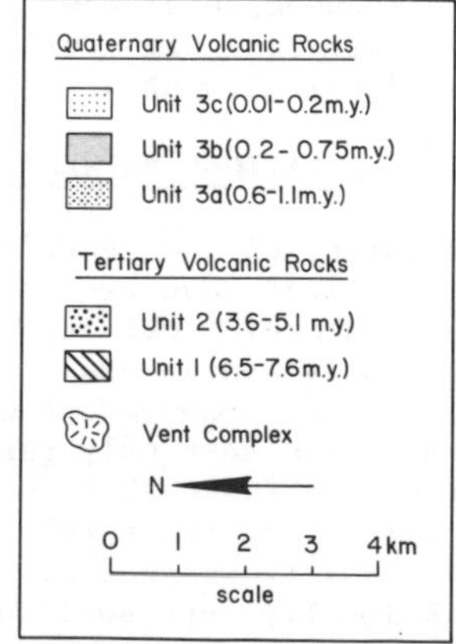

Fig. 4. Generalized geologic map of the southern part of the Cima volcanic field showing the approximate locations of specific site measurements and of longitudinal topographic profiles on pediment remnants buried by the Cima lavas. Numbers refer to the specific site locations listed in Table 1. Letters refer to the longitudinal profiles presented in Figures 5, 6, and 7. Contour interval is 200 ft.

combination of circumstances has created an ideal situation for the study of pediment evolution, namely a series of pediment remnants that can be radiometrically dated and whose vertical positions, relative to modern pediment surfaces and to each other, record progressive downwasting through an extended period of geologic time.

The basaltic flows of the Cima field collectively cover an area of approximately 150 km^2 (Fig. 1). These lavas flowed down the flanks of the pediment domes and along major drainages between domes, and coalesced into caprock veneers across the crest and upper flanks of at least one dome (Fig. 2). Three principal periods of volcanic activity span the time from latest Miocene through latest Pleistocene (Dohrenwend et al., 1984; Turrin et al., 1984). During the initial period, from approximately 7.6 to 6.5 m.y.B.P., a small vent-and-flow complex formed on the northwest flank of Cimacita dome. During the intermediate period, from 5.1 to 3.3 m.y.B.P., a large pediment dome in the northern part of the field was veneered with a voluminous sequence of coalescing flows (Fig. 2). Lavas of this intermediate period also flowed down paleovalleys between the pediment domes immediately west and north of the field. During the latest period of activity, uniformly spanning the last million years, most of the southern half of the field was formed (Figs. 3 and 4). Lavas of this period flowed generally westward down the flanks of the Cimacita, Marl Mountains, and Cow Cove domes and across an irregular, partly dissected piedmont terrain of low to moderate relief.

The Cima lava flows, which are as much as 1.5 km wide and 9 km long, form a continuum between two distinct morphologic types: thin elongate flows having low gradients and low surface relief and thick equant flows having somewhat higher gradients and higher surface relief. Elongate flows, the most extensive flow type at Cima, are usually extruded during the earlier eruptive phases of a vent complex and commonly rest directly on nonvolcanic land surfaces. Consequently, these flows are the more useful for analyzing pediment evolution. Elongate flow thicknesses generally range between 2.5 and 4 m, except where locally ponded, and constructional relief on flows more than 0.25 m.y. old seldom exceeds one meter. Flow-surface gradients generally range between 1.5 and 3.5^{o} and are nearly parallel to the underlying pediment remnants which they bury.

Topographic relations demonstrate that the cones and flows of the Cima volcanic field have been erupted into a progressively downwasting erosional environment that has been active since before inception of the volcanism. The Cima lava flows form caprocks that protect the relatively nonresistant grus-forming quartz monzonite and poorly indurated terrigenous clastic rocks from fluvial erosion. With the eruption of each flow, part of an active pediment surface is buried and removed from the active erosional environment. Progressively younger flows bury progressively younger and lower surfaces; thus each caprock-protected remnant now stands at an average height above the modern pediment surface that is directly related to the age of its overlying basalt flow.

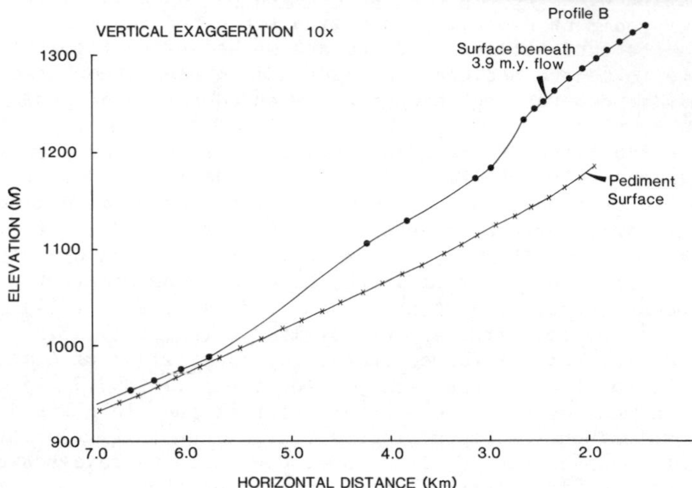

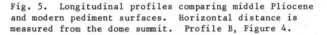

Fig. 5. Longitudinal profiles comparing middle Pliocene and modern pediment surfaces. Horizontal distance is measured from the dome summit. Profile B, Figure 4.

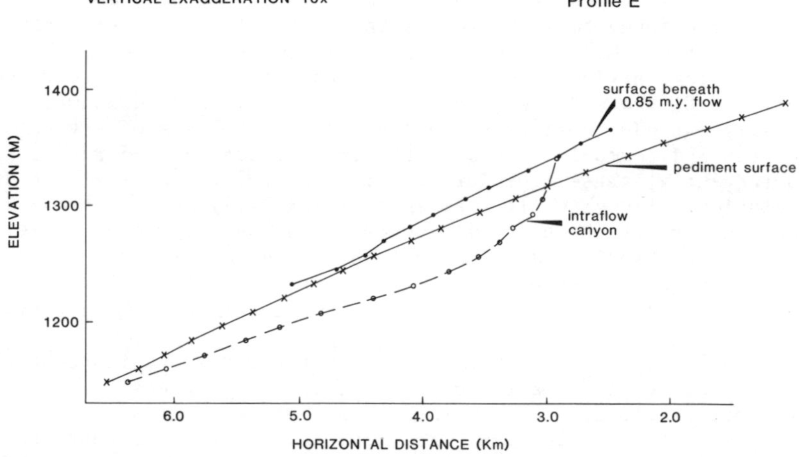

Fig. 6. Longitudinal profiles comparing early Pleistocene and modern pediment surfaces with intraflow canyon surfaces. Horizontal distance is measured from the dome summit. Profile E, Figure 4.

TOPOGRAPHIC PROFILES OF PEDIMENT REMNANTS

Longitudinal profiles were constructed along seven remnant pediment surfaces in the Cima volcanic field. Profiles were also constructed for modern pediment surfaces and fluvial valley floors immediately adjacent to these remnants. Profile locations are shown in Figure 4, and representative examples are presented in Figures 5, 6, and 7. To supplement these longitudinal profiles, pediment remnant heights were measured in the field at random locations along the profiles and at selected locations along other pediment remnants that were too small to be profiled.

Because elongate flows in the Cima field are uniformly thin and flow surfaces are nearly parallel to the remnant pediment surfaces which they bury, topographic profiling of these pediment remnants is facilitated. Longitudinal profiles of the surfaces of the overlying lava flows were first constructed, using measurements from 1:24,000-scale, 10-meter contour topographic maps. Profiles for the underlying pediment remnants were then approximated by subtracting 3 meters from each flow-surface profile. Flow thicknesses were measured at several locations along each flow-surface profile, and in all cases measured thicknesses were within one meter of the approximate three-meter average.

Comparison of the profiles of remnant and modern pediment surfaces indicates that downwasting has been the dominant mode of pediment modification over the past four million years. Although remnant surfaces are, in all cases, nearly parallel to modern surfaces, they slope more steeply than the modern surfaces. Differences in slope between Pleistocene and modern surfaces are 0.5° or less, and differences in slope between Pliocene and modern surfaces typically range between 1.0 and 1.5°.

AVERAGE DOWNWASTING RATES

Average rates of surface downwasting in the vicinity of each pediment remnant were determined by dividing the average height of each remnant above the adjacent modern surface by the K-Ar age of the basalt flow capping that remnant. These average rates are listed according to erosional environment in Tables 1 and 2.

On the flanks of the pediment domes, rates of average downwasting are clearly associated with distance from the dome summits. On upper flanks (2 to 4 km from dome summits), rates range between 1.2 and 2.8 $cm/10^3 yr$ and average approximately 2.0 $cm/10^3 yr$; whereas in mid-flank areas (5 to 12 km from dome summits), rates range between 0.0 and 0.4 $cm/10^3$ yr and average approximately 0.2 $cm/10^3 yr$ (Table 2). Along the boundaries between pediment domes, along flow margins oriented at high angles to pediment slopes, and within valleys cut by lava-flow-sourced drainage (all areas affected by concentrated fluvial erosion), erosion rates are very similar to the rates of downwasting on upper flanks of the pediment domes. Erosion rates in these areas range between 0.9 and 3.8 $cm/10^3 yr$ and average approximately 2.2 $cm/10^3 yr$ (Table 1). However, variations in these rates do not show any clear spatial patterns.

TABLE 1. Average downwasting rates in the Cima volcanic field

Map location	K-Ar sample number	Flow age (m.y.)[a]	Maximum height above modern surface (m)	Approximate downwasting rate ($cm/10^3 yr$)	Distance from dome summit (km)
Locations on the flanks of pediment domes					
1	C01	3.88 ± 0.09	8.5	0.2	6.0
2	C01	3.88 ± 0.09	110.0	2.8	2.0
3	C11	0.56 ± 0.08	0.0	0.0	12.0
4	C29	0.99 ± 0.07	3.5	0.3	5.5
5	C29	0.99 ± 0.07	18.0	1.8	3.5
6	C62	0.85 ± 0.05	3.5	0.4	5.0
7	C62	0.85 ± 0.05	24.0	2.8	2.5
8	C111	0.46 ± 0.05	0.0	0.0	7.0
Locations along the boundaries between pediment domes					
9	C41	0.27 ± 0.11	6.0	2.2	6.5
9	C40	0.67 ± 0.13	18.0	2.7	6.5
9	C40	0.67 ± 0.13	12.0(C41)[b]	3.0[b]	6.5
10	C59	3.64 ± 0.16	35.0	1.0	7.5
11	C59	3.64 ± 0.16	80.0	2.5	5.0
Locations within or adjacent to valleys cut by concentrated fluvial action					
12	C12	0.58 ± 0.16	7.0	1.2	
13	C25	0.39 ± 0.08	3.5	0.9	
14	C28	0.33 ± 0.03	4.0	1.2	
15	C51	0.27 ± 0.05	6.5	2.5	
15	C52	0.64 ± 0.05	17.0	2.6	
15	C52	0.64 ± 0.05	11.0(C51)[c]	3.0[c]	
16	C55	0.70 ± 0.06	27.0	3.8	

[a] Radiometric ages from Turrin et al. (1984); ± values are 2 sigma error ranges.
[b] Difference in height between the base of flow z_3 (sample C40) and the base of flow z_1 (sample C41); dissection rate for the period 0.67 to 0.27 m.y.B.P.
[c] Difference in height between the base of flow s_2 (sample C52) and the base of flow t_2 (sample 51); dissection rate for the period 0.64 to 0.27 m.y.B.P.

The average downwasting rates on the upper flanks of the Cima pediment domes and the average erosion rates in areas of concentrated fluvial erosion are similar to: (1) the 1.1 to 2.4 $cm/10^3yr$ erosion rates averaged over the last 10.8 m.y. for the east flank of the White Mountains of eastern California (Marchand, 1971), (2) the 2.5 $cm/10^3yr$ regional dissecton rate averaged over the last seven m. y. for the west margin of the Colorado Plateau (Hamblin et al., 1981), and (3) the 1.8 to 3.5 $cm/10^3yr$ erosion rates averaged over the last six m.y. for crestal and upper flank areas of the Reveille Range of south-central Nevada (Dohrenwend et al., 1985). However, the Cima rates are significantly less (two to four times) than average dissection rates for more tectonically active areas such as the Grand Wash-Hurricane fault area of the western Colorado Plateau (Hamblin et al., 1981) or the Mount Taylor area along the western margin of the Rio Grande Rift (Grimm,1982); and they are significantly greater (two to ten times) than average rates in tectonically quiescent areas such as the east flank of the Great Dividing Range in New South Wales, Australia (Young, 1983).

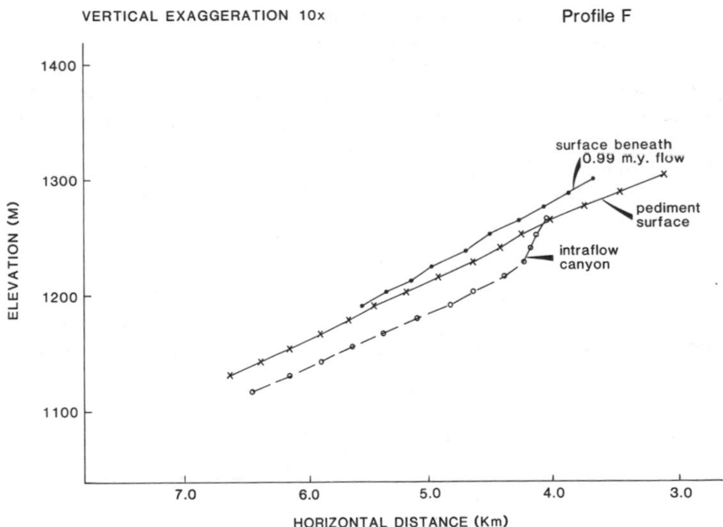

Fig. 7. Longitudinal profiles comparing early Pleistocene and modern pediment surfaces with intraflow canyon surfaces. Horizontal distance is measured from the dome summit. Profile F, Figure 4.

PEDIMENT DOME EVOLUTION

Relations between the Cima lava flows, the pediment surfaces underlying these flows, and the rock units truncated by these pediment surfaces indicate that pediment domes have been a dominant landscape element in the Cima area since at least late Miocene time. The late Tertiary lava flows of the Cima field cap extensive pediment surfaces cut across both deeply weathered Cretaceous quartz monzonite and Tertiary terrigenous clastic rocks. The contact between these two older rock units is characterized by high relief and substantial irregularity, yet the pediment surfaces pass smoothly and uninterruptedly across this contact. These relations indicate that a substantial period of time, probably more than a million years, elapsed between deposition of the Tertiary sediments and formation of the pediment surfaces, and a significant part of that time was required for pediment formation. However, absence of precise chronologic control prevents detailed reconstruction of the actual process of dome formation or of early dome evolution.

After emplacement of the late Tertiary lava flows, rates and trends of pediment dome evolution are well documented (Table 2, Fig. 8.). Degradation has followed a general pattern of crestal lowering, upper slope decline, and mid-slope stability. Downwasting has been greatest in crestal and upper flank areas and has progressively decreased downslope to mid-flank areas. Crestal and upper flank areas have downwasted at nearly constant rates ranging between 1.2 and 3.3 cm/10^3 yr. Mid-flank areas have remained in a state of approximate erosional equilibrium with little or no net downwasting or aggradation (less than 0.5 cm/10^3yr), and lower flanks have probably aggraded to some extent.

TABLE 2. Average downwasting rates versus distance from dome crests

	Downwasting Rates (cm/10^3yr)			
Distance from dome crest (km)	Profile B 3.88 m.y.	Profile E 0.99 m.y.	Profile F 0.85 m.y.	Average
2.0	2.8	-	-	2.8
2.5	2.6	2.8	-	2.7
3.0	1.5	2.1	-	1.8
3.5	1.3	1.7	1.8	1.6
4.0	1.2	1.2	1.2	1.2
5.0	0.5	0.4	0.7	0.5
6.0	0.2	-	0.3	0.2
7.0	0.0	-	0.0	0.0
12.0	-	-	0.0	0.0

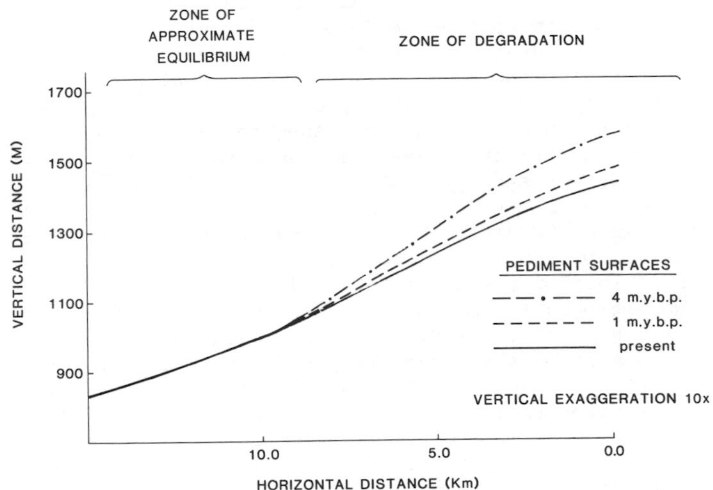

Fig. 8. Empirical model of pediment dome evolution synthesized from the longitudinal profiles shown in Figures 5 through 7 and the data presented in Tables 1 and 2. The present surface is a generalized composite of two representative profiles of the modern surface of Cimacita dome. The Pliocene and Pleistocene surfaces were reconstructed using a smoothed plot of average downwasting rates versus distance from dome summits. Crestal and upper flank areas have downwasted, mid-flank areas have remained in a state of approximate equilibrium (i.e. little or no downwasting or aggradation over time), and lower flank areas (not shown) have probably aggraded. Horizontal distance is measured from the dome summit.

A similar pattern of landscape evolution occurs in the Reveille Range of south-central Nevada (Dohrenwend et al., 1985). Since latest Miocene time, denudation rates along the crest and upper flanks of the Reveille Range have ranged between 1.8 and 3.5 $cm/10^3 yr$. Downwasting of lower flank and upper piedmont areas has ranged between 0.4 and 1.8 $cm/10^3 yr$; and most middle and lower piedmont areas have remained in a state of approximate erosional equilibrium for practically all of the last six m.y. This general pattern of maximum erosion in crestal and upper flank areas and approximate erosional equilibrium in middle and lower piedmont areas may well apply to many, if not most, upland areas flanking closed basins in arid regions of the American southwest.

CANYON FORMATION BY STREAMS DRAINING LAVA FLOW SURFACES

An instructive contrast to the general scenario of uniform downwasting on the Cima pediment domes is provided by canyon downcutting in areas capped by the Cima lava flows. Deep steep-sided and locally flat-floored valleys have been cut into several of the older lava-flow-covered pediment surfaces. These canyons, carved by surface drainage with headwaters on the lava flows, are as much as 400 m wide and 65 m deep. The floors of these canyons lie as much as 45 m below the modern pediments surrounding the lava flows, and longitudinal gradients along the lower reaches of these canyons are significantly less than the gradients of the modern pediments (Figs. 6 and 7).

An insight into the development of these canyons is provided by dated lava flows and associated remnant pediment surfaces in the southeast corner of the Cima field. In this area, a 0.99 ± 0.07 m.y. flow has been incised by a 60-m-deep canyon. This canyon is partly filled by a 0.33 ± 0.03 m.y. flow, the base of which stands only 4 to 5 m above the modern canyon floor. Lack of dissection of the younger lava flows at Cima indicates that at least 0.2 to 0.3 m.y. are required to initiate canyon cutting. In addition, soils developed in boulder-protected alluvium on canyon side slopes suggest that these slopes have been stable for probably more than 0.2 m.y. (These soils are characterized by 2-m-thick Bk horizons (7.5 YR 5/4 dry) with abundant thin clay films and stage II to III carbonate accumulations). These relations indicate that the rate of canyon downcutting varies significantly through time (Fig. 9). After extrusion of a lava flow, little downcutting occurs for at least 0.2 to 0.3 m.y. Then an intense pulse of canyon downcutting is initiated and proceeds at rates as high as 20 $cm/10^3 yr$, almost an order of magnitude faster than the average rate of pediment downwasting. After a period of probably not more than 0.2 to 0.3 m.y., downcutting slows abruptly to the average downwasting rate of the surrounding pediment and canyon side slopes stabilize beneath a veneer of boulder talus.

Canyon downcutting within areas capped by lava flows probably occurs because of differences in sediment load between streams draining the pediments and streams draining the lava flows and concentration of lava-flow-sourced drainage between caprock protected valley walls. Pediment drainage is highly charged with grus and widely dispersed in large numbers of laterally migrating braided drainageways. In contrast, streams draining the lava flows are confined and relatively starved for sediment. The lava flows, serving as caprocks for the underlying pediment surfaces, inhibit general erosion by concentrating drainage between canyon walls while yielding little sediment of their own to feed the streams that drain them. The resulting excess energy of lava-flow-sourced streams is expended in downcutting. This downcutting continues until the canyons are deep enough to equilibrate with drainage from the surrounding pediment surfaces.

SUMMARY AND CONCLUSIONS

Lava flows of the Cima volcanic field, superimposed upon several evolving pediment domes, provide precise time control for documentation of four million years of pediment downwasting. Progressively younger flows spread across and partly bury progressively younger and lower pediment surfaces. Comparison of topographic profiles of remnant and modern pediment surfaces indicates that downwasting has been the dominant mode of pediment modification over the past four million years. Downwasting in crestal and upper flank areas has ranged between 1.2 and 3.3 cm/10^3 yr and has proceeded through time at an average rate of approximately 2.0 cm/10^3 yr. Degradation of the pediment domes has been greatest in crestal areas and has progressively decreased downslope. Mid-flank areas have remained in a state of approximate equilibrium, and lower flanks have probably aggraded to some extent. This pattern of downwasting has also been documented for the Reveille Range of central Nevada and probably represents a general erosional pattern in upland areas flanking closed basins. In contrast, erosion by concentrated fluvial activity, such as canyon downcutting in areas capped by lava flows, has proceeded in pulses and at maximum rates as high as 20 cm/10^3 yr, approximately one order of magnitude faster than the average rate of pediment downwasting.

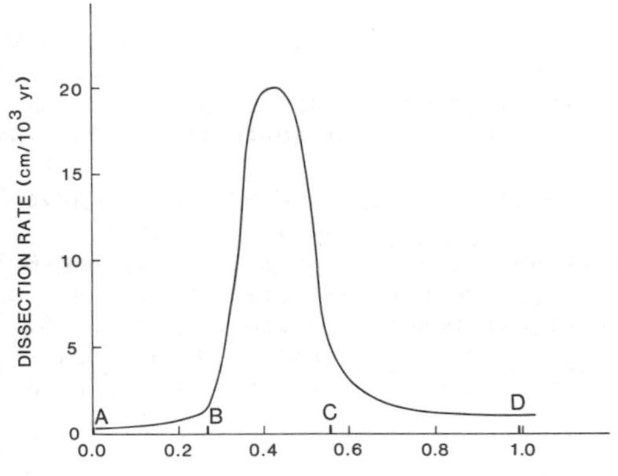

Fig. 9. Canyon development through time in areas capped by lava flows. Canyons have been formed by relatively rapid and short-lived pulses of erosion. After emplacement of the lava flows (time A), little dissection occurs for at least 0.2 to 0.3 m.y. Canyon cutting then proceeds at rates 10 to 20 times greater than the average rates of pediment downwasting (time B to time C). As equilibrium is approached (time C), dissection rates drop precipitously to the average downwasting rate of the surrounding pediment. Time D is 1985.

REFERENCES

Cooke, R. U., and Warren A. 1973. Geomorphology in deserts, Berkeley and Los Angeles, University of California Press.

Davis, W. M. 1933. Granitic domes of the Mohave Desert, California. San Diego Society Natural History Transactions, 7, 211-258.

Dohrenwend, J. C., McFadden, L. D., Turrin, B. D., and Wells, S. G. 1984. K-Ar dating of the Cima volcanic field, eastern Mojave Desert, California: late Cenozoic volcanic history and landscape evolution. Geology, 12, 163-167.

Dohrenwend, J. C., Turrin, B. D., and Diggles, M. F. 1985. Topographic distribution of dated basaltic lava flows in the Reveille Range, Nye County, Nevada: Implications of late Cenozoic erosion of upland areas in the Great Basin. Geol. Soc. Am. Abstracts with Programs, 17, 352.

Grimm, J. P. 1982. Base-level changes and incision rates for canyons draining the Mount Taylor volcanic field, New Mexico, in Grambling, J. A. and Wells, S. G., eds., Albuquerque country II, New Mexico Geological Society, 33rd Annual Field Conference Guidebook, 60-61.

Hamblin, W. K., Damon, P. E., and Bull, W. B. 1981. Estimates of vertical crustal strain rates along the western margins of the Colorado Plateau. Geology, 9, 293-298.

Hewett, D. F. 1956. Geology and mineral resources of the Ivanpah quadrangle, California and Nevada. U S. Geol. Survey Prof. Paper 275.

Marchand, D. E. 1971. Rates and modes of denudation, White Mountains, eastern California. Am. J. Sci., 270, 109-135.

Oberlander, T. M. 1974. Landscape inheritance and the pediment problem in the Mojave Desert of southern California. Am. J. Sci., 274, 849-875.

Sharp, R. P. 1957. Geomorphology of Cima Dome, Mojave Desert, California. Geol. Soc. Am. Bull., 68, 273-290.

Turrin, B. D., Dohrenwend, J. C., Wells, S. G., and McFadden, L. D. 1984. Geochronology and eruptive history of the Cima volcanic field, eastern Mojave Desert, California, in J. C. Dohrenwend, ed., Surficial geology of the eastern Mojave Desert, California: Geol. Soc. Am. 1984 Annual Meeting Guidebook, Reno, Nevada, 88-100.

Warnke, D.A. 1969. Pediment evolution in the Halloran Hills, central Mojave Desert, California. Z. Geomorph., 13, 357-389.

Young, R. W. 1983. The tempo of geomorphological change: evidence from southeastern Australia. J. Geol., 91, 221-230.

International Geomorphology 1986 Part II
Edited by V. Gardiner

TERTIARY GEOMORPHOLOGICAL EVOLUTION OF THE MARGINAL BULGE OF THE NORTH-WEST OF THE IBERIAN PENINSULA, AND LITHOSTRATIGRAPHY OF THE GRABENS OF THE NORTH-EAST OF GALICIA (SPAIN)

C. Vergnolle

CIMA - LA 366 C.N.R.S., Institut de Géographie, Université de Toulouse - Le Mirail, 5 Allées Antonio Machado, 31 058 Toulouse Cedex, France.

ABSTRACT

During the Tertiary era, the marginal bulge of the North-West of the Iberian Peninsula settled into place in relation to the structural development of the north-Spanish continental margin, its coastal relief taking on an individual form before those of the interior. The latter appeared during a period of intense tectonic activity which took the form of vertical movements. In the north-east of Galicia these dissected a vast planed surface and gave rise to three morphological units. To the west, where the metamorphic series are cut by numerous batholiths, the initial surface has been little deformed. To the east, on the contrary, where only metamorphic rocks outcrop, it has been violently raised and fragmented. Between these two units of differing mobility there opened an alignment of grabens. These are endoreic basins filled in with two superimposed deposits which correspond to two successive morphogenetic systems: the first is contemporary with the formation of the basins and the relief which borders them; the second with the end of the positioning of the volumes. This evolution reveals that the structural development of the interior side of the marginal bulge led to the individualisation of small regions, the evolution of which was autonomous.

INTRODUCTION

In the North-West of the Iberian Peninsula (Fig.1), parallel to the southern edge of the Bay of Biscay, stretches out a marginal bulge, the particular features of which are linked to the structural development of the continental margin of Northern Spain. In this vast swelling, numerous deposits, trapped in the grabens or conserved on the surface, give us an idea of the evolution from the coastline towards the interior. The basins studied (Villalba, Lugo, Sarria, Monforte) open onto the interior side of the marginal bulge and run north-south along the points of contact (Fig.2) between the first plutons (of granite or granodiorite) of Galicia and the folded metamorphic series (schists and quartzites) of Asturias (Matte, 1968). The grabens are filled in with Tertiary deposits the age of which could not be determined because of a lack of fossils. These sediments have given rise to numerous studies which come to divergent conclusions. Several authors (Schulz, 1934; Birot and Solé, 1954;

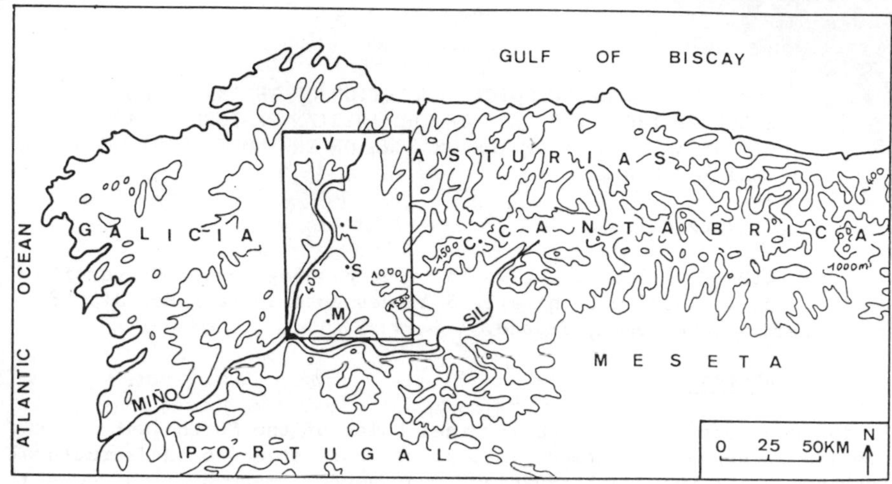

Fig.1. Situation of the region studied
V: Villalba; L: Lugo; S: Sarria; M: Monforte

Nonn, 1966; Brell and Doval, 1974; Brell, 1975; Virgili et al., 1975) consider that the filling of these different basins was not quite contemporary. On the contrary Martin Serrano (1979, 1980, 1982) believes that these grabens appeared individually at the same time and that they were filled up with two series of superimposed sediments ("red deposits" and then "arkosic deposits") separated by an hiatus. This lithostratigraphy was not taken up by Olmo (1986) for whom the Monforte basin is filled in with a single series, the facies of which vary from the periphery towards the centre. The analysis of the surface outcrops, electric soundings made by the I.G.M.E. (1979, 1980, 1981) and the borings executed by Promotora de Recursos Naturales (1977, 1978), have led to another interpretation of the sedimentary filling of the basins which, when related to the morphology of the relief around them, makes it possible to put forward an hypothesis concerning the Tertiary geomorphological evolution of the marginal bulge of the North-West of the Iberian Peninsula.

THE GEOMETRY OF THE GRABENS AND THE MORPHOLOGY OF THE RELIEFS WHICH BORDER THEM

To the west of the grabens studied (Fig.2) where batholiths cut the metamorphic series, a vast erosional surface has developed. Its average height is about 600m, but locally it is dominated by ridges which reach up to 800m. Most of these correspond to bars of quartzite. However, to the west of the Sarria and Monforte basins, certain of these are made up of small horsts bounded by faults running NNE-SSW and NE-SW. On this large surface there are some alluvial veneers, often much reworked, under which weathering profiles are sometimes conserved. They reveal that, whatever its nature, the parent rock is little weathered: its minerology has been little modified, and it is only weakened and rubified down to a few metres.

To the east of the basins, where only the Pre-Cambrian and Palaeozoic metamorphic series outcrop (Fig.2) this vast planed surface gives way to a mountainous massif subdivided into blocks separated by NE-SW fractures. The summits of these different compartments are very planate and correspond to surface fragments sometimes dominated by little quartzite ridges. The height to which these fragments have been carried increases from north to south: to the east of the Villalba basin they are to be found at 900m, while at the latitude of the Monforte basin they are at more than 1400m. Near to the Sarria graben they are at 1200m and lean against a higher zone, the massif of the Los Ancares which, considerably dissected by erosion, corresponds to the western end of the Cordillera Cantabrica (Fig.1). On these various surface fragments perched above deep, narrow valleys, no weathering profiles or alluvial coverings have survived.

The Villalba, Lugo, Sarria and Monforte grabens open out between these two morphological units (Fig.2). They are bounded by normal faults resulting from the vertical play of certain late Hercynian strike-slip faults (Arthaud and Matte, 1975). Those which border the basins to the west are long NNE-SSW fractures between the plutons and the metamorphic series and which are related to the great fractures of the North of Portugal (Vila Real, Chaves, Bragança). To the east of the basins, most of the border faults are to be found in the prolongation of the fractures which separate the eastern mountainous massifs. The geometry of these basins varies according to their surface area (about 150 square kilometres for the largest, Villalba and Monforte, but only 12 square kilometres for the Sarria basin) and according to the height of their bottoms. This decreases from north to south: the bottoms of the Villalba and Lugo basins are situated between 340 and 400m above sea level, the lowest points of the Sarria graben are at about 295m while the lower ones of the Monforte graben reach only 180m. But in these last two basins the topography of the bottoms is very broken (Fig.3) by block faulting in the Sarria basin and by the Appalachian relief which is particularly clear in the western part of the Monforte basin. Whatever their height or their topography, the bottoms of these grabens are always covered with a thin mantle of reddish, little weathered regolith.

Although there are no convincing correlation criteria, it is possible to suppose that the western surface, the bottoms of the basins and the fragments from the planed surface which crown the eastern massifs constitute the fragments of a peneplain which extended to the foot of a little chain of hills foreshadowing the Cordillera Cantabrique. This relief, which is not very contrasted, has been violently dislocated, especially in the axis of this Cordillera near Sarria and Monforte, by vertical movements which can be studied through deposits trapped in the collapsed zones.

THE DEPOSITS CONTEMPORARY WITH THE OPENING OF THE GRABENS

These constitute the essential part of the infilling and consequently their thickness varies according to the depth of each graben: about 150m in the Villalba and Lugo basins, and 400m in the Sarria basin and between 400 and 450m in the two parts of the Monforte basin. But

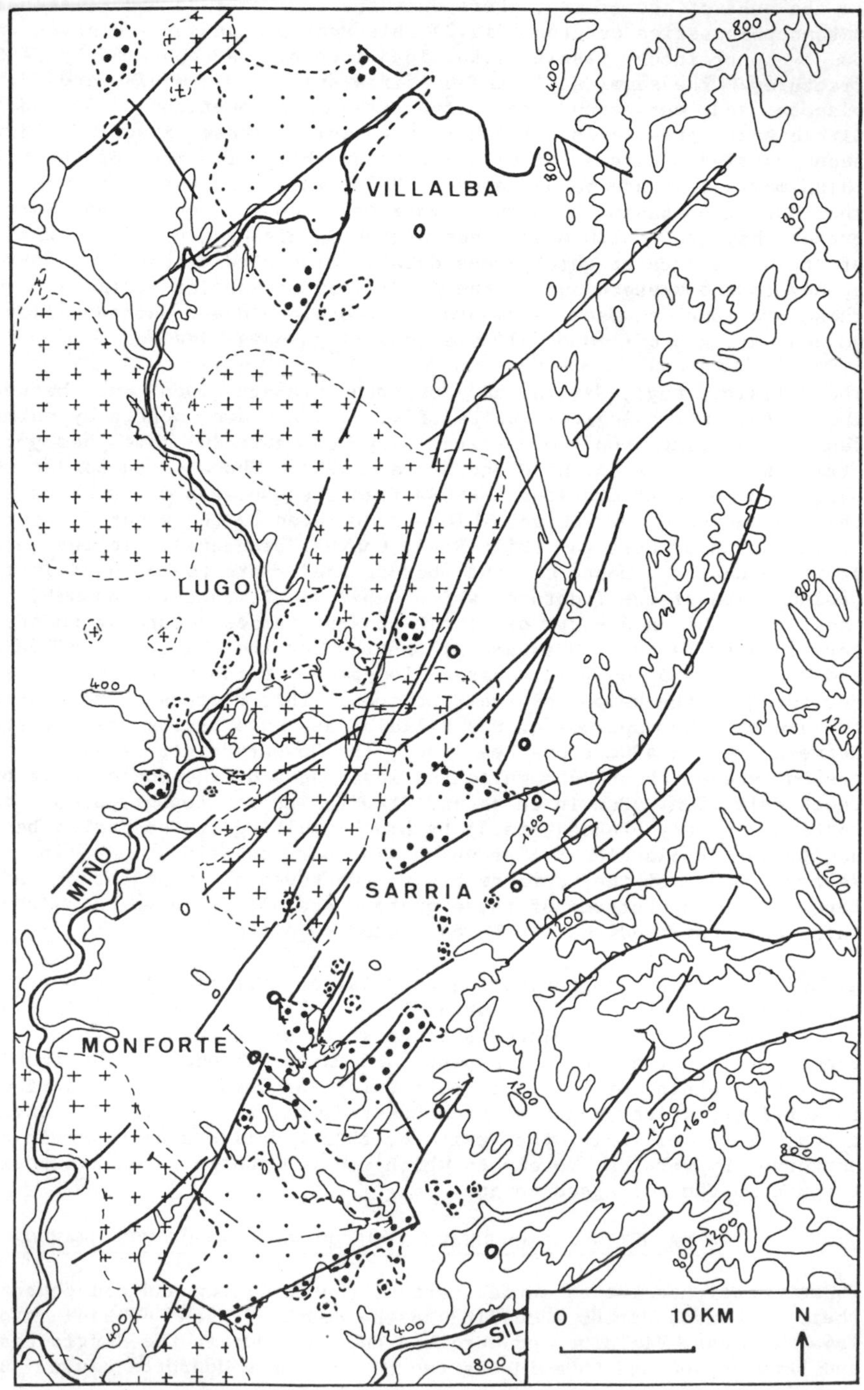
VILLALBA
LUGO
SARRIA
MONFORTE
MIÑO
SIL
0 10KM
N
400
800
1200
1600

whatever their thickness, these deposits always present the same facies.

Small alluvial fans have built up at the foot of the metamorphic rock massifs (Figs.2 and 3). They are clearly structured when the blocks from which they are derived are elevated and when the basins in which they are deposited are deep. These fans are thus thick and short. Conglomeratic channels predominate in the proximal facies. They are made up of heterometric material, badly blunted, poorly sorted and laid down in a not very abundant, gravelly matrix. They erode to form red silty-sand banks containing illite and kaolinite. Downstream, the flow and depositional conditions are rapidly transformed: sandy clay levels rich in smectite and long, thin, gravelly lenses alternate in the distal facies. Between these permanent fans, small alluvial cones considerably less well structured and connected to short, temporary valleys have been built.

Massive banks of slightly blunted, coarse feldspathic sands have been laid down along the edge of the plutonic rock massifs (Figs.2 and 3). These levels are separated by reddish clay beds containing a large proportion of smectite. No lateral variation can be observed in these deposits because the morphology (tabular) and the lithology (granite and grandiorite) of the catchment basins do not favour the establishment of permanent valleys or the supply of coarse elements.

The materials which come out of the plutonic rock massifs and those which come from the schisto-quartzitic blocks all converge towards the centre of the basins where they mix into a swampy environment. In the deeper basins (Fig.3) the appearance of these deposits has completely changed with the course of time. The base is comprised of a thin breccia bed partly cemented with carbonates. This is surmounted by a thick layer of compact clay rich in smectite laid down in alternately red and green beds. Towards the top, the concentration of carbonates increases and indurations appear. In the centre of the Sarria basin and in certain parts of the Monforte basin (close to the fault intersection) these beds are composed of dolomitic flagstones with attapulgite and sepiolite (Lucas et al., 1963).

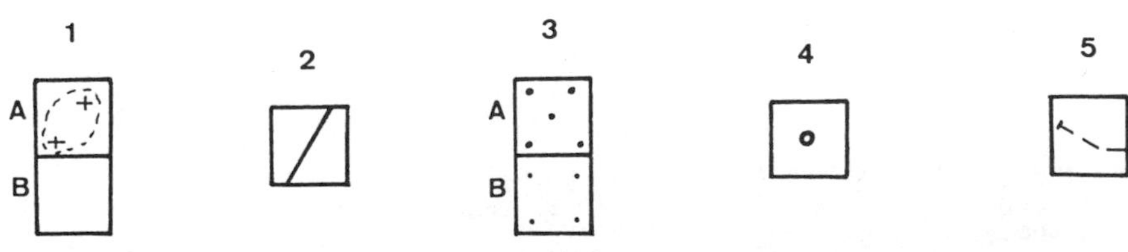

Fig.2. (FACING PAGE). Geological diagram of the basement (according to the 1/50,000 and 1/200,000 geological maps) and of the Tertiary deposits studied. 1. Basement A: Plutonic rocks B: Metamorphic rocks. 2. Faults. 3. Deposits contemporary with the opening of the basins A: Conglomerate B: Sand and clay. 4. Deposits marking the end of the filling of the basins. 5. Lines of the sections in Fig.3.

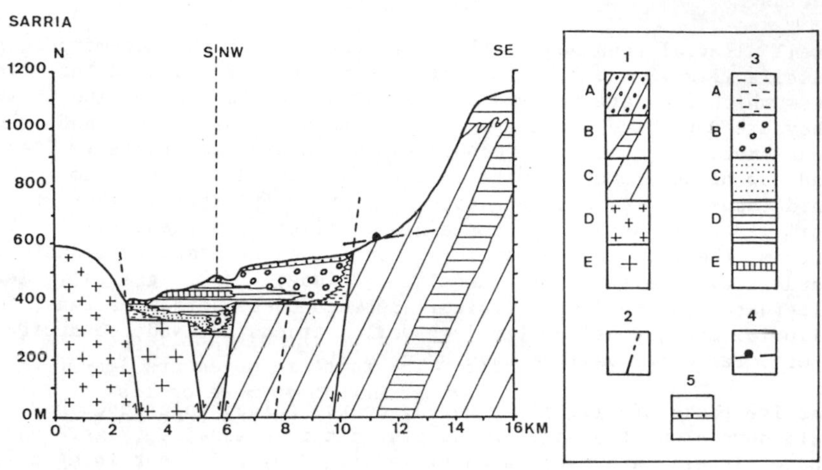
SARRIA
N
S NW
SE
1200
1000
800
600
400
200
OM
0
2
4
6
8
10
12
14
16KM
1
A
B
C
D
E
3
A
B
C
D
E
2
4
5

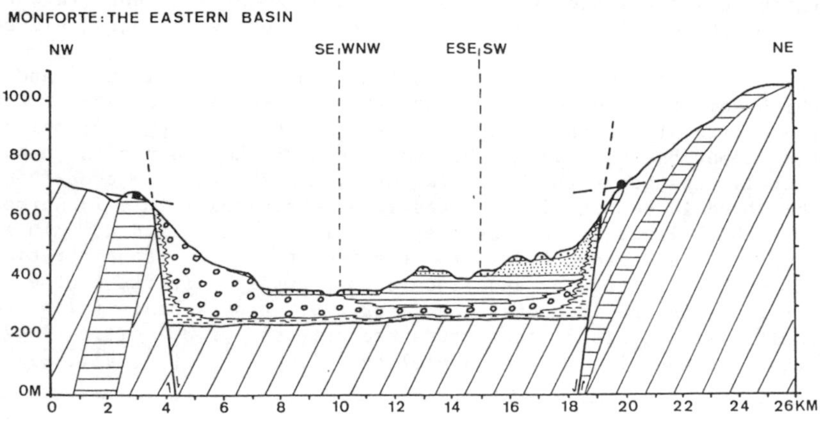
MONFORTE: THE EASTERN BASIN
NW
SE WNW
ESE SW
NE
1000
800
600
400
200
OM
0
2
4
6
8
10
12
14
16
18
20
22
24
26KM

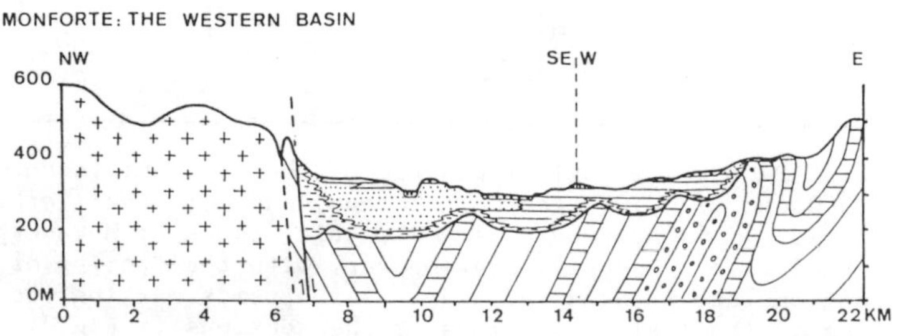
MONFORTE: THE WESTERN BASIN
NW
SE W
E
600
400
200
OM
0
2
4
6
8
10
12
14
16
18
20
22KM

Progressively, without erosional or angular discordance, this maximal level concentrated with carbonates was covered by coarser and coarser material (sand then gravels and sometimes pebbles) disposed in an illitic clay matrix. These elements come either from the plutonic or metamorphic rock massifs.

The deposits are slightly deformed. Beside the faulted escarpments, the sediments are sometimes affected by little oblique faults parallel in direction to the nearest main fracture. In the centre of the deepest basins, when the outcrops are well exposed, the beds may be seen to have been tilted several degrees to the west.

The distribution of the facies and the types of deformation observed indicate that these deposits settled gradually with the collapse of the basins and that they result from the dismantling of the border relief during relative uplift. The tectonic movements leading to the formation of these sediments must have been more or less continuous, because no discordance or sedimentary hiatus has been recognised. The dislocation of the vast surface of eastern Galicia came about therefore - in the region studied - in a single phase. The progressive appearance of this varied relief brought about a modification of the environment reflected in the progressive vertical evolution of the sedimentary filling (Vergnolle, 1985a).

THE DEPOSITS MARKING THE END OF THE INFILLING OF THE BASINS

The infilling is sealed by a conglomerate layer several metres thick (Figs.2 and 3). It is made up of pebbles and often well blunted blocks of quartz and quartzite. They are poorly sorted and disposed in an abundant matrix of reddy sands and fragments of schist, but more often than not the latter have been carried away by erosion.

The deposits are very localised. Sometimes they cover the tops of the small alluvial fans built up against the fault scarps, but much more frequently they rest on benches cut into the escarpments bordering the most uplifted blocks. These little planed surfaces sometimes penetrate several hundreds of metres towards the interior of the massifs permitting a reconstruction of the trace of the old valleys. These short, narrow, slightly encased dales come out of the highest relief (principally the eastern massifs) towards the grabens. Unfortunately, no trace of the paths of these flows after their arrival in the basins remains. They must have traversed the surface, then flowed westward before spreading out on the western planed surface

---

Fig.3. (FACING PAGE). Lithostratigraphic sections interpreting the infilling of the Sarria and Monforte grabens. 1. Basement A:Pre-Cambrian metagraywacke B: Palaeozoic quartzite C: Palaeozoic schist D: Granodiorite E: Granite. 2. Faults. 3. Deposits contemporary with the opening of the basins A: Breccia B: Conglomerate of schist and quartzite C: Feldspatic sand D: Clay E: Dolomitic flagstone. 4. Conglomerate marking the end of the infilling of the basins. 5: Quaternary terrace.

which was at that time at approximately the same level as the top of the infilling (Fig.3).

The characteristics of these flows vary slightly according to the height of the reliefs from which they originate. The higher they are, the greater the number and the size of the well blunted blocks. Equally, it is possible to relate the lowering of the eastern relief towards the north to the fact that from north to south these water courses arrived into the basins at a lower and lower level: at more than 700m in the Monforte basin and in the southern part of the Sarria basin, at about 600m in the north of this latter basin and in the Lugo basin, and finally at scarcely 500m in the Villalba basin.

These deposits and these relief forms could only have appeared as the result of an attenuation of the vertical movements. They mark therefore the end of the deepening of the basins and the relative uplifting of the border relief. Concomitant with this return to a greater tectonic stability the drainage pattern was modified. It was no longer conditioned by current deformations, but became adapted to the most elevated blocks which these deformations brought about. Thus, the most uplifted massifs gave rise to very concentrated and very competent torrents. Through their hydrodynamic characteristics as well as through their respective courses, they foreshadowed the actual Quaternary water courses which partially drain the basins and dissect the surrounding massifs.

CONCLUSION : GEOMORPHOLOGICAL EVOLUTION OF THE MARGINAL BULGE OF NORTH-WESTERN SPAIN

The infilling of the grabens of the North-East of Galicia was contemporary with the formation of the relief and the deposits comprise two superimposed sedimentary units corresponding to two successive morphogenetic systems. The latter are controlled by a single tectonic phase linked to the uprising of the Cordillera Cantabrica. Indeed, the uplifting of the Cordillera Cantabrica brought about the dislocation of a vast planed surface which stretched to its foot: the eastern part, the closest to this orogen, has been considerably raised, while towards the west the deformations came to abut up against a region stabilised by the presence of numerous plutons. Situated at the limit between these last two zones, the grabens studied took on the role of a hinge and absorbed the deformations.

These events fit into the framework of the Tertiary evolution of the whole of the marginal bulge of the North-West of the Iberian Peninsula and the margin to which it is joined. Indeed, the coastal massifs were the first to take on individual form (Vergnolle, 1985b) following upon the deformation of the continental edge of Northern Spain acting as an active margin during the paroxysmal phase previous to the Middle Eocene (Temime, 1984; CNEXO, 1984). In the interior, the relief conserved little energy and streams became established at the foot of a slightly elevated and hardly mobile massif which prefigured the Cordillera Cantabrica. The materials carried away from the vast surface of Galicia were collected by large regularly flowing streams which ran out towards the Meseta endoreic basin (Fig.2), passing by a lower zone which corresponded approximately to the path of the actual Sil valley (Hérail, 1984). Then, while the continental margin and the coastal zone remained relatively stable, the internal regions became the centre of important vertical movements (Vergnolle, 1985b), the intensity of which varied from east to west as a function of the nature of the basement. The bulge was divided into compartments based on the structures inherited from the Hercynian orogeny. Thus, this phase of intense, tectonic activity led to the fragmentation of the interior slope and to the isolation of regions, the evolution of which became autonomous: the intra-montane basins of the North-East of Galicia, those of the Sil valley and, in particular, the Bierzo graben (Hérail, 1984) and the Meseta basin (Perez Garcia, 1977).

REFERENCES

Arthaud, F., and Matte, Ph., 1975. Les décrochements tardi-hercyniens du Sud-Ouest de l'Europe. Géométrie et essai de reconstitution des conditions de déformation. Tectonophysics, 25, 139-171.

Birot, P., and Sole Sabaris, L., 1954. Recherches morphologiques dans le Nord-Ouest de la Péninsule ibérique. Mem. et Doc. du CNRS, T.IV, 9-61.

Brell, J. M., 1975. Applicación de las correlaciones al estudio del Terciario continental. Trabajos de Congresos y Reuniones, Primero y secundo ciclos de correlaciones estratigráficas, I.N.I. - ADARO, Serie 7, No.2, 123-130.

Brell, J. M., and Doval, M., 1974. Un ejemplo de correlación litoestratigráfica aplicado a las cuencas terciarias del N.W. de la Península. Est. Geol., VXXX, 631-638.

CNEXO, 1984. La marge déformée du Nort-Ouest de l'Espagne. Campagne cybère du submersible Cyana, Aôut 1982. Publications du Centre National pour l'Exploitation des Océans, Résultats des campagnes à la mer, 26.

Espinosa Godoy, J., and Rey de la Rosa, J., 1983. Caracterizaciones geológicas de las cuencas terciarias gallegas y interés económico. Tecniterrae, 52, 58-70.

Hérail, G., 1984. Géomorphologie et gîtologie de l'or détritiqu. Piémonts et bassins intramontagneux du Nord-Ouest de l'Espagne. Toulouse.

Ibergesa, 1978. Informe de los materiales arcillosos de las cuencas de Monforte, Sarria, Meira, Chantada, Páramo, Visantona (Galicia). (Unpublished).

I.G.M.E., 1979; 1980; 1981. Proyecto de exploración de lignito en la región gallega. (Unpublished).

Lucas, J., Nonn, H., and Paquet, H., 1963. Présence de niveaux à sépiolite et attapulgite dans les sédiments tertiaires de Galice (Espagne). Bul. Serv. Géol. Alsace et Lor., 26, 227-232.

Martín Serrano, A., 1979. El conocimiento del lignito y del Terciario en Galicia. Exposición y critica. Tecniterrae, 31, 1-8.

Martín Serrano, A., 1980. Nouvelles hypothèses concernant la signification géologique du lignite de Galice (Nord-Ouest de l'Espagne). Industries Minérales, Les techniques, Juin, 249-258.

Martín Serrano, A., 1982. El Terciario de Galicia. Significado y posición cronoestratigráfica de sus yacimientos de lignitos. Tecniterrae, 48, 19-41.

Matte, Ph., 1968. La structure de la virgation hercynienne de Galice (Espagne). Trav. Lab. Géol. Univ. Grenoble, 44, 153-281.

Nonn, H., 1966. Les régions côtières de Galice (Espagne). Etude géomorphologique. Strasbourg.

Olmo, A. del., 1986. Estudio sedimentológico de las cuencas post-orogénicas de Monforte de Lemos y Quiroga. Personal communication of a work included in a programme of thc I.G.M.E., (to appear).

Perez García, L. C., 1977. Los sedimentos auriferos del Noroeste de la cuenca del Duero (Provincia de León, Espana). Thesis Univ. Oviedo. (Unpublished).

Promotora de Recursos Naturales, 1977. Informes de exploración de los permisos de Monforte de Lemos, Páramo, Puertomarín, Sarria, Meira, Villalba-Lugo. (Unpublished).

Promotora de Recursos Naturales, 1978. Campaña de testificación de los sondeos realizados en las cuencas terciarias de Monforte de Lemos, Sarria, Páramo, Meira, Villalba (Lugo). (Unpublished).

Schulz, G., 1834. Descripción geognóstica del Reino de Galicia, Madrid.

Temime, D., 1984. Contribution à l'étude géologique de la marge au Nord-Ouest de l'Espagne. Thesis Univ. Paris VI. (Unpublished).

Vergnolle, C., 1985. Géométrie du remplissage sédimentaire des bassins de Sarria et de Monforte (Galice, Espagne) et évolution géomorphologique régionale. Mélanges de la Casa de Velázquez, XXI (to appear).

Vergnolle, C., 1985. Lithostratigraphie des dépôts tertiaires du Nord-Est de la Galice (Espagne) et évolution géomorphologique régionale. Estudios geológicos, (to appear).

Virgili Rodón, C., and Brell Parlade, J. M., 1975. Algunas características de la sedimentación durante el Terciario en Galicia. Bol. Real Soc. Esp. Hist. Nat., centenary volume, 515-523.

KARST GEOMORPHOLOGY

International Geomorphology 1986 Part II
Edited by V. Gardiner

KARST GEOMORPHOLOGY

V. Gardiner

Department of Geography, University of Leicester, Leicester, UK

The karst symposium at the First International Conference on Geomorphology was attended by delegates from most parts of the world, including a particularly large delegation of distinguished Chinese scientists. The papers included here, which have been edited by Dr. M. Sweeting and myself, include a substantial contribution from Chinese work on karst. The first paper sets these into a context by outlining the major mechanisms operating in the karst regions of China, and proposing a typology of Chinese karst. Papers by He Caihua and Yuan Daoxian go on to describe karst landscapes in two regions, stressing how the interplay between lithology and morpho-climatic evolution produces a variety of karst landform assemblages. At a more detailed scale the evolutions of individual cave systems are discussed by Liu Zechun and Lin Junshu et al. As the final Chinese contribution Lu Yaoru's second paper demonstrates that karst studies are not only academic exercises, but have great practical value in considering karst waters for water supply, electricity generation and as a source of minerals.

The paper by Marker demonstrates that there is much potential for work on karst regions of Southern Africa, despite the progress which has already been made. Russell shows how this progress has allowed the formulation of a simple model for karst development. Sjoberg's contribution is very different, in that it describes an invaluable inventory of caves in part of Sweden, these arising by a variety of processes. Finally, Castellani and Dragoni report on detailed observations of some of the processes operating in a desert limestone area.

The papers, although only a small sample from those read at Manchester, give a fair impression of the symposium as a whole. Above all they emphasize the importance of karst in some landscapes, as in China, and the great range of techniques demanded for karst studies. These range from mapping to detailed process measurement, often underground in very difficult conditions. The papers and the discussions which followed them conveyed a great sense of enthusiasm, which karst geomorphologists undoubtedly need - and possess.

International Geomorphology 1986 Part II
Edited by V. Gardiner

KARST GEOMORPHOLOGICAL MECHANISMS AND TYPES IN CHINA

Lu Yaoru

Institute of Hydrogeology and Engineering Geology,
Ministry of Geology and Mineral Resources
Zhengding, Hebei, The People's Republic of China

ABSTRACT

The development of karst geomorphological landscapes in the vast karst regions of China is closely related to lithological character and is controlled by many natural conditions, in particular structural events in the Cenozoic era and the five-fold climatic changes in the Pleistocene have interacted to influence and to control karst development. Many uplift, subsidence, dissolution and deposition rates in typical karst regions have been calculated and are examined in order to discuss the karst geomorphologic mechanisms; polyphyletic evolution in karst regions may be summarized in eight models. Karts geomorphologic landscapes as well as karst types in China have been delimited, and a map in this paper simply expresses their distribution.

INTRODUCTION

Karst with its various magnificent landscapes is distributed widely in China. Bare and semi-bare carbonate bed areas (the pure kind, mostly of carbonate strata), and the intercalated kind (of carbonate layers intercalated with non-carbonate beds), together occupy about 1.2 million km in China; the interbedded kind (that is carbonate and non-carbonate layers as interbedding), and intermediated kind (that is less carbonate beds with intermediate layers of non-carbonate strata), are also distributed over 1.2 million km^2. Therefore the four kinds with bare and semi-bare carbonate beds occupy about one-fourth of the total area of China (about 9.6 million km^2). As to buried carbonate layers, their areas are even more vast. Including buried karst, the total karst area is over 70 percent of the whole territory of China. The accumulated thickness of carbonate beds in many regions reaches 2-3000 metres, and the greatest thickness is about 19000 metres.

MAJOR CONDITIONS INFLUENCING THE DEVELOPMENT OF KARST GEOMORPHOLOGICAL LANDSCAPES

The development of karst geomorphologic landscapes is related firstly to lithologic characters and composite layer types; the different microscale to macroscale karst phenomena together make up the local character of karst development. Obviously typical karst

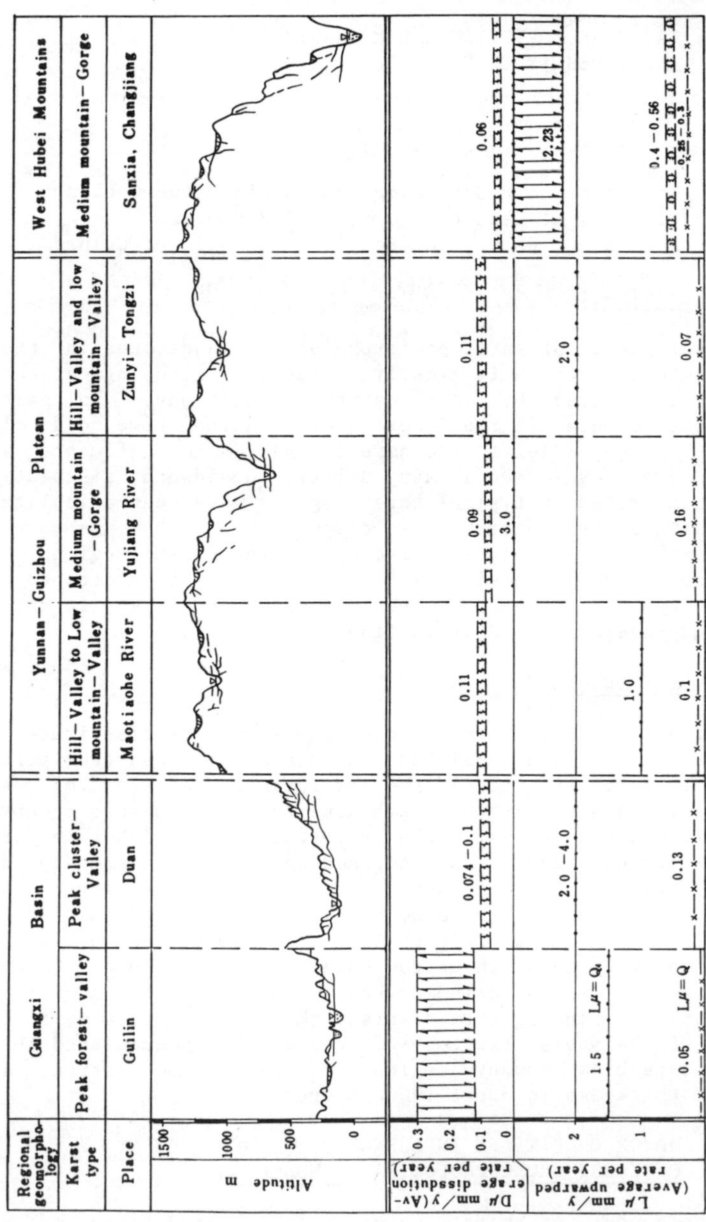

Fig.1. Analysis of dissolution rates and structural upwarping rates in some typical karst regions of South China.

landscapes are mostly developed on and in pure and intercalated kinds of carbonate strata, but the landscapes developed on and in interbedded and intermediated kinds are usually similar to those of non-carbonate beds. Except for this internal factor, the development of karst landscapes has been closely controlled by geologic structure, climate and hydrological systems as well as biogenic process and other natural conditions and factors. The influences of former structural and climatic conditions to karst development are most important [1], [2], [3].

The Yan Shan Movement, from the beginning of the Jurassic period to the end, has widely influenced the karst and its landscapes; later geologic structures in the Cenozoic era have continued the early results. Particularly in the later stages, one of the major structural events was the strongest uplift of the Himalayas and upwarping of the Qinghai-Xizang Plateau, that was caused by the collision between the Indian Plate and the Eurasian Continent. Another structural event was the formation of a series of island arcs in the eastern part of Asia and the opening of the marginal sea basins made by the subduction of the Pacific Plate. Both events (Li Chunyu et al., 1982) have closely influenced tectonic and karst development in mainland China to form new regional uplift and subsidence on the bases of palaeo-structures in east and west, south and north as well as northeast directions. The results changed the circumstances of bare and buried conditions of carbonate beds in many regions and offered the structural foundations for karst development. The Himalayas, with Mount Qomolangma at an altitude of over 8848 metres above sea level, have rapidly uplifted since the Pliocene at an average rate of about 0.98mm per year; at present the rate for Mount Qomolangma is still as much as 3.2 - 12.7mm per year [19].

Summarizing the chronologic evidence from cave deposits and/or from surface Quaternary layers, obtained by using uranium series, palaeomagnetism, fission track, thermoluminescence, amino-acid and isotope dating, as well as geologic-geomorphologic analytic methods, the upwarp rates of several typical karst regions can be listed in Table 1. Structural lifting and upwarping have always been accompanied by subsidence, and the mountains or plateaux are usually therefore adjacent to basins or subsided plains. The depositional rates of beds which were deposited in the basins can also be calculated by using the above mentioned methods. Structural movements are mostly related to igneous and hydrothermal processes and lead to hydrothermal karstification from deep to shallow zones, towards the surface, and influencing even the karst landscapes.

Different climatic conditions with unequal precipitation, rain intensity, and atmospheric and water temperatures, have closely influenced the surface run-off; genetic conditions of carbon dioxide and other acids control the intensity and features of karst processes, as well as river erosion, mechanical weathering and biogenic processes. Generally, the intensities of karstification under hotter and rainy tropical and subtropical climatic conditions are stronger than those in temperate and frigid zones. The

TABLE 1. Upwarping rates of several regions in China.

Region or belt	Karst (geomorphologic) type	Upwarping rate (mm/yr) Early Pleistocene	Middle Pleistocene	Late Pleistocene	Holocene
Zhoukoudian in Beijing	Low mountain - valley	0.02	0.02	0.03	0.5
Guilin in Guangxi	Peak forest - valley	0.038	0.05-0.07	0.05	1.5
Maotiao He River, Guizhou	Corroded hill - Dale	0.08	0.1-0.5	0.33	1.0
Wujiangdu in Guizhou	Medium - high Mountain - Gorge	0.05	0.33	0.8	3.0

TABLE 2. The sedimentary rates of travertines in several caves.

Region	Travertines in cave	Chronologic age (yrs BP)	Sampling distance (mm)	Sedimentary rate (mm/yr)	Palaeo-climatic condition
Maotiaoh of Guizhou	Stalagmite in K18 cave	126000 - 100000 (U)	132.5	0.0233	Warm - semitropic
West Hubei	Stalagmite in Huangjin cave	84000 - 74000 (U)	85	0.0085	Semitropic
Xingwen of Sichuan	Stalagmite in Tianquan cave	60000 - 53000 (U)	111.0	0.011	Rather warm - dry
Beijing	Tower-shaped travertine in Yunshui cave	330000 - 300000 (U)	1500	0.05	Warm
Guilin of Guangxi	Stalagmite in Zhenpiyan cave	14600 - 11300 (^{14}C)	4	0.001	Rather warm - dry to rather cold
Guilin of Guangxi	Plate-shaped travertine in Zhenpiyan cave	6600 - 3370 (^{14}C)	32	0.01-0.009	Semitropic rather dry

dissolution rates under damp and hot climatic conditions are larger than that of the warm to cold climatic conditions in north China by ten to over a hundred times [3], [4], [13].

Usually, the relationship between structural movement and climate will be expressed as an inverse relationship between structural upwarp rate and dissolution rate. The strong uplift led to changes of local climatic conditions. The real values of both rates in several typical karst regions are compared in Fig.2. The upwarp rates in Guilin of Guangxi in some stages of the Quaternary are lower than those of the Yunnan-Guizhou Plateau and the west Hubei mountain lands, but the dissolution rates are reversed (Corbel, 1959), [2], [3]. According to spore-pollen and mineral analyses of cavern and related surface deposits, and dates obtained by the methods previously listed, there are five stages of changing climate from damp-hot or warm into cold or glacial conditions. These changes influence the karst development and may be divided and summarized as in Fig.2.

In China the above-mentioned structural events and five times of climatic change have combined to influence and control karst development. Several depositional rates calculated from chronologic data of travertines in some typical caves are summarized in Table 2.

The results of the combined influences of uplift, structure and climate may be expressed in the following functions:

$$H = Hs - (Hc+Hm) \quad \ldots . (1)$$

$$Hi = (Hi-Hi-1) = [Lui - (Dui+Mui)] \; ti \quad \ldots . (2)$$

Hi - Increase (+) or decrease (-) in height for time interval i, mm/yr
Hs - average change in height per year, mm
Lui - average upwarp rate in time interval i, mm/yr
Dui - average dissolution rate in time interval i, mm/yr
Mui - average mechanical erosion rate in time interval i, mm/yr

$$Mu = Qs \times Ps/A \times 1000 \quad \ldots . (3)$$

Qs - total surface run-off, per annum, m^3/yr
Ps - coefficient of solid run-off in surface stream
A - catchment area to calculated position, m^2

$$Hc = Hc_1 + Hc_2 \quad \ldots . (4)$$

Hc - average decreased height of rock surface by corrosion, per year, mm
Hc - average decreased height of rock surface by corrosion of surface streams, per year, mm
Hc_1 - average decreased height of rock surface by corrosion of subsurface water before sinking into ground, per year, mm
Hm_2 - average decreased height of rock surface by mechanical erosion, per year, mm

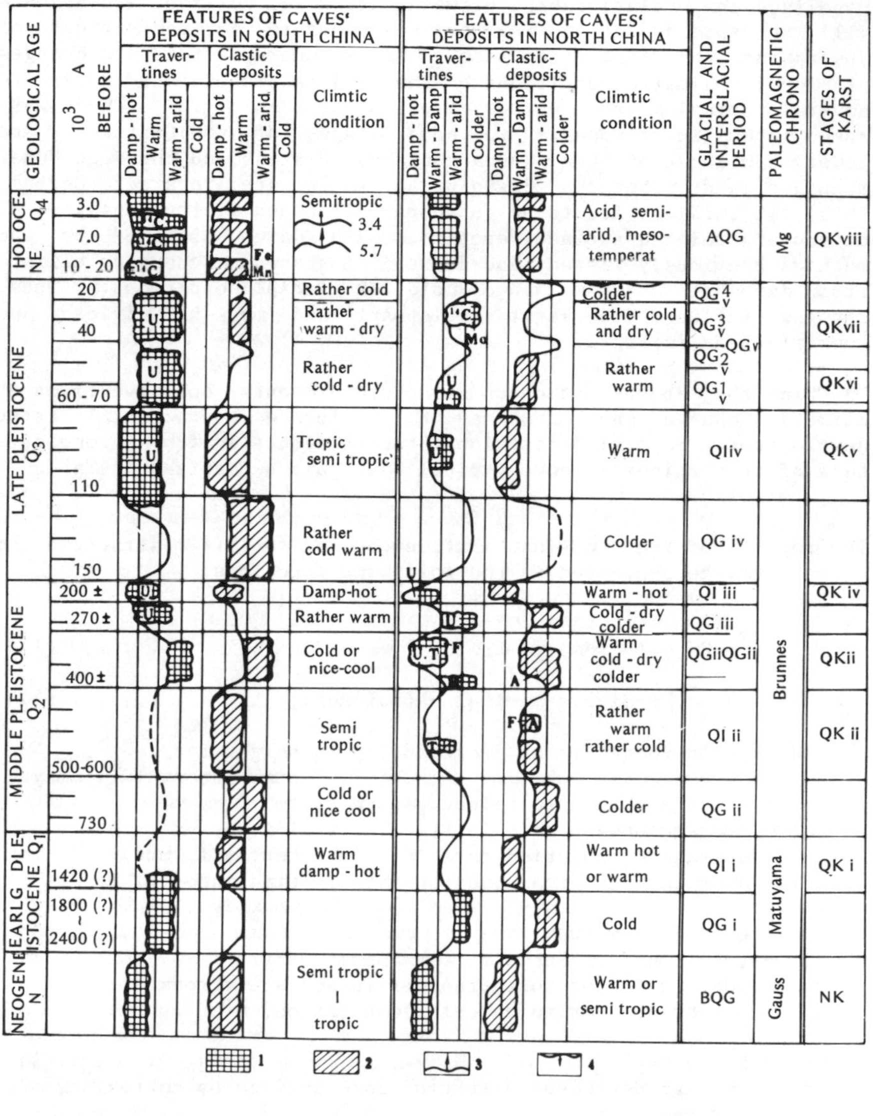

GEOLOGICAL AGE	10^3 A BEFORE	FEATURES OF CAVES' DEPOSITS IN SOUTH CHINA — Climtic condition	FEATURES OF CAVES' DEPOSITS IN NORTH CHINA — Climtic condition	GLACIAL AND INTERGLACIAL PERIOD	PALEOMAGNETIC CHRONO	STAGES OF KARST
HOLOCENE Q_4	3.0, 7.0, 10 - 20	Semitropic 3.4 5.7	Acid, semi-arid, meso-temperat	AQG	Mg	QKviii
LATE PLEISTOCENE Q_3	20	Rather cold	Colder	QG_v^4	Brunnes	
	40	Rather warm - dry	Rather cold and dry	QG_v^3 QG_v		QKvii
	60 - 70	Rather cold - dry	Rather warm	QG_v^2 QG_v^1		QKvi
	110	Tropic semi tropic	Warm	QIiv		QKv
	150	Rather cold warm	Colder	QG iv		
MIDDLE PLEISTOCENE Q_2	200 ±	Damp-hot	Warm - hot	QI iii		QK iv
	270 ±	Rather warm	Cold - dry colder	QG iii		
	400 ±	Cold or nice-cool	Warm cold - dry colder	QGiiQGii		QKii
	500-600	Semi tropic	Rather warm rather cold	QI ii		QK ii
	730	Cold or nice cool	Colder	QG ii		
EARLG PLEISTOCENE Q_1	1420 (?)	Warm damp - hot	Warm hot or warm	QI i	Matuyama	QK i
	1800 (?) ~ 2400 (?)		Cold	QG i		
NEOGENE N		Semi tropic I tropic	Warm or semi tropic	BQG	Gauss	NK

Travertines — Damp - hot, Warm, Warm - arid, Cold; Clastic deposits — Damp - hot, Warm, Warm - arid, Cold. (North China): Travertines — Damp - hot, Warm - Damp, Warm - arid, Colder; Clastic-deposits — Damp - hot, Warm - Damp, Warm - arid, Colder.

1 2 3 4

THE MECHANISMS OF KARST GEOMORPHOLOGICAL DEVELOPMENT

The complex changes of both conditions, the geological structure and climate, have found expression in the various mechanisms of karst geomorphological development in the vast area of China. It is necessary however to emphasize the inlay-combined patterns. The meaning of this is that positive and negative karst phenomena and landscapes, such as hill or peak/depression, dale or valley, faulted mountain/faulted basin, uplifted mountain/subsided basin or plain etc., are linked together in either their development or distribution. These relationships are controlled by structural events.

Fig.2. [FACING PAGE] A comprehensive analysis of palaeoclimates and karstified stages from deposits in some Quaternary caves in China.

Legend: 1. travertines in caves 2. clastic deposits in caves
3. high sea water level 4. lower sea water level.

Chronologic results [5], [6], [7], [14]: ^{14}C - radiocarbon U - uranium series
T - thermoluminescence M - palaeomagnetism F - fission track A - amino acid
Mg - Gothenburg event Mc - Biwa event mo - samples of moonmilk.
AQG - after Quaternary glacial stages
BQG - before Quaternary glacial stages;
QG_{I} to QG_{V} - first to fifth glacial or cold stages in Quaternary
Q_{I} to Q_{IV} - first to fourth interglacial stages in Quaternary
QK_{I} to QK_{VIII} - main karst development stages in Quaternary

Q_{I} - 1420-730( ) x 1000 yrs BP; Q_{II} - 600-400( ) x 1000 yrs BP
Q_{III} - 380-270( ) x 1000 yrs BP; Q_{IV} - 200-150( ) x 1000 yrs BP
Q_{V} - 110-70( ) x 1000 yrs BP; Q_{VI} - 66-50( ) x 1000 yrs BP
Q_{VII} - 40-22( ) x 1000 yrs BP; Q_{VIII} - Holocene
NK - karst development stages in Neogene

Explanations and assumptions:

A. The representative chemical deposits in caves under damp-hot climatic condition are stalactites, stalagmites and current-shaped travertines with larger dimension.

B. Typical deposits under warm-arid or cold-dry climatic conditions are plate-shaped travertines etc., deposited from sheet or floor flows.

C. Moonmilk, powdery sponge-shaped and honeycome-shaped etc., are the typical travertines deposited under colder or rather cold climatic conditions.

D. The chronologic features of mechanical deposits in caves are referenced to the results of travertines, spore-pollens and their mineral components.

E. The varieties of spore-pollens have been found either in travertines or in clastic deposits of caves under different climatic conditions, for examples the spores of *Lygodium*, *Polypodiaceae*, *Cyatheaceae* and of other tropic to semi-tropic plants [8] deposited under damp to hot climatic conditions; deciduous-broadleaf trees and coniferous-broadleaf trees' spores deposited under warm-humid climatic conditions; thermophyte, cold resisting plants, *Botrychiaceae* and most of the herbaceous and woody plants deposited their spores under warm-dry climatic conditions; the spores of mixed coniferous-broadleaf trees and of distal coniferous trees were deposited under colder or rather cold-dry climatic conditions.

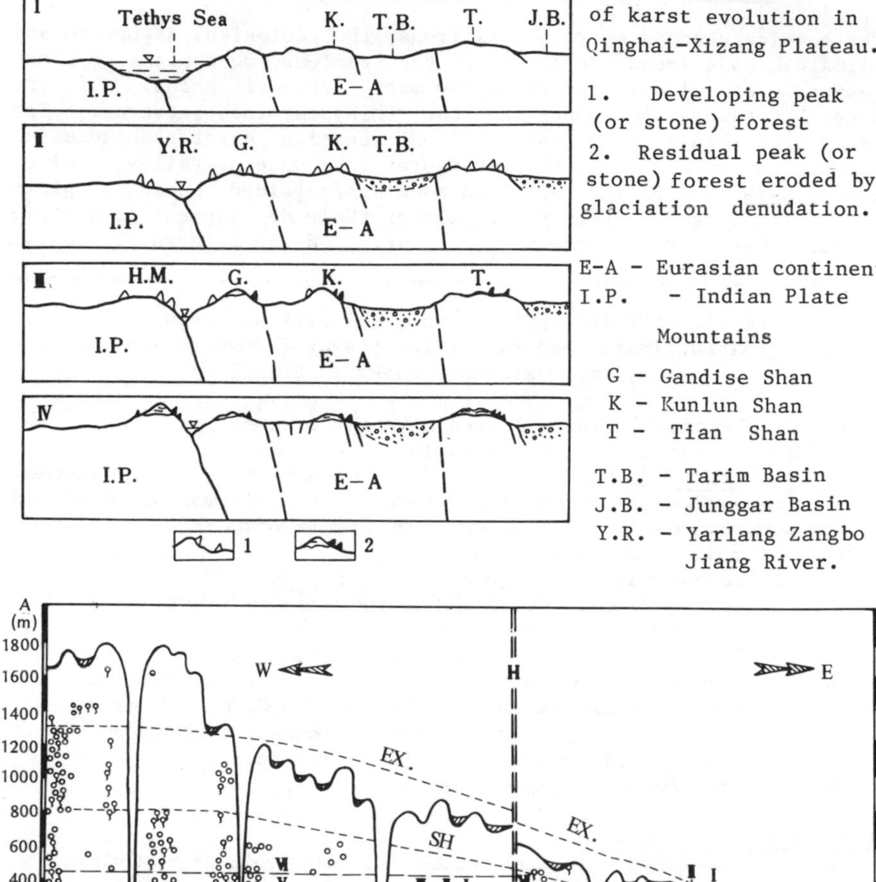

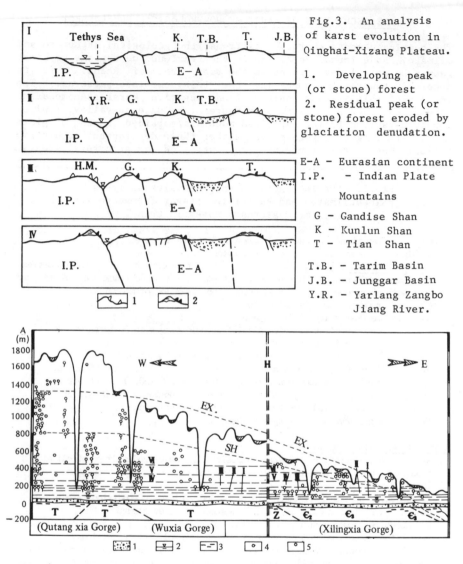

Fig.3. An analysis of karst evolution in Qinghai-Xizang Plateau.

1. Developing peak (or stone) forest
2. Residual peak (or stone) forest eroded by glaciation denudation.

E-A - Eurasian continent
I.P. - Indian Plate

Mountains

G - Gandise Shan
K - Kunlun Shan
T - Tian Shan

T.B. - Tarim Basin
J.B. - Junggar Basin
Y.R. - Yarlang Zangbo Jiang River.

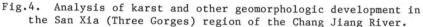

Fig.4. Analysis of karst and other geomorphologic development in the San Xia (Three Gorges) region of the Chang Jiang River.

H - Huangling Dome; Ex - The strong karstified zone of E'Xi Qi (west Hubei) stage in late Cretaceous-Eogene period; SH - Strong karstified zone of Shanyuan Qi stage in Neogene to early Quaternary period; I, II, III, IV, V, VI - River terraces developed in the Quaternary; 1 - Sand and gravel; 2 - Chang Jiang River water level; 3 - Caverns developed under river bed; 4 - Projection of caves in two banks; 5 - Projection of underground river or larger karst spring in two banks.

From Qinghai-Xizang Plateau to Xinjiang the upwarped mountains with an east-west direction alternate with larger subsided basins, where the structures were developed by the collision of plates to bring the tectonic pressure from south to north. In later stages its secondary stress led early fault blocks and massive mountains to be sheared, compressed and tensioned to form the inlay-combined patterns but mostly gave rise to the alternating structures in a west-east direction. For the east part of China, where the structures have been mostly influenced and controlled by the subduction of the Pacific Plate with the tectonic pressure in an east-west direction, the related forms of positive mountains and negative plains or basins, either in the mainland or in the marine area, are mainly expressing the alternation as an east-west direction.

Based on the above discussion, the polyphyletic evolution of karst regions of China may be summarized into many evolutionary models, but in this paper only several of the more important ones are introduced:

Strong uplift model. In Qinghai-Xizang Plateau and Xinjiang many mountains still retain the residual corroded peak forest or stone forest as well as other karst phenomena at altitudes of 4000-5000 metres above sea level; these were developed at lower altitudes during the damp-hot or warmer climatic stages [16] and eroded in later stages by strong uplift. The palaeoloxodon fossil of Pleistocene age in the north piedmont of the Tian Shan Mountains and spore-pollens and Hipparion Fauna of Pliocene age in Qinghai-Xizang Plateau [17] together illustrate the evolution; the climatic conditions during the early Pleistocene or Neogene were rather damp and hot or warm and the area was at lower altitude (Fig.3 I, II). There then resulted strong tectonic pressure in a north-south direction, from the collision between the Indian Plate and Eurasian Continent, uplifting old structural zones such as the Tanggula Shan Mountains, Bayan Har Shan Mountains and Kunlun Shan Mountains as the first stage (Fig.1 III). Secondly the Himalayas were uplifted strongly with upwarped rates from lower than 1mm per year to about 50mm per year [18], [19], [20], accompanied by upwarping of the Qinghai-Xizang Plateau, where secondary upwarping and subsidence occurred in the second or third stage (Fig.1 IV).

Continuous upwarping model. The Yunnan-Guizhou Plateau is formed by continuous upwarping, where the corrosion phenomena still occupy important situations; the upwarped rates and the dissolution rates are approximately equal, which means that the values of erosional index (Pg = Hc/Hm) are usually equal to 1. Owing to the limited uplift the gentle plateau surfaces are mostly at 1000-2000 metres above sea level, and the climatic conditions have been changed to a lesser degree and accompanied by limited uplift. The high mountains in west Yunnan and the sloping lands of the Qinghai-Xizang Plateau, are however always favourable for karst development (Fig.1).

Differential upwarping model. Because of unequal upwarping rates the development of karst geomorphologic landscapes always brought about differential evolution; for example the San Xia (Three Gorges) region of the Chang Jiang (Yangtze) River, is the result of the upwarped Huangling Dome, the uplift rates being different in its west and east limbs; the karst landscapes in the west limb are tending to change from corrosion types into erosion forms, but that in the east limb is mostly overlapping karstification forms and transitional features between mountain land and downwarped plain. The comparisons between karst process and related karst phenomena are briefly expressed in Fig.4. The climatic changes associated with the upwarping to some degree are usually linked to the karst development, but the differences of local climatic conditions between the two limbs are lesser at present (Table 3).

TABLE 3. Comparison of terraces in the San Xia (Three Gorges) region of Changjiang (Yangtze River) [Altitude of local terraces in metres]

Position	Fengjie (entrance of Quiang Xia Gorge - first gorge)	Wushan (near the entrance of Wu Xia Gorge - second gorge)	Nanjinguan (mouth of Xiling Xia Gorge - third gorge)
I	60	67	--
II	95	101	55
III	125	124	65
IV	190	192	110
V	245	270	--
VI	335	362	--

Slowly upwarping model. In slowly upwarped regions, or in repeatedly uplifted-downwarped regions with smaller upwarping rates, the evolution of karst geomorphologic landscapes is determined by major climatic conditions. For example, in the south China Basin the climates belonged to tropical or subtropical conditions for a long period, with large dissolution rates occurring. Therefore, most of the landscape evolution in areas such as Guangxi and Guangdong are overprinted by the strong development of corrosion phenomena in different stage of the Quaternary. But for the regions with temperate or cold climatic conditions in north and northeast China, or with less carbonate rocks in some zones of east and south China, other processes are the main causes of their karstic evolution.

Widespread subsidence model. In these areas the bare karst landscapes formed in early stages have subsided widely and rapidly. This results in a buried palaeo-relief of mountains and valleys covered by the Tertiary and/or Quaternary layers, their thicknesses ranging from several tens to over several thousand metres. The depositional rates of two typical kinds of this evolutionary model as indicated by the chronologic data, have been calculated to indicate the subsidence features and the changes of karst geomorphologic landscapes. Some results ([9], [12]) are summarized in Table 4. The strong uplift of the Qinghai-Xizang Plateau at high

TABLE 4. Several deposition rates in typical regions of China

Region	Kunming Basin	North China Plain	Tarim Basin
Geomorphological features	Fault Basin	Subsided Plain	Interior Basin
Maximum thickness of covering stratum	About 1000m	Over 2400m	Over 10000m
Deposition rates in Quaternary period (mm/yr)	0.11 - 0.28 0.017 (average)	0.19 - 0.207	0.02 - 1.5
Deposition rates in Tertiary period	0.11 (N)	0.02 - 0.038	0.14 - 0.098

altitude to form the first plane of geomorphologic landscapes in China was closely accompanied by rapid subsidence to form the larger Tarim Basin, where the climatic conditions changed from warm and damp into interior aridity. The Taihang Shan Mountains and Shangxi Plateau at medium-high altitude, as a part of the second plane, are joined with the formation of the North China Plain in middle depositional rates, where the climatic conditions have obviously changed in several stages from warm or hot to cold, and from damp to semi-arid. The profile (Fig.5) reflects the relationship between the upwarped Taihang Shan Mountains as fault blocks and the subsided North China Plain where the palaeo-karstified landscapes of a deeply buried zone are shown in Fig.6, as drawn by electronic computer.

Restricted subsidence model. As a result of secondary tectonic pressure in vast upwarping plateaux there are narrow subsidences in several belts, closely related to faults. Such evolution will be summarized as two important kinds: one type is the basins formed in

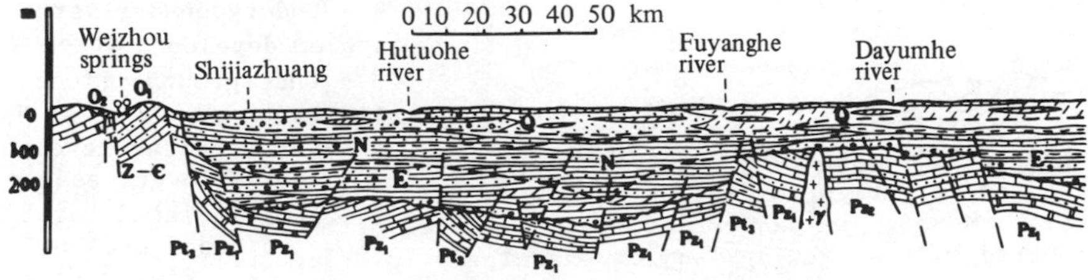

Fig.5. A profile from Taihang Shan Mountain Lands with bare karst to the North China Plain with buried palaeo-karst.

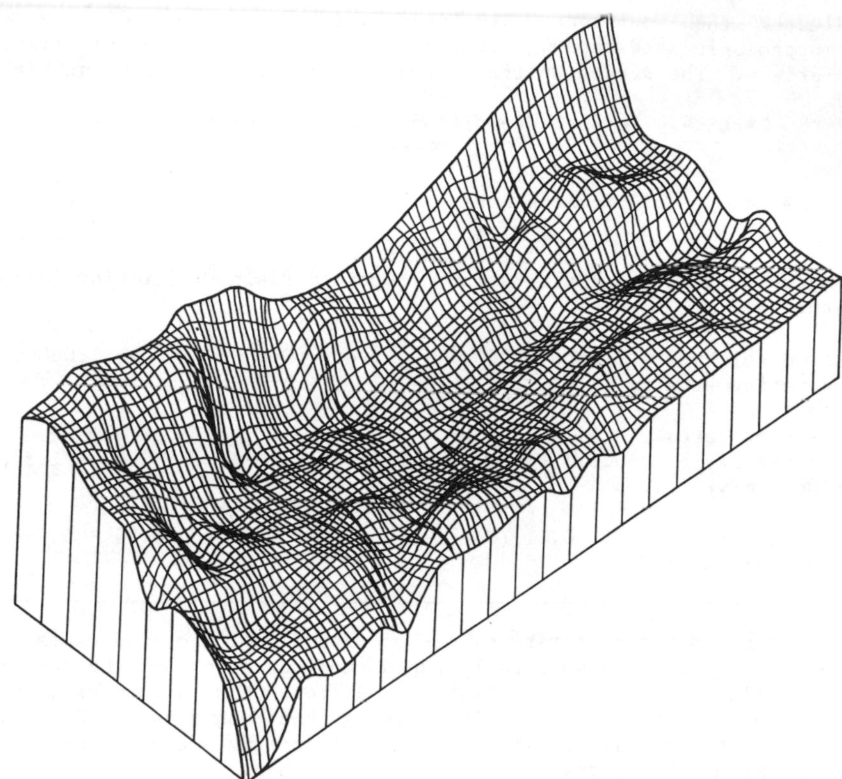

Fig.6. Buried palaeo-karstified landscapes in a zone of the North China Plain, as drawn by electronic computer (from Lin Jinxuan).

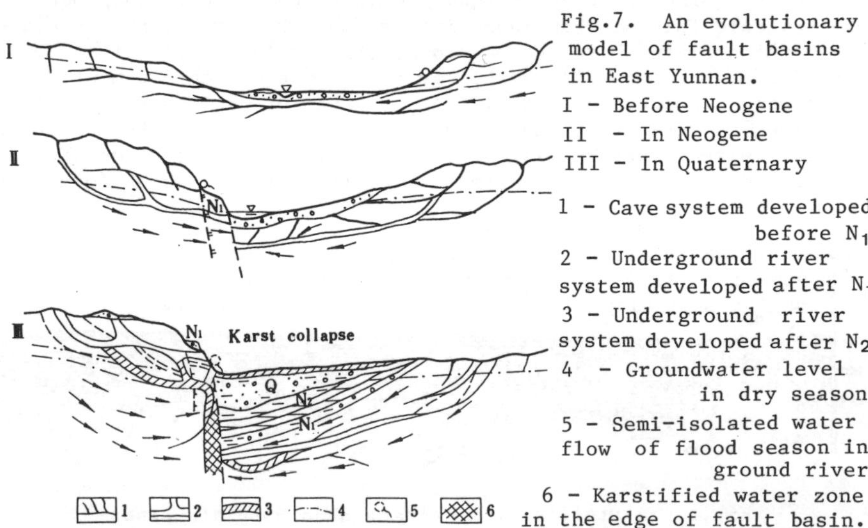

Fig.7. An evolutionary model of fault basins in East Yunnan.
I - Before Neogene
II - In Neogene
III - In Quaternary
1 - Cave system developed before N_1
2 - Underground river system developed after N_1
3 - Underground river system developed after N_2
4 - Groundwater level in dry season
5 - Semi-isolated water flow of flood season in ground river
6 - Karstified water zone in the edge of fault basin.

Qinghai-Xizang Plateau, where the climates changed from damp-hot into dry-cold interior conditions as the uplift of the plateau occurred, to form the salt lakes commonly found there. Another type is the fault basins in Yunnan-Guizhou Plateau, where in several belts the surface and subsurface karst streams are still collecting as fresh water lakes. For instance, the Kunming Basin is formed by secondary geopressure with low thickness of overlying beds and a smaller area of about 825 km^2, including Dian Chi Lake of about 340 km^2, compared with the basin and plain belonging to the above model. The climates of these basins changed as the uplift of the plateau made differences between basins and their surrounding mountain lands. These basins formed by such evolution with several steps in altitude are similar to those formed in the Dinaric regions of Yugoslavia (Herak and Stringfield, 1972; Milanovic, 1979). The evolution of such fault basins may be diagrammatically expressed as in Fig.7.

Exposing Model. The carbonate beds had been covered by non-carbonate layers in early stages: as a result of uplift and other structural movement in later stages the overlying layer has undergone erosion to reduce its thickness gradually or rapidly, then the carbonate layers are exposed. The evolution of geomorphologic landscapes is mostly turned from that of a non-carbonate system into a karstified one. This develops new karst phenomena and landscapes or re-activates palaeo-karstification.

Marine model. This model will be sub-divided into three kinds according to the changes of sea level, which are:

A. the result of upwarping of the seaboard or sea platform, decreasing the sea level. Then the carbonate beds with developed karst buried under marine conditions will be turned into the exposing evolutionary model above sea water.

B. As the sea water intrudes because of the settlement of the coast zone, the karst process may be changed from a land model into the marine condition.

C. multiple rises and falls in sea level because of climatic conditions in the glacial periods; the coast zone will develop karst and its geomorphologic evolution will be influenced by the fluctuating sea level.

KARST (GEOMORPHOLOGICAL) TYPES IN CHINA

Based on the above discussion, the karst geomorphological landscapes may be classified into seven types in the regions where the carbonate rocks are mainly distributed.

Broad corrosion karst types (K). The vast regions with bare carbonate rocks are usually favourable to the broad corrosion and percolation of rain water, to develop typical karst phenomena and landscapes strongly.

Limited corrosion karst types (KS). The thick carbonate beds intercalated with more non-carbonate layers have been influenced in their entire distribution by folds and/or faults; the impermeable

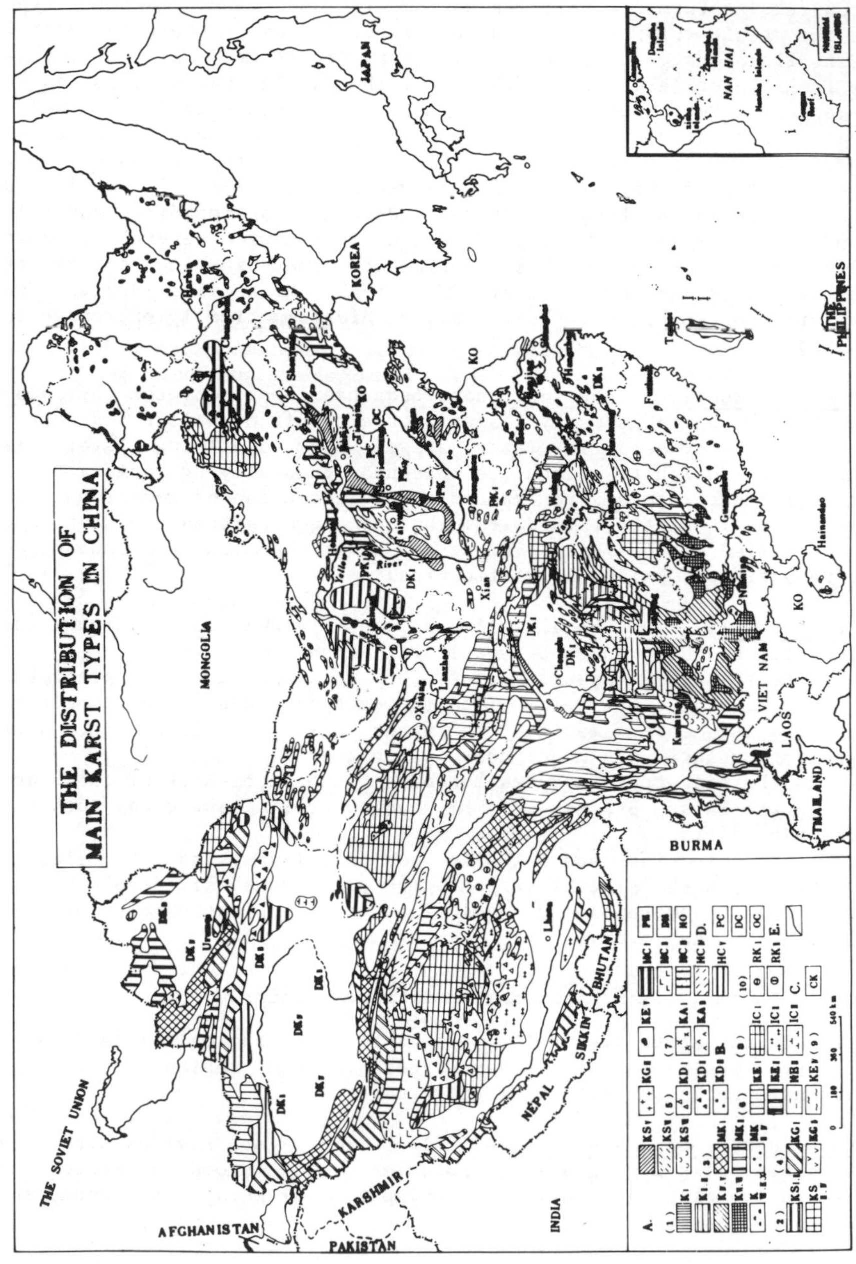
THE DISTRIBUTION OF
MAIN KARST TYPES IN CHINA
THE SOVIET UNION
MONGOLIA
KOREA
JAPAN
AFGHANISTAN
PAKISTAN
KARSHMIR
INDIA
NEPAL
SIKKIM
BHUTAN
BURMA
THAILAND
LAOS
VIET NAM
THE PHILIPPINES
NAN HAI
Yellow River
Xian
Chengdu
Kunming
Lhasa
Xining
Lanzhou
Haimandao
Taiyuan

strata and faults may obviously control the karst water movement and karst processes, allowing development of corrosion phenomena and landscapes under their limitations.

Corrosion-resorption karst types (MK). In regions with wide distribution of igneous rocks the carbonate beds have undergone dissection, resorption and metamorphism; correspondingly, the hydrothermal processes will be activated from the deep towards the

---

Fig.8. [FACING PAGE] The distribution of main karst types in China.

A. BARE CARBONATE KARST
(1) BROAD CORROSION KARST TYPES: K_{i} - stone forest-depression type; K_{ii} - hill depression type; K_{iii} - hill-dale type; K_{iv} - peak cluster-depression type; K_{v} - peak cluster-dale type; K_{vi} - peak forest-valley type; K_{vii} - peak forest-peneplain type; K_{viii} - residual peak-rolling land type; K_{ix} - isolated peak-rolling land type; K_{x} - isolated peak-peneplain type.
(2) LIMITED CORROSION KARST TYPES: KS_{i} - ridge-trough type; KS_{ii} - mountain-dale type; KS_{iii} - ridge-slope land type; KS_{iv} - fold basin type; KS_{v} - block mountain type; KS_{vi} - fault mountain type; KS_{vii} - fault basin type.
(3) CORROSION-RESORPTION KARST TYPES: MK_{i} - upwarped alteration mountain range type; MK_{ii} - upwarped alternation mountain land type; MK_{iii} - tattered fault mound type; MK_{iv} - tattered fault basin type;
(4) CORROSION-GLACIATION KARST TYPES: KG_{i} - high mountain-glacial valley type; KG_{ii} - glaciated residual corroded peak type; KG_{iii} - glaciated intermontane lake type.
(5) CORROSION-DENUDATION KARST TYPES: KD_{i} - inland mountain range type; KD_{ii} - inland mountain type; KD_{iii} - inland hill type.
(6) CORROSION-EROSION KARST TYPES: KE_{i} - high mountain-deep valley type; KE_{ii} - medium-high mountain-gorge type; KE_{iii} - low mountain-valley type; KE_{iv} - medium-high mountain-lake type; KE_{v} - massif-lake type.
(7) CORROSION-ABRASION KARST TYPES: KA_{i} - coastal karst type; KA_{ii} - reef karst type.

B. BARE CARBONATE-SULPHATES-HALIDES COMPOUND KARST
(8) INTERIOR BASIN COMPOUND KARST TYPES: IC_{i} - interior lake depression type; IC_{ii} - interior lake basin type; IC_{iii} - interior dry lake type.
(9) INTERIOR HIGHLAND COMPOUND KARST TYPES: HC_{i} - interior high land-lake depression type; HC_{ii} - interior mountain-lake depression type; HC_{iii} - interior mountain-lake basin type; HC_{iv} - interior mountain-lake-valley type; HC_{v} - interior high land-dry lake type.
(10) ORIGINAL ROCKS COMPOUND KARST TYPES: RK_{i} - mountain range-original rocks type; RK_{ii} - mountain lands-original rocks type.

C. COVERED AND BURIED CARBONATE KARST
(1) QUATERNARY COVERED KARST TYPES (CK); (2) POLYCOVERED-SETTLED TYPES (PK); (3) DEPRESSED-SEG KARST TYPES (DK); (4) MARINE BURIED KARST TYPES (KO).

D. BURIED CARBONATE-SULPHATES-HALIDES COMPOUND KARST
(5) SETTLED COMPOUND KARST TYPES (PC); (6) SEG COMPOUND KARST TYPES (SC); (7) MARINE COMPOUND KARST TYPES (OC).

---

shallow zone, giving rise to hydrothermal-mixed corrosion. Therefore in such regions, where no typical karst landscapes appear on the surface, underground karst phenomena have developed with a close relation to the igneous activity and hydrothermal processes.

Corrosion-glaciation karst types (KI). By undergoing glacial and periglacial activity for long periods the karst landscapes developed during early stages at lower altitudes with damp/hot or warm climatic conditions have obviously been changed; nowadays they are still proceeding, but with weak intensity on the base of early karstified conditions in upwarped high-cold mountains.

Corrosion-denudation karst types (KD). Karst phenomena and landscapes developed under early damp/hot or warm climatic conditions have been denuded by mechanical weathering to remain as residual phenomena and landscapes. Owing to climatic change into interior dry conditions in later stages the karst is still developing, but with weak intensity; karst fissures and small passage-ways are its main character, whereas most of the surface landscapes are similar to the denuded regions of non-carbonate rocks.

Corrosion-erosion karst types (KE). The regions where distributed carbonate beds exist have undergone not only the obvious corrosion but also stronger erosion from water flow. The surface landscapes developed are usually similar to those of the non-carbonate rock regions, with characteristics of water erosion, where the larger cave systems, karst springs and water erosion-corrosion landscapes are all closely controlled by local river water level and/or local lake water level.

Corrosion-abrasion karst types (KA). Except for corrosion from rain water, the karst developed in littoral regions and reef islands are mostly controlled by the sea level. The differential height between the minimum level in the last glacial stage of the late Pleistocene and the maximum value in the Holocene ranges from several tens to over 100 metres. Such changes have closely influenced the sea water corrosion-abrasion, karst water hydrodynamic conditions, as well as karst development in the coast zone and under the sea bottom.

As the interior basis subsides further the water flow from carbonate, halide and sulphate rocks, which are distributed in and around the basin, accumulates in the lowermost areas as an interior lake, to form compound karst. By evaporation under the arid climate the interior basins always appear as salt lakes, where a series of carbonate, borate, sulphate and halide minerals is deposited from the lake water.

CONCLUSIONS

The distribution of major karst (and geomorphologic) types is expressed in Fig.8, simplified from the coloured national map (The Distributions of Main Karst Types in China) in an important new scientific atlas (CHINA KARST), the compilation of which the author has just completed.

The different karst types have different landscapes and various hydrogeological-engineering, geological and minerogenetic conditions; these problems are omitted from this paper because of limited space but are discussed in other papers and reflected in the atlas.

In concluding this paper, it must be emphasized that the development of karst phenomena and related landscapes is controlled and influenced by many natural factors and conditions; in particular the structure and climate, the most important conditions, are usually linked together to influence and control the karst process closely.

REFERENCES

〔1〕卢耀如，1965，中国南方喀斯特发育基本规律的初步研究。地质学报，第45卷第1期。P.108－128。

〔2〕卢耀如 赵成梁 刘福灿，1966，初论喀斯特的作用过程及其类型。第一届全国水文地质工程地质学术会议论文选编。中国工业出版社。P.1－28

〔3〕卢耀如 杰显义 张上林 赵成梁 刘福灿，1973，中国岩溶(喀斯特)发育规律及其若干水文地质工程地质条件。地质学报，第一期。P.121－136。

〔4〕中国科学院地质研究所岩溶研究组，1979，中国岩溶研究。科学出版社。P.315－330。

〔5〕钱方等， 1979， 周口店猿人洞堆积物磁性地层的研究。科学通报。 24期。科学出版社。 P.192。

〔6〕郭士伦等，1980，裂变径迹法测定北京猿人时代。同上。P.1137－1139。

〔7〕裴静娴，1980，热发光年龄测定在"北京人"遗址文化层中的应用。中国第四纪研究，第五卷，1期。科学出版社。P.87－95。

〔8〕林钧枢等，1982，广西武鸣盆地岩溶发育的古地理因素分析。 地质学报，第37卷，第2期。科学出版社。 P.128－134。

〔9〕徐世浙，1982，古地磁学概论。地震出版社。P.165－175。

〔10〕《第二届岩溶学术会议论文选集》编辑组，1982，中国地质学会第二届岩溶学术会议论文选集。科学出版社。

〔11〕中国地质学会岩溶地质专业委员会编，1982，中国北方岩溶和岩溶水。地质出版社。

〔12〕罗建宁等，1983，滇池湖盆第四系沉积相古地磁和孢粉的初步研究。中国地质科学院院报，第6号。地质出版社。P.65－78。

〔13〕任美锷 刘振中，1983，岩溶学概论。 商务印书馆。P.52－59。

〔14〕赵树森 刘明林，1984，洞穴堆积物铀系测定数据报导。科学通报。16期。科学出版社。P.1004－1006。

〔15〕中国科学院西藏科学考察队，1974，珠穆朗玛峰地区科学考察报告（1966－1968）。科学出版社。

〔16〕崔之久，1981，古岩溶与青藏高原抬升。青藏高原隆起时代、幅度和形式问题。科学出版社。

〔17〕中国科学院青藏高原综合科学考察队，1980，西藏古生物，（第一分册）。科学出版社。

〔18〕地质矿产部青藏高原地质文集编委会， 1982，青藏高原地质文集（1）。地质出版社。

〔19〕中国科学院青藏高原综合科学考察队，1982，青藏高原地质构造。科学出版社。P.56。

〔20〕地质矿产部青藏高原地质文集编委会，1984，青藏高原地质文集(15)岩石、构造地质。地质出版社。

[21] Corbel, J., 1959. Erosion en terrain calcaire. Annals. de Geogr, 68, 97-120.

[22] Herak, K. and Stringfield, V.T., 1972. Karst - Important Karst Regions of The Northern Hemisphere. Amsterdam, Elsevier.

[23] Li Chunyu, Wang Quan, Liu Xueya and Tang Yaoqing, 1982. Explanatory Notes to the Tectonic Map of Asia. Cartographic Publishing House.

[24] Milanovic, P. T., 1979. Hidrogeologiza Karata l Metode Istrazivanza.

[25] Song Lin Hua, 1981. Progress of Karst Hydrology in China. Progress in Physical Geography, 5, 563-574.

International Geomorphology 1986 Part II
Edited by V. Gardiner

THE CHARACTERISTICS OF KARST GEOMORPHOLOGY IN GUIZHOU

He Caihua

Geography Department,
Guizhou Teachers' University, Guizhou, China.

ABSTRACT

Karst is well developed in Guizhou Province, and is controlled by lithology, structural characteristics and neotectonics. Major caves exist, of fairly recent formation. Waterfalls often result from the differential down-cutting rates of surface and underground streams. The Guizhou karst developed in three stages, from Cretaceous, through Tertiary to Pleistocene, and has largely humid-tropical karst landforms. The rate of karst denudation by corrosion is estimated at 33.8-77.6mm/1000 years.

INTRODUCTION

Guizhou Province, 176,400 square kilometres in area, is situated on a karst plateau in the subtropical zone of South China, in the slope zone of the transitional area from the western plateau to the eastern hilly zones of China, and forms the divide between the Changjiang River System and the Pearl River System.

Guizhou, some 1000m above sea level on average, has a rolling relief and a dense network of rivers,which dissect the land severely. Since the Quaternary it has been uplifted and tilted from West to East over large areas and the relief appears to be a step-shaped major slope, descending from the Weining Plateau more than 2,200m above sea level in the West, to the hilly surface of denudation in the centre (1,000-1,400m), and to the hills of Tongren and Yuping (300-500m) in the east of the province. From the middle of the province the land slopes down toward South and North; thus another two major slopes are formed which control the directions of flow of Wujiang, Chishui, Beipan, Qingshui and Duliu Rivers.

There are also mountain ranges in Guizhou. In the North, Dalou Mountain (1,000-1,500m) is the divide between the Wujiang and Chishui Rivers. In the East, Wuling Mountain, whose highest peak Fanjing Mountain is as high as 2,572m, is the divide between the Wujiang and Yuanjiang Rivers. In the middle, Miao Mountain Range runs across the province, at 1,200-1,800m above sea level; its highest peak, Leigong Mountain, reaches as high as 2,168m, and is the divide between the Changjiang and Pearl Rivers. In the West, Wumeng Mountain (more than 2,500m, with Jiucaiping reaching 2,900m) stretches from Yunnan Province into Guizhou Province. All the mountains above except Miao Mountain Range, which runs East West, are characterized by a structurally-controlled North-South trend.

As a result of continuous strong headward erosion toward the middle of the plateau from the North, East and South by the Wujiang, Beipan, Chishui, Qingshui, Duliu, Zhangjiang and Pearl River systems, the plateau is mountainous, and the rivers cut deeply, with rugged valley relief. The farmland is situated above the valleys. The deep groundwater is difficult to exploit, but water power resources are abundant. Those parts of the plateau near the divides which are not affected by headward erosion are quite smooth, with broad valleys and sluggish water flow. Shallow groundwater occurs there, and may be exploited.

The great thickness and large exposed areas of carbonate rocks allow the full development of karst landforms. Because of recent intermittent uplift of the earth's crust, the water table is declining, encouraging karst development. Typical and widespread are dolines, funnels, depressions, caves, underground rivers, lakes, peak clusters, peak forests, pinnacles, etc.

THE GEOLOGICAL BASIS OF KARST DEVELOPMENT

The Guizhou strata are mainly of shallow sea facies. Carbonate rocks make up 70% of the whole area of Guizhou Province. Their vertical thickness totals over 8,500m. The lower part of each sequence is formed by clayey clastic rocks or coal formations, the middle part mainly by carbonate rocks, and the upper part also by clastic and clayey rocks. In the strata of the same period there are often alternating layers of insoluble and soluble rocks, which make up 70-80% of the total thickness of the cap-strata in central and southern Guizhou, 33% or less in the North-east, and 55% in the rest of the province (Table 1, Fig.1). The CaO:MgO ratio generally tends to increase with increasing age (Table 1)

The Guizhou folded mountains were formed during the Yenshan movement and are the basic skeleton of the present landforms. Eight varied tectonic systems can be recognised, based on strong foldings, the development of fractures and clear structural features. The features and patterns of the different structural systems greatly affect the distribution and development of karst landforms. The alternation of carbonate and clastic strata gives rise to multiple water-bearing layers and karst development in belts, as a result of folding and fracturing (Fig.2).

The frequent intermittent uplifts have brought about six erosional unconformities in Guizhou. The upheaval of the central part of Guizhou caused the direct contact of the Middle and Upper Cambrian with the Lower Permian and Middle and Upper Carboniferous, giving widespread karst planes of denudation.

The depositional structure of the carbonate rocks produces a great impact upon karst forms. The thick-bedded limestone and dolomitic limestone take the shape of fenglin (peak forest), fengtsung (peak cluster) and depressions in the South of Guizhou, and a variety of karst landform types. Thin-bedded limestone and dolomitic limestone take the shape of karst hills, depressions, karren, and caves, which are small in scale.

TABLE 1. Percentage of the total thickness, and chemical composition of carbonate rocks of different periods in Guizhou.

Period	Thickness	CaO	MgO	SiO_2
Sinian (Pre-Cambrian)	87	27	20	11.3
Lower Cambrian	15	10	-	-
Middle and Upper Cambrian	81	32.3	17.3	1.6
Ordovician	35	40.1	5	2.6
Silurian	2.5	-	-	-
Devonian	65	45.2	7.9	1
Lower Carboniferous	76	43.1	9.8	3.6
Middle and Upper Carboniferous	98	47.3	6.2	1.6
Lower Permian	90	46.1	4.4	2.4
Upper Permian	30	35	2.3	4.5
Lower Triassic	86	45.8	3.2	2.6
Upper Triassic	57	39.2	12.5	12

The joints and bedding planes provide channels along which the infiltrating water runs, with corrosive and erosive abilities; in this way corrosive cracks and fissures, karren, caves, etc., are formed. The Caihua Cave on the Huanghou underground river in Dushan County has developed along bedding planes.

THE CHARACTER OF KARST DEVELOPMENT

There is a variety of karst features. On the surface there are clints, karren, funnels,avens, shafts, box-shaped valleys, natural bridges, waterfalls, depressions, poljes, trough valleys, peak forests and peak clusters. Under the surface there are karst caves, subterranean streams, subsidence and multi-tide springs, etc. They vary in shape. For example, karst depressions may be saucer, cylindrical- or trough-shaped.

Karst caves can be classified as corrosive caves and cyclic caves, and are typified by their storey-like distribution, usually of 3-4 storeys.

Guizhou peak forests, 10-100m high, are mainly cone shaped, and often accompanied by karst depressions, hence the typical peak-cluster depressions; some peak forests are also scattered on the plateau surface.

The negative landforms of Guizhou appear to be karst basins or poljes. Some develop along fracture zones and some along contacts between soluble and insoluble rocks. Those close to the watersheds (the plateau surface) are vast, those near the valleys are small.

Owing to the lithological character, structure, neotectonic movements and the regional individuality of the development process, Guizhou karst landforms have their own regional characteristics of shape, degree and regularities of karst development.

Karst undergoes a "deepward" development and is thus characterized by superposed features, as during karst development the rising crust and falling groundwater-table control karst development. This downward hydrodynamic progress results in continuous thickening of the vertical water circulation zone, represented, as landforms by incised valleys, deep enclosed rounded depressions, deep funnels, and dolines; the funnels and dolines in the major depressions lead to deep-seated rivers. Karst water often occurs at great depths.

Horizontal karst water movement is accompanied by distinct changes in underground gradient. To keep pace with the descending base level, in the belts near valley slopes the hydraulic gradient suddenly increases, forming "underground knick points". This enables the vertical circulation zones to thicken. Several kilometres away from the valleys of the main streams some tributaries become disappearing streams, and pour into the main streams with very steep gradients. At the confluences, natural bridges and waterfalls occur where the knick points exist. On the banks below the knick points are a great number of dolines and funnels, with increasing density and depth, as along the Wujiang, Maotiao, Liuchong, Nanming and Beipan Rivers. The gradients of these rivers are low in the upper reaches, but high elsewhere, forming "profiles of anti-equilibrium", which are widespread in Guizhou (Fig.3).

Karst landform types are continuously evolving. Karst's physical features possess their own universal but unique spatial regularity, which is the geomorphic expression of the hydrodynamic characteristics controlled by neotectonic movements. It shows not only the adaptation but also the "backwardness" of karst development. That is to say, from the major watersheds to the incised valleys, peak forests are gradually replaced by peak cluster depressions and peak cluster valleys. This regularity is obvious in the areas of Wujiang, Maotiao, Sancha and Peipan Rivers (Fig.4).

THE DEVELOPMENT OF KARST CAVES

Karst caves are one of the most important physical features of karst regions. In Guizhou there are many caves of diverse types and scales, with rich calcium deposition. Some caves are as long as 10km or more, with heights and widths over 80-100m. The multi-layered caves usually have two or three storeys, and some have five, connected by shafts. The bottom caves are often occupied by underground streams. For instance, the Monk Cave in Permian limestone on the left bank of Yuliang River in Wudang district is five-storeyed. Rich chemical deposition in the cave forms various dripstones. Fossils of giant pandas of Old Stone Age and Middle and Later Pleistocene are also found in the cave - the fossils of stegodon faunas. In the Erge, Juhua and Xiaobi Caves in Guiyang the fossils of deer, rhinoceros, pig, horse, bear, porcupine and tapir, etc., are found.

The Guiyang Underground Park (White Dragon Cave, 550m long and 1,090-1,100m high) runs in a zig-zag fashion. The cave floor is flat and the cross sections of the cave are roughly rectangular or flattened in shape. On the cave floor are gravel deposits and dolines. One shaft, open to the underground river, is 19.6m deep and has a flow capacity of some 25 litres per second. The calcium deposits are dazzling. Besides stalagmites, stalacto-stalagmites, stalactites, curtains, there are rimstones, stone shields, stone waterfalls and stone flowers, as well as scallops and cave wall troughs.

The Daji Cave in Zhiji County is 5,400m long. It contains an enormous cave chamber, the maximum width being 110m, and height 90m. The cave has developed in the Triassic limestone of the Yulong Mountains. It is situated in the transition area between the Mountain Basin Period erosional surface and the Wujang Period gorge. Caused by subsurface drainage, it was improved by permeating streams. Its speleothems are the most abundant of all caves in China (Fig.5).

From field survey we can draw the following conclusions:
(1) The evolution of the caves is in accord with the concentrated progress of subsurface runoff of underground drainage, i.e. water flows develop from fissure to vein, to tubular, to cave flow. Karst caves develop from crack to solution fissure to solution tube to large-scale karst cave.
(2) The geological activity of underground water consists of corrosion, mechanical erosion, accumulation and collapse, of which corrosion is effective all the time. Many caves undergo the process of desiccation due to alternate filling and partial emptying by water.
(3) Zhao Shu-sen (1984) points out that Guizhou cave calcium deposits are fairly new, being less than 300,000 years old. Between 70,000 and 120,000 years B.P., cave deposition was well developed, while between 150,000 and 160,000 years B.P. deposition was at a standstill.
(4) The major regularities of karst cave development are:
 (a) Controlled by lithological character: karst caves mainly develop in pure and dolomitic limestones. For example, caves in the North of Guizhou develop mostly in Sinian strata. Those in the middle part develop mostly in

Permo-Triassic strata, those in the South develop mostly in Carboniferous and Permo-Triassic strata, and those in the West develop mostly in Carboniferous and Permo-Triassic strata.

(b) Related to structure: In the folded areas the cave generally runs parallel with synclinal axes. Both major and branch caves, influenced by break and crack structures, have reticulated shapes. At the anticlinal axes caves develop along the bedding planes, faults and joints. At the flanks they develop along the bedding planes and fault belts, or the contact zones between permeable and impermeable beds.

(c) Affected by physical features: On the plateau surfaces, horizontal caves develop, while in the valley regions storeyed karst caves develop.

TABLE 2. Uranium age determination in Guizhou caves.

Place	Sample	Radioactive Age ($x10^3$ year)
Xiuwei K 18 Cave	Stalagmite	91 ± 5
Guiyang Underground Park	Stalagmite	50 ± 3
Zhenning Huoniu Cave	Calcium plate	75 ± 4
Anshun Zhenjia Cave	Stalagmite	109 ± 9
Dushan Shenxian Cave	Stalagmite	$230 \pm ^{60}_{40}$
Dushan Chuan Cave	Stalagmite	$220 \pm ^{20}_{15}$
Maotiao River K 18 Cave	Travertine	130 ± 7

DEVELOPMENT OF WATERFALLS

There are many waterfalls and steep steps on the rivers' longitudinal profiles, being related to the rivers' development stages, and having diverse complex causes. Most occur in relation to the subsurface rivers. They often develop where there are definite structures, active fractures and physiographic boundaries. Waterfalls on steep cliffs at the outlets of underground rivers and giant cave springs have been found in 88 places in Guizhou. These are caused by the down-cutting rate of the surface rivers being greater than that of karst downward development; this may be due to insoluble strata lying under soluble strata, which prevents underground rivers from developing downward. Structural waterfalls are conspicuous on the banks of incised valleys of main rivers. For example, hanging waterfalls such as the synclinal Dayanmen Waterfall at Qingping in Wuchuang County and Naoshui Cave Waterfall of Dejiang, are more than 200m high, and the waterfall of Yangpi Cave on the left bank of the Maotiao River in Qingzhen is 20m high, (Fig.6).

In Guizhou karst regions, due to clean water and low sand load, the wash-out pits of the waterfalls are not easily filled. Several wash-out pits along a river are evidence of the distance and rate of waterfall recession. The Orange Waterfall in Zhenning has four pits, the total recession being 400m. It is at the junction of the Guizhou "Shanpen" (Mountain Basin) and "Xiagu" (Gorge) stages of geomorphic development, where typical knick points of the river exist (Fig.6).

In Guizhou karst regions remarkable tufa deposits exist on many waterfalls and the surface of rocks and gravel on rapids. Tufa deposition also occurs in the water, often in the high velocity flows of hydraulic structures and machinery. For example, near Leigong Beach of Nanpan River, there is a water turbine pump, which ceased to work because of tufa sediment. Tufa sediment commonly accumulates on the overflow surfaces of dams.

THE THREE STAGES OF GUIZHOU KARST DEVELOPMENT

During the Yenshan Movement (Cretaceous) of strong folding and fracturing, the framework of the landforms of Guizhou was built. Since then they have undergone three different chronological stages, under the interaction of neotectonic movements and active exogenic forces.

Dalou Mountain Stage. Since the end of the Mesozoic the Yenshan folded mountains have been denuded and worn away. Conglomerate of the Older Tertiary was accumulated in some sunken basins. In such places as Qingzhen, Wudan, Huishui, and Jiuzhou, the plane of denudation can now, because of repeated destruction and deformation in later stages, be identified only from peak surfaces, or from grave-shaped karst hills on the highest plateau surface.

Mountain Basin Stage. The time from Miocene to early Pleistocene is represented by the plateau surface composed of large karst basins and peak forests, with wide, shallow valleys, and low, attenuated mountain ranges, often accompanied by thick lateritic crust. Karst pools, lakes and horizontal caves occur and water is at shallow depth. Typical of such areas are Weining, Qingzhen, Pingba, Anlong, Xingyi and Meitang.

Wujiang Stage. Since the middle of the Pleistocene crust has been uplifted drastically. As a result of the erosion and rejuvenation of the main rivers and strong down-cutting, the landsurface of the Mountain Basin Stage has been re-dissected and destroyed, forming deeply-cut gorges. Intermittent uplifts have lead to several minor developmental stages, represented by river terraces and associated caves. For example, Wujiang, Nanpan, Peipan and Qingshui Rivers all have 4-5 terraces as well as 3-4 layered caves. Wujiang River has 4 terraces: the second is 30-35 m high, the third about 60m, the fourth 80-90m, and the fifth about 190m, accompanied by caves (Fig.7).

THE TENDENCY OF KARST DEVELOPMENT

Neotectonic movements bring about intense crustal uplift and as a result the main drainage paths are drastically cut down; the tributaries and karst horizontal caves and underground rivers, due to the small amount of water, have a slower rate of downcutting than that of the main rivers. Hence there are many discordant phenomena and many hanging springs and dry valleys. At the same time, because of different magnitude of each confluent, the degree of discordance differs. So karst developments in the same period are different in height.

In the upper course of Wujiang River and other main drainage paths and the middle-upper course of tributaries, involving no Wujiang headward erosion, dissection is slight. Karst landforms remain intact, and development is now mainly horizontal.

In Wujiang main streams and below the knick points of tributaries, because of the lowering of base level, the rivers cut down deeply and river beds, being deep troughs and rapids, are still in the course of downcutting. Karst is dominated by vertical downward development.

Guizhou karst landform types show basic regularities imposed by the rising landsurface, and reflect climatic changes since the Tertiary. In general, peak forest landforms are dominant in the South, while Qiufeng (hill peak) landforms are very common in the North. The relict peak forests originated under the wet tropical climate.

During the Quaternary period, because climate changed into a subtropical one and crustal uplift caused rivers to dissect greatly, fluvial erosion played an important part. The rate of fluvial erosion was 310-980mm/1,000 years, whereas the rate of corrosion was 33.8-77.6mm/1,000 years.

CONCLUSIONS

Thick carbonate rocks are widely distributed in Guizhou with strong folding and fracturing. Karst water circulation is proceeding actively. Karst landforms, which develop perfectly, have a great variety of types and shapes, and caves have developed under ideal conditions.

Guizhou karst development is greatly controlled by structure, lithological character and neotectonic movements, which result in banded spatial distributions, and downward development.

Guizhou karst is typified by humid tropical karst landforms. Since the Pleistocene, flowing water action has had an increased impact upon the development of karst landforms.

Since the Quaternary, neotectonic movements have become stronger and the crust has uplifted severely. As a result, gorges of Wujiang, Beipan, Maotiao and Caodu Rivers and deep karst landforms have been formed. During this time there have been at least four temporary still-stands, which brought about four steps of terraces and caused horizontal karst caves and subsurface rivers to develop. The rate of karst denudation is preliminarily estimated at 33.8-77.6mm/1,000 yr.

REFERENCES AND FURTHER INFORMATION

He Caihua. Guizhou Karst Landforms. A Treatise of National Karst Conference.

Kao Ping Xiong shuyi He Caihua, 1964. The Karst Geomorphology Study of Guizhou Plateau.

Mo Zhongdo, 1982. The characters of modern river fall Geomorphology development in Guizhou Karst Regions.

Qin Qiwan. The accumulation of Guizhou Caves and the regularities of the landform development. A Treatise of National Karst Conference.

Yang Mingde, 1982. The Geomorphological regularities of karst water occurrences in Guizhou Plateau. Carsologica Sinica, 1, (2).

Zhang Dian, 1983. Conditions of groundwater movement for calcium carbonate deposition in cave system. Carsologica Sinica, 2, (1).

Zhang Yingjun and Mo Zhongda, 1982. The Origin and evolution of Orange Fall. Acta Geographica Sinica. Sept.

Zhao Shu-sen and Liu Minglin, 1984. Uranium System dating data of cave accumulation. Science Bulletin, 016.

Zhou Zheng, 1979. The Karst rock series characters in Guizhou. A Treatise of National Karst Conference.

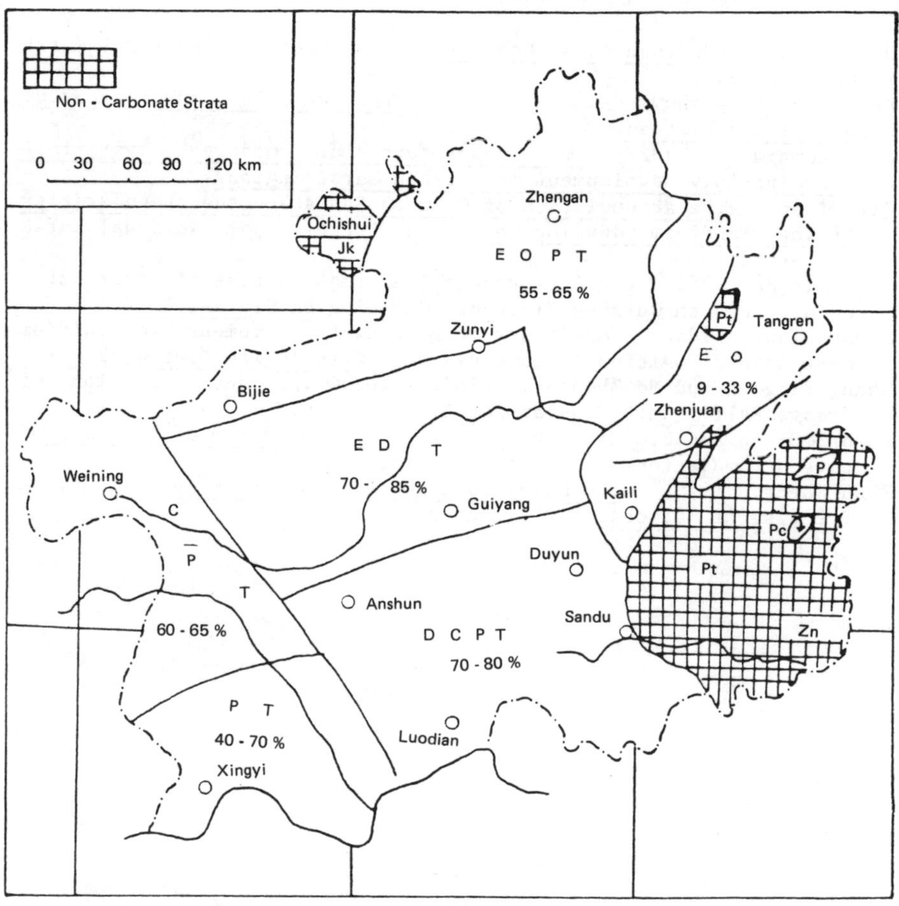

Fig.1. The percentage of the thickness of carbonate rocks in cap strata in regions of Guizhou.

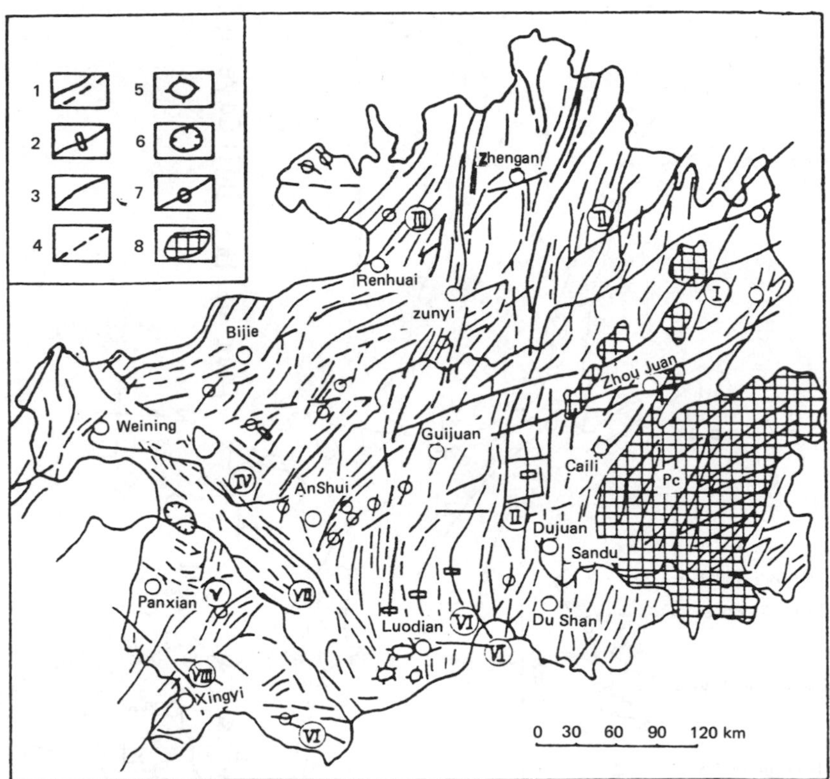

Fig.2. Structural systems of Guizhou 1. fault and inferred fault. 2. box-shaped anticline. 3. anticline. 4. syncline. 5. dome structure. 6. structural basin. 7. branch axis anticline. 8. basement. I. Cathaysian shear structure zone. II. Meridinal structure zone. III. Loushan arc united structure zone. IV. ⊥ -shaped structure. V. United structure. VI. Latitudinal structure zone. VII. N-W structure zone. VIII. Turbine-like structure.

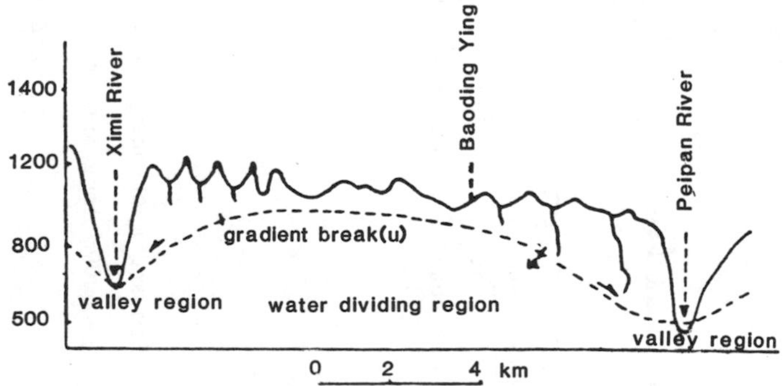

Fig.3. Profile of anti-equilibrium of karst ground water (underground knick point).

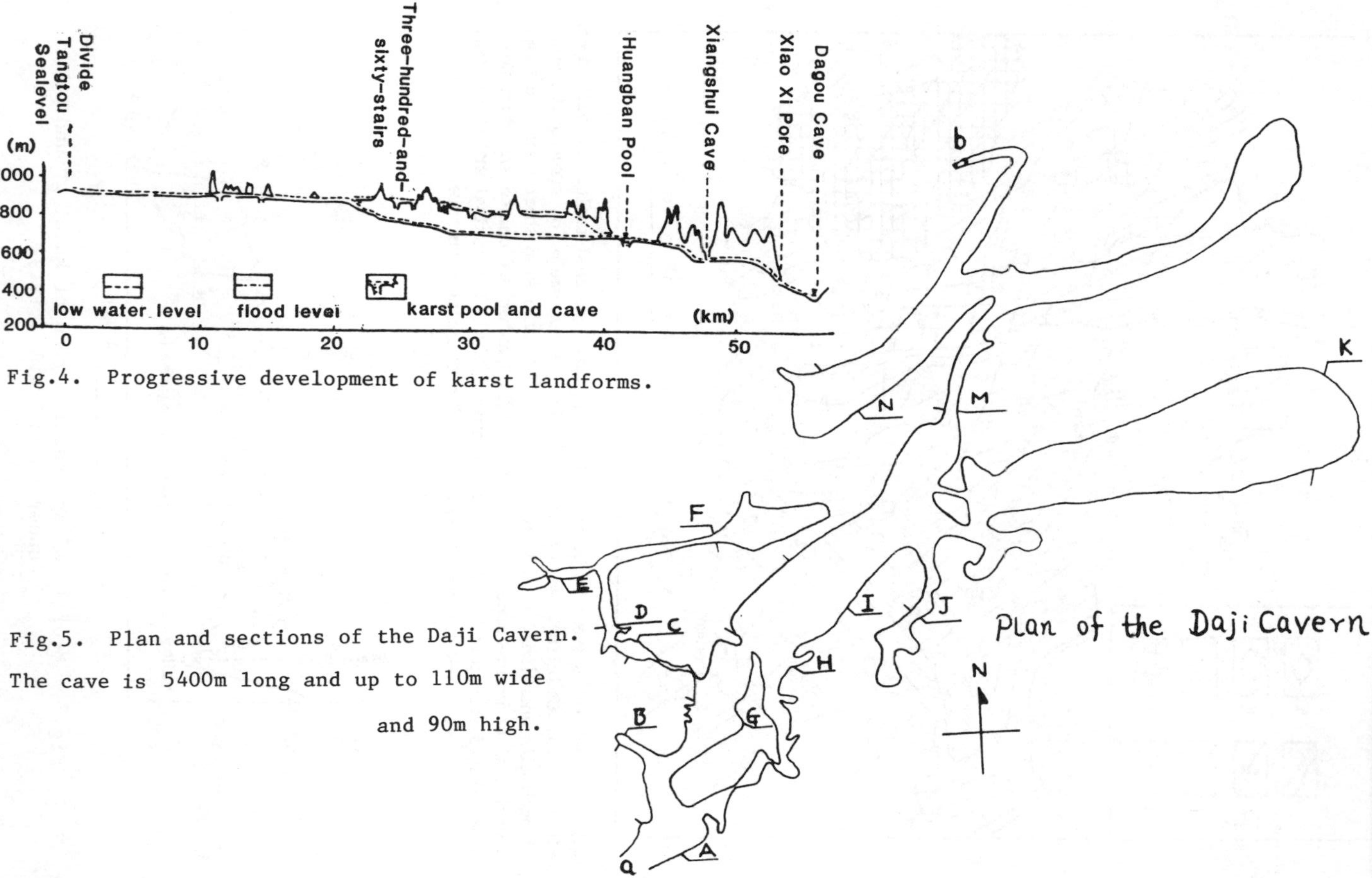

Fig.4. Progressive development of karst landforms.

Fig.5. Plan and sections of the Daji Cavern. The cave is 5400m long and up to 110m wide and 90m high.

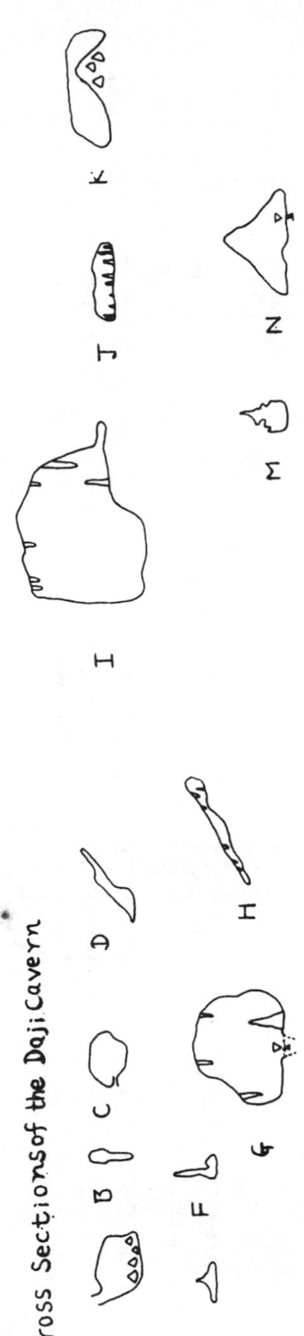
Cross Sections of the Daji Cavern
A
B
C
D
E
F
G
H
I
J
K
M
N

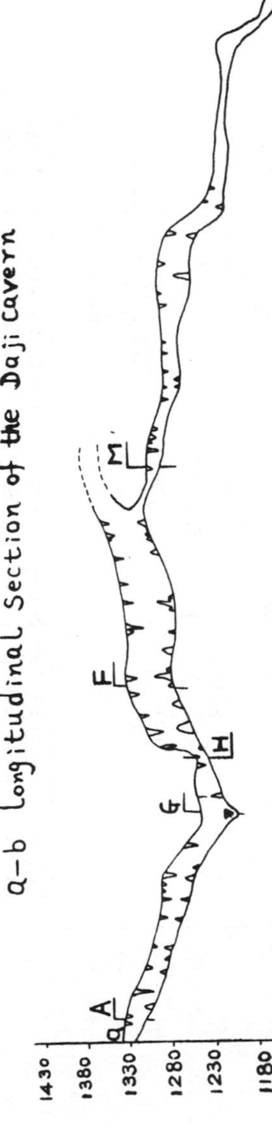
a–b Longitudinal Section of the Daji Cavern
1430
1380
1330
1280
1230
1180
a
A
G
F
H
M
b

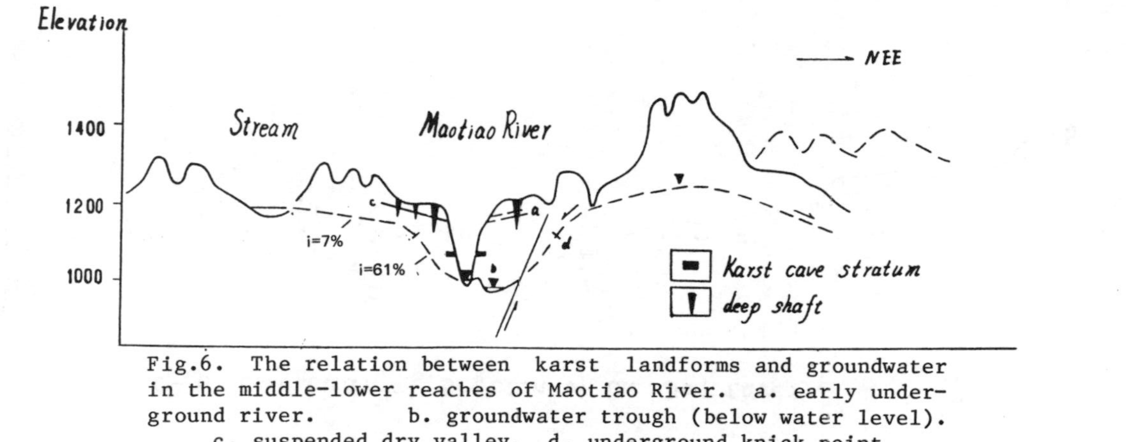

Fig.6. The relation between karst landforms and groundwater in the middle-lower reaches of Maotiao River. a. early underground river. b. groundwater trough (below water level). c. suspended dry valley. d. underground knick point.

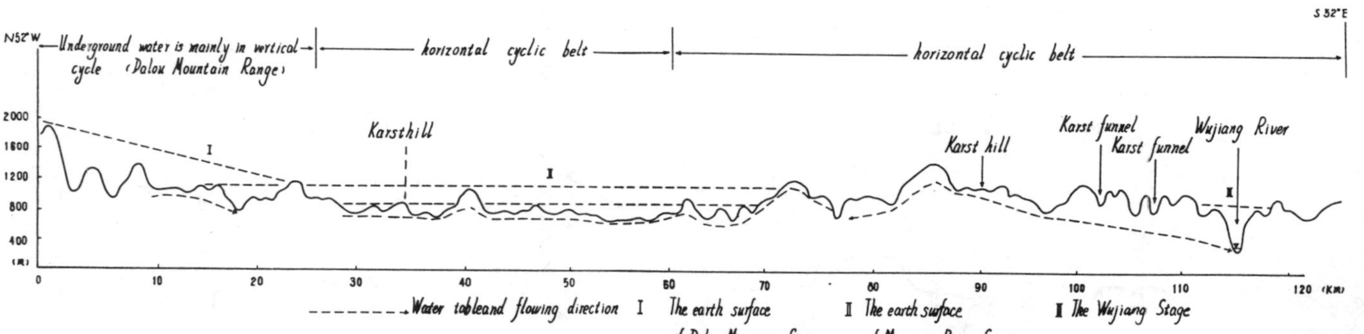

Fig.7. Stages of landscape evolution.
I. Dalou Mountain Stage. II. Mountain Basin Stage.
III. Wujiang Stage.

International Geomorphology 1986 Part II
Edited by V. Gardiner

NEW OBSERVATIONS ON TOWER KARST

Yuan Daoxian

Institute of Karst Geology,
Ministry of Geology and Mineral Resources
Guilin, Guangxi, China

ABSTRACT

Two basic varieties of tower karst, namely, the peak cluster and the peak forest, and the transitional types between them, are recognizable in South China. From recent field observations on their distribution, and its relation with geology and hydrology, the traditional views on zoning and evolutionary sequence from peak cluster to peak forest are examined, and the origins of the two basic types of tower karst are discussed. In addition, the possible influences on modern tower karst evolution of Pre-Cretaceous landforms and protection provided by Cretaceous red beds are discussed in the light of the discovery of the Late Cretaceous *Atopochara* flora in reddish calcareous breccia. This is widely distributed in the South China karst, both on the bottom parts of karst depressions and on the top of some stone peaks.

INTRODUCTION

Southwest China has the largest extent of tower karst in the world. Moreover, in contrast to the similar landforms of such well-known karst areas as Southeast Asia, New Guinea, or Central America, which are mainly developed in Cenozoic poorly diagenized carbonates, and hence usually have a lower and more or less rounded peak shape, the tower karst in the mainland part of South China is developed in compact, pre-Triassic carbonates. It has a more conspicuous form, towering up abruptly from a karstified plain or hillslope (Plates 1 and 2). This is why South China tower karst attracts interest from karstologists of many countries, and is sometimes called the China Type Karst.

The Chinese have long paid attention to this kind of unique landform. Xu Xiake (1587-1641 A.D.), the great geographer and traveller of the Ming dynasty, noticed the difference between tower karst and other types of karst landform, and pointed out in his book "Travels of Xu Xiake" that tower karst extends from Daoxian County (Hunan Province) westward to Luoping County (Yunnan Province); he was the first scholar to describe tower karst [8]. During the past three decades, in accordance with the economic need to link studies of the South China Karst with water and mineral resources exploration, many karstologists have been involved in the study of karst geomorphology, and a lot of papers dealing with tower karst were published.

The main points included could be summarized as follows:

1. Many varieties of tower karst are recognizable, but the two most essential types are peak cluster (Fengcon) and peak forest (Fenglin) [9].
2. Different types of tower karst show spatial zonation [1]. For instance, four karst landform belts were distinguished for the karst area of Guangxi Province, which totals about 90,000 square kilometres. From west to east they are: the peak cluster belt; the peak forest belt; the isolated peak belt; and the residual karst knob belt [12]. In Guilin area, an altitudinal zonation was claimed such that the peak cluster zone is above 360m in height, the peak forest zone is between 360 and 150m; and the isolated peak zone is lower than 150m. Moreover, the zones can be ascribed to the Cretaceous, Tertiary, and Quaternary geological period, from the uppermost to the lowermost one respectively [10].
3. Different explanations exist for the origin of peak cluster and peak forest karst. Many consider that they reflect different stages of karst landform development in humid tropical climates, i.e. there is an evolutionary sequence from peak cluster to peak forest, and then from peak forest to isolated peak [4, 12]. This view is also shared by some foreign publications (Sweeting, 1973), although different geomorphological terms such as mogotes, cone karst, cockpit karst, tower karst, or polygonal karst are given to the corresponding forms. Definite conditions are considered to be necessary for such evolution. For example, some believe that neotectonic movement plays an important part, so that the peak cluster type will develop in uplift areas, whereas long term stability will give rise to the peak forest type [4, 5, 6, Zhang Zhigan, 1980]. Others consider that the shifting of drainage base level is more essential, i.e. peak cluster will follow lowering of the base level and increasing of the vadose zone, and the peak forest will develop where the water table is near the surface for a long time [6]. Moreover, some geomorphologists treat this issue on the basis of the relation between groundwater and surface water processes, and consider the peak forest type a result of prevailing river processes [9]; or that the peak cluster type is mainly developed in the upper reaches of a river, and the peak forest type in its lower reaches (Zhang Zhigan, 1980). There are also climatological explanations of the peak shape, for example some hold that peak forest karst is a tropical type, whereas subtropical karst is characterized by knob (low peak) and depressions [12].

After recent field observations on tower karst of typical karst regions in Guangxi, Hunan, Guizhou, and Yunnan Provinces, the author considers that it is pertinent in this paper to review and make some comments on such problems as the horizontal and altitudinal distributions of peak cluster and peak forest karst, their evolutionary processes and conditions of formation. In this paper, peak cluster karst is defined as a group of stone peaks with a common stony basement (Plate 2); there are usually closed depressions among the peaks, so the combination is sometimes called peak cluster-depression. Peak forest karst is defined as stone peaks that are isolated from each other by a flat stony surface, generally covered by a thin layer of loose sediments (Plates 1 and 10); the peaks are usually surrounded by a plain, so the combination is sometimes called peak forest-plain.

CHARACTERISTICS OF PEAK CLUSTER AND PEAK FOREST DISTRIBUTION

The peak cluster and peak forest types are mixed with each other on the map. It is difficult to establish a clear zoning in spite of observations on a macroscopic (e.g. zoning for karst region of Guangxi Province), or mesoscopic scale (e.g. zoning for Guilin karst area). On the contrary, they are usually mixed with each other. For example, the region between Hechi and Shangchao was considered to be all occupied by peak cluster landforms [12]. However, this is only true for its northern and southern end. Marvellous peak forest very similar to that which is near Guilin occurs in quite a few points in the middle part of this region, such as Yong'an (Plate 7), Gubin, and Guangnan (Plate 8; Fig.1). The landforms between Guilin

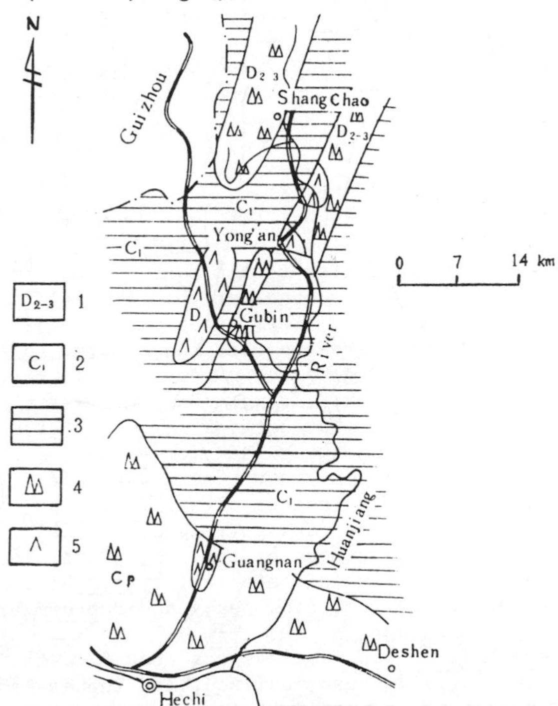

Fig.1. Distribution of peak cluster and peak forest in the region between Shangchao and Hechi

1. Middle and Upper Devonian limestone
2. Lower Carboniferous sandstone
3. Non-karst area 4. peak cluster 5. peak forest

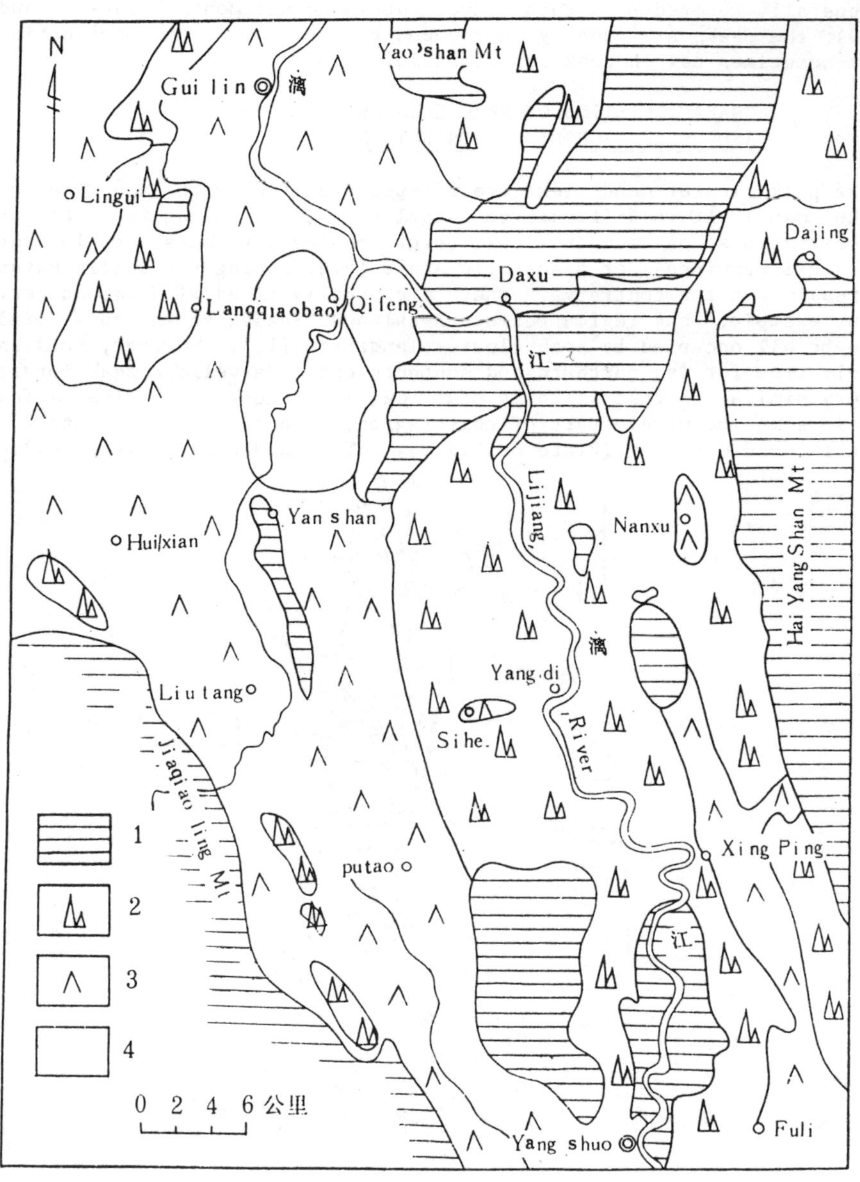

Fig.2. Distribution of peak cluster and peak forest in Guilin-Yangshuo area. 1. Non-karst area; 2. Peak cluster; 3. Peak forest; 4. Karst area.

Plate 1. Duxiufeng, Guilin urban area.

Plate 2. Peak cluster along Lijiang River, Guilin.

Plate 3. Cuesta type peak near Dafengshan, south-west Guilin.

Plate 4. Peak cluster - Depression near West Hill, Western Guilin.

Plate 5. Rounded isolated low knobs developed on Dolomite, Near Ertang, Lingui County.

Plate 6. Peak forest near Mawei, South Guizhou.

and Yangshuo were previously classified as peak forest, but actually peak forest only makes one half of this area, while the other is peak cluster. Moreover, the two types are mixed with each other in a complicated fashion. Near Guilin urban area, patches of peak cluster are enclosed in peak forest (Plate 4), whereas on both flanks of Lijiang river small patches of peak forest appear within a peak cluster region (Plate 10; Fig.2). If an evolutionary sequence from peak cluster to peak forest is proposed, it should explain why, within a small area, some small parts changed, while the majority remained unchanged, or vice versa. If the evolution is considered to be linked with relative stability of neotectonic movement, the possibility of differential uplift or subsidence for a small area totalling less than one square kilometre is a challenge.

Both peak forest and peak cluster can occur at different altitudes. At a regional scale, the peak cluster not only occurs in Northwest Guangxi, with altitudes of 600-1,000m, but also at altitudes of about 100m, such as Hechi suburb and both flanks of Lijiang Gorges (Plate 2). On the other hand, the peak forest is not only distributed at lower altitudes, such as Guilin urban area (150m), but also at higher altitudes, such as the Mawei (Plate 6) area on the boundary between Guangxi and Guizhou Provinces (800-900m). At a local scale, the peak cluster around Tangjiawan area, six kilometres to the south-west of Guilin, and the peak forest of Guilin urban area have the same basal level (150m). Besides, the peak forest

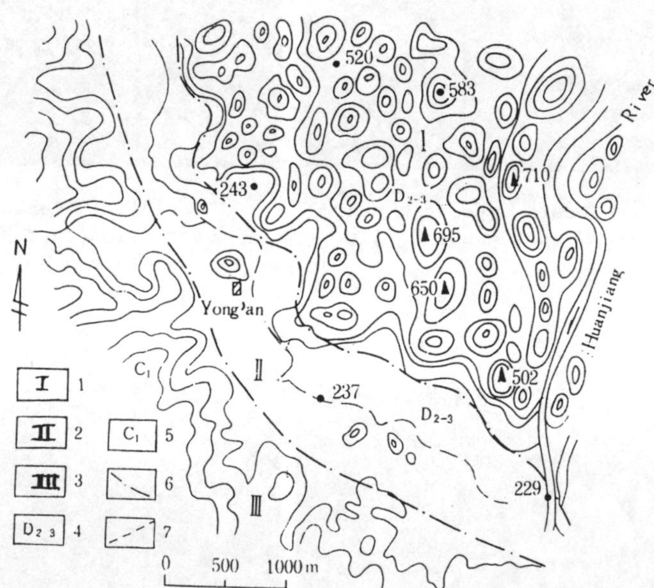

Fig.3. The distribution of peak cluster and peak forest near Yong'an, Huanjiang County, Guangxi
1. Peak cluster; 2. Peak forest; 3. Non-karst knobs; 4. Middle and Upper Devonian; 5. Lower Carboniferous; 6. Landform type boundary; 7. Intermittent stream.

Plate 7. Peak forest near Yong'an, Huanjiang, with peak cluster in the background.

Plate 8. Peak forest near Guangnan, Hechi County.

Plate 9. Peak forest near Daoxian County, South Hunan.

Plate 10. Peak forest of Nanxu, Southeast Guilin, enclosed in a major peak cluster area.

Plate 11. Conical peak forest on Guizhou Plateau.

Plate 12. Old Man Hill, Guilin, showing a cap of red breccia.

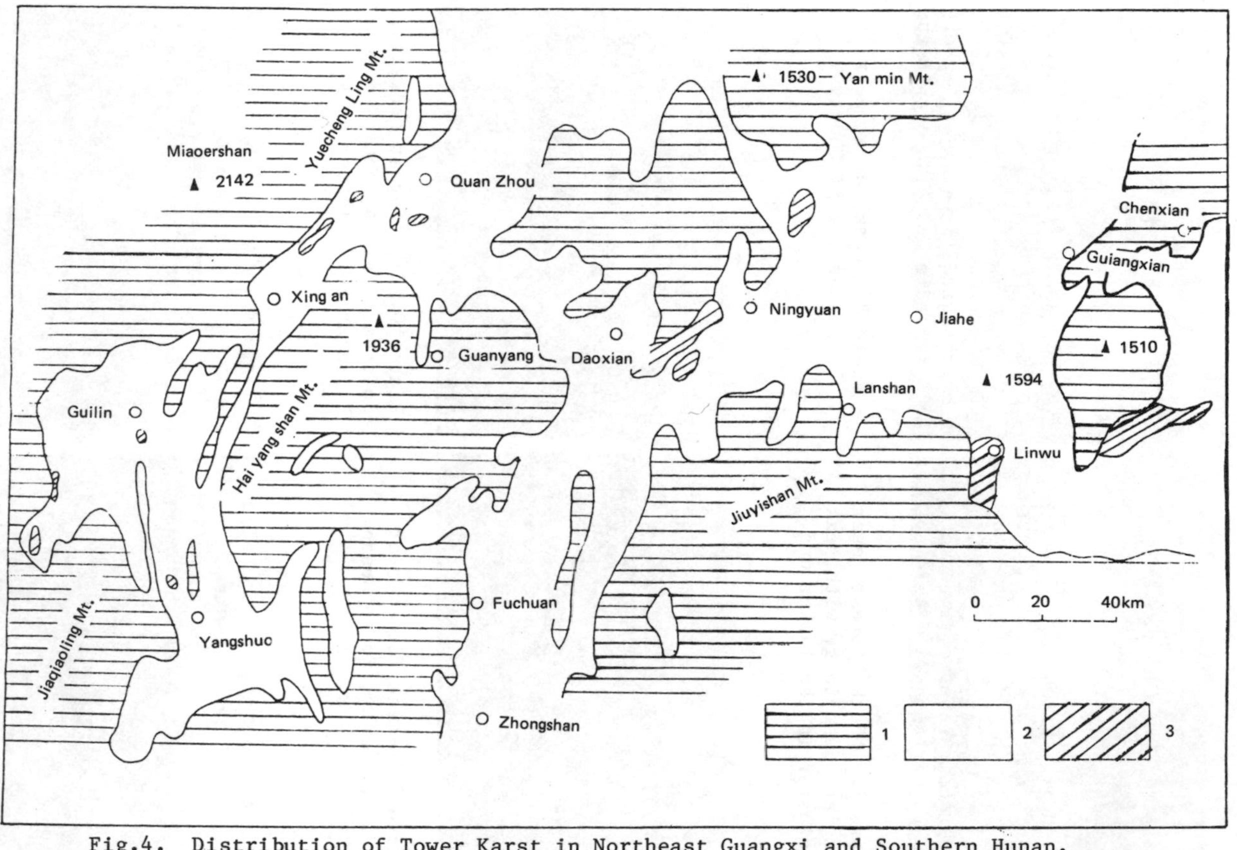

Fig.4. Distribution of Tower Karst in Northeast Guangxi and Southern Hunan. 1. Non-Karst; 2. Karst Area; 3. Red Bed (Cretaceous).

occurring near Yong'an village, Huanjiang County, is at a similar basal level to the peak cluster nearby (Fig.1; Fig.3; Plate 7). Therefore, it is difficult to assert that the peak cluster to the east of Yong'an is an uplift region, while the area near to Yong'an is relatively stable. The peak cluster is not necessarily in an uplifted area, and the peak forest is not necessarily developed in a relatively stable one. There must be some other factors to control the development of peak cluster and peak forest.

The peak forest areas are mainly developed where adequate surface runoff is available, whereas the peak cluster type is in regions lacking surface runoff, or where it is very weak. Although karst regions are generally characterized by no surface drainage network, under certain special geological and hydrological situations, there are strong surface fluvial processes. According to evidence in the South China karst, these situations could be summarized as follows:

(1) Karst areas in the neighbourhood of non-karst regions, so that there are strong allogenic surface water inputs.
(2) Carbonate rock intercalated with non-carbonate or impure carbonates, hence an underground drainage system develops with difficulty.
(3) Soluble rock covered by relatively impermeable overburden, so percolation of surface water is detained. The majority of typical peak forest regions in South China could be ascribed to these settings.

Peak forest near Guilin is predominantly distributed in those parts that are influenced by allogenic water from the Yuechengling Jiaqiaoling and Haiyangshan Mountains, to north, west and east respectively; moreover, it is noticed that the larger the catchment area of the allogenic water, the broader the relevant peak forest patches (Fig.2). Observing on a larger scale the region between Guangxi and Hunan Provinces, the peak forest near Daoxian (Plate 9), Zhongshan, and Xing'an Counties is also related to allogenic water from the Haiyangshan, Dupangling, Minzhuling, and Jiu-yi-shan Mountains (Fig.4). Near Yong'an (Fig.3), the small patch of peak forest developed in massive Upper Devonian limestone occurs near its boundary with Lower Carboniferous sandstone and shale, so it could be imagined that the allogenic water from the west has played an important part in its development. These points are supported by the widespread distribution of clay and gravel with such allogenic material as sandstone and vein quartz - its main composition in the peak forest plain of Guangxi and Southern Hunan. Besides, cave sediment and scallop investigations show most caves to be of an inflow nature, thereby offering additional evidence [11].

The peak forest landscape in Mawei (Plate 6), a part of a strongly uplifted region near the boundary between Guangxi and Guizhou Provinces, is an example of the role of impure beds intercalating the carbonates, and making surface flow available to some extent. The small patch of peak forest form in Sihe Polje, being enclosed by a peak cluster region on the west side of the Lijiang river between Guilin and Yangshuo (Fig.2), is considered to be linked with an anticline at the west end of the polje, where a small patch of Lower Middle Devonian sandstone is exposed.

Because of the limitation of the influence of surface flow on the plain, small patches of peak cluster may remain in a large peak forest area. This is the case in Southwest Guilin (Plate 4) and Northwest Yangshuo (Fig.2). The distribution of peak forest may also be limited if there is a lower drainage base level in the karstified system; in this case the allogenic water, after flowing a certain distance on the surface, may be drained rapidly by the underground system. For instance, on both flanks of Huanjiang river to the east of Yong'an (Fig.3), and of Lijiang Gorge between Guilin and Yangshuo (Fig.2), there are peak cluster landscapes (Plate 2), despite there being a perennial river.

So peak forest is the result of predominant surface fluvial processes, whereas peak cluster is the landform developed in those regions where the surface fluvial processes are rather weak. Karst regions with stronger surface runoff occur in accordance with certain geological settings. Peak forest is not necessarily a landform of a certain stage of geomorphological evolution [9]; peak cluster is not necessarily an initial form, but is itself a result of karst landform evolution following a different route.

TYPES AND ORIGIN OF PEAK SHAPE

Much work has been done on peak shape, both in China and elsewhere. For instance, a peak shape classification was given as follows: the cylindrical or Duxiufeng (the peak of unique beauty) type; the gastropoda shape type; the tower shape type; and the Cuesta type or the Old Man Hill type [2, 8]. It is generally considered that peak shapes are controlled by geological factors such as lithology and structure; however, some karstologists try to explain them with an evolutionary or climatological approach.

The difference between peak shapes on Guizhou Plateau and Guilin karst has long been noticed. The former has usually a gentle slope, and is conical in shape, whereas the latter is characterized by steep slopes and cylindrical features. This difference is considered to be a result of geological factors, i.e. the conical peak forest on Guizhou Plateau (Plate 11) is developed on gently-dipping Triassic limestones, usually intercalated with argillaceous beds, and the cylindrical peak forest of urban Guilin (Plate 1) is based on low-dip Devonian massive limestones. Some geomorphologists suggest that the peak forest on Guizhou Plateau is a relict of Tertiary tower karst, meeting the challenge of available palaeoclimate data claiming that the Tertiary had a semi-arid environment in this region [7]. Moreover, the Quaternary here is generally accepted as being humid and subtropical or temperate. Recent limestone denudation measurement gives a rather high corrosion rate, so how and why Tertiary karst features could remain for millions or even tens of millions of years is another question needing explanation.

Various peak shapes occur in a small area with more or less uniform climate conditions, such as Guilin. This supports the argument that the peak shape is mainly controlled by geological factors, and is not necessarily determined by climatic differences. For example, in some parts of Guilin, the orientations of peaks show clear structural lineament. Near Putao, 40km to the south of Guilin, the peak surface declines eastward, coinciding with the bedding plane dip. The cylindrical Duxiufeng type peaks occur in urban Guilin where Upper Devonian massive limestone strata are nearly horizontal. In the western part of Guilin, from West Hill to the Dafengshan Hill, the limestone beds are generally dipping 20-30° eastward, so stone peaks here are characterized by cuesta shapes with asymmetrical profiles (Plate 3), which have a gentle slope on the eastern side, and a steep slope on the west. Cuesta peaks are sometimes called the Old Man Hill type, but actually the Old Man Hill is capped by red breccia (Plate 12), and is not as typical as the cuesta peaks distributed from the West Hill to Dafengshan Hill. The gastropoda-shaped peaks in Guilin are found to be developed in Carboniferous limestone with argillaceous intercalations, and gently dipping bedding planes are also required. Near Ertang, Lingui County, 15km to the west of Guilin, well-rounded isolated low knobs are developed on Lower Carboniferous medium to coarse grained dolomites (Plate 5).

All these facts indicate that if one wants to discuss the evolution of tower karst on the basis of some morphometric indices of peak shape, such as the ratios of diameter to height, or the long axis to short axis plan lengths, one should first analyse carefully the geological background of the particular examples.

NEW EVIDENCE FOR STUDYING THE EVOLUTIONARY HISTORY OF TOWER KARST

On the basis of analysis of the physiographic cycles of peak surfaces at different altitudes, some karstologists have discussed the evolution of tower karst. For instance, four epochs of karst evolution are distinguished in Guangxi, namely, the Peak Surface Epoch (Cretaceous; 1500m in western Guangxi, 350-400m in eastern Guangxi); the West Guangxi Epoch (Early Tertiary; 300-1500m, rising westward); the Peak Forest Epoch (Late Tertiary; peak surface 200-1000m); and the Hongshuihe Epoch (Quaternary, the down cutting of gorges) [4]. The main difficulty in understanding tower karst history is lack of contemporaneous sedimentary sequences. For a long time, Quaternary sediments were only found covering the peak forest plain, and a lot of Quaternary Mammalia fossils have been unearthed from caves at lower altitudes. This information is important to the study of the history of low altitude caves and tower karst, mainly after the Pleistocene. However, little of importance has been found relevant to high altitude tower karst and its caves. Moreover, difficulty also comes from the limitation of available dating techniques for Cenozoic sediments, which are effective only in the range of 50,000, 350,000, or about one million years B.P. according to the different techniques.

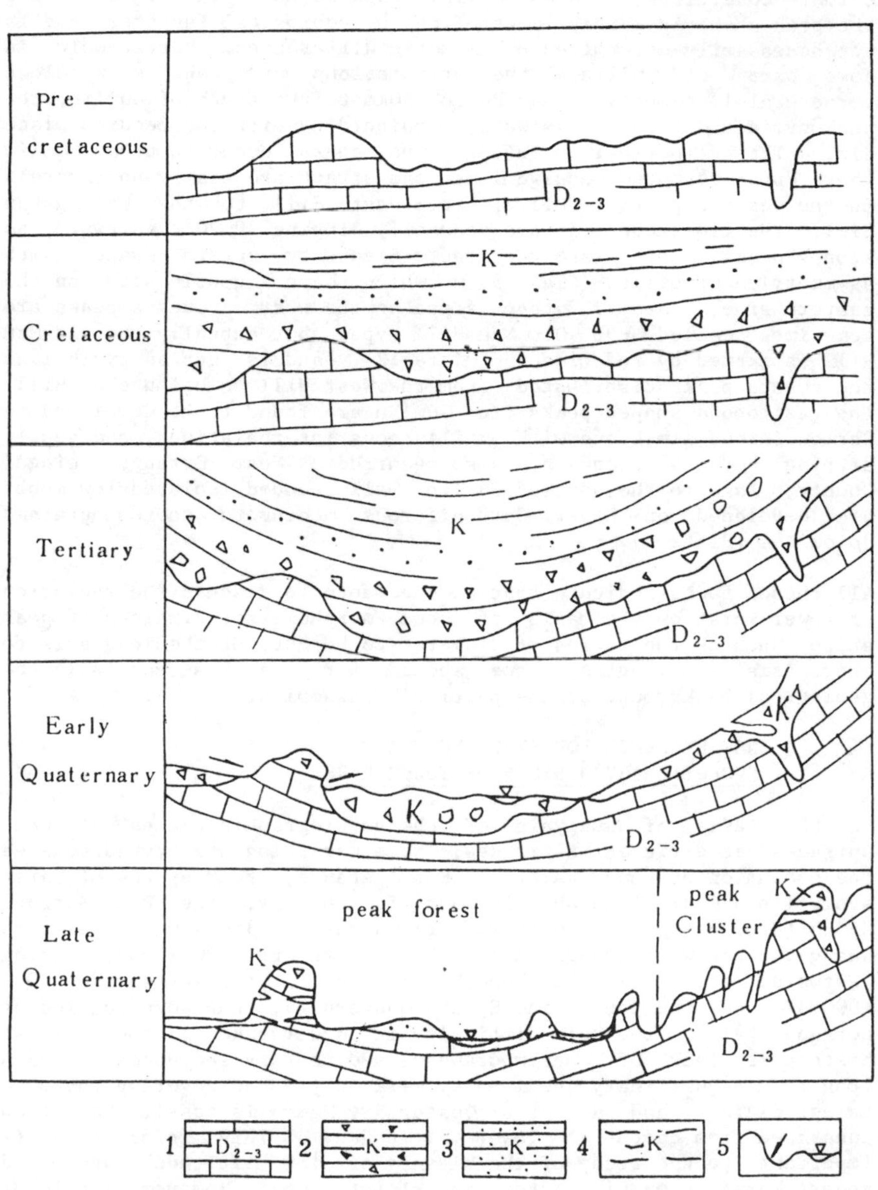

Fig.5. A simplified model for the evolution of Guilin Karst. 1. Middle and Upper Devonian limestone; 2. Cretaceous Red Breccia; 3. Cretaceous Red siltstone; 4. Cretaceous Red Mudstone; 5. Surface stream and swallet.

Between the towns of Qifeng and Yanshan, 12km and 24km to the south of Guilin, there is a plain totalling about 100 square kilometres without any peaks (Fig.2). Visiting geomorphologists often wondered about the origin of this plain: why are there no peaks at all? What is its relation with the surrounding peak forest and peak cluster? Is it the final result of tower karst denudation? On the margin of this plain, for example near Langqiaobao village, a spectrum of landform transitional sequences from the outside to the centre part of the plain appears to exist. This gives an impression of gradual reduction of the peak density, and the evolution from peak cluster to peak forest, and then from peak forest to plain without peaks. However, this has long been questioned because there is a patch of Red Beds just at the western side of Guilin Airport on the plain, although it totals only about one square kilometre; no conclusions have been reached owing to uncertainty concerning the origin and geological age of the Red Bed. According to recent drilling, the Red Bed is more than 100m thick, and is composed of red mudstone, siltstone and conglomerate, with greenish grey mottles, so it is suggested to be a lacustrine deposit. Moreover, abundant fossils of _Atopochara_ flora are found in it, among them are _Atopochara_ sp., _Charites_ sp. and so on; accordingly, the Red Beds are identified as Late Cretaceous [13]. From these new findings it appears that this plain without peak is not the final result of gradual reduction of peak density, or the denudation process of tower karst. There has never been tower karst at this part of Guilin after the Cretaceous. Contemporary with the tower karst evolution in its surrounding area, here there has been denudation and removal of the Red Beds, with their thickness reduced gradually.

When dealing with the evolution of tower karst, it is necessary to know first at what time did the soluble rock at different parts of the studied area begin to be exposed, and when the process of karst landform evolution started; nevertheless, there is an increasing emphasis on the importance of subsoil corrosion (Gams, 1981). For the Guilin karst, it is important to reconstruct the area that was once covered by Cretaceous Red Beds. Thanks to the work of Lin Yushi, Wong Jintao, Deng Ziqiang et al., [14], a red breccia contemporary with the Late Cretaceous Red Beds occurring near Guilin Airport is found scattered widespread on the top of peaks, such as Houshan Hill (the Monkey Hill, 580m), and Old Man Hill (320m; Plate 12), or in the high altitude karst depressions. This red breccia is composed predominantly of carbonate gravel, and its cementation, where the _Atopochara_ flora is also found, is reddish, argillaceous and calcareous; sorting and lamination are often seen. From this it is reasonable to consider that the region that had been concealed by Late Cretaceous Red Beds was much greater than is seen today around Guilin Airport. The area with altitudes 300-600m was very likely to be included. Owing to the upwarping movements that happened during the Tertiary, it is impossible to estimate the thickness of the Red Beds for different parts of Guilin on only the basis of the relative height of the points where the breccia is found; the time for exposure of limestone still remains a question, depending on the denudation rate of the Red Beds. However, bearing in mind the available corrosion rate data for this area (100-300mm/thousand years, or possibly 200-600m during the Quaternary), it is difficult to imagine that a peak capped by red breccia which is also soluble, like the Old Man Hill (Plate 12), with altitude 320m and relative height 170m, could have started evolving from the Early Tertiary. It is possible that the varieties of tower karst which the available evolution models are based upon, are all different forms developed on carbonate rocks exposed at different times, and under different geological and hydrological settings. A simplified model for such a consideration is summarized in Fig.5.

More elaborate work is necessary to reach a conclusion for this hypothesis, but the red breccia is widespread over the peak forest or peak cluster regions of Guilin, Liuzhou (on top of the E-shan Hill or Goose Hill), Litang, Huanjiang, as well as Southern Hunan. In dealing with the problem of tower karst evolution, both the Pre-Cretaceous landforms, and the protection provided by Cretaceous Red Beds on limestone corrosion are not negligible.

CONCLUSIONS

The climatic control of karst landforms, as verified by the great contrast between the normal shaped limestone mountains in Northern China and the marvellous tower karst in Southern China, is still convincing.

Tower karst has many varieties, but the essential types are the peak forest and the peak cluster. The two types are mixed with each other, both horizontally or vertically. It is difficult to establish a clear zoning or evolutionary sequence for tower karst varieties. The peak forest is developed in those areas where the surface fluvial processes are rather strong, whereas the peak cluster occurs where there is weak or no surface runoff. Whether a karst area enjoys strong surface runoff or not depends on climatic geological settings.

The peak shapes are mainly controlled by structural and lithological features. If one wants to investigate the problems of climatic influence on peak shape, or its evolution, the same geological background is a pre-requisite.

The Pre-Cretaceous landforms, and the protection provided by Cretaceous Red Beds on limestone denudation, are two factors that should be taken into account for a better understanding of the evolution of parts of the South China tower karst. For studying the evolution of tower karst, and its controlling factors, it is necessary to consider the climatic factors, as well as the geological factors comprehensively. Moreover, emphasis should be placed on the differences of hydrologic conditions, which is a combination of both climatic and geological conditions, as well as their historic change.

Every natural condition is changing with space and time. Any attempt to establish an evolutionary sequence for karst landforms under the same conditions, and merely changing with time, could only be done in a theoretical way. It is not appropriate to summarize into an evolutionary sequence different karst landforms that developed under different conditions. Two basic questions should be answered before finding an evolutionary sequence by comparing present surface karst features: "Are these forms developed under the same conditions?" and "What happened to the old forms to stop their development, so that they are still now visible?"

Acknowledgements

The author wishes to express his sincere thanks to Mr. Zhu Xuwen, Mr. Lin Yushi, Mr. Wong Jintao, Mr. Deng Ziqiang, Mr. Jiang Lideng, Dr. M. M. Sweeting, Dr. Tony Waltham and Dr. P. Smart for helpful discussions. Special thanks should extend to Prof. P. Williams for his helpful comments on some points in this paper through correspondence. The author is also indebted to Mrs. Li Jian for drawing illustrations, and Mrs. Song Ailing for typing the manuscript.

REFERENCES

〔1〕李梓中等，1962，广西喀斯特发育的基本规律。《全国喀斯特研究会议论文选集》，科学 出 版 社，P. 25—44.

〔2〕祁延年，1962，广西中部及东北部地区喀斯特地貌。《全国喀斯特研究会议论文选集》，科学出版社，P. 69—84.

〔3〕黄万波，1979，华南动物群的性质时代。《古脊椎动物与古人类》，第17卷，第4期。P. 325—343.

〔4〕中国科学院地质研究所岩溶研究组，1979，《中国岩溶研究》。科学出版社。

〔5〕陈治平、刘金荣，1980，桂林盆地岩溶发育史探讨。《地理学报》，第35卷，第4期。

〔6〕朱学稳等，1980，桂林地区的岩溶峰林地貌及其发育。《国际交流地质学术论文集》，地质出版社，P. 98—108

〔7〕吴萍、杨振强等，1979，《中南地区白垩—第三纪岩相古地理及含矿性》，地质出版社，P. 1—75.

〔8〕任美锷，1982，中国岩溶发育规律初步研究。《中国地质学会第二届岩溶学术会议论文选集》，科学出版社，P. 1—4

〔9〕曾昭璇，1982，论我国南部喀斯特地形的特征。《中国岩溶》第一卷第一期，P. 27—31

〔10〕陈文俊，1982，桂林岩溶地质概况，《中国地质学会第二届岩溶学术会议论文选集》，科学出版社，P. 30—37.

〔11〕朱学稳，1982，桂林地区灰岩洞穴的溶蚀形态。《中国岩溶》，第一卷，第二期，P. 93—103

〔12〕任美锷、刘振中等，1983，《岩溶学概论》。商务印书馆

〔13〕刘功余，1984，桂林岩溶区红色钙泥质岩中发现白垩纪轮藻化石。《中国岩溶》，第三卷，第一期．P. 14.

〔14〕林玉石、邓自强等，1984，岩溶交[illegible]岩。《中国岩溶》，第三卷，第一期，P. 1—13.

Gams, I., 1981. Comparative Research of Limestone Solution by Means of Standard Tablets. Proceedings of the Eighth International Congress of Speleology, 273-275.

Sweeting, M. M., 1973. Karst Landforms. Columbia University.

Williams, P., 1973. Variation in Karst Landforms with Altitude in New Guinea. Geographische Zeitschrift, 25-33.

Zhang Zhigan, 1980. Karst Types in China. Geojournal, 4, 541-570.

International Geomorphology 1986 Part II
Edited by V. Gardiner

DEVELOPMENT AND FILLING OF CAVES IN DRAGON BONE HILL AT ZHOUKOUDIAN,BEIJING

Liu Zechun

Department of Geography,
Nanjing Normal University, Nanjing, China.

ABSTRACT

Dragon Bone Hill has limestone caves, in which many animal fossils, including those of Peking Man, have been found. The cave morphology is controlled by geological structure, with both vertical and horizontal types occurring; the caves formed during the Middle Miocene, Pliocene and Quaternary. Sediments in the caves include old river sediments, limestone breccia and travertines. Habitation of the caves by early man depended partly on the pattern of sedimentation and stage of cave development; this is illustrated by detailed consideration of the development of Peking Man's Cave, Upper Cave and New Cave.

INTRODUCTION

The Dragon Bone Hill (Longgu Shan), is a small piedmont hill of Middle Ordovician carbonate rocks, some 150m above sea level, and 70m above the nearby river bed; it has an area of less than one square kilometre, and is located at Zhoukoudian, in front of the Beijing West-Mountains, about 50km southwest of Beijing (Fig.1). A great many animal fossils, especially the famous fossils of Peking Man (Sinanthropus Pekinensis, i.e. erectus pekinensis), were discovered in the caves of the hill, from which its name has been gained (Fig.2).

There are over ten limestone caves in the Hill and eight contain mammalian fossils; in four of them the old hominid fossils (from Homo erectus to Homo Sapiens) and stone tools have been found. This paper deals mainly with the types of caves in Dragon Bone Hill, the times of cave formation and the cave development and filling processes in relation to the life activities of early man, and the comparison of stratigraphy during the Quaternary.

THE TYPE OF CAVES

The development of the caves in the Dragon Bone Hill is mostly controlled by geological structure and lithology. The hill is in a limb of a syncline, but the limestone strata of the hill do not display the simple monoclinal structure of a fold limb. They were secondarily deformed and consequently planes of the limestone strata are wavy, with dip angles of from 10° to 70°. Several sets of nearly vertical joints are well developed.

The caves in the hill are mostly developed along bedding planes or joints of limestone, so that their morphology is closely related to the geological structure; they can be grouped into two categories: the vertical type and the horizontal type. The former was formed by corrosion along surfaces of the strata in steeply dipping zones, for instance, the Peking Man's Cave (Locality 1) and the cave of Locality 12; the latter type are mainly along strata in gently dipping zones, for example New Cave (Fig.3).

In addition, there are a few small caves and fissures formed along joints of limestone. However, the Upper Cave above Peking Man's Cave is one of a transitional type. Its upper part is mainly developed along the gently inclined surface of the strata (dip angle 20°-30°) and has a morphology quite like that of horizontal caves; the lower part of the cave appears to be a vertical cave, because there the dip of the strata steepens to 50°-60° and shear joints exist. This causes corrosion to penetrate deeply downward, forming a vertical shaft once connected with Peking Man's Cave (Plate 1).

Most of the caves in the region are filled with sediments, which include breccia fragments resulting from the collapse of the cave roofs and overhangs and the degradation of the cave walls. They also include deposits brought into caves by sheet flow; fluvial fine-gravelly sand; sand, silt and clay; ash layers resulting from fires; and chemically deposited travertines.

THE AGE OF CAVE FORMATION

The general causes and ages of cave development in the area have been established according to stratigraphical comparison, mammalian fauna and recent dating with several methods. The summit level of the Dragon Bone Hill is an old erosional surface, called Tang Xian surface, on which there are Miocene beds containing fossils of the *Hipparion* fauna of North China. The elevation and dissection of the surface caused the formation of Dragon Bone Hill, and the development of the caves.

On top of the hill there are deposits named "Upper Gravel" and "Cap travertine", with *prosiphneus* sp. and *postchinotherium* sp., of the Late Miocene. The cave deposits of fine sand-silt at Locality 14, just under a hill top, contain fish fossils (*Barbus*) belonging to the Late Miocene. This shows that some caves in the region were developed during the Middle Miocene. This kind of cave generally lacks travertines or chemical deposits, and is more than 100m above sea level. It is possible that the caves were formed in the saturated zone before the old surface had been dissected by rivers.

Caves in the region at a lower level and developed largely vertically, were formed during the Pliocene and Quaternary; for instance, the caves of locality 12 and 1. In the former, early Pleistocene mammalian fossils have been found. The latter, i.e. Peking Man's Cave, was dated at one million years B.P., for its lower layers of deposits were magneostratigraphically laid down before the Brunhes epoch. This type of cave is caused by erosion of underground water along the surfaces of the limestone strata with steep dips and vertical joints, as uplift occurred. The bottom of the cave is lower than the underground water level of that time and the recent river bed, at less than 80m above sea level; this is believed to be so because the terrace deposits outside the Peking Man's Cave, called "Lower Gravel", at 90m above sea level, belong to the Middle Pleistocene. These facts support the interpretation that the development of the caves in the region was mainly in the vadose zone, and mostly accomplished by the Early Pleistocene (some one million yrs B.P.).

THE FILLING PROCESSES OF THE CAVE

Cave fillings in Dragon Bone Hill are varied and depend on the conditions of the cave and changes in environment. They have had a large influence on the living activities of early man. The filling processes of caves can be divided as follows:

Old river sediments. These cave deposits are chiefly silty sand and fine gravel, sometimes bedded with a small amount of limestone breccia and travertine. They were formed during the Late Pliocene and early Pleistocene; where there is little limestone breccia this indicates that the climate during the sedimentation was warmer than today. The cave deposits at locality 14, about one kilometre southwest of Longgu Shan, may be cited as an example. They are a silty and fine sand, 7m thick, in which many fish fossils (_Barbus_) have been discovered. Around Longgu Shan (including the southeast slope of the Hill), several caves also have similar deposits. It is possible that the old river on the Tong Xian landsurface (or underground river) transported a large load into the caves through karst fissures, and filled them (Plate 2, Fig.5). A similar phenomenon can be observed in some small limestone caves and fissures in quarries near the Longgu Shan, but these deposits are formed as the present weathering product from the coal series of the Permian Period, which lie above.

Limestone breccia sediments. Limestone breccias of the cave deposits are so thick that they almost fill up the caves. They are often interbedded with non-breccia layers, (deposits of ash, silt, sand and travertine, etc.), but the former dominate over the latter. A lot of fossils of mammals and man, and stone implements, have been excavated in these cave deposits which often belong to the Middle and Late Pleistocene.

The changes from cave excavation to filling processes are usually caused by changes in landform and environment. The evolution of Peking Man's Cave is a typical example of cave filling. The cave is a vertical fissure cave measuring some 107 x 25m. It is filled with at least 40m of sediments, which can be divided into 17 layers, numbered from the top downwards. According to lithological and sedimentological studies, the silts and sands of layers 12 and 15-17 are due to fluvial deposition by the old Zhoukou River; the gravels of layer 14 are due to the bringing up of Upper Gravel into the cave. They have blocked the drainage of the cave bottom, so that the development of the cave turned into a filling process (Fig.4).

In relation to the life of early man, the evolution of the Peking Man's Cave can be divided into five stages (Fig.6):
(1) stage of formation, deeply buried, at some one million years B.P.
(2) stage when the cave was enlarged, with an east entrance, at some 700,000 years B.P.
(3) stage when Peking Man lived in the cave, at some 500,000 years B.P.
(4) stage of partial collapse of the cave roof and westward migration of Peking Man, at some 300,000 years B.P.
(5) stage when the cave was filled up, at some 230,000 years B.P.

Cave sediments with erosional unconformity. In some caves of the region, there is a plane of erosional unconformity in the depositional sequence, which indicates that the cave filling was sometimes interfered with by erosion and corrosion. For example, Upper Cave seems to be a lateral branch cave, above Peking Man's Cave. It was filled up while layer 4 of Peking Man's Cave was being laid down. After that, about 320,000 years B.P., the deposits of the upper cave were eroded; stalagmites were formed on the erosion surface in the cave. However, no early man had chosen the cave as his habitat, and it is clear that the cave was not fit for habitation during that time. We assume that this is because the cave floor was then a steeply dipping surface. In the Late Pleistocene, about 50,000 years B.P., limestone breccia was deposited, which overlies the layer of light red clay and travertines in Upper Cave. In the lower part of the layer of limestone breccia a whole mammalian bone has been found, derived from an animal falling into the cave. With continuous filling of the cave with limestone breccias, the floor of the cave was gradually flattened, and Upper Cave Man (late Homo Sapiens) came into the cave at some 10,000-20,000 years B.P. Thus, a large number of stone and bone implements have been excavated from the middle to upper part of the breccia layer.

The development of Upper Cave may be subdivided into four stages (Fig.7):
(1) stage of corrosion and erosion penetrating deeply downward, at some 320,000 years B.P.
(2) stage of cave filling, with light red clay corresponding to layer 4 in Peking Man's Cave, at some 300,000 years B.P.
(3) stage of erosion of deposits in the cave, and stalagmites forming on the cave floor, at some 300,000-50,000 years B.P. (Plate 3)
(4) stage of a second filling of the Upper Cave, with breccias and inhabitation by early man (late Home Sapiens), at some 10,000-20,000 years B.P.

In the west of Longgu Shan, at Locality 2, the fissure cave deposits are similar to those in Upper Cave, though the upper deposits belong to the Late Pleistocene and they overlie unconformably lower ones which may be Late Pliocene.

Chemically deposited travertines. Sediments of stalagmite and stalactite have been found in the New Cave, which is on the south slope of Longgu Shan. The stalagmite is dated as 680,000-750,000 years B.P., and the stalactite 980,000 years B.P. The chemical sediments are some 2m thick, and have only filled a small part of the cave so there is still a space left 2.5-4m high, 15-20m long and 10-15m wide (Plate 5).

New Cave was favourable as a habitat for early man: a stable roof and wall, enough space, an even floor and an entrance facing south. Based on the correlation of stratigraphy and the principles of karst evolution, New Cave could not be formed later than the Peking Man's Cave and Upper Cave. But, why did neither Peking Man nor Upper Cave Man not live there? The reason for this, we suppose, is that the entrance of the cave faced a quite wide and deep fissure, which was filled up with red clay at the time of the Late Pliocene or the Early Pleistocene; the fissure was expanded and deepened by corrosion and erosion until the Late Pleistocene, and was difficult for Peking Man to enter. Later, much limestone breccia was deposited, and filled up the fissure, so it became a road to the cave. Then, New Cave Man (early Homo Sapiens) could live in it. Yet the limestone breccias raised the floor continually until they blocked the entrance of the cave, and as a result early man was made to leave the cave (Plate 5). This can be proved by the fact that the thin layer of ash, left in the base of travertine layers in the cave, is only 4-12cm thick. This is indicative of early man living there for only a short time. The ash layer is between the talus from the near slope, and dips towards the entrance of the cave, showing man's activity outside of it. Thus, because the entrance of the cave was blocked, man had no chance to live in the cave. In fact, the cave was only found in the 1970's, and hence it is called "New Cave".

The development of New Cave can be subdivided into three stages (Fig.7): (1) stage of continual expansion, with a wide deep fissure in front of the entrance, at some 200,000 years B.P. (2) stage of filling of the fissure in front of the cave entrance, at about 120,000 years B.P., forming a passage leading to the early man's habitat, although that entrance was soon blocked with breccia (3) stage of sedimenting travertines, at some 100,000-70,000 years B.P.

CONCLUSIONS

From the above, it can be seen that in Longgu Shan, the limestone caves have developed under a great influence of geological structure; they can be broadly grouped into two types, the vertical and horizontal types. The former is corroded along the steep surfaces of limestone strata and vertical joints, the latter along the surfaces of limestone strata which are gently inclined.

The caves developed mainly in two geological periods. The caves that formed during the Miocene and the Pliocene were under the peneplain of the Tongxian stage near the saturated zone of that time; they are now filled with fine gravel-sand, silt-sand and clay with only a few deposits of limestone breccia and travertine. In the Pleistocene, with the uplift of the Tongxian surface, the downcutting of rivers and falling of the groundwater level, a cave system was hollowed out in the vadose zone, and this has many deposits of limestone breccia, travertine and ash.

The caves in certain areas have their own process of development, so there are chronologically various sequences of sediment in caves, even at the same time. Only by understanding the processes of cave development can we answer when and where early man lived in the cave, correlate the cave deposits inside and outside, and ascertain the sequences of cave deposits in a temporal context.

REFERENCES AND FURTHER INFORMATION

Chia, L. P., 1978. A note on the weather conditions in Choukoudian area of Peking Man's time. Acta Stratigraphica Sinica, 2, 53-56 (in Chinese).

Chou Min-chen, 1955. Sinanthropus living environment inferred from vertebrate fossils. Kexue Tongbao, 1, 15-22.

Hsu, J., 1965. The climatic condition in North China during the time of Sinanthropus. Scientia Sinica, 15, 410-414 (in Chinese).

Kahlke, H. D., and Chow, P. S., 1961. A summary of stratigraphic and palaeontological observations in the lower layer of Choukoutian, locality 1, and on the chronological position of the site. Vert. Palas, 3, 212-200 (in Chinese).

Kung, Z. C., 1981. Discussion on the environmental changes during Peking Man's time and earlier or later than it as viewed from the analysis of pollen. Kexue Tongbao. 25, 65.

Liu, Zechun, 1982. The character of the formation deposited in a limestone cave. Journal of Nanjing University (Natural Sciences) 1, 163-178 (in Chinese).

Liu, Zechun, 1982. Climotostratigraphy of the Sediments in the Peking Man's Cave, in Quaternary Geology and environment of China pp25-31. China Ocean Press, Beijing.

Liu, Zechun, 1983. Peking Man's cave yields new finds. The Geographical Magazine, LV(6), 297-300.

Liu, Zechun, 1983. Le Remplissage de la Grothe de L'Homme de Pékin Choukoutien-Localité 1. L'Anthropologie, (Paris), 87, 163-176.

Liu, Zechun, 1985. Sequence of sediments at locality 1 in Zhoukoudian and correlation with loess stratigraphy in Northern China and with the chronology of Deep-sea cores. Quaternary Research, 23, 139-153.

Pei, J., and Sun, J., 1979. Thermoluminescence ages of quartz in ash materials from Homo erectus Pekinensis site and its geological implication. Kexue Tongbao, 24, 849.

Pei, W. C., 1939. New fossil materials and artifacts collected from Choukoutian during 1927-1939. Bull.Geol.Soc.China, 19, 147-188.

Qian, F., Zhang, J., and Yin, W., 1980. Magnetostratigraphic study of the cave deposit containing fossil Peking Man at Zhoukoudian. Kexue Tongbao, 25, 359.

Ren Meie, Liu Zechun et al., 1981. Evolution of limestone caves in relation to the life of early man at Zhoukoudian, Beijing. Scientia Sinica, 24, 843-851.

Yang, Z., and Mou, Y., 1982. Latest conception of the late Cenozoic strata in Zhoukoudian. Kexue Tongbao, 27, 55-61.

Zhu, S., Xia, M., and Wan, S., 1980. Uranium-Series dating of Peking Man. Kexue Tongbao, 25, 447.

Plate 1. Part of Upper Cave developed along knee-like bending of limestone.

Plate 2. Fine sand and silt in a cave on the East slope of Dragon Bone Hill.

Plate 3. Old stalagmites and breccias left after excavation of a shaft in Upper Cave.

Plate 5. Wide and deep fissure in front of the entrance to New Cave after excavation.

Platé 4. Travertines and stalagmites in New Cave and the ash layer in the lower part of the cave deposits (pointed to by hammer).

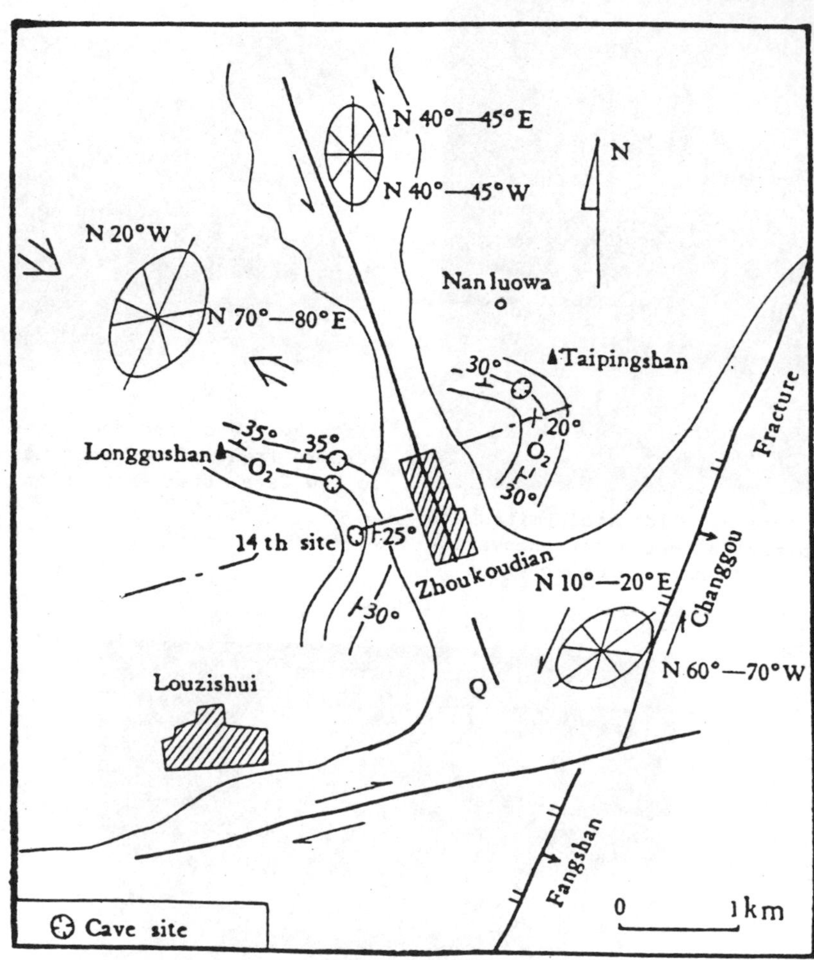

Fig.1. Sketch of geological structure at Zhoukoudian.

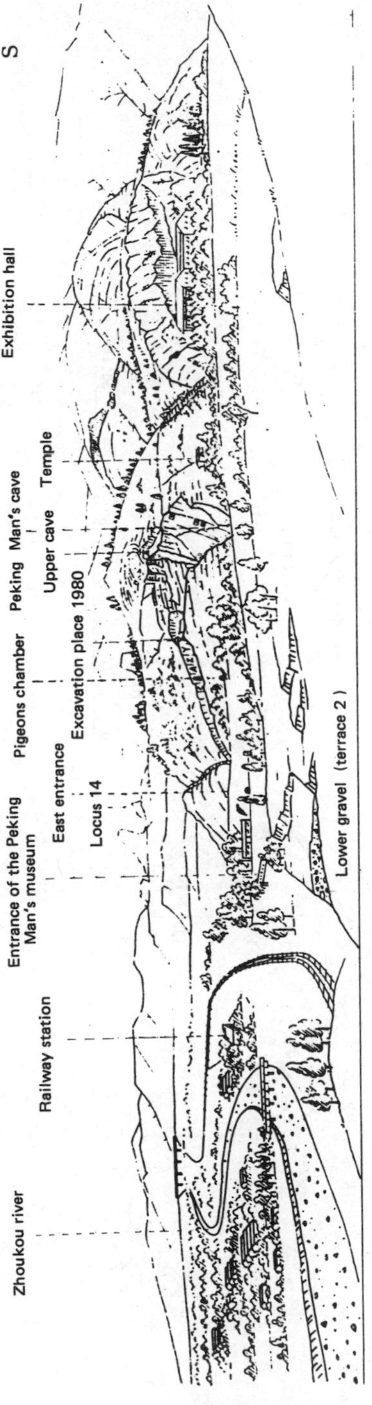

Fig.2. Sketch showing view of Dragon Bone Hill where fossils of early man (from Homo erectus to Homo Sapiens) were discovered.

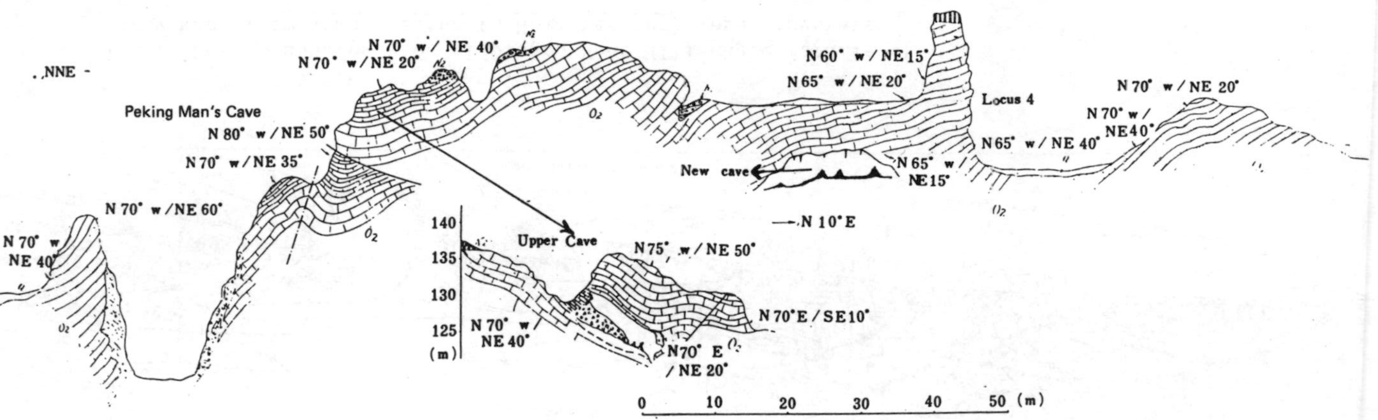

Fig.3. Geologic section of Dragon Bone Hill at Zhoukoudian.

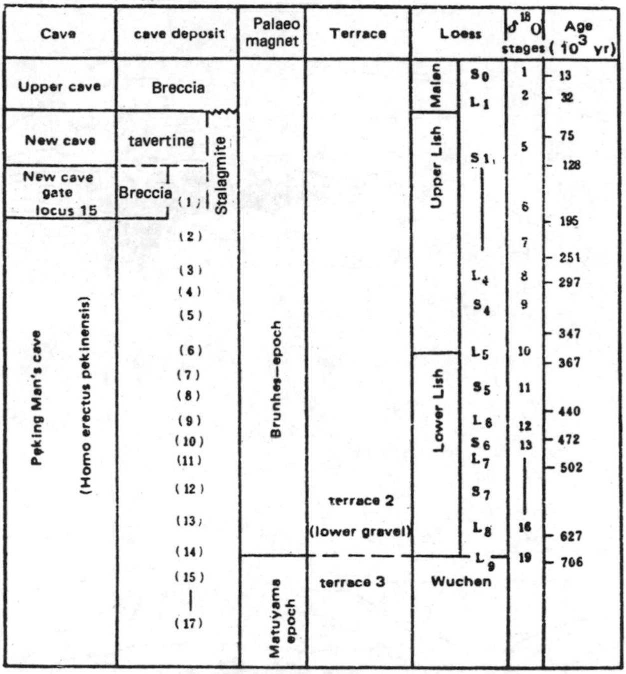

Cave	cave deposit		Palaeo magnet	Terrace	Loess		$\delta^{18}O$ stages	Age (10^3 yr)
Upper cave	Breccia		Brunhes–epoch		Malan	S_0	1	- 13
						L_1	2	- 32
New cave	tavertine	Stalagmite			Upper Lish			- 75
						S_1	5	- 128
New cave gate locus 15	Breccia	(1)					6	- 195
Peking Man's cave (Homo erectus pekinensis)	(2)						7	- 251
	(3)					L_4	8	- 297
	(4)					S_4	9	
	(5)							- 347
	(6)				Lower Lish	L_5	10	- 367
	(7)					S_5	11	
	(8)							- 440
	(9)					L_6	12	
	(10)					S_6	13	- 472
	(11)					L_7		- 502
	(12)			terrace 2		S_7		
	(13)			(lower gravel)		L_8	16	- 627
	(14)					L_9	19	- 706
	(15) – (17)		Matuyama epoch	terrace 3	Wuchen			

Fig.4. Stratigraphic Sequence of deposits, inside and outside the caves in Zhoukoudian.

Fig.5. Block diagram showing land surface of the Tongxian stage and silt-sand of cave sediments.

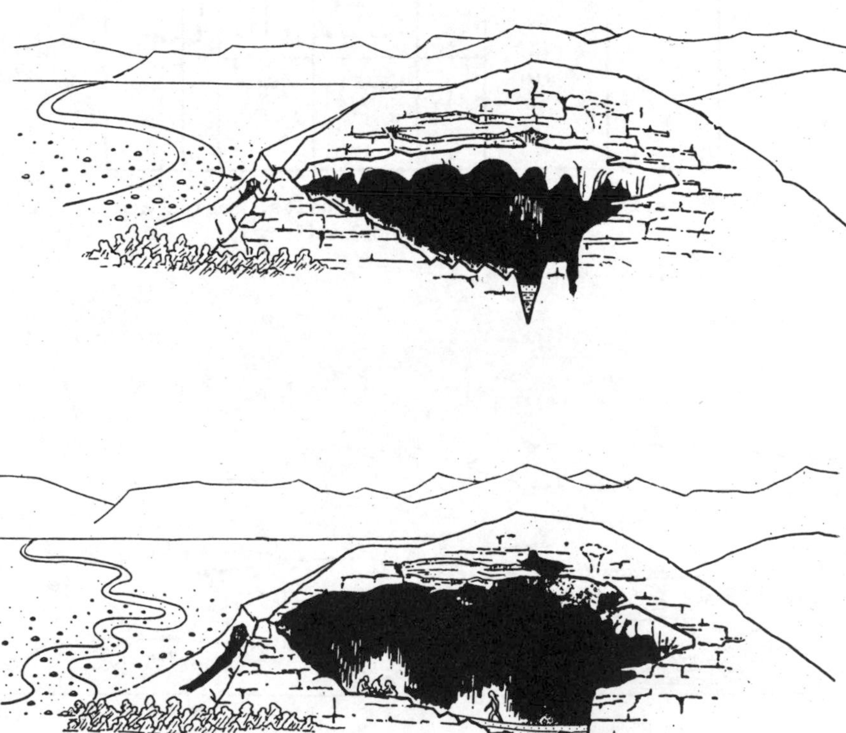

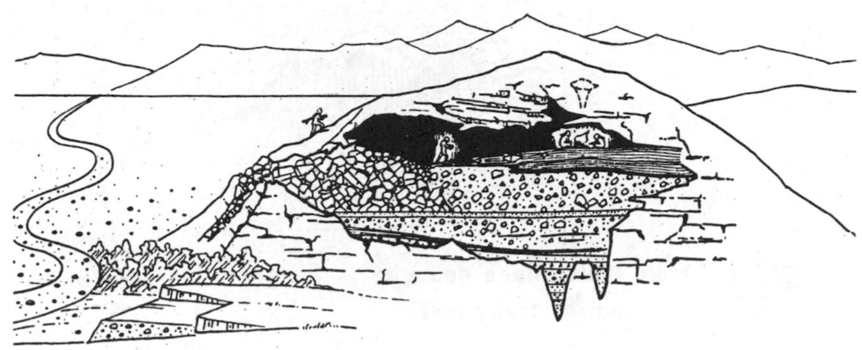

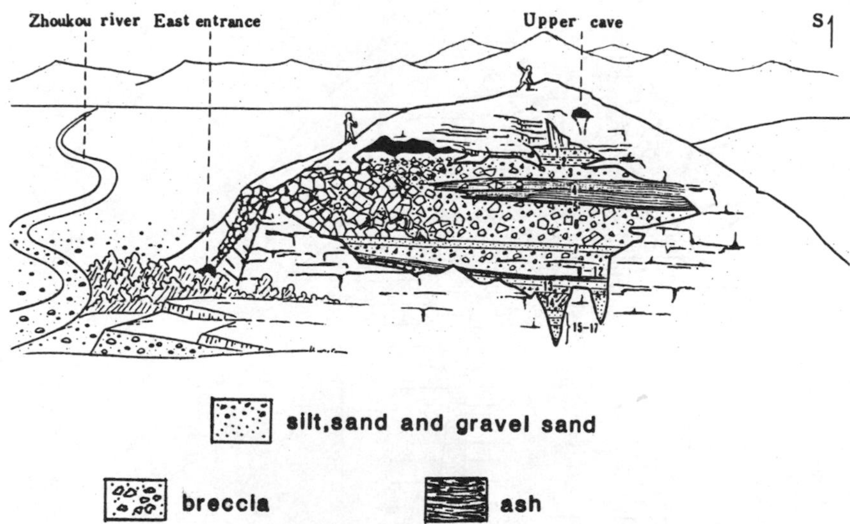

Fig.6. Block diagrams showing five stages in the evolution of Peking Man's Cave.

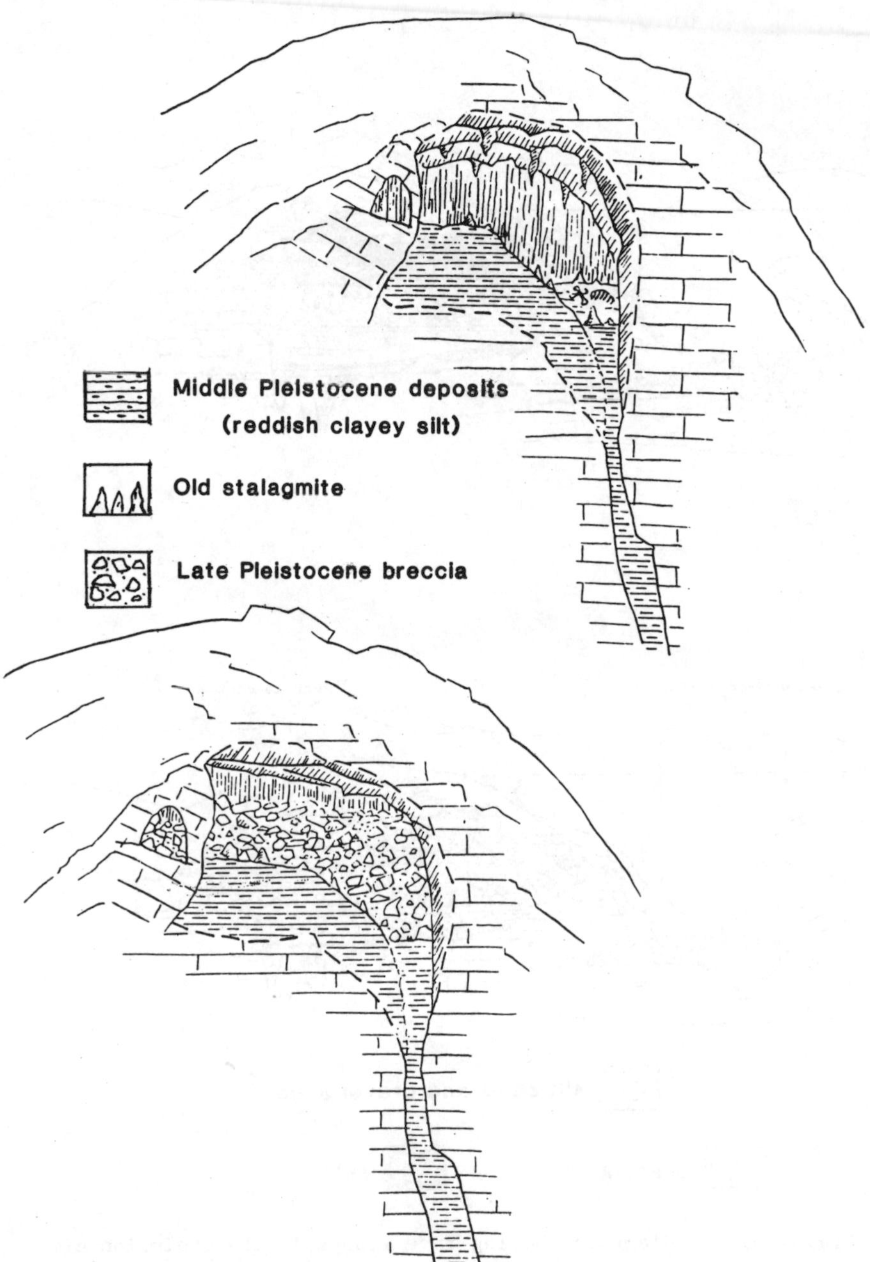

Fig.7. Block diagrams showing stages in the filling of Upper Cave. Top : stage of erosion of cave deposits (about 300,000 years ago). Bottom : stage of another filling of the cave and Upper Cave Man living in the cave (about 10-50,000 years ago).

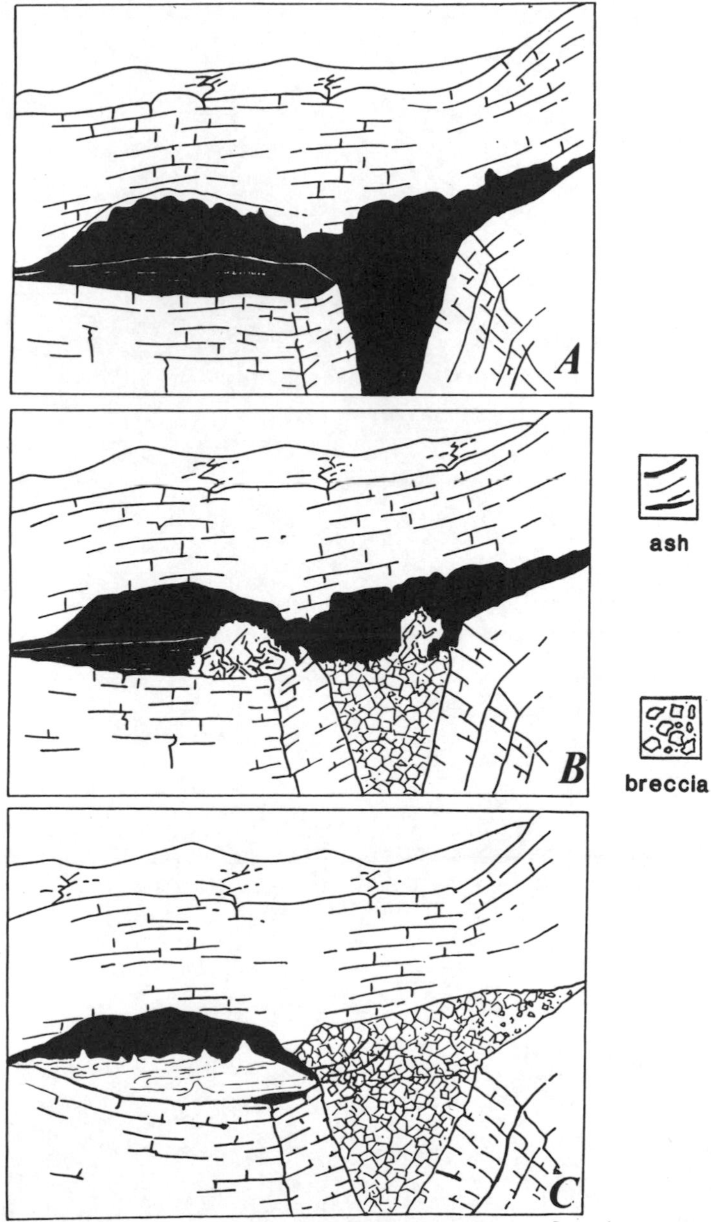

Fig.8. Block diagrams showing the filling processes of New Cave. A. Stage of continual expansion, with a wide, deep fissure in front of the cave entrance (about 200,000 years ago). B. Stage of filling up with limestone breccia and New Cave Man living in the cave (about 120,000 years ago). C. Stage of sedimenting travertines (about 70-100,000 years ago).

International Geomorphology 1986 Part II
Edited by V. Gardiner

A PRELIMINARY STUDY OF DEVELOPMENT PROCESSES AND DYNAMIC CONDITIONS IN THE YAOLIN CAVE IN EAST CHINA

Lin Junshu, Zhang Yaoguang and Huang Yunlin

Institute of Geography, Academia Sinica, Beijing, China.

ABSTRACT

The Yaolin Cave, in an area of subtropical climate and pure Carboniferous limestones, is one of the largest caves in East China. It has developed in four stages at four levels since the Middle Pleistocene, under the control of intermittently falling base level.

INTRODUCTION

The Yaolin Cave, situated in Tongluo county, Zhejiang Province, 85km from the famous picturesque city of Hangzhou, is one of the largest caves in East China and attracted explorers as early as the Sui-Tang dynasties (581-907 A.D.), as evidenced by poems written on the cave walls at that time. Geomorphologically, the region belongs to the unit of low mountains and hills in the west Zhejiang (Fig.1). It is in a subtropical climate, with the annual average temperature about 16°C, and mean annual precipitation 1500mm at Tongluo meteorologic station. The present karst denudation rate estimated by Corbel's formula is roughly 50mm per 1000 years.

Geologically, carbonate rocks outcrop widely on the southeast limb of the Bipu Syncline. The main karstic formations are the Middle and Upper Carboniferous Huanglong (C_2h) and Chuanshan (C_3c) limestones, and the Lower Permian Qixie (P_1q) limestones. They are pure and thick limestones, interbedded with a few dolomites, and all together about 400m thick. The chemical compositions of limestones in the west Zhejiang are shown in Table 1.

The Chuanshan and Huanglong limestones provide favourable conditions for karst and cave development and are widely distributed in the Yaolin Cave area. The Yaolin Cave is principally controlled by a normal fault zone of NE-SW alignment. A main cavern with six chambers has developed, with a length of about one kilometre and an area of 30,000 square metres; the cave system has been formed under a favourable palaeogeographic environment.

Table 1. Chemical composition of the Carboniferous limestones in the West Zhejiang, based on data from the Second Geological Team, Geological Bureau, Zhejiang Province.

Formations	Composition						(%)	
	CaO	MgO	SiO_2	Al_2O_3	Fe_2O_3	S	P	Loss
Chuanshan	54.68	0.40	1.25	0.26	0.10	0.04	0.004	43.22
Huanglong	55.68	0.06	0.22	0.05	0.04	0.004	0.003	43.65

DEVELOPMENT OF THE CAVE

The development of the Yaolin Cave is closely related to geomorphic processes in the area, in which the Shangyaoping planation surface at an altitude about 200m above sea level was formed in the Pliocene Epoch. According to the rounded gravels and the features of quartz grain surface textures seen with the scanning electron microscope, it can be recognized that water flowed over the ground surface at that time (Fig.2). Information we have obtained so far has shown that the Yaolin Cave did not develop at shallow depths (up to 150m) below the planation surface, but was probably formed beneath the karstified base-level.

Based on regional geomorphic processes and speleochronology of the speleothems dated by uranium series, radiocarbon and so on, we suggest that the Yaolin Cave has been gradually developing since the Middle Pleistocene. Based on the comparison of clastic sedimentary sequences and their petrologic properties, including chemical and mineral composition, between the caves and the terraces; the levels of the cave and the elevations of terraces; and the dating of speleothems, it has been suggested that the cave has undergone four development stages (Fig.3):
(1) The first, high level cave stage (the highest cave) was formed in the Middle Pleistocene. The altitude of the bottom is about 47-50m or more above sea level.
(2) The second level cave stage was formed in the Middle to Late Pleistocene. The altitude is about 40-47m.
(3) The third level cave stage was chiefly developed in the Late Pleistocene, the altitude being about 35-38m.
(4) The present underground passage stage (the lowest cave) was dominantly formed in the Holocene. The altitude of the channel is about 28-33m above sea level.

By comparing cave levels with those of the terraces, the altitudes of the first two levels of caves are seen to be 10-20m lower than those of the corresponding terraces. After that the differences tend to be gradually reduced. As the base level was lowered intermittently, the caves were first developed under a phreatic environment; they then became in the vadose zone and were modified.

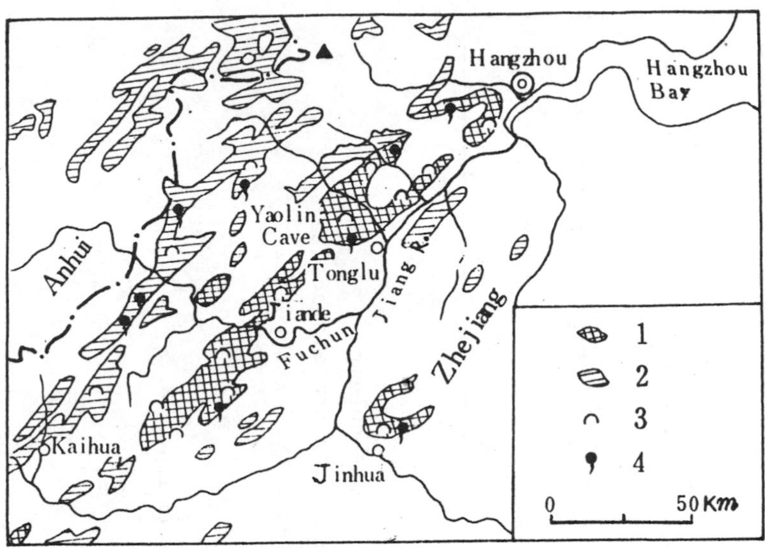

Fig.1. Location map of the Yaolin Cave and karst distribution in the West Zhejiang Province, China. 1. Pure Permian and Carboniferous carbonate rocks. 2. Impure Cambrian carbonate rocks. 3. Cave 4. Karst spring

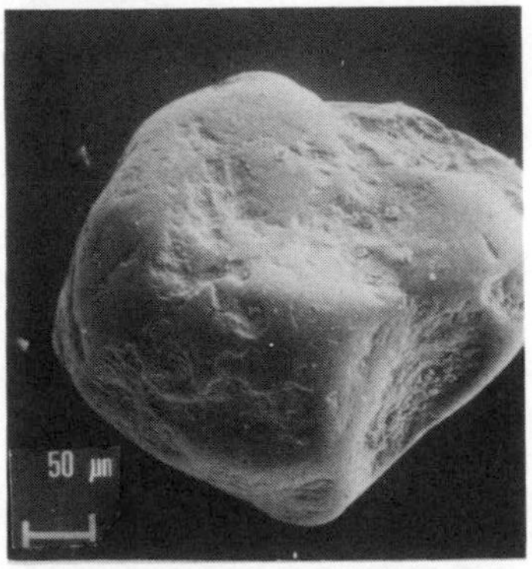

Fig.2. SEM image of a rounded quartz grain from the laterite on the Shangyaoping denudation surface. There are many striations and shallow dish-like pits on the surface of the grain.

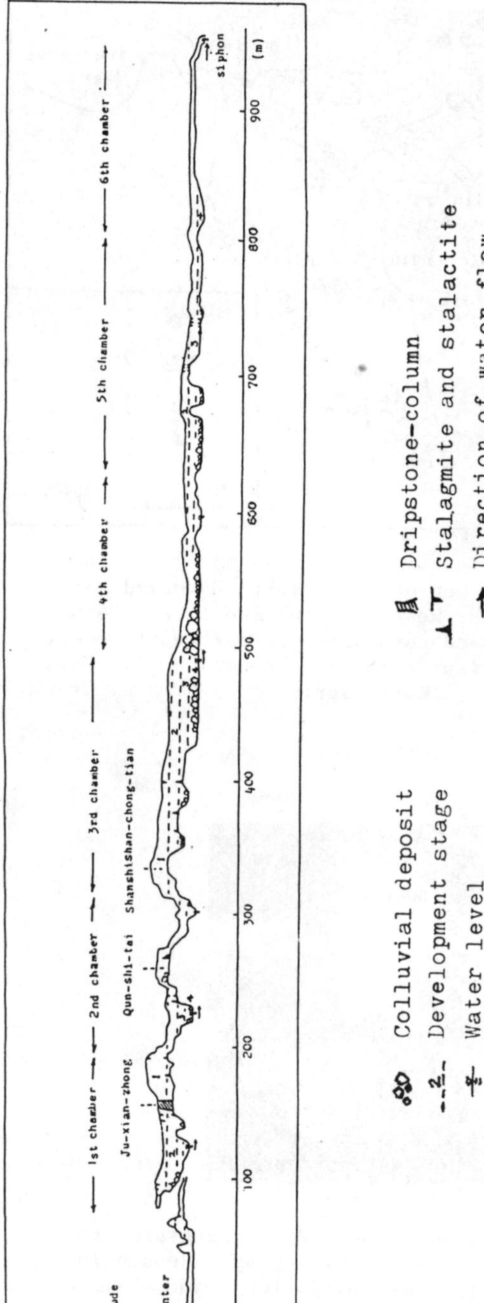

Fig.3. Longitudinal profile and development stages of the Yaolin Cave.

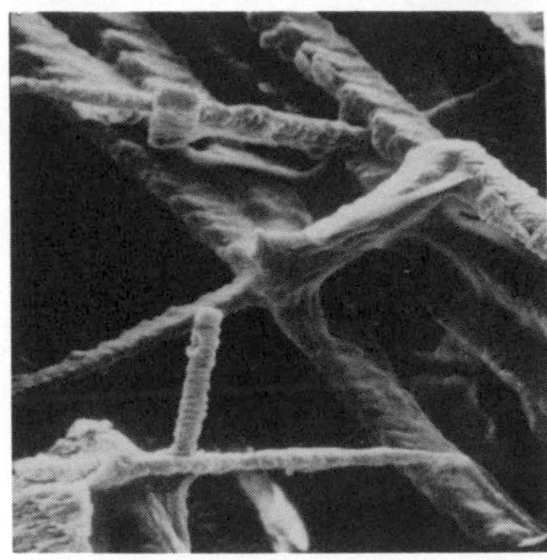

Fig.4. The Ju-xian-zhon speleothem.
(A) Part of a cross-section at the base. The black tone and pale grey-white tone represent dark red-brown and yellowish-white thin layers, deposited in wet and dry periods respectively.
(B) Sample from dark layer, of crystal with clay and acicular fibrocrystalline minerals (SEM image X 240).
(C) Sample from light-coloured sinter, of crystalline calcium carbonate, with less clay, and with dense and short bar-like crystal aggregates.

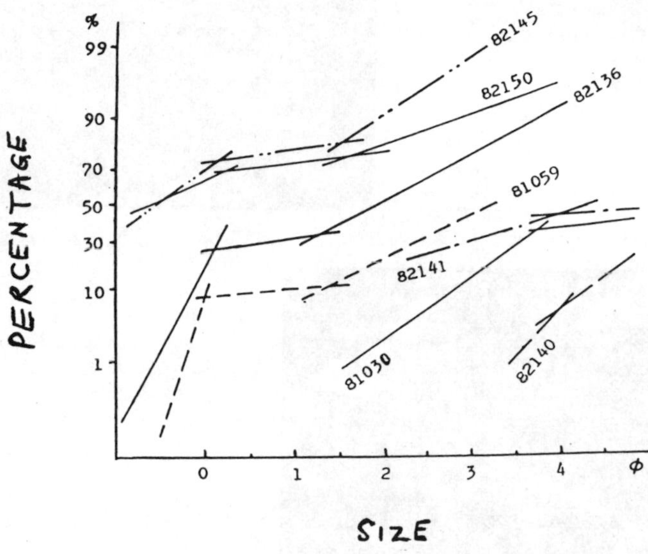

Fig.5. Probability curves of grain sizes of sediments in the Yaolin Cave.

In the latest period the modern underground passages are directly connected with the surface rivers through the ponors, and cut down together with the surface rivers as the base-level lowers.

DYNAMIC ENVIRONMENT AND CONDITION

Geochemically, the dynamic environment has evolved in several stages, especially in the Middle Pleistocene, in which the "Zijiang Formation", of gravels and reticulated laterite, was widely developed in this area. It has a low SiO_2/Al_2O_3 ratio of about 2.4 to 2.6, and low Sr/Ba ratio, as well as an assemblage of kaolinite-illite multiple clay minerals, the kaolinites occupying a dominant position. It is very clear that the environment was, in a dynamic geochemical condition, in the siallitic weathering stage, and the climate was hot and humid. Undoubtedly, it was favourable to solution, and karst denudation was higher than at present. Thus the first and second level caves have a huge original space, and the cave systems have abundant chemical deposits, such as colossal bell-shaped stalagmites and tall columns. For example, the Ju-xian-zhong (meaning the bell for calling gods) speleothem is about 18m high, and the radius of annual ring-like deposits, which recorded the periodic oscillations of the sedimentary environment, is more than 170cm at the bottom (Fig.4).

Hydraulically, the granularity analyses of clastic sediments in the second level of caves suggests that they are turbulent fluvial deposits; there also exist subterranean lacustrine deposits and mud debris-like deposits of a high density medium. By determining the interception points in the grain size probability curve (Fig.5), the velocity of underground flow is indicated to be about 7-57cm/sec when deposited. That is to say, the dynamic condition was similar to the present condition, and it fluctuated with higher velocities for long periods and lower velocities for short periods. Analyses of heavy minerals, spore-pollen and $\delta^{18}O$ of calcium sinters give similar results. The level of oxygen isotopes and the palaeotemperatures fluctuate around the level of the present mean annual temperature; the mean annual palaeotemperature amplitude might have been as much as 7°C at that period.

At the bottom and middle of the third level cave pebbles may be found. These indicate that since that time the cave developed into the stage of opening and direct connection with surface rivers. Many V-shaped pits were formed on the quartz sand surfaces by ground flow collision. This indicates that mechanical erosion increased with the base level intermittently descending and subterranean river downcutting gradually, and the roof of the lower-level cave collapsed. Thus, caves with a maximum height of 37m, and widths of over 70m, together with underground drainage passages were progressively developed.

CONCLUSION

Generally speaking, with the base level lowering relatively and intermittently and the periodic variation of palaeoclimate, leading to dynamic oscillations of corrosion and erosion, the whole cave systems developed in multiple stages, so that the cave pattern was characterized by multisteps or levels. In addition, at the same time as the main passage of the Yaolin Cave developed, on the planation surface a series of sinkholes, dolines, shafts and corrosion fissures connected with them, as indicated by tracing tests with environmental isotope 3_H , and tributary passages connected by shafts, such as the Yepan Cave, were formed. Thus a unified ground and underground karst morphology, and an integrated Yaolin Cave system were formed.

Acknowledgements

Grateful acknowledgements are made to Prof. Chen Zhiping, Drs. Huang Cixuan, Ye Xiangqing, Leng Daming, Jing Desheng and Song Linhua of the Institute of Geography, Academia Sinica; to Prof. Zhang Shuyue, Drs. Zhao Shusen, Ye Sujuan, Pei Jingxian and Tian Xingyou of the Institute of Geology, Academia Sinica; to Drs. Zhou Xuansen and Yu Jiansheng of Hangzhou University, for many valuable comments and data. We also thank the Sediment Laboratory of the Institute of Geography, for analyses.

REFERENCES AND SOURCES OF INFORMATION*

Institute of Geography, Academia Sinica, 1959. Geomorphologic regionalism of China. Science Press.

Group of karstology, Institute of Geology, Academia Sinica, 1979. Karst research of China. Science Press.

Zhou Xuansen, 1981. The Yaolin Cave in Zhejiang and its deposits. Journal of Hanzhou University, 8, (1).

"Physical Geography of China" Editorial Committee, Academia Sinica, 1984. Palaeogeography, Science Press.

Gu Chiliang et al., 1964. On the genesis, types and its age of Zijiang Formation. Journal of Hangzhou University, 2, (2).

Lin Junshu, Zhang Yaoguang et al., 1982. An analysis of the palaeogeographic elements of karst development in the Wumin Basin, Guangxi. Acta Geographica Sinica, 37, (2).

Zhou Xuansheng, 1985. Preliminary study of karst caverns in West Zhejiang. Karst Geomorphology and Speleology. Science Press.

Lin Junshu, Zhang Yaoguang, 1985. Progress of karst speleology study in China. Karst Geomorphology and Speleology. Science Press.

Huang Yunlin 1985. Application of the study of quartz sand grain surface textures with scanning electron microscope in karst researches. Karst Geomorphology & Speleology. Science Press.

* All published in Chinese

International Geomorphology 1986 Part II
Edited by V. Gardiner

WATER RESOURCES IN KARST REGIONS AND THEIR COMPREHENSIVE EXPLOITATION AND HARNESSING

Lu Yaoru

Institute of Hydrogeology and Engineering Geology,
Ministry of Geology and Mineral Resources,
Zhengding, Hebei, The People's Republic of China.

ABSTRACT

In the vast karst areas of China surface and underground water are intimately connected and together make up the water resources. The first part of this paper deals with the general situation of karst water resources in China, which are sources of electric energy, water supply and minerals. The chronologic nature of karst water is an important problem, discussed in part two; their different ages are caused and affected by many factors such as the kind of medium, mode of percolation, differential temperature effects, glacial influence, karstification, hydrothermal processes, overlying strata and diagenesis. A simple classification for the chronologic features of karst water is put forward. The third part centres on the discussion of the types of water resources in karst regions, based on karst types, karst hydrodynamic systems and chronologic features. Finally, the comprehensive exploitation and harnessing of water resources in karst regions is discussed.

INTRODUCTION

Water resources in karst regions usually include surface streams and subsurface water, which are intimately related to each other, and are very important in large areas of China. To study the formation, distribution and characteristics of these resources we have to base research on the fundamentals of karst development; karst (and their geomorphologic) types, karst hydrogeological engineering, and geological and minerogenetic conditions. All these are the premises for projecting the water resources in karst regions, which need to be exploited and harnessed comprehensively. This paper discusses several problems related to this.

THE SIGNIFICANCE OF WATER RESOURCES IN KARST REGIONS AND THEIR DISTRIBUTION IN CHINA

Surface and groundwaters in karst regions are not only significant to water supply but are also sources of electric energy and mineral resources [1].

Electric energy. Surface and subsurface sources of water power and sources of thermal water energy are included in this context. In the large karst regions, particularly in middle-south China and southwest China, there are many deep valleys with large gradients, formed by structural uplift. There is strong dissection, with large river flows caused by abundant rainfall in vast catchment areas. These are favourable for the construction of water conservancy schemes and hydro-electric stations, with benefits to irrigation, water supply, and flood control as well as navigation. They are also beneficial to forestry, animal husbandry and fishery. For example, the upper and middle streams of the Changjiang (Yangtze) River and its tributaries, the Hongshuihe River system and other river basins, have steep gradients (Table 1) and average flows from 400 to 14,300m^3/s. Water power stations will be constructed here on a large scale, with capacities of commonly several hundreds to several millions Kw. The hydroelectric station will be located in the main river course in San Xia (Three Gorges) of the Changjiang River. Main streams which have well developed knick points are very favourable, as they provide the high heads needed. The hydroelectric resources in southwest China and in middle-south China total about 83.3% of the total for China.

TABLE 1. Gradients of some rivers in karst regions

River	Length (km)	Gradient (°/oo)
Three Gorges of Changjiang	193	0.2
Qingjian	423	3.3
Wujiang	1050	1.65
Maotiao He	180	3.05
Dadu He	1062	3.9
Jinshajiang (reach in Sichuan)	1545	1.4
Nanpanjiang	936	1.98
Hongshuihe (Shuangjiangkou to Shanjiangkou)	659	0.3
Hong He (reach in China)	677	3
Lancangjiang (reach in China)	2153	2
Huang He (Hekou-Longmen)	725	0.84

Over 3000 large underground rivers with average discharge above 0.5m^3/s are known in southwest and middle-south China. Their flood discharges are mostly between 5 and 300m^3/s; therefore these underground rivers may be used to construct subsurface reservoirs or underground reservoirs connected with surface reservoirs, for the best regulation of all water resources. It is necessary to utilize topography in drawing the flow from subsurface reservoirs, for getting greater head. Most underground water power stations may be constructed on only a small scale, with installed capacity from several hundreds to several thousands Kw. A series of subsurface water conservation projects and water power stations have been constructed in many karst regions of China, and have already supplied important local electric sources [2], (Fig.1).

There are many karst thermal groundwater and karst thermal springs in Sichuan, Yunnan, Guizhou, Xizang and in north and east China. Many zones 1000-3000 metres deep have been exploited for confined thermal karst water with temperatures about 50-80°C and yields of one well typically about 2000-3000m^3 /day. These may be used directly for electric power, and indirectly reduce the demand for consumable electric sources by exploiting them for special water supply in industries, agriculture and medicine.

<u>Sources of water supply</u>. Surface and subsurface karst water, with lower total dissolved solids content and ordinary temperatures, recharged by natural recharge from rain, have significance for common water supply. For example over 100 larger karst springs with average flows over 1m^3/s and instability coefficients 1-5 have been found and exploited in north China. The larger discharges reach 5-13.7m^3/s, and most are about 1-10 l/s/km^2. The larger karst springs and underground rivers in south China are fine sources for water supply, where underground runoffs are mainly 5 to 55 l/s/km^2.

Basing our primary calculations on Yunnan, Guizhou, Sichuan, Guangxi, Hunan and Hubei, the karst water resources in bare karstified regions form 16.9-18.8% of the total surface runoff of each province or autonomous region; in the semi-arid north China, including Shanxi, Hebei, Shandong, Beijing and Tianjin, they are about 18.9 per cent. These water resources have filled important roles in water supplies for local industries and agriculture in intermontance basins and piedmont belts.

To exploit surface water in karst regions usually involves groundwater problems. For example, most of the larger karst springs and underground rivers are the main heads of surface rivers, and development of groundwater resources always influences surface runoff (Table 2).

There are no problems of water quantity in exploiting surface water for water supply in south China, but the key issues are related to the method of development, protection of water quality and utilization in a comprehensive manner.

<u>Liquid mineral resources</u>. Salt water, brine and mineralized water as well as groundwater containing special mineral elements may be exploited in karst regions as liquid mineral resources. These include the karst water of some karstified beds buried deeply in several zones of Sichuan. These waters were formed in the Triassic

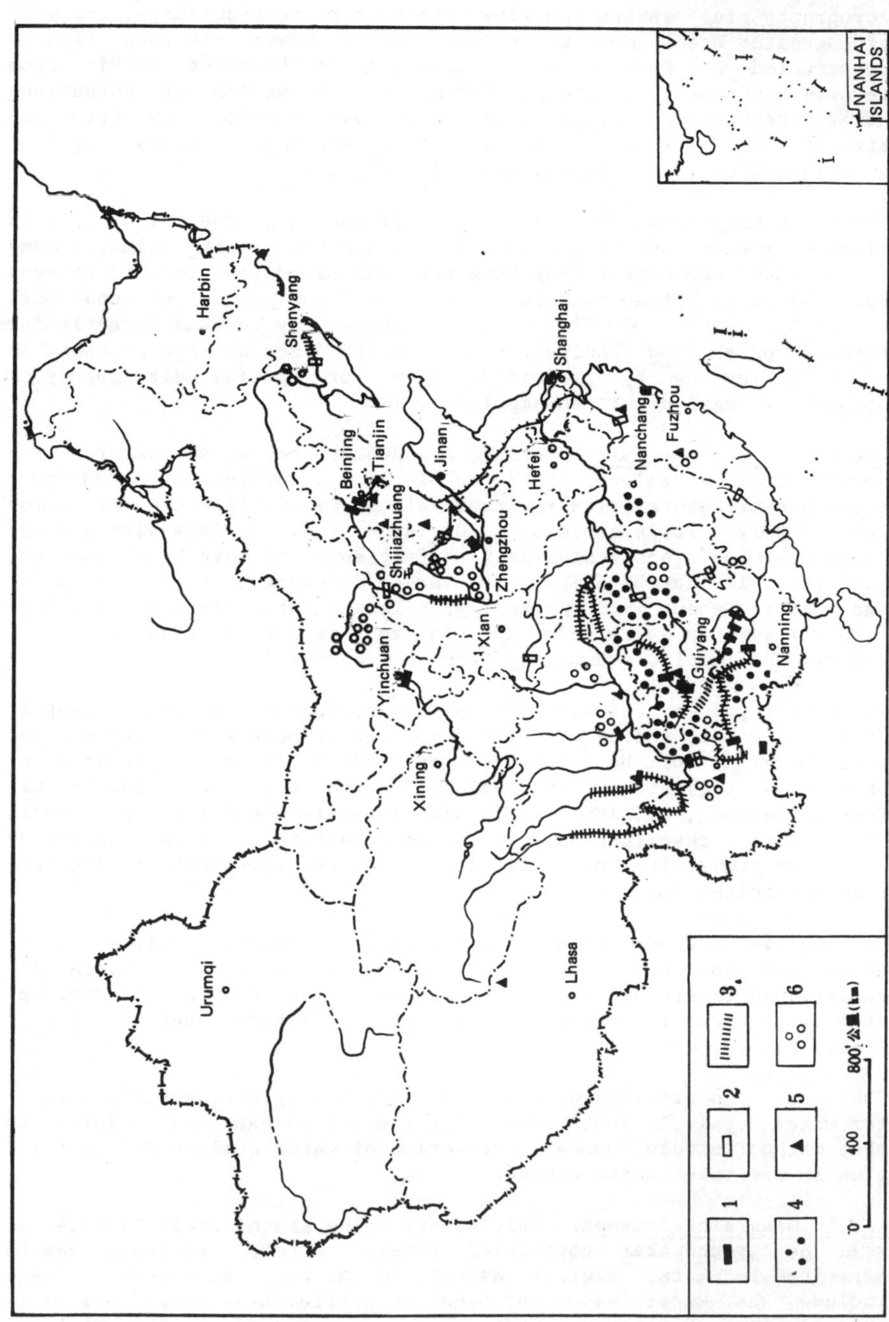
NANHAI ISLANDS
Harbin
Shenyang
Beinjing
Tianjin
Jinan
Shanghai
Fuzhou
Nanchang
Hefei
Shijiazhuang
Zhengzhou
Xian
Yinchuan
Xining
Guiyang
Nanning
Urumqi
Lhasa
1
2
3
4
5
6
0 400 800

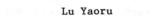

period and enclosed by later overlying beds and/or by the calcareous deposits from karst processes to make up the enclosed lixiviation type of liquid mineral resource enriched in elements Br, I, K, Li, Sr, Ba, B, U, Ra and others. The origins of brines and mineralized waters have certain similarities to hydrothermal processes. According to conditions of enclosure and total dissolved solids (commonly 5-250 g/l), karst water as a mineral resource may be separated into salt-brine, brackish water and mineralized water.

TABLE 2. Discharges in several rivers

River	Hydrological profile	Catchment area A (km^2)	Average flow quantity Q (m^3/s)	Average runoff per annum Qy (10^8)
Changjiang (Yangtse)	Yichang	1000,000	14,300	4509
Hongshuihe	Datengxia Gorge	190,400	4,110	1,296
Nanpanjiang	Tianshengqiao	50,194	6.5	193
Liujiang	Liuzhou	45,800	1,292	407
Qingjiang	Geheyan	14,430	380	120
Daduhe	Leshan	76,400	1,500	473
Wujiang	Pengshui	69,920	1,300	410

The salt lakes, salt water lakes and brackish water lakes formed in interior basins have accumulated karst streams from halide, carbonate and sulphate rocks, to produce enrichment in elements B, K, Li, Sr, Ba, Rb, Cs, U, Th, Ag, Hg, Cu, Zn, Pb, and so on. Owing to changes of geological-geographical environments and climatic conditions, strong evaporation in the arid basins has caused many series of borate, sulphate, halide and carbonate minerals to be deposited from these lake waters. In many inland basins the salt water occupies from about 86% to 100% of the lake water. The content of tritium ^3H in several lake waters with total dissolved solids about 49-365 g/L is from ten to hundreds T.U.; the ratio of 32s/34s is about 22 (data from Qinghai Salt Lake Institute, Academia Sinica, etc.). These

Fig.1. [FACING PAGE]. Electricity resources based on karst water in China.
1. Typical water power station and water conservancy projects, with dams and reservoirs on karstified beds. 2. Typical water power or water conservancy projects with karstified beds in the reservoir zone. 3. Rivers in typical karst regions with potentially rich water power exploitation in the near future. 4. Typical karst region with rich subsurface water power. 5. Typical place with karst water resources to be developed. 6. Regions with rich coal energy sources yet to be exploited, and with a relationship to karst water resources in water supply and mine drainage.

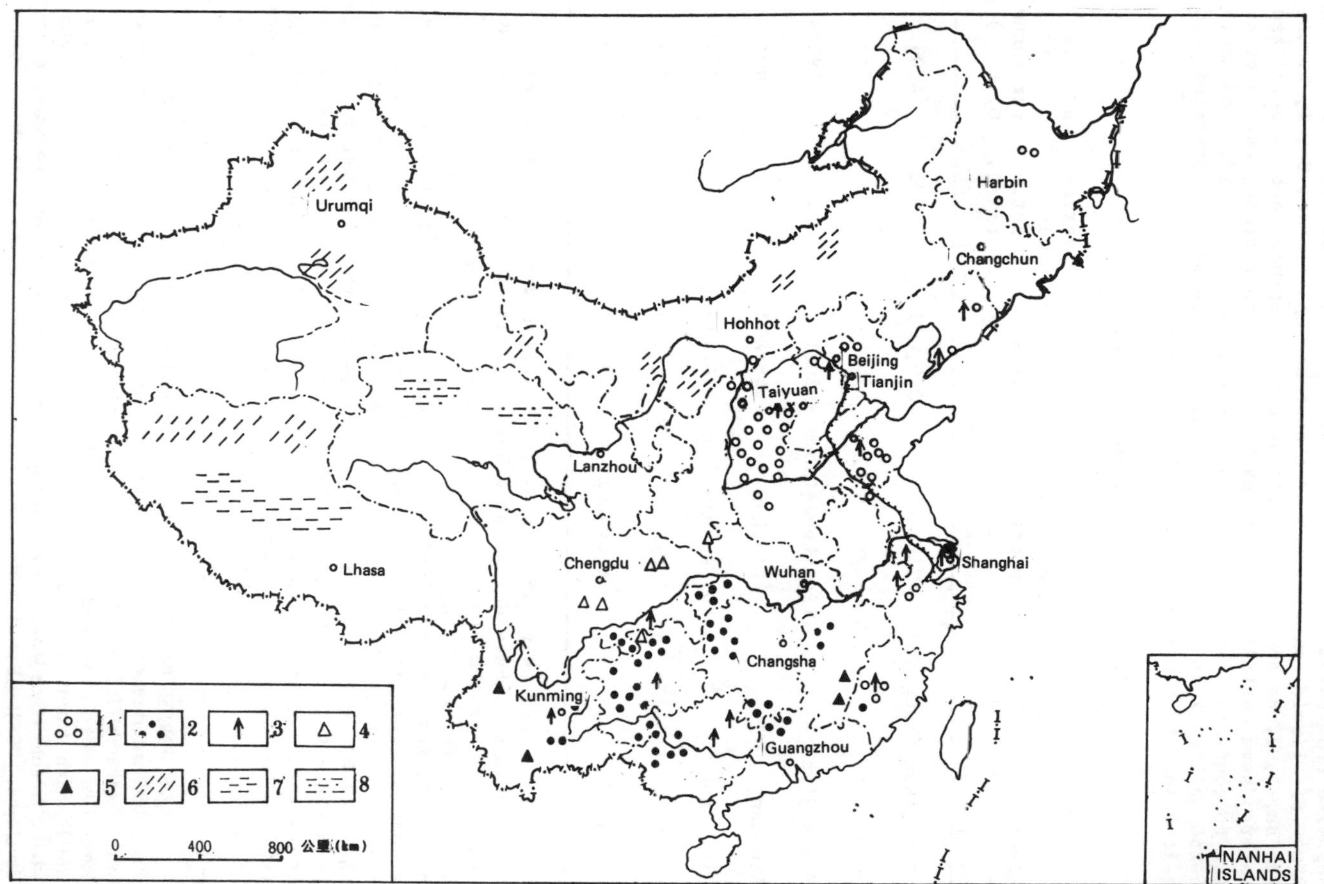
Urumqi
Harbin
Changchun
Hohhot
Beijing
Tianjin
Taiyuan
Lanzhou
Lhasa
Chengdu
Wuhan
Shanghai
Changsha
Kunming
Guangzhou
NANHAI ISLANDS
1
2
3
4
5
6
7
8
0
400
800
公里(km)

results, and other studies, indicate that the sources of salt lakes water are in many ways related to surface and sub-surface streams and, in early stages, to hydrothermal processes. In later periods, these liquid mineral resources are always influenced by atmospheric rainfall, melting snow runoff and the activities of fresh karst groundwater. The general situation concerning exploitation of karst water liquid mineral resources in regions of China is expressed in Fig.2.

Certain karst regions allow all three modes of exploitation of water resources, but many regions have prospects for only one or two. An important problem is the connection between these three purposes; generally, water supply and hydroelectric power will be exploiting the same sources, either on surface rivers or in subsurface rivers. Therefore, it is necessary to establish an equilibrium between the development of industry and of agriculture, as well as establishing their economic benefits.

FEATURES AND CLASSIFICATION OF KARST WATER CHRONOLOGY

The evaluation of water resources in karst regions must rest on the basis of their chronologic characteristics; thus their genetic mechanisms and recharge, runoff and discharge characteristics will be more fully understood. Building the reservoirs will also need to take chronologic data into account in order to study the existence and distributions of seepage ways from reservoirs [4], [5]. It is also necessary to obtain chronologic data in order to grasp the genetic and storage characteristics of karst water as a liquid mineral resource.

Chronologic features of karst water. To study the chronologic features of karst water three dating methods are usually adopted: tritium, radiocarbon and uranium series. These chronologic results are useful for studying flow velocity, flow direction and flow net as well as the zoning of karst water.

As pores, fissures and caverns develop in the same karstified bed they make the karst water unhomogeneous. Therefore, the three kinds of karst water may occur in the same bed and in the same aquifer; with accumulation, movement and storage the result is complex chronologic evidence. In addition to the influences of cosmic radioactivity and artificial nuclear explosions, latitudinal and seasonal effects also must be considered. For example, the tritium contents

---

Fig.2. [FACING PAGE] Exploitation of karst water in China
1. Typical regions for exploiting karst water resources by wells, tunnels and guiding karst springs. 2. Typical regions for exploiting karst water resources by constructing underground dams and guiding karst springs. 3. Important cities exploiting karst water resources for water supply. 4. Typical region exploiting subsurface salt water and brine. 5. Typical region with rock salts. 6. Typical region with salt water - brackish water lakes for extracting salts and other minerals. 7. Compound karst region enriched in Lithium and Boron in salt lakes. 8. Compound karst region enriched in Potassium, Magnesium and Boron in salt lakes.

of rain in Beijing was about 30 T.U. in 1953; influenced by nuclear explosions, the maximum value reached thousands of T.U. in 1963; then the average value decreased to about 92 T.U. in 1979, and about 64 T.U. in 1982. In spite of that, several results of chronological tests have been used to bring together studies of groundwater regime, hydrodynamic condition and water quality, and approximate ages have been determined.

Some tritium results from karst groundwaters and karst spring waters, sampled by the author in typical regions of China in 1980-83, and other related hydrogeological results and their approximate chronologic features, are expressed in Table 3. These indicate how karst waters in different hydrodynamic conditions in several typical karst regions have different values of tritium contents.

Analyses of the formation of karst water of different ages. The karst waters in different aquifers or in different zones of an aquifer or hydrodynamic system, ordinarily have different chronologic features, caused by many factors. These are:

(1) Medium effect: the different media in the same karst aquifer have been formed by unhomogeneous karstification; water percolates at different speeds, and chronologic ages will differ by from half a year over several years.
(2) Percolating effect: under the control of the same flow net in an aquifer or hydrodynamic system, percolating karst water flows may nevertheless differ in either their velocities or in their circulating distances; their chronologic ages will therefore have values, differing by several months to over ten years.
(3) Temperature effect: surface flows may sink or percolate into subsurface karstified beds and meet the karst groundwater with different temperatures. This temperature effect may enhance corrosion to form larger caverns, or may lead to calcareous deposition outside the caves.
(4) Karstification effect: the normal karst process of corrosion is always accompanied by chemical and mechanical deposition which may enclose or block several early karst caverns and corroded galleries in the same karstified system. This causes the early trapped karst water to have an older chronologic age.
(5) Glacial periodic effect: during the glacial periods the percolating water had a lower temperature and contained a greater concentration of dissolved carbonates. This water moved toward deep zones, where their increased temperature enhanced the diffusion of carbon dioxide and promoted re-deposition of carbonates from karst groundwater; such processes over a long period had the result of enclosing the karst water at depth. Researches illustrate that the last glacial stage of the Quaternary was characterised by a lower sea level, which influenced the lake and river levels in typical karst regions of China. Early results suggest that the chronologic ages of the karst water which was enclosed in the last glacial stage is about 10500-30500 years old (Geyh, 1972; Smith et al., 1976), but the semi-enclosed karst water is only several thousand years old, owing to the later mixture with new water.

TABLE 3. Contents of ^3H and approximate ages of karst water from typical regions of China.

Place	Flow feature	Quantity of ^3H T.U.	Approximate age in years
Bohai Sea water	Influenced by karst spring	32.1± 5	
Dalian coastal	Karst spring of Sinian	32.1± 5	c.8 - 10
Huang hai Sea water in Dalian	Normal sea water	12.5± 5	
Dalian	Karst groundwater	59.5± 5	c.2 - 3
Benxi, Liaoning	Ground river water	66.7± 5	c.several
Taizihe river water	Converging karst springs	68.7± 6	Influenced by karst groundwater
Benxi region	Karst spring in O bed	70.4± 5	c.1 (influenced by subsurface stream)
	Karst stream in C bed	98.1± 6	c.1 - 2
Niangzi Guan, Shanxi	Karst springs in O bed	up to 7	c.10 to several tens
Xingtai, Hebei	Karst springs in O bed	50.7± 6	several to over ten
Yongan, Fujian	Karst spring water in C-P bed	24.6± 5	several
Beijing	Ground pond in cave	117.4± 7	c.1 - 2
	Ground pond water	67.7± 6	several to 18
	Ground river water	62.2± 6	over ten
	Thermal karst water 1000m deep	2.16± 2.1	c.several tens
	Karst thermal spring	up to 2	c.several tens
Shanghai	Karst groundwater 200m deep	up to 7	over ten
Xingzi, Jiangxi	Similar karst thermal spring	up to 7	c.20 - 30
Shanggao, Jiangxi	Karst ground river	23± 6	several
Hangzhou, Zhejiang	Karst spring	26± 6	several
Jinan, Shandong	Karst spring	83.5± 7	c.3 - 4
Zunyi, Guizhou	Karst flow Karst fissure flow	40± 10-20	1 - 4 or more

(6) Hydrothermal effect: hydrothermal water in deep zones will develop karst caverns and passage-ways and will become enriched with dissolved carbonates and other minerals; then it may rise to shallow zones to meet ordinary karst water percolating from the surface. The results of this usually cause mixture corrosion and differential

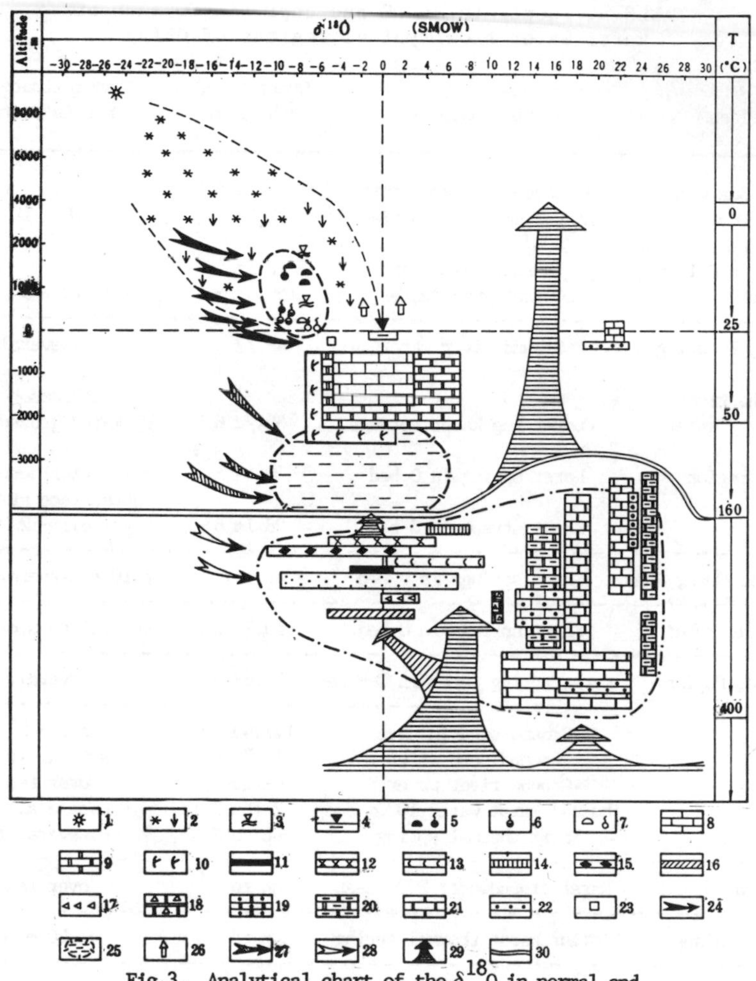

Fig.3. Analytical chart of the $\delta^{18}O$ in normal and hydrothermal karst water of some regions in China.

1. new snow on Mount Qomolangma. 2. snow and rain water of regions from Xizang to Beijing and Shanghai. 3. river water. 4. normal sea water. 5. underground river and karst springs in E'mei Mountain region of Sichuan. 6. karst springs in Jinan etc. regions of Shandong. 7. underground river and karst springs in Hongzhou-Tonglu region of Zhejiang. 8. normal limestones. 9. normal dolomites. 10. coral beds etc. 11. magnetic iron ore. 12. mineral liquid. 13. hydrothermal of minerogenetic stage. 14. calcite near iron ore. 15. lead-zinc ore. 16. calcite near uranium deposit. 17. dolomite-magnetite. 18. ore dolomite. 19. mineralized dolomite. 20. country rock of dolomite. 21. country rock of limestone. 22. sparry rock. 23. shallow buried karst water in Shanghai. 24. flow direction of shallow karst water. 25. karst salt water and brine in deep zones. 26. karst thermal spring and brine spring. 27. movement directions of karst water flow in deep zones. 28. percolation direction of upper karst water at minerogenetic stage. 29. hydrothermal injective direction and the limited line of mineralized zone. 30. distributive line of karst hydrothermal mineral deposits without the level limitation.

TABLE 4. Simple classification for chronologic features of karst water.

	Chronologic features of karst groundwater	Approximate age (years)	Measurement method
Annual water (A)	Rain karst (R)	up to 0.5	^{3}H and artificial trading
	Basic karst (B)	about 1	^{3}H
New water (N)	Multi-yrs. karst (M)	1 - 10	^{3}H
	Long term karst (L)	10 - 100	^{3}H, ^{14}C
Early water (E)	100s of yrs. karst (H)	100 - 1000	^{3}H, ^{14}C
	1000s of yrs. karst (T)	1000 - 10000	^{14}C, uranium series
Older water (O)	Late Pleistocene older (OQ_3)	10000 - 150000	^{14}C and other data comprehensive analyses
	Early-mid Pleistocene (OQ_{1-2})	150000 - 2400000	^{18}O, ^{2}H and other data comprehensive analyses
Palaeo water (P)	Tertiary palaeo karst (PR)	2.4×10^6 - 67×10^6	^{18}O, ^{2}H, $^{32}S/^{34}S$, He/Ar etc isotopic methods and comprehensive analyses
	Pre Tertiary palaeo karst (PK) to (PZ)	67×10^6 - 850×10^6	
Diagenetic water	Each sedimentary period of soluble rocks (DQ) to DP_t)	more than 850×10^6	^{18}O, ^{2}H, in carbonate rocks and water etc. isotopic methods and analyses

temperature effects in dissolution, as well as deposition of calcite and other minerals to fill the pores, fissures and caverns. Obviously, karst water which has been closed or semi-enclosed always has the older chronologic ages. Using the stable isotopes ^{18}O and ^{2}H will help in the study of the genetic environment and the features of enclosed karst water and minerals, formed by early hydrothermal processes (O'Neil, 1977; McCrea, 1950).

(7) Overlying strata effect: karst water and gas masses formed in early karstified stages and enclosed by later overlying strata, will be transported upward and accumulated in suitable zones. For example, the karst water gas systems in some regions of the Sichuan Basin have been tested for their chronologic ages by the He/Ar method (for instance the data from Lanzhou Geological Institute of Academia Sinica). If we analyse these chronologic data and related karst hydrodynamic results we may reach several conclusions. First, the ages of gas masses stored in Permian carbonate beds in some regions are 500-600 million years old; such results indicate that the water-gas masses were formed in the Sinian system and were transported upward and trapped in Permian layers at a later stage. Secondly, the

gas masses stored in Triassic carbonate beds have ages over 200 million years old, and resulted from early water-gas masses in the original beds of the Permian system, which rose up and were trapped again.
(8) Diagenesis effect: carbonate deposits covered by clastic or other carbonate materials after their sedimentation immediately undergo diagenesis; they are compressed by gravity and quantities of water and gas from pores and fissures are released and are trapped in suitable strata. The gas masses stored in Permian layers in some regions have ages about 200 million years old, corresponding to the age of their source beds.

Results of isotopic studies of several typical karst regions, that were processed by the author, and also incorporating other related results [16], [17], [18], are summarized in Fig.3. This shows the value of $\delta^{18}O$ of snow from Mount Qomolangma, which is -25.76 (Zhang Rongshen, [16]), and from the Qinghai-Xizang Plateau; it also shows figures from Beijing and Shanghai, both located in plains at lower altitudes and near seas; the main values of $\delta^{18}O$ in snow and/or rain water are obviously increased, but those of both regions are still smaller than -1. Most brine and mineral liquids have been enclosed and have values of $\delta^{18}O$ between -12 and +8. Values of ore bodies and their carbonate country rocks with mineral deposits are between -11 to +26. Ordinary values are positive, but negative values express the possibility that mixed processes from shallow karst flow had occurred. Several calcites and dolomites sampled from a mercury ore zone in a karst region have had their genetic temperatures (about 88°C - 246°C) determined by testing their liquid-gas inclusions [11]. All these results clearly indicate that hydrothermal processes are also important in making up the liquid minerals or brines that are enclosed, with older chronologic ages.

Based on the above discussions, the chronologic features of karst groundwater may be classified as in Table 4. Most of the karst waters are supplemented mainly by rain water directly or indirectly, and belong to annual water or new water; this is exploited for water power and water supply. However, karst water as a liquid mineral resource belongs mostly to the enclosed or semi-enclosed earlier karst waters, as well as diagenetic water.

TYPES OF WATER RESOURCES IN KARST REGIONS AND THEIR COMPREHENSIVE EXPLOITATION

The classification of water resources in karst regions may be based on different principles; valuable discussions of the related problems are [12], [13] and [14].

Principles for the classification of water resources in karst regions

The reason for classifying water resources in karst regions is to stress the formation, storage, runoffs and discharges, in advance of the comprehensive exploitation and harnessing of karst water. Therefore:

(1) Classifying water resources in karst regions must consider their significance and exploitation values for electric energy, water supply and liquid mineral resources.
(2) Both surface and subsurface waters are linked together, and should be considered together in a classification for their integrated management, exploitation and harnessing.
(3) Chronologic features of karst groundwater, an important factor for classification, may also be important in exploitation of karst water resources.
(4) The discharges from karst springs and the flows in surface rivers are all natural resources; we need to consider their basic flow quantities in the natural ecosystem. Thus flows must not be less than 30 - 50% in streams when exploited, in order to protect the ecological environment.

TABLE 5. Simplified types of natural water resources in karst regions.

	Karst water resource type	SURFACE WATER				KARST GROUNDWATER					
		River	Lake	Salt lake	Sea	Annual (A)	New (N)	Early (E)	Older (O)	Palaeo (P)	Diagenetic (D)
	Corroded peak-hill (PH)	PHR	PHL			PHA	PHN	PHE			
B	Corroded valley-plane (PL)	PLR				PLA	PLN	PLE			
A	Eroded high to mid-height gorge (HM)	HMR				HMA	HMN	HME			
R	Eroded lwr. mountain river, lake (VL)	VLR	VLL			VLA	VLN	VLE			
E	Coastal and reef island (CI)	CIR	CIL	CISL	CIS	CIA	CIN	CIE			
K	Anticline & dome mountain land (AN)	ANR	ANL			ANA	ANN	ANE			
A	Syncline structural mountain land (SY)	SYR	SYL			SYA	SYN	SYE			
R	Fault basin and mountain (FA)	FAR	FAL			FAA	FAN	FAE	FAO		
S	Complex structural high/mid mtn. (SH)	SHR	SHL			SHA	SHN	SHE	SHO		
T	Scattered structural low mtn./hill (TA)	TAR	TAL			TAA	TAN	TAE			
	Structural island lake (IN)	INR	INL	INSL		INA	INN	INE	INO		
B	Covered by Quaternary beds (PE)	PER				PEA	PEN	PEE	PEO	PEP	
U R	Covered by Tertiary beds (SL)	SLR					SLN	SLE	SLO	SLP	
I E	Covered by pre-Tertiary beds (ST)	STR						STE	STO	STP	STD
D	Buried in sea bottom (CL)			CLSL	CLS			CLE	CLO	CLP	CLD

(5) Apart from evaporation losses, streams in valleys or underground may be used as hydroelectric sources and, after generating electricity, they can be wholly or partly used in other ways, such as in water supply.
(6) Different karst types have different hydrodynamic conditions, which is an important factor in classifying water resources in karst regions.

Classisification of natural water resources in karst regions. For the above reasons, no unified evaluation of the potential of karst surface and subsurface waters for exploitation has been developed. For comprehensive exploitation, karst types and their hydrodynamic systems may be the major principles for classifying the water resources in karst regions; these have been divided by the author into a series of types and sub-types, but this paper only lists the simplified classification as in Table 5. Certainly, not all types of water resources exist in every karst region. In using the electronic computer for storage of the related data, the types are expressed in letter combinations (Table 5).

Comprehensive exploitation and harnessing of water resources of karst regions. It is essential to involve the rational distribution of industries and agriculture as well as the benefits to them in considering the comprehensive exploitation and harnessing of water resources in karst regions. This paper summarizes the following experiences:
(1) Surface water and shallow groundwater are recharged by rain water as renewable water resources; these quantities are closely determined by precipitation and the quantities of percolation. But most of the older or palaeo-water can not be treated as a renewable resource. Therefore, it is necessary to consider limitations to regeneration of water resources.
(2) Continual changes related to the formation, storage, movement and discharge of karst water occur normally; therefore, understanding changes of karst water characteristics in karstification is very important when considering their comprehensive utilization.
(3) Exploitation and harnessing of the karst water will rapidly influence the local or zonal natural environment and geological conditions. For example, drainage of karst flows from mining pits for the exploitation of solid mineral resources may quickly influence karst water in adjacent zones and may even lead the surface or ground to collapse; large or middle scale reservoirs will influence the storage, movement, discharge and re-distribution of local or zonal karst resources; harnessing karst water in railways or other industrial or civil engineering constructions may also influence the natural state of karst water flows.

Primarily, therefore, the comprehensive utilization and harnessing of karst water resources lies in the most effective exploitation, with the best economic benefits; harmful phenomena must be avoided in order to protect the environment. Thus, using experience from different regions, the comprehensive exploitation and harnessing of water resources in karst regions may be summarized in the following ways:

(1) Broad corrosion karst regions (PH, PL):

(a) storage and guiding by constructing subsurface reservoirs to store groundwater and then to guide the water into electric stations for generating electricity; or guiding into canals or pipes for water supply and irrigation.

(b) storing-pumping-guiding: building subsurface reservoirs to store water for pumping up to mountain lands and for guiding into lower hydroelectric stations.

(c) storing-accumulating-guiding: constructing subsurface reservoirs to store groundwater and surface reservoirs to accumulate surface streams; then both resources may be used for water power and irrigation.

(d) blocking-draining-guiding: draining water in mining regions and blocking outside surface water by dams to avoid more percolating flows from surface valleys into mining pits; surface water and pumped groundwater can then be guided for water supply and/or for generating electricity.

(2) Limited corrosion karst regions (AN, SY, FA):

(a) storage-guiding: constructing surface reservoirs to store surface streams for water power and guiding flows for water supply.

(b) storing-collecting-guiding: constructing surface and subsurface reservoirs to store and collect the streams for water power and water supply.

(c) pouring-guiding-pumping: artificial pouring of surface water into the ground and guiding or pumping water from down stream in the same hydrogeologic system for water supply.

(3) Corrosion-erosion or denudation karst regions (HM, VL, CI):

(a) collecting-guiding: constructing surface reservoirs for guiding flows for water power and water supply.

(b) collecting-pumping-guiding: collecting surface streams in surface reservoirs and storing groundwater in subsurface reservoirs for water power and water supply by guiding.

(c) storing-guiding-pumping: pumping or guiding water for water supply or water power from subsurface reservoirs.

(d) pouring-guiding: artificially pouring water into the ground or constructing leakage reservoirs. Water from certain surface reservoirs will seep into the ground by natural karst caverns or corroded fissures. Recharge will flow to larger karst springs that are located downstream of the same hydrogeological system which is being exploited for water supply or power.

(4) Water resources in corrosion-glaciation or corrosion-resorption karst regions (SH, TA):

(a) collecting-guiding: collecting the melting snow water or flood water flows into reservoirs in deep valleys and then guiding them for water supply and/or for water power.

(b) storing-guiding: storing the melting snow water in karstified beds, with enclosed structural conditions, as underground reservoirs at high altitude for use as a source of water supply.

(c) pouring-pumping: to pour surface water into the ground, where the complete structural basin will store a large quantity of flow for pumping.

(5) Water resources in compound karst regions (TA):

(a) enclosing-pumping: enclosing fresh karst water and pumping for water supply; for this it is necessary to avoid the natural percolation and diffusion of salt water from salt lakes which deteriorates the fresh karst water.

(b) drawing-storing-guiding: drawing surface water into reservoirs for storing and guiding them for water supply; for this the salt lake water has to be desalinated.

(6) Water resources in buried karst regions (PE, SL, ST, CL):

(a) pouring-pumping: artificial recharge water will recharge the buried karst aquifers in shallow to medium deep zones, where the groundwater may be pumped for water supply. This can take place in two circumstances - pouring water in summer for winter pumping and the contrary by pouring water in winter for use in summer.

(b) cutting-pouring-pumping: cutting and enclosing groundwater or pouring surface water into the ground will also recharge the groundwater. Water may be poured in summer, for use in winter, or in winter for use in summer.

(c) collecting-pouring-pumping: using surface reservoirs to collect surface streams and then to recharge into the ground is another way to utilize the karst water, by later pumping in a suitable season.

(d) pressing-pumping-guiding: pressing surface water into the ground under high pressure will help confined special karst water. Liquid mineral sources in deep zones are pumped easily from artesian flows from exploited wells.

Most of the aforesaid methods are successful, but some problems remain. Each method must be selected according to local karst geological conditions and needs.

There are rich water resources in the karst regions of China. In water conservancy and water power constructions a series of successful experiences have been accumulated, relating to foundation treatments and prevention of karst leakage [2], [15]. However, hydroelectric energy sources that have been developed only occupy a small part of the total reserves. Exploitation of karst water for water supply has great benefits in many ways in north and south China; on the other hand, there have been harmful results in several karst regions, where the comprehensive utilization as well as reasonable uses of karst groundwater had not been studied beforehand. As to liquid mineral resources, the related investigations have only proceeded in some typical regions, where the rich sources are not well exploited. Therefore, in the study of the water resources in karst regions for reasonable comprehensive exploitation and harnessing, and also for protection and control, there are still important problems, which this paper has discussed.

Acknowledgements

The author expresses his deep gratitude to many geologists and karst investigators in many karst regions of China; they have given great assistance and co-operation to the author either by studying and sampling in the field or in laboratory tests.

REFERENCES

〔1〕卢耀如，1982，略谈岩溶(喀斯特)及其研究方向。自然辩证法通讯，第四卷第一期。

〔2〕卢耀如，1982，岩溶地区主要水利工程地质问题与水库类型及其防渗处理途径。水文地质工程地质，第四期。

〔3〕地质矿产部水文地质工程地质研究所编，1982，深层卤水形成问题及其研究方法。地质出版社。

〔4〕ErguvanLi K. Yuzer，1978，E.KarstificationProblems and Their Effects on Dam Foundation and Reservoir。International Association of Engineering Geology，Vol 1 Madrid，Spain。

〔5〕卫克勤　刘邦良等，1983，乌江渡水电站深部岩溶地下水中氚含量测定的初步结果。水力发电。第3期。

〔6〕藤井厚　仓克干・中山康，1974，冲永良部岛にすは为琉球曾群下の埋没段丘と第四纪海水准变动について。地质学志，第80卷第一期。

〔7〕Geyh M.A.，1972，Basic Studies in Hydrology and ^{14}C and ^{3}H Measurements。International Geology Congress，24 th.Sect.11。

〔8〕Smith D. B. et al，1976，The Age of Groundwater in the Chalk of the London Basin。Water Resources Research，Vol.12，No.3.

〔9〕O'Neil J. R.，1977，Stable Isotopes in Mineralogy。Physics and Chemistry of Minerals，Vol.2 No 1－2。

〔10〕McCrea J. M.，1950，On the Isotopic Chemistry of Carbonates and a Paleotemperature Scale。Journal of Chemical Physics，Vol.18，No.8.

〔11〕花永丰，1982，中国汞矿成因及其找矿预测。贵州人民出版社。

〔12〕王兆馨，1982，不同类型地区地下水资源形成和评价方法的若干问题。地下水资源评价理论与方法的研究。地质出版社。

〔13〕袁道先，1982，岩溶水资源评价的几个问题。同上。

〔14〕钱学溥，1982，娘子关泉水流量的相关分析。同上。

〔15〕刘邦良等，1983，对乌江渡水电站岩溶地基渗漏问题的初步认识。水力发电，第3期。

〔16〕丁悌平，1980，氢氧同位素地球化学。地质出版社。

〔17〕王英华　刘立本，1981，氧、碳同位素组成在研究碳酸盐岩成岩作用中的意义。北京大学。

〔18〕中国矿物岩石地球化学学会同位素地球化学委员会，1982，第二届全国同位素地球化学学术讨论会论文(摘要)汇编。

International Geomorphology 1986 Part II
Edited by V. Gardiner

KARST AREAS OF SOUTHERN AFRICA

Margaret E. Marker

University of Fort Hare,
Private Bag X1314,
Alice 5700,
CISKEI,
SOUTH AFRICA

ABSTRACT

Since 1967 detailed studies of karst landforms in various areas of southern Africa have demonstrated the existence of complex and varied assemblages of karst landforms. Karst is associated with outcrops inland of sparitic Proterozoic dolomitic limestones and Cambrian limestones in the Cape fold belt and Namibia, and with micritic limestones of the Tertiary coastal limestones and Namib Tsondab Limestones. Not all these assemblages have as yet been investigated in detail.

A systems model to explain the local variation of karst assemblage characteristics incorporates the use of variables. The most important appear to be total available precipitation, rock lithology, depth to the saturation zone and degree of relief. Most of southern Africa has a seasonal growing season which affects availability of biotic carbon dioxide but temperature never acts as a constraint

INTRODUCTION

Although the area covered by potential karst rocks in southern Africa is considerable, knowledge of southern Africa's karst areas is still largely restricted to those working in the area. Even twenty years ago Sweeting (1972), with first hand knowledge, could write 'Karsts are relatively rare in Central and Southern Africa but there are significant though scattered examples in the dolomites of the western and northern Transvaal'. The dolomites, more accurately dolomitic limestones, of the Transvaal are by no means the only karst host rocks nor is karst confined to their outcrops. As knowledge of southern African karst is increasing all the time, this paper sets out to survey the extent and variation of southern Africa's karst areas.

THE KARST ROCKS

Potential karst host rocks cover over 50 000 km^2. (Fig. 1, Table 1).

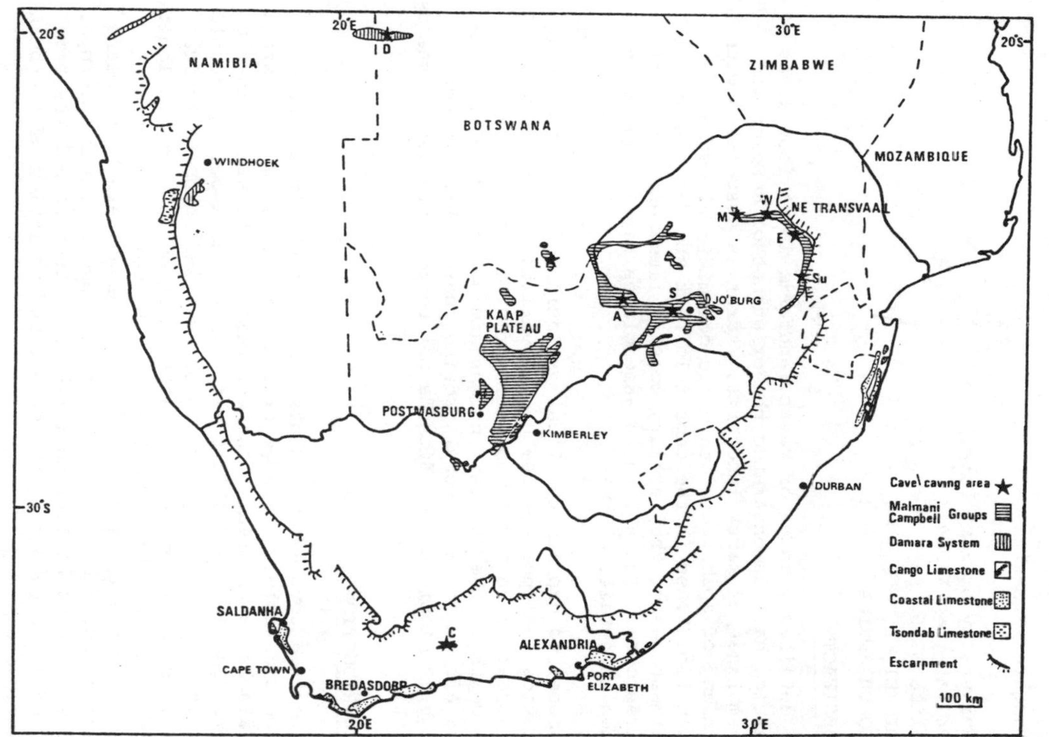

Figure 1 Karst areas of southern Africa (A = Apocalypse, C = Cango, D = Drotsky's Caves, E = Echo, L = Lobatse Caves, M = Makapansgat, S = Sterkfontein, Su = Sudwala, W = Wolkberg)

CLASSIFICATION	LOCALITY	ROCK FORMATION	AREA (km²)	ANNUAL RAINFALL (mm)	TOTAL AREA (km²)
SPARITES					
	NE Transvaal	Transvaal Sequence	3025	550 - 1200	
	Marble Hall	Malmani and Duitschland Groups	450	600	
	Central and N Transvaal	(dolomitic limestones)	10450	450 - 600	
	S Transvaal		350	650	
	Botswana		1000	200 - 400	
	Kaap Plateau	Griqualand West Sequence	17500	350	
	Postmasburg-Sishen	Campbell Group	900	300	
	Prieska	(dolomitic limestones)	300	300	
	Botswana	Damara System	115	450	
	Namibia	(marble and limestone)	7775	200 - 400	
	Cango	Cango Limestone	175	350	
					42040
MICRITES	W Cape	Coastal Limestone	525	400	
	S Cape	Longebaan Fm	2575	500 - 700	
	E Cape	Bredasforp Fm	2300	500 - 700	
	Zululand	Alexandria Fm	1850	1000	
	Namibia	Tsondab Limestone	450	250	
					7700
					49740

Table 1 Karst rocks of southern Africa

The most important of the karst rocks are the Proterozoic dolomitic limestones of the Transvaal and Griqualand West Sequences. Though deposited in separate basins, the Chuniespoort Formation and the less extensive Duitschland Formation of the Malmani Group of the Transvaal Sequence are very similar litholologically to the Campbell Group of the Griqualand West Sequence of northern Cape Province (Brink,1979). The dolomitic limestones are all resistant, sparitic, well-jointed and highly lithified ancient rocks. In contrast, the other extensive karst hosting rocks are the informally grouped. Tertiary Coastal Limestones which are soft, of varying purity, weakly jointed and dominantly micritic limestones. These Coastal Limestones outcrop discontinuously around the coast of southern Africa but on the continental shelf are more continuous (Siesser, 1972) (Fig.1). In addition to these extensive series of outcrops, there are outcrops of Palaeozoic Damara marble and associated limestones in Namibia and in western Botswana (Fig.1). Palaeozoic Cango Limestones outcrop in the Cape fold belt and Tertiary calcretic limestones have isolated occurrences along the inner margin of the Namib desert. There is thus a basic division between the old hard limestones above the Great Escarpment and the Tertiary micritic limestones below it.

VARIATION WITHIN THE KARST

Southern African karst is essentially a soil covered fluvio-karst (Marker,1971). Karst surface landforms have a localised distribution and overall their density is low. Nevertheless distinct karst provinces can be identified. Major cave systems exist, though not of world rank (Fig.1). These systems are often blocked by clay, colluvium and collapse material but are gradually being extended by exploration. Large dimension, joint-controlled phreatic systems, very similar to other mesothermal karst environments with clay fills, large dimension speleothems and long evolutionary histories, are characteristic. There are indications that most caves originated as shallow phreatic systems since their gross altitudinal distribution has a clear relationship with major planation surfaces. Locally the gross pattern is affected by structural and lithological controls (Moon,1972). In areas of high relief where rejuvenation has been active, subsequent vadose modification is important (Marker & Brook,1970; Marker,1974). Even in caves within limited outcrops phreatic origins can be detected even though the major development has been vadose. This is true of the Cango valley, parts of the Bredasdorp cave area and in parts of the northeastern Transvaal.

Surface karst assemblages are also variable. On the dolomitic limestones of the mountainous Eastern Transvaal, in the vicinity of the Blyde valley, is a karst plain-cone karst. Small dolines occur rarely. Incision permits relatively fast through-flow and the permanent flow of the Blyde river is maintained by springs.

In the northern sector of the same area where relief is also pronounced, and the cycle of incision is less advanced, karst plains and cones are virtually absent. Surface karst forms are scarcer as surface runoff is rapid. By contrast the central and western Transvaal is an area of small amplitude, low relief. Where rejuvenation has occurred, either naturally, as in the vicinity of Sterkfontein near Johannesburg, or artificially by pumping as in the gold mining Far West Rand, collapse sinkholes are common. Elsewhere uvalas and small poljes are aligned along former drainage lines (Cooks, 1968). Farther west beyond the mining areas, in the Lichtenburg-Koster district, large shallow poljes, uvalas and occasional cenotes are found, all partially concealed by redistributed terra rossa. It would appear that rainfall seasonality with low and erratic totals, usually as high intensity falls, results preferentially in solution from below and subsequent collapse following upward piping. Many of the collapse dolines in the mining area appear to be re-activated into much older cavities (Brink, 1979).

In the northern Cape Province, the extensive dolomitic limestones incorporate some actual limestone beds, form the Kaap Plateau and are virtually undissected. There, where so far little detailed investigation has been carried out, shallow pan dolines and mound tufa springs aligned along dykes are the dominant forms. On the Postmasburg-Sishen outcrop, dolines, interstratal large dimension collapse sinkholes and giant karren appear to predominate, concealed beneath metalliferous rubble or planed by later pediments.

On the Tertiary Coastal Limestones a much higher density of karst forms occurs; they are predominantly enclosed hollows of various dimensions. The eastern Cape Alexandria area is essentially a doline karst. Shallow pan forms predominate in the east where the soft calcarenite is underlain by impermeable planed Palaeozoic rocks; whereas deep funnel form dolines, uvalas and poljes showing a much greater overall assemblage complexity, occur where the substrate is a permeable Cretaceous sandstone (Marker & Sweeting, 1983) (Fig. 2). The southern Cape karst areas host dolines and uvalas whose surface dimensions relate depth to saturation level. Elongated rand poljes have developed along the inland contact with the Palaeozoic Cape System shales and quartzites. This karst is an aligned karst, and whether it developed syngenetically or as a function of periodic marine withdrawal is not yet known. Research is still in progress there and in the less spectacular western Saldanha and northern Zululand outcrops (Russell, pers. comm.).

Palaeozoic Damara marble and limestone strata have limited surface relief being usually planed and overlain by Kalahari sands. The actual outcrops are resistant to surface solution in the arid environments in which they occur. Linear caves associated with collapse dolines and shafts appear to be the chief karst manifestations (Cooke, 1975; Heine & Gehr, 1985). The Palaeozoic Cango limestones of the Cape fold belt have a narrow and fault-displaced outcrop.

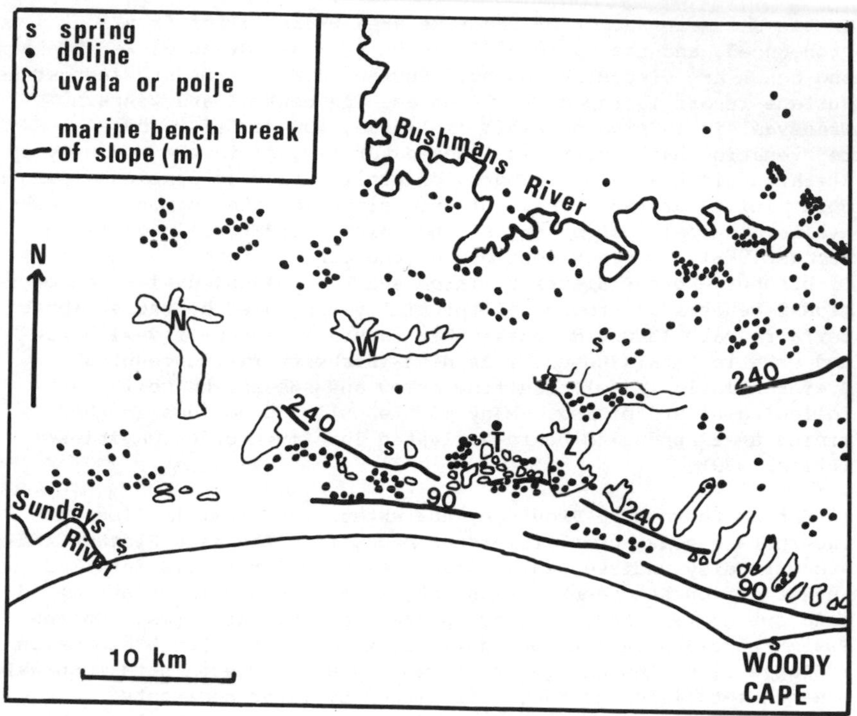

Figure 2 The western portion of the Eastern Cape (Alexandria) Karst area where limestone overlies permeable Cretaceous sandstones (N = Nanaga, T = Thornhill, W = Wycombe Vale and Z = Zuney)

Table 2 Karst publications

Period	Total Publications	(Overseas)	Theses (unpub.)
Pre 1959	2	-	-
1960-1964	3	-	-
1965-1969	10	(1)	-
1970-1974	12	(5)	2
1975-1979	13	(4)	1
1980-1984	14	(7)	2
TOTAL	54	17	5

Surface karst is limited to occasional dolines despite the relatively high cave density (Marker,1975). The Tertiary Tsondab Limestone of the Namib margin supports a locally high doline density (Marker,1982). A shallow doline karst exists on this thin, impure limestone which is restricted to low plateau interfluves in an arid environment; the karst is clearly the result of slow drainage to slowly incising main channels and to former wetter conditions.

KARST EXPLANATION

A systems model for karst development was evolved after working in these southern African highly varied karst environments (Marker, 1980). Although water availability must be of importance and could be expected to lead to differentiation between the high rainfall areas (up to 1 000 mm p.a.) of the northeastern and central Transvaal and the more arid remainder (less than 400 mm p.a.) of the other inland karst areas, little difference can be detected. It may, however, explain the greater density of karst forms in the central and northeastern Transvaal as opposed to the northern Cape. Probably more significant is the limited availability of biotic CO_2 which is strongly constrained by rainfall seasonality and erratic annual rainfall even in the higher rainfall regions. Karst processes therefore operate very slowly.

The most significant controls are probably first, the thickness of the karst rock whether of the limestone itself or of limestone including underlying permeable substrate, and secondly the degree of dissection which governs the rate of through-flow and thereby promotes solution. Much of the regional karst variation can be explained in terms of this factor. The lithology of the limestone, micritic or sparitic, hard or soft, is another major control. Density of jointing, and the quantity and size range of insoluble residuum have been found to be important in the development of micro karst solution features (Marker,1981; 1985). Landscape history, by which is implied planation and periodic rejuvenation whether fluvially controlled inland or by sea level change along the coast, has exerted a strong influence through its control of water levels. But to date there is little conclusive evidence to explain precisely the distribution of specific variants in karst assemblages. There are many variables interacting. Furthermore the present day variables identified on a regional basis may not, in fact, be those which existed at the time of karst development.

Much of the karst of southern Africa is relict. Its present distribution is a function of preservation in sites of less vigorous erosion within the karst hostrock outcrops. But dating of the period of major karst development is difficult. All karst evolution inevitably post-dates the age of the host rock itself. If karst areas are more or less contemporaneous in development light may be shed by investigation of karst on the Tertiary Coastal Limestones. In the southern and eastern Cape the Tertiary limestones range in age from at least Palaeocene to late Pliocene, although some authors (Rogers,1985) would include within the grouping many Pleistocene aeolianites.

Dating even of the earlier Coastal Limestones is difficult as they are essentially beach deposits (Siesser,1972) and macro fossils are rare and non specific. Large karst features associated with a high degree of landform assemblage complexity, hall-marks of a more advanced karst, are associated only with limestones older than Pliocene (Marker,1984). Pliocene and younger strata support only shallow pan dolines.

Inland on the Proterozoic dolomitic limestones, solution enlargement of major joints to create giant karren has occurred in many areas. At Irene south of Pretoria sinkholes associated with these giant karren contain Ecca sandstone (Marker,1974). At Rietfontein on the West Rand, Karoo age deposits have also been preserved in a similar context (Brink,1979). The presence of these inliers within karst sinkholes suggest that the overlying Jurassic Karoo rocks were gradually let down and preserved in the cavities during the evolution of the African planation surface which was simultaneously a period of active karst solution. This erosion phase was initiated at the coast in the mid Cenozoic period but reached inland sites later (King,1963). Giant karren and associated sinkhole deposits also occur in the Postmasburg area but there contain mineralised breccia of probable Palaeozoic age. Although likely to be contemporaneous this locality sheds little light on the date of karst formation. The active phase of karst development is thus attributed to the mid to late Tertiary with periods of reactivation during Quaternary wetter climatic oscillations. Palaeontological and palynological evidence suggest that the Miocene was the last period of widespread forest cover which is further support for the probable period of karst development.

The terminal date of the period of major karst development can be approached via cave deposits. Large dimensions $CaCO_3$ deposition is the product of active surface solution. In most Transvaal caves and also in Cango Cave, at least two periods of large volume speleothem deposition exist, interrupted by phases of high cave water levels, all preceding the present small scale speleothem formation. The present speleothem deposition in caves on the dolomitic limestones is very largely in the form of aragonite, known to be preferentially deposited in the presence of high magnesium: calcium ratios, themselves a product of drought conditions. Analysis of some of the larger speleothems indicates that even those of the second period are older than 41 000 years BP. In Wolkberg a date of 43 500 years BP has been published (Talma & Vogel,1974) and more recent work in the Namib gives dates earlier than 41 5000 years BP (Heine & Gehr,1984). On the surface large tufa carapace fans were also formed at several discrete periods, the oldest being very strongly lithified and containing Australopithecine fossils tentatively dated to 0.8 million years BP (Partridge,1973). More precise dating of the major period of karst formation is constrained by lack of fossil evidence and by the dating methods available.

It lies between mid Tertiary and mid Quaternary with a strong probability of a Miocene initiation. Aridity has prevailed to varying degrees since then, with only shorter periods when karst processes were reactivated.

CONCLUSION

The relatively limited knowledge of southern Africa's extensive and varied karst assemblages must be attributed to a late start in the research, to a paucity of specialised researchers and to the remoteness of many of the karst areas. The number of publications has risen steadily over the past 30 years but of the total of 60 publications including 5 karst theses, only 19 have appeared in overseas journals (Table 2). Karst in southern Africa deserves greater publicity. Its great age and complexity demands detailed studies which need experienced workers with time to spend in the field. Less spectacular than many of the classical karst areas, southern Africa holds considerable potential for future work on karst.

REFERENCES

Brink, A.B.A., 1979. Engineering geology of southern Africa Vol. I Building Publications

Cooke, H.J., 1975. The palaeoclimatic significance of caves and adjacent landforms in western Ngamiland, Botswana. Geog. J. 141, 430 - 44.

Cooks, J., 1968. Palaeodreineringslyne in die Wes Transvaalse Karst gebied SA Geog. J. 50, 101 - 9.

Heine, K., & Gehr, M.A., 1984. Radiocarbon dating of speleothems from the Rossing cave, Namib desert and palaeoclimatic Implications. Late Cainozoic Palaeoclimates of S. Hemisphere 465 - 70 (Ed Vogel, J.C.)

King, L.C., 1963. South African Scenery 3rd Ed. Oliver & Boyd pp 302

Marker, M.E., 1971. Karst landforms of the northeastern Transvaal PhD thesis Witwatersrand (unpub).

1974. A note on the occurrence of Karroo sediments near Pretoria and its relevance to the dating of karst weathering. Trans. Geol. Soc. SA 77, 69 - 70.

1975. The development of the Cango cave system. Proc. Int. Symp. Cave Biol. & Palaeont. 1 - 6.

1980. A systems model for karst development with relevance for southern Africa. SA Geog. J. 62, 151 - 63.

1981. Aspects of the geology of two contrasted South African Karst areas. Br. Cave Res. Asso. 8, 43 - 51.

1982. Aspects of Namib geomorphology : a doline karst Palaeoec. Africa 15, 187 - 99.

1984. Marine benches of the eastern Cape, South Africa. Trans. Geol. Soc. SA 87, 11 - 18.

1985. A note on factors controlling micro-solutional karren and carbonate rocks of the Griqualand West Sequence Br. Cave Res. Ass, 12, 61 - 65.

Marker, M.E.,& Brook, G.A., 1970. Echo Cave : a tentative Quaternary chronology for the Eastern Transvaal. Env. St. Occ. Paper 3 Witwatersrand Univ. pp 38.

Marker, M.E.,& Sweeting,M.M., 1983. Karst development on the Alexandria Coastal Limestone, Eastern Cape Province, South Africa. Zeit. f. Geomorph. 27, 21 - 38.

Moon, B.P., 1972. Factors controlling the development of caves in the Sterkfontein area. SA Geog. J. 54, 145 - 51.

Partridge, T.C., 1973. Geomorphological dates of cave opening at Makapansgat Sterkfontein Swartkrans and Taung. Nature 246 5428, 75 - 79.

Rogers, J., 1985. Cenozoic geology of the southern Cape coastal plain between Cape Agulhas and Mossel Bay focussing on the area between Kafferkuils and Gouritz rivers. Tech Rept 15 Marine Report Unit 173 - 199.

Siesser, W.G., 1972. Petrology of the Cainozoic Coastal Limestones of Cape Province, SA Trans. Geol. Soc. SA 75, 178 - 185.

Sweeting, M.M., 1972. Karst landforms Macmillan

Talma, A.S., Vogel, J.C., Patridge, T.C., 1974. Isotopic contents of some Transvaal speleothems and their palaeoclimatic significance. SA J. Sc. 70, 135 - 40.

Acknowledgements : Funding from CSIR for much of the fieldwork on which is based, is gratefully acknowledged. Support for attendance at the First Internation Conference on Geomorphology, at which the paper was presented, was also given.

International Geomorphology 1986 Part II
Edited by V. Gardiner

KARST DEVELOPMENT : THE APPLICATION OF A SYSTEMS MODEL

Lin Russell
University of Fort Hare,
Private Bag X1314,
Alice 5700,
CISKEI,
SOUTH AFRICA

ABSTRACT

A systems model has been proposed as an effective tool for karst explanation. The model was tested empirically on the Agulhas area, a region of high density surface Karst characterised by dolines, uvalas, dry valleys and poljes. These landforms are developed on calcareous Tertiary and Quaternary beach and dune deposits derived from the reworking of marine limestones.

Distinct topographic surfaces, related to Tertiary and Quaternary sea levels, are characterised by specific karst landform assemblages. Various factors may be implicated in the control of karst development on these surfaces : these include time, climate, limestone lithology structure and thickness, topography. The systems model, allowing for the interaction of several variables was therefore applied in order to determine the relative importance of these factors. No single factor could be described as dominant in the control of karst development in this region.

THE MODEL

A systems approach to karst explanation is more widely applicable than former explanations based on time, morphoclimatic or geological factors alone, since a model of this type allows for greater flexibility in the interpretation of variable interactions. The general model comprises 7 variables acting on a central open system within which are 2 subsystems : karst system and assemblage (Figure 1; Marker 1980). This model allows more than one control to account for differences in landform development and hence does not place dominance on a single factor.

TESTING THE MODEL : (i) The Study area

Although there are extensive outcrops of carbonate rocks in South Africa, (Marker, 1985) the occurence of high frequency and well developed karst landforms is rare. Most research to date has been carried out on the Protozoic Malmani Dolomite of the Transvaal (Marker, 1971; 1980) and few studies have been made of karst landforms along almost 1 200 km of the Cape Province coastline where Tertiary and Quaternary limestones, informally termed the Coastal

Limestones, outcrop. A small area of this Coastal Limestone outcrop, the De Hoop Nature Reserve located between Bredasdorp and the Potberg Range south of Swellendam (Figure 2), was selected as a suitable test site for the model. This area supports a large number of karst landforms, although their type and distribution is by no means uniform. Specific karst landforms characterise different topographic areas; testing of the model would show whether this non-random distribution is a result of the dominance of one or several of the controlling variables.

(ii) Landform distribution

The topography of the area consists of several distinct subregions (Figure 3). From inland southwards to the coast there is first a border depression occupying the junction between impermeable shales and sandstone and the limestone. This area comprises a narrow, elongated valley with discontinuous drainage in the west and a large rand polje, 5 000 m long and 2 500 m wide, in the east. The land rises from this depression southwards to a bevelled ridge (called locally 'Die Duine') at 180 m then drops sharply seawards.

Deep dolines (average depth 8.9m), uvalas and rocky dry valleys characterise the 180 m surface. However, although landforms are well developed vertically, they only occupy 27% of the surface area. At the foot of this ridge is a smooth footslope sloping at between 2 and 5°. Topographically the footslope is almost featureless (only 2% of the surface area is occupied by depressions) except for shallow stream valleys, now dry. The area between the base of the footslope and the sea comprises 4 distinct surfaces at 90 - 100 m, 60 m, 30 - 40 m and 15 - 20 m. These surfaces are characterised by dolines and uvalas; as surface height increases so does the extent and complexity of landform development. The lowest surface at 15 - 20 m supports small, shallow pan dolines (average depth 1.6 m) and occasional shallow uvalas. Depressions account for 16% of the total surface area. The 30 - 40 m surface has 25% of its surface area occupied by deeper dolines (average depth 5.6 m) and uvalas. The 60 m surface comprises more complex uvalas, often with dolines developed in their base (average depth is 7.3 m). 50% of the surface area is occupied by depressions. Lastly, the 90 - 100 m surface supports similar depressions to the 60 m surface although their extent has increased and they occupy 60% of the total surface and have an average depth of 7.8 m. (Table 2).

Thus each topographic area supports different landform assemblages and this non-random distribution is the result of the interaction of definite controlling factors. The relative importance of these factors and hence the applicability of the systems model is now assessed.

(iii) The Variables

Variables affect both the rate and type of landform development and may be listed as follows : climate (past and present), ratio of non karst to karst process, biotic dioxide, time (the amount of time available for landform development), hydrology, geology (limestone structure, thickness and lithology) and topography.

Climate

The De Hoop area receives an annual rainfall of 350 mm. This is far too low to account for the large solution and fluvial features occurring on the 180 m surface. Therefore past, wetter conditions were responsible for the development of these landforms on the slopes and summit of this surface. During this wet period possibly of high intensity storms the drainage on the 180 m surface was a flash-flood fluvial type. It resulted in valleys running in an east-west direction draining into what is now the discontinuous Potberg River and the rand polje occupying the border depression. Runoff from these limestone valleys and the adjacent impermeable rocks accounts for the considerable width and depth of the polje.

The relatively short duration of these wetter conditions, estimated as occurring during the Pliocene (Russell, 1981), is shown by the failure of the gorges on the slopes of Die Duine to cut very far back into the surface. The return to drier conditions is shown by the fragmentation of the east-west valleys into long uvalas and by the abrupt termination of steep sided gorges. It is believed that the remaining surfaces were formed after this wet period. Although these areas have also been subjected to climatic fluctuations during the Quaternary conditions were never wet enough to allow fluvial features to develop on any of the level surfaces.

Therefore climate as a control does not account for differences in size or type of depression on these lower surfaces. However, it is an important control when accounting for firstly the development of fluvial features localised on the 180 m surface and secondly when accounting for the formation of the border depression and rand polje.

Non-karst Ratio

As all landforms are developed wholly on limestone with the exception of the border depression non karstic effects are minimal. However, in the case of the border depression shales and sandstones outcrop to the north. Surface runoff at the junction of the limestone and these impermeable beds (focused on a fault intersection) has led to active lateral corrosion and the consequent development of a rand polje. A period of past wetter climate has also aided its development.

Biotic Carbon Dioxide

Due to the limited extent of the study area soils and vegetation and hence available biotic carbon dioxide are similar for all topographic areas. Therefore this factor could not account for the differences in landform assemblage.

Time:

Topographic surfaces become younger in a seawards, southwesterly direction and are related to Tertiary and Quaternary sea levels with the exception of the border depression which was formed under sub aerial conditions. Thus, with this exception, age of surface correlates with surface height.

The highest surface at 180 m is the oldest one. It is believed to have been initiated by a marine transgression in Oligocene-early Miocene times. The limestones forming and underlying this surface were probably deposited in late Eocene times, and later reworked into beach and dune deposits as seas regressed. The footslope and 90 - 100 m surfaces are believed to have been initiated in late Pliocene times, limestones underlying them having a mid-Miocene to mid-Pliocene age. The remaining surfaces at 60 m, 30 - 40 m and 15 - 20 m were cut in Pleistocene times during high interglacial sea levels. (Russell, 1981)

The age gap between the 180 m surface and the next lowest one is greater than the range of age of all other surfaces combined; this surface is therefore considerably older than all others. This helps to explain the greater depression depth developed on this surface. However, it does not explain the fact that this surface has the second lowest total area of depressions.

Although the range in time between all the other lower surfaces is only estimated at 3 million years this factor could still account for the progressive increase in percentage area of depression shown to occur on progressively older (higher) surfaces. It does not explain why although depression area significantly increases with surface age, depression depth does not show this significant increase statistically. Although depths do increase progressively with surface age, there appears to be some other control acting as a constraint on vertical solution and limiting the effectiveness of the time factor.

Hydrology

Effectiveness of depth of solution depends very much upon the distance below the ground surface of the water table. When a land surface is first exposed the piezometric surface will be near the land surface and will then fall progressively as the level falls. Provided there is sufficient available thickness of limestone, a large fall in sea level should be accompanied by a large drop in the piezometric surface. The further this surface is below the ground surface the greater will be the hydraulic gradient and hence the greater the development of vertical solution.

After the initiation and exposure of the present 180 m surface, sea level fell by approximately 90 m to the present 90 - 100 m surface. This would be accompanied by a rapid fall in water level, and the consequent development of a steep hydraulic gradient, would favour the development of the deep depressions found on this surface. Subsequent falls of sea level, causing the exposure of all the lower surfaces, became less sharp. Between the 90 - 100 m and 60 m surface the drop is 30 m, between the 60 m and 30 - 40 m surface it is approximately 20 m and between the 30 - 40 m and 15 - 20 m surface it is only 10 m. Hence accompanying vertical solution becomes progressively less effective.

Therefore the hydrological control is reflected in the shape of depressions found on these surfaces. However, piezometric surface altitudes are related to climate and limestone thickness as well as to former sea levels. Climate has already been discussed; the importance of the control of limestone thickness follows.

Geology : Limestone thickness

Limestone thickness has a direct influence on the depth to which a depression is able to develop unless the limestone is underlain by a semi-permeable material. In the study area the limestone is underlain by impermeable rock. Average limestone thickness for each surface are given in Table 1. Limestone thickness therefore accounts for differences in the shape of depressions on the 2 lowest surfaces when compared particularly to the shape of those on the 180 m surface. The wide, shallow nature of depression on the 15 - 20 m and 30 - 40 m surfaces suggests that vertical solution is limited by thin limestone and that most enlargement is taking place in a lateral direction. On the 180 m surface limestone thickness does not act as a constraint on the vertical development of depressions, hence deep dolines and uvalas are formed.

However, the limestone thickness control does not explain why depressions have tended to develop vertically on the 180 m surface but show a greater horizontal development on the 60 m and 90 - 100 m surfaces. On these 3 areas there is a large available thickness. Water table control must be important here.

Geology : Limestone Structure

The limestones on all surfaces are almost horizontal which has favoured the development of horizontal planation surfaces with the consequent development of symmetrical depressions. Both horizontal surfaces and a predominance of symmetrical depressions occur on all topographic surfaces with the exception of the footslope. Therefore since all limestone in the area is approximately horizontally bedded, this factor could not account for any difference in land-forms on specific surfaces but could explain the lack of depressions on the footslope. This surface has slopes varying between 2 and 5°. Such angles are sufficient to prevent water remaining on the surface and having time to infiltrate.

Instead, it runs off, cutting shallow valleys. On a large scale, the presence of lineaments trending in an east-west direction influences the location of the deep dolines and rocky uvalas of the 180 m surface and of shallow, enclosed depressions on the lower surfaces. The development of these large scale joint patterns on all surfaces and their consistent influence on landform location precludes their accounting for differences in landform development on the various topographic areas.

Geology : Limestone Lithology

Siesser (1972) describes the Coastal limestones as stranded beach and dune deposits from regressive seas, consisting mainly of comminuted shells mixed with terrigenous grains, which, when subject to diagenesis in a subaerial environment, resulted in a dominantly sparitic rock of low magnesian calcite. Specimens from the study area showed a micritic texture.

A comparison of the solubility, porosity and density of surface limestone samples is shown in Table 3. The 2 lowest surfaces are seen to have more soluble limestone whereas the 90 - 100 m surface and the 180 m surface have limestone of lower solubility.

A low limestone solubility is believed to account for the low percentage area of depressions on the 180 m surface. Although it has been shown that this surface has had the longest time available for landform development, and also has been subjected to a wetter period before other surfaces were exposed, these favourable factors have been offset by a low limestone solubility preventing the expansion of depressions. Thus solution is not as effective here as on the lower surfaces and has remained in the fluvial cut valleys due to a concentration of runoff here. The initial location of these valleys is largely structurally controlled, being aligned in an east-west direction. Hence depression and valleys have only developed where weaknesses in the rock favour solution.

Both the 15 - 20 m and the 30 - 40 m surfaces consist of highly soluble limestone. This factor explains the large number of depressions initiated on the 15 - 20 m surface which is the youngest surface. There were probably just as many on the 30 - 40m surface but because this surface is older they have now enlarged and coalesced to form uvalas. The high solubility of the footslope and its lack of enclosed depressions suggests that another control is dominant and has counteracted the initiation of depressions favoured by the high solubility.

The 60 m and 90 - 100 m surfaces show a higher percentage of insoluble residue and thus lower solubility than all other surfaces except for the 180 m surface. The higher solubility of the 60 m surface might explain why there is little difference in landform development on the 2 surfaces even though the 90 - 100 m surface is the older one and hence has had more time for depressions to enlarge.

The fact that landforms on these 2 surfaces are more developed than those on the 15 - 20 m and 30 - 40 m surfaces shows that the age factor is relatively more important in this case. In other words, the shorter time available for landform development on these lower surfaces has offset their more favourable solubility.

CONCLUSION

The dominant controls of karst landform development were found to be time, former climates, limestone thickness, limestone lithology and topography. The controls of time and limestone thickness were particularly important in accounting for the degree of development of depressions on the 15 - 20 m, 30 - 40 m, 60 m and 90 - 100 m surfaces.

Past climatic conditions, wetter than at present, plus a large available thickness of limestone were largely responsible for the development of deep enclosed depressions and dry valleys on the 180 m surface. This period of wetter climate plus the presence of adjacent impermeable strata accounted for the development of a large rand polje in the border depression. Differences in limestone lithology are sufficient to explain differences in the extent of aerial development of depressions on the lower surfaces compared with the 180 m surface. The latter surface is composed of limestone that is less soluble than that underlying all other surfaces, hence depression location is limited to places where prominent joints intersect the surface. Finally, topography was found to be the most important factor controlling the lack of depressions developed on the footslope.

When these findings are related to the various models of karst landform development that have been suggested to account for differences in karst assemblages found in various areas of the world it is seen that the systems model is the only one that is applicable to this particular area. This model, unlike others, allows for the interaction of more than one variable to explain differences in landform development. Other models place dominance on a single factor (time, climate or geology) to account for all differences in landform development.

REFERENCES

Marker, M.E., 1971. Karst landforms of the northeastern Transvaal PhD thesis Witwatersrand (unpub)

1980. A systems model for karst development with relevance for southern Africa. SA Geog. J. 62, 151 - 63.

1985. Karst areas of Southern Africa. Proc. First. Internat. Conf. on Geomorph.

Russell, L., 1981. Karst surface landforms of the De Hoop Nature Reserve MSc thesis Univ. Fort Hare (unpub)

Siesser, W.G., 1972. Petrology of the Cainozoic Coastal Limestones of Cape Province, SA Trans. Geol. Soc. SA 75, 178 - 185.

ACKNOWLEDGEMENTS

Funding from CSIR towards the cost of the fieldwork is acknowledged.

TABLE 1. Limestone Thickness

Surface	Average Limestone Thickness
180 m	160 - 180 m
Footslope	90 m (top) 20 m (base)
90 - 100 m	85 m
60 m	60 m
30 - 40 m	30 m
15 - 20 m	20 m

TABLE 2. Karst characteristics

Land Unit	Dolines-uvalas	
	Depth (m)	Area covered (%)
180 m	8.9	27
Footslope	Minimal	2
90 - 100 m	7.8	60
60 m	7.3	50
30 m	5.6	25
15 m	1.6	16

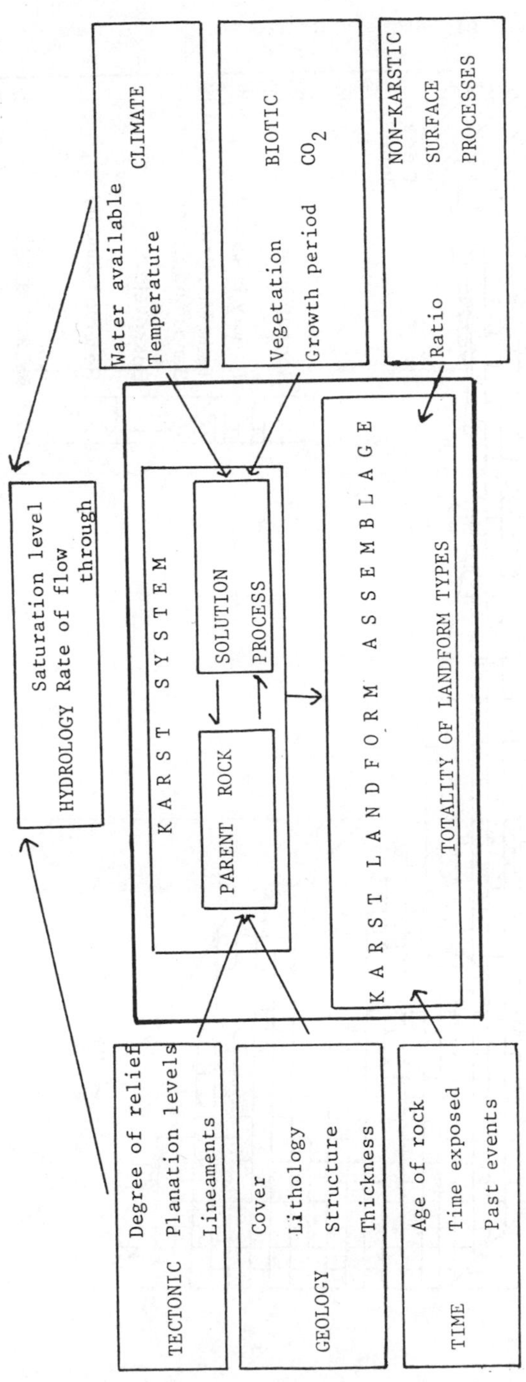

Fig.1. A Systems Model of Karst Development (Marker, 1980).

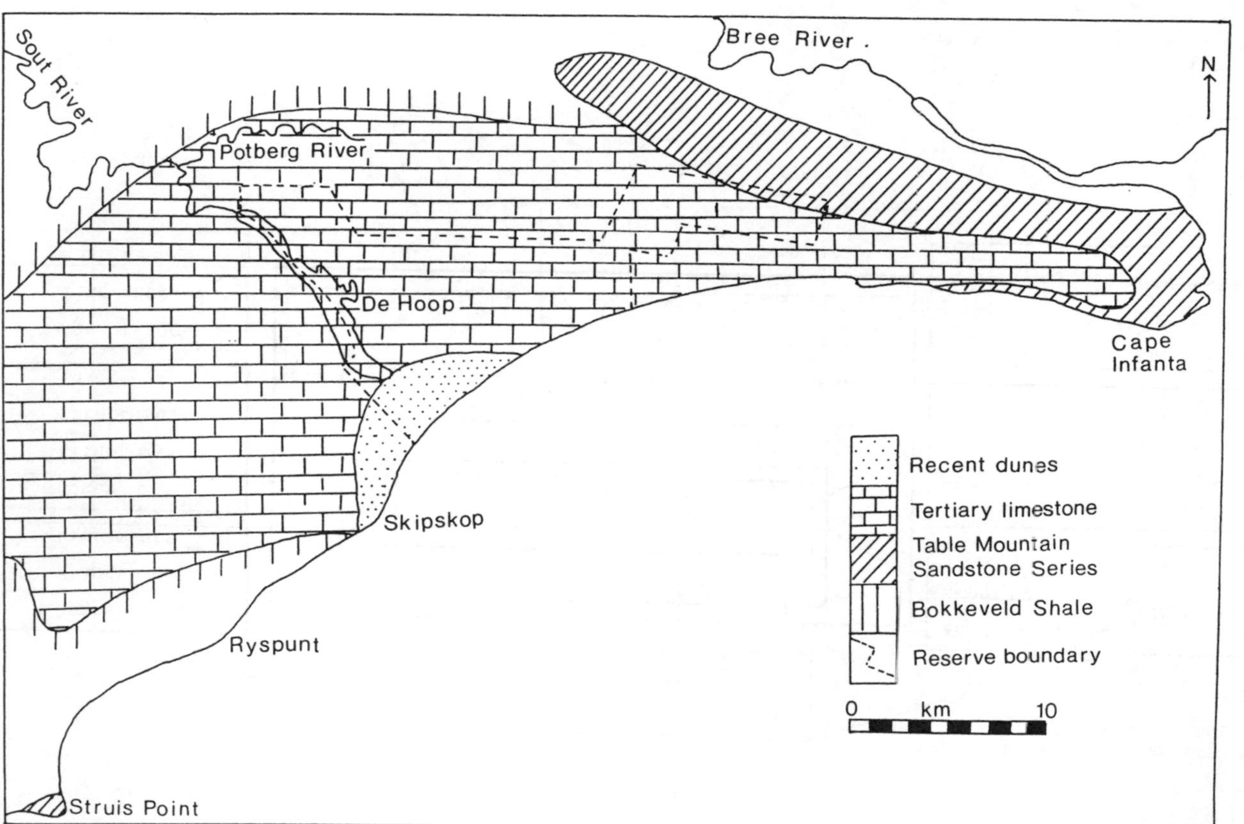

Fig. 2. Geology of the Study Area

TABLE 3. Comparison of the solubility, porosity and density of limestone samples taken from each topographic area.

	Surface						Average percentage for all surfaces
	15-20m	30-40m	60m	90-100m	Footslope	180m	
Percentage insoluble residue	3.14	3.23	11.95	19.82	3.21	17.89	9.87
Percentage solubility	96.86	96.77	88.05	80.18	96.79	82.11	90.13
Percentage porosity	5.06	8.08	14.92	19.90	6.5	8.75	10.53
Density	1.39	1.38	1.25	1.14	1.40	1.70	1.37

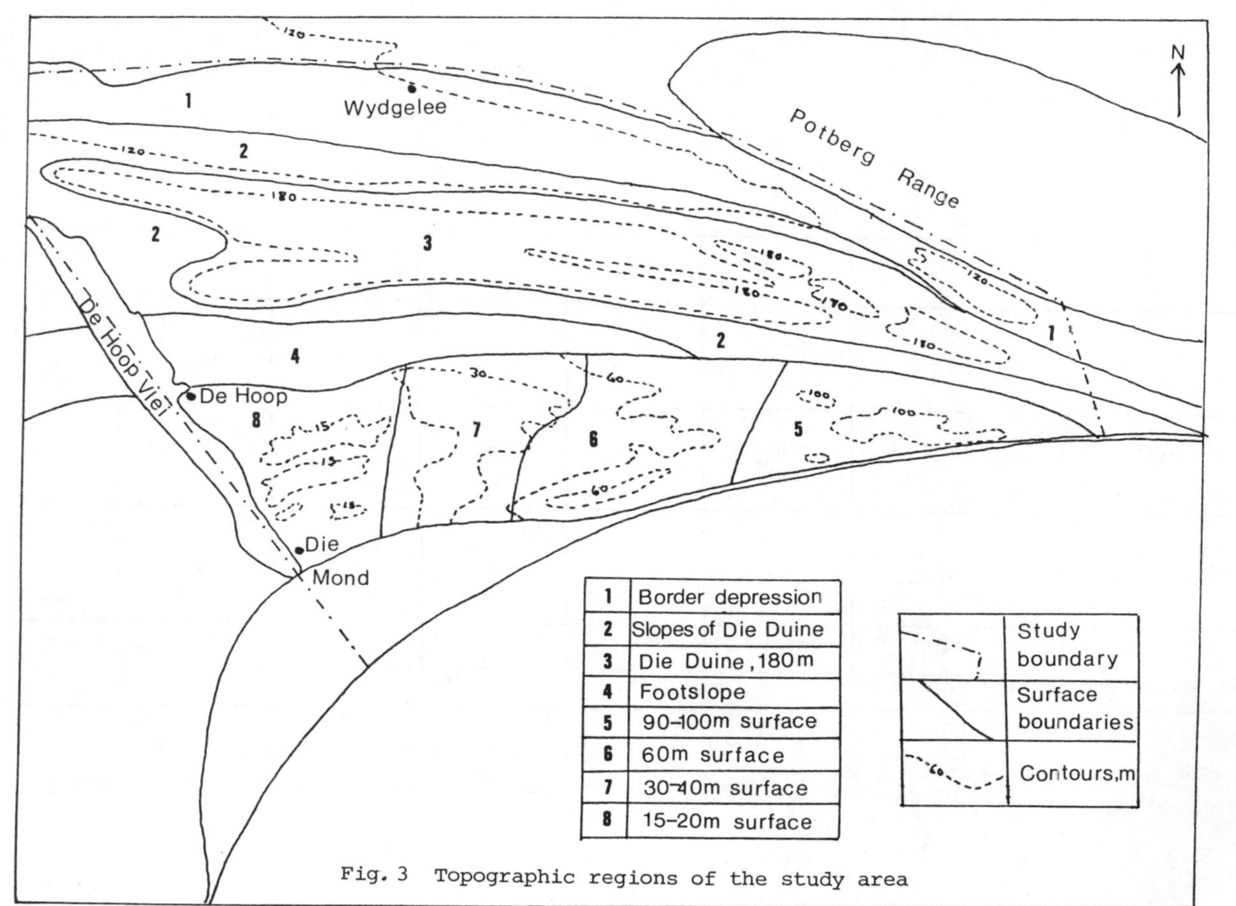

Fig. 3 Topographic regions of the study area

International Geomorphology 1986 Part II
Edited by V. Gardiner

AN INVENTORY OF CAVES IN THE COUNTY OF VÄSTERNORRLAND, N. SWEDEN.

R. Sjöberg

Department of Physical Geography,
University of Umeå, Sweden.

INTRODUCTION

The study region has an area of 21 780 km² and is situated in NE Sweden between the latitudes of 62° and 64°N. Within this area more than 300 caves have been surveyed and valued by scientific, ethnographical and recreational criteria (Fig. 1), using a three degree scale, where III is low and I is a high value (Table 1).

TABLE 1.

Value	Scientific value		Ethnographical value		Recreational value	
I	29	16,6%	10	5,6 %	27	15,2%
II	70	40,o	35	19,6	89	51,8
III	76	43,4	133	74,7	59	33,7

GEOLOGICAL SITUATION AND GEOMORPHOLOGICAL EVOLUTION

The study region consists of sedimentary rocks deposited in a primary geosyncline and later metamorphosed to metasediments; through these, magmatic rocks such as granites, anorthosites, dolerites etc. have penetrated during different periods. The magmatic rocks are mostly more resistant to mechanical physical erosion than are the usually more schisty sedimentary rocks.

During several billions of years the area has been affected by weathering; this weathering seems to have reached a depth of several hundreds of meters in parts of the region. During the last 1,5 millions of years repeated glaciations have passed over the region and cleaned the deeply weathered areas to form the broken topography and almost fjordlike landscape termed the "High Coast of Sweden".

During these periods of evolution a tectonic shearzone along the northern Baltic coast of Sweden has been active at various times together with tectonic zones striking more SE-NW (Strömberg, 1977). These tectonic movements probably activated by the melting of the great inland ice about 9 000 years ago, have accentuated the topography and caused major fault-zones striking N-S and E-W, which are dominant in the region. Glacial deposition in the valleys and fluvial erosion,

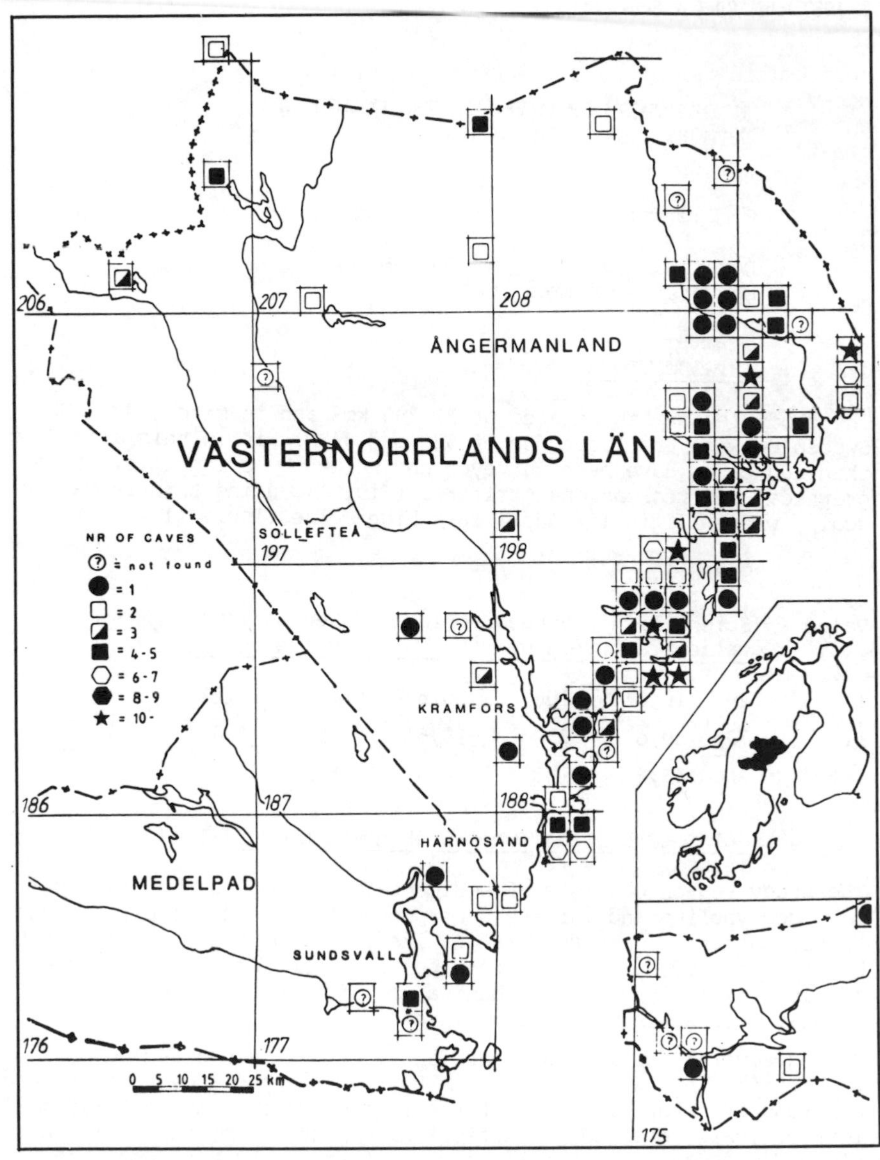

Fig. 1. The distribution of caves in the county of Västernorrland, N. Sweden. The area of the small squares is 25 km².

together with marine abrasion, have been major landscape forming agents during the last 10 000 years.

The highest post-glacial coastline in the region is 285 m above the present level of the sea, and is the highest in Sweden. The present mean uplift is 0,8 cm per year; this means that, with exception of occasional high mountains in the NE, the eastern parts of the region and the valleys of the major rivers are situated below the highest coastline.

CLASSIFICATION OF DIFFERENT TYPES OF CAVES IN THE REGION

The structure of the region´s caves vary considerably. Thus a classification of the caves has been made, based primarily on their morphology. Three main groups have been recognized:

Fracture caves (I)
Other types of caves in solid rocks (II)
Boulder caves (III).

These types have secondly been subdivided according to the morphogenetic agents such as tectonics, glaciation, marine abrasion and weathering (Fig. 2). When different types of caves can be recognized within these subgroups, a tertiary division is made.

DIFFERENT TYPES OF CAVES IN THE REGION

I. Fracture caves.

I.1. Tectonic fracture caves. The cave forming movements in the rocks are mostly of a very great age, but in many cases there are caves formed by neotectonics, caused by the isostatic rise of the land during the melting of the last inland ice. These purely tectonic caves occur mainly in the inner parts of the region and above the highest coastline.

I.2. Glacial fracture caves. This type of cave is found in the inner parts of the region and above the highest coastline. The fractured rock is mostly covered by till, leaving visible only a small vertical cave-entrance.

I.3. Abrasions crevice caves. Most of the caves in the region are found below the highest coastline, thus they are, to a higher or lower degree, affected by coastal processes, of which abrasion is found to be most important.

I.3.1. Tunnel-caves (Fig. 3). A tunnel-cave is defined as a cave formed along a crevice of tectonic origin in non-calcareous rocks. The tunnel-cave has a pear-shaped cross-section which is often repeated in the cliff outside the cave mouth (Sjöberg, 1981, 1982). This type of cave is very rare outside the Scandinavian peninsula. In Sweden 60 caves of this type are known, 29 of which are found in this region (Sjöberg, 1983, 1985).

1.3.2. Bedding-plane caves (Fig. 4). These are often low caves with huge floor area, formed by the excavation of one or more, generally horizontal sheets divided by fractures. Abrasion has affected the lower fracture causing the sheet above to drop down. Most of these caves have polished walls and roofs.

I.3.3. Abrasion-formed cave-shelters. This type of cave is closely

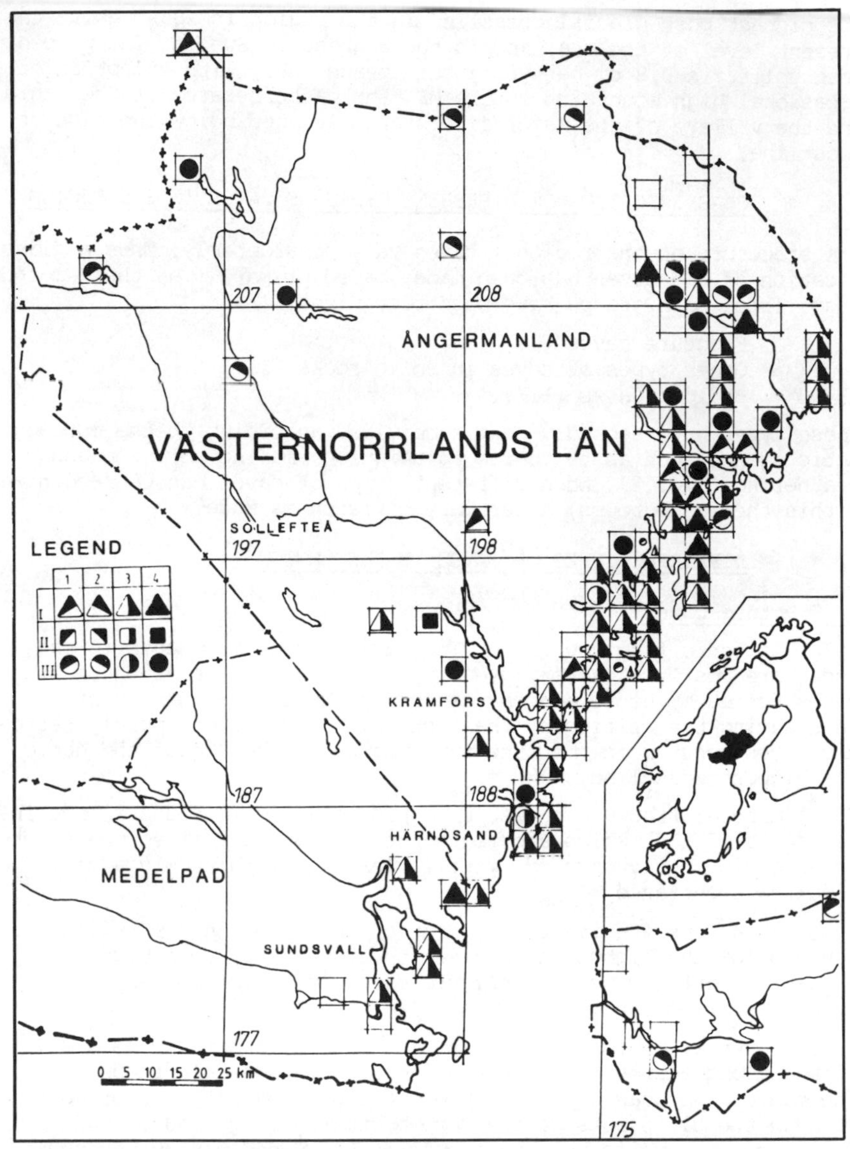

Fig. 2. The dominant type of cave per 25 km² in the study region. Legend: I. Fracture caves, II. Other types of caves in solid rocks, III. Boulder caves, 1. tectonic origin, 2. glacial origin, 3. abrasive origin, 4. formed by frost-wedging.

related to the above-mentioned. These cave-shelters are formed by a) the excavation of a great number of boulders from the rock by breakers, or b) by the excavation of one single giant boulder by the

Fig. 3. A couple of tunnel-caves on the island of Härnön. Square 188. (Photo. R. Sjöberg, 1984).

widening of vertical and horizontal fractures in the rock. These caves are mostly polished by abrasion.

II. Other types of caves in solid rocks.

II.2. Fluvial and glaciofluvial caves. In a glaciofluvially-formed canyon in the northernmost part of the region a giant half-kettle is big enough to be called a cave.

II.4. Frost-wedging caves. Caves affected by secondary frost-wedging are quite common in the region, but no examples are known of caves formed solely by frost-wedging.

III. Boulder caves.

III.1. Glacial boulder caves (Fig. 5). This group contains caves in coarse moraine and caves in erratics. Caves of this type are found both inland and in the coastal parts of the region. Caves in erratics can be formed in boulder-piles as well as in single split blocks which, primarily, might be core-stones (Twidale, 1982).

III.2. Abrasive boulder caves. These caves are formed under boulders excavated by abrasive processes along steep coasts and former coasts, and they are only found below the highest post-glacial shoreline in the eastern parts of the region.

III.3. Boulder caves formed by frost-wedging. Only one definite example of this type is found. The big sheet-like boulder forming the

Fig. 4. A bedding-plane cave in the mountains N of the city of Örnsköldsvik. Square 199. (Photo. R. Sjöberg, 1980).

roof of this cave in the eastern part of the region slipped down from the steep mountain-side less than one hundred years ago. Other examples of this type are caves formed in crevices, more or less covered by boulders. These are very common in the eastern parts of the region.

III.4. Neotectonic boulder caves (Fig. 6). Caves caused by neotectonic movements (Sjöberg, 1985b) appears as collapsed mountain-sides, and broken roches moutonnées. Caves of this type are found in both the coastland and the interior of the region.

CAVE FORMING ROCKS

The rocks in the region are pre-Cambrian, ranging from 2 000 ma greywackes to 1 215 ma dolerites (Table 2).

TABLE 2. Cave forming rocks

Type of cave Type of rock	I	II	III	Total
Metagreywackes and sediment-gneisses	42	4	21	65
Older granites	18	4	26	48
Anorthosite, labradorite	18	-	2	20
Younger granites	37	4	20	61
Dolerites	6	4	2	12

Fig. 5. A glacial boulder cave in a split erratics NW of the city of Örnsköldsvik. Square 208. (Photo. R. Sjöberg, 1981).

Table II shows that metagreywackes/ sedimentgneisses and younger granites are the dominant cave-forming rocks. The older magmatic rocks, such as granites, anorthosites and labradorites, are considerably fractured, thus forming fracture caves and boulder caves. The dolerites are intersected by vertical fractures caused by the cooling of the rock and, especially in areas near the coast there are large numbers of smaller caves - here omitted - in the fractures and in the boulder piles within them.

ANALYSIS OF DIRECTION OF THE CAVE MOUTHS

Only caves with definite direction of the cave mouth are discussed in this analysis. The main exposures of the caves are between N61°E - S60°E (49%). Within the interval S59°E - S another 34% is found. The remaining 20% are found in the sectors N - N60°E and S - S30°W.

This analysis shows clearly that most of the caves are exposed toward the present Baltic sea and its postglacial stages.

THE ALTITUDE OF THE CAVES

The caves were found at altitudes between 0 - 500 m a.s.l. Only 12 caves (7%) were situated above the highest coastline. The distribution of the remaining caves is fairly regular from the present sea level up to the highest postglacial shoreline at 285 m a.s.l.

Fig. 6. A neotectonic boulder cave formed by the collapse of a mountainside. On the coast NE of the city of Härnösand. Square 188. (Photo. R. Sjöberg, 1981).

CONCLUSIONS

The many caves in the region with a clear abrasive origin and the main alignments of the cave mouths, as well as the altitude of the caves, proves that even small seas such as the Baltic and its predecessors have a great abrading capacity, and that caves can be numerous also in non-limestone areas.

REFERENCES

Sjöberg, R., 1981. Tunnelcaves in Swedish Archean Rocks. Transactions British Cave Research Association, 8, 159-167.

Sjöberg, R., 1982. Tunnelgrottor i södra Västerbotten, morfografiska och morfogenetiska studier. Gerum report A:31, 1-68. University of Umeå, Sweden.

Sjöberg, R., 1983. Tunnelgrottor i södra och mellersta Sverige, morfografiska studier. Gerum report C:59, 1-65. University of Umeå, Sweden.

Sjöberg, R., 1985a. Tunnelgrottor i Västernorrlands län, morfografiska och morfogenetiska studier. Länsstyrelsen i Västernorrlands läns rapportserie; 1985:2 NE, 1-77. Härnösand.

Sjöberg, R., 1985b. Caves as indicators of neotectonics in Sweden. Proc. II. Symposium o Pseudokrasu Janovicky u Broumova, CSSR. (in print).

Strömberg, A., 1978. Early tectonic zones in the Baltic shield. Precambrian Research, 6, 217-222.

Twidale, C.R., 1982. Granite landforms. Elsevier, Amsterdam.

International Geomorphology 1986 Part II
Edited by V. Gardiner

SOME CONSIDERATIONS REGARDING KARSTIC EVOLUTION OF DESERT LIMESTONE PLATEAUS

V. Castellani* and W. Dragoni**

* Università La Sapienza, Roma, Italy

** C.N.R. - I.R.P.I., Località Madonna Alta, Perugia, Italy

ABSTRACT

When karst features are present in desert areas, they are generally considered to be the remains of karst evolution under ancient more humid climatic conditions. In Southern Morocco the authors found some field evidences suggesting that also in arid environments there may be considerable karst development. Such karst features consist of vertical and cylindrical tubes (about 0.5 m in diameter and 10-15 m deep) formed in horizontal-bedded limestone cliffs; the bottom of such tubes are connected to a thin network of horizontal karstic channels, resting on a marly bed. The tubes stop abruptly before reaching the surface and are present where two or more vertical fractures cross. Anywhere over the surface examined there are no surface karst features comparable in size with the tubes; this suggests that the karstic mechanism forming the tubes has little to do with water percolating from the surface, water that, in any case, in the area is rare. A simple quantitative model of formation has been made, which assumes that the tubes were and are formed by dew condensation coming from air circulating in the fractures.

KARSTIC STRUCTURE IN THE MOROCCAN "HAMADA DU GUIR"

The "Hamada du Guir" is a flat limestone plateau in the south east region of Morocco, near the town of Erfoud. Surface karstic forms were recognized early on under the form of small "dayas". Fig. 1 shows the surface of the Hamada with "dayas" scattered all over. A field investigation revealed that the Hamada edge is intensely perforated by karstic features consisting of nearly cylindrical, vertical tubes of about 0.5 m in diameter and 10-15 m deep. The bottom of such tubes are connected to a thin network of horizontal karstic channels, resting on a marly bed. The tubes stop abruptly before reaching the surface and they seem to be at the crossing of two or more vertical fractures (Figs. 2-3-4).

Fig. 1. Satellite photograph of a portion of Hamada of Guir. One side of the photo is 30 km long.

Fig. 2. Karst tubes on the edge of a butte, detached part of main Hamada.

Fig. 3. Close up view of a tube.

Fig. 4. Close up view of tubes.

DEW CONDENSATION AND THE ORIGIN OF TUBE KARST

We do not think the tubes originated by percolating water for the following reasons: i) the lack of channeled features along the wall, ii) the present aridity of the region (less than 50 mm/year of rain) and iii) the lack of surface features indicating a much more humid climate in the past. Thus we investigated in some detail the possible role of dew condensation as possibly produced by air circulation in the limestone mass. As it is well known, consistent phenomena of "breathing" are indeed common in karstic areas, though their effect on the speleogenesis is not yet clear (see Castellani and Dragoni, 1982). Our attention has been in particular raised by some experimental data by Schoeller (1953), who described in Tunisia two "breathing holes" ("trous sufflant") which may be regarded as karst tubes at work. According to this Author, winds of some m/s should blow from those holes; from this and from an analysis of the meteorological conditions in the region, he gives a figure of 35 m^3/year of water condensed in the limestone through each hole.

UNDERGROUND AIR CIRCULATION

Various situations can be envisaged for which air is forced to circulate through a network of fractures. Both differences in temperature and/or in humidity (see: e.g., Boegli, 1980) act inducing a continuous circulation of air and a continuous exchange with the atmosphere. The quoted data by Schoeller indicate that, at least in that case, aerial movements can sensitively contribute to karst. Modelling this phenomenon in a general case is rather a complex task, being a rather sophisticated multiparameter problem. For this reason we started the production of a computer program, simulating the response of an underground network of openings to various physical conditions. In the following we report some simplified cases which can instructively illustrate the potential power of condensation.

SIMPLE MODELS

Let us consider the schematic situation in Fig. 5. In absence of water, the air circulation follows the well-known rules early on described by Trombe (1952), shown in Figs. 5a and 5b. If water is present at the bottom of the vertical shaft (Fig. 5c), as given by a thin water table, the shaft tends to be occupied by saturated air, which is less dense (at the same conditions of pressure and temperature), than dry air. This, by itself, tends to induce a circulation of air as indicated by the dotted arrow, shifting the circulation toward "winter" conditions.

Let us indicate with ρ the density, T the temperature and μ the

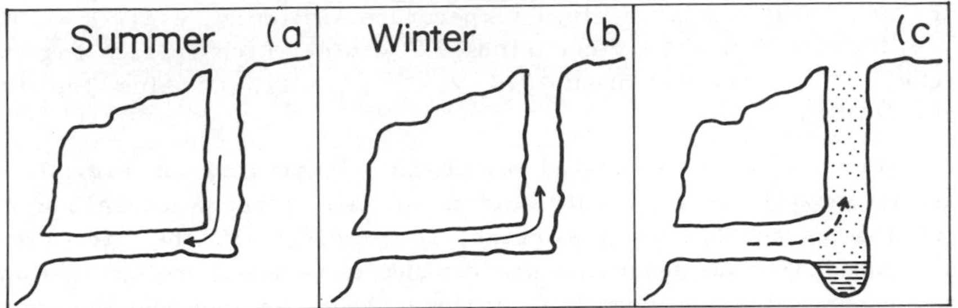

Fig. 5. Air circulation in karst systems according to thermal and humidity effects.

molar mass of the air. Since $d\rho/\rho = - dT/T + d\mu/\mu$ one can easily evaluate that the saturation can be regarded as an "equivalent dT" of only a few degrees, in the sense that circulation will follow a "winter" path unless the interior is cooler than the atmosphere by such an amount.

In spite of its smallness such an effect can play a rather interesting role when no other driving forces are present. This is sketched in Fig. 6 in a rather crude and strongly idealized model. A circulation is now induced, whose velocity can be derived from:

$$V^2 = 2000 \times D \times (P_2 - P_1) / (L \times \rho)$$

where V = wind speed (m/s), D = pipe diameter (m); P_1 and P_2 = internal and external pressure (kg/m^2), L = pipe lenghth (m), ρ = average gas density (kg/m^3).

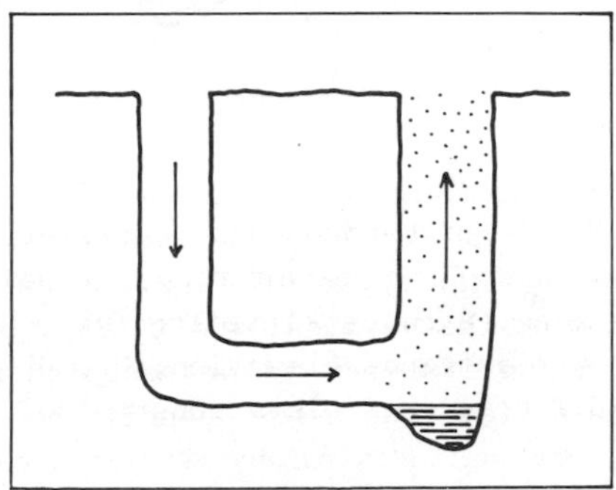

Fig. 6. Air circulation in karst system according to humidity effects.

The difference P_2-P_1 can be easily computed through the density of dry and moist air at a given temperature. Assuming, e.g., T = 20°C, L = 100 m and D = 1 m, one finds $V \cong 4$ m/s, which implies that the tube will circulate about 3.5×10^5 m^3/day, a value far from negligible.

A rather more sophisticated model is illustrated in Fig. 7. We assume now that the distribution of air temperature along the vertical tube follows the adiabatic gradient of the circulating saturated air. As saturated air expands and cools, one can evaluate the amount of water ready-to-condense along the tube and the amount of carbonate that can be eroded by such an aggressive water. Figs. 8 and 9 report some data about the efficiency of the process.

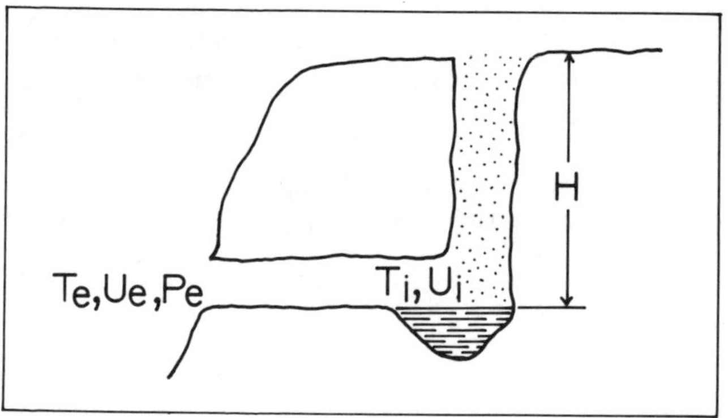

Fig. 7. Model for condensation of water through adiabatic expansion. T_e, T_i = external, internal air temperature (°C) at lower entrance elevation; P_e = external pressure (mb) at lower entrance elevation; U_e, U_i = humidity of air (%), H = difference of elevations (m) of the entrances.

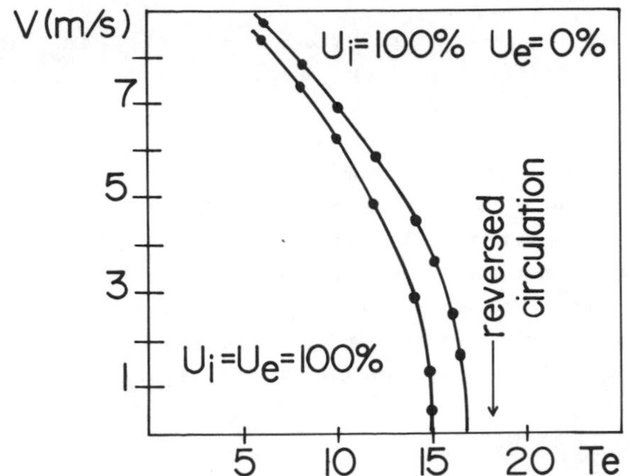

Fig. 8. Wind speed in a tube according to the model of Fig. 7 as a function of external temperature T_e. H = 50 m, P_e = 1000 mb, total lenght of cavity 100 m. T_i = 15°C.

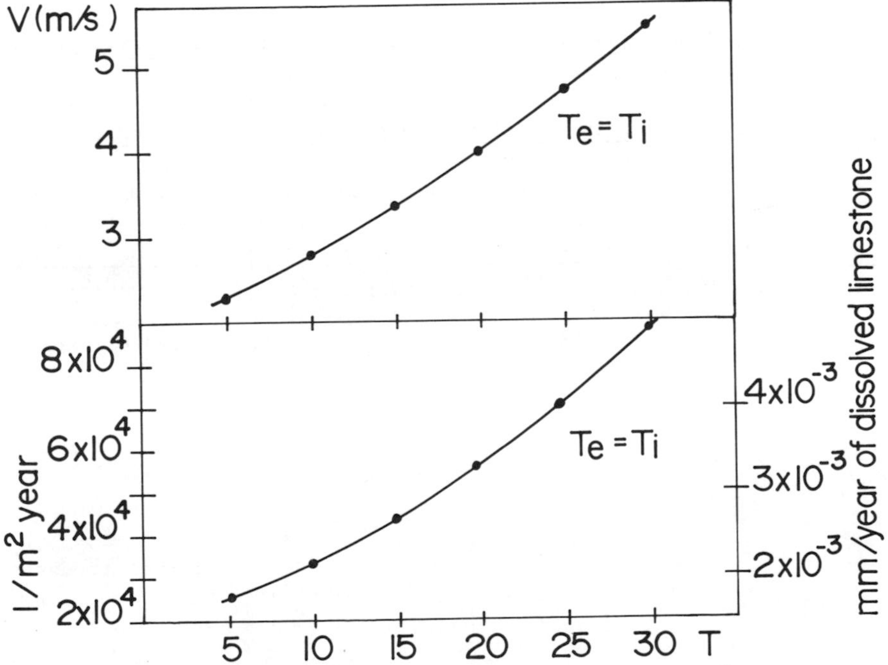

Fig. 9. Wind speed, water condensation (l/m^2 year) and limestone dissolution in the situation of Fig. 7, as a function of temperature T = T_i = T_e. U_i = 100%, U_e = 0%.

CONCLUSIONS

We found evidence that deep karst does exist in the limestone plateau of the arid Hamada du Guir. Observations by Schoeller support the idea that tubes can be originated by condensation. Numerical experiments, though rather crude, give results pretty close to Schoeller's data. Assuming these results as an order of magnitude, one would derive that tubes as those observed in the Hamada du Guir need some 10^5 years to develop. One would finally notice that the numerical experiments we are referring to show how small differences in temperature and/or humidity appear able to produce not negligible winds, which in general tend to be in the range of some m/s.

ACNOWLEDGEMENTS

The authors wish to thank prof. Marjorie Sweeting for the critical reading of the manuscript.

REFERENCES

Boegli, A., 1980. Karst hydrology and physical speleology. Springer Verlag Ed., pp. 214-224.

Castellani, V., and Dragoni, W., 1982. About the genesis of karstic cavities, in II International Symposium on Utilization of Karstic Areas, pp. 29-35. Bari.

Schoeller, H., 1953. Condensations occultes, en particulier dans les affleurements de terrains calcaires ou greseux de l'Afrique du Nord, in Actions eoliennes. Phenomenes d'evaporation et d'hydrologie superficielle dans les regions arides, Ed. C.N.R.S., pp. 348-358. Paris.

Trombe, F., 1952. Traité de spéléologie. Payot, pp. 1-376. Paris.

ARID LAND AND AEOLIAN GEOMORPHOLOGY

International Geomorphology 1986 Part II
Edited by V. Gardiner

ARID LAND AND AEOLIAN GEOMORPHOLOGY : INTRODUCTION

Andrew Warren

Department of Geography, University College,
London, U.K.

The range of modern and ancient geomorphological processes that has shaped the dry world is no less diverse than those that have shaped other parts of the earth, and each process has a distinctive mode of operation and expression in the arid zone. Only unhampered aeolian processes are really characteristic of these parts of the globe, even if they are not entirely restricted to them. With the exception of the work of the wind, therefore, we should not expect a coherent group of papers on the geomorphology of arid lands from the global geomorphological community, and coherence is unfortunately not a property of the few papers in the following section.

Although small, the collection does represent some of the current range in the interests and methods of geomorphologists in the dry world. There are three papers on the work of the wind: one on dunes in the very arid parts of China; one on erosive and depositional processes on the semi-arid high plains in Algeria; and one which, by describing dunes on the Chilean coast, reminds us that aeolian processes are also common on many non-arid coasts. All three demonstrate the economic hazards of aeolian landforms, but each has a different scientific approach, a useful lesson that the understanding of wind-generated forms needs many different kinds of effort: the cataloguing and classification of form; deductions about processes from field description of forms; and the development of concepts about what is happening. The gap in this very small collection is itself representative of the field: in the literature at large there are very few reports of detailed quantitative observations of aeolian processes; and there is none here.

The remaining three papers could not hope to cover the scope of the current geomorphological effort in deserts, but they too are examples of different approaches in use. The first is another illustration of the basic need to classify and catalogue landforms, in this case volcanic forms in arid Saudi Arabia. The second, on slope processes in the central Sudan, makes use of a range of simple and sometimes effective field techniques. And the last, on the effects of salts on rock-weathering, makes use of more complex observational equipment and some fundamental physico-chemical concepts.

The very sparsity of this little geomorphological harvest from the arid lands actually highlights a general conclusion about this field of research: much remains to be done. There is more to be understood about these landforms than those of any other major environment, because inaccessibility has protected them from study, but this is to the detriment of the geomorphological community. Some of our most fundamental concepts were derived from the landforms of the dry American West through the eyes of Gilbert and Davis, and many others since. Some of these insights are illustrated in the other sections of these Proceedings. The arid lands still hold secrets that will give us new and basic insights into landforms in general.

International Geomorphology 1986 Part II
Edited by V. Gardiner

THE CHARACTERISTICS OF AEOLIAN LANDFORMS AND THE CONTROL OF MOBILE DUNES IN CHINA

Zhu Zhenda, Zou Bengong and Yang Youlin

Institute of Desert Research,
Academia Sinica, Lanzhou, Gansu, China

ABSTRACT

Aeolian landforms cover 8.4% of China. In the arid areas they are the result of long-term aridity. Aeolian landforms occur typically in basins of internal drainage, and aeolian sands are derived from river alluvium, lacustrine deposits or diluvium. Arid mobile dunes have complex shapes. In semi-arid areas vegetation fixes dunes, which are often honeycomb-shaped, or barchanoid where vegetation has been destroyed. Uni-directional winds produce barchanoid dunes, two dominant directions produce longitudinal dunes, multi-directional winds produce complex forms. Dune movement is influenced by wind velocity, dune height, moisture content and vegetation cover. Erosional aeolian landforms also occur. Measures for controlling mobile dunes are suggested.

INTRODUCTION

Aeolian landforms cover 8.4% of China's surface or approximately 803,000 square kilometres, of which inland dunes occupy 76.8%, vegetated dunes occupy 21.9%, in the semi-arid zone, and coastal and river dunes occupy 1.3% in the sub-humid and humid zones.

AEOLIAN LANDFORMS OF THE ARID ZONE

The aeolian landforms of the arid areas have formed under conditions of long-term aridity. Aridity was progressively intensified by the rise of the Qinghai-Tibet Plateau from the end of the Tertiary. The rise of the Plateau, (Ling Zhengyao et al., 1978), which reached 3,000m in the Meso-Pleistocene, reinforced the formerly weak zone of high atmospheric pressure and moved its centre north to the eastern Tarim Basin at 40°N. Most of the lakes in this Basin gradually dried up, and the sandy sediments of the rivers from the surrounding mountain ranges began to be moved by the wind and formed dunes.

The Qingai-Tibet Plateau rose to 4,000m in the Late Pleistocene and Early Recent, and the centre of the winter high pressure zone was moved even further north, to 55°N. The winter monsoon winds now blew outward from the high pressure zone to the north-east and north-west at 96°-97°E. A north-west wind blew over the south-east of the desert and formed the aeolian landforms of the Axla and

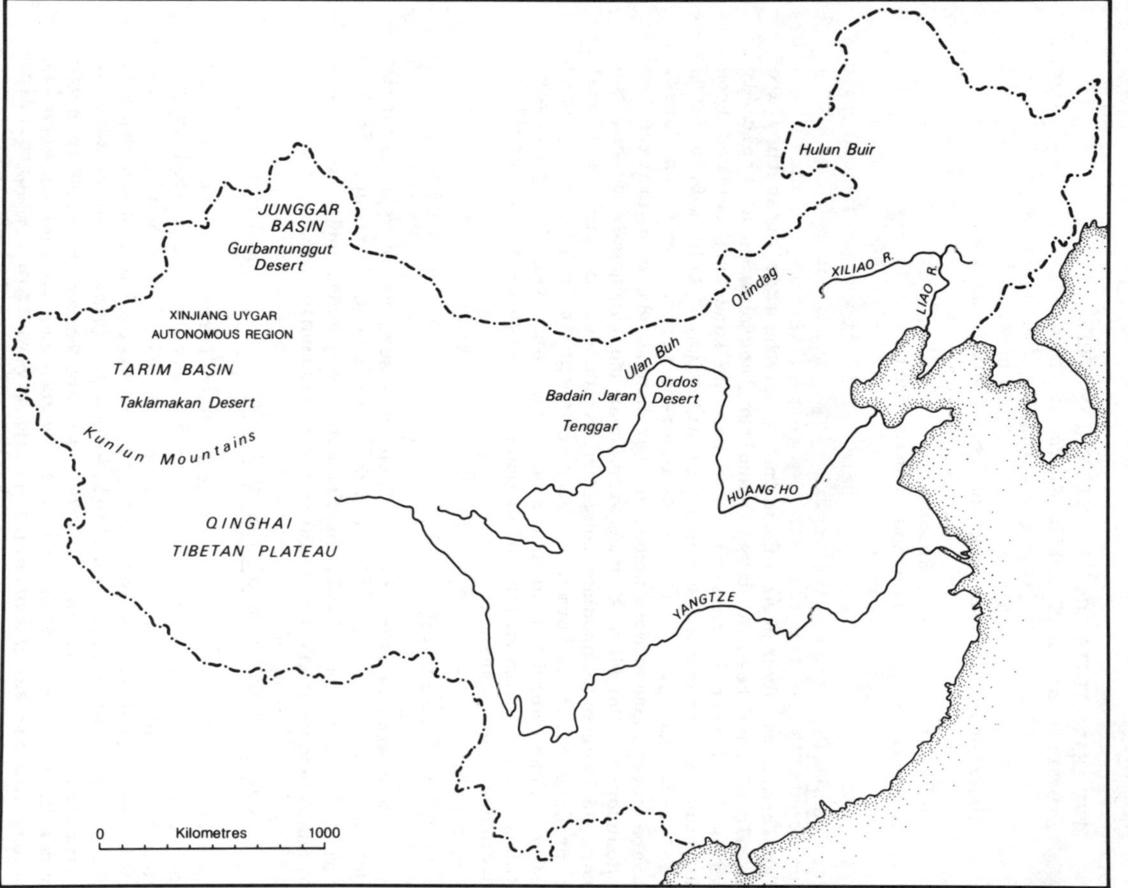

Fig.1. Deserts of China and locations in text.

Ordos. A south-east wind blew in the south-western desert and moulded the landforms of the western Hexi Corridor and the Tarim Basin. In the western part of the Tarim Basin the wind was changed to north-westerly.

The aeolian landforms of China are found typically in basins of internal drainage. Among these are the Taklimakan Sand Desert of the Tarim Basin, the Gurbantunggut in the Junggar Basin, and the Badain Jaran and Tenggar Sand Deserts in the south-west of the Hanhai Basin (Fig.1).

Sand that had accumulated along river channels and in lakes provided the source areas for these dune fields. We have shown, by heavy mineral analysis, that the southern part of the Taklimakan Sand Desert was derived from alluvial-diluvial fans on the borders of the Kunlun Mountains. Forty to fifty per cent of the heavy minerals are hornblende, with epidote and metal minerals also abundant. The northern part, on the other hand, derived from the fans of the Tianshan Mountains. Here mica forms 43.8% of the sands, with epidote and hornblende also abundant.

Studies of Quaternary palaeogeography and heavy mineral analysis (Zhu Zhenda and Wu Zhen, 1979) show that the origins of aeolian sands in China can be summarised as follows:

1. River alluvium in the northern and southern Taklimakan, the Ulan Buh and the southern Gurbantunggut.
2. Alluvium and lacustrine deposits dominate the Badain Jaran, the Tenggar and the eastern Taklimakan.
3. Alluvium and diluvium in the southern central Taklimakan.

AEOLIAN LANDFORMS OF THE SEMI-ARID ZONE

In the past, the fragile ecosystems of the sandy steppes were misused and largely destroyed. Here the main sources of sand were the deposits of rivers such as the Xiliao in north-eastern China, and lacustrine sediments as in the Hulun Buir and Otindag Sandy Lands.

DUNE FORMS

In the arid lands, sandy deserts are characterised by mobile dunes, which cover 85% of their surface. The forms of these dunes include complex mega-dunes such as star-dunes, transverse dunes and longitudinal dunes. Isolated mega-dunes are commonly found in areas with residual hills. Where there are extensive sources of sand, as from river terraces, compound longitudinal dunes are common.

In the semi-arid zone, vegetation cover, often with such trees as *Pinus sylvestris var. mongolica*, has fixed the dunes. The dune forms are often honeycomb-longitudinal or honeycombs on sandy ridges. Barchans and barchanoid dunes are only found where the vegetation has been destroyed. River and coastal dune forms are often linear, and their form is influenced by the trends of the river or coast. Most coastal dunes are fixed or partially fixed by vegetation.

Some general statements can be made about the relations between dune forms and their environment:

A. Different wind regimes produce different dune patterns (Zhu Zhenda et al., 1980; Wasson and Hyde, 1983): Uni-directional winds produce barchans and barchanoid ridges; examples are found in the south-western Taklimakan, the south-eastern Axla, and the south-western Ordos Plateau. Where there are two dominant weak winds, longitudinal dunes are found, as in the central Junggar Basin. Where there are multi-directional winds of similar wind velocity, pyramidal or star dunes are found, as in the south-western Axla and the Taklimakan.

B. Windward slip-faces are orientated to the resultant of the sand-drifting winds. The relationship is not exact; the angle between the resultant and the dominant direction of slip-faces (theta) is related to the complexity of the sand-drifting regime. In uni-directional winds, theta is 10° to 50°. Where there are multi-directional winds, the angle varies between 90° and 150°.

DUNE MOVEMENT

In the eastern part of northern China dunes generally encroach from south-west to north-east. In the Axla of northern China and in the western Tarim and Qadaim Basins, the dunes move from north-west to south-east. In the eastern Tarim Basin and in the western Hexi Corridor the dunes move from north-east to south-west.

In addition to wind velocity, the rate of dune movement is also influenced by the height of the dune and by the moisture content of the sand and vegetative cover. Dunes can be classified according to celerity into four groups:

1. slow dunes, moving less than 1m annually;
2. moderately fast dunes moving at between 1 and 5m annually;
3. fast dunes, moving 6 to 20m annually; and
4. very fast dunes, moving over 20m per annum.

LANDFORMS OF WIND EROSION

Where there are soft rocks, as with the mudstones, clay-sandstones and sandstones of the north-western Qadaim Basin, the wind has eroded low hills parallel to the wind direction. In the eastern Xinjiang Uygar Autonomous Region, violent storms can form yardangs from river and lacustrine deposits of salt, clay-sandstone and mudstone. Where the rocks are largely clay and siltstone, the landscape is dominated by depressions and low hills.

THE CONTROL OF MOVING SANDS

Mobile dunes threaten farmland, roads, railways and settlements, and so must be controlled. In arid zones, moving dunes that threaten oases can be controlled in the following ways:

1. The establishment of sand-breaks around oases;
2. The planting of trees within oases;
3. The building of grass fences around oases and in dune fields;
4. The construction of straw and clay chequer-board grids on dune slopes to check the wind velocity;
5. The planting of sand-holding species within the grids to stabilize the dunes.

As to the control of dunes that threaten roads and railways, vertical barriers made of trees and shrubs should be erected next to the road or railway, and these should be protected by vertical barriers in chequer-board patterns, within which sand-holding tree or shrub species should be planted.

To control mobile dunes in semi-arid areas, the following methods can be used:

1. Nature reserves;
2. Wood-lot planting in inter-dune areas;
3. Stabilization with shrubs or grasses;
4. The avoidance of disturbance.

REFERENCES

Ling Zhengyao et al., 1978. The uplift for the influence of the weather in the Qinhai-Tibet Plateau, in The Collected Works of the Scientific Symposium on Qinhai-Tibet Plateau. Science Press, Beijing.

Wasson, R. J., and Hyde, R., 1983. Factors determining desert dune type. Nature, 304, 337-339.

Zhu Zhenda and Wu Zhen, 1979. Aeolian Landforms in China (Physical Geography of China). Science Press, Beijing.

Zhu Zhenda, Liu Shu and Wu Zhen, 1980. An Introduction to Chinese Deserts. Science Press, Beijing.

International Geomorphology 1986 Part II
Edited by V. Gardiner

A COMPARATIVE STUDY OF THE LUNETTES IN THE TARF AND HODNA BASINS (ALGERIA)

M T Benazzouz

Institut des Sciences de la Terre, Université de Constantine, Algérie

ABSTRACT

The formation of an endoreic drainage system in the High Plains of Eastern Algeria is linked to the appearance of substrates following disruption of the drainage pattern at the end of the mid-Quaternary.

There appear to be two causes for endoreism in this region:

a) accentuated tectonic movement from late Pliocene times causing active subsidence leading to the endoreic isolation of the Chott el Hodna basin.

b) climatic endoreism furthur east where the Tarf Sebkha is typical of sebkhas produced by Late Quaternary aridity. Conditions in the Tarf sebkha, at 830 metres altitude with 400 to 500 mm annual rainfall favour the formation of the pelletal aggregates of which the present active lunette is being made.

The lower Chott el Hodna (400 m altitude) which has a Saharan-type continental climate (annual rainfall of 200 to 250 mm) lacks an actively forming lunette.

The Tarf Sebkha experienced two phases of lunette generation in the upper Pleistocene and in the Holocene. The single lunette formation of the Hodna comprises longitudinal clay-dunes separated by flat depressions which are virtually fossilised beneath gypsum encrusted silty-clay deposits derived from the erosion of longitudinal dunes. The significant water deficit restricts degradation of the lunette in the Hodna-basin so preserving the original longitudinal dunes which are separated by flat floored depressions in direct contact with the base of the sebkhas.

INTRODUCTION

The geomorphological expression of the endoreic hydrologic regime of the high plains of Algeria is a series of seasonally inundated marshy depressions. Research on the origin and fluvial regimes of these depressions reveals two ways in which these sebkhas develop:

a) development through tectonic acvitity

b) a progressive evolution towards a sebkha regime linked to increasing aridity.

On the other hand the sebkha landscape of the Southern Constantine High Plains can be shown to be but a stage in the disruption of an old drainage network which ceased to function at the beginning of the Late Quaternary. Lunettes are a characteristic aeolian landform of crescent shape on the eastern edge of the sebkha with the concavity opening on to the sebkha, as defined by Hills (1940). The clay dunes studied in eastern Algeria lack the characteristic crescent form, but are closely linked in their origin and form to the adjacent sebkhas. These aeolian features might be more appropriately termed ribbon lunettes (lunette laniérée) in the cases, at Tarf, where the two points of the crescent are found at the north eastern and south eastern extremes of the sebkha. In the Hodna, however, there are individual dunes which can be called aeolian mounds (bourrelet éolien).

THE CAUSES OF ENCLOSED DRAINAGE IN THE HIGH PLAINS OF EASTERN ALGERIA

Tectonic factors in drainage development in the Hodna basin. The enclosed Hodna basin is essentially a corridor, marked with a series of sebkhas, closed off by two mountain barriers, the Tellien Atlas, rising to 1900 m in the Djebel Bou-Taleb to the north and the Saharan Atlas, to the south, with a general elevation of 700 to 1000 m. A former graben, the Hodna basin still shows some signs of tectonic instability (the tremors of M'Sila 1960-65) at present. The geological history involves three episodes.

- the basin developed early in the Lutetian folding, becoming a zone of continental sedimentation until the end of the Oligocene. The Burdigalian subsidence later led to 1000 - 2500 m of marine sediments.

- subsequent evolution witnessed the final retreat of the sea at the end of the Miocene with infilling by conglomerates until the end of the Pliocene.

- the Quaternary began with a phase of relative tectonic stability during which debris was evacuated southwards, probably towards the Chott Melhrir, across the M'doukal divide separating that basin from the Hodna depression (Fig. 1).

These processes led to the development of a series of glacis on the northern side of the Hodna basin, converging to the south east at the height of the M'doukal divide. The thin detrital covering of these glacis is further evidence of the lack of subsidence in the Hodna during the early Quaternary.

The formation of the sebkha proper, the Chott El-Hodna, resulted from the mid Quaternary subsidence of the Hodna basin, creating 100 m of detrital deposits, and uplift of the M'doukal divide, finally closing off the depression. Since then, through the Late Quaternary, the basin has operated as an enclosed depression with a sebkha in the centre. Thus subsidence led to the formation of the depression and to its thick layers of detrital sediments, with differential uplift of the divide cutting off the external drainage.

THE TARF COMPLEX OF INTERNAL DRAINAGE

At the eastern extremity of the High Plains, dominated to the south by the Aures, is a large topographic elongated hollow marked by small sebkhas. In contrast to the Hodna, the structure here is a more complex sequence of tectonic fractures with a series of horsts and grabens, leading a set of small plains, many containing a sebkha. The area's geological history reflects tectonic events in the Aures.

During the Miocene, the Tarf basin was subsiding actively and accumulating several hundred metres of gypsiferous marls. Later, Pliocene tectonic stability led to the formation of lakes in which a sequence of conglomerates, sands and clays accumulated, terminating in sediments whose chemical character indicates stable conditions.

Quaternary processes led to the degradation of the Neogene deposits by the development of a large drainage network extending northwards towards the sea from the Aures piedmont. The existence of this drainage axis is indicated by a sequence of glacis mantled with thick layers of alluvial deposits. The deposits were probably laid down by powerful wadis under a more humid climate than the present. However, dried, mid-Quaternary conditions reduced wadi discharges and led to a series of small alluvial fans breaking up the main valley into a series of small depressions in which sebkhas formed. In the Late Quaternary, intensive wind action led to a creation of the lunette on the eastern side of the Sebkhet Tarf. The formerly integrated exoreic drainage system became broken up into a series of small endoreic catchments (Fig.2).

While in the Tarf, enclosed depression formation is related to increasing aridity and dismemberment of an ancient drainage system, in the Hodna, tectonic events are the prime cause, as the Hodna mountains always ensure an adequate water supply.

EVOLUTION OF THE TARF AND HODNA LUNETTES

THE HODNA AEOLIAN COMPLEX AND THE ACCENTUATION OF ARIDITY

On the eastern edge of the Chott el Hodna has developed an aeolian gypsiferous clay complex of longitudinal dunes, several km long, separated by broad corridors linked to the main sebkha. As with other north African sebkhas, southern margin of the sebkha is marked by a massive, mainly sandy, unbroken aeolian complex terminating in a small chott at its south eastern end. This sebkha has a single generation of Late Quaternary dunes, the chains, or ribbons, of which stand 10-20 m above the sebkha. On the flank of one of these elongated dunes, the Draa Souid, two distinct stratigraphic horizons were recognised with multicoloured gypsiferous clays containing gypsum crystals arranged in horizontal beds at the base and grey to beige gypsiferous sands above characterised by bedding. On the dune surface, a slight concentration of gypsum forms a two centimetre thick layer.

Prehistoric relics in these deposits and the general stratigraphy indicate that the base of the sebkha was originally a few metres above its present level. The aeolian complex essentially developed from these dunes and lies directly on the multicoloured gypsiferous clays of the former sebkha floor (Fig.3).

THE EVOLUTION OF CHOTT EL HODNA AEOLIAN COMPLEX

The Hodna aeolian complex developed in one morphogenetic phase producing a single generation of landforms. The dune chains are not derived from previous aeolian or fluvial features. The original cross-bedded dune stratigraphy is found through the complex without any more clayey B horizon and no surface development of gypsum. Present-day processes may be deduced from three groups of features: a) one metre long, 50 cm high nebkhas of clay and gypsum aggregates on the northern, windward flanks of the dune chains; b) wind-blown corrasion striae between the nebkhas on the summits of the dunes exposed to the northwest; and c) water-eroded features of the aeolian complex, where the main dune features are being dissected.

THE RIBBON LUNETTE OF THE SEBKHET TARF

The Sebkhet Tarf ribbon lunette is much larger than the Hodna complex, being 5 km long and 20 m high. It has developed among parallel rows of longitudinal dunes and has its greatest extent in the central zone of the eastern margin. Although the north eastern end of the lunette forms a half-crescent, the south eastern end has remarkable transverse dunes. The complexity of this lunette reflects Holocene climatic fluctuations in the High Plains of eastern Algeria.

SEDIMENTOLOGICAL ASPECTS OF THE TARF LUNETTE

The Tarf dunes have flatter, whiter summits than those of Hodna, the whiteness being due to a 50 to 80 cm thick gypsum crust, of over 90% gypsum, protecting the dune surface. The floors of the entirely enclosed inter-dune depressions are covered with halophytic steppe vegetation. These depressions lead towards the Sebkhet Tarf.

Beneath the gypsum crust, the dunes are composed of 10 m or more of gypsiferous sands in an orange to dark brown clay loam matrix. A borehole through the dune revealed the following sequence:

0-20 cm yellowish loams with gypsum sand-sized concretions, becoming more clayey towards the base with veins of gypsum;

1 m - 3 m or more fine sand-sized concretions, consolidated towards the top, in a clayey matrix with gypsum veins, more friable towards the base; composed of gypsum, quartz and calcite with poorly crystallline detrital kaolinitic instratified clays, the aridity not permitting neoformation of clays. The grain sizes of the material reflect wind action on aggregates of the cracked floor of the sebkha, not the original composition of the sedimentary particles. Salt encrustations and cements give particles a low density which favours their transport by wind. Further salt weathering processes modify the sediments once they are accumulated in dunes, which contain up to 50% clay. Ancient sebkha floors are often found in interdune depressions where the multicoloured clays are bruied beneath several metres of sandy-clay fill.

The main sedimentological features of the Tarf lunette are: a) lack of preservation of original stratification throughout the lunette profile; b) post-depositional leaching of aeolian formations, with the development of an illuviated clay horizon; c) protective surface gypsum crust and d) an infill on the floor of the interdune depressions. These findings confirm the existence of humid conditions to the north of the Aures during the Holocene period.

THE SUBSIDIARY FEATURES OF THE TARF LUNETTE

A west-east section from the sebkha to the summit of the lunette reveals two subordinate phases of lunette development at the foot of the major complex (Fig.4): a recent lunette and an actively forming lunette. The recent lunette forms an extensive 2 to 4 km high feature 30-50 m wide following the edge of the present sebkha sitting on the former sebkha floor. Containing a high proportion of dark brown aeolian clay with lenses of gypsiferous concretions the whole formation is very friable when dry. A borehole shows the following sequence:

0 - 70 cm	clay loam, with weak loamy matrix (7.5 YR, 4/3)
70 - 95 cm	slightly sandy clay with gypsum crystals (7.5 YR, 5/4)
95 - 130 cm	fine loam (10 YR, 6/4)
130 - 140 cm	sandy clay (10 YR, 6/3)
140 - 260 cm	loamy sand (10 YR, 6/3)
260 - 280 cm	gypsiferous sand in a clayey matrix
280 - 300 cm	fine, siliceous sand (2.5 UR, 6/2)

This arrangement encourages concentrated runoff which imprints a distinctive relief pattern on the lunette. Where degradation is most developed a system of rills has broken the lunette into a series of isolated remnants.

The active lunette is made up a field of nebkhas of dark brown aeolian clay piled up against the recent lunette. The nebkha sediments are largely gypsum granules and crystals, with some poorly preserved, highly soluble whitish salt crystals. The fresh surface sediments are small aggregates, several millimetres in diameter comprising salt and gypsum crystals with clay and fine sand particles. Halophytic sebkha vegetation gradually colonises and fixes this last generation of lunette forms.

EVIDENCE FOR A CHRONOLOGY

Few prehistoric remains are available to reconstruct a precise chronology of landform development. Capsian snails on the chains of the major dune system suggest a minimum age of 10,000 BP for the first generation of lunette features. Such landforms have been pecisely dated in south eastern Australia as forming between 17,000 and 15,000 BP during a phase of extreme aridity, using human artifacts and midden shells in several horizons in the lunettes for radiocarbon dates (Bowler, 1971).

A section along the Wadi El Meniri exposes the components of the interdune fills where a depth of a metre is palaeosol with 40-50 cm of dark brown clay (10 YR, 4/4) with snail shells. An alluvial formation has cut into the surface of this palaeosol also has snail shells which have been C_{14}dated by J C Fontes at the hydrology and isotope geochemistry laboratory of Orsay - Paris Sud. Two main humid phases since the development of the major dune formation are identified, corresponding to palaeosol development at 3680 $\pm$ 100 BP and deposition of the overlying alluvial formation around 1495 $\pm$ 100 BP. The soil formation period was more humid than the latter, with the second humid phase predating the development of the active lunette.

DISCUSSION

Increasing climatic aridity has been a fundamental factor in sebkha development in eastern Algeria, unlike the general pattern of sebkhas described by Bowler (1973) which involved the derivation of dunes by the drying out of ancient lakes whose wave created sand beach dunes become buried by clay dunes due to aeolian action on the former lake bed. On the eastern Algerian high plains however, sebkhas have almost unique aeolian complexes lying on the former sebkha floor on their eastern margins. Sebkha development in both cases stems from disruption of an ancient drainage network, but in the Hodna and Tarf basins the modification occurred later in the Quaternary, with more rapid dessication which must have eliminated any traces of intermediate lacustrine phases of which there are no vestiges on the margins of the sebkhas. Moreover, sedimentological and morphological analysis has failed to reveal any evidence of lakes in these two basins.

CONCLUSION

Studies of the Tarf lunette revealed two humid periods in the mid- and late- Holocene. Lunettes can provide much palaeoclimatic evidence for the last few thousand years, as at Tarf, but at Hodna such evidence was not found. If there were any such indications, they could have been removed by intense fluvial erosion by wadi sheetfloods descending from the Belezma range to the east and flowing on to the sebkha through the interdune corridors of the lunette.

The 3680 ± 100 BP date for the lower palaeosol confirms the existence of a mid- Holocene pluvial period corresponding roughly to the date of 4830 ± 120 BP given by Ballais (1981) for the Ain-Touta palaeosol in the Aures mountains. Possibly the humid phase was slightly later in the High Plains than further south. The humid conditions persisted longer further east in the Tarf area, whereas conditions became drier earlier in the Hodna area.

Today the Hodna area suffers from a Saharan continental climate with irregular rainfalls averaging 200 to 250 mm y^{-1}. This aridity favours the development of a saline crust on the surface of the sebkha, thereby reducing the efficiency of wind erosion. However, the eastern extremity of the High Plains at 830 m altitude has a fresher, moister climate with 400 - 500 mm average annual rainfall in the Tarf basin. Such moisture keeps the sebkha surface sufficiently humid for the development of miniature polygonal crack structures from which material is readily supplied to active fields of nebkhas.

REFERENCES

Ballais, J. L., 1981. Recherches géomorphologiques dans les Aurès (Algérie). Thèse Doctorat ès Lettres, Paris I, 566 pp.

Bowler, J. M., 1971. Pliestocene salinities and climatic change: evidence from lakes and lunettes in south eastern Australia, in: Aboriginal Man and Environment in Australia, (eds Mulvaney and Gotson), pp 47-65. Australian National University Press, Canberra.

Bowler, J. M., 1973. Clay dunes: their occurrence, formation and environmental significance. Earth Science Reviews, 9, 315-338.

Hills, E. S., 1940. The lunette; a new landform of aeolian origin. Australian Geographer, 3, 15-21.

Fig.1. (Facing) The Hodna Basin.

- Mountain massif
- Sands and dunes
- former drainage system to the Chott Melrhir
- aeolian complex
- sebkha

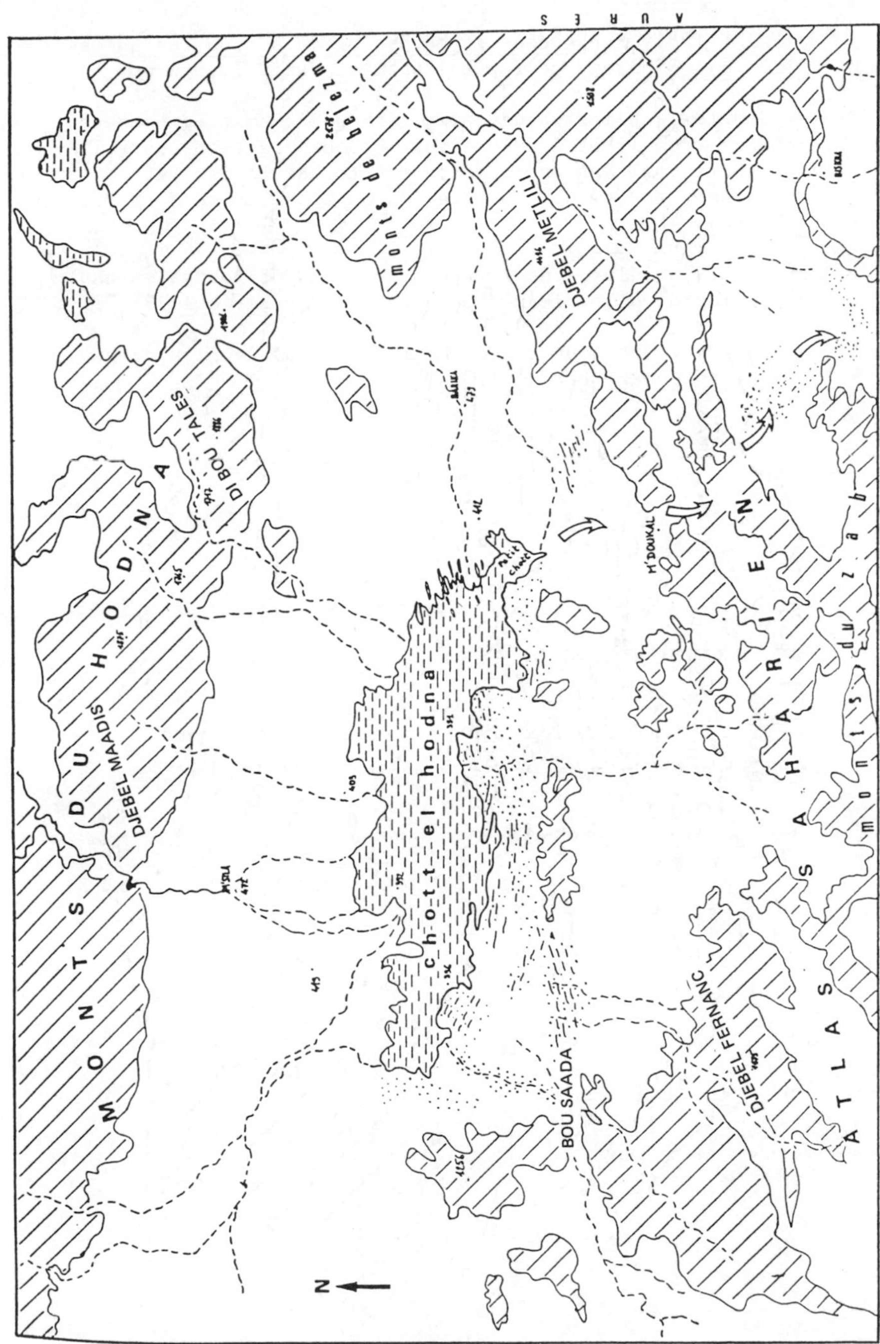
AURES
MONTS DU HODNA
DJEBEL MAADIS
DJ BOU TALES
monts de belezma
DJEBEL METLILI
chott el hodna
Petit chott
M'DOUKAL
BISKRA
BOU SAADA
DJEBEL FERNANC
ATLAS SAHARIEN
monts du zab
N

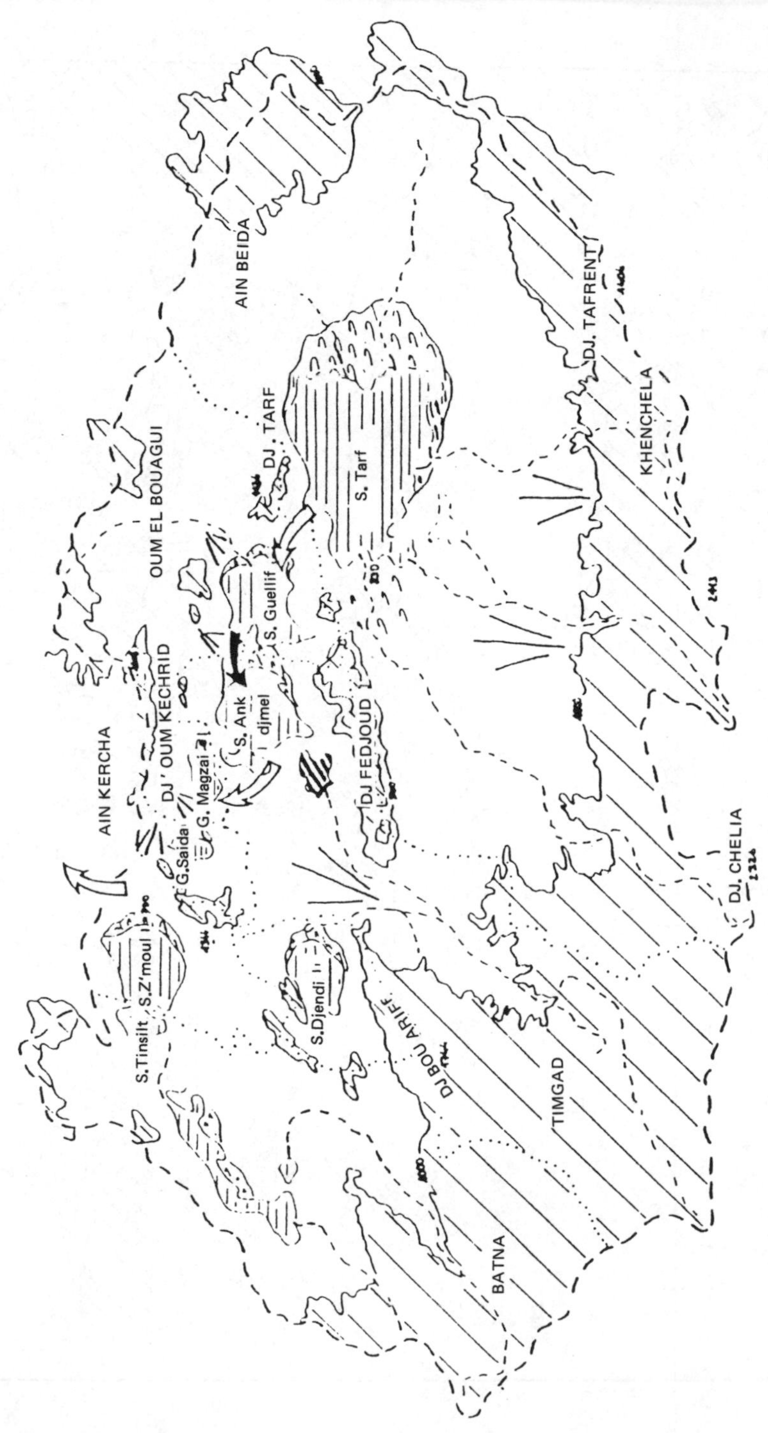
AIN BEIDA
OUM EL BOUAGUI
DJ. TARF
S. Tarf
DJ. TAFRENT
KHENCHELA
S. Guellif
DJ OUM KECHRID
AIN KERCHA
S. Ank djmel
G. Magzai
G.Saida
DJ FEDJOUD
S.Tinsilt
S.Z'moul
S.Djendi
DJ BOU ARIF
TIMGAD
BATNA
DJ. CHELIA

Fig.2. Factors in the development of internal drainage in the Constantine High Plains.

1000m contour

major upland areas surrounding the region

divide separating the internal and external drainage

divides between individual sebkhas

sebkha or salt flat

Wadi

former abandoned drainage line to the north

drainage by sebkha overlow

alluvial fans

lunettes

irregular clay dunes

longitudinal dunes

artificial lake

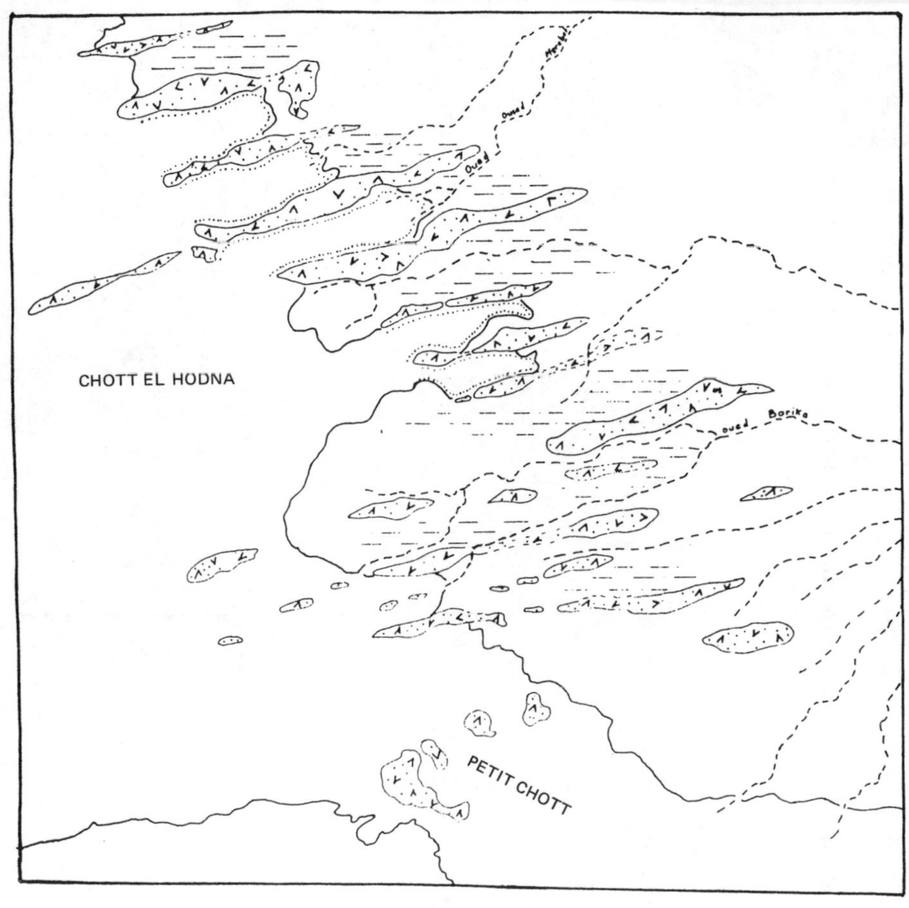

ribbon or string lunette of sand-sized saline nodules

dissected ribbon dunes

interdune corridors separating the ribbon dunes

extensions of the sebkha onto the interdune corridors

outwash plain

wadi

Fig.3. Relief features of the eastern Chott el Hodna lunette

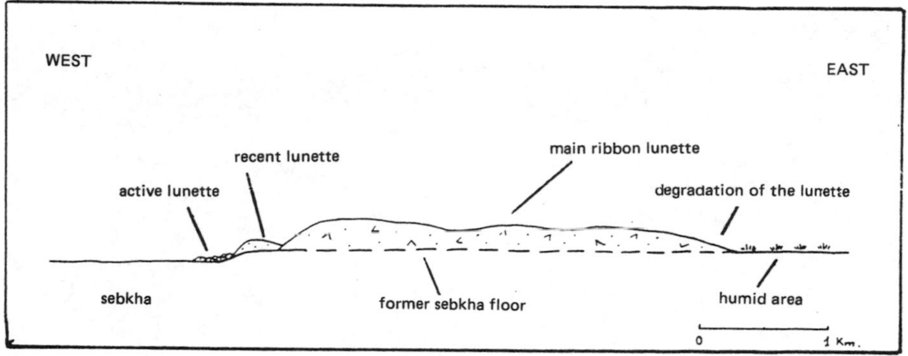

Fig.4. Section through the eastern dune of the Tarf Sebkhet

International Geomorphology 1986 Part II
Edited by V. Gardiner

THE EVOLUTION OF MODERN COASTAL DUNE SYSTEMS IN CENTRAL CHILE

J. F. Araya-Vergara

Department of Geography, University of Chile,
Cas. 3387, Santiago de Chile.

ABSTRACT

Changes in central Chilean dune systems over the period 1955 to 1978 were studied in order to discover: (a) transformations of one dune class to another; (b) structural changes in dune systems; and (c) trends. These observations were used to test the "transmudación" theory. Different types of transformation occurred on shorelines with different orientation. On wind-parallel shorelines, small coalescent barchans and transverse ridges suffered splash and runoff erosion and then erosive aeolian thresholds were surpassed; the dune pattern lost compactness and some new barchans appeared. On shorelines oblique to the wind, large coalescent barchans and transverse patterns also lost compactness of pattern, and in places broke into individual barchans after splash and runoff erosion. On quasi-wind-transverse shorelines, transverse dunes and massive dune ridges remained in a quasi-steady state, but the management of *Ammophila* induced blow-outs. On the leeward side of these coasts blowouts and parabolic structures appeared. Thus the orientation of a beach is an important element in the control of the morphology and evolution of dune patterns. Dune patterns lose their compactness in rainy years, and some individual dunes may move. Small dunes, and especially barchans reach their greatest geomorphological activity. The "transmudación" theory was most applicable where processes were most effective and where forms were best defined.

INTRODUCTION

Theories of dune-change. Walther, as long ago as 1900, observed that a transverse ridge was formed by the coalescence of a number of barchans. In contrast, Bagnold (1941) believed that transverse dunes were inherently unstable and would break-up either into barchans or, if winds blew at a narrow angle to each other, into longitudinal ('seif') dunes. Bloom (1978) agreed that transverse dunes appeared where there are large amounts of sand and relatively ineffective winds. This is confirmed by the recent survey of Wasson and Hyde (1983). De Martonne (1926), basing his hypotheses partly on classic German work, believed that parabolic dunes resulted from the erosion of a transverse ridge, into longitudinal ridges. Verstappen (1968) observed that parabolic dunes were elongated by downwind migration to form "hairpin" or "upsiloidal" dunes, which might themselves be transformed into longitudinal ridges. Vanney et al., (1979) studied the association of transverse and parabolic dunes, and Norrman (1981) described systems that comprised a single continuous foredune ridge backed by partly active parabolic dunes. Goldsmith et al., (1977) described large fields of parabolic dunes behind foredunes, formed from "médanos" or transverse dunes. These parabolics were about one kilometre long and differed in origin from smaller parabolic forms.

As for barchans, Hastenrath (1967, p.329) observed that "they grow to a maximum size, and then shrink .... and vanish .... downwind." Finkel (1959) and Hastenrath (1967) correlated the rate of displacement with the crest height, and confirmed Bagnold's (1941) theory that small dunes travelled faster than large ones. Tricart (1977) believed that barchans were the most active forms and that other dune systems were relatively inactive. Wasson and Hyde (1983) confirmed that barchans occurred where there was very little sand and almost unidirectional winds; transverse dunes occurred where sand was abundant; longitudinal dunes where there was little sand; and star dunes occurred where sand was abundant.

<u>Objectives</u>. This paper seeks to classify changes in dune forms systematically. In particular it is a test for the "transmudación" hypothesis, first briefly formulated by Araya-Vergara (1981). The Spanish word "transmudación" (transmutative transference; <u>verwandelnde Ubertragung</u>) may be used when transport of mass produces transformation of form. The theory is used below to produce a classification of changes in dune-form.

The hypothesis was tested on the fundamentally different forelands of Putú-Quivolgo and Chanco by Araya-Vergara (1977), using SKYLAB panchromatic and infra-red images (Fig.1). The difference between these forelands is due mainly to their orientation with respect to the dominant southwesterly winds. Putú-Quivolgo is a rounded foreland with wind-parallel and wind-oblique shorelines, whereas Chanco is a quasi-transverse shoreline. The local mean annual rainfall is about 1000mm.

METHODS

<u>Categories and Concepts</u>. Categories of dune-transformation were based on the dune classification of Smith (1954, in Davies) and developed by Davies (1977). The basis of the categorization is the transformation of one dune type into another. Dunes are classed as primary and secondary, according to the interpretations of the authorities mentioned above. "Primary" dunes are barchans and transverse dunes. "Secondary" dunes are parabolic, upsiloidal and longitudinal dunes.

Empirical data, collected from the north-central Chilean coast, allowed the following system of categories to be developed:

1. Quasi-steady state: primary and secondary.
2. Changes between primary dune types: e.g. barchans transformed to transverse dunes and vice versa; barchan advance.

3. "Secondarization": e.g. barchans transformed to longitudinal dunes; barchans coalescing to produce blowouts or parabolic forms; transverse dunes transformed to parabolic dunes.
4. Changes between secondary types: blowouts changing to parabolic dunes; elongation of parabolic dunes into upsiloidal dunes or longitudinal dunes.
5. Human influence: changes that can disturb other processes, such as pine plantations; psamoseres; and urbanisation.

Following the observation of barchans on a wind-parallel coastline in north-central Chile, and the movement of its upwind side, the concept of group celerity was developed. On the windward edge of this dune field the group velocity was 60m per year. This group velocity was considerably less than that of isolated barchans in the same field which were slowed down to a celerity of 10 to 20m per year when they entered the field.

Data collection. These concepts were applied to a study of change over 23 years. This was based on three air-photographic surveys for the Putú-Quivolgo foreland (1955, scale 1:70,000; 1963 at 1:30,000; and 1978 at 1:30,000) and three for the Chanco dunes (1955 at 1:70,000; 1970 at 1:30,000; and 1978 at 1:30,000).

RESULTS

Validation of the "transmudación" hypothesis. Observations on coastal dunes in north-central Chile showed that the character of dune transformation differed according to the orientation of the shoreline.

(a) On shorelines transverse to the wind, dune systems have a homogeneous pattern of change. Dune processes continue without fundamental structural changes, i.e. there is a steady-state. In general, secondary dunes do not appear. Some transformation of transverse dunes to parabolics occurs, but rarely.
(b) On wind-parallel shorelines in contrast, the proportion of stable primary dunes is small, and transformation into secondary dunes is common (Araya-Vergara, 1985).
(c) On wind-oblique shorelines the principal pattern is of parabolic dunes in transverse groups (Araya-Vergara, 1985). The system appears to be in a quasi-steady state, but parabolic dunes do become elongated as the first stage in their evolution to upsiloidal and longitudinal dunes (seifs) (Araya-Vergara, 1985). Of the three groups, this one is the only one with a mobile distal edge (moving 2m per year along 600m of edge). Parabolic dunes appear to be characteristic of oblique shorelines.

These observations give general support to the "transmudación" hypothesis as explained above, but further detailed observations can be used to validate the concept.

The characteristics of the dune patterns on the two forelands. Putú-Quivolgo. The rounded Putú-Quivolgo foreland (Fig.1), is a broad convex progradation of the Chilean coast, 24km long by 5km wide, which is built of a series of beach ridges. The highest inner ridges reach 10m above sea level and the swales reach 5 to 6m at Putú. The ridges converge towards the mouth of the Maule River, from which their sediments are derived. There are at least three different generations of these ridges. Marshes fill the areas between some of the ridges, the central marsh being the largest. Substantial dunes cover the most recent ridges; there are longitudinal dunes in the south, long coalescent barchans in the central area, and short coalescent barchans in the north (Fig.1). Some short transverse dunes at the outermost convexity exist temporarily on an inundatable depression.

Active modern dunes overlie ancient, stable and vegetated parabolic and upsiloidal dunes on the outer foreland and ancient, vegetated marginal ridges inland. Because the southern shore of the foreland is oblique to the swell while its northern shoreline is parallel, south-west swell produces drift alignment along the shore and beach sands are delivered from South to North.

Chanco. The Chanco system, unlike the Putú-Quivolgo system, is not associated with a major river mouth. It occurs inland from a concave beach which is about 17km long. The dune field is about 5km wide at maximum, with a festooned inner boundary formed by upsiloidal and longitudinal dunes (Fig.1). In the dune-field itself, however, the dunes are mostly transverse ridges associated with massive quantities of sand. They are active and overlie a system of low terraces and ancient upsiloidal and longitudinal dunes which are stable and outcrop only at the boundaries of the field. Because the shore is transverse to the south-west swell, the delivery of sand to the beach is controlled by swash alignment. South-west winds predominate both on the rounded foreland and on the transverse massive.

The relationship of shoreline orientation and dune transformation. On the rounded foreland.

(I) On the wind-parallel shorelines.

The January 1955 aerial photographs show a south-to-north gradation from large coalescent barchans in the south, through short coalescent barchans to transverse ridges in the north. These northern dunes grade to coalescent barchans downwind. Downwind of the mouth of a small river there is a similar truncated sequence.

The 1963 photography (Fig.2a) shows a similar situation to that in 1955, although the downwind barchans are less coalescent and more isolated. The northern downwind sequence remained in a steady state.

The August 1978 photographs (taken in the rainy season), show that the river mouth and the littoral plain at the windward end of the wind-parallel sector, which had formerly been buried by coalescent barchans, had been exhumed. A kilometre-long upsiloidal dune had virtually disappeared and been replaced by some isolated or semi-coalescent barchans. The inundated depressions between the dunes gave the impression that the dune topography had become disaggregated (Fig.2b). Blowouts and parabolic dunes were apparent. The transverse dune pattern had become less compact, and appeared to be like a series of coalescent barchans.

Field observations in 1983 and 1984 suggest that the pattern seen in 1978 had not changed. Following the very wet season in 1982, there were a large number of lagoons in the inter-dune depressions. The barchans had advanced in the manner described by Finkel (1959) and Hastenrath (1967). The horns of the larger barchans were covered by smaller ones and because of their different celerities, the smaller ones were catching up with the bigger.

(II) <u>On the Oblique Shoreline</u>

In 1955 two different zones could be distinguished: the shoreward section was occupied by longitudinal dunes which began from the foredune and partially covered the plain behind; inland were large coalescent barchans, some merging downwind into transverse ridges. The highest coalescent barchans were found between these two zones. In the north, a longitudinal dune had grown downwind from one of these. Windward and leeward of this intermediate zone the dunes were smaller. The leeward boundary was festooned with barchans.

Only minor changes had taken place by 1963 (Fig.3a). Some barchans had moved and changed the type of coalescence. The landward boundary of the dunefield was practically unaltered, perhaps because the barchans here were coming up against an inundatable marsh over which advance was impossible.

The August 1978 cover again showed little change in the shoreward part of the dune field, where flooding picked out the dune pattern; a barchan in this zone had advanced (Fig.3b). In the landward zone the pattern appeared to be more dispersed than in 1963, perhaps because of the flooding of the interdune depressions. More barchans had moved than between 1955 and 1963, and this process had also changed the texture of the pattern. Several changes in the shapes of barchans, including some coalescence, were difficult to explain and may have been the result of splash and runoff erosion. The eroded barchans, though small in area, could have been easily rejuvenated as active dunes, and as such would have moved more quickly than other dunes. Transverse dunes had been replaced by blowouts. The inner boundary remained immobile, presumably for the same reasons as before.

At the outermost point of the foreland, where on the 1955 and 1963 photographs there was a cover of short transverse ridges, a lagoon had appeared in 1978, suggesting that the short transverse ridges were impermanent features (see Fig.3a and 3b).

The transverse system at Chanco.

The 1955 air photographs show two systems of transverse dunes at Chanco. The first was parallel to the shoreline, and therefore at right angles to the south-west winds. The second had an east-west orientation, evidently transverse to southerly winds. The first system had a similar wavelength to the second in its upwind portion, but a shorter wavelength in the downwind zone. The windward part of the field comprised a sand plain with smooth undulating forms. Pine had been planted in some peripheral areas.

The massive pattern remained on the 1970 photography (Fig.4a), and there had been no major changes in the dune pattern, but there had been extensive planting with Ammophila. Planting had been in systematic rows on the shoreward smooth dunes transverse to southerly winds.

By 1978 this type of management had evidently reduced the compactness of the dune pattern in the northern part of the field (near Pahuil, see Fig.4b). Blowouts had followed the lines of planting, and had exhumed the underlying plain in places. The leeward edge of the dune-field had advanced by a few metres in places, and it is possible that these advances were associated with the new blowouts.

There had been more significant transformations south of Pahuil and the Reloca River in the northern Chanco dune field. Some of them, as at La Puntiaguda, seem to have been associated with dune management as at Pahuil, although there had not been as appreciable advance as at Chanco. Other changes seem to have happened independently of management. For example, short transverse ridges which were sub-parallel to the shore had broken into a zig-zag pattern containing blowouts, leading to a loss in compactness of the pattern. The windward boundary had advanced in one place (Figs.5a and b).

Comparisons with the northern central Chilean coast.
Observations of the effects of shoreline orientation on dune changes in southern central Chile can be compared to the ones reported above for north-central Chile. On the south-central coast, as on the north-central coast, a homogeneous pattern of transformation occurs on shorelines transverse to the dominant winds. In the absence of major disturbances, dune-forming mechanisms continue to operate unchanged, and this retains the compactness of the pattern. In some cases, transverse dunes are transformed into parabolic dunes, and some of these transformations occur after management.

On shorelines parallel to the dominant winds, the south-central Chilean examples show some differences to those in north-central Chile. On the northern, as on the north-central coast the pattern is very different from that on wind-transverse shorelines, but in the south there is little "secondarization", and more change between primary forms as when coalescent barchans grade into transverse dunes, as in the zeta-form bays of north-central Chile. Both in north and south-central Chile there is a small proportion of isolated barchans, whose celerity is measurable.

On shorelines oblique to the dominant wind there are more similarities in dune transformation between the southern and north-central shorelines. Both contain secondary longitudinal and upsiloidal dunes in a quasi-steady state. In the south however, there are some coalescent barchans at the distal end of the dune field, some of which are then transformed into longitudinal dunes.

DISCUSSION AND CONCLUSIONS

Categories of change. Of the 11 categories of change outlined, five are clearly demonstrated for the north-central Chilean coast, namely transverse dunes to barchans and *vice versa*, transverse dunes to blowouts, barchans to blowouts, barchan advance and psamoseres. The classification of change therefore appears to be a valuable research tool.

Three of these categories of change are significant: the formation of transverse dunes from barchans; the passage of transverse dunes into coalescent barchans; and the transformation of transverse dunes into blowouts. The first two were observed on the rounded foreland and the third on the Chanco massif. A genetic sequence can be suggested for the dunes on the rounded Putú-Quivolgo foreland:

1. Coalescent barchans form a grid. This arises partly because of the differential celerities of barchans of different sizes, as observed by Finkel (1959) and Hastenrath (1967), and explained by Bagnold (1941).

2. High dunes formed by coalescence slow down and are overtaken by smaller ones which then merge with them. Maximum coalescence creates transverse dunes, and this takes place especially in the downwind zone where group celerities are low. However, some small barchans can reach to the windward edge of the dune field intact because they are more active than the transverse ridges.

3. If some threshold of wind energy or surface stability is passed, instability may reappear, and transverse ridges may break into coalescent or isolated barchans.

On the transverse Chanco massif on the other hand, if a wind threshold is passed, the transverse dunes are transformed into blowouts and a zig-zag pattern appears.

It is believed that most of the thresholds that must be surpassed to produce changes are extrinsic. Wind-speed variations are the most likely, although the study of the changes between the various air photographic coverages shows that splash and runoff erosion may have been important in some instances. This process probably restructures the dune surfaces, reducing the compactness of the pattern and facilitates later re-working by the wind. Flooding and perhaps the resulting wave action appear to widen the inter-dune depressions. Winter storms can therefore be seen as preparing the dunes for reworking by the wind. It may indeed be that the compactness of the dune pattern is related to mean annual rainfall. However, the relationships are probably complex, and one should at least consider also the seasonal patterns of wind velocity and rainfall, as did Park and You (1979). The most effective winds blow after there has been erosion associated with rainfall.

In general, Tricart's (1977) observations on the contrast between the effectiveness of a process and the clarity of forms it produces is nicely demonstrated on the rounded foreland of Putú-Quivolgo. Here only the smallest dunes are moving at a significant rate, and the process as a whole only speeds up after a loss of massiveness in the pattern when small dunes are produced again. The more massive transverse dunes of Chanco are in more of a steady-state, in spite of management. Therefore, in spite of the analysis of Wasson and Hyde (1983), it seems that sand supply is a more important factor than wind direction variability in the explanation of dune types and their changes.

The effect of vegetation. It appears that in some circumstances planting of the dunes with grasses can produce a loss of compactness in the dune pattern, under the influence of runoff erosion and high wind velocities.

The transmudación hypothesis. The sequences of dune transformation described here validate the transmudación hypothesis: the transformations are all changes of form associated with a transport of mass. Barchans have been seen to be "transmuted" into transverse dunes and *vice versa*. The transmudación concept is similar to Tricart's (1965) notion of "morphogenetic sequences". Both contain the idea of a period of preparation and a threshold of change. For example, the grid of barchans is in preparation for change in the period when it increases its compactness of pattern as the barchans attract accretion; the equilibrium is broken in periods of high rainfall and high winds; a relaxation path follows until the system again prepares itself by increasing its compactness, and the whole process begins again.

REFERENCES

Araya-Vergara, J. F., 1977. Análisis geomorfológico de la costa central sur de Chile en imàgenes de Skylab 3 - Experimento 190 - A. *Informaciones Geográficas, Chile*, 24, 37-47.

Araya-Vergara, J. F., 1981. El concepto de delta en ria y su significado en la evolución litoral (Ej. Chile Central). *Informaciones Geográficas, Chile*, 28, 71-102.

Araya-Vergara, J. F., 1985. Trend analysis of shoreline changes and coastal management in Central Chile, in *Acta, Excursion -Symposium No. 9*. (Ed. Paskoff), pp99-110. Union Géographique International Comm. Fhvir. Côtiers.

Bagnold, R. A., 1941. *The physics of blown sand and desert dunes*. Methuen & Co., London.

Bloom, A. L., 1978. *Geomorphology, A Systematic Analysis of Late Cenozoic Landforms*. Prentice Hall, New Jersey.

Davies, J. L., 1977. *Geographical variations in coastal development*. Longman, London.

De Martonne, E., 1926. *Traité de Géographie Physique*, t 2: *Le Rélief du Sol*. Colin, Paris.

Finkel, H. J., 1969. The barchans of southern Perú. *Journal of Geology*, 67, pp614-647.

Goldsmith, V., Henniger, H. F., and Gutman, A. L., 1977. The "VAMP" coastal dune classification, in Coastal Processes and Resulting Forms of Sediment Accumulation. (Ed. Goldsmith), pp26.1-26.20. SHAMSOE 143, Virginia Inst. Marine Sci., Gloucester Point, Virginia.

Hastenrath, S. L., 1967. The barchans of the Arequipa Region, Southern Peru. Zeitschrift für Geomorphologie, N.F., 11, 300-331.

Norrman, J. O., 1981. Coastal Dune Systems, in Coastal Dynamics and Scientific Sites. (Eds. Bird and Koike), pp119-157. I.G.U., Komazawa University.

Park, D. W., and You, K. B., 1979. A study of the morphology of the coastal dunes of western coast of Korea. Journal of Geography (Seoul), 6, 1-10. (In Korean with English summary).

Tricart, J., 1965. Principes et méthodes de la gémorphologie. Masson, Paris.

Tricart, J., 1977. Précis de Géomorphologie, t 2: Géomorphologie dynamique Géneral. C.D.U. - SEDES, Paris.

Wasson, R. J., and Hyde, R., 1983. Factors determining desert dune type. Nature, 304, 337-339.

International Geomorphology 1986 Part II
Edited by V. Gardiner

A SURVEY OF CENOZOIC BASALT OUTCROPS (AL-HARRA'AT) IN SAUDI ARABIA AND YEMEN

M. R. Mohammad

Department of Geology, Faculty of Science,
Alexandria University, Alexandria, Egypt.

ABSTRACT

The term "Harra" or "Harrah", or "Harrat" if followed by locality or other name, may be generalized to denote any localised outcrop of rocks formed by Cenozoic basaltic or basic volcanic eruptions, especially in arid regions. A survey is presented of the distribution and planimetric areas of Al-Harra'at (plural of Harrah) in Saudi Arabia and Yemen territory.

INTRODUCTION

Cenozoic volcanic activity is usually localised. The Arabian Peninsula has many outcrops of rocks formed by this activity. Volcanic activity in the Peninsula (Fig.1) increased through the Tertiary-Quaternary periods (Beydoun, 1966; Brown, 1970; Coleman et al., 1970-75). The central part of the Arabian Shield, occupied by the provinces of Al-Madinan, Makkah and Ha'il, is covered by extensive areas of Tertiary-Quaternary eruptions (Fig.2), some of which occurred in historic times (Abed, 1977). The eruptions formed scattered outcrops of varying dimensions ranging from small flows to basalt plateaus, occupying more than 150,000 square kilometres. Unlike Pre-Cambrian volcanoes in the Peninsula, they do not form rugged mountainous belts but flat to moderately undulating hilly country. They consist mostly of alkaline rocks, mainly olivine basalts (Al-Egl and Abd-Al-Rahman, 1974; Coleman et al., 1970-75; Abed, 1977; Al-Sayari and Zotl, 1978; Abed, 1982). The rocks are almost fresh to weakly altered, dark grey to black and fine grained. The original rugged surfaces are maintained in some places (Abo-Al-Haggagg, 1982). The term "Harra" or "Harrah", meaning "hot" in Arabic, is used for about thirteen of these occurrences in Saudi Arabian regional maps, for example Harrat Rabat, Harrat Uwayrid, ..., or simply Harrah or Al-Harrah (Figs.2 and 3). More than 80 uses of the term Al-Harra'at are mentioned without definite locations in geographical and historical books and dictionaries (Al-Andalusy, 1947; Al-Balady, 1978; Al-Gasir, 1981).

It is proposed here that the term "Harra" or Harrah", if followed by locality or other name, be generalized to denote any localised outcrop of rocks formed by Cenozoic basaltic or basic volcanic

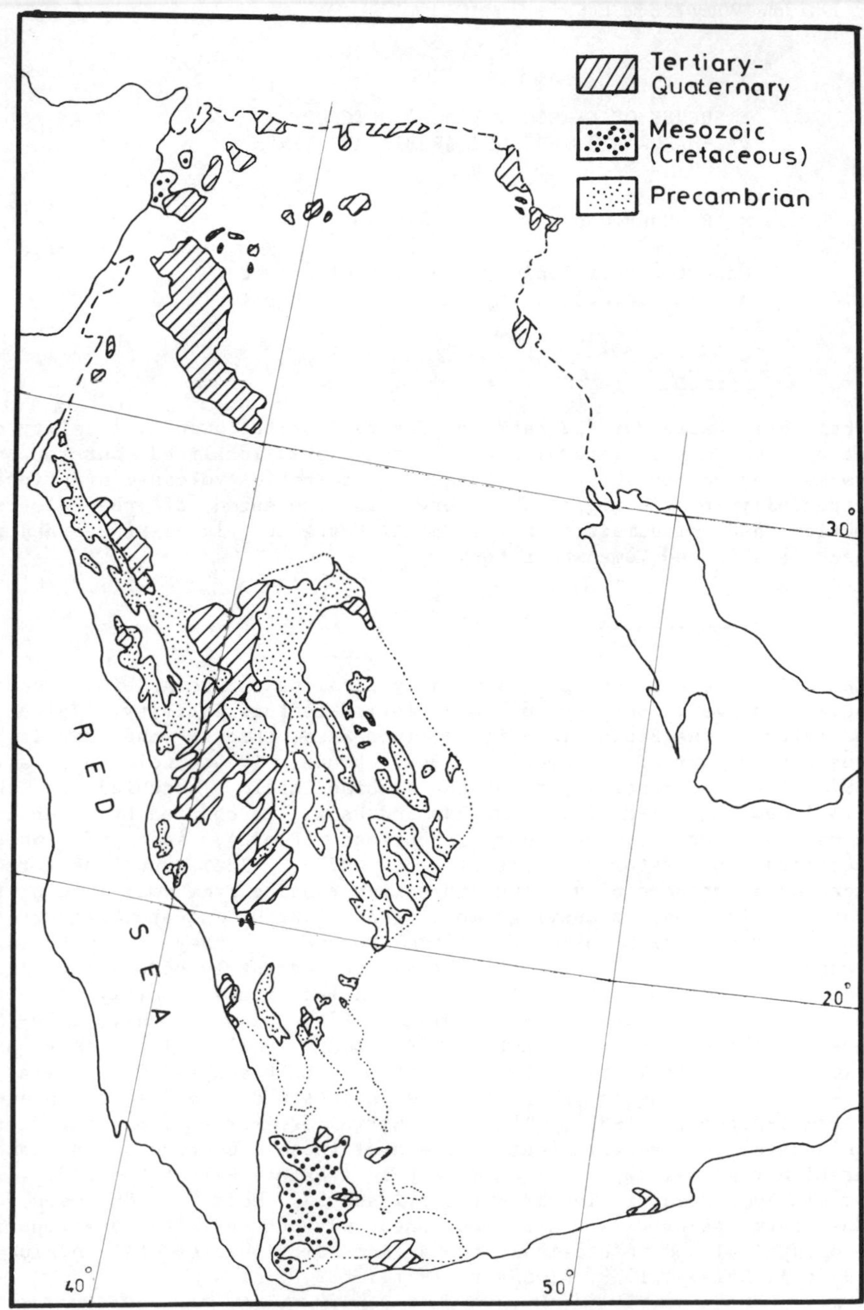

Fig.1. Volcanic rocks in the Arabian Peninsula (compiled from Brown et al., 1963; Coleman et al., 1970-5; Al-Egl and Abd-Al-Rahman, 1974; Abed, 1982).

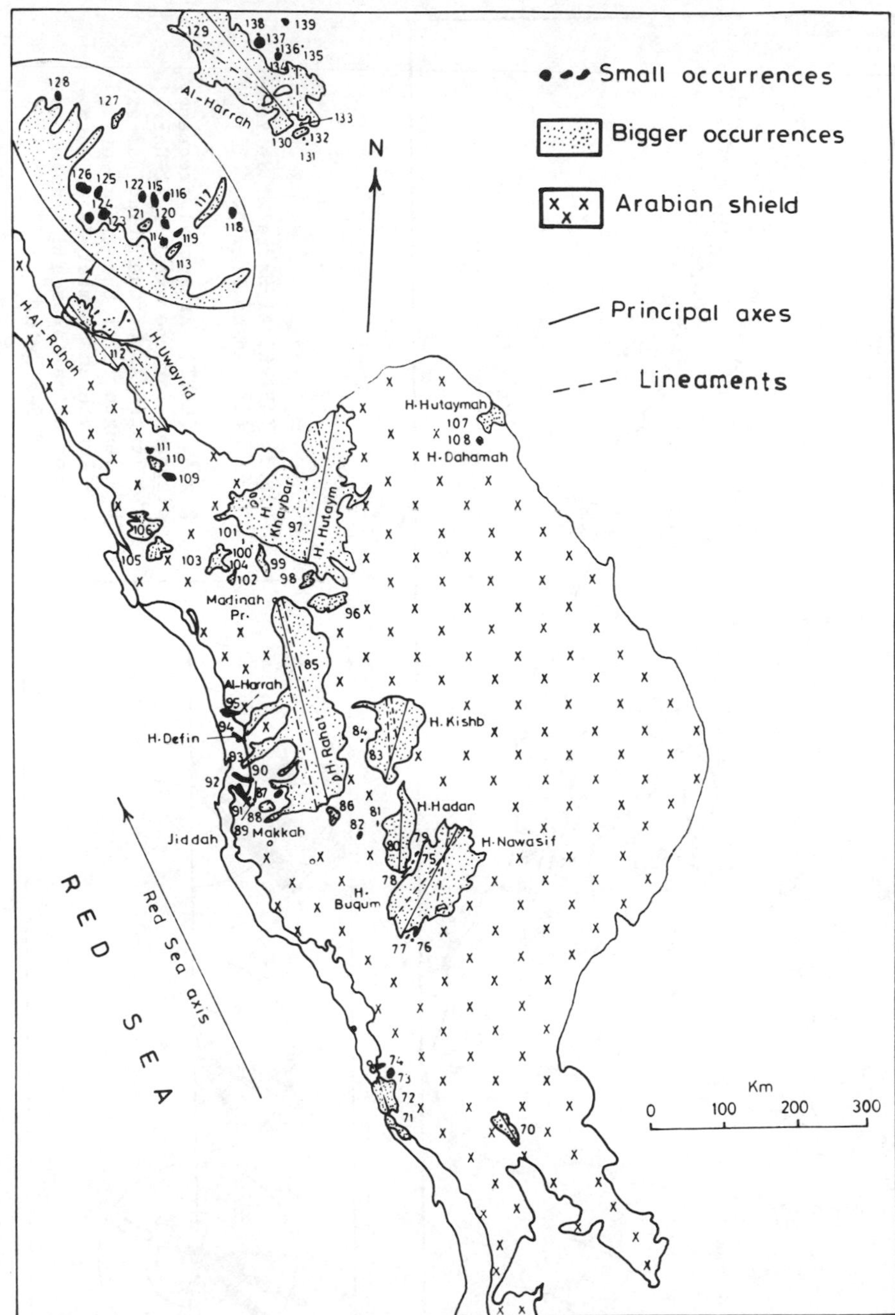

Fig.2. Al-Harra'at in Saudi Arabia. H - Abbreviation of Harrat; 1, 2, 3, etc. - Serial numbers of Al-Harra'at (see Table 1).

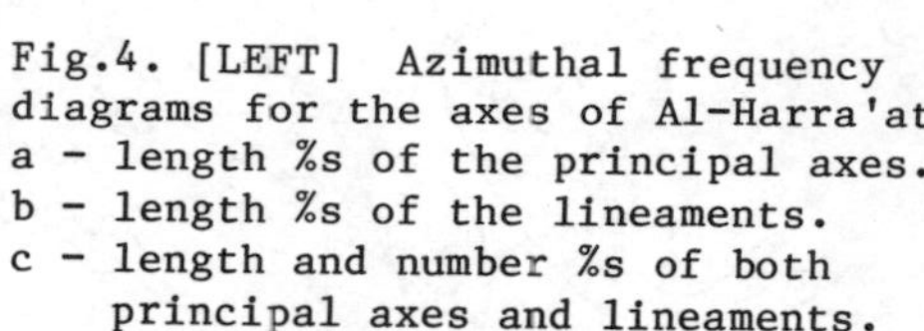

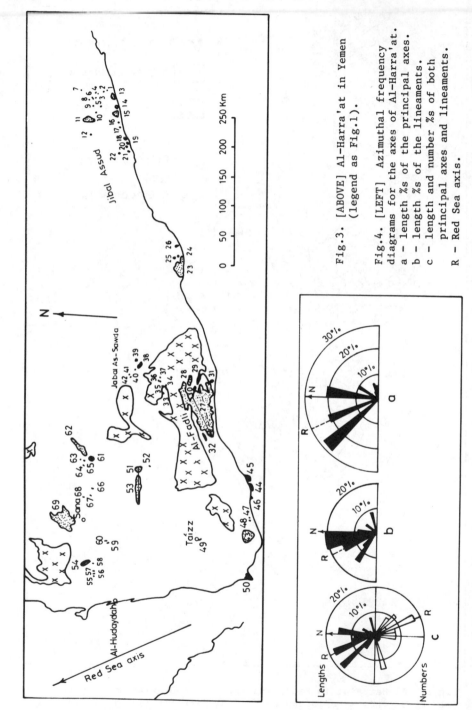

Fig.3. [ABOVE] Al-Harra'at in Yemen (legend as Fig.1).

Fig.4. [LEFT] Azimuthal frequency diagrams for the axes of Al-Harra'at.
a - length %s of the principal axes.
b - length %s of the lineaments.
c - length and number %s of both principal axes and lineaments.
R - Red Sea axis.

TABLE 1. The areas of Al-Harra'at (square kilometres).

A. Yemen

Ser. No.	Area	Ser. No.	Area	Ser. No.	Area	Ser. No.	Area
1	7.64	19	14.96	37	16.20	55	6.30
2	13.40	20	9.68	38	55.80	56	7.10
3	2.55	21	9.04	39	22.60	57	3.10
4	3.12	22	10.92	40	6.88	58	4.50
5	2.03	23	357.60	41	4.80	59	6.70
6	3.12	24	11.04	42	9.36	60	19.50
7	2.65	25	10.52	43	230.80	61	67.60
8	3.40	26	14.00	44	51.60	62	321.20
9	5.92	27	2664.00	45	45.60	63	4.68
10	8.10	28	40.00	46	11.40	64	3.25
11	113.20	29	71.60	47	12.50	65	5.00
12	9.20	30	27.00	48	6.20	66	6.40
13	73.20	31	47.44	49	20.00	67	3.80
14	22.76	32	38.20	50	58.00	68	10.69
15	80.80	33	8.56	51	84.00	69	850.40
16	11.68	34	2.70	52	5.50		
17	3.60	35	10.44	53	338.80		
18	18.84	36	24.80	54	106.80		

A. Saudi Arabia

Ser. No.	Area	Ser. No.	Area	Ser. No.	Area	Ser. No.	Area
70	688.00	88	176.80	106	1165.84	124	8.72
71	476.40	89	26.36	107	853.20	125	7.64
72	1152.40	90	9.04	108	41.60	126	29.00
73	125.60	91	3.74	109	64.80	127	37.70
74	120.80	92	121.20	110	269.60	128	4.00
75	10832.50	93	125.60	111	44.80	129	15227.60
76	26.80	94	106.00	112	7152.00	130	194.68
77	15.60	95	79.20	113	9.00	131	5.76
78	20.12	96	1056.40	114	5.20	132	14.80
79	44.00	97	21662.20	115	6.60	133	12.92
80	3544.40	98	280.00	116	4.20	134	74.00
81	26.20	99	311.20	117	36.00	135	25.88
82	64.40	100	27.44	118	10.12	136	10.76
83	6264.00	101	30.88	119	5.28	137	118.80
84	13.24	102	121.20	120	5.60	138	15.60
85	20240.80	103	552.40	121	25.00	139	113.60
86	267.60	104	14.64	122	5.30		
87	90.80	105	652.40	123	7.60		

eruptions, especially in the dry regions. A study of these occurrences in Saudi Arabia and Yemen is presented here.

AL-HARRA'AT IN SAUDI ARABIA AND YEMEN

One hundred and thirty nine occurrences of Al-Harra'at have been traced from the geologic and topographic maps of the Peninsula, especially the geological map at 1:2,000,000 (Brown et al., 1963). They are numbered serially in Figs.2 and 3 and their planimetric areas have been measured using the Leitz A.S.M. "Image Analysis System". The occurrences are concentrated mainly in the western sector of the territory, near the Red Sea coast and Gulf of Aden.

The planimetric areas of these Al-Harra'at range from a few to more than 20,000 square kilometres (Table 1). They may be classified into six categories:

1. Very small; comprising areas of less than 10 square kilometres, and including 44 occurrences.
2. Minor; less than 100 to 10 square kilometres, including 58 occurrences.
3. Small; less than 1,000 to 100 square kilometres, including 26 occurrences.
4. Medium; less than 5,000 to 1,000 square kilometres, including 5 occurrences.
5. Large; less than 10,000 to 5,000 square kilometres, including 2 occurrences.
6. Very large; areas of 10,000 square kilometres and more; four occurrences belong to this category.

The shapes of Al-Harra'at are mostly modified by erosion and controlled by the pre-existing topography. Being mostly fissure eruptions, they are elongated, especially in the 4th, 5th and 6th categories. The longest axes of the elongated occurrences are shown as the principal axis in Figs.2 and 3. Some lineaments are also identifiable within the medium, large and very large occurrences, formed by alignment of the volcanic peaks as seen on the topographic maps (shown also in Figs.2 and 3). The Azimuth-frequency diagram (Fig.4) for the principal axes and the lineaments shows that they are mostly parallel or subparallel (by acute angle up to 40°) to the Red Sea axis. This indicates interdependence between the Al-Harra'at volcanoes and the tectonics of the Red Sea. Only 16 such lineaments can be distinguished, having a mean length of about 95km. There are 18 principal axes with average length of about 105km.

REFERENCES

Abed, A. A., 1982. Geology of Jordan. Maktabat Al-Nahda Al-Eslamiya, Amman.

Abed, Abdulkader M., 1977. Al-Madinah harra (basalts), petrology and geochemistry. Bull.Fac.Sci., King Abdulaziz Univ., Jeddah, 1, 119-128.

Abo-Al-Haggag, Yousef, 1982. Geographical observations in western Arabian Peninsula. Dep. Geog., Kuwait Univ. Pap.48. (in Arabic).

Al-Andalusy, Abeid, 1947. Dictionary of vaguenesses. Writing--Translating-Publishing Committee Press, Cairo, Part 1, (in Arabic)

Al-Balady, Atiq Ibn-Gheith, 1978. Dictionary of Al-Higaz features. Dar Makkah for Printing and Publishing, Makkah, Saudi Arabia. (in Arabic).

Al-Egl, F., and Abd-Al-Rahman, A. H., 1974. Geology of Syria. Dar Al-Fikr, Beirut. (in Arabic).

Al-Gasir, Hamad, 1981. In northwest of the Peninsula. Dar Al-Yamama, Ar-riyad. (in Arabic).

Al-Hamawy, Yakout, 1956. Dictionary of countries. Dar Sadir for Printing and Publishing, Beirut. (in Arabic).

Al-Sayari, S. S., and Zotl, J. G., (Eds.) 1978. Quaternary Period in Saudi Arabia. Springer Verlag, Wein, New York.

Beydoun, Z. R., 1966. Geology of the Arabian Peninsula, Eastern Aden Protectorate and parts of Dhufar. U.S. Geol. Surv. Prof. Pap. 560-11.

Brown, G. F., 1970. Eastern margins of the Red Sea and coastal structures in Saudi Arabia. Phil. Trans. Roy. Soc. Lond. Series A 267, 75-87.

Brown, G. F. et al., 1963. Geologic map of the Arabian Peninsula. U.S. Geol. Surv. Misc. Geol. Inv. Map 1-270 A, (1:2,000,000).

Coleman, R. G., Fleck, R. R., Hedge, C. E., Ghent, E. D., 1970-75. The volcanic rocks of southwest Saudi Arabia and the opening of the Red Sea. Red Sea Research Bull., 1970-75.

International Geomorphology 1986 Part II
Edited by V. Gardiner

THE DISTRIBUTION OF WEATHERING AND EROSION ON AN INSELBERG-PEDIMENT SYSTEM IN SEMI-ARID SUDAN

M.D. Campbell, R.A. Shakesby and R.P.D. Walsh

Department of Geography, University College of Swansea, UK

ABSTRACT

Field and laboratory techniques are used to assess spatial patterns in degree of weathering, weathering potential and current erosion in the Jebel Arashkol inselberg-pediment system of gneissic lithology in semi-arid Sudan. Simple weathering indices are developed using selected quartz grain surface features obtained using SEM analysis ; these proved useful in characterising regoliths in different parts of the inselberg system. Schmidt hammer 'R' values were similar for all exposed bare rock surfaces regardless of position on the inselberg complex. Weight losses from weathering tablets (gypsum cubes placed 25 cm deep in regolith) over a two year period were greatest on the inselberg slopes. This is attributed to the concentration of runoff from the mainly boulder and bedrock surfaces on the slope into the small patches of regolith. On the pediment and at the break of slope, infiltration is both less and not so localised, leading to reduced tablet solution. The concentration of runoff and solution on the slopes is thought to enhance the slope microrelief. Erosion measurements from 1983-1985 using simple repeat survey techniques showed that significant erosion is occurring in the upper pediment/piedmont area, mainly through headward growth of and lateral erosion by ephemeral gullies. Vertical incision and planar wash erosion are insignificant in comparison.

INTRODUCTION

The numerous approaches to the study of inselbergs (reviewed by Thomas (1978)) have generally tended to be static and theoretical in nature, attempting to explain inselberg form and distribution with reference to factors such as climate, lithology and tectonic uplift history with little regard to or empirical backing for the details of the processes involved. Studies of current processes on inselbergs have been few, yet knowledge of them is essential if our understanding of the evolution of inselberg-pediment systems is to be furthered. Although it is clearly arguable that _present_ processes may be of little relevance in explanations of what may be old or relict features produced by perhaps very different _past_ processes, it is nevertheless important to assess the role and significance of current processes and in particular whether they are either actually actively modifying (and if so in what way) or having no discernible effect upon an inselberg-pediment system.

In this paper some results are presented of a research programme into weathering and erosion patterns and processes on the Jebel Arashkol inselberg-pediment complex in the White Nile Province of semi-arid Sudan (Fig.1). An attempt is made to assess spatial patterns in (a) degree of weathering, (b) weathering potential and (c) current erosion within the inselberg-pediment system. The possibility of enhanced footslope weathering and erosion, a feature of inselbergs emphasized by several workers (e.g. Ruxton, 1958 ; Twidale and Bourne, 1975) and considered responsible for maintaining the sharp break in slope typical of inselberg-pediment systems, is particularly investigated. A range of techniques are employed, including SEM (Scanning Electron Microscope) quartz grain analysis, X-ray diffraction analysis of clay minerals, Schmidt hammer hardness tests, weathering tablet experiments and repeat erosion surveys.

THE JEBEL ARASHKOL AREA

The Jebel Arashkol inselberg complex is situated 10 km west of the White Nile and 30 km northwest of Ed Dueim, the White Nile Province capital. It consists (Fig.1) of two major ridges separated by a dissected irregular basin, all aligned SSW/NNE in accordance with bedrock structure. About 8 km long and 1.5 km wide, it rises about 120 m above the surrounding pediment. The Precambrian Basement Complex rocks of the inselberg are composed of well-foliated quartzofelspathic gneiss with secondary hornblende, biotite and magnetite. Vail (1982) has suggested that the inselberg rises above the plain because of the greater resistance of these rocks to erosion, but, in the absence of information on the rocks of the surrounding pediment, this must remain conjecture. The bare rock surfaces of the inselberg are extensively weathered, with exfoliation slabs up to 1 m thick, weathering rinds and 'rotting' of boulders all common features.

The present climate is hot and semi-arid with a low and highly variable annual rainfall averaging 256 mm at Ed Dueim in 1951-80 (Walsh, 1983). The rainfall is concentrated into a short summer wet season from June to September, with on average 20 rain-days and 5 days with 10 mm or more rain per annum (Hulme and Walsh, in press). During the Quaternary, both wetter and drier phases than at present are known to have occurred (Warren, 1970 ; Williams and Adamson, 1980), but little is known of pre-Pleistocene climatic history. Current vegetation cover is sparse, mainly semi-arid scrub and wet season grasses, degraded by human interference in recent years.

Drainage (Fig.1) consists of a series of major SE-flowing ephemeral watercourses, which originate on the higher pediment surface to the northwest and cut through the inselberg range with steep channel gradients often exceeding 1°. These become diffuse washes soon after emerging on the gently sloping clay plain. Structurally-guided tributaries, mostly aligned SSW-NNE, drain into these watercourses from the inselberg complex and there are numerous gullies on the

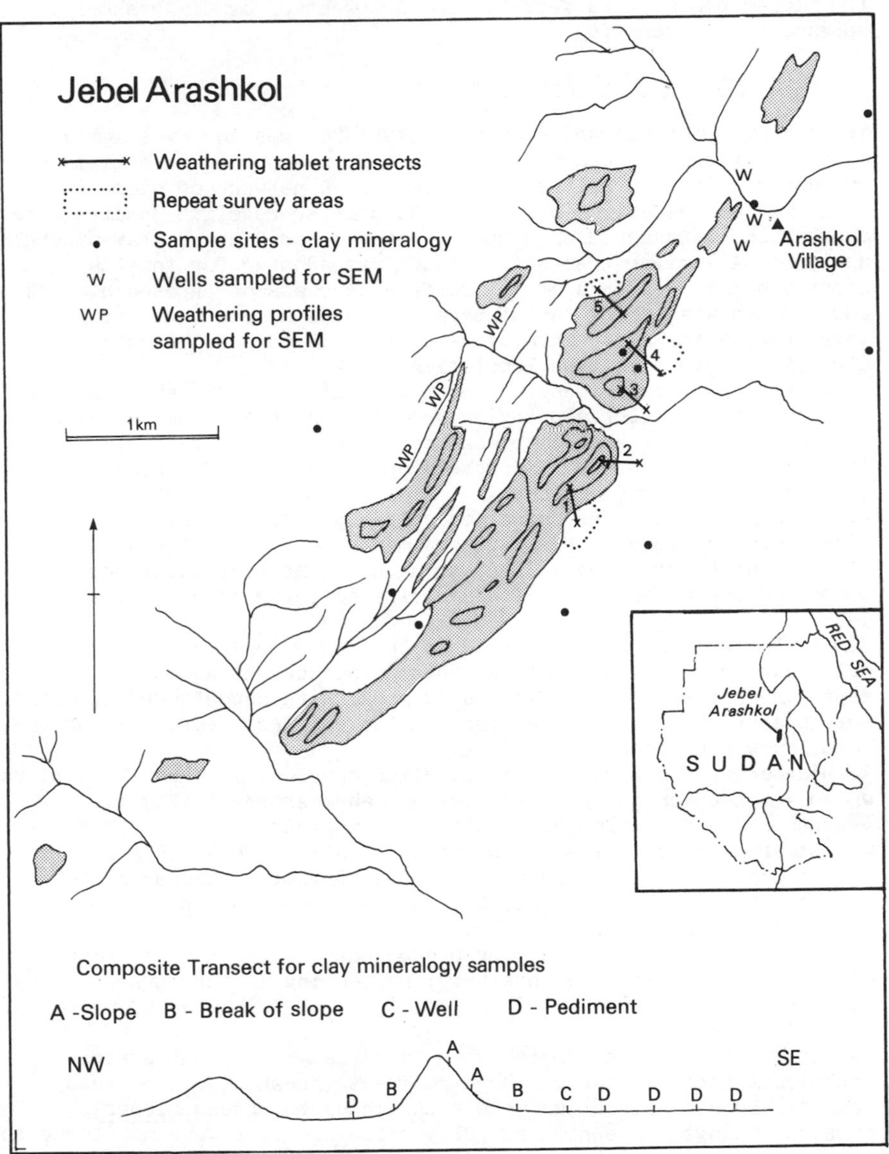

<u>Fig. 1</u> The Jebel Arashkol inselberg complex and location of sampling and monitoring sites.

pediment surfaces around each inselberg slope. Active incision of the higher pediment surface to the northwest of Jebel Arashkol appears to be occurring.

THE FIELDWORK PROGRAMME

The aim of the fieldwork programme (1983-85) was to investigate spatial patterns in degree of weathering, weathering potential and current erosion on the inselberg complex. A network of slope profile transects was established (Fig.1). Subsoil or regolith samples were systematically taken along these transects for SEM and clay mineral analysis. A similar transect approach was adopted for *in situ* Schmidt hammer readings of the surface hardness of exposed bedrock and for the weathering tablet experiments to assess weathering potential. Sites along inselberg-pediment slope profiles were classified into five type localities :

1. Inselberg slopes. These are mostly rectilinear and about 30° and are composed of material ranging from bedrock bosses and domes and boulders up to 5 m across through to gravel, silt and clay, the finer material occurring in pockets up to 30 cm thick between bedrock outcrops and boulders.
2. Bases of slopes. The characteristically sharp transition from the 30° inselberg slope to the 1-2° pediment angle generally occurs within a distance of only 30-40 m, although at some locations dissected colluvial material gives rise to intermediate slopes of 5-7°.
3. Shallow well locations. Close to the inselberg front, shallow wells dug in weathered bedrock (which commences at about 1.5 m depth) reveal rotted, buff-coloured rock, with unweathered minerals set in a friable matrix that retains the original bedrock structure. Corestones are common.
4. Weathering profiles exposed by river erosion. In channel sections up to 4 m deep on the western side of Jebel Arashkol (Fig.1), weathering is irregular with little horizonization visible and, whilst similar to that seen in the well walls, there is less alteration to clay. The material is less friable and quartzitic corestones are more frequent. Sampling was from the upper loose disaggregated material.
5. Pediment. Here silt and clay predominate, with particle size decreasing away from the inselberg, reflecting in part transport by wash.

Assessment of current erosion using a variety of repeat survey techniques concentrated on three pediment/ break of slope areas (Fig.1). Erosion pin grids were established to assess *planar* erosion. Changes in ephemeral gully cross-sections were monitored to assess *linear* fluvial erosion of the pediment, differentiating as far as possible between headward, lateral and vertical components.

TABLE 1 SEM quartz grain surface features (after Culver et al., 1983)

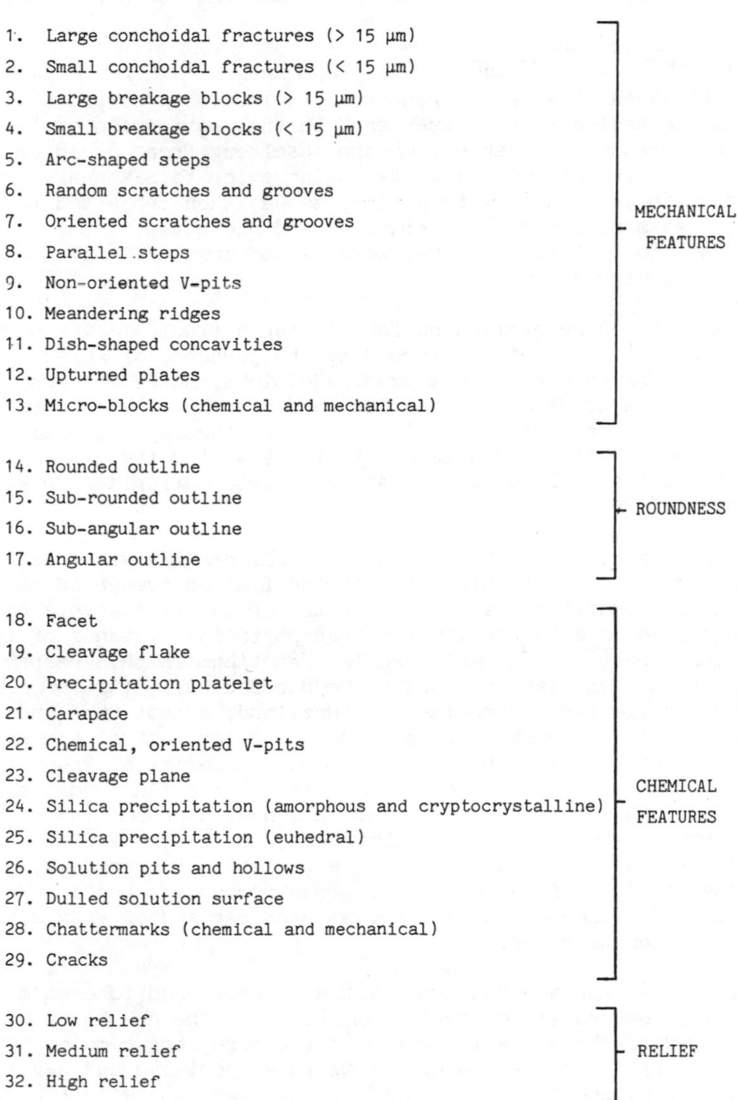

1. Large conchoidal fractures (> 15 µm)	MECHANICAL FEATURES
2. Small conchoidal fractures (< 15 µm)	
3. Large breakage blocks (> 15 µm)	
4. Small breakage blocks (< 15 µm)	
5. Arc-shaped steps	
6. Random scratches and grooves	
7. Oriented scratches and grooves	
8. Parallel steps	
9. Non-oriented V-pits	
10. Meandering ridges	
11. Dish-shaped concavities	
12. Upturned plates	
13. Micro-blocks (chemical and mechanical)	
14. Rounded outline	ROUNDNESS
15. Sub-rounded outline	
16. Sub-angular outline	
17. Angular outline	
18. Facet	CHEMICAL FEATURES
19. Cleavage flake	
20. Precipitation platelet	
21. Carapace	
22. Chemical, oriented V-pits	
23. Cleavage plane	
24. Silica precipitation (amorphous and cryptocrystalline)	
25. Silica precipitation (euhedral)	
26. Solution pits and hollows	
27. Dulled solution surface	
28. Chattermarks (chemical and mechanical)	
29. Cracks	
30. Low relief	RELIEF
31. Medium relief	
32. High relief	

WEATHERING PATTERNS OVER THE INSELBERG-PEDIMENT SYSTEM

Three techniques were used to investigate spatial patterns in degree and nature of weathering : SEM analysis of quartz grain surface texture of regolith samples ; clay mineralogy analysis of regolith samples using X-ray diffraction techniques ; and Schmidt hammer surface hardness investigations of weathered bedrock outcrops.

Quartz grain surface texture

In all 24 samples were taken from the inselberg complex, with more from the pediment than elsewhere, so that patterns over the pediment with increasing distance from the inselberg front could be assessed. Operator bias was minimized by coding prior to SEM analysis. For each sample, sand 200-500 µm in size was then subjected to standard pre-treatment procedures (Krinsley and Doornkamp, 1973) and 30 grains, controlled for size, were viewed on a JEOL 35C Scanning Electron Microscope.

A table of 32 recognized surface textures (Table 1) was used to create a tally sheet for recording the presence or absence of each feature for each grain analyzed. Additional notes concerning the general character of the samples were also made, with particular attention being paid to the criteria outlined by Doornkamp and Krinsley (1971) and Goudie and Bull (1984) for the analysis of tropical chemical weathering and semi-arid colluvial processes respectively.

The most striking feature of the overall results was their similarity. In part this reflects the limited number of different textures identified, a result perhaps of the first-cycle nature (weathered *in situ*) or limited transportation distance of the grains, apart from a small aeolian fraction. In this respect the results are similar to those of Goudie and Bull (1984) for a somewhat similar environment in Swaziland, except that chemical modification appears to be more advanced for most of the samples at Jebel Arashkol. In part, however, the similarity of the results may simply reflect the greater suitability of the standard table of surface features to the differentiation of *transporting* rather than *weathering* environments. Nevertheless, there were differences between the samples as regards their degree of breakdown, as evidenced by differences in the gross morphology and a few specific surface features of the grains, as a comparison of Figs 2 and 3 clearly demonstrates.

These 'key' surface features (Table 2) were used to create a 'Weathering Index' to enable comparison of the degree of weathering at the different type locations on the inselberg complex. The index is calculated as the sum of the weighted positive and negative values assigned to the key surface features, the weighting being dependent on the inferred degree of weathering (see Table 2). The higher the positive value of the sum (or index), the greater is deemed the degree of weathering, and *vice versa*.

TABLE 2 Quartz grain surface feature criteria used in the derivation of the Weathering and Intensity Indices

CATEGORY 1 : Features of fresh, unweathered grains

Feature	Score
Meandering ridges	- 1
Steps	- 1
Ribbed feature	- 2
Subangularity	- 1
Angularity	- 2

CATEGORY 2 : Features of old, thoroughly weathered grains

Feature	Score
Partly dulled surface	+ 1
Very dulled surface	+ 2
Silica precipitation	+ 1
Carapace	+ 1

CATEGORY 3 : Features of grains subject to intense, concentrated chemical action

Feature	Score
Pits and hollows	+ 1
Chattermarks	+ 1
V-pits	+ 1
Crevasses and lids and rims	+ 2

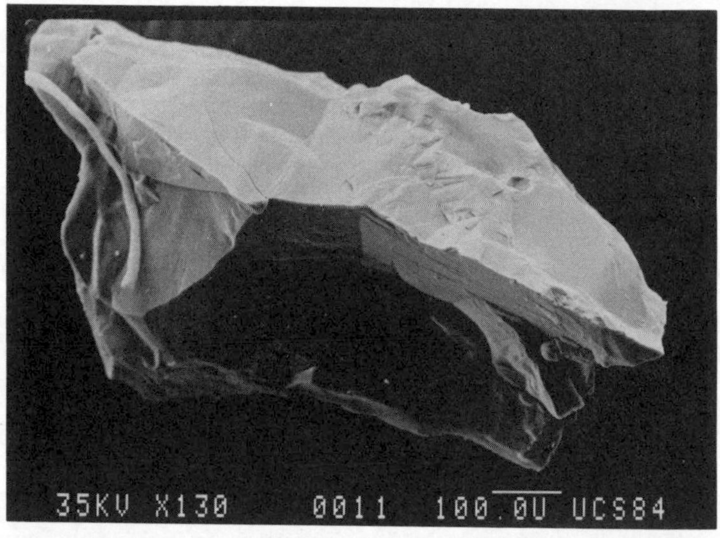

Fig. 2 A relatively 'fresh' unweathered quartz grain from slightly weathered gneissic bedrock. Note angularity and meandering ridges.

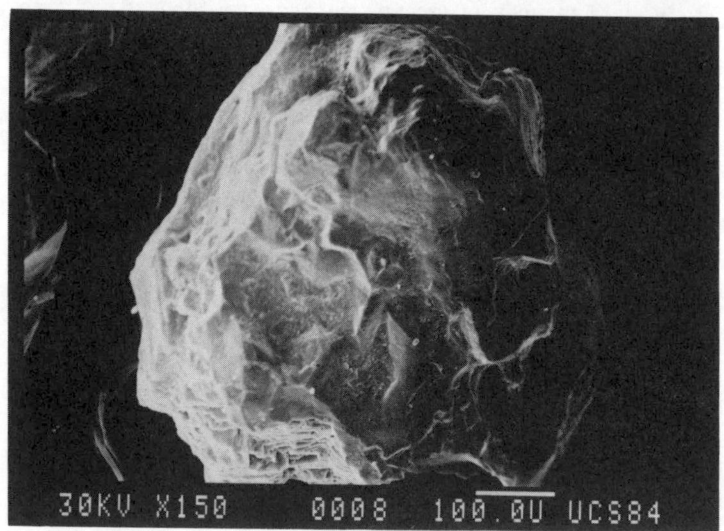

Fig. 3 A quartz grain from regolith on the inselberg slope. Note the effect of chemical activity superimposed on an original blocky 'source rock' type grain.

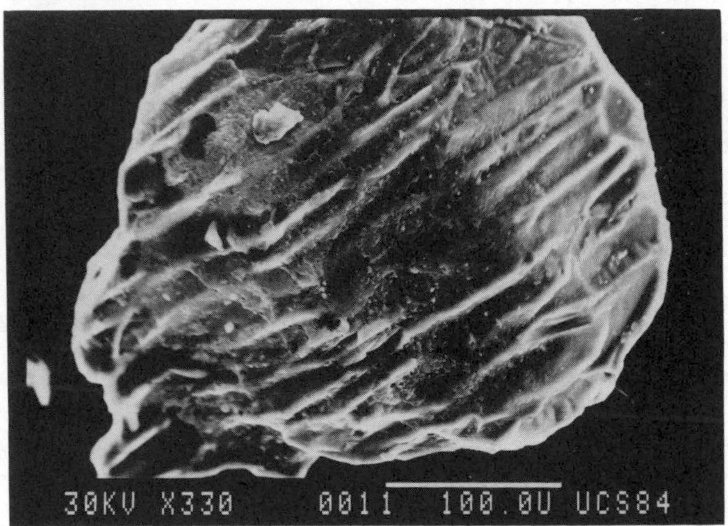

Fig. 4 Parallel rib-like features on a quartz sand grain. These are thought to reflect the structure of an adjoining crystal in the original bedrock. The presence of such features is thus considered to indicate a relatively unweathered grain surface.

Meandering ridges and parallel and arc-shaped steps are often regarded as indicating a high energy transporting environment, but here they were felt to be relatively unweathered source-rock features and thus were assigned a negative weighting. Likewise, parallel rib-like features (Fig.4), not previously reported in the literature, appear to have undergone little alteration since removal from bedrock and therefore are also given a negative weighting. Angularity also appears to be a characteristic of fresh grains and is accorded a negative weighting (-1 for sub-angularity, -2 for high angularity).

Positive scores were given to features generally attributed to chemical modification (Bull, 1981). These chemical features have here been subdivided into two categories. Features in Category 2 in Table 2 are thought to represent long-term weathering with general solution and silica reprecipitation leading to a dulled and subdued surface (e.g. Fig.5). As many grains were affected by such dulled solution surfaces to some extent, mere presence or absence was not a useful diagnostic measure, so that weighting (+1 or +2) was introduced to indicate the extent of dulled surfaces. On the other hand, features in Category 3, such as pits and hollows and crevasses with lids and rims (Fig.6) are considered to result from (and therefore to indicate) solution acting more rapidly and in a more concentrated manner on the grain surface.

Mean Weathering Index values for the five type localities described earlier are shown in Fig. 7a. Also shown are the scores derived solely from the Category 3 features of 'intense' weathering and the ratio of these 'intense' weathering features to all weathering features ; this latter ratio is termed the 'Intensity Index'. The Weathering Index results show that material on the pediment is the most weathered, followed by the inselberg slope, the break of slope (including wells) and finally the weathering profiles. According to this index, therefore, there is little evidence of enhanced weathering at the base of the inselberg slopes, though the high values for the pediment material may be inferred to relate to the greater age of the material there. This is supported by the values for the Intensity Index and the scores given by the 'intense' weathering features. On the basis of frequency of 'intense' weathering features, the slope stands out as intensely weathered, with the weathering profiles again showing relatively little chemical modification, whilst the other locations are intermediate between the two. Examination of the Intensity Index results shows a roughly similar pattern , with the slope and base of slope having higher values than those of the pediment and weathering profiles.

Taking all these results into consideration, it appears that the inselberg slopes have the highest weathering rates (indicated by their high Intensity Index values), followed by the base of slope, whereas the highly weathered nature of the pediment material appears to be a function of time rather than speed of weathering (as indicated by the high values of the Weathering Index but low values of the Intensity Index). Furthermore there is no evidence that the

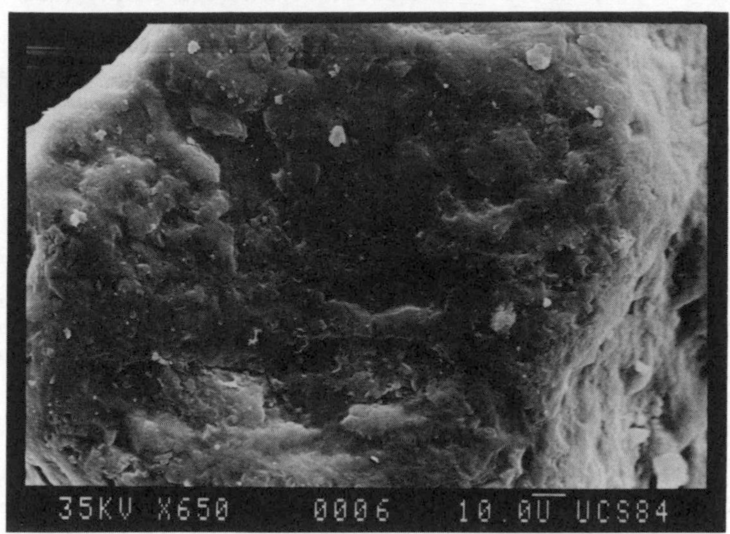

Fig. 5 Dulled solution surface on a quartz grain. This is thought to represent a very long period of chemical weathering.

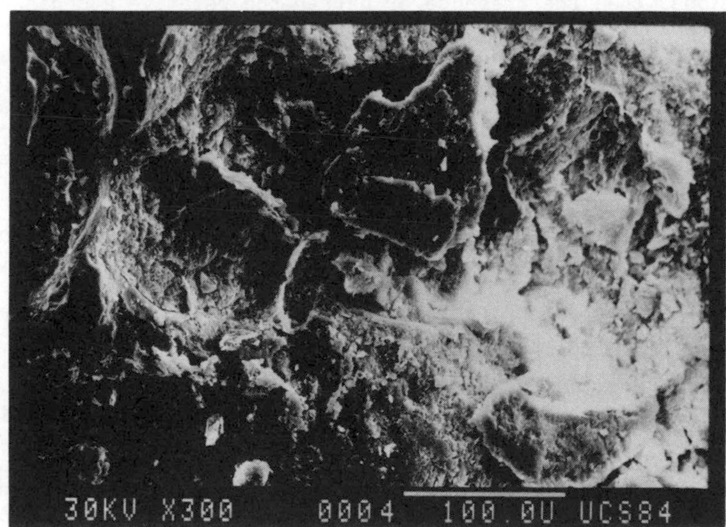

Fig. 6 Chemically produced pits and hollows on a quartz sand grain. This is considered indicative of particularly intense chemical weathering and should be contrasted with the grain of Fig. 5.

highly weathered nature of the pediment material is the result of transport processes, as there were no surface features typical of of transport and grains had been controlled for size anyway. In the weathering profile locations exposed by stream incision of the pediment there has been little alteration of the quartz component (indicated by the low Weathering Index values). However, the feldspars and micas have been substantially altered to give intergranular clays and a weathered friable structure to the rock. It would appear, therefore, that at this latter type location only slow degradation of quartz is taking place through gradual solutional rounding, whilst more susceptible minerals serve to promote the actual gradual breakdown of the bedrock. However, only at locations where there is increased concentration and throughput of subsurface or percolating water (as in the pockets of regolith on the slopes and at the base of slope) do more pronounced features and higher rates of chemical weathering occur.

Clay mineralogy

Clay mineralogy of regolith is generally regarded as important in diagnosing the nature and intensity of chemical weathering processes. X-ray diffraction analysis of clay samples from the Jebel Arashkol inselberg complex was undertaken (Table 3). The limited data do not permit firm conclusions, but several points of note emerge. If kaoloinite is considered indicative of enhanced leaching conditions and montmorillonite is considered the 'expected' residual clay mineral produced from igneous and metamorphic rocks under arid conditions, then the presence or better the relative abundance of kaolinite in a well crystallized form may be taken as an index of weathering conditions. However, it should be noted that whether it is an index of present or past weathering conditions and, if past, at what time in the past is unknown ; there is no *a priori* reason in view of the history of major climatic change during the Quaternary in the region to expect residual products to be in equilibrium with present climate.

Using abundance of kaolinite as a percentage of total clay minerals as a criterion, the sample for Group 2 (base of slope) seems to represent conditions of enhanced leaching. This sample, however, also has substantial proportions of illite and montmorillonite (and possibly some chlorite and mixed-layer clays) with both kaolinite and montmorillonite having low crystallinity indices. This combination may indicate early stages of weathering, with the influence of bedrock still strong prior to the regolith reaching an equilibrium composition. The sample in Group 3 (a shallow well weathering profile) shows a contrasting clay mineralogy with a predominance of well crystallized montmorillonite. This difference perhaps reflects the rather contrasting hydrological conditions at the well site, where local groundwater plays a major role and where a longer time period has elapsed in which equilibrium conditions could be achieved. The clays on the inselberg slopes and on the pediment have similar mineralogies to the well profile, although the slopes have a slightly higher kaolinite content which is also better

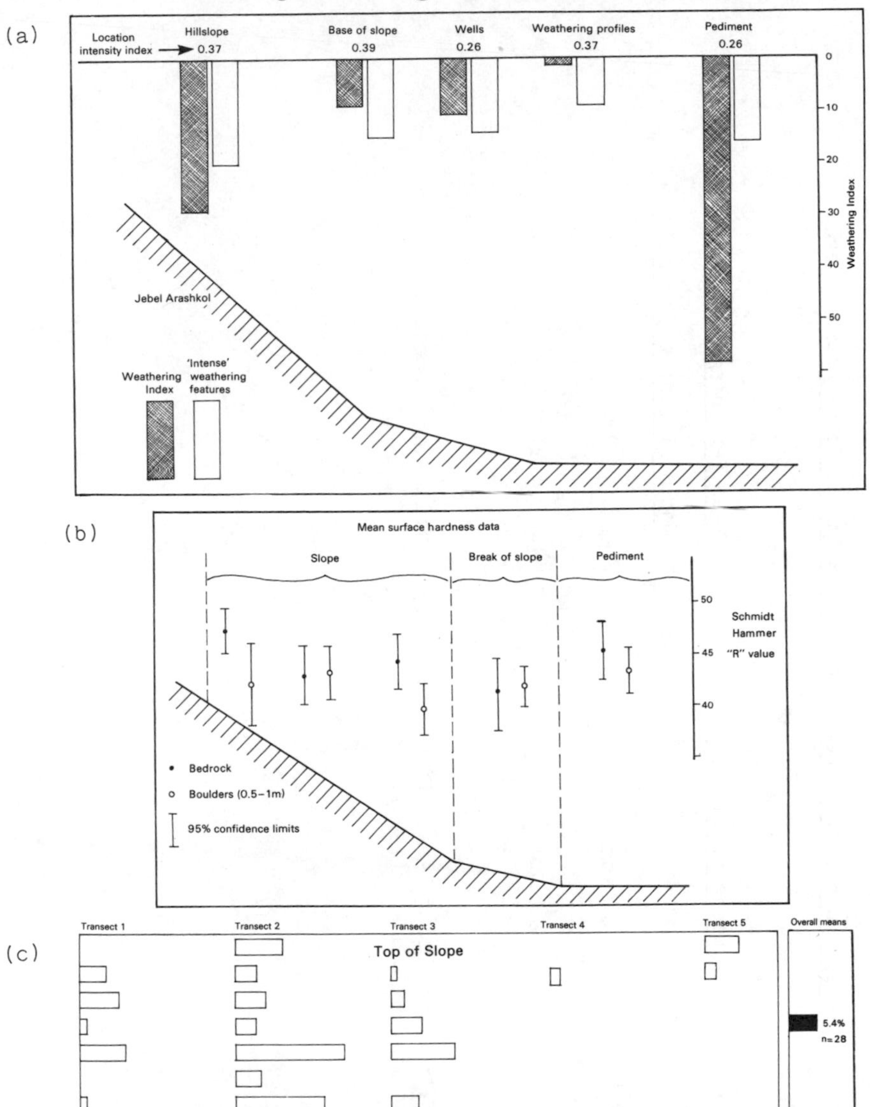

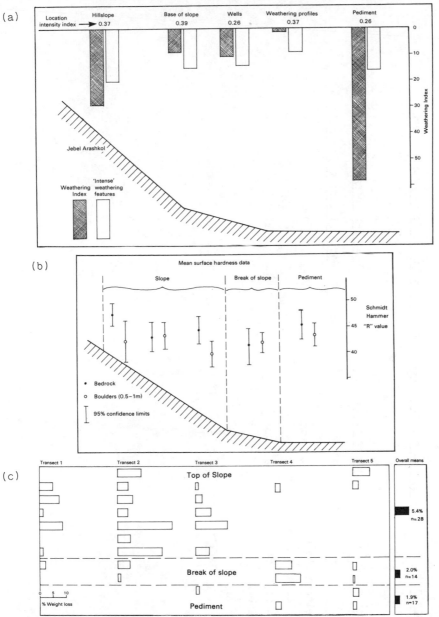

Fig. 7 (a) Mean values for the Weathering Index, the Frequency of 'Intense' Weathering Surface Features, and the Intensity Index along inselberg slope profiles.

(b) Schmidt hammer 'R' readings on inselberg slope profiles.

(c) Weathering tablet percentage weight losses along slope transects over the period March 1983 to January 1985.

TABLE 3 Clay mineralogy of regolith samples in the Jebel Arashkol area

Location	Number of samples	Mean Percentage		
		Illite	Montmorillonite	Kaolinite
Inselberg slopes	2	5	73	22
Base of slopes	1	23	27	50
Well profiles	1	7	91	2
Pediment	6	7	74	19

* Data obtained using X-ray diffraction analysis following procedures described by Biscaye (1965).

crystallized. This would support the hypothesis that the slopes are subject to a somewhat higher degree of leaching, but the evidence is not of itself conclusive.

That kaolinite is present in most samples, regardless of location, is surprising in view of the low rainfall (and hence low leaching potential) of the semi-arid environment. Chemical weathering, therefore, either is or in the past has been rather more active than might be assumed under the present alkaline, low leaching potential, weathering environment.

Schmidt hammer 'R' values

Bare rock surfaces in the form of bedrock and boulders constitute about 25 % of the inselberg slope area ; thus possible differences in the degree of weathering of such surfaces, in addition to soil and regolith, need to be considered. At Jebel Arashkol, bare rock surfaces exhibited signs of both mechanical breakdown, in the form of exfoliation at a range of scales from flakes less than 1 mm in thickness to sheeting slabs over 1 m thick, and chemical weathering, indicated by surface discoloration to reddish brown and black and the presence of discoloured bands or rinds on boulders parallel to the surface.

Recently the Schmidt hammer 'R' value has been shown to be a useful index of degree of surface weathering in the context of boulder surfaces of varying age (Matthews and Shakesby, 1984). They used the Schmidt hammer as an aid in dating boulder moraines of different ages ; more recently exposed and less weathered boulders yielded significantly higher average 'R' readings than older, more weathered rocks. In the present study, Schmidt hammer 'R' values were recorded on bedrock and for a range of boulder size categories at different points along slope profiles on the inselberg complex. The number of sample locations (see Table 4) varied with each slope location/rock surface category, with fewest samples in the >3 m boulder categories, as very large boulders were rare. Each individual sample value was itself the mean 'R' value of 20 readings after the two highest and lowest had been discounted. Some 8620 readings were taken in all.

Results are summarized in Table 4. The 95 % confidence limits for the mean 'R' values for all location/rock surface categories overlap (Fig. 7b), implying no significant systematic differences in means either between locations or between rock surface/boulder size types sampled. Mean 'R' values all fell within the narrow range 39.7 to 47.8. From these results it would appear that either (1) the degree of weathering of exposed rock surfaces does not vary with position on the inselberg-pediment slope, or (2) any differences in degree of weathering are not reflected in differences in surface hardness. If (1) is true, then the variations in degree of weathering already noted from the SEM quartz grain analysis and clay mineralogy results over the slope system apply only to material within (and possibly beneath) a regolith. The rock surfaces could be of similar degrees of weathering either because of similar weathering processes and periods of exposure, or, alternatively, because all the rock surfaces have been exposed for so long that they have reached a weathering equilibrium where further weathering leads to removal of weathered material and exposure of fresher rock beneath.

It is difficult in view of the ambiguity of the above results to draw any firm conclusions about inselberg evolution. The similarity in surface hardness (and, it is inferred, degree of weathering) over all parts of the inselberg slope could be interpreted as evidence of a similar time of exposure of the whole slope and hence evidence of parallel retreat ; however, if surface weathering is rapid relative to therate of (gradual or episodic) exposure of bedrock and boulders on the slope, then the similarity in surface hardness may simply represent equilibrium weathering conditions and thus could also accord with inselberg evolution theories which involve progressive downwearing of the pediment or exhumation of deep weathering surfaces.

WEATHERING POTENTIAL OVER THE INSELBERG SYSTEM : EVIDENCE FROM WEATHERING TABLET EXPERIMENTS

Weathering tablets, weighing about 11 g and with a surface area of 37 cm^2, were made from Plaster of Paris (gypsum), a highly soluble and easily worked substance suited to field solution experiments (Crabtree and Trudgill, 1984). They were each numbered, air-dried, weighed and then buried in March 1983 at depths of about 25 cm along 5 transects (Fig. 1) involving locations on the inselberg slope, at the break of slope and on the pediment. After two summer wet seasons the tablets were retrieved in January 1985 and reweighed. Of 140 buried tablets, 67 were recovered, the remainder having been prematurely excavated by local children.

Results (Fig. 7c) indicate clearly greater weight loss on the slopes than at the break of slope or on the pediment. Weathering and solution of gypsum tablets are effectively proportional to the quantity of water passing through the tablet site (Trudgill, 1975). The technique, therefore, is really assessing <u>potential</u> rather than <u>actual</u> weathering and solution, as the latter will clearly be dependent on the weatherability of the regolith, which varies greatly with location. Thus, material which is already thoroughly weathered (as on the pediment) will undergo relatively little further change, whatever the leaching conditions. Nevertheless, that weathering potential rates are highest on the inselberg slopes rather than at the break of slope or on the pediment probably results from the fact that regolith occupies only a small proportion of the inselberg slope. These patches of regolith within which the tablets were buried tend to receive disproportionately large water inputs because of runoff from the adjacent bare rock boulder and bedrock surfaces as well as direct rainfall. Weathering and solution on the slopes are therefore concentrated in the crevices, cracks and hollows, including the small regolith pockets, arguably thereby also maintaining or enhancing the bouldery microrelief of the slopes. The

TABLE 4 Summary of Schmidt hammer 'R' readings at different locations and for different types of rock surface in the Jebel Arashkol inselberg-pediment complex

Location on inselberg-pediment slope profile	Type of rock surface											
	Bedrock			Boulders 0.5-1m			Boulders 1-3m			Boulders >3m		
	R	n	s	R	n	s	R	n	s	R	n	s
Top of inselberg slope	45.0	23	6.2	43.1	33	6.2	44.1	37	5.5	38.8	8	5.8
Middle of inselberg slope	41.4	12	6.4	41.7	36	5.5	42.6	31	5.7	44.2	13	4.8
Lower part of slope	44.2	15	5.0	39.7	31	6.7	43.3	28	5.5	44.3	8	5.2
Break of slope	43.0	18	5.9	43.3	29	6.8	45.9	26	4.9	47.8	9	3.0
Pediment	47.4	16	4.0	42.3	15	7.3	44.6	22	5.5	44.3	3	3.0

R = Mean Schmidt hammer 'R' reading ; n = number of sample sites (20 readings per site) ; s = standard deviation of readings.

lower rates of weathering potential recorded at the break of slope do not necessarily imply that weathering is less significant there than on the inselberg slope, as regolith is more widespread and infiltration is spatially less concentrated than on the slopes. Weathering data are needed for the bare rock surfaces of the inselberg complex, not just the regolith areas assessed in the experiments, in order to make valid comparisons of overall weathering potential rates between the slope and break of slope areas.

CURRENT EROSION IN THE PEDIMENT AND BREAK OF SLOPE AREAS

Process studies are notoriously difficult in semi-arid areas because of the episodic nature of much of the geomorphic activity. At Jebel Arashkol, the practical difficulties of measurement are compounded by two further major factors. First, present climatic conditions (and hence present erosion patterns and rates) are arguably very different from those operative in wetter phases not only in the Pleistocene (Warren, 1970 ; Williams and Adamson, 1980), but also earlier in this century (Trilsbach and Hulme, 1984 ; Hulme and Walsh, in press). Second, Man's impact on the environment has been particularly significant in Sahelian Sudan during the last 30 years with a progressively denuded vegetation cover and accompanying desertification problems.

Despite these problems of interpretation, an attempt was made to monitor between March 1983 and January 1985 patterns of erosion in gullied areas around the base of inselberg slopes at three locations (Fig. 1) in order to investigate how the break of slope area is currently evolving. Simple, robust techniques were preferred to more elaborate, but vulnerable equipment. The three sites investigated were each about 0.025 km^2 in area and dissected by ephemeral gullies starting at or just below the break of slope and extending up to 50 m across the pediment. These dissected areas were considered representative of the intermediate (3-7°) slope zone.

Six types of measurement were made :

(i) Lateral markers. Permanent markers (either plastic pegs or marked rock outcrops) were established near to the courses of incised channels in order to measure lateral fluvial erosion. The distance and bearing of the incised bank from the marker were noted in April 1983 and remeasured after two wet seasons in January 1985. Changes of at least 2 cm were taken to indicate lateral erosion or deposition.

(ii) Head markers. A measurement procedure similar to (i), except that it was made from a marker to the incised head of a channel, was used to assess headward erosion of channels.

(iii) Head-cut markers. Where the incised head of a channel corresponded with a bedrock outcrop or a large boulder, the head of the channel could be marked (using paint) in outline on the rock. Exposure of additional rock upstream of the painted line after the two wet seasons clearly indicated headward erosion.

TABLE 5 Erosion and deposition during the period March 1983 to January 1985 in gullied areas of the Break of Slope/Upper Pediment zone at Jebel Arashkol

Type of measurement*	Number monitored	Number showing		Mean erosion (mm)	
		Erosion	Deposition	Of those changed	Overall
Lateral markers (L)	174	34	6	51	12
Head markers (H)	56	25	5	76	41
Head-cut markers (H)	12	4	0	78	27
Cross-sections (V)	38	7	12	- 3	- 1
Bedrock markers (V/L)	31	14	2	Lateral : 39 Vertical : 2	20 1

Notes :
L = assessing lateral erosion
H = assessing headward erosion
V = assessing vertical erosion
* For explanation see text

(iv) Channel cross-sections. Depth to the channel (and banks) was measured every 25 cm along a tape stretched between two marked points aligned across a channel in order to assess vertical incision or infill. For each cross-section, mean erosion or deposition over the survey period was calculated by averaging change at each depth profile. Changes of less than 1 cm were deemed zero. For each ephemeral channel monitored, a series of cross-sections at increasing distance from the inselberg front were measured.
(v) Bedrock markers. These were similar to the head-cut markers but located in the sides and bottom of the channels. Again erosion or deposition could be clearly seen at the time of resurvey with respect to the painted line, but interpretation is more difficult. At many markers, the amount of change varied with upstream and downstream orientation. Also it was often difficult to differentiate between vertical and lateral change on markers that were in both bank and bed of the gullies.
(vi) Erosion pins. Networks of erosion pins (rigid metal rods of about 5 mm diameter) were established on the inter-gully and pediment areas in order to assess planar erosion and deposition as opposed to the linear/fluvial component being assessed by the other techniques. The relatively few pins that survived human interference and removal indicated very little change in surface level compared with changes in the gullies over the same period.

The most striking feature of the results (Table 5) is the substantial erosion indicated, despite the dryness of the two wet seasons (213 mm in 1983 and 51 mm in 1984 compared with a 1921-80 mean of 294 mm). Considerable aeolian infill occurred in the lower reaches of the channels, presumably partly because of the lack of stream incision there as the streams rapidly lost their low discharge through bed infiltration. Headward erosion is clearly important, thus implying that the gullies are extending on the pediment towards the break of slope and in some cases extending into weathered material on the inselberg slope itself. Lateral erosion, indicated the lateral marker and 'lateral' bedrock marker results in Table 5, is also important, particularly when compared to the small degrees of vertical change indicated by the cross-section and 'vertical' bedrock marker results. Vertical incision into weathered material at the break of slope appears therefore to be unimportant. The combination of lateral and headward gully erosion would, if continued , arguably tend to (1) cause slope retreat whilst maintaining the sharp break of slope and (2) lower the general pediment surface via replacement through lateral shifting of the gullies. However, it would be unwise to draw firm conclusions about long-term inselberg-process relationships from the results of just two, arguably unrepresentatively dry years.

CONCLUSIONS

1. The diagnostic capabilities of standard techniques of interpretation of quartz grain surface features using SEM analysis are greatest where high energy transporting environments are being considered, but are far less rewarding where in situ weathered products are being investigated, as in this study. Nevertheless, the

two weathering indices (assessing degree and intensity of weathering respectively) developed in this paper using 'key' surface grain features have proved capable of distinguishing _in situ_ regoliths at different locations on the Jebel Arashkol inselberg-pediment system. Further work is necessary to refine such an approach, however.

2. Differences in Schmidt hammer 'R' values between different parts of the inselberg-pediment slope system proved insignificant. This implies that the rock surfaces are equally weathered either because they have been exposed for similar time periods (as opposed to progressive exhumation or exposure) or because surface weathering on exposed boulders and bedrock rapidly reaches an equilibrium, whereby further weathering merely leads to spalling and flaking and exposure of underlying less weathered rock.

3. At Jebel Arashkol, weathering potential, as measured by weathering tablet solution loss beneath regolith, was greatest at inselberg slope locations and much lower at the break of slope and on the pediment. Lower infiltration capacities of the more clay-rich pediment compared with the high permeability of the bouldery slopes may be partly responsible, but concentration of runoff (and hence solution) into the small pockets of regolith on the slopes is probably the main cause. This concentration of weathering and solution in the regolith pockets, together with low weathering rates on the bare rock surfaces, is probably responsible for the maintenance and enhancement of the micro-relief of the inselberg slopes, with development of bedrock domes, knobs and castellated outcrops.

4. Footslope erosion on the upper pediment and at the break of slope is currently active and concentrated in the ephemeral gullies. Headward and lateral erosion by the gullies appear dominant, with vertical gully incision and planar erosion of the pediment relatively insignificant.

ACKNOWLEDGEMENTS

Campbell's research was supported by an NERC research studentship (GT4/82/GS/103). Shakesby and Walsh would like to thank The British Council for financial support for their field research in the Sudan. Thanks are also due to Mr Guy Lewis for drawing the maps and diagrams and to Mr Alan Cutliffe for the photographs and glossy print production. All authors are also greatly in debt to the immensely kind hospitality of Sheikh Wasila and the people of Arashkol.

REFERENCES

Biscaye, P.E., 1965. Mineralogy and sedimentation of Recent Deep-sea clay in the Atlantic Ocean and adjacent seas and oceans. _Geological Society of America Bulletin_, 76, 803-832.

Bull, P.A., 1981. Environmental reconstruction by scanning electron microscopy. _Progress in Physical Geography_, 5 (3), 368-397.

Crabtree, R.W., and Trudgill, S.T., 1984. The use of gypsum spheres for identifying water flow routes in soils. _Earth Surface Processes and Landforms_, 9, 25-34.

Culver, S.J., Bull, P.A., Whalley, W.B., Campbell, S., and Shakesby, R.A., 1983. Operator varaince in quartz grain surface studies. Sedimentology, 30, 129-136.

Doornkamp, J.C., and Krinsley, D.H., 1971. Electron microscopy applied to quartz grains from a tropical environment. Sedimentology, 17, 89-101.

Goudie, A.S., and Bull, P.A., 1984. Slope Process Change and Colluvium Deposition : and SEM Analysis. Earth Surface Processes and Landforms, 9, 289-300.

Hulme, M., and Walsh, R.P.D., In Press. Hydrological consequences of recent climatic change in the west-central Sudan and some suggestions for future monitoring and assessment, in Proceedings of the Workshop on Monitoring and Controlling Desertification in the Sudan : 20-24th February 1983, Khartoum (Eds Johnson, D. and Khogali, M.), Clark University, Worcester, Massachussetts.

Krinsley, D.H., and Doornkamp, J.C., 1973. An Atlas of Quartz Sand Surface Textures. Cambridge University Press, 91 pp.

Matthews, J.A., and Shakesby, R.A., 1984. The status of the 'Little Ice Age' in southern Norway : relative age dating of Neoglacial moraines with Schmidt hammer and lichenometry. Boreas, 13, 333-46.

Ruxton, B.P., 1958. Weathering and subsurface erosion in granite at the piedmont angle, Balos, Sudan. Geological Magazine, 95, 353-77.

Thomas, M.F., 1978. The study of inselbergs. Zeitschrift für Geomorphologie, N.F., Supplement-Band, 31, 1-41.

Trilsbach, A., and Hulme, M., 1984. Recent rainfall changes in central Sudan and their physical and human implications. Transactions of the Institute of British Geographers, New Series, 9, 280-298.

Trudgill, S.T., 1975. Measurement of erosional weight loss of rock tablets. British Geomorphological Research Group, Technical Bulletin, 17, 13-19.

Twidale, C.R., and Bourne, J.A., 1975. Episodic exposure of inselbergs. Geological Society of America Bulletin, 86, 1473-1481.

Vail, J.R., 1982. Geology of the central Sudan, in A Land Between Two Niles : Quaternary Geology and Biology of the Central Sudan (Eds.Williams, M.A.J. and Adamson, D.A.), pp 51-63. Balkema, Rotterdam.

Walsh, R.P.D., 1983. Observations on hydrological problems of the Nuba Mountains area of the Sudan. Swansea Geographer, 20, 16-25.

Warren, A., 1970. Dune trends and their implications in the central Sudan. Zeitschrift für Geomorphologie, N.F., Supplement-Band, 10, 154-180.

Williams, M.A.J., and Adamson, D.A., 1980. Late Quaternary depositional history of the Blue and White Nile rivers in central Sudan, in The Sahara and the Nile (Eds. Williams, M.A.J., and Faure, H.), pp 281-304. Balkema, Rotterdam.

International Geomorphology 1986 Part II
Edited by V. Gardiner

TEMPERATURE MEASUREMENTS OF ROCK SURFACES IN HOT DESERTS (NEGEV, ISRAEL)

K. Rögner

University of Paderborn
Paderborn, West Germany

ABSTRACT

Rock temperature variations are described which cannot be explained by factors such as insolation, shadow effects or changing air temperature. These include warming within the rock which coincides with the rise of air temperature after sunrise and the simultaneous reduction of relative humidity. These anomalous variations are attributed to the process of salt crystallisation out of highly concentrated solutions. Salt crystallisation is the main process in flaking, scaling and the origin of the tafoni in Machtesh HaGadol.

INTRODUCTION

Over several decades there has been a large number of studies concerning the forms and processes of weathering in arid areas (Barton, 1916; Futterer, 1902; Knetsch, 1960; Knetsch and Refai, 1955; Mortensen, 1930; Ollier, 1963; Schattner, 1961; Walther, 1888, 1924; Wilhelmy, 1964; for a review: Evans, 1970). That the discussion is by no means complete is demonstrated by the increase during the last twenty years of papers dealing with this subject (Besler, 1979; Brettschneider, 1980, 1981; Dragovich, 1967, 1969; Eichler, 1982; Goudie, 1977; Jäkel and Dronia, 1976; Kaiser, 1972; Kerr, Smith, Whalley and McGreevy, 1984; Klaer, 1970; Peel, 1974; Roth, 1965; Smith, 1977; Wellmann and Wilson, 1965). Field mapping and the description of forms and products of weathering from which weathering mechanisms have been deduced have been accompanied from a relatively early stage by laboratory simulation (Blackwelder, 1925, 1933; Bonnell and Nottage, 1939; Cooke and Smalley, 1969; Correns and Steinborn, 1939; Goudie, 1974; Goudie, Cooke and Evans, 1970; Goudie, Cooke and Doornkamp, 1979; Griggs, 1936; McGreevy and Smith, 1984; Smith and McGreevy, 1983; Tarr, 1915) and theoretical computation (Mortensen, 1933).
Nevertheless there has been little measurement of rock surface and subsurface temperatures under natural conditions in arid regions until 1950 and the temperature recordings carried out between 1950 and 1970 were concerned primarily with daily temperature ranges and not seasonal or yearly ones (Smith, 1977). Seasonal differences are reported by Smith (1977) and Brettschneider (1980), where account is also taken of aspect related variations.
Independent of these reports H. Eichler and the author began in 1976

to measure the values of rock temperatures in the Negev (Israel). Serial recordings were made in connection with studies of recent arid morphodynamics for both dry (August 1979) and humid seasons (March 1982). The measurements were set out to identify the microclimatological conditions of weathering processes.

Location and climate. The Negev in southern Israel is part of the subtropical desert belt (Table 1) but borders the mediterranean climatic zone; the steppe zone between them often extends only for a few kilometers.
The following measurements have been carried out in Machtesh HaGadol (Fig. 1) at one of the numerous tafoni in the surrounding cliffs.

TABLE 1. Climatic data of Machtesh HaGadol (according to Atlas of Israel, 1971; own measurements from August 1979 and March 1982)

Mean annual temperature:		21°C
Mean temperature of hottest month:		26 - 28°C
Mean temperature of coldest month:		10°C
Mean annual rainfall:		90 mm (includes 30 mm as dew)
Evaporation	- annual:	1 700 mm
	- July:	220 - 260 mm
	- January:	60 mm
Mean annual humidity:		50 - 55%
Climatic region according to Thornthwaite:		E
Climatic region according to Koeppen:		BWh
Own results:		
Highest value of air temperature:		39°C
Lowest value of air temperature:		9°C
Highest value of rock surface temperature:		51°C
Lowest value of rock surface temperature:		10°C
Highest value recorded within the rock:		47°C
Lowest value recorded within the rock:		10°C
Air humidity according to own measurements:		18 - 100%

METHODS

Recordings in 1976 and 1979 were carried out using small hand-held equipment, but in March 1982 an automatic temperature recorder with printer was used. In this way it was possible to measure twenty

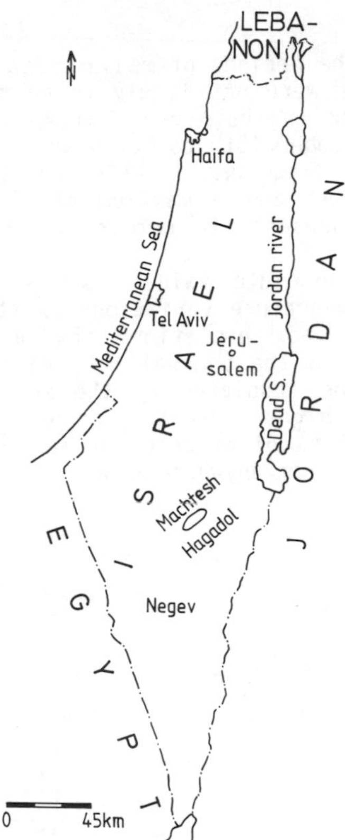

Fig. 1. Location map.

different positions simultaneously at 15 and 30 minute intervals. While in 1976 only rock surface temperatures were measured, in 1979 subsurface temperatures to a depth of 5 cm were recorded using thermocouples set in small holes (3.5 mm diameter). In March 1982 seventeen additional thermocouples were added to the measurement positions of 1979 in an attempt to identify minor temperature variations in time and space.

The drilling of these holes followed along a transect from the rock surface (position 0) to the roof of the cavern (position 11) and then from the backwall surface (position 12) with increasing depth into the backwall (position 16 was the deepest one). The depth of the different measurement positions is given in Fig. 4.

The indicated values of air temperature and relative humidity in Figures 3, 5 and 6 are recorded by means of a thermohygrograph.

Samples of rock pieces and salt horizons were collected from different positions (latin numbers I, II, III, IV in Fig. 4) at tafoni 'two'; the salts and the other minerals (Table 2) were identified by X-ray analyses.

RESULTS

It was clear that during the periods of measurement (August 1979 and March 1982) extreme values were not likely to be registered, since maximum temperatures appear in the Negev from April 15th to August 30th, the minimum from December 15th to March 1st.

Rock surface temperatures from August 1979 revealed a minimum of 19.2°C (5.40 a.m.; 23.8.79) and a maximum of 50.8°C (1.10 p.m.; 22.8.79). These values represent the temperature ranges during the hot, dry season.

More important than the absolute values is the observation that surface and subsurface temperature variations exist which cannot be explained by factors such as insolation, shadow effects and air temperature. At one point on the backwall of the tafoni (C 2, C 3; Fig. 2) which could not be insolated by the sun, the temperature rose to a level which was higher than the surrounding air (C 1) and since all the other positions of measurement (A 1-3, B 1-3) stayed cooler, a transfer of heat from anywhere else could not have existed (Fig. 3).

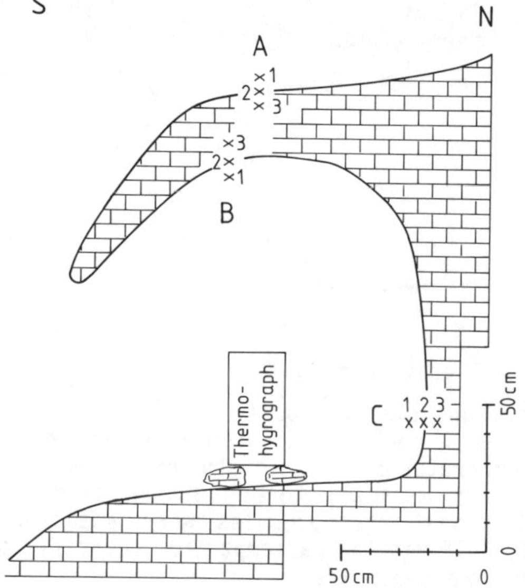

Fig. 2. Positions of measurement from August 1979 at tafoni 'two' in Machtesh HaGadol (Negev, Israel). Position 1: Air temperature 5 cm above surface; position 2: Rock surface temperature; position 3: Subsurface temperature from a depth of 5 cm.

The primary hypothesis is that the temperature rise was triggered by internal factors (for example by the crystallisation of salts out of highly concentrated solutions). The relative humidity of 100% (6 a.m., 23.8.1979; Fig. 3) and its decrease after 6 a.m. (6.30 a.m.: 95%; 7 a.m.: 92%) seem to support this hypothesis.

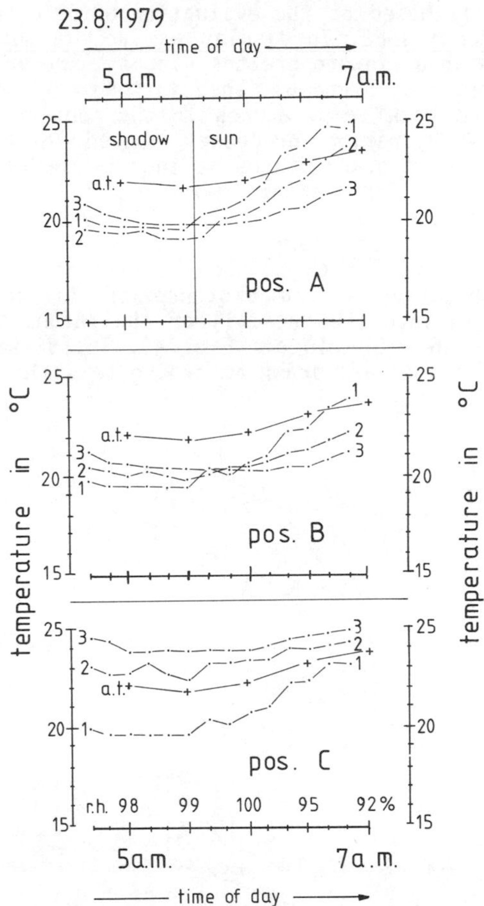

Fig. 3. Temperature values at tafoni 'two' during the early morning of the 23th of August 1979. a.t. = air temperature; r.h. = relative humidity, both recorded by means of a thermohygrograph. For the positions of measurement (A 1 - C 3) see Fig. 2.

The second law of thermodynamics states that heat cannot be transferred from a colder (C 1 or C 2) to a warmer rock part (C 3) without an inside source of energy. The heat must be produced in the rock itself.

Measurements collected during March 1982 can thus be regarded first as an attempt to measure the subsurface heat producing processes again, and second as a supplementary to published values of rock temperatures (c.f. Kerr et al. 1984).

If the hypothesis is correct, that heat producing process occurs and that this process is caused by salt crystallisation together with changes in moisture during the dry season, then there should be a greater possibility of recording such phenomena during the rainy season.

The following is based on the evaluation of 5000 temperature values measured in March 1982. To display as large a quantity of data as possible and with a view to greater clarity, the values are given by thermoisopleths (Fig. 5 and 6). This is then followed by an analyses of the recorded rock temperatures in the context of the values of the hydration/dehydration processes quoted by Mortensen (1933). Finally these analyses are discussed in relation to salts identified with X-ray analyses from tafoni 'two'.

Thermoisopleths of rock temperatures. Temperatures were measured along a transect from rock surface (position 0) to the surface of the cavern (position 11; a distance of 30 cm) and then with increasing depth into the backwall of the tafoni at the lower edge (position 12 - 16 = 0 - 10 cm; Fig. 4). The distances between the measurement positions are drawn according to scale in Fig. 5 and 6.

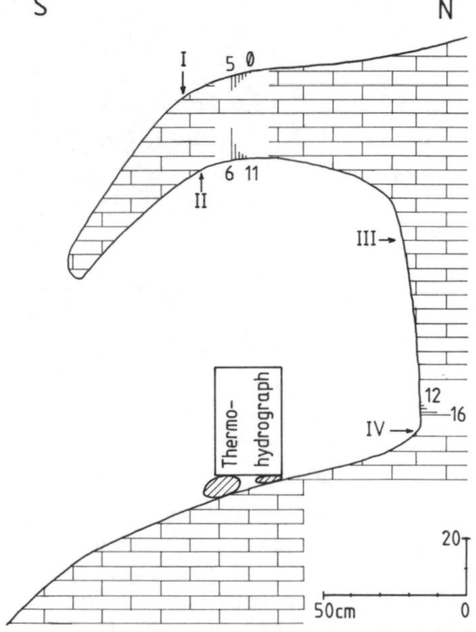

Fig. 4. The arrangement of the thermocouples at tafoni 'two' during March 1982. The drilling of the holes for the thermocouples follows along a transect from the rock surface (position 0) to the roof of the cavern (position 11) and then with increasing depth into the backwall (position 12 is the surface and 16 the deepest one).
The depths of the different positions: 0 = 0.2 cm; 1 = 1.0 cm; 2 = 2.0 cm; 3 = 3.0 cm; 4 = 4.0 cm; 5 = 5.0 cm; 6 = 11 cm; 7 = 5.7 cm; 8 = 3.0 cm; 9 = 2.0 cm; 10 = 1.3 cm; 11 = 0.2 cm; 12 = 0.2 cm; 13 = 1.3 cm; 14 = 2.0 cm; 15 = 5.0 cm; 16 = 10 cm. The latin numbers indicate the sampling points of the material analysed by X-ray analysis (Table 2).

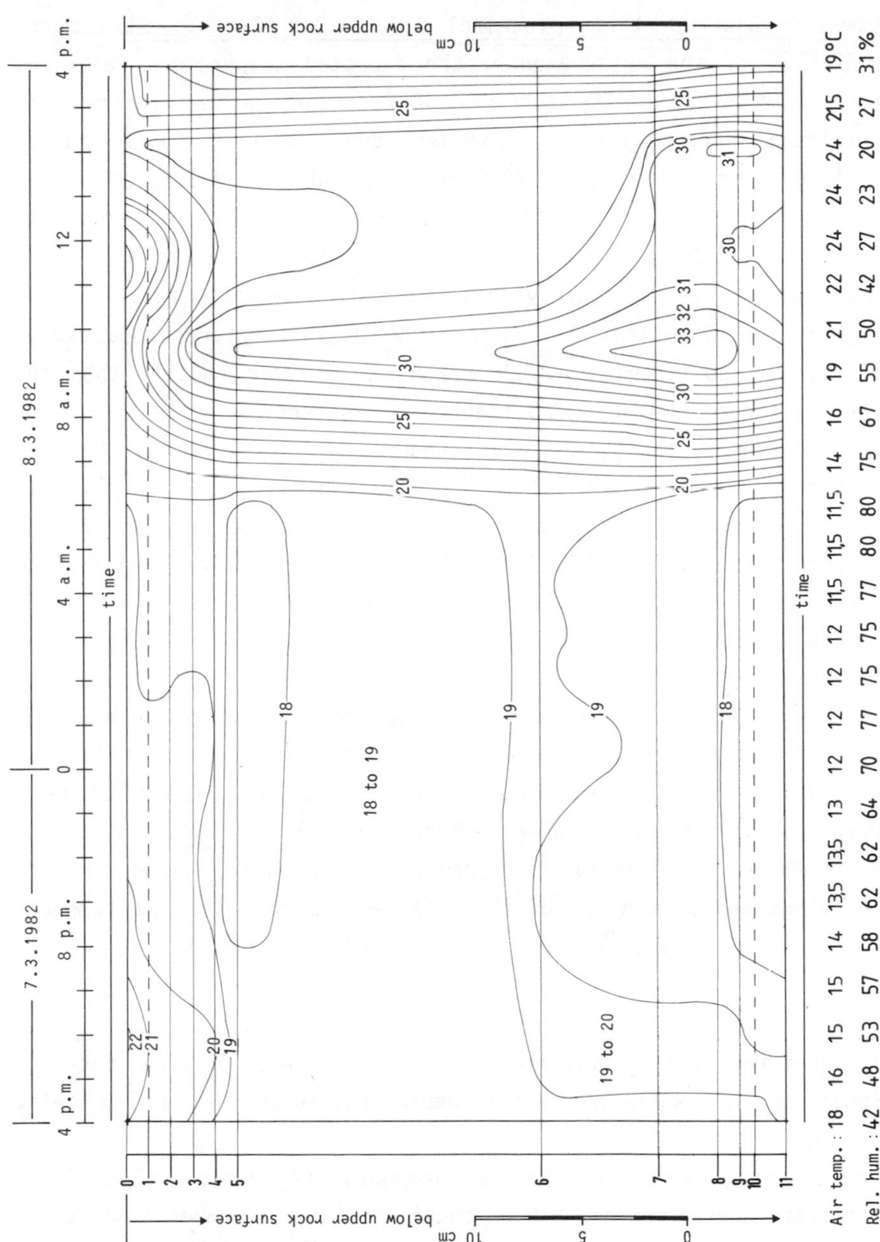

Fig. 5. Thermoisopleths (°C) of rock temperatures recorded in different depth at the roof of tafoni 'two'. Position 0 is the upper rock surface and position 11 the surface of the roof of the cavern.

Example: Roof of tafoni (period: 7.3. - 8.3.82; Fig. 5). At the beginning of the measurements, the expected temperature relations were substantiated. The surface and near-surface temperatures for insolated surfaces were the highest (position 0, 1, 2). A gradual cooling started at 6 p.m. (7.3.82) and ended on 8.3.82 at 5 a.m. From 8 p.m. (7.3.82) to 5 a.m. (8.3.82) temperatures differed only by 3°C (20.4°C at 8 p.m. at position 0 and 17.4°C at 5 a.m. at position 11; all the other values fluctuate between these).

From 6 a.m. (sunrise: 5.30 a.m.) onwards temperatures rose everywhere, but positions 6, 7, 8 and 9 heated more rapidly than the others. The maximum of more than 33°C was recorded at 9.30 a.m. at positions 7, 8 and 9; while air temperature was 19°C and the rock surface temperature was only 23°C (position 0). The warming within the rock eventually influenced temperatures above and underneath. Between 10 a.m. and 12 o'clock strong winds cooled the surface temperature from 22.9°C (9.30 a.m.) to 19.9°C (11.30 a.m.). That cooling effect influenced the subsurface temperatures (positions 1-4) too.

From 12 a.m. to 2 p.m. surface temperatures and those near the surface rose to 26°C as a result of insolation. Positions 7, 8, 9 and 10 had at the same time higher values again (30-31°C) as a result of the previous higher temperatures (33°C at 9.30 a.m.).

After 2 p.m. a general and rapid cooling was recorded at all positions and at the end of the measurements at 4 p.m. the temperatures were between 25°C (position 0) to 22°C (positions 5, 6, 7, 8).

Example: Backwall of tafoni (period: 7.3. - 8.3.82; Fig. 6).The temperatures were measured from the surface of the backwall (position 12) with increasing depth (position 16: 10 cm below surface).

At the beginning of the measurements (4 p.m.; 7.3.82) the temperature at a depth of 10 cm (position 16) was higher than at all other depths. The cooling continued until 6 a.m., but position 16 remained always the warmest. From 6 a.m. onwards rock temperature at all depths rose more quickly than air temperature, although the sun could never insolate this part of the tafoni. After 9 a.m. warming

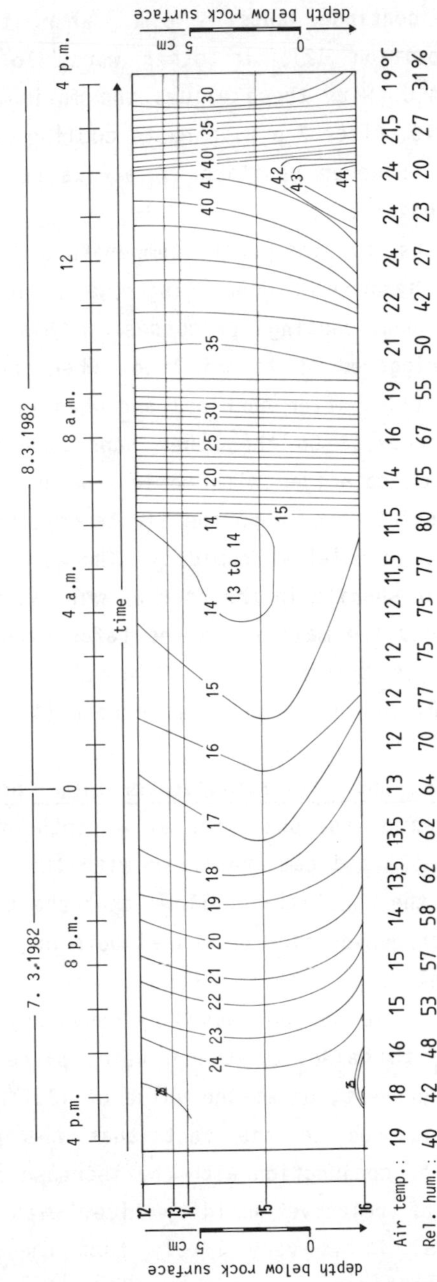

Fig. 6. Thermoisopleths (°C) of rock temperatures recorded in different depths at the backwall of tafoni 'two'. Position 12 is the surface of the backwall and position 16 is the deepest one.

slowed down, but continued until 2 p.m., when it was 40°C on the rock surface (position 12), in other words 16°C above the air temperature of 24°C. With these values the maximum temperature was 44°C at position 16. After 2 p.m. a rapid cooling process took place at all positions. However position 16 always remained warmer than the neighbouring positions.

The thermoisopleths of the rock temperatures (Fig. 5 and 6) substantiate the hypothesis concerning the values of August 1979 which indicated heat-causing processes within the rock. Such processes were triggered at tafoni 'two' when the air temperature rose considerably (i.e. after sunrise) and when in addition relative humidity also decreased. On the other hand, the moments of rapid cooling of the rock coincided with periods when the surface of the tafoni was not insolated, or when the air temperature fell without a corresponding rise in relative humidity. The warming process within the rock started at sunrise in the roof as well as in the back wall. Temperatures reached the maximum in the tafoni roof (positions 0 - 11) during the late morning (9 - 10 a.m.) and in the backwall (positions 12 - 16) during the early afternoon (1 p.m. - 2 p.m.).

Rock temperatures and hydration/dehydration processes. If we evaluate the curves of the positions at a depth of 5 cm (positions 5, 7 and 15; Fig. 7) and compare these with the results quoted by Mortensen (1933), the conclusion follows that the temperatures would allow the hydration/dehydration processes both of Na_2SO_4 and of Na_2CO_3. Since the dehydration process needs energy, the heat producing process could not be caused by dehydration. The hydration process of Na_2SO_4 to Na_2So_4 x 10 H_2O takes place at the value of 17.9°C if NaCl is present; or at the value of 13.7°C if $MgSO_4$ and K_2SO_4 are present. Due to the fact that the process of heat production begins in conjunction with the increase of air temperature and decrease of relative humidity (i.e. with the decrease of available moisture) it is very likely that the observed warming within the rock was based on warming caused by crystallisation and not by hydration. A gradient of moisture pressure between the moist rock and the dryer air is created by the decrease of humidity. The

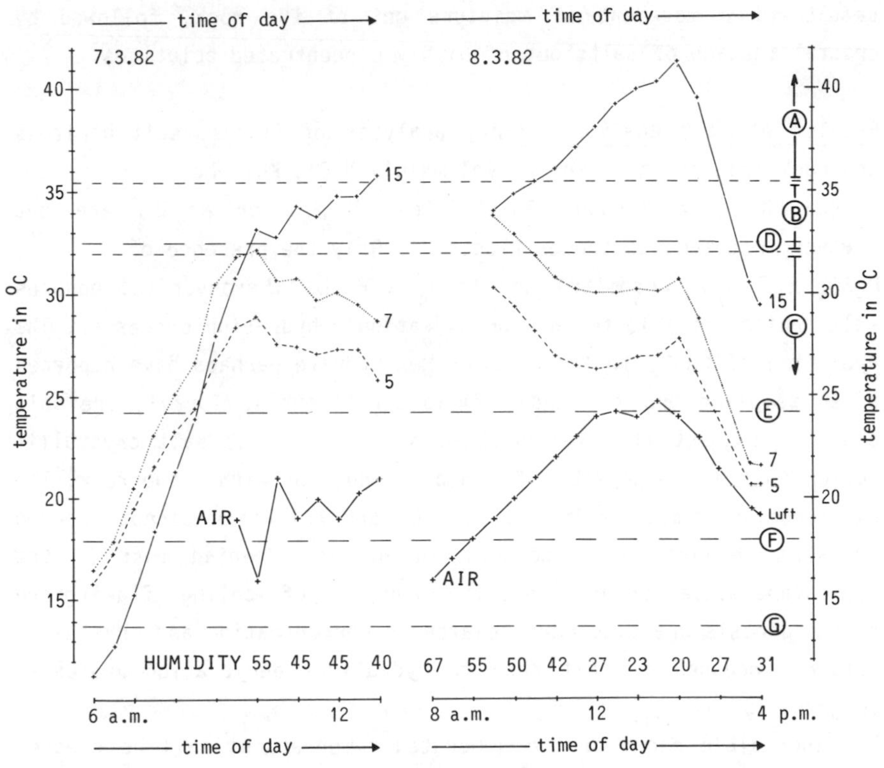

Fig. 7. Temperature curves in a depth of 5 (positions 5, 15) or 5.7 cm (position 7) and values of hydration/dehydration transition temperatures (after Mortensen 1933). The figure reports the possibility of hydration/dehydration processes under the recorded temperatures. But X-ray analyses of the salts reveal no existence of Na_2CO_3 and Na_2SO_4. And at least hydration depends also upon vapour pressure of the salt/hydrate system (i.e. upon moisture conditions).

A = $Na_2CO_3 \times 1\ H_2O$ above 35.4°C;

B = $Na_2CO_3 \times 7\ H_2O$ between 35.4 and 32°C;

C = $Na_2CO_3 \times 10\ H_2O$ below 32.0°C;

D = $Na_2SO_4 \rightleftharpoons Na_2SO_4 \times 10\ H_2O$ at 32.4°C;

E = $Na_2SO_4 \rightleftharpoons Na_2SO_4 \times 10\ H_2O$ at 24.4°C if $NaNO_3$ is present;

F = $Na_2SO_4 \rightleftharpoons Na_2SO_4 \times 10\ H_2O$ at 17.9°C if NaCl is present;

G = $Na_2SO_4 \rightleftharpoons Na_2SO_4 \times 10\ H_2O$ at 13.9°C if $MgSO_4$ and K_2SO_4 are present.

result is a movement of moisture out of the rock, followed by crystallisation of salts out of highly concentrated solutions.

Results of salt analyses. X-ray analyses of flakes, salt horizons and rock from tafoni 'two' reveal mainly NaCl, KCl and $K_2MgCa_2(SO_4)_4$ x 2 H_2O (Polyhalite). Na_2CO_3 or Na_2SO_4 are not present, not even as traces (Table 2). Only the presence of $MgSO_4$ x 7 H_2O (Epsomite) and $MgSO_4$ x 6 H_2O (Hexahydrite) ensures salts which are able to undergo hydration/dehydration processes. The hydration of $MgSO_4$ to Epsomite or Hexahydrite perhaps have happened after sampling and not under field conditions in Israel. The salt analyses suggest that rather than hydration it is salt crystallisation that is responsible for heat production within the rock. The existence of common salts (like NaCl and KCl which cannot undergo hydration/dehydration processes) in horizons behind most of the flakes and scales ensures that the processes of scaling, flaking and tafoni genesis are also more related to concentration and crystallisation processes of salt than to hydration/ dehydration processes (Rögner, 1985).

The incredible range of the generated temperatures will be checked by laboratory experiments in 1986. However, it is worth noting that Bonnell and Nottage observed "a sudden evolution of heat" (1939, p. 18) at the range of 6 - 8°C (1939, Fig. 4) within porous material while the surrounding air was cooling. They combined the heat evolution and the immediately following expansion with the hydration process of Na_2SO_4.

DISCUSSION

In August 1979 as well as in March 1982 internal rock temperatures were measured which cannot be explained by exogenous factors such as insolation and changes in air temperature. The origin of these anomalously high temperatures is therefore attributed to endogenous processes. The start of the heat-producing processes at tafoni 'two' coincided with the rise of air temperature after sunrise and the simultaneous reduction of relative humidity.

It is suggested that the warming could be caused mainly by the crystallisation of salts out of highly concentrated solutions. Salts are very frequent in the tafoni of Machtesh HaGadol forming segregated horizons parallel to the surface of the cavern indepen-

Table 2. X-ray analyses of salts and minerals of tafoni 'two' (Machtesh HaGadol, Negev, Israel)

Position in Fig.4	Specification of the sampling point:		Determined salts and minerals:
I	surface of the tafoni	analysed material from a depth of 2 - 3 cm (position 2)	Dolomite: $CaMg(CO_3)_2$ Albite: $Na(AlSi_3O_8)$
II	surface of the roof of the cavern	analysed material from a depth of 2 - 3 cm (position 8)	Dolomite Halite: NaCl Calcite: $CaCO_3$ Quartz: SiO_2
		analysed material rock chips from drilling, depth 5 - 8 cm (position 6)	Dolomite Calcite Halite Sylvite: KCl
III	backwall of the cavern	crust of salts from a depth of 2 cm	Halite Dolomite Calcite Polyhalite: $K_2MgCa_2(SO_4)_4 \times 2\ H_2O$
		analysed material from a depth of 4 cm	Dolomite Halite Calcite Quartz Gypsum: $CaSO_4 \times 2\ H_2O$ Goethite: FeO(OH) Epsomite: $MgSO_4 \times 7\ H_2O$ Hexahydrite: $MgSO_4 \times 6\ H_2O$
IV	lower edge of the tafoni	analysed material from a depth of 2 cm (positions 13 - 14)	Dolomite Halite Calcite Quartz Kaolinite Illite

dent of the rock structure.

The rock surface outside the tafoni neither contains salts nor shows flaking or scaling, although this is the position with the highest temperature ranges caused by insolation. At this exposed rock surface (positions 0 - 5) temperatures are recorded which can be related solely to insolation and atmospheric cooling processes. These normal temperatures contrast with the 'anomalous' values measured in the roof of the cavern (positions 6 - 11) and in the backwall (positions 12 - 16). These are also the positions where salts occur. From the results we conclude that flaking, scaling and ultimately the origin of the tafoni is not caused by insolation processes alone. The dominant process in cavernous weathering in Machtesh Hagadol is salt weathering mainly by crystal growth (and not by hydration). And this is based on the interaction of changes in relative humidity, in temperature and in moisture conditions between rock and air. Most of the salts which occur in the tafoni of Machtesh HaGadol are derived from seepage and are locally concentrated in different segregation horizons. These horizons possibly occur in relation to different wetting/drying depths and/or different heating/ cooling boundary planes.

The hygroscopic effect of the salts influences the direction of the seepage to the positions where salts already occur and promotes the penetration of air humidity into the rock. This continuing enrichment of salts in the smallest joints and cracks enhances the process of salt crystallisation which is the main cause of scaling, flaking and the genesis of the tafoni.

REFERENCES

Atlas of Israel, 1970. Survey of Israel, Ministery of Labour, Jerusalem and Elsevier Publishing Company, Amsterdam.

Barton, D.C., 1916. Notes on the disintegration of granite in Egypt. Journal of Geology, 24, 382 - 393.

Besler, H., 1979. Feldversuche zur aktuellen Granitverwitterung und Rindenbildung in der Namib. Stuttgarter Geographische Studien, 93, 95 - 106.

Blackwelder, E., 1925. Exfoliation as a phase of rock weathering. Journal of Geology, 33, 793 - 806.

Blackwelder, E., 1933. The insolation hypothesis of rock weathering. American Journal of Science, Series 5, 26, No. 152, 97 - 113.

Bonell, D.G.R. and Nottage, M.E., 1939. Studies in porous materials with special reference to building materials - I. The crystallisation of salts in porous materials. Journal of the Society Chemical Industry, 58A, 16 - 21.

Brettschneider, H., 1980. Mikroklima und Verwitterung an Beispielen aus der Sierra Nevada Spaniens und aus Nordafrika mit Grundlagenstudien zur Glatthanggenese. Münstersche Geographische Arbeiten, 9, 65 - 141.

Brettschneider, H., 1981. Interpretation mikroklimatischer Daten in Hinsicht auf das Verwitterungsgeschehen an Beispielen aus dem Hohen Atlas. Zeitschrift für Geomorphologie, N.F., Suppl. Bd. 39, 81 - 94.

Cooke, R.U. and Smalley, I.J., 1968. Salt weathering in deserts. Nature, 220, 1226 - 1227.

Correns, C.W. and Steinborn, W., 1939. Experimente zur Messung und Erklärung der sogenannten Kristallisationskraft. Zeitschrift für Kristallographie, A, 101, 117 - 133.

Dragovich, D., 1967. Flaking, a weathering process operating on cavernous rock surfaces. Geological Society of America, Bulletin, 78, 801 - 804.

Dragovich, D., 1969. The origin of cavernous surfaces (tafoni) in granitic rocks of southern South Australia. Zeitschrift für Geomorphologie, N.F., 13, 163 - 181.

Eichler, H., 1982. Flechten, biotische Verwitterung und ökologische Wertung. Wissen Heute, 9, 68 - 75.

Evans, I.S., 1970. Salt crystallisation and rock weathering: a review. Revue de Geómorphologie dynamique, 19, Nr. 4, 153 - 177.

Futterer, K., 1902. Der Pe-schan als Typus der Felswüste. Geographische Zeitschrift, 8, 249 - 266 and 323 - 339.

Goudie, A., 1974. Further experimental investigation of rock weathering by salt crystallisation and other mechanical processes. Zeitschrift für Geomorphologie, N.F., Suppl. Bd. 21, 1 - 12.

Goudie, A., 1977. Sodium sulphate weathering and the disintegration of Mohenjo-Daro, Pakistan. Earth Surface Processes, 2, 75 - 86.

Goudie, A., Cooke, R. and Evans, I., 1970. Experimental investigation of rock weathering by salts. Area, 4, 42 - 48.

Goudie, A., Cooke, R. and Doornkamp, I.C., 1979. The formation of silt from quartz dune sand by salt-weathering processes in deserts. Journal of Arid Environments, 2, 105 - 112.

Griggs, D.T., 1936. The factor of fatique in rock exfoliation. Journal of Geology, 44, 7833- 796.

Jäkel, D. and Dronia, H., 1976. Ergebnisse von Boden- und Gesteinstemperaturmessungen in der Sahara. Berliner Geographische Abhandlungen, 24, 55 - 64.

Kaiser, Kh., 1972. Prozesse und Formen der ariden Verwitterung am Beispiel des Tibesti-Gebirges und seiner Rahmenbereiche. Berliner Geographische Abhandlungen, 16, 49 - 80.

Kerr, A., Smith, B.J., Whalley, W.B. and McGreevy, J.P., 1984. Rock temperatures from southeast Marocco and their significance for experimental rock weathering studies. Geology, 12, 306-309.

Klaer, W., 1970. Formen der Granitverwitterung im ganzjährig ariden Gebiet der östlichen Sahara (Tibesti). Tübinger Geographische Studien (Festschrift für H. Wilhelmy), 71 - 78.

Knetsch, G., 1960. Über aride Verwitterung unter besonderer Berücksichtigung natürlicher und künstlicher Wände in Ägypten. Zeitschrift für Geomorphologie, N.F., Suppl. Bd. 1, 191 - 205.

Knetsch, G. and Refai, E., 1955. Über Wüstenverwitterung, Wüstenfeinrelief und Denkmalzerfall in Ägypten. Neues Jahrbuch für Geologie und Paläontologie, Abhandlungen 101, 227 - 256.

Mc Greevy, J.P. and Smith, B.J., 1984. The possible role of clay minerals in salt weathering. Catena, 11, 169 - 175.

Mortensen, H., 1930. Die Wüstenböden, in Handbuch der Bodenlehre, 3, 437 - 490; Berlin (Springer).

Mortensen, H., 1933. Die "Salzsprengung" und ihre Bedeutung für die regional-klimatische Gliederung der Wüsten. Petermanns Geographische Mitteilungen, 79, 130 - 135.

Ollier, C.D., 1963. Insolation weathering: Examples from Central Australia. American Journal of Science, 261, 376 - 381.

Peel, R.F., 1974. Insolation weathering: Some measurements of diurnal temperature changes in exposed rocks in the Tibesti region, Central Sahara. Zeitschrift für Geomorphologie, N.F., Suppl. Bd. 21, 19 - 28.

Roth, E.S., 1965. Temperature and water content as factors in desert weathering. Journal of Geology, 73, 454 - 468.

Rögner, K., 1985. Geomorphologische Untersuchungen in Negev und Sinai - Ein Beitrag zur rezenten Morphodynamik, zur Quantifizierung arid-morphodynamischer Prozesse und Prozeßkombinationen sowie zur Landschaftsgenese. Habil. thesis, Universität Paderborn, 218 + 127 p.

Schattner, I., 1961. Weathering phenomena in the crystalline of the Sinai in the light of current notions. Bulletin of the Research Council of Israel. 10 G, 247 - 266.

Smith, B.J., 1977. Rock temperature measurements from the northwest Sahara and their implications for rock weathering. Catena, 4, 41 - 63.

Smith, B.J. and McGreevy, J.P., 1983. A simulation study of salt weathering in hot deserts. Geografiska Annaler, 65A, 127 - 133.

Tarr, W.A., 1915. A study of some heating tests, and the light they throw on the cause of the disintegration of granite. Economic Geology, 10, 348 - 367.

Walther, J., 1888. Über Ergebnisse einer Forschungsreise auf der Sinaihalbinsel und in der Arabischen Wüste. Verhandlungen der Gesellschaft für Erdkunde zu Berlin, 15, 244 - 255.

Walther, J., 1924. Das Gesetz der Wüstenbildung, 4; Leipzig.

Wellmann, H.W. and Wilson, A.T., 1965. Salt weathering, a neglected geological erosive agent in coastal and arid environment. Nature, 205, 1097 - 1098.

Wilhelmy, H., 1964. Cavernous rock surfaces (tafoni) in semiarid and arid climates. Pakistan Geographical Review, 19, 9 - 13.

Author Index

Subject Index